Edited by
Jürgen Popp, Valery V. Tuchin,
Arthur Chiou, and
Stefan Heinemann

Handbook of Biophotonics

Related Titles

Popp, J., Tuchin, V. V., Chiou, A., Heinemann, S. H. (eds.)

Handbook of Biophotonics

3 Volume Set
approx. 2012
ISBN: 978-3-527-40728-6

Popp, J., Tuchin, V. V., Chiou, A., Heinemann, S. H. (eds.)

Handbook of Biophotonics

Vol. 1: Basics and Techniques
2011
ISBN: 978-3-527-41047-7

Popp, J., Tuchin, V. V., Chiou, A., Heinemann, S. H. (eds.)

Handbook of Biophotonics

Vol. 3: Photonics in Pharmaceutics, Bioanalysis and Environmental Research
approx. 2012
ISBN: 978-3-527-41049-1

Edited by
*Jürgen Popp, Valery V. Tuchin, Arthur Chiou,
and Stefan Heinemann*

Handbook of Biophotonics

Vol. 2: Photonics for Health Care

WILEY-VCH Verlag GmbH & Co. KGaA

The Editors:

Prof. Jürgen Popp
Institute of Physical Chemistry
Friedrich-Schiller-University of Jena
Jena, Germany

Prof. Valery V. Tuchin
Department of Optics and Biophotonics
Saratov State University
Saratov, Russia

Prof. Arthur Chiou
Department of Biophotonics Engineering
National Yang Ming University
Taipei, Taiwan, China

Prof. Stefan Heinemann
Institute of Biochemistry and Biophysics
Friedrich-Schiller-University of Jena
Jena, Germany

Editorial Assistance/Coordination:

Dr. Marion Strehle
Marion.strehle@uni-jena.de

Dr. Thomas Mayerhöfer
Thomas.Mayerhoefer@uni-jena.de

Cover:
Background: Frozen sections of mammacarcinoma (Tissue: Brigitte Mack, LMU; Immunohistochemistry: Sabine Sandner, LMU; TKTL1-antibody: R-Biopharm, Darmstadt)
Small image: Safe and ergonomic work environment with OR1, KARL STORZ GMBH & CO. KG

All books published by **Wiley-VCH** are carefully produced. Nevertheless, authors, editors, and publisher do not warrant the information contained in these books, including this book, to be free of errors. Readers are advised to keep in mind that statements, data, illustrations, procedural details or other items may inadvertently be inaccurate.

Library of Congress Card No.: applied for

British Library Cataloguing-in-Publication Data
A catalogue record for this book is available from the British Library.

Bibliographic information published by the Deutsche Nationalbibliothek
The Deutsche Nationalbibliothek lists this publication in the Deutsche Nationalbibliografie; detailed bibliographic data are available on the Internet at http://dnb.d-nb.de.

© 2012 Wiley-VCH Verlag & Co. KGaA, Boschstr. 12, 69469 Weinheim, Germany

All rights reserved (including those of translation into other languages). No part of this book may be reproduced in any form – by photoprinting, microfilm, or any other means – nor transmitted or translated into a machine language without written permission from the publishers. Registered names, trademarks, etc. used in this book, even when not specifically marked as such, are not to be considered unprotected by law.

Composition   Thomson Digital, Noida, India

Printing and Binding   betz-druck GmbH, Darmstadt

Cover Design   Adam Design, Weinheim

Printed in the Federal Republic of Germany
Printed on acid-free paper

ISBN: 978-3-527-41048-4

Contents

Preface *XXXIII*
List of Contributors *XXXV*

Part 1 Laboratory Medicine *1*

1 Microscopic and Spectroscopic Methods *3*
Brent D. Cameron
1.1 Bright Field Microscopy *3*
1.2 Dark Field Microscopy *3*
1.3 Phase Contrast Microscopy *4*
1.4 Differential Interference Contrast Microscopy *4*
1.5 Confocal Microscopy *5*
1.6 Fluorescence Spectroscopy *5*
1.7 Infrared Spectroscopy *6*
1.8 Raman Spectroscopy *7*
References *8*

2 Glucose Sensing and Glucose Determination Using Fluorescent Probes *11*
Mark-Steven Steiner, Axel Duerkop, and Otto S. Wolfbeis
2.1 Why Glucose Sensing? *11*
2.2 Classification of Sensors According to the Recognition Scheme *12*
2.3 Glucose Sensing via the Optical Properties of Oxidative or Reductive Enzymes *13*
2.3.1 Nonlabeled Enzymes *13*
2.3.2 Labeled Enzymes *13*
2.4 Fluorescent Sensing of Glucose via Measurement of the Consumption of Oxygen Caused by the Action of GOx *14*
2.4.1 Planar and Fiber-Optic Sensors *14*
2.4.2 Sensors Based on the Use of Microparticles (μPs) and Nanoparticles (NPs) *16*

2.5	Fluorescent Sensing of Glucose via Measurement of the Formation of Hydrogen Peroxide Caused by the Action of GOx 18
2.6	Enzymatic Sensing of Glucose via Changes of pH 20
2.7	Fluorescent Sensing of Glucose via Synthetic Boronic Acids 20
2.8	Fluorescent Sensing of Glucose via Concanavalin A 23
2.9	Fluorescent Sensing of Glucose using Glucose-Specific Apoenzymes 26
2.10	Fluorescent Sensing of Glucose via Glucose-Binding Proteins 27
2.11	Conclusion and Outlook 30
	References 30
3	**Biochips as Novel Bioassays** 35
	Bettina Rudolph, Karina Weber, and Robert Möller
3.1	Introduction 35
3.2	Types of Microarrays 36
3.2.1	DNA Microarrays 36
3.2.2	Protein Microarrays 37
3.2.3	Tissue Microarrays 38
3.2.4	Cell-Based Microarrays (Transfection Microarrays) 39
3.3	Principle Steps of an Experimental Microarray Process 40
3.3.1	Basics 41
3.3.1.1	Surface Biofunctionalization 41
3.3.2	Preparation of Biomolecules (Sample Preparation) 42
3.3.2.1	DNA 42
3.3.2.2	RNA 43
3.3.2.3	Proteins 43
3.3.3	Target Amplification and Labeling 43
3.3.3.1	PCR – Polymerase Chain Reaction 44
3.3.3.2	*In Vitro* Transcription – cDNA/IVT 44
3.3.3.3	Bio-Barcode Amplification of Nucleic Acids and Proteins 44
3.3.3.4	Radioactive and Fluorophore Labeling 45
3.3.3.5	Enzymes 46
3.3.3.6	Nanoparticles 47
3.3.4	Readout/Detection 47
3.3.4.1	Radioactivity 48
3.3.4.2	Fluorescence/Absorbance 49
3.3.4.3	Surface Plasmon Resonance 51
3.4	Examples of Biochip Applications in Biology and Medicine 51
3.4.1	Genomic Research 51
3.4.1.1	SNP Detection 52
3.4.1.2	Gene Expression Profile Analysis 54
3.4.1.3	Pharmacogenomics and Toxicogenomics 55
3.4.2	Pathogen Detection and Clinical Diagnosis 56
3.4.2.1	Pathogen Detection 56
3.4.2.2	Clinical Diagnosis 57

3.5	Trends in Biochip Research 58
3.5.1	Metal Nanoparticles 58
3.5.2	Quantum Dots 59
3.5.3	Enzymes 61
3.6	Conclusion 63
	References 64

4	**Microstructure Fibers in Biophotonics** 77
	Aleksei M. Zheltikov
4.1	Introduction: From Early Concepts of Fiber-Based Bioimaging to Biophotonics with Specialty Fibers 77
4.2	Microstructure Fiber Sensors 78
4.3	Designing Specialty Fibers for High-Resolution Imaging 80
4.4	Fiber-Based Imaging of Diffusively Scattering Tissues 82
4.5	Enhancement of Guided-Wave Two-Photon-Excited Luminescence Response with a Photonic-Crystal Fiber 84
4.6	Microstructure Fibers for Nonlinear Optical Microspectroscopy 90
4.7	Microstructure Fibers for Neurophotonics 94
4.8	Toward Fiber-Based Multicolor Imaging 98
	References 100

5	**Identification and Characterization of Microorganisms by Vibrational Spectroscopy** 105
	Stephan Stöckel, Angela Walter, Anja Boßecker, Susann Meisel, Valerian Ciobotă, Wilm Schumacher, Petra Rösch, and Jürgen Popp
5.1	Introduction 105
5.2	Sample Preparation Techniques 106
5.3	Statistical Analysis of Vibrational Spectroscopic Data 107
5.4	Monitoring of Biofilms 116
5.5	Role of Vibrational Spectroscopy in Epidemiological Studies 120
5.6	Effect of Antibiotics on Microorganisms 127
5.7	Detection and Identification of Food-Borne Pathogens 131
5.8	Conclusion 135
	References 136

Part 2 Pathology 143

6	**Cell Sorting** 145
	Jochen Guck
6.1	Conventional Techniques in Cell Sorting 145
6.2	Photonic Techniques in Single-Cell Sorting 145
6.3	Photonic Techniques in Bulk Sorting 147
	References 148

7	**Laser Microtomy** *151*
	Holger Lubatschowski, Fabian Will, Sabine Przemeck, and Heiko Richter
7.1	Cutting by Optical Breakdown *151*
7.2	The Laser Microtome at a Glance *151*
7.3	Preparing and Sectioning the Tissue Samples *152*
7.4	OCT Controlled 3D Processing *154*
7.5	Tissue Samples *155*
7.6	Not for Routine but for Special Cases *155*
	References *157*

8	**Classical Microscopy** *159*
	Kerstin Roeser, Dietlind Zuehlke, Martin Gluch, Gernoth Grunst, Markus Sticker, Klaus Bendrat, and Axel Niendorf
8.1	Appearance *159*
8.2	Application *159*
8.3	Pathology *159*
8.4	Materials *160*
8.5	Methods (Tissue Processing) *160*
8.6	Histological Staining *163*
8.7	Microscopy *165*
8.8	Diagnostics *167*
8.8.1	Context-Dependent Aspects in Morphological Diagnostics *167*
8.8.2	Cognitive Aspects of Digital Microscopy *167*
	References *174*

9	**Virtual Microscopy** *175*
	Thorsten Heupel
9.1	Introduction *175*
9.2	Principles *176*
9.3	Applications *177*
9.3.1	Telepathology *177*
9.3.2	Routine Pathology *177*
9.3.3	Education *178*
9.3.4	Tissue Microarrays *178*
9.3.5	Research *179*
9.4	Outlook *179*
	References *180*

10	***In Vitro* Instrumentation** *181*
	Sergio Coda, Rakesh Patalay, Christopher Dunsby, and Paul M.W. French
10.1	Introduction *181*
10.2	Autofluorescence of Biological Tissue *181*
10.2.1	Fluorescence Contrast *183*
10.2.2	Spectroscopic Techniques for Fluorescence Imaging and Metrology *184*
10.2.2.1	Spectral Techniques *184*

10.2.2.2	Fluorescence Lifetime Techniques	184
10.2.3	Single-Point Spectroscopic Techniques	185
10.2.4	Fluorescence Microscopy	186
10.2.4.1	Laser Scanning Confocal/Multiphoton Microscopy	186
10.2.4.2	Wide-Field Fluorescence Imaging	187
10.2.4.3	Multidimensional Fluorescence Microscopy	187
10.3	*Ex Vivo* Tissue Spectroscopy and Imaging: Application to Cancer	189
10.3.1	Oral Cavity	190
10.3.2	Gastrointestinal Tract	190
10.3.3	Bladder	191
10.3.4	Breast	192
10.3.5	Brain	192
10.3.6	Skin	193
10.4	Conclusion	194
	References	194

11 **In Vivo Instrumentation** *201*
Herbert Schneckenburger, Michael Wagner, Petra Weber, and Thomas Bruns

11.1	Introduction	201
11.2	Microscopy	201
11.3	Endoscopy	202
11.4	Imaging Techniques	203
	References	205

12 **Fluorescent Probes (Fluorescence Standards, Green Protein Technology)** *207*
Markus Sauer
References *212*

13 **Optical Coherence Imaging for Surgical Pathology Assessment** *215*
Wladimir A. Benalcazar and Stephen A. Boppart

13.1	Structural Contrast Imaging	217
13.2	Molecular Contrast Imaging	218
13.3	Conclusion	222
	References	222

14 **Diffuse Optical Imaging** *225*
Jeremy C. Hebden

14.1	Introduction	225
14.2	Optical Topography	226
14.2.1	Evoked Response to Sensory Stimulation	227
14.2.2	Cognitive Activity and Language Processing	229
14.2.3	Epilepsy	229
14.3	Optical Tomography	229
14.3.1	Optical Mammography	230

14.3.2	Optical Tomography of the Brain 230
14.4	Conclusion 232
	References 232

15	**Raman Microscopy** 235
	Christoph Krafft and Jürgen Popp
15.1	Introduction 235
15.2	Application to Cell Characterization 236
15.3	Application to Tissue Characterization 239
15.4	Conclusions 240
	References 241

16	**CARS Microscopy** 243
	Andreas Zumbusch
16.1	Introduction 243
16.2	Technical Background and Experimental Realizations 244
16.2.1	Excitation Geometry 245
16.2.2	Excitation Lasers 245
16.3	Applications 246
16.4	Outlook 248
	References 248

17	**Detection of Viral Infection in Epithelial Cells by Infrared Spectral Cytopathology** 251
	Max Diem, Nora Laver, Kristi Bedrossian, Jennifer Schubert, Kostas Papamarkakis, Benjamin Bird, and Miloš Miljković
17.1	Introduction 251
17.2	Methods 252
17.3	Results and Discussion 252
17.4	Conclusions 256
	References 256

Part 3 Oncology 259

18	**Clinical and Technical Aspects of Photodynamic Therapy – Superficial and Interstitial Illumination in Skin and Prostate Cancer** 261
	Katarina Svanberg, Niels Bendsoe, Sune Svanberg, and Stefan Andersson-Engels
18.1	Background – Cancer and Its Treatment 261
18.2	Phototherapy 261
18.3	Phototoxicity 263
18.4	Photodynamic Procedure 265
18.5	Photosensitizers 265
18.5.1	Laser-Induced Fluorescence for Tumor Detection 267
18.6	Light Sources 269

18.7	Photodynamic Therapy as a Clinical Procedure	271
18.8	Superficial PDT in Skin Lesions	271
18.9	Interstitial PDT for Prostate Cancer	274
18.10	Discussion	282
	References	283

19 Molecular Targeting of Photosensitizers   289
Hasan Tayyaba, Prakash R. Rai, Sarika Verma, and Yiang Zheng

19.1	Introduction	289
19.2	Molecular Modifications of Photosensitizers	290
19.2.1	Small Molecules, Peptides, and Aptamers	290
19.2.2	Proteins	291
19.3	Delivering Photosensitizers via Targeted Nanoconstructs	291
19.4	Target Modification/Cellular Activation	292
19.5	Conclusion	292
	References	293

20 Photodynamic Molecular Beacons   295
Pui-Chi Lo, Jonathan F. Lovell, and Gang Zheng

References   303

21 Oxygen Effects in Photodynamic Therapy   305
Asta Juzeniene

21.1	Introduction	305
21.2	The Role of Oxygen in Photodynamic Therapy: $^1O_2$ Generation	305
21.3	Dependence of the Photosensitizing Effect on O_2 Concentration	306
21.4	The Oxygenation Status in Tumors and Normal Tissues	306
21.5	PDT-Induced Reduction of Tumor Oxygenation	307
21.5.1	O_2 Consumption (Primary Reduction)	307
21.5.2	Vascular Damage (Secondary Reactions)	307
21.5.3	Photosensitizer Photobleaching (Secondary Reactions)	307
21.6	Methods to Reduce Tumor Deoxygenation During PDT	308
21.6.1	Low Fluence Rates	308
21.6.2	Fractionated Light Exposure	308
21.6.3	Other Methods	309
21.7	Changes of Quantum Yields Related to Photosensitizer Relocalization	309
21.8	Changes of Optical Penetration Caused by Changes in O_2 Concentration	309
21.9	Conclusion	309
	References	310

22		**Organic Light-Emitting Diodes (OLEDs): Next-Generation Photodynamic Therapy of Skin Cancer** *315*
		Ifor D.W. Samuel, James Ferguson, and Andrew McNeill
22.1		Introduction *315*
22.2		PDT of Skin Cancer: Current Approach *315*
22.3		Advances in Light Sources: Organic Light-Emitting Diodes *316*
22.4		Transforming PDT with Wearable Light Sources *317*
22.4.1		Developing OLEDs for PDT *317*
22.4.2		Testing OLEDs for PDT *318*
22.4.3		Ambulatory PDT: a New Paradigm *319*
22.4.4		Commercialization of Ambulatory PDT *319*
22.5		Conclusion *320*
		References *320*
23		**Nanoparticles for Photodynamic Therapy** *321*
		Wen-Tyng Li
23.1		Introduction *321*
23.2		Active Nanoparticles for PDT *323*
23.3		Nanoparticles as Photosensitizer Carriers for PDT *326*
23.3.1		Biodegradable Nanoparticles for PDT *326*
23.3.2		Nonbiodegradable Nanoparticles for PDT *330*
23.4		Conclusion *332*
		References *333*
24		**Optical Coherence Tomographic Monitoring of Surgery in Oncology** *337*
		Elena V. Zagaynova, N.D. Gladkova, N.M. Shakhova, O.S. Streltsova, I.A. Kuznetsova, I.A. Yanvareva, L.B. Snopova, E.E. Yunusova, E.B. Kiseleva, V.M. Gelikonov, G.V. Gelikonov, and A.M. Sergeev
24.1		Introduction *337*
24.2		OCT Technique *339*
24.3		Urology *342*
24.3.1		Accuracy of OCT in Endoscopic Detection of Early Cancer *343*
24.3.2		Combined Use of OCT and Fluorescence Imaging for Cancer Detection *345*
24.3.3		OCT Guided Surgery in Bladder Cancer (Open Surgery and Transurethral Resections) *346*
24.3.3.1		Transurethral Resections *346*
24.3.3.2		Open Surgery *348*
24.4		Gastroenterology *348*
24.4.1		OCT for Endoscopic Monitoring of Barrett's Esophagus Malignancy *348*
24.4.1.1		Differentiation Between Normal Esophagus and Benign BE *350*
24.4.1.2		Differentiation Between HGD and Benign BE *351*

24.4.1.3	Differentiation Between IMAC and Benign BE 352
24.4.1.4	Differentiation Between Invasive Carcinoma and Benign BE 352
24.4.1.5	Malignant BE Versus Benign BE 353
24.4.2	OCT for Surgical Guidance of Colon Polyps 355
24.4.3	OCT in Guided Surgery of Gastrointestinal Cancer (Esophagectomy and Rectoectomy) 357
24.5	Gynecology 358
24.5.1	Methodology and Patient Selection 358
24.5.2	Guidance Biopsy During Colposcopy 359
24.6	Cross-Polarization OCT in Oncology 363
24.6.1	CP-OCT in Oncogynecology 364
24.6.2	CP-OCT in Oncourology 366
24.6.3	CP-OCT in Oncogastroenterology 369
24.7	Conclusion 370
	References 371

25	**Intraoperative Optical Coherence Tomographic Monitoring** 377
	Alexandre Douplik
25.1	Optical coherence tomography as a Surgery-Assisting Tool 377
25.2	OCT Modalities and Implementations for Intraoperative Assistance 378
25.3	Advantages, Limitations, and Reproducibility of Intraoperative OCT Monitoring 379
25.4	Specificity and Sensitivity of OCT for Recognizing Normal and Pathologic Tissue 379
25.5	Normal Tissue Differentiation with OCT 380
25.6	OCT Biopsy Guidance 381
25.7	Identifying Malignancy 381
25.8	Identifying Atherosclerotic Plaque Composition 381
25.9	Identifying Necrotic Tissue 382
25.10	Resection Delineation 382
25.11	Particular Intraoperative OCT Applications 383
25.12	Intraoperative OCT Monitoring in Ophthalmology 383
25.13	Intraoperative OCT Monitoring in Cardiovascular Surgery 385
25.14	Intraoperative OCT Monitoring in Dermatology 386
25.15	Intraoperative OCT Monitoring in Gastroenterology 387
25.16	Intraoperative OCT Monitoring in Urology 387
25.17	Intraoperative OCT Monitoring in Bone and Joint Surgery 388
25.18	Intraoperative OCT Monitoring in Neurology 389
25.19	Intraoperative OCT Monitoring in Breast Surgery 390
25.20	Intraoperative OCT Monitoring in ENT 391
25.21	Intraoperative OCT Monitoring in Gynecology 392
25.22	Prediction of Future Developments in Intraoperative OCT Monitoring 393
	References 393

Part 4 Cardiology, Angiology 401

26 Microvascular Blood Flow: Microcirculation Imaging 403
Martin J. Leahy
26.1 Introduction 403
26.2 Nailfold Capillaroscopy 406
26.2.1 Measurements 407
26.2.2 Experimental Considerations 409
26.2.3 Limitations and Improved Analysis 409
26.2.4 Conclusion 410
26.3 Laser Doppler Perfusion Monitoring 410
26.3.1 Light Sources 411
26.3.2 Laser Doppler Perfusion Imaging (LDPI) 411
26.3.3 Point Raster Technique 412
26.3.4 Line Scanning Technique 413
26.3.5 Full-Field Scanning Technique 414
26.3.6 Applications 414
26.4 Laser Speckle Perfusion Imaging 415
26.4.1 Full-Field Measurements 416
26.4.2 Limitations and Improved Analysis 417
26.4.3 Microcirculation 418
26.4.4 Agreement with Laser Doppler 420
26.5 Polarization Spectroscopy 421
26.5.1 Current Technology 422
26.5.2 Image and Data Processing 423
26.6 Photoacoustic Tomography 426
26.7 Optical Microangiography (OMAG) 427
26.8 Comparison of Planar Microcirculation Imagers TiVi, LSPI, and LDPI 429
26.8.1 Laser Doppler Imaging – Moor Instruments moorLDLS 430
26.8.2 Laser Speckle Imaging – Moor Instruments moorFLPI 431
26.8.3 Polarization Spectroscopy – WheelsBridge TiVi Imager 431
26.8.4 Results 432
26.8.5 Discussion 433
26.8.5.1 Depth of Measurement 436
26.8.6 Conclusions 437
References 438

27 Optical Oximetry 445
Stephen J. Matcher
27.1 Introduction 445
27.2 Co-Oximetry 446
27.3 Pulse Oximetry 448

27.4	Near-Infrared Spectroscopy and Imaging	*449*
27.5	New Approaches	*452*
27.6	Conclusion	*453*
	References	*454*

28 Photonic Therapies *457*
Igor Peshko

28.1	Atrial Fibrillation	*457*
28.1.1	Fiber Photocatheters	*458*
28.1.1.1	What is a Fiber Photocatheter?	*458*
28.1.1.2	Fiber Photocatheter Design	*459*
28.2	All-in-One Endoscope Devices	*460*
	References	*461*

Part 5 Pulmonology *463*

29 Bronchoscopy *465*
Lutz Freitag, Kaid Darwiche, and Dirk Hüttenberger

29.1	Diagnostics	*465*
29.1.1	Cancer Detection	*466*
29.1.2	Filter-Based Techniques for Tumor Detection	*467*
29.1.3	Fluorescence Techniques, Drug Induced Fluorescence	*469*
29.1.4	Fluorescence Techniques, Auto-Fluorescence	*471*
29.1.5	Commercially Available Systems	*474*
29.1.6	Remaining Problems	*477*
29.2	Therapy	*477*
29.2.1	Lasers in Bronchology	*477*
29.2.2	Thermal Laser Applications	*478*
29.2.3	Photodynamic Therapy	*479*
29.3	New Techniques: Fluorescence Microimaging, Optical Coherence Tomography	*486*
	References	*486*

30 Laser Therapy *491*
Anant Mohan

30.1	Diagnostic Applications of Lasers	*491*
30.1.1	Autofluorescence Bronchoscopy (AFB)	*491*
30.2	Fibered Confocal Fluorescence Microscopy (FCFM)	*494*
30.3	Therapeutic Uses of Laser	*495*
30.3.1	Photodynamic Therapy (PDT)	*495*
30.3.2	Laser Therapy	*496*
30.4	Contraindications	*498*
30.5	Laser Bronchoscopy Procedure	*498*

30.5.1	Anesthesia	498
30.5.2	Technique	498
30.6	Complications	499
30.7	Outcome	499
	References	501

Part 6 Urology and Nephrology 503

31 Bladder Biopsy with Optical Coherence Tomography 505
Alexandre Douplik

31.1	Acknowledgments	507
	References	507

32 Advanced Laser Endoscopy in Urology: Laser Prostate Vaporization, Laser Surgery, and Laser Removal of Renal Calculi 509
Ronald Sroka, Michael Seitz, and Markus Bader

32.1	Introduction	509
32.2	High-Power Diode Laser for BPH-Treatment	509
32.2.1	Preclinical Investigations	510
32.2.2	Clinical Laser Application for BPH	511
32.2.3	Summary	512
32.3	Laser Therapy for the Upper Urinary Tract Tumors	512
32.3.1	Laser Interventions	513
32.3.2	Complications	514
32.4	Laser-Assisted Lithotripsy	514
32.4.1	Experience from Preclinical Investigations	514
32.4.2	Laser Lithotripsy from the Clinical Point of View	515
32.5	Conclusion	516
	References	516

Part 7 Gastroenterology 521

33 Barrett's Esophagus and Gastroesophageal Reflux Disease – Diagnosis and Therapy 523
Jason M. Dunn, Steven G. Bown, and Laurence B. Lovat

33.1	Introduction	523
33.2	Diagnosis	524
33.2.1	Spectroscopy	524
33.2.1.1	Scattering (Reflectance) Spectroscopy	524
33.2.1.2	Raman Spectroscopy	527
33.2.1.3	Absorption Spectroscopy	527
33.2.1.4	Fluorescence Spectroscopy	527
33.2.1.5	Summary	529
33.3	Therapy	529

33.3.1	Argon Plasma Coagulation	*529*
33.3.2	Photodynamic Therapy	*529*
33.3.2.1	Photofrin	*530*
33.3.2.2	Levulan	*531*
33.3.2.3	Foscan	*533*
33.3.3	Radiofrequency Ablation	*534*
33.3.3.1	Clinical Studies	*535*
33.3.3.2	Adverse Events	*535*
33.3.4	Subsquamous specialized intestinal metaplasia SSIM	*535*
33.4	Conclusion	*536*
	References	*536*

Part 8 Rheumatology *545*

34 Noninvasive Fluorescence Imaging in Rheumatoid Arthritis – Animal Studies and Clinical Applications *547*
Andreas Wunder and Bernd Ebert

34.1	Introduction	*547*
34.2	Non-Invasive Fluorescence Imaging Techniques and Fluorescent Imaging Agents	*547*
34.3	Experimental Studies in Arthritis Models	*548*
34.3.1	Nonspecific Fluorescent Imaging Agents in Arthritis Models	*549*
34.3.2	Targeted Fluorescent Imaging Agents in Arthritis Models	*551*
34.3.3	Activatable Fluorescent Imaging Agents in Arthritis Models	*553*
34.3.4	Monitoring Cell Migration in Arthritis Models	*554*
34.4	Clinical Applications	*555*
34.5	Critical Overview of Fluorescence Imaging in Arthritis	*555*
	References	*558*

35 Diffuse Optical Tomography of Osteoarthritis *561*
Zhen Yuan and Huabei Jiang

35.1	Motivation	*561*
35.2	Diffuse Optical Tomography Imaging of Osteoarthritis in the Hands	*562*
35.2.1	DOT and Hybrid X-Ray–DOT Systems	*562*
35.2.2	DOT Reconstruction Methods	*564*
35.2.3	Reconstructed Results	*564*
35.2.4	Diffuse Model Versus Transport Model	*566*
35.3	Image-Guided Optical Spectroscopy in Diagnosis of OA	*567*
35.4	Future Directions	*568*
	References	*569*

36 Laser Doppler Imaging *571*
Jamie Turner, Bernat Galarraga, and Faisel Khan

36.1	Development and Use of Laser Doppler Flowmetry	*571*

36.2	Laser Doppler Imaging	572
36.3	Endothelial Dysfunction	574
36.4	LDI and Iontophoresis	574
36.5	Studies Using LDI in RA Patients	575
36.6	Conclusion	576
	References	577

Part 9 Ophthalmology and Optometry 581

37 Ocular Diagnostics and Imaging 583
Michael Larsen

37.1	Ocular Diagnosis, Imaging and Therapy	583
37.2	Outline of Anatomy and Physiology	583
37.3	Basic Optical Instrumentation for Examination of the Eye	586
37.4	Fundus Photography	587
37.5	Fluorescence Imaging	592
37.6	Photocoagulation Therapy	593
37.7	Photodisruption	598
37.8	Photodynamic Therapy	600
37.9	Optical Coherence Tomography	601
37.10	Adaptive Optics Fundus Photography	606
37.11	Refractive Surgery	606
37.12	Intraocular Lens Implantation	609
37.13	Photoablation	609
37.14	Thermal Keratoplasty	610
37.15	Blood Flow Measurement	610
37.16	Perspectives	611
	References	611

38 Glaucoma Diagnostics 613
Marc Töteberg-Harms, Cornelia Hirn, and Jens Funk

38.1	Introduction	613
38.2	Optical Coherence Tomography	614
38.3	Scanning Laser Tomography – Heidelberg Retina Tomography	617
38.4	Scanning Laser Polarimetry – GDx	620
	References	621

39 Early Detection of Cataracts 629
Rafat R. Ansari and Manuel B. Datiles III

39.1	Visual Acuity Testing	629
39.2	Ophthalmologic Clinical Examinations	630
39.2.1	Hand-Held Light Examination	630
39.2.2	Ophthalmoscopy	630
39.2.3	Slit-Lamp Examination	630
39.2.4	Slit-Lamp Photography of the Lens and Grading of Cataracts	632

39.2.5	Modified Slit-Lamp Photography	*632*
39.2.6	Scheimpflug Cameras	*633*
39.2.7	Retroillumination	*633*
39.3	New Methods Under Development	*633*
39.3.1	Optical Coherence Tomography	*634*
39.3.2	Raman Spectroscopy for the Detection of Cataract	*635*
39.3.3	Autofluoresence for the Detection of Cataract	*636*
39.3.4	Wavefront in Cataract Detection	*636*
39.3.5	Magnetic Resonance Imaging and Nuclear Magnetic Resonance Spectroscopy of the Lens	*637*
39.3.6	Dynamic Light Scattering (DLS)	*639*
39.4	Conclusion	*640*
	References	*640*

40	**Diabetic Retinopathy**	*645*
	Adzura Salam, Sebastian Wolf, and Carsten Framme	
40.1	Introduction	*645*
40.2	Diagnosis of DR and DME	*646*
40.3	Spectral-domain OCT in DME	*648*
40.4	How OCT Changes Our View in Management of DR	*653*
40.5	SD-OCT Scans as a Prognostic Features for DME	*654*
40.6	Summary for the Clinician	*657*
	References	*657*

41	**Glaucoma Laser Therapy**	*661*
	Marc Töteberg-Harms, Peter P. Ciechanowski, and Jens Funk	
41.1	Introduction	*661*
41.2	Laser Trabeculoplasty (LTP) and Selective Laser Trabeculoplasty (SLT)	*661*
41.3	Excimer Laser Trabeculotomy (ELT)	*662*
41.4	Cyclophotocoagulation (CPC)	*663*
	References	*663*

42	**Ocular Blood Flow Measurement Methodologies and Clinical Application**	*665*
	Yochai Z. Shoshani, Alon Harris, and Brent A. Siesky	
42.1	Background and Significance	*665*
42.2	Blood flow Methodologies	*665*
42.2.1	Color Doppler Imaging	*665*
42.2.2	Laser Doppler Flowmetry and Scanning Laser Flowmetry	*665*
42.2.3	Retinal Vessel Analyzer	*666*
42.2.4	Ocular Pulse Amplitude and Pulsatile Ocular Blood Flow	*666*
42.2.5	Laser Speckle Method (Laser Speckle Flowgraphy)	*666*
42.2.6	Digital Scanning Laser Ophthalmoscope Angiography	*666*
42.2.7	Doppler Optical Coherence Tomography	*667*

42.2.8	Retinal Oximetry	667
42.2.9	Newly Developed Techniques	668
42.3	Relationship Between Blood Flow and Visual Function	668
42.4	Relationship Between Blood Flow and Optic Nerve Structure	668
	References	670

43 Photodynamic Modulation of Wound Healing in Glaucoma Filtration Surgery 673
Salvatore Grisanti

43.1	Introduction	673
43.2	Photodynamic Therapy	673
43.3	Photodynamic Modulation of Wound Healing in Glaucoma Filtration Surgery	674
43.3.1	Preclinical Studies	674
43.3.2	Clinical Studies	676
	References	681

44 Correction of Wavefront Aberrations of the Eye 685
Dan Reinstein and Manfred Dick

44.1	Introduction	685
44.2	Excimer Laser-Based Refractive Surgery for Correction of Wavefront Aberrations	686
	References	691

Part 10 Otolaryngology (ENT) 693

45 Optical Coherence Tomography of the Human Larynx 695
Marc Rubinstein, Davin Chark, and Brian Wong

45.1	Anatomy	695
45.2	Clinical Problems	695
45.2.1	Laryngeal Cancer	695
45.2.2	Importance of Laryngeal Optical Coherence Tomography	696
45.3	Clinical *In Vivo* Applications	697
45.3.1	OCT During Microsurgical Endoscopy	697
45.3.2	Office-Based OCT	699
45.3.3	OCT Integrated with a Surgical Microscope	700
45.4	Other Preclinical Applications of OCT in the Larynx	700
	References	702

46 Optical Coherence Tomography of the Human Oral Cavity and Oropharynx 705
Marc Rubinstein, Davin Chark, and Brian Wong

46.1	Anatomy	705
46.2	Progression of Disease	705

46.3	Usefulness of OCT *707*	
	References *710*	

Part 11 Neurology *713*

47	***In Vivo* Brain Imaging and Diagnosis** *715*	
	Christoph Krafft and Matthias Kirsch	
47.1	Introduction *715*	
47.2	Basic Principles *719*	
47.2.1	Two-Dimensional Camera-Based Imaging: Absorption *719*	
47.2.2	Two-Dimensional Camera Based Imaging: Fluorescence *720*	
47.2.3	Two-Dimensional Camera-Based Imaging: Laser Speckle *720*	
47.2.4	Two-Dimensional Fiber Bundle-Based Imaging *721*	
47.2.5	Three-Dimensional Imaging: Optical Coherence Tomography *721*	
47.2.6	Confocal Microscopy *722*	
47.2.7	Two-Photon Microscopy *722*	
47.2.8	Second-Harmonic Generation *723*	
47.2.9	Inelastic Light Scattering: Raman Spectroscopy *723*	
47.2.10	Noninvasive Optical Brain Imaging: Topography, Tomography *724*	
47.3	Applications in Neuro-Oncology and Functional Imaging *725*	
47.3.1	Absorption-Based Imaging *726*	
47.3.2	Fluorescence-Based Imaging *727*	
47.3.2.1	5-ALA *727*	
47.3.2.2	Nanoparticles *729*	
47.3.2.3	Quantum Dots *729*	
47.3.3	Bioluminescence Imaging *730*	
47.3.4	OCT-Based Imaging *730*	
47.3.5	Raman-Based Techniques *731*	
47.4	Applications in Neurovascular Pathologies *732*	
47.4.1	Absorption-Based Imaging *733*	
47.4.2	Thermography *734*	
47.5	Applications to Regeneration, Transplantation, and Stem Cell Monitoring *734*	
47.5.1	Identification of Cellular Fractions and Monitoring of Implants in the CNS *734*	
47.6	Conclusions *735*	
	References *737*	
48	**Assessment of Infant Brain Development** *743*	
	Nadege Roche-Labarbe, P. Ellen Grant, and Maria Angela Franceschini	
48.1	Introduction *743*	
48.2	Techniques *744*	
48.2.1	Measurement of Local Oxy- and Deoxyhemoglobin Concentrations *744*	
48.2.2	Local Cerebral Blood Flow Measurement *745*	

48.2.3	Estimates of Cerebral Metabolic Rate of Oxygen	746
48.3	Functional and Cognitive Studies	747
48.4	Baseline Hemodynamic Assessment and Clinical Applications	748
48.5	Conclusion and Perspectives	749
	References	750

49 Revealing the Roles of Prefrontal Cortex in Memory 759
Hui Gong and Qingming Luo

49.1	Introduction	759
49.2	Working Memory Modality	760
49.3	Long-Term Memory Modality	763
	References	765

50 Cerebral Blood Flow and Oxygenation 767
Andrew K. Dunn

50.1	Introduction	767
50.2	Laser Speckle Contrast Imaging of Cerebral Blood Flow	767
50.3	Multispectral Reflectance Imaging of Hemoglobin Oxygenation	769
50.4	Imaging of pO_2 with Phosphorescence Quenching	770
	References	772

Part 12 Dermatology 775

51 Skin Diagnostics 777
Martin Kaatz, Susaane Lange-Asschenfeldt, Joachim Fluhr, and Johannes Martin Köhler

51.1	Introduction	777
51.2	Dermoscopy	777
51.2.1	Technical Principle and Development	777
51.2.2	Applications	778
51.2.2.1	Pigmented Skin Lesions	778
51.2.2.2	Non-Pigmented Skin Lesions	779
51.2.3	Conclusion	780
51.3	Optical Coherence Tomography (OCT)	780
51.3.1	Technical Principle and Development	780
51.3.2	Applications	781
51.3.2.1	Skin Tumors	781
51.3.2.2	Inflammatory Skin Diseases	781
51.3.2.3	Wound Healing	782
51.3.2.4	Bullous Skin Disorders	783
51.3.2.5	Other	783
51.3.3	Conclusion	783
51.4	Laser Doppler Imaging	784
51.4.1	Technical Principle and Development	784
51.4.2	Applications	784

51.4.2.1	Skin Tumors	*784*
51.4.2.2	Inflammatory Skin Diseases	*784*
51.4.2.3	Microcirculation of Chronic Wounds	*784*
51.4.2.4	Collagenosis and Rheumatic Diseases	*785*
51.4.3	Conclusion	*786*
51.5	Confocal Laser Scanning Microscopy (CLSM)	*786*
51.5.1	Technical Principle and Development	*786*
51.5.2	Applications	*787*
51.5.2.1	Topography of Normal Skin	*787*
51.5.2.2	Skin Tumors	*787*
51.5.2.3	Vascular Skin Lesions	*788*
51.5.2.4	Inflammatory Skin Diseases	*788*
51.5.2.5	Infections	*789*
51.5.2.6	Drug Delivery	*789*
51.5.3	Conclusion	*789*
51.6	Multiphoton Laser Imaging	*790*
51.6.1	Technical Principle and Development	*790*
51.6.2	Applications	*791*
51.6.2.1	Topography of Normal Skin	*791*
51.6.2.2	Skin Tumors	*791*
51.6.2.3	Skin Aging	*791*
51.6.2.4	Drug Delivery	*794*
51.6.3	Conclusion	*795*
51.7	Raman Spectroscopy	*795*
51.7.1	Technical Principle and Development	*795*
51.7.2	Applications	*795*
51.7.2.1	Molecular Structure of Normal Skin	*795*
51.7.2.2	Inflammatory Skin Diseases	*796*
51.7.3	Conclusion	*798*
	References	*799*

52	**Non-Melanoma Skin Cancer**	*805*
	Martin Kaatz, Susanne Lange-Asschenfeldt, Martin Johannes Koehler, and Uta Christina Hipler	
52.1	Introduction	*805*
52.2	Dermoscopy	*806*
52.2.1	Basal Cell Carcinoma	*806*
52.3	Optical Coherence Tomography (OCT)	*806*
52.3.1	Basal Cell Carcinoma	*806*
52.3.2	Actinic Keratosis and Squamous Cell Carcinoma	*807*
52.4	Confocal Laser Scanning Microscopy (CLSM)	*808*
52.4.1	Basal Cell Carcinoma	*808*
52.4.2	Actinic Keratoses and Squamous Cell Carcinoma	*809*
52.5	Multiphoton Tomography (MPT)	*810*
52.5.1	Basal Cell Carcinoma	*810*

52.5.2	Actinic Keratoses and Squamous Cell Carcinomas *811*
52.6	Laser Doppler Imaging *812*
52.6.1	Basal Cell Carcinoma *812*
52.7	Raman Spectroscopy *813*
52.7.1	Basal Cell Carcinoma *813*
52.7.2	Actinic Keratoses and Squamous Cell Carcinomas *814*
	References *814*

53 Pigmented Skin Lesions *817*
Martin Kaatz, Susanne Lange-Asschenfeldt, Peter-Elsner, and Sindy Zimmermann

53.1	Introduction *817*
53.2	Dermoscopy *817*
53.3	Optical Coherence Tomography (OCT) *819*
53.4	Confocal Laser Scanning Microscopy (CLSM) *821*
53.5	Multiphoton Laser Tomography (MPT) *822*
53.6	Raman Spectroscopy *824*
53.7	Laser Doppler Imaging *824*
	References *825*

54 Monitoring of Blood Flow and Hemoglobin Oxygenation *827*
Sean J. Kirkpatrick, Donald D. Duncan, and Jessica Ramella-Roman

54.1	Introduction *827*
54.2	Laser Doppler Approaches to Quantifying Cutaneous Blood Flow *828*
54.3	Laser Speckle Methods for Blood Flow Estimations *829*
54.4	Oxygenation *830*
	References *831*

55 Sensing Glucose and Other Metabolites in Skin *835*
Elina A. Genina, Kirill V. Larin, Alexey N. Bashkatov, and Valery V. Tuchin

55.1	Introduction *835*
55.2	Light–Tissue Interaction *836*
55.3	Fluorescence Measurements *837*
55.4	IR Spectroscopy *839*
55.5	Photoacoustic Technique *841*
55.6	Raman Spectroscopy *842*
55.7	Occlusion Spectroscopy *843*
55.8	Reflectance Spectroscopy *844*
55.9	Polarimetric Technique *845*
55.10	OCT Technique *846*
55.11	Conclusion *847*
	References *849*

56 Psoriasis and Acne 855
Bernhard Ortel and Piergiacomo Calzavara-Pinton

56.1 Psoriasis 855
56.1.1 General Principles of Phototherapies 855
56.1.2 Ultraviolet B Phototherapy 856
56.1.3 Photochemotherapy 858
56.1.4 Photodynamic Therapy 859
56.1.5 Laser Therapy 860
56.1.6 Mechanisms of Action 861
56.1.7 Adverse Effects 861
56.1.8 Summary 862
56.2 Acne 862
56.2.1 General Principles of Phototherapies 863
56.2.2 Ultraviolet Phototherapy and PUVA 863
56.2.3 Visible Light Phototherapy 864
56.2.4 Photodynamic Therapy 864
56.2.5 Mechanisms and Adverse Effects 865
56.2.6 Summary 865
References 866

57 Wound Healing 869
Bernard Choi

57.1 Introduction 869
57.2 Wound-Healing Response to Selective Optical Injury to the Microvasculature 869
57.3 Recent Applications of Optical-Based Methods in Wound Healing 872
57.3.1 Low-Level Laser Therapy (LLLT) – Wound Healing 872
57.3.2 LLLT to Aid Cerebral Nervous System Repair Processes 872
57.3.3 Wound Healing after Laser Cartilage Reshaping 873
57.3.4 Novel Diagnostic Applications of Biophotonics to Study Wound Healing 873
57.4 Summary 874
References 874

Part 13 Gynecology and Obstetrics 877

58 Diagnosis of Neoplastic Processes in the Uterine Cervix 879
Natalia Shakhova, Irina Kuznetsova, Ekaterina Yunusova, and Elena Kiseleva

58.1 Introduction 879
58.2 Materials and Methods 880
58.3 Results 880
58.4 Discussion and Conclusion 882
References 883

Part 14 Reproductive Medicine 885

59 **The Use of Single-Point and Ring-Shaped Laser Traps to Study Sperm Motility and Energetics** 887
Linda Z. Shi, Bing Shao, Jaclyn Nascimento, and Michael W. Berns
59.1 Introduction 887
59.2 Single-Point Laser Trap 887
59.3 Ring-Shaped Laser Traps 889
59.4 Biological Studies 890
59.4.1 Sperm Competition 890
59.4.2 Sperm Energetics 891
59.5 Conclusion 892
References 892

60 **Laser Thinning of the Zona Pellucida** 895
Hanna Balakier
60.1 The Zona Pellucida 895
60.2 Laser Assisted Hatching 896
60.3 Laser-Assisted ICSI 897
60.4 Safety Aspects of ZPT 897
References 898

Part 15 Genetics 901

61 **Transfection of Cardiac Cells by Means of Laser-Assisted Optoporation** 903
Alena V. Nikolskaya, Vladimir P. Nikolski, and Igor R. Efimov
61.1 Introduction 903
61.2 Methods 903
61.2.1 Primary Cultures of Neonatal Rat Cardiac Cells 903
61.3 Experimental Design 904
61.4 Results 905
61.5 Conclusion 908
References 908

Part 16 Laser Surgery 911

62 **Laser Surgery: an Overview** 913
Roberto Pini, Francesca Rossi, and Fulvio Ratto
62.1 Introduction 913
62.2 Laser–Tissue Interactions 914
62.2.1 Photothermal Effects 918
62.2.1.1 Low-Temperature Effects (43–100 °C) 919
62.2.1.2 Medium- and High-Temperature Effects ($\geq$100 °C) 920
62.2.1.3 Controlling Heating Effects 921

62.2.2	Photomechanical Effects	*923*
62.2.2.1	Photoablation with Nanosecond Pulses	*924*
62.2.2.2	Photodisruption	*927*
62.2.2.3	Photoablation with Femtosecond Pulses	*928*
62.3	Notes on Laser Systems Suitable for Surgical Applications	*930*
	References	*935*

63 Endovenous Laser Therapy – Application Studies, Clinical Update, and Innovative Developments *939*
Ronald Sroka and Claus-Georg Schmedt

63.1	Introduction	*939*
63.2	Clinical Application of ELT	*940*
63.3	Clinical Studies on ELT	*942*
63.4	Analysis of Comparison Studies	*943*
63.5	Investigations to Improve Light Application	*945*
63.5.1	Experiments	*945*
63.5.2	Experimental Results	*947*
63.6	Discussion	*949*
63.7	Conclusion	*950*
	References	*951*

64 Laser Treatment of Cerebral Ischemia *955*
Ying-Ying Huang, Vida J. Bil De Arce, Luis De Taboada, Thomas McCarthy, and Michael R. Hamblin

64.1	Introduction	*955*
64.2	The Problem of Cerebral Ischemia	*955*
64.3	Mechanisms of Brain Injury After Cerebral Ischemia	*956*
64.4	Current Treatment for Cerebral Ischemia	*956*
64.5	Investigational Neuroprotectants and Pharmacologic Intervention	*956*
64.6	Mechanism of Low-Level Laser Therapy	*957*
64.7	LLLT on Neuronal Cells	*958*
64.8	LLLT for Stroke in Animal Models	*958*
64.9	TLT in Clinical Trials for Stroke	*960*
64.10	Conclusions and Future Outlook	*960*
	References	*961*

65 Laser Vision Correction *963*
Diogo L. Caldas and Renato Ambrósio Jr.

65.1	Introduction	*963*
65.2	Current Lasers Used in Refractive Surgery	*964*
65.2.1	The Excimer Laser	*964*
65.2.2	The Femtosecond Laser	*964*

65.3	Refractive Surgery Techniques	966
65.3.1	Surface Ablation	967
65.3.1.1	PRK	968
65.3.1.2	Transepithelial PRK	968
65.3.1.3	LASEK	969
65.3.1.4	EpiLASIK	969
65.3.2	LASIK	970
65.3.2.1	Mechanical microkeratome LASIK	971
65.3.2.2	Femtosecond laser LASIK	971
65.4	Custom Corneal Ablations	972
	References	972

66 Laser Trabeculoplasty 975
Stephan Eckert

66.1	Laser Systems	975
66.2	Mechanisms of Action	975
66.3	Methods of Treatment	976
66.4	Indications for Laser Trabeculoplasty	978
66.5	Contraindications for Laser Trabeculoplasty	979
66.6	Effectiveness of Laser Trabeculoplasty	979
66.7	Complications After Laser Trabeculoplasty	980
66.8	Conclusion	980
	References	980

67 Laser Photocoagulation 983
Bettina Fuisting and Gisbent Richard

67.1	Introduction	983
67.2	Retinal Degeneration	984
67.2.1	Retinal Tears, Retinoschisis, Retinal Detachment	984
67.3	Diabetic Retinopathy (DRP)	985
67.3.1	Diabetic Macular Edema (DME)	985
67.3.2	Proliferative Diabetic Retinopathy	986
67.4	Retinal Vein Occlusions	987
67.4.1	Central Retinal Vein Occlusion (CRVO)	987
67.4.2	Branch Retinal Vein Occlusion (BRVO)	987
67.5	Macular Diseases and Choroidal Neovascularization	988
67.5.1	Age-Related Macular Degeneration (ARMD)	988
67.5.2	Transpupillary Thermotherapy (TTT) for the Treatment of CNV	988
67.5.3	Photodynamic Therapy (PDT)	989
67.5.4	Central Serous Chorioretinopathy (CSC)	989
67.6	Choroidal Melanoma	990
	References	991

Part 17 Dentistry 993

68 Diagnostic Imaging and Spectroscopy 995
Daniel Fried
68.1 Caries Detection 995
68.2 Optical Properties of Dental Hard Tissue 996
68.2.1 Optical Properties of Dental Hard Tissue in the Visible and Near-Infrared Regions 996
68.2.2 Optical Properties of Sound Enamel and Dentin 996
68.2.3 Optical Property Changes of Demineralized Enamel 998
68.2.4 Optical Properties of Dental Hard Tissue in the IR region 1000
68.3 Reflectance Spectroscopy 1002
68.4 Optical Transillumination 1002
68.5 Optical Coherence Tomography 1004
68.6 Fluorescence Imaging 1010
68.6.1 Quantitative Light Fluorescence (QLF) 1010
68.6.2 DIAGNOdent (Porphyrin Fluorescence) 1012
68.6.3 Other Fluorescence-Based Imaging Methods 1013
68.7 Thermal Imaging 1013
68.8 Infrared Spectroscopy 1015
68.9 Raman Spectroscopy and Imaging 1016
68.10 Ultrasound and Terahertz Imaging 1018
References 1019

69 Optical Coherence Tomography in Dentistry 1029
Natalia D. Gladkova, Yulia.V. Fomina, Elena B. Kiseleva, Maria M. Karabut, Natalia S. Robakidze, Alexander A. Muraev, Stefka G. Radenska-Lopovok, and Anna V. Maslennikova
69.1 Introduction 1029
69.2 Hard Tooth Tissues 1030
69.3 Periodontal Tissues 1030
69.4 Oral Cavity Mucosa 1032
69.5 OCT Images of Labial Salivary Glands in Norm and Pathology 1034
References 1037

70 Fluorescence Spectroscopy 1041
Adrian Lussi
70.1 Introduction 1041
70.2 Laser/Light-Induced Fluorescence 1041
70.3 DIAGNOdent 1042
70.3.1 Quantitative Light-Induced Fluorescence (QLF) 1043
References 1045

71	**Photothermal Radiometry and Modulated Luminescence: Applications to Dental Caries Detection** *1047*	
	Jose A. Garcia, Andreas Mandelis, Stephen H. Abrams, and Anna Matvienko	
71.1	Introduction *1047*	
71.2	Theoretical Modeling *1048*	
71.3	Depth Profilometry *1048*	
71.4	Pits and Fissures Caries Detection *1049*	
71.5	Interproximal Caries Detection *1049*	
71.6	Early Lesion Detection *1050*	
71.7	Recent Developments *1050*	
71.8	Conclusion *1051*	
	References *1051*	
72	**Lasers in Restorative Dentistry** *1053*	
	Anil Kishen	
72.1	Introduction *1053*	
72.2	Classification of Lasers in Restorative Dentistry *1054*	
72.3	Lasers and Delivery Systems *1055*	
72.4	Laser–Dental Tissue Interaction *1056*	
72.5	Laser Effects on Dental Tissues *1059*	
72.5.1	Laser Effect on Enamel *1059*	
72.5.2	Laser Effects on Dentin *1060*	
72.5.3	Laser Effects on Dental Pulp *1061*	
72.6	Lasers in Operative Dentistry *1062*	
72.6.1	Laser-Assisted Tooth Bleaching *1062*	
72.6.2	Laser-Assisted Cavity Preparation and Caries Removal *1063*	
72.6.3	Laser-Assisted Adhesion in Tooth Color Restoration *1065*	
72.6.4	Laser-Activated Polymerization of Composite Resin *1065*	
72.6.5	Laser-Assisted Removal of Restorative Materials and Metal Dowels *1066*	
72.6.6	Laser-Based Management of Dentin Hypersensitivity *1066*	
72.7	Application of Lasers in Endodontics *1067*	
72.7.1	Application of Lasers to Diagnose Dental Pulp Health *1067*	
72.7.2	Laser-Assisted Pulp Capping and Pulpotomy *1068*	
72.7.3	Laser-Assisted Root Canal Disinfection and Shaping *1070*	
72.7.4	Laser-Assisted Root Canal Obturation *1072*	
72.7.5	Laser-Assisted Root Canal Retreatment *1073*	
72.7.6	Laser-Assisted Apical Surgery *1074*	
72.8	Lasers in Periodontal Therapy *1075*	
72.8.1	Laser-Assisted Periodontal Soft Tissue Management *1075*	
72.8.2	Laser-Assisted Management of Periodontal Pocket *1076*	
72.8.3	Bactericidal Effect of Lasers on Periodontal Pocket *1077*	
72.8.4	Laser-Assisted Calculus Removal *1077*	
72.8.5	Laser-Assisted Management of Oral Hard Tissues *1077*	
72.8.6	Laser-Assisted Management of Dental Hard Tissue *1078*

72.9	Conclusion *1079*	
	References *1079*	
73	**Laser Ablation of Dental Restorative Materials** *1085*	
	Ernst Wintner and Verena Wieger	
73.1	Introduction *1085*	
73.2	Dental Restorative Materials *1086*	
73.3	Ablation by Erbium Laser *1086*	
73.3.1	Results for Dentin and Enamel *1087*	
73.3.2	Results for Composites *1089*	
73.4	Ablation by Ultra-Short Pulse Laser *1090*	
73.5	Conclusion *1091*	
	References *1092*	
74	**Laser Ablation of Hard Tissues** *1095*	
	Gregory B. Altshuler, Andrey V. Belikov, and Felix I. Feldchtein	
74.1	Introduction *1095*	
74.2	Tooth Structure *1095*	
74.3	Laser Ablation *1096*	
74.4	Conclusion *1105*	
	References *1108*	
	Index *1111*	

Preface

Biophotonics as part of photonics has proved to be an important enabling technology for accelerated progress in medicine and biotechnology. It can do so, because it originated at the interface of the most innovative academic disciplines of the last century, i.e. photonics, biotechnology and nanotechnologies. This new field of research brings together scientists from different professions, such as physicists, biologists, pharmacists, medical doctors etc. To work on common projects, they need to understand the ideas and the techniques of their counterparts - physicians have to learn about laser and microscopy, physicists have to learn about metabolism and cell structures etc. So far, they have research papers or text books from their own subject written for the students and scientists of their limited field of research. Cross disciplinary information is rare but urgently needed. To solve this task, the "Handbook of Biophotonics" is divided into three volumes: the first volume will offer introductory chapters to the fundamental aspects in the different fields of research i. e. photonics and biology; the subsequent two volumes will focus on applications and techniques in various fields of health care, industry and research. These application-related volumes will follow a new application-related approach, starting with the application in particular the different medical disciplines and then citing the various photonic tools to solve the scientific task. This approach improves the value of the handbook for all readers, in particular medical doctors.

Building up on the technical and methodical biophotonic background presented in the first volume, the starting points of volume two are the different medical specialties, which, at the same time, determine the overarching structure of the volume. It is the goal of this volume to provide an overview of the medical applications of biophotonics techniques and methods in an as comprehensive way as this is possible, given the fast development of biophotonics and the fact that a single book, despite containing more than 70 contributions, can never incorporate everything that is already mature or starts out developing in the field.

In the case of some of the medical specialties, like e.g. ophthalmology or oncology, the application of biophotonic technologies and methods is obviously better established or more advanced than in others. Accordingly, larger overview articles are presented, often flanked by shorter contributions. Other specialties are covered only by a single overview article or a short contribution, often reflecting the present stage of development.

We want to express our sincere gratitude to all of the more than 200 colleagues who have been occupied in authoring this multitude of single chapters of volume two and thereby its content. Consequently, a tremendous amount of work was put on the shoulders of the team of Wiley Publishers, who never lost their confidence and patience for which we thank them most gratefully. We would also like to thank all the people who further contributed to this volume and the many helping and supporting hands that were involved in the different stages of putting it together.

Last but not least we want to express a special word of thanks to our editorial assistant Dr. Marion Strehle and to PD Dr. Thomas Mayerhöfer for their strong support and their help in getting this volume organized.

Jürgen Popp

List of Contributors

Stephen H. Abrams
University of Toronto
Center for Advanced
Diffusion Wave Technologies
Department of Mechanical and
Industrial Engineering
5 King's College Road
Toronto, ON M5S 3G8
Canada

Gregory B. Altshuler
Dental Photonics, Inc.
1600 Boston-Providence Highway
Walpole, MA 02081
USA

Renato Ambrósio Jr.
Instituto de Olhos Renato Ambrósio
Oftalmologia, Rio de Janeiro Corneal
Tomography and Biomechanics Study
Group
Rua Conde de Bonfim 211/712
Tijuca, Rio de Janeiro, RJ 20520-050
Brazil

Stefan Andersson-Engels
Lund University
Department of Physics
Solvegatan 14A
221 00 Lund
Sweden

Rafat R. Ansari
National Aeronautics and Space
Administration (NASA)
John H. Glenn Research Center
at Lewis Field
21000 Brookpark Road
Cleveland, OH 44135
USA

Susaane Lange-Asschenfeldt
Department of Dermatology
Venerology and Allergology
Charite University Medicine Berlin
Chariteplatz 1
10117 Berlin
Germany

Markus Bader
Ludwig Maximilians University Munich
Department of Urology
Marchioninists 15
81377 Munich
Germany

Hanna Balakier
CReATe
Suite 1100, 790 Bay Street
Toronto, ON M5G 1N8
Canada

Alexey N. Bashkatov
Saratov State University
Research-Educational Institute
of Optics and Biophotonics
83 Astrakhanskaya Street
410012 Saratov
Russia

Kristi Bedrossian
Tufts Medical Center
Department of Pathology
800 Washington Street
Boston, MA 02111
USA

Andrey V. Belikov
Saint Petersburg State University of
Information Technologies, Mechanics
and Optics
Sablinskaya Street 14
197101 Saint Petersburg
Russia

Wladimir A. Benalcazar
University of Illinois at Urbana-
Champaign
Beckman Institute for Advanced
Science and Technology
405 North Mathews Avenue
Urbana, IL 61801
USA

Klaus Bendrat
University Medical Center
Hamburg-Eppendorf
Martinistrasse 52
20246 Hamburg
Germany

Niels Bendsoe
Lund University
Department of Dermatology
Barngatan 2B
221 00 Lund
Sweden

Michael W. Berns
University of California at San Diego
Department of Bioengineering and
Department of Electrical and Computer
Engineering
9500 Gilman Drive
La Jolla, CA 92093
USA
and
University of California at Irvine
Beckman Laser Institute
Department of Biomedical Engineering
1002 Health Sciences Road
Irvine, CA 92612
USA

Vida J. Bil De Arce
Massachusetts General Hospital
Wellman Center for Photomedicine
40 Blossom Street
Boston, MA 02114
USA

Benjamin Bird
Northeastern University
Department of Chemistry
and Chemical Biology
Laboratory for Spectral Diagnosis
360 Huntington Avenue
Boston, MA 02115
USA

Stephen A. Boppart
University of Illinois at
Urbana-Champaign
Beckman Institute for Advanced
Science and Technology
405 North Mathews Avenue
Urbana, IL 61801
USA

Anja Boßecker
Friedrich Schiller University Jena
Institute of Physical Chemistry
Helmholtzweg 4
07743 Jena
Germany

Steven G. Bown
University College London
National Medical Laser Centre
Charles Bell House
67-73 Riding House Street
London W1W 7EJ
UK

Sylvia Brockmöller
Gemeinschaftspraxis für Pathologie
Lornsenstra. 4
22767 Hamburg
Germany

Thomas Bruns
Hochschule Aalen
Institut für Angewandte Forschung
Beethovenstr. 1
73428 Aalen
Germany

Diogo L. Caldas
Instituto de Olhos Renato Ambrósio
Oftalmologia, Rio de Janeiro Corneal
Tomography and Biomechanics Study
Group
Rua Conde de Bonfim 211/712
Tijuca, Rio de Janeiro, RJ 20520-050
Brazil

Piergiacomo Calzavara-Pinton
University of Brescia
Department of Dermatology
Piazzale Spedale Civile
25123 Brescia
Italy

Brent D. Cameron
University of Toledo
Department of Bioengineering
5030 Nitschke Hall
Toledo, OH 43606
USA

Davin Chark
Beckman Laser Institute
1002 Health Sciences Road
Irvine, CA 92612
USA

Bernard Choi
Beckman Laser Institute
Surgery and Biomedical Engineering
1002 Health Sciences Road
Irvine, CA 92697
USA

Peter P. Ciechanowski
University Hospital Zurich
Division of Ophthalmology
Frauenklinikstr. 24
8091 Zurich
Switzerland

Valerian Ciobotă
Friedrich Schiller University Jena
Institute of Physical Chemistry
Helmholtzweg 4
07743 Jena
Germany

List of Contributors

Sergio Coda
Imperial College London
Department of Physics
Photonics Group
South Kensington Campus
606 Blackett Building
London SW7 2AZ
UK
and
Imperial College London
Department of Medicine
Division of Experimental Medicine
Hammersmith Campus
Commonwealth Building
Du Cane Road
London W12 0NN
UK

Kaid Darwiche
Lungenklinik Hemer
Pneumologie
Theo-Funccius-Str. 1
58675 Hemer
Germany

Manuel B. Datiles III
National Institutes of Health (NIH)
National Eye Institute
31 Center Drive
Bethesda, MD 20892
USA

Annegrct Debus
Gemeinschaftspraxis für Pathologie
Lornsenstr. 4
22767 Hamburg
Germany

Luis De Taboada
PhotoThera Inc.
5925 Priestly Drive, Suite 120
Carlsbad, CA 92008
USA

Manfred Dick
London Vision Clinic
138 Harley Street
London W1G 7LA
UK

Max Diem
Northeastern University
Department of Chemistry
and Chemical Biology
Laboratory for Spectral Diagnosis
360 Huntington Avenue
Boston, MA 02115
USA

Alexandre Douplik
Friedrich-Alexander University
Erlangen-Nuremberg
Chair of Photonics Technologies
Medical Photonics Engineering Group
Paul-Gordan-Stra. 3
91052 Erlangen
Germany
and
Friedrich-Alexander University
Erlangen-Nuremberg
Erlangen Graduate School in Advanced
Optical Technologies (SAOT)
Clinical Photonics Laboratory
Paul-Gordan-Str. 6
91052 Erlangen
Germany

Axel Duerkop
University of Regensburg
Institute of Analytical Chemistry,
Chemo- and Biosensors
Universitätsstr. 31
93040 Regensburg
Germany

Donald D. Duncan
Portland State University
Department of Electrical
and Computer Engineering
1900 SW 4th Avenue
Portland, OR 97201
USA

Andrew K. Dunn
University of Texas at Austin
Institute for Cellular and Molecular
Biology
Department of Biomedical Engineering
1 University Station
Austin, TX 78712
USA

Jason M. Dunn
University College London
National Medical Laser Centre
Charles Bell House
67-73 Riding House Street
London W1W 7EJ
UK

Christopher Dunsby
Imperial College London
Department of Physics
Photonics Group
South Kensington Campus
606 Blackett Building
London SW7 2AZ
UK
and
Imperial College London
Department of Medicine
Division of Experimental Medicine
Hammersmith Campus
Commonwealth Building
Du Cane Road
London W12 0NN
UK

Bernd Ebert
Physikalisch-Technische
Bundesanstalt (PTB)
Department of Biomedical Optics
Abbestr. 2-12
10587 Berlin
Germany

Stephan Eckert
Universitätsklinik Ulm
Augenklinik
Prittwitzstr. 43
89075 Ulm
Germany

Igor R. Efimov
Washington University
Department of Biomedical Engineering
One Brookings Drive
St. Louis, MO 63130-4899
USA

Peter-Elsner
Department of Dermatology
University Hospital Jena
Erfurter Str. 35
07743 Jena
Germany

Felix I. Feldchtein
Dental Photonics, Inc.
1600 Boston-Providence Highway
Walpole, MA 02081
USA

James Ferguson
Ambicare Health Ltd.
Kinburn Castle, Doubledykes Road
St Andrews KY16 9DR
UK
and
University of Dundee
Ninewells Hospital and Medical School
Photobiology Unit
Ninewells Avenue
Dundee DD1 9SY
UK

Joachim Fluhr
Department of Dermatology
Venerology and Allergology
Charite University Medicine Berlin
Chariteplatz 1
10117 Berlin
Germany

Yulia .V. Fomina
Nizhny Novgorod State Medical Academy
10/1 Minina Square
603005 Nizhny Novgorod
Russia

Carsten Framme
University of Bern
Inselspital, Universitätsklinik für Augenheilkunde
Freiburgstr.
3010 Bern
Switzerland

Maria Angela Franceschini
Athinoula A. Martinas
Center for Biomedical Imaging
Massachusetts
General Hospital
Boston, MA 0244
USA

Lutz Freitag
Lungenklinik Hemer
Pneumologie
Theo-Funccius-Str. 1
58675 Hemer
Germany

Paul M.W. French
Imperial College London
Department of Physics
Photonics Group
South Kensington Campus
606 Blackett Building
London SW7 2AZ
UK

Daniel Fried
University of California
Department of Preventive and Restorative Dental Sciences
Dentistry 2234
707 Parnassus Avenue
San Francisco, CA 94143
USA

Bettina Fuisting
University Medical Center Hamburg-Eppendorf
Department of Ophthalmology
Martinistr. 52
20246 Hamburg
Germany

Jens Funk
University Hospital Zurich
Division of Opthalmology
Frauenklinikstr. 24
8091 Zurich
Switzerland

Bernat Galarraga
University of Dundee
School of Medicine
Ninewells Hospital and Medical School
Ninewells Avenue
Dundee DD1 9SY
UK

Jose A. Garcia
University of Toronto
Center for Advanced
Diffusion Wave Technologies
Department of Mechanical and
Industrial Engineering
5 King's College Road
Toronto, ON M5S 3G8
Canada

Grigory V. Gelikonov
Institute of Applied Physics
Russian Academy of Sciences
Division of Nonlinear Dynamics
and Optics
46 Ul'yanov Street
603950 Nizhny Novgorod
Russia

Valentin M. Gelikonov
Institute of Applied Physics
Russian Academy of Sciences
Division of Nonlinear Dynamics
and Optics
46 Ul'yanov Street
603950 Nizhny Novgorod
Russia

Elina A. Genina
Saratov State University
Research-Educational Institute of Optics
and Biophotonics
83 Astrakhanskaya Street
410012 Saratov
Russia

Natalia D. Gladkova
Nizhny Novgorod State
Medical Academy
Institute of Applied and
Fundamental Medicine
10/1 Minina Square
603005 Nizhny Novgorod
Russia

Martin Gluch
Carl Zeiss MicroImaging GmbH
Market Intelligence
Carl-Zeiss-Promenade 10
07745 Jena
Germany

Hui Gong
Huazhong University
of Science and Technology
Wuhan National Laboratory
for Optoelectronics
Britton Chance Center
for Biomedical Photonics
Wuhan 430074
China

P. Ellen Grant
Athinoula A. Martinas
Center for Biomedical Imaging
Massachusetts
General Hospital
Boston, MA 0244
USA
and
Center for Fetal and Neonatal
Neuroradiology
Childeren's Hospital Boston
300 Longwood Avenue
Boston, MA 0245
USA

Salvatore Grisanti
University of Lübeck
Department of Ophthalmology
Ratzeburger Allee 160
23538 Lübeck
Germany

Gernoth Grunst
Fraunhofer Institute for
Applied Information Technology FIT
Schloss Birlinghoven
53754 Sankt Augustin
Germany

Jochen Guck
University of Cambridge
Department of Physics
Cavendish Laboratory
J.J. Thomson Avenue
Cambridge CB3 0HE
UK

Michael R. Hamblin
Massachusetts General Hospital
Wellman Center for Photomedicine
40 Blossom Street
Boston, MA 02114
USA
and
Harvard Medical School
Department of Dermatology
25 Shattuck Street
Boston, MA 02115
USA
and
Harvard-MIT Division of Health
Sciences and Technology
77 Massachusetts Avenue
Cambridge, MA 02139
USA

Klaw Hamper
Gemeinschaftspraxis für Pathologie
Lornsenstr. 4
22767 Hamburg
Germany

Alon Harris
Indiana University School of Medicine
Glaucoma Research
and Diagnostics Center
Department of Ophthalmology
702 Rotary Circle
Indianapolis, IN 46202
USA

Jeremy C. Hebden
University College London
Department of Medical Physics and
Bioengineering
Malet Place Engineering Building
Gower Street
London WC1E 6BT
UK

Thorsten Heupel
Carl Zeiss MicroImaging GmbH
MIB-MP
Kistlerhoftstr. 75
81379 Munich
Germany

Christina Hipler
Department of Dermatology
University Hospital Jena
Erfurter Str. 35
07743 Jena
Germany

Cornelia Hirn
University Hospital Zurich
Division of Opthalmology
Frauenklinikstr. 24
8091 Zurich
Switzerland

Ying-Ying Huang
Massachusetts General Hospital
Wellman Center for Photomedicine
40 Blossom Street
Boston, MA 02114
USA
and
Harvard Medical School
Department of Dermatology
25 Shattuck Street
Boston, MA 02115
USA
and
Guangxi Medical University
Aesthetic Plastic Laser Center
22 Shuangyong Road
Nanning 530021, Guangxi
China

Dirk Hüttenberger
Lungenklinik Hemer, Pneumologie
Theo-Funccius-Str. 1
58675 Hemer
Germany

Huabei Jiang
University of Florida
Department of Biomedical Engineering
University Avenue
Gainesville, FL 32611
USA

Christian Jüngst
Department of Chemistry
POB 722
Universitaet Konstanz
78457 Konstanz
Germany

Asta Juzeniene
Norwegian Radium Hospital Oslo
University Hospital
Department of Radiation Biology
Montebello
0310 Oslo
Norway

Martin Kaatz
Friedrich Schiller University
Hospital Jena
Department of Dermatology
and Allergology
Bachstr. 18
07740 Jena
Germany

Maria M. Karabut
Nizhny Novgorod State Medical
Academy
10/1 Minina Square
603005 Nizhny Novgorod
Russia

Faisel Khan
University of Dundee
School of Medicine
Division of Medical Sciences
Centre for Cardiovascular
and Lung Biology
Ninewells Hospital and Medical School
Ninewells Avenue
Dundee DD1 9SY
UK

Sean J. Kirkpatrick
Michigan Technological University
Department of Biomedical Engineering
1400 Townsend Drive
Houghton, MI 49931
USA

Matthias Kirsch
Dresden University of Technology
Department of Neurosurgery
Carl Gustav Carus University Hospital
Fetscherstr. 74
01307 Dresden
Germany

Elena B. Kiseleva
Nizhny Novgorod State
Medical Academy
Institute of Applied and
Fundamental Medicine
10/1 Minina Square
603005 Nizhny Novgorod
Russia

Anil Kishen
Discipline of Endodontics
Faculty of Dentistry
124 Edward Street
Toronto, ON M5G 1G6
Canada

Martin Johannes Koehler
Department of Dermatology
University Hospital Jena
Erfurter Str. 35
07743 Jena
Germany

Christoph Krafft
Institute of Photonic Technology eV
(IPHT)
Albert-Einstein-Str. 9
07745 Jena
Germany

Irina A. Kuznetsova
Nizhny Novgorod Regional Hospital
603129 Nizhny Novgorod
Russia

Stefan Langenegger
University Hospital Zurich
Division of Opthalmology
Frauenklinikstr. 24
8091 Zurich
Switzerland

Kirill V. Larin
Saratov State University
Research-Educational Institute of
Optics and Biophotonics
83 Astrakhanskaya Street
410012 Saratov
Russia
and
University of Houston
Department of Biomedical Engineering
4800 Calhoun Road
Houston, TX 77204
USA

Michael Larsen
University of Copenhagen
Department of Ophthalmology
Nordre Ringvej 57
2600 Glostrup
Denmark

Nora Laver
Tufts Medical Center
Department of Pathology
800 Washington Street
Boston, MA 02111
USA

Martin J. Leahy
Chair of Applied Physics School of
Physics
National University of Ireland
Galwaly
Ireland

Wen-Tyng Li
Chung Yuan Christian University
Department of Biomedical Engineering
R&D Center for Biomedical Microdevice
Technology
Center for Nano-Technology
200 Chung Pei Road
Chung Li 32023
Taiwan
Republic of China

Pui-Chi Lo
Chinese University of Hong Kong
Department of Chemistry
Shatin
NT Hong Kong
China

Laurence B. Lovat
University College London
National Medical Laser Centre
Charles Bell House
67-73 Riding House Street
London W1W 7EJ
UK

Jonathan F. Lovell
University of Toronto
Institute of Biomaterials and Biomedical
Engineering
Toronto, ON M5G 1L7
Canada

Holger Lubatschowski
Universität Hannover
Institut für Quantenoptik
Welfengarten 1
30167 Hannover
Germany

Qingming Luo
Huazhong University
of Science and Technology
Wuhan National Laboratory
for Optoelectronics
Britton Chance Center
for Biomedical Photonics
Wuhan 430074
China

Adrian Lussi
University of Bern
School of Dental Medicine
Department of Preventive, Restorative
and Pediatric Dentistry
Freiburgstr. 7
3010 Bern
Switzerland

Andreas Mandelis
University of Toronto
Center for Advanced
Diffusion Wave Technologies
Department of Mechanical and
Industrial Engineering
5 King's College Road
Toronto, ON M5S 3G8
Canada

Anna V. Maslennikova
Nizhny Novgorod State
Medical Academy
10/1 Minina Square
603005 Nizhny Novgorod
Russia

Stephen J. Matcher
University of Sheffield
The Kroto Institute
Broad Lane
Sheffield S3 7HQ
UK

Anna Matvienko
University of Toronto
Center for Advanced
Diffusion Wave Technologies
Department of Mechanical and
Industrial Engineering
5 King's College Road
Toronto, ON M5S 3G8
Canada

Thomas McCarthy
PhotoThera Inc.
5925 Priestly Drive, Suite 120
Carlsbad, CA 92008
USA

Andrew McNeill
University of St. Andrews
School of Physics and Astronomy
Organic Semiconductor Center and
Biophotonics Collaboration
North Haugh
St. Andrews KY16 9SS
UK
and
Ambicare Health Ltd.
Kinburn Castle, Doubledykes Road
St Andrews KY16 9DR
UK

Susann Meisel
Friedrich Schiller University Jena
Institute of Physical Chemistry
Helmholtzweg 4
07743 Jena
Germany

Miloš Miljković
Northeastern University
Department of Chemistry and Chemical Biology
Laboratory for Spectral Diagnosis
360 Huntington Avenue
Boston, MA 02115
USA

Anant Mohan
All India Institute of Medical Sciences
Department of Medicine
Aurobindo Marg
Ansari Nagar
New Delhi 110029
India

Robert Möller
Institute of Photonic Technology
Albert-Einstein-Str. 9
07745 Jena
Germany

Alexander A. Muraev
Nizhny Novgorod State Medical Academy
10/1 Minina Square
603005 Nizhny Novgorod
Russia

Amit Kumar Murar
All India Institute of Medical Sciences
Department of Medicine
Aurobindo Marg
Ansari Nagar
New Delhi 110029
India

Jaclyn Nascimento
University of California at San Diego
Department of Bioengineering
9500 Gilman Drive
La Jolla, CA 92093
USA

Axel Niendorf
Pathologie Hamburg-West
Lornsenstrasse 4
22767 Hamburg
Germany

Alena V. Nikolskaya
Washington University
Department of Biomedical Engineering
One Brookings Drive
St. Louis, MO 63130-4899
USA

Vladimir P. Nikolski
Washington University
Department of Biomedical Engineering
One Brookings Drive
St. Louis, MO 63130-4899
USA

Bernhard Ortel
University of Chicago Medical Center
Dermatology
5841 South Maryland Avenue
Chicago, IL 60637
USA

Kostas Papamarkakis
Northeastern University
Department of Chemistry and Chemical Biology
Laboratory for Spectral Diagnosis
360 Huntington Avenue
Boston, MA 02115
USA

Rakesh Patalay
Imperial College London
Department of Physics
Photonics Group
South Kensington Campus
606 Blackett Building
London SW7 2AZ
UK

and

Imperial College London
Department of Medicine
Division of Experimental Medicine
Hammersmith Campus
Commonwealth Building
Du Cane Road
London W12 0NN
UK

Igor Peshko
Wilfrid Laurier University
Department of Physics and Computer Science
75 University Avenue West
Waterloo, ON N2L 3C5
Canada

Roberto Pini
Consiglio Nazionale delle Ricerche
Istituto di Fisica Applicata
"Nello Carrara"
Via Madonna del Piano 10
50019 Sesto Fiorentino
Italy

Jürgen Popp
Institute of Photonic Technology eV (IPHT)
Albert-Einstein-Str. 9
07745 Jena
Germany

and

Friedrich Schiller University Jena
Institute of Physical Chemistry
Helmholtzweg 4
07743 Jena
Germany

Sabine Przemeck
Rowiak GmbH
Garbsener Landstr. 10
30419 Hannover
Germany

Stefka G. Radenska-Lopovok
Russian Medical Academy of Sciences
Institute of Rheumatology
Kashirskoe Road 34A
155522 Moscow
Russia

Prakash R. Rai
Harvard Medical School
Massachusetts General Hospital
Wellman Center for Photomedicine
Department of Dermatology
40 Blossom Street
Boston, MA 02114
USA

Jessica Ramella-Roman
Catholic University of America
Department of Biomedical Engineering
620 Michigan Avenue
Washington, DC 20064
USA

Fulvio Ratto
Consiglio Nazionale delle Ricerche
Istituto di Fisica Applicata
"Nello Carrara"
Via Madonna del Piano 10
50019 Sesto Fiorentino
Italy

Dan Reinstein
London Vision Clinic
138 Harley Street
London W1G 7LA
UK

Heiko Richter
Rowiak GmbH
Garbsener Landstr. 10
30419 Hannover
Germany

Gisbert Richard
University Medical Center
Hamburg-Eppendorf
Department of Ophthalmology
Martinistr. 52
20246 Hamburg
Germany

Natalia S. Robakidze
St. Petersburg Medical Academy of
Postgraduate Studies
Kirochnaya Street 41
191015 St. Petersburg
Russia

Nadege Roche-Labarbe
Athinoula A. Martinas
Center for Biomedical Imaging
Massachusetts
General Hospital
Boston, MA 0244
USA

Kerstin Roeser
Department of Biochemistry and
Molecular Biology II: Molecular
Cell Biology
University Medical Center
Hamburg-Eppendorf
Martinistrasse 52
20246 Hamburg
Germany

Petra Rösch
Friedrich Schiller University Jena
Institute of Physical Chemistry
Helmholtzweg 4
07743 Jena
Germany

Francesca Rossi
Consiglio Nazionale delle Ricerche
Istituto di Fisica Applicata
"Nello Carrara"
Via Madonna del Piano 10
50019 Sesto Fiorentino
Italy

Marc Rubinstein
Beckman Laser Institute
1002 Health Sciences Road
Irvine, CA 92612
USA

Bettina Rudolph
Institute of Photonic Technology
Albert-Einstein-Str. 9
07745 Jena
Germany

Adzura Salam
University of Bern
Inselspital, Universitätsklinik für
Augenheilkunde
Freiburgstr.
3010 Bern
Switzerland

Ifor D.W. Samuel
University of St. Andrews
School of Physics and Astronomy
Organic Semiconductor Center and
Biophotonics Collaboration
North Haugh
St. Andrews KY16 9SS
UK
and
Ambicare Health Ltd.
Kinburn Castle, Doubledykes Road
St Andrews KY16 9DR
UK

Markus Sauer
Bielefeld University
Applied Laser Physics and Laser
Spectroscopy
Universitätsstr. 25
33615 Bielefeld
Germany

Claus-Georg Schmedt
Klinikum Stuttgart - Katharinenhospital
Department of Vascular Surgery
Kriegsbergstr. 60
70174 Stuttgart
Germany

Herbert Schneckenburger
Hochschule Aalen
Institut für Angewandte Forschung
Beethovenstr. 1
73428 Aalen
Germany

Jennifer Schubert
Northeastern University
Department of Chemistry and Chemical
Biology
Laboratory for Spectral Diagnosis
360 Huntington Avenue
Boston, MA 02115
USA

Wilm Schumacher
Friedrich Schiller University Jena
Institute of Physical Chemistry
Helmholtzweg 4
07743 Jena
Germany

Michael Seitz
Ludwig Maximilians University Munich
Department of Urology
Marchioninistr. 15
81377 Munich
Germany

List of Contributors

Romedi Selm
Department of Chemistry
POB 722
Universitaet Konstanz
78457 Konstanz
Germany

Alexander M. Sergeev
Institute of Applied Physics
Russian Academy of Sciences
Division of Nonlinear Dynamics and Optics
46 Ul'yanov Street
603950 Nizhny Novgorod
Russia

Natalia M. Shakhova
Russian Academy of Sciences
Institute of Applied Physics
Division of Nonlinear Dynamics and Optics
46 Ul'yanov Street
603950 Nizhny Novgorod
Russia

Bing Shao
University of California at San Diego
Department of Electrical and Computer Engineering
9500 Gilman Drive
La Jolla, CA 92093
USA

Linda Z. Shi
University of California at San Diego
Department of Bioengineering
9500 Gilman Drive
La Jolla, CA 92093
USA

Yochai Z. Shoshani
Indiana University School of Medicine
Glaucoma Research and Diagnostics Center
Department of Ophthalmology
702 Rotary Circle
Indianapolis, IN 46202
USA

Brent A. Siesky
Indiana University School of Medicine
Glaucoma Research and Diagnostics Center
Department of Ophthalmology
702 Rotary Circle
Indianapolis, IN 46202
USA

Ludmila B. Snopova
Nizhny Novgorod State Medical Academy
Institute of Applied and Fundamental Medicine
603005 Nizhny Novgorod
Russia

Ronald Sroka
Ludwig Maximilians University Munich
LIFE Center
Laser Research Laboratory
Geschwister-Scholl-Platz 1
80539 Munich
Germany
and
Klinikum der Universität München
LIFE-Zentrum, Laser-Forschungslabor
Marchioninistr. 23
81377 Munich
Germany

Mark-Steven Steiner
University of Regensburg
Institute of Analytical Chemistry,
Chemo- and Biosensors
Universitätsstr. 31
93040 Regensburg
Germany

Markus Sticker
Carl Zeiss MicroImaging GmbH
Advanced Development Microscopy
Carl-Zeiss-Promenade 10
07745 Jena

Stephan Stöckel
Friedrich Schiller University Jena
Institute of Physical Chemistry
Helmholtzweg 4
07743 Jena
Germany

Olga S. Streltsova
Nizhny Novgorod State
Medical Academy
Institute of Applied and Fundamental
Medicine
603005 Nizhny Novgorod
Russia

Katarina Svanberg
Lund University
Department of Oncology
Paradisgatan 5
221 00 Lund
Sweden

Sune Svanberg
Lund University
Department of Physics
Solvegatan 14A
221 00 Lund
Sweden

Hasan Tayyaba
Harvard Medical School
Massachusetts General Hospital
Wellman Center for Photomedicine
Department of Dermatology
40 Blossom Street
Boston, MA 02114
USA

Marc Töteberg-Harms
University Hospital Zurich
Medical Faculty
Division of Ophthalmology
Frauenklinikstr. 24
8091 Zurich
Switzerland

Valery V. Tuchin
Saratov State University
Research-Educational Institute of Optics
and Biophotonics
83 Astrakhanskaya Street
410012 Saratov
Russia
and
Russian Academy of Sciences
Institute of Precise Mechanics and
Control
24 Rabochaya Street
410028 Saratov
Russia

Jamie Turner
University of Dundee
School of Medicine
Ninewells Hospital and Medical School
Ninewells Avenue
Dundee DD1 9SY
UK

Sarika Verma
Harvard Medical School
Massachusetts General Hospital
Wellman Center for Photomedicine
Department of Dermatology
40 Blossom Street
Boston, MA 02114
USA

Michael Wagner
Hochschule Aalen
Institut für Angewandte Forschung
Beethovenstr. 1
73428 Aalen
Germany

Angela Walter
Friedrich Schiller University Jena
Institute of Physical Chemistry
Helmholtzweg 4
07743 Jena
Germany

Karina Weber
Institute of Photonic Technology
Albert-Einstein-Str. 9
07745 Jena
Germany

Petra Weber
Hochschule Aalen
Institut für Angewandte Forschung
Beethovenstr. 1
73428 Aalen
Germany

Fabian Will
Rowiak GmbH
Garbsener Landstr. 10
30419 Hannover
Germany

Ernst Wintner
Vienna University of Technology
Photonics Institute
Gusshausstr. 27-29
1040 Vienna
Austria

Martin Winterhalder
Department of Chemistry
POB 722
Universitaet Konstanz
78457 Konstanz
Germany

Sebastian Wolf
University of Bern
Inselspital, Universitätsklinik für Augenheilkunde
Freiburgstr.
3010 Bern
Switzerland

Otto S. Wolfbeis
University of Regensburg
Institute of Analytical Chemistry, Chemo- and Biosensors
Universitätsstr. 31
93040 Regensburg
Germany

Brian Wong
Beckman Laser Institute
1002 Health Sciences Road
Irvine, CA 92612
USA

Andreas Wunder
Charité - University of Medicine Berlin
Center for Stroke Research Berlin (CSB)
Small Animal Imaging Center (SAIC)
Department of Experimental Neurology
Charitéplatz 1
10117 Berlin
Germany

I.A. Yanvareva
Nizhny Novgorod State
Medical Academy
Institute of Applied and Fundamental
Medicine
603005 Nizhny Novgorod
Russia

Zhen Yuan
University of Florida
Department of Biomedical Engineering
University Avenue
Gainesville, FL 32611
USA

Ekaterina E. Yunusova
Nizhny Novgorod State
Medical Academy
10/1 Minina Square
603005 Nizhny Novgorod
Russia

Elena V. Zagaynova
Nizhny Novgorod State
Medical Academy
Institute of Applied and
Fundamental Medicine
10/1 Minine Square
603005 Nizhny Novgorod
Russia

Aleksei M. Zheltikov
M.V. Lomonosov Moscow State
University
International Laser Center
Physics Department
Leninskie Gory
119991 Moscow
Russia

Gang Zheng
University of Toronto
Institute of Biomaterials and Biomedical
Engineering
Toronto, ON M5G 1L7
Canada
and
University of Toronto
Ontario Cancer Institute
Department of Medical Biophysics
Toronto, ON M5G 1L7
Canada

Xiang Zheng
Harvard Medical School
Massachusetts General Hospital
Wellman Center for Photomedicine
Department of Dermatology
40 Blossom Street
Boston, MA 02114
USA

Sindy Zimmermann
Department of Dermatology
SRH Hospital Gera
Stiaße des Friedens 122
D 07548 Gera
Germany

Dietlind Zuehlke
Fraunhofer Institute
for Applied Information Technology
FIT
Schloss Birlinghoven
53754 Sankt Augustin
Germany

Andreas Zumbusch
University of Konstanz
Department of Chemistry
Universitätsstr. 10
78457 Konstanz
Germany

Part 1
Laboratory Medicine

1
Microscopic and Spectroscopic Methods

Brent D. Cameron

Microscopy is a term used to refer to methods that allow for the visual examination of the physical properties of a sample on a micron-level scale. Although there are a wide variety of microscopic methods ranging from optical to electron microscopy, the focus of this chapter will concentrate on light-based methods. Although microscopy by itself sheds insight into the physical structure of a sample, optical spectroscopic methods, which concentrate on how photons of light interact with a sample (e.g., absorbed or scattered), can be used to examine a sample's chemical structure. Spectroscopic methods can be performed using single-point measurements or integrated with microscopic methods to allow for chemical imaging; also known as microspectroscopy.

1.1
Bright Field Microscopy

One of the most basic microscopy techniques is known as bright field microscopy. In this approach, sample illumination occurs in transmission mode through the bottom of the sample under investigation. Visual observation of the sample occurs from the top through standard microscope optics to allow for image magnification up to 1000×. Although this is one of the most common microscopy methods, bright field microscopy suffers from contrast limitations and often has difficulty with thick samples [1], as a considerable amount of the sample under investigation will be out of focus, which hinders the image resolution. Bright field microscopy often requires samples to be stained or have sufficient natural pigmentation to allow for proper viewing [1, 2]. This approach is also well suited for visualization of living preparations, such as single-cell organisms.

1.2
Dark Field Microscopy

To deal with some of the contrast issues associated with bright field imaging, especially when dealing with unstained samples, dark field microscopy is commonly employed. For this approach, the configuration uses an opaque disc placed in front of the light source and behind the condenser lens [1, 2]. At this point, the light interacts with the sample and only the scattered light is collected due to a direct light block. This method can significantly improve image contrast and does not require sample staining [1]. This technique, however, is limited in terms of the intensity of the image and therefore often requires very intense illumination, which may degrade or damage the sample. Dark field configurations are commonly used as low-cost alternatives to phase contrast approaches.

1.3
Phase Contrast Microscopy

When working with biological specimens, a significant amount of the detail is undetectable when using bright field microscopy due to contrast limitations, especially for microstructures with similar pigmentation or transparencies. In 1934, a Dutch physicist, Frits Zernike, described a technique known as phase contrast microscopy. This approach can greatly enhance the contrast of transparent samples, such as microstructures contained in living cells, microorganisms, and thin tissues [3]. The phase contrast method translates minute variations in a specimen's refractive indices, which alters the phase relationships within the electric field of light into corresponding changes in amplitude [1–4]. These changes can then be visualized as differences in image contrast, and therefore details that would normally be transparent or invisible in bright field imaging can now be easily observed. Furthermore, living cells can be examined in their natural state without being fixed and/or stained. Such ability allows the dynamics of various biological processes to be observed and recorded in significant detail.

In terms of implementation, light is focused on to a specialized annulus positioned in the condenser prior to the sample (i.e., front focal plane) [1]. The light then illuminates the sample, at which point it will either pass straight through (i.e., unaltered) or will be diffracted and/or retarded in phase by structures in the specimen. Both the undeviated and diffracted light collected by the objective is then segregated at the back focal plane by a phase plate and focused on to an intermediate image plane to form the final phase contrast image. It should be noted that without the phase plate, there would be no destructive interference and therefore no significant improvement in contrast. The phase plate allows for the slight phase variations to be converted into visible amplitude changes, which provide for the contrast enhancement.

1.4
Differential Interference Contrast Microscopy

Differential interference contrast (DIC) microscopy is another technique used to improve the contrast of highly transparent or unstained samples. DIC microscopy is somewhat similar to phase contrast microscopy in that it converts changes in the phase shift of the electric field of light, due to variations in refractive index, into detectable intensity or amplitude changes. The resulting image is similar; however, it is not plagued by the sometimes distracting bright diffraction halo [3, 5]. DIC microscopy also depicts differences in optical density as something like a three-dimensional relief image, although this sometimes may not entirely correctly represent actuality, as the occurrence is due to an optical effect instead of physical geometry.

With regard to implementation, the illumination source is initially linearly polarized at an angle of 45° and enters an optical component known as a Nomarski-modified Wollaston prism [5]. This component separates the light into its two orthogonal polarization components. One ray will be used as the sampling ray and the other as a reference. The microscope condenser focuses the rays through the sample; however, they will be slightly spatially offset to each other. Therefore, the rays may experience slightly different refractive indices and/or sample thicknesses, resulting in different optical pathlengths. The resulting rays are then focused by an objective lens and travel through a second Nomarski–Wollaston prism, which recombines the beams. At this point, the rays will interfere with each other, either constructively or destructively, leading to either bright or dark enhancements, respectively, depending on the corresponding optical path difference.

1.5
Confocal Microscopy

One of the most significant advances in the field of optical microscopy was the development of confocal microscopy. The popularity of this approach lies in its ability to examine the three-dimensional structure of thick biological samples. Confocal microscopy produces an optical image using an alternative approach to the traditional wide-field microscope. This method relies on the use of a scanning point approach instead of full sample illumination, which provides slightly higher resolution and considerable enhancement in the optical sectioning of the sample by eliminating the influence of out-of-focus light that would otherwise distort the image [6, 7]. In applications where it is desired to investigate the three-dimensional structure of a sample, confocal microscopy is commonly employed. Lastly, in terms of implementation, confocal microscopy may be combined with fluorescence spectroscopy, which allows for further exploration of biological samples.

1.6
Fluorescence Spectroscopy

Fluorescence spectroscopy is a popular research method used in biomedical research and clinical pathology. Applications include organic compound analysis,

DNA sequencing, the study of protein conformational changes, medical diagnostics, and so on. Although fluorescence measurements can be performed using single-point detectors, image-based analysis is very commonplace today. Fluorescence imaging/microscopy can provide for localization of intracellular compounds and in certain implementations detection at the level of single molecules.

Fluorescence is related to the luminescence family of processes in which certain molecules have the ability to emit light when they are in an excited state. The two categories of luminescence include fluorescence and phosphorescence, which specifically depend on the electronic configuration of the excited state and the emission pathway [8, 9]. Specifically, fluorescence is a property of certain molecules (e.g., aromatic molecules) that absorb light at a given wavelength which excites the species from its ground state to one of the various vibrational levels in an excited electronic state [8]. When the molecule falls back to a specific vibrational level of the ground state, a photon of light will be emitted in the process. Depending on the energy level in the drop, this will determine the frequency or wavelength of light emitted. Therefore, by measuring the emitted wavelengths (i.e., emission spectra) along with their respective intensities, significant information about the molecule under investigation can be determined.

In fluorescence microscopy, the fluorescence emission can be characterized not only by intensity and spatial position, but also by lifetime [10], polarization, and wavelength. Therefore, fluorescence microscopy is commonly performed using a variety of configurations. Some of these configurations include epifluorescence or incident light fluorescence [8], laser scanning fluorescence [6], confocal fluorescence [6], fluorescence lifetime imaging [10], and two-photon excitation fluorescence microscopy [11].

1.7
Infrared Spectroscopy

Infrared (IR) spectroscopy is a well-established technique for the identification of organic molecules. Specifically, it is a subcategory of spectroscopy which employs the IR region of the electromagnetic spectrum, which is normally broken up into the near-infrared (NIR) (0.78–2.5 µm), mid-infrared (MIR) (2.5–50 µm), and far-infrared (FIR) (50–1000 µm) spectral regions [12]. The method is based on vibrations of atoms within a molecule. For analysis, an IR spectrum is normally obtained by passing IR radiation through a sample and then determining which wavelengths are absorbed and to what extent. From this analysis, considerable information on the molecular and bonding structure of the sample can be determined. For biological analysis, IR spectroscopy allows for detailed examination of complex biomolecules, including proteins, lipids, carbohydrates, and nucleic acids [13].

In terms of implementation, the two main methods include a dispersive approach and the more common Fourier transform infrared (FTIR) spectrometer configuration, which is based on optical interference [12]. Detection can be made as single-point, line, or image-based measurements. MIR spectroscopy can distinguish very subtle changes in chemical structure and therefore often this range is used in the identification of unknown substances. However, since this region of the electromagnetic spectrum is highly subject to water absorption, analysis of aqueous samples can prove extremely difficult and often requires alternative sample preparation (e.g., desiccation) or alternative sampling methods such as attenuated total reflectance (ATR). NIR spectroscopy is better suited to handle aqueous samples and is more commonly employed for applications in medical diagnostics, pharmaceutical analysis, and biological investigations [13–15]. NIR spectra consist of molecular overtones and combination bands of the fundamental molecular absorptions found in the MIR region. Therefore, NIR spectra generally consist of overlapping vibrational bands that may appear nonspecific and/or difficult to resolve. However, through the use of chemometric multivariate processing techniques, both qualitative and quantitative analyses have proven fairly successful [16, 17].

1.8
Raman Spectroscopy

Raman spectroscopy has become an important analytical tool in a multitude of research disciplines. It can be used for applications ranging through pharmaceuticals, forensics, DNA analysis, and medical diagnostics [18, 19]. The technique can provide significant information that may be used for chemical identification, bonding effect analysis, and characterization of molecular structures. Raman spectroscopy is a powerful light-scattering technique, which in simple terms can be thought of as a process in which a photon of light at a specific wavelength interacts with a sample and scattered light at different wavelengths result. This scattered light can then be analyzed to characterize the internal structure of molecules.

More specifically, Raman emissions are generated based on the concept of inelastic scattering of monochromatic light. The source excitation is usually provided by a laser source in the visible to NIR region of the electromagnetic spectrum [20]. In the case of inelastic scattering, depending on the molecular and bonding structure of the sample, a small amount of the light will experience a change in energy. In the case of reduced energy, longer wavelengths will be emitted that are unique to the sample. These are referred to as the Stokes bands and are more commonly used for analysis. In the case of higher energy (due to the molecule already being in an excited vibrational state), shorter wavelengths will be emitted, also unique to the sample. These are commonly referred to as the anti-Stokes bands. Inelastic scatter is very uncommon and accounts for only about 0.001% of the scattered light coming from the sample; most of the light is elastically scattered at the same wavelength of the excitation source.

As with IR spectroscopy, in terms of implementation, the two main methods include a dispersive approach and a Fourier transform (FT) spectrometer configuration [20]. The information contained in Raman spectra is commonly complementary to MIR absorption spectroscopy and both organic and inorganic materials possess Raman spectra. These normally include sharp bands that are representative of the chemical structure of the sample under investigation. It should be noted, however, that since Raman measurements are commonly collected in the visible range of the electromagnetic spectrum, sample autofluorescence can sometimes mask the weak Raman signature, especially for biological samples [21]. Moving the excitation source out to the NIR region can normally help minimize this effect. A main benefit of Raman measurements compared with MIR spectroscopy is that the sample requires little to no preparation and water absorption is not a problem (i.e., sample desiccation is not required), hence the approach is well suited to biological samples.

It should be noted that several variations of Raman spectroscopy are used in practice. These variants are usually implemented to improve sensitivity or spatial resolution, or to acquire specific types of information. Some of these techniques include surface-enhanced Raman spectroscopy (SERS), Raman optical activity (ROA), coherent anti-Stokes Raman spectroscopy (CARS), and resonance Raman spectroscopy [21].

References

1 Leng, Y. (2008) *Materials Characterization: Introduction to Microscopic and Spectroscopic Methods*, John Wiley & Sons (Asia) Pte. Ltd., Singapore.
2 Bradbury, S. and Bracegirdle, B. (1998) *Introduction to Light Microscopy*, BIOS Scientific Publishers, Oxford.
3 Ruzin, S.E. (1999) *Plant Microtechnique and Microscopy*, Oxford University Press, New York.
4 Bennett, A., Osterberg, H., Jupnik, H., and Richards, O. (1951) *Phase Microscopy: Principles and Applications*, John Wiley and Sons, Inc., New York.
5 Murphy, D. (2001) *Fundamentals of Light Microscopy and Digital Imaging*, Wiley-Liss, New York.
6 Pawley, J.B. (ed.) (1995) *Handbook of Biological Confocal Microscopy*, Plenum Press, New York.
7 Wilson, T. (ed.) (1990) *Confocal Microscopy*, Academic Press, London.
8 Lakowicz, J.R. (2006) *Principles of Fluorescence Spectroscopy*, Springer, New York.
9 Rost, F. (1991) *Quantitative Fluorescence Microscopy*, Cambridge University Press, Cambridge.
10 Suhling, K., French, P.M., and Phillips, D. (2005) Time-resolved fluorescence microscopy. *Phtochem. Photobiol. Sci.*, 4 (1), 13–22.
11 So, P.T., Dong, C.Y., Masters, B.R., and Berland, K.M. (2000) Two-photon excitation fluorescence microscopy. *Annu. Rev. Biomed. Eng.*, 2, 399.
12 Hsu, C.P.S. (1997) Infrared spectroscopy, in *Handbook of Instrumental Techniques for Analytical Chemistry* (ed. F. Settle), Prentice Hall, Englewood Cliffs, NJ, pp. 247–283.
13 Manitsch, H.H. and Casal, H.L. (1986) Biological applications of infrared spectrometry. *Fresenius' Z. Anal. Chem.*, 324, 655–661.
14 Burns, D. and Ciurczak, E. (2001) *Handbook of Near-Infrared Analysis*, Marcel Dekker, New York.
15 Ciurczak, E. and Drennen, J. (2002) *Near-Infrared Spectroscopy in Pharmaceutical and Medical Applications*, Marcel Dekker, New York.
16 Martens, H. and Naes, T. (1992) *Multivariate Calibration*, John Wiley and Sons, Ltd., Chichester.
17 Mark, H. and Workman, J. (2003) *Statistics in Spectroscopy*, Elsevier, Amsterdam.

18 Choo-Smith, L.P., Edwards, H.G., Endtz, H.P., Kros, J.M., Heule, F., Barr, H., Robinson, J.S. Jr., Bruining, H.A., and Puppels, G.J. (2002) Medical applications of Raman spectroscopy: from proof of principle to clinical implementation. *Biopolymers*, **67** (1), 1–9.

19 Shih, W., Bechtel, K.L., and Feld, M.S. (2008) Quantitative biological Raman spectroscopy, in *Handbook of Optical Sensing of Glucose in Biological Fluids and Tissues* (ed. V.V. Tuchin), Taylor & Francis, Abingdon, pp. 354–381.

20 Colthup, N.B., Daly, L.H., and Wiberley, S.E. (1990) *Introduction to Infrared and Raman Spectroscopy*, Academic Press, New York.

21 Laserna, J.J. (ed.) (1996) *Modern Techniques in Raman Spectroscopy*, John Wiley and Sons, Ltd., Chichester.

2
Glucose Sensing and Glucose Determination Using Fluorescent Probes
Mark-Steven Steiner, Axel Duerkop, and Otto S. Wolfbeis

2.1
Why Glucose Sensing?

The measurement of glucose is among the most important analytical tasks. It has been estimated that about 40% of all blood tests are related to it. In addition, there are numerous other situations where glucose is to be determined, for example, in biotechnology, the production and processing of all kinds of food, in biochemistry in general, and in numerous other areas. The continuing interest in sensing glucose, mainly in blood, is one result of the increasing age and (alarming) size of the world's population, and the fact that about 4% of its (Caucasian) population suffers from diabetes. Normal blood glucose levels range from 3 to 10 mM (54–180 mg dl^{-1}).

The market for glucose sensors is probably the biggest single one in the diagnostic field, its size being about US$40 billion per year at present. Given this size, it does not come as a surprise that any true improvement in sensing glucose represents a major step forward. The greatest need at present is, however, for continuous sensors.

Both optical and electrochemical sensors can be manufactured at costs that are so low that they can also be of the disposable type. Clinical multiparametric instrumentation (e.g., for sensing glucose and blood parameters such as pO$_2$, pH, Na$^+$, K$^+$, Cl$^-$, lactate, and urea) relies on reusable sensors. In this case, a blood sample is inserted into the instrument, a reading is made, the surface of the sensor is washed, and the sensor is recalibrated before the next sample is introduced. Electrochemical methods are mainly used in stand-alone instruments in clinical laboratories and in near-patient testing. Millions of disposable electrochemical (mediator-based) blood glucose meters are used in homecare devices that enable glucose to be determined within less than 1 min in blood samples as small as 0.3 μl [1]. Electrical wiring of enzymes [2] has led to smaller quantities of blood being taken, thus leading to almost painless sampling in a new generation of glucose sensors that have had tremendous commercial success.

Optical methods are based on the measurement of photons rather than of electrons. The absence of electrodes can be advantageous in the case of patients with heart pacemakers. Although optical schemes for sensing glucose have not had

Handbook of Biophotonics. Vol.2: Photonics for Health Care, First Edition. Edited by Jürgen Popp, Valery V. Tuchin, Arthur Chiou, and Stefan Heinemann.
© 2012 Wiley-VCH Verlag GmbH & Co. KGaA. Published 2012 by Wiley-VCH Verlag GmbH & Co. KGaA.

the success of electrochemical schemes, they still are a topic of highly active research. In addition to color test strips, a couple of reversible optical sensors rely on the measurement of the consumption of oxygen. Among the optical methods, photometry (and reflectometry), fluorescence, and surface plasmon resonance (SPR) have had the greatest success. Almost all optical sensors for continuous monitoring rely on either fluorescence or SPR. Not a single reflectometric method is known for use in continuous sensing. This chapter is restricted to luminescence-based methods.

2.2
Classification of Sensors According to the Recognition Scheme

This review is subdivided into sections according to the method for recognition of glucose. Once selective recognition of glucose has occurred, it has to be transduced into (optical) information. The first fundamental type of glucose sensor is based on the recognition of glucose by certain enzymes (or coenzymes) that subsequently undergo changes in their intrinsic absorption and/or fluorescence, or carry a label placed near the site of interaction and capable of reporting the interaction with glucose. The second class of sensors measure (kinetically) the formation/consumption of metabolites of enzyme activity, mainly of glucose oxidase (GOx). One scheme is based on the measurement of the quantity of oxygen consumed according to Eq. (2.1). Alternatively, the hydrogen peroxide (HP) formed according to Eq. (2.1) may be determined by electrochemical or optical means. A third option consists in the determination of the quantity of protons formed (i.e., the decrease in pH) [Eq. (2.2)].

$$\beta\text{-D-glucose} + O_2 \rightarrow \text{D-glucono-1, 5-lactone} + H_2O_2 \qquad (2.1)$$

$$\text{D-glucono-1, 5-lactone} + H_2O \rightarrow \text{gluconate} + H^+ \qquad (2.2)$$

The enzyme glucose dehydrogenase has also been used. It catalyzes the conversion of glucose to form NADH according to

$$\beta\text{-D-glucose} + NAD^+ \rightarrow \text{D-glucono-1, 5-lactone} + NADH \qquad (2.3)$$

The amount of NADH formed according to Eq. (2.3) may be measured, for example, by photometry at 345 nm or by measuring its fluorescence peaking at 455 nm, but this reaction is not reversible and comes to an end once all the NAD^+ has been consumed. Hence it is less suited (and less elegant) in terms of continuous sensing.

The third large group of sensors relies on organic boronic acids that act as molecular receptors for 1,2-diols and saccharides. The high affinity of boronic acids towards saccharides facilitates recognition of glucose levels in blood samples (3–50 mM). When coupled to a conjugated π-system (a dye), these show a change in the optical properties as a result of binding glucose [3, 4]. The fourth large group of receptors exploits the capability of glucose-binding proteins (GBPs) to bind glucose with high specificity. This includes concanavalin A (ConA) based sensor schemes and apo-GOx, a glucose oxidase whose coenzyme has been removed. On binding glucose, some proteins undergo changes in their intrinsic optical properties in the ultraviolet (UV) region. In order to shift analytical wavelengths into the (longwave) visible region, they have been labeled with fluorophores. Methods based on labeled proteins are preferred because the choice of a proper label enables the optical properties of the system to be fine-tuned.

2.3
Glucose Sensing via the Optical Properties of Oxidative or Reductive Enzymes

2.3.1
Nonlabeled Enzymes

The first group of recognition schemes comprises enzymes and coenzymes that undergo optical changes in their spectral properties (e.g., GOx [5, 6], glucose dehydrogenase, glucokinase [7], and apo-GOx). The optical changes usually occur in the UV region (based on the intrinsic tryptophan fluorescence of GOx) and in the visible region [the coenzyme flavin adenine dinucleotide (FAD) [8] of GOx]. These respective enzymes have also been labeled with fluorophores in order to shift the analytical wavelength into the visible region. Longwave sensing is highly desirable in view of the reduced absorbance and fluorescence of blood and serum at >600 nm. Labeling usually does not strongly affect the binding constants of the enzymes.

$$\beta\text{-D-glucose} + FAD \rightarrow \text{D-glucono-1,5-lactone} + FADH_2 \quad (2.4)$$

$$FADH_2 + O_2 \rightarrow FAD + H_2O_2 \quad (2.5)$$

$$\text{D-glucono-1,5-lactone} + H_2O \rightarrow \text{gluconate} + H^+ \quad (2.6)$$

Changes of the intrinsic fluorescence may involve (a) the UV fluorescence of tryptophan groups [excitation/emission maxima ($\lambda_{ex}/\lambda_{em}$) at 295/330 nm] in the protein [5], (b) the fluorescence of FAD (Figure 2.1) ($\lambda_{ex}/\lambda_{em}$ 450/520 nm) [8], or (c) NAD$^+$ ($\lambda_{ex}/\lambda_{em}$ 340/460 nm) [9]. An interesting competitive FRET (Förster resonance energy transfer) assay was presented by D'Auria et al. [10]. They used the intrinsic tryptophan fluorescence of glucose kinase as donor whereas glucose derivatized with o-nitrophenyl-β-D-glucose-co-pyranoside is the acceptor. Addition of glucose increases donor emission and decreases FRET. The analytical range of these sensors is typically from 0.5 to 20 mM, and rarely up to 200 mM [7].

2.3.2
Labeled Enzymes

In addition to the intrinsic optical properties of glucose-specific enzymes, the optical properties of labels of the enzyme may be exploited. Fluorescein [11, 12] and coumarin [13] are among the most commonly used labels. Excitation at above 400 nm

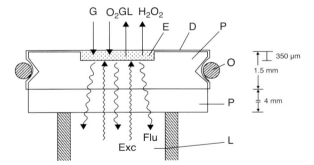

Figure 2.1 Cross-section through the sensing platelet of a fiber-optic glucose sensor. P, Plexiglas; D, dialyzing membrane; E, enzyme solution; O, O-ring; L, light guide. The arrows indicate the diffusion processes involved (G, glucose; GL, gluconolactone). The directions of the exciting light (Exc) and fluorescence (Flu) are also shown. The platelet has o.d. 20 mm; diameter of cavity, 4 mm [8]. Reprinted from W. Trettnak and O.S. Wolfbeis, Fully reversible fiber-optic glucose biosensor based on the intrinsic fluorescence of glucose oxidase, *Anal. Chim. Acta*, 1989, **221**, 195–203, with permission from Elsevier.

is more adequate for determination of glucose in real samples because of the ubiquitous UV absorption of proteins and other compounds in these samples.

2.4
Fluorescent Sensing of Glucose via Measurement of the Consumption of Oxygen Caused by the Action of GOx

The measurement of the consumption of oxygen [Eq. (2.1)] via quenchable probes (luminescent complexes of ruthenium, platinum, or palladium) immobilized in a sensor layer together with the enzyme is a fairly successful approach. The sensors described in the literature differ from each other mainly in the kind of fluorescent probe and the type of polymer matrix [hydrogels, sol–gels, polystyrene (PS)]. Among the technical layouts, optodes (optical fibers with the sensor layer fixed at the tip) and planar sensors in microwells or in a microfluidic flow cell have been reported. Continuous sensing using such enzymes and reversible oxygen sensors became available in the 1980s.

2.4.1
Planar and Fiber-Optic Sensors

The luminescence of probes such as decacyclene [14], complexes of Pt(II) [15, 16] or Pd(II) with porphyrins, Ru(phen) [17], Ru(bpy) [18] and Ru(dpp) [19], and Al-Ferron [20] have been reported to be quenched by molecular oxygen. The metal complexes can be excited in the visible spectrum, show large Stokes shifts, and have comparatively long decay times and good photostability. The signal increases strongly

 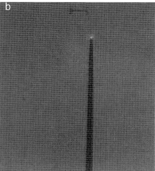

Figure 2.2 Photographs of fiber-optic glucose biosensors. (a) Sensor prepared from an unpulled, single-mode, 3–5 μm core optical fiber. The scale bar represents 50 μm. (b) Sensor prepared from a pulled, micrometer-sized optical fiber tip. The scale bar represents 10 μm [17]. Reprinted with permission from Z. Rosenzweig and R. Kopelman, Analytical properties and sensor size effects of a micrometer-sized optical fiber glucose biosensor, Anal. Chem., 1996, **68**, 1408–1413. Copyright 1996 American Chemical Society.

on addition of glucose (because of the consumption of oxygen). Both fluorescence intensity and lifetime increase. Most common GOx–O_2 sensors consist of the enzyme immobilized on, or in a polymer film (2–5 μm), mostly a hydrogel or polyacrylamide (PAA) [17]. The oxygen-sensitive indicator dye is immobilized in a second polymer, preferably absorbed on silica gel beads, and incorporated in a silicone film [21]. This type of sensor with Ru(dpp) as the probe for oxygen is the only optical sensor produced commercially in very large numbers. Sensors are also known where the enzyme and dye are embedded in the same layer. The sensing layers are then attached to a support such as PS, controlled-pore glass (CPG) [15], or an optical fiber (Figure 2.2), and are often shielded by a diffusion barrier layer [22].

A fiber-optic dual sensor for the continuous and simultaneous determination of glucose and oxygen with Ru(bpy)$_3$ as O_2 transducer was described in 1995 [18]. The sensor comprises two sensing sites in defined positions on the distal end of an imaging fiber (Figure 2.3). Each sensing site contains an individual polymer cone covalently attached to the activated fiber surface using localized photopolymerization. The fluorescence images of both sensing sites are captured with a charge-coupled device (CCD) camera and oxygen quenching data for both sensing sites are fitted by a two-site Stern–Volmer quenching model. Within response times from 9 to 28 s, glucose can be detected in the range 0–20 mM with this self-referenced scheme.

Further glucose biosensors have used unusual supports (eggshell membranes or swim bladder membranes) on which GOx was immobilized [23, 24]. The glucose biosensor has a long shelf-life and was successfully applied to the determination of glucose in a beverage sample in a flow cell.

A limitation of oxygen-based detection schemes in general and oxygen-depleting detection schemes in particular is based on the varying background of oxygen in samples. This may be overcome by using a large excess of air-saturated buffer

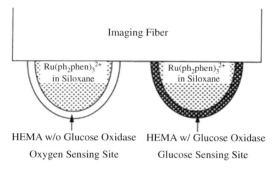

Figure 2.3 Schematic showing a cross-sectional view of the glucose and oxygen sensing sites of a fiber-optic dual sensor for the continuous and simultaneous determination of glucose [18]. Reprinted with permission from L. Li and D.R. Walt, Dual-analyte fiber-optic sensor for the simultaneous and continuous measurement of glucose and oxygen, *Anal. Chem.*, 1995, **67**, 3746–3752. Copyright 1995 American Chemical Society.

solution, or by compensating the variable oxygen background by determining it independently, for example, with a second sensor. This kind of dual sensor was reported first by Li and Walt [18], subsequently by Wolfbeis *et al.* [22], and more recently by Klimant's group [25, 26]. They used a dual sensor, consisting of two commercially available oxygen optodes in close proximity, where one had been modified with GOx and the other served as a reference (Figure 2.4). This sensor is unaffected by fluctuations of the oxygen concentration in the sample and also allows for the compensation of slight temperature fluctuations. Comparable sensor schemes have been patented [27–29].

2.4.2
Sensors Based on the Use of Microparticles (μPs) and Nanoparticles (NPs)

Particle-based sensors have attracted substantial interest for intracellular glucose testing in recent years. They may be applied in the future in the bloodstream as molecular analytical machines reporting blood glucose levels, provided that the optical signal they are giving can be interrogated. Xu *et al.* incorporated GOx, sulfo-Ru (dpp)$_3$ and Oregon Green 488–dextran (as a reference-dye) in 45 nm diameter PAA nanoparticles to obtain so-called PEBBLE sensors for self-referenced ratiometric measurements [30]. The PEBBLEs were implanted into living cells with minimal physical and chemical perturbations to their biological functions. Another implantable microsensor was reported by Brown and co-workers (Figure 2.5) [31, 32].

GOx and horseradish peroxidase (HRP) conjugated CdSe/ZnS core–shell quantum dots (QDs) were used for FRET-based glucose sensing. Upon irradiation, the QDs shuttle electrons to GOx and HRP, upon which they rapidly oxidize glucose to gluconic acid and reduce H_2O_2 (or O_2). This nonradiative energy transfer results in quenching of the QD emission proportional to the concentration of glucose up to 28 mM with a 30 min response time [33].

2.4 Fluorescent Sensing of Glucose via Measurement of the Consumption of Oxygen

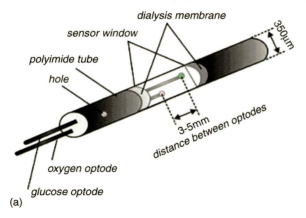

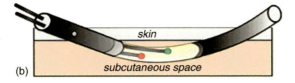

Figure 2.4 Schematic representation of (a) hybrid sensor and (b) implanted hybrid sensor [26]. Reprinted from A. Pasic, H. Koehler, I. Klimant, and L. Schaupp, Miniaturized fiber-optic hybrid sensor for continuous glucose monitoring in subcutaneous tissue, *Sens. Actuators B*, 2007, **122**, 60–68, with permission from Elsevier.

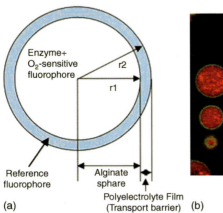

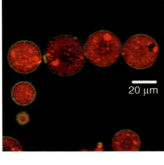

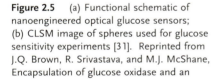

Figure 2.5 (a) Functional schematic of nanoengineered optical glucose sensors; (b) CLSM image of spheres used for glucose sensitivity experiments [31]. Reprinted from J.Q. Brown, R. Srivastava, and M.J. McShane, Encapsulation of glucose oxidase and an oxygen-quenched fluorophore in polyelectrolyte-coated calcium alginate microspheres as optical glucose sensor systems, *Biosens. Bioelectron.*, 2005, **21**, 212–216, with permission from Elsevier.

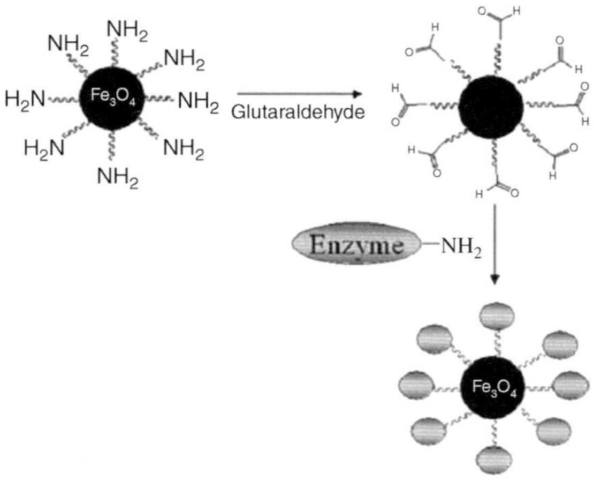

Figure 2.6 Covalent conjugation of glucose oxidase to amino-functionalized magnetic nanoparticles [34]. Reprinted from L.M. Rossi, A.D. Quach, and Z. Rosenzweig, Glucose oxidase–magnetite nanoparticle bioconjugate for glucose sensing, *Anal. Bioanal. Chem.*, 2004, **380**, 606, with permission from Springer Science + Business Media.

GOx-coated magnetite (Fe_3O_4) nanoparticles with $Ru(phen)_3$ can be used as nanometric glucose sensors in glucose solutions (Figure 2.6) [34]. This allows the magnetic separation of the nanoparticles from the analyte sample, permitting reuse in multiple samples.

The O_2 transducer is often influenced by temperature. This drawback can be compensated for by dual sensors that sense both oxygen and temperature [35]. Here, the oxygen probe Pt(porph) in a layer of PS and the europium complex $Eu(tta)_3(dpbt)$ in a layer of poly(vinyl methyl ketone) as the temperature probe allow dual lifetime determination to monitor the consumption of oxygen at varying temperatures. The layers lie upon one another on a polyester support.

2.5
Fluorescent Sensing of Glucose via Measurement of the Formation of Hydrogen Peroxide Caused by the Action of GOx

HP produced in the enzymatic reaction can be linked to the concentration of glucose, as shown in Eq. (2.1). Monitoring of HP has the advantage of a measurement against an almost zero background. Only few sensors for continuous mode operation have been reported.

A probe for HP was reported consisting of a conjugated cationic polymer incorporated with boronate-protected peroxyfluorescein. HP deprotects the fluorescein to create anionic fluorescein, which interacts electrostatically with the cationic fluorene. The FRET between fluorene (donor) and fluorescein (acceptor) can be used to monitor glucose in the nanomolar [36] to micromolar range [37].

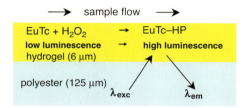

Figure 2.7 Cross-section of the membrane of a glucose biosensor using luminsecent EuTc–HP.

An enzymatic assay for glucose based on luminescent europium(III) tetracycline (EuTc) was described [38]. Weakly luminescent EuTc and enzymatically generated HP form a strongly luminescent complex (EuTc–HP). The probe can be excited at 400 nm and emits at 616 nm. It responds to HP by a 15-fold increase in luminescence and a strong increase in lifetime. This permits lifetime-based measurements of glucose with a substantially reduced luminescence background (Figure 2.7).

The reaction of EuTc with HP is fully reversible but takes about 10 min in both directions and is strongly pH dependent. The complex was used for time-resolved imaging of glucose [39]. Furthermore, GOx was covalently labeled to Mn-doped ZnS QDs. HP produced in the presence of glucose quenches the emission of the dots. The nanosensors were successfully applied to serum samples (Figure 2.8) [40].

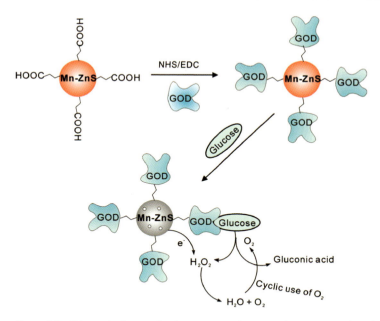

Figure 2.8 Schematic diagram for the preparation and application of GOx–Mn-doped ZnS QD bioconjugate for glucose sensing [40]. Reprinted with permission from P. Wu, Y. He, H.-F. Wang, and X.-P. Yan, Conjugation of glucose oxidase onto Mn-doped ZnS quantum dots for phosphorescent sensing of glucose in biological fluids, *Anal. Chem.*, 2010, **82**, 1427–1433. Copyright 2010 American Chemical Society.

2.6
Enzymatic Sensing of Glucose via Changes of pH

Changes of pH may also serve as analytical information to sense glucose according to Eq. (2.2). However, this approach is limited because often the initial pH of the sample and its buffer capacity are unknown. Therefore, only a few optical biosensors make use of pH transducers.

McCurley reported on a cadaverine unit linked to rhodamine that was incorporated into a hydrogel (along with GOx and catalase) placed at the end of an optical fiber [41]. The cadaverine moiety is protonated on the enzymatic reaction of glucose, which causes swelling of the hydrogel. This, in turn, decreases the fluorescence intensity of the rhodamine in the 0–1.6 mM glucose range.

Tohda and Gratzl [42] covalently immobilized GOx on cellulose acetate beads and embedded a pH-sensitive phenoxazine derivative in polymer microbeads. These beads, along with white reference beads, were arranged in a microarray to read the reversible color change caused by pH with a CCD camera. Another scheme comprises internally cationically charged biocompatible capsules containing the pH probe 8-hydroxypyrene-1,3,6-trisulfonic acid (HPTS) and electrostatically adsorbed GOx. pH changes induce ratiometric changes of emission spectra and are monitored as a function of glucose concentration in the physiological range [43]. Two different sensors with a pH-sensitive azlactone and GOx embedded in a sol–gel were reported [44]. A setup with immobilization of GOx and fluorescein isothiocyanate (FITC) as pH reporter in a polymer at the end of an optical fiber was patented by Applied Research Systems [45].

2.7
Fluorescent Sensing of Glucose via Synthetic Boronic Acids

Boronic acids can reversibly react with 1,2- or 1,3-diols in aqueous solution to form five- or six-membered cyclic esters. The state of the art in optical boronate probes for saccharides has been reviewed recently [3]. The rigid *cis*-diols found in many saccharides generally form stronger complexes than acyclic diols such as ethylene glycol and *trans*-diols. On binding a saccharide, the neutral trigonal boronic acid transforms into the anionic tetrahedral form and releases a proton (Figure 2.9). This change in geometry is accompanied by a reduction in the pK_a from $\sim$9 to $\sim$6 because the electrophilicity of the boronic acid group is enhanced on interaction with a diol. Hence the trigonal form is present at pH values below the pK_a but in the presence of a saccharide it is converted into its tetrahedral anionic form. The pK_a of the boronic acid is tunable to lower values with electron-withdrawing groups, whereas electron-donating groups increase the pK_a. When attached to an appropriate fluorophore, the changes in the symmetry of the boronic acid also induce changes in fluorescence emission (intensity, lifetime, and polarization). One fundamental disadvantage of most boronic acid-based probes for glucose sensing is either a higher selectivity for fructose than for glucose or/and a stronger response to fructose than to glucose.

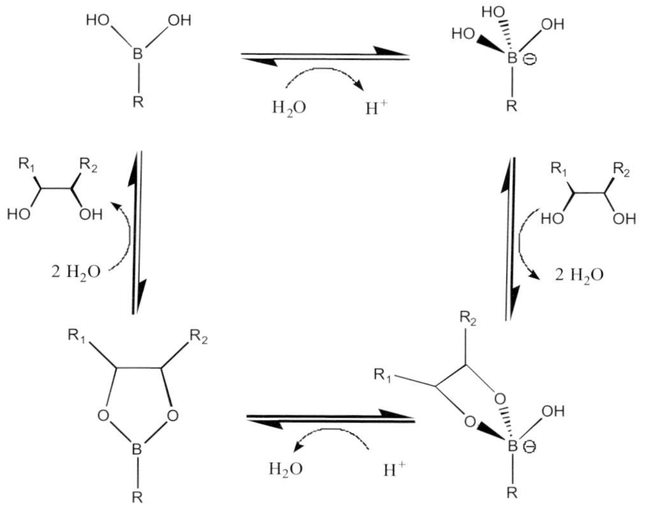

Figure 2.9 Geometries of reaction products of boronic acids with water or 1,2-diols (e.g., glucose).

Three quinoline-based probes for the determination of glucose in tear fluid were presented [46, 47]. The probes were structurally modified to allow compatibility with disposable plastic contact lenses. The sugar-bound pK_a is reduced, favorable with the mildly acidic lens environment. One probe shows a similar affinity for both glucose and fructose from the emission decrease at 450 nm. Glucose and fructose were determined at sub-millimolar levels with a response time of 10 min (Figure 2.10). The probe showed negligible leaching and was quenched only modestly by chloride.

Competitive assay schemes use the reversible interaction between Ru(2,2'-bipyridine)$_2$-(5,6-dihydroxy-1,10-phenanthroline) and arylboronic acid derivatives at pH 8 [48, 49]. Complexation is accompanied by a strong increase in fluorescence intensity at 620 nm. Addition of glucose reduces emission intensity because of intercepting the binding between boronic acid and the metal complex. The ruthenium complex may also be incorporated into a glucose-permeable and implantable polymer [50, 51].

Singaram and co-workers [52] reported on a two-component system comprising an anionic fluorescent dye and a cationic viologen quencher containing a bisboronic acid functionality to form a nonfluorescent ion pair. Their electrostatic interaction is reduced when a saccharide binds to the viologen to yield a negatively charged boronate ester and the fluorescence intensity is increased (Figure 2.11). A modular ensemble composed of three different viologen quenchers was also shown. The increase in luminescence of 11 anionic fluorescent dyes (e.g., HPTS) on addition of 12 saccharides in the presence of these three quenchers was determined and evaluated via two-dimensional linear discriminant analysis. Differentiation between glucose, fructose, mannose, and galactose at a concentration of 2 mM was accomplished [53].

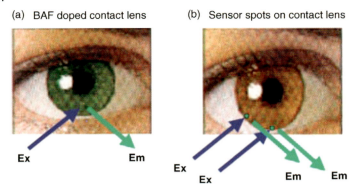

Figure 2.10 Potential methods for noninvasive continuous tear glucose monitoring.
(a) Contact lens doped with optical probe for glucose. (b) Sensor spots on the surface of the lens to monitor additionally other analytes in addition to glucose, such as chloride or oxygen. Sensor spot regions may also allow for ratiometric, lifetime, or polarization based fluorescence glucose sensing [47]. Reprinted from R. Badugu, J.R. Lakowicz, and C.D. Geddes, A glucose sensing contact lens: a non-invasive technique for continuous physiological glucose monitoring, *J. Fluoresc.*, 2003, **13**, 371, with permission from Springer Science + Business Media.

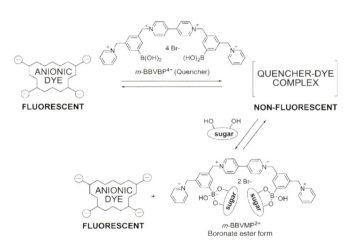

Figure 2.11 General sensing mechanism with the boronic acid-functionalized viologen quencher m-BBVBP^{4+} used as the saccharide receptor unit. Viologens quench the luminescence of dyes through an ionic attraction to form a nonfluorescent complex. Upon saccharide binding to the boronic acid of the viologen quencher, the negatively charged boronate ester causes a diminished electrostatic attraction resulting in a fluorescence increase [52]. Reproduced from D. B. Cordes, A. Miller, S. Gamsey, S. Zach, P. Thoniyot, R. Wessling, and B. Singaram, Optical glucose detection across the visible spectrum using anionic fluorescent dyes and a viologen quencher in a two-component saccharide sensing system, *Org. Biomol. Chem.*, 2005, **3**, 1708–1713, with permission from The Royal Society of Chemistry.

CdSe QDs were utilized as fluorescent reporters in combination with a viologen quencher with a boronic acid functionality as the glucose receptor. The fluorescence of the reporter is quenched by the viologen before addition of glucose. When glucose is added to the solution, the emission at 604 nm recovers. Both carboxy- and amino-substituted QDs are quenched by a viologen in aqueous solution at pH 7.4 [54].

2.8
Fluorescent Sensing of Glucose via Concanavalin A

ConA is a plant lectin protein originally extracted from jack beans. The ConA tetramer consists of two dimers and has four binding sites for glucose. Its specific function involves the agglutination of biologically relevant complexes such as glycoproteins, starches, and erythrocytes. ConA is used traditionally in a competitive assay where glucose and another carbohydrate such as dextran mannoside or glycated proteins compete for the lectin-binding sites. The protein and the competitor are generally labeled with appropriate fluorescent dyes. An inherent but often disregarded problem with ConA-based sensors regarding sensor stability is that unbound lectin tends to aggregate irreversibly over a period of some hours.

Schultz *et al.* [55] described a glucose sensor in which ConA is immobilized inside a hollow dialysis fiber connected to a fluorescence detection system by a single optical fiber. FITC-labeled dextran was the competing ligand to pass in and out of the fiber. Glucose displaces the competitor from ConA and increases the concentration of free dextran and, consequently, the fluorescence. Optimization steps comprised immobilization of ConA on Sepharose [56] and optical readout to reduce response times to 5 to 7 min [57].

Transdermal glucose monitoring was achieved [58] by confining Alexa488-labeled ConA and colored macroporous beads inside a sealed, small segment of a hollow-fiber dialysis membrane. Immobilized pendant glucose moieties inside the colored beads compete with glucose for binding of ConA. In the absence of glucose, the excitation and emission of Alexa488 is blocked. Glucose diffuses through the membrane into the sensor chamber and competitively displaces Alexa 488–ConA from the glucose residues of the beads. Consecutively, Alexa488–ConA is fully exposed to the excitation light, and the increase in fluorescence at 520 nm is measured (Figure 2.12). The sensor exhibits the strongest signal change between 0.2 and 30 mM, with a response time of around 4 min. *In vivo* tests were conducted with a ratiometric FRET system consisting of Cy7-labeled agarose-immobilized ConA as acceptor/quencher and unbound Alexa647–dextran as donor. In the absence of glucose, dextran is bound to ConA, therefore both are in close proximity, enabling a FRET. Glucose displaces dextran from ConA, which results in decreased FRET and an increase in Alexa647 fluorescence at 670 nm. *In vivo* tests with a small hollow fiber implanted in dermal skin tissue revealed a delay of 5–15 min in actual blood glucose concentrations [59]. PreciSense has patented a similar but lifetime-based sensor setup [60].

A hollow-fiber setup [61] uses FRET between rhodamine-labeled ConA and fluorescein-labeled dextran. The emission of dextran is restored on addition of

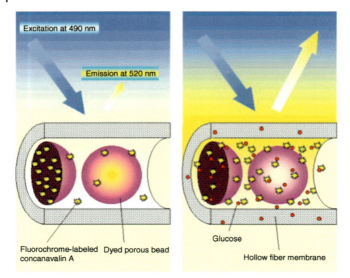

Figure 2.12 In the absence of glucose, fluorochrome-labeled ConA is bound to fixed glucose residues inside the porous beads (a) in a hollow fiber. The beads are colored with dyes to prevent the excitation light from penetrating into them and inducing ConA to fluoresce, thus keeping the fluorescence emission at 520 nm. After diffusion of glucose through the hollow-fiber membrane, ConA is displaced from the beads and diffuses out of them, and hereby fluorochrome-labeled ConA becomes exposed to excitation light, resulting in a strong increase in fluorescence (b) [58]. Reprinted with permission from R. Ballerstadt and J.S. Schultz, A fluorescence affinity hollow fiber sensor for continuous transdermal glucose monitoring, *Anal. Chem.*, 2000, **72**, 4185–4192. Copyright 2000 American Chemical Society.

glucose due to competitive displacement. Novartis has patented a related sensor that can be incorporated into a contact lens and an apparatus to irradiate the sensor dyes and detect FRET [62]. Cheung *et al.* [63] reported on a similar system in microplates (MTP). Donor and acceptor were incorporated into hydrogel pads that were then attached to hydrophobic surfaces of the wells of a microtiter plate (Figure 2.13). A layer-by-layer (LbL) self-assembly process was used to coat the surface of the hydrogel pads with polyelectrolyte multilayers with the aim of creating a permeability-controlled membrane with nanometer thickness. LbL hydrogel pads allow regeneration within 17 min with a signal reproducibility of more than 90% to yield a reusable and reagentless optical bioassay platform.

Birch and co-workers developed a non-radiative FRET sensing setup [64]. Con A was labeled with the NIR fluorescent protein allophycocyanin (APC) as donor and dextran labeled with Malachite Green (MG) as acceptor, which shields APC emission as long as dextran is bound to the lectin. Glucose competitively displaces dextran–MG and leads to restoration of the APC fluorescence at 663 nm, quantified by time-domain fluorescence lifetime measurements. Another setup exploits the luminescence decay time of Cy5-labeled ConA [65] as a FRET donor. The acceptor consisted of

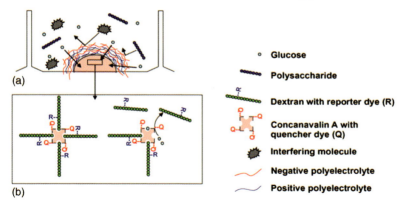

Figure 2.13 Schematic diagram of a LbL hydrogel pad (a) and the entrapped Q–ConA/R–dextran glucose sensing system (b) [63]. Reprinted from K.Y. Cheung, W.C. Mak, and D. Trau, Reusable optical bioassay platform with permeability-controlled hydrogel pads for selective saccharide detection, Anal. Chim. Acta, 2008, **607**, 204–210, with permission from Elsevier.

insulin covalently linked to MG and to maltose (MIMG) to provide binding affinity for ConA. MIMG was displaced competitively from ConA in the presence of glucose, and the Cy5 emission and lifetime increased. This sensor has the advantage of a reduced rate of aggregation and better reversibility in comparison with other setups by using a protein as glucose competitor instead of dextran. A further benefit is its NIR emission, allowing measurement of decay times through skin. This suggests the possibility of application as an implantable glucose sensor. Further experiments were carried out with the sensing setup, where the Cy5 label of ConA was replaced with Ru (bpy)$_3$ [66]. The advantages here are the large Stokes shift and long decay times, which simplifies instrumentation for phase-modulation lifetime measurements.

Another sensor based on FRET was described by Liao *et al.* [67]. They used QDs labeled with ConA as donor. Rhodamine-labeled β-cyclodextrin served as acceptor, showing a lower affinity to ConA to improve the signal detection sensitivity 60-fold while preserving glucose sensitivity in the physiological range. ConA and the dextrin were photopolymerized in a hydrogel at the end of an optical silica fiber with a diameter of only 250 μm (Figure 2.14). This permitted interstitial and minimal invasive *in vivo* glucose determination. The ratio of the emission at 525 and 570 nm correlated linearly with the glucose concentration up to 28 mM.

A related sensing scheme by Tang *et al.* [68] comprised ConA-labeled CdTe QDs and β-cyclodextrin-modified gold nanoparticles (AuNPs). In absence of glucose, FRET is observed between QDs as donors and AuNPs as acceptors. In the presence of glucose, the AuNP–CDs are displaced competitively by glucose, resulting in recovery of the QD fluorescence. The setup was also applied to direct glucose determination in human serum with satisfactory results.

The first solution-phase nanotube affinity sensor is based on dextran-modified single-walled carbon nanotubes (SWNTs) [69]. These were aggregated by ConA,

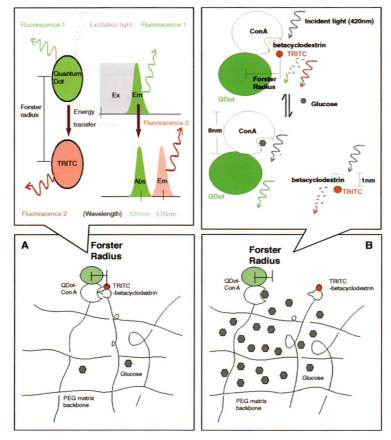

Figure 2.14 Changes in FRET between fluorophores covalently immobilized on the flexible PEG matrix depend on changes in the distance between them, which in turn depends on the competitive natural affinity between Con A and various saccharides such as betacyclodextrin and glucose.

resulting in quenched SWNT photoluminescence ($\lambda_{ex}/\lambda_{em}$ 633/>900 nm). Addition of glucose restored the initial luminescence.

2.9
Fluorescent Sensing of Glucose using Glucose-Specific Apoenzymes

In this (non-kinetic) scheme, the enzyme acts as both a molecular receptor and transducer. It represents an attractive alternative to sensing schemes utilizing the specific binding of glucose to ConA or to a bacterial or engineered glucose-binding protein. Typical enzymes used in this format include GOx and glucose dehydrogenase (GDH). In order to prevent the oxidation or reduction of glucose but to retain enzyme specificity, the cofactors FAD or NAD, respectively, have to be removed from the holoenzyme, resulting in apo-GOx or apo-GDH. Two major sensing schemes have been reported for apo-enzyme affinity assays:

1) labeling of apo-enzyme with a polarity-sensitive dye and utilization of conformational changes on binding glucose
2) competitive assay similar to ConA using a labeled dextran as a competitor.

Lakowicz and co-workers [70] found that on binding glucose to the apo-GOx, a decrease of up to 18% of its intrinsic tryptophan fluorescence occurs. As UV wavelengths are not useful for routine diagnostic purposes, apo-GOx was non-covalently labeled with the environment sensitive dye 8-anilino-1-naphthalenesulfonate (ANS). In the presence of glucose, both the fluorescence intensity and the mean lifetime of the bound ANS decreased due to conformational changes of apo-GOx. These results suggest that apo-glucose oxidase can be used as a reversible sensor for glucose which – unlike in the case of functional enzymes – is not consumed [71]. A further approach was described using apo-GDH with similar sensitivity [72].

Chinnayelka and McShane [73] described an example of a competitive FRET assay scheme comprising tetramethylrhodamine isothiocyanate (TRITC)-labeled apo-GOx and FITC-labeled dextran. On competitive binding of glucose, a reduction in the efficiency of FRET occurs when the apo-GOx–dextran complex dissociates. A glucose concentration range from 0 to 90 mM can be monitored. The system was applied in an implantable minimally invasive sensor using the LbL self-assembly technology of the former ConA sensor [74]. TRITC–apo-GOx and FITC–dextran were encapsulated in LbL-assembled semipermeable microcapsules (Figure 2.15). These capsules showed five times better specificity for glucose over other saccharides [75]. The sensor design was later extended to longer wavelength emission [76].

2.10
Fluorescent Sensing of Glucose via Glucose-Binding Proteins

The periplasm of Gram-negative bacteria such as *Escherichia coli* comprises a family of proteins [referred to as glucose-binding proteins (GBPs)] with highly specific binding of sugars and other ligands. These proteins are primary receptors for transport and have a very similar structure in common. The monomeric arrangement consists of two globular domains linked by a hinge. The GBPs also show a minute affinity for galactose. The binding constant of native GBP for glucose is in the micromolar range and therefore not very appropriate for sensing in the physiological range. Hence engineered GBPs were developed with mutations of the binding pocket for reduced affinity to glucose. GBP-based sensing schemes are superior to enzyme-based glucose sensors in that they are not limited by enzyme substrates such as

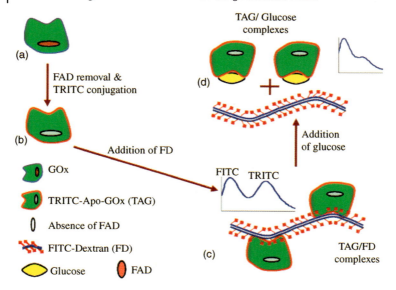

Figure 2.15 Schematic of the RET glucose assay based on competitive binding between dextran and glucose for binding sites on apo-GOx.5 [75]. Reprinted with permission from S. Chinnayelka and M.J. McShane, Microcapsule biosensors using competitive binding resonance energy transfer assays based on apoenzymes, Anal. Chem., 2005, 77, 5501–5511. Copyright 2005 American Chemical Society.

oxygen or potentially toxic reaction products such as HP. Therefore, they are often referred to as reagentless sensors. All sensing schemes utilizing native or engineered GBPs are based on the large change in conformation that occurs on binding of glucose. Hence the binding event can be transduced by site-specifically attached and polarity-sensitive luminophores. Two basic approaches are reported, that is, either one fluorophore labeled on GBP or two fluorophores for ratiometric measurements. The latter have the advantage of being unaffected by variations in excitation light intensity, light path, sample positioning, reagent concentration, and so on.

In the first report on a genetically engineered GBP for use in a glucose sensor [77], two fluorophores were covalently attached to cysteines next to the ligand-binding site. The fluorescence of the first was quenched, whereas the fluorescence of the other was enhanced on addition of glucose due to changes in the polarity of the microenvironment. A series of patents cover this sensing scheme [78–80]. GBP is most often labeled with dyes such as Cy3, Cy5 [81], IANBD, Nile Red derivatives [82], and dyes with a coumarin or benzodioxazole nucleus. The conjugate is then embedded in a polymer and often attached to the end of an optical fiber.

Ye and Schultz [83] were among the first to report on a continuous dialysis hollow-fiber microsensor based on an engineered GBP showing a FRET. The so-called glucose indicator protein (GIP) contains a UV-excitable green fluorescent protein (GFPuv) and a yellow fluorescent protein (YFP) fused to the GBP. The reporter permits efficient FRET in the absence of glucose, whereas on addition of glucose the distance between the two fluorescent proteins increases accompanied by a decreased FRET (Figure 2.16). A similar setup was used to monitor intracellular glucose levels. GBP was labeled with cyano fluorescent protein and YFP to create a FRET system. Again, addition of glucose decreased the emission of YFP at 535 nm. The developed GBP mutant showed a K_d of ~600 μM, which allowed specific monitoring of the glucose distribution and levels in cells [84]. An implantable glucose-sensing device based on the GIP–FRET sensing scheme was patented [85].

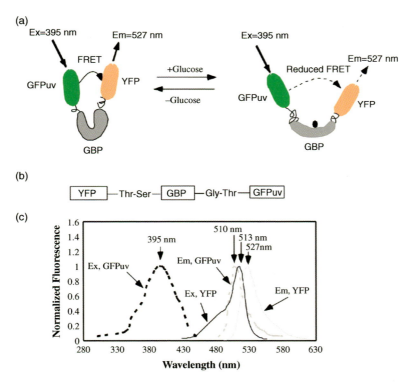

Figure 2.16 Design of a GIP (glucose indicator protein) for sensing glucose based on FRET. (a) Diagram of the GIP structure showing how FRET between two GFPs can measure glucose concentration. The GBP adopts an "open" form in the presence of the glucose, which triggers a conformation change, causing two GFPs apart from the center of GBP leading to the change in FRET. The large dot at the GBP on the right panel represents one molecule of glucose bound to the binding cleft of the GBP. (b) Domain structure of the GIP. GFPuv, green fluorescent protein with several mutations to enhance the excitation by UV light; YFP, yellow fluorescent protein; GBP, glucose-binding protein. (c) Spectral overlap of GFPuv and YFP. The absorbance spectra are denoted by black lines and the emission spectra by gray lines [83]. Reprinted with permission from K. Ye and J.S. Schultz, Genetic engineering of an allosterically based glucose indicator protein for continuous glucose monitoring by fluorescence resonance energy transfer, *Anal. Chem.*, 2003, **75**, 3451–3459. Copyright 2003 American Chemical Society.

2.11
Conclusion and Outlook

Fluorescent technologies, many involving FRET, provide attractive methods for sensing glucose. They have specific features that make them applicable to challenges not covered by electrodes, for example, intracellular sensing and imaging. Clinical sensing of glucose is the largest market and is dominated by electrochemical sensors. All the optical sensors used for routine clinical assays are based on the use of the (very robust) enzyme GOx and fluorimetric measurement of the oxygen consumed. The other schemes presented here still await commercialization. Notwithstanding this, they represent valuable tools for research and, predictably, will find their use in clinical situations where existing sensors cannot be applied.

References

1 Kondepati, V.R. and Heise, H.M. (2007) Recent progress in analytical instrumentation for glycemic control in diabetic and critically ill patients. *Anal. Bioanal. Chem.*, **388**, 545–563.

2 Heller, A. (1999) Implanted electrochemical glucose sensors for the management of diabetes. *Annu. Rev. Biomed. Eng.*, **1**, 153–175.

3 Mader, H.S. and Wolfbeis, O.S. (2008) Boronic acid based probes for microdetermination of saccharides and glycosylated biomolecules. *Microchim. Acta*, **162**, 1–34.

4 James, T.D., Phillips, M.D., and Shinkai, S. (2006) Boronic acids in saccharide recognition, in *Monographs in Supramolecular Chemistry* (ed. J.F. Stoddard), Royal Society of Chemistry, Cambridge.

5 De Luca, P., Lepore, M., Portaccio, M., Esposito, R., Rossi, S., Bencivenga, U., and Mita, D.G. (2007) Glucose determination by means of steady-state and time-course UV fluorescence in free or immobilized glucose oxidase. *Sensors*, **7**, 2612–2625.

6 Sierra, J.F., Galbán, J., and Castillo, J.R. (1997) Determination of glucose in blood based on the intrinsic fluorescence of glucose oxidase. *Anal. Chem.*, **69**, 1471–1476.

7 Hussain, F., Birch, D.J.S., and Pickup, J.C. (2005) Glucose sensing based on the intrinsic fluorescence of sol–gel immobilized yeast hexokinase. *Anal. Biochem.*, **339**, 137–143.

8 Trettnak, W. and Wolfbeis, O.S. (1989) Fully reversible fibre-optic glucose biosensor based on the intrinsic fluorescence of glucose oxidase. *Anal. Chim. Acta*, **221**, 195–203.

9 Narayanaswamy, R. and Sevilla, F. III (1988) An optical fibre probe for the determination of glucose based on immobilized glucose dehydrogenase. *Anal. Lett.*, **21**, 1165–1175.

10 D'Auria, S., Di Cesare, N., Siaano, M., Gryczynski, Z., Rossi, M., and Lakowicz, J.R. (2002) A novel fluorescence competitive assay for glucose determinations by using a thermostable glucose cokinase from the thermophilic microorganism *Bacillus stearothermophilus*. *Anal. Biochem.*, **303**, 138–144.

11 De Marcos, S., Galindo, J., Sierra, J.F., Galban, J., and Castillo, J.R. (1999) An optical glucose biosensor based on derived glucose oxidase immobilised onto a sol–gel matrix. *Sens. Actuators B*, **57**, 227–232.

12 Sanz, V., Galban, J., de Marcos, S., and Castillo, J.R. (2003) Fluorometric sensors based on chemically modified enzymes: glucose determination in drinks. *Talanta*, **60**, 415–423.

13 Sierra, J.F., Galban, J., de Marcos, S., and Castillo, J.R. (1998) Fluorimetric–enzymatic determination of glucose based on labelled glucose oxidase. *Anal. Chim. Acta*, **368**, 97–104.

14 Trettnak, W., Leiner, M.J., and Wolfbeis, O.S. (1988) Fibre optic glucose biosensor with an oxygen optrode as the transducer. *Analyst*, **113**, 1519–1523.

15 Papkovsky, D.B., Ovchinnikov, A.N., Ogurtsov, V.I., Ponomarev, G.V., and Korpela, T. (1998) Biosensors on the basis of luminescent oxygen sensor: the use of microporous light-scattering support materials. *Sens. Actuators B*, **51**, 137–145.

16 Collins, T.C., Munkholm, C., and Slovacek, R.E., assigned to Bayer Corporation (2000) Optical oxidative enzyme-based sensors. PCT Int. Appl. WO 2000011205.

17 Rosenzweig, Z. and Kopelman, R. (1996) Analytical properties and sensor size effects of a micrometer-sized optical fiber glucose biosensor. *Anal. Chem.*, **68**, 1408–1413.

18 Li, L. and Walt, D.R. (1995) Dual-analyte fiber-optic sensor for the simultaneous and continuous measurement of glucose and oxygen. *Anal. Chem.*, **67**, 3746–3752.

19 Neubauer, A., Pum, D., Sleytr, U.B., Klimant, I., and Wolfbeis, O.S. (1996) Fibre-optic glucose biosensor using enzyme membranes with 2-D crystalline structure. *Biosens. Bioelectron.*, **11**, 317–325.

20 Valencia-Gonzalez, M.J., Liu, Y.M., Diaz-Garcia, M.E., and Sanz-Medel, A. (1993) Optosensing of D-glucose with an immobilized glucose oxidase minireactor and an oxygen room-temperature phosphorescence transducer. *Anal. Chim. Acta*, **283**, 439–446.

21 Moreno-Bondi, M.C., Wolfbeis, O.S., Leiner, M.J., and Schaffar, B.P. (1990) Oxygen optrode for use in a fiber-optic glucose biosensor. *Anal. Chem.*, **62**, 2377–2380.

22 Wolfbeis, O.S., Oehme, I., Papkovskaya, N., and Klimant, I. (2000) Sol–gel based glucose biosensors employing optical oxygen transducers, and a method for compensating for variable oxygen background. *Biosens. Bioelectron.*, **15**, 69–76.

23 Choi, M.M.F., Pang, W.S.H., Wu, X., and Xiao, D. (2001) An optical glucose biosensor with eggshell membrane as an enzyme immobilisation platform. *Analyst*, **126**, 1558–1563.

24 Zhou, Z., Qiao, L., Zhang, P., Xiao, D., and Choi, M.M.F. (2005) An optical glucose biosensor based on glucose oxidase immobilized on a swim bladder membrane. *Anal. Bioanal. Chem.*, **383**, 673–679.

25 Pasic, A., Koehler, H., Schaupp, L., Pieber, T.R., and Klimant, I. (2006) Fiber-optic flow-through sensor for online monitoring of glucose. *Anal. Bioanal. Chem.*, **386**, 1293–1320.

26 Pasic, A., Koehler, H., Klimant, I., and Schaupp, L. (2007) Miniaturized fiber-optic hybrid sensor for continuous glucose monitoring in subcutaneous tissue. *Sens. Actuators B*, **122**, 60–68.

27 Slate, J.B., and Lord, P.C., assigned to Minimed Inc. (1996) Optical glucose sensor. Can. Pat. Appl. CA 2152862.

28 Curry, K.M., assigned to Baxter International, Inc. (1989) Fiber optic biosensor containing oxygen-quenching fluorescent dyes and gas-permeable polymers. Eur. Pat. Appl. EP 309214.

29 Wagner, D.B., assigned to Becton, Dickinson and Co. (1988) Method and apparatus for monitoring glucose. Eur. Pat. Appl. EP 251475.

30 Xu, H., Aylott, J.W., and Kopelman, R. (2002) Fluorescent nano-PEBBLE sensors designed for intracellular glucose imaging. *Analyst*, **127**, 1471–1477.

31 Brown, J.Q., Srivastava, R., and McShane, M.J. (2005) Encapsulation of glucose oxidase and an oxygen-quenched fluorophore in polyelectrolyte-coated calcium alginate microspheres as optical glucose sensor systems. *Biosens. Bioelectron.*, **21**, 212–216.

32 Brown, J.Q., Srivastava, R., Zhu, H-g., and McShane, M.J. (2006) Enzymatic fluorescent microsphere glucose sensors: evaluation of response under dynamic conditions. *Diabetes Technol. Ther.*, **8**, 288–295.

33 Duong, H.D. and Rhee, J.I. (2007) Use of CdSe/ZnS core–shell quantum dots as energy transfer donors in sensing glucose. *Talanta*, **73**, 899–905.

34 Rossi, L.M., Quach, A.D., and Rosenzweig, Z. (2004) Glucose oxidase–magnetite nanoparticle bioconjugate for glucose sensing. *Anal. Bioanal. Chem.*, **380**, 606–613.

35 Nagl, S., Stich, M.I.J., Schäferling, M., and Wolfbeis, O.S. (2009) Method for simultaneous luminescence sensing of two species using optical probes of different decay time, and its application to an enzymatic reaction at varying temperature. *Anal. Bioanal. Chem.*, **393**, 1199–1207.

36 He, F., Tang, Y., Yu, M., Wang, S., Li, Y., and Zhu, D. (2006) Fluorescence-amplifying detection of hydrogen peroxide with cationic conjugated polymers, and its application to glucose sensing. *Adv. Funct. Mater.*, **16**, 91–94.

37 He, F., Feng, F., Wang, S., Li, Y., and Zhu, D. (2007) Fluorescence ratiometric assays of hydrogen peroxide and glucose in serum using conjugated polyelectrolytes. *J. Mater. Chem.*, **35**, 3702–3707.

38 Wolfbeis, O.S., Schaeferling, M., and Duerkop, A. (2003) Reversible optical sensor membrane for hydrogen peroxide using an immobilized fluorescent probe, and its application to a glucose biosensor. *Microchim. Acta*, **143**, 221–227.

39 Schäferling, M., Wu, M., and Wolfbeis, O.S. (2004) Time-resolved fluorescent imaging of glucose. *J. Fluoresc.*, **5**, 561–568.

40 Wu, P., He, Y., Wang, H.-F., and Yan, X.-P. (2010) Conjugation of glucose oxidase onto Mn-doped ZnS quantum dots for phosphorescent sensing of glucose in biological fluids. *Anal. Chem.*, **82**, 1427–1433.

41 McCurley, M.F. (1994) An optical biosensor using a fluorescent, swelling sensing element. *Biosens. Bioelectron.*, **9**, 527–533.

42 Tohda, K. and Gratzl, M. (2006) Micro-miniature autonomous optical sensor array for monitoring ions and metabolites 2: color responses to pH, K^+ and glucose. *Anal. Sci.*, **22**, 937–941.

43 Nayak, S.R. and McShane, M.J. (2006) Fluorescence glucose monitoring based on transduction of enzymatically-driven pH changes within microcapsules. *Sensor Lett.*, **4**, 433–439.

44 Ertekin, K., Cinar, S., Aydemir, T., and Alp, S. (2005) Glucose sensing employing fluorescent pH indicator: 4-[(p-N,N-dimethylamino)benzylidene]-2-phenyloxazole-5-one. *Dyes Pigm.*, **67**, 133–138.

45 Attridge, J.W. and Robinson, G.A., assigned to ARS Holding 89 (1993) Enzyme biosensor with optical waveguide for optical assay. PCT Int. Appl. WO 9325892.

46 Badugu, R., Lakowicz, J.R., and Geddes, C.D. (2005) Boronic acid fluorescent sensors for monosaccharide signalling based on the 6-methoxyquinolinium heterocyclic nucleus: progress toward non-invasive and continuous glucose monitoring. *Bioorg. Med. Chem.*, **13**, 113–119.

47 Badugu, R., Lakowicz, J.R., and Geddes, C.D. (2003) A glucose sensing contact lens: a non-invasive technique for continuous physiological glucose monitoring. *J. Fluoresc.*, **13**, 371–374.

48 Murtaza, Z., Tolosa, L., Harms, P., and Lakowicz, J.R. (2002) On the possibility of glucose sensing using boronic acid and a luminescent ruthenium metal–ligand complex. *J. Fluoresc.*, **12**, 187–192.

49 Lakowicz, J.R. and Murtaza, Z., assigned to the Regents of the University of Maryland (2000) Glucose biosensor using fluorescent metal–ligand complexes. PCT Int. Appl. WO 2000043536.

50 Satcher, J.H. Jr, Lane, S.M., Darrow, C.B., Cary, D.R., and Tran, J.A., assigned to the Regents of the University of California and Minimed Inc. (2001) Glucose sensing molecules having selected fluorescent properties. PCT Int. Appl. WO 2001020334.

51 Satcher, J.H., Lane, S.M., Darrow, C.B., Cary, D.R., and Tran, J.A., assigned to University of California and Minimed Inc. (2004) Saccharide sensing molecules having enhanced fluorescent properties. U.S. Pat. 6673625.

52 Cordes, D.B., Miller, A., Gamsey, S., Zach, S., Thoniyot, P., Wessling, R., and

Singaram, B. (2005) Optical glucose detection across the visible spectrum using anionic fluorescent dyes and a viologen quencher in a two-component saccharide sensing system. *Org. Biomol. Chem.*, **3**, 1708–1713.

53 Schiller, A., Wessling, R.A., and Singaram, B. (2007) A fluorescent sensor array for saccharides based on boronic acid appended bipyridinium salts. *Angew. Chem. Int. Ed.*, **46**, 6457–6459.

54 Cordes, D.B., Gamsey, S., and Singaram, B. (2006) Fluorescent quantum dots with boronic acid substituted viologens to sense glucose in aqueous solution. *Angew. Chem. Int. Ed.*, **45**, 3829–3832.

55 Schultz, J.S., Mansouri, S., and Goldstein, I.J. (1982) Affinity sensor: a new technique for developing implantable sensors for glucose and other metabolites. *Diabetes Care*, **5**, 245–253.

56 Schultz, J.S., assigned to United States Department of Health, Education, and Welfare (1980) Optical sensor for blood plasma constituents. U. S. Pat. 4344438.

57 Mansouri, S. and Schultz, J.S. (1984) A miniature optical glucose sensor based on affinity binding. *Bio/Technology*, **2**, 885–890.

58 Ballerstadt, R. and Schultz, J.S. (2000) A fluorescence affinity hollow fiber sensor for continuous transdermal glucose monitoring. *Anal. Chem.*, **72**, 4185–4192.

59 Ballerstadt, R., Evans, C., Gowda, A., and McNichols, R. (2006) *In vivo* performance evaluation of a transdermal near-infrared fluorescence resonance energy transfer affinity sensor for continuous glucose monitoring. *Diabetes Technol. Ther.*, **8** (3), 296–311.

60 Kristensen, J.S., Gregorius, K., Struve, C., Frederiksen, J.M., and Yu, Y., assigned to Precisense A/S (2006) Sensor for detection of carbohydrate. PCT Int. Appl. WO 2006061207.

61 Meadows, D. and Schultz, J.S. (1993) Design, manufacture and characterization of an optical fiber glucose affinity sensor based on an homogeneous fluorescence energy transfer assay system. *Anal. Chim. Acta*, **280**, 21–30.

62 March, W.F., assigned to Novartis AG and Novartis-Erfindungen Verwaltungsgesellschaft mbH (2002) Apparatus for measuring blood glucose concentrations. PCT Int. Appl. WO 2002087429.

63 Cheung, K.Y., Mak, W.C., and Trau, D. (2008) Reusable optical bioassay platform with permeability-controlled hydrogel pads for selective saccharide detection. *Anal. Chim. Acta*, **607**, 204–210.

64 McCartney, L.J., Pickup, J.C., Rolinski, O.J., and Birch, D.J.S. (2001) Near-infrared fluorescence lifetime assay for serum glucose based on allophycocyanin labelled concanavalin A. *Anal. Biochem.*, **292**, 216–221.

65 Tolosa, L., Malak, H., Rao, G., and Lakowicz, J.R. (1997) Optical assay for glucose based on the luminescnence decay time of the long wavelength dye Cy5. *Sens. Actuators B*, **45**, 93–99.

66 Tolosa, L., Szmacinski, H., Rao, G., and Lakowicz, J.R. (1997) Lifetime-based sensing of glucose using energy transfer with a long lifetime donor. *Anal. Biochem.*, **250**, 102–108.

67 Liao, K.-C., Hogen-Esch, T., Richmond, F.J., Marcu, L., Clifton, W., and Loeb, G.E. (2008) Percutaneous fiber-optic sensor for chronic glucose monitoring *in vivo*. *Biosens. Bioelectron.*, **23**, 1458–1465.

68 Tang, B., Cao, L., Xu, K., Zhuo, L., Ge, J., Li, Q., and Yu, L. (2008) A new nanobiosensor for glucose with high sensitivity and selectivity in serum based on fluorescence resonance energy transfer (FRET) between CdTe quantum dots and Au nanoparticles. *Chem. Eur. J.*, **14**, 3637–3644.

69 Barone, P.W. and Strano, M.S. (2006) Reversible control of carbon nanotube aggregation for a glucose affinity sensor. *Angew. Chem. Int. Ed.*, **45**, 8138–8141.

70 D'Auria, S., Herman, P., Rossi, M., and Lakowicz, J.R. (1999) The fluorescence emission of the apo-glucose oxidase from *Aspergillus niger* as probe to estimate glucose concentrations. *Biochem. Biophys. Res. Commun.*, **263**, 550–553.

71 Lakowicz, J.R. and D'Auria, S., assigned to the Regents of the University of Maryland

(2001) Inactive enzymes as non-consuming sensors. PCT Int. Appl. WO 2001018237.

72 D'Auria, S., Di Cesare, N., Gryczynski, Z., Gryczynski, I., Rossi, M., and Lakowicz, J.R. (2000) A thermophilic apoglucose dehydrogenase as nonconsuming glucose sensor. *Biochem. Biophys. Res. Commun.*, **274**, 727–731.

73 Chinnayelka, S. and McShane, M.J. (2004) Resonance energy transfer nanobiosensors based on affinity binding between apo-GOx and its substrate. *Biomacromolecules*, **5**, 1657–1661.

74 Chinnayelka, S. and McShane, M.J. (2004) Glucose-sensitive nanoassemblies comprising affinity-binding complexes trapped in fuzzy microshells. *J. Fluoresc.*, **5**, 585–595.

75 Chinnayelka, S. and McShane, M.J. (2005) Microcapsule biosensors using competitive binding resonance energy transfer assays based on apoenzymes. *Anal. Chem.*, **77**, 5501–5511.

76 Chinnayelka, S. and McShane, M.J. (2006) Glucose sensors based on microcapsules containing an orange/red competitive binding resonance energy transfer assay. *Diabetes Technol. Ther.*, **8**, 269–278.

77 Marvin, J.S. and Hellinga, H.W. (1998) Engineering biosensors by introducing fluorescent allosteric signal transducers: construction of a novel glucose sensor. *J. Am. Chem. Soc.*, **120**, 7–11.

78 Hellinga, H.W., assigned to the Regents of Duke University (1998) Biosensor. US Pat. 6277627.

79 Jacobson, R.W., Weidemaier, K., Alarcon, J., Herdman, C., and Keith, S., assigned to Becton Dickinson (2005) Fiber optic device for sensing analytes and method of making same. US Pat. Appl., US 20050113658 A1.

80 Pitner, B.J., Sherman, D.B., Ambroise, A., and Thomas, K.J., assigned to Becton Dickinson (2006) Long wavelength thiol-reactive fluorophores. US Pat. Appl., US 20060280652 A1.

81 Tian, Y., Cuneo, M.J., Changela, A., Höcker, B., Beese, L.S., and Hellinga, H.W. (2007) Structure-based design of robust glucose biosensors using a *Thermotoga maritima* periplasmic glucose-binding protein. *Protein Sci.*, **16**, 2240–2250.

82 Thomas, J.K., Sherman, D.B., Amiss, T.J., Andaluz, S.A., and Pitner, J.B. (2006) A long-wavelength fluorescent glucose biosensor based on bioconjugates of galactose/glucose binding protein and nile red derivatives. *Diabetes Technol. Ther.*, **8**, 261–268.

83 Ye, K. and Schultz, J.S. (2003) Genetic engineering of an allosterically based glucose indicator protein for continuous glucose monitoring by fluorescence resonance energy transfer. *Anal. Chem.*, **75**, 3451–3459.

84 Fehr, M., Lalonde, S., Lager, I., Wolff, M.W., and Frommer, W.B. (2003) In vivo imaging of the dynamics of glucose uptake in the cytosol of COS-7 cells by fluorescence nanosensors. *J. Biol. Chem.*, **278**, 19127–19133.

85 Gross, Y. and Hyman, T., assigned to Glusense Ltd. (2007) Implantable biosensor comprising live cells producing fluorescent proteins and an analyte-binding protein, for monitoring a body fluid analyte, such as glucose. PCT Int. Appl. WO 2007110867.

3
Biochips as Novel Bioassays

Bettina Rudolph, Karina Weber, and Robert Möller

3.1
Introduction

Through new methods in molecular biology, biochemistry, and the humane genome project, the amount of genetic information and the information on biomolecules and their interactions has risen exponentially in recent years. This new information allows the in-depth investigation of biological systems but also demand simple, fast, cheap, and miniaturized analytical tools, which can be produced in large numbers. Classical bioassays, which investigate the effect of a substance on a biomolecule, cell, or even organisms, were normally performed in batch experiments, meaning that the interactions between the investigated substance and a particular biomolecule of interest were performed as one experiment. To investigate the interactions of the same substance with another biomolecule, a new experiment had to be conducted. This experimental setup was time consuming and also needed large amounts of reagents. To save time and reagents, bioassays were miniaturized, for example, using microtiter plates, and molecules were immobilized to surfaces to allow the parallel testing of multiple molecules in one experiment. These approaches for the development of paralleled and miniaturized bioassays led to the development of biochips or microarrays.

Biochips normally consist of a planar solid substrate, on which known biomolecules are immobilized in an ordered fashion. The immobilized biomolecules, which are also referred to as capture molecules, are bound to the surface in spots and each spot contains molecules of the same kind. The biochip is then incubated with a solution containing target molecules that can bind specifically to the surface-immobilized capture molecules.

Because of sophisticated production processes, a high spot density can be achieved. Typical feature sizes for DNA chips produced by photolithography are $100\,\mu m^2$ per spot, which allows the immobilization of an enormous amount of different DNA sequences even on $1\,cm^2$. Hence a biochip permits miniaturized, highly parallel investigations of biomolecular interactions.

In this chapter, we introduce the basics of biochip technology and their use in pathogen detection and clinical diagnostics, and also innovative trends in this field aimed at increasing the applicability and efficiency of biochip-based detection of biomolecules.

3.2
Types of Microarrays

In molecular biology and medicine, miniaturized multiplexed systems are increasingly of importance for highly complex and simultaneous (parallel) testing of tens of thousands of probes. Microarrays, also called biochips, have evolved to a mainstream tool of life science research. Depending on the immobilized biological recognition systems, microarrays can be classified in different types. A classification by the type of transducer/detection method is another possibility, but in this chapter we describe types of microarrays depending on their immobilized receptor. The most common types of microarrays are DNA microarrays. Oligonucleotide arrays provide a powerful tool to study molecular basics and interactions on a scale that is impossible using conventional methods [1]. Further applications of microarrays based on immobilization of, for example, proteins, cells, or tissues are described in the following. Nowadays, many commercial standardized microarrays with different immobilized biological recognition systems, especially oligonucleotides or antibodies, are supplied by a variety of companies, for example, Affymetrix, Agilent, Illumina, GE Healthcare, Applied Biosystems, and Agendia. Most of these companies provide custom-made microarrays for individual research. Another common option is to manufacture specific microarrays directly in the laboratory by using different commercially available technologies for their individual production.

3.2.1
DNA Microarrays

Historically, microarray technology evolved from blotting methods. DNA microarrays are based on the Southern blot [2], which is used in medical diagnostics to control DNA sequences in a DNA sample. Instead of DNA fragments, RNA can be analyzed in a Northern blot. These conventional methods are often time consuming and not effective for investigating large numbers of genes. Hence DNA microarrays became a mainstream method in research for investigations of thousands of genes simultaneously.

Basically, there are three strategies for the development of a DNA chip for gene expression studies, which were described by Lemieux *et al.* [3]. First a strategically selection of known genes that play an important role in biological pathways is necessary. A second way uses clones from a library [4] restricted to cDNA microarrays. Third, the complete expressed sequence content of an organism is used for the production of a microarray [5]. For that purpose, a DNA microarray can consist of hundreds of thousands of spots of DNA oligonucleotides, called captures or probes. Each spot includes a specific DNA sequence, which binds complementarily to a target molecule in the investigated sample. Depending on the application, there exist different special arrays for genomic DNA, cDNA, RNA, plasmids, polymerase chain reaction (PCR) products, or long oligonucleotides. Usually complementary binding of capture and target molecules is detected and quantified by optical methods to determine the relative abundance of a nucleic acid target sequence in the sample.

Since the early 1990s, the field of DNA microarray technology has seen a constant technological evolution and intensive studies have been carried out. In 1991, Fodor and the company Affymetrix pioneered oligonucleotide arrays [6]. Numerous reviews and articles have reported and discussed the different possibilities and applications of DNA microarrays [7–25]. In 1999, *Nature Genetics* published a series of reviews about DNA microarrays and their fabrication and detection [26], gene expression profiling [27, 28], resequencing and mutational analysis [29], and also applications in drug discovery [30]. Furthermore, Southern *et al.* reported molecular interactions on microarrays [1]. In recent years, really intensive studies were carried out in the field of genomic research, pathogen detection, and clinical diagnostics, as described in Section 3.4.

In 1994, the HIV GeneChip of Affymetrix was the first commercially available DNA microarray, and is now the most widely known and applied microarray platform [31]. A good overview of the current status of used DNA microarrays was given by Shi *et al.* [15]. MammaPrint, a prognostic test for breast cancer from Agendia, became the first *in vitro* diagnostic multivariate index assay (IVDMIA) to acquire clearance from the US Food and Drug Administration (FDA) in 2007. Finally, the importance of DNA microarray technology is evidenced by its extensive literature and still increasing market share. Since the beginning of DNA microarray technology, these novel aspects of bioassays have become ubiquitous in a variety of applications in biological and medical research and drug discovery.

3.2.2
Protein Microarrays

DNA microarrays were the first described biochips applied in the investigation of the expression level of various genes of an organism. However, the evaluated results allow only a limited insight into the process of actual protein expression. Additionally, DNA microarrays provide no information about protein–protein interactions. Most of the protein monitoring on biochips is based on the analysis of the mRNA expression level. However, the data do not always correlate with the corresponding protein abundance. The main reasons for this effect are the post-translational protein modifications, which make the human proteome much more complex in composition than the coding portion of the genome. Therefore, protein microarrays notably increased the applications of biochips in many different fields, due to availability of much new and additional information.

Standard methods to analyze proteins and protein profiles are usually 2D gel electrophoresis and mass spectrometry. An alternative to standard methods is protein microarrays, which are utilized in research to analyze changes in the abundance and existence of proteins in a very high dynamic range in biological samples. Bilitewski reviewed protein-sensing assay formats and devices in general in 2006, based on affinity reactions between an immobilized capture agent and the target protein [32]. Nowadays, optimal ways to immobilize proteins to 2D surfaces to assure the retention of their biological activity and shape is a major challenge to create point-of-care devices for multianalyte diagnosis [33]. Due to their high specificity and affinity, antigen–antibody reactions are the most important tool for protein microarrays in diagnostics and biological studies. The development of protein microarrays and diagnostic applications has been reported in depth by various groups [34–42]. Critical points in antibody microarrays are the selection, production, and purification of suitable antibodies with high affinity and a reduced cross-reactivity to target molecules. Protein microarrays could be a powerful tool for the identification of biomarkers, for example, in tumor–antigen discovery and cancer diagnostics. For instance, Miller *et al.* [43] and Bouwman *et al.* [44] used antibody microarrays for the detection of prostate cancer. Other antibody applications examine the analysis of cytokine secretion by human dendritic cells with lipopolysaccharide or tumor necrosis factor-α [45] and the detection and quantification of cardiovascular risk markers [46]. Protein microarrays can also be used to identify specific targets of pharmaceutical intervention [47, 48].

A second important application of protein microarrays is to examine a multitude of protein–protein, receptor–ligand, protein–peptide [39, 49], protein–carbohydrate, and enzyme–substrate interactions and also protein interactions with small molecules. The knowledge of these interactions is important in applications of protein network profiling, drug discovery, and drug target confirmation [42, 50, 51]. Uttamchandani and Yao also gave an overview of peptide microarrays used in profiling, detection, or diagnostic applications to sense protein activity or act as ligands for potential therapeutic leads [52].

3.2.3
Tissue Microarrays

The understanding of the association between molecular changes and clinical pathology and the validation of new biomarkers are important in clinical research. Current tissue- and cell-based microarrays combine the analysis of tissue morphology with the identification, localization, and characterization of biomolecules in biological systems.

Tissue microarrays (TMAs) allow traditional tissue analysis of conventional histological paraffin blocks to become miniaturized and high throughput to study molecular alterations at the DNA, RNA, and protein level in multiple individual tissue types. Novel targets can be analyzed in hundreds to thousands of tissue samples with molecular pathology techniques in parallel on one slide.

For the construction of a tissue microarray, cylindrical core tissue biopsies from different donor paraffin-embedded tissue blocks were punched and precisely re-embedded in a blank paraffin block [53, 54]. From this new master block, several hundred to thousands of consecutive sections can be cut and molecular alterations can be detected in parallel in up to 500 or more different tissues. TMAs can be used in many fields such as cancer research and inflammatory, cardiovascular, and neurological diseases.

Today, in cancer research, different types of TMAs exist and allow the analysis of molecular alterations in multiple tumor types (multitumor TMA) or in different stages of one particular tumor (progression TMA). Also, (novel) prognostic parameters can be identified in TMAs containing tumor samples from patients with known clinical follow-up data and clinical endpoint or value testing for molecular alterations to predict chemotherapeutic agents.

The combination of tissue and cDNA microarrays will be a powerful approach for the rapid identification and evaluation of genes with clinical relevance in diagnostics and prognostics, or as therapeutic markers [55, 56].

Alterations in molecular targets such as DNA, RNA, and proteins can be visualized by classical histochemical and molecular detection techniques, for example, immunohistochemistry (IHC), *in situ* hybridization (ISH), fluorescent *in situ* hybridization (FISH), *in situ* PCR, RNA or DNA expression analysis, or TUNEL (terminal dUTP nick-end labeling) assay to detect apoptotic cells and for morphological and clinical characterization of patient tissues. Hence TMAs combine all the advantages of microarrays – low cost, high parallelism, and data collection with many other techniques of biomolecular diagnostics.

3.2.4
Cell-Based Microarrays (Transfection Microarrays)

Cell-based microarrays or so-called reverse transfection microarrays allow functional analysis in mammalian cells in parallel in a format without wells. In this array, nanoliter quantities of cDNA- or siRNA-expressing vectors and also small molecules (e.g., drug-like molecules) are printed in defined areas on a slide surface. Adherent cells are cultured on the top of the array and take up the printed molecules and create spots of localized transfections within a lawn of non-transfected cells. Each feature of 100–250 µm in diameter contains a cluster of 30–80 cells, for example, over- or underexpressing a special gene product, or are under the influence of a drug-like molecule [57]. Cell-based microarrays can be examined when the cells are still alive to visualize cellular processes in real time or after fixation. Also, a variety of detection assays can be applied such as *in situ* hybridization, immunofluorescence, autoradiography and also Western blot by transferring the cells on to nitrocellulose [51, 57]. A special form of cell-based microarrays is lentivirus-infected and *Drosophila* cell microarrays, which provide a platform for effective high-throughput loss-of-function screens through expression of short hairpin RNA/RNAi [58–61]. Lentiviruses infect a wide variety of mammalian cells but can also infect non-dividing cells, which is important for the study of primary cell types [14, 62].

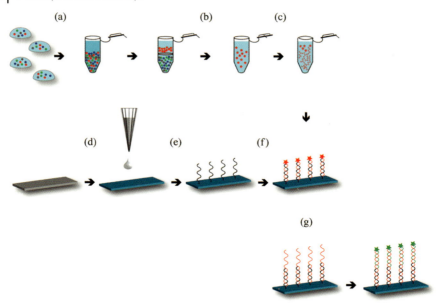

Figure 3.1 Principle process of a microarray experiment. The basic steps include sample preparation (a), purification (b), and target amplification and labeling (c). The functionalization of the solid substrate with capture molecules (e) needs a prior chemical modification of the biochip surface (d). The following hybridization of amplified and labeled target molecules (f) allows the final readout of the specific biomolecular interaction on the chip surface. Alternatively to the direct labeling of the target molecules during the amplification process, the necessary label can be introduced by a sandwich process (g).

3.3
Principle Steps of an Experimental Microarray Process

This section is focused on the microarray experimental setup process especially for DNA/RNA and protein probes. A typical microarray experiment is depicted schematically in Figure 3.1. It involves the basic steps of sample preparation (a), purification (b), target amplification, and labeling (c). The functionalization of the solid substrate with capture molecules (e) needs a prior chemical modification of the biochip surface (d). The subsequent hybridization of amplified and labeled target molecules (f) allows the final readout of the specific biomolecular interaction on the chip surface. Alternatively to the direct labeling of the target molecules during the amplification process, the necessary label can be introduced by a sandwich process (g). Finally, the huge amount of data must be analyzed by means of computational and statistical tools to generate a profile of, for example, expressed genes in a cell [63–67]. A variety of books and overviews have covered microarray analysis and different technical aspects [19, 28, 68–73].

3.3.1
Basics

The underlying principle of a microarray is based on a capture probe immobilized on a solid surface that can bind to a complementary target probe contained in a complex sample. Depending on the amount and density of spots, an array can be subdivided into low-, medium-, and high-density arrays, but overall have the same fundamental principle. Traditionally, an array consists of DNA fragments, but it also can contain other biological recognition units such as antigens or proteins. Often DNA microarrays, for example, in gene expression profiling, belong to high-density arrays due to the presence of hundreds of thousands of probes. It is possible to detect a small amount of target probes in the nanomolar to picomolar range.

3.3.1.1 Surface Biofunctionalization

Surface chemistry and modification are of great importance in microarray technology for binding capture molecules [74–76]. Especially protein immobilization is much more difficult than the immobilization of oligonucleotides or cDNA. Proteins have to be immobilized in their active form in order to bind specifically to their complementary target [33, 77, 78].

The requirements for optimal microarray performance are (a) the stable and covalent immobilization of biomolecules on the chip surface and (b) suitable mechanical and chemical properties of the substrate used [79]. Three different kinds of surfaces are used for the fabrication of microarrays: first, 2D solid structures, such as glass, silicon, silicon nitride, or gold layers, all of which have chemical modifications located on the surface, for instance thiol [80–82], amine [83], epoxy [84], or aldehyde [85] groups; second, 3D coated structures based on the gelation or polymerization of monomers, such as agarose gels [86, 87] or thin layers of polyacrylamide [88, 89], which are commonly immobilized on solid structures for support (e.g., glass); and thirdly, 3D solid structures based on mechanically stable polymeric materials such as poly(methyl methacrylate) [75] and polydimethylsiloxane [90], which do not require a mechanical support.

The advantages of the 2D solid structures are the ease and low cost of fabrication. In addition, some 2D solid structures are transparent towards different light wavelengths, hence making them the substrates of choice for optical detection. Furthermore, their mechanical rigidity allows the deposition of electrodes on their surface, making them compatible with electrochemical detection. Nevertheless, the surface of 2D solid structures is chemically inert. Therefore, these biochip substrates need chemical modification for the stable binding of biomolecules, hence the need to apply chemical reagents or coatings with gels or non-rigid polymers, increasing their fabrication costs.

Therefore, despite the advantages provided by the 2D solid structures and 3D coated structures, it is evident that the development of new materials is required. The 3D solid structures based on polymeric materials can be functionalized during their fabrication procedure. These polymeric materials can be functionalized for biomolecule immobilization by already embedded functional groups in their monomers, gaining the advantages of 3D coated substrates, but retaining the desired mechanical properties (e.g., transparency and rigidity) of 2D solid structures without reducing the ease or increasing the costs of their fabrication.

Depending on surface modifications, the array can be prepared with capture probes, for example, oligonucleotides with defined length [normally between 20 and 60 base pairs (bps)] or proteins. Oligonucleotides bind on surfaces using combinatorial chemistry and photolithographic techniques. Affymetrix pioneered the fabrication of microarray platforms with high-density arrays. The Affymetrix GeneChip arrays use photolithography techniques for the *in situ* synthesis of oligonucleotides. Semiconductor fabrication, solid-phase chemical synthesis, combinatorial chemistry, and molecular biology are integrated in a robotic manufacturing process [6, 15, 31, 91]. Agilent used Hewlett-Packard's inkjet printing technology for array preparation [92]. Analogous to printing a document, nucleotides are printed on glass slides coated with a hydrophobic surface [15, 93].

Instead of synthesizing capture molecules directly on the surface, another method is the spotting of presynthesized oligonucleotide probes (e.g., GE Healthcare's CodeLink system, Applied Biosystems' Genome Survey Microarrays, custom microarrays printed with Operon's oligonucleotide sets) on modified surfaces. An approach to immobilization of modified oligonucleotides even on unmodified glass surfaces can be made by UV cross-linking [94]. This well known method is widely used in different applications for the attachment of DNA or proteins on membranes, solid surfaces, and polymers [95, 96]. However, the exposure to UV light can damage the DNA [97].

An innovative tool for DNA or RNA studies in genetic analysis is the Illumina bead-based microarray (BeadChip microarray), where presynthesized oligonucleotide probes are deposited on beads. The highest density array platform which is currently available is the Sentrix BeadChip from Illumina, containing 6.5 million features that are just 3 µm wide [98]. The bead-based technology represents an alternative technique that is an accurate, scalable, and flexible approach to a microarray [99].

A detailed comparison of different platforms as tools for functional genomics based on microarray technology was made by Wick and Hardiman [100] and Shi *et al.* also gave a good overview of the current status of DNA microarrays [15].

3.3.2
Preparation of Biomolecules (Sample Preparation)

3.3.2.1 **DNA**
The ability to extract pure DNA from a variety of sources is an important step in many molecular biological assays. Many different protocols exist for the preparation of genomic DNA. All methods include three steps: cell lysis, deproteination, and concentration/precipitation of DNA. Commonly used protocols are alkaline lysis of cells and phenol–chloroform extraction followed by ethanol precipitation of the nucleic acid. In recent years, traditional purification methods have been replaced by a variety of commercial kits available from many companies (e.g., Qiagen, Invitrogen,

Roche Applied Science, Promega). These kits include necessary solutions for preparation and solubilization of the DNA and columns for anion-exchange chromatography or glass beads. The DNA binds on the silica membranes or glass beads of the spin columns during the separation from contaminating RNA, proteins, and metabolites and can subsequently be easily eluted with high-salt buffer and should be desalted and concentrated before use. A further approach is adsorption of DNA on carboxylated magnetic beads after cell lysis. Subsequently the DNA–magnet particle complex is separated from any residual contaminants by applying a magnetic field. After separation, the purified DNA is released by heating the complex. In this way, a high yield of pure DNA can be produced quickly and conveniently [101, 102].

3.3.2.2 RNA

The preparation of high-quality, intact RNA is the first and often the most critical step in performing many molecular biology experiments. This procedure is complicated by the ubiquitous presence of ribonuclease enzymes in cells and tissues, which can rapidly degrade RNA.

The cell probes are first lysed in the presence of a highly denaturing guanidinium isothiocyanate buffer, which immediately inactivates ribonucleases. Then RNA is fractionated from other cellular components via phenol extraction and under conditions that suppress any RNase activity and subsequently concentrated by alcohol precipitation [103].

Several kits for RNA isolation are commercially available, including guanidinium thiocyanate/phenol–chloroform mixtures or silica membrane/glass bead spin columns for RNA extraction without organic solvents. The use of oligo[dT]-cellulose columns or oligo[dT]-magnetic beads selects mRNA from total RNA. Magnetic bead separation allows mRNA to be purified in a single tube by application of a magnetic field, thus eliminating potential sample loss during liquid handling [101, 103].

RNA extraction kits for different starting materials and applications are offered by various companies such as Qiagen, Invitrogen, and Promega.

3.3.2.3 Proteins

The first step in protein analysis for any cell line of interest is the lysis of cells and the extraction of total proteins. Typically, a detergent-based method is applied. The lysis method should be mild enough to deliver active and native conformation proteins. Additives such as protease inhibitors protect the proteins from enzymatic degradation whereas nucleases digest DNA. Cell debris and total protein fraction are separated by centrifugation [104].

3.3.3
Target Amplification and Labeling

Because of minimal amounts of starting material, for example from needle biopsies, cell sorting, and laser capture microdissection, an amplification step before use in analytical assays is required in most applications.

3.3.3.1 PCR – Polymerase Chain Reaction

The PCR is a rapid procedure for *in vitro* amplification of a specific DNA region. The process based on the sequence-specific annealing and extension of two oligonucleotide primers which bracket the target DNA region. Three steps, denaturation of the double-stranded target DNA by heating, specific annealing of the primer during temperature reduction, and polymerase-catalyzed extension of the primer sequences in 5′–3′ orientation, define one PCR cycle. The target DNA is enriched exponentially during each PCR cycle and the amplicons produced are also used as templates for subsequent cycles [105, 106].

For the detection of RNA, the PCR procedure is adapted to RNA templates. In this reaction, a defined primer complementary to the polyadenylated tail on the 3′ end of most eukaryotic mRNAs was used and also the reverse transcriptase substitutes/replaces the polymerase. The so-called reverse transcriptase polymerase chain reaction (RT-PCR) produces a cDNA copy of the RNA which can be used in the same context as DNA [107, 108].

During the amplification processes, the amplicon can be modified via the incorporation of altered primers or dNTPs. Hence the introduction of labels such as fluorescent dyes, radioactivity, digoxigenin (DIG), or biotin and also mutations or restriction sites are possible.

3.3.3.2 *In Vitro* Transcription – cDNA/IVT

The so-called Eberwine method and its modifications is a strategy for mRNA amplification by template-directed *in vitro* transcription. In the first step, the poly(A) + RNA pool is reverse transcribed using an oligo-dT primer bearing a specific promoter site for phage RNA polymerases. After the second strand synthesis, cRNA can be amplified by RNA polymerase using the double-stranded cDNA as template. The common RNA polymerases used in *in vitro* transcription reactions are SP6, T7, and T3 polymerases [109].

3.3.3.3 Bio-Barcode Amplification of Nucleic Acids and Proteins

The bio-barcode technology does not include enzymatic target or signal amplification, but is based on nanoparticle technology. In this two-sided sandwich system, the target is immobilized on a solid surface capture probe and is detected through binding a detection probe.

Magnetic-based microparticles of one to several diameters interact with the target molecule through biological recognition (DNA–DNA interaction or antigen–antibody interaction) and allow the separation of the sandwich from unreacted material and medium. Amplifier nanoparticles recognize and bind the target–magnetic particle complex through immobilized detection probes (Figure 3.2). Additionally bound bio-barcode oligonucleotides specific for each target can be released for representing and amplifying the target in a subsequent detection assay [110–113]. Hundreds to thousands of barcodes of one nanoparticle provide a 2–3 log amplification of a target that has been successfully captured and sandwiched. Typical materials for amplifier nanoparticles are gold and polystyrene [114].

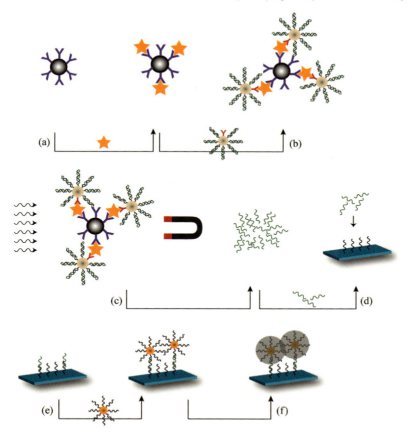

Figure 3.2 Principle of bio-barcode nanoparticle-based PCR-less DNA amplification. An antibody-labeled nanoparticle binds to the analyte (a). A second DNA-labeled magnetic nanoparticle can also bind via antibody–antigen interaction to the analyte (b). In a magnetic separation process (c) followed by barcode DNA denaturation (d), the free amplified bio-barcode DNA can hybridize to a biochip (e). After nanoparticle labeling and silver enhancement, the signal can be optically detected (f).

3.3.3.4 Radioactive and Fluorophore Labeling

The visualization of biomolecules can be mediated by labeling with radioactive, fluorescent, luminescent, or other chemical modifications. In nucleic acids, the incorporation of radioactive and nonradioactive labels is common. The label can be attached to one of the dNTPs or the 5′ end of one or both primers, which will be enzymatically incorporated in the amplicon [115, 116].

The efficiency of enzymatic incorporation of labeled nucleotides may be impaired due to steric effects of the bulky label. Also, the use of labeled nucleotides is more costly per reaction than 5′-labeled oligonucleotides. An indirect labeling, for example, with aminoallyl-dNTPs or DIG, allows subsequent conjugation of a fluorescent dye to the purified amplicon, but it is more time consuming because incubation and washing steps are required.

Proteins commonly will be covalently linked to a nucleoside or nucleotide which carries radioactivity. The radioactivity can be anywhere on the nucleoside (C, N, H, S, I, Hg). Labeling of proteins is performed by reactions of the label with the N- and C-terminus or functional groups of side chains on the surface such as an amino group of lysine, sulfhydryl group of cysteine, or carbohydrate group of the protein [117].

Radioactive labeling provides a highly sensitive method for detection. However, radioisotopes are expensive, have a short half-life, and require a long exposure time before detection. Radioactive materials are also potentially hazardous and their use requires adherence to strict safety precautions. The majority of radioisotopes used in biological experiments emit ionizing radiation and impart energy in living cells. In large enough doses, this energy can damage cellular structures, such as chromosomes and membranes. Such damage can kill the cell or impair its normal functionality.

Nonradioactive labels such as fluorophores become more important due to their sensitivity and easy handling compared with radioactive labels. A wide variety of fluorophores in many different colors are available and compatible with the usual excitation/emission systems. Current fluorescent dyes achieve levels of sensitivity comparable to radioactivity and can be detected immediately, in contrast to isotope labels. Disadvantages of fluorophores are photobleaching and quenching of the fluorescence signal. Additionally, fluorescence detection requires extensive equipment such as an excitation energy source (laser beam) and can be performed only with specialized scanners and image acquisition systems.

3.3.3.5 Enzymes

Enzyme assays represent a special group in the field of bioassays [118]. In this labeling method, the substrate of the enzyme is often used as the analyte to be detected. However, also cofactors or even the enzyme itself, for example in clinical diagnostics, can be analyzed by enzyme assays. The advantage of this method is the high specificity. Hence there is no need for a sophisticated sample preparation to eliminate negative matrix effects. Even the investigation of complex probe material can be managed rapidly and easily by enzyme assays, without the use of expensive chromatographic processes. In most cases these assays are based on redox-active enzymes that catalyze the creation of a soluble dye in the presence of specific cofactors (e.g., nicotinamide adenine dinucleotide (NAD) or by co-products of the enzymatic reaction (e.g., H_2O_2). From the analytical point of view, the oxidoreductases and also the hydrolases are of main interest. For example, glucose oxidase and horseradish peroxidase (both belong to the group of oxidoreductases) or alkaline phosphatase and acetylcholinesterase (both belong to the group of hydrolases) are typical representatives and widely used in enzyme-based bioanalytical investigations. A further benefit of enzymes is the wide knowledge about their reactions, kinetics, structure, substrate specificity, and inhibition of catalytic processes. Nowadays, traditional systems such as enzyme-linked immunosorbent assay (ELISA) [119] and the Western blot [120, 121] are still used as standard analytical methods in biomolecular and immunological laboratories.

3.3.3.6 Nanoparticles

Metallic nanoparticles have attracted increasing interest as alternative labels in analytical biochip-based assays. Due to their unique properties, metal nanoparticles allow the development of various novel approaches for the detection of biomolecules. Their attributes depend strongly on the size, shape, and composition of the single particles [122]. A large variety of fabrication methods have already been described in the literature. Different metals can be used to produce nanoparticles, such as Au, Ag, Cu, Pd, and Pt. Since the properties of metallic nanoparticles depend on their shape, a variety of shapes, such as prisms, rods, and cubes, have been synthesized in addition to the common spherical structure. A further important factor for their use in biochip and bioassay applications is the stable and reliable biofunctionalization of the metallic nanoparticles [123, 124]. The coupling of suitable ligands allows the subsequent specific interaction with complementary biomolecules. Functionalization has been studied with high intensity especially on gold surfaces, because this material provides the simplest and most stable conjugation [125].

Because of their high extinction coefficients and also the large scattering coefficients, metal nanoparticles can be used in optical setups based on absorption or scattering processes [126, 127].

In addition to their special optical properties, metal nanoparticles can be used further in electrical, electrochemical, and electromechanical setups [128–130]. By measuring the electrical conductivity of metallic nanoparticles between an electrode gap, a robust and simple approach for the detection of biomolecules can be realized [131].

Furthermore, these metallic nanoscopic structures can be used as reaction seeds for a specific, reductive metal deposition process [132]. The additional metal deposition leads to so-called "core–shell" structures. During the "core–shell" process, the metal nanoparticles increase in size, which leads in many different systems to an improvement of the sensitivity.

3.3.4
Readout/Detection

A key step in the microarray experimental process is the detection method or readout after the hybridization event. Therefore, optical detection methods remain the preferred technique in genomic and proteomic chip-based applications. For the evaluation of detection techniques, some criteria must be considered [23, 133].

Figure 3.3 illustrates the most frequent detection methods for optical biochips. In addition to fluorescence (a), metal nanoparticles can be used in absorbance (light microscopy in transmission mode) (b) and light-scattering setups (c). The approach to detect the scattered light of metal nanoparticles enables a multilabeling system comparable to modern fluorescence setups to be developed.

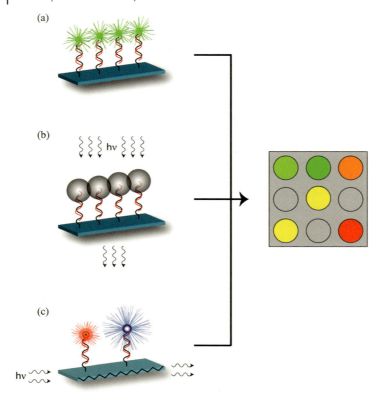

Figure 3.3 The optical detection on biochips is mainly dominated by fluorescence (a). This readout technology allows highly sensitive multilabeling of biomolecules. An alternative approach is the utilization of metal nanoparticles. In both absorbance (b) and scattering setups (c), metal nanoparticles showed the potential to replace traditional markers in biochip technology.

3.3.4.1 Radioactivity

Radioactive signals can be captured using autoradiography films, storage phosphor screens, and image acquisition systems, depending on the labeling method and the required levels of sensitivity and resolution. Autoradiography films are commonly used; they are inexpensive, reliable, and provide a good level of sensitivity and resolution. Results obtained with this method can be scanned and analyzed using image analysis software. Scanner-based image acquisition systems are able to detect radioactive signals using a storage phosphor screen, which is about 10 times faster than regular autoradiography films [134].

For the detection of radioactive spots on a microarray, the array is coated with a photographic emulsion. After drying, the sealed slides are stored for a few days in a radiation box. Subsequently the signals are developed with a photographic emulsion.

The reduced silver ions then can be detected and imaged with a scanning light microscope [GE Healthcare (Amersham Biosciences)].

3.3.4.2 Fluorescence/Absorbance

Fluorescence is the most commonly applied detection technology in biological and clinical research and drug discovery. Therefore, most of the current commercial microarrays use fluorescence signals for their readout. Reviews have already surveyed fluorescence-based nucleic acid detection methods and microarrays [19, 135–137]. Bally et al. described different optical sensing techniques based on microarrays and their applications [133]. Fluorescence immunoassays have largely replaced radioimmunoassay, and fluorescence dominates other detection techniques because of its high sensitivity, dynamic range, and high spatial resolution.

Briefly, the fluorescence process involves the emission of visible light by electronic excitation of molecules, called a fluorophore or fluorescent dye. These molecules have a specialized conjugated π-electron system so that they are capable of absorbing light of different, shorter wavelength than the emission wavelength. Detailed descriptions of the principle of fluorescence have been given in a variety of books and reviews and will not be repeated in this chapter [135, 138]. Microarray technology often uses organic fluorophores as labels due to their characteristics of functionalization, high stability, solubility, and biocompatibility in aqueous solutions [137, 139, 140]. Primary factors limiting fluorescence detectability are photobleaching and quenching processes. Photobleaching is a dynamic process, in which excited fluorophores undergo irreversible destruction under high-intensity illumination conditions and thus lose their ability to emit light [141]. Also, the fluorescence signal depends strongly on the environmental conditions.

In addition to the relatively small size of the microarray, a large periphery is necessary for detection and data interpretation, for instance, light sources, optical instruments, photomultipliers, or a charged-coupled device (CCD) camera. Conventional instruments include scanners and imagers – the two main designs of detection instruments. Scanners are distinguished from imagers in moving their position from the substrate or optics as a function of spatial coordinates in two dimensions. Scanner techniques include laser scanning methods and often photomultipliers for detection. Imagers often use white light sources and CCD technology to achieve images in a single step. A powerful analytical method is required for rapid and automatic data acquisition. Subsequently, the huge amount of data must be analyzed and interpreted by means of bioinformatics [142, 143].

Based on fluorescence, there are different parameters that characterize fluorescent molecules, such as fluorescence quantum yield, wavelength, and lifetime. For instance, information about the lifetime and decay of fluorescence signals is used in DNA microarray [144, 145] and protein array technology [146]. Cubeddu et al. presented an overview of time-resolved fluorescence in biology and medicine [147].

Förster resonance energy transfer (FRET), a special form of fluorescence quenching, is often used in biophysics and biochemistry to quantify molecular interactions such as protein–protein or DNA–protein interactions and also protein conformational changes. FRET also plays an important role in microarray analysis. However, FRET depends strongly on the distance between two molecules, a donor and an acceptor molecule. It depends on the spectral overlap between the emission spectrum of the donor and the absorption spectrum of the acceptor. The fluorescence signals are generated or quenched due to the spatial separation of these two fluorophores. A pair of interactive fluorophores are attached to the ends of two different oligonucleotides or to the two ends of the same oligonucleotide probe. There are two possibilities for detecting the hybridization between capture and target molecules that are used in microarray technology. One possibility is a nonradiative energy transfer from donor fluorophore to acceptor fluorophore if they are in close proximity to each other. If the acceptors are quenchers, the fluorescence emission intensity of the donor fluorophore is decreased. However, if the acceptors are also fluorescent molecules, which are activated by the emission of the donor, the binding event leads to an increase in the emission intensity of the acceptor fluorophore. The other method is to separate the two fluorophores from each other to prevent the energy transfer and generate the donor characteristic emission spectrum uninfluenced by the acceptor fluorophore [11, 136, 148–152]. For instance, a microarray analysis of protein–protein interactions based on FRET using the fluorescence lifetime as an analytical parameter was investigated in the model system of bovine serum albumin and streptavidin [152].

Bio- and chemiluminescence are also useful techniques in biomedical research, diagnostics, and drug discovery and development. Chemiluminescence microscope imaging is a valuable tool for localizing and quantitating enzymes, metabolites, antigens, and nucleic acids in many kinds of specimens [153–157]. In comparison with fluorescence light emission, chemiluminescence is generated by a chemical reaction under exclusion of light. This is advantageous for this technique owing to a lower nonspecific signal and absence of light scattering. However, a disadvantage is that the light-emitting chemical reaction could be uncontrollably inhibited, enhanced, or triggered by sample matrix constituents [158]. A new chemiluminescence-based Ziplex gene expression array technology was evaluated based on Affymetrix GeneChip profiles and applied in ovarian cancer research [159].

Similarly to other methods such as localized surface plasmon resonance (LSPR) and surface-enhanced Raman spectroscopy (SERS), a surface-enhanced fluorescence effect can occur when fluorophores are in close proximity to metal nanoparticles [160]. Lakowicz *et al.* reported on plasmon-controlled fluorescence and the properties of fluorophore–metal interaction bioassays [161].

Absorption measurements have also been used in microarray analysis. Although this method is less sensitive than fluorescence, its simplicity makes it interesting and gives advantages in certain applications. In principle, absorption techniques are based on attenuation of incident light as a function of wavelength. Microarray experimental processes often detect changes in the optical density or the color of probes [162]. Metal nanoparticles have been used in absorption measurements of DNA microarrays as a novel technique [163]. Thus, binding events of target DNA to capture DNA tagged with nanoparticles can be visualized using reflection or transmission of light depending on some characteristics of particles such size, shape, composition, and environment. Furthermore, nanoparticles are used in DNA microarray scattering detection based on by the extremely large scattering coefficient [125, 163].

3.3.4.3 Surface Plasmon Resonance

Apart from readout-based techniques such as radioactivity and fluorescence, only the development of surface plasmon resonance (SPR)-based biosensors has achieved commercial success (e.g., the Biacore system) [164]. SPR is an optical method that can be used for the real-time and label-free detection of biomolecular recognition reaction [165, 166]. This approach makes use of the interaction of plane polarized light with the free electrons at a metal surface [167]. Under certain conditions, the energy carried by the photons can be transferred to collective oscillations of the electrons, which are called plasmons [168]. The energy transfer occurs only at a specific resonance wavelength of light when the momentum of the photons and that of the plasmons are matched. To excite surface plasmons, in SPR systems, a laser beam is focused on the back of a prism covered with a thin film of metal, mostly gold or silver with a thickness of 40–50 nm. At a defined angle of incidence, the intensity of the reflected light decreases significantly. This effect appears to be due to energy absorption followed by the creation of surface plasmons at the metal surface. The surface plasmons can propagate along the metal surface, but they are not strictly located at the surface. The intensity of their electromagnetic field can be drastically decayed on going from the metal surface into the adjacent medium. Concerning the bioanalytical interest of SPR, the position of the critical angle and thereby the adsorption depend on the evanescent wave and therefore it is sensitive to the environment up to a distance close to the wavelength from the actual metal surface. When using SPR systems in bioanalytics, a major advantage is the label-free detection of biomolecules [169]. These detection units are ideally suited for the direct monitoring of binding events without the need for additional labeled substances. However, the additional use of metal nanoparticles increases the sensitivity up to the picomolar region [170, 171]. Nevertheless, the major problem in every analytical SPR-based system is the unspecific adsorption of the biological target components to the sensor surface. More recently, novel biochip assays have been combined with SPR technology to create a fast and sensitive detection system for different biomolecules [133, 172].

3.4
Examples of Biochip Applications in Biology and Medicine

3.4.1
Genomic Research

The aim of genomic research is the identification of new genes and the understanding of their function and expression levels under different conditions. Two main applications can be identified: (1) identification of gene mutation (sequence analysis) and (2) determination and monitoring of expression level (gene expression profile) for rapid mapping and identification of disease-related genes [3, 173]. The goal of this research is not only to catalog all genes and their characteristics but also to understand how the components work together to comprise functioning cells and organisms [13]. Pastinen *et al.* developed a powerful approach to scan all possible sequence variations provided by minisequencing assays based on high-density DNA–chip arrays [174]. Several reviews of microarray technologies and applications used in genomic research have been published [14, 175–182]. Grant and Hakonarson provided an overview of recent advances and further expectations within this field [182]. Comparative genome hybridization (CGH) has been mainly used to identify relatively large insertions and deletions in cancer cell lines, for example by cDNA expression arrays [183].

cDNA microarrays are a powerful technique to identify novel biomarkers for potential use as diagnostic, prognostic, and therapeutic tools in clinical diagnosis. They facilitate the analysis of the expression of thousands of genes simultaneously to provide static and dynamic information about expressed genes [91, 184]. It is based on the same principle as Northern blot and RT-PCR analysis. An overview of cDNA microarray technology was given by de Mello-Coelho and Hess [9]. Basically, cDNA microarrays are used for identifying the role of several genes involved in cellular processes. cDNA microarray techniques play an important role in research on human cancers and are used for studying distinct types of cancers, for example, lung cancer [185–187], prostate cancer [188, 189], breast cancer [183, 190–193], and ovarian cancer [194, 195], and also the pathogenesis of infectious diseases [196, 197]. Furthermore, differentiation in gene expression patterns of nondiseased and diseased tissues has contributed to a better understanding of the regulation of molecular mechanisms and potential for therapeutic use towards personalized medicine [198]. Parissenti *et al.* demonstrated gene expression profiles as biomarkers for the prediction of chemotherapy drug response in human tumor cells [179]. Different applications of cDNA microarrays include life science areas of endocrinology, microbiology, immunology, oncology, toxicogenomics, development, genetics, endocrinology, and gerontology [9, 199].

3.4.1.1 SNP Detection

A large number of variations in DNA sequence, mainly single nucleotide polymorphism (SNP), are revealed with the complete sequencing of the humane genome. Millions of SNPs are deposited in public databases [200, 201]. These variations in DNA sequence play a major role in the diagnosis of a variety of genetic diseases and cancer and in drug development. SNPs serve as genomic markers for assessing disease predisposition and prediction medication [202, 203]. The analysis of SNPs can help in understanding genetic differences between individuals and disease states and will be also useful in helping researchers to determine and understand why individuals differ in their response to absorb or clear certain drugs for the improvement of medical treatments [180, 204].

The application of oligonucleotide arrays for the detection of particular mutations or polymorphism was first reported in 1989 by Saiki *et al.* [205]. They described the genetic analysis of amplified DNA (HLA-DQA genotyping and 3-thalassemia mutations) with immobilized sequence-specific oligonucleotide probes on a nylon membrane.

Commercial SNP arrays are provided by numerous companies for numerous applications. For example, Illumina has developed a bead-based flexible, accurate, highly multiplexed SNP genotyping assay. Up to 1536 SNPs can be detected in a single DNA sample with this system [204]. Each bead contains hundreds of thousands of covalently attached oligonucleotide probes. For instance, in a cDNA-mediated annealing, selection, extension, and ligation assay, 1152 cSNPs derived from 380 cancer-related genes can be investigated. In this assay, information about both genomic DNA and RNA and also allele-specific expression is obtained, which is important for studying variations and origins of diseases. Owing to the variability and scalability of the Illumina platform, the user can perform small pilot studies or large-scale SNP genotyping studies for human genetic diseases, pharmacogenomic applications, or rapid development of molecular diagnostics [99, 204]. An automated magnetic bead-based platform for pretreatment of human leukocytes, gDNA extraction, and fast analysis of genetic genes was demonstrated by Lien *et al.* [206]. Furthermore, a comparison of genotyping kits and arrays from Applied Biosystems and Affymetrix was made by Selmer *et al.* [207].

A novel genotyping technique that employs a peptide nucleic acid (PNA) zip-code microarray [208] method was described by Mun *et al.* [209]. The zip-code array technique involves unique and distinct short oligonucleotides designed for the purpose of addressing complementary target sequences. It has become an interesting method for conducting routine genetic assays, for instance, investigating mutations in susceptibility gene BRCA1, which is considered to be involved in the development of breast and ovarian cancer. Based on a zip-code microarray mutation, detection of BRCA1 was described [195, 210, 211]. Further applications based on zip-code microarrays include investigations of diabetes caused by HNF-1α mutations [212].

Other applications have been reported for sensing of low-abundance DNA point mutation in K-*ras* oncogenes, which have a high diagnostic value for colorectal cancers [95, 148]. Generally, reviews and articles have focused on the wide range of applications of SNP microarray analysis in cancer research and potential applications in cancer risk assessment, diagnosis, prognosis, and treatment selection [12, 16, 198, 213–216]. Loss of heterozygosity (LOH), genomic copy number changes, and DNA methylation alterations [217, 218] of cancer cells can be determined by SNP array genotyping [216, 219]. LOH of chromosomal regions bearing mutated tumor suppressor genes involve genomic alterations [185, 186, 189, 220]. Also in molecular karyotyping, SNP arrays are used for genome-wide genotyping [221]. Bruno *et al.* studied the feasibility of replacing targeted testing of patients with mental retardation and/or congenital abnormalities using SNP arrays [222].

3.4.1.2 Gene Expression Profile Analysis

Point mutations in the DNA sequence are mainly identified by using SNP analysis. In gene expression profiling, the analysis of point mutations is not preferentially considered. Gene expression profile analysis is used for the monitoring and determination of expressed genes. Depending on their environmental conditions, different expression patterns of genes can be obtained. SNPs are genetic variations and almost unaffected by their environment.

High-throughput molecular technologies are reshaping our understanding of diseases, and microarray-based gene expression has attracted the most attention for performing massive parallel profiling of gene expression from a single sample. General reviews of DNA microarrays include gene expression profiling by Hughes and Shoemaker [223] and also by Deyholos and Galbraith [224]. A technology review about navigating gene expression using microarrays was presented given by Schulze and Downward [225]. Gene expression profile analysis has been successfully used to derive a molecular taxonomy to gain biological insights into basic biochemical pathways and molecular mechanisms of diseases and their regulatory circuits to generate a multitude of prognostic/predictive signatures. Cancer research is one of the most important clinical fields based on microarray technology [175]. The application of biochip technologies in cancer research as a basis for individualized treatment of cancer patients was reported by Kallioniemi [198]. Polyak and Riggins included in a review the serial analysis of gene expression techniques and the implications for cancer research [226]. Other reviews have described the contribution of gene expression profiling to our understanding of breast cancer [227] and its clinical management and what still remains be done for these classifiers to be incorporated in clinical practice [192, 228].

The first customized microarray-based gene expression diagnostic breast cancer array, accepted by US FDA in 2007, called MammaPrint, was developed based on the Agilent two-color platform and method [229]. Breast cancer is the most common research field for gene expression profile analysis and the application of microarrays. Another review covered genome-wide profiling of genetic alterations in acute lymphoblastic leukemia [230]. Also, results of genomic analysis revealed few genetic alterations in pediatric acute myeloid leukemia [231]. Nannini et al. analyzed the state of the art of gene expression profiling in colorectal cancer using microarray technology [232]. In addition to gene expression profiling in cancer research, other exciting examples covering fields such as cardiovascular diseases [233, 234] and psychiatric disorders [222, 235] are being researched. Furthermore, gene expression analysis plays an important role in human embryonic stem cell research and the differentiated progeny [236]. It is useful for the treatment of a variety of inherited and acquired diseases. Gene expression profiles can also serve as transcriptional "fingerprints," for example to determine the target of a drug [237, 238]. Another field of expression profiling analysis is the investigation of proteins. However, achieving the equivalent genome-wide profiling of proteins is technically more difficult than investigations of DNA or RNA, due to the already described problems with the complexity of capture molecules and their immobilization [239, 240].

In addition to cancer analysis by microarray techniques, several other diseases have also been investigated, including muscular dystrophy [241], Alzheimer's disease [242–244], schizophrenia [245, 246], and HIV infection [247, 248].

3.4.1.3 Pharmacogenomics and Toxicogenomics

Pharmacogenomics means the application of genomic technologies to drug discovery and development. Usually, drug discovery started with a biochemical pathway implicated in a pathophysiological process, and an appropriate enzyme commonly from the rate-limiting step in the pathway or a receptor was characterized, purified, and screened against a variation of structurally diverse small molecules. Today, the monitoring of genetic polymorphisms and differential expression patterns are the most biologically informative application of microarray technology and will be an important improvement in drug discovery [249].

Genetic polymorphisms (SNPs, cSNPs) represent inherited differences in drug-metabolizing capacity. Most drugs interact with specific target proteins to exert their pharmacological effects, such as receptors, enzymes, or proteins involved in signal transduction, cell cycle control, and many other cellular events. In many cases, genetic polymorphisms are associated with altered activity of encoded proteins which influence pharmacokinetics and pharmacologic effects, and as a consequence the sensitivity of specific medications. In drug discovery, genetic polymorphism analyses will be the initiating step followed by biochemical and clinical studies to evaluate the phenotypic consequences of these polymorphisms [250].

Another approach for the discovery of potential drug targets is the investigation of differential expression patterns. The gene expression in a variety of tissues can be compared: normal, diseased, and also drug-affected tissues, in addition to model organisms, examining the effects of selected overexpressed genes or drug treatment. Yeast and mouse models are commonly used. Drug-induced gene products can be used as surrogate markers to follow readily the effect and dose of a drug in the clinical setting. Also, a highly selective tissue expression of a drug target is attractive, as the potential for unwanted side effects may be more restricted [30].

The study of pathogen gene expression during the time of acute infection or during latency and also the identification of genes and gene products turned on by the response of the host against the pathogen is a further approach to establishing new therapeutic targets.

Microarrays are a powerful tool for *in vitro* target identification. They determine drug efficacy and toxicity and improve the understanding of the mechanism of drug action. Various companies are involved in the application of pharmacogenomic technology [238] for drug target validation and identification of secondary drug target effects facilitated by DNA microarrays [251].

The field of toxicogenomics contains global mRNA, protein, and metabolite analysis to study molecular and cellular effects in relation to exposure to toxic elements. The prediction of possible side effects of pharmaceuticals is an important part of the preclinical drug development process. The comparison of gene expression patterns from toxicant-treated organisms versus a control sample, healthy versus diseased tissue, and susceptible versus resistant tissue helps us to understand the influence of hazards and environmental stressors in biological systems.

There are increasing online resources for biological data and information for toxicogenomic studies (e.g., TOXNET, NIEHS, NCT, TRC, CEBS, and other) [252]. Toxicology microarray data sets comprising genes of different subject areas such as apoptosis, DNA repair, DNA damage, inflammation, and cell proliferation and also cell response to oxidative stress and xenobiotic metabolism have been reported. For the selection of genes from the knowledge-based microarrays, the ratio of the intensities is calculated and the induction and repression of genes inferred. Optimal microarray measurements can detect differences as small as a 1.2-fold increase/decrease in gene expression. Gene expression changes are more sensitive and comprehensive markers of toxicity than typical toxicological endpoints such as morphological changes, reproductive toxicity, and carcinogenicity [253].

Toxicology arrays, for example, the ToxBlot [254] and ToxChip [255] from the National Institute of Environmental Health Sciences Microarray Group, include gene classes related to cancer, immunology, endocrinology, neurobiology, investigative toxicology, predevelopment toxicology, and safety assessment. ToxChips will also be developed to monitor drinking water quality and identify microorganisms [256]. Toxicology biochips determine the relative security of natural and synthesized compounds and chemical mixtures to which humans are exposed.

3.4.2
Pathogen Detection and Clinical Diagnosis

3.4.2.1 Pathogen Detection

The detection and identification of pathogens are essential for ensuring the safety of humans [257]. The difficulty in implementing pathogen detection in biochip and bioassay technology lies in the complexity of most of the samples, making it hard to ensure the sensitivity and specificity of the method. Nowadays, many microbiological laboratories working on pathogen detection have to use microbiological culturing techniques to identify the pathogens reliably. After the culturing process, biochemical and immunological methods are used to detect specific antigens of the pathogen [258]. However, this procedure is very time consuming, and additionally the traditional tests do not provide information about the potential pathogenicity of the organisms.

New biochip approaches facilitate pathogen analysis. The main factors influencing this development are the assay speed, information content, and costs. Further promising opportunities such as miniaturization and integration allow the construction of "lab-on-a-chip" devices for on-site testing [259]. These point-of-care systems permit the analysis of pathogens even for slow-growing organisms. Pathogen detection at the bedside will lead to faster diagnosis and initiation of an appropriate therapy.

In recent years, many groups made remarkable progress in biochip-based pathogen detection [260]. In 2002, Wang et al. demonstrated the detection of viral pathogens [261]. They successfully developed a biochip platform to detect PCR products, labeled with a fluorescent dye, of 140 different viruses. Garaizar et al. listed a number of bacterial pathogens investigated by biochip studies [262]. A further point of interest is the analysis of host–pathogen interactions. Such an interaction causes substantial changes of the host and also the pathogen. The biochip-based detection of host–pathogen interactions was demonstrated by Diehn and Relman [263]. Especially new spectroscopic methods have the potential for the sensitive and specific detection of pathogens. Identification at the subspecies/strain level on the basis of the SERS fingerprint was reported in 2003 by Grow et al. [264]. The highly specific SERS fingerprint could also reflect the physiological state of the bacterial pathogens. A proof-of-principle for the point-of-care-analysis of *Escherichia coli*- and *Bacillus subtilis*-infected samples was developed by Yeung et al. in 2006, including sample preparation, amplification, and specific detection [265].

3.4.2.2 Clinical Diagnosis

In clinical diagnosis, novel biochip systems have been used in gene analysis for prognosis and treatment. The innovative biochip technology showed the ability to change the current molecular diagnosis to point-of-care analytics that allow the integration of diagnostics with therapeutics and the development of personalized medicine [213, 266]. Personalized medicine will improve the precision and outcome of treatment. Tailor-made therapy for individual patients is based on transcriptional profiling. Additionally, biochips can investigate specific factors of a disease, such as the presence of specific cell types (e.g., white blood cells), proteins, and different kinds of enzymes, in a rapid and efficient manner. For this purpose, the body fluids (e.g., blood), exhaled air, and tissue samples from the patient serve as probe materials for the final biochip analysis. Especially in cancer diagnosis, research in recent years has led to considerable progress in prognosis and tumor classification, where so-called methylation chips served for the diagnosis and prognosis of different cancer types [267].

Aberrant DNA methylation has frequently been found in cancer cells, which silenced different tumor suppressor genes and genes important for cell cycle function and DNA repair. Therefore, various groups tried to detect tumor tissue based on the investigation of DNA methylation. Yan et al. presented a tumor classification of breast cancer patients on the basis of their hormone receptor status and clinical aggressiveness [304]. Shi et al. detected and identified unique genes preferentially methylated in particular tumors [268]. DNA microarrays were used to analyze DNA methylation. Therefore, the authors analyzed methylated and unmethylated DNA targets on a chip surface after amplification (bisulfite-PCR) and hybridization of the amplified targets to the immobilized capture molecules. In 2006, Schumacher et al. developed a technology for unbiased, high-throughput DNA methylation profiling of large genomic regions [269]. Uunmethylated and methylated DNA fractions were enriched using a series of treatments with methylation-sensitive restriction enzymes, and interrogated on microarrays. Estécio et al. introduced the methylated CpG island amplification (MCAM) technique in 2007 [270]. This technique was stated to be a suitable method to discover methylated genes and to profile methylation changes in clinical samples in a high-throughput fashion. Further array-based detection of DNA methylation in cancer diagnosis was described by Melnikov et al. [271] and Kimura et al. [272]. Other studies targeted the detection of special proteins for cancer diagnosis. Thus, in 2001 Askari et al. introduced the anti-p53 tumor suppressor gene antibody in human serum on a biochip [273]. A review by Pollard et al. addressed the development and application of protein microarrays for cancer diagnosis and clinical proteomics [34].

Of course, cancer diagnosis and prognosis represent one of the largest fields in clinical diagnosis, although smaller research topics include the investigation and detection of Alzheimer's disease [244] and autism spectrum disorders (ASDs) [180]. However, the application of microarrays in clinical diagnosis is constantly growing.

3.5
Trends in Biochip Research

A wide variety of novel exciting technologies are becoming available with potential use in analytical applications. These innovative, novel sensing techniques have great promise in the analysis of a wide range of targets and various properties in biological, chemical, and immunological problems. In this connection, most notably nanoparticles, quantum dots, and novel enzymatic technologies show the potential to improve or to replace current analytical methods. Especially with respect to the development of point-of-care systems, the new labeling methods offer robust and easy-to-use detection units that can be applied outside specialized laboratories. Another important point is the stability of the signals, which is of interest for comparisons of the received data and storage.

These methods form the basis for a new generation of biochip platforms that allow fast and sensitive on-site detection of biomolecules.

3.5.1
Metal Nanoparticles

The main reasons for the development of innovative bioanalytical approaches based on metal nanoparticles include the stability of the signals, the high signal intensities, and harmless reagents. Additionally, the utilization of metal nanoparticles allows the construction of robust, cost-effective, and portable read-out devices to realize easy-to-use point-of-care systems.

Using the scattered light of metal nanoparticles permits sensitive imaging and quantification of biomolecules. The spectrum of the scatted light is dependent on the attributes of the nanoparticles [126, 127]. As the spectrum depends on particle size, shape, and composition, a multi labeling system based on metal nanoparticles has been demonstrated [163]. A common glass slide, functionalized with oligonucleotides, was used as a planar wave guide. After specific hybridization with target oligonucleotides, labeled with different sizes of metal nanoparticles, the size-selective scattered light could be used for sequence specific differentiation between the immobilized oligonucleotides.

Furthermore, Quinten demonstrated the influence on plasmon resonance of the interparticle spacing [274]. With this effect, a simple colorimetric DNA detection using gold nanoparticles could be realized [275, 276]. Mirkin's group modified different batches of gold nanoparticles with different sized oligonucleotides [275]. Due to mixing of the batches and by adding a single-stranded target DNA, which was complementary to the two oligonucleotides bound to the different sized particles, the particles formed DNA-linked aggregates. Thereby, the color of the colloidal nanoparticle solution shifted from red to blue. Since this color change was caused by DNA–DNA interactions, heating of the solution above the melting temperature of the double-stranded DNA removed the single-stranded target DNA. Consequently, the nanoparticle aggregate was resolved, resulting in a color shift back to red. By transferring the hybridized particle aggregate to a chromatography plate, the test could be permanently recorded [275].

In addition to the detection of light scattered directly from metal nanoparticles, metal surfaces can enhance electromagnetic fields, which has been exploited in SERS [277, 278], which is an emerging technology in the field of analytics [279]. In comparison with other spectroscopic detection techniques such as fluorescence, Raman spectroscopy is hampered by the inherently small scattering cross-section, which prevents the detection of low analyte concentrations [280, 281]. The huge SERS enhancement of more than 10^6 times the weak Raman signal was considered to be caused by interactions between the analyte molecule and a metallic surface having nanostructured morphology. Since the discovery of the SERS technique, it has become a versatile tool for chemical analysis, environmental monitoring, and biomedical applications, applying the unique properties of Raman spectroscopy, for example, the highly specific molecular fingerprint spectrum [282, 283]. Wabuyele and Vo-Dinh demonstrated the use of plasmonics-based nanoprobes that act as molecular sentinels for DNA diagnostics [284]. The plasmonics nanoprobe consists of a metal nanoparticle and a stem–loop DNA molecule tagged with a Raman label. As long as the hairpin loop is active, the Raman label is in close proximity to the metal nanoparticle, so an intense SERS signal can be detected. In the presence of a complementary target DNA, the hairpin loop will be disrupted. Thereby, the SERS signal will be decreased by removing the Raman label form the metal nanoparticle (Figure 3.4).

3.5.2
Quantum Dots

Quantum dots are most useful in biomedical research and for *in vitro* and *in vivo* clinical diagnostics. These monodisperse semiconductive nanocrystals have dimensions

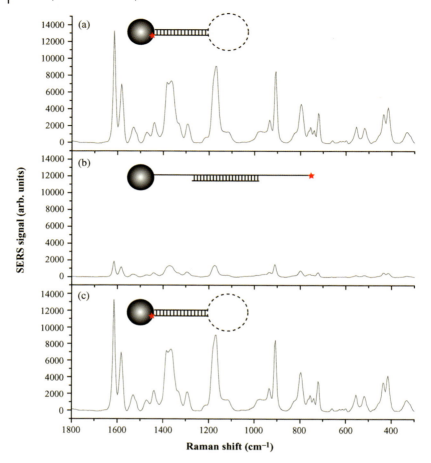

Figure 3.4 SERS intensity dependent on the presence of a plasmonic-based nanoprobe, consisting of a metal nanoparticle and a stem–loop DNA molecule tagged with a Raman label. With a closed hairpin loop the SERS signal can be detected (a). Due to the hybridization of a complementary target DNA, the hairpin loop will be disrupted (b). The SERS signal will be automatically decreased. Heating above the melting temperature of the DNA construct will rearrange the hairpin loop and consequently the SERS signal (c).

between 1 and 10 nm. Because of the small dimensions, the physicochemical properties of quantum dots depend strongly on their size. Colloidal semiconductor quantum dots are single crystals whose size can be controlled by the synthesis conditions. They have very high extension coefficients, much higher than those of organic fluorophores, and these particles adsorb light over a wide range of wavelengths. Quantum dots represent an alternative to organic or natural fluorophores with a variety of advantages, for instance, their large quantum yields, ability to excite a wide range of sizes, with emission ranging from the visible to the near-infrared region with one laser frequency, and longer fluorescence lifetimes. Furthermore, these nanocrystals offer high photostability, in contrast to organic fluorophores there is no bleaching effect, and their optical attributes allow their use in multiplexed applications for biological microarrays [285].

Especially for *in vivo* applications, significant concerns about the use quantum dots limit their utilization in various applications. The toxicology of quantum dots is determined by their chemical make-up. An inorganic coating of quantum dots insulates the core, amplifies the optical properties, and provides chemical stability. Then an organic coating is necessary for coupling biomolecules [11]. Quantum dots can also be incorporated into latex particles (quantum beads), making them more useful to those attempting antibody conjugation. Basically, chemical coatings (silica and other materials) should provide solubility and the possibility of functionalizing the particles with biomolecules, including DNA, proteins, and antibodies. The toxicology of quantum dots is a combination of the metal, the chemical coating, and the functionalization.

Smith and Nie in 2004 reported details of surface chemistry and bioconjugation, current applications in bioanalytical chemistry and cell biology, and future research directions of quantum dots [286]. Quantum dots are utilized in fluorescence *in situ* hybridization, DNA and protein analysis, and targeting a variety of cell components. In 2005, the Ting's group demonstrated the quantum dot-based detection of proteins via biotin–streptavidin interactions [287]. Additionally, Roullier *et al.* showed the potential of the direct visualization of protein–protein interactions at the single molecule level [288]. Recent reviews have underlined the applicability of quantum dots and the challenges to be overcome to make them commercially available [137, 289–291]. Nowadays, several quantum dot antibody and secondary detection reagents, labels, and labeling kits are commercially available from different companies, for example, Invitrogen and Evident Technologies.

Figure 3.5 illustrates selected applications of quantum dots in biomolecule diagnostics: fluorescence *in situ* hybridization, donors in fluorescence resonance energy transfer, labels for microarray hybridization, and for spectral bar coding [292].

3.5.3
Enzymes

In recent years, much effort has been directed at using enzymes as catalysts to generate metal nanoparticles. The combination aimed at the fusion of the fast and highly specific enzymatic reaction with the unique properties of metal nanoparticles. This technology extended the field of application of enzymes in bioanalytics because of the generation of insoluble stable products. An enzyme-induced growth of metal nanoparticles leads to stable signals that do not fade or bleach.

The use of enzymatic reactions to generate metal nanoparticles was described for the first time in the field of immunohistology and *in situ* hybridization [293, 294].

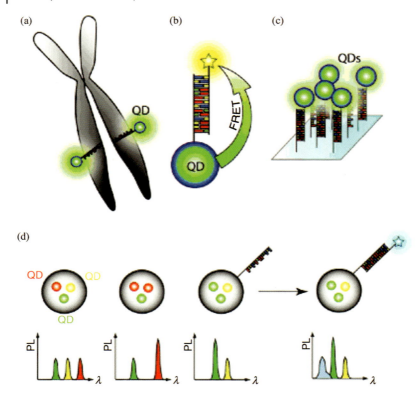

Figure 3.5 Selected applications of quantum dots in nucleic acid diagnostics: (a) quantum dots as labels for fluorescence *in situ* hybridization (FISH); (b) quantum dots as donors in hybridization-mediated fluorescence resonance energy transfer (FRET); (c) quantum dots as labels for biochips/solid-phase hybridization; (d) encapsulated combinations of quantum dots for spectral barcoding. From Algar *et al.* [292].

Both methods used enzymes as labels for the production of specific dyes. However, the enzymatically converted molecules started to diffuse after a short time into neighboring tissue, which led to a decrease in sensitivity and a reduction in resolution. Based on the demonstrated advantages such as the efficient reaction, stability, and reproducibility, different enzymes were investigated in terms of their ability to generate metal nanoparticles. The electrochemical detection of silver nanoparticles deposited by alkaline phosphatase is one of the first described reactions for a chip-based investigation of biomolecules [295–297]. In 2005, Fritzsche's group introduced the horseradish peroxidase-induced growth of silver nanoparticles [298]. The particle growth and the kinetics of the reaction were further investigated at the single-particle level to illustrate the reaction mechanism [299]. Further enzymes, widely used in molecular biological tests, such as glucose oxidase and acetylcholinesterase, were found to be possible initiators for the enzyme-induced growth of nanoparticles [300, 301]. Silver nanoparticles produced by alkaline phosphatase were used to produce nanoscopic metallic conductor paths by Dip-Pen nanolithography [302]. In addition to nanoparticles, enzymes can generate insoluble colored products that also can be exploited to detect different microorganisms [303]. This detection method is used in biochip applications by CLONDIAG Chip Technologies (Jena, Germany).

The successful combination of enzymes and metal nanoparticles represents a powerful technique that can be used to improve analytical methods in terms of sensitivity and specificity.

3.6 Conclusion

In the last decade, biochips have revolutionized analytical diagnostics. Because of their high throughput ability, biochips save time and costs in experiments. Advantages such as miniaturization and automation led to the further development of this technology. For this purpose, different scientific and engineering fields worked closely together. Due to the collaboration of workers involved in microfabrication, microelectronics, the computer industry, molecular biology, and so on, biochip technology is rapidly growing, with a large number of applications in pathogen detection, clinical diagnosis, genomic research, gene expression profiling, toxicogenomics, and pharmacogenomics.

The method of choice for chip-based biomolecule detection is fluorescence. This technique is well established and widely used in scientific laboratories. Despite all their advantages and the wide applicability of biochips, the technology is still limited to research applications. This is generally caused by a lack of reproducibility and reliability. The rapid development of the technology led to a large number of different biochip systems using various chip platforms, materials, labeling techniques, immobilization strategies, and detection units. Unfortunately, the comparability between these different systems, in terms of sensitivity and specificity, is uncertain. Nevertheless, the development of a new biochip generation, the so-called point-of-care systems, is promising for on-site testing and will continue to advance the technology. Due to the use of alternative labels such as metallic nanoparticles, quantum dots, and enzymes, biochips can be turned into robust and mobile analytical systems. Compared with fluorescence, the nanoparticle-based automation of biochips (e.g., integration in microfluidic systems) will significantly improve the reproducibility and reliability of these on-site diagnostic devices. The required high robustness and short analysis times of point-of-care systems allow their use outside specialized laboratories.

Finally, the biochip technology has already prepared the ground for a new class of bioanalytical tools in order to realize fast, sensitive, easy-to-use, mobile, and reliable diagnostics.

References

1. Southern, E., Mir, K., and Shchepinov, M. (1999) Molecular interactions on microarrays. *Nat. Genet.*, **21**, 5–9.
2. Southern, E.M. (1975) Detection of specific sequences among DNA fragments separated by gel electrophoresis. *J. Mol. Biol.*, **98** (3), 503–517.
3. Lemieux, B., Aharoni, A., and Schena, M. (1998) Overview of DNA chip technology. *Mol. Breed.*, **4** (4), 277–289.
4. Schena, M. (1996) Genome analysis with gene expression microarrays. *Bioessays*, **18** (5), 427–431.
5. DeRisi, J.L., Iyer, V.R., and Brown, P.O. (1997) Exploring the metabolic and genetic control of gene expression on a genomic scale. *Science*, **278** (5338), 680–686.
6. Fodor, S.P., Read, J.L., Pirrung, M.C. et al. (1991) Light-directed, spatially addressable parallel chemical synthesis. *Science*, **251** (4995), 767–773.
7. Green, N.S. and Pass, K.A. (2005) Neonatal screening by DNA microarray: spots and chips. *Nat. Rev. Genet.*, **6** (2), 147–151.
8. Greenberg, S.A. (2001) DNA microarray gene expression analysis technology and its application to neurological disorders. *Neurology*, **57** (5), 755–761.
9. de Mello-Coelho, V. and Hess, K.L. (2005) A conceptual and practical overview of cDNA microarray technology: implications for basic and clinical sciences. *Braz. J. Med. Biol. Res.*, **38** (10), 1543–1552.
10. Stoughton, R.B. (2005) Applications of DNA microarrays in biology. *Annu. Rev. Biochem.*, **74**, 53–82.
11. Ng, J.H. and Ilag, L.L. (2003) Biochips beyond DNA: technologies and applications. *Biotechnol. Annu. Rev.*, **9**, 1–149.
12. Heller, M.J. (2002) DNA microarray technology: devices, systems, and applications. *Annu. Rev. Biomed. Eng.*, **4** (1), 129.
13. Lockhart, D.J. and Winzeler, E.A. (2000) Genomics, gene expression and DNA arrays. *Nature*, **405** (6788), 827–836.
14. Aburatani, H. (2004) Functional genomics analysis on DNA microarray. *Tanpakushitsu Kakusan Koso*, **49** (11 Suppl), 1853–1858.
15. Shi, L., Perkins, R.G., and Tong, W. (2009) The current status of DNA microarrays, in *Microarrays Integrated Analytical Systems* (eds K. Dill, R. Liu, and P. Grodzinsky), Springer, New York, pp. 3–24.
16. Guo, Q.M. (2003) DNA microarray and cancer. *Curr. Opin. Oncol.*, **15** (1), 36–43.
17. Hardiman, G. (2008) Applications of microarrays and biochips in pharmacogenomics. *Methods Mol. Biol.*, **448**, 21–30.
18. Jordan, B.R. (2001) *DNA Microarrays: Gene Expression Applications*, Springer, Heidelberg.
19. Schena, M. (2003) *Microarray Analysis*, John Wiley & Sons, Inc., Hoboken, NJ.
20. Watson, S.J. et al. (2000) The "chip" as a specific genetic tool. *Biol. Psychiatry*, **48** (12), 1147–1156.
21. Ehrenreich, A. (2006) DNA microarray technology for the microbiologist: an overview. *Appl. Microbiol. Biotechnol.*, **73** (2), 255–273.
22. Wang, J. (2000) From DNA biosensors to gene chips. *Nucleic Acids Res.*, **28** (16), 3011–3016.
23. Schena, M. (ed.) (1999) *DNA Microarrays: a Practical Approach*, Oxford University Press, Oxford.
24. Ramsay, G. (1998) DNA chips: state-of-the-art. *Nat. Biotechnol.*, **16** (1), 40–44.
25. Southern, E.M. (1996) DNA chips: analysing sequence by hybridization to oligonucleotides on a large scale. *Trends Genet.*, **12** (3), 110–115.
26. Cheung, V.G. et al. (1999) Making and reading microarrays. *Nat. Genet.*, **21**, 15–19.
27. Brown, P.O. and Botstein, D. (1999) Exploring the new world of the genome with DNA microarrays. *Nat. Genet.*, **21**, 33–37.
28. Duggan, D.J. et al. (1999) Expression profiling using cDNA microarrays. *Nat. Genet.*, **21**, 10–14.
29. Hacia, J.G. (1999) Resequencing and mutational analysis using

oligonucleotide microarrays. *Nat. Genet.*, **21**, 42–47.

30 Debouck, C. and Goodfellow, P.N. (1999) DNA microarrays in drug discovery and development. *Nat. Genet.*, **21**, 48–50.

31 Dalma-Weiszhausz, D.D. et al. (2006) The Affymetrix GeneChip(R) platform: an overview. *Methods Enzymol.*, **410**, 3–27.

32 Bilitewski, U. (2006) Protein-sensing assay formats and devices. *Anal. Chim. Acta*, **568** (1–2), 232–247.

33 Rusmini, F., Zhong, Z.Y., and Feijen, J. (2007) Protein immobilization strategies for protein biochips. *Biomacromolecules*, **8** (6), 1775–1789.

34 Pollard, H.B. et al. (2007) Protein microarray platforms for clinical proteomics. *Proteomics Clin. Appl.*, **1** (9), 934–952.

35 Schena, M. (ed.) (2005) *Protein Microarrays*, Jones & Bartlett, Sudbury, MA.

36 Zhu, H. and Snyder, M. (2003) Protein chip technology. *Curr. Opin. Chem. Biol.*, **7** (1), 55–63.

37 Templin, M.F. et al. (2002) Protein microarray technology. *Trends Biotechnol.*, **20** (4), 160–166.

38 Emili, A.Q. and Cagney, G. (2000) Large-scale functional analysis using peptide or protein arrays. *Nat. Biotechnol.*, **18** (4), 393–397.

39 Panicker, R.C., Huang, X., and Yao, S.Q. (2004) Recent advances in peptide-based microarray technologies. *Comb. Chem. High Throuhput Screen.*, **7** (6), 547–556.

40 Liotta, L.A. et al. (2003) Protein microarrays: meeting analytical challenges for clinical applications. *Cancer Cell*, **3** (4), 317–325.

41 MacBeath, G. and Schreiber, S.L. (2000) Printing proteins as microarrays for high-throughput function determination. *Science*, **289** (5485), 1760–1763.

42 Tomizaki, K.Y., Usui, K., and Mihara, H. (2005) Protein-detecting microarrays: current accomplishments and requirements. *ChemBioChem*, **6** (5), 783–799.

43 Miller, J.C. et al. (2003) Antibody microarray profiling of human prostate cancer sera: antibody screening and identification of potential biomarkers. *Proteomics*, **3** (1), 56–63.

44 Bouwman, K. et al. (2003) Microarrays of tumor cell derived proteins uncover a distinct pattern of prostate cancer serum immunoreactivity. *Proteomics*, **3** (11), 2200–2207.

45 Schweitzer, B. et al. (2002) Multiplexed protein profiling on microarrays by rolling-circle amplification. *Nat. Biotechnol.*, **20** (4), 359–365.

46 Gul, O. et al. (2007) Sandwich-type, antibody microarrays for the detection and quantification of cardiovascular risk markers. *Sens. Actuators B Chem.*, **125** (2), 581–588.

47 Wilson, D.S. and Nock, S. (2003) Recent developments in protein microarray technology. *Angew. Chem. Int. Ed.*, **42** (5), 494–500.

48 Kumble, K.D. (2007) An update on using protein microarrays in drug discovery. *Expert Opin. Drug Discov.*, **2** (11), 1467–1476.

49 Falsey, J.R. et al. (2001) Peptide and small molecule microarray for high throughput cell adhesion and functional assays. *Bioconjug. Chem.*, **12** (3), 346–353.

50 Uttamchandani, M. et al. (2005) Small molecule microarrays: recent advances and applications. *Curr. Opin. Chem. Biol.*, **9** (1), 4–13.

51 Bailey, S.N., Wu, R.Z., and Sabatini, D.M. (2002) Applications of transfected cell microarrays in high-throughput drug discovery. *Drug Discov. Today*, **7** (18), S113–S118.

52 Uttamchandani, M. and Yao, S.Q. (2008) Peptide microarrays: next generation biochips for detection, diagnostics and high-throughput screening. *Curr. Pharm. Des.*, **14** (24), 2428–2438.

53 Kononen, J. et al. (1998) Tissue microarrays for high-throughput molecular profiling of tumor specimens. *Nat. Med.*, **4** (7), 844–847.

54 Battifora, H. (1986) The multitumor (sausage) tissue block: novel method for immunohistochemical antibody testing. *Lab. Invest.*, **55** (2), 244–248.

55 Bubendorf, L. et al. (1999) Survey of gene amplifications during prostate cancer

progression by high throughput fluorescence in situ hybridization on tissue microarrays. *Cancer Res.*, **59** (4), 803–806.
56. Moch, H. et al. (1999) High-throughput tissue microarray analysis to evaluate genes uncovered by cDNA microarray screening in renal cell carcinoma. *Am. J. Pathol.*, **154** (4), 981–986.
57. Ziauddin, J. and Sabatini, D.M. (2001) Microarrays of cells expressing defined cDNAs. *Nature*, **411** (6833), 107–110.
58. Bailey, S.N. et al. (2006) Microarrays of lentiviruses for gene function screens in immortalized and primary cells. *Nat. Methods*, **3** (2), 117–122.
59. Wheeler, D.B. et al. (2004) RNAi living-cell microarrays for loss-of-function screens in *Drosophila melanogaster* cells. *Nat. Methods*, **1** (2), 127–132.
60. Wheeler, D.B., Carpenter, A.E., and Sabatini, D.M. (2005) Cell microarrays and RNA interference chip away at gene function. *Nat. Genet.*, **37**, 25–30.
61. Kulkarni, M.M. et al. (2006) Evidence of off-target effects associated with long dsRNAs in *Drosophila melanogaster* cell-based assays. *Nat. Methods*, **3** (10), 833–838.
62. Federico, M. (ed.) (2003) *Lentivirus Gene Engineering Protocols, Methods in Molecular Biology*, vol. **229**, Humana Press, Totawa, NJ
63. Bono, H. and Nakao, M.C. (2003) Introduction to the practical analysis of DNA microarray data. *Tanpakushitsu Kakusan Koso*, **48** (2), 167–172.
64. Do, J.H. and Choi, D.K. (2008) Clustering approaches to identifying gene expression patterns from DNA microarray data. *Mol. Cells*, **25** (2), 279–288.
65. Fan, J. and Ren, Y. (2006) Statistical analysis of DNA microarray data in cancer research. *Clin. Cancer Res.*, **12** (15), 4469–4473.
66. Hanai, T., Hamada, H., and Okamoto, M. (2006) Application of bioinformatics for DNA microarray data to bioscience, bioengineering and medical fields. *J. Biosci. Bioeng.*, **101** (5), 377–384.
67. Hackl, H. et al. (2004) Analysis of DNA microarray data. *Curr. Top. Med. Chem.*, **4** (13), 1357–1370.
68. Barbulovic-Nad, I. et al. (2006) Bio-microarray fabrication techniques – a review. *Crit. Rev. Biotechnol.*, **26** (4), 237–259.
69. Thompson, M. and Furtado, L.M. (1999) High density oligonucleotide and DNA probe arrays for the analysis of target DNA. *Analyst*, **124** (8), 1133–1136.
70. Pirrung, M.C. (2002) How to make a DNA chip. *Angew. Chem. Int. Ed.*, **41** (8), 1277.
71. Dill, K., Liu, R.H., and Grodzinski, P. (eds.) (2009) *Microarrays: Preparation, Microfluidics, Detection Methods, and Biological Applications (Integrated Analytical Systems)*, Springer Science + Business Media, New York.
72. Nguyen, D.V. et al. (2002) DNA microarray experiments: biological and technological aspects. *Biometrics*, **58** (4), 701–717.
73. Pasquarelli, A. (2008) Biochips: technologies and applications. *Mater. Sci. Eng. C Biomimet. Supramol. Syst.*, **28** (4), 495–508.
74. Oh, S.J. et al. (2006) Surface modification for DNA and protein microarrays. *OMICS*, **10** (3), 327–343.
75. Henry, A.C. et al. (2000) Surface modification of poly(methyl methacrylate) used in the fabrication of microanalytical devices. *Anal. Chem.*, **72** (21), 5331–5337.
76. Pirrung, M.C., Davis, J.D., and Odenbaugh, A.L. (2000) Novel reagents and procedures for immobilization of DNA on glass microchips for primer extension. *Langmuir*, **16** (5), 2185–2191.
77. Kohn, M. (2009) Immobilization strategies for small molecule, peptide and protein microarrays. *J. Pept. Sci.*, **15** (6), 393–397.
78. Uttamchandani, M., Wang, J., and Yao, S.Q. (2006) Protein and small molecule microarrays: powerful tools for high-throughput proteomics. *Mol. Biosyst.*, **2** (1), 58–68.
79. Schena, M. et al. (1998) Microarrays: biotechnology's discovery platform for functional genomics. *Trends Biotechnol.*, **16** (7), 301–306.
80. Lamture, J.B. et al. (1994) Direct detection of nucleic acid hybridization on the

81 Rogers, Y.-H. et al. (1999) Immobilization of oligonucleotides onto a glass support via disulfide bonds: a method for preparation of DNA microarrays. *Anal. Biochem.*, **266** (1), 23–30.

82 Chrisey, L.A., Lee, G.U., and Oferrall, C.E. (1996) Covalent attachment of synthetic DNA to self-assembled monolayer films. *Nucleic Acids Res.*, **24** (15), 3031–3039.

83 Diehl, F., Beckmann, B., Kellner, N. et al. (2002) Manufacturing DNA microarrays from unpurified PCR products, *Nucleic Acids Res.*, **30** (16), e79.

84 Lamture, J.B. et al. (1994) Direct detection of nucleic acid hybridization on the surface of a charge coupled device. *Nucleic Acids Res.*, **22** (11), 2121–2125.

85 Schena, M. et al. (1996) Parallel human genome analysis: microarray-based expression monitoring of 1000 genes. *Proc. Natl. Acad. Sci. U. S. A.*, **93** (20), 10614–10619.

86 Dufva, M. et al. (2004) Characterization of an inexpensive, nontoxic, and highly sensitive microarray substrate. *Biotechniques*, **37** (2), 286–296.

87 Belleville, E. et al. (2003) Quantitative assessment of factors affecting the sensitivity of a competitive immunomicroarray for pesticide detection. *Biotechniques*, **35** (5), 1044–1051.

88 Pemov, A. et al. (2005) DNA analysis with multiplex microarray-enhanced PCR. *Nucleic Acids Res.*, **33** (2), e11.

89 Miller, J.C. et al. (2003) Antibody microarray profiling of human prostate cancer sera: antibody screening and identification of potential biomarkers. *Proteomics*, **3** (1), 56–63.

90 Liu, D. et al. (2004) Immobilization of DNA onto poly(dimethylsiloxane) surfaces and application to a microelectrochemical enzyme-amplified DNA hybridization assay. *Langmuir*, **20** (14), 5905–5910.

91 Lockhart, D.J. et al. (1996) Expression monitoring by hybridization to high-density oligonucleotide arrays. *Nat. Biotechnol.*, **14** (13), 1675–1680.

92 Hughes, T.R. et al. (2001) Expression profiling using microarrays fabricated by an ink-jet oligonucleotide synthesizer. *Nat. Biotechnol.*, **19** (4), 342–347.

93 Wolber, P.K. et al. (2006) The Agilent *in situ*-synthesized microarray platform. *Methods Enzymol.*, **410**, 28–57.

94 Duroux, M. et al. (2007) Using light to bioactivate surfaces: a new way of creating oriented, active immunobiosensors. *Appl. Surf. Sci.*, **254** (4), 1126–1130.

95 Soper, S.A. et al. (2005) Fabrication of DNA microarrays onto polymer substrates using UV modification protocols with integration into microfluidic platforms for the sensing of low-abundant DNA point mutations. *Methods*, **37** (1), 103–113.

96 Nahar, P., Naqvi, A., and Basir, S.F. (2004) Sunlight-mediated activation of an inert polymer surface for covalent immobilization of a protein. *Anal. Biochem.*, **327** (2), 162–164.

97 Sinha, R.P. and Hader, D.P. (2002) UV-induced DNA damage and repair: a review. *Photochem. Photobiol. Sci.*, **1** (4), 225–236.

98 http://www.illumina.com/Documents/products/techbulletins/techbulletin_rna.pdf; Gene expression profiling with Sentrix Focused Arrays; 2005 Illumina, Inc. 2011/07/20.

99 Fan, J.B. et al. (2006) Illumina universal bead arrays. *Methods Enzymol.*, **410**, 57–73.

100 Wick, I. and Hardiman, G. (2005) Biochip platforms as functional genomics tools for drug discovery. *Curr. Opin. Drug Discov. Dev.*, **8** (3), 347–354.

101 Ausubel, F.M., Brent, R., Moore, D.D., Seidman, J.G., Smith, J.A., and Stuhl, K. (eds) (2010) Current Protocols in Molecular Biology, Chapter 4: Preparation and Analysis of RNA, John Wiley & Sons Inc.

102 Sambrook, J. and Russell, D. (2001) *Molecular Cloning: a Laboratory Manual*, 3rd edn, Cold Spring Harbor Laboratory Press, Cold Spring Harbor, NY.

103 Stull, D. and Pisano, J.M. (2001) Purely RNA: new innovations enhance the quality, speed, and efficiency of RNA isolation techniques. *Scientist*, **15** (22), 29–31.

104 Qiagen (2006). Qproteome Mammalian Protein Preparation Handbook, Qiagen, Hilden.

105 Saiki, R.K. et al. (1985) Enzymatic amplification of beta-globin genomic sequences and restriction site analysis for diagnosis of sickle cell anemia. *Science*, **230** (4732), 1350–1354.

106 Saiki, R.K. et al. (1988) Primer-directed enzymatic amplification of DNA with a thermostable DNA polymerase. *Science*, **239** (4839), 487–491.

107 Frohman, M.A., Dush, M.K., and Martin, G.R. (1988) Rapid production of full-length cDNAs from rare transcripts: amplification using a single gene-specific oligonucleotide primer. *Proc. Natl. Acad. Sci. U. S. A.*, **85** (23), 8998–9002.

108 Berchtold, W. (1989) A simple method for direct cloning and sequencing cDNA by the use of a single specific oligonucleotide and oligo(dT) in a polymerase chain reaction (PCR). *Nucleic Acids Res.*, **17** (1), 453.

109 van Gelder, R. et al. (1990) Amplified RNA synthesized from limited quantities of heterogeneous cDNA. *Proc. Natl. Acad. Sci. U. S. A.*, **87** (5), 1663–1667.

110 Nam, J.-M., Stoeva, S.I., and Mirkin, C.A. (2004) Bio-bar-code-based DNA detection with PCR-like sensitivity. *J. Am. Chem. Soc.*, **126** (19), 5932–5933.

111 Nam, J.-M., Thaxton, C.S., and Mirkin, C.A. (2003) Nanoparticle-based bio-bar codes for the ultrasensitivedetection of proteins. *Science*, **301** (5641), 1884–1886.

112 Stoeva, S.I. et al. (2006) Multiplexed DNA detection with biobarcoded nanoparticle probes. *Angew. Chem. Int. Ed.*, **45** (20), 3303–3306.

113 Oh, B.-K. (2006) A fluorophore-based bio-barcode amplification assay for proteins. *Small*, **2** (1), 103–108.

114 Muller, U.R. (2006) Protein detection using biobarcodes. *Mol. Biosyst.*, **2** (10), 470–476.

115 Brown, T.A. (2002) DNA labeling, in *Genomes*, 2nd edn, Wiley-Liss, Oxford, p. 163.

116 Zhu, Z. et al. (1994) Directly labeled DNA probes using fluorescent nucleotides with different length linkers. *Nucleic Acids Res.*, **22** (16), 3418–3422.

117 Chen, J., Strand, S.-E., and Sjögren, H.-O. (1996) Optimization of radioiodination and biotinylation of monoclonal antibody chimeric BR96: an indirect labeling using N-succinimidyl-3-(tri-n-butylstannyl) benzoate conjugate. *Cancer Biother. Radiopharm.*, **11** (3), 217–226.

118 Goddard, J.-P. and Reymond, J.-L. (2004) Enzyme assays for high-throughput screening. *Curr. Opin. Biotechnol.*, **15** (4), 314–322.

119 Engvall, E. and Perlmann, P. (1971) Enzyme-linked immunosorbent assay (ELISA). Quantitative assay of immunoglobulin G. *Immunochemistry*, **8** (9), 871–874.

120 Renart, J., Reiser, J., and Stark, G.R. (1979) Transfer of proteins from gels to diazobenzyloxymethyl-paper and detection with antisera: a method for studying antibody specificity and antigen structure. *Proc. Natl. Acad. Sci. U. S. A.*, **76** (7), 3116–3120.

121 Towbin, H., Staehelin, T., and Gordon, J. (1979) Electrophoretic transfer of proteins from polyacrylamide gels to nitrocellulose sheets: procedure and some applications. *Proc. Natl. Acad. Sci. U. S. A.*, **76** (9), 4350–4354.

122 Liz-Marzan, L.M. (2006) Tailoring surface plasmons through the morphology and assembly of metal nanoparticles. *Langmuir*, **22** (1), 32–41.

123 Nuzzo, R.G. and Allara, D.L. (1983) Adsoption of bifunctional organic disulfides on gold surfaces. *J. Am. Chem. Soc.*, **105**, 4481–4483.

124 Xiao, S. et al. (2002) Selfassembly of metallic nanoparticle arrays by DNA scaffolding. *J. Nanopart. Res.*, **4**, 313–317.

125 Fritzsche, W. and Taton, T.A. (2003) Metal nanoparticles as labels for heterogeneous, chip-based DNA detection. *Nanotechnology*, **14** (12), R63–R73.

126 Yguerabide, J. and Yguerabide, E.E. (1998) Light-scattering submicroscopic particles as highly fluorescent analogs and their use as tracer labels in clinical and biological applications. I. Theory. *Anal. Biochem.*, **262** (2), 137–156.

127 Yguerabide, J. and Yguerabide, E.E. (1998) Light-scattering submicroscopic particles as highly fluorescent analogs and their use as tracer labels in clinical and biological applications. II. Experimental

characterization. *Anal. Biochem.*, **262** (2), 157–176.

128 Wang, J., Liu, G., and Merkoci, A. (2003) Electrochemical coding technology for simultaneous detection of multiple DNA targets. *J. Am. Chem. Soc.*, **125** (11), 3214–3215.

129 Dequaire, M., Degrand, C., and Limoges, B. (2000) An electrochemical metalloimmunoassay based on a colloidal gold label. *Anal. Chem.*, **72** (22), 5521–5528.

130 Okahata, Y. *et al.* (2000) Quantitative detection of binding of PCNA protein to DNA strands on a 27MHz quartz-crystal microbalance. *Nucleic Acids Symp. Ser.* (44), 243–244.

131 Möller, R. *et al.* (2001) Electrical classification of the concentration of bioconjugated metal colloids after surface adsorption and silver enhancement. *Langmuir*, **17**, 5426–5430.

132 Steinbrück, A. *et al.* (2006) Preparation and optical characterization of core–shell bimetal nanoparticles. *Plasmonics*, **1** (1), 79–85.

133 Bally, M. *et al.* (2006) Optical microarray biosensing techniques. *Surf. Interface Anal.*, **38** (11), 1442–1458.

134 Johnston, R.F., Pickett, S.C., and Barker, D.L. (1990) Autoradiography using storage phosphor technology. *Electrophoresis*, **11** (5), 355–360.

135 Lakowicz, J.R. (2006) *Principles of Fluorescence Spectroscopy*, 3rd edn, Springer, Berlin.

136 Epstein, J.R., Biran, I., and Walt, D.R. (2002) Fluorescence-based nucleic acid detection and microarrays. *Anal. Chim. Acta*, **469** (1), 3–36.

137 Nagl, S., Schaeferling, M., and Wolfbeis, O.S. (2005) Fluorescence analysis in microarray technology. *Microchim. Acta*, **151** (1), 1–21.

138 Valeur, B. (2001) *Molecular Fluorescence Principles and Applications*, Wiley-VCH Verlag GmbH, Weinheim, p. 388.

139 Sebat, J. *et al.* (2004) Large-scale copy number polymorphism in the human genome. *Science*, **305** (5683), 525–528.

140 Ekins, R.P. and Chu, F.W. (1991) Multianalyte microspot immunoassay – microanalytical compact-disk of the future. *Clin. Chem.*, **37** (11), 1955–1967.

141 Song, L.L. *et al.* (1995) Photobleaching kinetics of fluorescein in quantitative fluorescence microscopy. *Biophys. J.*, **68** (6), 2588–2600.

142 Timlin, J.A. (2006) Scanning microarrays: current methods and future directions. *Methods Enzymol.*, **411**, 79–98.

143 Huang, G.L. *et al.* (2008) Digital imaging scanning system and biomedical applications for biochips. *J. Biomed. Opt.*, **13** (3), 034006.

144 Valentini, G. *et al.* (2000) Time-resolved DNA-microarray reading by an intensified CCD for ultimate sensitivity. *Opt. Lett.*, **25** (22), 1648–1650.

145 Waddell, E. *et al.* (2000) High-resolution near-infrared imaging of DNA microarrays with time-resolved acquisition of fluorescence lifetimes. *Anal. Chem.*, **72** (24), 5907–5917.

146 Striebel, H.M. *et al.* (2004) Readout of protein microarrays using intrinsic time resolved UV fluorescence for label-free detection. *Proteomics*, **4** (6), 1703–1711.

147 Cubeddu, R. *et al.* (2002) Time-resolved fluorescence imaging in biology and medicine. *J. Phys. D Appl. Phys.*, **35** (9), R61–R66.

148 Wabuyele, M.B. *et al.* (2003) Approaching real-time molecular diagnostics: single-pair fluorescence resonance energy transfer (spFRET) detection for the analysis of low abundant point mutations in K-ras oncogenes. *J. Am. Chem. Soc.*, **125** (23), 6937–6945.

149 Cardullo, R.A. *et al.* (1988) Detection of nucleic-acid hybridization by nonradiative fluorescence resonance energy-transfer. *Proc. Natl. Acad. Sci. U. S. A.*, **85** (23), 8790–8794.

150 Mere, L. *et al.* (1999) Miniaturized FRET assays and microfluidics: key components for ultra-high-throughput screening. *Drug Discov. Today*, **4** (8), 363–369.

151 Marras, S.A., Kramer, F.R., and Tyagi, S. (2002) Efficiencies of fluorescence resonance energy transfer and contact-mediated quenching in oligonucleotide probes. *Nucleic Acids Res.*, **30** (21), e122.

152 Nagl, S. *et al.* (2008) Microarray analysis of protein–protein interactions based on FRET using subnanosecond-resolved

fluorescence lifetime imaging. *Biosens. Bioelectron.*, **24** (3), 397–402.
153 Fan, A.P., Lau, C.W., and Lu, J.Z. (2005) Magnetic bead-based chemiluminescent metal immunoassay with a colloidal gold label. *Anal. Chem.*, **77** (10), 3238–3242.
154 Li, H.G. and He, Z.K. (2009) Magnetic bead-based DNA hybridization assay with chemiluminescence and chemiluminescent imaging detection. *Analyst*, **134** (4), 800–804.
155 Marquette, C.A. and Blum, L.J. (2006) Applications of the luminol chemiluminescent reaction in analytical chemistry. *Anal. Bioanal. Chem.*, **385** (3), 546–554.
156 Roda, A. et al. (2003) Analytical bioluminescence and chemiluminescence. *Anal. Chem.*, **75** (21), 463A–470A.
157 Seidel, M. and Niessner, R. (2008) Automated analytical microarrays: a critical review. *Anal. Bioanal. Chem.*, **391** (5), 1521–1544.
158 Roda, A. et al. (2005) Bio- and chemiluminescence imaging in analytical chemistry. *Anal. Chim. Acta*, **541** (1–2), 25–35.
159 Quinn, M.C.J. et al. (2009) The chemiluminescence based Ziplex(R) automated workstation focus array reproduces ovarian cancer Affymetrix GeneChip(R) expression profiles. *J. Translat. Med.*, **7**, 53.
160 Zhang, J. et al. (2008) Plasmon-coupled fluorescence probes: effect of emission wavelength on fluorophore-labeled silver particles. *J. Phys. Chem. C*, **112** (25), 9172–9180.
161 Lakowicz, J.R. et al. (2001) Intrinsic fluorescence from DNA can be enhanced by metallic particles. *Biochem. Biophys. Res. Commun.*, **286** (5), 875–879.
162 Myers, F.B. and Lee, L.P. (2008) Innovations in optical microfluidic technologies for point-of-care diagnostics. *Lab Chip*, **8** (12), 2015–2031.
163 Taton, T.A., Lu, G., and Mirkin, C.A. (2001) Two-color labeling of oligonucleotide arrays via size-selective scattering of nanoparticle probes. *J. Am. Chem. Soc.*, **123** (21), 5164–5165.
164 Karlsson, R. (2004) SPR for molecular interaction analysis: a review of emerging application areas. *J. Mol. Recognit.*, **17** (3), 151–161.
165 Wegner, G.J. et al. (2004) Real-time surface plasmon resonance imaging measurements for the multiplexed determination of protein adsorption/desorption kinetics and surface enzymatic reactions on peptide microarrays. *Anal. Chem.*, **76** (19), 5677–5684.
166 Liedberg, B., Nylander, C., and Lundstrom, I. (1995) Biosensing with surface plasmon resonance – how it all started. *Biosens. Bioelectron.*, **10** (8), i–ix.
167 Peterlinz, K.A. and Georgiadis, R.M. (1996) In situ kinetics of self-assembly by surface plasmon resonance spectroscopy. *Langmuir*, **12**, 4731–4740.
168 Georgiadis, R.M., Peterlinz, K.A., and Peterson, A.W. (2000) Quantitative measurements and modeling of kinetics in nucleic acid monolayer films using SPR spectroscopy. *J. Am. Chem. Soc.*, **122** (13), 3166–3173.
169 Perkins, E.A. and Squirrell, D.J. (2000) Development of instrumentation to allow the detection of microorganisms using light scattering in combination with surface plasmon resonance. *Biosens. Bioelectron.*, **14** (10–11), 853–859.
170 Stuart, D.A. et al. (2005) Biological applications of localised surface plasmonic phenomenae. *IEE Proc. Nanobiotechnol.*, **152** (1), 13–32.
171 Lyon, L.A., Musick, M.D., and Natan, M.J. (1998) Colloidal Au-enhanced surface plasmon resonance immunosensing. *Anal. Chem.*, **70** (24), 5177–5183.
172 Hoa, X.D., Kirk, A.G., and Tabrizian, M. (2007) Towards integrated and sensitive surface plasmon resonance biosensors: a review of recent progress. *Biosens. Bioelectron.*, **23** (2), 151–160.
173 Vo-Dinh, T. (2003) Biochips and microarrays: tools for the new medicine, in *Biomedical Photonics Handbook* (ed. T. Vo-Dinh), CRC Press, Boca Raton, FL, p. 51-1.
174 Pastinen, T. et al. (1997) Minisequencing: a specific tool for DNA analysis and

diagnostics on oligonucleotide arrays. *Genome Res.*, **7** (6), 606–614.
175 Macgregor, P.F. (2003) Gene expression in cancer: the application of microarrays. *Expert Rev. Mol. Diagn.*, **3** (2), 185–200.
176 Meldrum, D. (2000) Automation for genomics, part two: sequencers, microarrays, and future trends. *Genome Res.*, **10** (9), 1288–1303.
177 Mockler, T.C. and Ecker, J.R. (2005) Applications of DNA tiling arrays for whole-genome analysis. *Genomics*, **85** (1), 1–15.
178 Morozova, O. and Marra, M.A. (2008) Applications of next-generation sequencing technologies in functional genomics. *Genomics*, **92** (5), 255–264.
179 Parissenti, A.M. et al. (2007) Gene expression profiles as biomarkers for the prediction of chemotherapy drug response in human tumour cells. *Anticancer Drugs*, **18** (5), 499–523.
180 Shen, Y. and Wu, B.L. (2009) Microarray-based genomic DNA profiling technologies in clinical molecular diagnostics. *Clin. Chem.*, **55** (4), 659–669.
181 Waddell, N. (2008) Microarray-based DNA profiling to study genomic aberrations. *IUBMB Life*, **60** (7), 437–440.
182 Grant, S.F.A. and Hakonarson, H. (2008) Microarray technology and applications in the arena of genome-wide association. *Clin. Chem.*, **54** (7), 1116–1124.
183 Pollack, J.R. et al. (1999) Genome-wide analysis of DNA copy-number changes using cDNA microarrays. *Nat. Genet.*, **23** (1), 41–46.
184 Schena, M. et al. (1995) Quantitative monitoring of gene expression patterns with a complementary DNA microarray. *Science*, **270** (5235), 467–470.
185 Janne, P.A. et al. (2004) High-resolution single-nucleotide polymorphism array and clustering analysis of loss of heterozygosity in human lung cancer cell lines. *Oncogene*, **23** (15), 2716–2726.
186 Lindblad-Toh, K. et al. (2000) Loss-of-heterozygosity analysis of small-cell lung carcinomas using single-nucleotide polymorphism arrays. *Nat. Biotechnol.*, **18** (9), 1001–1005.
187 Santos, E.S., Perez, C.A., and Raez, L.E. (2009) How is gene-expression profiling going to challenge the future management of lung cancer? *Future Oncol.*, **5** (6), 827–835.
188 Bubendorf, L. et al. (1999) Hormone therapy failure in human prostate cancer: analysis by complementary DNA and tissue microarrays. *J. Natl. Cancer Inst.*, **91** (20), 1758–1764.
189 Lieberfarb, M.E. et al. (2003) Genome-wide loss of heterozygosity analysis from laser capture microdissected prostate cancer using single nucleotide polymorphic allele (SNP) arrays and a novel bioinformatics platform dChipSNP. *Cancer Res.*, **63** (16), 4781–4785.
190 Brennan, D.J. et al. (2005) Application of DNA microarray technology in determining breast cancer prognosis and therapeutic response. *Expert Opin. Biol. Ther.*, **5** (8), 1069–1083.
191 Cheang, M.C.U., van de Rijn, M., and Nielsen, T.O. (2008) Gene expression profiling of breast cancer. *Annu. Rev. Pathol. Mech. Dis.*, **3**, 67–97.
192 Geyer, F.C. and Reis, J.S. (2009) Microarray-based gene expression profiling as a clinical tool for breast cancer management: are we there yet? *Int. J. Surg. Pathol.*, **17** (4), 285–302.
193 Lacroix, M. et al. (2002) A low-density DNA microarray for analysis of markers in breast cancer. *Int. J. Biol. Markers*, **17** (1), 5–23.
194 Fehrmann, R.S.N. et al. (2007) Profiling studies in ovarian cancer: a review. *Oncologist*, **12** (8), 960–966.
195 Lux, M.P., Fasching, P.A., and Beckmann, M.W. (2006) Hereditary breast and ovarian cancer: review and future perspectives. *J. Mol. Med.*, **84** (1), 16–28.
196 Bednar, M. (2000) DNA microarray technology and application. *Med. Sci. Monit.*, **6** (4), 796–800.
197 Yoo, S.M. and Lee, S.Y. (2008) Diagnosis of pathogens using DNA microarray. *Recent Pat. Biotechnol.*, **2** (2), 124–129.
198 Kallioniemi, O.P. (2001) Biochip technologies in cancer research. *Ann. Med.*, **33** (2), 142–147.

199 Anisimov, S.V. (2007) Application of DNA microarray technology to gerontological studies. *Methods Mol. Biol.*, **371**, 249–265.

200 Sachidanandam, R. *et al.* (2001) A map of human genome sequence variation containing 1.42 million single nucleotide polymorphisms. *Nature*, **409** (6822), 928–933.

201 NCBI. Single Nucleotide Polymorphism, http://www.ncbi.nlm.nih.gov/SNP/., National Center for Biotechnology Information, Bethesda, MD (last accessed 26 March 2011).

202 Hirschhorn, J.N. and Daly, M.J. (2005) Genome-wide association studies for common diseases and complex traits. *Nat. Rev. Genet.*, **6** (2), 95–108.

203 Collins, F.S. *et al.* (2003) A vision for the future of genomics research. *Nature*, **422** (6934), 835–847.

204 Shen, R. *et al.* (2005) High-throughput SNP genotyping on universal bead arrays. *Mutat. Res.*, **573** (1–2), 70–82.

205 Saiki, R.K. *et al.* (1989) Genetic-analysis of amplified DNA with immobilized sequence-specific oligonucleotide probes. *Proc. Natl. Acad. Sci. U. S. A.*, **86** (16), 6230–6234.

206 Lien, K.Y. *et al.* (2009) Extraction of genomic DNA and detection of single nucleotide polymorphism genotyping utilizing an integrated magnetic bead-based microfluidic platform. *Microfluidics Nanofluidics*, **6** (4), 539–555.

207 Selmer, K.K. *et al.* (2009) Genome-wide linkage analysis with clustered SNP markers. *J. Biomol. Screen.*, **14** (1), 92–96.

208 Zabarovsky, E.R. *et al.* (2003) Restriction site tagged (RST) microarrays: a novel technique to study the species composition of complex microbial systems. *Nucleic Acids Res.*, **31** (16), e95.

209 Mun, H.Y. *et al.* (2009) SNPs detection by a single-strand specific nuclease on a PNA zip-code microarray. *Biosens. Bioelectron.*, **24** (6), 1706–1711.

210 Girigoswami, A. *et al.* (2008) PCR-free mutation detection of BRCA1 on a zip-code microarray using ligase chain reaction. *J. Biochem. Biophys. Methods*, **70** (6), 897–902.

211 Jung, C. *et al.* (2008) Microarray-based detection of Korean-specific BRCA1 mutations. *Anal. Bioanal. Chem.*, **391** (1), 405–413.

212 Song, J.Y. (2005) *et al.* Diagnosis of HNF-1 alpha mutations on a PNA zip-code microarray by single base extension. *Nucleic Acids Res.*, **33** (2), e19.

213 Anderson, J.E. *et al.* (2006) Methods and biomarkers for the diagnosis and prognosis of cancer and other diseases: towards personalized medicine. *Drug Resist. Updat.*, **9** (4–5), 198–210.

214 Zhao, X.J. *et al.* (2004) An integrated view of copy number and allelic alterations in the cancer genome using single nucleotide polymorphism arrays. *Cancer Res.*, **64** (9), 3060–3071.

215 Ohira, M. *et al.* (2005) A review of DNA microarray analysis of human neuroblastomas. *Cancer Lett.*, **228** (1–2), 5–11.

216 Dutt, A. and Beroukhim, R. (2007) Single nucleotide polymorphism array analysis of cancer. *Curr. Opin. Oncol.*, **19**, 43–49.

217 Gebhard, C. *et al.* (2006) Genome-wide profiling of CpG methylation identifies novel targets of aberrant hypermethylation in myeloid leukemia. *Cancer Res.*, **66** (12), 6118–6128.

218 Adrien, L.R. *et al.* (2006) Classification of DNA methylation patterns in tumor cell genomes using a CpG island microarray. *Cytogenet. Genome Res.*, **114** (1), 16–23.

219 Mao, X.Y., Young, B.D., and Lu, Y.J. (2007) The application of single nucleotide polymorphism microarrays in cancer research. *Curr. Genomics*, **8** (4), 219–228.

220 Wang, Z.G.C. *et al.* (2004) Loss of heterozygosity and its correlation with expression profiles in subclasses of invasive breast cancers. *Cancer Res.*, **64** (1), 64–71.

221 Rauch, A. *et al.* (2004) Molecular karyotyping using an SNP array for genomewide genotyping. *J. Med. Genet.*, **41** (12), 916–922.

222 Bruno, D.L. *et al.* (2009) Detection of cryptic pathogenic copy number variations and constitutional loss of heterozygosity using high resolution SNP microarray analysis in 117 patients referred for cytogenetic analysis and impact on clinical practice. *J. Med. Genet.*, **46** (2), 123–131.

223 Hughes, T.R. and Shoemaker, D.D. (2001) DNA microarrays for expression profiling. *Curr. Opin. Chem. Biol.*, **5** (1), 21–25.

224 Deyholos, M.K. and Galbraith, D.W. (2001) High-density microarrays for gene expression analysis. *Cytometry*, **43** (4), 229–238.

225 Schulze, A. and Downward, J. (2001) Navigating gene expression using microarrays – a technology review. *Nat. Cell Biol.*, **3** (8), E190–E195.

226 Polyak, K. and Riggins, G.J. (2001) Gene discovery using the serial analysis of gene expression technique: implications for cancer research. *J. Clin. Oncol.*, **19** (11), 2948–2958.

227 Stadler, Z.K. and Come, S.E. (2009) Review of gene-expression profiling and its clinical use in breast cancer. *Crit. Rev. Oncol. Hematol.*, **69** (1), 1–11.

228 Geyer, F.C., Decker, T., and Reis-Filho, J.S. (2009) Genome-wide expression profiling as a clinical tool. *Pathologe*, **30** (2), 141–146.

229 Glas, A.M. *et al.* (2006) Converting a breast cancer microarray signature into a high-throughput diagnostic test. *BMC Genomics*, **7**, 278.

230 Mullighan, C.G. and Downing, J.R. (2009) Genome-wide profiling of genetic alterations in acute lymphoblastic leukemia: recent insights and future directions. *Leukemia*, **23** (7), 1209–1218.

231 Radtke, I. *et al.* (2009) Genomic analysis reveals few genetic alterations in pediatric acute myeloid leukemia. *Proc. Natl. Acad. Sci. U. S. A.*, **106** (31), 12944–12949.

232 Nannini, M. *et al.* (2009) Gene expression profiling in colorectal cancer using microarray technologies: results and perspectives. *Cancer Treat. Rev.*, **35** (3), 201–209.

233 Boerma, M. *et al.* (2005) Microarray analysis of gene expression profiles of cardiac myocytes and fibroblasts after mechanical stress, ionising or ultraviolet radiation. *BMC Genomics*, **6** (1), p. 6.

234 Yanagawa, B. *et al.* (2005) Affymetrix oligonucleotide analysis of gene expression in the injured heart. *Methods Mol. Med.*, **112**, 305–320.

235 Ducray, F., Honnorat, J., and Lachuer, J. (2007) DNA microarray technology: principles and applications to the study of neurological disorders. *Rev. Neurol.*, **163** (4), 409–420.

236 Bhattacharya, B., Puri, S., and Puri, R.K. (2009) A review of gene expression profiling of human embryonic stem cell lines and their differentiated progeny. *Curr. Stem Cell Res. Ther.*, **4** (2), 98–106.

237 Marton, M.J. *et al.* (1998) Drug target validation and identification of secondary drug target effects using DNA microarrays. *Nat. Med.*, **4** (11), 1293–1301.

238 Jain, K.K. (2000) Applications of biochip and microarray systems in pharmacogenomics. *Pharmacogenomics*, **1** (3), 289–307.

239 Zhu, H., Bilgin, M., and Snyder, M. (2003) Proteomics. *Annu. Rev. Biochem.*, **72**, 783–812.

240 Boguski, M.S. and McIntosh, M.W. (2003) Biomedical informatics for proteomics. *Nature*, **422** (6928), 233–237.

241 Chen, Y.W. *et al.* (2000) Expression profiling in the muscular dystrophies: identification of novel aspects of molecular pathophysiology. *J. Cell Biol.*, **151** (6), 1321–1336.

242 Galvin, J.E. and Ginsberg, S.D. (2004) Expression profiling and pharmacotherapeutic development in the central nervous system. *Alzheimer Dis. Assoc. Disord.*, **18** (4), 264–269.

243 Ginsberg, S.D. *et al.* (2006) Single cell gene expression profiling in Alzheimer's disease. *NeuroRx*, **3** (3), 302–318.

244 Georganopoulou, D.G. *et al.* (2005) Nanoparticle-based detection in cerebral spinal fluid of a soluble pathogenic biomarker for Alzheimer's disease. *Proc. Natl. Acad. Sci. U. S. A.*, **102** (7), 2273–2276.

245 Mirnics, K., Levitt, P., and Lewis, D.A. (2004) DNA microarray analysis of postmortem brain tissue. *Int. Rev. Neurobiol.*, **60**, 153–181.

246 Mirnics, K. *et al.* (2001) Analysis of complex brain disorders with gene expression microarrays: schizophrenia as a disease of the synapse. *Trends Neurosci.*, **24** (8), 479–486.

247 Geiss, G.K. et al. (2000) Large-scale monitoring of host cell gene expression during HIV-1 infection using cDNA microarrays. *Virology*, **266** (1), 8–16.

248 van't Wout, A.B. et al. (2003) Cellular gene expression upon human immunodeficiency virus type 1 infection of CD4(+)-T-cell lines. *J. Virol.*, **77** (2), 1392–1402.

249 Braxton, S. and Bedilion, T. (1998) The integration of microarray information in the drug development process. *Curr. Opin. Biotechnol.*, **9**, 643–649.

250 Evans, W.E. and Relling, M.V. (1999) Pharmacogenomics: translating functional genomics into rational therapeutics. *Science*, **286** (5439), 487.

251 Marton, M., Derisi, J., and Bennett, H. (1998) Drug target validation and identification of secondary drug target effects using DNA microarrays. *Nat. Med.*, **4**, 1293–1301.

252 Lee, K.M., Kim, J.H., and Kang, D. (2005) Design issues in toxicogenomics using DNA microarray experiment. *Toxicol. Appl. Pharmacol.*, **207** (2 Suppl), 200–208.

253 Hamadeh, H.K. et al. (2002) An overview of toxicogenomics. *Curr. Issues Mol. Biol.*, **4**, 45–56.

254 Pennie, W.D. (2000) et al. The principles and practice of toxicogenomics: applications and opportunities. *Toxicol. Sci.*, **54** (2), 277–283.

255 Nuwaysir, E.F. et al. (1999) Microarrays and toxicology: the advent of toxicogenomics. *Mol. Carcinog.*, **24** (3), 153–159.

256 Lemarchand, K., Masson, L., and Brousseau, R. (2004) Molecular biology and DNA microarray technology for microbial quality monitoring of water. *Crit. Rev. Microbiol.*, **30** (3), 145–172.

257 Uttamchandani, M. et al. (2009) Applications of microarrays in pathogen detection and biodefence. *Trends Biotechnol.*, **27** (1), 53–61.

258 Kostrzynska, M. and Bachand, A. (2006) Application of DNA microarray technology for detection, identification, and characterization of food-borne pathogens. *Can. J. Microbiol.*, **52** (1), 1–8.

259 Schulze, H. et al. (2009) Multiplexed optical pathogen detection with lab-on-a-chip devices. *J. Biophotonics*, **2** (4), 199–211.

260 Garaizar, J. et al. (2006) Use of DNA microarray technology and gene expression profiles to investigate the pathogenesis, cell biology, antifungal susceptibility and diagnosis of *Candida albicans*. *FEMS Yeast Res.*, **6** (7), 987–998.

261 Wang, D. et al. (2002) Microarray-based detection and genotyping of viral pathogens. *Proc. Natl. Acad. Sci. U. S. A.*, **99** (24), 15687–15692.

262 Garaizar, J., Rementeria, A., and Porwollik, S. (2006) DNA microarray technology: a new tool for the epidemiological typing of bacterial pathogens? *FEMS Immunol. Med. Microbiol.*, **47** (2), 178–189.

263 Diehn, M. and Relman, D.A. (2001) Comparing functional genomic datasets: lessons from DNA microarray analyses of host–pathogen interactions. *Curr. Opin. Microbiol.*, **4** (1), 95–101.

264 Grow, A.E. et al. (2003) New biochip technology for label-free detection of pathogens and their toxins. *J. Microbiol. Methods*, **53** (2), 221–233.

265 Yeung, S.W. et al. (2006) A DNA biochip for on-the-spot multiplexed pathogen identification. *Nucleic Acids Res.*, **34** (18), e118.

266 de Leon, J. et al. (2009) DNA microarray technology in the clinical environment: the AmpliChip CYP450 test for CYP2D6 and CYP2C19 genotyping. *CNS Spectr.*, **14** (1), 19–34.

267 Hatada, I. et al. (2002) A microarray-based method for detecting methylated loci. *J. Hum. Genet.*, **47** (8), 448.

268 Shi, H. et al. (2003) Oligonucleotide-based microarray for DNA methylation analysis: principles and applications. *J. Cell Biochem.*, **88** (1), 138–143.

269 Schumacher, A. et al. (2006) Microarray-based DNA methylation profiling: technology and applications. *Nucleic Acids Res.*, **34** (2), 528–542.

270 Estécio, M.R.H. et al. (2007) High-throughput methylation profiling by MCA coupled to CpG island microarray. *Genome Res.*, **17** (10), 1529–1536.

271 Melnikov, A.A. et al. (2008) Array-based multiplex analysis of DNA methylation in breast cancer tissues. *J. Mol. Diagn.*, **10** (1), 93–101.

272 Kimura, N. et al. (2005) Methylation profiles of genes utilizing newly developed CpG island methylation microarray on colorectal cancer patients. *Nucleic Acids Res.*, **33** (5), e46.

273 Askari, M. et al. (2001) Application of an antibody biochip for p53 detection and cancer diagnosis. *Biotechnol. Prog.*, **17** (3), 543–552.

274 Quinten, M. (2001) Local fields close to the surface of nanoparticles and aggregates of nanoparticles. *Appl. Phys. B*, **73** (3), 245–255.

275 Elghanian, R. et al. (1997) Selective colorimetric detection of polynucleotides based on the distance-dependent optical properties of gold nanoparticles. *Science*, **277** (5329), 1078–1081.

276 Storhoff, J.J. et al. (1998) One pot colorimetric differentiation of polynucleotides with single base imperfections using gold nanoparticle probes. *J. Am. Chem. Soc.*, **120** (9), 1959–1964.

277 Fleischmann, M., Hendra, P.J., and McQuillan, A.J. (1974) Raman spectra of pyridine adsorbed at a silver electrode. *Chem. Phys. Lett.*, **26** (2), 163–166.

278 Freeman, R.G. et al. (1995) Self-assembled metal colloid monolayers: an approach to SERS substrates. *Science*, **267** (5204), 1629–1632.

279 Petry, R., Schmitt, M., and Popp, J. (2003) Raman spectroscopy – a prospective tool in the life sciences. *ChemPhysChem*, **4** (1), 14–30.

280 Schmitt, M. and Popp, J. (2006) Raman spectroscopy at the beginning of the twenty-first century. *J. Raman Spectrosc.*, **37** (1–3), 20–28.

281 Haynes, C.L., McFarland, A.D., and Van Duyne, R.P. (2005) Surface-enhanced Raman spectroscopy. *Anal. Chem.*, **77** (17), 338A–346A.

282 Hering, K. et al. (2008) SERS: a versatile tool in chemical and biochemical diagnostics. *Anal. Bioanal. Chem.*, **390** (1), 113–124.

283 Cao, Y.C., Jin, R., and Mirkin, C.A. (2002) Nanoparticles with Raman spectroscopic fingerprints for DNA and RNA detection. *Science*, **297** (5586), 1536–1540.

284 Wabuyele, M.B. and Vo-Dinh, T. (2005) Detection of human immunodeficiency virus type 1 DNA sequence using plasmonics nanoprobes. *Anal. Chem.*, **77** (23), 7810–7815.

285 Han, M. et al. (2001) Quantum-dot-tagged microbeads for multiplexed optical coding of biomolecules. *Nat. Biotechnol.*, **19**, 631–635.

286 Smith, A.M. and Nie, S.M. (2004) Chemical analysis and cellular imaging with quantum dots. *Analyst*, **129** (8), 672–677.

287 Howarth, M. et al. (2005) Targeting quantum dots to surface proteins in living cells with biotin ligase. *Proc. Natl. Acad. Sci. U. S. A.*, **102** (21), 7583–7588.

288 Roullier, V. et al. (2009) High-affinity labeling and tracking of individual histidine-tagged proteins in live cells using Ni^{2+} tris-nitrilotriacetic acid quantum dot conjugates. *Nano Lett.*, **9** (3), 1228–1234.

289 Jain, K.K. (2007) Applications of nanobiotechnology in clinical diagnostics. *Clin. Chem.*, **53**, 2002–2009.

290 Michalet, X. et al. (2005) Quantum dots for live cells, *in vivo* imaging, and diagnostics. *Science*, **307** (5709), 538–544.

291 Lucas, L.J., Chesler, J.N., and Yoon, J.-Y. (2007) Lab-on-a-chip immunoassay for multiple antibodies using microsphere light scattering and quantum dot emission. *Biosens. Bioelectron.*, **23** (5), 675–681.

292 Algar, W.R., Massey, M., and Krull, U.J. (2009) The application of quantum dots, gold nanoparticles and molecular switches to optical nucleic-acid diagnostics. *TrAC Trends Anal. Chem.*, **28** (3), 292–306.

293 Hainfeld, J.F. et al. (2002) Enzymatic metallography: a simple new staining method, in *Proceedings of Microscopy and Microanalysis 2002* (eds. E. Voekl et al..), Cambridge University Press, New York, p. 916CD.

294 Mayer, G. et al. (2000) Introduction of a novel HRP substrate–Nanogold probe for signal amplification in

immunocytochemistry. *J. Histochem. Cytochem.*, **48** (4), 461–469.
295 Hwang, S., Kim, E., and Kwak, J. (2005) Electrochemical detection of DNA hybridization using biometallization. *Anal. Chem.*, **77**, 579–584.
296 Fanjul-Bolado, P. et al. (2007) Alkaline phosphatase-catalyzed silver deposition for electrochemical detection. *Anal. Chem.*, **79** (14), 5272–5277.
297 Zhang, P. et al. (2008) Electrochemical detection of point mutation based on surface ligation reaction and biometallization. *Biosens. Bioelectron.*, **23** (10), 1435–1441.
298 Möller, R. et al. (2005) Enzymatic control of metal deposition as key step for a low-background electrical detection for DNA chips. *Nano Lett.*, **5** (7), 1475–1482.
299 Schüler, T. et al. (2008) Enzyme-induced growth of silver nanoparticles studied on single particle level. *J. Nanopart. Res.*, **11**, 939–946.
300 Willner, I., Basnar, B., and Willner, B. (2007) Nanoparticle–enzyme hybrid systems for nanobiotechnology. *FEBS J.*, **274**, 302–309.
301 Willner, I., Baron, R., and Willner, B. (2006) Growing metal nanoparticles by enzymes. *Adv. Mater.*, **18** (9), 1109–1120.
302 Basnar, B., Willner, I., and Willner, B. (2006) Synthesis of nanowires using Dip-Pen nanolithography and biocatalytic inks. *Adv. Mater.*, **18** (6), 713–718.
303 Felder, K.M. et al. (2009) A DNA microarray facilitates the diagnosis of *Bacillus anthracis* in environmental samples. *Lett. Appl. Microbiol.*, **49** (3), 324–331.
304 Yan, P.S., Chen, C.H., Shi, H.D. (2001) Dissecting complex epigenetic alternations in breast cancer using CpG island microarrays. *Cancer Research*, **61** (23), 8375–8380.

4
Microstructure Fibers in Biophotonics
Aleksei M. Zheltikov

4.1
Introduction: From Early Concepts of Fiber-Based Bioimaging to Biophotonics with Specialty Fibers

The development of fiber-based methods of biomedical imaging spans more than eight decades [1]. The capability of optical fibers to transfer images was first realized in the late 1920s. In 1927, Clarence Hansell, an electronic engineer at the Radio Corporation of America, filed a patent on fiber-based imaging [2], which covered the use of optical cables consisting of many fibers. In 1929, Heinrich Lamm, a medical student at the University of Munich, reported the first successful experiments on image transfer through a bundle of optical fibers [3], motivated by the development of a flexible gastroscope (see [4] for a detailed account of the early days of the history of fiber optics). These early inventions and pioneering concepts had to wait for several decades, however, before fiber fabrication technologies and low-loss fiber materials became mature enough to support the creation of practical fiber imaging systems, leading to revolutionary breakthroughs in biophotonics.

Modern fiber-optic technologies [5] provide an advanced platform for the creation of novel fiber-based imaging systems [6, 7], optical endoscopes [8, 9], biophotonics-oriented fiber laser sources [10], and optical neural interfaces [11, 12]. Although standard single-mode fibers (SMFs) can help implement many of the biophotonic protocols in the endoscopic mode, guiding continuous-wave radiation to the area of interest in living organisms, such fibers are much less efficient in delivering ultrashort laser pulses, needed for multiphoton bioimaging, and in collecting the incoherent optical response from biotissues. The group-velocity dispersion of such fibers tends to stretch ultrashort light pulses, while the numerical aperture of SMFs is typically too low to permit efficient collection of incoherent radiation.

Novel micro- and nanostructured optical fibers, also referred to as photonic-crystal fibers (PCFs) [13, 14], have been shown to offer attractive solutions to these problems [15]. Both group-velocity dispersion and numerical aperture can be tailored in PCFs through fiber structure engineering [16]. Fibers of this class have been used to improve substantially the signal collection efficiency in nonlinear microscopy [17] and have allowed the creation of nonlinear optical fiber endoscopes [18, 19].

Handbook of Biophotonics. Vol.2: Photonics for Health Care, First Edition. Edited by Jürgen Popp,
Valery V. Tuchin, Arthur Chiou, and Stefan Heinemann.
© 2012 Wiley-VCH Verlag GmbH & Co. KGaA. Published 2012 by Wiley-VCH Verlag GmbH & Co. KGaA.

Advanced PCF lasers [10, 20], and also PCF frequency converters [16] and supercontinuum sources [21, 22], are gaining growing applications in bioimaging based on coherent anti-Stokes Raman scattering (CARS) [23]. Hollow-core PCFs [24, 25] allow the delivery of high-energy, ultrashort laser pulses for biomedical applications [26]. These fibers also offer much promise for the creation of tunable wavelength shifters for multiphoton microscopy and CARS microspectroscopy [27, 28]. The low nonlinearity of the gas filling the core of the fiber and the choice of the central wavelength of laser pulses to the zero group velocity dispersion (GVD) wavelength of the fiber help to reduce temporal envelope distortions of the transmitted light signal. In this chapter, we give a brief overview of recent developments in biophotonics based on the use of microstructure fibers

4.2
Microstructure Fiber Sensors

Microstructure fibers suggest a variety of attractive strategies for optical sensing. The most common approach involves using the evanescent field of waveguide modes, confined to a high refractive index material in a waveguide, to produce a fluorescent signal, which is employed to detect species of a certain type in the gas, liquid, or solid phase. When implemented with PCFs [29–38], this sensing strategy allows the creation of compact fiber format sensors of gas-phase media [36–38] and biomolecules in aqueous solution [31]. Another widespread approach to sensing with advanced waveguide components is based on interferometric techniques [39], most often the Mach–Zehnder interferometer configuration, intended to detect small changes in the phase of an optical signal transmitted through a waveguide sensor. Devices based on this principle can sense small changes in the refractive index of an analyte or detect the formation of thin molecular layers on a waveguide cladding.

Owing to the remarkable flexibility of their core–cladding structure (Figure 4.1), PCFs offer attractive and often unique options for biosensing. For example, double-clad PCFs, as shown elegantly by Myaing et al. [40], can substantially improve the efficiency of two-photon fluorescence detection of biomolecules with the fluorescence signal delivered to the detector in the backward direction through the same fiber. The possibility of extending the concept of fiber grating-based sensing to PCFs has been demonstrated by Eggleton et al. [41]. Rindorf et al. [33] developed PCF-based sensors with long-period gratings for the detection of molecular monolayers containing immobilized DNA. Advanced fiber micro- and nanostructuring technologies have been shown to permit the creation of high-performance DNA sensors for gene-expression analysis [42], and also specific genomics- and proteomics-oriented sensor systems designed to detect specific genes and proteins (see [43] for a review).

Air-guided modes of hollow-core PCFs (Figure 4.1e) have been shown to be ideally suited for gas-phase sensing using linear [44] and nonlinear [45] optical methods. The biosensing ability of ring-cladding hollow-core PCFs has been demonstrated [46]. The ring cladding in such waveguides, conveniently implemented in a PCF design, serves as a built-in Fabry–Pérot interferometer, allowing the detection of thin

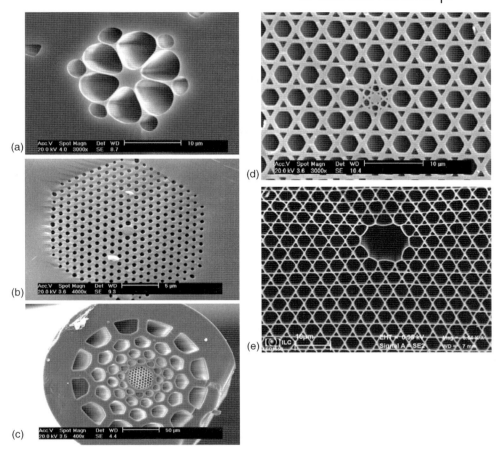

Figure 4.1 SEM images of PCFs: (a) solid-core high refractive index-step silica PCFs with a few-airhole-ring cladding; (b) periodic cladding PCF; (c) solid-core silica PCF with a dual microstructure cladding; (d) multicomponent glass PCF with a nanohole array-modified core; (e) hollow-core PCF.

molecular layers and ensuring a high sensitivity of transmission spectra of waveguide modes to small changes in the refractive index of an analyte filling the hollow core of the waveguide and air holes in the fiber cladding. We show that waveguides of this type offer a platform for the creation of compact and efficient biochemical sensors and sensor arrays. Components of this type can sense biolayers a few nanometers thick and provide a high sensitivity with respect to refractive index changes of an analyte filling the air holes of the waveguide structure.

A diversified architecture of photonic-crystal fibers (Figure 4.1) suggests various approaches to optical sensing. Experiments reported by Konorov et al. [32] provide a quantitative comparison of two protocols of optical sensing with PCFs. In the first protocol, diode-laser radiation is delivered to a sample through the central core of a dual-cladding PCF with a diameter of a few micrometers, while the large-diameter fiber cladding serves to collect the fluorescent response from the sample and to guide it to a detector in the backward direction. In the second scheme, liquid sample is collected by a microcapillary array in the PCF cladding and is interrogated by laser radiation guided in PCF modes. For sample fluids with refractive indices exceeding the refractive index of PCF material, fluid channels in PCFs can guide laser light by total internal reflection, providing an 80% overlap of interrogating radiation with sample fluid. Specially designed PCFs have been shown to integrate an optical sensor with a micropipette, a sample collector, and a microfluidic polycapillary array in a single fiber-optic component, which is ideally suited for "lab-on-a-chip" applications, offering an attractive platform for fiber-based sensors and probes for genomics and proteomics.

4.3
Designing Specialty Fibers for High-Resolution Imaging

The development of subdiffraction microscopy techniques is one of the key tendencies in modern optical physics. Rapidly progressing methods of near-field microscopy [47] provide an unprecedentedly high spatial resolution on specific samples, giving an access to a unique information on the structure of matter and physical processes in subsurface layers. Methods of near-field microscopy, however, face serious difficulties when applied to a broad class of problems related to processes and details of structure in the bulk of a material and in living systems. Problems of this class can be attacked by using novel physical principles of far-field microscopy, such as multiphoton microscopy [48, 49], microscopy based on CARS [50], and stimulated Raman scattering [51], and also microscopy using stimulated emission depletion (STED) [52]. New methods of far-field microscopy have been shown to provide spatial resolution as high as a few tens of nanometers (in certain cases better than 20 nm [53]). These techniques find growing applications in studies of intracellular processes in living organisms, single-molecule detection, and analysis of a broad class of nanostructures and nano-objects.

Modern fiber-optic technologies allow the creation of optical micro- and nanowaveguide systems, including submicrometer-core optical fibers [16]. Such components open up wide opportunities for the implementation of novel methods of microscopy in the fiber format. Rapidly growing research in this field includes the development of fiber-optic sources and systems for the delivery of ultrashort light pulses for nonlinear optical endoscopy and mapping of neuronal activity in the living brain [54, 55]. Unique opportunities offered by fiber-format ultrahigh-resolution microscopy call for a detailed understanding of fundamental physical factors controlling the confinement of electromagnetic field in waveguide systems and analysis of methods of transmission of ultracompact light beams in fiber-optic networks.

The spatial resolution of fiber-based optical imaging is controlled by the confinement of the light field in the fiber core, which can be quantified in terms of the effective mode radius:

$$w = (S/\pi)^{\frac{1}{2}} \tag{4.1}$$

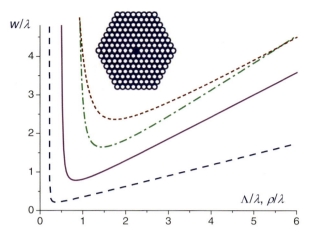

Figure 4.2 Effective radius of the PCF mode as a function of the pitch Λ of the hexagonal fiber cladding for $\lambda = 1\,\mu\text{m}$, $d/\Lambda = 0.3$ (dashed-dotted line), 0.5 (solid line), and 0.9 (dashed line). The dotted line shows the effective radius of the fundamental mode of a standard fiber with $n_1 - n_2 = 0.01$ as a function of the core size ϱ. A PCF with a hexagonal-lattice cladding is shown in the inset.

where

$$S = 2\pi \left[\int_0^\infty |F(r)|^2 r dr \right]^2 \left[\int_0^\infty |F(r)|^4 r dr \right]^{-1} \quad (4.2)$$

is the effective mode area, $F(r)$ being the transverse field profile in the fiber mode. The mode radius w is sensitive to the refractive index contrast, $\Delta = n_1 - n_2$, where n_1 and n_2 are the refractive indices of the fiber core and fiber cladding, respectively, and generally differs from the physical size of the fiber core ϱ. This difference may become especially significant in the case of fibers with very small, subwavelength cores, where diffraction arrests a tight localization of the light field in the fiber core, making a significant fraction of the mode power propagate as the evanescent field outside the fiber core [56, 57].

Figure 4.2 compares the typical behavior of the effective mode radius w as a function of the physical radius of the fiber core ϱ in the case of a conventional step-index fiber (dotted line) and a generic-type PCF with a hexagonal structure of the cladding with a lattice constant Λ (shown in the inset). Calculations presented in Figure 4.2 suggest [58] that the light confinement in the PCF core and, hence, the spatial resolution provided by a PCF probe can be controlled through fiber structure engineering. Larger d/Λ ratios, as can be seen from Figure 4.2, provide a tighter localization of the light field in the fiber core, translating into a higher spatial resolution of a fiber probe. With $d/\Lambda = 0.9$, the minimum mode radius, achieved when $\Lambda \approx 0.4\lambda$, is $w_{\min} \approx 0.2\lambda$ (the dashed line in Figure 4.2), allowing fiber-based optical probing of areas adjacent to the fiber end with subwavelength resolution. Interestingly, and quite importantly, for smaller core sizes, the field confinement in the fiber core becomes weaker, because of strong diffraction, leading to larger effective mode sizes (see Figure 4.2), with a significant fraction of mode power guided outside the fiber core. All these predictions are fully confirmed by fiber-probe experiments on test objects [59].

4.4
Fiber-Based Imaging of Diffusively Scattering Tissues

To illustrate the significance of the fiber structure for the regime of collection of a fluorescence signal from a diffusively scattering tissue, we consider a generic fiber (Figure 4.3a) consisting of a core with a diameter d_c, which is used to deliver the laser light to a tissue, and a cladding that includes a ring (similar to a ring seen in the PCF structure in Figure 4.1c) with an inner diameter d_r, which serves to pick up the fluorescent response from the tissue within its larger aperture. The case of $d_c = d_r$ corresponds to a standard optical fiber. In a PCF, a cross-section structure can be designed in such a way as to isolate a signal-collecting cladding ring from the central core of the fiber. When a fiber probe with such a design is used to detect the fluorescence response from a uniformly scattering biotissue, the probability of the ring in the fiber cladding to pick up a fluorescent photon generated at a depth z (measured from the fiber tip, Figure 4.3a) peaks, due to the general properties of diffusive scattering of optical signals in random media [60, 61], at $z = L_0$. The full width at half-maximum, δz, of the axial profile $I_{fl}(z)$ of fluorescence collected by the cladding ring is controlled by scattering and absorption of the tissue, and also the diameter of the ring d_r, its thickness t_r, and the refractive index step $\delta n = n_r - n_{eff}$, where n_r is the refractive index of the material of the ring ($n_r \approx 1.45$ for a silica ring in our experiments) and n_{eff} is the effective refractive index of the microstructure surrounding this ring. A ring with a larger diameter d_r will probe tissue layers lying at larger distances L_0 from the fiber tip at the cost of lower axial resolution (larger δz). The thickness of the tissue layer providing the main contribution to the signal detected by the outer ring of the PCF can be reduced by lowering the refractive index contrast δn and decreasing the thickness of the ring t_r. For the PCF shown in Figure 4.1c with $d_r \approx 300\,\mu m$ and $t_r \approx 25\,\mu m$, numerical simulations for typical brain-tissue parameters (reduced scattering coefficient $\mu'_s \approx 1\,mm^{-1}$ and the absorption coefficient $\mu_a \approx 0.1\,cm^{-1}$ [62]) give $L_0 \approx 250\,\mu m$ with $\delta z \approx 160\,\mu m$. The lateral dimensions of the interrogated volume are determined by the outer diameter of the signal-collecting aperture of the fiber probe. In our experiments, this diameter was about 325 μm.

Special measurements were performed to test how well the fluorescence-collecting ring in the PCF cladding is isolated from the central fiber core for PCFs with different cladding-ring diameters. To this end, the intensity I_s of the signal collected by PCFs of different types was measured as a function of the displacement ϱ of the fiber core with respect to a 2 μm spot of a tightly focused laser beam, intended to mimic a point light source. Results of these measurements are presented in Figure 4.3b. For a PCF

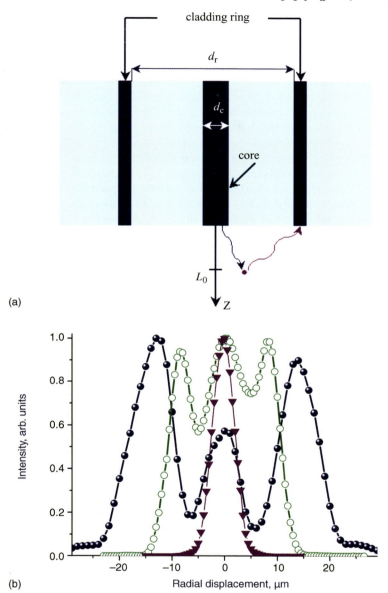

Figure 4.3 (a) Fiber probe with a fluorescence-collecting ring in the cladding. (b) Intensity of the signal collected by a PCF with a cross-section structure shown in Figure 4.1a (filled and open circles) and Figure 4.1d (triangles) measured as a function of the displacement of the fiber core with respect to a 2 μm radiation source. The diameter of the outer light collecting ring in the core of PCF shown in Figure 4.1a is 18 μm (open circles) and 28 μm (filled circles).

without a signal-collecting cladding ring (Figure 4.1a), the measured radial profile $I_s(\varrho)$ features only one maximum (triangles in Figure 4.3b), corresponding to the center of the fiber core. For fibers with signal-collecting cladding rings (Figure 4.1c), the $I_s(\varrho)$ profile displays three peaks (solid and open circles in Figure 4.3b), corresponding to the signal delivered through the central core and the cladding ring. Whereas for fibers with $d_r \leq 20$ μm, the signal transmitted through the cladding ring suffers from interference due to radiation transmitted through the core and the inner part of the microstructure cladding (open circles in Figure 4.3b), PCFs with $d_r \geq 25$ μm show well-resolved peaks in the $I_s(\varrho)$ profile (solid circles in Figure 4.3b), indicating almost no interference between the signals transmitted through the central core and the signal-collecting cladding ring. The experimental results thus support numerical simulations, confirming that properly designed fiber probes can effectively resolve the fluorescence response from a brain tissue along the axial coordinate even in the regime of single-photon excitation, where high laser intensities, typically required for multiphoton methods, are not needed.

4.5
Enhancement of Guided-Wave Two-Photon-Excited Luminescence Response with a Photonic-Crystal Fiber

Two-photon absorption (TPA) is the backbone of a broad variety of bioimaging techniques and biomedical strategies, including high-resolution microscopy [48, 49, 63], photodynamic therapy [64, 65], and drug-delivery monitoring and drug activation [66]. Luminescence of TPA-excited molecules provides a measurable optical response that allows imaging of TPA-excited regions in two-photon microscopy and helps to visualize tiny objects and identify fine details of structure and morphology inside biological tissues [48, 49]. A combination of TPA approaches with capabilities of fiber-optic probes [54] offers numerous advantages [67], suggesting a convenient format for beam delivery, facilitating manipulation of excitation radiation, and allowing this excitation to be applied locally and selectively. Although many attractive fiber solutions have been developed to meet the needs of TPA-based technologies, collection of the two-photon-excited luminescence (TPL) response still leaves much room for optimization, with the main source of problems being related to the small aperture of fiber probes, making it difficult to collect efficiently the TPL signal, which is uniformly emitted in a 4π solid angle.

PCFs can substantially improve the collection efficiency of the TPL response from TPA-excited molecules in waveguide modes [68], thus allowing the sensitivity of TPA-based techniques to be radically improved. In the earlier relevant work, PCFs of various types were shown to offer numerous advantages as efficient optical sensors [30–32] and compact wavelength-tunable sources for nonlinear microspectroscopy [16, 23]. Dual-cladding PCFs have been employed to enhance the efficiency of two-photon biosensing [40]. As shown by Fedotov et al. [68], with only a few nanoliters of TPA-excited molecules filling air holes in a specifically designed PCF, the guided-wave TPL signal can be enhanced by two orders of magnitude relative to

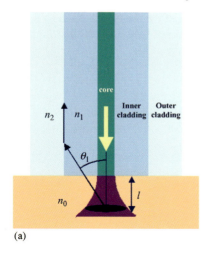 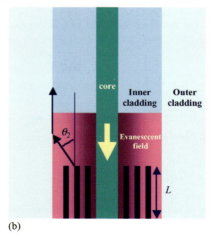

(a) (b)

Figure 4.4 TPA excitation and TPL response collection with a dual-cladding fiber probe: (a) excitation light is delivered through the fiber core to an object adjacent to the fiber end, while the TPL signal is collected by a high numerical aperture inner part of the fiber cladding; (b) an evanescent field of the mode guided along the fiber core generates the TPL signal through a TPA excitation of dye molecules in a solution filling the air holes in the PCF cladding, with the inner part of the fiber cladding capturing the TPL response into a waveguide mode and delivering this signal to a detection system.

the maximum TPL signal attainable from a cell with the same dye excited and collected by the same PCF.

We compare two arrangements for the collection of the TPL response with a dual-cladding fiber. In the first scheme (Figure 4.4a), excitation light is delivered along the fiber core to an object adjacent to the fiber end. The TPL signal is then collected by the inner part of the fiber cladding. As shown by Myaing et al. [40], this method of TPA excitation using dual-cladding PCF probes offers important advantages for TPA bioimaging. In the second approach (Figure 4.4b), we let a tiny amount of the liquid- or gas-phase substance fill the air holes of the PCF. The TPL signal is then excited by the evanescent field of the mode guided along the fiber core and is captured into a waveguide mode inside the inner part of the fiber cladding.

The total TPL signal collected by the fiber probe in the first experimental arrangement (Figure 4.4a) is

$$P_1 = 2\pi \int_0^a \sigma_1 I_1(r) r dr \qquad (4.3)$$

where

$$I_1(r) = \mu \beta l I_0^2 [f_1(r)]^2 \qquad (4.4)$$

is the intensity of the TPL signal, σ_1 is the TPL signal collection efficiency, μ is the quantum yield of TPA-induced luminescence, β is the TPA coefficient, l is the length of the interaction region, I_0 is the on-axis intensity of the waveguide mode, a is the radius of the fiber core, and $f_1(r)$ is the radial profile of field intensity distribution.

The TPL signal collection efficiency σ_1 is given by

$$\sigma_1 = 2\pi \int_0^{\theta_1} \sin\xi \, d\xi \tag{4.5}$$

where θ_1 is the maximum acceptance angle of the high numerical aperture inner part of the fiber cladding:

$$\sin\theta_1 = \frac{n_1}{n_0}\left[1 - \left(\frac{n_2}{n_1}\right)^2\right]^{\frac{1}{2}} \tag{4.6}$$

where n_0 is the refractive index of the medium adjacent to the fiber probe end and n_1 and n_2 are the refractive indices of the inner and outer sections of the fiber cladding, respectively.

With $(n_1^2 - n_2^2)/n_0^2 \ll 1$, integration of Eq. (4.5) yields

$$\sigma_1 \approx \pi\theta_1^2 \approx \pi\left(\frac{NA_1}{n_0}\right)^2 \tag{4.7}$$

where $NA_1 = (n_1^2 - n_2^2)^{1/2}$ is the numerical aperture of the inner cladding of the fiber probe.

In the case of a Gaussian field intensity profile, $f_1(r) = \exp(-r^2/r_1^2)$, where r_1 is the effective field intensity radius, Eqs. (4.3), (4.4), and (4.7) give

$$P_1 \approx \pi\mu\beta l I_0^2 \frac{S_1}{2} \theta_1^2 \left[1 - \exp\left(-2\frac{a^2}{r_1^2}\right)\right] \tag{4.8}$$

where $S_1 = \pi r_1^2$ is the effective mode area.

For a large-core fiber with $(a/r_1)^2 \gg 1$, Eq. (4.8) reduces to

$$P_1 \approx \pi\mu\beta l I_0^2 S_1 \theta_1^2 \tag{4.9}$$

For the second experimental scheme (Figure 4.2b), the total TPL signal collected by the fiber probe is

$$P_2 = 2\pi \int_a^R I_2 \left[f_2(r)\right]^2 r \, dr \tag{4.10}$$

where $f_2(r)$ is the radial profile of field intensity distribution in a liquid-filled fiber, R is the radius of the inner part of the cladding, and the TPL intensity I_2 can be found from the equation

$$I_2 = \mu\beta\sigma_2 \int_0^L \frac{I_0^2 \, dx}{(1 + \beta I_0 x)^2} \tag{4.11}$$

where L is the length of the interaction region, σ_2 is the TPL signal collection efficiency defined by

$$\sigma_2 \approx \pi\theta_2{}^2 \approx \pi\left(\frac{NA_2}{n_{co}}\right)^2 \qquad (4.12)$$

θ_2 is the maximum acceptance angle, $NA_2 = (n_{co}{}^2 - n_{cl}{}^2)^{1/2}$, n_{co} is the effective refractive index of the inner part of the liquid-filled fiber bounded by the outer cladding, and n_{cl} is the refractive index of the outer part of the cladding.

Integration of Eq. (4.11) yields

$$I_2 = \mu\beta\sigma_2 \frac{I_0{}^2 L}{1 + \beta I_0 L} \qquad (4.13)$$

with $\beta I_0 L \ll 1$ and a Gaussian field intensity profile, $f_2(r) = \exp(-r^2/r_2{}^2)$, r_2 being the effective field intensity radius in a liquid-filled fiber, Eqs. (4.10), (4.12), and (4.13) lead to

$$P_2 \approx \pi\mu\beta\theta_2{}^2 I_0{}^2 L \frac{S_2}{2} \exp\left(-2\frac{a^2}{r_2{}^2}\right) \qquad (4.14)$$

where $S_2 = \pi r_2{}^2$ and we assume that $r_2 \ll R$.

The length of the interaction region in the first experimental scheme (Figure 4.4a) is dictated by diffraction and can be estimated as $l \approx 2\pi a^2/\lambda$, λ being the radiation wavelength, we find that the second experimental arrangement provides the following TPL signal collection enhancement with respect to the first scheme implemented with a large-core fiber with $(a/r_1)^2 \gg 1$:

$$\eta = \frac{P_2}{P_1} \approx \frac{\theta_2{}^2}{\theta_1{}^2} \frac{S_2}{S_1} \frac{L\lambda}{2\pi a^2} \exp\left(-2\frac{a^2}{r_2{}^2}\right) \qquad (4.15)$$

from which it can be seen that the enhancement of the TPL response provided by a fiber with air holes partially filled with a liquid under study scales linearly with the length L of the interaction region and rapidly decreases [$\propto a^{-2}\exp(2a^2/r_2{}^2)$] with an increase in the fiber core radius a because of the evanescent character of the field inducing the TPL response in the second experimental arrangement. For typical parameters of experiments described below, with $\lambda \approx 0.8\,\mu m$, $a \approx 2.3\,\mu m$, $r_2 \approx 2.5\,\mu m$, and $L \approx 1.5\,cm$, we find $L/l \approx 360$, and Eq. (4.15) gives $\eta \approx 200$. Waveguide collection of the luminescence signal using an appropriately designed fiber probe can thus provide orders of magnitude enhancement of the TPL response relative to the maximum TPL signal attainable from a cell with the same TPA medium.

Fiber probes used in experiments [68] were based on suitably designed fused-silica PCFs, produced by means of standard PCF fabrication technology. A Ti:sapphire laser, delivering 40 fs, 0.5 W pulses of 805 nm radiation at a pulse repetition rate of 90 MHz, was employed as a source of light inducing two-photon luminescence in solutions of various dyes. Light pulses with an average power of ~50 mW were coupled into the fiber probe with a micro-objective. Because of the nonlinear nature of the TPA process, efficient TPL generation requires excitation of molecules by light pulses with a sufficiently high peak power. It is therefore critical for highly efficient TPL excitation to minimize temporal spreading of femtosecond light pulses transmitted through the fiber. To this end, the PCF was designed in such a way as to provide a low GVD at the central laser wavelength. In experiments, a laser pulse with an input pulse width of 40 fs transmitted through a 50 cm long piece of PCF with a GVD parameter $\beta_2 \approx 5 \times 10^{-3}\,\mathrm{ps}^2\,\mathrm{m}^{-1}$ was stretched up to 75 fs. This pulse width was still short enough to provide a sufficient laser peak power for efficient TPA excitation of Rhodamine 6G molecules.

In the first scheme of TPL detection, laser pulses were delivered to a 10^{-3} M solution of Rhodamine 6G in a cell through the central core of the fiber with a

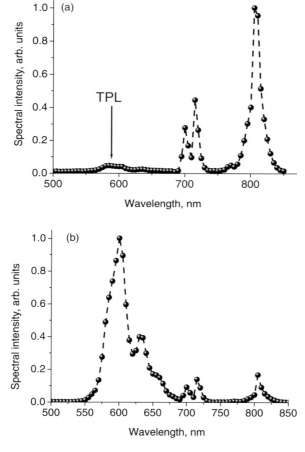

Figure 4.5 Spectra of the optical response of Rhodamine 6G solution excited by femtosecond Ti:sapphire laser pulses with a central wavelength of 805 nm detected (a) from dye solution in a cell (as sketched in Figure 4.4a) and (b) from dye solution filling the air holes in the PCF cladding (as shown in Figure 4.4b).

diameter of 2.3 μm in accordance with the generic approach illustrated in Figure 4.4a. The inner fiber cladding serves to collect the TPL response from the sample and to guide it in the backward direction, towards the input end of the fiber. Outside the fiber, the TPL signal is separated from the 800 nm light and is sent to a detection system by a dichroic mirror. In this regime, the TPL signal is reliably detectable, but is much weaker than the background signal in the spectral range 690–810 nm (Figure 4.5a), which includes back-reflected laser light centered at 805 nm and the nonlinear signal in the spectral region of 700–715 nm generated in the PCF through dispersive-wave emission by optical solitons. This type of emission originates from the instability of solitons induced by higher order dispersion [69]. Its central frequency is controlled by phase matching between the soliton and dispersive waves [70]. The dispersion profile of the PCF used in our experiments dictates emission of dispersive waves that are blue shifted relative to the central wavelength of a soliton. The ratio of the amplitude of the TPL peak in the spectrum of optical response in Figure 4.5a to the amplitude of the peak corresponding to the back-reflected laser light at 805 nm is $\xi_1 \approx 0.04$.

In the second scheme of TPL detection, sketched in Figure 4.4b, a PCF is brought in contact with a liquid-phase sample (a 10^{-3} M solution of Rhodamine 6G in our experiments) so that a small amount of liquid fills the air holes in the fiber cladding due to sample flows induced by surface tension. Dye molecules in the liquid filling the air holes in the fiber cladding are interrogated with the evanescent field of PCF waveguide modes. The TPL signal generated by dye molecules is detected in the backward direction using a dichroic mirror, a monochromator, and a photomultiplier. The fiber infiltration length was estimated from an abrupt change in the level of scattered continuous-wave laser light detected with a standard fiber probe scanned along the infiltrated fiber. With such an experimental arrangement, the ratio of the amplitude of the TPL peak in the spectrum of an optical response from Rhodamine 6G dye penetrating into the air holes inside the PCF over $L \approx 1.5$ cm is $\xi_2 \approx 6.3$ times higher than the intensity of the back-reflected laser signal at 805 nm (Figure 4.5b). The overall volume of interrogated dye solution in this experiment is about 4 nl. Using the back-reflected signal at 805 nm as a reference, we then find that the enhancement of the TPL response provided by the second regime of TPL fiber probing is $\eta = (\xi_2/\xi_1)(P_{02}/P_{01})$, where P_{02} and P_{01} are the powers of the back-reflected signal at 805 nm in the first and second schemes, respectively. With $\xi_1 \approx 0.04$, $\xi_2 \approx 6.3$, and $P_{02}/P_{01} \approx 0.85$, we find $\eta \approx 130$. This estimate agrees well with the analysis presented in the previous section. As a demonstration of a biophotonic application of the developed approach, the dual-cladding PCF shown in Figure 4.1c was used to perform TPL measurements on the Alexa Fluor® family of dyes, which are commonly used as labels in cell biology and fluorescence microscopy, and on green fluorescent protein (GFP), which has gained wide acceptance as a reporter of gene expression [71, 72].

A PCF is thus shown to increase substantially the guided-wave luminescent response from molecules excited through TPA by femtosecond near-infrared laser pulses. With only a few nanoliters of TPA-excited molecules filling air holes in a specifically designed PCF, the guided-wave two-photon-excited luminescence signal is enhanced by more than two orders of magnitude relative to the maximum TPL signal attainable from a cell with the same dye excited and collected by the same PCF. Biophotonic implications of this waveguide TPL response enhancement include fiber-format solutions for online monitoring of drug delivery and drug activation, interrogation of neural activity, biosensing, endoscopy, highly parallelized gene and protein detection in genomics and proteomics, and locally controlled singlet oxygen generation in photodynamic therapy.

4.6
Microstructure Fibers for Nonlinear Optical Microspectroscopy

CARS [73] has long been established as a powerful tool for time-resolved studies of ultrafast molecular dynamics and diagnostics of excited gases, combustion, flames, and plasmas [74–77]. Because of the nonlinear nature of this type of scattering, the CARS field-interaction region is strongly confined to the beam-waist area, suggesting an attractive approach for microscopy and high-spatial-resolution imaging [50, 78]. Recent advances in femtosecond laser technologies have given a powerful momentum to the development of CARS imaging techniques, making CARS a practical and convenient instrument for biomedical imaging and visualization of processes inside living cells [78, 79].

In the early days of CARS, stimulated Raman scattering (SRS) was successfully used to generate the Stokes field for time-resolved measurements [74] of vibrational lifetimes of molecules in liquids and phonons in solids. Dye lasers and optical parametric sources later proved to be convenient sources of wavelength-tunable Stokes radiation [80, 81], giving a powerful momentum to CARS as a practical spectroscopic tool. The rapid progress of femtosecond lasers and the advent of novel highly nonlinear fibers put a new twist on the strategy of Stokes field generation in CARS, allowing the creation of compact and convenient SRS-based wavelength-tunable fiber-format ultrashort-pulse Stokes sources for a new generation of CARS spectrometers and microscopes [82, 83]. Below in this section, we discuss a compact CARS apparatus that relies on the use of an unamplified femtosecond laser output and two types of PCFs optimized for the generation of wavelength-tunable Stokes field and spectral compression of the probe pulse.

CARS involves coherent excitation of Raman-active modes with a frequency Ω_p by pump and Stokes fields, whose central frequencies, ω_1 and ω_2, are chosen in such a way as to meet the condition of a two-photon, Raman-type resonance, $\omega_1 - \omega_2 \approx \Omega_p$. The third (probe) field with a central frequency ω_3 is then scattered off the coherence induced by the pump and Stokes field to give rise to the anti-Stokes signal at the central frequency $\omega_a = \omega_1 - \omega_2 + \omega_3$, thus reading out the information on the physical properties of the medium projected on the Raman mode. Experiments described below in this section employed a two-color version of CARS with the pump and probe fields delivered by the same light source, implying that $\omega_1 = \omega_3$ and $\omega_a = 2\omega_1 - \omega_2$.

The CARS system that we developed recently [84, 85] (Figure 4.6) is based on a specifically designed long-cavity Cr:forsterite laser oscillator [86]. With a 10 W pump

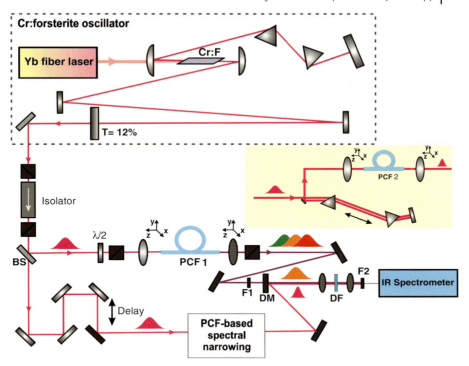

Figure 4.6 Diagram of the experimental setup: Cr:F, Cr:forsterite crystal; BS, beam splitter; PCF1, PCF2, PCFs; F1, F2, filters; DM, dichroic mirror; DF, diamond film. The PCF-based scheme of spectral narrowing is shown in the inset.

provided by an ytterbium fiber laser, the Cr:forsterite oscillator delivered 70 fs, 340 mW light pulses with a central wavelength of 1.25 m at a repetition rate of 18 MHz. A beam splitter (BS in Figure 4.6) was used to divide the Cr:forsterite laser output into two beams. The first beam was launched into a first PCF (PCF1) for the generation of a frequency-tunable Stokes field through soliton self-frequency shift. CARS with a PCF-based soliton frequency shifter as a Stokes source was demonstrated earlier by Andresen *et al.* [82] and Sidorov-Biryukov *et al.* [83]. By varying the power of laser pulses launched into the PCF, we were able to tune the soliton PCF output within the range of wavelengths from 1.35 to 1.75 μm (Figure 4.7a). The smoothly tunable wavelength-shifted soliton output of the PCF was characterized by means of cross-correlation frequency-resolved optical gating (XFROG), implemented in our experiments by mixing the PCF output with reference pulses from the Cr:forsterite laser. Typical XFROG traces of a frequency-shifted PCF1 output measured in experiments with different input peak powers are presented on one map in Figure 4.7a, with the pulse width of the PCF output varying within the range 55–65 fs for different output wavelengths. As can be seen from these XFROG traces, the PCF1 output is nearly transform limited, owing to the soliton regime of wavelength shift,

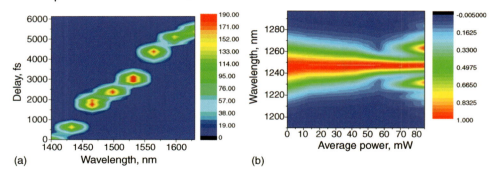

Figure 4.7 (a) A map showing XFROG traces of the frequency-tunable PCF1 output. Traces with different central wavelengths and different delay times were measured in experiments with different input peak powers. (b) Spectra of the output of the photonic crystal fiber (PCF2) used for the spectral narrowing of Cr:forsterite laser pulses measured as a function of the average power of input light pulses.

which helps to avoid unwanted oscillations in CARS spectra due to the interference of resonant and nonresonant parts of the nonlinear signal.

The second beam was intended to provide the pump and probe fields for the CARS process. To improve the spectral resolution and to enhance nonresonant background

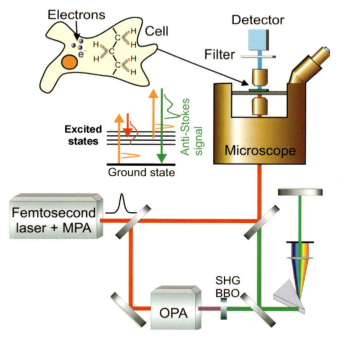

Figure 4.8 Diagram of CARS apparatus: MPA, multipass amplifier; OPA, optical parametric amplifier; SHG BBO, second-harmonic generation in a BBO crystal. A diagram of photons and quantum states involved in CARS is also shown.

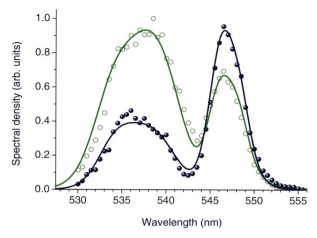

Figure 4.9 Typical CARS spectra from two different regions of mouse brain with a high (open circles) and low (filled circles) content of water versus the results of simulations (solid lines) using the model of the Raman response as described in [90].

suppression in CARS spectra, we used a second PCF (PCF2), which was adjusted to narrow the spectrum of laser pulses through a self-phase modulation process [87–89]. For efficient spectral compression, the light pulses were negatively prechirped with a prism pair, which stretched the pulses to ∼700 fs, before they were launched into the PCF. Self-phase modulation in a medium with a positive nonlinear refractive index tends to red shift the leading edge of the pulse and blue shift its trailing edge. For a pulse with an initial negative chirp, such a temporally nonuniform spectral shifting transfers the energy from the wings of the field spectrum toward its central part, thus narrowing the spectrum. Figure 4.7b displays the spectra of the PCF output measured as a function of the input average power of light pulses. Spectral compression ratios of 3.5–4.0 provided an adequate level of spectral resolution in our CARS measurements with the output of the PCF spectral compressor used as a source of pump and probe fields in CARS. Although the spectral compression improves the spectral resolution and helps to suppress the nonresonant background in CARS spectra, a strong chirp of pump and probe pulses should be avoided, as it induces unwanted oscillations due to the interference of resonant and nonresonant parts of the nonlinear signal.

Typical CARS spectra from a coronal section of mouse brain measured [90] with a broadband Stokes field, as shown in Figure 4.8, are presented in Figure 4.9. These spectra are dominated by the lipid symmetric CH stretch vibrational modes (the peak centered at 547 nm in Figure 4.9), and also by the symmetric and asymmetric H_2O stretch vibrational modes of liquid water (merging into the peak centered at 537 nm in Figure 4.9). The ratio of the amplitudes of the 547 and 537 nm peaks is controlled (Figure 4.9) by the content and orientation of lipids and water in the brain tissue region probed by the focused laser beams. Myelin sheaths are especially rich in CH bonds, allowing white matter to be readily distinguished by their high CARS contrast from gray matter in CARS images of brain tissues [91]. CARS spectra from dehydrated brain samples did not display any feature at 537 nm, confirming the assignment of this peak to vibrational modes of liquid water.

4.7
Microstructure Fibers for Neurophotonics

Advanced fiber-optic components open up a variety of new opportunities in neuroscience, including *in vivo* brain imaging, functional studies of neurons in the brain [92, 93], and the creation of unique optical neural interfaces [11, 12], allowing control over the activity of neurons with light and leading to revolutionary breakthroughs in cognitive sciences. The recently developed advanced fiber tools for high-speed fluorescence microscopy of freely moving animals [54] help to confront the long-standing challenges in experimental studies of neuronal bases of cognition and memory. Specifically designed PCFs have been shown to enhance the two-photon-excited luminescence response from fluorescent protein biomarkers and neuron activity reporters [55].

In vivo work typically imposes demanding requirements on fiber components with regard to their flexibility, compactness, and the ability to integrate multiple functions, such as an optimal geometry local optical excitation of biomolecules, pickup of optical response, and low-loss pump and signal delivery. Several attractive and elegant fiber-format solutions have been developed [5], based on advanced fiber technologies, to meet the needs of biophotonics. A combination of the capabilities of fiber-optic probes with approaches based on multiphoton absorption (MPA) techniques [48, 49, 63] offers numerous advantages, suggesting a convenient format for beam delivery, facilitating manipulation of excitation radiation, and allowing this excitation to be applied locally and selectively. However, three serious problems need to be adequately addressed to make fiber solutions fully compatible with MPA-based technologies. The first difficulty is related to fiber dispersion, which stretches ultrashort laser pulses, reducing the MPA efficiency. The second problem originates from a small aperture typical of a fiber probe, which makes it difficult to collect efficiently the MPA-excited luminescence response, which is uniformly emitted in a 4π solid angle. Finally, the third issue is the laser damage of biotissues, which imposes strict limitations on laser intensity in *in vivo* work and calls for solutions enabling the enhancement of the MPA-excited luminescence response of biomarker molecules. In this section, we show that a new generation of optical fibers offers a promising platform for the creation of wavelength-tunable sources for MPA-based imaging of stained cells and biotissues, including neural activity visualization and brain imaging applications.

In a fused-silica PCF used in experiments [55] (Figure 4.1c), a 2.7 μm diameter core serves to deliver ultrashort laser pulses to the region of TPL interrogation. The inner part of the microstructure cladding with a diameter of 20 μm is designed to provide a high optical nonlinearity and the desired dispersion profile of the fundamental waveguide mode. Along with its outer ring boundary with a diameter of 24 μm, this section of the fiber serves as a high numerical aperture ($NA \approx 0.55$) multimode waveguide for the collection and guiding of the TPL response signal from dye molecules excited via TPA. Advantages of dual-core PCFs with a high-NA inner cladding for luminescent response collection have been demonstrated earlier in experiments on two-photon biosensing [40] and remote analyte detection [32].

The design of the central fiber core and the inner part of the microstructure cladding of the PCF is targeted towards achieving high nonlinearity and a dispersion profile providing (i) anomalous dispersion at the central wavelength of laser radiation used in experiments (800 nm), (ii) soliton self-frequency shift (SSFS) to a desired wavelength needed for efficient TPL excitation, and (iii) two zero points of GVD for SSFS stabilization. As a short-pulse Ti:sapphire-laser output with a central wavelength of 800 nm enters this dual-cladding two zero GVD point PCF (with zero GVD points at 690 and 1050 nm), it tends to evolve toward a soliton. In Figure 4.10, we present the results of numerical analysis of this soliton dynamics for femtosecond 800 nm input laser pulses mimicking the output of the mode-locked Ti:sapphire laser used in our experiments. Simulations were performed by numerically solving the generalized nonlinear Schrödinger equation (GNSE) [22, 94]. High-order fiber dispersion induces soliton instabilities with respect to the emission of dispersive

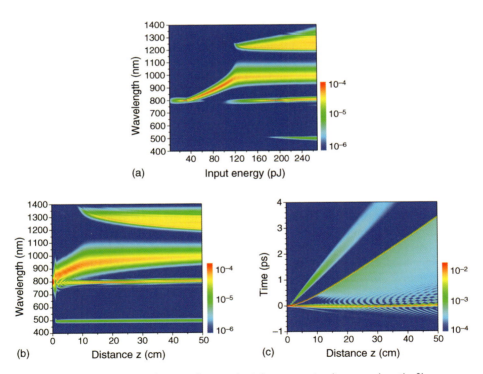

Figure 4.10 Spectrally (a, b) and temporally (c) resolved output of the dual-cladding, two zero GVD point PCF calculated with the use of the GNSE as a function of (a) the input energy and (b, c) the propagation distance z along the fiber: (a) the fiber length is 50 cm; (b, c) the input pulse energy is 265 pJ.

waves – a phenomenon that is often referred to as Cherenkov radiation by optical solitons [69]. At the initial stage of pulse evolution in the PCF (propagation distance z ranging from 1 to 10 cm in Figure 4.10a), phase matching between the soliton and a dispersive wave dictates a predominant emission of blue-shifted dispersive waves, observed in GNSE simulations (Figure 4.10a and b) as an intense spectral component centered at the wavelength $\lambda_b \approx 490$ nm. In the time domain, this blue-shifted dispersive-wave radiation is seen as a clearly resolved feature that branches off the soliton part of the field around $z \approx 1$ cm and becomes dispersed by normal fiber dispersion in the process of field evolution (Figure 4.10c).

The Raman effect induces a continuous red shift of a soliton. This effect is clearly seen in Figure 4.10a and b as a well-resolved feature that sweeps over the range of wavelengths from 800 to 990 nm as the pump pulse energy changes from 30 to 120 pJ and the propagation distance increases from 1 to 12 cm. Due to fiber dispersion, the red-shifted soliton accumulates a time delay of about 3.4 ps at the output of a 50 cm piece of PCF (Figure 4.10c). As the central wavelength of a soliton is shifted to the region of a negative slope of the fiber dispersion profile ($z > 10$ cm in Figure 4.10b and c), phase matching between a soliton and a dispersive wave [95] allows efficient generation of red-shifted dispersive waves, observed as a spectral component centered at $\lambda_r \approx 1300$ nm in Figure 4.10a and b. In the time domain, the red-shifted dispersive-wave radiation becomes clearly visible for $z > 10$ cm as a dispersed part of the field adjacent to the red-shifted soliton branch (Figure 4.10c). For input pulse energies exceeding 120 pJ, the central wavelength of the red-shifted soliton becomes insensitive to variations in the input peak energy (Figure 4.10a and b) and stays constant within a broad range of input pulse energies at least up to 800 pJ. This stabilization of the SSFS relative to variations in the pump power, verified by experimental results presented in the inset in Figure 4.11a, and also by earlier experiments [96], is due to the spectral recoil of a soliton by the red-shifted dispersive wave, which exactly compensates [95] for the SSFS induced by the Raman effect.

With its dispersion and nonlinearity designed to generate a stabilized frequency-shifted soliton centered at a desired wavelength, the PCF offers an ideal source for the excitation and detection of the TPL response from a variety of dyes, including fluorescent proteins used as markers of neuron activity, and also brain tissue-labeling dyes. With appropriate modifications of the PCF structure, frequency-shifted ultrashort pulses optimized for the detection of the TPL response of such dyes have been generated. In particular, the fiber design presented in Figure 4.1c, providing a stabilized soliton output with a central wavelength of 980–990 nm (inset in Figure 4.11b), is ideally suited for efficient TPA excitation of Alexa Fluor 488 dye, which is commonly used as a label in cell biology and fluorescence microscopy. Filled circles in the inset in Figure 4.11b display the spectrum of the TPL response of Alexa Fluor 488 dye, measured for a dye solution in a quartz cell with the TPL response collected by the inner, 24 μm diameter part of the microstructure cladding. Our measurements suggest that the use of a frequency-shifted PCF output enhances the TPL response power collected by the fiber by more than an order of magnitude relative to the level of TPL response attainable with a frequency-unshifted, 800 nm, 40 fs output of the Ti:sapphire laser with the same field intensity. This not only opens

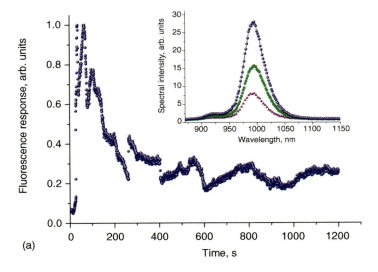

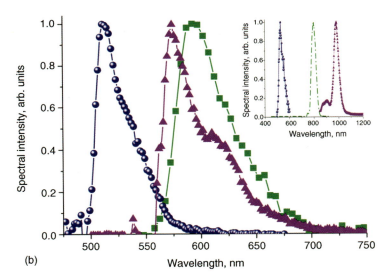

Figure 4.11 (a) Decay kinetics of the fluorescence response of EGFP from the brain of a transgenic mouse in open-skull experiments. The inset shows the spectrum of the PCF output measured for three different input energies of Ti:sapphire laser pulses: triangles, 150 pJ; rectangles, 250 pJ; and filled circles, 400 pJ. (b) Spectrally resolved fluorescent response of EGFP in the brain of a transgenic mouse (filled circles), DsRed2 fluorescent protein in the tail of a *Xenopus laevis* tadpole (rectangles), and Alexa Fluor 555 dye in stained mouse brain (triangles). The inset shows the fluorescent response of Alexa Fluor 488 dye (filled circles) TPA-excited by the frequency-shifted soliton output of a highly nonlinear PCF (triangles). The spectrum of the laser pulse is shown by the dashed-dotted line in the inset.

the ways for a higher speed and improved quality of neuron activity mapping, but also helps to maintain moderate levels of laser intensity in *in vivo* work, making it possible to reduce the laser-induced heat load and avoid radiation damage to biotissues.

The rectangles in Figure 4.11b present the results of spectrally resolved *in vivo* measurements performed on the optical response of DsRed2 fluorescent protein in the muscle tissue of a *Xenopus laevis* tadpole. In these experiments, the DsRed2 reporter gene is controlled by an actin promoter gene. Actin protein produced as a result of promoter gene expression helps build a scaffold, necessary for muscle contraction due to the force generated by myosin proteins. *In vivo* work on neuron activity mapping in brain was performed on transgenic mice, in which the expression of enhanced green fluorescent protein (EGFP) is controlled by an immediate early gene promoter, encoding the *zif*268 mammalian transcription factor. Optical detection of EGFP can thus visualize an accumulation of *zif*268, which is coordinated by a Ca^{2+} influx into a cell and indicates activation of neurons [71, 97]. For the maximum efficiency of TPL detection of EGFP, the absorption of which peaks at $\lambda_{EGFP} \approx 480$ nm, a PCF was designed to deliver a wavelength-shifted soliton centered at $2\lambda_{EGFP} \approx 960$ nm. The pulse width of this wavelength-shifted soliton PCF output was estimated as 25 fs. The spectrum of the TPL response of EGFP expressed in the brain of a transgenic mouse is shown by filled circles in Figure 4.11b. The overall fall-off kinetics of this response (Figure 4.11a), measured in experiments with an open-skull transgenic mouse, reveal the decaying activity of neurons in a fading brain. The maximum intensity of light pulses used for TPL excitation in our *in vivo* experiments (on the order of $10\,GW\,cm^{-2}$) was well below the level of light intensities where photodamage of biotissues and photobleaching of dye or fluorescent protein biomarkers became noticeable. With slight modifications of the fiber structure, a wavelength tunability range as broad as 600 nm (from 800 to 1400 nm) could be achieved for the soliton output of the type of PCF considered, suggesting a strategy for multiplex multicolor bioimaging with an appropriate combination of dye and fluorescent-protein biomarkers.

Dye–cell measurements and *in vivo* experiments [55] demonstrated that a dual-cladding PCF with two zero GVD points and properly engineered dispersion profile can substantially enhance the TPL response of fluorescent protein biomarkers and neuron activity reporters. The SSFS, stabilized against laser power fluctuations in a two zero GVD point PCF, has been shown to allow the wavelength of the soliton PCF output to be accurately matched with the TPA spectrum of dye or fluorescent protein biomarker molecules, enhancing their TPL response and allowing laser damage of biotissues to be avoided.

4.8
Toward Fiber-Based Multicolor Imaging

The use of fluorescent proteins for the simultaneous detection of multitudes of neurons in a living brain is one of the most exciting recent achievements of optical technologies in neuroscience [98]. Implementation of this technique in the

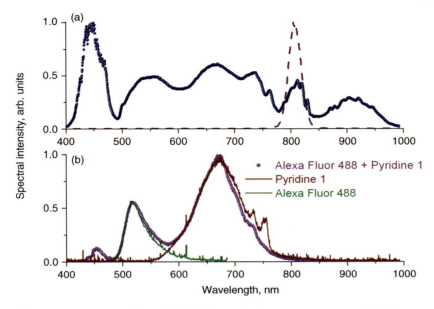

Figure 4.12 Spectrum of the mode-locked Ti:sapphire laser output (a, dashed line) and the spectrum of supercontinuum (a, filled circles) from the highly nonlinear PCF used for the simultaneous measurement of the fluorescent response of Alexa Fluor 488 and Pyridine 1 dyes in a mixture. Also shown are (b) the individual fluorescent responses of the dyes.

endoscopic mode is a challenging task and a highly promising direction for biophotonics. We show below that PCFs offer an attractive platform for the integration of fiber-optic light delivery and fluorescent response pickup with a compact fiber-format multicolor light source for a multiplex optical interrogation of a variety of fluorescent-protein reporters.

The key technology behind the integration of a multicolor light source into a fiber endoscope for neuroscience applications is supercontinuum generation [21, 22] – dramatic spectral broadening of optical waveforms induced by an involved combination of nonlinear optical effects, including self- and cross-phase modulation, pulse self-steepening, wave-mixing phenomena, and solitonic transformations of optical fields assisted by the Raman effect. In experiments [59], efficient supercontinuum generation for multicolor imaging was provided by a highly nonlinear PCF with a core diameter less than 2 μm pumped by a Ti:sapphire laser output with a typical energy of a few nanojoules, a central frequency of 810 nm, and a pulse width of 50 fs at a pulse repetition rate of 90 MHz. The spectrum of the Ti:sapphire laser output is shown by the dashed line in Figure 4.12a. The highly nonlinear PCF is optimized to transform Ti:sapphire laser pulses into a broadband supercontinuum radiation with a spectrum spanning over more than an octave from 420 to 960 nm (filled circles in Figure 4.12a), with a radiation power contained in the 400–700 nm spectral region estimated as approximately 3 mW. The supercontinuum PCF output was focused with an objective into a beam spot with an area of about $10\,\mu m^2$. With a supercontinuum pulse width on the order of 100 fs and a pulse repetition rate of 90 MHz, such a focusing corresponds to a maximum field intensity of about $3\,GW\,cm^2$. A Keldysh-type formalism of ionization in biotissues [99] estimates the electron density induced by a 100 fs pulse with an intensity in a water-type dielectric such as $n_e \approx 10^3\,cm^{-3}$, indicating that ionization effects are negligible in this regime. The situation becomes radically different, however, in the range of intensities needed for multiphoton brain-tissue excitation and probing. In the case of two-photon excitation of EGFP with 960 nm femtosecond pulses, more than one free electron is generated per laser pulse per each photon emitted through two photon absorption-induced fluorescence (with the cross-section of TPA estimated as $\sigma_{TPA} \approx 6.5\,GM = 6.5 \times 10^{-50}\,cm^4\,s^{-1}$ per photon for wild-type GFP) for sub-$TW\,cm^{-2}$ field intensities [100]. In this regime, special care should be taken in choosing the repetition rate of excitation pulses in order to prevent the accumulation of free electrons in the laser–tissue interaction region.

The supercontinuum PCF output is used to induce fluorescence in an aqueous solution of Alexa Fluor® 488 and Pyridine 1 dyes, intended to mimic a two-component mixture of fluorescent biomarkers. When this mixture of dyes was irradiated with the entire supercontinuum spectrum, a fluorescent response from both dyes was observed in the spectrum of the fluorescence signal collected with an additional standard optical fiber and studied with a spectrum analyzer. To provide individual excitation of the dyes in the solution, appropriate filters were inserted into the beam of supercontinuum in front of the cell. The spectra of the individual fluorescence responses of Alexa Fluor 488 and Pyridine 1 are shown by the solid lines in Figure 4.12b. The spectrum of the fluorescence response from the mixture of these dyes collected with a standard optical fiber is represented by the bold line in Figure 4.12b. The spectral brightness and the overall intensity of the supercontinuum output of the PCF used in our experiments were found to be ideally suited for experiments on living brain, providing a high efficiency of fluorescence response excitation and producing no photodamage of brain tissues.

References

1 Hecht, J. (1999) *City of Light: the Story of Fiber Optics*, Oxford University Press, Oxford.
2 Hansell, C.W. (1927) Picture transmission, US Patent 1,751,584, filed 13 August 1927, issued 25 March 1930.
3 Lamm, H. (1930) *Z. Instrumentenkd.*, **50**, 579.
4 Hecht, J. (1999) *Opt. Photon. News*, **10** (11), 26.
5 Flusberg, B.A., Cocker, E.D., Piyawattanametha, W., Jung, J.C., Cheung, E.L.M., and Schnitzer, M.J. (2005) *Nat. Methods*, **2**, 941.
6 Lin, C.H. and Webb, R.H. (2000) *Opt. Lett.*, **25**, 954.
7 Bird, D. and Gu, M. (2002) *Opt. Lett.*, **27**, 1031.
8 Jung, J.C. and Schnitzer, M.J. (2003) *Opt. Lett.*, **28**, 902.
9 Liang, C., Sung, K.B., Richards-Kortum, R., and Descour, M.R. (2002) *Appl. Opt.*, **41**, 4603.
10 Limpert, J., Röser, F., Schreiber, T., and Tünnermann, A. (2006) *IEEE*

J. Sel. Top. Quantum Electron., **12**, 233.

11. Boyden, E.S., Zhang, F., Bamberg, E., Nagel, G., and Deisseroth, K. (2005) *Nat. Neurosci.*, **8**, 1263.
12. Aravanis, A.M., Wang, L.P., Zhang, F., Meltzer, L.A., Mogri, M.Z., Schneider, M.B., and Deisseroth, K. (2007) *J. Neural Eng.*, **4**, S143.
13. Russell, P.St.J. (2003) *Science*, **299**, 358.
14. Knight, J.C. (2003) *Nature*, **424**, 847.
15. Zheltikov, A.M. (2000) *Phys. Usp.*, **43**, 1125.
16. Zheltikov, A.M. (2007) *Phys. Usp.*, **50**, 705.
17. Myaing, M.T., Ye, J.Y., Norris, T.B., Thomas, T., Baker, J.R. Jr., Wadsworth, W.J., Bouwmans, G., Knight, J.C., and Russell, P.St.J. (2003) *Opt. Lett.*, **28**, 1224.
18. Flusberg, B., Jung, J., Cocker, E., Anderson, E., and Schnitzer, M. (2005) *Opt. Lett.*, **30**, 2272.
19. Fu, L., Jain, A., Xie, H., Cranfield, C., and Gu, M. (2006) *Opt. Express*, **14**, 1027–1032.
20. Fang, X.-H., Hu, M.-L., Liu, B.-W., Chai, L., Wang, C.-Y., and Zheltikov, A.M. (2010) *Opt. Lett.*, **35**, 2326.
21. Dudley, J.M., Genty, G., and Coen, S. (2006) *Rev. Mod. Phys.*, **78**, 1135.
22. Zheltikov, A.M. (2006) *Phys. Usp.*, **49**, 605.
23. Zheltikov, A.M. (2007) *J. Raman Spectrosc.*, **38**, 1052.
24. Cregan, R.F., Mangan, B.J., Knight, J.C., Birks, T.A., Russell, P.St.J., Allen, D., and Roberts, P.J. (1999) *Science*, **285**, 1537.
25. Smith, C.M., Venkataraman, N., Gallagher, M.T., Muller, D., West, J.A., Borrelli, N.F., Allan, D.C., and Koch, K. (2003) *Nature*, **424**, 657.
26. Konorov, S.O., Fedotov, A.B., Mitrokhin, V.P., Beloglazov, V.I., Skibina, N.B., Shcherbakov, A.V., Wintner, E., Scalora, M., and Zheltikov, A.M. (2004) *Appl. Opt.*, **43**, 2251.
27. Ivanov, A.A., Podshivalov, A.A., and Zheltikov, A.M. (2006) *Opt. Lett.*, **31**, 3318.
28. Zheltikov, A.M. (2005) *Nat. Mater.*, **4**, 267.
29. Monro, T.M., Richardson, D.J., and Bennett, P.J. (1999) *Electron. Lett.*, **35**, 1188.
30. Monro, T.M., Belardi, W., Furusawa, K., Baggett, J.C., Broderick, N.G.R., and Richardson, D.J. (2001) *Meas. Sci. Technol.*, **12**, 854.
31. Jensen, J.B., Pedersen, L.H., Hoiby, P.E., Nielsen, L.B., Hansen, T.P., Folkenberg, J.R., Riishede, J., Noordegraaf, D., Nielsen, K., Carlsen, A., and Bjarklev, A. (2004) *Opt. Lett.*, **29**, 1974.
32. Konorov, S., Zheltikov, A., and Scalora, M. (2005) *Opt. Express*, **13**, 3454.
33. Rindorf, L., Jensen, J.B., Dufva, M., Pedersen, L.H., Høiby, P.E., and Bang, O. (2006) *Opt. Express*, **14**, 8224.
34. Cordeiro, C.M.B., Franco, M.A.R., Chesini, G., Barretto, E.C.S., Lwin, R., Brito Cruz, C.H., and Large, M.C.J. (2006) *Opt. Express*, **14**, 13056.
35. Jensen, J., Hoiby, P., Emiliyanov, G., Bang, O., Pedersen, L., and Bjarklev, A. (2005) *Opt. Express*, **13**, 5883.
36. Hoo, Y.L., Jin, W., Ho, H.L., Wang, D.N., and Windeler, R.S. (2002) *Opt. Eng.*, **41**, 8.
37. Hoo, Y.L., Jin, W., Shi, C., Ho, H.L., Wang, D.N., and Ruan, S.C. (2003) *Appl. Opt.*, **42**, 3509.
38. Pickrell, G., Peng, W., and Wang, A. (2004) *Opt. Lett.*, **29**, 1476.
39. Alfernes, R.C. (1988) in *Guided Wave Optoelectronics* (ed. T. Tamir), Springer, Berlin, p. 145.
40. Myaing F M.T., Ye, J.Y., Norris, T.B., Thomas, T., Baker, J.R. Jr., Wadsworth, W.J., Bouwmans, G., Knight, J.C., and Russell, P.S.J. (2003) *Opt. Lett.*, **28**, 1224.
41. Eggleton, B., Kerbage, C., Westbrook, P., Windeler, R., and Hale, A. (2001) *Opt. Express*, **9**, 698.
42. Ferguson, J.A., Christian Boles, T., Adams, C.P., and Walt, D.R. (1996) *Nat. Biotechnol.*, **14**, 1681.
43. Deiss, F., Sojic, N., White, D.J., and Stoddart, P.R. (2010) *Anal. Bioanal. Chem.*, **396**, 53.
44. Ritari, T., Tuominen, J., Ludvigsen, H., Petersen, J., Sørensen, T., Hansen, T., and Simonsen, H. (2004) *Opt. Express*, **12**, 4080.
45. Fedotov, A.B., Konorov, S.O., Mitrokhin, V.P., Serebryannikov, E.E., and Zheltikov, A.M. (2004) *Phys. Rev. A*, **70**, 045802.

46 Zheltikov, A.M. (2008) *Appl. Opt.*, **47**, 474.
47 Kawata, S., Ohtsu, M., and Irie, M. (2002) *Nano-Optics*, Springer, Berlin.
48 Zipfel, W.R., Williams, R.M., and Webb, W.W. (2003) *Nat. Biotechnol.*, **21**, 1369.
49 Helmchen, F. and Denk, W. (2005) *Nat. Methods*, **2**, 932.
50 Evans, C.L. and Sunney Xie, X. (2008) *Annu. Rev. Anal. Chem.*, **1**, 883.
51 Freudiger, C., Min, W., Saar, B., Lu, S., Holtom, G., He, C., Tsai, J., Kang, J., and Sunney Xie, X. (2008) *Science*, **322**, 1857.
52 Hell, S.W. (2009) *Nat. Methods*, **6**, 24.
53 Bates, M., Huang, B., Dempsey, G.P., and Zhuang, X. (2007) *Science*, **317**, 1749.
54 Flusberg, B.A., Nimmerjahn, A., Cocker, E.D., Mukamel, E.A., Barretto, R.P.J., Ko, T.H., Burns, L.D., Jung, J.C., and Schnitzer, M.J. (2008) *Nat. Methods*, **5**, 935.
55 Doronina, L.V., Fedotov, I.V., Voronin, A.A., Ivashkina, O.I., Zots, M.A., Anokhin, K.V., Rostova, E., Fedotov, A.B., and Zheltikov, A.M. (2009) *Opt. Lett.*, **34**, 3373.
56 Snyder, A.W. and Love, J.D. (1983) *Optical Waveguide Theory*, Chapman and Hall, London.
57 Zheltikov, A.M. (2005) *J. Opt. Soc. Am. B*, **22**, 1100.
58 Zheltikov, A.M. (2010) *JETP Lett.*, **91**, 378.
59 Doronina-Amitonova, L.V., Fedotov, I.V., Ivashkina, O.I., Zots, M.A., Fedotov, A.B., Anokhin, K.V., and Zheltikov, A.M. (2010) *J. Biophoton.*, **3**, 660.
60 Farrell, T.J., Patterson, M.S., and Wilson, B.C. (1992) *Med. Phys.*, **19**, 879.
61 Hull, E.L., Nichols, M.G., and Foster, T.H. (1998) *Appl. Opt.*, **37**, 2755.
62 Bevilacqua, F., Piquet, D., Marquet, P., Gross, J.D., Tromberg, B.J., and Depeursinge, C. (1999) *Appl. Opt.*, **38**, 4939.
63 Denk, W., Strickler, J.H., and Webb, W.W. (1990) *Science*, **248**, 73.
64 Bhawalkar, J.D., Kumar, N.D., Zhao, C.-F., and Prasad, P.N. (1997) *J. Clin. Lasers Med. Surg.*, **15**, 201.
65 Wilson, B.C., Olivo, M., and Singh, G. (1997) *Photochem. Photobiol.*, **65**, 166.
66 Fisher, W.G., Partridge, W.P. Jr., Dees, C., and Wachter, E.A. (1997) *Photochem. Photobiol.*, **66**, 141.
67 Ye, J.Y., Myaing, M.T., Norris, T.B., Thomas, T., and Baker, J. Jr. (2002) *Opt. Lett.*, **27**, 1412.
68 Fedotov, I.V., Fedotov, A.B., Doronina, L.V., and Zheltikov, A.M. (2009) *Appl. Opt.*, **48**, 5274.
69 Akhmediev, N. and Karlsson, M. (1995) *Phys. Rev. A*, **51**, 2602.
70 Herrmann, J., Griebner, U., Zhavoronkov, N., Husakou, A., Nickel, D., Knight, J.C., Wadsworth, W.J., Russell, P.St.J., and Korn, G. (2002) *Phys. Rev. Lett.*, **88**, 173901.
71 Chalfie, M., Tu, Y., Euskirchen, G., Ward, W.W., and Prasher, D.C. (1994) *Science*, **263**, 802.
72 Tsien, R.Y. (2005) Breeding molecules to spy on cells, in *The Harvey Lectures, Series 99*, John Wiley & Sons, Inc., Hoboken, NJ, pp. 77–93.
73 Eesley, G.L. (1981) *Coherent Raman Spectroscopy*, Pergamon Press, Oxford.
74 von der Linde, D., Laubereau, A., and Kaiser, W. (1971) *Phys. Rev. Lett.*, **26**, 9541971.
75 Eckbreth, A.C. (1988) *Laser Diagnostics for Combustion Temperature and Species*, Abacus, Cambridge, MA.
76 Zheltikov, A.M. and Koroteev, N.I. (1999) *Phys. Usp.*, **42**, 321.
77 Zheltikov, A.M. (2000) *J. Raman Spectrosc.*, **31**, 653.
78 Zumbusch, A., Holtom, G.R., and Xie, X.S. (1999) *Phys. Rev. Lett.*, **82**, 4142.
79 Evans, C.L., Potma, E.O., Poureshagh, M., Coté, D., Lin, C.L., and Xie, X.S. (2005) *Proc. Natl. Acad. Sci. U. S. A.*, **102**, 16807.
80 Akhmanov, S.A., Dmitriev, V.G., Kovrigin, A.I., Koroteev, N.I., Tunkin, V.G., and Kholodnykh, A.I. (1972) *JETP Lett.*, **15**, 425.
81 Levenson, M.D., Flytzanis, C., and Bloembergen, N. (1972) *Phys. Rev. B*, **6**, 3962.
82 Andresen, E.R., Birkedal, V., Thøgersen, J., and Keiding, S.R. (2006) *Opt. Lett.*, **31**, 1328.

83 Sidorov-Biryukov, D.A., Serebryannikov, E.E., and Zheltikov, A.M. (2006) *Opt. Lett.*, **31**, 2323.

84 Mitrokhin, V.P., Fedotov, A.B., Ivanov, A.A., Alfimov, M.V., and Zheltikov, A.M. (2007) *Opt. Lett.*, **32**, 3471.

85 Savvin, A.D., Lanin, A.A., Voronin, A.A., Fedotov, A.B., and Zheltikov, A.M. (2010) *Opt. Lett.*, **35**, 919.

86 Fedotov, A.B., Voronin, A.A., Fedotov, I.V., Ivanov, A.A., and Zheltikov, A.M. (2009) *Opt. Lett.*, **34**, 851.

87 Oberthaler, M. and Hopfel, R.A. (1993) *Appl. Phys. Lett.*, **63**, 1017.

88 Planas, S.A., Pires Mansur, N.L., Brito Cruz, C.H., and Fragnito, H.L. (1993) *Opt. Lett.*, **18**, 699.

89 Sidorov-Biryukov, D.A., Fernandez, A., Zhu, L., Pugžlys, A., Serebryannikov, E.E., Baltuška, A., and Zheltikov, A.M. (2008) *Opt. Express*, **16**, 2502.

90 Voronin, A.A., Fedotov, I.V., Doronina-Amitonova, L.V., Ivashkina, O.I., Zots, M.A., Fedotov, A.B., Anokhin, K.V., and Zheltikov, A.M. (2010) Presented at the European Conference on Nonlinear-Optical Spectroscopy, Bremen, June 2010.

91 Evans, C.L., Xu, X., Kesari, S., Sunney Xie, X., Wong, S.T.C., and Young, G.S. (2007) *Opt. Express*, **15**, 12076.

92 Mehta, A.D., Jung, J.C., Flusberg, B.A., and Schnitzer, M.J. (2004) *Curr. Opin. Neurobiol.*, **14**, 617.

93 Jung, J.C., Mehta, A.D., Aksay, E., Stepnoski, R., and Schnitzer, M.J. (2004) *J. Neurophysiol.*, **92**, 3121.

94 Agrawal, G.P. (2001) *Nonlinear Fiber Optics*, Academic Press, San Diego, CA.

95 Skryabin, D.V., Luan, F., Knight, J.C., and Russell, P.St.J. (2003) *Science*, **301**, 1705.

96 Liu, B.-W., Hu, M.-L., Fang, X.-H., Li, Y.-F., Chai, L., Wang, C.-Y., Tong, W., Luo, J., Voronin, A.A., and Zheltikov, A.M. (2008) *Opt. Express*, **16**, 14987.

97 Dynes, J.L. and Ngai, J. (1998) *Neuron*, **20**, 1081.

98 Livet, J., Weissman, T.A., Kang, H., Draft, R.W., Lu, J., Bennis, R.A., Sanes, J.R., and Lichtman, J.W. (2007) *Nature*, **450**, 56.

99 Vogel, A., Noack, J., Huttman, G., and Paltauf, G. (2005) *Appl. Phys. B*, **81**, 1015.

100 Voronin, A.A. and Zheltikov, A.M. (2010) *Phys. Rev. E*, **81**, 051918.

5
Identification and Characterization of Microorganisms by Vibrational Spectroscopy

Stephan Stöckel, Angela Walter, Anja Boßecker, Susann Meisel, Valerian Ciobotă, Wilm Schumacher, Petra Rösch, and Jürgen Popp

5.1
Introduction

Vibrational spectroscopy has been successfully applied to the characterization, identification, and classification of microorganisms [1–5]. In this chapter, the applicability of vibrational spectroscopy to clinically relevant microorganisms is demonstrated.

Infrared (IR) absorption spectroscopy and Raman spectroscopy – the two main vibrational spectroscopic techniques – are based on energetic characteristics of molecular vibrations of major biopolymeric constituents within microorganisms, such as nucleic acids, proteins, lipids, and carbohydrates. Hence specific spectral information within the low-wavenumber region of the IR absorption spectra and the Raman scattering spectra can be achieved, since microorganisms exhibit complex substance compositions of varying concentrations, which depend on phylogenetic characteristics and growth conditions, for example, age, temperature, and nutrition availability [6–8]. Therefore, the fingerprint region of IR absorption and Raman spectra contains specific information, which functions as a molecular fingerprint for classification and identification. In addition to high specificity, IR absorption and Raman spectroscopy exhibit further advantages which are complementary to both techniques [9].

The investigation of microorganisms by means of IR absorption spectroscopy has been increasingly performed since the 1950s [10]. Fourier transform infrared (FTIR) spectrometers have improved the acquisition of IR absorption spectra and by introducing micro-FTIR spectroscopy – the combination of FTIR absorption spectroscopy with a microscope – spatially resolved detection of certain microcolonies was achieved [11]. FTIR absorption analyses can be performed in either transmission or reflectance mode with the transmission mode resulting in high-quality spectra provided that homogeneity and a small sample thickness are maintained. The signal intensity within the IR spectrum correlates with the polarity of particular bonds. This is extremely high for water, which consequently exhibits intense signals and masks

Handbook of Biophotonics. Vol.2: Photonics for Health Care, First Edition. Edited by Jürgen Popp, Valery V. Tuchin, Arthur Chiou, and Stefan Heinemann.
© 2012 Wiley-VCH Verlag GmbH & Co. KGaA. Published 2012 by Wiley-VCH Verlag GmbH & Co. KGaA.

other relevant peaks within the IR absorption spectrum. This characteristic of IR absorption spectroscopy limits the application to biological samples since water is omnipresent in microorganisms, cells, and tissues.

In contrast to the IR absorption process, the scattering mechanism of Raman spectroscopy correlates with the polarizability of molecular bonds. Thus, water causes no intense signals within the fingerprint region of Raman spectra. This facilitates the analysis of microorganisms under *in vivo* conditions. Additionally, the drying step can also be omitted, which simplifies the sample preparation tremendously [12]. Micro-Raman spectroscopy, which results in better spatial resolution than micro-IR absorption spectroscopy, is capable of focusing on single bacterial cells. A preceding sample cultivation step to generate sufficient biomass for the measurements is therefore dispensable. Raman spectroscopy has been proven to be suitable for the fast and reliable investigation of microorganisms with high throughput providing real-time monitoring [13, 14].

Raman spectroscopy is limited by the weak intrinsic scattering, which has been overcome by two enhancement techniques: UV–resonance Raman (UVRR) spectroscopy and surface-enhanced Raman spectroscopy (SERS) [15, 16]. For resonance Raman spectroscopy, the excitation wavelength is shifted to an electronic absorption of the target chromophores to enhance selectively the related chromophore vibrations. An intensity increase of up to 10^6 compared with the non-resonant Raman scattering can be achieved. Hence certain biomolecules can be detected selectively and sensitively in complex biological environments even if present only in low concentrations. Within UVRR spectra excited, for example, at 244 or 257 nm, DNA and proteins are selectively enhanced, where the former functions potentially as a taxonomic marker in genotypic studies.

SERS is another approach to increase the weak Raman signal in which either silver or gold substrates are applied to accomplish an electronic enhancement in the order of 10^5–10^{14} [17].

The combination of SERS with atomic force microscopy (AFM) results in tip-enhanced Raman spectroscopy (TERS). TERS combines the high specificity of Raman spectroscopy with the high spatial resolution of AFM [16].

To illustrate the application of vibrational spectroscopy in medicine and medical-related fields, several examples especially concerning the investigation of biofilms, influence of antibiotics on microorganisms, food-borne pathogens, and epidemiological studies are introduced later in the chapter. Additionally, commonly applied sample preparation procedures are presented in the next section. Since vibrational spectroscopy of microorganisms always deals with huge databases, the necessary preprocessing and data evaluation are also discussed.

5.2
Sample Preparation Techniques

Each vibrational spectroscopic method requires its own characteristic sample preparation technique and specific sample substrates. Since both IR absorption and

Raman spectra are sensitive towards changes in environmental parameters such as different nutrition media, temperature, light, or cultivation age, these parameters should be considered for each study and implemented within databases. For bulk measurements, which require pre-cultivation, this can be achieved by rigid standardization of the cultivation conditions. For single cell measurements, all relevant parameters have to be included in the database [4].

Since the identification of bacteria was first established by means of IR absorption spectroscopy, all further preparation methods are variations of the original protocol established by Naumann [18]. Here, bacterial strains are grown on agar plates applying the so-alled four-quadrant streak pattern. For the IR measurements, a small amount of a colony is removed from the third quadrant. The bacteria are suspended in 80 μl of distilled water and an aliquot of 30 μl is then transferred to a ZnSe optical plate. Subsequently,, the bacterial suspension is dried under moderate vacuum conditions (2.5–7.5 kPa) to form a transparent film [19]. In addition to ZnSe, Si plates are also utilized as substrate for IR absorption spectroscopy [2].

The sample preparation for UVRR spectroscopy [20–22] and Raman spectroscopy with excitation wavelengths in the visible [7] and near-infrared (NIR) regions [23] is similar to that described for IR absorption spectroscopy. For Raman spectroscopic studies, either fused-silica slides or CaF_2 are utilized as substrates. Alternatively, for Raman excitation in the visible and NIR regions, bacterial smears can also be used for investigation [24, 25].

For the application of SERS, the sample preparation varies slightly. The collected bacteria are directly suspended in the corresponding colloid and not in water as described above. For the measurements, both bacteria and colloids are dried on the substrates, either CaF_2 or fused-silica slides [26, 27].

Whenever a microscope is included in the experimental setup, IR and Raman spectroscopy are additionally capable of investigating small micro-colonies [24]. For IR microscopy, a special stamping device is applied which transfers the micro-colonies from the agar plates on to the ZnSe substrate [9]. Using micro-Raman spectroscopy, it is even possible to measure the micro-colonies from the agar plates directly and the sample preparation step can therefore be omitted. Whenever micro-colonies are too small, the medium makes a contribution to the Raman spectrum, which has to be removed [24, 28].

In addition to bulk measurements, investigations of single bacterial or yeast cells can also be performed by means of micro-Raman spectroscopy. For the investigation of single bacterial cells, bacterial smears with well-separated cells are prepared [4, 25, 29]. For the localization of single bacteria inside heterogeneous sample matrices, even conventional fluorescence staining methods can be applied [30].

5.3
Statistical Analysis of Vibrational Spectroscopic Data

Spectroscopic investigations of microorganisms result in datasets which cannot be handled without statistical evaluation. First, the datasets are usually too extensive, so an intuitive registration is impossible. Second, the chemical variations between different species or strains are so minor that the impact on the spectra cannot be seen with the naked eye. These are the two main reasons why investigations of microorganisms cannot be discussed separately from statistical analysis. Statistical analysis is able to handle huge datasets and is very sensitive to small spectral changes. Therefore, statistical methods offer immense possibilities for the evaluation of biological data but also bear the risk of interpretations induced by artifacts. Artifacts can derive from several sources, for example, sample preparation, measurement procedures, and the measurement setup. Therefore, sample preparations and measurement procedures have to be standardized and the spectrometer must be carefully calibrated. Since variances due to varying laser power intensity and varying sample thickness cannot be completely avoided, careful preprocessing has to be implemented within the data analysis. The preprocessing usually involves, for example, cosmic spike removal, baseline correction, normalization, and dimension reduction. The preprocessed dataset can then be subjected to the statistical methods, which can be divided into supervised and unsupervised methods. For supervised methods, such as artificial neural networks (ANNs), a training procedure is carried out aiming for the optimal output of the algorithm regarding the given labels. Hence, *a priori* knowledge is induced which enables the classification to be validated. Unsupervised methods, such as hierarchical cluster analysis (HCA) and principal component analysis (PCA), instead cluster the data regarding their position within the feature space and not because of their group membership. For unsupervised methods, the actual dataset is used without further information.

A crucial part of the data mining is the preprocessing step. Therefore, before the actual statistical methods are presented, the steps for preprocessing will be discussed.

One very important part is the removal of cosmic spikes. Cosmic spikes are the result of photons which are not caused by the scattering of the laser light of the sample, but derive from outer space. These signals distort the measured spectrum and change the results of the chemometric analysis, so their removal is essential. Because of their origin, the cosmic spikes are detected randomly, and as they are not correlated in either time or space they can therefore be localized by repeating the measurement were measured time or space. If, for example, two Raman spectra in at the same measuring position, a cosmic spike would increase the counts in one channel. The intensity difference between the two spectra would increase abruptly in this channel compared with all other channels of the spectra.

This property enables the algorithm to find spikes by spectra subtraction. In Figure 5.1, the result for this cosmic spike removal procedure is visualized for two Raman spectra, which are depicted in black and red. The black spectrum contains a cosmic spike and the green line is the absolute value of the difference between the black and the red spectra. The distinct green peak determines the localization of a cosmic spike within the black spectrum. With several statistical techniques, it is possible to detect the outlier (very high count) in the green spectrum and so it is possible to detect the cosmic spikes. This method is generalizable for more spectra or spectra of the same sample for different points in space.

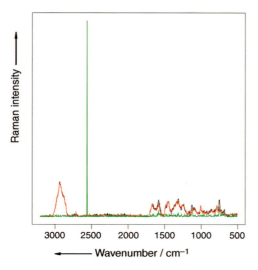

Figure 5.1 Visualization of the cosmic spike removal procedure: two Raman spectra of P. aeruginosa were measured at exactly the same position. The first spectrum is represented by the black line and the second by the red line. The green line is the absolute value of the difference of both spectra and clearly helps to identify the channels containing a spike.

The next important preprocessing step is background removal. Not all measured counts result from Raman scattering; other effects such as fluorescence also contribute to the spectra. Fluorescence is a much more likely process than Raman scattering and overlaps or even obscures the Raman signal. Because fluorescence changes over time and may not be constant over the whole sample, it would distort the learning procedure of the supervised method. This has an undesired impact on the outcome of the chemometric analysis and has to be minimized. Characteristically, fluorescence signals vary more slowly than the desired Raman signal over the wavenumber region. This characteristic offers the possibility to detect and remove the spectral background. There are several algorithms to accomplish this task. The most common ones are the SNIP algorithm and Lieber's algorithm [31, 32]. The SNIP algorithm is a composite of a low statistic filter and a peak clipping algorithm. The low statistic filter smoothes domains low statistics (low information) containing with a wide smoothing window. The peak clipping algorithm localizes peaks in order to remove them from the spectra. Both algorithms combined are able to estimate the background. A "background-free" spectrum is achieved by subtracting the estimated background from the measured spectrum. Exemplarily, the result of the SNIP algorithm performed on a spectrum is plotted in Figure 5.2, where the original spectrum is depicted in black, the estimated background is represented by the red line, and the green line is referred to as the preprocessed spectrum.

Lieber's algorithm is based on the assumption that the background changes much more slowly than the actual spectrum. The measured spectrum is a superposition of a

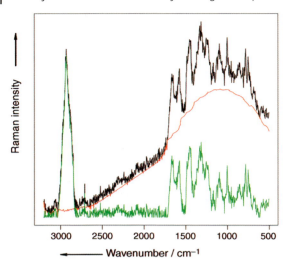

Figure 5.2 Background removal procedure: the black line is the original micro-Raman spectrum of *P. aeruginosa* and the red line shows the estimated background by the SNIP algorithm. The green line presents the difference of the original spectrum (black) and the background (red), which is then referred to as the background corrected spectrum.

rapidly changing positive spectrum and a slowly varying irrelevant background. The spectrum is fitted with a polynomial function of appropriately choser order. Every channel that is bigger than the fit is decreased to the level of the polynomial function and the algorithm starts again. This procedure is repeated until no channel is manipulated any longer.

A further preprocessing step is the normalization of the spectra. The absolute intensity values of the spectra depend on the measurement time, laser intensity, and the scattering properties of the sample. Consequently, it is important to achieve intensity-comparable spectra. There are several normalization techniques available. The first is normalization referring to one band. If all spectra share a common band, it is possible to divide the spectra by the maximum value or the area of this band. This method suffers from the presupposition that one common band is required in all spectra. Another, more general method, the so-called vector normalization, involves normalization by the area of the complete spectrum. For this method, the spectrum is divided by the area of the spectrum, which is calculated as the Euclidean distance of the spectrum to the zero spectrum (2-norm). This procedure has no further prerequisites but is not as selective as the previously mentioned procedure.

Most often the number of spectra is much smaller than or comparable to the number of spectral channels. Therefore, the dimension of the space of the spectra has to be reduced. There are several methods to reduce the dimension of the problem without losing relevant spectral information. The two most common techniques are the independent component analysis (ICA) and the PCA [33, 34]. The latter

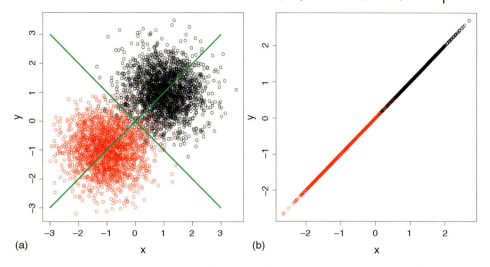

Figure 5.3 Classification of a dataset via PCA: (a) an artificial two-dimensional dataset, where the black and red points represent different classes. The green lines are the axes of the rotated feature space (PCA loadings). (b) the points in (a) projected on the first axis of the new space to achieve a group discrimination based on reduced dimensions of the dataset.

searches for a new orthogonal vector system reconstructing the spectra space. The new coordinates are sorted with falling diversity over the coordinates for the data cloud. High dimensions containing white noise can only be removed without loss of significant spectral information.

An artificial example of a PCA is shown in Figure 5.3, where Figure 5.3a contains the original positions of the elements belonging to the red and the black classes, which are depicted in a two-dimensional space. The green lines represent the new coordinate system in which the two classes are transformed to by the PCA. In Figure 5.3b, the projection of the data points to the first axis of the new space is shown and the dataset is separated into the red and the black groups. A PCA allows classification based on dimensionally reduced data. The ICA works similarly with the difference that no assumption of orthogonality between the new coordinates is required.

The reduction of the dimension has several advantages. First, it decreases the computational power needed, second, it reduces the white noise, and finally, it avoids overfitting. Overfitting is one of the major problems in chemometrics. The learning algorithms of the classifier are very powerful and are capable of learning ever white noise, which increases the accuracy of the classifiers on the training set but decreases the accuracy on unknown data. This problem is called overfitting.

The evaluation of extended datasets containing spectra with high similarities, which cannot be differentiated by the naked eye, is usually accomplished by statistical analysis. One of the most commonly applied statistics is unsupervised HCA [35]. The dataset is analyzed without any *a priori* information. The grouping is based on the similarities of the spectra – the spectral distances. The spectral distance determines how the similarities of two elements (spectra) are calculated and depends on the applied distance measurements. Two examples are the Euclidian and Pearson correlation coefficient distance measurements. The Euclidian distance measurement refers to the geometric distance, which can corresponds to the area of the difference spectrum. Two different spectra result in a difference spectrum with intense signals and a large area below the curve. Two similar spectra create a difference spectrum with minimal spectral features and a small area below the curve. The Pearson correlation coefficient is the multiplication of two preprocessed spectra. The internal preprocessing includes the subtraction of a spectral mean and the normalization within the spectrum. This preprocessing results in spectra that contain minimized element values. By multiplying those spectral intensities, only values between 1 and -1 can be obtained. Whenever two spectra are very similar, the size of the Pearson coefficient, which is also described as degree of dissimilarities, is close to 1.

After determining the spectral distances, a clustering algorithm follows to depict the results in a tree-like structure called a dendrogram. The x-axis of a dendrogram represents the heterogeneity between the spectra and visualizes the spectral distances of all elements from each other by the characteristic of the dendrogram paths. The length of a dendrogram path along the x-axis symbolizes the heterogeneity of the connected spectra. The y-axis of the dendrogram does not contain any further relevant information apart from the sectioning of the cluster. The most similar spectra within a dataset are connected by short paths along the x-axis, which represent low heterogeneity and therefore short spectral distances. Spectra of low heterogeneity are connected via a node and are collected in subclusters. The subclusters are combined in higher level clusters via longer paths along the x-axis which represent higher spectral heterogeneity. The intra-cluster heterogeneity is therefore lower than the inter-cluster heterogeneity. The clusters and subclusters contain spectra of certain spectral distances which are determined by a pure mathematical procedure and based on spectral features. That does not necessarily result in classes, which can be compared with the results of microbiological methods such as phenotypic and genotypic identification. Usually the spectral distance does not exhibit "pure" genotypic or phenotypic features; more often a combination of both is expected because the complete microbes with all cell components and metabolites are measured.

The HCA is an attractive analysis method in biological and biomedical applications, since dendrograms, representing phylogenetic trees, are commonly used in "biological-related" sciences. Phylogenetic trees represent the degree of relationship between bacterial strains, species, genera, and other compared classes. The presentation of HCA in the form of a dendrogram offers easily accessible and intuitive comprehension of the results for fast visualization and interpretation. However, besides the advantages of the HCA, an objective application is challenging. The existing risks are outlined in the following after the theoretical background of supervised chemometric methods is presented.

There are numerous supervised classifiers and some of them exhibit very interesting and unique properties. Here only one will be discussed exemplarily. We

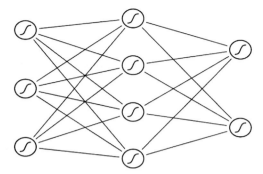

Figure 5.4 Schematic diagram of a feed-forward ANN: the input becomes first mapped to the excitation by the activation function, depicted by the line in the neurons. Subsequently, the excitation becomes mapped weighted by the weights of every connection and the procedure starts again. In this case, it is done twice. The output of the last layer is the output of the whole net.

chose the ANN, because it is one of the strongest classifiers and there are some examples of its application present in the literature [36].

ANNs comprise supervised and unsupervised statistical methods, depending on the used topology of the applied neural network. Most commonly feed-forward neural networks, a supervised technique, are applied. An example is given in Figure 5.4. The ANN is in some aspects modeled after the human brain. ANNs are basically a net of neurons with weighted connections between them. Every neuron contains a function, the so called activation function, which determines. At one step every, neuron collects the excitations of all neurons which are connected to it and sums over the neurons output for a given input weighted with the weights of every connection. The neuron reacts to this weighted sum with excitation or input disexcitation, depending on the activation function. Which activation function is used for in particular depends on the applied net.

For feed-forward nets, several layers exist and every neuron of one layer is connected to every neuron of the next layer. The excitation is propagated forwards layer by layer, giving rise to the net's name. It is obvious that the critical point of the ANNs is the right choice of the connections. Finding the best connections is called training, where several methods exist. The most prominent method for feed-forward nets is the so-called back-propagation algorithm. A systematic introduction to this field of research was given by Rojas and Feldman [36].

To emphasize the different advantages of unsupervised HCA and supervised ANNs, two examples are discussed in the following, focusing on the statistical approach.

Maquelin *et al.* presented results for the identification of pathogenic microorganisms by employing statistical analysis [24]. Figure 5.5a shows Raman spectra of clinically relevant bacterial strains: (a) *Staphylococcus aureus* ATCC 29213, (b) *Staphylococcus aureus* UHR 28624, (c) *Staphylococcus epidermidis* UHR 29489, (d) *Escherichia*

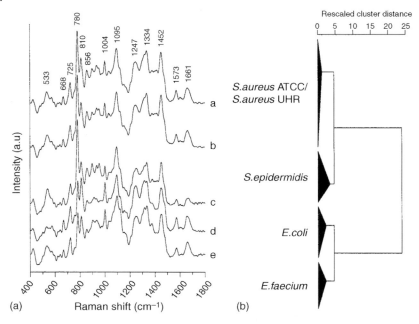

Figure 5.5 (a) Representative micro-Raman spectra of (trace a) *S. aureus* ATCC 29213, (trace b) *S. aureus* UHR 28624, (trace c) *S. epidermidis* UHR 29489, (trace d) *E. coli* ATCC 25922, and (trace e) *E. faecium* BM 4147. (b) Dendrogram as a result of the HCA performed on the principal component scores of the vector-corrected Raman spectra [24].

coli ATCC 25922, and (e) *Enterococcus faecium* BM 4147. The bacterial spectra were measured by micro-Raman spectroscopy directly from the agar plate after a cultivation time of 6 h. The spectra were recorded with an excitation wavelength of 830 nm. The Raman spectra of *Staphylococcus* spp. (Figure 5.5a–c) exhibit an intense band at 780 cm^{-1} assigned to uracil, whereas the representative spectrum of *E. coli* and *E. faecium* (Figure 5.5d and e) feature a more intense Raman signal at 1004 cm^{-1} originating from the phenylalanine vibration. With the application of an HCA, the classification of the spectra was successfully carried out and is depicted as a dendrogram in Figure 5.5b. The dendrogram is divided into two main branches: one cluster contains all *Staphylococcus* species and the other cluster combines the *E. coli* strain and the *E. faecium* strain, which are clearly divided into two clusters on a lower level. Within the *Staphylococcus* cluster, a distinct division between the *Staphylococcus* species *S. aureus* and *S. epidermidis* is obtained, but the two *S. aureus* strains cannot be differentiated and are unified in one subcluster. In the case discussed here, the HCA is capable of differentiating bacterial genera and *Staphylococcus* spp. by means of their Raman spectra, but fails to distinguish between the two *S. aureus* strains.

The phylogenetic interpretation of a dendrogram resulting from statistical analysis is limited. It seems to be certain that the *Staphylococcus* spp., which are combined in

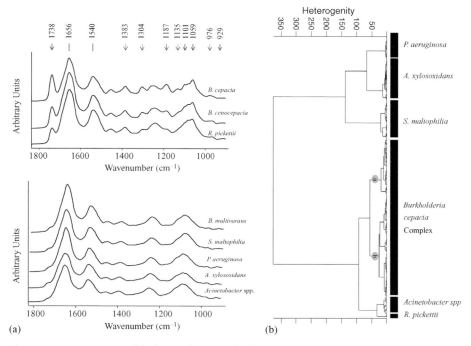

Figure 5.6 (a) IR spectra of the bacterial groups which were isolated from CF patients. (b) Dendrogram of the HCA clustering of the first derivative of the spectra within three spectral windows [38].

one main cluster, are most closely related to each other. It is also reasonable to interpret the dendrogram structure by concluding that the heterogeneity between the two main clusters is higher compared to each intra-cluster heterogeneity. However, it cannot be argued that *S. epidermidis* is genetically more similar to the closer neighbor *E. coli* than to *E. faecium* from the dendrogram, since the clusters can be reflected at their nodes, resulting in opposite relations.

Providing the phenotypic variance or heterogeneity of the culture condition is one of the challenges for the application of statistics such as HCA. Willemse-Erix *et al.* applied hierarchical cluster analysis to a Raman dataset of 785 nm excitation to differentiate between five isolates of *S. aureus* [37]. To ensure that the classification of the five isolates is based on strain specificities and not on differing growth behavior, the influence of growth was included by sampling at 18, 20, 22, and 24 h after incubation. Again, the application of the HCA is suitable to a certain extent, since two isolates could not be separated from each other.

The second example is a study by Bosch *et al.* to differentiate bacterial groups isolated from cystic fibrosis (CF) patients: *Pseudomonas aeruginosa*, *Ralstonia pickettii*, *Achromobacter xylosoxidans*, *Actinobacter* spp., *Stenotrophomonas maltophilia*, and

Burkholderia cepacia complex (BCC) by means of IR spectroscopy in combination with statistical analysis [38]. At first the sample preparation was carefully standardized and the HCA and ANN methods were applied. Figure 5.6 shows (a) the IR spectra and (b) the dendrogram resulting from the HCA. The dendrogram shows clearly six defined groups representing the six bacterial genera. However, a clear distinction of the different BCC strains could not be achieved, but a separation into two groups appeared. One subcluster contained the reference strains, whereas the second subcluster combined the clinical isolates. Although several approaches were implemented to investigate the BCC bacteria, all resulted in division of the clinical and reference isolates. The HCA model discriminated successfully between the six groups (*P. aeruginosa*, *R. pickettii*, *A. xylosoxidans*, *Actinobacter* spp., *S. maltophilia*, and BCC bacteria) but failed to differentiate between the BCC species. The presented misidentifications are considered to be caused by the insufficient phenotypic variance presented by the measured strains within the bacterial species.

The spectral data were also subjected to a two-level ANN model. Both ANN levels were independent and had different tasks that demanded individual preprocessing steps and net characteristics. The first level was applied to identify whether the spectrum originated from one of the six Gram-negative groups. The second level was switched on whenever the bacterium was identified as belonging to the BCC group to continue the identification on the species level. The complete dataset contained around 2000 spectra comprising clinical isolates and reference strains. The results of the ANN model were verified twice, by internal and external validation. For this, the complete dataset was divided into a training dataset and a test dataset. The model was exclusively trained by the training dataset and the internal validation was performed by removing spectra randomly and testing for correct identification, which resulted in 100% correct identification. The external validation was carried out by a completely independent training dataset containing spectra of clinical isolates. The combined ANN model was able to classify correctly 98.1% at the genus level and 93.8% at the BCC species level. Hence the ANN was able to differentiate on BCC species level, which could not be accomplished by the unsupervised HCA method described earlier.

5.4
Monitoring of Biofilms

The self-organization of various microorganisms in biofilms leads, especially in the field of healthcare, to serious problems. The potential of such well-organized organisms to cause infections in patients with indwelling medical devices increases in proportion to their resistance to antimicrobial agents. The process of biofilm formation takes place on either living surfaces such as tooth enamel, heart valves, the lung, and middle ear, or on inanimate surfaces such as catheters, any kind of artificial prosthesis, and intrauterine devices. Biofilm-associated microorganisms commonly isolated from selected indwelling medical devices are, for example, *Candida albicans*, *S. aureus*, *P. aeruginosa* and *E. coli* [39]. This section introduces the examination of biofilm models by different vibrational spectroscopic techniques, which offer a beneficial approach in medical microbiology.

The multifarious aggregations of different groups of attached microbes with their excretory products have different phenotypes and antibiotic susceptibility compared with their planktonic (suspended) organisms [40]. The nondestructive (i.e., leaving the sample intact during the analysis) and direct analysis of such a multilateral matrix is essential to understand the role of biofilms in infections. Fingerprint spectra display the chemical composition of cells and allow the rapid characterization of microbial strains. The application of vibrational spectroscopic techniques is suitable for fundamental biofilm research and also for the monitoring of biofilm formation [41]. Of these techniques, IR and Raman spectroscopy are complementary concerning spectral information [9].

Investigation of biofilms that develop directly on IR-transparent materials is possible in the transmission mode. Nevertheless, this sample preparation is difficult to standardize and to integrate into a routine protocol. It also limits the possibility of analyzing intact larger samples and the utilization of thick or opaque surfaces. Recent advances in detector sensitivity allow the use of reflectance micro-FTIR spectroscopy for the analysis of raw materials without the need for previous treatment. The penetration depth of the IR beam ranges from 2 to 5 μm. Ojeda *et al.* showed that the reflectance mode is a suitable technique for studying the functional groups in biofilms in opaque or thick samples, such as stainless steel, without sample preparation [42]. The characterization of the biofilm formation of *Aquabacterium commune* was carried out by applying reflectance micro-FTIR spectroscopy. Sterile stainless-steel slides were left on the culture medium with *A. commune* to be colonized at pH 7 for 100 h. Figure 5.7 shows an optical image of stainless steel and an *A. commune* biofilm. The false color images (b–f) of the scanned sample depict the location of different molecules. Thus, the reflectance micro-FTIR analysis showed reliable results for the location of different functional groups in *A. commune* biofilms on stainless steel. According to "free" (not attached) bacteria from planktonic suspension, the carboxylate and phosphoryl groups in the biofilm could play an important role in the adhesion to stainless steel [42].

Attenuated total reflection (ATR) offers a further possibility for FTIR techniques for the observation of biofilms formed directly on the surface of an ATR crystal. Investigations can be carried out without sample preparation and spectra can be obtained nondestructively, *in situ*, and in real time. The ATR crystal is the so-called internal reflection element (IRE). It has a higher refractive index than the sample or the surrounding medium. The beam penetrates through the interface into the sample, which is adjacent to the crystal surface. If the sample absorbs part of the radiation, the reflection will be attenuated. The penetration depth of the light depends on the wavelength of the light, the incident angle, and the refractive indices of the sample and crystal [43]. Immersion of the IRE in a liquid allows the observation of the subsequent colonization of the surface by bacteria and other microorganisms. With this setup, biofilm studies can be carried out in a flow-through cell, where a biofilm grows directly on the crystal surface. Observation of biofilm formation over many days can thus be performed with such an experiment by recording absorption spectra, which depend on the chemical nature of the cells attached to the crystal surface, which was explored by Delille *et al.* [43]. Here *Pseudomonas fluorescens* bacteria were pumped as a suspension in Luria–Bertani (LB) medium through the cell and the appropriate biofilm was formed on a germanium crystal. Figure 5.8 illustrates the accumulation of biomass on the IRE surface depicted by the acquired spectra over this period and the microscope image shows an increase in the IRE surface coverage by bacteria. After the initial bacterial attachment and biofilm detachment, the authors observed regrowth kinetics depending on the dissolved organic carbon level determined from changes in the areas of bands associated with proteins and polysaccharides [43].

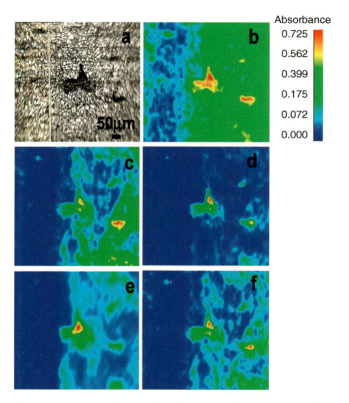

Figure 5.7 (a) Optical image of stainless steel and *Aquabacterium commune* biofilm. (b) Full-range false color image (4000–700 cm^{-1}). (c) Location of molecules with intrinsic absorption bands of amides (N−H stretching at 3550–3200, amide I at 1690, and amide II at 1540 cm^{-1}). (d) Location of molecules with absorption bands intrinsic of C−H, C=O, and C−O. (e) Location of molecules with absorption bands intrinsic of phosphorylated proteins, polyphosphates, and phosphodiester groups (1270–1220, 1100–1070, 1000–950 cm^{-1}). (f) Location of molecules with absorption bands intrinsic of hydroxyl groups (3600–3200 cm^{-1}) [42].

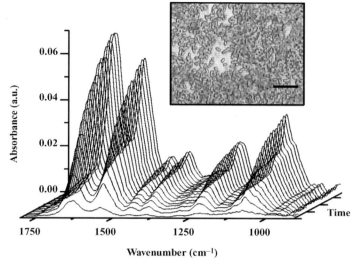

Figure 5.8 Temporal evolution of ATR–FTIR spectra during initial attachment of P. fluorescens to the germanium IR [43].

In fact, various groups were able to investigate the development of biofilms by means of IR absorption spectroscopy correlated with the corresponding changes in carbohydrate and protein expression or changes in persistence [42, 43]. This analytic approach could also be used to address the role of individual virulence factors, resistance against antimicrobial agents, and shear stress in the development and adhesion of pathogens grown on surfaces.

Complementary to IR absorption, a Raman "fingerprint" of the sample can be measured. This facilitates an additional analysis of hydrated and living bacterial biofilms under *in vitro* conditions. Therefore, Raman spectroscopy proved to be a suitable technique to investigate the effects of various additives and environmental factors on biofilm growth. The excitation with visible light (e.g., 532 nm) allows the use of common standard optics. Under these conditions, the sample can be visualized by a bright field image and spectra can be processed at different locations within the biofilm. Additionally, it is possible to record vertical and horizontal maps to show the spatial distribution of molecular or cellular species. The nondestructive character of Raman spectroscopy permits the observation of biofilm development and its spatial heterogeneities over a period of time. Raman spectra are furthermore considered to be more specific than IR spectra as they generate sharper bands. The spatial distribution of the biofilm biomass, for example, was analyzed by means of micro-Raman spectroscopy. The chemical mapping was successfully employed to probe directly the water and the biomass distribution within biofilms after only minimal sample treatment. The authors suggested that with this method, the penetration of solutes in biofilms can be determined [44, 45].

The composition of different microorganisms and other substances (minerals) inside an intact multispecies biofilm was characterized by Pätzold et al. [46]. The analyses were based on the identification of specific bands in correlation with reference materials. Ivleva et al. demonstrated that Raman spectroscopy can correlate various structural appearances within the biofilm to variations of its chemical composition and provide detailed chemical information about the different constituents in this complex matrix [47].

Biofilm analysis by means of SERS is characterized by a higher number of distinguishable peaks in the spectra, suggesting a promising potential of SERS to probe components of the biofilm matrix even at low concentrations. Ivleva et al. reported on reproducible *in situ* SERS analysis of biofilm. A heterotrophic biofilm was cultivated on a glass slide and immersed in a silver colloid suspension. SERS spectra from different constituents of multispecies biofilm were recorded [48].

The application of TERS to alginates for analyzing nanostructures in biofilm was reported by Schmid et al. [49]. A problem with the interpretation of TERS spectra are contaminated tips since they have a significant influence on the spectrum in contrast to weak analyte signals. An appropriate background subtraction and a careful selection of reference materials are indispensable.

5.5
Role of Vibrational Spectroscopy in Epidemiological Studies

In this section, an overview is given of recent developments in optical typing methods, namely IR and Raman spectroscopy. Here, vibrational spectroscopic methods are successfully applied as identification modalities in clinical environments. This outline mainly comprises investigations on clinical isolates of medically relevant microorganisms to validate the discriminatory power of the spectroscopic approaches and further to perform epidemiological analysis.

While micro-Raman spectroscopic identification at the single cell level allows direct presumptive analysis of isolates, in most of the publications cited here the investigated strains were precultured prior to the actual measurement. Thus, not only enough biomass is generated to perform measurements on bulk material and micro-colonies, but also possible background interferences due to the matrices, from which the microorganisms were isolated, are minimized. Although the mandatory culturing steps contradict the time efficiency of the spectroscopic methods to a certain extent, the achievable gain in time is still significant compared to the most preferred genotyping techniques up to now. In fact, in the case of epidemiological surveillance of microorganisms, pulsed field gel electrophoresis (PFGE) and multi-locus sequence typing (MLST) are mostly applied in retrospective analysis of outbreak situations due to their long turnaround times (48–72 h when starting from a pure culture) and low sample throughput.

Vibrational spectroscopy meets the requirements for a high-throughput, fast, and reliable typing method to provide real-time monitoring of the spread of bacterial isolates, as demonstrated in several publications. The common strategy in those studies was to classify and consequently identify medically relevant microorganisms originating from clinical isolates by means of vibrational spectroscopic measurements combined with chemometric analyses. The identification results were accomplished by comparing them with typing results obtained via common established daily routine techniques, including DNA sequencing or PFGE [50].

For example, Willemse-Erix et al. [37] employed a custom-built Raman setup with 785 nm laser excitation to validate spectroscopic typing using the methicillin-sensitive *Staphylococcus aureus* (MSSA) and methicillin-resistant *Staphylococcus aureus* (MRSA) isolates of four different collections (118 isolates). MRSA is known to be a leading cause of hospital-acquired infections and therefore a candidate for rapid epidemiological surveillance in clinics. The reproducibility of Raman spectroscopy was tested and also its discriminatory power in comparison with established typing techniques, such as PFGE and *spa* typing. Furthermore, the concordance of the Raman spectroscopic results with epidemiological data was checked. Thus, 20 well-characterized HRSA isolates were classified by performing an HCA on their Raman spectra. Isolates with different PFGE patterns were distinguished by using Raman spectroscopy, whereas PFGE-identical isolates were not. In another retrospective study, a collection of 78 MRSA isolates, retrieved from patients and hospital staff members collected in a Dutch hospital during five outbreaks in 2002, were evaluated by two PFGE analyses and Raman spectroscopy. Raman spectroscopy differentiated 72 out of 78 isolates with an accuracy of 92–95% identical with the two PFGE results and identified the majority of unique PFGE isolates as unique Raman types. In total, only three discrepancies were documented between the three typing methods – the HCA clustering of the Raman spectra thus attained good agreement with the epidemiological data [37]. In the same study, Willemse-Erix et al. collected four MRSA isolates (one from an index patient and three from a colonized staff member) during a contact screening at a Dutch hospital [37]. Based on the Raman spectroscopic data, it was possible to confirm within 1 day that no transmission between the index patient and the staff member occurred. Four days later, PFGE results reaffirmed these results. This significant gain of time is based upon the protocol employed: starting from positive microbial cultures, only 45 min are needed to prepare 24 isolates for Raman measurement, which usually takes 10 s–1 min per sample or 1 s per spectrum. This rapidity of signal collection is a crucial requirement for preliminary real-time identification of clinical isolates and is achieved due to constant improvement of the Raman equipment, which entails both simplification and improved sensitivity and robustness of the spectrometer combined with a powerful laser (220 mW with 785 nm excitation).

One drawback of non-resonant Raman spectroscopy is the low probability of Raman scattering, with typically only 1 in 10^6 photons Raman scattered, which leads to rather long Raman spectra collection times. An alternative to the classical non-resonant Raman scattering is UVRR scattering, which was described earlier.

An application of this resonant Raman technique in the field of identifying pathogenic organisms is in the discrimination of urinary tract infection (UTI) bacteria [21]. The typical causal agents of UTI are Enterobacteriaceae, especially

E. coli and *Klebsiella* spp. in addition to Gram-positive enterococci [51]. Twenty well-characterized and identified clinical bacterial isolates from patients with UTI from a hospital were obtained belonging to *E. coli*, *Klebsiella oxytoca*, *Klebsiella pneumoniae*, *Enterococcus* spp., and *Proteus mirabilis*. All isolates were cultured for 12 h and were analyzed by means of UVRR spectroscopy with 0.1 mW laser power on the sample and a 120 s integration time for a single spectrum of the bacterial bulk material. This rather long accumulation time despite the aforementioned enhancement mechanism is due to the photochemical destruction or "burning effects" in the sampled microorganisms because of the high energetic nature of UV photons. The applicable laser power on the samples and thus the minimal integration time are therefore limited for the sake of the spectral quality.

The measured and preprocessed resonance Raman spectra of the UTI bacteria were analyzed chemometrically with respect to similarities among them with numerous clustering algorithms. Each of the four classes/genera analyzed was correctly resolved into four separate groups, which leads to the assumption that discrimination between these main causal agents of UCI was achieved. However, because only four *a priori* groups were fed into the cluster algorithms, the resolution at strain level was still low.

Research has also been carried out recently on the spectroscopic identification of nonfermenting Gram-negative rods isolated from sputum samples of cystic fibrosis (CF) patients. Although the major causes of the morbidity and mortality in patients with CF are chronic lung infections caused by *P. aeruginosa*, *S. aureus*, and *Haemophilus influenzae*, an increased fraction of CF patients has been colonized by other bacilli, such as members of the Burkholderia cepacia complex (BCC) [52]. Because of their high transmissibility and multiple antibiotic resistances, BCC infections are difficult to treat once a patient has become infected. Currently, the BCC comprises a group of genetically distinct but phenotypically similar bacteria and contains 17 recognized species. Despite the importance of the rapid and accurate identification of this group, only a few studies using vibrational spectroscopy have been published.

In 2008, Bosch *et al.* employed FTIR spectroscopy to identify bacteria isolated from sputum samples of CF patients [38], including *P. aeruginosa*, *R. pickettii*, *A. xylosoxidans*, *Actinobacter* spp., *S. maltophilia* and BCC bacteria in combination with HCA and ANNs. From 150 CF patients attending three different hospitals in Argentina, 169 clinical isolates of nonfermenting Gram-negative bacteria were studied in combination with 15 reference strains. The clinical isolates were characterized according to guidelines for clinical microbiology practices for respiratory tract specimens from CF patients. In particular, BCC bacteria were further identified by *recA*-based PCR followed by restricted fragment length polymorphism (RFLP), and 16S rRNA gene sequencing was performed on some bacterial species belonging to genera different from BCC. FTIR spectra have been acquired of all strains to establish a database containing 2000 IR spectra, which were fed into the statistical algorithms. Bosch *et al.* [38] managed to discriminate successfully between the six groups with the help of an HCA model. However, differentiation within the BCC was not accomplished [38]. The two-level ANN model proved to be superior to the HCA in this respect, since a differentiation at the BCC species level was eventually achieved with a classification rate of 93.8%.

However, not only chemical modifications of the bacteria due to cultivation, sampling, and measurement parameters can be a limiting factor for some of the spectroscopic typing methods, which probe the complete biochemical makeup of the microorganisms: chemical modifications resulting from adaptation strategies to the CF lung stress environment, for example, aggressive and prolonged antibiotic therapy, can also affect the discriminatory power. The influence of the phenotypic variability of BCC isolates, for example caused by CF therapeutic procedures, on IR spectroscopy-based models for species and ribopattern discrimination was studied by Coutinho et al. [53]. A collection of 185 BCC isolates, representing four different BCC species, from respiratory secretions of 35 patients with CF from a Portuguese hospital, gathered from January 1995 to March 2006, was analyzed. All were previously classified by established molecular methods at species and ribopattern levels. On the basis of discriminant partial least-squares (DPLS) calculations performed on the FTIR absorption spectra, misclassification error rates at the ribopattern level of 2–8% and at the species level of 10% for the hospital isolates and 32% for the reference strains were achieved. This result indicates that the model calibrated with the hospital strains cannot be used to predict the reference strains. Whereas the latter originated from different sources, the isolates obtained in the hospital were sequential isolates from a limited number of different patients, several of them exhibiting the same ribopattern. This demonstrates the restricted generalization ability of a model calibrated with only a limited number of samples. Also, an important degree of variability for isolates belonging to the same species, same ribopattern, and collected from the same patient during prolonged colonization was registered. This may, on the other hand, indicate the possibility of revealing changes in the biochemical composition of the microorganisms due to adaptation strategies to the CF lung stressing environment by means of FTIR spectroscopy.

Another widespread cause of hospital-acquired sepsis is *Candida* spp. This serious noscomial pathogen strikes preferentially in surgical services, and especially in non-neutropenic critically ill patients in intensive care units (ICUs). *Candida*-associated infections exhibit high mortality rates of up to 46–75% and they represent the fourth most commonly recovered organisms from blood cultures [54]. Hence early and systematic antifungal therapy has a tremendously positive effect on the outcome of patients with invasive candidiasis in respect of morbidity and (attributable) mortality. Fluconazole followed by amphotericin B is often the first choice for prophylactic or empirical antifungal therapy. Unfortunately, there exist a number of fluconazole-resistant non-*albicans Candida* spp. and also the fluconazole-resistant *Candida albicans*. A prompt identification of significant isolates to the species, or even strain level, is therefore imperative to be able to decide which initial regimen for prompt antifungal therapy is appropriate, since susceptibility data for isolates may not be available immediately. However, at present the conventional yeast identification is based on an extensive series of biochemical assays, following an obligatory culture period sufficient to obtain a biomass of 10^6–10^8 cells and thus leading to up to 4 days before a definitive report reaches the clinician. Vibrational spectroscopy might render assistance in this context: a clinical study was reported by Ibelings et al., where Raman spectra of microcolonies 10–100 μm in diameter were obtained after only a 6 h incubation period [55]. Upon arrival in the microbiology laboratory, each specimen was divided into two groups, with one part used for conventional microbiological identification (via a commercially available yeast identification panel with a turnaround time of 48–96 h), and the other for identification by Raman microspectroscopy (within 1 day). A collection of 93 reference *Candida* strains, comprising 10 different *Candida* spp., backed up the Raman spectroscopy-based identification of 88 pertioneal specimens obtained from peritonitis patients from a surgical ICU and a general surgical ward. Spectra obtained from patient specimens were analyzed with a sequential identification model based on six linear discriminant models (LDAs). In three of the 29 *Candida*-positive cases, a difference between the microbiological and Raman identifications occurred, giving a prediction accuracy of 90% by Raman spectroscopy, if the conventional microbiological method is interpreted as the reference method. In a similar study by Maquelin et al., an accuracy of even 97–100% for the Raman identification of *Candida* spp. was obtained with the same method but including 42 reference strains [56].

Similar approaches concerning epidemiological investigations of *Candida* have also been performed with FTIR spectroscopy. Essendoubi et al. studied 29 *Candida glabrata* isolates from patients within a two-step procedure [57], as described in the following. First, an internal validation phase with 16 strains was carried out to determine the statistical parameters which result in the highest discriminatory power. A genotypic technique based on randomly amplified polymorphic DNA analysis (RAPD) confirmed the outcomes of this validation step. Second, the 13 remaining clinical strains of *C. glabrata*, isolated from multiple sites in four ICU patients, were tested by FTIR and HCA in order to eliminate the possibility of inter-human infection by comparing the spectral signatures of all patient strains. In the end, six spectral groups (four from the patients and two from reference strains) led to the conclusion that no cross-infection had occurred. Again, the saving of time compared with conventional approaches is striking, since the FTIR spectra were recorded after only 24 h of cultivation of the isolates. Also, a decrease in growth time to 10–18 h seems within reach to gain enough microcolonies for FTIR microspectroscopy.

Recent studies regarding spectroscopic species identification in a clinical diagnostic setting were also focused on cultured mycobacteria. The genus *Mycobacterium* comprises more than 80 species, some of which are potentially pathogenic to humans and animals. Whereas the classically defined lung tuberculosis is predominantly caused by *M. tuberculosis* complex (TB), an increasing number of incidents of pulmonary disease are caused by nontuberculous *Mycobacteria* (NTM) [58, 59]. The clinical features of the NTM-derived disease are in some cases indistinguishable from those of tuberculosis. To stipulate an appropriate therapy and to offer comprehensive infection control, a timely and correct discrimination between TB and NTM and also an identification of causative NTM at the species and strain levels is mandatory. A large number of commercial techniques are available for a limited number of frequently encountered species, which are rapid but expensive. For rarely emerging *Mycobacterium* isolates, DNA sequencing of the 16S rRNA gene is usually performed in reference laboratories. However, the implementation of these techniques in peripheral laboratories has various drawbacks, such as high cost, complexity, and the absence of clear, unambiguous interpretations and peer-reviewed databases. The spectroscopic data for TB and NTM isolates suggest so far that vibrational spectroscopy may provide a practical solution to identify unknown isolates rapidly in routine clinical–microbiological laboratories.

Buijtels *et al.* presented the first study on the use of Raman spectroscopy for the identification of *M. tuberculosis* and the most frequently encountered NTM species with the objective of assessing the reproducibility of the Raman spectra of inactivated *Mycobacteria* versus viable *Mycobacteria* and to compare the performance of this method with that of 16S rRNA sequencing [60]. The pilot inactivation study in Buijtels *et al.*'s publication on 12 level two NTM species revealed that heat inactivation is superior to formalin inactivation. Raman spectra of heat-inactivated samples showed the highest resemblance to those of viable cells. Heat inactivation did not decrease either the classification rate or the intra-species reproducibility substantially. In a subsequent identification study, a collection of 63 *Mycobacterium* strains, comprising eight different *Mycobacterium* species, was identified species-wise first with 16S rRNA sequencing before the Raman measurements were performed. The main differences in the Raman spectra of the eight *Mycobacterium* species were found in the intensity of peaks due to carotenoid vibrations. The similarity between the Raman spectra of the samples was calculated using a squared Pearson correlation coefficient (R^2). Species identification was performed via a leave-one-out approach, by which the R^2 value of one sample relative to all the other measured samples was calculated. The predicted species of a sample was assumed to be identical with the species of the sample with which the highest correlation occurred. This leave-one-out approach simulated the situation in a diagnostic setting where a new measurement is compared with an existing database.

In this study, the classification results of Raman spectroscopy versus 16S rRNA sequencing predicted 60 of 63 strains correctly at the species level, which represents an overall classification rate of 95.2%. These outcomes were achieved within 3 h after a positive signal from the automated culture system. For all strains, the differentiation between *M. tuberculosis* and NTM was correct. Since the model in this work was based on only a limited number of species, future research is suggested in order to extend the spectral database with more spectra of other TB and NMT strains with a larger number of isolates per species.

A promising approach to achieve cultivation-independent measurements is the analysis of single bacterial cells directly after their isolation from a patient's sample. Simple and fast enrichment and purification steps remain obligatory, but a prolonged preanalytical cultivation to gain sufficient biomass becomes obsolete, when techniques such as micro-Raman spectroscopy need hundreds of intact bacterial cells only,

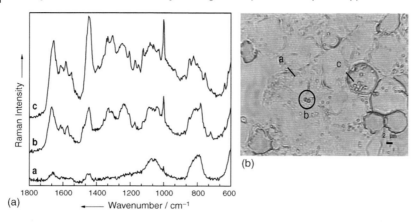

Figure 5.9 (a) Raman spectra of meningococcal patient's sample sedimented via cytocentrifugation: (trace a) background, (trace b) bacteria, and (trace c) leukocyte. (b) Microscopic image of a meningococcal patients sample: (a) background, (b) bacteria, and (c) leukocyte [61].

instead of billions in the case of measurements on bulk samples and microcolonies, as is the case in, for example, FTIR microspectroscopy.

Harz et al. proved in a pilot study the capability of micro-Raman spectroscopy to classify bacterial contaminants directly in the cerebrospinal fluid (CSF) of a meningococcal patient [61]. The sample was obtained from a patient with bacterial meningitis caused by *Neisseria meningitides*, who was already being treated with antibiotics. The preparation consisted simply of sedimentation of the cellular components onto fused silica slides by cytocentrifugation followed by a hot-air fixation (10 min at 110 °C) to inactivate the pathogens.

Direct targeting of individual cells, which were localized in the microscope's brightfield image as shown in Figure 5.9, was ensured due to the micro-Raman setup, which provided a focused laser beam approximately 0.8 μm in diameter. Thus, only an individual bacterial cell was probed and delivered a Raman spectrum after 30–60 s, as shown exemplarily in spectrum b in Figure 5.9. The Raman spectra a and c originated from the CSF matrix and a neighboring leukocyte, respectively. It is obvious that, except for contributions of the fused-silica substrate, the CSF did not give any strong Raman signals, which could have a serious impact on the bacterial Raman spectra. The intense Raman spectrum of the leukocyte (c) posed no problem since by targeting only the single bacterial cell any spectral disturbance of the blood cell could be avoided.

The measured spectra from the patient's sample were compared via hierarchical cluster analysis with spectra obtained from bacteria that had been cultivated *in vitro*. Among them are typical causative agents of bacterial meningitis such as *Neisseria meningitis*, *Streptococcus pneumoniae*, and *Listeria monocytogenes*. All of the patient's sample bacteria spectra were correctly placed in the *N. meningitis* cluster; the overall classification rate at the species level was 95.7% (121/127 spectra). It could be shown that the CSF matrix did not mask the signals of bacterial cells. Furthermore, other cellular components such as leukocytes also did not seriously hamper the detection of individual meningococci cells. In cases with lower cell numbers, where the localization time might increase to impractical dimensions, direct fluorescence staining of the bacteria can be advantageous, requiring an appropriate dye [62]. With a fluorescence detection modality it is then possible to find stained bacteria selectively.

Overall, it can justifiably be concluded that vibrational spectroscopy is suitable for real-time application in medical microbiology. In short, clinicians can benefit from spectroscopy based typing methods because of several advantages: Recording vibrational spectra does not require qualified specialists, the number of essential or expensive chemicals is limited, and, above all, microbial fingerprints are generated at tremendously high speed while still allowing a good discriminatory power. An appropriately equipped medical diagnostic laboratory therefore has the means to identify microbial cultures rapidly, preferably on the same day that an initial culture is obtained from patient material. Additionally, as the spectroscopic methods rely on the identification of variable biological and biochemical features of bacterial strains and thus are phenotypic methods, they supplement perfectly molecular methods, such as sequence-based identification schemes, for example, pulsed-field gel electrophoresis (PFGE) and multilocus sequence typing (MLST), which are mostly applied in retrospective analyses of outbreak situations due to their long turnaround times (48–72 h when starting from a pure culture) and low sample throughput [24, 63].

5.6
Effect of Antibiotics on Microorganisms

During the last two decades, the number of newly discovered antimicrobial agents has decreased drastically with a concomitant increase in multidrug-resistant pathogens. This alarming development is partially due to the prescription frequency of antibiotics and the use of a broad spectrum of antibacterial agents. The only solution for the increasing bacterial resistance is to continuously develop new drugs. However, the development of new antibiotics is affected by the lack of detailed information about the modes of action of various antibacterial agents [64].

To establish a detailed understanding of bacterial metabolism and the effectiveness of various antibacterial agents, noninvasive, highly sensitive techniques are required. In recent years, different vibrational spectroscopic methods have been applied to characterize the interaction between various bacteria and drugs. The vibrational spectroscopic approaches were applied to unravel the mode of action of the antibiotics.

For example, the investigation of the effects of β-lactam antibiotics on microorganisms can be performed successfully by means of the SERS technique [65]. β-Lactam substances inhibit the biosynthesis of the cell wall. The SERS enhancement originates from the strong optical intensity localized within 10 nm from the surface of metallic nanostructures. Hence SERS is particularly suitable for investigating the interactions of the antibiotics with the microbial cell wall. For example, oxacillin, a β-lactam agent, was found to affect the cell wall of the Gram-positive (*S. aureus*) and Gram-negative (*E. coli*) bacteria differently due to the diverse cell wall composition. The concentration of oxacillin used in the experiment was five times the minimal inhibitory concentration (20 mg l^{-1}). After the initial SERS perturbation of oxacillin, the Gram-positive *S. aureus* bacteria exhibited a transitory SERS recovery. The compensation of synthesis inhibition by the thick peptidoglycan layer is discussed as a possible explanation. Gram-negative bacteria do not possess any reserve of peptidoglycan and their response to oxacillin was found to be irreversible. The time required by β-lactam agents to affect the cell wall integrity was found to be ∼1 h. The changes in the SERS spectra of the investigated microorganisms were related to the destruction of the cell wall. SERS spectra of the Gram-positive *S. aureus* treated with antibiotics that inhibit protein synthesis, for example, gentamicin and tetracycline, exhibited changes after relatively long treatment times of about 9–12 h. This is due to the maintenance of the cell wall integrity during the inhibition of the protein synthesis [65].

Huleihel *et al.* investigated the effect of the commonly applied β-lactam antibiotic ampicillin (100 μg l^{-1}) on Gram-negative bacteria (*E. coli*, *Serratia marcescens*, *P. aeruginosa*, *H. influenza*, and *Salmonella enteridis*) and also on Gram-positive bacteria (*S. aureus*, *Staphylococcus olysgalacticae*, *Staphylococcus mitis*, *Bacillus subtilis*, *Bacillus cereus*, *Bacillus megaterium*, and *Bacillus thuringiensis*) by means of FTIR spectroscopy [66]. In contrast to SERS, which provides chemical information only about the bacterial cell wall, FTIR spectroscopy gives information about all the bacteria cell components. In contrast to the previously presented study, no spectral changes were detected 2 h after the antibiotic (β-lactam) treatment of a bacterial culture. This outcome was attributed by the authors to the relatively low sensitivity to cell wall composition of the FTIR technique compared with SERS. Moreover, no differences in the spectral response between Gram-positive and Gram-negative bacteria treated with antibiotics were observed on applying FTIR spectroscopy.

Huleihel *et al.* also compared the impact of ampicillin on the effect of caffeic acid phenylethyl ester (CAPE) on bacteria [66]. CAPE is a protein synthesis inhibitor and therefore changes of the relative proteins to nucleic acids ratios within the cells were expected. Significant spectral changes were observed 2 h after treatment. Interestingly, Gram-positive and Gram-negative bacteria presented distinctive changes in the biochemical composition. Considerable increases in band areas assigned to proteins and sugars were detected for Gram-positive bacteria, concomitant with major reductions in band areas arising from nucleic acid vibrations. These alterations were considered to be due to the reduced activity of the treated microorganisms, thereby causing an accumulation of carbohydrates in the cells. The increase in the protein level was reasoned to result from an overexpression of other bacterial genes. The IR spectra of Gram-negative bacteria showed an increase in the relative sugar to protein content. The reported results suggest that FTIR spectroscopy can be used as an effective tool to evaluate rapidly the efficiency of the antibacterial effect of various protein synthesis inhibitors.

In contrast, investigations focused on fluoroquinolones targeting the bacterial enzyme gyrase were mainly performed by means of UVRR spectroscopy [67–70]. This technique was preferred due to the abundant information supplied regarding the DNA/RNA and aromatic amino acids within the investigated microorganisms. Gyrase catalyzes the introduction of negative coils into the DNA structure. Thus, the DNA transcription and replication is blocked by binding the fluoroquinolones to the gyrase–DNA complex. UVRR spectroscopy is applied to study minor modifications of the nucleic acid and protein contents within bacteria. The effect of ciprofloxacin (0.9 and 5 µg ml^{-1}) as a representative of the fluorquinolones on *Bacillus pumilus* was investigated by Neugebauer *et al.* by means of UVRR spectroscopy in combination with statistical methods (PCA and HCA) [67, 69]. By increasing the concentration of ciprofloxacin, notable changes in the band intensity of nucleic acids and proteins were detected. The main changes in the spectra were assigned to nucleic acids, especially guanine and adenine, to proteins represented by tyrosine and tryptophan, and to a CC-stretching vibration assigned to the drug ciprofloxacin itself. For more precise assignment of the detected changes, the Raman spectrum of a relaxed plasmid was compared with that of a superhelical plasmid and also with a spectrum recorded from a mixture of a relaxed plasmid with a subunit of gyrase. The outcome suggests that the Raman spectra of the bacterial cells contain valuable information about the DNA topology and the impact of antibiotics on the DNA structure can be detected by means of UVRR spectroscopy.

The effect of various concentrations of moxifloxacin, another fluoroquinolone drug, on *Streptococcus epidermidis* bacteria was investigated using complementary spectroscopic techniques including FTIR absorption, nonresonant Raman and UVRR spectroscopy [71]. Although the IR spectra of *S. epidermidis*, recorded 80 min after the drug addition appeared to be very similar, quantitative differences according to the drug concentration could be found. HCA was used successfully to differentiate between untreated bacteria and bacteria exposed to small amounts of moxifloxacin (0.08 and 0.16 µg ml^{-1}). The Raman spectra of the bacteria exposed to higher doses of the drug (0.27 and 0.63 µg ml^{-1}) were combined in one cluster and were therefore not differentiated. The results suggest that the antibiotic applied at low concentrations (around the minimal inhibition concentration) did not reach the saturation level and the effect of the drug on the bacteria was still increasing. At high concentrations, saturation was achieved and no further changes in the chemical composition of the cells were detected with increasing amount of drug.

The underlying variances in the IR spectra were elucidated by means of the PCA. Moxifloxacin targets mainly the DNA and protein structures. The drug causes structural changes in the DNA backbone, which suggested to involve reorientation of the phosphate groups, and an additional decrease in the phosphate bindings was observable. Also, an increase of intermolecular β-sheet aggregates (characteristic of denatured proteins) concomitant with a decrease in ordered α-helical and β-sheet proteins was deduced from the IR absorption spectra. Neugebauer *et al.* interpreted this as a consequence of protein (gyrase)–drug interaction. The nonresonant Raman measurements were performed using an excitation laser in the green region at 532 nm. HCA applied to IR absorption spectra resulted in clear separation of untreated and treated bacteria. The bacterial spectra formed individual clusters dependent on moxifloxacin concentration, exhibiting the inner-class variation. To identify the main changes in the bacterial spectra, PCA was additionally applied. Drug treatment induces an intensity increase in Raman bands assigned to guanine and adenine, to adenosine and thymidine, to glycosidic linkages, and to the phosphate stretching vibration of the DNA backbone. Additionally, an increase in random coil structure upon drug addition was obtained within the Raman spectra. A simultaneous interaction of moxifloxacin with DNA and the protein gyrase was concluded to be the reason.

The results obtained by means of UVRR spectroscopy are in good agreement with the FTIR results, since similar bacteria clusters were obtained using HCA. The PCA outcomes supported the proposed mechanism of interaction between drug and bacteria by focusing on DNA and protein moieties.

The mode of action of amikacin, an aminoglycoside drug, on *P. aeruginosa* bacteria was monitored using UVRR spectroscopy [64]. In contrast to the above-mentioned studies, the antibiotic was provided in various sub-inhibitory concentrations. The spectral changes were ascribed to tryptophan and tyrosine. To study the dynamic nature of the spectral changes over the full antibiotic range below the inhibitory concentrations, the authors applied two-dimensional (2D) correlation analysis.

Generalized 2D correlation spectroscopy is a cross-correlation method that provides information regarding the dynamic spectral changes due to an external perturbation of the system such as increasing antibiotic concentration. The synchronous 2D correlation spectra are characterized by autopeaks along the diagonal and cross-peaks off the diagonal for bands that exhibit similarities in the dynamic behavior. In turn, the asynchronous 2D correlation spectrum contains only cross-peaks that reveal out-of-phase (dissimilar) behavior of two bands. Peaks that do not change their intensities throughout the series will not appear in the 2D spectra. Thus, 2D correlation asynchronous analysis provided information about the order of the changes registered in the Raman spectra [72].

The 2D correlation synchronous spectra obtained by Lopez-Diez *et al.* [64] indicated a decrease in protein signals with increasing amikacin concentration, whereas DNA bands increased. Also, a shift of the protein band was observed from the 2D correlation analysis. The shift of the protein band is consistent with the fact that, as the amount of newly synthesized proteins decreases and amino acids are misincorporated into proteins, the molecular environment of the proteins changes. The 2D correlation asynchronous spectra indicated that the nucleic bands are affected at lower concentration than the protein-related peaks. From the reported Raman results, the mode of action of the amikacin could clearly be described. The drug targeted the ribosomal RNA and thereby inhibited new protein synthesis and induced misincorporation of amino acids in proteins [64].

5.7
Detection and Identification of Food-Borne Pathogens

In 1999, Mead *et al.* calculated that ~76 million illnesses were caused by spoiled food in the United States each year [73]. Some of the main causes are pathogens (bacteria, viruses, and parasites), toxins, and metals. In the same context, Mead *et al.* defined also that *Salmonella, Listeria,* and *Toxoplasma* bacteria are responsible for about 1500 deaths each year. Additionally, other illnesses-causing bacteria including *E. coli, Shigella* spp., *Brucella* spp., *Campylobacter* spp., and *B. cereus* exist.

In most cases, medical diagnostics is highly correlated with the knowledge of the anamnesis for distinct identification. This includes the origin, the outbreak, and the spread of a disease, especially when pathogenic microorganisms are involved in this process. The demand on the scientists is to link a pathogenic isolate from a patient to the source of infection to decrease the risk of similar outbreaks. Since microorganisms are ubiquitous, contact cannot be avoided and risk of infections is relatively high. In this context, sources of infection include contaminated food products such as meat, fruits, milk, and juices.

Identification is always a time-sensitive matter. Protecting the public against disease-causing bacteria depends on the efficiency and reliability of the methods designed to detect these pathogens, requiring rapid, sensitive, and accurate procedures. Almost all conventional techniques used today to identify specific pathogens in foods need from not less than 48 h up to 1 week [74]. In addition to immunological and molecular genetic approaches, modern detection systems based on the analysis of a specific spectroscopic signature allow potentially faster identification with enhanced sensitivity and detection limits down to the single-cell level [75]. For that reason, investigations on food-borne pathogens by means of vibrational spectroscopy are considered below.

One method that is used for the classification of various pathogens is FTIR spectroscopy. In this context, Al-Holy *et al.* published a classification of *B. cereus, Salmonella enterica, E. coli,* and *Listeria* spp. at genus and strain levels based on pattern recognition. The separation of bacterial strains was considered to be due to differences in the structure and quantity of cell wall polysaccharides, lipids, and proteins, which are reflected in the IR spectra between 1700 and 700 cm^{-1} [76]. Furthermore, clear distinctions between different genera, species, and strains of bacteria were observed by PCA. The same method was used by Lin *et al.* to show discrimination between intact and sonication-injured *L. monocytogenes* and to distinguish this strain from other selected uninjured *Listeria* strains [77].

A more complex study based on FTIR spectroscopy was presented by Yu *et al.* [78]. The analysis was applied to the differentiation and quantification of microorganisms in apple juice in comparison with plate counting methods. Eight different pathogenic strains of bacteria (e.g., *Salmonella* spp., *Enterobacter* spp., and *Pseudomonas vulgaris*), which are known to cause gastroenteric diseases, were investigated. A concentration series was generated by diluting a bacterial starting concentration of 10^9 CFU ml^{-1} (CFU, colony-forming units) with autoclaved juice to six different concentrations from 10^8 to 10^3 CFU ml^{-1}. Exactly 2 min after sample preparation, measurements were made to assure that the number of cells settled on the ATR crystal was proportional to the concentration of the microbial suspension. By using a PCA in combination with canonical variate analysis (CVA), it could be shown that differentiation of microorganisms in apple juice is feasible down to a low concentration level of 10^3 CFU ml^{-1} [78]. CVA, performed on the spectral dataset within the concentration range mentioned above, resulted in consistent and satisfactory separation. Only the three lowest concentrations (10^3–10^5 CFU ml^{-1} apple juice) of *P. vulgaris* were difficult to separate in the chemometric studies (PCA and CVA). To validate the discriminant model, spectra of unknown samples were obtained and their concentrations were predicted, resulting in good agreement with the plate counting method.

Al-Holy *et al.* also published results concerning the identification of pathogens from juice by means of FTIR spectroscopy, especially focusing on the separation of *E. coli* O175:H7 leerzeichen from others. In contrast to the aforementioned method, the inoculated juices were treated with several purification steps, including centrifugation and two washing steps with 0.9% saline solution, to remove juice components and concentrate bacteria. Thereby a pure bacterial pellet was prepared before the spectroscopic analysis was commenced [79] and separation of *E. coli* at the strain level was achieved.

Whittaker *et al.* studied the identification of bacterial pathogens by ATR–FTIR spectroscopy basing on cellular fatty acid methyl esters (FAMEs) [80]. Lipid cellular components, which are responsible for bacterial speciation, are commonly analyzed only by using capillary gas chromatography with flame ionization or mass spectrometric detection. Whittaker *et al.* showed that the identification of FAME mixtures from pathogenic bacteria such as *B. cereus*, *Shigella sonnei*, and *E. coli* is feasible by applying FTIR spectroscopy in combination with data analysis such as PCA and SIMCA.

The described studies emphasize the potential of FTIR spectroscopy to identify pathogenic microorganisms in liquid foods such as juices, and similar food systems, after extraction. Nevertheless, direct measurements on the food matrices are desirable. In this context, food-borne microorganisms have been analyzed directly on food surfaces by means of Fourier transform Raman spectroscopy.

Yang and Irudayaraj demonstrated the characterization and classification of six different bacteria including the pathogenic *E. coli* O157:H7 on whole apples by means of FT-Raman spectroscopy [74]. The apples were carefully covered with the bacterial suspension and then analyzed. The spectrum of the apple skin and the spectra of the apples skin smeared with the different microorganisms were very similar, as shown in Figure 5.10.

Minor spectral differences were localized within the fingerprint region between 1800 and 600 m^{-1} which are due to varying protein and lipid contents within the

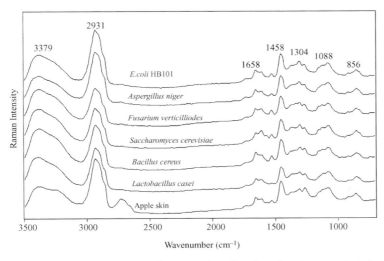

Figure 5.10 FT-Raman spectra of uncontaminated apple and apple contaminated with different types of microorganisms [74].

microbial cells. In this context, spectral analyses were performed with focus on differentiation between various *E. coli* strains. By using multivariate statistics (PCA and CVA), *E. coli* strains were successfully classified and separated into pathogens and nonpathogens according to Figure 5.11. Whereas the pathogenic strain *E. coli* O157:H7 was classified separately from the five different nonpathogenic *E. coli* strains by means of the first two CV scores according to Figure 5.11,

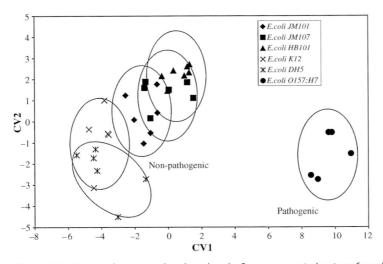

Figure 5.11 Canonical variate analysis based on the first two canonical variates from the spectra of whole apple surface contaminated with nonpathogenic and pathogenic strains of *E. coli* [74].

the nonpathogenic strains overlapped in their clusters. By considering only the nonpathogenic *E. coli* strains in a separate classification by using the first two scores again, different clusters for the strains JM101, JM107 and HB101 without overlapping were also achieved. The clusters of the two other strains, K12 and DB5alpha, still overlap.

The results indicate that Raman spectroscopy is a promising technique, because the sample or the food product can be placed directly under the laser focus and any further sample preparation can be omitted [74].

In 2008, Luo and Lin reported an analysis for discrimination of food-borne pathogens with a dispersive Raman system by using SERS on the basis of silver colloidal nanoparticles for enhanced signals. Spectra of three food-borne pathogens were processed, resulting in clear identification on the basis of their unique spectral bands [81].

All the investigations discussed above have not been introduced in the analytical industry so far, because some obstacles concerning sample preparation and handling practices still have to be overcome. Batt *et al.* pointed out that locating pathogens in complex matrices is not possible by simple physical means, for example, centrifugation [75]. The extraction procedure for bacteria in water differs from that for the same bacteria from juices or milk, depending on the complexity of the matrix composition. Advances in techniques regarding those complex issues will play an important role in food safety and, along with this, in the detection and reduction of illnesses caused by food-borne pathogens.

Ravindranath *et al.*, developed a pathogen sensor for food matrices addressing the above problems regarding the complexity of food matrices. Biofunctionalized magnetic nanoparticles in combination with mid-IR analysis were employed. The nanoparticles were functionalized with antibodies of two well-known food-borne pathogens: *E. coli* 0175:H7 and *Salmonella typhimurium* [82]. The separate isolation of both species from a mixture of other bacteria was achieved and these microorganisms were detected in food matrices by means of FTIR spectroscopy. This advanced technology can also be integrated in a portable device, which enables this approach to be used as a point-of-care technology. Ravindranath *et al.* reported that highly selective detection was achieved in less than 30 min with both investigated strains in complex matrices, such as milk and spinach extract, with a detection limit of 10^4–10^5 CFU ml^{-1}. On-site food-borne pathogen monitoring, with an in-built extraction step using magnetic nanoparticles combined with the specificity of IR spectra, offers the possibility of complementing the conventional and still established microbiological methods with molecular innovative methods [82].

Recent outbreaks of food-borne illnesses were associated with bacterial contamination. Therefore, not only the identification of bacterial cells but also processes of food spoilage by microorganisms concerning, for example, meat or dairy products are investigated. Especially pathogenic microorganisms, such as *Listeria* spp., *Salmonella* spp., *Campylobacter jejuni*, and *Yersinia enterolytica*, trigger food-borne illnesses. It is generally accepted that organoleptic changes result from microbial spoilages and that the compounds responsible for these changes are various metabolites produced by the enzymatic activity and growth of microorganisms [83]. For example, milk is certainly an ideal medium for microbial growth because of the high water content, the large variety of available nutrients, and the almost neutral pH. Nicolaou and Goodacre investigated spoilage in milk using FTIR spectroscopy and chemometrics [83]. They tested ATR and HT (high-throughput) FTIR techniques in combination with principal component–discriminant function analysis (PCA–DFA) and partial least-squares regression (PLSR) in comparison with the total viable cell counts by plating with three different kinds of milk (whole, skimmed, and semi-skimmed). The FTIR data for the different milk samples showed reasonable results for bacterial loads above 10^5 CFU ml^{-1}. However, HT FTIR spectroscopy provided higher accuracy for lower viable bacterial counts down to 10^3 CFU ml^{-1} for whole milk and 4×10^2 CFU ml^{-1} for semi-skimmed and skimmed milk [83]. Summarizing, they pointed out that by using these techniques in combination with PLSR, rapid acquisition of a metabolic fingerprint and accurate quantification of the microbial load of milk samples with reduced sample preparation are feasible.

FTIR spectroscopy was also applied by Ellis *et al.* to detect microbial spoilage of meat [84]. Organoleptic changes, including discoloration, slime formation, changes in taste or other attributes, lower muscle food quality to an unacceptable limit. These changes result from the decomposition and the formation of metabolites caused by bacterial growth. In this context, it is known that spoilage organisms belong primarily to the genus *Pseudomonas* and other major members of spoilage flora, such as *Moraxella* spp., *Psychrobacter* spp., and *Acinetobacter* spp. Ellis *et al.* exploited the information about microbial spoilage of chicken breast by spectroscopic analysis [84]. The spectra were recorded directly from the meat surface and integrated for 60 s. The authors demonstrated that the application of HATR (horizontal attenuated total reflectance) FTIR and PLS analysis are able to acquire a metabolic snapshot and quantify noninvasively the microbial load of food samples accurately and rapidly in 60 s, directly from the sample surface.

5.8 Conclusion

The application of different vibrational spectroscopic techniques such as IR absorption and Raman spectroscopy allows the rapid and reliable identification of microorganisms. Depending on the application different specialized setups can be applied in order to meet the specific requirements. For FTIR absorption spectroscopy either a fast identification routine of bulk samples can be established or by means of an ATR setup or an IR microscope the heterogeneity of a sample e.g. a biofilm can be analysed. Using Raman spectroscopy with excitation wavelength from the ultraviolet over the visible up to the infrared region allows for a screening and identification of microbial bulk samples or single cells which are related to epidemiological studies or food safety. Even the impact of antibiotics on bacteria and their internal mode of action can be addressed.

The variety of applications allows an identification of microorganisms within hours and will therefore play a central role in future bacterial identification routines when a fast diagnosis is required.

References

1 Helm, D., Labischinski, H., Schallehn, G., and Naumann, D. (1991) Classification and identification of bacteria by Fourier-transform infrared spectroscopy. *J. Gen. Microbiol.*, **137**, 69–79.

2 Naumann, D. (2001) FT-infrared and FT-Raman spectroscopy in biomedical research, in *Infrared and Raman Spectroscopy of Biological Materials* (eds. H.-U. Gremlich and B. Yan), Marcel Dekker, New York, pp. 323–378.

3 Kirschner, C., Maquelin, K., Pina, P., Thi, N.A.N., Choo-Smith, L.P., Sockalingum, G.D., Sandt, C., Ami, D., Orsini, F., Doglia, S.M., Allouch, P., Mainfait, M., Puppels, G.J., and Naumann, D. (2001) Classification and identification of Enterococci: a comparative phenotypic, genotypic, and vibrational spectroscopic study. *J. Clin. Microbiol.*, **39**, 1763–1770.

4 Harz, M., Rösch, P., and Popp, J. (2009) Vibrational spectroscopy – a powerful tool for the rapid identification of microbial cells at the single-cell level. *Cytometry A*, **75**, 104–113.

5 Rösch, P., Harz, M., Schmitt, M., and Popp, J. (2005) Raman spectroscopic identification of single yeast cells. *J. Raman Spectrosc.*, **36**, 377–379.

6 Escoriza, M.F., Vanbriesen, J.M., Stewart, S., and Maier, J. (2006) Studying bacterial metabolic states using Raman spectroscopy. *Appl. Spectrosc.*, **60**, 971–976.

7 Harz, M., Rösch, P., Peschke, K.-D., Ronneberger, O., Burkhardt, H., and Popp, J. (2005) Micro-Raman spectroscopical identification of bacterial cells of the genus *Staphylococcus* in dependence on their cultivation conditions. *Analyst*, **130**, 1543–1550.

8 Schuster, K.C., Urlaub, E., and Gapes, J.R. (2000) Single-cell analysis of bacteria by Raman microscopy: spectral information on the chemical composition of cells and on the heterogeneity in a culture. *J. Microbiol. Methods*, **42**, 29–38.

9 Choo-Smith, L.P., Maquelin, K., Van Vreeswijk, T., Bruining, H.A., Puppels, G.J., Thi, N.A.N., Kirschner, C., Naumann, D., Ami, D., Villa, A.M., Orsini, F., Doglia, S.M., Lamfarraj, H., Sockalingum, G.D., Manfait, M., Allouch, P., and Endtz, H.P. (2001) Investigating microbial (micro)colony heterogeneity by vibrational spectroscopy. *Appl. Environ. Microbiol.*, **67**, 1461–1469.

10 Heber, J.R.S. and Orvil, E.A.B. (1952) Infrared spectrophotometry as a means for identification of bacteria. *Science*, **116**, 111–113.

11 Naumann, D. (1998) Infrared and NIR Raman spectroscopy in medical microbiology, infrared spectroscopy: new tool in medicine. *Proc. SPIE*, **3257**, 245–257.

12 Petry, R., Schmitt, M., and Popp, J. (2003) Raman spectroscopy – a prospective tool in the life sciences. *ChemPhysChem*, **4**, 14–30.

13 Rösch, P., Harz, M., Peschke, K.-D., Ronneberger, O., Burkhardt, H., Schuele, A., Schmauz, G., Lankers, M., Hofer, S., Thiele, H., Motzkus, H.-W., and Popp, J. (2006) On-line monitoring and identification of bioaerosols. *Anal. Chem.*, **78**, 2163–2170.

14 Kalasinsky, K.S., Hadfield, T., Shea, A.A., Kalasinsky, V.F., Nelson, M.P., Neiss, J., Drauch, A.J., Vanni, G.S., and Treado, P.J. (2007) Raman chemical imaging

spectroscopy reagentless detection and identification of pathogens: signature development and evaluation. *Anal. Chem.*, **79**, 2658–2673.

15 McHale, J.L. (2002) Resonance Raman spectroscopy, in *Handbook of Vibrational Spectroscopy* (eds. J. Chalmers and P. Griffiths), John Wiley & Sons, Ltd., Chichester, pp. 534–556.

16 Hering, K., Cialla, D., Ackermann, K., Dörfer, T., Möller, R., Schneideweind, H., Mattheis, R., Fritzsche, W., Rösch, P., and Popp, J. (2008) SERS: a versatile tool in chemical and biochemical diagnostics. *Anal. Bioanal. Chem.*, **390**, 113–124.

17 Nie, S. and Emory, S.R. (1997) Probing single molecules and single nanoparticles by surface-enhanced Raman scattering. *Adv. Drug Deliv. Rev.*, **275**, 1102–1106.

18 Naumann, D. (2000) Infrared spectroscopy in microbiology, in *Encyclopedia of Analytical Chemistry* (ed. R.A. Meyers), John Wiley & Sons, Ltd., Chichester, pp. 102–131.

19 Naumann, D., Keller, S., Helm, D., Schultz, C., and Schrader, B. (1995) FT-IR spectroscopy and FT-Raman spectroscopy are powerful analytical tools for the noninvasive characterization of intact microbial-cells. *J. Mol. Struct.*, **347**, 399–405.

20 Gaus, K., Rösch, P., Petry, R., Peschke, K.-D., Ronneberger, O., Burkhardt, H., Baumann, K., and Popp, J. (2006) Classification of lactic acid bacteria with UV–resonance Raman spectroscopy. *Biopolymers*, **82**, 286–290.

21 Jarvis, R.M. and Goodacre, R. (2004) Ultraviolet resonance Raman spectroscopy for the rapid discrimination of urinary tract infection bacteria. *FEMS Microbiol. Lett.*, **232**, 127–132.

22 Lopez-Diez, E.C. and Goodacre, R. (2004) Characterization of microorganisms using UV resonance Raman spectroscopy and chemometrics. *Anal. Chem.*, **76**, 585–591.

23 Maquelin, K., Choo-Smith, L.-P.i., Endtz, H.P., Bruining, H.A., and Puppels, G.J. (2000) Raman spectroscopic studies on bacteria. *Proc. SPIE*, **4161**, 144–150.

24 Maquelin, K., Choo-Smith, L.-P.i., Van Vreeswijk, T., Endtz, H.P., Smith, B., Bennett, R., Bruining, H.A., and Puppels, G.J. (2000) Raman spectroscopic method for identification of clinically relevant microorganisms growing on solid culture medium. *Anal. Chem.*, **72**, 12–19.

25 Rösch, P., Harz, M., Schmitt, M., Peschke, K.-D., Ronneberger, O., Burkhardt, H., Motzkus, H.-W., Lankers, M., Hofer, S., Thiele, H., and Popp, J. (2005) Chemotaxonomic identification of single bacteria by micro-Raman spectroscopy: application to clean-room-relevant biological contaminations. *Appl. Environ. Microbiol.*, **71**, 1626–1637.

26 Jarvis, R.M., Brooker, A., and Goodacre, R. (2006) Surface-enhanced Raman scattering for the rapid discrimination of bacteria. *Faraday Discuss.*, **132**, 281–292.

27 Jarvis, R.M., Brooker, A., and Goodacre, R. (2004) Surface-enhanced Raman spectroscopy for bacterial discrimination utilizing a scanning electron microscope with a Raman spectroscopy interface. *Anal. Chem.*, **76**, 5198–5202.

28 Maquelin, K., Kirschner, C., Choo-Smith, L.P., Ngo-Thi, N.A., van Vreeswijk, T., Stammler, M., Endtz, H.P., Bruining, H.A., Naumann, D., and Puppels, G.J. (2003) Prospective study of the performance of vibrational spectroscopies for rapid identification of bacterial and fungal pathogens recovered from blood cultures. *J. Clin. Microbiol.*, **41**, 324–329.

29 Rösch, P., Harz, M., Peschke, K.-D., Ronneberger, O., Burkhardt, H., and Popp, J. (2006) Identification of single eukaryotic cells with micro-Raman spectroscopy. *Biopolymers*, **82**, 312–316.

30 Krause, M., Rösch, P., Radt, B., and Popp, J. (2008) Localizing and identifying living bacteria in an abiotic environment by a combination of Raman and fluorescence microscopy. *Anal. Chem.*, **80**, 8568–8575.

31 Lieber, C.A. and Mahadevan-Jansen, A. (2003) Automated method for subtraction of fluorescence from biological Raman spectra. *Appl. Spectrosc.*, **57**, 1363–1367.

32 Ryan, C.G., Clayton, E., Griffin, W.L., Sie, S.H., and Cousens, D.R. (1988) SNIP,

a statistics-sensitive background treatment for the quantitative analysis of PIXE spectra in geoscience applications. *Nucl. Instrum. Methods B*, **34**, 396–402.

33 Comon, P. (1994) Independent component analysis, a new concept? *Signal Process.*, **36**, 287–314.

34 Pearson, K. (1901) Principal components analysis. *Philos. Mag.*, **2**, 559–572.

35 Fraley, C. and Raftery, A.E. (1998) How many clusters? Which clustering method? Answers via model-based cluster analysis. *Comput. J.*, **41**, 578–588.

36 Rojas, R. and Feldman, J. (1996) *Neural Networks: a Systematic Introduction*, Springer, Berlin.

37 Willemse-Erix, D.F.M., Scholtes-Timmerman, M.J., Jachtenberg, J.-W., van Leeuwen, W.B., Horst-Kreft, D., Bakker Schut, T.C., Deurenberg, R.H., Puppels, G.J., van Belkum, A., Vos, M.C., and Maquelin, K. (2009) Optical fingerprinting in bacterial epidemiology: Raman spectroscopy as a real-time typing method. *J. Clin. Microbiol.*, **47**, 652–659.

38 Bosch, A., Minan, A., Vescina, C., Degrossi, J., Gatti, B., Montanaro, P., Messina, M., Franco, M., Vay, C., Schmitt, J., Naumann, D., and Yantorno, O. (2008) Fourier transform infrared spectroscopy for rapid identification of nonfermenting Gram-negative bacteria isolated from sputum samples from cystic fibrosis patients. *J. Clin. Microbiol.*, **46**, 2535–2546.

39 Donlan, R.M. (2001) Biofilm formation: a clinically relevant microbiological process. *Clin. Infect. Dis.*, **33**, 1387–1392.

40 Donlan, R.M. (2001) Biofilms and device-associated infections. *Emerg. Infect. Dis.*, **7**, 277–281.

41 Heitz, E., Flemming, H.-C., and Sand, W. (eds.) (1996) *Microbially Influenced Corrosion of Materials: Scientific and Engineering Aspects*, Springer, Berlin.

42 Ojeda, J.J., Romero-Gonzalez, M.E., and Banwart, S.A. (2009) Analysis of bacteria on steel surfaces using reflectance micro-Fourier transform infrared spectroscopy. *Anal. Chem.*, **81**, 6467–6473.

43 Delille, A., Quiles, F., and Humbert, F. (2007) *In situ* monitoring of the nascent *Pseudomonas fluorescens* biofilm response to variations in the dissolved organic carbon level in low-nutrient water by attenuated total reflectance–Fourier transform infrared spectroscopy. *Appl. Environ. Microbiol.*, **73**, 5782–5788.

44 Marcotte, L., Barbeau, J., and Lafleur, M. (2004) Characterization of the diffusion of polyethylene glycol in *Streptococcus mutans* biofilms by Raman microspectroscopy. *Appl. Spectrosc.*, **58**, 1295–1301.

45 Huang, W.E., Ude, S., and Spiers, A.J. (2007) *Pseudomonas fluorescens* SBW25 biofilm and planktonic cells have differentiable Raman spectral profiles. *Microb. Ecol.*, **53**, 471–474.

46 Pätzold, R., Keuntje, M., and Anders-von Ahlften, A. (2006) A new approach to non-destructive analysis of biofilms by confocal Raman microscopy. *Anal. Bioanal. Chem.*, **386**, 286–292.

47 Ivleva, N., Wagner, M., Horn, H., Niessner, R., and Haisch, C. (2009) Towards a nondestructive chemical characterization of biofilm matrix by Raman microscopy. *Anal. Bioanal. Chem.*, **393**, 197–206.

48 Ivleva, N.P., Wagner, M., Horn, H., Niessner, R., and Haisch, C. (2008) In situ surface-enhanced Raman scattering analysis of biofilm. *Anal. Chem.*, **80**, 8538–8544.

49 Schmid, T., Messmer, A., Yeo, B.-S., Zhang, W., and Zenobi, R. (2008) Towards chemical analysis of nanostructures in biofilms II: tip-enhanced Raman spectroscopy of alginates. *Anal. Bioanal. Chem.*, **391**, 1907–1916.

50 Maquelin, K., Cookson, B., Tassios, P., and van Belkum, A. (2007) Current trends in the epidemiological typing of clinically relevant microbes in Europe. *J. Microbiol. Methods*, **69**, 222–226.

51 Bitsori, M., Maraki, S., Raissaki, M., Bakantaki, A., and Galanakis, E. (2005) Community-acquired enterococcal urinary tract infections. *Pediatr. Nephrol.*, **20**, 1583–1586.

52 Hutchison, M.L. and Govan, J.R.W. (1999) Pathogenicity of microbes associated with cystic fibrosis. *Microb. Infect.*, **1**, 1005–1014.

53 Coutinho, C.P., Sá-Correia, I., and Almeida Lopes, J. (2009) Use of Fourier

transform infrared spectroscopy and chemometrics to discriminate clinical isolates of bacteria of the *Burkholderia cepacia* complex from different species and ribopatterns. *Anal. Bioanal. Chem.*, **394**, 2161–2171.

54 Pfaller, M.A. and Diekema, D.J. (2007) Epidemiology of invasive candidiasis: a persistent public health problem. *Clin. Microbiol. Rev.*, **20**, 133–163.

55 Ibelings, M.S., Maquelin, K., Endtz, H.P., Bruining, H.A., and Puppels, G.J. (2005) Rapid identification of *Candida* spp. in peritonitis patients by Raman spectroscopy. *Clin. Microbiol. Infect.*, **11**, 353–358.

56 Maquelin, K., Choo-Smith, L.P., Endtz, H.P., Bruining, H.A., and Puppels, G.J. (2002) Rapid identification of *Candida* species by confocal Raman microspectroscopy. *J. Clin. Microbiol.*, **40**, 594–600.

57 Essendoubi, M., Toubas, D., Lepouse, C., Leon, A., Bourgeade, F., Pinon, J.-M., Manfait, M., and Sockalingum, G.D. (2007) Epidemiological investigation and typing of *Candida glabrata* clinical isolates by FTIR spectroscopy. *J. Microbiol. Methods*, **71**, 325–331.

58 Martin-Casabona, N., Bahrmand, A.R., Bennedsen, J., Ostergaard Thomsen, V., Curcio, M., Fauville-Dufaux, M., Feldman, K., Havelkova, M., Katila, M.L., Koeksalan, K., Pereira, M.F., Rodrigues, F., Pfyffer, G.E., Portaels, F., Rossell-Urgell, J., Ruesch-Gerdes, S., Tortoli, E., Vincent, V., and Watt, B. (2004) Non-tuberculous mycobacteria: patterns of isolation; a multi-country retrospective survey. *Int. J. Tuberc. Lung Dis.*, **8**, 1186–1193.

59 Griffith, D.E., Aksamit, T., Brown-Elliott, B.A., Catanzaro, A., Daley, C., Gordin, F., Holland, S.M., Horsburgh, R., Huitt, G., Iademarco, M.F., Iseman, M., Olivier, K., Ruoss, S., von Reyn, C.F., Wallace, R.J. Jr., Winthrop, K., ATS Mycobacterial Diseases Subcommittee, American Thoracic Society, and Infectious Disease Society of America (2007) An official ATS/IDSA statement: diagnosis, treatment, and prevention of nontuberculous mycobacterial diseases. *Am J. Respir. Crit. Care Med.*, **175**, 367–416.

60 Buijtels, P.C.A.M., Willemse-Erix, H.F.M., Petit, P.L.C., Endtz, H.P., Puppels, G.J., Verbrugh, H.A., van Belkum, A., van Soolingen, D., and Maquelin, K. (2008) Rapid identification of mycobacteria by Raman spectroscopy. *J. Clin. Microbiol.*, **46**, 961–965.

61 Harz, M., Kiehntopf, M., Stöckel, S., Rösch, P., Straube, E., Deufel, T., and Popp, J. (2009) Direct analysis of clinical relevant single bacterial cells from cerebrospinal fluid during bacterial meningitis by means of micro-Raman spectroscopy. *J. Biophotonics*, **2**, 70–80.

62 Krause, M., Radt, B., Rösch, P., and Popp, J. (2007) The investigation of single bacteria by means of fluorescence staining and Raman spectroscopy. *J. Raman Spectros.*, **38**, 369–372.

63 Maquelin, K., Kirschner, C., Choo-Smith, L.P., van den Braak, N., Endtz, H.P., Naumann, D., and Puppels, G.J. (2002) Identification of medically relevant microorganisms by vibrational spectroscopy. *J. Microbiol. Methods*, **51**, 255–271.

64 Lopez-Diez, E.C., Winder, C.L., Ashton, L., Currie, F., and Goodacre, R. (2005) Monitoring the mode of action of antibiotics using Raman spectroscopy: investigating subinhibitory effects of amikacin on *Pseudomonas aeruginosa*. *Anal. Chem.*, **77**, 2901–2906.

65 Liu, T.-T., Lin, Y.-H., Hung, C.-S., Liu, T.-J., Chen, Y., Huang, Y.-C., Tsai, T.-H., Wang, H.-H., Wang, D.-W., Wang, J.-K., Wang, Y.-L., and Lin, C.-H. (2009) A high speed detection platform based on surface-enhanced Raman scattering for monitoring antibiotic-induced chemical changes in bacteria cell wall. *PLoS ONE*, **4**, e5470.

66 Huleihel, M., Pavlov, V., and Erukhimovitch, V. (2009) The use of FTIR microscopy for the evaluation of anti-bacterial agents activity. *J. Photochem. Photobiol. B*, **96**, 17–23.

67 Neugebauer, U., Schmid, U., Baumann, K., Holzgrabe, U., Ziebuhr, W., Kozitskaya, S., Kiefer, W., Schmitt, M., and Popp, J. (2006) Characterization of bacterial growth and the influence of antibiotics by means of UV resonance Raman spectroscopy. *Biopolymers*, **82**, 306–311.

68 Neugebauer, U., Schmid, U., Baumann, K., Simon, H., Schmitt, M., and Popp, J. (2007) DNA tertiary structure and changes in DNA supercoiling upon interaction with ethidium bromide and gyrase monitored by UV resonance Raman spectroscopy. *J. Raman Spectrosc.*, **38**, 1246–1258.

69 Neugebauer, U., Schmid, U., Baumann, K., Ziebuhr, W., Kozitskaya, S., Holzgrabe, U., Schmitt, M., and Popp, J. (2007) The influence of fluoroquinolone drugs on the bacterial growth of *S. epidermidis* utilizing the unique potential of vibrational spectroscopy. *J. Phys. Chem. A*, **111**, 2898–2906.

70 Neugebauer, U., Szeghalmi, A., Schmitt, M., Kiefer, W., Popp, J., and Holzgrabe, U. (2005) Vibrational spectroscopic characterization of fluoroquinolones. *Spectrochim Acta A Mol. Spectrosc.*, **61**, 1505–1517.

71 Neugebauer, U., Schmid, U., Baumann, K., Ziebuhr, W., Kozitskaya, S., Deckert, V., Schmitt, M., and Popp, J. (2007) Towards a detailed understanding of bacterial metabolism: spectroscopic characterization of *Staphylococcus epidermidis*. *ChemPhysChem*, **8**, 124–137.

72 Noda, I. (1993) Generalized two-dimensional correlation method applicable to infrared, Raman, and other types of spectroscopy. *Appl. Spectrosc.*, **47**, 1329–1336.

73 Mead, P.S., Slutsker, L., Dietz, V., McCaig, L.F., Bresee, J.S., Shapiro, C., Griffin, P.M., and Tauxe, R.V. (1999) Food-related illness and death in the United States. *Emerg. Infect. Dis.*, **5**, 607–625.

74 Yang, H. and Irudayaraj, J. (2003) Rapid detection of foodborne microorganisms on food surfaces using Fourier transform Raman spectroscopy. *J. Mol. Struct.*, **646**, 35–43.

75 Batt, C.A. (2007) Materials science: food pathogen detection. *Science*, **316**, 1579–1580.

76 Al-Holy, M.A., Lin, M., Cavinato, A.G., and Rasco, B.A. (2006) The use of Fourier transform infrared spectroscopy to differentiate *Escherichia coli* O157:H7 from other bacteria inoculated into apple juice. *Food Microbiol.*, **23**, 162–168.

77 Lin, M., Al-Holy, M., Al-Qadiri, H., Kang, D.-H., Cavinato, A.G., Huang, Y., and Rasco, B.A. (2004) Discrimination of intact and injured *Listeria monocytogenes* by Fourier transform infrared spectroscopy and principal component analysis. *J. Agric. Food Chem.*, **52**, 5769–5772.

78 Yu, C., Irudayaraj, J., Debroy, C., Schmilovtich, Z., and Mizrach, A. (2004) Spectroscopic differentiation and quantification of microorganisms in apple juice. *J. Food Sci.*, **69**, 268–272.

79 Al-Holy, M.A., Lin, M., Al-Qadiri, H., Cavinato, A.G., and Rasco, B.A. (2006) Classification of foodborne pathogens by Fourier transform infrared spectroscopy and pattern recognition techniques. *J. Rapid Methods Autom. Microbiol.*, **14**, 189–200.

80 Whittaker, P., Mossoba, M.M., Al-Khaldi, S., Fry, F.S., Dunkel, V.C., Tall, B.D., and Yurawecz, M.P. (2003) Identification of foodborne bacteria by infrared spectroscopy using cellular fatty acid methyl esters. *J. Microbiol. Methods*, **55**, 709–716.

81 Luo, B.S. and Lin, M. (2008) A portable Raman system for the identification of foodborne pathogenic bacteria. *J. Rapid Methods Autom. Microbiol.*, **16**, 238–255.

82 Ravindranath, S.P., Mauer, L.J., Deb-Roy, C., and Irudayaraj, J. (2009) Biofunctionalized magnetic nanoparticle integrated mid-infrared pathogen sensor for food matrixes. *Anal. Chem.*, **81**, 2840–2846.

83 Nicolaou, N., and Goodacre, R. (2008) Rapid and quantitative detection of the microbial spoilage in milk using Fourier transform infrared spectroscopy

and chemometrics. *Analyst*, **133**, 1424–1431.
84 David I. Ellis, David Broadhurst, Douglas B. Kell, Jem J. Rowland, and Royston Goodacre (2002) Rapid and quantitative detection of the microbial spoilage of meat by fourier transform infrared spectroscopy and machine learning. *Appl. Environ. Microbiol.*, **68**, 2822–2828.

Part 2
Pathology

6
Cell Sorting
Jochen Guck

6.1
Conventional Techniques in Cell Sorting

The standard technique for the sequential characterization and sorting of cells is fluorescence-activated cell sorting (FACS) [1]. A cell suspension, in which cells of interest are labeled with one or several fluorescent antibodies against specific surface markers, is flowed through a nozzle such that the resulting droplets are charged and each contains a single cell. This stream of droplets then passes a detector that analyzes each drop for the presence of fluorescence. Triggered by a fluorescence signal, the drop is deflected by electric fields and collected. Current high-speed sorters are capable of reliably sorting on the order of 10^4 cells s^{-1} [2].

A variant of fluorescent antibodies is the use of antibodies conjugated with paramagnetic beads. Cells labeled with such beads can then be sorted from a mixed population by magnetic gradients. This magnetic-activated cell sorting (MACS) can process on the order of 10^6 cells s^{-1} [3]. In contrast to FACS, which analyzes many single cells sequentially, MACS is a bulk technique, where cell classification and sorting are achieved in a single step. Bulk sorters generally have a higher throughput rate but are limited in the number of different parameters that can be used simultaneously.

6.2
Photonic Techniques in Single-Cell Sorting

The basis for all photonic techniques used to sort cells is the fact that light carries momentum and can exert forces. In contrast to electric and magnetic forces, optical forces are much weaker because the momentum of photons, and hence their ability to exert forces, are comparatively small. Therefore, lasers are required to achieve photon fluxes sufficient for the manipulation of micron-sized objects, such as cells, with light. A pioneer of such optical manipulation was Arthur Ashkin, who showed in

Handbook of Biophotonics. Vol.2: Photonics for Health Care, First Edition. Edited by Jürgen Popp, Valery V. Tuchin, Arthur Chiou, and Stefan Heinemann.
© 2012 Wiley-VCH Verlag GmbH & Co. KGaA. Published 2012 by Wiley-VCH Verlag GmbH & Co. KGaA.

the early 1970s that small glass beads can be moved by laser beams and that two counterpropagating laser beams can trap small objects in their middle [4]. The optical force has two components: the scattering force pushes the object away from the light source, while the gradient force acts along the gradient of light intensity – in a nonfocused laser beam with Gaussian intensity distribution towards the axis of the beam (see Figure 6.1a).

The most common trap built with optical forces is optical tweezers, which are ideal for single-molecule manipulations and have been used extensively in biophysical studies [5, 6]. In the context of cell characterization and sorting, these optical forces can be used to hold single cells in place during investigation or to move them about for sorting [7]. Here, nonfocused laser beams, such as in dual-beam laser traps, are often better suited than optical tweezers because they allow manipulation of cells of arbitrary size and with reduced light intensities, which could otherwise endanger the integrity of the cell [4, 8–10]. Such optical traps are usefully combined with microfluidics into "lab-on-a-chip" systems for the efficient transport, analysis, and sorting of cells [10–13]. In such lab-on-a-chip systems, cells in suspension are serially delivered into the trapping region where the optical trap (e.g., two counterpropagating laser beams perpendicular to the flow direction) attracts the cell, centers it automatically, and holds it in place for any characterization desired (see Figure 6.1b). While trapped, the cell can be tested for an overall fluorescence signal in analogy to normal FACS machines. In contrast to FACS, however, the cell can be held stationary for an arbitrary number of other assays until a final verdict about the nature of the cell is reached. For example, the cell can be analyzed with microscopy for size, morphology, and structure [9] or with spectroscopy for biochemical information [14]. Optical forces can also be used to rotate the cell [15], so that tomographic information can be obtained, or even to deform the cell in a controlled way to obtain mechanical information [16]. Since the mechanical properties of cells are intricately linked to cell function, this mechanical phenotyping of cells with optical forces represents a powerful new option for cancer diagnosis [17, 18] and stem cell identification [19], and suggests the development of elasticity-activated

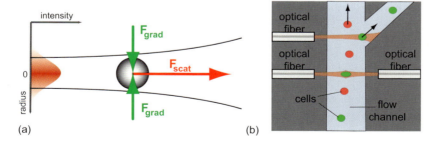

Figure 6.1 Cell sorting with optical forces. (a) Illustration of the direction of scattering and gradient forces acting on a cell in a Gaussian laser beam. (b) Combination of optical forces with a microfluidic system for analysis and sorting. A pair of counterpropagating laser beams, emerging from optical fibers, center and trap single cells sequentially for analysis. A third laser beam then deflects cells of interest into a sorting channel.

cell sorting (EACS) [10]. Once the cell has been identified, it can be released from the optical trap and flowed to the sorting region. Here, another laser beam perpendicular to the flow direction can be used to deflect cells into a sorting channel [20] (see Figure 6.1b). Once sorted, the cells can then be subjected to further analysis such as single-cell polymerase chain reaction (PCR) on the same chip [11] or collected for culture and analysis off-line.

There is also another approach for isolating cells with photonic techniques, called laser microdissection and pressure catapulting (LMPC) [21]. Rather than dissolving tissue to yield a cell suspension, tissue slices can be used directly as a source of cells. Once a cell in the tissue has been identified, for example, using standard histological staining, it is cut out of the tissue slice using a tightly focused UV or IR laser as an optical scalpel. The cut-out cell is then catapulted into a collection reservoir by a laser pulse. LMPC can be used to obtain cells, or even the DNA from within cells, with no contamination because the sample is never in contact with any physical surface – only with light.

6.3
Photonic Techniques in Bulk Sorting

Optical forces can also be used for bulk sorting, where cells are not analyzed serially but analysis is intimately coupled to the sorting process and many cells are sorted simultaneously. MACS is one such bulk technique. One purely optical approach to bulk sorting analyzes the size or optical properties of cells. A cell in an optical lattice, for example, generated by interfering two laser beams, feels a higher deflection force when its refractive index is higher or when its size is comparable to that of the interference pattern spacing. Larger cells, or cells with a lower refractive index, are less affected. When a cell suspension is flowed through such an extended lattice, cells are sorted based on these parameters [22]. The intensity pattern can also be generated by the use of a spatial light modulator [23] or nondiffracting Bessel beams [24]. The throughput of this optical fractionation technique can exceed that of FACS sorters, with a sorting efficiency approaching 100% [22], but obviously at the cost of the amount of specific information used for the distinction. One application of bulk optical fractionation is the sorting of white blood cells from whole blood as they have different size and refractive index to red blood cells. Although sorting by size can also be achieved with micropillar arrays [25], which let smaller cells pass while larger cells are retained, there is an obvious advantage to the optical technique in that it can never clog up. Once the light has been turned off, the sorting mechanism is gone.

In summary, optical forces have their place in both single-cell and bulk sorting applications. In sequential single-cell sorting, the main advantage over conventional sorting approaches such as FACS is that an individual cell can be held stationary by optical forces for high-content analysis, leading to very high purity of the sorted cells at the cost of high throughput. In bulk sorting, optical techniques can operate at sorting rates even in excess of other standard techniques and offer, in addition to size, new sorting parameters such as refractive index that have not yet been explored.

In both cases, optical forces are generally very mild as there is no physical contact involved that could cause unwanted cell activation or alteration. Some 50 years after the invention of the laser, optical cell sorting is another field where its application is set to increase.

References

1 Orfao, A. and Ruiz-Arguelles, A. (1996) General concepts about cell sorting techniques. *Clin. Biochem.*, **29**, 5–9.
2 Ibrahim, S.F. and van den Engh, G. (2003) High-speed cell sorting: fundamentals and recent advances. *Curr. Opin. Biotechnol.*, **14**, 5–12.
3 Miltenyi, S., Müller, W., Weichel, W., and Radbruch, A. (1990) High gradient magnetic cell separation with MACS. *Cytometry*, **11**, 231–238.
4 Ashkin, A. (1970) Acceleration and trapping of particles by radiation pressure. *Phys. Rev. Lett.*, **24**, 156–159.
5 Ashkin, A. (1986) Observation of a single-beam gradient force optical trap for dielectric particles. *Opt. Lett.*, **11**, 288–290.
6 Kuo, S.C. (2001) Using optics to measure biological forces and mechanics. *Traffic*, **2**, 757–763.
7 Jonas, A. and Zemanek, P. (2008) Light at work: the use of optical forces for particle manipulation, sorting, and analysis. *Electrophoresis*, **29**, 4813–4851.
8 Constable, A., Kim, J., Mervis, J., Zarinetchi, F., and Prentiss, M. (1993) Demonstration of a fiberoptic light-force trap. *Opt. Lett.*, **18**, 1867–1869.
9 Grover, S.C., Skirtach, A.G., Gauthier, R.C., and Grover, C.P. (2001) Automated single-cell sorting system based on optical trapping. *J. Biomed. Opt.*, **6**, 14–22.
10 Lincoln, B., Schinkinger, S., Travis, K., Wottawah, F., Ebert, S., Sauer, F., and Guck, J. (2007) Reconfigurable microfluidic integration of a dual-beam laser trap with biomedical applications. *Biomed. Microdevices*, **9**, 703–710.
11 Sims, C.E. and Allbritton, N.L. (2007) Analysis of single mammalian cells on-chip. *Lab Chip*, **7**, 423–440.
12 Eriksson, E., Scrimgeour, J., Graneli, A., Ramser, K., Wellander, R., Enger, J., Hanstrop, D., and Goksor, M. (2007) Optical manipulation and microfluidics for studies of single cell dynamics. *J. Opt. A Pure Appl. Opt.*, **9**, S113–S121.
13 Bellini, N., Vishnubhatla, K.C., Bragheri, F., Ferrara, L., Minzioni, P., Ramponi, R., Cristiani, I., and Osellame, R. (2010) Femtosecond laser fabricated monolithic chip for optical trapping and stretching of single cells. *Opt. Express*, **18**, 4679–4688.
14 Jess, P.R.T., Garcés-Chávez, V., Smith, D., Mazilu, M., Paterson, L., Riches, A., Herrington, C.S., Sibbett, W., and Dholakia, K. (2006) Dual beam fibre trap for Raman micro-spectroscopy of single cells. *Opt. Express*, **14**, 5779–5791.
15 Kreysing, M.K., Kiessling, T., Fritsch, A., Dietrich, C., Guck, J.R., and Käs, J.A. (2008) The optical cell rotator. *Opt. Express*, **16**, 16984–16992.
16 Guck, J., Ananthakrishnan, R., Mahmood, H., Moon, T.J., Cunningham, C.C., and Käs, J. (2001) The optical stretcher: a novel laser tool to micromanipulate cells. *Biophys. J.*, **81**, 767–784.
17 Guck, J., Schinkinger, S., Lincoln, B., Wottawah, F., Ebert, S., Romeyke, M., Lenz, D., Erickson, H.M., Ananthakrishnan, R., Mitchell, D., Käs, J., Ulvick, S., and Bilby, C. (2005) Optical deformability as an inherent cell marker for testing malignant transformation and metastatic competence. *Biophys. J.*, **88**, 3689–3698.
18 Remmerbach, T.W., Wottawah, F., Dietrich, J., Lincoln, B., Wittekind, C., and Guck, J. (2009) Oral cancer diagnosis by mechanical phenotyping. *Cancer Res.*, **69**, 1728–1732.
19 Lautenschläger, F., Paschke, S., Schinkinger, S., Bruel, A., Beil, M., and Guck, J. (2009) The regulatory role of cell mechanics for migration of

differentiating myeloid cells. *Proc. Natl. Acad. Sci. U. S. A.*, **106**, 15696–15701.

20 Wang, M.M., Tu, E., Raymond, D.E., Yang, J.M., Zhang, H., Hagen, N., Dees, B., Mercer, E.M., Forster, A.H., Kariv, I., Marchand, P.J., and Butler, W.F. (2005) Microfluidic sorting of mammalian cells by optical force switching. *Nat. Biotechnol.*, **23**, 83–87.

21 Schütze, K., Niyaz, Y., Stich, M., and Buchstaller, A. (2007) Noncontact laser microdissection and catapulting for pure sample capture. *Methods Cell Biol.*, **82**, 649–673.

22 MacDonald, M.P., Spalding, G.C., and Dholakia, K. (2003) Microfluidic sorting in an optical lattice. *Nature*, **426**, 421–424.

23 Ladavac, K., Kasza, K., and Grier, D.G. (2004) Sorting mesoscopic objects with periodic potential landscapes: optical fractionation. *Phys. Rev. E*, **70**, 010901.

24 Paterson, L., Papagiakoumou, E., Milne, G., Garces-Chavez, V., Briscoe, T., Sibbett, W., Dholakia, K., and Riches, A.C. (2007) Passive optical separation within a 'nondiffracting' light beam. *J. Biomed. Opt.*, **12**, 054017–054013.

25 Panaro, N.J., Lou, X.J., Fortina, P., Kricka, L.J., and Wilding, P. (2005) Micropillar array chip for integrated white blood cell isolation and PCR. *Biomol. Eng.*, **21**, 157–162.

7
Laser Microtomy

Holger Lubatschowski, Fabian Will, Sabine Przemeck, and Heiko Richter

7.1
Cutting by Optical Breakdown

The key technology of a laser microtome is an ultrashort (femtosecond) pulsed laser, emitting light in the near-infrared (NIR) range. Due to the rapid development of commercially available turn-key laser systems in recent years, the use of ultrafast laser technology has become much easier and has opened up new applications in the life sciences field.

Laser light in the NIR range around 1000 nm is well suited for the processing of biological material, since most biological tissues have a very low absorption coefficient at this wavelength. Hence manipulation of tissue is not just limited to the surface, but can even be performed inside the material.

While cutting, the laser beam of a laser microtome is tightly focused into the specimen by a high numerical aperture objective. Due to the extreme intensities of up to $1\,TW\,cm^{-2}$ inside the focus, multi-photon absorption causes ionization of the tissue. This process is called optical breakdown and leads to the formation of plasma [1, 2]. The rapid expansion of the plasma causes disruption of the tissue and is responsible for the cutting process (Figure 7.1).

If the pulse duration is sufficiently short (100–500 fs) and the diameter of the focal spot is nearly diffraction limited (1 µm), only a very low pulse energy of ~10–100 nJ is needed to induce the optical breakdown. This limits the interaction range to diameters below 1 µm. The material separation only takes place within the focal region. Outside this region, no thermal or mechanical damage can be detected.

7.2
The Laser Microtome at a Glance

The main component of a typical laser microtome (Figure 7.2) is a high-power femtosecond oscillator emitting ultrashort laser pulsing with a wavelength of 1030 nm, a pulse duration of about 300 fs, and a pulse repetition rate variable

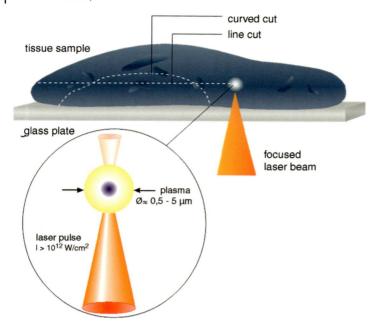

Figure 7.1 Working principle of the laser microtome. A laser-induced plasma is used to cut the tissue without any mechanical forces being applied. The NIR radiation of the laser penetrates the tissue up to 1 mm.

ranging from some kHz to 10 MHz. The maximum average power is ∼2.5 W, where laser pulse energies of up to 100 nJ were focused into the sample with a high numerical aperture (>0.5).

The laser beam is deflected by a fast scanner device for lateral deflection whereas the z-direction is controlled by a piezo-driven positioning stage. Parameters such as pulse overlap, slice thickness, and the size of the cutting area can be set by the user. Typically, an area of up to 14×14 mm can be processed. Larger areas are possible. The cutting speed depends on the properties of the sample, $1\,\text{mm}^2\,\text{s}^{-1}$ being typical. At present slices of 5–100 μm thickness are generated. Thinner slices are also possible, but handling of such tissue slices following removal of the sample bulk is more difficult. For light microscopic examinations, a thickness of 5–10 μm is in most cases sufficient. However, thicker slices are of interest for, for example, primary culture, neurobiological experiments, and pretreatment of samples for ultra microtomy.

7.3
Preparing and Sectioning the Tissue Samples

Because no mechanical forces are applied to the sample, tissue can be processed in its native state and does not need prior fixation, embedding, or freezing. The tissue

7.3 Preparing and Sectioning the Tissue Samples

Figure 7.2 The laser microtome as a stand-alone system.

sample is placed on a conventional glass microscope slide. In order to achieve good results, the tissue surface has to be plane with good optical contact with the glass slide. If needed, a small drop of liquid is helpful to assist refractive index matching. Saline solutions are as good as most of the media used for cell culture (Figure 7.3).

The software allows three cutting modes: cutting of the whole field (maximum size and maximum time), of a defined rectangular field (e.g., 5×6 mm), or of a customized field by setting position markers. According to the precision of the z-axis, the thickness of the slices can be controlled with an accuracy of better than 3 µm.

During the cutting process, a live video of the sample, that is, the cutting plane, is shown on the control screen. This allows the user to observe whether the cutting procedure is successful or not. A rapidly growing field of microbubbles is a typical indication of the occurrence of photodisruption and therefore a sign of correct cutting. After the cutting process, the tissue section is separated from the

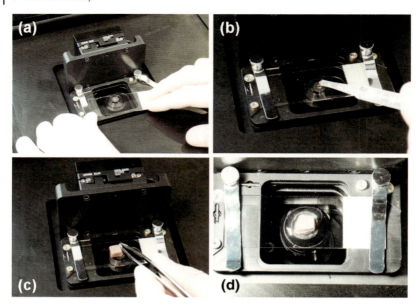

Figure 7.3 Preparation of the sample. (a) A conventional microscopic glass slide acts a tissue holder. (b) A small drop of liquid is helpful to assist refractive index matching. (c) The fresh tissue is placed on the glass slide and (d) a built-in camera of the laser microtome helps to find the area of interest.

bulk with tweezers and placed on a glass slide. Now it can be stained and covered if necessary.

7.4
OCT Controlled 3D Processing

Cutting of flat sections by femtosecond laser technology is only the first step in a new direction in laser microtomy. As a next step, Optical coherence tomography (OCT) imaging was implemented in the system. OCT [3, 4] is a method for imaging different layers of transparent or scattering tissue by scanning the area of interest with a low coherent light source. The measuring principle is similar to that of ultrasound imaging. As a typical light source, superluminescent diodes or even the femtosecond laser pulses themselves with a broad emission band and a corresponding low coherence length can be used. The penetration depth of the NIR radiation is up to 2 mm into the tissue. Due to differences in their optical pathlengths, photons reflected from different layers inside the tissue can be distinguished by interference measurements with an external reference plane. Figure 7.4 shows an OCT image of corneal tissue where the cut was navigated right below the epithelium.

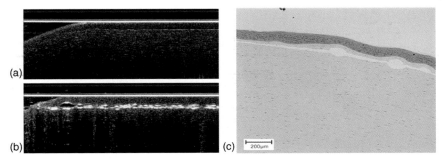

Figure 7.4 Example of an OCT-guided cut below the epithelium of a porcine cornea. (a) OCT image before the cut. The picture is upside down. The bright horizontal line represents the surface of the glass plate. (b) The laser-induced gas bubbles can be monitored as the cutting line right below the epithelium. The bubbles disappear within minutes. (c) Histological section of the cornea after the cut.

Having such a tomographic imaging system implemented, real 3D cutting near the surface of tissue samples is possible, targeting specific areas or volumes of interest.

7.5
Tissue Samples

In principle, every soft tissue with the exception of melanin-containing tissue can be processed, since the 1040 nm radiation is barely absorbed by tissue chromophores. Hard tissues such as bones or of dental origin are also suitable for cutting without decalcifying them. For hard tissues, embedding may be necessary, especially if the tissue is very brittle. The time-consuming preparation of ground sections of embedded hard tissues is therefore shortened. In Figure 7.5, sections of different hard and soft tissue are shown as an example of laser-processed sectioning. The samples show no thermal or mechanical alterations and are well suited for further light microscopic examination. Typical thickness of these slices was 20 μm. Even plastic material can be cut, provided that it has a bright color (Figure 7.6).

7.6
Not for Routine but for Special Cases

Of course, the laser microtome is not a device for routine pathology, as the procedure of tissue separation is too slow. However, it has its clear advantages over standard procedures when preparing tissue in its native state. For investigations based on immunostaining, it can be desirable to keep tissue alive. Furthermore, there are

7 Laser Microtomy

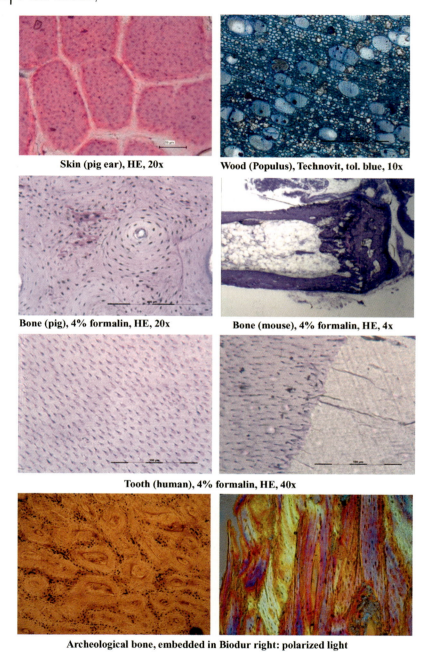

Skin (pig ear), HE, 20x

Wood (Populus), Technovit, tol. blue, 10x

Bone (pig), 4% formalin, HE, 20x

Bone (mouse), 4% formalin, HE, 4x

Tooth (human), 4% formalin, HE, 40x

Archeological bone, embedded in Biodur right: polarized light

Figure 7.5 Sections of different hard and soft tissues are shown as an example of laser-processed sectioning. The samples show no thermal or mechanical alterations and are well suited for further light microscopic examination. Typical thickness of these slices was 20 μm.

Figure 7.6 Example of Teflon blocks, which were cut into 20 μm slices.

several applications which are, at present, difficult or impossible to realize with conventional sectioning methods, including the cutting of hard tissue, plants, wood, and other materials. The laser microtome might be a good option for these applications. It has the potential to solve different problems in various research fields from plant biology to regenerative medicine.

References

1 Vogel, A. and Venugopalan, V. (2003) Mechanisms of pulsed laser ablation of biological tissues. *Chem. Rev.*, **103**, 577–644.

2 Vogel, A., Noack, J., Hüttman, G., and Paltauf, G. (2005) Mechanisms of femtosecond laser nanosurgery of cells and tissues. *Appl. Phys. B*, **81**, 1015–1047.

3 Huang, D., Swanson, E.A., Lin, C.P., Schuman, J.S., Stinson, W.G., Chang, W., Hee, M.R., Flotte, T., Gregory, K., Puliafito, C.A., and Fujimoto, J.G. (1991) Optical coherence tomography. *Science*, **254** (5035), 1178–1181.

4 Drexler, W. and Fujimoto, J.G. (2008) *Ultrahigh Resolution Optical Coherence Tomography*, Springer, Berlin.

8
Classical Microscopy

Kerstin Roeser, Dietlind Zuehlke, Martin Gluch, Gernoth Grunst, Markus Sticker, Klaus Bendrat, and Axel Niendorf

8.1
Appearance

Classical microscopy relates to light microscopy, that is, incident or transmitted light with a magnification range from $10\times$ to $1000\times$. This technique has a wide range of applications in the life and material sciences. Here we focus on the role played by classical microscopy in medicine, especially in surgical pathology. However, it is recognized that there is considerable overlap between pathology, anatomy, and the broad field of cell biology.

8.2
Application

Pathology is one of the fundamental diagnostic disciplines. Tissues and cells can be viewed under a microscope and thereby be classified in an order that discriminates between normal, disorders of metabolism, malformation, inflammation, and tumor. In the case of tumor diagnosis, a subclassification into benign and malignant groups is of crucial importance for therapeutic procedures.

8.3
Pathology

Today, more than 90% of a pathologist's work load concerns living patients. One of the most important tasks is to diagnose and classify tumors, which apart from few exceptions is still done by microscopy. The basic requirement is a representative small tissue specimen followed by a optimal processing with the option of subsequent use. After applying selected visualization techniques to give the tissue some contrast or to mark specific molecules, specimens are all examined by means of light microscopic methods. The diagnosis will then be the basis for the risk assessment and also any therapeutic decisions. However, taking into account the fact that a biopsy always

Handbook of Biophotonics. Vol.2: Photonics for Health Care, First Edition. Edited by Jürgen Popp, Valery V. Tuchin, Arthur Chiou, and Stefan Heinemann.
© 2012 Wiley-VCH Verlag GmbH & Co. KGaA. Published 2012 by Wiley-VCH Verlag GmbH & Co. KGaA.

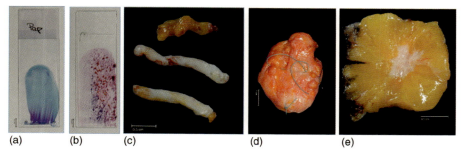

Figure 8.1 Biopsy techniques: (a) cytologic smear; (b) fine needle aspirate (FNA); (c) core biopsies; (d) biopsy; (e) wide excision section.

represents a sample and the microscopic image shows only a section, the need for contextual information leading to a general finding is obvious [1].

This chapter addresses primarily the nonmedical reader and is intended to inform about techniques and procedures in the field of classical microscopy. This is demonstrated exemplarily for breast cancer diagnostics.

8.4
Materials

For diagnostic purposes, tissue samples are used for microscopic investigations. These samples are taken by *biopsy*, which is the removal of cells, tissue, or fluid for diagnosis of diseased tissue.

There are two main techniques to be distinguished: the closed techniques, including *fine needle aspiration* (FNA) and *core needle biopsy*, and the open technique. FNA is a type of biopsy that removes cells trough suction using a syringe. A core needle biopsy is done with a larger needle. Biopsies in the classical sense are the removal of tissue. When a malignancy is suspected, then the tissue specimen is marked to define the margins. However, all techniques correspond to common principles, namely obtaining sufficient tissue for diagnosis and using the least traumatic method (see Figure 8.1).

8.5
Methods (Tissue Processing)

Tissue processing in diagnostic pathology comprises fixation, dehydration, embedding, sectioning, staining, and cover slipping.

Chemical fixatives are used to preserve tissue from degradation and to maintain the structure of cells and their subcellular components. There are five groups of fixatives based on different mechanisms (Table 8.1). The choice of fixative is based on the type of tissue and the histological details to be demonstrated.

Table 8.1 Fixatives used in histology.

Substance	Application	Concentration/content	Advantages/disadvantages	Mechanism of action
Aldehydes	Surgical pathology; autopsy tissue; routine light microscopy	4% formaldehyde in phosphate-buffered saline	Penetrate tissue well, act relatively slowly	Cross-linking proteins
Glutaraldehydes	Electron microscopy	2.5% glutaraldehyde in phosphate-buffered saline	Penetrate poorly; act very quickly; give best overall cytoplasmic and nuclear detail	Deformation of the α-helix structure in proteins
Mercurials	Reticuloendothelial tissue (lymph nodes, spleen, thymus, bone marrow)	Mercuric chloride in Zenker's or B-5 fixative	Penetrate relatively poorly, act fast, give excellent nuclear detail	Unknown
Alcohols	Cytologic smears	95% ethanol	Penetrate cells well, act quickly, give good nuclear detail	Protein denaturation
Oxidizing agents	Electron microscopy; only for special applications	Permanganate, dichromate, osmium tetraoxide	Penetrate relatively poorly	Cross-linking proteins
Picrates	Testis, gastrointestinal tract, endocrine tissue, hematopoietic tissue	Picric acid in Bouin's solution	Penetrate relatively poorly, give good nuclear detail	Unknown

In addition to the nature of the fixatives, there are some critical steps that influence the quality of fixation. First, there must be buffering capacity in the fixative to prevent excessive acidity. Then, the thickness of the tissue sample should be adjusted to the penetration properties of the particular fixative. The fixative will be exhausted during the process, so a 10:1 volume ratio of fixative to tissue is required. Alternatively, interval-type changes of the fixative can solve the problem. By increasing the temperature, the tissue will be fixed faster, but excessive heat will cause artifacts. The concentration of the fixative should be adjusted to the lowest possible level because too high concentrations may adversely affect the tissue and produce artifacts. Also important is the time interval from removal of tissue to fixation. A prolonged dwell time causes artifacts by drying and also the loss of cellular organelles, more nuclear shrinking, and artifactual clumping.

In order to produce microscopic specimen slides, the fixed tissue has to be embedded. Paraffin wax is most frequently used. It is similar in density to tissue, allowing sufficiently thin sections to be cut. Since water and paraffin are immiscible, all traces of water must be removed from tissue. This *dehydration* process is done by passing the tissue through a series of increasing alcohol concentrations. This is followed by a hydrophobic clearing agent (xylene) to remove the alcohol, and finally the infiltration agent (molten paraffin wax), which replaces the xylene. The tissue samples are now ready for external embedding.

During the *embedding* process, the tissue samples are placed in molds along with liquid embedding material, which is then hardened by cooling. The hardened blocks containing the tissue samples are then ready to be sectioned.

Formalin-fixed, paraffin-embedded (FFPE) tissues may be stored at room temperature, and nucleic acids (both DNA and RNA) may be recovered from them decades after fixation, making FFPE tissues an important resource for retrospective studies in medicine. Paraffin-embedded tissues are routinely stored in pathology archives.

Cutting the embedded tissue into sections is done with a microtome. Microtomes have a mechanism for either advancing the block across the knife (rotary microtome) or the knife across the block (sled microtome). *Sectioning* takes much skill and practice and it is important to have properly fixed and embedded tissue to avoid artifacts (tearing, ripping, holes, folding). Sections are floated on warm water to remove wrinkles and mounted on glass microscope slides.

Direct sectioning of a frozen tissue is performed with a cryomicrotome. This method does not need tissue fixation and embedding and therefore has the advantages of speed, maintenance of most enzymes and immunological functions, and relative ease of handling. It has the disadvantage that ice crystals formed during the freezing process will distort the image of the cells. Frozen sections are necessary to obtain rapid diagnosis of a pathological process during surgical procedures (see Figure 8.2).

Figure 8.2 Tissue processing.

8.6 Histological Staining

Staining is employed in both to give contrast to the tissue and to highlight particular features of interest.

To allow water-soluble dyes to penetrate the sections, the paraffin wax has to be removed from the tissue. This is done by running the sections through xylene to alcohol to water. Various dyes are available with different abilities to stain cellular and tissue components. The choice of stain is based on the diagnostic need.

Routine staining is carried out with hematoxylin and eosin (H&E), which is the most commonly used stain in light microscopic histology and histopathology. Hematoxylin, a basic dye, stains nuclei blue, and eosin, an acidic dye, stains the cytoplasm pink.

For special staining, various techniques have been used to stain cells and cellular components selectively; for more details, see Sternberg *et al.* [2]. Stains that are commonly used in breast cancer diagnosis are listed in Table 8.2.

For *immunohistochemistry*, antibodies have been used to visualize specifically proteins, carbohydrate, and lipids. This technique allows the identification of categories of cells under a microscope; more details can be found elsewhere [3–6]. A list of antibodies is given in Table 8.3.

In situ hybridization is used to identify specific DNA or RNA molecules either with fluorescence probes or otherwise labeled nucleic acid probes. To detect fluorescent signals, fluorescence microscopy (see the next section) and confocal microscopy are used, giving good intracellular details [7, 8]; for more details, see Table 8.4.

Table 8.2 Histological stainings used in breast cancer diagnostics.

H&E	General overview staining for histological specimens
	Nucleus, blue. Cytoplasm, pink. Red blood cell, orange/red. Collagen fiber, pink
Giemsa	General overview staining for cytological specimens and used to differentiate cells present in hematopoietic tissue (lymph nodes)
	Nucleus, purple–red. Cytoplasm, light blue
EvG (Elastica van Gieson)	Staining to differentiate different types of connective tissue in histological sections
	Elastic fibers, blue. Nucleus, red. Collagen, light pink. Muscle fibers, light red
PAS (periodic acid–Schiff reaction)	Common staining technique in histology for localizing glycogens and carbohydrates
	Membranes and mucus, magenta. Nucleus, blue. Cytoplasm, light pink
Iron	Staining to demonstrate ferric iron in tissue sections
	Iron, blue. Nucleus, red. Background, pink

Table 8.3 Antibodies used for immunohistological staining in breast cancer diagnostics.

Cytokeratins	Monoclonal anti-cytokeratins are specific markers of epithelial cell differentiation and have been widely used as tools in tumor identification and classification. They facilitate typing of normal, metaplastic, and neoplastic cells. They are also useful in detecting micrometastases in lymph nodes, bone marrow, and other tissues and for determining the origin of poorly differentiated tumors
BCL2	BCL2 is an integral outer mitochondrial membrane protein that suppresses apoptosis in a variety of cell systems, including factor-dependent lymphohematopoietic and neural cells. It regulates cell death by controlling the mitochondrial membrane permeability. Heterodimerization with BAX requires intact BH1 and BH2 domains, and is necessary for anti-apoptotic activity
HER2	ErbB 2 is a receptor tyrosine kinase of the ErbB 2 family. It is closely related in structure to the epidermal growth factor receptor. ErbB 2 oncoprotein is detectable in a proportion of breast and other adenocarcinomas, and also transitional cell carcinomas. In the case of breast cancer, expression determined by immunohistochemistry has been shown to be associated with poor prognosis
ER (estrogen receptor)	The estrogen receptor (ER) is a 66 kDa protein that mediates the actions of estrogens in estrogen-responsive tissues. It is a member of a large superfamily of nuclear hormone receptors that function as ligand-activated transcription factors. The ER is an important regulator of growth and differentiation in the mammary gland. The presence of ER in breast tumors indicates an increased likelihood of response to anti-estrogen (e.g., tamoxifen) therapy
PR (progesterone receptor)	The human progesterone receptor (PR) is a member of the steroid family of nuclear receptors. It has been proposed that expression of PR determination indicates a responsive ER pathway, and therefore may predict likely response to endocrine therapy in human breast cancer. A number of studies have shown that PR determination provides supplementary information to ER, both in predicting response to endocrine therapy and estimating survival. PR has proved superior to ER as a prognostic indicator in some studies

Table 8.4 Selection of probes used for *in situ* hybridization in cancer diagnostics.

ERBB2	Epidermal growth factor (EGF) receptor that enhances kinase-mediated activation of downstream signaling pathways. Amplification and/or overexpression of this gene has been reported in numerous cancers, including breast and ovarian tumors
TOP2A	DNA topoisomerase, involved in processes such as chromosome condensation, chromatid separation, and the relief of torsional stress that occurs during DNA transcription and replication. The TOP2A gene functions as the target for several anticancer agents and a variety of mutations in this gene have been associated with the development of drug resistance
cMYC	cMYC functions as a transcription factor that regulates transcription of specific target genes. Mutations, overexpression, rearrangement, and translocation of this gene have been associated with a variety of hematopoietic tumors, leukemias, and lymphomas, including Burkitt lymphoma
CCND1	Cyclins function as regulators of CDK kinases. Cyclin D1 has been shown to interact with tumor suppressor protein Rb. Mutations, amplification, and overexpression of this gene, which alters cell cycle progression, are observed frequently in a variety of tumors and may contribute to tumorigenesis
EGFR	EGFR is a cell surface protein that binds to epidermal growth factor. Binding of the protein to a ligand induces receptor dimerization and tyrosine autophosphorylation and leads to cell proliferation. Mutations in this gene are associated with lung cancer
ESR1	The estrogen receptor, a ligand-activated transcription factor, essential for sexual development and reproductive function, but also plays a role in other tissues such as bone. Estrogen receptors are also involved in pathological processes including breast cancer, endometrial cancer, and osteoporosis
MDM2	MDM2 is a target gene of the transcription factor tumor protein p53. Overexpression of this gene can result in excessive inactivation of tumor protein p53, diminishing its tumor suppressor function. This protein also affects the cell cycle, apoptosis, and tumorigenesis through interactions with other proteins

8.7 Microscopy

The microscope is the enabling tool for the pathologist when it comes to morphological analysis, as it allows viewing of the patient sample from overview up to subcellular resolution. The basic equipment has remained the same for more than 100 years: the compound microscope comprises sample illumination with transmitted light, an objective revolver allows the primary magnification range to be selected between $1\times$ and $100\times$ and the ocular lens delivers the magnified image to the observer. Together with the ocular lens, the total magnification range extends from $10\times$ to $1000\times$, which corresponds to a field of view between 2.3 mm and 23 µm. In addition to the magnification, the resolution is also important for seeing small object details. The resolution is dependent on the numerical aperture of the objective and illumination, and object structures down to 0.25 µm can be resolved.

8 Classical Microscopy

In order to make the microscope workplace for the pathologist an efficient tool, many technological improvements have been made over time:

- automatic alignment of the microscope to maintain the highest image fidelity over time
- ocular lenses with a large field of view which allow one to look at large sample areas at high magnification
- ergonomic improvements to minimize stress on neck and hands during day to day operation
- LED illumination with long lifetime and high stability (Figure 8.3).

In addition to these ergonomic additions there are two major additions:

1) With the growing importance of molecular markers, fluorescence has entered routine pathology. It allows visualization of multiple biomarkers on the same sample region, which increases the information density. However, the downside compared with colorimetric staining techniques is that the tissue morphology

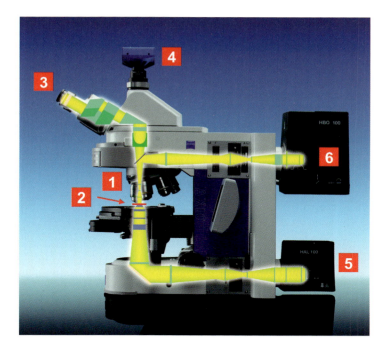

Figure 8.3 Course of beam in light microscopy. In the bright field beam path the objective (1) magnifies the object (2), which can be viewed through the ocular lens (3) or imaged on to a camera (4). The object is illuminated from a white light source (5). In the fluorescence beam path, the sample is illuminated through the objective with a certain wavelength (6), which is absorbed by the fluorophore. The fluorophore emits a longer wavelength which is collected from the objective. After the objective, a dichroic mirror reflects the excitation light and transmits the emission. Only the molecular marker becomes visible for the pathologist.

is not available. So far, fluorescence has been deployed mostly in areas where this is not of crucial importance (e.g., chromosome analysis).
2) Documentation with a digital camera is now an instant easy-to-use option to capture a region of interest. Most of the laboratory information systems today support image attachments, which makes it easy to share images between pathologists for a second opinion.

In the case of unstained samples the pathologist has some contrast techniques that permit the examination of the morphology without staining the sample:

- Dark field contrast shows sample structures where light is scattered or diffracted.
- Polarization contrast shows sample structures due to differences in the birefringent and depolarizing properties of the sample.
- Differential interference contrast and phase contrast make visible small structures that have different refractive indices.

8.8 Diagnostics

8.8.1 Context-Dependent Aspects in Morphological Diagnostics

Diagnostic procedures in surgical pathology are based on the total clinical information and morphological examination at the gross and microscopic levels. The microscopic examination starts with a low magnification to obtain an overview of the whole specimen followed by a selection of areas of interest to be examined in more detail at a higher magnification. Histopathology is a "precision tool" because pathologists are confined to certain (morphological) criteria to define entities. These entities are listed in monographs or, for instance in the case of tumor diagnostics, in series such as the WHO and the Armed Forces Institute of Pathology series, which again are based on the scientific literature. In a practical sense, the pathologist addresses a case with knowledge about the possible single criteria and also the context in which these criteria are likely to occur. In modern pathology, in addition to morphological criteria, biochemical, immunological, and molecular findings become an essential part of the diagnosis. Furthermore, there has been growing interest in tools able to reduce human subjectivity and improve workload. Whole slide scanning technology combined with automated image analysis can offer the capacity to generate fast and reliable results (Figures 8.4–8.7).

8.8.2 Cognitive Aspects of Digital Microscopy

Digital microscopy (DM) can enhance the potential to derive light-based information about reality. Combinable IT modules enable support of human information processing.

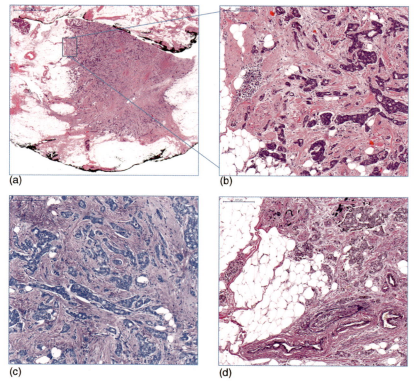

Figure 8.4 Histological analysis. (a) H&E staining is used to identify tumor. Here is shown a classical invasive ductal carcinoma of the breast (G1pT1cpN0V0R0) (×1); (b) H&E staining of the same section (x10); (c) Giemsa staining for improved demonstration of cytological characteristics (×10); (d) EvG staining for more detailed differentiation of connective tissue (×10).

Human experts such as pathologists are forced to focus on the visual data overflow in order to derive relevant information. In contrast, computer-based data handling can take into account extensive and vastly more heterogeneous data. For instance, it is technically possible to merge various images of tissue slices with different staining and to use the integrated information in a common coordinate system.

The amount of data that is produced in this way is huge. Taking into account four different stainings with about 1.5 billion pixels per image (which is the average size of a digital slide with an optical resolution of 40), and relate for image analysis purposes each pixel to a surrounding neighborhood of about 1 cm^2, the resulting data volume is approximately 22 million terabytes! On the other hand, without integration of the analyses into human comprehension processing, the impact of this kind of technical support is without effect in life science research and pathological routine (Figure 8.8).

Digital images in this respect provide a very promising starting point. They can be condensed into semantic units through the application of machine learning-based

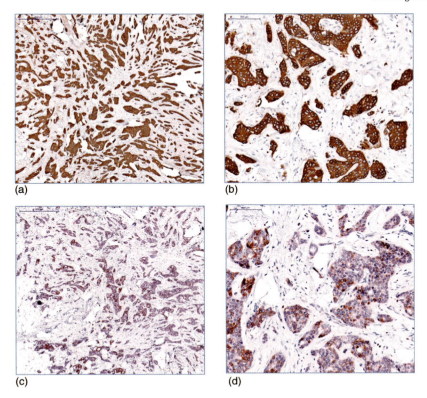

Figure 8.5 Immunohistochemical analysis of tumor marker in breast cancer. (a) Epithelial tumor component demonstrated by expression of cytokeratins (mAB, mouse anti-human pan cytokeratin) (5%). (b) Medium-sized tumor cell islands surrounded by a cell-rich connective tissue ($\times 20$); (c) tumor cells of this particular case expresses express bcl-2 in the cytoplasm (mAB, mouse anti-human bcl-2) ($\times 5$); (d) stained cells show a distinctive expression level that become obvious at higher magnification by different staining intensities ($\times 20$).

image analyses (for an overview, see [9]). The integration process requires mutual "understanding" of humans and machine concerning their respective contributions. Even image experts such as pathologists lack an understanding of what type of additional information can be expected from machine-made image analyses. Therefore, when starting a research program they are unable to give the system cues about important and technically processable image details. Procedures that start with expert labeling of images are called supervised learning methods. From experience concerning DM in life science projects, this is not the most efficient starting point for the integration of human and technical cognition (Figure 8.9).

An alternative is unsupervised learning methods that exploit cluster analyses in order to identify technically manageable image features that can, for example, discriminate various tissue types in pathological slices. The results can be mapped on to the images in the form of color codings for different regions. Visual exploration

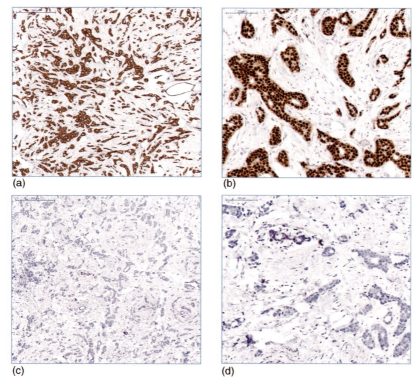

Figure 8.6 Immunohistochemical analysis of steroid receptors in breast cancer. (a) Epithelial tumor component is highly positive (100%) for the estrogen receptor (ER) (mAB, mouse anti-human estrogen receptor) (×5); (b) ER is expressed in the nucleus, in this particular case with a slight cytoplasmic staining (×20); (c) the tumor sample demonstrated here is negative for the progesterone receptor (PR) (mAB, mouse anti-human progesterone receptor) in the epithelial tumor component (×5); (d) only a few cells lining nontimorous ducts show a positive nuclear staining (×20).

and control systems can support the human expert in the necessary validation and adjustment of these findings. If the results are validated by the expert, there is a considerable advantage of technical image analyses. They are comprehensive and therefore applicable to quantitative evaluations of tissue slices that are notoriously problematic for human experts (Figure 8.10).

Results that are obtained in this way can be seen as expedient findings in a diagnostic pattern if they can trigger further experimental steps such as microscopy records applying additional structural and functional staining. Even if senseless "data cemeteries" can be avoided through such interlocking of human–computer interaction, there will arise a multilayered complexity of image-related information that will soon become confusing. In order to support navigation in the growing information space and even the introduction of technical data mining, it is necessary to introduce an image database with pertinent metadata. This combination of machine

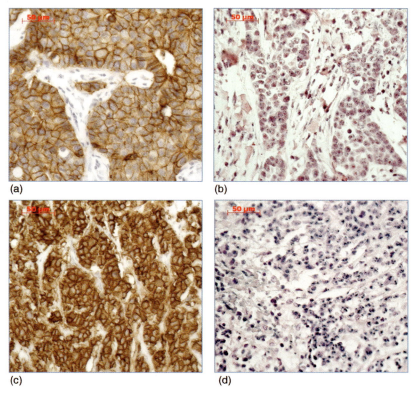

Figure 8.7 Immunohistochemical analysis and *in situ* hybridization for Her-2/neu gene amplification in breast cancer for evaluating Her-2/neu status. (a) Her-2 staining pattern in immunohistochemistry refer to a score of 2 + according to the guideline recommendations for HER2 testing in breast cancer [10] (mAB, mouse anti-human erbB2) (×40); (b) CISH, using dual-color probes against Her-2 (blue) located on chromosome 17 and a probe against the centromeric region of chromosome 17 (red) (ZytoDot 2C Spec HER2/CEN 17) reveals no amplification for Her-2 (×40); (c) another case shows a Her-2 staining pattern that refers to a score of 3 + in immunohistochemistry [10] (×40); (d) CISH reveals an amplification for Her-2 as demonstrated by large dots for the Her-2 signal (blue) and two separate signals for Cen 17 (×40).

learning-based analyses and an image database can considerably enhance the value of DM, especially for the life sciences. Such an enabling environment supports the gradual extension and validation of complex and multilayered knowledge that is impossible without the mutual completion and control of human and machine-made analytical findings.

Digital microscopy renders images of heterogeneous resolution scales and staining. Pathological tissue images can show overviews, cells, and even functional molecules within cells. The spatially correlated evaluation of this heterogeneous information yields a potential for biological and pathological insights that could

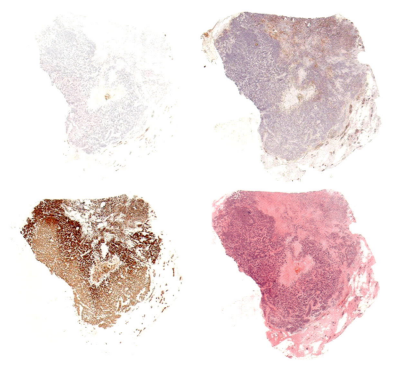

Figure 8.8 Four different histological stainings (CD45, VIM, AE13, HE) for registration into one synthetic image stack.

not be exploited without the described integration of DM and IT modules. For example, it is possible to register different images with structural and functional staining of tissue slices on various magnification levels into one virtual image space that can be analyzed via machine learning algorithms. The combination of identified

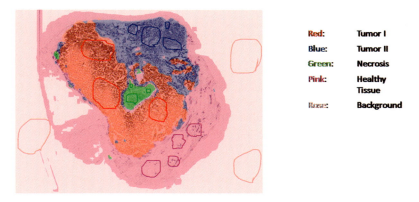

Figure 8.9 Resulting tissue characterization for supervised learning given expert annotations.

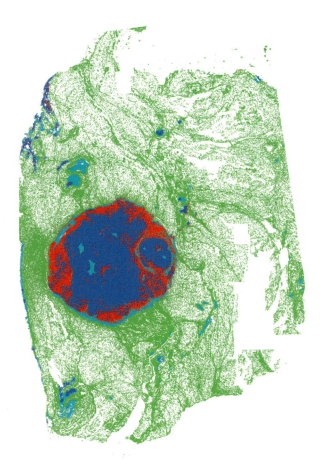

Figure 8.10 Resulting tissue characterization for unsupervised learning with highlighted CD45 (red).

features cannot be "seen" by humans, whereas the resulting classification can be mapped on to color codings of tissue areas and thus show and measure the extension of tumors.

A technical success model for such IT-based integration and graspable presentation of spatially encoded information in a common coordinate system include geographical information environments such as Google Maps.

In conclusion, pathologists and light microscopic analysis of tissue specimens have an integral role in modern cancer units. The information that they provide about tumor type, grade, stage, completeness of excision, and prognosis are derived mainly from traditional methods of histopathology. The information value can be improved as these methods are linked together with advanced techniques such as molecular analyses and automated image analyses. Having said that, we are in accordance with the tenet "pathology is far from dead" [1].

References

1 Lakhani, S.R. and Ashworth, A. (2001) Microarray and histopathological analysis of tumours: the future and the past? *Nat. Rev. Cancer*, **1**, 151–157.
2 Sternberg, S.S. (ed.) (1997) *Histology for Pathologists*, Lippincott Williams and Wilkins, Philadelphia.
3 Hayat, M.A. (2004) *Handbook of Immunohistochemistry and In Situ Hybridization of Human Carcinomas: Molecular Genetics – Lung and Breast Carcinomas*, **1**, Academic Press, New York.
4 Pearse, A.G.E. (1980) *Histochemistry: Theoretical and Applied*, **1**, Churchill Livingstone, Oxford.
5 Pearse, A.G.E. (1985) *Histochemistry: Theoretical and Applied*, **2**, Churchill Livingstone, Oxford.
6 Polak, J.M. and Van Noorden, S. (2002) *Introduction to Immunocytochemistry*, Bios Scientific Publishers, Oxford.
7 Gray, J.W., Pinkel, D., and Brown, J.M. (1994) Fluorescence *in situ* hybridization in cancer and radiation biology. *Radiat. Res.*, **137**, 275–289.
8 Jin, L. and Lloyd, R.V. (1997) *In situ* hybridization: methods and applications. *J. Clin. Lab. Anal.*, **11**, 2–9.
9 Caelli, T. and Bischof, W.F. (2008) *Machine Learning and Image Interpretation*, Springer, Berlin.
10 Wolff, A.C., Hammond, M.E., Schwartz, J.N., Hagerty, K.L., Allred, D.C., Cote, R.J., Dowsett, M., Fitzgibbons, P.L., Hanna, W.M., Langer, A., McShane, L.M., Paik, S., Pegram, M.D., Perez, E.A., Press, M.F., Rhodes, A., Sturgeon, C., Taube, S.E., Tubbs, R., Vance, G.H., van de Vijver, M., Wheeler, T.M., Hayes, D.F., American Society of Clinical Oncology , and College of American Pathologists (2007) American Society of Clinical Oncology/College of American Pathologists guideline recommendations for human epidermal growth factor receptor 2 testing in breast cancer. *J. Clin. Oncol.*, **25**, 118–145.

9
Virtual Microscopy

Thorsten Heupel

9.1
Introduction

The basic concept of virtual microscopy is to distribute microscopic images via computer networks. Based on this definition, there are three evolutionary steps along which this technology has developed:

(a) **Static remote microscopy:** From a technology standpoint, this is the easiest way to transmit microscopic images, because only a microscope, a digital camera, and capture software are needed. The operator makes single snapshots from regions of interest and subsequently they can be transferred for example via e-mail to another person for review. The limitations are obvious, such as the possibility of a misaligned microscope, an operator and the physical specimen is always needed, and, the most prominent, only a small part of the specimen can be transferred.

(b) **Dynamic remote microscopy:** Here a motorized microscope is remotely controlled by an operator and a live image from the camera is broadcasted (e.g., via satellite, Internet). This allows one to pan the specimen in the x and y directions, to focus, and to change objectives. Thus the remote person can investigate a complete specimen; some advanced systems even allow switching between slides with a robotic system. This technology was mainly driven by pathologists in order to make remote diagnosis possible, where presence on-site was not possible. Systems were deployed by the US Army to provide expert diagnosis even close to the battlefield. Another application area was instant section diagnosis, allowing remote hospitals with no resident pathologist to obtain the required support during surgery. The term telepathology is a combination of telemedicine and pathology. Telemedicine has been used for several decades (e.g., on-line monitoring of health status by NASA). Telepathology was introduced in 1968 in Boston [1]. In the past, the deployment of telepathology was limited to regions with a mature telecommunication infrastructure. In underdeveloped countries there is a need for telepathology, but so far the infrastructure has mostly been lacking. However, now the increasing coverage of GSM networks in Africa has triggered the development of telepathology solutions using smartphones [2].

(c) **Static–dynamic hybrid remote microscopy (virtual microscopy):** Virtual microscopy (VM) is a combination of the two technologies described above. The entire sample is automatically scanned at high resolution without user intervention. The user can access the image via the Internet or locally. The remote navigation within the large dataset can be realized with the same technology as for Google Earth. The digitization of the complete sample leads to the so-called virtual slide (VS). VM is becoming the most popular technology, as it paves the way to applications beyond remote viewing of samples. In recent years, the driving forces for VM were the increasing performance of scanners [3] and the exponential growth of information technology (IT), namely cheaper storage, faster processors, and faster Internet connections. In addition to telepathology, new applications are emerging from the biomedical research field. Therefore, it is fair to say that VM has the characteristics of a disruptive technology that will radically change the way in which we use microscopes in the future.

9.2
Principles

A VS is a digital representation of a microscopic specimen. All object regions on a slide are imaged at high magnification and captured with a digital sensor. Scanners use microscope objectives with a high numerical aperture (e.g., 0.8) and a magnification of mainly 20×. The digital sensors used can be divided into two main categories: line sensors (with stripe-wise scanning) and area sensors (with tile-based scanning). After scanning, the stripes or tiles have to be merged together to form a large montage image of the entire specimen. Consequently, most scanners are nowadays closed-box systems with limited functionality, in contrast to research microscopes, which allow for example the selection of multiple magnifications and contrast modes.

Quality of focusing is a critical issue with scanners, because the VS depicts a single focus plane of the specimen and for the viewer the focus dimension is lost. To overcome this drawback, many systems offer the possibility of digitizing the sample at different z levels. Depending on the need, the different z levels are stored or merged into a single plane.

The image size of VS depends on the specimen area on the slide and the scan resolution. The typical scan resolution for VS is between 0.5 and 0.2 µm per pixel, which gives sufficient detail to perform most diagnoses.

Based on the above parameters, the size of the resulting VS is in the range of megabytes to gigabytes. It is obvious that the falling costs of hard disk storage were key for the success of VS. Within 10 years time, the cost of storage space has dropped by a factor of 1000.

The majority of scanners use brightfield illumination as the imaging modality. The reason for this comes from the needs of pathology with standard stains such as H&E (Hematoxylin&Eosin) and PAS (Periodicacid-Schiff stain), where this contrast mode is demanded. With the evolution of new molecular techniques for pathology (such as fluorescence *in situ* hybridization for HER2/neu) and research applications, the scanning of fluorescently labeled specimens became of increasing importance. This represents a new challenge for scanners also, because scanning of fluorescent samples is much more critical in terms of focus and speed.

9.3 Applications

As already mentioned, telepathology is only one of several application areas for VM technology. The following sections summarize the important areas to date. As the technology is still under development, there will certainly be more applications to follow.

9.3.1 Telepathology

So far, remote access to VS is still a dominant application. In the clinical area, the need for remote consultation is due to the lack of pathologists in certain regions. With VS, their workflow can be optimized, because travel time reduces the effective time of a pathologist. It is as simple as accessing the complete specimen with high resolution from another location. In the case of intraoperative sectioning, it is important to scan with the appropriate resolution very rapidly to obtain a quick feedback from the pathologist (less than 10 min). Outside the clinical environment, peer review is also a well-accepted application field for pharmaceutical companies, because the global distribution of such companies makes it necessary to connect research and validation from different continents (e.g., for toxicology studies) for peer reviews. Another advantage of VS is that once the specimen has been digitized it is preserved (in contrast to physical immunostains, which fade over time). This allows compliance with regulatory procedures which demand that results need to be archived for re-examination if required.

9.3.2 Routine Pathology

In addition to telepathology, the use of VS in routine clinical practice was always considered a major application. However, as of today, the VS has not yet become an integral part of the routine clinical workflow. There are several reasons why this adoption is only slowly taking place. In many cases, pathologists can make their diagnoses faster using glass slides instead of loading and navigating a VS on a PC. The graphical user interface has to be improved, regarding speed and usability, and the software needs an interface to the clinical IT [e.g., PACS (Picture Archiving and Communication System)]. Routine pathology needs faster scanners to handle several hundred slides per day, which means slide processing times of the order of 1 min. Additionally in some countries there are regulatory questions how to use/validate this technology in routine diagnosis which are not fully answered so far. Combined with image analysis this technology will gain strength in the computer-assisted diagnosis.

9.3.3
Education

In a typical education scenario, there is a lecture room with a set of microscopes for students. One or more students use one microscope to investigate a set of specimens. To make sure that every student can benefit from the tutorial, it is important that all microscopes are perfectly adjusted and every student can investigate the same specimen. This is not trivial as sectioned samples are never identical, and there are rare specimens for which it is impossible to produce the number of samples required for the course. During the course, slides get broken, and microscopes have to maintained and serviced; in summary, the amount of work to keep a microscope training course at a high quality level is fairly high.

Using computer workstations and VS overcomes these problems. Once a specimen has been scanned and the VS put on a server, every student can explore exactly the same specimen. In addition, the lecturer can provide additional information via annotations, or links to notes. Instead of the physical microscope seminar, where the number of participants is limited by the available microscopes and the seminar schedule, VS seminars can be undertaken by students from home. Owing to these clear benefits, implementation of VM in education is already fairly widespread.

9.3.4
Tissue Microarrays

Tissue microarrays (TMAs) are collections of small tissue samples (each tissue is a spot with a diameter of around 1 mm) which are arranged on a slide to form an array. This setup enables the user to place up to several hundred different tissue samples on a slide to perform comparative studies. The first reference to this technique appeared in 1987 [4].

TMA is an enabling tool for clinical research (in particular cancer research). Many clinics have large collections of tissue/tumors built up over the years. These samples originated from patients whose disease record is well known. This allows searching for molecular tissue markers that are specific to known disease patterns. It is possible to investigate these samples for common features, and then correlate these features with characteristics of the disease [5].

The tissue spots are taken from standard tissue blocks, by means of a fine hollow needle. As the spot(s) represent only a small part of the whole block, it is necessary that the researcher/pathologist first investigates the specimen and marks the regions to be used for a TMA. Selecting these regions on a VS allows direct control of the needle for core extraction. This procedure optimizes the whole workflow.

The cores from the several donor blocks are finally transferred to an acceptor (TMA) block. From this acceptor block, several cuts are made for processing with different stains (e.g., a counterstain and immunohistochemical tumor markers).

In many cases, it is required to quantify the relative stain intensity and covered area over an ensemble of spots (scoring). This can be done by means of image analysis software or manually. Based on the patients' information, the spots can now be filtered and grouped in a gallery, which helps in making side-by-side comparisons.

9.3.5
Research

The number of publications on VM is increasing year by year whereas the number of publications on telepathology is decreasing [6]. With the increased quality and speed of scanning fluorescent probes, this technology is becoming even more interesting for research use. The rapid fading of fluorescent signals makes VS a perfect tool to archive a specimen digitally and to perform off-line image analysis to extract object and pattern information without harming the sample. With this technology also less experienced users can quickly obtain high-quality images. As the VS represents the complete specimen, it is now easy to perform image analysis on larger areas of the specimen. Especially for neurobiology, VM is a technology to acquire a more detailed insight. Research on Alzheimer disease is already utilizing this technology to analyze for example the influence of certain substances or parameters on the formation of plaques [7].

With VM, not only is the complete specimen available, but also together with serial sections it is possible to obtain a three-dimensional representation of the specimen to analyze the structures.

The VS technique is a also a perfect tool to generate high resolution representations of the whole specimen as source for further investigations and combination with other imaging modalities.

9.4
Outlook

It is obvious that future improvements in scanner technology (especially regarding scanning speed with brightfield illumination) and IT will pave the way towards the routine deployment of VS in pathology. In terms of clinical standardization, there is working group 26 within the DICOM organization to define the interface for VS within a standard. This is finalized and will support the implementation into the clinical workflow.

In addition to this prospect, we will also see penetration of this technology in research (universities), especially if the next generations of instruments exhibit increased speed and quality for fluorescence applications.

In the research environment, we will see a fusion between VS and other technologies. The pathologist/researcher will merge other modalities with their classical view of the stained specimen. An example is the combination with MALDI (matrix-assisted laser desorption/ionization) imaging mass spectrometry. This technology is a very important tool for protein profiling of tissue sections. However, interpretation of the results can only be done in the context of the histological sample. The future will bring more combinations of different technologies besides other imaging modalities with VS, such as proteohistography, tissue microdissection, and mass spectrometry providing data of higher quality and quantity.

VM will support users in the future with more reproducible and quantitative results and this will make it an ideal tool for quality control in the field of material applications but also for food inspection.

Acknowledgment

The author would like to thank Dr. Martin Gluch for the critical review of this chapter and his valuable contribution.

References

1 Kayser, K. (1999) History of Telemedicine and Telepathology, in *Telepathology: Telecommunication, Electronic Education in Pathology* (eds K. Kayser, J. Szymas, and R. Weinstein), Springer, Berlin, pp. 24ff.
2 Zimic, M., Coronel, J., Gilman, R.H., Luna, C.G., Curioso, W.H. *et al.* (2009) Can the power of mobile phones be used to improve tuberculosis diagnosis in developing countries? *Trans. R. Soc. Trop. Med. Hyg.*, **103**, 638–640.
3 Rojo, M.G., García, G.B., Mateos, C.P., García, J.G., and Vicente, M.C. (2006) Critical Comparison of 31 Commercially Available Digital Slide Systems in Pathology. *Int. J. Surg. Pathol.*, **14** (4), 1–21.
4 Wan, W.H., Fortuna, M.B., and Furmanski, P. (1987) A rapid and efficient method for testing immunohistochemical reactivity of monoclonal antibodies against multiple tissue samples simultaneously. *J. Immunol. Methods*, **103**, 121–129.
5 Simon, R., Mirlacher, M., and Sauter, G. (2003) Tissue microarrays in cancer diagnosis. *Expert Rev. Mol. Diagn.*, **3** (4), 421–430.
6 Kayser, K., Molnar, B., and Weinstein, R.S. (2006) *Virtual Microscopy*, VSV Interdisciplinary Medical Publishing, Berlin.
7 Scheffler, K., Stenzel, J., Krohn, M., Lange, C., Hofrichter, J., Schumacher, T., Brüning, T., Plath, A.S., Walker, L., and Pahnke, J. (2011) Determination of spatial and temporal distribution of microglia by 230nm-high-resolution, high-throughput automated analysis reveals different amyloid plaque populations in an APP/PS1 mouse model of Alzheimer's disease. *Curr. Alzheimer Res.*, [Epub ahead of print].

10
In Vitro Instrumentation
Sergio Coda, Rakesh Patalay, Christopher Dunsby, and Paul M.W. French

10.1
Introduction

This chapter describes the application of multidimensional fluorescence imaging techniques to biological tissue, with particular emphasis on fluorescence lifetime imaging. It reviews the main fluorescence measurement and imaging modalities and describes their application to label-free imaging of tissue autofluorescence (AF), concentrating on *ex vivo* measurements, although some of the technology presented has also been applied to *in vivo* imaging.

10.2
Autofluorescence of Biological Tissue

Fluorescence provides a powerful means of achieving optical molecular contrast [1] in single-point (cuvette-based or fiber-optic probe-based) fluorimeters, in cytometers and cell sorters, and in microscopes, endoscopes and multiwell plate readers. Typically for cell biology, fluorescent molecules (fluorophores) are used as "labels" to tag specific molecules of interest. For clinical applications, it is possible to exploit the AF of target molecules themselves to provide *label-free molecular contrast*, which is the main focus of this chapter, although there is an increasing trend to investigate the clinical use of exogenous labels to detect diseased tissue, for example, using photodynamic diagnosis (PDD) [2].

When studying AF of biological tissue, the molecular composition is often unknown and so tissue AF can present a complex signal resulting from an unknown number of endogenous fluorophores present with unknown relative concentrations. Interpretation of AF data is therefore highly challenging, typically requiring *a priori* knowledge of tissue structure and/or physiological function to yield quantitative information. Nevertheless, the potential for label-free molecular contrast as a diagnostic tool is compelling and empirical AF contrast is finding its way into clinical practice while an increasing number of *ex vivo* and *in vivo* studies are directed at understanding the molecular origins of tissue AF and the contrast available.

In biological tissue, AF can provide a source of label-free optical molecular contrast, offering the potential to discriminate between healthy and diseased tissue. The prospect of detecting molecular changes associated with the early manifestations of diseases such as cancer is particularly exciting. For example, accurate and early detection of cancer allows earlier treatment and significantly improves prognosis [3]. Although the exophytic tumors can often be visualized, the cellular and tissue perturbations of the peripheral components of neoplasia may not be apparent by direct inspection under visible light and are often beyond the discrimination of conventional noninvasive diagnostic imaging techniques. In general, label-free imaging modalities are preferable for clinical imaging, particularly for diagnosis, as they avoid the need for administration of an exogenous agent with associated considerations of toxicity and pharmacokinetics. A number of label-free modalities based on the interaction of light with tissue have been proposed to improve the detection of malignant change. These include fluorescence, elastic scattering, Raman, infrared absorption, and diffuse reflectance spectroscopy [4–7]. To date, most of these techniques have been limited to point measurements, where only a very small area of tissue is interrogated at a time, for example, via a fiber-optic contact probe. This enables the acquisition of biochemical information, but provides no spatial or morphological information about the tissue. AF can provide label-free "molecular" contrast that can be readily utilized as an imaging technique, allowing the rapid and relatively noninvasive collection of spatially resolved information from areas of tissue up to tens of centimeters in diameter. In order to exploit AF for clinical applications, it is first necessary to investigate the AF "signatures" of normal and diseased tissue states, and this has led to a number of *in vitro* studies. The instrumentation applied to such investigations is often similar to that envisaged for *in vivo* clinical applications, although the nature of *in vitro* experiments clearly relax many of the constraints associated with *in vivo* studies and permit the use of more sophisticated (and time consuming) measurements.

The principal endogenous tissue fluorophores include collagen and elastin cross-links, reduced nicotinamide adenine dinucleotide (NADH), oxidized flavins (FAD and FMN), lipofuscin, keratin, and porphyrins. As shown in Figure 10.1 [7], these fluorophores have excitation maxima in the UV-A or blue (325–450 nm) spectral regions and emit Stokes-shifted fluorescence in the near-UV to visible (390–520 nm) region of the spectrum. The actual AF signal excited in biological tissue will depend on the concentration and the distribution of the fluorophores present, on the presence of chromophores (principally hemoglobin) that absorb excitation and fluorescence light, and on the degree of light scattering that occurs within the tissue [6]. AF therefore reflects the biochemical and structural composition of the tissue, and consequently is altered when the tissue composition is changed by disease states such as atherosclerosis, cancer, and osteoarthritis.

Although AF can be observed using conventional "steady-state" imaging techniques, it is challenging to make sufficiently quantitative measurements for diagnostic applications since the AF intensity signal may be affected by fluorophore

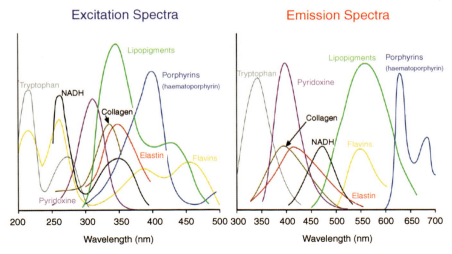

Figure 10.1 Excitation and emission spectra of main endogenous tissue fluorophores. Reproduced with permission from [7].

concentration, variations in temporal and spatial properties of the excitation flux, the angle of the excitation light, the detection efficiency, attenuation by light absorption and scattering within the tissue, and spatial variations in the tissue microenvironment altering local quenching of fluorescence. The remainder of this chapter introduces some of the increasingly sophisticated detection and analysis techniques that aim to overcome these problems.

10.2.1
Fluorescence Contrast

More robust measurements can be made using *ratiometric* techniques, since spectrally and time-resolved measurements can also be sensitive to variations in the local fluorophore environment. In the spectral domain, one can assume that unknown quantities such as excitation and detection efficiency, fluorophore concentration, and signal attenuation will be approximately the same in two or more spectral windows and may be effectively "canceled out" in a ratiometric measurement. This approach is used, for example, with excitation [8] and emission [9] ratiometric calcium sensing dyes and the approach has been demonstrated to provide useful contrast between, for example, malignant and normal tissue [10]. Fluorescence lifetime measurement is also a ratiometric technique as it is assumed that the various unknown quantities do not change significantly during the fluorescence decay time (typically nanoseconds) and the lifetime determination effectively compares the fluorescence signal at different delays after excitation. The fluorescence lifetime is the average time that a fluorophore takes to decay radiatively after having been excited from its ground energy level, and, like the quantum efficiency, it is also a function of the radiative and nonradiative decay rates and so can provide quantitative fluorescence-based molecular contrast.

10.2.2
Spectroscopic Techniques for Fluorescence Imaging and Metrology

10.2.2.1 Spectral Techniques

The most common and experimentally straightforward approach to tissue spectroscopy has been the acquisition of information about the AF emission spectrum for a fixed excitation wavelength. In its simplest form, this can be implemented using dichroic beamsplitters and multiple detectors, as is often applied to fluorescence microscopy for ratiometric imaging or colocalization studies. This *multispectral* approach is useful when the spectral profiles of fluorophores are known *a priori* and can be used with spectral unmixing techniques to separate contributions from different fluorophore labels or to classify different types and states of tissue using known spectral components.

Without *a priori* knowledge, it is useful to learn as much about the tissue AF as possible and so *hyperspectral* measurements that acquire the full (emission) spectral profile of a fluorescent sample are desirable [6]. For single-channel measurements, for example, made using a spectrofluorimeter, it is straightforward to use a spectrograph that can be read out using a CCD (*charge-coupled device*). Alternative approaches include scanning Fabry–Pérot interferometers, which can be implemented using mechanical scanning or by taking advantage of the electro-optic effect to sweep the refractive index of an étalon. Acousto-optic tunable filters (AOTFs) also provide a means to rapidly scan spectral profiles. Both scanning étalons and AOTFs effectively sample the fluorescence signal, rejecting the "out-of-band" light and so are lossy compared with spectrographs. A more photon-efficient approach is that of a scanning Michelson interferometer, from which the spectrum is obtained by Fourier transformation of the acquired interferogram. Hyperspectral imaging can be readily implemented in laser scanning microscopes by utilizing single-pixel detection techniques, although the rapid raster scanning rates can be a challenge for the read-out rates of the spectral detectors. With no *a priori* knowledge of tissue properties, such rich hyperspectral sets can be analyzed using techniques such as principal component analysis (PCA) that can extract common spectral components from "training sets" of tissue fluorescence data and subsequently be applied for diagnostic purposes, for example. Fluorescence emission data can also be complemented by excitation data.

10.2.2.2 Fluorescence Lifetime Techniques

Increasingly, there is interest in exploiting fluorescence lifetime contrast to analyze tissue AF signals, since fluorescence decay profiles depend on relative (rather than absolute) intensity values and fluorescence lifetime imaging (FLIM) is therefore largely unaffected by many factors that limit steady-state measurements [11]. The principal tissue fluorophores exhibit characteristic lifetimes (ranging from hundreds to thousands of picoseconds) [6, 12] that can enable spectrally overlapping fluorophores to be distinguished. Furthermore, the sensitivity of fluorescence lifetime to changes in the local tissue microenvironment (e.g., pH, [O_2], [Ca^{2+}]) [1] can provide a readout of biochemical changes indicating the onset or progression of disease. FLIM is now being actively investigated as a means of obtaining or enhancing intrinsic AF contrast in tissues [13, 14].

In general, fluorescence lifetime measurement and fluorescence lifetime imaging techniques are categorized as time-domain or frequency-domain techniques, according to whether the instrumentation measures the fluorescence signal as a function of time delay following pulsed excitation or whether the lifetime information is derived from measurements of phase difference between a sinusoidally modulated excitation signal and the resulting sinusoidally modulated fluorescence signal. In principle, frequency and time domain approaches can provide equivalent information, but specific implementations present different trade-offs with respect to cost, complexity, performance, and acquisition time. In general, the most appropriate method should be selected according to the target application. A further categorization of fluorescence lifetime measurement techniques can be made according to whether they are sampling techniques, which use gated detection to determine the relative timing of the fluorescence compared with the excitation signal, or photon counting techniques, which assign detected photons to different time bins. In general, wide-field FLIM is usually implemented with gated or modulated imaging detectors that sample the fluorescence signal whereas photon counting techniques have been widely applied to single-point lifetime measurements. For laser scanning FLIM, single-point fluorescence lifetime measurement techniques can be readily implemented in laser scanning microscopes as modern electronics makes it straightforward to assign detected photons to their respective images pixels. There are extensive reviews of the various techniques [1, 15–17].

10.2.3
Single-Point Spectroscopic Techniques

Single-point spectroscopic measurements are typically made in spectrofluorimeters that are usually designed for cuvette-based solution measurements, although some can be adapted for solid samples. For studying tissue samples, it is common to use a fiber-optic-based probe to deliver the excitation light and collect the resulting fluorescence from a sample. Fiber-optic probes provide experimental convenience, particularly for clinical applications. Such probes are frequently combined with spectrographs that may be used to capture fluorescence emission spectra and also reflected light spectra. With broadband illumination, reflected light spectra can provide information concerning the elastic scattering properties of tissue samples, and this has been used to identify diseased tissue [5]. AF spectra are frequently analyzed using methods such as PCA in order to distinguish different tissue components. Spectral measurements have also been combined with fluorescence lifetime measurements in spectrofluorimeters and fiber-optic probe instruments. Multispectral fluorescence lifetime measurements are often realized using one or more dichroic filters with multiple time-resolved detection channels or using a scanning spectrometer with a single time-resolved detector [18], and spectrographs have been used with array detectors, including gated intensified CCDs [19], streak cameras [20], and time-correlated single-photon counting (TCSPC) [21, 22].

10.2.4
Fluorescence Microscopy

10.2.4.1 Laser Scanning Confocal/Multiphoton Microscopy

Confocal or multiphoton laser scanning microscopes are used widely for biological imaging because they provide optical sectioning and improved contrast compared with wide-field microscopes. In general, multiphoton excitation permits imaging at greater depths than confocal microscopy due to the reduced scattering and absorption experienced at the longer excitation wavelengths [23], although the longer wavelength leads to a slightly degraded image resolution and multiphoton excitation cross-sections are typically significantly lower than their single-photon counterparts, resulting in increased image acquisition times. Multiphoton excitation confers the benefit of reduced out of focal plane photobleaching compared with confocal microscopy, which is important when acquiring dense z-stacks for 3D imaging. The photobleaching process associated with multiphoton excitation is nonlinear, however, and is more severe in the focal volume.

Multichannel spectral detection is routinely implemented in laser scanning microscopes using dichroic beamsplitters and filters, and hyperspectral imaging is realized by directing the collected fluorescence to a scanning spectrometer or to a spectrograph readout using a CCD or multianode photomultiplier array. FLIM is conveniently achieved on laser scanning microscopes using TCSPC [24], photon binning [25] or frequency domain techniques [26].

For all laser scanning microscopes, the sequential pixel acquisition means that increasing the imaging speed requires a concomitant increase in excitation intensity, which can be undesirable due to photobleaching and phototoxicity considerations. In general, parallel pixel excitation and detection are applicable to all laser scanning microscopes and have been extended to optically sectioned line-scanning microscopy, for example, by using line illumination or by using a rapidly scanned multiple beam array to produce a line of fluorescence excitation. The resulting emission can be relayed to the input slit of a spectrograph to facilitate "push-broom" hyperspectral imaging, as discussed below, or to the input slit of a streak camera to implement FLIM [27]. It is also possible to scan multiple excitation beams rapidly in parallel to produce a 2D optically sectioned fluorescence image that can be recorded using wide-field detectors. This has been implemented with single-photon excitation in spinning Nipkow disc microscopes that have been combined with wide-field time-gated detection for FLIM [28–30], and with multibeam multiphoton microscopes that have also been adapted for FLIM [31–33].

10.2.4.2 Wide-Field Fluorescence Imaging

The parallel nature of wide-field imaging techniques can support FLIM imaging rates of tens to hundreds of hertz [34, 35], although the maximum acquisition speed is inevitably limited by the number of photons/pixels available from the (biological) sample. Wide-field FLIM is most commonly implemented using modulated image intensifiers with frequency or time domain approaches proving successful. The frequency domain approach initially utilized sinusoidally modulated laser excitation and the resulting fluorescence was analyzed using a microchannel plate (MCP) image intensifier with a sinusoidally-modulated gain function by acquiring a series of gated "intensified" images at different relative phases between the MCP modulation and the excitation signal [36–38]. In the time domain, FLIM is also realized using modulated MCP image intensifiers to sample the fluorescence decays, but for this approach the MCP image intensifier gain is gated for short periods (pico- to nanoseconds) after excitation [39–41]. The use of gated image intensifiers with a wire mesh proximity-coupled to the MCP photocathode has led to devices with sub-100 ps resolution [42], although it is usually preferable to use longer time gates to increase the detected signal [43].

Although wide-field FLIM can achieve frame rates of around tens of hertz in both the frequency [44] and time [34] domains under the assumption of monoexponential fluorescence decay profiles and has been demonstrated in endoscope systems [34, 45], sampling and fitting of complex decay profiles inevitably increase the FLIM acquisition time since they require significantly more detected photons for accurate lifetime determination. Nevertheless, wide-field FLIM can still be faster than TCSPC implemented in a laser scanning microscope. If high-speed optically sectioned FLIM is required, the fasted systems reported to date have been based on multibeam confocal (Nipkow disc) microscopes with wide-field gated image intensifier detection [28].

10.2.4.3 Multidimensional Fluorescence Microscopy

It can be useful to obtain similar multidimensional fluorescence information from fluorescence imaging experiments, so that functional spectroscopic information can be correlated with morphology. This can be useful for imaging Förster resonance energy transfer (FRET) experiments, where spectral and lifetime measurements can give complementary information [46], for spatially resolved assays of chemical reactions [47], and for studying tissue AF, particularly when investigating samples with no *a priori* knowledge. For example, FLIM can be combined with multispectral imaging, for which time-resolved images are acquired in a few discrete spectral windows [48], or it can be combined with hyperspectral imaging, for which the full time-resolved excitation or emission spectral profile is acquired for each image pixel.

There are several established approaches for implementing hyperspectral imaging. In single-beam scanning fluorescence microscopes, the fluorescence radiation can be dispersed in a spectrometer and the spectral profiles acquired sequentially, pixel by pixel. This is readily combined with FLIM in laser scanning microscopes, for example, using TCSPC systems that incorporate a spectrometer and multi-anode photomultiplier [22, 49]. To increase the imaging rate, it is necessary to move to parallel pixel acquisition, for example, by combining wide-field FLIM with hyperspectral imaging

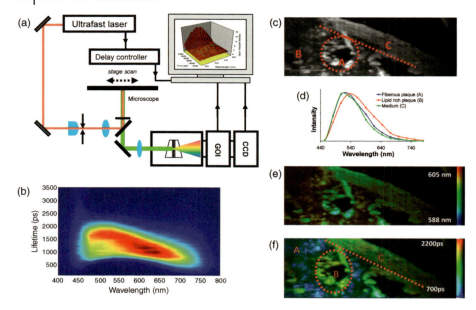

Figure 10.2 (a) Experimental set-up for line-scanning hyperspectral FLIM; (b) integrated intensity image of sample of frozen human artery exhibiting atherosclerosis; (c) time-integrated spectra of sample regions corresponding to medium and fibrous and lipid-rich plaques; (d) AF lifetime-emission matrix; (e) map of time-integrated central wavelength; (f) spectrally integrated lifetime map of sample AF. Adapted from [54].

implemented using filter wheels or acousto-optic [50] or liquid crystal [51] tunable filters to acquire images sequentially in different spectral channels. This approach is not photon efficient because the "out-of-band" light is rejected and the acquisition time will increase with the number of spectral channels. Alternative approaches that can be more efficient include Fourier transform spectroscopic imaging [22] and encoding Hadamard transforms using spatial light modulator technology [52]. In general, however, any approach that requires the final image to be computationally extracted from acquired data can suffer from a reduction in signal-to noise ratio compared with direct image detection. A compromise between photon economy and parallel pixel acquisition that directly detects the spectrally resolved image information is the so-called "push-broom" approach to hyperspectral imaging [53] implemented in a line-scanning microscope. For hyperspectral FLIM, the fluorescence resulting from a line excitation is imaged to the entrance slit of a spectrograph to produce an $(x–\lambda)$ "sub-image" that can be recorded on a wide-field FLIM detector. Stage scanning along the y-axis then sequentially provides the full $(x–y–\lambda–\tau)$ data set. This approach is photon efficient and the line-scanning microscope configuration provides "semi-confocal" optical sectioning [54]. This instrument was developed to study AF contrast in diseased tissue and Figure 10.2 shows the experimental configuration and presents multidimensional fluorescence data from a hyperspectral FLIM acquisition of an unstained frozen section of human artery exhibiting atherosclerosis of human cartilage.

This multidimensional fluorescence imaging (MDFI) approach can be further extended if we employ a tunable excitation laser also to resolve the excitation wavelength. Using spectral selection of a fiber laser-pumped supercontinuum source to provide tunable excitation from 390 to 510 nm with this line-scanning hyperspectral FLIM microscope, we demonstrated the ability to acquire the fluorescence excitation–emission–lifetime (EEL) matrix for each image pixel of a sample. Such a multidimensional data set enables us subsequently to reconstruct conventional "excitation–emission matrices" excitation–emission matrix (EEM) for any image pixel and to obtain the fluorescence decay profile for any point in this EEM space. Although this can provide exquisite sensitivity to perturbations in fluorescence emission and can enhance the ability to unmix signals from different fluorophores, we note that the size and complexity of such high content hyperspectral FLIM data sets is almost beyond the scope of *ad hoc* analysis by human investigators. Sophisticated bioinformatics software tools are required to analyze automatically and present such data to identify trends and fluorescence "signatures." The combination of MDFI with image segmentation for high content analysis should provide powerful tools, for example, for histopathology and for screening applications.

10.3
Ex Vivo Tissue Spectroscopy and Imaging: Application to Cancer

In the era of minimally invasive technology, laser-induced fluorescence microscopy and spectroscopy provide label-free highly discriminative intrinsic contrast between benign and neoplastic tissue. This potential has been demonstrated extensively in a multitude of *ex vivo* studies, and nowadays *in vivo* applications of different techniques are opening up a new concept of clinical imaging that can provide both histological and physiological information label-free and in real time. It is not possible to provide a comprehensive review of *ex vivo* studies of tissue AF within this chapter, so this section aims to review some representative examples of the application of fluorescence microscopy, spectroscopy, and endoscopy to a variety of *ex vivo* human tissues. The discussion will be limited to cancer, which is the disease that has attracted the most attention using optical techniques, although we note that multidimensional fluorescence imaging techniques have also been applied to atherosclerosis [55, 56] and other diseases. A review of MDFI applications by tissue type will be given, with a focus on studies using freshly resected biopsy specimens rather than fresh frozen or paraffin-embedded specimens. Important reviews of this field can be found in the literature [4, 5, 57, 58] and the reader is referred to Chapter 11 for a review of *in vivo* studies using fluorescence spectroscopy and imaging.

10.3.1
Oral Cavity

Evaluation of the oral cavity, pharynx, and larynx using AF imaging systems has now become an essential tool for the screening of head and neck squamous cell cancers as a result of pioneering *in vitro* studies [59, 60]. Furthermore, fluorescence spectroscopy and imaging of oral mucosa have been evaluated by several groups with the aim of detecting malignant or premalignant conditions using excitation wavelengths in the UV region [60, 61] and for a wide range of UV and visible wavelengths (250–500 nm) [59].

Uppal and Gupta [62] carried out enzymic measurements of NADH concentrations in samples from nine patients with oral squamous cell carcinoma. All measurements were carried out on malignant tissue and normal-appearing mucosa adjacent to the lesion from the same patient. They showed that the concentrations of NADH in malignant sites of oral cavity tissue were significantly lower than those in the healthy mucosa.

10.3.2
Gastrointestinal Tract

Chwirot and co-workers [63, 64] reported fluorescence imaging of 21 resected specimens of stomach cancers and normal tissues *in vitro* at six visible emission wavelengths and with 325 nm excitation. They observed a significant difference in the fluorescence intensities measured at 440 and 395 nm, both normalized to intensity measured at 590 nm, and used that difference as a diagnostic parameter to classify malignant tissues with high sensitivity and low predictive value.

Abe *et al.* [65] applied an AF endoscopic imaging system [light-induced fluorescence endoscopy (LIFE)-GI system] immediately after surgery to 61 gastric cancers resected from 50 patients. A comparison with conventional histopathology showed that the detection rates increased significantly as the depth of invasion and the mucosal thickening increased (differentiated cancers). Detection rates for undifferentiated cancers were disappointing due to the characteristic dispersion in the modes of invasion, most of which did not alter the mucosal thickness.

Silveira *et al.* [66] used fluorescence spectroscopy with excitation at 488 nm to detect benign and malignant lesions of *ex vivo* human gastric mucosa. Biopsies with endoscopic diagnosis of gastritis and gastric cancer of the antrum were collected from 35 patients during the endoscopic examination. On each biopsy fragment, the AF spectrum was collected at two random points via a fiber-optic catheter coupled to the excitation laser. Analysis of fluorescence spectra was able to identify the normal tissue from neoplastic lesions with 100% sensitivity and specificity.

Xiao and co-workers [67, 68] defined the color characteristics of gastric cancer in AF images using a double-channel laser scanning confocal microscope with an argon ion laser (excitation wavelength 488 nm) and a helium–neon laser (excitation wavelength 543 nm) to detect AF from 16 gastric cancer tissue specimens and corresponding normal gastric tissue. AF from normal gastric tissue produced a green image whereas a reddish brown image was found to be characteristic of AF in gastric cancer.

Kapadia et al. [69] obtained fluorescence emission spectra from 35 normal specimens and 35 adenomatous polyps of the colon with 325 nm excitation. They used multivariate linear regression analysis of the spectra and a binary classification scheme to develop and optimize an algorithm that differentiates adenomatous polyps from normal mucosa and hyperplastic polyps, retrospectively. The algorithm prospectively discriminated 16 adenomatous polyps from 16 hyperplastic polyps and 34 normal tissues, with a sensitivity of 100% and a specificity of 98%.

Richards-Kortum et al. [70] measured fluorescence emission spectra from normal colon tissues and adenomatous colonic polyps *in vitro* at a range of excitation wavelengths, spanning the ultraviolet and visible spectrum. They found that the spectral differences between normal and adenomatous colon tissues were greatest with 330, 370 and 430 nm excitation. Furthermore, at an excitation wavelength of 370 nm, fluorescence intensities at 404, 480 and 680 nm were found to be most useful for differentiating adenomas from normal colon tissues.

Yang et al. [71] evaluated the ratio of the fluorescence emission intensities at 360 and 380 nm with excitation at 325 nm, and the ratio of the fluorescence emission intensities measured at 450 nm with excitation at 290 and 340 nm. An algorithm based on the ratio of these fluorescence intensities separated adenocarcinomas from normal tissues with 94% sensitivity and 92% specificity.

Chwirot et al. [72] investigated whether digital imaging of AF could be applied in the detection of colonic malignancies in a study of 50 resected specimens. AF was excited with 325 nm radiation from an He–Cd laser and images were recorded *in vitro* in six spectral bands. The main result was the observation that for a majority of malignant and premalignant lesions the intensity of AF was lower than for the corresponding normal mucosa in all of the spectral bands selected for imaging. The spectral bands centered around 440 and 475 nm seemed to be most promising in terms of possible future clinical applications.

10.3.3
Bladder

In the bladder, Demos et al. [73] imaged a series of fresh surgical specimens obtained following cystectomy or transurethral resection using near-infrared AF under long-wavelength laser excitation in combination with cross-polarized elastic light scattering. The intensity of the near-infrared emission, and also that of the cross-polarized backscattered light, was significantly different in cancerous tissue than the contiguous normal tissue.

Cicchi et al. [74] used a combination of four different systems – two-photon excitation intrinsic fluorescence (TPE), second harmonic generation (SHG) microscopy, FLIM, and multispectral two-photon emission (MTPE) detection – to investigate different states of *ex vivo* fresh biopsies of bladder. In particular, they focused on normal mucosa and carcinoma *in situ* (CIS), reporting significant morphological and spectroscopic differences, in both spectral emission and fluorescence lifetime distribution.

10.3.4
Breast

AF has been evaluated for breast cancer diagnosis using cell cultures, animal models, and human *ex vivo* and *in vivo* tissue. Although results using different models can suggest contradictory conclusions [75, 76], the majority of research using human tissue has confirmed a role for AF in the diagnosis of breast cancer.

Yang *et al.* excited fresh benign and malignant lesions using a range of excitation (275–450 nm) and emission (340–630 nm) wavelengths [77]. The ratios of 289/300 to 289/268 nm in excitation peak intensities for emission at 340 nm were found to be most specific and sensitive for distinguishing benign from malignant pathology.

Breslin *et al.* excited AF in 56 samples of *ex vivo* breast tissue at several wavelengths from 300 to 460 nm [78]. Diffuse reflectance and AF emission spectra were collected between 300 and 600 nm and reduced in dimension using PCA. A sensitivity of 70.0% and a specificity of 91.7% were achieved for distinguishing benign from malignant tissue.

Demos *et al.* imaged *ex vivo* high-grade ductal carcinomas using 532 and 632.8 nm laser excitation and with near-infrared polarized light between 700 and 1000 nm [79]. AF and reflectance images were collected in the 700–1000 nm spectral range. They concluded that there was a significant difference in the tissue AF in this spectral range between cancerous and contiguous non-neoplastic human tissue.

Keller *et al.* used fluorescence spectroscopy and reflectance microscopy on *ex vivo* samples during surgery. An 85% sensitivity and 96% specificity were reported for assessing positive tumor margins [80].

A number of groups have assessed AF imaging in breast ductoscopy [81, 82]. Douplik *et al.* illuminated fresh *ex vivo* breast tissue with blue light (390–450 nm) and collected AF images and point measurements in the green spectral range (490–580 nm). The processed AF images and spectra detected a decrease in green AF from malignant tissue compared with normal breast tissue [82].

Contrast between normal and dysplastic tissue has also been observed using AF lifetime contrast in *ex vivo* specimens with point measurements [83] and lifetime imaging instrumentation [84].

10.3.5
Brain

AF lifetime studies of *ex vivo* brain samples have shown contrast between healthy and normal tissue using both single-point measurements [85, 86] and multiphoton fluorescence imaging has been used in combination with intraperitoneal administration of 5-aminolevulinic acid to identify cancerous tissue [87].

10.3.6
Skin

AF of the skin has been used for diagnosis since the introduction of the Woods lamp (a UV light source) in the earlier part of the twentieth century. Due to its accessibility and the emergence of a clinically licensed and commercially available two-photon microscope [88], there is a rapidly increasing literature concerning the application of MDFI to skin. Although most research has focused on skin malignancies, there is also interest in investigating non-cancer applications such as skin aging [89], inflammatory dermatoses such as eczema [90], and hypertrophic scars [91]. The application of fluorescent precursors to the skin, such as protophorphyrin IX, to aid diagnosis is also being increasingly investigated.

Using a single-point spectral probe, Brancaleon *et al.* measured the emission spectra *in vivo* and from fresh frozen samples of basal cell carcinomas (BCCs)/squamous cell carcinomas (SCCs) [92]. The samples were excited at 350, 360, 390 and 420 nm and fluorescence was collected at 390, 400 and 450 nm. They recorded a loss of AF from a region surrounding the malignant tissue that extended up to three times the size of the tumor. This corresponded to a histologically visible loss of collagen and elastin in 78% of tumors that were studied.

Using two-photon microscopy (TPM) at 810 nm excitation, Eichhorn *et al.* [93] measured the spectra from fresh skin ($n = 150$) and formalin-fixed, unstained sections ($n = 27$) of melanocytic lesions. They observed a single spectral peak at 490 nm in benign nevi and at 600 nm in nodular melanomas, with both peaks appearing in dysplastic nevi. These observations were also reported to be found *in vivo*.

TPM has also been used to study the morphological features of skin malignancies and been reported to show good correlation with histology. Studies of BCC/SCC *ex vivo* samples excited at 780 nm [94] showed irregularly distributed keratinocytes in SCC *in situ* with widened intercellular spaces. In pleomorphic cells and dyskeratotic cells, the cytoplasmic fluorescence intensity was found to be higher than in the surrounding cells. In BCC, nests of BCC cells were found in only one-third of BCCs, although the imaging depth was reported to be insufficient for complete assessment of these tumors. Melanomas have been studied both *ex vivo* and *in vivo* with excitation at 760 nm [95] and were observed to exhibit architectural disarray of the epidermis, poorly defined keratinocyte cell borders, and the presence of pleomorphic or dendritic cells.

In recent years, as FLIM instrumentation has improved, fluorescence lifetime contrast of skin lesions has also been reported. Galletly *et al.* [96] excited freshly excised BCCs at 355 nm and collected fluorescence above 375 nm using a wide-field FLIM system. The observed a significant lifetime difference between BCCs and normal perilesional skin (1.40 ns versus 1.55 ns, respectively) when fitting to a single exponential decay model. In contradiction, however, lifetime data from three out of four BCCs imaged using TPM with excitation at 740 nm by De Giorgi *et al.* [97] demonstrated a shift of 91 ps towards longer lifetimes.

Fluorescence lifetimes obtained *in vivo* and freshly excised *ex vivo* benign nevi and melanomas were recently reported by Dimitrow *et al.* using TPM with 760 nm excitation [98]. Although a difference was seen in the lifetimes (fitted to a double exponential decay model) between keratinocytes and melanocytes, no difference could be detected between benign and malignant tissue.

10.4 Conclusion

A wide range of different types of instrument has been developed for multidimensional fluorescence imaging and many of these have been applied to the study of tissue AF, including *ex vivo* studies of endogenous and exogenous fluorophores. A key outcome of the fluorescence microscopy and spectroscopy of *ex vivo* tissues has been the identification of the main endogenous fluorophores: the cellular components, tryptophan, tyrosine, NADH and FAD, the structural proteins, collagen and elastin, and the porphyrins. These results have provided a sound basis for the application of fluorescence techniques to distinguish between normal and neoplastic tissue *in vivo*.

Acknowledgments

The authors acknowledge funding in part from the UK Engineering and Physical Sciences Research Council (EPSRC), the UK Biotechnology and Biological Sciences Research Council (BBSRC), the Department of Trade and Industry (DTI)/Technology Strategy Board (TSB), the Wellcome Trust, and the European Commission (HEALTH-F5-2008-201577). Paul French acknowledges a Royal Society Wolfson Research Merit Award.

References

1 Lakowicz, J.R. (1999) *Principles of Fluorescence Spectroscopy*, 2nd edn, Kluwer Academic/Plenum Publishers, New York.
2 Borisova, E.G., Vladimirov, B., Terziev, I., Ivanova, R., and Avramov, L. (2009) 5-ALA/PpIX fluorescence detection of gastrointestinal neoplasia, *Proc. SPIE*, **7368**, 241–246.
3 American Cancer Society (2006) *Cancer Facts and Figures 2006*, American Cancer Society, Atlanta, GA.
4 Sokolov, K., Follen, M., and Richards-Kortum, R. (2002) Optical spectroscopy for detection of neoplasia. *Curr. Opin. Chem. Biol.*, **6**, 651–658.
5 Bigio, I.J. and Mourant, J.R. (1997) Ultraviolet and visible spectroscopies for tissue diagnostics: fluorescence spectroscopy and elastic-scattering spectroscopy. *Phys. Med. Biol.*, **42**, 803–814.
6 Richards-Kortum, R. and Sevick-Muraca, E. (1996) Quantitative optical spectroscopy for tissue diagnosis. *Annu. Rev. Phys. Chem.*, **47**, 555–606.
7 Wagnieres, G.A., Star, W.M., and Wilson, B.C. (1998) *In vivo* fluorescence

spectroscopy and imaging for oncological applications. *Photochem. Photobiol.*, **68**, 603–632.

8 Smith, M.W., Phelps, P.C., and Trump, B.F. (1991) Cytosolic Ca^{2+} deregulation and blebbing after $HgCl_2$ injury to cultured rabbit proximal tubule cells as determined by digital imaging microscopy. *Proc. Natl. Acad. Sci. U. S. A.*, **88** (11), 4926–4930.

9 Millot, J.M., Pingret, L., Angiboust, J.F., Bonhomme, A., Pinon, J.M., and Manfait, M. (1995) Quantitative determination of free calcium in subcellular compartments, as probed by Indo-1 and confocal microspectrofluorometry. *Cell Calcium*, **17** (5), 354–366.

10 Andersson-Engels, S., Johansson, J., Stenram, U., Svanberg, K., and Svanberg, S. (1990) Malignant tumour and atherosclerotic plaque diagnosis using laser induced fluorescence. *IEEE J. Quantum Electron.*, **26** (12), 2207–2217.

11 Chen, H.M., Chiang, C.P., You, C., Hsiao, T.C., and Wang, C.Y. (2005) Time-resolved autofluorescence spectroscopy for classifying normal and premalignant oral tissues. *Lasers Surg. Med.*, **37**, 37–45.

12 Elson, D., Requejo-Isidro, J., Munro, I., Reavell, F., Siegel, J., Suhling, K., Tadrous, P., Benninger, R., Lanigan, P., McGinty, J., Talbot, C., Treanor, B., Webb, S., Sandison, A., Wallace, A., Davis, D., Lever, J., Neil, M., Phillips, D., Stamp, G., and French, P. (2004) Time-domain fluorescence lifetime imaging applied to biological tissue. *Photochem. Photobiol. Sci.*, **3**, 795–801.

13 Das, B.B., Liu, F., and Alfano, R.R. (1997) Time-resolved fluorescence and photon migration studies in biomedical and model random media. *Rep. Prog. Phys.*, **60**, 227–291.

14 Dowling, K., Dayel, M.J., Lever, M.J., French, P.M.W., Hares, J.D., and Dymoke-Bradshaw, A.K.L. (1998) Fluorescence lifetime imaging with picosecond resolution for biomedical applications. *Opt. Lett.*, **23** (10), 810–812.

15 Cubeddu, R., Comelli, D., D'Andrea, C., Taroni, P., and Valentini, G. (2002) Time-resolved fluorescence imaging in biology and medicine. *J. Phys. D Appl. Phys.*, **35**, R61–R66.

16 Gadella, T.W. Jr. (ed.) (2008) *FRET and FLIM Imaging Techniques, Laboratory Techniques in Biochemistry and Molecular Biology*, vol. 33, Elsevier, Amsterdam.

17 Esposito, A., Gerritsen, H.C., and Wouters, F.S. (2007) Optimizing frequency-domain fluorescence lifetime sensing for high-throughput applications: photon economy and acquisition speed. *J. Opt. Soc. Am. A*, **24** (10), 3261–3273.

18 Fang, Q., Papaioannou, T., Jo, J.A., Vaitha, R., Shastry, K., and Marcu, L. (2004) Time-domain laser-induced fluorescence spectroscopy apparatus for clinical diagnostics. *Rev. Sci. Instrum.*, **75** (1), 151–162.

19 Pitts, J.D. and Mycek, M.A. (2001) Design and development of a rapid acquisition laser-based fluorometer with simultaneous spectral and temporal resolution. *Rev. Sci. Instrum.*, **72** (7), 3061–3072.

20 Glanzmann, T., Ballini, J.P., van den Bergh, H., and Wagnieres, G. (1999) Time-resolved spectrofluorometer for clinical tissue characterization during endoscopy. *Rev. Sci. Instrum.*, **70**, 4067–4077.

21 O'Connor, D.V. and Phillips, D. (1984) *Time-Correlated Single-Photon Counting*, Academic Press, London.

22 Becker, W. (2005) *Advanced Time-Correlated Single Photon Counting Techniques*, Springer, Berlin.

23 Denk, W., Strickler, J.H., and Webb, W.W. (1990) Two-photon laser scanning fluorescence microscopy. *Science*, **248**, 73–76.

24 Bugiel, I., Konig, K., and Wabnitz, H. (1989) Investigation of cells by fluorescence laser scanning microscopy with subnanosecond time resolution. *Lasers Life Sci.*, **3** (1), 47–53.

25 Gerritsen, H.C., Asselbergs, M.A.H., Agronskaia, A.V., and Van Sark, W.G.J.H.M. (2002) Fluorescence lifetime imaging in scanning microscopes: acquisition speed, photon economy and lifetime resolution. *J. Microsc.*, **206** (3), 218–224.

26 Carlsson, K. and Liljeborg, A. (1998) Simultaneous confocal lifetime imaging of multiple fluorophores using the intensity-modulated multiple-wavelength scanning (IMS) technique. *J. Microsc.*, **191** (2), 119–127.

27 Krishnan, R.V., Saitoh, H., Terada, H., Centonze, V.E., and Herman, B. (2003) Development of a multiphoton fluorescence lifetime imaging microscopy system using a streak camera. *Rev. Sci. Instrum.*, **74** (5), 2714–2721.

28 Grant, D.M., Elson, D.S., Schimpf, D., Dunsby, C., Requejo-Isidro, J., Auksorius, E., Munro, I., Neil, M.A.A., French, P.M.W., Nye, E., Stamp, G.W., and Courtney, P. (2005) Optically sectioned fluorescence lifetime imaging using a Nipkow disc microscope and a tunable ultrafast continuum excitation source. *Opt. Lett.*, **30**, 3353–3355.

29 van Munster, E.B., Goedhart, J., Kremers, G.J., Manders, E.M., and Gadella, T.W. (2007) Combination of a spinning disc confocal unit with frequency-domain fluorescence lifetime imaging microscopy. *Cytometry A*, **71**, 207–214.

30 Grant, D.M., McGinty, J., McGhee, E.J., Bunney, T.D., Owen, D.M., Talbot, C.B., Zhang, W., Kumar, S., Munro, I., Lanigan, P.M.P., Kennedy, G.T., Dunsby, C., Magee, A.I., Courtney, P., Katan, M., Neil, M.A.A., and French, P.M.W. (2007) High speed optically sectioned fluorescence lifetime imaging permits study of live cell signaling events. *Opt. Express*, **15**, 15656–15673.

31 Straub, M. and Hell, S.W. (1998) Fluorescence lifetime three-dimensional microscopy with picosecond precision using a multifocal multiphoton microscope. *Appl. Phys. Lett.*, **73**, 1769–1771.

32 Benninger, R.K.P., Hofmann, O., McGinty, J., Requejo-Isidro, J., Munro, I., Neil, M.A.A., deMello, A.J., and French, P.M.W. (2005) Time-resolved fluorescence imaging of solvent interactions in microfluidic devices. *Opt. Express*, **13** (16), 6275–6285.

33 Leveque-Fort, S., Fontaine-Aupart, M.P., Roger, G., and Georges, P. (2004) Fluorescence-lifetime imaging with a multifocal two-photon microscope. *Opt. Lett.*, **29**, 2884–2886.

34 Requejo-Isidro, J., McGinty, J., Munro, I., Elson, D.S., Galletly, N., Lever, M.J., Neil, M.A.A., Stamp, G.W.H., French, P.M.W., Kellet, P.A., Hares, J.D., and Dymoke-Bradshaw, A.K.L. (2004) High-speed wide-field time-gated endoscopic fluorescence lifetime imaging. *Opt. Lett.*, **29** (19), 2249–2251.

35 Agronskaia, A.V., Tertoolen, L., and Gerritsen, H.C. (2003) High frame rate fluorescence lifetime imaging. *J. Phys. D Appl. Phys.*, **36**, 1655–1662.

36 Morgan, C.G., Mitchell, A.C., and Murray, J.G. (1990) Nanosecond time-resolved fluorescence microscopy: principles and practice. *Trans. R. Microsc. Soc.*, **1**, 463–466.

37 Gratton, E., Feddersen, B., and van de Ven, M. (1990) Parallel acquisition of fluorescence decay using array detectors, *Proc. SPIE*, **1204**, 21–25.

38 Gadella, T.W.J., Jovin, T.M., and Clegg, R.M. (1993) Fluorescence lifetime imaging microscopy (FLIM) – spatial resolution of structures on the nanosecond timescale. *Biophys. Chem.*, **48**, 221–239.

39 Oida, T., Sako, Y., and Kusumi, A. (1993) Fluorescence lifetime imaging microscopy (flimscopy). Methodology development and application to studies of endosome fusion in single cells. *Biophys. J.*, **64**, 676–685.

40 Scully, A.D., MacRobert, A.J., Botchway, S., O'Neill, P., Parker, A.W., Ostler, R.B., and Phillips, D. (1996) Development of a laser-based fluorescence microscope with subnanosecond time resolution. *J. Fluoresc.*, **6** (2), 119–125.

41 Dowling, K., Dayel, M.J., Lever, M.J., French, P.M.W., Hares, J.D., and Dymoke-Bradshaw, A.K.L. (1998) Fluorescence lifetime imaging with picosecond resolution for biomedical applications. *Opt. Lett.*, **23** (10), 810–812.

42 Hares, J.D. (1987) Advances in sub-nanosecond shutter tube technology and applications in plasma physics, *Proc. SPIE*, **831**, 165–170.

43 Munro F I., McGinty, J., Galletly, N., Requejo-Isidro, J., Lanigan, P.M.P., Elson, D.S., Dunsby, C., Neil, M.A.A., Lever, M.J., Stamp, G.W.H., and French, P.M.W. (2005) Toward the clinical application of time-domain fluorescence lifetime imaging. *J. Biomed. Opt.*, **10**, 051403.

44 Holub, O., Seufferheld, M.J., Gohlke, G., and Clegg, R.M. (2000) Fluorescence lifetime imaging (FLI) in real-time – a new technique in photosynthesis research. *Photosynthetica*, **38** (4), 581–599.

45 Mizeret, J., Stepinac, T., Hansroul, M., Studzinski, A., van den Bergh, H., and Wagnieres, G. (1999) Instrumentation for real-time fluorescence lifetime imaging in endoscopy. *Rev. Sci. Instrum.*, **70** (12), 4689–4701.

46 Becker, W., Bergmann, A., Haustein, E., Petrasek, Z., Schwille, P., Biskup, C., Kelbauskas, L., Benndorf, K., Klocker, N., Anhut, T., Riemann, I., and Konig, K. (2006) Fluorescence lifetime images and correlation spectra obtained by multidimensional time-correlated single photon counting. *Microsc. Res. Tech.*, **69**, 186–195.

47 Robinson, T., Valluri, P., Manning, H.B., Owen, D.M., Munro, I., Talbot, C.B., Dunsby, C., Eccleston, J.F., Baldwin, G.S., Neil, M.A.A., de Mello, A.J., and French, P.M.W. (2008) Three-dimensional molecular mapping in a microfluidic mixing device using fluorescence lifetime imaging. *Opt. Lett.*, **33** (16), 1887–1889.

48 Siegel, J., Elson, D.S., Webb, S.E.D., Parsons-Karavassilis, D., Lévêque-Fort, S., Cole, M.J., Lever, M.J., French, P.M.W., Neil, M.A.A., Juskaitis, R., Sucharov, L.O., and Wilson, T. (2001) Whole-field five-dimensional fluorescence microscopy combining lifetime and spectral resolution with optical sectioning. *Opt. Lett.*, **26**, 1338–1340.

49 Bird, D.K., Eliceiri, K.W., Fan, C.H., and White, J.G. (2004) Simultaneous two-photon spectral and lifetime fluorescence microscopy. *Appl. Opt.*, **43**, 5173–5182.

50 Wachman, E.S., Niu, W.H., and Farkas, D.L. (1996) Imaging acousto-optic tunable filter with 0.35-micrometer spatial resolution. *Appl. Opt.*, **35**, 5220–5226.

51 Farkas, D.L., Du, C.W., Fisher, G.W., Lau, C., Niu, W.H., Wachman, E.S., and Levenson, R.M. (1998) Non-invasive image acquisition and advanced processing in optical bioimaging. *Comput. Med. Imaging Graph.*, **22**, 89–102.

52 Hanley, Q.S., Verveer, P.J., and Jovin, T.M. (1999) Spectral imaging in a programmable array microscope by Hadamard transform fluorescence spectroscopy. *Appl. Spectrosc.*, **53**, 1–10.

53 Schultz, R.A., Nielsen, T., Zavaleta, J.R., Ruch, R., Wyatt, R., and Garner, H.R. (2001) Hyperspectral imaging: a novel approach for microscopic analysis. *Cytometry*, **43**, 239–247.

54 De Beule, P., Owen, D.M., Manning, H.B., Talbot, C.B., Requejo-Isidro, J., Dunsby, C., McGinty, J., Benninger, R.K.P., Elson, D.S., Munro, I., Lever, M.J., Anand, P., Neil, M.A.A., and French, P.M.W. (2007) Rapid hyperspectral fluorescence lifetime imaging. *Microsc. Res. Tech.*, **70**, 481–484.

55 Marcu, L., Jo, J.A., Fang, Q.Y., Papaioannou, T., Reil, T., Qiao, J.H., Baker, J.D., Freischlag, J.A., and Fishbein, M.C. (2009) Detection of rupture-prone atherosclerotic plaques by time-resolved laser-induced fluorescence spectroscopy. *Atherosclerosis*, **204** (1), 156–164.

56 Talbot, C.B., McGinty, J., McGhee, E., Owen, D., Grant, D., Kumar, S., De Beule, P., Auksorius, E., Manning, H., Galletly, N.P., Treanor, B., Kennedy, G.T., Lanigan, P.M.P., Munro, I., Elson, D.S., McGee, A., Davis, D., Neil, M.A.A., Stamp, G.W.H., Dunsby, C., and French, P.M.W. (2010) Fluorescence lifetime imaging and metrology for biomedicine, in *Handbook of Photonics for Biomedical Science* (ed. V. Tuchin), Taylor and Francis/CRC Press, Boca Raton, FL, pp. 159–196

57 Andersson-Engels, S., Johansson, J., Svanberg, K., and Svanberg, S. (1991) Fluorescence imaging and point measurements of tissue: applications to the demarcation of malignant tumors and atherosclerotic lesions from normal tissue. *Photochem. Photobiol.*, **53** (6), 807–814.

58 Ramanujam, N. (2000) Fluorescence spectroscopy of neoplastic and non-neoplastic tissues. *Neoplasia*, **2**, 89–117.

59 Ingrams, D.R., Dhingra, J.K., Roy, K., Perrault, D.F. Jr., Bottrill, I.D., Kabani, S., Rebeiz, E.E., Pankratov, M.M., Shapshay, S.M., Manoharan, R., Itzkan, I., and Feld, M.S. (1997) Autofluorescence characteristics of oral mucosa. *Head Neck*, **19**, 27–32.

60 Majumder, S.K., Gupta, P.K., and Uppal, A. (1999) Autofluorescence spectroscopy of tissues from human oral cavity for discriminating malignant from normal. *Lasers Life Sci.*, **8**, 211–227.

61 Wang, C.Y., Chiang, H.K., Chen, C.T., Chiang, C.P., Kuo, Y.S., and Chow, S.N. (1999) Diagnosis of oral cancer by light-induced autofluorescence spectroscopy using double excitation wavelengths. *Oral Oncol.*, **35**, 144–150.

62 Uppal, A. and Gupta, P.K. (2003) Measurement of NADH concentration in normal and malignant human tissues from breast and oral cavity. *Biotechnol. Appl. Biochem.*, **37**, 45–50.

63 Chwirot, B., Jedrzejczyk, W., Chwirot, S., Michniewicz, Z., and Redzinski, J. (1996) Tissue spectroscopy. New generation of optical methods for cancer detection. *Pol. Merkur. Lekarski.*, **1**, 355–358 (in Polish).

64 Chwirot, B.W., Chwirot, S., Jedrzejczyk, W., Jackowski, M., Raczynska, A.M., Winczakiewicz, J., and Dobber, J. (1997) Ultraviolet laser-induced fluorescence of human stomach tissues: detection of cancer tissues by imaging techniques. *Lasers Surg. Med.*, **21**, 149–158.

65 Abe, S., Izuishi, K., Tajiri, H., Kinoshita, T., and Matsuoka, T. (2000) Correlation of *in vitro* autofluorescence endoscopy images with histopathologic findings in stomach cancer. *Endoscopy*, **32**, 281–286.

66 Silveira, L. Jr., Betiol Filho, J.A., Silveira, F.L., Zangaro, R.A., and Pacheco, M.T. (2008) Laser-induced fluorescence at 488 nm excitation for detecting benign and malignant lesions in stomach mucosa. *J. Fluoresc.*, **18**, 35–40.

67 Xiao, S.D., Ge, Z.Z., Zhong, L., and Luo, H.Y. (2002) Diagnosis of gastric cancer by using autofluorescence spectroscopy. *Chin. Dig. Dis.*, **3**, 99–102.

68 Xiao, S.D., Zhong, L., Luo, H.Y., Chen, X.Y., and Shi, Y. (2002) Autofluorescence imaging analysis of gastric cancer. *Chin. Dig. Dis.*, **3**, 95–98.

69 Kapadia, C.R., Cutruzzola, F.W., O'Brien, K.M., Stetz, M.L., Enriquez, R., and Deckelbaum, L.I. (1990) Laser-induced fluorescence spectroscopy of human colonic mucosa. Detection of adenomatous transformation. *Gastroenterology*, **99**, 150–157.

70 Richards-Kortum, R., Rava, R.P., Petras, R.E., Fitzmaurice, M., Sivak, M., and Feld, M.S. (1991) Spectroscopic diagnosis of colonic dysplasia. *Photochem. Photobiol.*, **53**, 777–786.

71 Yang, Y., Tang, G.C., Bessler, M., and Alfano, R.R. (1995) Fluorescence spectroscopy as a photonic pathology method for detecting colon cancer. *Lasers Life Sci.*, **6**, 259–276.

72 Chwirot, B.W., Michniewicz, Z., Kowalska, M., and Nussbeutel, J. (1999) Detection of colonic malignant lesions by digital imaging of UV laser-induced autofluorescence. *Photochem. Photobiol.*, **69**, 336–340.

73 Demos, S.G., Gandour-Edwards, R., Ramsamooj, R., and deVere White, R. (2004) Spectroscopic detection of bladder cancer using near-infrared imaging techniques. *J. Biomed. Opt.*, **9**, 767–771.

74 Cicchi, R., Crisci, A., Nesi, G., Cosci, A., Giancane, S., Carini, M., and Pavone, F.S. (2009) Time- and spectral-resolved multiphoton imaging of fresh bladder biopsies, *Proc. SPIE*, **7367**, 73670M–73679M.

75 Palmer, G.M., Keely, P.J., Breslin, T.M., and Ramanujam, N. (2003) Autofluorescence spectroscopy of normal and malignant human breast cell lines. *Photochem. Photobiol.*, **78**, 462–469.

76 Conklin, M.W., Provenzano, P.P., Eliceiri, K.W., Sullivan, R., and Keely, P.J. (2009) Fluorescence lifetime imaging of endogenous fluorophores in histopathology sections reveals differences between normal and tumor epithelium in carcinoma *in situ* of the breast. *Cell Biochem. Biophys.*, **53**, 145–157.

77 Yang, Y., Katz, A., Celmer, E.J., Zurawska-Szczepaniak, M., and Alfano, R.R. (1997) Fundamental differences of excitation spectrum between malignant and benign breast tissues. *Photochem. Photobiol.*, **66**, 518–522.

78 Breslin, T.M., Xu, F., Palmer, G.M., Zhu, C., Gilchrist, K.W., and Ramanujam, N. (2004) Autofluorescence and diffuse reflectance properties of malignant and benign breast tissues. *Ann. Surg. Oncol.*, **11**, 65–70.

79 Demos, S.G., Gandour-Edwards, R., Ramsamooj, R., and White, R. (2004) Near-infrared autofluorescence imaging for detection of cancer. *J. Biomed. Opt.*, **9**, 587–592.

80 Keller, M.D., Majumder, S.K., Kelley, M.C., Meszoely, I.M., Boulos, F.I., Olivares, G.M., and Mahadevan-Jansen, A. (2010) Autofluorescence and diffuse reflectance spectroscopy and spectral imaging for breast surgical margin analysis. *Lasers Surg. Med.*, **42**, 15–23.

81 Jacobs, V.R., Paepke, S., Schaaf, H., Weber, B.C., and Kiechle-Bahat, M. (2007) Autofluorescence ductoscopy: a new imaging technique for intraductal breast endoscopy. *Clin. Breast Cancer*, **7**, 619–623.

82 Douplik, A., Leong, W.L., Easson, A.M., Done, S., Netchev, G., and Wilson, B.C. (2009) Feasibility study of autofluorescence mammary ductoscopy. *J. Biomed. Opt.*, **14**, 044036.

83 Pradhan, A., Das, B.B., Yoo, K.M., Cleary, J., Prudente, R., Celmer, E., and Alfano, R.R. (1992) Time-resolved UV photoexcited fluorescence kinetics from malignant and non-malignant human breast tissues. *Lasers Life Sci.*, **4**, 225–234.

84 Provenzano, P.P., Eliceiri, K.W., Yan, L., Ada-Nguema, A., Conklin, M.W., Inman, D.R., and Keely, P.J. (2008) Nonlinear optical imaging of cellular processes in breast cancer. *Microsc. Microanal.*, **14**, 532–548.

85 Marcu, L., Jo, J.A., Butte, P.V., Yong, W.H., Pikul, B.K., Black, K.L., and Thompson, R.C. (2004) Fluorescence lifetime spectroscopy of glioblastoma multiforme. *Photochem. Photobiol.*, **80**, 98–103.

86 Butte, P.V., Pikul, B.K., Hever, A., Yong, W.H., Black, K.L., and Marcu, L. (2005) Diagnosis of meningioma by time-resolved fluorescence spectroscopy. *J. Biomed. Opt.*, **10** (6), 064026–064029.

87 Kantelhardt, S.R., Diddens, H., Leppert, J., Rohde, V., Huttmann, G., and Giese, A. (2008) Multiphoton excitation fluorescence microscopy of 5-aminolevulinic acid induced fluorescence in experimental gliomas. *Lasers Surg. Med.*, **40**, 273–281.

88 König, K. (2008) Clinical multiphoton tomography. *J. Biophotonics*, **1** (1), 13–23.

89 Koehler, M.J., Konig, K., Elsner, P., Buckle, R., and Kaatz, M. (2006) In vivo assessment of human skin aging by multiphoton laser scanning tomography. *Opt. Lett.*, **31**, 2879–2881.

90 Lee, J.H., Chen, S.Y., Yu, C.H., Chu, S.W., Wang, L.F., Sun, C.K., and Chiang, B.L. (2009) Noninvasive in vitro and in vivo assessment of epidermal hyperkeratosis and dermal fibrosis in atopic dermatitis. *J. Biomed. Opt.*, **14**, 014008.

91 Chen, G., Chen, J., Zhuo, S., Xiong, S., Zeng, H., Jiang, X., Chen, R., and Xie, S. (2009) Nonlinear spectral imaging of human hypertrophic scar based on two-photon excited fluorescence and second-harmonic generation. *Br. J. Dermatol.*, **161** (1), 48–55.

92 Brancaleon, L., Durkin, A.J., Tu, J.H., Menaker, G., Fallon, J.D., and Kollias, N. (2001) In vivo fluorescence spectroscopy of nonmelanoma skin cancer. *Photochem. Photobiol.*, **73**, 178–183.

93 Eichhorn, R., Wessler, G., Scholz, M., Leupold, D., Stankovic, G., Buder, S., Stucker, M., and Hoffmann, K. (2009) Early diagnosis of melanotic melanoma based on laser-induced melanin fluorescence. *J. Biomed. Opt.*, **14**, 034033.

94 Paoli, J., Smedh, M., Wennberg, A.M., and Ericson, M.B. (2008) Multiphoton laser scanning microscopy on non-melanoma skin cancer: morphologic features for future non-invasive diagnostics. *J. Invest. Dermatol.*, **128**, 1248–1255.

95 Dimitrow, E., Ziemer, M., Koehler, M.J., Norgauer, J., Konig, K., Elsner, P., and Kaatz, M. (2009) Sensitivity and specificity of multiphoton laser tomography for

in vivo and *ex vivo* diagnosis of malignant melanoma. *J. Invest. Dermatol.*, **129**, 1752–1758.

96 Galletly, N.P., McGinty, J., Dunsby, C., Teixeira, F., Requejo-Isidro, J., Munro, I., Elson, D.S., Neil, M.A., Chu, A.C., French, P.M., and Stamp, G.W. (2008) Fluorescence lifetime imaging distinguishes basal cell carcinoma from surrounding uninvolved skin. *Br. J. Dermatol.*, **159**, 152–161.

97 De Giorgi, V., Massi, D., Sestini, S., Cicchi, R., Pavone, F.S., and Lotti, T. (2009) Combined non-linear laser imaging (two-photon excitation fluorescence microscopy, fluorescence lifetime imaging microscopy, multispectral multiphoton microscopy) in cutaneous tumours: first experiences. *J. Eur. Acad. Dermatol. Venereol.*, **23**, 314–316.

98 Dimitrow, E., Riemann, I., Ehlers, A., Koehler, M.J., Norgauer, J., Elsner, P., Konig, K., and Kaatz, M. (2009) Spectral fluorescence lifetime detection and selective melanin imaging by multiphoton laser tomography for melanoma diagnosis. *Exp. Dermatol.*, **18**, 509–515.

11
In Vivo Instrumentation
Herbert Schneckenburger, Michael Wagner, Petra Weber, and Thomas Bruns

11.1
Introduction

Optical analysis of human tissue requires homogeneous illumination with thermally tolerable, non-phototoxic light doses, together with high-resolution and sensitive detection. For cell cultures and biopsies, microscopic methods are commonly used, including transillumination, reflection, and fluorescence techniques. For patients, more specific systems have to be applied for various organs, which are based on backscattering or fluorescence measurements. Again, microscopic techniques, but also various endoscopic systems, are appropriate for high-resolution imaging. In addition to these incoherent systems, optical coherence tomography [1] and also interferometric and holographic methods [2] are used increasingly, if high axial resolution is required. These techniques are based on interference of an electromagnetic wave with a reference wave, and only in a well-defined depth of the sample is positive interference attained. In spite of important achievements with these coherent techniques, this chapter focuses on incoherent methods and instrumentation, including microscopy, endoscopy, and imaging equipment.

11.2
Microscopy

Only for thin samples, for example, cell monolayers or thin biopsies, can conventional transillumination microscopy be used. Köhler's illumination and additional equipment for phase contrast or interference contrast microscopy provide optimum image quality. Most *in vivo* samples, however, require epi-illumination either by darkfield equipment (e.g., a darkfield objective lens) for reflection measurements or by a dichromatic mirror in combination with appropriate filters for fluorescence experiments. Fluorescence microscopy currently is the most widespread method, since many cellular compartments or organelles can be stained selectively with various fluorescent dyes or nanoparticles [3], and a large number of fluorescent

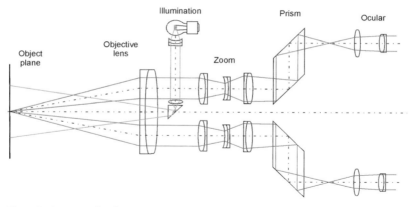

Figure 11.1 Principle of an operation microscope.

proteins can be encoded genetically [4]. Application of fluorescent probes for humans, however, is restricted to very few dyes, for example, fluorescein or indocyanine green used for fluorescence angiography or blood volume determination [5]. Therefore, intrinsic fluorescence, for example, of the free and protein-bound coenzyme nicotinamide adenine dinucleotide (NADH) [6], plays an increasing role in the diagnosis of patients. Whereas conventional fluorescence microscopy generally does not provide axial resolution, laser scanning microscopy [7] and some wide-field techniques using structured [8] or selective plane [9] illumination permit measurements of well-defined layers of a sample and 3D image reconstruction.

In recent years, the operation microscope has become an essential tool in microsurgery with application in many clinical disciplines, for example, ophthalmology, ear, nose, and throat (ENT), gynecology, and neurosurgery. In comparison with *in vitro* microscopy, ultimate resolution is not required, but a large field of vision, high focal depth, long working distance, and stereoscopic observation. Commonly, a binocular tube with an angle ranging from 3 to 10° between individual light paths is used, as depicted in Figure 11.1. An achromatic objective lens with a numerical aperture A_N between 0.01 and 0.05 creates two parallel light beams which are focused to an image plane and further visualized by an ocular. A zoom system may be used to provide variable magnification, which for the whole microscope is between 5× and 20×. With the comparably low A_N, a resolution between 6 and 30 μm and a focal depth up to a few millimeters are attained. Epi-illumination systems including various kinds of light sources, for example, light-emitting diodes (LEDs), halogen or xenon lamps, fiber-optics and specific focusing devices, are commonly used.

11.3
Endoscopy

Endoscopy is the method of choice for measurements inside a body or a specific organ, often in combination with surgery or certain therapies, for example, photo-

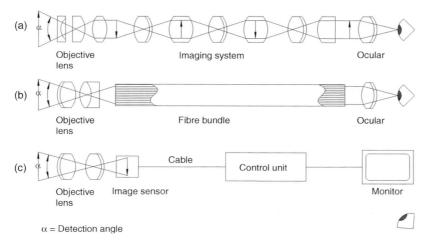

Figure 11.2 Types of endoscopes: (a) rigid endoscope with a system of lenses; (b) flexible endoscope with an oriented fiber bundle; (c) flexible endoscope with electronic image transfer.

dynamic therapy [10]. Different light paths are used for illumination and image detection, in either a reflection or a fluorescence mode. Depending on the specific application, either rigid or flexible endoscopes are preferred, as depicted in Figure 11.2. In rigid endoscopes, the detection path consists of a system of lenses generating an image of the object either in an ocular or on a camera chip. These endoscopes include arthroscopes, laryngoscopes, bronchoscopes, laparoscopes, cystoscopes, and rectoscopes. In contrast, flexible endoscopy is based either on optical image transfer by an oriented fiber bundle or on electronic transfer with an image sensor being located close to the object. Generally, image quality and resolution are better in the latter case. Like rigid endoscopes, flexible endoscopes can be used as bronchoscopes, cystoscopes, or rectoscopes, but there are additional applications in, for example, gastroscopy, choledochoscopy, and coloscopy. For all optical imaging systems, specific techniques can be used, for example, laser scanning endoscopy for high-resolution imaging [11] and integration of grin lenses to keep the diameter of the detection channel small at relatively high numerical aperture [12].

11.4
Imaging Techniques

In modern microscopy and endoscopy visual observation by an ocular is increasingly being replaced by high-performance imaging systems with charge-coupled device (CCD) or electron-multiplying charge-coupled device (EMCCD) cameras whose sensitivities range between about 10^{-13} and 10^{-18} W per pixel. Alternatively, the complementary metal oxide semiconductor (CMOS) detector camera and the image-intensified charge-coupled device (ICCD) camera are becoming more and more popular.

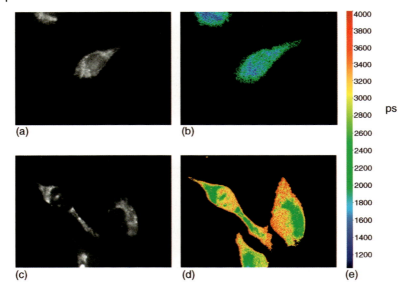

Figure 11.3 Fluorescence intensity (a, c) and lifetime images (b, d) of U251-MG glioblastoma cells (a, b) and U251-MG cells with an activated suppressor gene PTEN (c, d). (e) Scale of fluorescence lifetimes. Excitation wavelength, 375 nm; images measured at $\lambda \geq 420$ nm; image size, $150 \times 200\,\mu m$. Data from BMBF report No. 1792C08 (2010). Cells kindly provided by Professor Jan Mollenhauer, Department of Molecular Oncology, University of South Denmark, Odense, Denmark.

In addition to conventional imaging methods, some special techniques, including spectral imaging, polarization, and fluorescence lifetime imaging (FLIM) are increasingly being used. Together with spatial distribution, spectral or lifetime parameters can give valuable information about molecular or cellular interactions of a metabolite or a dye molecule with its microenvironment. This may include pH values, microviscosities, aggregation, or transfer of optical excitation energy between adjacent molecules. An example of FLIM is given in Figure 11.3 for intrinsic fluorescence of U251-MG human glioblastoma cells (a, b) and U251-MG cells with an activated tumor suppressor gene PTEN. Whereas the fluorescence intensities (a, c) show similar patterns of distribution, the fluorescence lifetimes are generally longer for the less malignant cells (d) than for the tumor cells (b), possibly due to different metabolic pathways. Fluorescence lifetime images can be deduced either from fluorescence decay kinetics recorded for each pixel of a laser scanning microscope [13] or from fluorescence intensities (I_1, I_2) recorded within two sub-nanosecond time gates shifted by a time interval Δt between one another according to the equation $\tau = \Delta t/\ln(I_1/I_2)$, where τ corresponds to the real fluorescence lifetime in the case of a monoexponential decay and to an "effective fluorescence lifetime" in the case of a multi-exponential decay.

References

1. Fujimoto, J.G. (2003) Optical coherence tomography for ultrahigh resolution *in vivo* imaging. *Nat. Biotechnol.*, **21**, 1361–1367.
2. Kemper, B. and von Bally, G. (2008) Digital holographic microscopy for live cell applications and technical inspection. *Appl. Opt.*, **47**, A52–A61.
3. Bruchez, M., Moronne, M., Gin, P., Weis, S., and Alivisatos, A.P. (1998) Semiconductor nanocrystals as fluorescence biological labels. *Science*, **281**, 2013–2016.
4. Rizzuto, R., Brini, M., Pizzo, P., Murgia, M., and Pozzan, T. (1995) Chimeric green fluorescent protein as a tool for visualizing subcellular organelles in living cells. *Curr. Biol.*, **5**, 635–642.
5. Brancato, R. and Trabucchi, G. (1998) Fluorescein and indocyanine green angiography in vascular chorioretinal diseases. *Semin. Ophthalmol.*, **13**, 189–198.
6. Galeotti, T., van Rossum, G.D.V., Mayer, D.H., and Chance, B. (1970) On the fluorescence of NAD(P)H in whole cell preparations of tumours and normal tissues. *Eur. J. Biochem.*, **17**, 485–496.
7. Pawley, J. (1990) *Handbook of Biological Confocal Microscopy*, Plenum Press, New York.
8. Neil, M.A.A., Juskaitis, R., and Wilson, T. (1997) Method of obtaining optical sectioning by structured light in a conventional microscope. *Opt. Lett.*, **22**, 1905–1907.
9. Huisken, J., Swoger, J., del Bene, F., Wittbrodt, J., and Stelzer, E.H.K. (2004) Optical sectioning deep inside live embryos by SPIM. *Science*, **305**, 1007–1009.
10. Dougherty, T.J. and Marcus, S.L. (1992) Photodynamic therapy. *Eur. J. Cancer*, **28**, 1734–1742.
11. de Palma, G.D. (2009) Confocal laser endomicroscopy in the *"in vivo"* histological diagnosis of the gastrointestinal tract. *World J. Gastroenterol.*, **15**, 5770–5775.
12. Fu, L., Gan, X., and Gu, M. (2005) Characterization of gradient-index lens-fiber spacing towards application in two-photon fluorescence endoscopy. *Appl. Opt.*, **44**, 7270–7274.
13. Becker, W. (2005) *Advanced Time-Correlated Single Photon Counting Techniques*, Springer, Berlin.

12
Fluorescent Probes (Fluorescence Standards, Green Protein Technology)
Markus Sauer

Molecules displaying strong fluorescence possess delocalized electrons formally present in conjugated double bonds. Most proteins and all nucleic acids are colorless in the visible region of the spectrum. However, they exhibit absorption and emission in the ultraviolet (UV) region. Natural fluorophores in tissue include nicotinamide adenine dinucleotide (NADH) and flavin adenine dinucleotide (FAD), structural proteins such as collagen, elastin, and their cross-links, and the aromatic amino acids, each of which has a characteristic wavelength for excitation with an associated characteristic emission. A serious limitation of native fluorescence detection of, for example, aromatic amino acids is their low photostability under one- and two-photon excitation conditions, which renders difficult the application of native fluorescence for highly sensitive detection schemes, for example, single-molecule fluorescence spectroscopy [1, 2].

Phycobiliproteins derived from cyanobacteria and eukaryotic algae belong to a class of relatively bright water-soluble natural fluorophores [3]. They absorb light in the visible wavelength range that is poorly utilized by chlorophyll and, through fluorescence energy transfer, convey the energy to the membrane-bound photosynthetic reaction centers where fast electron transfer occurs with high efficiency, converting solar energy to chemical energy. Phycobiliproteins are classified on the basis of their color into phycoerythrins (red) and phycocyanins (blue), with absorption maxima lying between 490 and 570 nm and between 610 and 665 nm, respectively. For example, B-phycoerythrin is comprised of three polypeptide subunits forming an aggregate containing a total of 34 bilin chromophores. It exhibits an absorption cross-section equivalent to that of ~20 rhodamine 6G chromophores and the highest fluorescence quantum yield of all phycobiliproteins of 0.98 with a fluorescence lifetime of 2.5 ns [4–6]. B-phycoerythrin was the first species to be detected at the single-molecule level using laser-induced fluorescence [6–8]. However, single allophycocyanin molecules with a smaller fluorescence quantum yield of 0.68 [9] can also be easily visualized immobilized on a cover-glass surface under aqueous buffer [10].

Organic fluorophores or fluorescent dyes are characterized by a strong absorption and emission band in the visible region of the electromagnetic spectrum (i.e., between 400 and 700 nm). The long-wavelength absorption band of a fluorophore is attributed to the transition from the electronic ground state S_0 to the first excited singlet state S_1. Since the transition moment for this process is typically very large, the corresponding absorption bands exhibit oscillator strengths on the order of unity. The reverse process $S_1 \rightarrow S_0$ is responsible for spontaneous emission known as fluorescence and for stimulated emission [11]. Since the earlier use of organic fluorescent dyes for qualitative and quantitative determination of analyte molecules (especially for automated DNA sequencing in the 1980s), their importance for bioanalytical applications has increased considerably [12–15]. Owing to the enhanced demand for fluorescent markers, the development of new fluorescent dyes has increased notably in recent years [16–21]. Using fluorescent dyes, extremely high sensitivity down to the single-molecule level can be achieved. Furthermore, time- and position-resolved detection without contacting the analyte is possible. Typically, the fluorescent probes or markers are identified and quantified by their absorption and fluorescence emission wavelengths and intensities. The sensitivity achievable with a fluorescent label is directly proportional to the molar extinction coefficient for absorption and the quantum yield of fluorescence. The extinction coefficient typically has maximum values of about $10^5 \, l \, mol^{-1} \, cm^{-1}$ in organic fluorophores and the quantum yield may approach values close to 100%. Absorption and emission spectra and also the fluorescence quantum yield and lifetime are dependent on environmental factors. Furthermore, for labeling of biological compounds, for example, antibodies and DNA/RNA, the dye must carry a functional group suitable for a mild covalent coupling reaction, preferentially with free amino or thiol groups of the analyte. In addition, the fluorophore should be as hydrophilic as possible to avoid aggregation and nonspecific binding in aqueous solvents. Today, a broad palette of fluorescent dyes (see Figure 12.1) activated as N-hydroxysuccinimide (NHS) ester, sulfo-NHS ester, isothiocyanates, or maleimides for mild covalent coupling to amino or thiol groups have been commercialized by several companies (see, e.g., www.invitrogen.com, www.atto-tec.com, www.dyomics.com).

Suitable organic fluorescent dyes are distinguished by a high fluorescence quantum yield. That is, upon excitation into the excited singlet state and subsequent relaxation to the lowest vibronic level of S_1, the radiative decay to the singlet ground state S_0 is the preferred deactivation pathway. However, there are many nonradiative processes that can compete efficiently with light emission and thus reduce fluorescence efficiency to an extent that depends in a complicated fashion on the molecular structure of the dye. Internal conversion, that is, the nonradiative decay of the lowest excited singlet state S_1 directly to the ground state S_0, is mostly responsible for the loss of fluorescence efficiency of organic dyes. In addition to fluorescence quenching via photoinduced electron transfer (PET) or Förster resonance energy transfer (FRET), a molecule excited to S_1 may enter a triplet state and relax to the lowest level T_1. The occupation of triplet states is undesirable with respect to highly sensitive fluorescence applications such as single-molecule fluorescence spectroscopy [23].

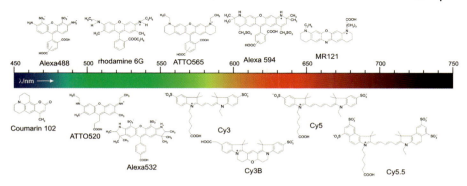

Figure 12.1 Molecular structures of some prominent commercially available organic fluorophores.

Figure 12.1 gives the molecular structures and rough absorption and emission properties of some of the most widely used organic fluorophores. For example, Alexa Fluor 350 and 430 and the two dyes ATTO 390 and ATTO 425 belong to the class of commercially available coumarin derivatives. Their absorption maxima are reflected in their names, that is, the absorptions maxima are at ~350, 430, 390, and 425 nm, respectively. The emission maxima of the three coumarins are located around 445, 545, 480, and 485 nm, respectively. They exhibit extinction coefficients in the range $20\,000-50\,000\,l\,mol^{-1}\,cm^{-1}$ with fluorescence quantum yields of up to 0.90 and lifetimes between 3 and 4 ns. Today, most (bio)analytical applications requiring high sensitivity use, in addition to carbocyanines, rhodamine dyes that absorb and emit in the wavelength range 500–700 nm. Owing to their structural rigidity, rhodamine dyes show high fluorescence quantum yields. They exhibit a small Stokes shift of about 20–30 nm and are applied as complementary probes together with other fluorophores in double-label staining, as energy donors and acceptors in FRET experiments, and as fluorescent markers in DNA sequencing and immunoassays. To increase the water solubility of the fluorophores for bioanalytical applications, the chromophores are often modified by the attachment of sulfonate groups. There are several functional rhodamine derivatives commercially available for biological labeling: ATTO 488, ATTO 520 and ATTO 532 related to rhodamine 6G, ATTO 550 related to rhodamine B, ATTO 565 related to rhodamine 630, ATTO 590, and ATTO 594. The molecular structures of the Alexa derivatives Alexa 488, Alexa 532, Alexa 546, Alexa 568, Alexa 594, and Alexa 633 are also based on rhodamines (Figure 12.1) [24]. Typically, rhodamine dyes exhibit fluorescence quantum yields close to unity (100%) and a fluorescence lifetime of ~4 ns.

Owing to the small number of compounds that demonstrate intrinsic fluorescence above 600 nm, the use of near-infrared (NIR) fluorescence detection in bioanalytical samples is a desirable alternative to visible fluorescence detection. This has prompted current efforts to use NIR dyes for bioanalytical applications. However, there are few fluorophores that show sufficient fluorescence quantum yields in the NIR region, especially in aqueous surroundings, and that can be coupled covalently to analyte molecules [25]. Otherwise, there are several advantages of using fluorescent dyes that absorb in the red region over those that absorb at shorter blue and green wavelengths. The most important of these advantages is the reduction in background that ultimately improves the sensitivity achievable. There are three major sources of background: (i) elastic scattering, that is, Rayleigh scattering, (ii) inelastic scattering, that is, Raman scattering, and (iii) fluorescence from impurities. The efficiency of both Rayleigh and Raman scattering are dramatically reduced by shifting to longer wavelength excitation (scales with $1/\lambda^4$). Likewise, the number of fluorescent impurities is significantly reduced with longer excitation and detection wavelengths.

One type of red-absorbing fluorophores is cyanine dyes. Cyanine derivatives belong to the class of polymethine dyes, that is, planar fluorophores with conjugated double bonds where all atoms of the conjugated chain lie in a common plane linked by σ-bonds. Figure 12.1 also gives the molecular structures of some commercially available symmetric cyanine derivatives (Cy3, Cy5, Cy5.5, and the bridged Cy3B). The molar extinction coefficients of cyanine derivatives are comparably high and lie between 1.2×10^5 and $2.5 \times 10^5 \, l\,mol^{-1}\,cm^{-1}$ [26–30]. On the other hand, the fluorescence quantum yield varies only between 0.04 and 0.4 with lifetimes in the range of a few hundred picoseconds (0.2–1.0 ns).

Another class of fluorescent probes, namely fluorescent nanoparticles associated with great promise, is semiconductor nanocrystals (NCs) such as core-shell CdSe/ZnS NCs [31]. Their unique optical properties – tunable narrow emission spectrum, broad excitation spectrum, high photostability, and long fluorescence lifetime (on the order of tens of nanoseconds) – make these bright probes attractive in experiments involving long observation times and multicolor and time-gated detection. Furthermore, the relatively long fluorescence lifetime of CdSe nanocrystals, in the range of several tens of nanoseconds, can be used advantageously to enhance fluorescence biological imaging contrast and sensitivity by time-gated detection [32].

For the investigation of biologically relevant samples, target molecules have to be labeled *in vivo* with a fluorescent tag. Fluorescent labels useful for labeling of biomolecules in living cells have to fulfill special requirements, such as high biocompatibility, high photostability, and retention of biological function. In addition, the observation of the fluorescence signal of a fluorophore is more complicated than *in vitro*, primarily due to strong autofluorescence, especially in the blue/green wavelength region, and concentration control is generally difficult to perform. The first problem, however, is site-specific labeling inside a living cell and some procedures have been described in the literature. A very promising method for *in vivo* labeling with fluorescent probes is protein transduction. Several naturally occurring proteins have been found to enter cells easily, including the TAT protein from HIV [33, 34]. Furthermore, so-called hybrid systems composed of a small molecule that can covalently bind to genetically specified proteins inside or on the surface of living cells have been developed. One of these methods for covalent labeling of proteins in living cells is the tetracysteine–biarsenical system [35], which requires incorporation of a 4-cysteine α-helical motif – a 12-residue peptide sequence that includes four cysteine residues – into the target protein. The tetracysteine motif

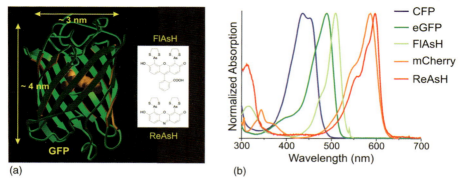

Figure 12.2 (a) Comparison (to scale) of images of the green fluorescent protein (GFP) and the biarsenical fluorophores FlAsH and ReAsH. (b) Normalized absorption spectra of CFP, eGFP, mCherry, and two biarsenical fluorophores in aqueous solvents.

binds membrane-permeable biarsenical molecules, notably the green and red fluorophores FlAsH and ReAsH, with picomolar affinity (Figure 12.2).

On the other hand, the protein of interest can be expressed fused to a protein tag that is capable of binding a small fluorescent ligand [36, 37]. A prominent example uses the enzymatic activity of human O^6-alkylguanine-DNA-alkyltransferase (hAGT) that irreversibly transfers a substrate alkyl group, an O^6-benzylguanine (BG) derivative, to one of its cysteine residues [38]. Kits for genetic labeling of proteins with hAGT are commercially available as SNAP-tags. Alternatively, trimethoprim (TMP) derivatives can be used to selectively tag *Escherichia coli* dihydrofolate reductase (eDHFR) fusion proteins in wild-type mammalian cells with minimal background and fast kinetics [39]. However, one should be aware that the labeling specificity and efficiency of tags in living cells are often reduced due to the attachment of organic fluorophores. Hence all tag technologies require sensitive controls to ensure specific labeling of the desired target protein with minimal perturbation.

The most elegant way to label proteins *in vivo* specifically is direct genetic labeling with fluorescence. The fluorophore used is a genetically encoded protein such as the green fluorescent protein (GFP) or its relatives. GFP from the bioluminescent jellyfish *Aequorea victoria* has revolutionized many areas of cell biology and biotechnology because it provides direct genetic encoding of strong visible fluorescence [40, 41]. GFP can function as a protein tag, as it tolerates N- and C-terminal fusion to a broad variety of proteins, many of which have been shown to retain native function [40–42]. According to this method, the DNA sequence coding for GFP is placed immediately adjacent to the sequence coding for the protein of interest. During biosynthesis, the protein will be prepared as a GFP-fusion protein. GFP is comprised of 238 amino acids and exhibits a barrel-like cylindrical structure in which the fluorophore is highly protected, located on the central helix of the geometric center of the cylinder. These cylinders have a diameter of about 3 nm and a length of about 4 nm, that is, significantly larger than common fluorophores with a size of $\leq$1 nm (Figure 12.2). The fluorophore is a *p*-hydroxybenzylideneimidazolinone formed from residues 65–67, which are Ser–Tyr–Gly in the native protein. The fluorophore appears to be self-catalytic, requiring proper folding of the entire structure. The protein is relatively stable, with a melting point above 65 °C.

Wild-type GFP has an extinction coefficient of $9500\,l\,mol^{-1}\,cm^{-1}$ at 475 nm and a fluorescence quantum yield Φ_f of 0.79 compared with $53\,000\,l\,mol^{-1}\,cm^{-1}$ at 489 nm in enhanced GFP (EGFP) with $\Phi_f = 0.60$. Further, several spectral GFP variants with blue, cyan, and yellowish green emission have been successfully generated [43, 44]. Proteins that fluoresce at red or far-red wavelengths are of specific interest because cells and tissues display reduced autofluorescence at longer wavelengths. Furthermore, red fluorescent proteins (RFPs) can be used in combination with other fluorescent proteins that fluoresce at shorter wavelengths for both multicolor labeling and fluorescence resonance energy transfer measurements. Here, the discovery of new fluorescent proteins from nonbioluminescent *Anthozoa* species, in particular the red-shifted fluorescent protein DsRed, is of general interest [45]. DsRed (drFP583) has absorption and emission maxima at 558 and 583 nm, respectively. However, several major drawbacks, such as slow maturation and residual green fluorescence, need to be overcome for the efficient use of DsRed as an *in vivo* reporter, especially in Single-molecule fluorescence spectroscopy (SMFS) applications. To improve the maturation properties, and to reduce aggregation, a number of other red fluorescent proteins and variants of DsRed have been developed, for example, DsRed2, DsRed-Express, eqFP611, and HcRed1. The oligomerization problem of DsRed has been solved by mutagenetic means (mRFP1) [46].

In addition to their use for selective fluorescence labeling for dynamic studies, some optical highlighters (photoconvertible or photoactivatable fluorescent proteins such as PA-GFP, Dronpa, and EosFP) and organic fluorophores are also valuable tools in super-resolution fluorescence imaging based on single-molecule detection and precise localization of the position of each fluorophore, for example, PALM, STORM, and dSTORM [47–50]. These methods rely on the light-induced activation of only a small subset of fluorophores per image spaced further apart than the diffraction limit.

References

1 Lippitz, M., Erker, W., Decker, H., van Holde, K.E., and Basché, T. (2002) *Proc. Natl. Acad. Sci. U. S. A.*, **99**, 2772.
2 Schüttpelz, M., Müller, C., Neuweiler, H., and Sauer, M. (2006) *Anal. Chem.*, **78**, 663.
3 MacColl, R. and Guard-Friar, D. (1987) *Phycobiliproteins*, CRC Press, Boca Raton, FL.
4 Mathies, R.A. and Peck, K. (1990) *Anal. Chem.*, **62**, 1786.
5 Wehrmeyer, W., Wendler, J., and Holzwarth, A.R. (1985) *Eur. J. Cell. Biol.*, **36**, 17.
6 Wu, M., Goodwin, P.M., Ambrose, W.P., and Keller, R.A. (1996) *J. Phys. Chem.*, **100**, 17406.
7 Nguyen, D.C., Keller, R.A., Jett, J.H., and Martin, J.C. (1987) *Anal. Chem.*, **59**, 2158.
8 Peck, K., Stryer, L., Glazer, A.N., and Mathies, R.N. (1989) *Proc. Natl. Acad. Sci. U. S. A.*, **86**, 4087.
9 Oi, V., Glazer, A.N., and Stryer, L. (1982) *J. Cell Biol.*, **93**, 981.
10 Ying, L. and Xie, X.S. (1998) *J. Phys. Chem. B*, **102**, 10399.

11 Schäfer, F.P. (ed.) (1973) *Dye Lasers, Topics in Applied Physics*, vol. **1**, Springer, Berlin.

12 Smith, L.M., Fung, S., Hunkapillar, M.W., and Hood, L.E. (1985) *Nucleic Acids Res.*, **13**, 2399.

13 Prober, J.M., Trainer, G.L., Dam, R.J., Hobbs, F.W., Robertson, C.W., Zagursky, R.J., Cocuzza, A.J., Jensen, M.A., and Baumeister, K. (1987) *Science*, **238**, 336.

14 Smith, L.M. (1991) *Nature*, **349**, 812.

15 Ansorge, W., Sproat, B., Stegemann, J., Schwager, C., and Zenke, M. (1987) *Nucleic Acids Res.*, **15**, 4593.

16 Hemmilä, I.A. (1989) *Appl. Fluoresc. Technol.*, **1**, 1.

17 Whitaker, J.A., Haugland, R.P., Ryan, D., Hewitt, P.C., and Prendergast, F.G. (1992) *Anal. Biochem.*, **207**, 267.

18 Zen, J.M. and Patonay, G. (1991) *Anal. Chem.*, **63**, 2934.

19 Ernst, L.A., Gupta, R.K., Mujumdar, R.B., and Waggoner, A.S. (1989) *Cytometry*, **10**, 3.

20 Sauer, M., Han, K.T., Müller, R., Schulz, A., Tadday, R., Seeger, S., Wolfrum, J., Arden-Jacob, J., Deltau, G., Marx, N.J., and Drexhage, K.H. (1993) *J. Fluoresc.*, **3**, 131.

21 Sauer, M., Han, K.T., Müller, R., Nord, S., Schulz, A., Seeger, S., Wolfrum, J., Arden-Jacob, J., Deltau, G., Marx, N.J., Zander, C., and Drexhage, K.H. (1995) *J. Fluoresc.*, **5**, 247.

22 Coons, H., Creech, H.J., Jones, R.N., and Berliner, E. (1942) *J. Immunol. Methods*, **45**, 159.

23 Tinnefeld, P., Hofkens, J., Herten, D.P., Masuo, S., Vosch, T., Cotlet, M., Habuchi, S., Müllen, K., De Schryver, F.C., and Sauer, M. (2004) *ChemPhysChem*, **5**, 1786.

24 Panchuk-Voloshina, N., Bishop-Stewart, J., Bhalgat, M.K., Millard, P.J., Mao, F., Leung, W.L., and Haugland, R.P. (1999) *J. Histochem. Cytochem.*, **47**, 1179.

25 Soper, S.A. and Mattingly, J. (1994) *J. Am. Chem. Soc.*, **116**, 3744.

26 Boyer, A.E., Lipowska, M., Zen, J.M., Patonay, G., and Tsung, V.C.W. (1992) *Anal. Lett.*, **25**, 415.

27 Southwick, P.L., Ernst, L.A., Tauriello, E.V., Parker, S.R., Mujumdar, R.B., Mujumdar, S.R., Clever, H.A., and Waggoner, A.S. (1990) *Cytometry*, **11**, 418.

28 Mujumdar, R.B., Ernst, L.A., Mujumdar, S.R., and Waggoner, A.S. (1989) *Cytometry*, **10**, 11.

29 Terpetschnig, E., Szmacinski, H., Ozinskas, A., and Lakowicz, J.R. (1994) *Anal. Biochem.*, **217**, 197.

30 Mujumdar, R.B., Ernst, L.A., Mujumdar, S.R., Lewis, C.J., and Waggoner, A.S. (1993) *Bioconjug. Chem.*, **4**, 105.

31 Bruchez, M., Moronne, M., Gin, P., Weiss, S., and Alivisatos, A.P. (1998) *Science*, **281**, 2013.

32 Dahan, M., Laurence, T., Pinaud, F., Chemla, D.S., Alivisatos, A.P., Sauer, M., and Weiss, S. (2001) *Opt. Lett.*, **26**, 825.

33 Torchilin, V.P., Rammohan, R., Weissig, V., and Levchenko, T.S. (2001) *Proc. Natl. Acad. Sci. U. S. A.*, **98**, 8786.

34 Schwarze, S.R., Ho, A., Vocero-Akbani, A., and Dowdy, S.F. (1999) *Science*, **285**, 1569.

35 Griffin, B.A., Adams, S.R., and Tsien, R.Y. (1998) *Science*, **281**, 269.

36 Müller, L.W. and Cornish, V.W. (2005) *Curr. Opin. Chem. Biol.*, **9**, 56.

37 Chen, I. and Ting, A.Y. (2005) *Curr. Opin. Biotechnol.*, **16**, 35.

38 Keppler, A., Gendreizig, S., Gronemeyer, T., Pick, H., Vogel, H., and Johnsson, K. (2003) *Nat. Biotechnol.*, **21**, 86.

39 Miller, L.W., Cai, Y., Sheetz, M.P., and Cornish, V.W. (2005) *Nat. Methods*, **2**, 255.

40 Tsien, R.Y. (1998) *Annu. Rev. Biochem.*, **67**, 509.

41 Baird, D., Zacharias, A., and Tsien, R.Y. (2000) *Proc. Natl. Acad. Sci. U. S. A.*, **97**, 11984.

42 Moores, S.L., Sabry, J.H., and Spudich, J.A. (1996) *Proc. Natl. Acad. Sci. U. S. A.*, **93**, 443.

43 Wiedenmann, J., Oswald, F., and Nienhaus, G.U. (2009) *IUBMB Life*, **61**, 1029.

44 Shaner, N.C., Patterson, G.H., and Davidson, M.W. (2007) *J. Cell Sci.*, **120**, 4247.

45 Matz, M.V., Fradkov, A.F., Labas, Y.A., Savitsky, A.P., Zaraisky, A.G., Markelov, M.L., and Lukyanov, S.A. (1999) *Nat. Biotechnol.*, **17**, 969.
46 Campbell, R.E., Tour, O., Palmer, A.E., Steinbach, P.A., Baird, G.S., Zacharias, D.A., and Tsien, R.Y. (2002) *Proc. Natl. Acad. Sci. U. S. A.*, **99**, 7877.
47 Hell, S.W. (2007) *Science*, **316**, 1153.
48 Betzig, E., Patterson, G.H., Sougrat, R., Lindwasser, O.W., Olenych, S., Bonifacino, J.S., Davidson, M.W., Lippincott-Schwartz, J., and Hess, H.F. (2006) *Science*, **313**, 1642.
49 Rust, M.J., Bates, M., and Zhuang, X. (2006) *Nat. Methods*, **3**, 793.
50 Heilemann, M., van de Linde, S., Mukherjee, A., and Sauer, M. (2009) *Angew. Chem. Int. Ed.*, **48**, 6903.

13
Optical Coherence Imaging for Surgical Pathology Assessment
Wladimir A. Benalcazar and Stephen A. Boppart

Surgical pathology is the field that specializes in assessing surgically removed tissue for disease in order to provide important information to guide surgical intervention and postoperative treatment. In surgical oncology, this includes confirmation of the preoperative diagnosis of the disease, staging the extent or spread of the malignancy, establishing whether the entire affected area was surgically removed, identifying the presence of any unsuspected concurrent diseases, and integrating this information to determine optimal strategies for postoperative treatment.

The entire assessment by a surgical pathologist (so-called intraoperative consultation) requires a combination of macroscopic and microscopic examinations. Whereas a macroscopic examination of tissue specimens can be performed in the operating room by gross examination, a microscopic examination requires that the tissue specimen be frozen or fixed, sectioned into micron-thick slices, placed on microscope slides, and stained for assessment by brightfield microscopy. Frozen sectioning (cryosectioning) can usually be performed within 30 min to provide a fair assessment of the specimen, and has been used most often to confirm if the surgical margins are free of tumor cells. Unfortunately, frozen sectioning has been largely unreliable, and is being used less frequently. More advanced molecular-based immunohistochemical analysis of tissue sections is performed in the pathology laboratory, after the surgical procedure has ended, and often days later. To address some of these limitations, studies have investigated the use of cytology preparations (e.g., touch imprints), polymerase chain reaction (PCR) analysis, and flow cytometry in an effort to provide more information about the disease based on cellular and molecular composition.

Intraoperative consultation is widely used in oncologic surgery. The increase in cancer screening rates in recent years has led to the detection of more tumors at earlier stages of development, which subsequently has resulted in eliminate, more breast-conserving (lumpectomy) surgical procedures. In lumpectomy procedures, the main goal of the surgical pathologist is to inform the surgeon if the surgical margin is clear of any tumor cells or residual cancer. Therefore, this poses a critical need for fast, high-resolution screening methods that can adequately sample the surgical margin and ensure that it is clear.

The use of frozen section analysis for tumor margin assessment in breast cancer has a sensitivity of 73% and a specificity of 98% compared with the more standard but time-consuming paraffin section analysis [1]. This relatively low quality of frozen section analysis has precluded it from becoming a widely accepted part of the standard of care. In particular, this technique presents problems in sectioning adipose tissue and cannot be done over the entire surface area of the tissue specimen, sampling at best only 10–15% of the surface area, just as with paraffin section analysis [2]. Additionally, this technique adds time (20–30 min) to the surgical procedure and increases costs.

Alternatively, touch preparation cytology can rapidly assess the entire surface area of a surgical specimen while preserving its physical integrity for later histopathologic sectioning and assessment. Although this may be suitable for identifying positive margins, the technique has a reported sensitivity of 75% and a specificity of 82% [2]. The major disadvantage is the requirement for tumor cells to detach from the surface of the examined area, making it unable to provide information about the presence of cancer cells immediately beneath the surface, which is used to determine close and negative margins.

Imaging techniques, such as X-ray mammography, diffuse optical tomography (DOT), ultrasound, and magnetic resonance imaging (MRI), have shown an ability to detect large tumor masses, but lack the resolution and sensitivity to detect the microscopic distribution of tumor cells at the periphery of solid tumors or those tumor cells metastasizing to other sites. Intraoperative radiography provides visualization of the margin in-depth by displaying two-dimensional X-ray projections. However, it does not have the ability to identify diffuse microscopic processes, especially where the tumor boundary is poorly defined. Intraoperative radiography has a sensitivity of only 49% and a specificity of 73% [3]. Alternatively, radiofrequency spectroscopy provides an average measurement over an area with a diameter of 0.7 cm and within a 100 μm depth. It has a sensitivity of 71% and a specificity of 68% [4].

In summary, no intraoperative method currently exists to nondestructively assess the microscopic status of lumpectomy margins in real time. However, high-speed optical imaging and spectroscopy could address this limitation. Optical techniques take advantage of a biological window that exists in the near-infrared region of the optical spectrum, where attenuation of light is governed more by scattering rather than by absorption, so that light at these wavelengths can penetrate into tissue to convey information about its structure or chemical composition. Intraoperative optical imaging with rapid feedback could eliminate the time spent in the surgical pathology laboratory when a fully staffed operating room is waiting for the pathologic assessment of tissue specimens, and could reduce or eliminate the need for reoperations when positive surgical margins are discovered in the pathology laboratory, after the surgical procedure has ended.

13.1
Structural Contrast Imaging

One optical biomedical imaging method that addresses this need for intraoperative, high-resolution imaging is optical coherence tomography (OCT) [5, 6], which interrogates the structure of tissues noninvasively by probing the subsurface refractive index up to 1–2 mm deep into scattering tissue. OCT is attractive because it allows real-time imaging of microstructure *in situ* while avoiding the need for excisions and histologic processing. Initial studies have shown favorable results towards the use of OCT as an intraoperative imaging technique for the assessment of tumor margins [7], lymph nodes [8], and optically guided needle biopsies in the treatment of breast cancer.

OCT is the optical analogue of ultrasound imaging. The contrast mechanism is the change in refractive index and density of tissue structures that cause backscattering of the incident light. Following the illumination of the sample by a focused optical beam, multiple back-reflections are collected and resolved using interferometric detection, which relies on the short coherence length of the incident source to achieve micron-scale axial resolution (i.e., it typically uses sources with a bandwidth of 90 nm centered at 1300 nm for an axial resolution of 6 μm). At a given position of the incident beam, an axial scan is acquired. Two- or three-dimensional images are formed by scanning the beam across the sample. The lateral resolution is limited by the diffraction of the focused incident beam. A system schematic and photograph of a portable OCT system used for intraoperative imaging are shown in Figure 13.1.

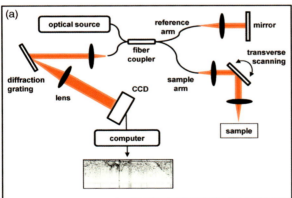

Figure 13.1 (a) Schematic of an OCT setup. A low-coherence optical source (superluminescent diode) is coupled to a fiber-based interferometer, in which a beam splitter (2 × 2 coupler) is used to direct 90% of the power to the sample (sample arm) and the remaining 10% to a mirror (reference arm). The back-reflected optical signals are combined and diffracted into a linear CCD camera to perform spectral interferometric detection. As a result, a depth-resolved profile of the refractive index variations of the sample is obtained. Scanning the beam across the sample allows image formation. (b) Typical implementation of a portable OCT system for intraoperative clinical use.

Since its introduction, advances in OCT technology have enabled significantly increased data acquisition rates to be achieved. Current systems can acquire 200 000 axial scans per second [9] or more, which permits acquisition of 400 frames per second in a range of 10 mm with a lateral resolution of 35 µm. The implementation of compact OCT systems facilitates their use inside the operating room, and the optical imaging beam can be delivered to the patient using hand-held surgical probes [10, 11]. This technique has already found clinical applications in ophthalmology, cardiology, and gastroenterology [12].

OCT imaging of breast tumors has shown that differences in image contrast are correlated with different cell and tissue structures. Adipose tissue, which comprises a large portion of the breast, is less dense and presents a more homogeneous structure than the more structured stromal or epithelial tissue. The same difference in contrast also exists between adipocytes and tumor cells, where the latter are typically densely packed, have an increase in their nuclear to cytoplasm ratio, and are more highly scattering. The large size and low scattering of adipocytes, relative to stromal and tumor cells, provides a method for differentiating these tissue types [13, 14]. In a study that used OCT to classify margins as positive or negative in 20 different lumpectomy specimens after resection but before margin assessment by the pathologist, an overall sensitivity of 100% and a specificity of 82% were achieved [7].

These findings suggest OCT as a potential method for intraoperative margin assessment in breast-conserving surgeries. In particular, following the resection of a tumor mass, OCT can be used to scan the walls of the tumor cavity in the search for tumor cells present at or just beneath the *in situ* surgical margin. Alternatively, resected masses can be imaged on all surfaces to verify a clean surgical margin (Figure 13.2). OCT can typically image to depths of 2 mm in tissue composed primarily of adipocytes and from 0.2 to 1 mm in cell-dense tumor tissue. These depths are comparable to the currently accepted margin widths that classify positive, close, and negative margins [7]. Additionally, although blood is highly scattering, intravascular blood in small vessels and capillaries constitutes a small percentage of the tissue volume and therefore has a minimal effect on the OCT penetration depth. OCT systems using optical sources with center wavelength around 1300 nm are also unaffected by the presence of dyes such as methylene blue and lymphazurin, which are used to map lymph drainage for sentinel and axillary lymph node dissections, because these dyes absorb in the spectral region below 700 nm.

13.2
Molecular Contrast Imaging

Although there is a promising future for OCT in assessing surgical pathology, important challenges and needs for optical microscopy remain, including the detection of tumors and tumor cells at the early stages of development, achieving good differentiation between benign fibrocystic changes and malignant lesions, and

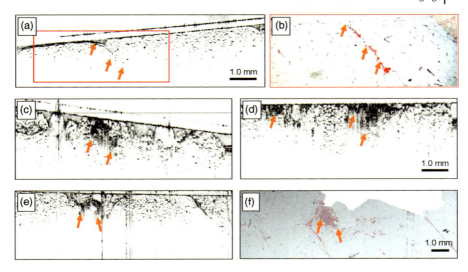

Figure 13.2 Intraoperative surgical margin assessment with OCT, and corresponding histology. (a) OCT image of a negative surgical margin composed primarily of adipose tissue (the small, dark, highly scattering point-like features correspond to individual nuclei of adipocytes, and arrows indicate vasculature) and (b) corresponding histology. (c) OCT image of a positive margin with ductal carcinoma *in situ* (arrows indicate no dense highly scattering features corresponding to tumor cells) and (d) OCT image of a positive margin with invasive ductal carcinoma (tumor cell features indicated by arrows). (e) OCT image and (f) corresponding histology of a small (∼1 mm) isolated focus of tumor cells at the surgical margin. Modified figure used with permission from [7].

characterizing tissue at the molecular level. For instance, OCT is limited by the fact that the optical refractive index does not differ significantly between stromal and tumor tissue [15].

In order to overcome these difficulties, OCT with new functional modalities [16, 17] and contrast agents [18, 19] is being investigated. In particular, magnetomotive optical coherence tomography (MM-OCT) has shown selective molecular imaging of tumors by the use of magnetic nanoparticles (MNPs) that target human epidermal growth factor receptor 2 (HER-2/neu) [19] (Figure 13.3).

Working towards more selective and noninvasive microscopy techniques, vibrational spectroscopy has been explored as a label-free method for assessing the molecular composition of tissue. The chemical constituents in breast tissue to which Raman spectroscopy is sensitive have been characterized, and Raman spectroscopy has shown the ability to differentiate normal, malignant (infiltrating ductal carcinoma), and benign (fibrocystic change) human breast tissue [20, 21]. Criteria for classification based on vibrational spectra rely mainly on the higher abundance of proteins in tumors, compared with normal tissue [22]. Other spectral differences suggest a higher presence of myoglobin and a lower degree of oxygenation in infiltrating ductal carcinomas. In specimens with fibrocystic changes, an indicator is a reduction in lipids and carotenoids and an absence of a heme-type signal

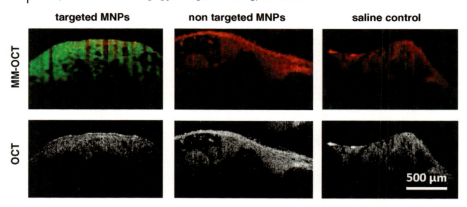

Figure 13.3 Targeted magnetomotive (MM) contrast from *in vivo* MM-OCT using magnetic nanoparticles conjugated to anti-HER-2/neu antibodies for site-specific molecular imaging. MM-OCT (green channel superimposed over red OCT channel, top row) and structural OCT images (bottom row) of tumors from a preclinical carcinogen-induced rat mammary tumor model that recapitulated the progression of human ductal carcinoma *in situ*. Tumors are shown for targeted MNP-injected (left column), nontargeted MNP-injected (center column), and saline-injected rats (right column). The scale bar applies to all six images. Modified figure used with permission from [19].

characteristic of porphyrins [21]. Finally, the degree of saturation in the hydrocarbon chain of lipids is also an important metric, as this varies in normal, benign, and malignant tumor tissues [23].

A study using Raman spectroscopy to identify carcinomas reported a sensitivity of 100%, a specificity of 100%, and an overall accuracy of 93% [24]. Despite these promising findings, Raman spectroscopy is limited by the weak Raman signal, which leads to long integration times (about 1 second per pixel) that make its use impractical for real-time imaging. Coherent anti-Stokes Raman scattering (CARS) microscopy can overcome this issue, but at the cost of generating an unwanted background that coherently interferes with the desired vibrational signal. Furthermore, conventional CARS techniques do not obtain broadband spectra, and the CARS signals are not linearly proportional to the concentration of the molecular species of interest. Despite these limitations, methods based on CARS are emerging, and have shown significant improvements in acquisition rates [25], rejection of background [26, 27], and broadband vibrational acquisition [28].

Techniques such as nonlinear interferometric vibrational imaging (NIVI), which use shaped pulses and interferometric detection, can overcome many of the drawbacks in conventional CARS. In particular, NIVI has been used to acquire background-free, broadband, vibrational spectra from mammary tissue at acquisition rates two orders of magnitude faster than Raman microscopy (Figure 13.4) [29, 30]. Recently, a study of NIVI spectra from various lipids showed its ability to reproduce Raman-like spectra 200 times faster than Raman spectroscopy, and the ability to distinguish (un)saturation levels in fatty acid chains because the NIVI signals are linearly dependent on the concentrations of the molecular bonds [31]. This type of fast

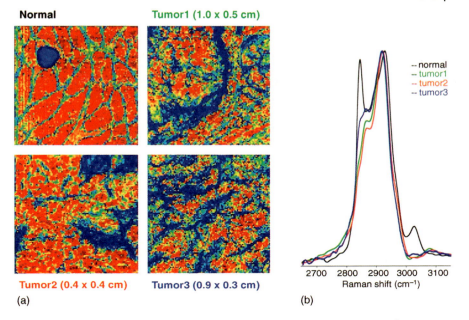

Figure 13.4 Nonlinear interferometric vibrational imaging. Spatial mapping of CH groups (present in different proportions in lipid and protein domains) in mammary tissue. (a) Endogenous-contrast NIVI images ($500 \times 500\,\mu m^2$) (color scale: blue to red, as percentage of CH vibrations) of one normal and three developed mammary tumors of various sizes. (b) Vibrational spectra for images in (a). Images and spectra reveal disruptions of lipid and collagen domains in adipose and tumor tissues, with a predominance of collagen-based structures in tumors.

vibrational spectroscopy could assist surgical procedures by taking spectra or even images from zones deep in the breast using minimally invasive methods involving needle-based probes, or in procedures similar to the commonly used fine-needle aspiration cytology or core-needle biopsies.

Since CARS uses the same contrast mechanism as spontaneous Raman spectroscopy, there is potential for techniques based on CARS to improve the optical diagnosis of breast disease further by adding molecular vibrational contrast to structural and spatial imaging capabilities. The integration of both molecular and morphologic information could contribute significantly to the analysis of disease states, in part because this builds upon traditional cancer histopathology methods that rely on molecularly staining and imaging tissue sections [22]. Currently, however, fast vibrational spectroscopy makes use of nonlinear processes that require two synchronized laser pulses, and the complex setups needed are neither compact nor portable enough for use today in an operating room. Advances are under way, including single-beam approaches to CARS [31], and methods to obtain ultrabroadband spectra free from background, particularly to target the fingerprint region of the vibrational spectrum, where a wealth of valuable molecular information used in spontaneous Raman spectroscopy is present.

13.3
Conclusion

Whether it is by means of providing structural or molecular contrast, coherent optical microscopy and imaging techniques have the potential to assist in advancing our ability to assess surgical pathology. In particular, OCT can be used for real-time identification of tumor margins, and new Raman-based spectroscopic imaging techniques can potentially aid in the fast confirmation of a diagnosis based on molecular signatures. Accompanying these technological advancements, as with all novel imaging techniques, classification and diagnosis will require the study of training and calibration sets by clinicians in order to establish the evaluation methodology and image feature criteria for correct identification. Complementary to this, the information provided by these techniques could eventually lead to automated tissue-type classification algorithms to improve further both the diagnosis at earlier stages and the identification of margins intraoperatively, thereby potentially reducing the need for additional surgical procedures and advancing the diagnostic capabilities in surgical pathology.

References

1 Olson, T.P., Harter, J., Muoz, A., Mahvi, D.M., and Breslin, T. (2007) Frozen section analysis for intraoperative margin assessment during breast-conserving surgery results in low rates of re-excision and local recurrence. *Ann. Surg. Oncol.*, **14** (10), 2953–2960.

2 Valdes, E.K., Boolbol, S.K., Cohen, J.-M., and Feldman, S.M. (2007) Intra-operative touch preparation cytology; does it have a role in re-excision lumpectomy? *Ann. Surg. Oncol.*, **14** (3), 1045–1050.

3 Goldfeder, S., Davis, D., and Cullinan, J. (2006) Breast specimen radiography: can it predict margin status of excised breast carcinoma? *Acad. Radiol.*, **13** (12), 1453–1459.

4 Karni, T., Pappo, I., Sandbank, J., Lavon, O., Kent, V., Spector, R., Morgenstern, S., and Lelcuk, S. (2007) A device for real-time, intraoperative margin assessment in breast-conservation surgery. *Am. J. Surg.*, **194** (4), 467–473.

5 Huang, D., Swanson, E.A., Lin, C.P., Schuman, J.S., Stinson, W.G., Chang, W., Hee, M.R., Flotte, T., Gregory, K., Puliafito, C.A., and Fujimoto, J. G. (1991) Optical coherence tomography. *Science*, **254**, 1178–1181.

6 Drexler W. and Fujimoto J.G. (eds.) (2008) *Optical Coherence Tomography*, Springer, Berlin.

7 Nguyen, F.T., Zysk, A.M., Chaney, E.J., Kotynek, J.G., Oliphant, U.J., Bellafiore, F.J., Rowland, K.M., Johnson, P.A., and Boppart, S.A. (2009) Intraoperative evaluation of breast tumor margins with optical coherence tomography. *Cancer Res.*, **69** (22), 8790–8796.

8 Nguyen, F.T., Zysk, A.M., Chaney, E.J., Adie, S.G., Kotynek, J.G., Oliphant, U.J., Bellafiore, F.J., Rowland, K.M., Johnson, P.A., and Boppart, S.A. (2010) Optical coherence tomography: the intraoperative assessment of lymph nodes in breast cancer. *IEEE Eng. Med. Biol.*, **29**, 63–70.

9 Huber, R., Adler, D.C., and Fujimoto, J.G. (2006) Buffered Fourier domain mode locking: unidirectional swept laser sources for optical coherence tomography imaging at 370,000 lines/s. *Opt. Lett.*, **31** (20), 2975–2977.

10 Boppart, S.A., Bouma, B.E., Pitris, C., Tearney, G.J., Fujimoto, J.G., and Brezinski, M.E. (1997) Forward imaging

instructions for optical coherence tomography. *Opt. Lett.*, **22** (21), 1618–1620.

11 Zysk, A.M., Nguyen, F.T., Chaney, E.J., Kotynek, J.G., Oliphant, U.J., Bellafiore, F.J., Johnson, P.A., Rowland, K.M., and Boppart, S.A. (2009) Clinical feasibility of microscopically-guided breast needle biopsy using a fiber-optic probe with computer-aided detection. *Technol. Cancer Res. Treat.*, **8** (5), 315–400.

12 Zysk, A.M., Nguyen, F.T., Oldenburg, A.L., Marks, D.L., and Boppart, S.A. (2007) Optical coherence tomography: a review of clinical development from bench to bedside. *J. Biomed. Opt.*, **12** (5), 051403.

13 Boppart, S.A., Luo, W., Marks, D.L., and Singletary, K.W. (2004) Optical coherence tomography: feasibility for basic research and image-guided surgery of breast cancer. *Breast Cancer Res. Treat.*, **84**, 85–97.

14 Hsiung, P.L., Phatak, D.R., Chen, Y., Aguirre, A.D., Fujimoto, J.G., and Connolly, J.L. (2007) Benign and malignant lesions in the human breast depicted with ultrahigh resolution and three-dimensional optical coherence tomography. *Radiology*, **244** (3), 865–874.

15 Zysk, A.M., Chaney, E.J., and Boppart, S.A. (2006) Refractive index of carcinogen-induced rat mammary tumours. *Phys. Med. Biol.*, **51** (9), 2165–2177.

16 de Boer, J.F., Milner, T.E., van Gemert, M.J.C., and Nelson, J.S. (1997) Two-dimensional birefringence imaging in biological tissue by polarization-sensitive optical coherence tomography. *Opt. Lett.*, **22** (12), 934–936.

17 Oldenburg, A.L., Crecea, V., Rinne, S.A., and Boppart, S.A. (2008) Phase-resolved magnetomotive OCT for imaging nanomolar concentrations of magnetic nanoparticles in tissues. *Opt. Express*, **16** (15), 11525–11539.

18 Lee, T.M., Oldenburg, A.L., Sitafalwalla, S., Marks, D.L., Luo, W., Toublan, F.J.-J., Suslick, K.S., and Boppart, S.A. (2003) Engineered microsphere contrast agents for optical coherence tomography. *Opt. Lett.*, **28** (17), 1546–1548.

19 John, R., Rezaeipoor, R., Adie, S.G., Chaney, E.J., Oldenburg, A.L., Marjanovic, M., Haldar, J.P., Sutton, B.P., and Boppart, S.A. (2010) In vivo magnetomotive optical molecular imaging using targeted magnetic nanoprobes. *Proc. Natl. Acad. Sci. U. S. A.*, **107** (18), 8085–8090.

20 Frank, C.J., Redd, D.C.B., Gansler, T.S., and McCreery, R.L. (1994) Characterization of human breast biopsy specimens with near-IR Raman spectroscopy. *Anal. Chem.*, **66**, 319–326.

21 Redd, D.C., Feng, Z.C., Yue, K.T., and Gansler, T.S. (1993) Raman spectroscopic characterization of human breast tissues: implications for breast cancer diagnosis. *Appl. Spectrosc.*, **47**, 787–791.

22 Kline, N.J. and Treado, P.J. (1997) Raman chemical imaging of breast tissue. *J. Raman Spectrosc.*, **28**, 119–124.

23 Liu, C.-H., Das, B., Glassman, W., Tang, G., Yoo, K., Zhu, H., Akins, D., Lubicz, S., Cleary, J., Prudente, R., Celmer, E., Caron, A., and Alfano, R. (1992) Raman, fluorescence, and time-resolved light scattering as optical diagnostic techniques to separate diseased and normal biomedical media. *J. Photochem. Photobiol. B*, **16**, 187–209.

24 Breslin, T. (2007) In vivo margin assessment during partial mastectomy breast surgery using Raman spectroscopy. *Breast Dis. Year Book Q.*, **17** (4), 379–380.

25 Evans, C.L., Potma, E.O., Puoris'haag, M., Côté, D., Lin, C.P., and Xie, X.S. (2005) Chemical imaging of tissue in vivo with video-rate coherent anti-Stokes Raman scattering microscopy. *Proc. Natl. Acad. Sci. U. S. A.*, **102**, 16807–16812.

26 Volkmer, A., Book, L.D., and Xie, X.S. (2002) Time-resolved coherent anti-Stokes Raman scattering microscopy: imaging based on Raman free induction decay. *Appl. Phys. Lett.*, **80**, 1505–1507.

27 Cheng, J.X., Book, L.D., and Xie, X.S. (2001) Polarization coherent anti-Stokes Raman scattering microscopy. *Opt. Lett.*, **26**, 1341–1343.

28 Wurpel, G.W.H., Schins, J.M., and Muller, M. (2002) Chemical specificity in

three-dimensional imaging with multiplex coherent anti-Stokes Raman scattering microscopy. *Opt. Lett.*, **27**, 1093–1095.

29 Benalcazar, W.A., Chowdary, P.D., Jiang, Z., Marks, D.L., Chaney, E.J., Gruebele, M., and Boppart, S.A. (2010) High-speed nonlinear interferometric vibrational imaging of biological tissue with comparison to Raman microscopy. *IEEE J. Sel. Top. Quantum Electron.*, **16** (4), 824–832.

30 Chowdary, P. D., Benalcazar, W. A., Jiang, Z., Chaney, E. J., Marks, D. L., Gruebele, M., Boppart, S. A. (2010) Molecular histopathology by nonlinear interferometric vibrational imaging. *Cancer Research*, **70**, 9562–9569.

31 Chowdary, P.D., Benalcazar, W.A., Jiang, Z., Marks, D.M., Boppart, S.A., and Gruebele, M. (2010) High speed nonlinear interferometric vibrational analysis of lipids by spectral decomposition. *Anal. Chem.*, **82** (9), 3812–3818.

32 von Vacano, B., Buckup, T., and Motzkus, M. (2006) Highly sensitive single-beam heterodyne coherent anti-Stokes Raman scattering. *Opt. Lett.*, **31**, 2495–2497.

14
Diffuse Optical Imaging
Jeremy C. Hebden

14.1
Introduction

When a pencil beam of visible or near-infrared light is incident upon biological tissue, profuse scattering ensures that the beam becomes diffuse (i.e., the initial directionality is lost) after penetrating a distance in excess of one transport scattering length, or about 1 mm in most tissues. As a consequence, the formation of images using light which has penetrated more than a few millimeters in tissue is known as diffuse optical imaging (DOI). The simplest and earliest example of DOI is transillumination, which involves viewing or otherwise forming an image when the tissue is illuminated from behind. In 1929, Cutler [1] reported a systematic attempt to detect cancer of the breast by transilluminating with a bright light source in a darkened room. While the study revealed the dominant role played by blood concentration in the degree of opacity of tissue, and showed that solid tumors often appeared to be opaque, it was not possible to differentiate between benign and malignant tumors. In this and subsequent studies, breast transillumination consistently exhibited poor sensitivity and specificity resulting from low inherent spatial resolution.

Interest in DOI has been revived during the past 20 years as a result of developments in two areas of technology. First is the availability of near-infrared sources and sensitive detectors, including low-cost portable devices. Second is the development of sophisticated algorithms for reconstructing images from measurements of diffuse light though tissue, associated with the availability of powerful computers able to perform modeling of photon migration through large thicknesses of biological tissues. Subsequently, two DOI approaches have emerged, known as optical topography and optical tomography (Figure 14.1). Although the distinction between them is somewhat arbitrary, the term optical topography is used to describe techniques which provide maps of changes in hemodynamic parameters (such as blood volume) close to the surface, with little or no depth resolution, whereas optical tomography involves reconstructing an image representing the 3D distribution of optical properties (or a transverse slice

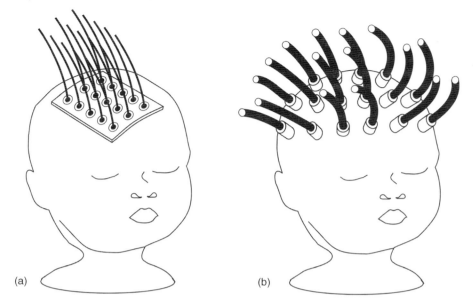

Figure 14.1 Arrangements of sources and detectors for performing a) optical topography, and b) optical tomography of the brain.

across the 3D volume). The distinction has evolved because of the differences in the complexity of the imaging problem and of the associated instrumentation. The two approaches are described in more detail below, with examples of clinical and scientific applications.

14.2
Optical Topography

The imaging technique known as optical topography involves measuring diffusely reflected light between multiple pairs of sources and detectors at discrete locations spread over the area of tissue of interest. The source–detector separations are small enough (a few centimeters or less) to ensure strong enough signals so that measurements can be acquired quickly, enabling images to be displayed in real time at a rate of a few hertz or faster. Fast acquisition is also facilitated by illuminating sources simultaneously, which requires that the detected signal from each source is uniquely identifiable, usually by modulating the intensity of each at a different frequency. The detectors then operate in parallel, employing either lock-in amplifiers or digital processing methods to isolate the signal from each source.

The principle motivation for the development of optical topography has been the display of hemodynamic activity in the cerebral cortex. Measurements of changes in attenuation of signals at two or more wavelengths can be converted into estimates of changes in concentrations of oxyhemoglobin ($[HbO_2]$) and deoxyhemoglobin ($[Hb]$), usually invoking some assumptions about the optical properties of the sampled tissue volume. In addition to monitoring hemodynamic activity, optical techniques have been used to map a so-called "fast" event-related optical signal, where changes observed over a few hundred milliseconds are considered to correspond directly to the neuronal activity associated with information processing.

The probes developed for optical topography systems typically consist of a flexible pad which supports an array of optical fibers, each coupled to either a source (e.g., a laser diode) or a detector (e.g., avalanche photodiode). Securing a probe safely, securely, and comfortably to the head, while minimizing the adverse effects of hair, has always been one of the greatest technical challenges in the implementation of DOI techniques, and a broad variety of solutions have been implemented. Several commercial optical topography systems are now available. For example, the ETG-4000 system developed by Hitachi Medical Corporation (Tokyo, Japan) samples data from up to 52 discrete source–detector pairs at two wavelengths (695 and 830 nm) at a rate of 10 Hz.

14.2.1
Evoked Response to Sensory Stimulation

The regions of the brain associated with processing of sensory information (e.g., touch, sound) are generally cortical areas located within 2–3 cm of the scalp in the adult head. Consequently, measurement of the evoked response of the brain to sensory stimuli has been a major focus of research with optical topography. The additional metabolic demand produced by the stimulus results in an increase in local blood volume and/or a change in blood oxygenation. The technique offers several advantages over BOLD fMRI: optical topography provides greater temporal resolution and signal-to-noise ratio, is portable, less expensive, and is sensitive to changes in both $[Hb]$ and $[HbO_2]$.

Imaging studies have been performed on the response to a variety of visual [2–5], tactile [6], and motor [6–9] stimuli. One of the most dramatic recent illustrations of the potential of optical topography has been presented by Culver and colleagues at Washington University (St. Louis, MO, USA) [5]. A high-density array has been employed to provide phase-encoded mapping of the retinotopic organization within the visual cortex of adult volunteers, allowing the study of subtle inter-subject differences. Figure 14.2a shows the placement of the array over the occipital cortex of an adult, and Figure 14.2b–j show the visual stimuli (a counter-phase flickering checkerboard in one quadrant of the visual field) and the corresponding maps of relative changes in $[HbO_2]$. All images are posterior coronal projections of a cortical shell, as if looking at the surface of the brain from behind the subject. In order to match the stimulus and response, a 6 s lag between stimulation and maximal response was used. The results show that the hemodynamic response is always maximal in the opposite visual quadrant from the stimulus.

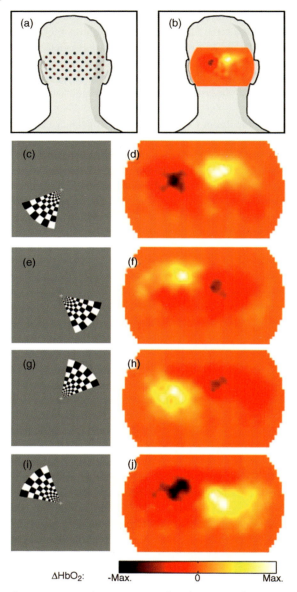

Figure 14.2 Visual activation stimuli and corresponding maps of changes in [HbO_2] the visual cortex. The stimulus is a rotating wedge-shaped flickering checkerboard (a, c, e, g), and the response is always in the opposite visual quadrant. See White and Culver [5] for more details.

14.2.2
Cognitive Activity and Language Processing

Optical topography has evolved into a popular tool for the study of cognitive and psychological processes, and especially for developmental studies on infants and children. The portability and silent function of the hardware make the technique particularly suitable for psychological studies and for use on awake infants. Published results include studies of face perception [10], language processing [11, 12], memory [13], object recognition [14], and post-traumatic stress disorder [15]. The specific application of optical imaging to studies of neurocognitive development in infants has been reviewed by Minagawa-Kawai *et al.* [16].

14.2.3
Epilepsy

Electroencephalography (EEG) is the tool routinely used to diagnose and monitor abnormal electrical activity in the brain, including epileptic seizure. Whereas EEG is sensitive to the electromagnetic dynamics of groups of active neurons, optical imaging provides a means of measuring associated changes in hemodynamic activity as a result of the neurons' increased metabolic demand. Optical topography has been used to display continuous changes in cerebral blood volume during seizure [17, 18]. It has also been combined with simultaneous EEG [19] and magnetoencephalography [20] as a means of studying neurovascular coupling.

14.3
Optical Tomography

Optical tomography involves generating three-dimensional (3D) volume or cross-sectional images using measurements of light transmitted across large thicknesses of tissue. The objective is to acquire maps representing either the spatial variation of optical properties (i.e., the absorption and scattering coefficients) at one or more wavelengths, or the variation in physiological parameters such as concentrations of specific chromophores, blood volume, and blood oxygenation. Typically, measurements are obtained of light transmitted between discrete sources and detectors located over as much of the available surface of the tissue as realistically achievable. Measuring light which is strongly attenuated over large thickness (up to 10 cm) requires powerful sources (consistent with patient safety) and sensitive detectors, and consequently optical tomography has a lower temporal resolution than optical topography. Furthermore, optical tomography requires a method to reconstruct the 3D distribution of optical properties or two-dimensional (2D) cross-section. Although various *ad hoc* back-projection methods have been proposed [21, 22], such approaches are not applicable to imaging arbitrary-shaped volumes. Instead, it is necessary to utilize a model of the propagation of light within the tissue which allows a set of predicted measurements on the surface to be derived for any given distribution of internal optical properties. Optical tomography is an inverse problem which is nonlinear, severely ill-posed, and generally underdetermined. Consequently, finding reliable and unique solutions for a given set of data involves considerable theoretical and practical difficulties, and various compromises and approximations are unavoidable. A comprehensive examination of the theory of optical tomography image reconstruction was presented by Arridge and Schotland [23]. In order to reduce the ill-posedness of the image reconstruction problem, the development of optical tomography systems has focused on increasing the amount of information available from light transmitted across the tissue. For example, imaging systems have been developed which measure the times-of-flight of photons traveling through the tissue, or measure the modulation and phase delay of light detected from sources modulated at megahertz frequencies. Devices that perform such measurements are known as time-domain and frequency-domain systems, respectively.

14.3.1
Optical Mammography

The principle clinical target of optical tomography has been imaging of the female breast as a means of detecting and characterizing tumors. Several commercial prototypes of so-called optical mammography systems have been produced [24–27], all requiring the patient to lie on a bed with one or both breasts suspended within a cavity surrounded by sources and detectors. Optical tomographic images of the breast typically demonstrate a strong heterogeneity, and significant sensitivity to many types of lesion [28–33]. However, spatial resolution remains low (1–2 cm) and a specificity sufficient for cancer screening remains elusive. The positive and negative aspects of optical mammography are both highlighted in Figure 14.3, which compares optical absorption and magnetic resonance imaging (MRI) images of the same breast cancer patient. The optical image of the right breast reveals the double tumor which is also present in the corresponding contrast agent-enhanced MRI image. However, the optical image of the right breast shows a feature of comparable contrast (lower left), almost certainly due to a surface blood vessel, which has no correlation in the MRI image. Further, optical mammography has demonstrated considerable promise as a mean of monitoring response to primary medical therapy. Since it does not involve ionizing radiation, patients can be scanned at regular intervals to monitor changes over a period of weeks. For example, optical images have exhibited sensitivity to changes in angiogenesis, apoptosis, and hypoxia resulting from neoadjuvant chemotherapy [34, 35].

14.3.2
Optical Tomography of the Brain

Although the attenuation of near-infrared light across the adult head is too great to image the center of the brain, 3D imaging of the entire infant brain is achievable.

14.3 Optical Tomography

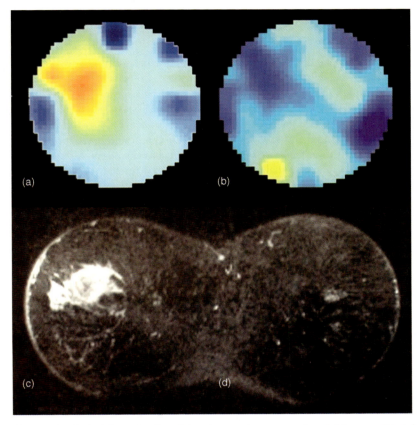

Figure 14.3 Optical absorption (a and b) and magnetic resonance (c and d) images of the left and right breasts of a patient with a malignant lesion in the right breast (a and c).

Optical tomography of the newborn infant brain was first demonstrated by researchers at Stanford University, who acquired images using a time-domain system which revealed intracranial hemorrhage [36], and focal regions of low oxygenation after acute stroke [37]. A more sophisticated time-domain system has been employed at University College London to reconstruct 3D images of the whole infant brain, which have revealed incidence of intraventricular hemorrhage [38] and changes in blood volume and oxygenation induced by small alterations to ventilator settings [39]. The system has also been used to generate 3D images of the entire neonatal head during motor-evoked response [40]. Partial tomographic reconstruction of the adult cortex has also been demonstrated. For example, the Stanford researchers successfully imaged localized contralateral oxygenation increases in the motor cortex of a healthy adult during hand movement [37], and Bluestone *et al.* [41] imaged hemodynamic changes within the frontal regions of the adult brain during the Valsalva maneuver.

14.4
Conclusion

Mapping of cerebral hemodynamic activity using optical topography has evolved into a powerful and widely used research tool, and there are excellent prospects for increased adoption of the technique within a clinical environment, particularly as imaging systems become more portable. The method provides complementary information to established tools such as EEG and functional MRI, with which they can be easily integrated. Optical tomography of the breast and brain has demonstrated the capacity to provide useful research information, and efforts to improve their clinical diagnostic potential are showing promising results.

References

1 Cutler, M. (1929) Transillumination as an aid in the diagnosis of breast lesions. *Surg. Gynecol. Obstet.*, **48**, 721–728.

2 Taga, G., Asakawa, K., Maki, A., Konishi, Y., and Koizumi, H. (2003) Brain imaging in awake infants by near-infrared optical topography. *Proc. Natl. Acad. Sci. U. S. A.*, **100**, 10722–10727.

3 Kusaka, T., Kawada, K., Okubo, K., Nagano, K., Namba, M., Okada, H., Imai, T., Isobe, K., and Itoh, S. (2004) Noninvasive optical imaging in the visual cortex in young infants. *Hum. Brain Mapp.*, **22**, 122–132.

4 Wolf, M., Wolf, U., Choi, J.H., Toronov, V., Paunescu, L.A., Michalos, A., and Gratton, E. (2003) Fast cerebral functional signal in the 100-ms range detected in the visual cortex by frequency-domain near-infrared spectrophotometry. *Psychophysiology*, **40**, 521–528.

5 White, B.R. and Culver, J.P. (2010) Phase-encoded retinotopy as an evaluation of diffuse optical neuroimaging. *Neuroimage*, **49**, 568–577.

6 Franceschini, M.A. and Boas, D.A. (2004) Noninvasive measurement of neuronal activity with near-infrared optical imaging. *Neuroimage*, **21**, 373–386.

7 Yamashita, Y., Maki, A., and Koizumi, H. (2001) Wavelength dependence of the precision of noninvasive optical measurement of oxy-, deoxy-, and total-hemoglobin concentration. *Med. Phys.*, **28**, 1108–1114.

8 Franceschini, M.A., Fantini, S., Thompson, J.H., Culver, J.P., and Boas, D.A. (2003) Hemodynamic evoked response of the sensorimotor cortex measured noninvasively with near-infrared optical imaging. *Psychophysiology*, **40**, 548–560.

9 Hintz, S.R., Benaron, D.A., Siegel, A.M., Zourabian, A., Stevenson, D.K., and Boas, D.A. (2001) Bedside functional imaging of the premature infant brain during passive motor activation. *J. Perinat. Med.*, **29**, 335–343.

10 Blasi, A., Fox, S., Everdell, N., Volein, A., Tucker, L., Csibra, G., Gibson, A.P., Hebden, J.C., Johnson, M.H., and Elwell, C.E. (2007) Investigation of depth dependent changes in cerebral haemodynamics during face perception in infants. *Phys. Med. Biol.*, **52**, 6849–6864.

11 Watanabe, E., Maki, A., Kawaguchi, F., Takashiro, K., Yamashita, Y., Koizumi, H., and Mayanagi, Y. (1998) Non-invasive assessment of language dominance with near-infrared spectroscopic mapping. *Neurosci. Lett.*, **256**, 49–52.

12 Pena, M., Maki, A., Kovacic, D., Dehaene-Lambertz, G., Koizumi, H., Bouquest, F., and Mehler, J. (2003) Sounds and silence: an optical topography study of language recognition at birth. *Proc. Natl. Acad. Sci. U. S. A.*, **100**, 11702–11705.

13 Tsujimoto, S., Yamamoto, T., Kawaguchi, H., Koizumi, H., and Sawaguchi, T. (2004) Prefrontal cortical

activation associated with working memory in adults and preschool children: an event-related optical topography study. *Cereb. Cortex*, **14**, 703–712.

14 Medvedev, A.V., Kainerstorfer, J., Borisov, S.V., Barbour, R.L., and VanMeter, J. (2008) Event-related fast optical signal in a rapid object recognition task: improving detection by the independent component analysis. *Brain Res.*, **1236**, 145–158.

15 Matsuo, K., Kato, T., Taneichi, K., Matsumoto, A., Ohtani, T., Hamamoto, T., Yamasue, H., Sakano, Y., Sasaki, T., Sadamatsu, M., Iwanami, A., Asukai, N., and Kato, N. (2003) Activation of the prefrontal cortex to trauma-related stimuli measured by near-infrared spectroscopy in posttraumatic stress disorder due to terrorism. *Psychophysiology*, **40**, 492–500.

16 Minagawa-Kawai, Y., Mori, K., Hebden, J.C., and Dupoux, E. (2008) Optical imaging of infants' neurocognitive development: recent advances and perspectives. *Dev. Neurobiol.*, **68**, 712–728.

17 Watanabe, E., Maki, A., Kawaguchi, F., Yamashita, Y., Koizumi, H., and Mayanagi, Y. (1998) Noninvasive cerebral blood volume measurement during seizures using multichannel near infrared spectroscopic topography. *J. Epilepsy*, **11** 335–340.

18 Gallagher, A., Lassonde, M., Bastien, D., Vannasing, P., Lesage, F., Grova, C., Bouthillier, A., Carmant, L., Lepore, F., Beland, R., and Nguyen, D.K. (2008) Non-invasive pre-surgical investigation of a 10 year-old epileptic boy using simultaneous EEG–NIRS. *Seizure*, **17**, 576–582.

19 Koch, S.P., Werner, P., Steinbrink, J., Fries, P., and Obrig, H. (2009) Stimulus-induced and state-dependent sustained gamma activity is tightly coupled to the hemodynamic response in humans. *J. Neurosci.*, **29**, 13962–13970.

20 Ou, W., Nissila, I., Radhakrishnan, H., Boas, D.A., Hamalainen, M.S., and Franceschini, M.A. (2009) Study of neurovascular coupling in humans via simultaneous magnetoencephalography and diffuse optical imaging acquisition. *Neuroimage*, **46**, 624–632.

21 Benaron, D.A., Ho, D.C., Spilman, S.D., van Houten, J.C., and Stevenson, D.K. (1994) Tomographic time-of-flight optical imaging device. *Adv. Exp. Med. Biol.*, **361**, 207–214.

22 Walker, S.A., Fantini, S., and Gratton, E. (1997) Image reconstruction by backprojection from frequency-domain optical measurements in highly scattering media. *Appl. Opt.*, **36**, 170–174.

23 Arridge, S.R. and Schotland, J.C. (2009) Optical tomography: forward and inverse problems. *Inverse Probl.*, **25** (12), 123010.

24 Colak, S.B., van der Mark, M.B., Hooft, G.W., Hoogenraad, J.H., van der Linden, E.S., and Kuipers, F.A. (1999) Clinical optical tomography and NIR spectroscopy for breast cancer detection. *IEEE J. Sel. Top. Quantum Electron.*, **5**, 1143–1158.

25 Grable, R.J. and Rohler, D.P. (1997) Optical tomography breast imaging. *Proc. SPIE*, **2979**, 197–210.

26 Intes, X. (2005) Time-domain optical mammography SoftScan: initial results. *Acad. Radiol.*, **12** (8), 934–947.

27 Schmitz, C.H., Klemer, D.P., Hardin, R., Katz, M.S., Pei, Y., Graber, H.L., Levin, M.B., Levina, R.D., Franco, N.A., Solomon, W.B., and Barbour, R.L. (2005) Design and implementation of dynamic near-infrared optical tomographic imaging instrumentation for simultaneous dual-breast measurements. *Appl. Opt.*, **44**, 2140–2153.

28 Grosenick, D., Moesta, K.T., Moller, M., Mucke, J., Wabnitz, H., Gebauer, B., Stroszczynski, C., Wassermann, B., Schlag, P.M., and Rinneberg, H. (2005) Time-domain scanning optical mammography: I. recording and assessment of mmammograms of 154 patients. *Phys. Med. Biol.*, **50**, 2429–2450.

29 Taroni, P., Torricelli, A., Spinelli, L., Pifferi, A., Arpaia, F., Danesini, G., and Cubeddu, R. (2005) Time-resolved optical mammography between 637 and 985 nm: clinical study on the detection and identification of breast lesions. *Phys. Med. Biol.*, **50**, 2469–2488.

30 Yates, T.D., Hebden, J.C., Gibson, A.P., Everdell, N.L., Arridge, S.R., and

Douek, M. (2005) Optical tomography of the breast using a multi-channel time-resolved imager. *Phys. Med. Biol.*, **50**, 2503–2517.

31 Boverman, G., Fang., Q., Carp, S.A., Miller, E.L., Brooks, D.H., Selb, J., Moore, R.H., Kopans, K.B., and Boas, D.A. (2007) Spatio-temporal imaging of the hemoglobin in the compressed breast with diffuse optical tomography. *Phys. Med. Biol.*, **52**, 3619–3641.

32 Tromberg, B.J., Pogue, B.W., Paulsen, K.D., Yodh, A.G., Boas, D.A., and Cerussi, A.E. (2008) Assessing the future of diffuse optical imaging technologies for breast cancer management. *Med. Phys.*, **35** (6), 2443–2451.

33 Wang, X., Pogue, B.W., Jiang, S., Dehgahni, H., Song, X., Srinivasan, S., Brooksby, B.A., Paulsen, K.D., Kogel, C., Poplack, S.P., and Wells, W.A. (2006) Image reconstruction of effective Mie scattering parameters of breast tissue *in vivo* with near-infrared tomography. *J. Biomed. Opt.*, **11** (4), 041106.

34 Choe, R., Corlu, A., Lee, K., Durduran, T., Konecky, S.D., Grosicka-Koptyra, M., Arridge, S.R., Czernieki, B.J., Fraker, D.L., DeMichele, A., Chance, B., Rosen, M.A., and Yodh, AG. (2005) Diffuse optical tomography of breast cancer during neoadjuvant chemotherapy: a case study with comparison to MRI. *Med. Phys.*, **32** (4), 1128–1139.

35 Jiang, S., Pogue, B.W., Carpenter, C.M., Poplack, S.P., Wells, W.A., Kogel, C.A., Forero, J., Muffly, L.S., Schwartz, G.N., Paulsen, K.D., and Kaufman, P.A. (2009) Evaluation of breast tumor response to neoadjuvant chemotherapy with diffuse optical spectroscopic tomography: case studies of tumor region-of-interest changes. *Radiology*, **252** (2), 330–331.

36 Hintz, S.R., Cheong, W.F., van Houten, J.P., Stevenson, D.K., and Benaron, D.A. (1999) Bedside imaging of intracranial hemorrhage in the neonate using light: comparison with ultrasound, computed tomography, and magnetic resonance imaging. *Pediatr. Res.*, **45**, 54–59.

37 Benaron, D.A., Hintz, S.R., Villringer, A., Boas, D., Kleinschmidt, A., Frahm, J., Hirth, C., Obrig, H., van Houten, J.C., Kermit, E.L., Cheong, W.F., and Stevenson, D.K. (2000) Noninvasive functional imaging of human brain using light. *J. Cereb. Blood Flow Metab.*, **20**, 469–477.

38 Hebden, J.C., Gibson, A., Yusof, R., Everdell, N., Hillman, E.M.C., Delpy, D.T., Arridge, S.R., Austin, T., Meek, J.H., and Wyatt, J.S. (2002) Three-dimensional optical tomography of the premature infant brain. *Phys. Med. Biol.*, **47**, 4155–4166.

39 Hebden, J.C., Gibson, A., Austin, T., Yusof, R., Everdell, N., Delpy, D.T., Arridge, S.R., Meek, J.H., and Wyatt, J.S. (2004) Imaging changes in blood volume and oxygenation in the newborn infant brain using three-dimensional optical tomography. *Phys. Med. Biol.*, **49**, 1117–1130.

40 Gibson, A.P., Austin, T., Everdell, N.L., Schweiger, M., Arridge, S.R., Meek, J.H., Wyatt, J.S., Delpy, D.T., and Hebden, J.C. (2006) Three-dimensional whole-head optical tomography of passive motor evoked responses in the neonate. *Neuroimage*, **30**, 521–528.

41 Bluestone, A.Y., Abdouleav, G., Schmitz, C.H., Barbour, R.L., and Hielscher, A.H. (2001) Three-dimensional optical tomography of hemodynamics in the human head. *Opt. Express*, **9**, 272–286.

15
Raman Microscopy

Christoph Krafft and Jürgen Popp

15.1
Introduction

Fluorescence microscopy is the most commonly biophotonic technique applied to assess cells and tissues. Because only a few biomolecules show inherent fluorescence, extrinsic fluorescent probes and molecular stains have been introduced as fluorophores. However, there are sometimes artifacts and drawbacks that have to be taken into account when using fluorophores. Along with the well-known problem of photobleaching, delivering labels can be a problem, particularly for whole organisms. Some labels work only on dead cells; others damage cells or perturb the very processes that they are intended to study. Other measurements destroy cells, creating isolated snapshots of different cells at given points in time. Such data are not particularly useful for heterogeneous cultures containing spontaneously differentiating cells. Label-free and nondestructive microscopy offers a way to investigate living cells while eliminating possible artifacts. Instead of detecting photons emitted from excited fluorophores, alternative techniques detect subtle changes in light as it is absorbed or altered by biological samples relying on linear or nonlinear optical phenomena. Nonlinear phenomena such as second harmonic generation, two-photon fluorescence, coherent anti-Stokes Raman scattering, and stimulated Raman scattering are observed when high-intensity light, in essence pulses of laser light, interacts with matter. They will not be discussed here as the sophisticated and expensive instrumentation is available in only a few specialized laboratories [1]. Raman spectroscopy is a linear optical phenomenon based on inelastic scattering of monochromatic light. It probes molecular vibrations, which are an inherent property of matter. As the spectrum of vibrations provides a specific fingerprint of the chemical composition and structure of samples, Raman spectroscopy has been introduced as a biophotonic tool to characterize cells and tissues. The coupling of a Raman spectrometer with a microscope offers two main advantages. First, lateral resolutions can be achieved down to Abbe's limit of diffraction below 1 µm. Second, maximum sensitivity can be achieved because the photon flux (photon density per unit area) of the focused laser beam on to the sample is at a maximum and the collection efficiency of scattered photons from the sample is at a maximum. User-friendly, commercial Raman microscopes are available from numerous companies [2]. Of course, Raman microscopy has limitations. Whereas fluorescence labeling can often allow discrimination of single molecules, label-free techniques are less sensitive and specific. To improve the sensitive and specific detection of biomolecules with low scattering intensities or low abundance, signal-enhancement techniques have been developed such as surface-enhanced Raman scattering (SERS) and resonance Raman scattering (RRS). This chapter briefly describes some biomedical examples. Recent reviews give a more comprehensive overview of Raman microscopy of cells and tissues [3] and disease recognition [4].

15.2
Application to Cell Characterization

Single cells are very suitable objects for Raman spectroscopy because of the high concentrations of biomolecules in their condensed volume. Protein concentration as high as $250\,\mu g\,\mu l^{-1}$ and DNA and RNA concentrations in the region of $100\,\mu g\,\mu l^{-1}$ have been reported. These values depend on the cell type, the phase of the cell cycle, and the location within the cell.

Figure 15.1 shows photomicrographs and Raman microscopic images of a dried Hep G2 cell and a fixed macrophage in medium. Hep G2 is a perpetual cell line which was derived from the liver tissue of a 15-year-old male with a well-differentiated hepatocellular carcinoma. The Raman image of 51×61 spectra (Figure 15.1b) was collected using a $100 \times$ /NA 0.9 objective with a step size of 500 nm and an exposure time of 6 s per spectrum. The numbers 51 and 61 indicate the spectra in x and y direction, respectively. After low-intensity spectra had been removed and the remaining spectra normalized, the data set was segmented by k-means clustering. This algorithm groups the spectra into k clusters (here seven) according to their similarity, which is calculated from the Euclidian distance in a given spectral range. The cluster memberships are color coded for display. Averaging all spectra within a cluster offers

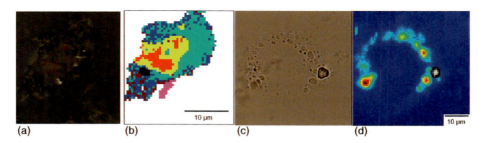

Figure 15.1 (a) Photomicrograph and (b) Raman image of a dried Hep G2 cell. The color code represents the membership of a cluster analysis. (c) Photomicrograph and (d) Raman image of a fixed macrophage in aqueous buffer. The color code indicates the intensity of CH_2 vibrations that corresponds to lipids. The gray scale code shows a foreign particle.

the advantages that pixel-to-pixel variations are reduced and the signal-to-noise ratio is improved compared with a single spectrum.

Macrophages comprise a special class of white blood cells (leukocytes). As phagocytes they engulf and digest cellular debris and pathogens. They are also involved in the uptake of lipids by various mechanisms that create the progressive plaque lesion of atherosclerosis. The Raman image of 78 × 72 spectra (Figure 15.1d) was registered using a 60 × /NA 1.0 water immersion objective with a step size of 700 nm. Typical features of the macrophage are lipid droplets that line the margin of the cell. The most intense Raman bands of lipids are CH_2 stretch vibrations near 2900 cm^{-1} (Figure 15.2). The chemical map in Figure 15.1d plots the intensity distribution of these bands. Under the experimental conditions, larger droplets are resolved; smaller droplets could not be resolved. Higher lateral resolution can be achieved by using a shorter excitation wavelength and a smaller step size.

Figure 15.2a shows five Raman spectra representing the clusters of the Hep G2 cell in Figure 15.1b (traces 1, 2, 3, 5, 6) and a difference spectrum (trace 4 = 2 − 1). Trace 1 represents the cytoplasm (cyan cluster). Spectral contributions of proteins dominate. Bands are assigned to the aromatic amino acid side chains (Phe, 622, 1003, 1032,

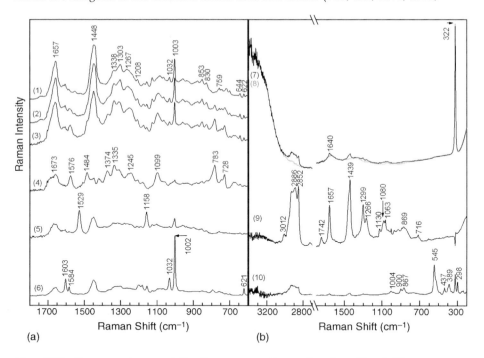

Figure 15.2 Raman spectra representing the clusters of the dried Hep G2 cell in Figure 15.1b (traces 1, 2, 3, 5, 6) and difference spectrum (trace 4 = 2 − 1). Unmanipulated Raman spectra of the macrophage in Figure 15.1d corresponding to the lipid droplets (trace 7) and surrounding medium (trace 8) and difference spectrum (trace 9 = 7 − 8). Raman spectrum of foreign particle (trace 10). The intensity scale was four fold compressed for the range 3400–2700 cm^{-1}.

1208 cm^{-1}; Tyr, 644, 830, 853, 1208 cm^{-1}; Trp, 759 cm^{-1}), aliphatic amino acid side chains (1338, 1448 cm^{-1}), and the peptide backbone (amide III at 1267 cm^{-1}, amide I at 1657 cm^{-1}). Spectral contributions of lipids are resolved at 1303, 1130, 1060, 716, and 700 cm^{-1}. The latter bands are not labeled. The other cytoplasm spectra (blue and brown clusters) deviate only slightly and will not be discussed here. Trace 2 represents the nucleus (yellow cluster). In addition to protein bands that are similar to those in the cytoplasm, spectral contributions of DNA are evident at 783 and 1099 cm^{-1}. As most DNA bands overlap with protein bands, a difference spectrum was calculated (trace 4). Then positive difference bands are assigned to nucleotides (adenine, 728, 1335, 1484, 1576 cm^{-1}; cytosine, 783, 1245 cm^{-1}; thymine, 783, 1374, 1673 cm^{-1}; guanine, 1484, 1576 cm^{-1}) and the phosphate backbone (1099 cm^{-1}). The intensities of DNA bands increase in trace 3, which represents nucleoli (orange cluster). Traces 5 and 6 represent special features (magenta and black clusters, respectively). Bands at 1158 and 1529 cm^{-1} are assigned to carotene. An inherent property of carotene compounds is their high Raman cross-section, which makes them easily detectable in Raman spectra even at low concentrations. Intense bands at 621, 1002, 1032, 1584 and 1603 cm^{-1} are typical of polystyrene. The detection of this polymer bead clearly demonstrates the power of Raman microscopic imaging. The location and distribution of numerous molecules can be assessed by their spectroscopic fingerprint without exogenous labels.

Unmanipulated Raman spectra of the macrophage represent the lipid droplets and the surrounding medium [Figure 15.2b, traces 7 (black) and 8 (gray), respectively]. Spectral contributions are assigned to the substrate calcium fluoride (322 cm^{-1}) and water (1640 and above 3000 cm^{-1}). The overlay of both spectra indicates that the spectral contributions of the lipid droplets are less intense. Subtracting the spectrum of the medium compensates bands of the substrate, water and also constant spectral contributions of optical elements in the light path. The resulting difference spectrum (trace 9) reveals pure lipid bands with a very low background. The bands are assigned to the choline group (716, 869 cm^{-1}), phosphate group (1080 cm^{-1}), ester group (1742 cm^{-1}), and C–C and C–H vibrations of fatty acids (1063, 1130, 1299, 1439, 2852, 2886 cm^{-1}). Bands at 1266, 1657, and 3012 cm^{-1} are indicative of unsaturated fatty acids. The Raman spectrum (trace 9) is consistent with the phospholipid phosphatidylcholine. The Raman spectrum of a foreign particle at the right part of the cell (trace 10) shows bands in the low-wavenumber region at 545, 437, 369, and 298 cm^{-1}. The distribution of the band at 545 cm^{-1} has been included in Figure 15.1d as a gray scale chemical map. This Raman image shows a macrophage at work engulfing a foreign particle.

To improve the sensitive and specific detection of biomolecules with low scattering intensities or low abundance, signal-enhancement techniques have been applied such as SERS and RRS. RRS is observed if the excitation wavelength is matched to an electronic transition of the molecule so that vibrational modes associated with the excited state are enhanced. As the enhancement is restricted to vibrational modes of the chromophore, the complexity of Raman spectra from large biomolecules is reduced. Examples of RRS of cells include the detection of heme and hemozoin in red blood cells (erythrocytes) using 785 nm excitation [5] and cytochrome b_{558} in neutrophilic granulocytes using 413 nm excitation [6]. SERS utilizes the optical excitation of surface plasmons, i.e., collective vibrations of the free electrons in metals. The result is an amplification of the electromagnetic field in the close vicinity of the metal surface (a few nanometers), which leads to the enhancement of Raman signals from adsorbed molecules. The first SERS experiments used colloidal gold nanoparticles that were taken up by cells. However, due nonspecific binding of biomolecules to these nanoparticles, every spectrum of a Raman image was different and the approach lacked reproducibility [7]. For specific recognition of antigens in cancer cells by SERS microscopy, gold nanoparticles have been functionalized with antibodies [8]. The advantage of the SERS approach compared with the detection of fluorescence labels is the multiplex capacity. The fingerprint of the SERS spectrum allows the simultaneous detection of numerous antigens.

15.3
Application to Tissue Characterization

Histology – the study of the anatomy of tissue – and histopathology – the study of diseased tissue – are performed by examining thin slices (sections) of tissue under a light microscope. The ability to visualize or differentially identify microscopic structures is frequently enhanced through the use of histologic stains because biological tissue has little inherent contrast. Hundreds of techniques exist that have been developed to stain cells and cellular components selectively. As Raman microspectroscopic imaging provides molecular contrast in tissues without stains, it offers prospects to complement established techniques. Instead of staining protocols, sophisticated chemometric tools are applied for the reconstruction and classification of Raman images from tissue sections. In principle, the concepts that have been described in the previous section can be transferred from cell characterization. For example, a SERS marker has been developed to detect the prostate-specific antigen in the context of prostate cancer tissue [9].

Figure 15.3 shows as an example a 70 × 70 Raman image with a 2.5 μm step size of a liver tissue section. A photomicrograph (a), three plots of a vertex component analysis (VCA) (b–d), and a composite image (e) are compared. VCA is a hyperspectral unmixing algorithm that projects the image raw data in a space of smaller dimensionality aiming to retain all information. The scope of VCA is that end-

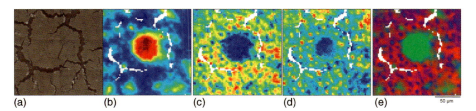

Figure 15.3 (a) Photomicrograph, (b–d) score plots of a vertex component analysis and (e) composite image of the central vein in a liver tissue section.

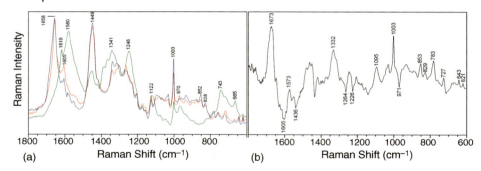

Figure 15.4 (a) Overlay of endmember spectra representing the score plots in Figure 15.3b–d and (b) difference spectrum (black trace = blue − red).

members represent spectra of most dissimilar constituents. It has been found that endmembers have high chemical relevance. Then, scores denote the concentration of the endmember spectra. Details in the context of Raman spectroscopy recently have been published [10]. Considering the bands in the endmember spectra (Figure 15.4) and the location and size of the morphologic features, the score plot in Figure 15.3b is assigned to the central vein and red blood cells, the score plot in Figure 15.3c to liver parenchyma, and the score plot in Figure 15.3d to cell nuclei. If the intensities of the plots in Figure 15.3b–d are converted to green, red, and blue channels, respectively, a composite image can be generated (Figure 15.3e). Such a composite image can be analyzed in a similar way to a histologically stained tissue section.

The endmember spectra are overlaid in Figure 15.4. The color code is analogous to the colors in the composite in Figure 15.3e. This means that the endmember spectrum representing the central vein and red blood cells is plotted green, liver parenchyma red, and cell nuclei blue. The green spectrum is characterized by intense contributions of the heme group at 665, 743, 1122, 1248, 1341, 1580, and 1619 cm^{-1} that are enhanced by a resonance effect. The red and blue spectra are dominated by spectral contributions of proteins near 828, 852, 1003, 1449, and 1658 cm^{-1}. The difference spectrum (black) between the blue and the red was calculated to improve visualization of the variations. Positive difference bands at 727, 783, 1095, and 1573 cm^{-1} are associated with DNA. Further positive difference bands at 621, 643, 829, 853, and 1003 cm^{-1} are assigned to proteins. Positive difference bands at 1332 and 1673 cm^{-1} contain overlapping contributions from proteins and DNA. Negative difference bands at 971, 1226, 1364, 1436, and 1605 cm^{-1} are consistent with bilirubin in bile acid. This result demonstrates the wealth of molecular information that can be obtained simultaneously from the analysis of Raman images of tissue sections.

15.4
Conclusions

Applications of Raman microscopy to cells and tissues are still in their infancy and are expected to grow during the years to come. A prohibitive factor in Raman microscopic imaging remains the acquisition time. Therefore, technical improvements aim to (i) increase the sensitivity to reduce the exposure time per spectrum and (ii) apply parallel data acquisition strategies to register multiple spectra simultaneously. Innovations include optical filters with higher transmission, spectrographs with reduced light losses, and detectors with optimized quantum efficiency. Modern Raman microscopic instruments are able to collect several hundred spectra per second. Another challenge is the automatic processing of extended Raman microscopic images because the spectral features are often small and distributed over a wide range. In addition to unsupervised segmentation and spectral unmixing algorithms such as cluster analysis, principal component analysis, and vertex component analysis, supervised algorithms such as artificial neural networks, linear discriminant analysis, and support vector machines have to be trained to classify cells and tissues objectively based on their Raman signatures. A full description of all technical and theoretical details is beyond the scope of this chapter. The intention of the chapter was to highlight the potential of Raman microscopy in biophotonics. Further progress requires interdisciplinary efforts of natural scientists, engineers, and physicians.

References

1 Müller, M. and Zumbusch, A. (2007) Coherent anti-Stokes Raman scattering microscopy. *ChemPhysChem*, **8**, 2156–2170.

2 Sage, L. (2009) Raman microscopes. *Anal. Chem.*, **81**, 3222–3226.

3 Krafft, C., Dietzek, B., and Popp, J. (2009) Raman and CARS microspectroscopy of cells and tissues. *Analyst*, **134**, 1046–1057.

4 Krafft, C., Steiner, G., Beleites, C., and Salzer, R. (2008) Disease recognition by infrared and Raman spectroscopy. *J. Biophotonics*, **2**, 13–28.

5 Bonifacio, A., Finaurini, S., Krafft, C., Parapini, S., Taramelli, D., and Sergo, V. (2008) Spatial distribution of heme species in erythrocytes infected with *Plasmodium falciparum* by use of resonance Raman imaging and multivariate analysis. *Anal. Bioanal. Chem.*, **392**, 1277–1282.

6 van Manen, H.J., Uzunbajakava, N., van Bruggen, R., Roos, D., and Otto, C. (2003) Resonance Raman imaging of the NADPH oxidase subunit cytochrome b558 in single neutrophilic granulocytes. *J. Am. Chem. Soc.*, **125**, 12112–12113.

7 Kneipp, K., Haka, A.S., Kneipp, H., Badizadegan, K., Yoshizawa, N., Boone, C., Shafer-Peltier, K.E., Motz, J.T., Dasari, R.R., and Feld, M.S. (2002) Surface-enhanced Raman spectroscopy in single living cells using gold nanoparticles. *Appl. Spectrosc.*, **56**, 150–154.

8 Lee, S., Chon, H., Lee, M., Choo, J., Shin, S.Y., Lee, Y.H., Rhyu, I.J., Son, S.W., and Oh, CH. (2009) Surface-enhanced Raman scattering imaging of HER2 cancer markers overexpressed in single MCF7 cells using antibody conjugated hollow gold nanospheres. *Biosens. Bioelectron.*, **24**, 2260–2263.

9 Schlücker, S., Küstner, B., Punge, A.R.B., Marx, A., and Ströbel, P. (2006) Immuno-Raman microspectroscopy: in situ detection of antigens in tissue specimens by surface-enhanced Raman scattering. *J. Raman Spectrosc.*, **37**, 719–721.

10 Miljković, M., Chernenko, T., Romeo, M.J., Bird, B., Matthaus, C., and Diem, M. (2010) Label-free imaging of human cells: algorithms for image reconstruction of Raman hyperspectral datasets. *Analyst*, **135**, 2002–2013.

16
CARS Microscopy
Andreas Zumbusch

16.1
Introduction

Optical microscopy is one of the major tools used in the biological sciences. The interaction between optics and biology has for centuries been enormously fruitful for both areas. In recent decades, especially fluorescence microscopic techniques have found very widespread applications. This is due to their high sensitivity, which can reach the single-molecule detection limit [1], their high specificity achieved by powerful labeling strategies, and the noninvasive character common to most optical microscopy techniques. Despite the great success of fluorescence microscopy, much effort has recently been invested in research on complementary label-free microscopy techniques. The motivation for these efforts is twofold: first, not all samples can be fluorescently labeled, and second, all fluorescent labels suffer from photobleaching, which severely limits the observation times. Ideally, new techniques should – despite working without labels – offer the above-mentioned advantages of fluorescence-based approaches.

In general, if the requirement of offering (molecular) specificity without the use of labels is to be fulfilled, contrast generation should not be based on electronic spectra. This is due to the large linewidths of electronic transitions with respect to the width of the relevant spectral detection window. Concerning this point, vibrational spectra offer much more information, since medium-sized molecules can clearly be identified based on their characteristic vibrational spectra, which typically feature rather narrow linewidths. Vibrational microscopy is mostly performed either by directly monitoring the absorption of infrared (IR) light or by monitoring Raman scattering. Apart from technical problems concerning the relatively low sensitivity and poor spatial resolution, biological applications of IR microscopy suffer especially from the broad and intense absorption bands of water. These two problems are avoided when using spontaneous Raman scattering microscopy. Here, however, low scattering cross-sections and an overwhelmingly strong fluorescence background, which is present in many samples, are new difficulties.

Handbook of Biophotonics. Vol.2: Photonics for Health Care, First Edition. Edited by Jürgen Popp, Valery V. Tuchin, Arthur Chiou, and Stefan Heinemann.
© 2012 Wiley-VCH Verlag GmbH & Co. KGaA. Published 2012 by Wiley-VCH Verlag GmbH & Co. KGaA.

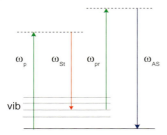

Figure 16.1 Energy level representation of the CARS process.

Despite these problems, Raman-type scattering can be exploited for generating images of unlabeled samples with high spatial resolution if nonlinear optical techniques are used. Since the advent of commercially available laser sources delivering ultrashort pulses, nonlinear optical effects such as two-photon fluorescence excitation [2] and second harmonic generation (SHG) and third harmonic generation (THG) microscopy [3] have been used extensively in optical microscopy. Whereas two-photon excitation as a variant of fluorescent microscopy relies on sample labeling, both SHG and THG microscopy generate contrast in unlabeled specimens. However, both techniques generate structural and not chemically specific contrast. Contrast with chemical selectivity can instead be generated by coherent anti-Stokes Raman scattering (CARS) microscopy [4, 5], a technique which is now commercially available and has found widespread applications.

16.2
Technical Background and Experimental Realizations

CARS is a third-order nonlinear optical process in which three excitation fields interact to produce a fourth field, which is detected. In general, two beams with frequencies ω_p and ω_{St} are tuned such that the frequency difference $\omega_p - \omega_{St}$ coincides with the frequency ω_{vib} of a vibrational transition of the sample molecules. Under this condition, a third beam with frequency ω_{pr} generates a fourth field with frequency $\omega_{AS} = \omega_p + \omega_{vib}$ in a resonance-enhanced manner (Figure 16.1). In most cases, the experiment is performed using a frequency-degenerate scheme with $\omega_p = \omega_{pr}$, such that only two laser beams are necessary for the excitation. The CARS signal intensity depends quadratically on the square modulus of the induced third-order polarization in the sample $I_{AS} \propto |\mathbf{P}^{(3)}|^2$. $\mathbf{P}^{(3)}$ itself depends on the third-order optical susceptibility $\chi^{(3)}$ as a material-specific parameter that can be decomposed into a resonant and a nonresonant term, and the amplitude of the exciting laser fields $\mathbf{E}$. One obtains

$$\mathbf{P}^{(3)} = \left[\chi_r^{(3)} + \chi_{nr}^{(3)}\right] \mathbf{E}_p \mathbf{E}_{pr} \mathbf{E}_{St}^* \tag{16.1}$$

This description already explains many of the features of CARS microscopy. (i) Image contrast is generated with chemical selectivity by tuning the frequency difference between the excitation lasers to ω_{vib}. (ii) Sample autofluorescence does not interfere with the CARS signal since the latter is blue shifted with respect to the excitation wavelengths. (iii) Three-dimensional images can be obtained without use of a confocal pinhole since CARS is a nonlinear optical process. However, because $\chi^{(3)}$ in all cases is very small, pulsed lasers with high pulse energies have to be used.

A CARS microscopy experiment then consists in tuning the laser wavelengths to the vibrational frequency of interest and then scanning the sample. Microscope setups similar to those used in confocal microscopy are easily adapted to CARS microscopy. This leads to images which represent the spatial distribution of compounds with a specific vibrational resonance. However, the nonresonant part of $\chi^{(3)}$ and the quadratic dependence of I_{AS} on $\chi^{(3)}$ in general make it difficult to extract quantitative information of local molecular concentrations.

16.2.1
Excitation Geometry

The first adaptation of CARS to microscopy was described in the early 1980s [6]. At that time, attempts were made to increase the efficiency of the signal generation process by matching the respective phases of the excitation beams through their geometric separation. However, this severely reduces the effective numerical aperture and therefore no high spatial resolution was achievable. Therefore, CARS microscopy initially did not find any applications. This changed, when it was realized that the reduction in the interaction volume by strong focusing leads to the generation of strong signals [7]. The subsequent development of CARS microscopy was mainly aimed at simplifying the experimental setup and at reducing the nonresonant background due to $\chi^{(3)}_{nr}$.

In most cases, CARS microscopy is carried out in a transmission-type geometry, because most of the signal is generated in the forward direction [8]. However, sample features which are of a size comparable to or smaller than the excitation wavelengths also lead to a strong CARS signal in the backward direction. It has been pointed out, however, that large parts of the strong forward-directed signal can be recovered in an epi-detection scheme due to CARS signal scattering [9]. CARS microscopy can therefore also be performed in a backscattering- or epi-detection scheme. With epi-detection, quantitative image interpretation is impossible, because the signal intensity shows an oscillatory behavior depending on the size of the sample features [10]. The epi-detection scheme is nevertheless of great importance since only one face of the sample has to be accessed for imaging. It is worth mentioning that CARS microscopy has also been performed in a wide-field geometry [11]. This allows very fast image acquisition over large sample areas, but decreases the axial resolution.

16.2.2
Excitation Lasers

Whereas well-established microscope setups can be used for CARS microscopy, much effort has been spent in order to improve the excitation sources used. The most important requirement is that the two (or three) excitation lasers are synchronized to an extent such that the jitter in the arrival time of the individual laser pulses is less than 300 fs. In addition, the laser system should offer fast tunability in order to address different vibrational resonances and a spectral linewidth of $10\,\mathrm{cm}^{-1}$ to ensure high spectral selectivity. At the same time, the pulse duration should be as short as possible for the CARS process to be driven efficiently. The last two requirements are contradictory, since short pulses come with broad spectra. The ideal compromise would consist in using transform-limited pulses with a duration of 1 ps. In practice, most experiments are currently performed by using 5–7 ps pulses available from an optical parametric amplifier system which is pumped by a high repetition rate pump laser at 532 nm [12]. Since in this experiment all excitation beams used are derived from one pump pulse, no timing jitter occurs. The performance of these systems is therefore much better than that of commercially available synchronized Ti:sapphire laser systems [13]. Common to all the laser sources discussed so far is that they are large and rather difficult to operate. Therefore, fiber lasers that are cheaper, more compact, more versatile, and more robust have recently been introduced as an alternative [14].

In contrast to the approach of imaging only one vibrational frequency at a time, so-called multiplexing experiments probe a whole range of frequencies with each excitation cycle. For this purpose, ω_{St} is provided not by a narrowband laser, but instead by a laser covering a broad spectral range [15]. Typically, short laser pulses are used for this purpose. Spectrally resolved detection is then used to gain information about vibrational resonances. Multiplex CARS microscopy is much faster than the single-frequency approach if spectroscopic information of a microscopic volume element is desired. It has therefore been employed successfully in microfluidic measurements [16, 17]. The integration times necessary for recording spectra, however, currently limit its suitability in imaging applications. Also in the field of multiplex CARS microscopy, fiber lasers have recently been used [18]. In this case, they offer the advantage of background-free imaging over the entire vibrational spectral region.

16.3
Applications

CARS microscopy has been developed in order to complement fluorescence microscopic techniques in experiments where fluorescent labeling is not possible. It turns out that especially C—H vibrational resonances lead to very strong CARS signals. Therefore, CARS microscopy has been used extensively to study the lipid distribution in a variety of live cells, live animals, and functional tissues [19]. Early applications of CARS experiments were aimed at following the distribution dynamics of exogenous compounds in live animals. This has the advantage that the vibrational spectra of the compounds of interest are well known. In this respect, mostly permeation studies of the skin were performed [9]. Increasingly, however, CARS microscopy of endogenous structures rich in C—H was performed. Examples are investigations related to atherosclerosis and neuronal damage [20, 21]. None of these experiments could be performed using the labeling strategies necessary for fluorescence imaging. It should be noted that the pulsed excitation employed in CARS microscopy can lead to sample

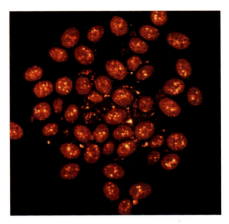

Figure 16.2 CARS images of lipid droplets in live yeast cells recorded at a resonance of 2845 cm^{-1}. Image dimensions are 150 × 150 μm (1024 × 1024 pixels). Integration time, 1 s. M. Winterhalder, unpublished results.

damage. In most cases, however, the excitation intensities necessary are below the threshold for the induction of noticeable damage even to sensitive organisms. As an example, high-resolution CARS microscopy of yeast cells is possible for extended periods without changing their normal behavior, that is, budding and similar processes are not affected (Figure 16.2). Likewise, several CARS microscopic studies on the lipid distribution in live *Caenorhabditis elegans* (Figure 16.3) have been published [22, 23].

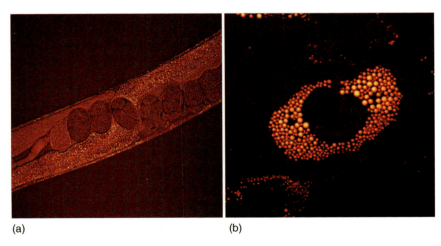

(a)   (b)

Figure 16.3 (a) CARS image of *C. elegans*. A maximum projection of 54 stacks clearly shows the intestine and fertilized eggs. Lipids mainly located in the intestine generate a strong CARS signal at 2845 cm^{-1}. The image dimensions are 150 × 150 μm (1024 × 1024 pixels). Integration time, 5 s. (b) CARS image of live human adipocytes, recorded at 2850 cm^{-1}. Image size, 84 × 84 μm (1024 × 1024 pixels). Line average, 2; integration time, 2 s. C. Jüngst, unpublished results.

The facts that CARS microscopy can be performed in a truly noninvasive manner, that is, no labeling is necessary, and microscopy with light levels below those causing photodamage to living animals is possible, hold great promise for future medical applications of CARS microscopy. Currently, the main areas of interest in this field are the diagnosis of malignant structures of the skin and intrasurgical application in brain surgery [24]. Although this chapter focuses on biological applications of CARS microscopy, it should nevertheless be mentioned that recently it has also found exciting applications in studies of problems related to material science, such as studies of catalytic processes [25].

16.4
Outlook

CARS microscopy is a new tool for truly noninvasive investigations. Although it does not reach the sensitivity of fluorescence microscopy, it does not require the use of labeling and it can be employed at light levels benign to living organisms, yet it delivers contrast which is molecularly specific. It therefore provides a complementary approach to other microscopy techniques, which is especially interesting for applications in the medical sciences.

References

1 Moerner, W.E. and Orrit, M. (1999) Illuminating single molecules in condensed matter. *Science*, **283** (5408), 1670–1676.
2 Denk, W., Strickler, J., and Webb, W. (1990) Two-photon laser scanning microscopy. *Science*, **248**, 73–76.
3 Olivier, N., Luengo-Oroz, M.A., Duloquin, L., Faure, E., Savy, T., Veilleux, I., Solinas, X., Debarre, D., Bourgine, P., Santos, A., Peyrieras, N., and Beaurepaire, E. (2010) Cell lineage reconstruction of early zebrafish embryos using label-free nonlinear microscopy. *Science*, **329** (5994), 967–971.
4 Müller, M. and Zumbusch, A. (2007) Coherent anti-Stokes Raman scattering microscopy. *ChemPhysChem*, **8** (15), 2156–2170.
5 Evans, C.L. and Xie, X.S. (2008) Coherent anti-Stokes Raman scattering microscopy: chemical imaging for biology and medicine. *Annu. Rev. Anal. Chem. (Palo Alto Calif)*, **1**, 883–909.
6 Duncan, M., Reintjes, J., and Manuccia, T. (1982) Scanning coherent anti-Stokes Raman microscope. *Opt. Lett.*, **7**, 350–352.
7 Zumbusch, A., Holtom, G., and Xie, X. (1999) Three-dimensional vibrational imaging by coherent anti-Stokes Raman scattering. *Phys. Rev. Lett.*, **82**, 4142–4145.
8 Cheng, J.X., Volkmer, A., and Xie, X. (2002) Theoretical and experimental characterization of coherent anti-Stokes Raman scattering microscopy. *J. Opt. Soc. Am. B*, **19**, 1363–1375.
9 Evans, C.L., Potma, E.O., Puoris'haag, M., Côté, D., Lin, C.P., and Xie, X.S. (2005) Chemical imaging of tissue *in vivo* with video-rate coherent anti-Stokes Raman scattering microscopy. *Proc. Natl. Acad. Sci. U. S. A.*, **102** (46), 16807–16812.
10 Cheng, J.X., Volkmer, A., Book, L.D., and Xie, X. (2001) An epi-detected coherent anti-Stokes Raman scattering (E-CARS) microscope with high spectral resolution and high sensitivity. *J. Phys. Chem. B*, **105**, 1277–1280.

11 Heinrich, C., Bernet, S., and Ritsch-Marte, M. (2006) Nanosecond microscopy with spectroscopic resolution. *New J. Phys.*, **8**, 36.
12 Ganikhanov, F., Carrasco, S., Xie, X., Katz, M., Seitz, W., and Kopf, D. (2006) Broadly tunable dual-wavelength light source for coherent anti-Stokes Raman scattering microscopy. *Opt. Lett.*, **31**, 1292–1294.
13 Jones, D., Potma, E., Cheng, J.X., Burfeindt, B., Pang, Y., and Xie, X. (2002) Synchronization of two passively mode-locked, picosecond lasers within 20 fs for coherent anti-Stokes Raman scattering microscopy. *Rev. Sci. Instrum.*, **73**, 2843–2848.
14 Krauss, G., Hanke, T., Sell, A., Träutlein, D., Leitenstorfer, A., Selm, R., Winterhalder, M., and Zumbusch, A. (2009) Compact coherent anti-Stokes Raman scattering microscope based on a picosecond two-color Er:fiber laser system. *Opt. Lett.*, **34** (18), 2847–2849.
15 Müller, M. and Schins, J. (2002) Imaging the thermodynamic state of lipid membranes with multiplex CARS microscopy. *J. Phys. Chem. B*, **106**, 3715–3723.
16 Schafer, D., Squier, J.A., van Maarseveen, J., Bonn, D., Bonn, M., and Müller, M. (2008) In situ quantitative measurement of concentration profiles in a microreactor with submicron resolution using multiplex CARS microscopy. *J. Am. Chem. Soc.*, **130**, 11592–11593.
17 Bergner, G., Chatzipapadopoulos, S., Akimov, D., Dietzek, B., Malsch, D., Henkel, T., Schlücker, S., and Popp, J. (2009) Quantitative CARS microscopic detection of analytes and their isotopomers in a two-channel microfluidic chip. *Small*, **5**, 2816–2818.
18 Selm, R., Winterhalder, M., Zumbusch, A., Krauss, G., Hanke, T., Sell, A., and Leitenstorfer, A. (2010) Ultrabroadband background-free coherent anti-Stokes Raman scattering microscopy based on a compact Er:fiber laser system. *Opt. Lett.*, **35** (19), 3282–3284.
19 Le, T.T., Yue, S., and Cheng, J.X. (2010) Shedding new light on lipid biology with coherent anti-Stokes Raman scattering microscopy. *J. Lipid Res.*, **51** (11), 3091–3102.
20 Wang, H.W., Langohr, I.M., Sturek, M., and Cheng, J.X. (2009) Imaging and quantitative analysis of atherosclerotic lesions by CARS-based multimodal nonlinear optical microscopy. *Arterioscler. Thromb. Vasc. Biol.*, **29** (9), 1342–1348.
21 Fu, Y., Sun, W., Shi, Y., Shi, R., and Cheng, J.X. (2009) Glutamate excitotoxicity inflicts paranodal myelin splitting and retraction. *PLoS One*, **4** (8), e6705.
22 Hellerer, T., Axäng, C., Brackmann, C., Hillertz, P., Pilon, M., and Enejder, A. (2007) Monitoring of lipid storage in *Caenorhabditis elegans* using coherent anti-Stokes Raman scattering (CARS) microscopy. *Proc. Natl. Acad. Sci. U. S. A.*, **104** (37), 14658–14663.
23 Yen, K., Le, T.T., Bansal, A., Narasimhan, S.D., Cheng, J.X., and Tissenbaum, H.A. (2010) A comparative study of fat storage quantitation in nematode *Caenorhabditis elegans* using label and label-free methods. *PLoS One*, **5** (9), e12810.
24 Evans, C.L., Xu, X., Kesari, S., Xie, X.S., Wong, S.T.C., and Young, G.S. (2007) Chemically-selective imaging of brain structures with CARS microscopy. *Opt. Express*, **15** (19), 12076–12087.
25 Kox, M.H.F., Domke, K.F., Day, J.P.R., Rago, G., Stavitski, E., Bonn, M., and Weckhuysen, B.M. (2009) Label-free chemical imaging of catalytic solids by coherent anti-Stokes Raman scattering and synchrotron-based infrared microscopy. *Angew. Chem. Int. Ed.*, **48**, 8990–8994.

17
Detection of Viral Infection in Epithelial Cells by Infrared Spectral Cytopathology

Max Diem, Nora Laver, Kristi Bedrossian, Jennifer Schubert, Kostas Papamarkakis, Benjamin Bird, and Miloš Miljković

17.1
Introduction

The Laboratory for Spectral Diagnosis (LSpD) at Northeastern University has been developing methodology for the automatic, infrared spectroscopy-based diagnosis of exfoliated cells [1–3]. These methods will be referred to henceforth as spectral cytopathology (SCP), since they deal with cytology – the "study of cells" using spectral methods. Whereas classical cytology renders a diagnosis based on cell morphology, SCP monitors changes in a cell's biochemical composition. This composition is determined via a global measurement, that is, all cellular biochemical components contribute to the observed infrared spectral pattern based on their abundance. Although infrared spectroscopy of isolated biochemical components can yield a variety of useful information, such as secondary and tertiary structure, degree of hydration and association, and the nature of counter ions, the global measurement of the infrared spectrum of a cell cannot be interpreted in similar detail, since the spectra of all components are superimposed and cannot be decomposed readily into component spectra. Nevertheless, recent advances in the use of multivariate methods of data analysis have produced distinct spectral patterns for a large number of tissue and cell types [4–11], and also for states of health and disease, cellular activity [12], and response to external stimuli [13].

Here, we summarize recent experimental evidence that SCP is able to detect viral infection in exfoliated epithelial cells. The results were unexpected, although some prior *in vivo* Raman results [14] can be interpreted in terms of sensitivity toward viral infections. In this chapter, we report preliminary results from a study of oral and cervical cells for which viral infection status was known *a priori*, or established via standard molecular biology/clinical testing procedures.

Handbook of Biophotonics. Vol.2: Photonics for Health Care, First Edition. Edited by Jürgen Popp, Valery V. Tuchin, Arthur Chiou, and Stefan Heinemann.
© 2012 Wiley-VCH Verlag GmbH & Co. KGaA. Published 2012 by Wiley-VCH Verlag GmbH & Co. KGaA.

17.2
Methods

Clinical samples of cervical and oral exfoliates were obtained from the Department of Pathology at Tufts Medical Center (TMC), Boston, MA. The fixation protocols, handling of samples, and the cell deposition methods have been reported previously. Classical cytopathology, and all testing for high-risk strains of human papillomavirus (HPV, including the most common oncogenic strains 16 and 18) were performed at TMC using the FDA-approved Digene Hybrid Capture II test (see, e.g., [15]).

Spectral data of epithelial cells were collected using a data acquisition procedure referred to as the PapMap approach [16] after deposition by cyto-centrifugation on to infrared reflective slides. In this approach, cellular spectra are reconstructed by averaging between 10 and 100 individual spectra collected from pixels $6.25 \times 6.25\,\mu m$ in size. Subsequent to spectral data acquisition, cells were stained and imaged. Spectral data collection and viral DNA testing were performed on parallel samples (i.e., a sample aliquot was tested for the virus, while another aliquot was used for spectroscopy). The sample of herpes simplex-infected oral cells was diagnosed by classical cytopathology only.

Raw spectral data were corrected via a universally applicable multivariate procedure known as extended multiplicative signal correction (EMSC) [17, 18] to minimize any residual water vapor spectral contribution, and also light scattered by the cells or subcellular entities such as the nucleus. The correction of raw spectral features for these effects has recently been discussed in a number of papers [19–21]. Spectral datasets were analyzed by principal component analysis (PCA) [22], an unsupervised method which does not require any training or validation datasets. PCA identifies correlated and reproducible spectral differences which may be used to partition data. Spectral types are visualized using scores plots, where correlated variance is indicated by clustering of spectra or items.

17.3
Results and Discussion

Recent results from cervical and oral cell samples diagnosed by classical cytology as "abnormal," but not cancerous [i.e., samples diagnosed as ASCUS (atypical squamous cell of unknown significance) or LGIL (low-grade intraepithelial lesion) in cervical cytology)] showed unexpected SCP results [2, 3]. Most cells from an abnormal sample are nondiagnostic in that they appear morphologically normal. However, these cells exhibit different spectral signatures to those from normal samples. This is shown in Figure 17.1, which shows a scores plot of cellular spectra collected from the tongues of normal subjects and from the tongues of patients with reactive abnormalities, and also cancer of the tongue. Whereas the cancerous cells are morphologically and spectrally very different from normal cells, it is the

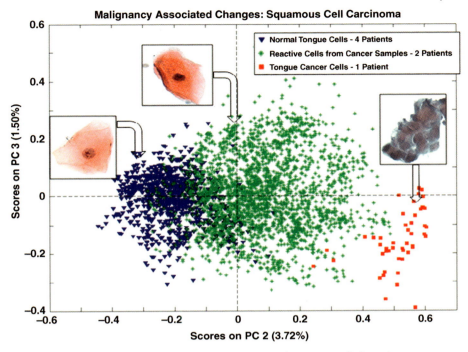

Figure 17.1 Scores plot of spectra from normal, reactive, and cancerous cells from the tongue.

nondiagnostic cells from the samples with reactive abnormalities that are most interesting: despite the fact that these cells exhibit normal morphology, they show spectral differences which range from near normal to near cancerous along the PC2 axis. This is in great contrast to classical cytology, where only 1% or fewer cells show morphologic abnormalities.

Completely analogous results were observed for cancer and premalignant states of other locations in the oral cavity [2]. Furthermore, cells from a patient with an active cold sore, caused by an outbreak of a viral infection (herpes simplex), showed that the infection was persistent in the majority of cells, and caused a gradual change in the spectral patterns between normal cells and cells that were so compromised by the viral infection that they could be diagnosed visually. The majority of cells, however, which exhibited normal morphology, showed spectral patterns that were different from those of normal cells. A comparison of spectral patterns observed for premalignant changes and virally infected cells revealed that there exists a similar change in both (Figure 17.2).

Although these observations do not prove that the precancerous conditions observed in the oral cavity are caused by viruses, recent oncologic studies have implicated both the human papillomavirus (HPV) and the Epstein–Barr (EB) virus as possible causes of head and neck malignancies [23, 24]. Further evidence of SCP

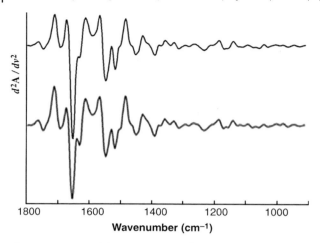

Figure 17.2 Second-derivative spectra of virally infected (top trace) and dysplastic (bottom trace) oral cells.

detecting viral infection (or their consequences) was recently described by preliminary cervical cancer screening efforts in the LSpD. These studies revealed that

- SCP can distinguish between normal [HPV(−)] and HPV-infected [HPV(+)] states.
- Cervical cell samples analyzed by SCP tend to follow the present or most recent HPV status.

Figure 17.3 shows a graphical representation of these two statements, displaying scores plots of cytologically normal cervical cells from HPV(+) samples (pale red) and HPV(−) samples (pale blue) from 10 subjects as a coarse indicator of the separability of these two states. The HPV status was established by DNA-based diagnostics, which detects oncogenic strains of HPV, including strains 16 and 18. Superimposed on this scores plot are samples with a normal cytologic and negative HPV result at the time of sample collection, but with recent histories of abnormal cytology. In Figure 17.3a, the sample had a history of ASCUS with a negative HPV result, and classified with the HPV(−) samples (pale blue). In Figure 17.3b, the sample had a history of HPV(+) ASCUS and LSIL, and classified with the HPV(+) samples (pale red). These results indicate that SCP has the sensitivity to detect the primary response to viral infection and also the effect of recent infection status.

In vivo Raman studies for cervical cancer screening [14, 25] have demonstrated very similar results, which were previously interpreted in terms of malignancy-associated changes (MACs) [26], a loosely defined term with usage similar to "field cancerization" [27], defined as follows: "Patients with a head and neck squamous cell carcinoma (HNSCC) often develop multiple (pre)malignant lesions. This finding led to the field cancerization theory, which hypothesizes that the entire epithelial surface of the upper aero-digestive tract has an increased risk for the development of (pre)malignant lesions because of multiple genetic abnormalities in the whole tissue

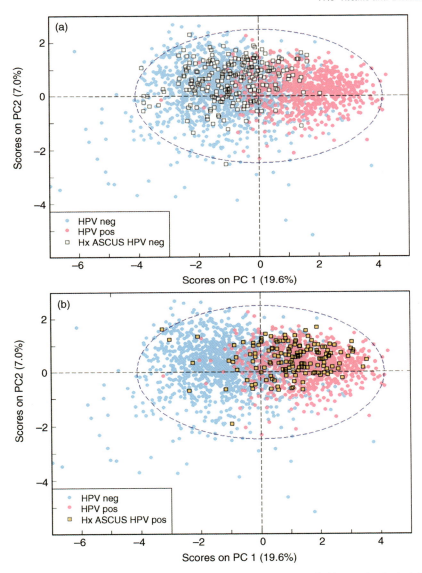

Figure 17.3 Scores plots of cervical cells diagnosed as HPV (−) (pale blue) and HPV (+) (pale red). Superimposed are the scores for an HPV(−) sample with a prior history of ASCUS (a), and a sample with a history of HPV(+), ASCUS, and LSIL (b).

region" [28]. Thus, both "field cancerization" and MACs describe very subtle cellular changes that help explain the occurrence of multiple lesions and the recurrence of HNSCC after treatment of a primary lesion. The results presented here suggest that these subtle morphologic changes could be due to persistent viral infections or their effects. Normally, viral infections by the HPV or the EB virus are cleared within months by the immune system; however, in cases when the immune system is not able to fight the virus, genital warts or squamous cell cancers may result, as either a direct or indirect result of the viral infection. In addition, recent studies have shown that certain viruses (notably the human herpes virus 6 and 7) can insert their DNA permanently into the human genome; these viruses have been implicated in a number of diseases.

17.4
Conclusions

The results presented here suggest that methods of vibrational (i.e., infrared and Raman) spectroscopy can detect spectral changes that are induced in cells by viral infection. At this point, the nature of the actual spectral change is not understood, and could be due to the presence of the viral genome or the viral proteome, or to the immune response of the body. Recent studies from the LSpD have shown that precancerous and virally infected cells show the largest changes in spectral regions associated with protein vibrations [29], pointing to the detection of either the viral proteome or the immune response. Infrared spectral changes due to the latter case scenario, the immune response, have been observed in serum by Lasch *et al.* [30], manifested by a large variation in the protein composition of the serum.

Acknowledgements

Partial support of this research by grant 090346 from the National Cancer Institute of the National Institutes of Health is gratefully acknowledged.

References

1 Romeo, M., Mohlenhoff, B., Jennings, M., and Diem, M. (2006) Infrared microspectroscopic studies of epithelial cells. *Biochim. Biophys. Acta*, **1758** (7), 915–922.

2 Papamarkakis, K., Bird, B., Schubert, J.M., Miljković, M., Wein, R., Bedrossian, K., Laver, N. and Diem M., (2010) Cytopathology by Optical Methods: Spectral Cytopathology of the Oral Mucosa. *Lab. Invest.*, **90** (4), 589–598.

3 Schubert, J.M., Bird, B., Papamarkakis, K., Miljković, M., Bedrossian, K., Laver, N., and Diem, M. (2010) Spectral cytopathology of cervical samples: detecting cellular abnormalities in cytologically normal cells. *Lab. Invest.*, **90** (7), 1068–1077.

4 Krafft, C. and Salzer, R. (2008) Neuro-oncological Applications of Infrared and Raman Spectroscopy, in *Vibrational Spectroscopy for Medical Diagnosis* (eds Diem, M., Griffith, P. and Chalmer, J.), John Wiley & Sons, Chichester, UK, pp. 231–259.

5 Steller, W., Einenke, J., Horn, L.C., Braumann, U.D., Binder, H., Salzer, R. and Krafft C. (2006) Delimitation of a squamous cell cervical carcinoma using infrared microspectroscopic imaging. *Anal Bioanal Chem.*, **384**, 145–154.

6 Lasch, P., Haensch, W., Naumann, D. and Diem, M., (2004) Imaging of colorectal adenocarcinoma using FT-IR microspectroscopy and cluster analysis. *Biochim Biophys Acta*, **1688** (2), 176–186.

7 Lasch, P., Diem, M., Hänsch, W. and Naumann, D. (2007) Artificial Neural Networks as Supervised Techniques for FT-IR Microspectroscopic Imaging. *JChemometrics.*, **20** (5), 209–220.

8 Gazi, E., Baker, M., Dwyer, J., Lockyer, N.P., Gardner, P., Shanks, J.H., Reeve, R.S., Hart, C.A., Clarke, N.W. and Brown, M.D. (2006) A Correlation of FTIR Spectra Derived from Prostate Cancer Tissue with Gleason Grade, PSA and Tumour Stage. *European Urology*, 750–61.

9 Ly, E., Piot, O., Wolthuis, R., Durlach, A., Bernard, P. and Manfait, M. (2008) Combination of FTIR spectral imaging and chemometrics for tumour detection from paraffin-embedded biopsies. *Analyst*, **133**, 197–205.

10 Wolthuis, R., Travo, A., Nicolet, C., Neuville, A., Gaub, M.-P., Guenot, D., Ly, E., Manfait, M., Jeannesson, P. and Piot, O. (2008) IR Spectral Imaging for Histopathological Characterization of Xenografted Human Colon Carcinomas. *Anal Chem.*, **80**, 8461–8469.

11 Romeo, M., Boydston-White, S., Matthäus, C., Miljkoviać, M., Bird, B., Chernenko, T., Diem M. (2008) Vibrational Microspectroscopy of Cells and Tissues, in *Biomedical Vibrational Spectroscopy* (eds P. Lasch and J. Kneipp), Wiley-Interscience, Hoboken, NJ, pp. 121–152.

12 Boydston-White, S., Romeo, M.J., Chernenko, T., Regina, A., Miljkovic, M., and Diem, M. (2006) Cell-cycle-dependent variations in FTIR micro-spectra of single proliferating HeLa cells: principal component and artificial neural network analysis. *Biochim. Biophys. Acta*, **1758** (7), 908–914.

13 Krishna, C.M., Kegelaer, G., Adt, I., Rubin, S., Kartha, V.B., Manfait, M., and Sockalingum, G.D. (2006) Combined FT-IR and Raman spectroscopic approach for identification of multidrug resistance phenotype in cancer cell lines. *Biopolymers*, **85** (5), 462–470.

14 Utzinger, U., Heintzelmann, D.L., Mahadevan-Jansen, A., Malpica, A., Follen, M., and Richards-Kortum, R. (2001) Near IR Raman Spectroscopy for in vivo detection of cervical precancers. *Appl. Spectroscopy*, **55** (8), 955–959.

15 Ronco, G., Segnan N., Giorgi-Rossi P., Zappa M., Casadei G.P., Carozzi F., Palma, P. D., Del Mistro, A., Folicaldi, S., Gillio-Tos, A., Nardo, G., Naldoni, C., Schincaglia, P., Zorzi, M.l., Confortini, M. and Cuzick, J. (2006) Human Papillomavirus Testing and Liquid-Based Cytology: Results at Recruitment From the New Technologies for Cervical Cancer Randomized Controlled Trial. *J. Natn Cancer Inst.*, **98** (11), 765–774.

16 Schubert, J.M., Mazur, A.I., Bird, B., Miljković, M. and Diem, M. (2010) Single Point vs. Mapping Approach for Spectral Cytopathology (SCP). *Biophotonics*, **3** (8–9), 588–596.

17 Kohler, A., Sule-Sosa, J., Sockalingum, G.D., Tobin, M., Bahramil, F., Yang, Y., Pijanka J., Dumas, P., Cotte, M., van Pittius, D.G., Parkes, G. and Martens, H. (2008) Estimating and Correcting Mie Scattering in Synchrotron-Based Microscopic Fourier Transform Infrared Spectra by Extended Multiplicative Signal Correction. *Appl. Spectroscopy*, **62** (3), 259–266.

18 Kohler, A., Kirschner, C., Oust, A. and Martens, H. (2005) Extended Multiplicative Signal Correction as a Tool for Separation and Characterization of Physical and Chemical Information in Fourier Transform Infrared Microscopy Images of Cryo-sections of Beef Loin. *Appl. Spectroscopy*, **59** (6), 707–716.

19 Bassan, P., Byrne, H.J., Bonnier, F., Lee, J., Dumas, P. and Gardner, P. (2009) Resonant Mie scattering in infrared spectroscopy of biological materials – understanding the 'dispersion artefact'. *Analyst*, **134**, 1586–1593.

20 Bassan, P., Byrne, H.J., Lee, J., Bonnier, F., Clarke, C., Dumas, P., Gazi, E., Brown, M.D, Clarke, N.W. and Gardner, P. (2009) Reflection contributions to the dispersion artefact in FTIR spectra of single biological cells. *Analyst*, **134**, 1171–1175.

21 Bassan, P., Kohler, A., Martens, H., Lee, J., Byrne, H.J., Dumas, P., Gazi, E., Brown, M., Clarke N. and Gardner, P. (2010) Resonant Mie Scattering (RMieS) correction of infrared spectra from highly scattering biological samples. *Analyst*, **135**, 268–277.

22 Adams, M.J. (2004) *Chemometrics in Analytical Spectroscopy*, 2nd edn., RSC Analytical Spectroscopy Monographs (series ed. N.W. Barnett), Royal Society of Chemistry, Cambridge.

23 Sugerman, P.B. and Shillitoe, E.J. (1997) The high risk human papillomaviruses and oral cancer: evidence for and against a causal relationship. *Oral Dis.*, **3**, 130–147.

24 Porter, S. and Waugh, A. (2000) Comment on: oral cancer in young adults. *Br. Dent. J.*, **188** (7), 366.

25 Robichaux-Viehoefer, A., Kanter E., Shappell, H., Billheimer, D., Jones III, H. and Mahadevan-Jansen, A. (2007) Characterization of Raman Spectra Measured in Vivo for the Detection of Cervical Dysplasia. *Appl. Spectroscopy*, **61** (9), 986–993.

26 Ogden, G.R., Cowpe, J.G., and Green, M.W. (1990) The effect of distant malignancy upon quantitative cytologic assessment of normal oral mucosa. *Cancer*, **65**, 477–480.

27 Ha, P.K. and Califano, J.A. (2003) The molecular biology of mucosal field cancerization of the head and neck. *Crit. Rev. Oral Biol. Med.*, **14** (5), 363–369.

28 van Oijen, M.G.C.T. and Slootweg, P.J. (2000) Oral field cancerization: carcinogen-induced independent events or micrometastatic deposits? *Cancer Epidemiol. Biomarkers Prev.*, **9**, 249–256.

29 Diem, M., Papamarkakis, K., Schubert, J.M., Bird, B., Romeo, M.J. and Miljković, M. (2009) The Infrared Spectral Signatures of Disease: Extracting the Distinguishing Spectral Features between Normal and Diseased States. *Appl. Spectroscopy*, **63** (11), 307A–318A.

30 Lasch, P., Beekes, M., Fabian, H. and Naumann D. (2008) Antemortem Identification of Transmissible Spongiform Encephalopathy (TSE) from Serum by Mid-infrared Spectroscopy, in *Vibrational Spectroscopy for Medical Diagnosis* (eds Diem, M., Griffith, P. and Chalmer, J.) John Wiley & Sons, Chichester, UK, pp. 97–122.

**Part 3
Oncology**

18
Clinical and Technical Aspects of Photodynamic Therapy – Superficial and Interstitial Illumination in Skin and Prostate Cancer

Katarina Svanberg, Niels Bendsoe, Sune Svanberg, and Stefan Andersson-Engels

18.1
Background – Cancer and Its Treatment

Worldwide, in almost all countries, the incidence of malignant tumor disease is growing. In developed countries, approximately every third person is diagnosed throughout his/her lifespan with one or more malignancies. With an aging population this figure is growing, as age is the paramount factor for presentation with a malignant tumor. Considering these facts, cancer is defined as one of the world's endemic diseases and the second highest killer in Western countries, out-numbered only by deaths due to heart and vascular diseases. Historically, the evidence for neoplasias of various kinds was described in Indian and Chinese scriptures and Egyptian papyrus rolls as early as 2000 BC. Also, in very old artwork, neoplasias can also be recognized in mummies from both the Old and New World. Already at that time treatment modalities were presented, such as burning sticks, herbal decoctions, and ointments. Surgery was the dominant treatment option until the beginning of the twentieth century, when ionizing radiation was discovered and introduced as another treatment modality. Interestingly, the inventor of ionizing radiation, Konrad Röntgen, became the first Nobel Laureate in 1901, motivated by his paper "Eine neue Art von Strahlen" [1], which was published in 1896. Just a few years later this new treatment modality was introduced in specialized hospitals in Germany, the United Kingdom, and France, and international conferences within this new field were organized and fascinating results presented.

18.2
Phototherapy

Ionizing radiation and also phototherapy (PT) and photochemotherapy (PCT) are treatment modalities within the group of therapies related to physical phenomena arising from the interaction of electromagnetic irradiation with biological tissue. In this chapter, we consider electromagnetic radiation within the visible range, that is, light. PT and PCT differ in the conceptual meaning, namely in PCT a photosensitizing

Handbook of Biophotonics. Vol.2: Photonics for Health Care, First Edition. Edited by Jürgen Popp, Valery V. Tuchin, Arthur Chiou, and Stefan Heinemann.
© 2012 Wiley-VCH Verlag GmbH & Co. KGaA. Published 2012 by Wiley-VCH Verlag GmbH & Co. KGaA.

agent is administered before the light exposure. Both therapies rely on a long tradition and date back to the Greeks, who practiced full-body sun exposure, later termed heliotherapy. A pioneer within the more modern clinical use of PT was Niels Finsen, who was also awarded the Nobel Prize for his discoveries using light in the treatment of cutaneous tuberculosis [2]. Finsen is often referred to as the founder of modern PT, a field which is still very relevant for dermatologic applications, such as for psoriasis and certain other types of dermatitis. Another example of PT is the treatment of juvenile jaundice utilizing UV light. Many mothers have met this type of PT when their newborn babies have been immunized with their own blood, causing a yellowish skin as a sign of jaundice [3]. Historically, these babies were placed in the window niche at the hospital for exposure to sunlight, nowadays more often under a UV lamp. This is the same type of UV lamp that has been used during many years for the diagnosis of cutaneous tinea, and also for the identification of false banknotes and antibacterial effects in laboratories, and in public toilets for camouflaging veins to prevent drug addict injections. A new and interesting field of PT is in psychiatry, where encouraging results have been achieved in the treatment of a particular type of depression called "seasonal affective disorder," as it appears more frequently during the dark season in Nordic countries, where daylight is restricted or absent during the autumn and winter periods [4]. The effect on depression is partly ascribed to a subsequent upgrading of the serotonin levels in the brain receptor transmitting system after light illumination [5].

Already more than 5000 years ago during the Egyptian dynasties and in ancient India, the first known examples of PCT in humans were reported when psoralen plants were used as sensitizing agents in combination with sunlight for the treatment of vitiligo (white/hypopigmented spots in the skin).

The term photodynamic therapy (PDT) refers to a variety of PCT when oxygen must be present in the tissue. There is a resemblance between PDT and photosynthesis of green plants, in which oxygen also is a prerequisite together with a ring-structured light-absorbing molecule, chlorophyll, although photosynthesis leads to build-up of organic material in contrast to tissue destruction in the case of PDT. PDT is a local treatment modality with the potential of being selective due to the tumor-preferred retention of the administered sensitizing agent. The first to demonstrate PDT action was the medical student Oscar Raab in 1898, when he showed the toxic action of acridine on paramecia (a unicellular microorganism) in conjunction with ambient light [6]. He worked in the laboratory of Herman von Tappenier in Munich, who reported in 1904 that the process that Raab had described was dependent on oxygen. Raab was the first scientist to coin the term *photodynamic therapy* to describe the phenomenon of oxygen-dependent photosensitization [7]. Subsequently, von Tappenier performed PDT on humans with skin cancer, cutaneous lupus erythematosis (a rheumatic and inflammatory disease), and genital condylomas (virus-induced warts) using eosin as a photosensitizing agent. Hence he was the first to introduce clinical PDT in modern times. The next step in PDT development occurred in 1908, when the physical properties of the sensitizer hematoporphyrin derivative (HpD) were described. The biological activity of the agent was shown a few years later in 1913 when a German physician, F. Meyer-Betz, injected himself with 200 mg of

HpD. He remained photosensitized for ambient light for 2 months and photographs of him show the typical edema occurring on the sun-exposed areas of his face and on the dorsal part of his hands [8]. The Meyer-Betz sensitization signs are similar to those shown by patients suffering from the inherited disorder porphyria, due to a deficiency of certain enzymes in the heme biosynthetic pathway. These patients accumulate various porphyrins and their precursors in the skin, and thus show the same type of skin sensitization. The anecdote about the mythological creatures called vampires, who were afraid of light as they were sensitized, had long teeth as the light destroyed the soft tissue, and were thirsting for people's blood due to anemia, seems to reflect the classical symptoms of porphyria disease. The term porphyrin comes from the Greek word *purphura*, which means purple pigment and refers to the fact that these patients excrete purple/brown urine. The selective retention of endogenous porphyrins in tumor tissue was first observed in Lyon, France, by Policard in 1924 using a Wood's lamp with UV light exciting fluorescence emission. The selective retention of HpD with fluorescence emission in tumor tissue was demonstrated in 1948 by Figge and Weiland [9, 10]. The modern era of PDT utilizing HpD was initiated in 1960 by Lipson and Baldes in the treatment of breast cancer [11], followed up by Bonnett *et al.* [12], who made a mixture of oligomeric porphyrins. Thomas Dougherty improved the porphyrin mixture mainly by linking it with ester and ether bridges and researched the potential of the PDT technique. Dougherty was instrumental in taking the treatment modality into clinical use [13]. Many review papers on PDT have now published; see, for example, [14–16]. In this chapter, we aim to provide clinical and technical perspectives of PDT and compare challenges in superficial and interstitial light administration.

18.3
Phototoxicity

PDT action relies on three components required in order to cause phototoxicity in targets such as tumors (Figure 18.1). These components are:

- **Photosensitizer (PS)**: A PS or a pro-drug to a PS that is administered (systemically or locally) to the subject. During a certain time interval, the PS is preferentially accumulated to a higher degree in the malignant than the normal surrounding tissue. The role of the PS is to capture/absorb the energy of a light photon and then transfer it to another molecule (the substrate or the tissue oxygen), resulting in tissue toxicity causing necrosis and apoptosis directly or by stimulating the immune system.
- **Light**: Light of an appropriate wavelength matching the peak in the absorption spectrum of the PS is required. Some PS agents have more than one absorption peak (usually in the blue–green and also in the red wavelength region). The depth of light treatment differs with wavelength and is increased in the red part of the spectrum, which is preferably used for deeper PDT action. Only for very thin lesions (less than 1 mm in depth) does blue–green light have therapeutic value.

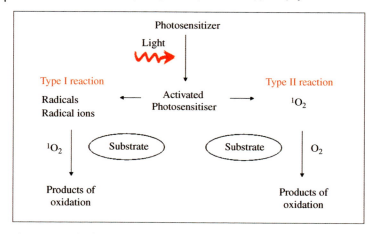

Figure 18.1 The three components of importance for photodynamic therapy: photosensitizer, light and molecular oxygen.

- **Oxygen**: The substrate in the PDT process is the oxygen in the tissue. The oxygen molecule accepts the excess energy gained in the light excitation of the PS. The excited substrate, singlet oxygen, causes chemical destruction of the tumor cells that involves oxidation.

The PDT process is illustrated in Figure 18.2. As light of an appropriate wavelength is sent towards (superficial illumination) or into (interstitially through optical fibers) PS-sensitized tissue, excitation of the PS molecule occurs. Thus, the PS molecule has

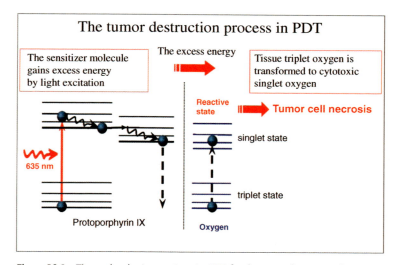

Figure 18.2 The molecular interactions in PDT for the case of protoporphyrin IX as a photosensitizing substance.

gained excess energy. However, as a general phenomenon in nature, nothing remains in an elevated energy level and the excess energy is transferred to the tissue oxygen, which normally resides in a triplet ground state. Cytotoxic singlet oxygen is generated and the PDT action occurs. The phototoxicity in PDT is either a direct momentarily appearing cell kill with breakage of the cell membranes, or a secondary effect after some days or weeks inducing damage to the vascular system or causing necrosis and apoptosis in the cells due to the involvement of the immune system. The action on the cell membrane is explained by the fact that some of the PSs are oligomers and thus lipophilic, and therefore attach to the cell membrane. In a second round, the lipophilic oligomers are split into hydrophilic monomers, which can easily pass through the membrane into the inner part of the cells. There they are mainly attached to the mitochondria or the lysosomes. The intracellular effect is then an action on the respiration of the cell [17–19].

18.4
Photodynamic Procedure

As a principle for cancer treatment, PDT is fairly simple. There are, however, certain important parameters that have to be taken into account for optimizing the therapy results. Examples of these parameters are:

- selection of the PS and mode of administration (local or systemic)
- drug dose for that particular tumor type (percentage of PS or mg/kg body weight)
- time interval between PS administration and light delivery (minutes to days)
- total light dose or light energy delivered (J/cm^{-2})
- light dose rate (W/cm^{-2}) and dose ($s \times W/cm^{-2} = J/cm^{-2}$)
- light fractionation (continuous or intermittent illumination)
- light wavelength (adjusted for the PS absorption property)
- mode of light administration (superficial, interstitial)
- light source (light-emitting diode (LED) or fiber-transmitted laser light).

18.5
Photosensitizers

An optimal photosensitizer for PDT should fulfill certain criteria, namely:

- nontoxic in the absence of light (no "dark" PDT effect)
- high efficiency in absorbing light energy and in transferring it to the substrate (tissue oxygen)
- able to absorb light at "longer" wavelengths, preferably above 600 nm for increased depth of the light penetration
- accumulated preferentially within tumors (high contrast between tumor and normal tissue)
- cleared rapidly from the normal surrounding tissue and organs at risk.

Table 18.1 List of photosensitizers with the major absorption peaks in the red part of the spectrum (names in square brackets are the commercial trade names).

Photosensitizer	Absorption peak wavelength (nm)
Hematoporphyrin derivative (HpD) [Photofrin]	630
δ-Aminolevulinic acid (ALA) [Metvix, Levulan]	635
meso-Tetra(hydroxyphenyl)chlorin (mTHPC) [Foscan]	652
Tin etiopurpurin [Pyrlytin]	660
Benzoporphyrin derivative (BPD) [Verteporfin, Visudyne]	690
Phthalocyanine	720
Lutetium texaphyrin [Lutrin]	732
Bacteriochlorophyll [Tookad]	760

Most probably there is no single sensitizer that fulfills the criteria for being the optimal PS. Different tumors will require specific capabilities of the PSs and substantial efforts have been devoted to developing ideal sensitizers. HpD has been described as belonging to the first generation of PSs and the others in the list above to the second generation.

Sensitizers and the red absorption wavelengths used are listed in Table 18.1. One of the agents in the table differs from the others as it is a precursor to a PS, namely δ-aminolevulinic acid (ALA), which is converted into an active PS following the heme cycle in the cells. All the other PSs are preformed chemical substances. If the wavelengths are translated into corresponding light penetration in the tissue, the depth at which light is attenuated to 37% (1/e) of its initial value, this means approximately a doubling from 630 to 720 nm and in terms of millimeters it corresponds to a change from about 3 to about 6 mm. Thin tumors can be treated efficiently with all these PSs using superficial illumination. Deeper tumors and tumors embedded in the body have to be treated with a different light administration mode — the interstitial mode, with the light transmitted through optical fibers inserted into the tumor mass.

There is a certain preferred time interval between the administration of a PS and the light illumination, and this time span varies from only minutes to days depending on which PS is used. One of the newer PSs, a bacteriochlorophyll (Tookad), exerts its action mainly on the endothelial cells in the vessels and therefore the illumination starts at the same time as the PS is injected. For skin cancer, an important discovery was reported when it was shown that topical application of (ALA) could be used to kill skin cancer and precancer cells [20–22]. In this way, it is possible to avoid one of the undesirable side effects with systemically administered PS, namely that in addition to sensitizing the tumor cells, the PS usually also causes light sensitivity in the skin over a certain time period. Most probably this has delayed the clinical acceptance to introduce PDT, as the patients have to be shielded from strong ambient light for several weeks when administering some of the PSs. However, it should be mentioned that drug-induced photosensitivity also occurs with many other pharmaceutical agents given to patients in daily practice, including some antibiotics (tetracyclines and sulfonamides), nonsteroidal anti-inflammatory drugs (e.g., naproxen, ketoprofen, and ibuprofen), diuretics (hydrochlorothiazide), and some neuroleptics. The use of some fragrances (e.g., musk perfume) and aspirin can also cause some skin sensitivity. The clinically reversible effects of these drugs and substances are similar to those of systemically administered PSs, with a dry bumpy or blistering rash on the skin.

There are several important physical parameters to take into account concerning the light illumination of PS-sensitized tissue. Most PSs exhibit several light absorption peaks. Usually the peak in the UV or blue part of the spectrum, (e.g., the Soret bands of porphyrins) has a higher amplitude. On the other hand, for therapy of tumors, red absorption is usually chosen owing to the deeper penetration of the activation light compared with UV or blue light.

18.5.1
Laser-Induced Fluorescence for Tumor Detection

The gold standard for diagnosis of a disease is histopathologic investigation of biopsy samples. This procedure is costly and time consuming as the tissue sample has to be fixed, cut, and stained. This conventional technique might be complemented by optical detection with the advantage of giving a diagnosis in real time. The technique is referred to as laser-induced fluorescence (LIF) and two different geometries can be utilized, either point monitoring with the recording of the full spectrum in a single point or an imaging mode over the whole area with a single or a few wavelength bands. The fluorescence excited from the endogenous chromophores is called the autofluorescence. The technique has shown great potential for use in early tumor detection and visualization of shallow lesions.

Blue or UV light is used for the fluorescence excitation. The autofluorescence is broadbanded as it is a composite of various chromophores, such as collagen, elastin, the coenzymes $NADH/NAD^+$, and others, as shown in Figure 18.3. The composite of the tissue endogenous chromophores differs depending on whether it is normal or malignant tissue and the fluorescence emission varies and can be used for tissue characterization. The endogenous fluorescence emission shows a drastic decrease in intensity in the tumor tissue and is usually also red shifted.

LIF can also be used to localize dysplasia and early cancer before it is possible to identify the affected areas with the naked eye. As an example, the detection of a carcinoma *in situ* in the cervical area in a young woman is shown in Figure 18.4. The five spectra with high-intensity peaking at about 410 nm were all recorded from the squamous and transformation zone of the cervical portio. The fifth spectrum shows a clear decrease in intensity and at the same time a red wavelength shift of approximately 60 nm. In this case, LIF guided the gynecologist to the right location for biopsy sampling and the histopathology showed noninvasive carcinoma of the cervix [23]. On visual inspection no sign of the early cancer was seen. The fluorescence signal corresponds to the early biochemical changes in the tissue. In the transformation process from normal to dysplasia, this status changes earlier than the

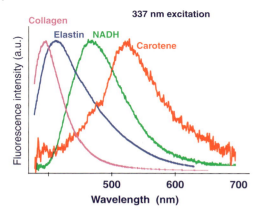

Figure 18.3 Autofluorescence emission spectra from the most important endogenous fluorophores in the tissue excited at 337 nm with a pulsed nitrogen laser. As can be seen, the autofluorescence is broadbanded as it is a composite of different chromophores with the maximum emission at about 470 nm.

cell architecture, which is observed with the naked eye. Therefore, there is a possibility to discover early cancer and precancer by using LIF and thus improve the prognosis for the patient.

As there is a certain preference in uptake of the sensitizers in cancerous and precancer tissue, the fluorescence signal in the red part of the spectrum from these

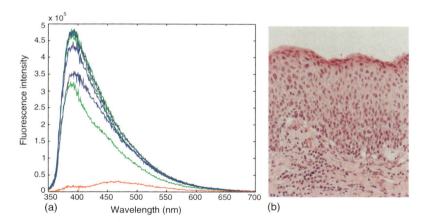

Figure 18.4 (a) Autofluorescence spectra recorded *in vivo* from a patient during an investigation of the cervical area. The fluorescence spectra with overall high intensity represent normal squamous epithelium while the spectrum with the low intensity was recorded from a spot of cancer *in situ* on the portio. In addition to showing a very low signal, the spectrum is shifted towards the red part of the spectrum, both being characteristic of precancer or cancer. (b) The corresponding histopathology from the suspicious area, verifying carcinoma *in situ*.

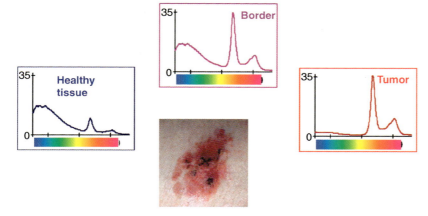

Figure 18.5 Fluorescence spectra from a basal cell carcinoma and the surrounding normal skin. The whole area was topically sensitized with ALA, which induced the protoporphyrin IX, which has an intense emission at about 635 nm. The spectrum from the normal skin shows a weak emission in the red part of the spectrum and a high intensity in the blue–green region. The tumor and the tumor border are characterized by low autofluorescence and high intensity corresponding to the PpIX peak.

agents can be used for enhanced tumor detection and delineation. Three LIF spectra recorded from healthy surrounding skin, tumor border, and tumor center are shown in Figure 18.5. Two main features can be recognized. The autofluorescence shows a decreased signal for the tumor. At the same time, the signal from the ALA-induced protoporphyrin at ∼635 nm shows high values for the border and the tumor. If the signal at 635 nm is divided by the signal at 470 nm, a clear demarcation ratio is formed and the tumor including the border is delineated.

The ultimate goal would be to combine LIF and PDT for better tumor target identification. The white light endoscopic image and the processed fluorescence image of the vocal cords are shown in Figure 18.6. The fluorescence image marks out the tumor area on the vocal cord all the way to the commissure between the two folds. The fiber transmits the red PDT light to the tumor area and the PDT treatment is performed guided by the fluorescence image.

18.6
Light Sources

Lasers with monochromatic light emission have been the natural choice for PDT. However, before realistic diode lasers became available, the systems were complicated and required strong support from collaborating physicists. The first lasers in PDT were the Au and Cu vapor lasers and argon ion pumped dye lasers emitting in the red spectral region. Another possibility emerged with the frequency-doubled neodymium:yttrium–aluminum–garnet (Nd:YAG) laser, emitting at 532 nm and

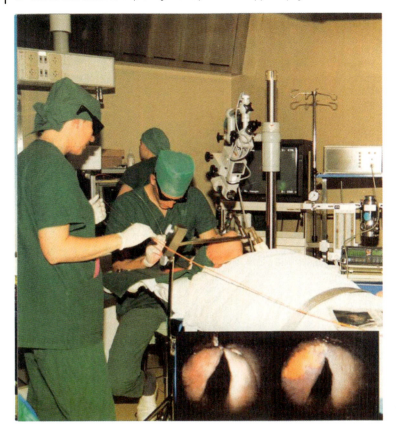

Figure 18.6 A photograph from the operating room at the ENT department at Lund University Hospital, where a patient is being treated with PDT. The white light image and the fluorescence image of the vocal cords were recorded in the pretreatment planning (inset). The tumor is marked out by the fluorescence image and the PDT procedure is guided to the target by the LIF image.

pumping a dye laser. Solid-state diode lasers were introduced in the clinic in the late 1990s. The advantage of lasers in PDT is the possibility of transmitting the light through optical fibers and thus opening up the option of treating tumors in hollow organs, such as the urinary bladder, bronchus, intestine, and esophagus. For superficial illumination, such as for skin, the female gynecologic tract, and the oral cavity, an array of LEDs can fulfill the task. As the PSs do not show very narrow absorption features, the LED with a bandwidth of ∼20 nm is advantageous as a treatment light source for superficial PDT. For interstitial PDT, lasers are convenient sources, as the output light can be coupled efficiently to optical fibers for guiding it to the target tissue.

18.7
Photodynamic Therapy as a Clinical Procedure

The most attractive feature of PDT as a local treatment modality is the selective action, causing minor damage to normal adjacent tissue with less scar formation than with alternative treatments. In particular in relation to ionizing radiation, PDT does not cause any DNA damage with the risk of developing secondary malignancies. PDT is characterized by:

- selective action for sensitized cancer/precancer tissue
- possibility of being repeated (in contrast to ionizing radiation)
- no accumulated toxicity
- fast healing with minimal scar formation and good cosmetics
- retained organ function.

18.8
Superficial PDT in Skin Lesions

The most common malignancy in the Western world is skin cancer and the diagnosis and treatment impose huge costs on the healthcare system. Also, the incidence of skin cancer is continuously increasing (10% per year) for several reasons, such as more leisure time with increased sun/UV exposure, an aging population, and for younger persons increased sunbed use. All classical skin cancers, such as basal cell carcinoma (BCC), squamous cell carcinoma (SqCC), and malignant melanoma are related to sun exposure. Also some of the precancerous skin lesions, such as actinic keratosis (AK) and Mb Bowen (SqCC *in situ*) are caused by UV light exposure.

Statistics for Sweden, with a population of approximately 9 million, show that the cost of skin tumor removal amounts to 125 000 000 Euro per year [24]. The corresponding cost for Germany, for example, is 1 000 000 000 Euro per year. As there are several treatment options for skin malignancies, such as surgery, cryosurgery, ionizing radiation, and topical cytotoxic drugs, a careful judgment has to be made regarding treatment decisions.

Even though it is a very common disease, the diagnosis of some kinds of skin malignancy is still uncertain and many benign lesions are unnecessarily surgically removed. A clinical estimate shows that approximately nine out of ten removed lesions are benign [25]. Therefore, LIF or other optical detection modalities should be further developed for more convenient and safe diagnosis.

There are several benign, premalignant, and malignant lesions of the skin which are suitable for PDT according to the list below:
Benign lesions

- psoriatic lesions
- acne vulgaris

- verucca vulgaris
- lichen ruber (mucosal).

Potentially malignant lesions

- actinic keratosis
- Mb Bowen (SqCC *in situ*)
- cutaneous lymphoma.

Malignant lesions

- BCC
- SqCC
- Mb Paget
- extramammary Mb Paget (adenocarcinoma *in situ*).

Malignant melanoma is the only primary skin malignancy which is currently not a candidate for PDT. This tumor type has to be surgically removed as soon as possible for extensive histopathologic examination for continued handling and prognostic evaluation. All the other types of skin malignancies can be considered for PDT. A single treatment session is usually sufficient for thin tumors with a thickness of up to 3 mm. Thicker tumors can be retreated after follow-up, or be pretreated with, for example, curettage, which means that some tumor tissue is removed mechanically and PDT is performed to the bottom of the tumor.

PDT has an important role in certain types of skin malignancies, such as:

- in elderly patients
- large tumors (diameter >3 cm)
- multiple tumor locations
- sensitive locations (pretibial, periocular, outer ear)
- recurrent tumors
- earlier radiation treatment
- high need for good cosmetics (face, thoracic wall).

When PDT was introduced clinically, the PS was administered systemically also for skin malignancies [26–29]. As mentioned, an unwanted side effect of systemically applied PS is skin sensitization for a certain period (varying from hours to months) depending of the type of PS given. When topical application of ALA was introduced, PDT for skin malignancies took a large leap forward and a new era within superficial PDT began. ALA is a naturally occurring five-carbon molecule and a straight-chain amino acid. It can easily be dissolved in water and thus mixed in water-based creams for topical use. ALA is the first step in heme synthesis in human cells and given to the organism it enters the endogenous heme cycle in the cells. In this cycle, through many enzymatically driven steps, protoporphyrin IX (Pp IX) is formed, which is a very potent PS. There is a transient build-up in the amount of PpIX in malignant cells due to lower levels of the enzyme ferrochelatase, which transfers PpIX to heme. The process can be followed by LIF, and thus the optimal time window when the contrast normal/tumor ratio is high can be chosen. LIF can also be used for tumor delineation with target definition to increase the treatment radicality.

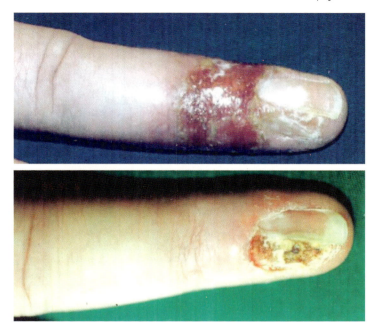

Figure 18.7 A squamous cell carcinoma *in situ* located on the finger before and after PDT utilizing topical ALA as a sensitizing agent. This tumor location is well suited for PDT as the conventional techniques, such as surgery and ionizing radiation, would not be preferred treatment modalities.

The clinical application of ALA was recognized by Kennedy *et al.*, who showed that ALA–PDT was an efficient treatment for human skin cancer, with the first result being presented in 1990 [21]. In 1991, the Lund group began clinical application of ALA–PDT for skin tumors and via clinical phase III studies introduced the treatment as a clinical routine modality for certain skin malignancies [30, 31]. Two clinical examples of ALA-induced PDT are shown in Figures 18.7 and 18.8. Both tumor locations are interesting for PDT as conventional modalities, such as surgery and ionizing radiation, would not be optimal in any of these cases. The tumors are shown before the ALA–PDT procedures and 3 months post-treatment. The tumor remission results are equal to those with other treatment modalities with a cure rate of ~85% and better cosmetic results compared with other methods [31]. The only modality that shows a lower recurrence rate is Moh's surgery, which is an interactive, time-consuming approach, often with very large areas of tissue being removed.

The light dose used for ALA–PDT of skin malignancies is typically 40–60 J/cm^{-2}. The fluence rate is usually kept below 150 mW/cm^{-2} in order not to heat the tissue. If higher fluence rates are applied, hyperthermic effects may occur, affecting the result of the PDT action with fibrosis induction, organ dysfunction with loss of tissue elasticity, and impaired cosmetic results [32].

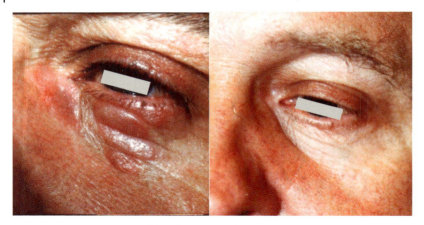

Figure 18.8 A cutaneous T-cell lymphoma tumor located below the margo border of the eye shown before and after ALA–PDT. The patient had gone through six cycles of cytotoxic treatment and had a remaining cutaneous tumor which was resistant to the systemic therapy.

18.9
Interstitial PDT for Prostate Cancer

Prostate cancer is the second most common malignant tumor in men, only outnumbered by skin cancer, and is strongly correlated with age. Prostate cancer is diagnosed based on the occurrence of urologic symptoms and an increased value of the prostate-specific antigen (PSA), which is over-expressed in prostate cancer. An elevated PSA value is a diagnostic indicator of a potential problem but can also be related to benign disorders, such as prostatitis or benign hyperplasia of the gland.

The treatment modality is chosen based on the stage and histopathology of the prostate cancer. For localized tumors, the preferred treatment is surgical prostatectomy. External or internal ionizing radiation is also considered. Hormone therapy is only used in more advanced disease or as a pretreatment to surgery in order to shrink the tumor size. Brachytherapy is administered by the insertion of trocars and sealed needles containing iridium-192 (high dose rate brachytherapy) for a short period (5–15 min), or by permanent implantation of ionizing iodine-125 or palladium-103 seeds (low dose rate brachytherapy). These conventional treatment modalities are all associated with undesired side effects such as urine incontinence, erectile dysfunction, and rectal wall damage with bleeding. These drawbacks open the way for alternative treatment modalities, such as interstitial PDT and the need for individualized dosimetry and treatment control.

Several groups, including our group in Lund, have been developing interstitial PDT to allow treatment of thicker and/or deeply located tumors. When light is delivered into the tissue by optical fibers, the light will be very intense close to the source and the fluence rate will decrease rapidly with distance from the optical fiber. This is one of the reasons why interactive online dosimetry is crucial in interstitial PDT as compared with surface illumination with a wide illumination field. Another reason is that organs at risk (in prostate cancer the sphincters, the venous/nerve plexus in the capsule, and the rectal wall) are located in the vicinity of the target tumor.

PDT of the prostate was first reported in 1990 by Windahl *et al.*, who successfully treated two patients with localized tumors using Photofrin (porfimer sodium) [33]. It took more than a decade before other groups followed this initial attempt and investigated the safety and feasibility of using ALA-induced PpIX for PDT of prostate cancer [34].

Despite ALA being an interesting candidate as a PS for PDT of malignant tumors, more potent PSs with absorption further towards the infrared region with better light penetration are of interest as the whole prostate gland has to be eradicated in prostate cancer therapy. This is a reason for investigating other PSs, such as *meso*-tetra (hydroxyphenyl)chlorin (mTHPC, FOSCAN, temoporfin) with an activation wavelength at 652 nm as reported by Nathan *et al.* at University College London [35] in patients with local recurrence after radiotherapy. This drug is attractive due to the slightly longer activation wavelength compared with Photofrin and ALA, and also to the higher efficiency of production of toxic reactive oxygen radicals. The same group applied mTHPC–PDT to untreated local prostate cancer in six patients [36]. The outcome was assessed by MRI examinations after the treatment. The treated prostate glands in some patients exhibited distinct features of devascularization, potentially indicating necrosis. The MRI images from some other patients showed patchy areas with reduced contrast uptake, indicating mild tumor response to the treatment. The volume of the prostate increased by $\sim$30% during the first week after treatment. This was interpreted as being caused by induced edema and inflammation. By 2–3 weeks post-treatment, the volume had decreased to $\sim$30% of the baseline volume. Reported complications following PDT were irritation of the urethra, which was cleared in all cases after 4 months. None of these patients needed a urinary catheter, which is common with other treatments.

Another PS investigated for prostate cancer treatment is Lutrin (lutetium texaphyrin), which was administered to patients with localized recurrent prostate cancer after radiotherapy in a trial at University of Pennsylvania [37, 38]. The primary goal of this trial was to assess the maximally tolerated dose of Lutrin–PDT. The drug–light interval between drug administration and the commencement of treatment varied between 3 and 24 h. Several properties, including measurements of the optical properties of the prostate gland [39], fluorescence spectroscopy of the PS [40], and spectroscopic assessment of tissue oxygenation [41], were investigated. Translatable spherical isotropic fiber-optic light diffusers were used for direct measurements of the light fluence rate. The optical measurements performed before, during, and after treatment all indicated substantial heterogeneity of the optical properties of the prostate and accumulation of the PS. The results also indicated that the tissue oxygenation was relatively constant in each patient, but that the total hemoglobin concentration decreased during treatment. The conclusion of the study was that Lutrin–PDT is an attractive alternative to conventional prostate cancer therapy even taking into account the mild, transient complications. Furthermore, the optical heterogeneity of the prostate gland must be considered to ensure correct light delivery. In a subsequent paper by the same group, short- and long-term effects on

PSA levels were described in relation to the PDT light dose [42]. This was defined as the product of the PS concentration and the *in situ* measured light dose. A prolonged delay to the time point when the PSA level began to increase was shown for high-dose as compared with low-dose PDT, 82 and 43 days, respectively. This group has also developed pretreatment dosimetry software intended to optimize various treatment parameters in order to tailor the emitted treatment light according to a predefined dose plan [43].

Trachtenberg and co-workers at the University of Toronto reported on vascular-targeted photodynamic therapy (VTP), using escalating drug doses of the photosensitizer WST09 [44, 45]. Spherical isotropic fiber tips were used as detector probes to study the fluence rate at selected sites during the treatment [46]. Complete response was achieved in 60% of the patients who received high-dose VTP. This assessment was judged using MRI investigation showing avascular areas 1 week after treatment [47]. Also, biopsies from these patients showed no viable cancer after 6 months. The software used in the dosimetry planning of these treatments was presented by Davidson *et al.* [48]. Inter-patient variations in the follow-up were seen. The most probable causes of these treatment response variations were the variability of the PS pharmacokinetics and the PS distribution as a function of time.

Our group started to consider interstitial PDT in localized prostate cancer in 2000 based on our experience of interstitial treatment of thick tumors using optical fibers (see, e.g., [49–53]). We also have long experience in diagnostic measurements, primarily in fluorescence spectroscopy, while dosimetric capability was being developed at that time. The first interstitial treatment system to be developed involved splitting the optical power from a treatment laser into six parts delivered through six individual fibers, which could be inserted into the tumor mass. The first clinical treatments with this system were performed on easily accessible thick skin tumors, intended to gain important clinical experience before moving on to treatment of prostate cancer. A unique feature with this system was that it could intermittently be used in a treatment and dosimetry measurement mode, as illustrated in Figure 18.9. For treatment, all the fibers were used as emitters, to allow the light to be distributed as efficiently and evenly as possible. In the measurement mode, only one fiber at a time was used as an emitter, while the other fibers were used as light sensors. Thus, it was possible to measure the light from one fiber tip to all the others. This dosimetry measurement mode made it possible to monitor the distribution throughout the tumor of fluence rates, PS concentration, and bleaching, and also the oxygenation of the hemoglobin present in the tissue. The system was evaluated in the treatment of experimental tumors in rats [49, 50], and was later used in early treatment of solid human tumors [51–53].

The interstitial PDT studies have contributed to a better understanding of the precise mechanisms of the PDT effects for the development of improved dosimetry models to plan and control the treatment. In order to be able to individualize the treatment as well as possibly, to account for inter-patient variations, it is very useful to be able to monitor all parameters of importance for the treatment outcome. With an improved understanding of the processes involved, would be valuable to know more parameters. It was in this context that the system shown in Figure 18.9 was

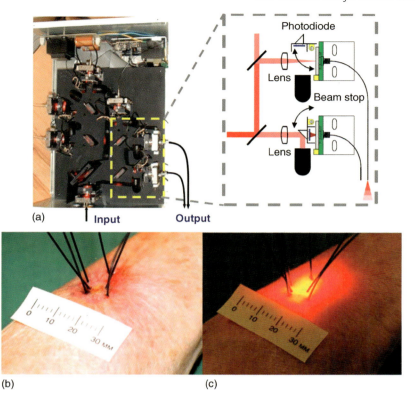

Figure 18.9 (a) Interstitial PDT of skin malignancies using a six-fiber system, where the intensity of light flowing between fibers can be measured by inserting a detector in front of a fiber, while at the same time blocking the treatment light intended for that particular fiber. The situation when one fiber is sending out treatment light and the other five are measuring the received light is shown in (b), and the situation when all six fibers are transmitting treatment light is shown in (c).

developed – a system that had capabilities both for delivering treatment light and also for measuring several parameters of importance for the treatment outcome. The basic elements of this instrument, developed for clinical trials, are presented in Figure 18.10 for the case of six interstitial fibers. Individually adjustable diode lasers were used to supply light to each of the treatment fibers. Photographs from a treatment session for a solid skin malignancy located on the ear are included in Figure 18.10.

Clearly, a system of the kind described above can be fully utilized only in conjunction with advanced dosimetric software, where measured diagnostic signals are fed into a dose distribution calculation program [52–56]. The most common approach is to define a treatment dose from a "threshold dose model" [57, 58]. This means ensuring that a minimum dose, that is, the dose required to induce direct cell

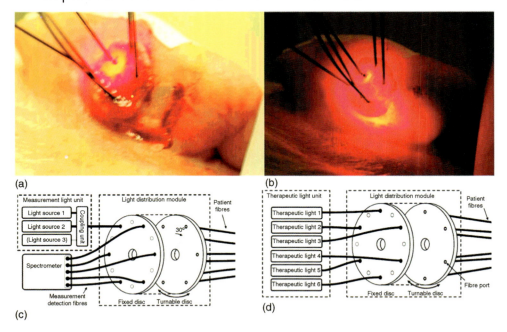

Figure 18.10 Schematic views of a six-fiber system with capability for treatment and also diagnostics of light flux, sensitizer concentration, and tissue oxygenation through the same fibers shown for the treatment situation (c) and for the diagnostic situation (d). Further, in the upper part of the figure a solid tumor is shown, (a) in a diagnostic measurement and (b) in treatment using all fibers Diagrams with fiber arrangement schemes from [54].

death, is delivered to all parts of the target volume, as depicted schematically in Figure 18.11, showing an interstitial PDT treatment session of a prostate with 18 fibers just after initiating the treatment (with small red spheres surrounding each fiber tip, indicating full treatment) and after almost full treatment (with the entire prostate, and some small regions of surrounding tissue being marked as receiving the threshold dose). In practice, the dose can be defined in several ways:

- One option is to multiply the light fluence by the PS concentration to obtain a value related to the amount of excited PS. One then assumes that this value is directly correlated with the amount of oxygen radicals produced and that this amount is related to the tissue damage potential. The idea is that this number should exceed a threshold value for the entire target volume. This means that the PS concentration and light fluence must be quantified in three dimensions.
- If one furthermore assumes that the PS concentration is homogeneously distributed throughout the target volume, it is sufficient to consider only the distribution of the light fluence within the target volume. This will clearly simplify the measurements required to control a treatment.

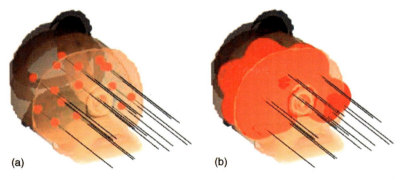

Figure 18.11 Treatment light delivered through multiple optical fibers positioned inside the prostate. Parts (a) and (b) show two different instances of the same treatment session, where the iso-surfaces mark tissue that has received at least the threshold dose.

The fluence rate can be assessed using calibrated optical probes, where integrating the signal over the treatment time yields the fluence dose [59, 60]. Such point measurements are valuable for providing representative values of the light dose delivered at one location. In order to obtain a spatial map of the fluence rate throughout the entire target volume, the photon propagation must be theoretically calculated. The optical properties of the tissue treated are then determined, followed by calculations using a photon propagation model. One example of a theoretical model is the analytical solution to the diffusion equation.

The optical properties in interstitial PDT are typically evaluated through steady-state, spatially resolved measurements [39, 46, 52, 56, 61–63], where absorption and scattering are assumed to be constant throughout the measured tissue. Recently, heterogeneous models of the tissue have been reported utilizing tomographic measurement schemes. The optical properties are then often reconstructed using structural information obtained from prior ultrasound examinations [64, 65] or MRI [66]. This is necessary as the reconstruction is an ill-conditioned and under-determined problem as the number of signals is limited for practical reasons. Many detectors are still essential in order to be able to assess the parameters as a function of position. It therefore quickly becomes a challenge to perform the measurements in the best way and to get the detectors into place. Several studies have been carried out in which the optical fibers delivering and collecting the light were placed transrectally.

Some type of model for the forward light problem calculating the light fluence distribution has to be used both for the evaluation of the optical properties of the tissue (the so-called inverse problem) and in the modeling of treatment light penetration (a true forward problem once the optical properties are known). A wide variety of models have been developed in interstitial PDT dosimetry. For example, accelerated Monte Carlo methods [67], higher order approximations of the radiative transport equation [68, 69], finite element methods [48], and homogeneous models of the diffusion equation [56, 70] have been utilized.

In the pretreatment planning, the issue is to optimize the treatment outcome in terms of destroying the target tissue, while avoiding or minimizing any side effects of the treatment. This requires modeling of the treatment parameters in every position of the target tissue, and in nearby tissue structures. The intention for the light delivery is to tailor the light distribution so that the entire target volume receives a fluence dose above the threshold. Organs at risk should receive as low a light dose as possible. For this optimization, not only the treatment time but also other parameters have to be considered and optimized. Such parameters are light source positions, shape, and power [43, 48, 55, 56, 70, 71]. The optimization has to take into account clinically realistic light delivery modes, meaning a limited number of optical fibers that can be inserted into the prostate gland.

Most groups utilize optical fiber diffusers to deliver the treatment light, whereas our rationale is to adopt bare-end optical fibers. These fibers allow well-defined positions when using them as sources or detectors, and thus provide a well-defined source–detector distance in any measurements conducted with the fibers. It should therefore be possible to obtain more accurate and robust values of the optical properties of the tissue. It has also been necessary to develop dosimetry software to calculate the optimum positioning of the fibers based on the geometry of the tumor and the nearby organs at risk. Predefined values for absorption and scattering can be assigned to different types of tissue in the pretreatment planning, when no values for the particular patient have yet been obtained. The fiber positions and irradiation times for the individual fibers can then be calculated. The aim of this calculation is to maximize the fluence dose to the prostate gland, while minimizing the dose to the organs at risk. In addition to the pretreatment planning, the instrument developed in our group performs dosimetry calculations during treatment. Thus, the irradiation times can be updated based on treatment-induced changes in the optical properties measured during the treatment. The general procedure is described in detail elsewhere [56, 72].

A clinical trial using an 18-fiber interstitial PDT system, constructed together with the spin-off company SpectraCure (Lund, Sweden) along the principles developed by the Lund University group, has so far involved four prostate cancer patients. The system, procedures, and clinical outcome of these activities have been described recently [73]. Photographs from the clinical procedures are shown in Figure 18.12. In addition to the light flux, the sensitizer concentration and level of oxygenation were also monitored in these treatments. These data, however, were not used to influence the treatment procedure. Incorporation of these data should help to achieve optimized, individually tailored treatments, while the parameters in such a model still have to be defined from real treatments. Several preclinical studies have reported less variation in PDT response when compensating for varying levels of sensitizer in the target medium [74, 75].

Photosensitizer concentration can be assessed through absorption [46] or fluorescence [40] point measurements. Fluorescence was employed in our measurements owing to its higher sensitivity. Such measurements could compensate for both intra- and inter-patient variations. Hence, analogously to the fluence rate calculations, the sensitizer distribution must be assessed in three dimensions. A full tomographic reconstruction of the PS concentration throughout the medium has been conducted, showing an interesting potential also in studying the rate of PS photobleaching as a

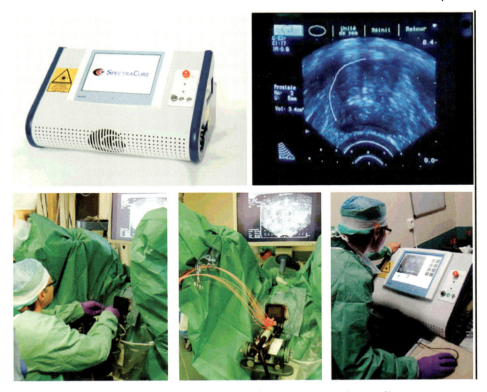

Figure 18.12 Photographs from a treatment session of prostate cancer with an 18-fiber system for interstitial PDT with integrated dosimetry feedback functions. Photographs courtesy of SpectraCure AB.

measure to predict treatment outcome. A method of doing this has been reported previously [76].

Interstitial PDT for the treatment of prostate cancer is being continuously developed. Several groups have stressed the need for more sophisticated dose planning approaches. Many aspects must be considered, such as individual variations in blood supply to the gland, type of PS, concentration and localization of PS, the oxygen concentration in and optical properties of the gland, and the size and pathology of the prostate gland. Risk organs in the close vicinity of the prostate must also be considered, for example, the urinary bladder sphincters, to avoid the risk of incontinence. Other potential side effects of the treatment also have to be carefully considered. Thus, the general health status of the patient, the treatment time, and light sensitivity of the patient during and following treatment have to be taken into account in the planning of the treatment. In our opinion, the results obtained so far indicate that this new modality for prostate cancer therapy is very promising. More studies are under way with the aim of getting the treatment modality approved for treatment of prostate cancer. Interstitial treatments are also under way for other tumors requiring interstitial light delivery, as mentioned above. These attempts to get interstitial PDT approved for these tumor types face very much the same challenges as described for the treatment of prostate cancer – a need for an improved dosimetry and measurements allowing more accurate planning for each individual treatment.

18.10
Discussion

PDT is gaining increased acceptance in the management of malignant disease. Its minimally invasive nature combined with the possibility of retreating without any memory effect are attractive features of this treatment modality. A serious draw-back with the technique is that many sensitizers leave a prolonged light sensitivity to the skin, requiring discipline in not being exposed to strong ambient light for periods which could extend to 1 month. This is not the case with ALA-induced PpIX sensitization, especially useful in the management of non-melanoma skin cancer, which therefore has gained wide application. Interstitial PDT of deep-lying lesions requires a detailed and individualized dosimetry taking spatially resolved light fluence, sensitizer concentration, and tissue oxygenation into account in order to achieve the full advantages of the modality. Several groups, including ours, are addressing this challenge. With more ideal sensitizers, having high tumor specificity, high quantum efficiency, long wavelength activation and fast skin clearance, PDT would gain a strong boost. In this context, extensive efforts along these lines include the development of functionalized nanoparticles.

The development of optimized PDT is a strongly multidisciplinary task where close collaboration between medical doctors of different specialities, physicists, and biochemists is needed. If pursued successfully, PDT might become a treatment of choice for patients with various malignant pathologies. Being minimally invasive and requiring limited infrastructure, this treatment modality could also prove particularly realistic in third-world countries with limited healthcare resources.

Acknowledgments

The authors gratefully acknowledge financial support from VINNOVA, the Swedish Strategic Research Foundation (SSF), the Swedish Research Council through direct grants, and through a Linnaeus Grant to the Lund Laser Center. Further, support from the Lund University Medical Faculty Funds and, the regional hospital organization (Region Skane) was very valuable. We thank a large number of graduate students and clinical collaborators who greatly contributed to our work on experimental and clinical PDT over a 20 year period. They include Göran Ahlgren, Johan Axelsson, Margret Einarsdottir, Ann Johansson, Tomas Johansson, Claes af Klinteberg, K. M. Kälkner, Sten Nilsson, Sara Pålsson, Jenny Svensson, Tomas

Svensson, Marcelo Soto Thompson, and Ingrid Wang. Fruitful collaboration with SpectraCure in developing clinically relevant instrumentation for interstitial PDT is also acknowledged, and thanks go especially to Kerstin Jakobsson, Masoud Khayyami, and Johannes Swartling.

References

1. Röntgen, W. (1895) Über eine neue Art von Strahlen (Vorläufige Mitteilung), in *Aus den Sitzungsberichten der Würzburger Physik.-Mediz. Gesellschaft 1895*, Verlag der Königlicher Hof- und Universitäts- Buch- und Kunsthandlung, Würzburg.
2. Finsen, N.R. (1901) *Phototherapy*, Edward Arnold, London.
3. Urbach, F., Forbes, P.D., Davies, R.E., and Berger, D. (1976) Cutaneous photobiology: past, present and future. *J. Invest. Dermatol.*, **67**, 209–224.
4. Lingjærde, O., Reichborn-Kjennerud, T., Haggag, A., Gärtner, I., Berg, E.M., and Narud, K. (1993) Treatment of winter depression in Norway. I. Short- and long-term effects of 1500-lux white light for 6 days. *Acta Psychiatr. Scand.*, **88**, 292–299.
5. Deguchi, T. (1979) Circadian rhythm of serotonin N-acetyltransferase activity in organ culture of chicken pineal gland. *Science*, **203**, 1245–1247.
6. Raab, O. (1899) Untersuchungen über die Wirkung fluorescierender Stoffe. *Z. Biol.*, **39**, 16.
7. von Tappeiner, H. and Jodlbauer, A. (1904) Über Wirkung der photodynamischen (fluorieszierenden) Stoffe auf Protozoan und Enzyme. *Dtsch. Arch. Klin. Med.*, **80**, 427–487.
8. Meyer-Betz, F. (1913) Untersuchungen über die biologische (photodynamische) Wirkung des Hämatoporphyrins und andere Derivate des Blut- und Gallenfarbstoffs. *Dtsch. Arch. Klin. Med.*, **112**, 476–450.
9. Policard, A. (1924) Etudes sur les aspects offerts par des tumeur experimentales examinée à la lumière de Wood's. *C. R. Soc. Biol.*, **91**, 1423–1423.
10. Figge, F.H.J. and Weiland, G.S. (1948) The affinity of neoplastic, embryonic, and traumatized tissue for porphyrins and metalloporphyrins. *Anat. Rec.*, **100**, 659.
11. Lipson, R.L. and Baldes, E.J. (1960) The photodynamic properties of a particular haematoporphyrin derivative. *Arch. Dermatol.*, **82**, 508–516.
12. Bonnett, R., Berenbaum, M.C., and Kaur, H. (1984) Chemical and biological studies on haematoporphyrin derivative: an unexpected photosensitisation in brain, in *Porphyrins in Tumor Phototherapy* (eds. A. Andreoni and R. Cubeddu), Plenum Press, New York, pp. 67–80.
13. Dougherty, T.J. (1979) Photoradiation in the treatment of recurrent breast carcinoma. *J. Natl. Cancer Inst.*, **62**, 231–237.
14. Huang, Z. (2005) A review of progress in clinical photodynamic therapy. *Technol. Cancer Res. Treat.*, **4**, 283–293.
15. Stylli, S.S. and Kaye, A.H. (2006) Photodynamic therapy of cerebral glioma – A review. Part II – Clinical studies. *J. Clin. Neurosci.*, **13**, 709–717.
16. Zeitouni, N.C., Oseroff, A.R., and Shieh, S. (2003) Photodynamic therapy for nonmelanoma skin cancers: current review and update. *Mol. Immunol.*, **39**, 1133–1136.
17. Kessel, D. (1982) Components of hematoporphyrin derivatives and their tumor localization capacity. *Cancer Res.*, **42**, 1703–1706.
18. Kessel, D. and Chou, T.H. (1983) Tumor-localizing components of the porphyrin preparation hematoporphyrin derivative. *Cancer Res.*, **43**, 1994–1999.

19 Bottiroli, G., Doccio, F., Freitas, I., Ramponi, R., and Sacchi, C. (1984) Hematoporphyrin derivative: fluorometric studies in solution and cells, in *Porphyrins in Tumor Phototherapy* (eds. A. Andreoni and R. Cubeddu), Plenum Press, New York, pp. 125–136.

20 Malik, Z. and Lugaci, H. (1987) Destruction of erythroleukaemic cells by photoinactivation of endogenous porphyrins. *Br. J. Cancer*, **56**, 589–595.

21 Kennedy, J.C., Pottier, R.H., and Pross, D.C. (1990) Photodynamic therapy with endogenous protoporphyrin IX: basic principles and present clinical experience. *J. Photochem. Photobiol. B*, **6**, 143–148.

22 Kennedy, J.C. and Pottier, R.H. (1992) Endogenous protoporphyrin IX, a clinically useful photosensitizer for photodynamic therapy. *J. Photochem. Photobiol. B*, **14**, 275–292.

23 Pålsson, S., Stenram, U., Soto Thompson, M., Vaitkuviene, A., Puskiene, V., Ziobakiene, R., Oyama, J., Gustafsson, U., DeWeert, M.J., Bendsoe, N., Andersson-Engels, S., Svanberg, S., and Svanberg, K. (2005) Methods for detailed histopathological investigation and localisation of cervical biopsies to improve the interpretation of autofluorescence data. *J. Environ. Pathol. Toxicol. Oncol.*, **25**, 321–340.

24 Tinghög, G., Carlsson, P., Synnerstad, I., and Rosdahl, I. (2007) *Samhällskostnader för Hudcancer Samt en Jämförelse Med Kostnaderna för Vägtrafikolyckor*, Linköping University Electronic Press, Linköping, http://www.ep.liu.se/ea/cmt/2007/005/.

25 Burton, R.C., Howe, C., Adamson, L., Reid, A.L., Hersey, P., Watson, A., Watt, G., Relic, J., Holt, D., Thursfield, V., Clarke, P., and Armstrong, B.K. (1998) General practitioner screening for melanoma: sensitivity, specificity, and effect of training. *J. Med. Screen.*, **5**, 156–161.

26 Dougherty, T.J., Kaufman, J.E., Goldfarb, A., Weishaupt, K.R., Boyle, D., and Mittleman, A. (1978) Photoradiation therapy for the treatment of malignant tumors. *Cancer Res.*, **38**, 2628–2635.

27 Bandieramonte, G., Marchesini, R., Melloni, E., Andreoli, C., di Pietro, S., Spinelli, P., Fava, G., Zunino, F., and Emanuelli, H. (1984) Laser phototherapy following HpD administration in superficial neoplastic lesions. *Tumori*, **70**, 327–334.

28 Wilson, B.D., Mang, T.S., Cooper, M., and Stoll, H. (1989) Use of photodynamic therapy for the treatment of extensive basal cell carcinomas. *Facial Plast. Surg.*, **6**, 185–189.

29 Tse, D.T., Kersten, R.C., and Anderson, R.L. (1984) Hematoporphyrin derivative photoradiation therapy in managing nevoid basal-cell carcinoma syndrome. *Arch. Ophthalmol.*, **102**, 990–994.

30 Svanberg, K., Andersson, T., Killander, D., Wang, I., Stenram, U., Andersson-Engels, S., Berg, R., Johansson, J., and Svanberg, S. (1994) Photodynamic therapy of non-melanoma malignant tumors of the skin using topical 5-amino levulinic acid sensitization and laser irradiation. *Br. J. Dermatol.*, **130**, 743–751.

31 Wang, I., Bendsoe, N., af Klinteberg, C., Enejder, A.M.K., Andersson-Engels, S., Svanberg, S., and Svanberg, K. (2001) Photodynamic therapy vs. cryosurgery of basal cell carcinomas; results of a phase III clinical trial. *Br. J. Dermatol.*, **144**, 832–840.

32 Cowled, P.A. and Forbes, I.J. (1985) Photocytotoxicity *in vivo* of haematoporphyrin derivative components. *Cancer Lett.*, **28**, 111–118.

33 Windahl, T., Andersson, S.O., and Lofgren, L. (1990) Photodynamic therapy of localised prostatic cancer. *Lancet*, **336**, 1139.

34 Zaak, D., Sroka, R., Höppner, M., Khoder, W., Reich, O., Tritschler, S., Muschter, R., Knüchel, R., and Hofstetter, A. (2003) Photodynamic therapy by means of 5-ALA induced PPIX in human prostate cancer – preliminary results. *Med. Laser Appl.*, **18**, 91–95.

35 Nathan, T.R., Whitelaw, D.E., Chang, S.C., Lees, W.R., Ripley, P.M., Payne, H., Jones, L., Parkinson, M.C., Emberton, M.,

Gillams, A.R., Mundy, A.R., and Bown, S.G. (2002) Photodynamic therapy for prostate cancer recurrence after radiotherapy: a Phase I study. *J. Urol.*, **168**, 1427–1432.

36 Moore, C.M., Nathan, T.R., Lees, W.R., Mosse, C.A., Freeman, A., Emberton, R., and Bown, S.G. (2006) Photodynamic therapy using meso tetra hydroxy phenyl chlorin (mTHPC) in early prostate cancer. *Lasers Surg. Med.*, **38**, 356–363.

37 Du, K.L., Mick, R., Busch, T., Zhu, T.C., Finlay, J.C., Yu, G., Yodh, A.G., Malkowicz, S.B., Smith, D., Whittington, R., Stripp, D., and Hahn, S.M. (2006) Preliminary results of interstitial motexafin lutetium-mediated PDT for prostate cancer. *Lasers Surg. Med.*, **38**, 427–434.

38 Verigos, K., Stripp, D., Mick, R., Zhu, T., Whittington, R., Smith, D., Dimofte, A., Finlay, J., Busch, T., Tochner, Z., Malkowicz, S., Glatstein, E., and Hahn, S. (2006) Updated results of a phase I trial of motexafin lutetium-mediated interstitial photodynamic therapy in patients with locally recurrent prostate cancer. *J. Environ. Pathol. Toxicol. Oncol.*, **25**, 373–387.

39 Zhu, T.C., Dimofte, A., Finlay, J.C., Stripp, D., Busch, T., Miles, J., Whittington, R., Malkowicz, S.B., Tochner, Z., Glatstein, E., and Hahn, S.M. (2005) Optical properties of human prostate at 732 nm measured *in vivo* during motexafin lutetium-mediated photodynamic therapy. *Photochem. Photobiol.*, **81**, 96–105.

40 Finlay, J.C., Zhu, T.C., Dimofte, A., Stripp, D., Malkowicz, S.B., Busch, T.M., and Hahn, S.M. (2006) Interstitial fluorescence spectroscopy in the human prostate during motexafin lutetium-mediated photodynamic therapy. *Photochem. Photobiol.*, **82**, 1270–1278.

41 Yu, G.Q., Durduran, T., Zhou, C., Zhu, T.C., Finlay, J.C., Busch, T.M., Malkowicz, S.B., Hahn, S.M., and Yodh, A.G. (2006) Real-time *in situ* monitoring of human prostate photodynamic therapy with diffuse light. *Photochem. Photobiol.*, **82**, 1279–1284.

42 Patel, H., Mick, R., Finlay, J., Zhu, T.C., Rickter, E., Cengel, K.A., Malkowicz, S.B., Hahn, S.M., and Busch, T.M. (2008) Motexafin lutetium-photodynamic therapy of prostate cancer: short- and long-term effects on prostate-specific antigen. *Clin. Cancer Res.*, **14**, 4869–4876.

43 Altschuler, M.D., Zhu, T.C., Li, J., and Hahn, S.M. (2005) Optimized interstitial PDT prostate treatment planning with the Cimmino feasibility algorithm. *Med. Phys.*, **32**, 3524–3536.

44 Trachtenberg, J., Bogaards, A., Weersink, R.A., Haider, M.A., Evans, A., McCluskey, S.A., Scherz, A., Gertner, M.R., Yue, C., Appu, S., Aprikian, A., Savard, J., Wilson, B.C., and Elhilali, M. (2007) Vascular targeted photodynamic therapy with palladium-bacteriopheophorbide photosensitizer for recurrent prostate cancer following definitive radiation therapy: assessment of safety and treatment response. *J. Urol.*, **178**, 1974–1979.

45 Trachtenberg, J., Weersink, R.A., Davidson, S.R.H., Haider, M.A., Bogaards, A., Gertner, M.R., Evans, A., Scherz, A., Savard, J., Chin, J.L., Wilson, B.C., and Elhilali, M. (2008) Vascular-targeted photodynamic therapy (padoporfin, WST09) for recurrent prostate cancer after failure of external beam radiotherapy: a study of escalating light doses. *BJU Int.*, **102**, 556–562.

46 Weersink, R.A., Bogaards, A., Gertner, M., Davidson, S.R.H., Zhang, K., Netchev, G., Trachtenberg, J., and Wilson, B.C. (2005) Techniques for delivery and monitoring of TOOKAD (WST09)-mediated photodynamic therapy of the prostate: clinical experience and practicalities. *J. Photochem. Photobiol. B*, **79**, 211–222.

47 Haider, M.A., Davidson, S.R.H., Kale, A.V., Weersink, R.A., Evans, A.J., Toi, A., Gertner, M.R., Bogaards, A., Wilson, B.C., Chin, J.L., Elhilali, M., and Trachtenberg, J. (2007) Prostate gland: MR imaging appearance after vascular

targeted photodynamic therapy with palladium-bacteriopheophorbide. *Radiology*, **244**, 196–204.

48 Davidson, S., Weersink, R.A., Haider, M.A., Gertner, M.R., Bogaards, A., Giewercer, D., Scherz, A., Sherar, M.D., Elhilali, M., Chin, J.L., Trachtenberg, J., and Wilson, B.C. (2009) Treatment planning and dose analysis for interstitial photodynamic therapy of prostate cancer. *Phys. Med. Biol.*, **54**, 2293–2313.

49 Stenberg, M., Soto Thompson, M., Johansson, T., Pålsson, S., af Klinteberg, C., Andersson-Engels, S., Stenram, U., Svanberg, S., and Svanberg, K. (2000) Interstitial photodynamic therapy – diagnostic measurements and treatment in malignant experimental rat tumors. *Proc. SPIE*, **4161**, 151–157.

50 Johansson, T., Soto Thompson, M., Stenberg, M., af Klinteberg, C., Andersson-Engels, S., Svanberg, S., and Svanberg, K. (2002) Feasibility study of a novel system for combined light dosimetry and interstitial photodynamic treatment of massive tumors. *Appl. Opt.*, **41**, 1462–1468.

51 Soto Thompson, M., Johansson, A., Johansson, T., Andersson-Engels, S., Bendsoe, N., Svanberg, K., and Svanberg, S. (2005) Clinical system for interstitial photodynamic therapy with combined on-line dosimetry measurements. *Appl. Opt.*, **44**, 4023–4031.

52 Johansson, A., Johansson, T., Thompson, M.S., Bendsoe, N., Svanberg, K., Svanberg, S., and Andersson-Engels, S. (2006) In vivo measurement of parameters of dosimetric importance during interstitial photodynamic therapy of thick skin tumors. *J. Biomed. Opt.*, **11**, 34029.

53 Johansson, A., Bendsoe, N., Svanberg, K., Svanberg, S., and Andersson-Engels, S. (2006) Influence of treatment-induced changes in tissue absorption on treatment volume during interstitial photodynamic therapy. *Med. Laser Appl.*, **21**, 261–270.

54 Johansson, A., Johansson, T., Thompson, M.S., Bendsoe, N., Svanberg, K., Svanberg, S., and Andersson-Engels, S. (2006) In vivo measurement of parameters of dosimetric importance during interstitial photodynamic therapy of thick skin tumors. *J. Biomed. Opt.*, **11**, 34029.

55 Johansson, A., Soto Thompson, M., Johansson, T., Bendsoe, N., Svanberg, K., Svanberg, S., and Andersson-Engels, S. (2005) System for integrated interstitial photodynamic therapy and dosimetric monitoring, *Proc. SPIE*, **5689**, 130–140.

56 Johansson, A., Hjelm, J., Eriksson, E., and Andersson-Engels, S. (2005) Pre-treatment dosimetry for interstitial photodynamic therapy. Presented at the European Conference on Biomedical Optics (ECBO), Munich, June 2005.

57 Johansson, A., Axelsson, J., Andersson-Engels, S., and Swartling, J. (2007) Realtime light dosimetry software tools for interstitial photodynamic therapy of the human prostate. *Med. Phys.*, **34**, 4309–4321.

58 Berenbaum, M.C., Bonnett, R., and Scourides, P.A. (1982) In vivo biological activity of the components of haematoporphyrin derivative. *Br. J. Cancer*, **45**, 571–581.

59 van Gemert, J.C., Berenbaum, M.C., and Gijsbers, G.H. (1985) Wavelength and light-dose dependence in tumor phototherapy with haematoporphyrin derivative. *Br. J. Cancer*, **52**, 43–49.

60 Dimofte, A., Zhu, T.C., Hahn, S.M., and Lustig, R.A. (2002) In vivo light dosimetry for motexafin lutetium-mediated PDT of breast cancer. *Lasers Surg. Med.*, **31**, 305–312.

61 Lilge, L., Pomerleau-Dalcourt, N., Douplik, A., Selman, S.H., Keck, R.W., Szkudlarek, M., Pestka, M., and Jankun, J. (2004) Transperineal in vivo fluence-rate dosimetry in the canine prostate during SnET2-mediated PDT. *Phys. Med. Biol.*, **49**, 3209–3225.

62 Zhu, T.C., Finlay, J.C., and Hahn, S.M. (2005) Determination of the distribution of light, optical properties, drug concentration, and tissue oxygenation *in-vivo* in human prostate during motexafin lutetium-mediated photodynamic therapy. *J. Photochem. Photobiol. B*, **79**, 231–241.

63 Wang, K.K.H. and Zhu, T.C. (2009) Reconstruction of *in-vivo* optical properties for human prostate using interstitial diffuse optical tomography. *Opt. Express*, **17**, 11665–11672.

64 Jiang, Z., Piao, D., Xu, G., Ritchey, J.W., Holyoak, G., Bartels, K.E., Bunting, C.F., Slobodov, G., and Krasinski, J.S. (2008) Trans-rectal ultrasound-coupled near-infrared optical tomography of the prostate. Part II: Experimental demonstration. *Opt. Express*, **16**, 17505–17520.

65 Xu, G., Piao, D., Musgrove, C., Bunting, C., and Dehghani, H. (2008) Trans-rectal ultrasound-coupled near-infrared optical tomography of the prostate. Part I: Simulation. *Opt. Express*, **16**, 17484–17504.

66 Li, C., Liengsawangwong, R., Choi, H., and Cheung, R. (2007) Using *a priori* structural information from magnetic resonance imaging to investigate the feasibility of prostate diffuse optical tomography and spectroscopy: a simulation study. *Med. Phys.*, **34**, 266–274.

67 Lo, W.C.Y., Redmond, K., Luu, J., Chow, P., Rose, J., and Lilge, L. (2009) Hardware acceleration of a Monte Carlo simulation for photodynamic therapy treatment planning. *J. Biomed. Opt.*, **14**, 014019.

68 Dickey, D.J., Moore, R.B., Rayner, D.C., and Tulip, J. (2001) Light dosimetry using the P3 approximation. *Phys. Med. Biol.*, **46**, 2359–2370.

69 Dickey, D.J., Partridge, K., Moore, R.B., and Tulip, J. (2004) Light dosimetry for multiple cylindrical diffusing sources for use in photodynamic therapy. *Phys. Med. Biol.*, **49**, 3197–3208.

70 Li, J., Altschuler, M.D., Hahn, S.M., and Zhu, T.C. (2008) Optimization of light source parameters in the photodynamic therapy of heterogeneous prostate. *Phys. Med. Biol.*, **53**, 4107–4121.

71 Rendon, A., Beck, J.C., and Lilge, L. (2008) Treatment planning using tailored and standard cylindrical light diffusers for photodynamic therapy of the prostate. *Phys. Med. Biol.*, **53**, 1131–1149.

72 Johansson, A., Axelsson, J., Swartling, J., Johansson, T., Pålsson, S., Stensson, J., Einarsdottir, M., Svanberg, K., Bendsoe, N., Kälkner, K.M., Nilsson, S., Svanberg, S., and Andersson-Engels, S. (2007) Interstitial photodynamic therapy for primary prostate cancer incorporating real-time treatment dosimetry. *Proc. SPIE*, **6427**, 64270.

73 Swartling, J., Axelsson, J., Ahlgren, G., Kälkner, K.M., Nilsson, S., Svanberg, S., Svanberg, K., and Andersson-Engels, S. (2010) System for interstitial photodynamic therapy with on-line dosimetry: first clinical experiences of prostate cancer. *J. Biomed. Opt.*, **15**, 058003.

74 Sheng, C., Jack Hoopes, P., Hasan, T., and Pogue, B.W. (2007) Photobleaching-based dosimetry predicts deposited dose in ALA–PpIX PDT of rodent esophagus. *Photochem. Photobiol.*, **83**, 738–748.

75 Zhou, X.D., Pogue, B.W., Chen, B., Demidenko, E., Joshi, R., Hoopes, J., and Hasan, T. (2006) Pretreatment photosensitizer dosimetry reduces variation in tumor response. *Int. J. Radiat. Oncol.*, **64**, 1211–1220.

76 Axelsson, J., Swartling, J., and Andersson-Engels, S. (2009) *In vivo* photosensitizer tomography inside the human prostate. *Opt. Lett.*, **34**, 232–234.

19
Molecular Targeting of Photosensitizers

Hasan Tayyaba, Prakash R. Rai, Sarika Verma, and Xiang Zheng

19.1
Introduction

Photodynamic therapy (PDT) is an emerging, externally activatable treatment modality. It involves the generation of toxic molecular species via excitation of chemicals called photosensitizers (PSs), by light of an appropriate wavelength, generally in the presence of oxygen. PDT is clinically approved for the treatment of neoplastic and non-neoplastic disease and offers several advantages over conventional chemotherapy by providing additional selectivity through the spatial confinement of light required for PS activation. A wide range of PSs have been evaluated and six of these have successfully transitioned from the bench to bedside applications [1]. Despite the regulatory approvals and the clinical success of PDT in oncology, a major limitation of all existing PSs is the lack of high selectivity for target tissue, which precludes its broad application to complex anatomic sites.

A key variable in ensuring favorable PDT outcome and a broader application is selective PS accumulation in the target tissue, thereby confining phototoxic effects to malignant cells. PS selectivity can be enhanced by several methods [2, 3] (Figure 19.1), and can be broadly classified into three categories: (i) molecular modification of the PS by using targeting moieties for improved recognition of target sites, (ii) delivering high PS payloads using passively or actively targeted nanoconstructs, and (iii) biological modulation of target cells to enhance PS concentration.

In this chapter, we discuss some recent advances, spanning the last 5 years, in developing such strategies for efficient PS targeting towards enhancing PDT efficacy against cancers.

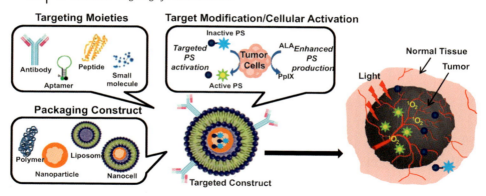

Figure 19.1 Schematic representation of molecular targeting of PS in PDT. Both targeting and packaging strategies can be employed to design efficient PSs, in combination with site-specific activation, to achieve optimum accumulation of PS in tumors.

19.2
Molecular Modifications of Photosensitizers

19.2.1
Small Molecules, Peptides, and Aptamers

Modifications can be made to the inherent structure of a PS by adding a side chain or functional group, or by conjugation of PS to small molecules for delivery such as folate [4], estradiol [5], carbohydrates [6], and fluorescent dyes that localize subcellularly [7]. Peptides have also been extensively used to target the PSs to surface receptors and to intracellular organelles including the nucleus [8]. Cell surface receptors are the most common molecular targets of peptide–PS conjugates and cause improved tumor selectivity over normal cells due to their overexpression on tumor cells. Some recent examples include vascular endothelial growth factor receptor (VEGFR) neuropilin-1, epidermal growth factor (EGF) [21], gastrin-releasing peptide receptor (GRPR), and $\alpha_v\beta_3$-integrin. Other organelles such as Golgi bodies, endoplasmic reticulum, and mitochondria have been explored to understand the effect of intracellular targeting on PDT [9]. As an example, an amphiphilic porphyrin derivative containing a mitochondrial localization sequences (MLS) peptide has been developed [10] and cell-penetrating peptide (CPP) has been used to deliver PSs intracellularly [11, 12]. These studies suggest that stability and affinity are the primary limitations with peptides. Some recent advances that may improve peptide-targeted PDT include the use of peptidomimetics to overcome *in vivo* stability and nanotechnology to enhance affinity to cancer cells. Aptamer–PS conjugates have also been developed for selective *in vitro* PDT targeting of a lymphoma cell line [13]. Thus such peptide– and aptamer–PS conjugates may provide an avenue for the development of highly efficient PDT regimens for clinical applications.

19.2.2
Proteins

Antibody-conjugated PSs were one of the earliest targeting strategies in PDT [14]. Despite all the efforts, antibodies have achieved limited success as targeting agents because of disadvantages such as nonspecific binding of Fc (fragment, crystallizable) receptors in normal cells and poor tumor penetration [15]. Efforts have been made to overcome these limitations using antibody fragments to direct PSs to the tumor [16]. scFv (single-chain fragment, variable) fragments provide efficient tumor penetration and the molecular specificity achieved by these scFvs has been found to vary for different PSs [17]. Some other advances include (i) production of a recombinant immunophotosensitizer [18] using a red fluorescent protein as PS, to overcome the synthetic limitations, such as purification, associated with antibody–PS constructs, and (ii) application of small immune proteins (SIPs) as antibody fragments [19]. SIPs may be preferable for many tumor targeting applications due to their excellent *in vivo* stability, long residence times in tumors, and good selectivity. *In vivo* PDT efficacy was enhanced with a SIP–PS conjugate as compared with an scFv conjugate [19]. Proteins such as transferrin [20] and albumin [22] have also been used for tumor targeting and can improve the specificity and efficiency of PDT [23]. In addition, antibodies have been used as targeting moieties in conjugation with other delivery constructs such as nanoparticles (NPs) [24]. Antibody-targeted PSs have rarely been used clinically, probably owing to the complicated and expensive regulatory process required for such transitions.

19.3
Delivering Photosensitizers via Targeted Nanoconstructs

Many PSs are hydrophobic, reconstitute poorly, have difficult transport in biological systems, and may be rapidly cleared by the reticuloendothelial system (RES) of the body. Preparation of pharmaceutical formulations for parenteral administration of such PSs is greatly hindered because of these limitations and, to overcome this, strategies that employ nanometer-sized delivery vehicles called NPs have been developed in order to permit the stable dispersion of these PSs into aqueous systems. NP encapsulation of PSs has broadened the spectrum of viable PSs for PDT due to the enhanced solubility and stability [24]. The leaky neovasculature of tumors favors accumulation and diminished clearing, often referred to as the enhanced permeability and retention (EPR) effect, helps passively target the tumor. PSs have been encapsulated by different strategies in different NPs, including liposomes, micelles, polymeric NPs, and dendrimers [24]. The use of nanoparticle-encapsulated two-photon dyes or upconversion NPs allows activation with low-energy light which can deeply penetrate tissues [25]. Instead of being mere carriers, these nanoconstructs participate in the actual mechanism of PDT, either by acting as PSs themselves or as transducers of X-rays or near-infrared radiation to emission wavelengths appropriate for excitation of attached PSs. Further, the NPs can have specific targeting agents such as monoclonal antibodies, aptamers, or peptides for greater selectivity and specificity [24]. Even additional imaging agents may be attached or encapsulated in the NPs to allow for drug/treatment monitoring by fluorescence optical imaging, computed tomographic scan, magnetic resonance imaging, or positron emission tomography [26–28]. It should be emphasized that by appropriate design of the NP, one can circumvent natural barriers to drug delivery, and nanotechnology provides a promising future direction with a platform for the development of multimodal PDT-based combinations for enhanced treatment effects [29, 30]. However, much work still needs to be done and PDT is now at an exciting stage of development. Very few clinical studies have evaluated the effect of the different delivery systems in terms of clinical efficiency. Following extensive characterization, the rapid translation of nanoparticle-encapsulated PSs into the clinic could help validate the potential of PDT towards impacting cancer patients.

19.4
Target Modification/Cellular Activation

Target cell biology has also been modulated to enhance PS accumulation in tumors. ALA (δ-aminolevulinic acid)-mediated treatment provides an excellent example of this technique, where cell differentiation agents are used to increase the conversion of PpIX (protoporphyrin IX) PS from ALA, resulting in increased PDT effectiveness. Differentiating agents that have been used include vitamin D [31], methotrexate [32], and iron chelators (CP94) [33], which are already in clinical use. These agents can induce tumor cell differentiation and upregulate key enzymes involved in PpIX synthesis, thereby rendering tumor cells more susceptible to PDT.

Another upcoming approach is the generation of "Smart PSs." These PSs are initially in a quenched state but can be activated for photokilling by specific components in the tumor region and can provide increased specificity with reduced toxicity. Such "smart PSs" can potentially provide a powerful tool for targeted PDT in cancer. Various activation mechanisms have been explored, including tumor microenvironment (pH, hydrophobicity), enzyme cleavage, and nucleic acid sequence displacement. Lovell *et al.* have reviewed the design and application of such "smart PSs" that provide additional specificity for cancer treatment [34]. Only at the tumor is PS converted from an inactive (quenched) form to an active phototoxic state, thus preventing any collateral damage to the normal tissue. Despite encouraging initial results, further studies are needed before the full potential of such "smart PSs" can be established.

19.5
Conclusion

PS conjugates and supramolecular delivery platforms can improve PDT selectivity by exploiting cellular and physiological markers of targeted tissue. Overexpression of receptors in cancer and angiogenic endothelial cells allows their targeting by affinity-based moieties for the selective uptake of PS conjugates and encapsulating delivery carriers, while the abnormal tumor neovascularization induces a specific accumulation of PS nanocarriers by the EPR effect. In addition, polymeric prodrug delivery platforms triggered by the acidic nature of the tumor environment or the expression of proteases can be designed. Promising results obtained with recent systemic carrier platforms will, in due course, be translated into the clinic for highly efficient and selective PDT protocols. The overall impact of the development of the more selective PDT agents will result in a broader applicability of this promising platform technology.

References

1 Triesscheijn, M., Baas, P., Schellens, J.H., and Stewart, F.A. (2006) *Oncologist*, **11**, 1034–1044.

2 Solban, N., Rizvi, I., and Hasan, T. (2006) *Lasers Surg. Med.*, **38**, 522–531.

3 Verma, S., Watt, G.M., Mai, Z., and Hasan, T. (2007) *Photochem. Photobiol.*, **83**, 996–1005.

4 Gravier, J., Schneider, R., Frochot, C., Bastogne, T., Schmitt, F., Didelon, J., Guillemin, F., and Barberi-Heyob, M. (2008) *J. Med. Chem.*, **51**, 3867–3877.

5 Swamy, N., Purohit, A., Fernandez-Gacio, A., Jones, G.B., and Ray, R. (2006) *J. Cell Biochem.*, **99**, 966–977.

6 Zheng, X. and Pandey, R.K. (2008) *Anticancer Agents Med. Chem.*, **8**, 241–268.

7 Ngen, E.J., Rajaputra, P., and You, Y. (2009) *Bioorg. Med. Chem.*, **17**, 6631–6640.

8 Schneider, R., Tirand, L., Frochot, C., Vanderesse, R., Thomas, N., Gravier, J., Guillemin, F., and Barberi-Heyob, M. (2006) *Anticancer Agents Med. Chem.*, **6**, 469–488.

9 Kessel, D. (2004) *J. Porph. Phthal.*, **8**, 1009–1014.

10 Sibrian-Vazquez, M., Nesterova, I.V., Jensen, T.J., and Vicente, M.G. (2008) *Bioconjug. Chem.*, **19**, 705–713.

11 Choi, Y., McCarthy, J.R., Weissleder, R., and Tung, C.H. (2006) *ChemMedChem*, **1**, 458–463.

12 Sehgal, I., Sibrian-Vazquez, M., and Vicente, M.G. (2008) *J. Med. Chem.*, **51**, 6014–6020.

13 Mallikaratchy, P., Tang, Z., and Tan, W. (2008) *ChemMedChem*, **3**, 425–428.

14 van Dongen, G., Visser, G.W., and Vrouenraets, M.B. (2004) *Adv. Drug Deliv. Rev.*, **56**, 31–52.

15 Carter, P. (2001) *Nat. Rev. Cancer*, **1**, 118–129.

16 Milgrom, L.R. (2008) *Sci. Prog.*, **91**, 241–263.

17 Bhatti, M., Yahioglu, G., Milgrom, L.R., Garcia-Maya, M., Chester, K.A., and Deonarain, M.P. (2008) *Int. J. Cancer*, **122**, 1155–1163.

18 Serebrovskaya, E.O., Edelweiss, E.F., Stremovskiy, O.A., Lukyanov, K.A., Chudakov, D.M., and Deyev, S.M. (2009) *Proc. Natl. Acad. Sci. U. S. A.*, **106**, 9221–9225.

19 Fabbrini, M., Trachsel, E., Soldani, P., Bindi, S., Alessi, P., Bracci, L., Kosmehl, H., Zardi, L., Neri, D., and Neri, P. (2006) *Int. J. Cancer*, **118**, 1805–1813.

20 Laptev, R., Nisnevitch, M., Siboni, G., Malik, Z., and Firer, M.A. (2006) *Br. J. Cancer*, **95**, 189–196.

21 Gilyazova, D.G., Rosenkranz, A.A., Gulak, P.V., Lunin, V.G., Sergienko, O.V., Khramtsov, Y.V., Timofeyev, K.N., Grin, M.A., Mironov, A.F., Rubin, A.B., Georgiev, G.P., and Sobolev, A.S. (2006) *Cancer Res.*, **66**, 10534–10540.

22 Anatelli, F., Mroz, P., Liu, Q., Yang, C., Castano, A.P., Swietlik, E., and Hamblin, M.R. (2006) *Mol. Pharmacol.*, **3**, 654–664.

23 Sharman, W.M., van Lier, J.E., and Allen, C.M. (2004) *Adv. Drug Deliv. Rev.*, **56**, 53–76.

24 Chatterjee, D.K., Fong, L.S., and Zhang, Y. (2008) *Adv. Drug Deliv. Rev.*, **60**, 1627–1637.

25 Zhang, P., Steelant, W., Kumar, M., and Scholfield, M. (2007) *J. Am. Chem. Soc.*, **129**, 4526–4527.

26 Reddy, G.R., Bhojani, M.S., McConville, P., Moody, J., Moffat, B.A., Hall, D.E., Kim, G., Koo, Y.E., Woolliscroft, M.J., Sugai, J.V., Johnson, T.D., Philbert, M.A., Kopelman, R., Rehemtulla, A., and Ross, B.D. (2006) *Clin. Cancer Res.*, **12** 6677–6686.

27 Koo, Y.E., Reddy, G.R., Bhojani, M., Schneider, R., Philbert, M.A., Rehemtulla, A., Ross, B.D., and Kopelman, R. (2006) *Adv. Drug Deliv. Rev.*, **58**, 1556–1577.

28 Spring, B., Mai, Z., Rai, P., and Hasan, T. (2010) Theranostic nanocells for simultaneous imaging and photodynamic therapy of pancreatic cancer. *Proc. SPIE*, **7551**, 755104.

29 Chen, W. and Zhang, J. (2006) *J. Nanosci. Nanotechnol.*, **6**, 1159–1166.

30 Rai, P., Chang, S., Mai, Z., Neuman, D., and Hasan, T. (2009) Nanotechnology-based combination therapy improves treatment response in cancer models. *Proc. SPIE*, **7380**, 73800W.

31 Ortel, B., Chen, N., Brissette, J., Dotto, G.P., Maytin, E., and Hasan, T. (1998) *Br. J. Cancer*, **77**, 1744–1751.

32 Sinha, A.K., Anand, S., Ortel, B.J., Chang, Y., Mai, Z., Hasan, T., and Maytin, E.V. (2006) *Br. J. Cancer*, **95**, 485–495.

33 Curnow, A. and Pye, A. (2007) *J. Environ. Pathol. Toxicol. Oncol.*, **26**, 89–103.

34 Lovell, J.F., Liu, T.W., Chen, J., and Zheng, G. (2010) Activatable photosensitizers for imaging and therapy. *Chem. Rev.*, **110**, 2839–2857.

20
Photodynamic Molecular Beacons
Pui-Chi Lo, Jonathan F. Lovell, and Gang Zheng

Photodynamic therapy (PDT) is a promising modality for cancer treatment [1, 2]. It involves the combined action of a photosensitizer, light, and molecular oxygen to generate reactive oxygen species (ROS), particularly singlet oxygen, to eradicate tumor cells. Although significant progress has been made over the last two decades in the development of highly efficient photosensitizers [3, 4], a major challenge is to enhance the selectivity of photosensitizers so that damage to surrounding tissues is minimized. To this end, various approaches have been explored to improve the tumor-targeting property of photosensitizers. One of the strategies involves conjugation with monoclonal antibodies or small-molecule ligands that have specific affinity for tumors [5, 6]. An alternative approach is through photosensitizer encapsulation in nanocarriers, such as liposomes, polymeric micelles, lipoproteins, and silica nanoparticles [7–9]. These nanomedicines can achieve long circulation times, allowing significant accumulation in tumors through the enhanced permeability and retention (EPR) effect [10, 11]. Rather than controlling the localization of photosensitizers, an alternative approach is to use activatable photosensitizers, which have recently attracted considerable attention [12–14]. These novel photosensitizers, upon interaction with various tumor-associated stimuli, can be switched "on" for photosensitization, offering a new level of selectivity for therapeutic applications. As shown in Figure 20.1, there are three steps to achieving targeted cell killing using activatable photosensitizers. First, the photosensitizers are delivered to the target cells and surrounding cells. This could be through systemic or topical routes. Next, the photosensitizers are activated only in the target tissue by enzymes or other factors associated with the target (and not the adjacent healthy cells). Finally, the target site is irradiated with light that is usually of a wavelength in the near-infrared range to maximize tissue penetration. Although the light is absorbed by the inactivated photosensitizers in the healthy tissue, the photosensitizers remain quenched and therefore no singlet oxygen is generated. The target cells activate the photosensitizers, and upon irradiation the photosensitizers generate singlet oxygen to kill those cells specifically.

To design an activatable photosensitizer targeting particular cells, several factors must be considered. The active linker, which regulates singlet oxygen

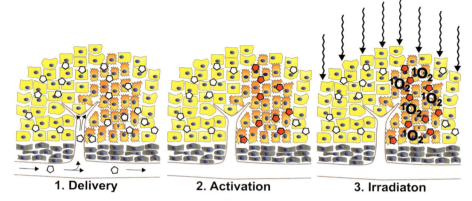

Figure 20.1 General principle of targeted cell killing by activatable photosensitizers.

production, is the centerpiece of the construct. Various activation mechanisms are possible, including the separation of photosensitizer and quencher through linker cleavage by target enzymes or linker elongation by a target nucleic acid. Selection of an appropriate photosensitizer is particularly important. It should absorb light at the appropriate wavelength (near-infrared wavelengths are usually ideal for *in vivo* applications) and have a high singlet oxygen quantum yield. Hydrophobicity should also be considered since very hydrophobic photosensitizers might affect the active linker control mechanism and could make purification more difficult. Finally, the quenching mechanism must be chosen carefully. The simplest quencher can be chosen based on FRET (Förster resonance energy transfer) efficiency with the fluorescence emission of the photosensitizers [15]. Alternatively, self-quenching mechanisms can be used, which often have the advantage of simpler synthesis and more photosensitizers per construct. However, it may be more difficult to predict a design that will ensure efficient self-quenching in the inactive state. Various activation mechanisms are summarized in Figure 20.2 and selected examples of these smart therapeutic agents are highlighted here, focusing on those which can be activated by the environment, enzymes, and nucleic acids.

It has well been documented that the extracellular pH in tumors is relatively low compared with that of normal tissues (about 6.8 versus 7.3) [16, 17]. This characteristic has been employed for the development of various pH-activatable fluorescence probes for cancer imaging [18, 19]. McDonnell *et al.* extended the study to prepare a series of amine-containing BF_2-chelated azadipyrromethenes (Figure 20.3), of which the singlet oxygen generation ability can be switched on and off in N,N-dimethylformamide (DMF) by changing the pH environment [20]. Under acidic conditions, the amino substituents are protonated, by which the photoinduced electron transfer (PET) process is inhibited. Compound **1**, for example, shows an approximately 10-fold enhancement in singlet oxygen generation efficiency. Compound **2** is a boron dipyrromethene derivative having an additional 15-crown-5 unit.

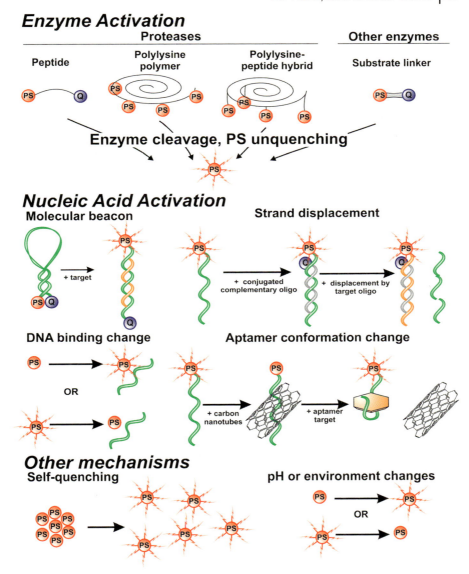

Figure 20.2 Different activation mechanisms of activatable photosensitizers for PDT, with PS representing a photosensitizer and Q representing a quencher.

It is responsive not only to the pH value, but also to the salt concentration, both of which are important physiological parameters [21]. Only at low pH and high salt concentration is this compound activated by hindering the PET process, showing an approximate sixfold increase in singlet oxygen production.

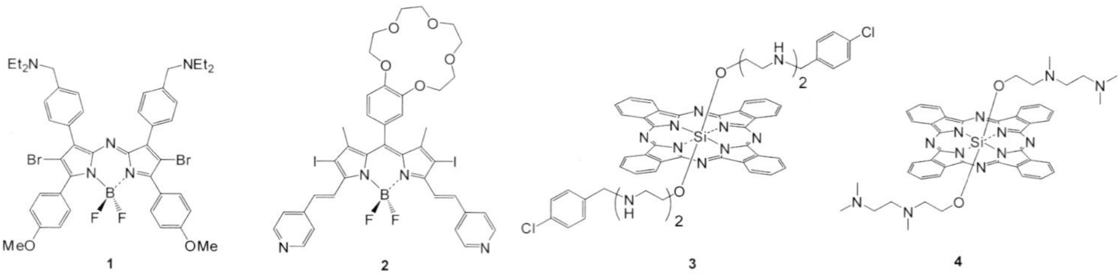

Figure 20.3 Selected examples of pH-activatable photosensitizers.

This approach has also been used by Jiang and co-workers, who synthesized a series of silicon(IV) phthalocyanines with two axial polyamine substituents (Figure 20.3) [22, 23]. The pH-dependent electronic absorption, fluorescence emission, and photosensitizing properties can be modulated by the axial substituents. Compounds **3** and **4** exhibit the most desirable changes in the pH range 5–7, which can roughly differentiate the tumor and normal tissue environments. Their intracellular fluorescence intensity and efficiency in generating singlet oxygen and superoxide radicals are greatly enhanced at lower pH, making them promising pH-controlled and tumor-selective fluorescence probes and photosensitizers for PDT.

Overexpression of specific enzymes is often associated with certain diseases. Proteases, for example, represent an important class of enzymes which are related to cancers, hypertension, AIDS, respiratory diseases, and so on. Therefore, they are the therapeutic targets of many clinically approved drugs for these diseases [24], and should be excellent candidates for activation of photosensitizers. In addition, due to their well-characterized and catalytic action, one enzyme molecule can in principle activate many molecules of the photosensitizer in a specific manner, resulting in high signal amplification. As shown in Figure 20.2, enzyme-activated photosensitizers have focused on enzymatic hydrolysis of covalent bonds that leads to a separation of photosensitizer and quencher. There have been several approaches to developing proteases beacons. Short peptides linking photosensitizer and quencher, polylysine backbones with self-quenched photosensitizers, and also peptide–polylysine hybrids have been described. Other substrates have been used to fuse photosensitizers so that quenching is impacted by enzymatic substrate cleavage. Chen *et al.* employed this concept to report a photodynamic molecular beacon (PMB), **5**, in which a pyropheophorbide *a* (Pyro) photosensitizer is linked to a carotenoid (CAR) quencher via a cleavable caspase-3 substrate, G<u>DEVD</u>GSGK (recognition site underlined; see Figure 20.4) [25]. Owing to the close proximity of Pyro and CAR, the excited state of Pyro is quenched by CAR, inhibiting the generation of singlet oxygen. However, in the presence of caspase-3, the peptide chain is cleaved and the detachment of Pyro and CAR leads to an increase in the singlet oxygen luminescence intensity (by about fourfold) and lifetime (by about 1.3-fold). In this particular PMB, the CAR not only deactivates the excited state of Pyro, but also directly scavenges singlet oxygen, which can minimize photodamage to nontargeted cells [26].

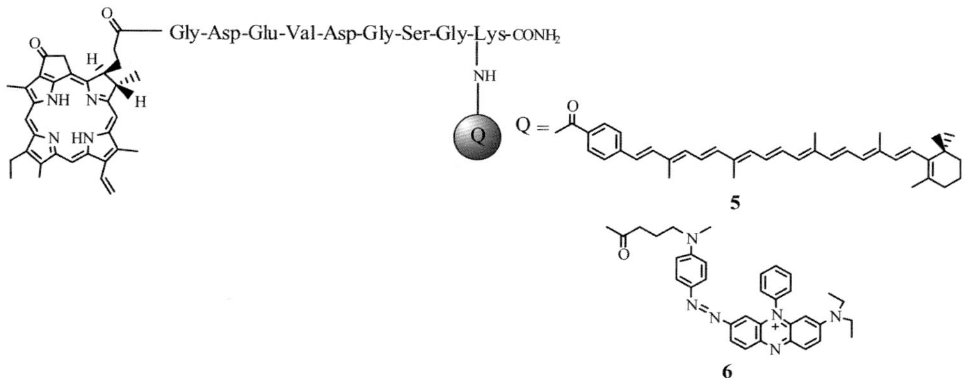

Figure 20.4 Caspase-3-activatable photosensitizers.

On the basis that caspase-3 is one of the executioner caspases involved in apoptosis, this kind of PMB can be used to indicate the occurrence apoptosis. PMB **6** is an analog of **5**, having a black hole quencher 3 (BHQ-3) instead of CAR as the quencher (Figure 20.4) [27]. This agent induces photodamage in irradiated cells by the sensitizing action of Pyro and simultaneously identifies apoptotic cells by the near-infrared fluorescence due to the detached Pyro. Hence this PMB can be considered as having a built-in apoptosis sensor for evaluating its therapeutic outcome *in situ*.

By using a similar strategy, PMBs triggered by matrix metalloproteinase-7 (MMP-7) [28] or fibroblast activation protein (FAP) [29] were also prepared and studied in detail, both *in vitro* and *in vivo*, by the same group. These PMBs are promising for the diagnosis and treatment of epithelial cancers. The studies validate the core principle of the PMB concept, showing that selective PDT-induced cell death can be achieved by controlling the singlet oxygen generation ability of the photosensitizer by responding to specific cancer-associated biomarkers.

One of the factors that affect the efficacy of PMBs is how close the photosensitizer and the quencher can be brought together by the peptide linkage. Different peptide sequences will give different background fluorescence as a result of their distinct natural folding. To overcome this problem, the concept of a "zipper" has been explored, which involves the introduction of two polycationic and polyanionic segments to hold the photosensitizer and the quencher in close proximity by electrostatic interactions. This results in silenced dye activity which is basically independent of the peptide linkage. The hairpin conformation of the substrate sequence can also facilitate the cleavage of the enzyme-specific linker. Chen *et al.* prepared a series of these zipper molecular beacons, in which Pyro connected to a segment of polyarginine and BHQ-3 connected to a segment polyglutamate are linked to two MMP-7-specific peptide sequences, GPLGLARK and RPLALWRSK [30]. It was found that the unsymmetrical beacon (BHQ-3)(E)$_5$GPLGLA(R)$_8$K(Pyro) can optimize the quenching efficiency and readiness of dissociation of the zipper arms after proteolysis. The polyanionic arm of the zipper can prevent the photosensitizer

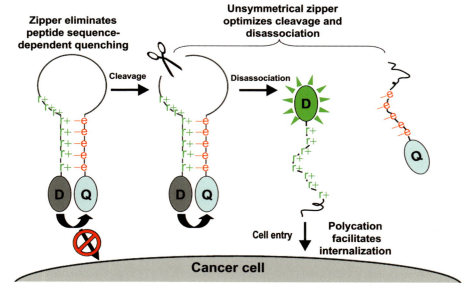

Figure 20.5 Schematic diagram showing the zipper molecular beacon design; D and Q represent the dye and the quencher, respectively. Reproduced with permission from Chen et al., Bioconjug. Chem., 2009, **20**, 1836.

from entering the cells, while the polycationic arm can enhance the cellular uptake of the dye after cleavage (Figure 20.5). This "zipper" concept can become a general approach to improve the functionality of a wide range of diagnostic and therapeutic probes through a simple modification of the substrate sequences.

In addition to these molecular beacons, polymeric protease-mediated photodynamic agents have also been reported [31–33]. Choi et al. developed a polylysine–chlorin e6 (Ce6) conjugate in which multiple Ce6 molecules are attached to a biodegradable poly-L-lysine (Lys) backbone grafted with monomethoxy poly(ethylene glycol) (MPEG) side chains (Figure 20.6) [31]. In this polymeric species, the Ce6 molecules are aggregated and the high local concentration of Ce6 causes self-

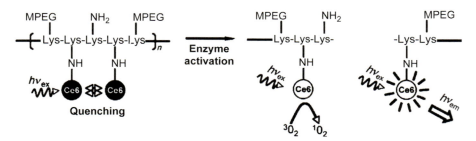

Figure 20.6 Activation of fluorescence emission and singlet oxygen generation of a polylysine–chlorin e6 conjugate by a protease. Reproduced with permission from Choi et al., Cancer Res., 2006, **66**, 7225.

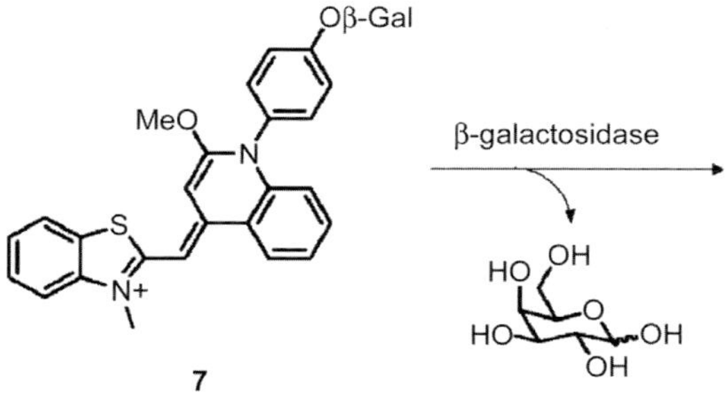

Figure 20.7 Activation of thiazole orange-based photosensitizer **7** by β-galactosidase. Reproduced with permission from Koide et al., *J. Am. Chem. Soc.*, 2009, **131**, 6058.

quenching, resulting in weak fluorescence and low singlet oxygen generation. The aggregation behavior of this polymeric species, however, can enhance the tumor-accumulating property by the EPR effect. In the presence of cathepsin B in buffer, the fluorescence of this polymeric beacon is activated and its singlet oxygen generation efficiency increases sixfold. A similar activation process was also demonstrated *in vivo* using a xenographic tumor model. An intensified fluorescence signal was observed in the animals injected with the agent, which also showed an improved antitumor efficacy compared with controls. Campo *et al.* further improved this system by using a derivatized nicotinic acid polymer with a hydrophilic cationic quaternary amine group, thus eliminating the interference of PEGylation with the photosensitizer interactions that give rise to self-quenching [34].

Nagano and co-workers recently reported another enzymatically activatable photosensitizer based on thiazole orange (**7**, Figure 20.7) [35]. The bulky and hydrophilic β-galactoside moiety of **7** reduces its intracellular fluorescence emission and photosensitizing ability. This may be a result of the blockage of the binding between the dye and biomolecules, especially DNA in the cells. Upon hydrolysis by β-galactosidase, **7** gives the thiazole orange derivative **8** (Figure 20.7), by which both the fluorescence intensity and phototoxicity in cultured cells are greatly increased. The action of **7** on *lacZ* gene-transfected HEK293 *lacZ*(+) cells and untransfected *lacZ*(−) cells has also been examined. β-Galactosidase is encoded by *lacZ* gene from *Escherichia coli*. Interestingly, the HEK293 *lacZ*(+) cells incubated with **7** show a bright fluorescence image and are killed upon irradiation, whereas the *lacZ*(−) cells are only weakly fluorescent and remain essentially unharmed. During the phototreatment, relocation of the dye to the nucleus was observed only in *lacZ*(+) cells. The results nicely demonstrate that **7** is activated specifically in β-galactosidase-expressing cells.

In addition to applications in cancer imaging and therapy, activatable photosensitizers are also potentially useful for the treatment of localized infections by drug-resistance bacteria. Zheng *et al.* reported a novel β-lactamase activated photosensitizer [36]. The short separation of the two photosensitizer moieties based on 5-(4′-carboxybutylamino)-9-diethylaminobenzo[*a*]phenothiazinium chloride leads to homo- or heterodimerization and eventually quenching of the fluorescence emission.

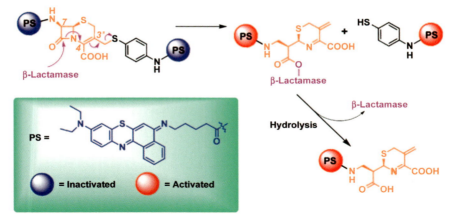

Figure 20.8 Mechanism of the cleavage of a β-lactamase activated photosensitizer. Reproduced with permission from Zheng *et al.*, *Angew. Chem. Int. Ed.*, 2009, **48**, 2148. Copyright Wiley-VCH Verlag GmbH & Co. KGaA.

In the presence of β-lactamase expressed by methicillin-resistant *Staphylococcus aureus* (MRSA), the β-lactam ring is cleaved, releasing activated photosensitizer moieties (Figure 20.8). The results show that photosensitizers can target MRSA selectively and cause less damage to host tissue than PDT with the free photosensitizer.

In addition to enzymes, nucleic acids can also be the targets for PMBs. On the basis that mRNAs have high tumor specificity and their hybridization to the complementary antisense oligonucleotide (AS-ON) sequences through Watson–Crick base pairing is highly selective and efficient, a novel mRNA-triggered PMB has been developed [37]. In this beacon, a Pyro photosensitizer and a CAR quencher are connected to a *c-raf-1* mRNA-targeted AS-ON as the loop sequence. The hairpin effect results in better control of the singlet oxygen production upon target–linker interactions compared with the peptide-based analog [26]. The signal-to-noise ratio is improved more than threefold. This PMB also shows efficient cellular uptake and PDT activation in *c-raf-1* expressing MDA-MB-231 cancer cells. By using a similar approach, Nesterova *et al.* recently prepared two dimer-based molecular beacons in

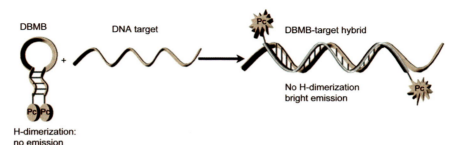

Figure 20.9 Schematic diagram showing the action of phthalocyanine dimer-based molecular beacons. Reproduced with permission from Nesterova *et al.*, *J. Am. Chem. Soc.*, 2009, **131**, 2432.

which two identical zinc(II) phthalocyanine moieties are linked together by an oligonucleotide sequence [38]. Owing to the intrinsic propensity of these phthalocyanines to form nonfluorescent H-type dimers, the beacons are in the "off" state. In the presence of the target DNA, the beacon's loop hybrids to its target, forcing the beacon to open. This disrupts the phthalocyanine dimer, restoring the fluorescence emission (Figure 20.9). Some attractive features of these dimer-based molecular beacons include their strong fluorescence emission in the open form due to the presence of two fluorescent reporter moieties and their relatively straightforward synthesis procedure.

This chapter has highlighted the design and effect of several classes of PMBs which can be activated by different tumor-associated stimuli. They have great potential in killing diseased cells effectively and specifically. Further development of this field of research should be intensified in the years ahead, particularly further improvement of their efficiency and selectivity, discovery of new activation mechanisms, and detailed *in vitro* and *in vivo* studies towards clinical implementation.

References

1 Allison, R.R., Bagnato, V.S., Cuenca, R., Downie, G.H., and Sibata, C.H. (2006) *Future Oncol.*, **2**, 53.
2 Wilson, B.C. and Patterson, M.S. (2008) *Phys. Med. Biol.*, **53**, R61.
3 Detty, M.R., Gibson, S.L., and Wagner, S.J. (2004) *J. Med. Chem.*, **47**, 3897.
4 Nyman, E.S. and Hynninen, P.H. (2004) *J. Photochem. Photobiol. B*, **73**, 1.
5 Sharman, W.M., van Lier, J.E., and Allen, C.M. (2004) *Adv. Drug Deliv. Rev.*, **56**, 53.
6 Verma, S., Watt, G.M., Mai, Z., and Hasan, T. (2007) *Photochem. Photobiol.*, **83**, 996.
7 Nishiyama, N., Jang, W.-D., and Kataoka, K. (2007) *New J. Chem.*, **31**, 1074.
8 Bechet, D., Couleaud, P., Frochot, C., Viriot, M.L., Guillemin, F., and Barberi-Heyob, M. (2008) *Trends Biotechnol.*, **26**, 612.
9 Zheng, G., Chen, J., Li, H., and Glickson, J.D. (2005) *Proc. Natl. Acad. Sci. U. S. A.*, **102**, 17757.
10 Ferrari, M. (2005) *Nat. Rev. Cancer*, **5**, 161.
11 Zhang, L., Gu, F.X., Chan, J.M., Wang, A.Z., Langer, R.S., and Farokhzad, O.C. (2008) *Clin. Pharmacol. Ther.*, **83**, 761.
12 Stefflova, K., Chen, J., and Zheng, G. (2007) *Frontiers Biosci.*, **12**, 4709.
13 Lovell, J.F. and Zheng, G. (2008) *J. Innov. Opt. Health Sci.*, **1**, 45.
14 Lovell, J.F., Liu, T.W., Chen, J., and Zheng, G. (2010) *Chem. Rev.*, **110**, 2839.
15 Lovell, J.F., Chen, J., Jarvi, M.T., Cao, W.-G., Allen, A.D., Liu, Y., Tidwell, T.T., Wilson, B.C., and Zheng, G. (2009) *J. Phys. Chem. B*, **113**, 3203.
16 Stubbs, M., McSheehy, P.M.J., Griffiths, J.R., and Bashford, C.L. (2000) *Mol. Med. Today*, **6**, 15–19.
17 Gerweck, L.E. (2000) *Drug Resist. Updates*, **3**, 49–50.
18 Tang, B., Yu, F., Li, P., Tong, L., Duan, X., Xie, T., and Wang, X. (2009) *J. Am. Chem. Soc.*, **131**, 3016.
19 Urano, Y., Asanuma, D., Hama, Y., Koyama, Y., Barrett, T., Kamiya, M., Nagano, T., Watanabe, T., Hasegawa, A., Choyke, P.L., and Kobayashi, H. (2009) *Nat. Med.*, **15**, 104.
20 McDonnell, S.O., Hall, M.J., Allen, L.T., Byrne, A., Gallagher, W.M., and O'Shea, D.F. (2005) *J. Am. Chem. Soc.*, **127**, 16360.
21 Ozlem, S. and Akkaya, E.U. (2009) *J. Am. Chem. Soc.*, **131**, 48.
22 Jiang, X.-J., Lo, P.-C., Yeung, S.-L., Fong, W.-P., and Ng, D.K.P. (2010) *Chem. Commun.*, 3188.

23 Jiang, X.-J., Lo, P.-C., Tsang, Y.-M., Yeung, S.-L., Fong, W.-P., and Ng, D.K.P. (2010) *Chem. Eur. J.*, **16**, 4777.

24 Turk, B. (2006) *Nat. Rev. Drug Discov.*, **5**, 785.

25 Chen, J., Stefflova, K., Niedre, M.J., Wilson, B.C., Chance, B., Glickson, J.D., and Zheng, G. (2004) *J. Am. Chem. Soc.*, **126**, 11450.

26 Chen, J., Jarvi, M., Lo, P.-C., Stefflova, K., Wilson, B.C., and Zheng, G. (2007) *Photochem. Photobiol. Sci.*, **6**, 1311.

27 Stefflova, K., Chen, J., Marotta, D., and Zheng, G. (2006) *J. Med. Chem.*, **49**, 3850.

28 Zheng, G., Chen, J., Stefflova, K., Jarvi, M., Li, H., and Wilson, B.C. (2007) *Proc. Natl. Acad. Sci. U. S. A.*, **104**, 8989.

29 Lo, P.-C., Chen, J., Stefflova, K., Warren, M.S., Navab, R., Bandarchi, B., Mullins, S., Tsao, M., Cheng, J.D., and Zheng, G. (2009) *J. Med. Chem.*, **52**, 358.

30 Chen, J., Liu, T.W.B., Lo, P.-C., Wilson, B.C., and Zheng, G. (2009) *Bioconjug. Chem.*, **20**, 1836.

31 Choi, Y., Weissleder, R., and Tung, C.-H. (2006) *Cancer Res.*, **66**, 7225.

32 Choi, Y., Weissleder, R., and Tung, C.-H. (2006) *ChemMedChem*, **1**, 698.

33 Gabriel, D., Busso, N., So, A., van den Bergh, H., Gurny, R., and Lange, N. (2009) *J. Control. Release*, **138**, 225.

34 Campo, M.A., Gabriel, D., Kucera, P., Gurny, R., and Lange, N. (2007) *Photochem. Photobiol.*, **83**, 958.

35 Koide, Y., Urano, Y., Yatsushige, A., Hanaoka, K., Terai, T., and Nagano, T. (2009) *J. Am. Chem. Soc.*, **131**, 6058.

36 Zheng, X., Sallum, U.W., Verma, S., Athar, H., Evans, C.L., and Hasan, T. (2009) *Angew. Chem. Int. Ed.*, **48**, 2148.

37 Chen, J., Lovell, J.F., Lo, P.-C., Stefflova, K., Niedre, M., Wilson, B.C., and Zheng, G. (2008) *Photochem. Photobiol. Sci.*, **7**, 775.

38 Nesterova, I.V., Erdem, S.S., Pakhomov, S., Hammer, R.P., and Soper, S.A. (2009) *J. Am. Chem. Soc.*, **131**, 2432.

21
Oxygen Effects in Photodynamic Therapy
Asta Juzeniene

21.1
Introduction

From the early days of photodynamic therapy (PDT), a combination of dyes (photosensitizers) and light has been used to kill cells. However, in 1902, the importance of a third element, molecular oxygen (O_2), was recognized by Ledoux-Lebards [1] and soon afterwards by von Tappeiner and Jesionek [2] and Straub [3]. They demonstrated that photodynamic action is inhibited in the absence of O_2 and concluded that O_2 is required in photosensitization reactions [1–3]. Several hypotheses have been proposed for the role of O_2 in these reactions. In 1931, Kautsky and de Bruijn proposed that O_2 in its singlet excited state could initiate photodynamic reactions [4], and Bäckström in 1934 suggested the involvement of free radicals [5]. Because of the difficulty in identifying and measuring the amounts of the photochemical reaction products *in vitro* and *in vivo*, the processes involved in photosensitization were not understood before the 1960s. In the presence of O_2, the photosensitized oxidation reactions can follow two pathways, called type I and type II reactions [6, 7]. In type I reactions, the triplet state photosensitizer can interact directly with biomolecules, by electron or hydrogen transfer, to form radicals and radical ions which react with O_2 to form oxidized products. In type II reactions (photodynamic action), the triplet state photosensitizer can transfer its energy directly to O_2, generating singlet molecular oxygen ($^1O_2$) [6–8]. Thus, the significance of O_2 in PDT was recognized, and later many important effects related to O_2 levels were observed. These are discussed in this chapter.

21.2
The Role of Oxygen in Photodynamic Therapy: $^1O_2$ Generation

$^1O_2$ is recognized to be the main mediator of photoinactivation of cells sensitized with most clinically used photosensitizers [8, 9]. In 2002, the production of $^1O_2$

during PDT was detected for the first time by direct measurements *in vivo* [10]. This highly reactive form of O_2 reacts with many biological molecules and causes damage that may be lethal for cells. The type II mechanism ($^1O_2$ mechanism) is generally considered to be dominant in PDT. The presence of sufficient O_2 is required for $^1O_2$ production in tissues.

21.3
Dependence of the Photosensitizing Effect on O_2 Concentration

When PDT became a cancer treatment, it was hoped that this treatment would not suffer from any O_2 effect, in comparison with radiation therapy or, at least, would be efficient down to very low O_2 levels. However, in the 1980s it was demonstrated that PDT was not efficient under anoxic conditions, either *in vitro* or *in vivo* [11–16]. The efficiency of photoinactivation of the sensitized cells decreases with decreasing O_2 concentration [12, 15]. The O_2 dependence of the photoinactivation of cells is affected by the type, intracellular concentration, and localization of the photosensitizer, the presence of $^1O_2$ quenchers, pH, and so on. For example, it has been demonstrated that the quantum yield of photoinactivation with hematoporphyrin derivative (HpD) was reduced by 50% when the partial pressure of oxygen (pO_2) was reduced to 7.6 mmHg in NHIK 3025 cells [15]. Similar results were later found *in vivo* by Henderson and Fingar [16].

21.4
The Oxygenation Status in Tumors and Normal Tissues

Oxygen concentrations are significantly lower in some solid tumors than in normal tissue. Up to half of locally advanced solid tumors may exhibit hypoxic and/or anoxic regions [17]. Hypoxic regions may be observed in breast, uterine cervix, vulva, head and neck, prostate, rectum, pancreas, brain tumors, soft tissue sarcomas, and malignant melanomas [17]. Tissue pO_2 is controlled by the O_2 delivery rate and by the O_2 consumption rate of vascular walls and tissues. An imbalance between O_2 supply and consumption in tumors is caused by abnormal structures and functions (tortuosity, leakiness, lack of pericytes, inhomogeneous distribution) of the microvessels, increased diffusion distances between the blood vessels and the tumor cells, and reduced O_2 transport capacity of the blood due to anemia [17]. Hypoxic areas in such tumors are insensitive to PDT [13, 16, 18]. Additionally, hypoxia may indicate an insufficient vascular network, likely to limit photosensitizer delivery.

PDT is now mainly used in dermatology to treat optically available skin cancer. The role of O_2 in cutaneous PDT has previously been described by Fuchs and Thiele [19].

21.5
PDT-Induced Reduction of Tumor Oxygenation

21.5.1
O$_2$ Consumption (Primary Reduction)

The majority of PDT-induced cytotoxic responses are due to type II reactions ($^1O_2$ generation) which require and consume O_2: one O_2 molecule is consumed for each $^1O_2$ that reacts with and oxidizes another molecule. The O_2 consumption rate depends on the concentration of photosensitizer, the fluence rate, the extinction coefficient of the photosensitizer, the quantum yield of photooxygenation, and so on [20–22]. The photochemical O_2 consumption rates under clinical conditions with Photofrin (porfimer sodium) at fluence rates greater than 100 mW cm^{-2} are higher than the metabolic oxygen consumption rates of most tissues and lead to O_2 depletion during exposure to light [23, 24].

21.5.2
Vascular Damage (Secondary Reactions)

Some photosensitizers, especially after shorter incubation times before light exposure, accumulate mostly in blood vessels, and PDT then mainly targets tumor vasculature. Thus, PDT can lead to vasoconstriction and vessel clogging, depending on the conditions such as fluence rate, fluence, drug concentration, and tissue type [25–30]. Vascular damage to the tumor microenvironment is serious, since it inevitably leads to changes in the O_2 supply to the tumor. For example, within minutes of light exposure acute hypoxia in tumors due to vascular damage has been demonstrated during PDT with Photofrin [31, 32]. This almost always limits direct tumor cell photodestruction [31].

21.5.3
Photosensitizer Photobleaching (Secondary Reactions)

In addition to $^1O_2$ reactions with cellular targets, $^1O_2$ may react with the photosensitizer itself. Generally, even under well-oxygenated conditions, the photobleaching reactions also consume O_2. Most photosensitizers used in PDT are photolabile [33, 34]. Photobleaching can decrease the effectiveness of PDT by reducing the photosensitizer and O_2 concentrations. However, it can also be useful for dosimetry [35–37]. Sometimes, when the O_2 concentration is reduced (due to PDT itself or due to hypoxic conditions), the O_2-independent photobleaching mechanism, caused by interactions between excited triplet-state photosensitizer and biological substrates, becomes dominant for tumor and tissue damage. Then, provided that the photosensitizer concentration is lower in the surroundings of the tumor than in the tumor itself, there will be a selective effect on the tumor and one does not need to worry about normal tissue damage, now matter how much light is delivered.

21.6
Methods to Reduce Tumor Deoxygenation During PDT

It is possible to minimize O_2 depletion during light exposure and to allow tumor reoxygenation.

21.6.1
Low Fluence Rates

Tumor responses to PDT with different photosensitizers depends on O_2 levels and on the fluence rate of the light [24, 38]. A reduction in fluence rate from high (75–200 mW cm^{-2}) to low (10–50 mW cm^{-2}) results in increased PDT efficiency [24, 38–42]. High fluence rate PDT is inefficient, since the consumption of O_2 exceeds the rate at which O_2 can be resupplied from the vasculature, leading to PDT-induced hypoxic tissue regions [24, 43]. O_2 depletion occurs on a time scale of seconds, and depends on the average fluence rate [22]. PDT carried out at a fluence rate of 50 mW cm^{-2} may consume O_2 at a rate of around 6–9 µM s^{-1} [20], whereas the rate is around 32 µM s^{-1} at 100 mW cm^{-2} [22]. There is a lower limit of efficient fluence rates (below 5.5 mW cm^{-2}) [24, 38]. A low fluence rate implies a long exposure time. If this time is too long, vessel damage will occur during exposure and, as a consequence, pO_2 will decrease. Thus, it has been found that low fluence rate PDT causes severe and longer lasting disruption of the microvascular perfusion [42, 44]. This may cause reduced PDT efficiency but may also lead to thrombosis and hemorrhage in tumor blood vessels that subsequently lead to tumor death via deprivation of O_2 and nutrients [42, 44, 45].

Recently, metronomic PDT has been proposed for delivery of ultra-low fluence rates (µW cm^{-2}) and low concentrations of a photosensitizer continuously over several hours, days, or weeks [22, 46].

21.6.2
Fractionated Light Exposure

O_2 consumption is more pronounced at higher fluence rates, and it may take only a few seconds to create anoxia [20]. This aspect is even more crucial when treating solid tumors, as they are often badly oxygenated. Fractionated light exposure regimes have been suggested to overcome this problem, because as long as the microvasculature remains functional, it is possible to stop light exposure for some time and allow reoxygenation. Effective reoxygenation of cells can occur in dark periods of around 30–150 s, depending on the degree of induced hypoxia and on the intactness of the microcirculation [20, 47]. Reperfusion injury can occur during PDT when blood flow is restored after a transient period of ischemia [48]. Furthermore, relocalization of the photosensitizer during the dark period may occur and improve the outcome [49].

21.6.3
Other Methods

Since most normal tissues contain about 5% oxygen, which is supplied by the circulation, they cannot be made more sensitive to PDT by increasing the oxygen tension because the relative quantum yield for $^1O_2$ formation is constant from 5 to 100% O_2, whereas it is halved at 1% oxygen [15, 16]. Hypoxic parts of tumors, however, can be made more sensitive to PDT by overcoming hypoxia. Several methods have been proposed to increase tumor oxygenation: (1) breathing hyperbaric oxygen, (2) using oxygen-carrying fluorocarbons combined with carbogen (95% oxygen) breathing, (3) using nicotinamide injection and carbogen breathing, (4) using oxygen-releasing substances, (5) modulating the oxygen binding capacity of hemoglobin, (6) decreasing the respiration rate, (7) increasing the oxygen solubility, (8) using blood flow modifiers, and (9) destroying hypoxic cells with bioreductive drugs or hypothermia [50]. Several of these methods have been exploited for PDT with promising experimental and clinical results [51–56].

21.7
Changes of Quantum Yields Related to Photosensitizer Relocalization

Under certain conditions, light exposure may result in a transient or permanent intracellular relocalization of photosensitizers (sulfonated *meso*-tetraphenylporphines, sulfonated aluminum phthalocyanines, etc.) [57–59], sometimes even to the nucleus [58–60]. This may lead to the DNA damage and sometimes increase the quantum yield of cell photodestruction [60, 61].

21.8
Changes of Optical Penetration Caused by Changes in O_2 Concentration

The O_2 concentration changes during PDT, as described above. This will lead to a change in the penetration spectrum of light into tissue since hemoglobin and oxyhemoglobin have different absorption spectra. This has to be considered in the choice of optimal wavelength. We have found that as O_2 depletion starts to manifest itself, and the hemoglobin/oxyhemoglobin ratio increases, the optical penetration will increase at 390–422, 455–500, 530–545, and 570–585 nm and decrease at 422–455, 500–530, 545–570, and 585–805 nm [62, 63].

21.9
Conclusion

The efficacy of PDT is crucially dependent on tissue oxygenation, which changes during PDT due to consumption in the photodynamic action itself, to photobleaching, and to vascular damage. The tissue oxygenation influences the penetration depth of light into tissue during PDT. Reliable, noninvasive methods for monitoring oxygenation during PDT are needed. Optical methods, exploiting differences in the absorption spectra of hemoglobin and oxyhemoglobin, are promising in view of their simplicity and precision.

References

1 Ledoux-Lebards, C. (1902) Action de la lumière sur la toxicité de l'éosine et de quelques autres substances pour les paramècies. *Ann. Inst. Pasteur*, **16**, 587–594.

2 von Tappeiner, H. and Jesionek, A. (1903) Therapeutische Versuche mit fluoreszierenden Stoffen. *Münch. Med. Wochenschr.*, **50**, 2042–2044.

3 Straub, W. (1904) Uber chemische Vorgänge bei der Einwirkung von Licht auf fluoreszierende Substanzen (Eosin und Chinin) und die Bedeutung dieser Vorgänge für die Giftwirkung. *Münch. Med. Wochenschr.*, **51**, 1093–1096.

4 Kautsky, H. and de Bruijn, H. (1931) Die Aufklärung der Photoluminescenztilgung fluorescierender Systeme durch Sauerstoff: die Bildung aktiver, diffusionsfähiger Sauerstoffmoleküle durch Sensibilisierung. *Naturwissenschaften*, **19**, 1043.

5 Bäckström, H.L.J.B. (1934) Der Kettenmechanismus bei der Autoxydation von Aldehyden. *Z. Phys. Chem. B*, **25**, 99–121.

6 Foote, C.S. (1991) Definition of type I and type II photosensitized oxidation. *Photochem. Photobiol.*, **54** (5), 659.

7 Foote, C.S. (1968) Mechanisms of photosensitized oxidation. *Science*, **29**, 963–970.

8 Moan, J., Pettersen, E.O., and Christensen, T. (1979) The mechanism of photodynamic inactivation of human cells *in vitro* in the presence of haematoporphyrin. *Br. J. Cancer*, **39** (4), 398–407.

9 Weishaupt, K.R., Gomer, C.J., and Dougherty, T.J. (1976) Identification of singlet oxygen as the cytotoxic agent in photoinactivation of a murine tumor. *Cancer Res.*, **36** (7 Pt. 1), 2326–2329.

10 Niedre, M., Patterson, M.S., and Wilson, B.C. (2002) Direct near-infrared luminescence detection of singlet oxygen generated by photodynamic therapy in cells *in vitro* and tissues *in vivo*. *Photochem. Photobiol.*, **75** (4), 382–391.

11 Gomer, C.J. and Razum, N.J. (1984) Acute skin response in albino mice following porphyrin photosensitization under oxic and anoxic conditions. *Photochem. Photobiol.*, **40** (4), 435–439.

12 Lee See, K., Forbes, I.J., and Betts, W.H. (1984) Oxygen dependency of photocytotoxicity with haematoporphyrin derivative. *Photochem. Photobiol.*, **39** (5), 631–634.

13 Freitas, I. (1985) Role of hypoxia in photodynamic therapy of tumors. *Tumori*, **71** (3), 251–259.

14 Mitchell, J.B., McPherson, S., DeGraff, W., Gamson, J., Zabell, A., and Russo, A. (1985) Oxygen dependence of hematoporphyrin derivative-induced photoinactivation of Chinese hamster cells. *Cancer Res.*, **45** (5), 2008–2011.

15 Moan, J. and Sommer, S. (1985) Oxygen dependence of the photosensitizing effect of hematoporphyrin derivative in NHIK 3025 cells. *Cancer Res.*, **45** (4), 1608–1610.

16 Henderson, B.W., and Fingar, V.H. (1987) Oxygen limitation of direct tumor cell kill during photodynamic treatment of a murine tumor model. *Cancer Res.*, **47** (12), 3110–3114.

17 Mees, G., Dierckx, R., Vangestel, C., and Van de Wiele, C. (2009) Molecular imaging of hypoxia with radiolabelled agents. *Eur. J. Nucl. Med. Mol. Imaging*, **36** (10), 1674–1686.

18 Henderson, B.W. and Fingar, V.H. (1989) Oxygen limitation of direct tumor cell kill during photodynamic treatment of a murine tumor model. *Photochem. Photobiol.*, **49** (3), 299–304.

19 Fuchs, J. and Thiele, J. (1998) The role of oxygen in cutaneous photodynamic therapy. *Free Radic. Biol. Med.*, **24** (5), 835–847.

20 Foster, T.H., Murant, R.S., Bryant, R.G., Knox, R.S., Gibson, S.L., and Hilf, R. (1991) Oxygen consumption and diffusion effects in photodynamic therapy. *Radiat. Res.*, **126** (3), 296–303.

21 Moan, J. and Juzeniene, A. (2008) The role of oxygen in photodynamic therapy, in *Advances in Photodynamic Therapy. Basic, Translational, and Clinical* (eds. M.R. Hamblin and P. Mroz), Artech House, Boston, pp. 135–149.

22 Wilson, B.C. and Patterson, M.S. (2008) The physics, biophysics and technology of photodynamic therapy. *Phys. Med. Biol.*, **53** (9), R61–R109.

23 Henderson, B.W., Busch, T.M., Vaughan, L.A., Frawley, N.P., Babich, D., Sosa, T.A., Zollo, J.D., Dee, A.S., Cooper, M.T., Bellnier, D.A., Greco, W.R., and Oseroff, A.R. (2000) Photofrin photodynamic therapy can significantly deplete or preserve oxygenation in human basal cell carcinomas during treatment, depending on fluence rate. *Cancer Res.*, **60** (3), 525–529.

24 Henderson, B.W., Busch, T.M., and Snyder, J.W. (2006) Fluence rate as a modulator of PDT mechanisms. *Lasers Surg. Med.*, **38** (5), 489–493.

25 Fingar, V.H., Wieman, T.J., Wiehle, S.A., and Cerrito, P.B. (1992) The role of microvascular damage in photodynamic therapy: the effect of treatment on vessel constriction, permeability, and leukocyte adhesion. *Cancer Res.*, **52** (18), 4914–4921.

26 Fingar, V.H. (1996) Vascular effects of photodynamic therapy. *J. Clin. Laser Med. Surg.*, **14** (5), 323–328.

27 Engbrecht, B.W., Menon, C., Kachur, A.V., Hahn, S.M., and Fraker, D.L. (1999) Photofrin-mediated photodynamic therapy induces vascular occlusion and apoptosis in a human sarcoma xenograft model. *Cancer Res.*, **59** (17), 4334–4342.

28 Fingar, V.H., Taber, S.W., Haydon, P.S., Harrison, L.T., Kempf, S.J., and Wieman, T.J. (2000) Vascular damage after photodynamic therapy of solid tumors: a view and comparison of effect in preclinical and clinical models at the University of Louisville. *In Vivo*, **14** (1), 93–100.

29 Krammer, B. (2001) Vascular effects of photodynamic therapy. *Anticancer Res.*, **21** (6B), 4271–4277.

30 Chen, B., Pogue, B.W., Hoopes, P.J., and Hasan, T. (2006) Vascular and cellular targeting for photodynamic therapy. *Crit. Rev. Eukaryot. Gene Expr.*, **16** (4), 279–305.

31 Henderson, B.W. and Dougherty, T.J. (1992) How does photodynamic therapy work? *Photochem. Photobiol.*, **55** (1), 145–157.

32 van Geel, I., Oppelaar, H., Rijken, P.F., Bernsen, H.J., Hagemeier, N.E., van der Kogel, A.J., Hodgkiss, R.J., and Stewart, F.A. (1996) Vascular perfusion and hypoxic areas in RIF-1 tumours after photodynamic therapy. *Br. J. Cancer*, **73** (3), 288–293.

33 Bonnett, R. and Martinez, G. (2001) Photobleaching of sensitisers used in photodynamic therapy. *Tetrahedron*, **57** (47), 9513–9547.

34 Moan, J., Juzenas, P., and Bagdonas, S. (2000) Degradation and transformation of photosensitisers during light exposure, in *Recent Research Developments in Photochemistry and Photobiology* (ed. S.G. Pandalai), Transworld Research Network, Trivandrum, pp. 121–132.

35 Finlay, J.C., Mitra, S., Patterson, M.S., and Foster, T.H. (2004) Photobleaching kinetics of Photofrin *in vivo* and in multicell tumour spheroids indicate two simultaneous bleaching mechanisms. *Phys. Med. Biol.*, **49** (21), 4837–4860.

36 Dysart, J.S. and Patterson, M.S. (2005) Characterization of Photofrin photobleaching for singlet oxygen dose estimation during photodynamic therapy of MLL cells *in vitro*. *Phys. Med. Biol.*, **50** (11), 2597–2616.

37 Weston, M.A. and Patterson, M.S. (2009) Simple photodynamic therapy dose models fail to predict the survival of MLL cells after HPPH-PDT *in vitro*. *Photochem. Photobiol.*, **85** (3), 750–759.

38 Veenhuizen, R.B. and Stewart, F.A. (1995) The importance of fluence rate in photodynamic therapy: is there a parallel with ionizing radiation dose-rate effects? *Radiother. Oncol.*, **37** (2), 131–135.

39 Gibson, S.L., VanDerMeid, K.R., Murant, R.S., Raubertas, R.F., and Hilf, R. (1990) Effects of various photoradiation regimens on the antitumor efficacy of photodynamic therapy for R3230AC mammary carcinomas. *Cancer Res.*, **50** (22), 7236–7241.

40 Feins, R.H., Hilf, R., Ross, H., and Gibson, S.L. (1990) Photodynamic therapy for human malignant mesothelioma in the nude mouse. *J. Surg. Res.*, **49** (4), 311–314.

41 Foster, T.H., Hartley, D.F., Nichols, M.G., and Hilf, R. (1993) Fluence rate effects in photodynamic therapy of multicell tumor spheroids. *Cancer Res.*, **53** (6), 1249–1254.

42 Sitnik, T.M., Hampton, J.A., and Henderson, B.W. (1998) Reduction of tumour oxygenation during and after photodynamic therapy *in vivo*: effects of fluence rate. *Br. J. Cancer*, **77** (9), 1386–1394.

43 Foster, T.H., Murant, R.S., Bryant, R.G., Knox, R.S., Gibson, S.L., and Hilf, R. (1991) Oxygen consumption and diffusion effects in photodynamic therapy. *Radiat. Res.*, **126** (3), 296–303.

44 Henderson, B.W., Sitnik-Busch, T.M., and Vaughan, L.A. (1999) Potentiation of photodynamic therapy antitumor activity in mice by nitric oxide synthase inhibition is fluence rate dependent. *Photochem. Photobiol.*, **70** (1), 64–71.

45 Castano, A.P., Demidova, T.N., and Hamblin, M.R. (2005) Mechanisms in photodynamic therapy: Part three – Photosensitizer pharmacokinetics, biodistribution, tumor localization and modes of tumor destruction. *Photodiagn. Photodyn. Ther.*, **2** (2), 91–106.

46 Bisland, S.K., Lilge, L., Lin, A., Rusnov, R., and Wilson, B.C. (2004) Metronomic photodynamic therapy as a new paradigm for photodynamic therapy: rationale and preclinical evaluation of technical feasibility for treating malignant brain tumors. *Photochem. Photobiol.*, **80**, 22–30.

47 Curnow, A., Haller, J.C., and Bown, S.G. (2000) Oxygen monitoring during 5-aminolaevulinic acid induced photodynamic therapy in normal rat colon. Comparison of continuous and fractionated light regimes. *J. Photochem. Photobiol. B*, **58** (2–3), 149–155.

48 Curnow, A. and Bown, S.G. (2002) The role of reperfusion injury in photodynamic therapy with 5-aminolaevulinic acid – a study on normal rat colon. *Br. J. Cancer*, **86** (6), 989–992.

49 Anholt, H. and Moan, J. (1992) Fractionated treatment of CaD2 tumors in mice sensitized with aluminium phthalocyanine tetrasulfonate. *Cancer Lett.*, **61** (3), 263–267.

50 Vaupel, P., Kelleher, D.K., and Thews, O. (1998) Modulation of tumor oxygenation. *Int. J. Radiat. Oncol. Biol. Phys.*, **42** (4), 843–848.

51 Fingar, V.H., Wieman, T.J., Park, Y.J., and Henderson, B.W. (1992) Implications of a pre-existing tumor hypoxic fraction on photodynamic therapy. *J. Surg. Res.*, **53** (5), 524–528.

52 Ma, L.W., Moan, J., Steen, H.B., and Iani, V. (1995) Anti-tumour activity of photodynamic therapy in combination with mitomycin C in nude mice with human colon adenocarcinoma. *Br. J. Cancer*, **71** (5), 950–956.

53 Tomaselli, F., Maier, A., Sankin, O., Anegg, U., Stranzl, U., Pinter, H., Kapp, K., and Smolle-Juttner, F.M. (2001) Acute effects of combined photodynamic therapy and hyperbaric oxygenation in lung cancer – a clinical pilot study. *Lasers Surg. Med.*, **28** (5), 399–403.

54 Chen, Q., Huang, Z., Chen, H., Shapiro, H., Beckers, J., and Hetzel, F.W. (2002) Improvement of tumor response by manipulation of tumor oxygenation

during photodynamic therapy. *Photochem. Photobiol.*, **76** (2), 197–203.
55. Huang, Z., Chen, Q., Shakil, A., Chen, H., Beckers, J., Shapiro, H., and Hetzel, F.W. (2003) Hyperoxygenation enhances the tumor cell killing of Photofrin-mediated photodynamic therapy. *Photochem. Photobiol.*, **78** (5), 496–502.
56. Skyrme, R.J., French, A.J., Datta, S.N., Allman, R., Mason, M.D., and Matthews, P.N. (2005) A phase-1 study of sequential mitomycin C and 5-aminolaevulinic acid-mediated photodynamic therapy in recurrent superficial bladder carcinoma. *BJU Int.*, **95** (9), 1206–1210.
57. Berg, K., Bommer, J.C., Winkelman, J.W., and Moan, J. (1990) Cellular uptake and relative efficiency in cell inactivation by photoactivated sulfonated *meso*-tetraphenylporphines. *Photochem. Photobiol.*, **52** (4), 775–781.
58. Berg, K. and Moan, J. (1994) Lysosomes as photochemical targets. *Int. J. Cancer*, **59** (6), 814–822.
59. Moan, J., Berg, K., Anholt, H., and Madslien, K. (1994) Sulfonated aluminium phthalocyanines as sensitizers for photochemotherapy. Effects of small light doses on localization, dye fluorescence and photosensitivity in V79 cells. *Int. J. Cancer*, **58** (6), 865–870.
60. Berg, K., Madslien, K., Bommer, J.C., Oftebro, R., Winkelman, J.W., and Moan, J. (1991) Light induced relocalization of sulfonated *meso*-tetraphenylporphines in NHIK 3025 cells and effects of dose fractionation. *Photochem. Photobiol.*, **53** (2), 203–210.
61. Wood, S.R., Holroyd, J.A., and Brown, S.B. (1997) The subcellular localization of Zn (II) phthalocyanines and their redistribution on exposure to light. *Photochem. Photobiol.*, **65** (3), 397–402.
62. Nielsen, K.P., Juzeniene, A., Juzenas, P., Stamnes, K., Stamnes, J.J., and Moan, J. (2005) Choice of optimal wavelength for PDT: the significance of oxygen depletion. *Photochem. Photobiol.*, **81** (5), 1190–1194.
63. Juzeniene, A., Nielsen, K.P., and Moan, J. (2006) Biophysical aspects of photodynamic therapy. *J. Environ. Pathol. Toxicol. Oncol.*, **25** (1–2), 7–28.

22
Organic Light-Emitting Diodes (OLEDs): Next-Generation Photodynamic Therapy of Skin Cancer

Ifor D.W. Samuel, James Ferguson, and Andrew McNeill

22.1
Introduction

The distinctive properties of modern light sources have opened up many uses of light in the diagnosis and treatment of disease. These light sources are particularly relevant in the treatment of skin diseases because the skin is so readily accessible, and as a result light is now used in the treatment of more than 30 skin diseases. A very important and rapidly growing family of skin diseases is skin cancers. Many of these cancers (notably basal cell carcinomas and Bowen's disease) can be treated by photodynamic therapy (PDT), in which a combination of light and a drug is used to cause a photochemical reaction to destroy the tumor.

Over the past 20 years, there has been considerable growth in PDT arising from a combination of advances in light sources and photosensitizers [1–5]. It is the treatment of choice for many skin cancers and is now widely used across Europe and Australia. Although light is used in the treatment of many skin diseases, its use is heavily restricted by constraints imposed by the hospital-based light sources that are typically used [6]. In this chapter, we review the strengths and weaknesses of PDT as currently practiced, introduce organic light-emitting diodes (OLEDs) as novel light sources, and show how wearable light sources inspired by them are transforming PDT.

22.2
PDT of Skin Cancer: Current Approach

Current treatments for skin cancers include surgical excision, cryotherapy, radiotherapy, and various creams. Surgery is very effective but can leave significant scarring and carries a risk of infection. PDT offers a gentler alternative, which can give excellent cosmetic outcomes. This is important as lesions are caused by sunlight

Handbook of Biophotonics. Vol.2: Photonics for Health Care, First Edition. Edited by Jürgen Popp,
Valery V. Tuchin, Arthur Chiou, and Stefan Heinemann.
© 2012 Wiley-VCH Verlag GmbH & Co. KGaA. Published 2012 by Wiley-VCH Verlag GmbH & Co. KGaA.

and therefore occur in places that can be seen. The current treatment cycle typically consists of an initial visit to a family doctor, followed by a visit to hospital for the lesion to be identified, two return visits for PDT, and a further return visit for follow-up.

On each of the visits for PDT, the lesion is initially abraded by light curettage (scraping) and a cream of a precursor to the photosensitizer [δ-aminolevulinic acid (ALA) or its methyl analog, Metvix] is applied. This precursor is metabolized to the photosensitizer protoporphyrin IX over the next few hours. Accumulation of protoporphyrin IX occurs preferentially in tumors and so strong selectivity for the tumor is achieved. Four hours after the cream has been applied, the lesion is illuminated under intense red light ($65\,\text{mW}\,\text{cm}^{-2}$) for 20 min, leading to a dose of $75\,\text{J}\,\text{cm}^{-2}$. The treatment is often painful and a variety of measures are used to manage the pain, including blowing cool air over the treatment area and in some cases local anesthetic. The procedure is labor intensive as the patient requires supervision to provide appropriate pain relief. The treatment is repeated 1 month later and the two sessions are normally sufficient to kill the tumor.

A variety of light sources are commonly used. These include solid-state diode lasers, filtered lamps, and large arrays of inorganic light-emitting diodes (LEDs). These light sources are expensive and cumbersome, requiring the patient to come to the hospital for treatment, and to lie still under the light source. The number of patients who can be treated is limited by the availability of light sources and supervising staff. From the patient's point of view, the treatment involves a considerable waiting time in hospital. The pain in conventional PDT is also a major concern as median pain scores are 6 on a scale where 10 is extremely severe pain [7]. Even so, the cosmetic results can be impressive.

22.3
Advances in Light Sources: Organic Light-Emitting Diodes

Organic semiconductors are plastic-like materials with semiconducting electronic properties. They combine favorable properties of plastics such as tunability by changing the structure, with scope for simple fabrication and manufacture, so enabling a new generation of semiconductor devices to be made. Many of these devices are related to their inorganic counterparts, and yet with distinctive aspects. For example, these materials can be dissolved and so deposited by simple processes such as ink-jet printing to make devices such as LEDs and transistors. The simplest organic light-emitting diode (OLED) structure consists of one or more ($\sim$100 nm) thin organic layers between contacts. When a voltage is applied, charges are injected and opposite charges can meet up, leading to emission of light.

The materials are amorphous so epitaxy is not required and different colors can be deposited side-by-side. As a result, most work on OLEDs has been directed towards making displays. Many mobile phones now contain OLED displays. Larger and even flexible displays are under development.

22.4 Transforming PDT with Wearable Light Sources

In this section, we show how OLEDs can be used for PDT and could lead to a transformation in the way in which PDT is delivered to patients.

22.4.1 Developing OLEDs for PDT

We recognized an opportunity to use OLEDs to perform PDT in a completely different way. OLEDs could be used to make a small, compact, wearable light source that would enable patients to undergo ambulatory treatment, that is, treatment in which they could move around (Figure 22.1). Displays are orders of magnitude dimmer than PDT light sources, so do not at first sight appear to be obvious candidates for PDT. However, as PDT works by a photochemical reaction, a lower intensity can be used for a longer time to achieve the desired result.

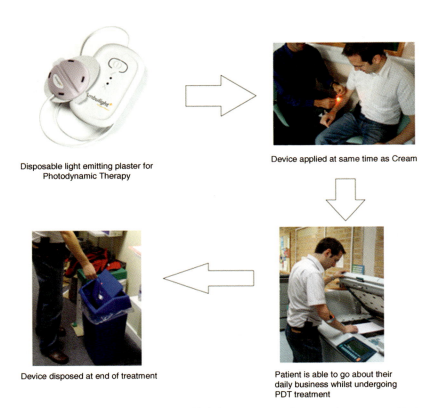

Figure 22.1 Ambulatory PDT. A wearable light source is used to treat skin cancer and is thrown away afterwards.

The demonstration of OLED-PDT required the development of suitable OLEDs. The requirements for OLED-PDT are very different from those for OLED displays. In OLED displays, important issues are achieving pure primary colors, making pixels, and ensuring a sufficiently long lifetime. For OLED-PDT, we envisage a disposable light source, so lifetime is not an issue. However, OLED-PDT requires uniform orange–red illumination over the area to be treated. Many lesions to be treated are small, so a light-emitting area 2 cm in diameter was sufficient for demonstration purposes. The device needs to run at much higher brightness than for a display, and in our initial studies we used OLEDs emitting 5 mW cm^{-2} in the orange–red region of the spectrum. An OLED is a very unusual light source because it emits light from an area, rather than a point. It also has a very distinctive shape – it is very thin. The device consists of two parts – a light-emitting head connected to a power supply containing batteries, that can be worn in a manner analogous to an iPod. Hence the patient can walk around or work during treatment. The light-emitting part of the device is attached to the area to be treated by adhesive tape, much like a sticking plaster.

22.4.2
Testing OLEDs for PDT

After obtaining ethical approval and initial studies on ourselves, we performed a pilot study at Ninewells Hospital, Dundee. The study involved following the same procedure as for conventional PDT, but applying light from an OLED for 3 hours instead of using one of the hospital fixed light sources. Patients could therefore move around during treatment.

Twelve patients with basal cell carcinoma or Bowen's disease were studied and received two treatments of OLED-PDT 1 month apart. At follow-up 6 months after treatment, seven of the 12 patients remained free from their lesions [7]. For the sample size, this is consistent with the normal (70–80%) success rate of PDT. In most cases, the cause of "failure" was a small region at the edge of the lesion in a crescent moon shape. The trial used OLED light sources that were 2 cm in diameter and treated lesions of up to 2 cm in size. It is therefore likely that the failures resulted from a slight misalignment of the light source over the lesion. This has been addressed in subsequent work by making the light source larger so that there is a significant margin of illumination around the lesion, and by introducing an alignment template for the clinician. These two measures now ensure that the entire lesion is treated evenly.

During the trial, patients were asked to rate the pain they felt on a scale from 0 to 10. For OLED-PDT, the median pain score was 1, whereas for conventional PDT the median pain score was 6. This is a remarkable difference and a major advantage of OLED-PDT. It is all the more remarkable as no pain relief was used with OLED-PDT, whereas a Cynosure (cool air) fan was used for all patients receiving conventional PDT. The reason for the reduced pain is that a much lower intensity of light is used for a much longer time.

22.4.3
Ambulatory PDT: a New Paradigm

The results of the pilot trial are extremely encouraging as they demonstrate ambulatory PDT with much reduced pain. There are many implications for treatment. One is that it is much more comfortable and convenient for the patient as they can move around during treatment and experience much less pain. Another is that the number of patients who can be treated is larger as there can be many of the light sources. It is also possible for the patient to return to work or home after the light source has been applied

The real scope for transformation of PDT comes from the possibility of avoiding a hospital visit altogether. Family doctors (general practitioners) could potentially administer the treatment, thereby removing the need for expensive and inconvenient hospital visits.

22.4.4
Commercialization of Ambulatory PDT

We are keen that the advantages of ambulatory PDT should be widely available. The path from concept to approved product is long and expensive and ambulatory PDT is now being commercialized via a spin-off company, Ambicare Health Ltd. The company raised money to enable a product to be designed that meets the requirements for medical CE marking. The medical CE mark has recently been awarded, meaning that the device can now be sold in the European Union. As OLED manufacture is still developing, the first product, Ambulight PDT (Figure 22.2), uses inorganic LEDs in combination with a diffuser as the light source. Future products

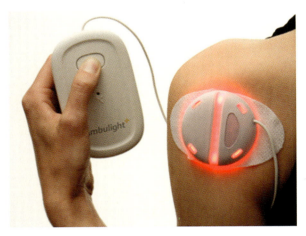

Figure 22.2 Ambulight PDT – a wearable light source for skin cancer treatment consisting of a self-adhesive light source and a separate battery-powered control unit.

may use OLEDs because they are area light sources by nature, matching the fact that a lesion is an area to be treated, and they are thin and can be flexible.

22.5
Conclusion

Advances in phototherapy and PDT have been closely linked to advances in light sources. PDT is an attractive treatment for many skin cancers, but its current use is limited by the need for cumbersome, specialized light sources, hospital visits, and many patients finding it painful. We have shown that OLEDs enable compact and convenient *wearable* light sources for PDT to be made, thereby making *ambulatory* treatment possible. This advance has the potential to transform PDT by enabling far more doctors to administer it, far more patients to receive it, and ultimately enabling treatment to take place without the need for hospital visits.

References

1 Babilas, P., Schreml, S., Landthaler, M., and Szeimies, R.M. (2010) *Photodermatol. Photoimmunol. Photomed*, **26**, 118.
2 Braathen, L.R., Szeimies, R.M. *et al.* (2007) *J. Am. Acad. Dermatol.*, **56**, 125.
3 Dougherty, T.J. (2002) *J. Clin. Laser Med. Surg.*, **20**, 3.
4 Ibbotson, S.H. (2010) *Photodiagn. Photodyn. Ther.*, **7**, 16.
5 Morton, C.A., McKenna, K.E., Rhodes, L.E., and The British Association of Dermatologists (2008) *Br. J. Dermatol.*, **159**, 1245.
6 Moseley, H., Allen, J.W. *et al.* (2006) *Br. J. Dermatol.*, **154**, 747.
7 Attili, S.K., Lesar, A. *et al.* (2009) *Br. J. Dermatol.*, **161**, 170.

23
Nanoparticles for Photodynamic Therapy
Wen-Tyng Li

23.1
Introduction

Photodynamic therapy (PDT) can be a curative modality for early or localized cancers and can improve quality of life and lengthen survival for advanced diseases. Owing to its minimal invasiveness and selective therapeutic effect, the applications of PDT have been expanded to treatment of precancerous conditions, infectious diseases, and age-related macular degeneration. Their advantages include cost effectiveness, highly localized treatments, sparing of extracellular matrix that allows regeneration of normal tissue, repetition of therapy without accumulation of toxicity, possibility of combining with chemotherapy, which leads to higher cure rates, and induction of immunity, which may contribute to long-term tumor control and suitability for outpatient therapy [1].

Depending on the part of the body being treated, the photosensitizer is administered by intravenous injection or local application. Light is applied to the area to be treated after the drug has been absorbed by the pathologic tissue. The photosensitizer activated by light forms reactive oxygen species (ROS) that kill the cancer cells directly by way of type I and type II reactions. Type I reaction involves the production of radicals resulting from the activated sensitizer reacting with plasma membrane or other intracellular molecules. Type II reaction involves generation of singlet oxygen ($^1O_2$) upon energy transfer from the activated sensitizer to oxygen. PDT may also work by destroying tumor-associated vasculature, leading to tumor infarction or by alerting the immune system to attack the cancer [2]. Responses to photodynamic treatment are dependent on the type of photosensitizer used, its extracellular and intracellular localization, the total dose administered, the total light exposure dose, light fluence rate, the time between the administration of the drug and light exposure, the oxygenation status of the tissue, and the type of cells involved [3].

Cancer cells respond to photodynamic damage by eliciting a rescue response and/or by undergoing cell death. Rescue responses often involve changes in gene and protein expression of stress proteins, allowing the cells to cope with the damage. PDT may result in cell death via apoptosis, necrosis, or autophagy, which can be affected by the cell type, the nature of photosensitizers, the incubation protocol, and the light dose [4, 5]. The physical and chemical natures of the photosensitizers, such as hydrophobicity and charge, are of great importance in determining their subcellular localization. The sensitivity of intracellular components to photooxidation via photodynamic action plays an important role in photocytotoxicity. Photosensitizers can localize in mitochondria, lysosomes, endoplasmic reticulum, Golgi apparatus, and plasma membranes, which leads to different signaling pathways involved in cell death.

The essential goal of PDT is to induce efficient damage to tumor tissue while sparing the surrounding tissue. A selective therapeutic effect of PDT is achieved from preferential accumulation of photosensitizers and from irradiation of the target tissue. Some photosensitizers can reach higher concentrations in tumor tissue than in surrounding healthy tissue due to the abnormal physiology of tumors, such as poor lymphatic drainage, leaky vasculature, decreased pH, increased number of receptors for low-density lipoprotein, and abnormal stromal composition [6]. However, most photosensitizers are hydrophobic and tend to aggregate easily in aqueous media, which causes a decrease in its quantum yield and problems for intravenous administration. Although many photosensitizers have been developed, few have made it to clinical trials owing to factors such as poor selectivity in terms of target tissue and healthy tissue, low extinction coefficients, absorption spectra at relatively short wavelengths, and high accumulation rates in skin [7].

Increasing the selective accumulation of the photosensitizers within the tumor tissue allows a lower effective dose of the PDT drug. One can take advantage of the intrinsic features of cancer cells, such as specific surface antigens, low-density lipoprotein (LDL) receptor, and oxidation state [8]. To enhance PDT efficacy, a photosensitizer can be bound to ligands such as monoclonal antibodies or LDL, or can be delivered via carrier systems such as liposomes and micelles [9, 10]. The use of nanoparticles to improve the efficiency of PDT is a promising approach because (1) their large surface area can be modified with functional groups for additional biochemical properties, (2) they have large distribution volumes and are generally taken up efficiently by cells, (3) controlled release of drugs is possible, (4) many synthetic strategies allow transportation of hydrophobic drugs in blood, and (5) preferential accumulation in the solid tumor site is easy due to the enhanced permeability and retention effect [11]. To satisfy the requirements for an ideal PDT drug, some nanoparticles comprising inorganic oxide, metallic, ceramic, and biodegradable polymer nanomaterials have emerged with potential application for PDT. Nanoparticles for PDT were classified into active and passive nanoparticles according to the functional roles in the enhancement of PDT efficacy (Table 23.1) [11]. The former was sub-classified by mechanism of activation into photosensitizer nanoparticles, upconversion nanoparticles, and self-lighting nanoparticles. The latter was sub-classified by material composition into biodegradable and nonbiodegradable nanoparticles. Here, approaches to improving the efficacy of PDT for cancers by using active and passive nanoparticles are described.

Table 23.1 Classification of nanoparticles used for PDT.

Category	Material	Mechanism
Active nanoparticles		
Photosensitizer	CdSe/CdS/ZnS	Indirect excitation of photosensitizer through a FRET mechanism from the nanoparticle to the photosensitizer
	Fullerene	Nanoparticles transfer energy from incident light directly to surrounding oxygen
	SWNTs	CNTs act as quenchers to control and regulate $^1O_2$ generation
Upconverting	$NaYF_4:Yb^{3+},Er^{3+}$	Low-energy light is transduced to higher energy emissions to activate associated photosensitizer
	$NaYF_4:Yb^{3+},Tm^{3+}$	
Self-lighting	$BaFBr:Eu^{2+},Mn^{2+}$	Scintillation on excitation by X-rays activates attached photosensitizer
	$LaF_3:Tb^{3+}$	
Passive nanoparticles		
Biodegradable	Alginate, chitosan, cyclodextrin, albumin	Drug is delivered by micelles, dendrimers, liposomes, or polymeric nanoparticles
	PLA, PLGA, PVAL	Controlled release of encapsulated photosensitizer through biodegradation
Nonbiodegradable	Polyacrylamide	Two-photon dye is encapsulated by microemulsion
	Silica	Photosensitizer is adsorbed/covalently bonded to porous shell
	Gold	Nanoparticles act as pure carrier, no release
	Magnetic iron oxide	Drug is carried directly or co-encapsulated in micelle or polymeric nanoparticle Targeted delivery is achieved by external magnetic field

23.2
Active Nanoparticles for PDT

Regarding active roles in PDT, nanoparticles may act like catalysts to produce free radicals from dissolved oxygen, and serve as ROS modulators or light sources for activating photosensitizers. Nanoparticles such as fullerenes can be generated upon light irradiation, hence they can serve as photosensitizers themselves. Carbon nanotubes (CNTs) are fluorescence quenchers, which can be engineered to modulate $^1O_2$ generation. Nanoparticles such as quantum dots (QDs), upconversion nanoparticles, and scintillation nanoparticles can be used as light sources to activate photosensitizers due to their fluorescence emission properties. To serve as a PDT light source, the nanoparticle selected for activating a photosensitizer must meet the following criteria: (1) its emission spectrum must match the photosensitizer's absorption spectrum, (2) it must have high luminescence efficiency, (3) it must be easily linked with photosensitizers, and (4) it must be nontoxic, water soluble, and stable in biological environments.

Fullerenes (C_{60}) are composed of 60 carbon atoms arranged in a soccer-ball structure. They absorb visible light, have a high triplet yield, and can generate ROS upon illumination due to their extended π-conjugation [12]. Functionalized fullerene has recently become an attractive tool for photodynamic applications. Depending on the functional groups introduced into the molecule, fullerenes can effectively photoinactivate cancer cells. The ability of fullerenes to generate reactive oxygen species and catalyze the phototoxic reaction after photoexcitation was first demonstrated on thymocytes and Ehrlich ascites cells [13]. One of the major problems in using fullerenes in biological systems is their poor solubility in aqueous solution, which may be overcome by encapsulation in special carriers, suspension with the help of co-solvents, introduction of hydrophilic appendages, or reduction to a water-soluble anion [14]. The photodynamic activity of the functionalized fullerenes as photosensitizers against mouse cancer cell lines was inversely proportional to the degree of substitution of the fullerene nucleus for both the neutral and the cationic groups [15]. Chemical modification of fullerene with poly(ethylene glycol) (PEG) not only made it soluble in water but also exhibited preferential accumulation and prolonged retention in the tumor tissue. PDT with fullerenes has been demonstrated in animal tumor models [16]. The major disadvantage of fullerenes is their optical absorption properties. The absorption spectrum of fullerenes is highest in the UVA and blue regions of the spectrum, where the tissue penetration depth of illumination is shortest due to a combination of light absorption by cellular chromophores and light scattering by cellular structures [12]. It is important to develop fullerenes that absorb red or near-infrared light for clinical application in PDT.

CNTs, which belong to the family of fullerenes, are made up of thin sheets of benzene ring carbons rolled up into the shape of a seamless tubular structure. Based on their structure, CNTs can be classified into single-walled nanotubes (SWNTs) and multi-walled nanotubes (MWNTs). Owing to their ultra-high surface area, molecules such as drugs, peptides, and nucleic acids can be integrated into their walls and tips, which enable them to cross the mammalian cell membrane by endocytosis or to recognize cancer-specific receptors on the cell surface. SWNTs are known to be efficient quenchers of fluorescence probes. A novel molecular complex of a photosensitizer (chlorin e6), an ssDNA aptamer, and SWNTs was engineered to control and regulate $^1O_2$ generation. Target binding with ssDNA aptamers to specific proteins could disturb the interaction between aptamers and SWNTs, resulting in the restoration of $^1O_2$ generation, whereby more selective PDT could be guaranteed [17]. Currently, the knowledge of CNT-based PDT is limited. Further research on the pharmacologic and toxicologic properties of CNTs is required before they could be recommended for routine clinical use [18].

QDs are semiconductor nanocrystals with high quantum yields, high photostability, and fluorescent emission properties which can be tuned from the ultraviolet to the infrared region by changing their size and composition. With a functionalized surface coating, they can be made water soluble and targeted to specific pathologic areas. Cadmium selenide (CdSe) is commonly contained in semiconductor QDs.

The emission of $^1O_2$ at 1270 nm generated by the energy transfer from CdSe QDs alone in toluene under excitation at 488 nm suggested the potential of QDs as possible therapeutic agents for PDT [19]. However, the yield of $^1O_2$ generation (~5%) of QDs was too low for practical applications. Because QDs exhibit broad absorption spectra, the combination of QDs with photosensitizers permits the use of variable excitation wavelengths to activate the photosensitizer. The interaction between CdSe QDs and photosensitizers was first studied using silicon phthalocyanine (Pc). The study revealed that the QDs could be used to sensitize the PDT agent through a Förster resonance energy transfer (FRET) mechanism, or interact directly with molecular oxygen via a triplet energy-transfer (TET) process, which resulted in the generation of ROS that can be used for PDT [20]. A study of the effect of linker chain length on the energy transfer from CdSe QDs to silicon Pc found that the energy-transfer efficiency increased as the chain length increased with maximum efficiency with seven bonds, following a non-Förster-type resonance energy-transfer behavior [21]. Applications of these conjugates may be limited due to the inherent cytotoxicity of core-only Cd-based QDs and the potential instability of charged-assembled complexes in the cellular environment. To enhance its quantum yield, CdSe–core–shell QDs were developed. ZnS shells were also shown to reduce the cytotoxicity of QDs greatly. Covalent conjugation of photosensitizers such as Rose Bengal (RB) and chlorin e6 to CdSe/CdS/ZnS QDs through phytochelatin-related peptides preserved the colloidal and photophysical properties of both the QDs and photosensitizers. Generation of $^1O_2$ was achieved via indirect excitation through FRET from the QDs to the photosensitizer [22]. The application of QDs for PDT has been demonstrated *in vitro*. However, the toxicity of QDs to living cells may limit its application *in vivo* due to release of Cd^{2+} with the protective shell deteriorating after prolonged circulation in the body. Its therapeutic effect for PDT for cancer remains to be determined *in vivo*.

Upconversion nanoparticles are modified nanocomposites which generate higher energy light from lower energy radiation. They are sometimes referred to as upconverting phosphors (UCPs), which are defined as lanthanide-containing, submicrometer-sized ceramic particles that can absorb near-infrared (NIR) or infrared light and emit visible light. $NaYF_4$ is often used as the core material of choice for UCPs. A versatile UCP, $NaYF_4:Yb^{3+},Er^{3+}$ nanoparticles coated with a porous, thin layer of silica doped with merocyanine-540 photosensitizer and functionalized with a tumor-targeting antibody, was reported to be a specific PDT agent against MCF-7/AZ breast cancer cells [23]. Another UCP, $NaYF_4$ nanocrystals coated with mesoporous silica-loaded zinc phthalocyanine (ZnPc), was able to convert NIR light to visible light upon excitation by an NIR laser, which further activated the photosensitizer to release $^1O_2$ to kill cancer cells. The photosensitizers loaded into the porous silica shell of the nanoparticles were not released from the silica while they continuously produced $^1O_2$ upon NIR irradiation [24]. To prevent clearance by the reticuloendothelial system, a three-layer UCP composed of an $NaYF_4:Yb^{3+},Er^{3+}$ nanoparticle interior, a coating of porphyrin photosensitizer, and a biocompatible PEG outer layer was constructed and millimolar amounts of $^1O_2$ was generated at NIR intensities far less than in two-photon techniques [25]. In addition to the resistance of UCPs to photobleaching, using UCPs in PDT has other advantages, for example, the excitatory NIR light allows deeper penetration into the soft tissue compared with visible light.

To enhance the effect of PDT without using an external light source, self-lighting PDT was proposed by using scintillation luminescent nanoparticle-attached photosensitizers. Scintillation nanoparticles such as $BaFBr:Eu^{2+},Mn^{2+}$ and $LaF_3:Tb^{3+}$ will emit luminescence to activate the photosensitizers upon exposure to ionizing irradiation; as a result, $^1O_2$ is produced to enhance the killing of cancer cells. In a pilot study, the energy transfer from $LaF_3:Tb^{3+}$ nanoparticles to *meso*-tetra(4-carboxyphenyl)porphyrin (mTCP) was demonstrated by fluorescence quenching techniques in water-soluble $LaF_3:Tb^{3+}$–mTCP nanoparticle conjugates after exposure to X-irradiation [26]. Persistent-luminescence nanoparticles consisting of luminescent materials with long decay lifetimes will further benefit the PDT method for deep cancer treatment by decreasing the ionization dose and prolonged photosensitizer excitation [27]. However, direct application in biological systems has not yet been reported.

23.3
Nanoparticles as Photosensitizer Carriers for PDT

Many clinically approved photosensitizers often suffer from poor bioavailability and unfavorable biodistribution. The enhanced permeability and retention (EPR) effect caused by the abnormal organization of the tumor neovasculature facilitates both diffusion and retention of photosensitizer carriers in tumors. Nanoparticles can serve as carrier platforms by taking advantage of the EPR effect to improve further the cellular uptake, biodistribution, bioavailability, and pharmacokinetics of the systemic delivery of the photosensitizers [28]. Biodegradable nanoparticles are made of polymers that are often enzymatically hydrolyzed in a biological environment and hence release the photosensitizers. Nonbiodegradable nanoparticles are used to protect the photosensitizers from the fluctuations of the environment, in which the release of the photosensitizers from the nanoparticle carriers is not necessary but oxygen diffusion freely in and out of the nanoparticles is essential [7].

23.3.1
Biodegradable Nanoparticles for PDT

Biodegradable polymer-based nanoparticles have attracted a lot of attention due to their advantages in the possibility of controlling drug release, versatility in material manufacturing processes, and high drug loading. The surface properties, morphologies, and composition of the polymers can be optimized to achieve the desired biocompatibility, controlled degradation rate, and drug release kinetics. The chemical composition and architecture of polymers can be tailored to accommodate photosensitizers with varying degrees of hydrophobicity, molecular weight, charge, and pH. Modifying the surface of nanoparticles with specific moieties also allows for active targeting to specific tissue sites [29].

Photosensitizers may be delivered by using micelles, liposomes, dendrimers, or nanoparticles. Micelles are spherical macromolecular complexes that form spontaneously when amphiphilic copolymers are mixed in an aqueous environment above the critical micelle concentration. The hydrophobic moieties of the polymer aggregate to form the micelle core, while the hydrophilic group forms a hydrated shell around the hydrophobic core. Polymeric micelles have been used to carry hydrophobic photosensitizers, which are physically entrapped in and/or covalently bound to the hydrophobic core. Polymers used for photosensitizer encapsulation include pluronics, PEG–lipid conjugates, pH-sensitive poly(N-isopropylacrylamide)-based micelles, and polyion complex micelles [30]. Liposomes are small unilamellar vesicles composed of a phospholipid bilayer membrane enclosing an inner aqueous environment. The phospholipid bilayer of liposomes can dissolve hydrophobic photosensitizers. They can also be used to encapsulate water-soluble molecules. The properties of liposomes can be altered by addition of different molecules to modify their surface to enhance cancer cell targeting specificity and the ability to control the release of photosensitizers [28]. Dendrimers are highly branched macromolecules composed of repetitive units branched on a multivalent core molecule. The photosensitizer can be attached at the periphery of the dendrimer branches or be encapsulated in the core of a dendrimer. The properties of dendrimers are dominated by their functional groups. Nanoparticles are made of natural or synthetic polymers. Photosensitizers can be adsorbed on the porous matrix of polymeric nanoparticles composed of natural or synthetic degradable polymers [28].

Natural degradable polymers including polysaccharides and proteins have been used in drug delivery systems. Commonly used natural biodegradable polymers are alginate, chitosan, dextran, albumin, gelatin, collagen, and agars. Modifications can be made easily to naturally occurring polymers to alter their physico-chemical properties. Alginates, polysaccharides prepared from brown algae, have been used extensively in drug delivery. A novel surfactant–polymer nanoparticle, formulated using the anionic surfactant dioctyl sodium sulfosuccinate [Aerosol-OT, (AOT)] and sodium alginate, was reported to be an efficient encapsulation and sustained delivery system for polar and weak bases such as methylene blue and doxorubicin. Increased ROS production and a higher incidence of necrosis were observed in breast cancer cell lines, MCF-7 and 4T1, following PDT using AOT–alginate nanoparticle-encapsulated methylene blue [31]. Using AOT–alginate nanoparticles for the simultaneous delivery of doxorubicin and methylene blue led to significant inhibition of drug-resistant tumor cell growth and improved survival of JC (mammary adenocarcinoma) tumor-bearing mice [32].

Chitosan is a modified natural carbohydrate polymer prepared by the partial N-deacetylation of the crustacean-derived natural biopolymer chitin. Chitosan-based nanoparticles can serve as drug carriers in cancer therapy because of their biocompatibility and biodegradability. Of chitosan derivatives with improved hydrophilicity, glycol chitosan is emerging as a novel carrier of drugs because of its solubility in water and biocompatibility. Hydrophobically modified glycol chitosans (HGCs) prepared by covalent attachment of 5β-cholanic acid to glycol chitosan can self-assemble into nanocarriers under aqueous conditions, that can efficiently take up various hydrophobic anticancer drugs into their hydrophobic inner cores. The amphiphilic chitosan-based nanoparticles are covered with biocompatible and biodegradable glycol chitosan shells and exhibit enhanced tumor target specificity and a prolonged blood circulation time. The water-insoluble photosensitizer protoporphyrin IX (PpIX) was efficiently encapsulated in chitosan-based nanoparticles (drug-loading efficiency >90%). The PpIX-nanoparticles exhibited phototoxicity against squamous cell carcinoma (SCC-7) cells and enhanced tumor specificity in SCC-7 tumor-bearing mice compared with free PpIX [33].

Cyclodextrins (CDs), produced from starch by means of enzymatic conversion, have wide application in the pharmaceutical industry. They are made of sugar molecules that are connected in a ring and have the ability to form water-soluble complexes encapsulating porphyrins in aqueous solution [34]. Heterotopic colloidal nanoparticles of a cationic amphiphilic cyclodextrin (CD) encapsulating anionic 5,10,15,20-tetrakis(4-sulfonatophenyl)-21H,23H-porphyrin (TPPS) were found to preserve the photodynamic properties of the entrapped photoactive agent in a range of TPPS:CD molar ratios between 1:10 and 1:50, with photodynamic efficacy proven by *in vitro* studies on tumor HeLa cells [35].

Albumin is a plasma protein responsible for the colloid osmotic pressure of the blood. It is nonantigenic and biodegradable, binds a great number of therapeutic drugs, and has been extensively investigated as a drug carrier. The average half-life of human serum albumin (HSA) is 19 days [36]. The photosensitizer pheophorbide was loaded on HSA nanoparticles with different cross-linked ratios by noncovalent adsorption. The $^1O_2$ quantum yield of pheophorbide-loaded HSA nanoparticles was low due to intramolecular interactions despite the fact that their final phototoxicity was at the same level as induced by free pheophorbide [37]. The results suggest that the photosensitizer loading strategies, biochemical stability, and *in vivo* PDT effect of pheophorbide-loaded HSA nanoparticles still need to be optimized.

Synthetic polymers such as aliphatic polylactide (PLA), polyglycolide (PGA), and their copolymer poly(D,L-lactide-co-glycolide) (PLGA) are often used as drug delivery carriers due to their favorable properties such as good biocompatibility, biodegradability, bioresorbability, and mechanical strength. For example, the photocytotoxicity was found to be more pronounced when EMT-6 mammary tumor cells were treated with PLGA (50:50) nanoparticles incorporated with *meso*-tetra(hydroxyphenyl)porphyrin (mTHPP), which allowed low drug doses and short drug administration–irradiation intervals [38]. To reduce the aggregation of the hydrophobic photosensitizer bacteriochlorophyll-*a* in solution, it was encapsulated into polymeric PLGA nanoparticles. Scanning electron microscopy analysis revealed that these nanoparticles were phagocytosed by P388-D1 cells within a 2 h incubation time [39]. In another study, internalization of PLGA encapsulated *meso*-tetraphenylporpholactol (mTPPL) nanoparticles resulted in 95% cell death of 9L glioblastoma, B16 melanoma, and BT breast carcinoma cells, respectively. *In vivo* experiments showed complete eradication of tumors in nude mice [40]. PLGA nanoparticles were also demonstrated to be a successful delivery system for improving the photodynamic activity of ZnPc in the target tissue in tumor-bearing mice [41]. In a comparative study, both PLGA and PLA were used to encapsulate hypericin. Hypericin-loaded

PLA nanoparticles exhibited a higher entrapment efficiency and photoactivity than PLGA nanoparticles and free drug on the NuTu-19 ovarian cancer cell model. Improved selective accumulation of hypericin in ovarian micrometastases was observed when hypericin-loaded PLA nanoparticles were injected intravenously into Fischer 344 rats bearing ovarian tumors [42, 43].

Disruption of the neovasculature is thought to play an important role in the eradication of tumors by PDT. An *in vitro* model cannot give any information regarding vascular photothrombosis, hence the well-vascularized chorioallantoic membrane (CAM) of the developing chick embryo was used to study the *in vivo* photodynamic activity of photosensitizer-loaded nanoparticles. The photosensitizer mTHPP was found to remain longer in the vascular compartment of CAM when incorporated into PLGA nanoparticles. The use of poly(vinyl alcohol) (PVAL) as stabilizing agent within mTHPP-loaded nanoparticles helped to control the mTHPP release and the photodynamic activity in the CAM model [44]. Analysis of the photodynamic activities of photosensitizers with different hydrophobicities entrapped in poly-D,L-lactic acid nanoparticles showed that the photosensitizer was more effectively entrapped in the polymeric matrix when its degree of lipophilicity increased. *meso*-Tetra-(4-carboxyphenyl)porphyrin (mTCPP) and chlorin e_6 are less hydrophobic than meso-tetraphenylporphyrin (mTPP) and pheophorbide-*a*. The nanoparticles loaded with the most lipophilic molecule (mTPP) induced the highest photothrombic efficiency in the CAM model compared with the other molecules tested [45].

Dendrimers are recognized as the most versatile, compositionally, and structurally controlled nanoscale building blocks. Thus, dendrimers have been investigated in the biomedical field for drug delivery. In 2001, a series of novel 5-aminolevulinate (ALA)-containing dendrimers were synthesized using a convergent growth approach [46]. The ability of the dendrimer conjugated with 18 ALA residues to deliver and release 5-ALA intracellularly for metabolism to the photosensitizer, PpIX, was demonstrated in tumorigenic PAM 212 keratinocytes and human epidermoid carcinoma A431 cells [47]. The dendrimer induced sustained porphyrin production for over 24 h and basal values were not reached until 48 h after systemic intraperitoneal administration, whereas the porphyrin kinetics from ALA exhibited an early peak between 3 and 4 h in most tissues in a murine tumor model [48]. Another group has developed two PEG-attached dendrimers derived from poly(amidoamine) (PAMAM) and poly(propylenimine) (PPI) dendrimers to encapsulate photosensitizers, RB and PpIX. Their results showed that PEG–PPI dendrimer held RB and PpIX in a more stable manner than PEG–PAMAM dendrimer because of their inner hydrophobicity. The complex of PpIX with PEG–PPI exhibited efficient photocytotoxicity in HeLa cells, compared with free PpIX [49]. A novel dendrimer phthalocyanine (DPc)-encapsulated polymeric micelle (DPc/m) was developed recently. The DPc/m might accumulate in the endo-/lysosomes. Both *in vitro* and *in vivo* PDT efficacy were observed in human lung adenocarcinoma A549 cells. Furthermore, DPc/m-treated mice bearing A549 cells did not show skin phototoxicity [50]. The development of dendrimers as nanocarriers with smart functions is expected to be a key to advance further the clinical applications of PDT.

Although synthetic polymers are preferable for use in drug delivery systems due to the ability to tailor their mechanical properties and degradation kinetics to suit various applications, natural polymers remain attractive because they are readily available, relatively inexpensive, and capable of a multitude of chemical modifications. The use of natural biodegradable polymers to deliver drugs will continue to be an area of active research despite the advances in synthetic biodegradable polymers.

23.3.2
Nonbiodegradable Nanoparticles for PDT

Nonbiodegradable nanoparticles play a different role in PDT due to their inability to degrade and release drugs in a controllable fashion. They are not destroyed by the treatment process; therefore, they can be used repeatedly with adequate activation. Nonbiodegradable nanoparticles have several advantages over biodegradable polymeric nanoparticles: (1) their particle size, shape, porosity, and monodispersibility can be easily controlled during the preparation process; (2) some of them are made of inert materials which are stable to environmental fluctuations; (3) they are not subject to microbial attack; and (4) exquisite control of the pore size allows oxygen diffusion in and out of the particles but not for the drug to escape [7, 11]. Similarly to biodegradable nanoparticles, nondegradable nanoparticles can serve as multifunctional platforms for drug delivery.

Polyacrylamide polymers can be used for the synthesis of nondegradable nanoparticles. The photosensitizer can be embedded in the nonporous core of polyacrylamide or sol–gel silica nanoparticles. A novel dynamic nanoplatform composed of polyacrylamide was designed based on a stable photosensitizer-loaded formulation and the concept of delivering $^1O_2$ from the outside of the tumor cells. The encapsulated methylene blue was able to generate sufficient $^1O_2$ upon light irradiation that could diffuse out of the matrix, reach the adjacent cell membranes, and kill rat C6 glioma tumor cells [51]. Furthermore, a polyacrylamide matrix could prevent the embedded methylene blue from being reduced by diaphorase enzymes, thereby retaining the photoactive form of methylene blue for efficient PDT treatment [52].

Most nondegradable nanoparticles are either silica or metallic based. The synthesis of silica nanoparticles has been extensively reported, but their application for drug delivery has not been fully exploited. Silica-based nanoparticles have successfully encapsulated photosensitizers such as mTHPC, Photolon and PpIX [53–55]. The uptake of PpIX-encapsulated colloidal mesoporous silica nanoparticles by tumor cells and the effect of photon-induced toxicity were demonstrated *in vitro* [54]. A highly stable aqueous formulation of organically modified silica-based (ORMOSIL) nanoparticles encapsulating 2-devinyl-2-(1-hexyloxyethyl)pyropheophorbide (HPPH) was developed. The encapsulated HPPH was able to generate $^1O_2$ upon irradiation with light of wavelength 650 nm and thereby caused significant damage to tumorigenic HeLa and UCI-107 cells [56]. The fluorescent dye 9,10-bis[4′-(4″-aminostyryl)styryl]anthracene (BDSA) and

HPPH were co-encapsulated in ORMOSIL nanoparticles with an aim of development for two-photon PDT. BDSA serves as an energy upconverting donor and HPPH serves as an acceptor. HPPH was indirectly activated through efficient two-photon excited intraparticle energy transfer from the BDSA aggregates in the intracellular environment of tumor cells and thereby led to cytotoxic effect on HeLa cells [57]. To prevent drug release from ORMOSIL nanoparticles in systemic circulation, the photosensitizer iodobenzylpyropheophorbide (IP) was covalently incorporated into ORMOSIL nanoparticles while retaining their spectroscopic and functional properties. The phototoxicity of IP–ORMOSIL nanoparticles was further demonstrated in RIF-1 tumor cells [58].

Metallic nanoparticles possess many fascinating properties which have been explored for their biological application in biochemistry as chemical and biological sensors, as systems for nanoelectronics and nanostructured magnetism, and in medicine as agents for drug delivery. Unlike silica-based nanoparticles, photosensitizers can be attached on the surface of the metallic nanoparticles. Because metallic nanoparticles can be confined to an extremely small size, a large dose of photosensitizer can be loaded due to the enormous surface area. Gold nanoparticles are well known for their chemical inertness and minimum acute cytotoxicity [59]. Pc-coated gold nanoparticles were shown to generate $^1O_2$ with enhanced quantum yields and to exhibit double the *in vitro* PDT efficiency on HeLa cells compared with the free drug [60]. To inhibit colloid aggregation in physiologic conditions, PEG was connected to the surface of gold nanoparticles, forming PEGylated gold nanoparticles. PEGylated gold nanoparticles were then conjugated with the photosensitizer Pc4. The dynamics of drug release *in vitro* and in cancer-bearing mice indicated that the process of drug delivery was highly efficient, the drug delivery time required for PDT being greatly reduced from 2 days to less than 2 h [61]. Recently, gold nanoparticles have been used as a vehicle to deliver 5-ALA. It was reported that PpIX accumulated preferentially in fibrosarcoma tumor cells treated with 5-ALA-conjugated nanoparticles while yielding significantly higher reactive oxygen species and 50% greater cytotoxicity than that with 5-ALA [62]. A novel development with gold nanoparticles is the use of intracellular glutathione as a trigger for drug release due to the higher glutathione levels found in cancerous and precancerous cells [63]. Although photosensitizer delivery using gold nanoparticles is still evolving, there is potential for developing multifunctional particles for imaging and drug delivery systems for PDT application. To extend the technique using metallic nanoparticles, magnetic nanoparticles such as iron oxide are used for selectively targeting to tumor tissue. Magnetic nanoparticles also possess other features, such as magnetic resonance imaging (MRI) visibility for MRI imaging and nanoparticle tracking. A nanocarrier consisting of polymeric micelles of diacylphospholipid–PEG co-loaded with the photosensitizer HPPH and magnetic Fe_3O_4 nanoparticles showed excellent stability and efficient uptake by HeLa cells. The magnetic response of the nanocarriers was demonstrated by their directed delivery to tumor cells *in vitro* upon exposure to an external magnetic field. The magnetophoretic control of the cellular uptake provided enhanced imaging and phototoxicity [64]. In another drug delivery system, a chitosan nanoparticle-containing magnetic core and the encapsulating photosensitizer 2,7,12,18-tetramethyl-3,8-di(1-propoxyethyl)-13,17-bis(3-hydroxypropyl)porphyrin (PHPP) was found to have excellent targeting and imaging ability. Non-toxicity and high photodynamic efficacy on SW480 carcinoma cells both *in vitro* and *in vivo* were achieved with these nanoparticles at the level of 0–100 µM. Photosensitivity and hepatotoxicity could also be attenuated [65]. Despite great enthusiasm in the field of metallic nanoparticle research, reproducible fabrication, aggregation, biocompatibility, nonspecific binding, and toxicity remain challenges to meet [66].

23.4
Conclusion

To improve PDT efficacy for cancers, nanoparticles with active and passive functional roles have been developed. Biodegradable nanoparticles are capable of controlling the drug release; therefore, they often serve as passive carriers for photosensitizers. Nonbiodegradable nanoparticles often resist the fluctuations of the environment and hence play active roles such as catalysts to produce free radicals from dissolved oxygen, ROS modulators, or light sources in PDT. The field of PDT is a quickly evolving technology with new approaches being tested constantly. Molecular strategies are developing to make PDT more efficient and selective. Innovative approaches including specific target moiety conjugates for active targeting, aptamer technology, and combination treatment regimens, which are not discussed here, have been brought into the field in recent years. Each has great advantages over conventional PDT but also comes with significant challenges in developing the process towards clinical application. For nanoparticles used in PDT, the focus is mainly on the enhancement of efficacy and reduction in phototoxicity obtained, whereas the possible toxicity of the residues of nanoparticles is less likely to be considered. Potential hazards with nanoparticles may arise and hence less toxic materials are of great interest for the synthesis of nanoparticles. A conceptual understanding of biological responses to nanoparticles is needed to develop and apply safe nanomaterials in PDT in the future. Long-term follow-up evaluation is necessary owing to the short history of clinical applications of nanoparticles in PDT. In conclusion, much more study remains to be done at the interface between PDT and nanoparticles. Because of its interdisciplinary nature, the application of nanoparticles in PDT is still in its infancy and presents numerous research opportunities to chemists, biologists, engineers, and physicians.

Acknowledgments

During the writing of this chapter, the author was supported by grant NSC 98-2218-E-033-010 from the National Science Council of Taiwan and grant CYCU-98-CR-BE from the project of specific research fields at Chung Yuan Christian University, Taiwan.

References

1 Triesscheijn, M., Baas, P., Schellens, J.H., and Stewart, F.A. (2006) Photodynamic therapy in oncology. *Oncologist*, **11** (9), 1034–1044.

2 Chen, B., Pogue, B.W., Hoopes, P.J., and Hasan, T. (2006) Vascular and cellular targeting for photodynamic therapy. *Crit. Rev. Eukaryot. Gene Expr.*, **16** (4), 279–305.

3 Dolmans, D.E., Fukumura, D., and Jain, R.K. (2003) Photodynamic therapy for cancer. *Nat. Rev. Cancer*, **3** (5), 380–387.

4 Moor, A.C. (2000) Signaling pathways in cell death and survival after photodynamic therapy. *J. Photochem. Photobiol. B*, **57** (1), 1–13.

5 Wyld, L., Reed, M.W., and Brown, N.J. (2001) Differential cell death response to photodynamic therapy is dependent on dose and cell type. *Br. J. Cancer*, **84** (10), 1384–1386.

6 Brown, S.B., Brown, E.A., and Walker, I. (2004) The present and future role of photodynamic therapy in cancer treatment. *Lancet Oncol.*, **5** (8), 497–508.

7 Bechet, D., Couleaud, P., Frochot, C., Viriot, M.L., Guillemin, F., and Barberi-Heyob, M. (2008) Nanoparticles as vehicles for delivery of photodynamic therapy agents. *Trends Biotechnol.*, **26** (11), 612–621.

8 Jori, G. (1996) Tumour photosensitizers: approaches to enhance the selectivity and efficiency of photodynamic therapy. *J. Photochem. Photobiol. B*, **36** (2), 87–93.

9 Solban, N., Rizvi, I., and Hasan, T. (2006) Targeted photodynamic therapy. *Lasers Surg. Med.*, **38** (5), 522–531.

10 Verma, S., Watt, G.M., Mai, Z., and Hasan, T. (2007) Strategies for enhanced photodynamic therapy effects. *Photochem. Photobiol.*, **83** (5), 996–1005.

11 Chatterjee, D.K., Fong, L.S., and Zhang, Y. (2008) Nanoparticles in photodynamic therapy: an emerging paradigm. *Adv. Drug Deliv. Rev.*, **60** (15), 1627–1637.

12 Mroz, P., Tegos, G.P., Gali, H., Wharton, T., Sarna, T., and Hamblin, M.R. (2007) Photodynamic therapy with fullerenes. *Photochem. Photobiol. Sci.*, **6** (11), 1139–1149.

13 Burlaka, A.P., Sidorik, Y.P., Prylutska, S.V., Matyshevska, O.P., Golub, O.A., Prylutskyy, Y.I., and Scharff, P. (2004) Catalytic system of the reactive oxygen species on the C_{60} fullerene basis. *Exp. Oncol.*, **26** (4), 326–327.

14 Doi, Y., Ikeda, A., Akiyama, M., Nagano, M., Shigematsu, T., Ogawa, T., Takeya, T., and Nagasaki, T. (2008) Intracellular uptake and photodynamic activity of water-soluble [60]- and [70] fullerenes incorporated in liposomes. *Chem. Eur. J.*, **14** (29), 8892–8897.

15 Mroz, P., Pawlak, A., Satti, M., Lee, H., Wharton, T., Gali, H., Sarna, T., and Hamblin, M.R. (2007) Functionalized fullerenes mediate photodynamic killing of cancer cells: Type I versus Type II photochemical mechanism. *Free Radic. Biol. Med.*, **43** (5), 711–719.

16 Liu, J., Ohta, S., Sonoda, A., Yamada, M., Yamamoto, M., Nitta, N., Murata, K., and Tabata, Y. (2007) Preparation of PEG-conjugated fullerene containing Gd^{3+} ions for photodynamic therapy. *J. Control. Release*, **117** (1), 104–110.

17 Zhu, Z., Tang, Z., Phillips, J.A., Yang, R., Wang, H., and Tan, W. (2008) Regulation of singlet oxygen generation using single-walled carbon nanotubes. *J. Am. Chem. Soc.*, **130** (33), 10856–10857.

18 Ji, S.R., Liu, C., Zhang, B., Yang, F., Xu, J., Long, J., Jin, C., Fu, D.L., Ni, Q.X., and Yu, X.J. (2010) Carbon nanotubes in cancer diagnosis and therapy. *Biochim. Biophys. Acta*, **1806** (1), 29–35.

19 Samia, A.C., Chen, X., and Burda, C. (2003) Semiconductor quantum dots for photodynamic therapy. *J. Am. Chem. Soc.*, **125** (51), 15736–15737.

20 Samia, A.C., Dayal, S., and Burda, C. (2006) Quantum dot-based energy transfer: perspectives and potential for applications in photodynamic therapy. *Photochem. Photobiol.*, **82** (3), 617–625.

21 Dayal, S., Lou, Y., Samia, A.C., Berlin, J.C., Kenney, M.E., and Burda, C. (2006) Observation of non-Förster-type energy-transfer behavior in quantum dot–phthalocyanine conjugates. *J. Am. Chem. Soc.*, **128** (43), 13974–13975.

22 Tsay, J.M., Trzoss, M., Shi, L., Kong, X., Selke, M., Jung, M.E., and Weiss, S. (2007) Singlet oxygen production by peptide-coated quantum dot–photosensitizer conjugates. *J. Am. Chem. Soc.*, **129** (21), 6865–6871.

23 Zhang, P., Steelant, W., Kumar, M., and Scholfield, M. (2007) Versatile photosensitizers for photodynamic therapy at infrared excitation. *J. Am. Chem. Soc.*, **129** (15), 4526–4527.

24 Qian, H.S., Guo, H.C., Ho, P.C., Mahendran, R., and Zhang, Y. (2009) Mesoporous-silica-coated up-conversion fluorescent nanoparticles for photodynamic therapy. *Small*, **5** (20), 2285–2290.

25 Ungun, B., Prud'homme, R.K., Budijon, S.J., Shan, J., Lim, S.F., Ju, Y., and Austin, R. (2009) Nanofabricated upconversion nanoparticles for photodynamic therapy. *Opt. Express*, **17** (1), 80–86.

26 Liu, Y., Chen, W., Wang, S., and Joly, A.G. (2008) Investigation of water-soluble X-ray luminescence nanoparticles for photodynamic activation. *Appl. Phys. Lett.*, **92** (043901), 1–3.

27 Chen, W. and Zhang, J. (2006) Using nanoparticles to enable simultaneous radiation and photodynamic therapies for cancer treatment. *J. Nanosci. Nanotechnol.*, **6** (4), 1159–1166.

28 Sibani, S.A., McCarron, P.A., Woolfson, A.D., and Donnelly, R.F. (2008) Photosensitiser delivery for photodynamic therapy. Part 2: systemic carrier platforms. *Expert Opin. Drug Deliv.*, **5** (11), 1241–1254.

29 Konan, Y.N., Gurny, R., and Allemann, E. (2002) State of the art in the delivery of photosensitizers for photodynamic therapy. *J. Photochem. Photobiol. B*, **66** (2), 89–106.

30 van Nostrum, C.F. (2004) Polymeric micelles to deliver photosensitizers for photodynamic therapy. *Adv. Drug Deliv. Rev.*, **56** (1), 9–16.

31 Khdair, A., Gerard, B., Handa, H., Mao, G., Shekhar, M.P., and Panyam, J. (2008) Surfactant–polymer nanoparticles enhance the effectiveness of anticancer photodynamic therapy. *Mol. Pharmacol.*, **5** (5), 795–807.

32 Khdair, A., Chen, D., Patil, Y., Ma, L., Dou, Q.P., Shekhar, M.P., and Panyam, J. (2010) Nanoparticle-mediated combination chemotherapy and photodynamic therapy overcomes tumor drug resistance. *J. Control. Release*, **141** (2), 137–144.

33 Lee, S.J., Park, K., Oh, Y.K., Kwon, S.H., Her, S., Kim, I.S., Choi, K., Lee, S.J., Kim, H., Lee, S.G., Kim, K., and Kwon, I.C. (2009) Tumor specificity and therapeutic efficacy of photosensitizer-encapsulated glycol chitosan-based nanoparticles in tumor-bearing mice. *Biomaterials*, **30** (15), 2929–2939.

34 Mazzaglia, A., Angelini, N., Lombardo, D., Micali, N., Patane, S., Villari, V., and Scolaro, L.M. (2005) Amphiphilic cyclodextrin carriers embedding porphyrins: charge and size modulation of colloidal stability in heterotopic aggregates. *J. Phys. Chem. B*, **109** (15), 7258–7265.

35 Sortino, S., Mazzaglia, A., Monsu Scolaro, L., Marino Merlo, F., Valveri, V., and Sciortino, M.T. (2006) Nanoparticles of cationic amphiphilic cyclodextrins entangling anionic porphyrins as carrier–sensitizer system in photodynamic cancer therapy. *Biomaterials*, **27** (23), 4256–4265.

36 Kratz, F. (2008) Albumin as a drug carrier: design of prodrugs, drug conjugates and nanoparticles. *J. Control. Release*, **132** (3), 171–183.

37 Chen, K., Preuss, A., Hackbarth, S., Wacker, M., Langer, K., and Roder, B. (2009) Novel photosensitizer–protein nanoparticles for photodynamic therapy: photophysical characterization and *in vitro* investigations. *J. Photochem. Photobiol. B*, **96** (1), 66–74.

38 Konan, Y.N., Berton, M., Gurny, R., and Allemann, E. (2003) Enhanced photodynamic activity of *meso*-tetra(4-hydroxyphenyl)porphyrin by incorporation into sub-200 nm nanoparticles. *Eur. J. Pharm. Sci.*, **18** (3–4), 241–249.

39 Gomes, A.J., Lunardi, C.N., and Tedesco, A.C. (2007) Characterization of biodegradable poly(D,L-lactide-co-

glycolide) nanoparticles loaded with bacteriochlorophyll-*a* for photodynamic therapy. *Photomed. Laser Surg.*, **25** (5), 428–435.

40 McCarthy, J.R., Perez, J.M., Bruckner, C., and Weissleder, R. (2005) Polymeric nanoparticle preparation that eradicates tumors. *Nano Lett.*, **5** (12), 2552–2556.

41 Fadel, M., Kassab, K., and Fadeel, D.A. (2010) Zinc phthalocyanine-loaded PLGA biodegradable nanoparticles for photodynamic therapy in tumor-bearing mice. *Lasers Med. Sci.*, **25** (2), 283–272.

42 Zeisser-Labouebe, M., Delie, F., Gurny, R., and Lange, N. (2009) Benefits of nanoencapsulation for the hypercin-mediated photodetection of ovarian micrometastases. *Eur. J. Pharm. Biopharm.*, **71** (2), 207–213.

43 Zeisser-Labouebe, M., Lange, N., Gurny, R., and Delie, F. (2006) Hypericin-loaded nanoparticles for the photodynamic treatment of ovarian cancer. *Int. J. Pharm.*, **326** (1–2), 174–181.

44 Vargas, A., Lange, N., Arvinte, T., Cerny, R., Gurny, R., and Delie, F. (2009) Toward the understanding of the photodynamic activity of m-THPP encapsulated in PLGA nanoparticles: correlation between nanoparticle properties and *in vivo* activity. *J. Drug Target.*, **17** (8), 599–609.

45 Pegaz, B., Debefve, E., Borle, F., Ballini, J.P., van den Bergh, H., and Kouakou-Konan, Y.N. (2005) Encapsulation of porphyrins and chlorins in biodegradable nanoparticles: the effect of dye lipophilicity on the extravasation and the photothrombic activity. A comparative study. *J. Photochem. Photobiol. B*, **80** (1), 19–27.

46 Battah, S.H., Chee, C.E., Nakanishi, H., Gerscher, S., MacRobert, A.J., and Edwards, C. (2001) Synthesis and biological studies of 5-aminolevulinic acid-containing dendrimers for photodynamic therapy. *Bioconjug. Chem.*, **12** (6), 980–988.

47 Battah, S., Balaratnam, S., Casas, A., O'Neill, S., Edwards, C., Batlle, A., Dobbin, P., and MacRobert, A.J. (2007) Macromolecular delivery of 5-aminolaevulinic acid for photodynamic therapy using dendrimer conjugates. *Mol. Cancer Ther.*, **6** (3), 876–885.

48 Casas, A., Battah, S., Di Venosa, G., Dobbin, P., Rodriguez, L., Fukuda, H., Batlle, A., and MacRobert, A.J. (2009) Sustained and efficient porphyrin generation *in vivo* using dendrimer conjugates of 5-ALA for photodynamic therapy. *J. Control. Release*, **135** (2), 136–143.

49 Kojima, C., Toi, Y., Harada, A., and Kono, K. (2007) Preparation of poly (ethylene glycol)-attached dendrimers encapsulating photosensitizers for application to photodynamic therapy. *Bioconjug. Chem.*, **18** (3), 663–670.

50 Nishiyama, N., Nakagishi, Y., Morimoto, Y., Lai, P.S., Miyazaki, K., Urano, K., Horie, S., Kumagai, M., Fukushima, S., Cheng, Y., Jang, W.D., Kikuchi, M., and Kataoka, K. (2009) Enhanced photodynamic cancer treatment by supramolecular nanocarriers charged with dendrimer phthalocyanine. *J. Control. Release*, **133** (3), 245–251.

51 Tang, W., Xu, H., Kopelman, R., and Philbert, M.A. (2005) Photodynamic characterization and *in vitro* application of methylene blue-containing nanoparticle platforms. *Photochem. Photobiol.*, **81** (2), 242–249.

52 Tang, W., Xu, H., Park, E.J., Philbert, M.A., and Kopelman, R. (2008) Encapsulation of methylene blue in polyacrylamide nanoparticle platforms protects its photodynamic effectiveness. *Biochem. Biophys. Res. Commun.*, **369** (2), 579–583.

53 Podbielska, H., Ulatowska-Jarza, A., Muller, G., Holowacz, I., Bauer, J., and Bindig, U. (2007) Silica sol–gel matrix doped with Photolon molecules for sensing and medical therapy purposes. *Biomol. Eng.*, **24** (5), 425–433.

54 Qian, J., Gharibi, A., and He, S. (2009) Colloidal mesoporous silica nanoparticles with protoporphyrin IX encapsulated for photodynamic therapy. *J. Biomed. Opt.*, **14** (1), 014012.

55 Yan, F. and Kopelman, R. (2003) The embedding of *meta*-tetra(hydroxyphenyl)-chlorin into silica nanoparticle platforms for photodynamic therapy and their singlet oxygen production and

pH-dependent optical properties. *Photochem. Photobiol.*, **78** (6), 587–591.

56 Roy, I., Ohulchanskyy, T.Y., Pudavar, H.E., Bergey, E.J., Oseroff, A.R., Morgan, J., Dougherty, T.J., and Prasad, P.N. (2003) Ceramic-based nanoparticles entrapping water-insoluble photosensitizing anticancer drugs: a novel drug-carrier system for photodynamic therapy. *J. Am. Chem. Soc.*, **125** (26), 7860–7865.

57 Kim, S., Ohulchanskyy, T.Y., Pudavar, H.E., Pandey, R.K., and Prasad, P.N. (2007) Organically modified silica nanoparticles co-encapsulating photosensitizing drug and aggregation-enhanced two-photon absorbing fluorescent dye aggregates for two-photon photodynamic therapy. *J. Am. Chem. Soc.*, **129** (9), 2669–2675.

58 Ohulchanskyy, T.Y., Roy, I., Goswami, L.N., Chen, Y., Bergey, E.J., Pandey, R.K., Oseroff, A.R., and Prasad, P.N. (2007) Organically modified silica nanoparticles with covalently incorporated photosensitizer for photodynamic therapy of cancer. *Nano Lett.*, **7** (9), 2835–2842.

59 Connor, E.E., Mwamuka, J., Gole, A., Murphy, C.J., and Wyatt, M.D. (2005) Gold nanoparticles are taken up by human cells but do not cause acute cytotoxicity. *Small*, **1** (3), 325–327.

60 Wieder, M.E., Hone, D.C., Cook, M.J., Handsley, M.M., Gavrilovic, J., and Russell, D.A. (2006) Intracellular photodynamic therapy with photosensitizer-nanoparticle conjugates: cancer therapy using a 'Trojan horse'. *Photochem. Photobiol. Sci.*, **5** (8), 727–734.

61 Cheng, Y., Samia, A.C., Meyers, J.D., Panagopoulos, I., Fei, B., and Burda, C. (2008) Highly efficient drug delivery with gold nanoparticle vectors for *in vivo* photodynamic therapy of cancer. *J. Am. Chem. Soc.*, **130** (32), 10643–10647.

62 Oo, M.K., Yang, X., Du, H., and Wang, H. (2008) 5-Aminolevulinic acid-conjugated gold nanoparticles for photodynamic therapy of cancer. *Nanomedicine (Lond.)*, **3** (6), 777–786.

63 Hong, R., Han, G., Fernandez, J.M., Kim, B.J., Forbes, N.S., and Rotello, V.M. (2006) Glutathione-mediated delivery and release using monolayer protected nanoparticle carriers. *J. Am. Chem. Soc.*, **128** (4), 1078–1079.

64 Cinteza, L.O., Ohulchanskyy, T.Y., Sahoo, Y., Bergey, E.J., Pandey, R.K., and Prasad, P.N. (2006) Diacyllipid micelle-based nanocarrier for magnetically guided delivery of drugs in photodynamic therapy. *Mol. Pharmacol.*, **3** (4), 415–423.

65 Sun, Y., Chen, Z.L., Yang, X.X., Huang, P., Zhou, X.P., and Du, X.X. (2009) Magnetic chitosan nanoparticles as a drug delivery system for targeting photodynamic therapy. *Nanotechnology*, **20** (13), 135102.

66 Juzenas, P., Chen, W., Sun, Y.P., Coelho, M.A., Generalov, R., Generalova, N., and Christensen, I.L. (2008) Quantum dots and nanoparticles for photodynamic and radiation therapies of cancer. *Adv. Drug Deliv. Rev.*, **60** (15), 1600–1614.

24
Optical Coherence Tomographic Monitoring of Surgery in Oncology

Elena V. Zagaynova, N.D. Gladkova, N.M. Shakhova, O.S. Streltsova, I.A. Kuznetsova, I.A. Yanvareva, L.B. Snopova, E.E. Yunusova, E.B. Kiseleva, V.M. Gelikonov, G.V. Gelikonov, and A.M. Sergeev

24.1
Introduction

Epithelial cancer is considered to account for the majority of cancer cases and deaths worldwide. The most common urologic cancers, those of the prostate, kidney, and urinary bladder, account for about 300 000 new cases of cancer per year in men with another 32 000 new renal and bladder cancer cases arising in women [1]. In the digestive system, most common are colon and esophageal cancer. Colorectal cancer is one of the leading causes of morbidity and mortality [2]. Colorectal cancer is the third most common cancer in the United States; in 2006, 148 610 new cases of colorectal cancer were registered According to the 2001 records, 13 200 cases of esophageal carcinoma were detected in the United States, of which 12 500 patients died. Adenocarcinoma accounts for about 90% of esophageal carcinoma in the Western hemisphere [3]. Malignant neoplasms of the female reproductive organs, including cervical cancer, are among the principal causes of death. About 510 000 new cervical cancer cases are registered worldwide every year; of these, 288 000 women die, and over US $5–6 billion are spent on patient treatment [4, 5]. A significant increase in the incidence of cervical cancer is observed for women in the early reproductive age group (by 2% yearly on average) [6]. This situation requires new approaches to cervical cancer prevention and diagnosis.

Endoscopic methods are used to diagnose different types of epithelial cancer. However, it is difficult to recognize early neoplasia such as dysplasia and carcinoma *in situ* by endoscopic visualization alone. Random biopsies during white light endoscopy have been used to identify and locate early neoplasia [7], but the sensitivity and specificity are poor. Numerous biopsies can provide histopathologic control only over a very small part of mucosa, for example, in Barrett's esophagus (BE) 3.5% of metaplastic mucosa are analyzed histologically [8].

Handbook of Biophotonics. Vol.2: Photonics for Health Care, First Edition. Edited by Jürgen Popp,
Valery V. Tuchin, Arthur Chiou, and Stefan Heinemann.
© 2012 Wiley-VCH Verlag GmbH & Co. KGaA. Published 2012 by Wiley-VCH Verlag GmbH & Co. KGaA.

Initial risk stratification for cervical neoplasia is performed using screening techniques, including Pap smear and HPV (human papilloma virus) testing. The next step in diagnosis is colposcopy guidance biopsy. Histopathologic evaluation of the biopsies serves as a "gold standard" for diagnosis. However, the diagnostic efficacy of colposcopy is limited. Similar colposcopic abnormalities may be observed for a variety of cervical conditions, benign and malignant, which reduces specificity and positive predictive value [9]. Hence the development of new techniques and the elaboration of modern protocols for management of Pap- and HPV-positive women are important for ensuring adequate diagnosis and, at the same time, for avoiding over-referral to colposcopy and over-treatment and to maintain sustainable costs.

In terms of diagnostic tasks, endoscopic methods may be classified into screening (assessment of the entire mucous surface aiming at detecting pathologic zones) and verification (assessment of structural changes with almost cellular resolution without damage to tissue integrity) techniques.

Chromoendoscopy, fluorescence endoscopy, and narrow-band imaging are the screening methods that improve the sensitivity of endoscopic diagnostics. Magnification endoscopy, optical coherence tomography, high-frequency ultrasound, confocal microscopy, and cytoendoscopy are among the verification techniques that enhance diagnostic specificity.

Chromoendoscopy has attracted considerable attention over the past 10 years. Methylene blue, Lugol's solution, and indigo carmine staining of the esophagus have been shown to highlight metaplastic epithelium, and may visualize areas of dysplasia and early adenocarcinoma. Several studies have demonstrated the high diagnostic accuracy of this method [10]. However, this technique is highly operator dependent, with low inter-observer agreement (kappa coefficient 0.36) [11].

Fluorescence imaging is known to have high sensitivity but limited specificity, thereby giving a low positive predictive value, especially for flat lesions. The rate of false-positive readings in fluorescence cystoscopy is high owing to simple hyperplasia of the epithelium, inflammation, and squamous metaplasia [12]. The use of fluorescence, both endogenous (auto) and exogenous, for visualization of high-grade dysplasia (HGD) in BE is particularly efficacious using integral analysis of several fluorescence spectra at several excitation wavelengths (337–620 nm) [13].

Modern verification endoscopic techniques, such as endoscopic confocal microscopy and multiphoton fluorescence microscopy, have cellular spatial resolution, whereas high-frequency ultrasound and optical coherence tomography (OCT), commonly used in clinical practice, have tissue-level resolution. According to some reports, high-frequency ultrasound can indirectly detect intestinal metaplasia by esophageal wall thickening and changed ratio between ultrasonic layers. It was shown that ultrasound is able to diagnose invasive cancer in 23% more patients and to detect lymph node involvement [14]. However, high-frequency ultrasound was shown to be ineffective in the detection of dysplasia or intramucosal carcinoma because its spatial resolution is comparable to the mucosal layer thickness.

24.2
OCT Technique

Since the first report on OCT in 1991 [15], different fundamental and applied studies of the OCT technique have been carried out, resulting in numerous publications. The foundations of OCT are described in a book [16]. A conventional time-domain OCT device uses near-infrared light to allow real-time cross-sectional imaging with high spatial resolution (10–20 µm) at a depth of ∼2 mm. When developing a device for clinical application, the fiber-optic modality is evidently to be preferred, allowing for miniaturization [17, 18]. The design of a compact device for effective scanning by Michelson interferometer arms mismatch became a key for such miniaturization. One of the most efficient devices is the controlled piezo-fiber delay line [19] with optical path modulation depth up to several millimeters. A scanning system with such an element can provide high accuracy in keeping a constant Doppler shift of optical frequency mismatch in the interferometer arms, which is necessary for narrow-band signal detection.

The advantages of the fiber-optic OCT modality have also been implemented in the design of endoscopic probes. Endoscopic OCT systems use different models of scanners: catheters with circumferential [20, 21], linear-transverse [22], or longitudinal scanning modality [23, 24], scanners with a built-in endoscope [25], and microelectromechanical systems (MEMS) [26].

Three types of scanning patterns are available for OCT imaging: radial [27], longitudinal [28], and transverse [22, 29]. The radial-scan probe directs the OCT beam radially, giving images that are displayed in a "radar-like" circular plot. Radial scanning can easily image large areas of tissue by moving the probe back and forth over the tissue surface, and has the highest definition when the probe is inserted within a small-diameter lumen, because the OCT images become progressively coarser when a large-diameter lumen is scanned, due to the progressive increase in pixel spacing with increasing distance between the probe and the tissue. The linear-transverse probes scan the longitudinal and transverse positions of the OCT beam at a fixed angle, generating rectangular images of longitudinal and transverse planes at a given angle with respect to the probe. The advantage of linear scanning is that the pixel spacing in the transverse direction is uniform and can better image a definite area of the scanned tissue, especially in the presence of large-diameter and noncircular lumens, where it may be impossible to maintain a constant distance from the probe to the surface over the entire circumferential scan. The transverse scanning modality provides a better depth of field. The depth of field is the range of distances from the probe over which optimal resolution of scanning can be obtained; current OCT scans permit imaging depths up to 1–2 mm in tissues, by using probes with different focuses. In our research, we use flexible endoscopic forward-looking probes 2.7 and 2.4 mm in diameter with linear-transverse scanning that may be introduced through biopsy channels of standard endoscopes to study mucous surfaces. We believe that such an approach to scanning has a number of advantages. The scanning technology implies contact of a probe with the studied object, which minimizes artifacts due to motion and ensures stable focusing of probing radiation.

The possibility of using the effect of tissue compression for obtaining additional information is also significant. The scanner described in this chapter is compatible with the majority of standard endoscopes, thus making OCT a useful technique for many clinical disciplines. Its design meets the requirements for instruments used in clinical practice.

Our previous research showed that conventional OCT technology is informative with regard to structural alterations in tissue, although it did not elucidate the origin of these alterations. Using OCT imaging, it is very difficult to differentiate inflammatory processes, papillomatosis, cancer, and scar changes. This problem may be solved by using polarization optics methods. Some structural components of biotissue can change the polarization state of backscattered radiation. Some visually similar pathologic cases may be differentiated by comparing co- and cross-polarized OCT images [30]. The polarization-sensitive (PS)- and cross-polarization (CP)-OCT techniques have undergone significant development in recent years [31–33]. In PS-OCT, the images obtained allow one to analyze birefringence in biotissues, thus permitting diagnostics, for example, of the state of burned tissues [32] or early glaucoma [34]. CP-OCT permits comparisons of co- and cross-polarization scattering from the sampling volume and, hence, differentiation of biotissue pathology [33]. This method was implemented almost simultaneously in open-space [33] and fiber-optics modalities [30, 35]. In this chapter, we present results of the application of a new tandem-polarization time-domain OCT device that allows one not only to obtain conventional OCT imaging but also to carry out polarization research.

The tandem-polarization time-domain OCT device developed at the Institute of Applied Physics of the Russian Academy of Science (Nizhny Novgorod) operates at 1300 nm with 3 mW power and 35–40 nm bandwidth (light source: SLD 561 HP2). A schematic of the CP-OCT system using two orthogonal polarization modes in an isotropic optical fiber (SMF-28) is shown in Figure 24.1. The full description of the optical scheme can be found elsewhere [36]. The CP-OCT scheme comprises a fiber-based Fizeau interferometer, a common path for signal and reference waves, and an autocorrelator based on a Michelson interferometer with Faraday mirrors [37] that is optimal when an isotropic optical fiber is used.

When operating in the simplest mode, the Fizeau interferometer provides heterodyne detection of weakly scattered light with maximum possible and stable visibility of the interference fringes. Indeed, when the directional pattern of the optical probe is identical for emitted and received light, a wave scattered from nondepolarizing structures and a wave reflected from the fiber tip have identical polarization states. The essential feature of the novel scheme is its ability for simultaneous independent detection of backscattered radiation in both initial and orthogonal polarizations at random polarization states of the probing radiation. The main idea underlying our CP-OCT system consists in creating at the entrance to the optical scheme two strictly orthogonally polarized waves with a predetermined time delay between coherent regions. In a general case, these waves in the probing beam may have arbitrary ellipticity under the condition of their strict mutual orthogonality.

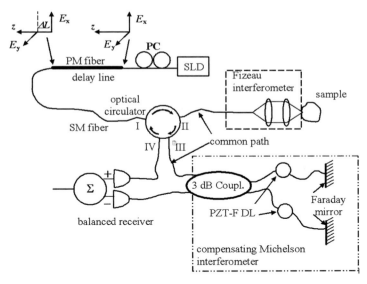

Figure 24.1 Optical scheme of tandem-polarization time-domain OCT device.

A forward-looking OCT probe (2.7 mm o.d.) was used in a routine endoscopic procedure: it was passed through the operating channel of the endoscope and placed in contact with mucosa. The OCT catheter has been developed as a miniaturized electromechanical unit controlling the lateral scanning process [22, 38]. This unit terminates the sampling arm of the fiber interferometer and has a size to fit the diameter and the curvature radius of the biopsy channel of endoscopes. The probe beam moves along the tissue surface within a distance of about 2 mm. The beam deviation system embodies the galvanometric principle, and voltage with a maximum of 5 V is supplied to the distal end of the endoscope. The distance between the output lens and the sample is 5–7 mm and the focal spot diameter is 20 mm. The scanning unit and the extended part of the fiber interferometer are sealed, which allows easy sterilization [22]. Each OCT image, with 200 × 200 pixels, 2 × 2 mm, 19 μm free space in depth resolution (13 μm in tissue), and 25 μm lateral resolution, was acquired in ∼2 s. Images are presented in a pseudo color, "positive" logarithmic palette, that is, the lighter area corresponds to the higher light scattering.

At the current stage of its development, OCT has rich capabilities in oncology. This technique is capable of solving a number of problems: early detection of cancer, targeting biopsy, more precise localization and spread of neoplastic processes, and monitoring results of treatment. There are some publications concerning the application of OCT in examination of the urinary bladder [39–42], digestive tract [43, 44], and cervix [45, 46].

In our recent work, we analyzed material from clinical studies of 895 patients: 175 patients with urinary bladder pathology, 390 with gastrointestinal tract pathology, and 330 with cervical pathology.

In spite of the different morphology of these organs, their study had similar problems that were solved by means of OCT. A multi-center study was carried out. The research was done in the Nizhny Novgorod Regional Hospital (Russia) and in three clinics abroad: Eppendorf Clinic, University of Hamburg (Germany), Cleveland Clinic Foundation (United States), and the George Washington University (GWU) Medical Center (United States). The different clinics used the same protocol, which permitted integration of data when analyzing results. Written informed consent was obtained from all patients.

Earlier studies made it possible to define OCT signs correlating with benign and malignant histopathology and to create a set of images to train observers performing OCT image recognition.

The following retrospectively tested criteria were used for OCT interpretation:

- Sharp, high-contrast horizontal layers through more than 75% of the lateral range. The upper layer (epithelium, glandular mucosa) has a lower signal intensity than the second (submucous) layer– benign images.
- Tissue layers are hardly differentiable, with increased intensity of the upper layer signal and a poorly defined border between the upper and second layers – suspicious images.
- No horizontal layers, or less than 75% (excluding evident motion artifacts) – neoplasia images.
- Very dark pattern of stripes associated with edema is not a diagnostic sign for neoplasia.

Hence the main OCT features characterizing images obtained by means of a transverse scanning probe are the presence or absence of high-contrast horizontal tissue layers and signal level from these layers.

24.3
Urology

The first OCT images of human urologic tissue *ex vivo* were demonstrated in 1997 [47], and the first *in vivo* OCT studies in urology were performed on the bladder urothelium in 1998 [39, 48]. Since then, a number of papers have been published exploring OCT for urologic clinical applications. In brief, the range of potential clinical use includes:

- early detection of bladder neoplasia, including visualization of carcinoma *in situ* (Tis)
- preoperative planning and intraoperative guidance of surgery for adequate resection with maximal bladder preservation
- differentiation of fluorescent bladder lesions
- identification and preservation of neurovascular tissue during prostatectomy
- detection of capsular penetration of prostate cancer during prostatectomy

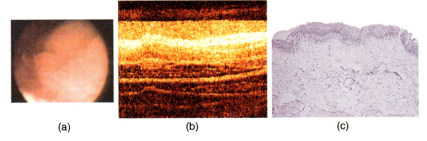

Figure 24.2 Urothelial hyperplasia. (a) Cystoscopic image; (b) OCT image (benign type); (c) corresponding histology (H&E stain).

- preservation of retroperitoneal nerves during lymph node dissection for testicular cancer
- differentiation of renal tumors.

The research done by our group concerns three possible applications of OCT in urologic oncology, outlined below.

24.3.1
Accuracy of OCT in Endoscopic Detection of Early Cancer

It is well known that early cancer in urinary bladder looks like a flat, suspicious lesion. We performed two clinical OCT studies of the urinary bladder: in the Nizhny Novgorod Regional Hospital (Russia) and in the GWU Medical Center (United States). We combined available flat lesion data from both the Russian and the GWU studies to include 80 patients with 114 sites (st) of OCT images with histopathologic and cystoscopic data. Different conditions were histologically diagnosed for each site: benign – urothelial hyperplasia (17 st) (Figure 24.2), von Brunn's nests (15 st), cystitis cystica (6 st), chronic cystitis with exudation (28 st), and chronic cystitis with infiltration (10 st) – and malignant – urothelial dysplasia (5 st), carcinoma *in situ* (10 st), and invasive cancer (2 st).

In both centers, we used a very similar procedure in which the OCT probe was introduced via the working channel of the rigid cystoscope and positioned on the mucosal area of interest under visual control. All images were obtained by placing the en-face OCT probe directly on the desired area normally to the bladder wall. Two or three OCT images were acquired from each site. Biopsies of all scanned areas were performed and submitted for histopathologic analysis. Two respondents made blind recognition of OCT images (Table 24.1).

All histologically confirmed invasive cancer sites had suspicious or abnormal OCT images; 17 of 20 cases of carcinoma *in situ* had suspicious or abnormal OCT images (Figure 24.3). The greatest false-negative error was in recognition of urothelial dysplasia. The cancerous potentiality of dysplasia is still under discussion worldwide.

Table 24.1 Results of blind recognition of OCT images of flat lesions in the urinary bladder.

Parameter	Russia	GWU	Total
Number of zones	61	53	114
Sensitivity (%)	81	88	82
Specificity (%)	78	92	85
Diagnostic accuracy (%)	79	92	85
Positive predictive value (%)	—	—	51
Negative predictive value (%)	—	—	96
Kappa	0.56	0.46	0.6

There are several possible reasons for false-positive OCT readings (benign conditions with suspicious or abnormal OCT image). Inflammatory processes and von Brunn's nest are the most common causes. Severe inflammation can partially or completely obscure the OCT contrast between the urothelium and connective tissue. The mechanism of these changes is still debated; however, it is hypothesized that optical scattering changes are associated with cell infiltration into the lamina propria.

OCT showed good sensitivity (82%) and good specificity (85%) in the identification of flat suspicious zones in the urinary bladder. Diagnostic accuracy at blind recognition of OCT images was higher in the GWU sampling (92%) compared with 79% in Nizhny Novgorod, which is due to different spectra of pathologic states. Based on the fact that OCT properly identified 165 out of 194 benign sites, we can hypothesize that when using white light cystoscopy with OCT, biopsy could have been safely avoided in 85% of cases.

Earlier experimental studies on animals demonstrated that OCT is potentially capable of detecting carcinogenesis in the urinary bladder. The difference between normal urothelium, inflammatory and proliferative changes, and early neoplasia was found [49]. Later, our group made the first clinical studies of the capabilities of OCT for the diagnosis of different clinical states of the urinary bladder [39]. Further, the accumulated clinical material enabled us to carry out preliminary statistical studies [50]. The sensitivity and specificity of OCT in detecting urinary bladder neoplasia were analyzed on the basis of different statistical samplings. However, statistical

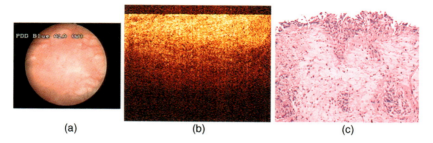

Figure 24.3 High-grade dysplasia (carcinoma *in situ*) in a flat suspicious zone. (a) Cystoscopic image; (b) OCT image (neoplasia type); (c) corresponding histology (H&E stain).

clinical studies in the urinary bladder are very sparse, perhaps owing to the poor adaptation of available probes (rotation probe or rigid OCT cystoscope) to a large, hollow organ. Wang *et al.* reported recent technical improvements of MEMS-based spectral-domain endoscopic OCT and applications for *in vivo* bladder imaging diagnosis. Preliminary clinical studies (20 cases) revealed high sensitivity (91%) and specificity (80%) in the diagnosis of neoplasia [26].

Recent publications confirmed our previous results that OCT was able to differentiate malignant from benign lesions with a positive predictive value of 89% and a negative predictive value of 100%. Benign lesions including cystitis cystica and hyperplasia, and also normal urothelium, had characteristic OCT findings that facilitated distinguishing them from malignant lesions in conjunction with white light cystoscopy [51].

24.3.2
Combined Use of OCT and Fluorescence Imaging for Cancer Detection

The OCT data obtained for bladder cancer suggest that OCT can increase the efficacy of diagnosis and possibly decrease the number of unnecessary biopsies.

Our current study of human fluorescence cystoscopy by means of OCT enrolled 26 patients. In preparation for fluorescence cystoscopy, 50 ml of 3% δ-aminolevulinic acid (ALA) solution was instilled in the bladder 2 h before the procedure. The urinary bladder was first examined in white light, followed by blue light illumination, and the red fluorescent zones were identified. The endoscopic OCT probe was placed through the cystoscope in direct contact with bladder mucosa and biopsy was performed for all fluorescent OCT zones.

A total of 107 fluorescent zones with 18 exophytic tumors and 89 flat suspicious areas were examined and compared with OCT and histology. All 18 cases of exophytic tumors were correctly detected by white light cystoscopy, OCT, and fluorescence cystoscopy. Fluorescence from 18 exophytic tumors verified correct photosensitizer administration. A total of 75 of 89 flat fluorescent zones (84%) had benign histology and, therefore, false-positive fluorescence. Of these 75 lesions, 59 (78.7%) had typical benign stratified OCT images and were considered true negatives (TN). In the remaining 16 histologically benign zones, the OCT images showed a bright or patchy urothelium with poor or no contrast with lamina propria; therefore, these zones had false-positive OCT. Interestingly, most of false-positive OCT cases contained von Brunn's nests and squamous metaplasia. Of the 14 malignant flat zones, OCT correctly revealed 12 areas with neoplastic pathology. OCT worked well in cases of invasive cancer and demonstrated satisfactory results in HGD (Figure 24.4); only two cases of HGD were erroneously recognized as false negative.

Specificity is frequently used to analyze the fluorescence imaging efficacy, but this is not truly accurate because this calculation depends on the arbitrary number of sites without fluorescence chosen for inclusion. Therefore, a positive predictive value (PPV) is a more relevant measure because it represents the ratio between false positives and true positives. In our series, the PPV for flat lesions with fluorescence cystoscopy alone was 16%, while the PPV for the fluorescence

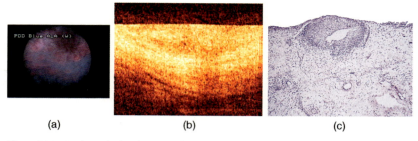

Figure 24.4 High-grade dysplasia. (a) Fluorescent cystoscopic image; (b) OCT image (suspicious type, true positive case); (c) corresponding histology (H&E stain).

cystoscopy combined with OCT increased to 43%, which is a very significant improvement. These data also suggest that 78.7% of the biopsies based on fluorescent positive findings could have been avoided based on OCT results.

Pioneering studies of the combined use of fluorescence and OCT were made on animals. Preliminary results based on rat bladder tumorogenesis studies demonstrate that ALA fluorescence imaging is highly effective and sensitive and covers a large area (diameter 30 mm) of the bladder surface, whereas OCT provides only a cross-sectional scan of 5 × 2.8 mm. However, because of the superior resolution and the ability to delineate bladder micromorphology, in addition to the detection of a neoplastic lesion, OCT can even detect precancerous lesions that fluorescence may sometimes miss and can differentiate inflammatory lesions that fluorescence may show positive [52]. Here, we present the results of the first clinical study of the combined application of fluorescence cystoscopy and OCT for the diagnosis of early cancer. We demonstrate that fluorescence cystoscopy and OCT do not add diagnostic information in zones of papillary tumors, but OCT greatly increases the specificity and PPV of fluorescence imaging tests in the diagnosis of early cancer in flat suspicious zones to 79% and 43%, respectively.

24.3.3
OCT Guided Surgery in Bladder Cancer (Open Surgery and Transurethral Resections)

The problem of an adequate tradeoff between radical tumor removal and maximum preservation of organs is very important for cancer management and patients' quality of life. We have evaluated OCT for pre- and intraoperative planning of the tumor resection margins on the examples of bladder, esophageal, and rectal carcinoma.

24.3.3.1 Transurethral Resections

OCT was used for intraoperative planning of 44 cases of transurethral resection of bladder tumor (TURBT). The OCT forward-looking probe was introduced through the operating channel of the cystoscope and anchored for TURBT. Under visual control, the probe was pressed directly against the mucosa. Two sites of the tumor and multiple sites along four directional paths around the visual tumor border (up to 2 cm in the 12, 3, 6, and 9 o'clock directions) were imaged.

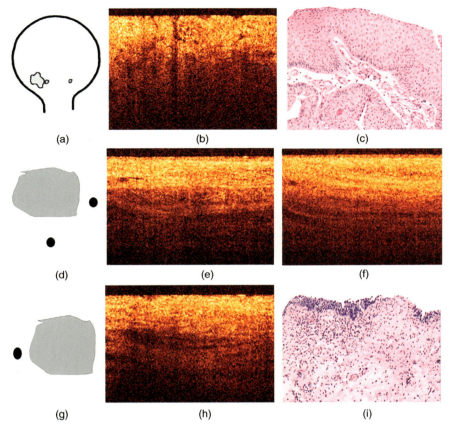

Figure 24.5 Intraoperative OCT planning of resection line during TURBT (T1G2). (a) Scheme of tumor localization; (b) OCT image of tumor focus; (d) scheme of OCT scanning along two directions of conventional clock-dial (6 o'clock); (e, f) OCT images of benign type at 6 o'clock; (g) scheme of localization of OCT image of suspicious type at 9 o'clock; (h) OCT image from the point at 9 o'clock.

To evaluate the postoperative resection margins following TURBT, OCT imaging was performed at the same four directional paths at the line of resection, 0.5 and 1 cm from the postoperative resection line.

All OCT image readings were compared with histologic data from the same site. The OCT and histologic borders coincided in 79% of cases. At the level of the traditional resection line (+0.5 cm from the visible tumor border along the perimeter), suspicious OCT images were observed in one-third of the cases (the sensitivity was 93% and specificity 74%). If an abnormal or suspicious OCT image was revealed along the line, then an additional resection was performed accordingly.

Based on OCT data, 14 of 44 TURBT patients with intraoperative resection line examination required additional resection (Figure 24.5). One TURBT patient with postoperative resection line examination required additional resection.

To determine the invasion depth, target investigation of the papillary tumor neck was undertaken in 28 cases. It was found that in the case of superficial tumor (Ta, T1 stages), sharp, high-contrast horizontal layers are visualized throughout the OCT image, whereas in the case of invasive tumor (T2 and higher) *no* horizontal layers were observed in the OCT image. The OCT sensitivity in determining tumor invasion depth by assessing neck image structure was 100% and the specificity was 77%.

24.3.3.2 Open Surgery

Using OCT intraoperatively in the open bladder, we can determine the condition of the inferior bladder at the bladder neck for possible preservation, which then may help to preserve normal urinary function for the patient [53]. This OCT study sample included 25 patients who underwent partial cystectomy. Three of 25 PC patients required intraoperative conversion to radical cystectomy as OCT revealed involvement near the bladder neck, while 67% of these patients could have had a more limited resection (Figure 24.6).

OCT can help to define the bladder tumor margins in real time and has the potential to improve the adequacy of bladder resection, optimize bladder sparing, and reduce the recurrence rate.

24.4
Gastroenterology

The history of using OCT in gastroenterology dates back to 1996, when animal organs were studied *ex vivo* and it was shown that OCT is capable of differentiating tissue layers of the wall of hollow organs [54]. The 13 year experience gained by different groups in the application of OCT in gastroenterology is sufficient to discuss its clinical value.

The spectrum of potential clinical use includes:

- OCT for endoscopic BE malignancy monitoring
- OCT for surgical guidance of colon polypscolitis
- OCT in guided surgery of gastrointestinal cancer (esophagectomy and rectoectomy)
- OCT for differential diagnosis of Crohn disease and ulcerative colitis
- OCT of the biliary tract.

Here we present the experience of our group in OCT monitoring surgery in gastroenterology.

24.4.1
OCT for Endoscopic Monitoring of Barrett's Esophagus Malignancy

A total of 78 patients had a complete qualifying set of data for inclusion in the statistical analysis. The protocol for this surveillance for BE was approved by the Cleveland Clinic Institutional Review Board. The protocol and the preliminary

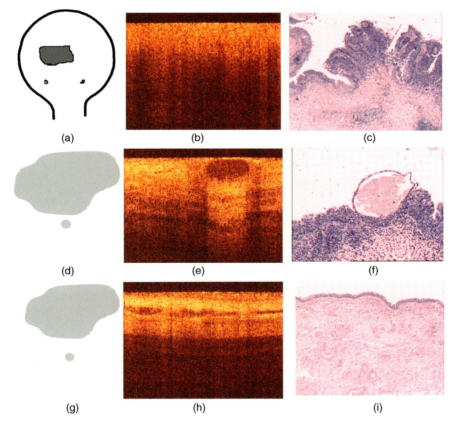

Figure 24.6 Intraoperative OCT planning of resection line during partial cystectomy. (a) Scheme of tumor localization; (b) OCT image of tumor focus; (c) H&E-stained histology section reveals transitional cell-invasive carcinoma (T2GIII); (d) scheme of OCT scanning along direction of conventional clock-dial (6 o'clock tumor border plus 3 mm); (e) OCT image of benign type from point (d); (f) H&E-stained histology section reveals cystitis cystica; (g) scheme of localization of OCT image of 6 o'clock tumor border plus 10 mm; (h) benign OCT image from the point (g); (i) histology section reveals chronic cystitis.

clinical results have been described earlier [55, 56]. Seventeen patients were histologically diagnosed with invasive adenocarcinoma (INVC) originating in the metaplastic epithelium, 17 with intramucosal adenocarcinoma (IMC), 11 with HGD, and 33 with benign BE. According to the Seattle protocol [57], OCT and morphologic analysis were performed on a per level of BE basis (rather than per biopsy site or per patient); 181 levels were analyzed. The analysis included comparison of OCT images and histopathology slides from the same level. For test purposes, histopathologic diagnosis of metaplastic epithelium and low-grade dysplasia (LGD) was considered to be benign (112 levels).

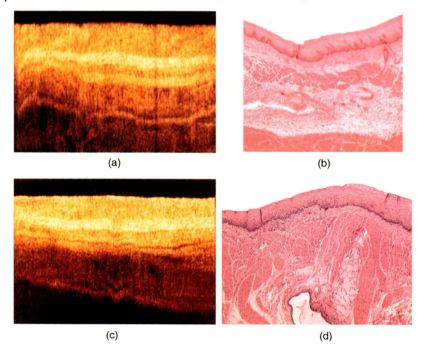

Figure 24.7 Images of normal esophagus: (a) OCT image with distinctive MM layer; (b) corresponding histology (H&E stain); (c) OCT image without distinctive MM layer; (d) corresponding histology (H&E stain). SSE, stratified squamous epithelium; LP, lamina propria; MM, muscularis mucosa; SM, submucosa; MP, muscularis propria (bars = 1 mm).

The following conditions were considered to be malignant: HGD (25 levels), IMC (27 levels), and adenocarcinoma invasive into muscularis propria (INVC) (17 levels). Altogether, 69 levels in malignant state were analyzed.

24.4.1.1 Differentiation Between Normal Esophagus and Benign BE

The OCT image of normal squamous esophageal mucosa has a five-layered (Figure 24.7a) or four-layered (Figure 24.7b) structure, depending on how distinctive the muscularis mucosa (MM) layer is.

In both cases, the upper layer with a moderate level of signal and thickness of ~100 μm corresponds to the stratified squamous epithelium (SSE). The second layer with the maximum level of backscattering corresponds to the lamina propria (LP). In the normal esophagus, the SSE layer is well contrasted from LP, and the basal membrane lies at the sharp interface between the two layers. If MM can be differentiated (Figure 24.7b), it will appear darker than the LP and submucosa (SM). In the absence of MM (Figure 24.7d), the contrast between the two connective tissue layers is less pronounced (Figure 24.7c). The fifth layer, muscularis propria (MP), always has low brightness.

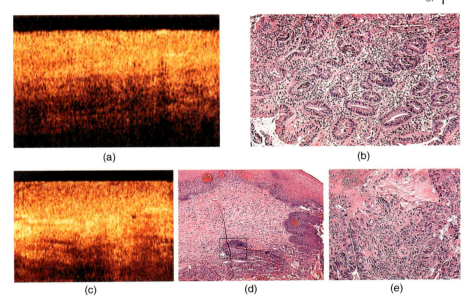

Figure 24.8 HGD of metaplastic epithelium. True-positive case: (a) absence of sharp interface between glandular mucosa and submucosa; (b) corresponding histology (H&E stain). False-negative case: (c) layered OCT image; (d) corresponding histology; (e) area of HGD delineated in a square (H&E stain).

Benign BE also has a layered structure. However, the BE OCT image is different, which is consistent with different histomorphology. Histologically, the upper layer, glandular mucosa (GM), is not a single layer and has a crypt structure with individual crypt size much higher than the resolution of our OCT system. However, the OCT contrast between crypts and mucus is insufficient at this resolution and the GM appears as a single layer.

The SM and MP are contrasted on OCT images of BE, but not so pronounced as in the normal squamous esophagus. As BE has an inflammatory origin, signs of inflammation are often present: dark areas correlating with exudate accumulations. Only three levels out of 112 with benign BE were false negative, the sensitivity of the method to detect BE versus normal esophagus was 97%, and the specificity was 100%.

24.4.1.2 Differentiation Between HGD and Benign BE

Loss of OCT contrast between epithelium and connective tissue is the main feature of HGD. This was observed for different types of epithelium, including stratified squamous, transitional, pseudostratified ciliated, and glandular epithelium [58]. OCT images of HGD were characterized by degradation of the sharp interface between GM and SM, and were considered to be positive for malignancy (Figure 24.8a).

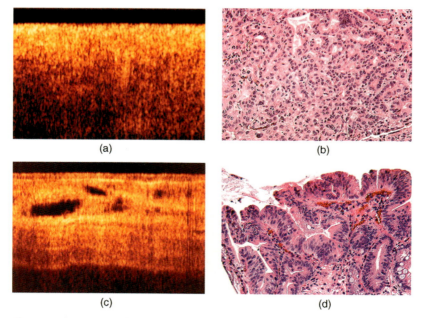

Figure 24.9 Intramucosal ADC. True-positive case: (a) OCT image; (b) corresponding histology (H&E stain). False-negative case: (c) OCT image; (d) corresponding histology (H&E stain).

For patients with BE, the sensitivity and specificity for OCT to detect HGD were 71% and 68%, respectively. Low contrast images of benign BE usually correlated with sites of scarring left from previous biopsies (false positive). Analysis of false negatives revealed that focal areas of SSE were overlapping underlying metaplastic epithelium with HGD (Figure 24.8c–e).

24.4.1.3 Differentiation Between IMAC and Benign BE

Adenocarcinoma is characterized by loss of layered structure on OCT images. The GM layer is indistinguishable from the underlying structures (Figure 24.9a).

The sensitivity for the detection of intramucosal adenocarcinoma (IMAC) in patients with BE was 85% and specificity 68%. Four cases out of 27 were false negative (Figure 24.9c), and represented a situation of focal invasion into the SM (when imaging and histology sites are not sufficiently correlated).

Endoscopic data showed that only 14 cases of IMAC were recognized visually by endoscopists. In the remainder of the cases, IMAC was detected in visually unchanged BE.

24.4.1.4 Differentiation Between Invasive Carcinoma and Benign BE

The main sign of invasive cancer on OCT images is loss of layered structure (Figure 24.10a). There were no false negatives during identification of invasive adenocarcinoma (INVAC) (sensitivity 100% and specificity 85%).

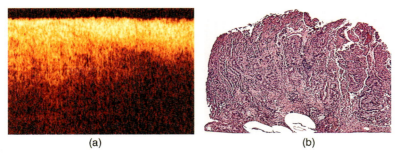

Figure 24.10 Invasive ADC: (a) true-positive OCT image; (b) corresponding histology (H&E stain).

24.4.1.5 Malignant BE Versus Benign BE

After combining all malignant states (HGD, IMAC, INVAC), the OCT sensitivity in detection of malignancy was 83% and specificity 68%.

The results of blind recognition of images are summarized in Table 24.2.

BE has been the focus of intense OCT research in the gastrointestinal tract. In Table 24.3, the diagnostic accuracy of OCT in detecting neoplasia in BE is presented. It must be emphasized that these results were obtained using different types of scanning and different criteria of OCT images [29, 55, 56, 59–61].

Evans *et al.* used the longitudinal scanning modality [29]. A catheter of this type exerts no pressure on mucous membrane surfaces, which enables one to observe surface features and differentiate enlarged glands. For diagnosis of IMC/HGD, the authors proposed to use criteria such as surface maturation and gland architecture. They hypothesized that incomplete surface maturation, indicative of dysplasia or regenerative changes, would be seen as a high surface OCT signal compared with the subsurface signal; gland irregularity by OCT might be characterized by irregular size, shape, and distribution of these architectural structures. A total of 177 biopsy correlated images were analyzed. Good correlation between histopathologic diagnosis of IMC/HGD and scores for each image feature [dysplasia index (Spearman correlation coefficient), $r = 0.50$, $p < 0.0001$], surface maturation ($r = 0.48$, $p < 0.0001$), and gland architecture ($r = 0.41$, $p < 0.0001$) was found. When a dysplasia index threshold of >2 was used, the sensitivity and

Table 24.2 Statistical parameters of blind comparison of OCT images and histopathology data.

Parameters	HGD + IMC + INVC	HGD + IMC	HGD	IMC
Sensitivity (%)	83	78	71	85
Specificity (%)	68	68	68	68
Diagnostic accuracy (%)	74	71	69	71
Positive predictive value (%)	62	53	33	39
Negative predictive value (%)	87	87	92	95
Kappa	0.626	0.585	0.562	0.512

Table 24.3 Diagnostic accuracy of OCT in detecting neoplasia in BE in blind recognition (according to data from different studies).

Study	No. of patients (images)	Sensitivity (%)	Specificity (%)	Diagnostic accuracy (%)	Ref.
1. BE/SE, Gastro, Cardia	121	97	92	–	Poneros et al., 2001 [59]
2. BE/LGD, HGD, IMC	23 (152)	69	71	70	Isenberg et al., 2003 [60]
3. BB/LGD, HGD, IMC	33 (314)	68	82	78	Isenberg et al., 2005 [61]
4. BB + LGD/HGD, IMC, INC	44	75	91	89	Zuccaro et al., 2001 [56]
5. BB/HGD, IMC					Evans et al., 2006 [29]
6. BB + LGD/HGD, IMC	67 (164)	78	68	70	Gladkova et al., 2007 [55]
7. BB + LGD/HGD, IMC, INC	78 (181)	83	68	74	Gladkova et al., 2007 [55]

specificity for diagnosing IMC/HGD were 83% and 75%, respectively. The authors concluded that an OCT image-scoring system based on histopathologic characteristics has the potential to identify IMC and HGD in BE. Note that these criteria cease to work for a dysplasia index >2 [29].

The team working at the Division of Gastroenterology in the Case Western Reserve University Hospitals of Cleveland [20, 62] employs endoscopic OCT with radial scanning. In 2005, a double-blinded endoscopic OCT study was performed by four endoscopists in 33 patients with documented BE. A total of 314 pairs of OCT images and jumbo biopsy specimens were analyzed. A cap-fitted, two-channel endoscope and rotation OCT probes were used that ensured target biopsy from the scanned regions. The formulated OCT criteria of BE for a rotation probe have much in common with those of ultrasonic imaging. Dysplasia and adenocarcinoma had the following OCT features: (1) focal areas of decreased light scattering and (2) focal loss of mucosal structure and organization. By using histology as the standard, the endoscopic OCT had sensitivity 68%, specificity 82%, PPV 53%, negative predictive value 89%, and diagnostic accuracy 78%. Based on these results, it was concluded that further modifications, including increased resolution and identification of further potential OCT characteristics of dysplasia, are needed before OCT can be used clinically.

Analysis of analogous groups of patients by our team enables us to argue that OCT is a promising technique for the diagnosis of BE neoplasia, which implies still further clinical experience and also further development of the technology. Making use of the specific OCT features of BE, we found that the sensitivity and specificity of endoscopic OCT in detecting malignization in BE (HGD, IMAC, INVAC) were 83% and 68%, respectively. However, the OCT sensitivity is rather low when detecting HGD only. Recent research in the field of Doppler OCT [63], ultra-high resolution OCT [64], high scanning speed OCT [65], *in vivo* 3D comprehensive microscopy [66], and computer-aided diagnosis for identification of dysplasia [67] demonstrated additional diagnostic features of the development of early neoplasia in BE, which are expected to increase the diagnostic value of OCT.

24.4.2
OCT for Surgical Guidance of Colon Polyps

The malignancy potential of colorectal polyps is well known in endoscopic practice [68].

Endoscopic OCT was tested as a possible tool for *in vivo* endoscopic differential diagnosis of colon polyps and for assessing the need for their removal during colonoscopy. The studies were undertaken at the Eppendorf Clinic, University of Hamburg (Germany) [69] and the Cleveland Clinic Foundation (United States) [70].

The studies revealed OCT features of adenomas and hyperplastic polyps based on which the diagnostic accuracy of OCT was statistically analyzed; 48 tubular adenomas, 12 tubulovillous adenomas, and 56 hyperplastic polyps were studied. The OCT features of different types of colon polyps were formulated:

1) The hyperplastic polyp is characterized by a three-layer "benign" OCT image, thickening of the upper layer (glandular mucous membrane), and a clear border between the glandular mucous membrane and submucous layer (Figure 24.11a–c).
2) Adenoma, independent of its size, is characterized by an OCT image without layers (Figure 24.11d–f).

Analysis of 116 polyps in the open test showed that OCT differentiates adenomas from hyperplastic polyps with good sensitivity (92%) and specificity (84%).

Our results on the differential diagnosis of colon polyps by the OCT technique agree with preliminary data obtained in clinical studies using an analogous OCT device [69]. Jaeckle *et al.* [69] identified OCT features of adenoma as an image without layers and of hyperplastic polyp as an image in which one can see a glandular mucous membrane, a submucous layer, and a muscular layer. Pfau *et al.* [27] used a rotation probe and also ascertained the capability of OCT to distinguish reliably between hyperplastic and adenomatous polyps. They investigated 44 polyps in 24 patients. Endoscopic optical coherence tomography (EOCT) showed that adenomas were rated significantly more disorganized, with less structure than hyperplastic polyps ($p = 0.0005$). Polyps that were subsequently found to be adenomas in histopathologic evaluation were judged during EOCT to exhibit a significant decrease in light scattering, appearing darker in EOCT images compared with hyperplastic polyps ($p = 0.0007$). It was found that the mean difference in light scattering was significantly greater between adenomas and normal tissue (mean difference = 45.81) than the mean difference in scattering between hyperplastic polyps and normal tissue (mean difference = 14.86) ($p = 0.0001$). The authors

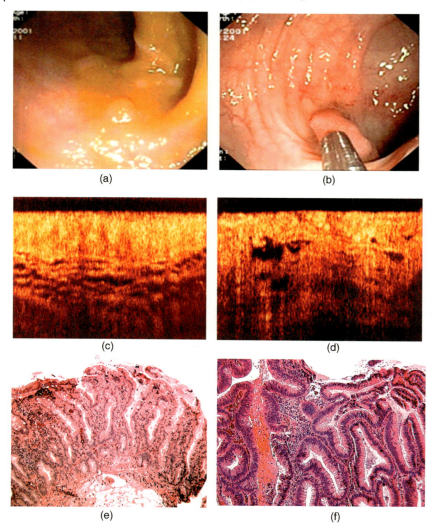

Figure 24.11 Endoscopic OCT and the corresponding histology images of (a–c) hyperplastic polyp and (d–f) adenomatous polyp.

considered that morphologic feature quantification of colonic crypt patterns using a microscope integrated OCT scanner [67] is highly promising for clinical applications.

We defined specific features of OCT images of adenoma and hyperplastic polyps obtained using the forward-looking probe on 116 polyps. It was found that OCT differentiates between adenoma and hyperplastic polyps with good sensitivity (92%) and specificity (84%).

Other groups have also reported on the potential capability of OCT to distinguish flat adenomas from normal mucous membrane of intestines, which is sometimes difficult in traditional endoscopy. Also reported was a preliminary study of a single case of flat adenoma using Doppler OCT. It was shown that the development of adenoma is accompanied, along with changes in structural features, by changes in the vascular pattern [71].

24.4.3
OCT in Guided Surgery of Gastrointestinal Cancer (Esophagectomy and Rectoectomy)

The problem of adequate tradeoff between radical tumor removal and maximum preservation of organs is very important for cancer management and patient quality of life.

We have evaluated OCT for pre- and intraoperative planning of the tumor resection margins on esophageal and rectal carcinoma. The standard EOCT was used for 19 patients with rectal adenocarcinoma and 24 patients with distal esophageal cancer (14 zsquamous cell, 10 adenocarcinoma, eight cardial type and two Barrett's adenocarcinoma).

This study was performed in the Nizhny Novgorod clinics (Russia). The samples included patients to be operated on for rectal or distal esophageal cancer. Examinations were performed for patients with esophageal cancer located not higher (no more proximal) than 5 cm from the Z-line and rectal adenocarcinoma with tumors in the upper and middle thirds. This inclusion criterion is associated with the possibility of performing organ-preserving surgery.

Taking into account that the endoscopic biopsy involves only the superficial esophageal layers (mucosal and submucosal), the OCT border was determined *ex vivo*, and histopathologic examination of the specimens was performed. We formulated the types of tumor growth along the esophagus and rectum. If the tumor spreads into the mucosa layer, the OCT image has no layers. If the tumor growth occurs along the submucous layer, OCT reveals a contrast border between the mucous and submucous layers, but does not show a contrast border between the submucous and muscle layers.

When the tumor had spread along the muscle layer, we acquired false-negative benign, very well-structured OCT images with two contrast borders. In the cases of mucosal growth, OCT correctly detected tumor borders in all patients with esophageal (4 pt) (Figure 24.12) and rectal cancer (16 pt). In the cases of submucosal tumor growth, OCT worked very well in the detection of tumor borders in all cases of esophageal (16 pt) and rectal cancer (9 pt). However, OCT could not determine tumor borders in patients with muscle cancer that had spread into either the esophagus (4 pt) or the rectum (2 pt).

Based on OCT *ex vivo* data, we developed a technique of endoscopic tumor border detection. Scanning was performed during the presurgery endoscopic examination. The OCT border was detected and marked for further surgery. The EOCT probe investigated the visual border of the tumor (distal for rectal carcinoma and proximal for esophageal carcinoma) along four directions of a conventional dial (12, 3, 6, and 9 o'clock). The above zones were OCT imaged in 0.3–0.5 cm steps until structural

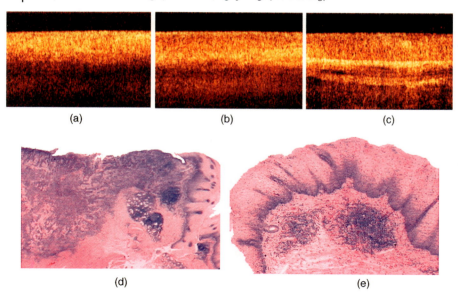

Figure 24.12 Finding the border of mucosal growth of squamous cell carcinoma of esophagus (*ex vivo*). OCT image of tumor focus (a) and the corresponding histology specimen (d); the OCT border (b, arrow) coincides with the histology border of squamous cell nonkeratinizing cancer (d); control OCT and histology study 1 cm from the border did not detect tumor growth (c, e). H&E staining of histology specimens.

images corresponding to normal mucosa were obtained. The OCT border was marked by electrocoagulator or methylene blue tattoo.

24.5
Gynecology

The first *in vivo* OCT images of the uterine cervix were obtained by the Nizhny Novgorod team in 1997 [22]. Later, the clinical feasibility of OCT in gynecology was discussed in a number of papers [72–76], describing the OCT features typical for a normal exocervix, endocervix, and endometrium. It was shown that OCT is capable of detecting neoplastic changes in stratified squamous epithelium. In this section, we demonstrate the potential of the clinical use of a new visualization method in gynecologic practice.

24.5.1
Methodology and Patient Selection

Development of the protocol for uterine cervix target biopsy was done in three stages: initial approbation of OCT for diagnosis of cervical pathology, study of the influence of acetic acid on OCT images, and development and approbation of the protocol based on OCT colposcopy. In all the stages of the investigation, the female patients had the following indications for colposcopy:

- history of abnormal colposcopic findings
- abnormal results of Pap smear (ASCUS, LSIL; HSIL)
- positive HPV test for oncogenic viral type.

The exclusion criterion was age under 18 years.

The study was approved by the Ethical Committee for scientific studies with human subjects. The patients had to sign an informed consent document stating that they understood the investigational nature of the proposed study.

A typical scenario of the study was similar in all the stages:

- review of patient's history, including cytology and HPV testing results
- colposcopy and identification of regions of interest (ROI)
- OCT of identified ROI
- biopsy of ROI, when clinically appropriate
- histopathology evaluation of biopsy and comparison with OCT and colposcopic findings.

The diagnostic efficiency of OCT in recognizing malignization was studied using blind recognition tests of OCT images.

24.5.2
Guidance Biopsy During Colposcopy

In the first stage, a preliminary investigation of OCT capabilities for detecting neoplasia-induced alterations in the uterine cervix was carried out; 120 patients took part in the study. It was revealed that the most typical OCT feature of malignization was loss of specific image structure, which allowed us to formulate features of "benign" and "malignant" OCT images. Using this approach, we assessed the diagnostic efficiency of OCT for the recognition of cervical malignization. The sensitivity was shown to be 82%, specificity 78%, error 19.2%, and kappa 0.65. It should be noted that the maximum percentage error (Table 24.4) was due to false-negative recognition of metaplasia whose optical images are low structured.

This led us to revise some of the OCT criteria of malignization, for classifying the interpreted images not only as "malignant" or "benign" but also as "suspicious" (Figure 24.13) [92], and for looking for ways to increase the diagnostic efficiency of the technique.

It is worth noting that OCT should not be regarded as an independent diagnostic technique. Research carried out in the first stage demonstrated that to increase the diagnostic quality of cervical pathology, OCT data should be used in combination with information about clinical symptoms, cytology data, and colposcopy results. Therefore, the development of the OCT-based diagnostic protocol demands, in the first place, maximum adaptation of the new method (OCT) to the basic procedure (colposcopy). For validating the OCT–colposcopy technique, in the second stage we

Table 24.4 Distribution of the indices of OCT diagnostic efficiency in recognizing malignization.

Condition	No. of recognitions	True positives	False positives	False negatives	True negatives	Error (%)
Invasive carcinoma	70	62	0	8	0	11.4
Microinvasive carcinoma	28	23	0	5	0	14.2
High-grade dysplasia (HGD)	91	70	0	21	0	23
Inflammation	14	0	0	0	14	0
Metaplasia	56	0	23	0	33	41
Low-grade dysplasia	42	0	2	0	40	4.7

studied the influence of traditional colposcopic tests (application of acetic acid) on the character of OCT images of the uterine cervix.

A total of 120 female patients took part in the study. The structure of the uterine cervix states is presented in Table 24.5.

It was found that local application of a 3% solution of acetic acid in the case of uterine cervical neoplasia reduces the depth of OCT scanning 2–3-fold. We formulated differential criteria for OCT visualization of metaplastic and neoplastic epithelia on local exposure to 3% acetic acid solution, taking into account that interpretation of immature metaplasia is the most difficult for diagnosis in both colposcopy and OCT. In the presence of immature metaplasia, the OCT image persists in being structureless and the depth of scanning does not change, in contrast to the case of severe dysplasia and cervical cancer, when the velocity of signal attenuation becomes very fast, which manifests itself as a pronounced decrease in the size of the image down to half to one-third frames (Figure 24.14).

Being the most difficult for traditional diagnosis and OCT study (the error rate in the previous stage was 41%), cervical metaplasia was chosen for assessing the diagnostic efficiency of the new approach in OCT diagnostics. Results of the "blind" recognition test (Table 24.6) showed that the use of the proposed interpretation criteria enabled the error rate to be reduced to 15%.

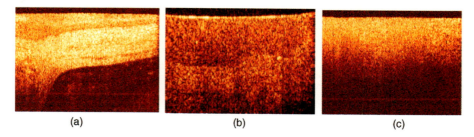

Figure 24.13 Types of OCT images of the cervix: (a) benign, (b) suspicious, and (c) malignant.

Table 24.5 The structure of cervical states.

Norm	21 (17.5%)		
Benign	56 (46.7%)		
Malignant	HGD 26 (21.6%)	Microinvasive carcinoma 5 (4.2%)	Invasive carcinoma 12 (10%)

This research allowed us to formulate the diagnostic protocol for examining the uterine cervix:

- colposcopy using acetic acid
- OCT study taking into account the impact of acetic acid
- target biopsy of the cervical mucous membrane by indications of OCT–colposcopy.

This protocol was tested in clinical conditions with the participation of 90 patients. The study was aimed at evaluating the efficiency of the developed protocol of OCT-colposcopy for taking a decision about uterine cervix biopsy. Examination was carried out following the proposed protocol, but the decision about biopsy was taken by a standard protocol based on colposcopic data. According to our data on abnormal

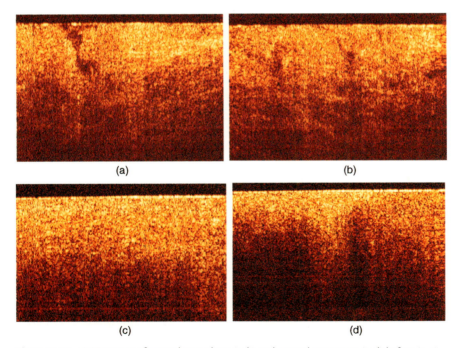

Figure 24.14 OCT images of cervical metaplasia (a, b) and cervical carcinoma (c, d) before (a, c) and after (b, d) application of a 5% solution of acetic acid.

Table 24.6 Diagnostic efficiency of the new approach in OCT diagnostics.

	True positives	False positives	False negatives	True negatives	Sensitivity (%)	Specificity (%)	Error (%)	Kappa
Total recognitions 430	217	20	43	150	83	88	15	0.67

colposcopic findings, 79% of cases had indications for biopsy and only in 36% of cases were malignant and suspicious types of images detected by OCT. Histology studies of biopsy material confirmed the presence of benign states in the benign types of OCT images (Figure 24.15).

We believe that inclusion of OCT in the protocol for uterine cervix examination allows more accurate assessment of the state of the object under study and reduces substantially (by more than 40%) the number of invasive procedures (biopsies) needed, which will have important medical, social, and economic impacts. When no biopsy is taken in the case of benign types of images, OCT may be an objective follow-up technique that will allow timely changes of therapeutic management in case of necessity.

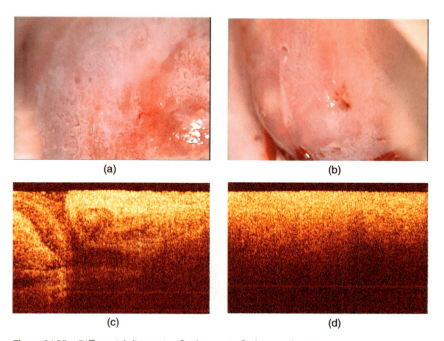

Figure 24.15 Differential diagnosis of colposcopic findings with OCT.

24.6
Cross-Polarization OCT in Oncology

Traditional OCT is most informative for the investigation of layered tissues covered by epithelium. The presence and loss of contrast between the epithelium and the underlying connective tissue are an important diagnostic feature recorded by means of OCT [77]. In some cases, traditional OCT images of benign and malignant tissues are both structureless, which makes them indistinguishable, thus reducing the specificity of the method. Loss of contrast between layers occurs as a result of changes in optical characteristics of the epithelium and also of connective stroma. Standard OCT is not intended for detecting changes in the connective stroma.

Structured components of connective tissue (collagen fibrils and fibers) are formed from collagen. High organization of collagen ensures its capability to change the polarization of light passing through biotissue. In the context of interest to us, collagen is an important birefringent molecule in the human body. Birefringence is the principal polarization characteristic of tissues that may be registered by means of an OCT modality known as polarization-sensitive OCT (PS-OCT) [78].

Another, no less important, polarization characteristic of biotissue is depolarization. Depolarization is the property of material to eliminate polarization in a sum scattered wave. If backscattered radiation contains radiation with polarization orthogonal to that of the probing wave, then it may be recorded by a PS-OCT modality referred to as cross-polarization OCT (CP-OCT) [36].

Neoplasia changes the properties and integrity of healthy tissues, hence it changes the collagen state and polarization features of tissues. That is why detection of birefringence and depolarization properties of tissues by means of PS-OCT may be a useful means for increasing the specificity of traditional OCT.

Elaboration of new polarization techniques is a promising trend in OCT development. Nearly all research groups involved in OCT studies have used polarization methods [79–81].

Histologic methods of assessing collagen and its birefringent properties have been developed, but only *ex vivo*. Of particular interest is the method of light microscopy in polarized light using PicroSirius Red (PSR) staining, which allows the reliable detection of collagen possessing high birefringence. The use of PSR staining in polarized light gives information about the size of the structures, their number, composition, and location, birefringence ability, and, as a consequence, the functional state of tissue [78]. The majority of authors working with PSR staining in polarized light are unanimous in reporting that staining confirms the collagen with enhanced linear birefringence. When stained by PSR, organized collagen shows up as bright regions whose yellow–orange and red colors are typical for thick fibers (type I collagen with fiber diameter 1.6–2.4 µm), and dark green for thin fibers (type III collagen, fiber diameter 0.8 µm) [82–84].

Most researchers have studied birefringence using CP-OCT systems [85, 86], which are a convenient tool for assessing biological tissues with high collagen content. These include tissues that are of low contrast for traditional OCT, such as tendon, articular cartilage [87], and derma [88], as they have a relatively uniform

Table 24.7 Volume of the studied material depending on biological tissue localization.

Object under study	No. of patients	No. of histology specimens	No. of CP-OCT images
Cervical tissue	68	78	148
Urinary bladder tissue	65	72	152
Esophagus tissue	17	23	143
Total	150	173	443

structure. Birefringence is visualized as contrast-alternating bright and dark stripes elongated primarily along the surface, whose spatial period is determined by the magnitude of birefringence. Several such bands may appear in the image at a strongly anisotropic refractive index in biotissue ($\Delta n = 10^{-3}$–10^{-2}).

Two polarization-conjugate images are demonstrated when assessing the depolarization properties of tissue [33, 35]. Comparison of scattering in initial and orthogonal polarizations gives detailed information about microstructures and biochemical alterations depolarizing tissue components, collagen in particular. Our preliminary studies have revealed that the CP-OCT technique may provide clinically important information about mucosa as connective tissue fibers are capable of scattering probing radiation not only to direct but also to orthogonal polarization. Comparison of the patterns of backscattering in direct and orthogonal polarizations proved to be most informative for differentiation between clinically similar cases of esophageal carcinoma and cicatrized esophagus tissue after chemical burning [30], and also for differentiation between cervical leukoplakia and cancer [77].

Thus, the possibility of *in vivo* (noninvasive) observation of collagen changes refines the treatment of suspicious OCT cases and opens up new diagnostic capabilities of OCT.

Cross-polarization investigation of urinary bladder, uterine cervix, and esophageal mucosas was carried out in the course of endoscopic procedures: cystoscopy, colposcopy, and fibrogastroscopy (Table 24.7).

24.6.1
CP-OCT in Oncogynecology

Figures 24.17–24.19 demonstrate two types of OCT images: in direct and in orthogonal polarization. OCT images of the unaltered uterine cervix are shown in Figure 24.16, where a layered horizontally organized mucous structure is visualized. In direct polarization (Figure 24.16a, lower image), the epithelium looks like a moderately uniformly scattering layer that has a sharp border with a more intensely scattering subepithelial tissue. In orthogonal polarization (Figure 24.16a, upper image), the epithelium has, primarily, a cellular structure and is visualized as a very weakly scattering layer. PSR staining of histology specimens in polarized light reveals thick, well-ordered, oriented along the tissue surface, bright-red collagen fibers (type I

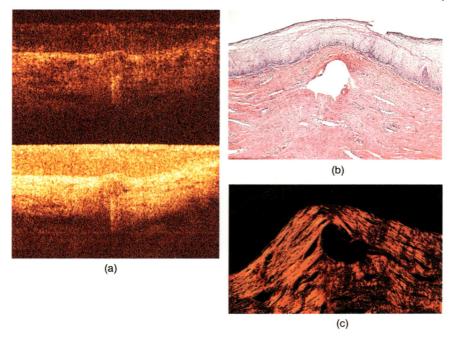

Figure 24.16 Mucous layer of normal uterine cervix. The layers were marked with Safil surgical suture 150 μm in diameter, transparent to OCT probing radiation. The suture was sewn immediately under the epithelium. (a) OCT image in direct (bottom) and orthogonal (top) polarization; (b) histologic specimen, H&E staining, frame size 1.7 × 1 mm; (c) histology specimen, PSR staining in polarized light, frame size 1.7 × 1 mm.

collagen). CP-OCT images of the mucous layer of the normal uterine cervix feature layered tissue structures. Visualization of images in direct and orthogonal polarization allow reliable differentiation of tissue layers with high contrast, and also various inclusions, blood vessels in particular. The signal intensity in orthogonal polarization reflects the presence of type I collagen in stroma.

Two cases where the colposcopic pattern is suspicious for cancer (stippling, mosaic, moderate aceto-white reaction) are illustrated in Figures 24.17 and 24.18. Figure 24.17 demonstrates a mucous layer of the uterine cervix with a growth site of squamous cell nonkeratinous carcinoma with an invasion depth of about 2–3 mm.

Figure 24.18 shows the site of the mucous layer of the uterine cervix (colposcopically suspicious for neoplasia) with histology features of cylindrical epithelium ectopia (immature metaplasia) (without epithelium epidermization and high-grade cellular infiltration of stroma) (benign state).

The presence or absence of a signal in orthogonal polarization may have differential-diagnostic value when traditional OCT images are structureless. Both OCT images are indistinguishable in direct polarization – they have no horizontal layered organization. The typical structureless OCT image depicted in Figure 24.19 may be interpreted as a sign of malignization, which is a false-positive result. In the case of

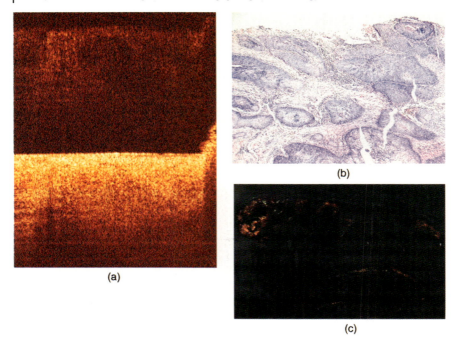

Figure 24.17 Mucous layer of the uterine cervix, growth site of squamous cell nonkeratinous carcinoma with an invasion depth of about 2–3 mm. (a) OCT image in direct (bottom) and orthogonal (top) polarization; (b) histology specimen, H&E staining, frame size 1.7 × 1 mm; (c) histology specimen, PSR staining in polarized light, frame size 0.75 × 0.5 mm. PSR staining of the histology specimen in polarized light shows that the area corresponding to cancer complexes is almost non-luminescent. Tumor stroma is represented by sporadic thin yellow–green fibers. OCT image: in direct polarization, the image field is uniform with a high-intensity signal, no layered structure is differentiated. In orthogonal polarization, the image field has no signal; there is signal at a site corresponding to stroma. Invasive squamous cell carcinoma is almost devoid of stroma containing organized collagen. Therefore, in orthogonal polarization, there is almost no signal at the site of cancer, which may be a differential feature of invasive tumor of arbitrary differentiation.

immature metaplasia, a signal is detected in the image in orthogonal polarization because the maturity and orientation of collagen fibers are conserved, in contrast to cancer when disorganized collagen has no polarization characteristics.

Thus, CP-OCT furnishes additional information about the states giving low-contrast images in direct polarization.

24.6.2
CP-OCT in Oncourology

CP-OCT images of normal mucosa of the urinary bladder in direct and orthogonal polarization feature a layered tissue structure. The signal intensity in orthogonal polarization reflects the presence of type I collagen in stroma. Visualization of images

24.6 Cross-Polarization OCT in Oncology

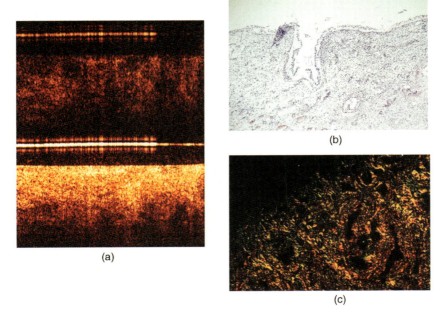

Figure 24.18 Mucous layer of the uterine cervix, cylindrical epithelium ectopia (immature metaplasia). (a) OCT image in direct (bottom) and orthogonal (top) polarization; (b) histology specimen, H&E staining, frame size 1.7 × 1 mm; (c) histology specimen, PSR staining in polarized light, frame size 0.75 × 0.8 mm.

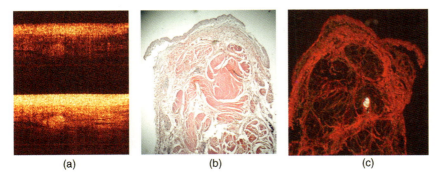

Figure 24.19 Urinal bladder wall specimen sewn with surgical suture transparent to OCT probing radiation. (a) OCT image in direct (bottom) and orthogonal (top) polarization; (b) histology specimen, H&E staining, frame size 1.8 × 2.3 mm; (c) histology specimen, PSR staining in polarized light, frame size 1.8 × 2.3 mm. The urinary bladder tissue contracts during OCT examination, hence layer thicknesses do not coincide in OCT and histology images.

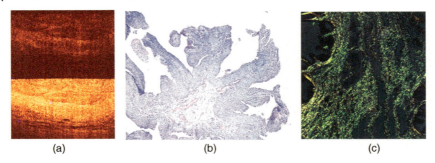

Figure 24.20 Simple urinary bladder papilloma. (a) OCT image in direct (bottom) and orthogonal (top) polarization; (b) histology specimen, H&E staining, frame size 1.8 × 2.3 mm; (c) histology specimen, PSR staining in polarized light, frame size 0.7 × 0.4 mm.

in direct and orthogonal polarizations allows reliable differentiation of all layers of bladder wall tissues with high contrast (Figure 24.19).

Benign states that look like zones of neoplasia at cyctoscopy [for example, papillomatous (exophytic) urinal bladder formation with histology features of simple papilloma without features of cell atypia] (Figure 24.20) give a signal in orthogonal polarization. Connective tissue of benign papilloma gives a signal in orthogonal polarization due to the unchanged mature type I and type III collagen (yellow and green luminescence in PSR-stained histology specimens, Figure 24.20c).

At the same time, exophytic papillomatous growth of urothelial tumors with high-grade differentiation but without signs of invasion into muscular layers (Figure 24.21) gives a structureless image in direct polarization and no signal in orthogonal polarization, because malignant papilloma is almost free of stroma containing organized collagen (Figure 24.21c).

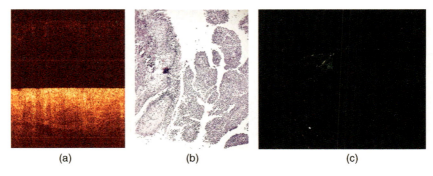

Figure 24.21 Exophytic papillomatous growth of urinary bladder mucosa – high-grade differentiation urothelial tumor without invasion into muscular layer. (a) OCT image in direct (bottom) and orthogonal (top) polarization; (b) histology specimen, H&E staining, frame size 1.7 × 1 mm; (c) histology specimen, PSR staining in polarized light, frame size 0.75 × 0.5 mm.

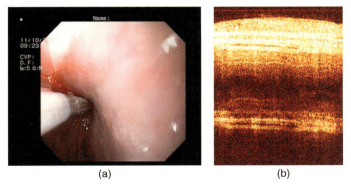

Figure 24.22 Normal squamous cell mucosa of esophagus. (a) Probe in esophagus lumen; (b) OCT image in direct (top) and orthogonal (bottom) polarization.

24.6.3
CP-OCT in Oncogastroenterology

Structured OCT images in standard and orthogonal polarization were also visualized in studying the tissues of the normal gastrointestinal tract (Figure 24.22). Verification of zones suspicious for cancer growth gave low-structured, non-layered cancer-type images (Figure 24.23). Malignant tumor does not depolarize radiation in orthogonal polarization and the OCT signal is minimal.

To conclude, when benign states are scanned, CP-OCT demonstrates structured layered images of benign type. Malignant neoplasms of different localizations are visualized as nonlayered images in standard polarization and are characterized by a weak OCT signal or no signal at all in orthogonal polarization. Some benign states are visualized as structureless images in direct polarization, whereas in orthogonal

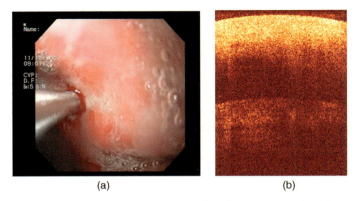

Figure 24.23 Tumor (adenocarcinoma) surface of the middle one-third of stomach body. (a) Probe in stomach lumen; (b) OCT image in direct (top) and orthogonal (bottom) polarization.

polarization a signal is detected, which enhances the specificity of CP-OCT for recognition of neoplasia.

24.7
Conclusion

Earlier pilot studies that used EOCT demonstrated an excellent ability to delineate clearly the layered structure of the esophagus, colon, bladder, vessels, and other biological structures into which the catheter is inserted. This feature of OCT images is universal for the great majority of EOCT devices [42, 75, 89]. Subsequent studies have validated the diagnostic accuracy of OCT for detecting specialized intestinal metaplasia, HGD, and intramucosal carcinoma of the esophagus and bladder [53] and dysplastic colonic polyps [90].

Our studies were aimed at statistical justification of using OCT in endoscopy. The clinical sampling included patients with urinary bladder (164 pt) and gastrointestinal tract (390 pt) pathology. The material was combined into one group as these organs are accessible to OCT. Further, in this fashion we demonstrated that the new diagnostic technique is unique and universal for organs of different morphology. As a result, we analyzed OCT images of pathologic states of the basic types of mucous membranes with transition, multilayer flat, and glandular epithelium.

In our research we used a handy en-face EOCT scanner. Its perpendicular position and moderate pressing on tissue ensure stable images with good tissue contrast; no additional fixing devices are required. The en-face design makes it convenient for investigation of both tubular (gastrointestinal tract) and large, hollow (urinal bladder) organs; further, it is useful for studying skin and mucous membranes of any localization. Its drawback, as in any other linear scanning technique, is the small surface area of a studied sample that is limited by the diameter of the scanner optical window.

Analysis of the results of the application of EOCT enables us to state that this novel visualization technique has found its niche in the diagnostic process. It is capable of solving several diagnostic tasks: early malignancy detection, tumor border marking, tumor removal control, and differential diagnosis of various pathologic forms. In this chapter, we have demonstrated the high diagnostic accuracy of EOCT in specific clinical tasks.

It is difficult to compare results of EOCT diagnostic efficiency obtained by different workers as they investigated different objects (different stages of the neoplastic process) and used different approaches to scanning. Nevertheless, the data presented demonstrate comparability of the independent results obtained. Statistical data for our study show that the diagnostic accuracy of the EOCT technique with criteria for image interpretation proposed by our team is comparable to results reported by other groups, and even better in some cases.

We intend to refine the technique further. The development of high-resolution fast scanning systems is known to be the main direction towards improving OCT. However, it should be taken into account that for promotion into clinical practice, characteristics such as handy and simple use, portability, and economy are very important. These properties are inherent in our device. We believe that the diagnostic accuracy of such an OCT modality may be enhanced as follows:

- by developing basically new schemes, for example, PS-OCT
- by creating a high-speed scanning system that would enlarge the area of the analyzed surface
- by further minimization of the en-face probe for better matching with standard endoscopic equipment.

Concerning trends in the development the OCT technique as a whole, we hope that a scanning system integrating all the advantages of available approaches will be created and universal criteria of image interpretation will be developed.

Acknowledgments

This work was supported by CRDF, State Contract of the Russian Federation No. 02.522.11.2002, and Presidium of RAS (program "Fundamental Science for Medicine"), Russian Foundation for Basic Research (projects 08.04.97098-r-a, 07-02-01146-a, 08-02-01152, and 08-02-99049-r_ofi), and IMALUX Corporation.

The authors are grateful to Professor M. J. Manyak and his team of the George Washington University Medical Center (United States), Dr. G. Zuccaro and his colleagues at the Cleveland Clinic Foundation (United States), Dr. S. Jäckle and his team at the Eppendorf Clinic, University of Hamburg (Germany), Dr. F. I. Feldchtein, Dr. A. I. Abelevich, Dr. A. N. Denisenko, Dr. V. E. Zagaynov, and A. G. Orlova for their participation in the clinical studies, and N. B. Krivatkina for technical support in this work.

References

1 Jemal, A., Siegel, R., Ward, E., Murray, T., Xu, J. *et al.* (2006) Cancer statistics, 2006. *CA Cancer J. Clin.*, **56**, 106–130.

2 Jemal, A., Siegel, R., Ward, E., Murray, T., Xu, J., and Thun, M. (2007) Cancer statistics, 2007. *CA Cancer J. Clin.*, **57**, 43–66.

3 Koshy, M., Esiashvilli, N., Landry, J., Thomas, C., and Matthews, R.J. (2004) Multiple management modalities in esophageal cancer: epidemiology, presentation and progression, work-up, and surgical approaches. *Oncologist*, **9**, 137–146.

4 Parkin, D.M., Bray, F., Ferlay, J., and Pisani, P. (2005) Global cancer statistics, 2002. *CA Cancer J. Clin.*, **55**, 74–108.

5 Anderson, P. and Runowicz, C. (2001) Beyond the PAP test. New techniques for cervical cancer screening minimizing false-negative results. *Women's Health Prim. Care*, **21**, 753–758.

6 Ronco, G. and Rossi, P. (2008) New paradigms in cervical cancer prevention: opportunities and risks. *BMC Women's Health*, **8**, 23.

7 Oosterlinck, W. (2004) Guidelines on diagnosis and treatment of superficial bladder cancer. *Minerva Urol. Nefrol.*, **56**, 65–72.

8 Tschanz, E.R. (2005) Do 40% of patients resected for Barrett esophagus with high-

grade dysplasia have unsuspected adenocarcinoma? *Arch. Pathol. Lab. Med.*, **129**, 177–180.

9 Burghardt, E., Pickel, H., and Girardi, F. (eds.) (2002) *Primary Care Colposcopy: Textbook and Atlas*, Thieme Medical Publishers, Stuttgart.

10 Sharma, P., Weston, A.P., Topalovski, M., Cherian, R., Bhattacharyya, A., and Sampliner, R.E. (2003) Magnification chromoendoscopy for the detection of intestinal metaplasia and dysplasia in Barrett's oesophagus. *Gut*, **52**, 24–27.

11 Meining, A., Rosch, T., Kiesslich, R., Muders, M., Sax, F., and Heldwein, W. (2004) Inter- and intra-observer variability of magnification chromoendoscopy for detecting specialized intestinal metaplasia at the gastroesophageal junction. *Endoscopy*, **36**, 160–164.

12 Baumgartner, R., Wagner, S., Zaak, D., and Knuchel-Klarke, R. (2000) *Fluorescence Diagnosis of Bladder Tumor by Use of 5-Aminolevulinic Acid: Fundamental and Results*, Endo-Press, Tutlingen.

13 Dacosta, R.S., Wilson, B.C., and Marcon, N.E. (2003) Photodiagnostic techniques for the endoscopic detection of premalignant gastrointestinal lesions. *Dig. Endosc.*, **15**, 153–173.

14 Scotiniotis, I.A., Kochman, M.L., Lewis, J.D., Furth, E.E., Rosato, E.F., and Ginsberg, G.G. (2001) Accuracy of EUS in the evaluation of Barrett's esophagus and high-grade dysplasia or intramucosal carcinoma. *Gastrointest. Endosc.*, **54**, 689–696.

15 Huang, D., Swanson, E.A., Lin, C.P., Schuman, J.S., Stinson, W.G. *et al.* (1991) Optical coherence tomography. *Science*, **254**, 1178–1181.

16 Bouma, B.E. and Tearney, G.J. (eds.) (2002) *Handbook of Optical Coherence Tomography*, Marcel Dekker, New York.

17 Gelikonov, V.M., Gelikonov, G.V., Gladkova, N.D., Kuranov, R.V., Petrova, GA. *et al.* (1995) Coherent optical tomography of microscopic inhomogenities in biological tissues. *Pis'ma Zh. Eksp. Teor. Fiz.*, **61**, 149–153.

18 Gelikonov, V.M., Gelikonov, G.V., Dolin, L.S., Kamensky, V.A., Sergeev, A.M. *et al.* (2003) Optical coherence tomography: physical principles and applications. *Laser Phys.*, **13**, 692–702.

19 Gelikonov, V.M., Gelikonov, G.V., Gladkova, N.D., Leonov, V.I., Feldchtein, F.I. *et al.* (1998) U.S. Pat. US 5835642.

20 Das, A., Sivak, M.V., Chak, A., Wong, R.C.K., Westphal, V. *et al.* (2001) High-resolution endoscopic imaging of the GI tract: a comparative study of optical coherence tomography versus high-frequency catheter probe EUS. *Gastrointest. Endosc.*, **54**, 219–224.

21 Sivak, M.V. Jr., Kobayashim, K., Izatt, J.A., Rollins, A.M., Ung-Runyawee, R. *et al.* (2000) High-resolution endoscopic imaging of the GI tract using optical coherence tomography. *Gastrointest. Endosc.*, **51**, 474–479.

22 Sergeev, A.M., Gelikonov, V.M., Gelikonov, G.V., Feldchtein, F.I., Kuranov, R.V. *et al.* (1997) *In vivo* endoscopic OCT imaging of precancer and cancer states of human mucosa. *Opt. Express*, **1**, 432–440.

23 Bouma, B.E., Tearney, G.J., Compton, C.C., Brand, S., Poneros, J.M. *et al.* (2000) Optical coherence tomography for upper gastrointestinal tract diagnosis. Presented at Biomedical Topical Meetings, Miami Beach, FL.

24 Evans, J.A. and Nishioka, N.S. (2005) The use of optical coherence tomography in screening and surveillance of Barrett's esophagus. *Clin. Gastroenterol. Hepatol.*, **3**, S8–S11.

25 Daniltchenko, D., Konig, F., Lankenau, E., Sachs, M., Kristiansen, G. *et al.* (2005) Utilizing optical coherence tomography (OCT) for visualization of urothelial diseases of the urinary bladder. *Radiologe*, **46**, 584–589.

26 Wang, Z., Lee, C.S., Waltzer, W.C., Liu, J., Xie, H. *et al.* (2007) *In vivo* bladder imaging with microelectromechanical-systems-based endoscopic spectral domain optical coherence tomography. *J. Biomed. Opt.*, **12**, 034009.

27 Pfau, P.R., Sivak, M.V., Chak, A., Kinnard, M., Wong, R.C.K. *et al.* (2003) Criteria for the diagnosis of dysplasia by endoscopic

optical coherence tomography. *Gastrointest. Endosc.*, **58**, 196–202.

28 Bouma, B.E. and Tearney, G.J. (2002) Clinical imaging with optical coherence tomography. *Acad. Radiol.*, **9**, 942–953.

29 Evans, J.A., Poneros, J.M., Bouma, B.E., Bressner, J., Halpern, E.F. *et al.* (2006) Optical coherence tomography to identify intramucosal carcinoma and high-grade dysplasia in Barrett's esophagus. *Clin. Gastroenterol. Hepatol.*, **4**, 38–43.

30 Kuranov, R.V., Sapozhnikova, V.V., Turchin, I.V., Zagainova, E.V., Gelikonov, V.M. *et al.* (2002) Complementary use of cross-polarization and standard OCT for differential diagnosis of pathological tissues. *Opt. Express*, **10**, 707–713.

31 Hee, M.R., Huang, D., Swanson, E.A., and Fujimoto, J.G. (1992) Polarization-sensitive low-coherence reflectometer for birefringence characterization and ranging. *J. Opt. Soc. Am.*, **9**, 903–908.

32 de Boer, J.F., Srinivas, S.M., Malekafzali, A., Chen, Z.P., and Nelson, J.S. (1998) Imaging thermally damaged tissue by polarization sensitive optical coherence tomography. *Opt. Express*, **3**, 212–218.

33 Schmitt, J.M. and Xiang, S.H. (1998) Cross-polarized backscatter in optical coherence tomography of biological tissue. *Opt. Lett.*, **23**, 1060–1062.

34 Cense, B., Chen, H.C., Park, B.H., Pierce, M.C., and de Boer, J.F. (2004) *In vivo* birefringence and thickness measurements of the human retinal nerve fiber layer using polarization-sensitive optical coherence tomography. *J. Biomed. Opt.*, **9**, 121–125.

35 Feldchtein, F.I., Gelikonov, G.V., Gelikonov, V.M., Iksanov, R.R., Kuranov, R.V. *et al.* (1998) *In vivo* OCT imaging of hard and soft tissue of the oral cavity. *Opt. Express*, **3**, 239–250.

36 Gelikonov, V.M. and Gelikonov, G.V. (2006) New approach to cross-polarized optical coherence tomography based on orthogonal arbitrarily polarized modes. *Laser Phys. Lett.*, **3**, 445–451.

37 Gelikonov, V., Gusovsky, D., Leonov, V., and Novikov, M. (1987) Birefringence compensation in single mode optical fibers. *Sov. Tech. Phys. Lett.*, **13**, 775–779.

38 Feldchtein, F.I., Gelikonov, V.M., and Gelikonov, V.M. (2002) Design of OCT scanners, in *Handbook of Optical Coherence Tomography* (eds. B.E. Bouma and G.J. Tearney), Marcel Dekker, New York, pp. 000–000.

39 Zagaynova, E.V., Strelzova, O.S., Gladkova, N.D., Snopova, L.B., Gelikonov, G.V. *et al.* (2002) *In vivo* optical coherence tomography feasibility for bladder disease. *J. Urol.*, **167**, 1492–1497.

40 Koenig, F., Tearney, G.J., and Bouma, B.E. (2002) Optical coherence tomography in urology, in *Handbook of Optical Coherence Tomography* (eds. B.E. Bouma and G.J. Tearney), Marcel Dekker, New York, pp. 000–000.

41 Manyak, M.J., Gladkova, N., Makari, J.H., Schwartz, A., Zagaynova, E. *et al.* (2005) Evaluation of superficial bladder transitional cell carcinoma by optical coherence tomography. *J. Endourol.*, **19**, 570–574.

42 Zagaynova, E.V., Gladkova, N.D., Streltsova, O.S., Gelikonov, G.V., Tresser, N. *et al.* (2008) Optical coherence tomography in urology, in *Optical Coherence Tomography*, Springer, Berlin, pp. 000–000.

43 Jackle, S., Gladkova, N.D., Feldchtein, F.I., Terentieva, A.B., Brand, B. *et al.* (2000) *In vivo* endoscopic optical coherence tomography of the human gastrointestinal tract – toward optical biopsy. *Endoscopy*, **32**, 743–749.

44 Poneros, J.M. and Nishioka, N.S. (2003) Diagnosis of Barrett's esophagus using optical coherence tomography. *Gastrointes.t Endosc. Clin. N. Am.*, **13**, 309–323.

45 Escobar, P.F., Belinson, J.L., White, A., Shakhova, N.M., Feldchtein, F.I. *et al.* (2004) Diagnostic efficacy of optical coherence tomography in the management of preinvasive and invasive cancer of uterine cervix and vulva. *Int. J. Gynecol. Cancer*, **14**, 470–474.

46 Ascencio, M., Collinet, P., Cosson, M., and Mordon, S. (2007) The role and value of optical coherence tomography in gynecology. *J. Gynecol. Obstet. Biol. Reprod.*, **36**, 749–755.

47 Tearney, G.J., Brezinski, M.E., Bouma, B.E., Boppart, S.A., Pitris, C. et al. (1997) In vivo endoscopic optical biopsy with optical coherence tomography. Science, 276, 2037–2039.
48 Feldchtein, F.I., Gelikonov, G.V., Gelikonov, V.M., Kuranov, R.V., Sergeev, A.M. et al. (1998) Endoscopic applications of optical coherence tomography. Opt. Express, 3, 257–269.
49 Pan, Y.T., Lavelle, J.P., Bastacky, S.I., Meyers, S., Pirtskhalaishvili, G. et al. (2001) Detection of tumorigenesis in rat bladders with optical coherence tomography. Med. Phys., 28, 2432–2440.
50 Zagaynova, E.V., Streltzova, O.S., Gladkova, N.D., Shakhova, N.M. and Feldchtein, F.I. et al. (2004) Optical coherence tomography in diagnostics of precancer and cancer of human bladder. Proc. SPIE, 5312, 75.
51 Lerner, S.P., Goh, A.C., Tresser, N.J., and Shen, S.S. (2008) Optical coherence tomography as an adjunct to white light cystoscopy for intravesical real-time imaging and staging of bladder cancer. Urology, 72, 133–137.
52 Pan, Y.T., Xie, T.Q., Du, C.W., Bastacky, S., Meyers, S., and Zeidel, M.L. (2003) Enhancing early bladder cancer detection with fluorescence-guided endoscopic optical coherence tomography. Opt. Lett., 28, 2485–2487.
53 Zagaynova, E., Gladkova, N., Shakhova, N., Gelikonov, G.V., and Gelikonov, V.M. (2008) OCT with forward looking probes: clinical studies in urology and gastroenterology. J. Biophoton., 1, 114–128.
54 Izatt, J.A., Kulkarni, M.D., Wang, H.-W., Kobayashi, K., and Sivak, M.V. (1996) Optical coherence tomography and microscopy in gastrointestinal tissues. IEEE J. Sel. Top. Quantum Electron., 2, 1017–1028.
55 Gladkova, N.D., Zagaynova, E.V., Zuccaro, G., Kareta, M.V. and Feldchtein, F.I. et al. (2007) Optical coherence tomography in the diagnosis of dysplasia and adenocarcinoma in Barret's esophagus. Proc. SPIE, 6432, 64320A.
56 Zuccaro, G., Gladkova, N.D., Vargo, J., Feldchtein, F.I., Zagaynova, E.V. et al. (2001) Optical coherence tomography of the esophagus and proximal stomach in health and disease. Am. J. Gastroenterol., 96, 2633–2639.
57 Levine, D., Haggitt, R.C., Blount, P., Rabinovitch, P., Rusch, V., and Reid, B.J. (1993) An endoscopic biopsy protocol can differentiate high-grade dysplasia from early adenocarcinoma in Barrett's esophagus. Gastroenterology, 105, 40–50.
58 Feldchtein, F.I., Gladkova, N.D., Snopova, L.B., Zagaynova, E.V., Streltzova, O.S. et al. (2003) Blinded recognition of optical coherence tomography images of human mucosa precancer. Proc. SPIE, 4956, 89–94.
59 Poneros, J.M., Brand, S., Bouma, B.E., Tearney, G.J., Compton, C.C., and Nishioka, N.S. (2001) Diagnosis of specialized intestinal metaplasia by optical coherence tomography. Gastroenterology, 120, 7–12.
60 Isenberg, G., Sivak, M.V., Chak, A., Wong, R., Willis, J. et al. (2003) Accuracy of endoscopic optical coherence tomography (EOCT) in the detection of dysplasia (D) in Barrett's esophagus (BE). Gastrointest. Endosc., 57, AB77–AB80.
61 Isenberg, G., Sivak, M.V. Jr., Chak, A., Wong, R.C., Willis, J.E. et al. (2005) Accuracy of endoscopic optical coherence tomography in the detection of dysplasia in Barrett's esophagus: a prospective, double-blinded study. Gastrointest. Endosc., 62, 825–831.
62 Qi, X., Sivak, M.V. Jr., Isenberg, G., Willis, J.E., and Rollins, M.A. (2006) Computer-aided diagnosis of dysplasia in Barrett's esophagus using endoscopic optical coherence tomography. J. Biomed. Opt., 11, 044010.
63 Standish, B.A., Yang, V.X., Munce, N.R., Wong, K., Song, L.M., Gardiner, G. et al. (2007) Doppler optical coherence tomography monitoring of microvascular tissue response during photodynamic therapy in an animal model of Barrett's esophagus. Gastrointest. Endosc., 66, 326–333.
64 Chen, Y., Aguirre, A.D., Hsiung, P.L., Desai, S., Herz, P.R. et al. (2007) Ultrahigh resolution optical coherence tomography of Barrett's esophagus: preliminary

descriptive clinical study correlating images with histology. *Endoscopy*, **39**, 599–605.
65 Chen, Y., Aguirre, A.D., Hsiung, P.L., Huang, S.W., Mashimo, H. *et al.* (2008) Effects of axial resolution improvement on optical coherence tomography (OCT) imaging of gastrointestinal tissues. *Opt. Express*, **16**, 2469–2485.
66 Melissa, J.S., Patrick, S.Y., Benjamin, J.V., Milen, S., Brett, E.B. *et al.* (2008) *In vivo* 3D comprehensive microscopy of the human esophagus for the management of Barrett's patients. *Gastrointest. Endosc.*, **67**, AB99.
67 Qi, X., Pan, Y., Hu, Z., Sivak, M.V., Willis, J. *et al.* (2008) T1853 morphological feature quantification of colonic crypt patterns using microscope-integrated OCT. *Gastroenterology*, **134** (4 Suuppl 1), A577.
68 Park, D.H., Kim, H.S., Kim, W.H., Kim, T.I., Kim, Y.H. *et al.* (2008) Clinicopathologic characteristics and malignant potential of colorectal flat neoplasia compared with that of polypoid neoplasia. *Dis. Colon Rectum*, **51**, 43–49.
69 Jaeckle, S., Gladkova, N.D., Feldchtein, F.I., Terentieva, A.B., Brand, B. *et al.* (2000) *In vivo* endoscopic optical coherence tomography of the human gastrointestinal tract – toward optical biopsy. *Endoscopy*, **32**, 743–749.
70 Zuccaro, G., Gladkova, N., Conwell, D., Feldchtein, F., Vargo, J. *et al.* (2002) Optical coherence tomography (OCT) of the colon in health and disease. *Gastrointest. Endosc.*, **55**, M1951.
71 Yang, V.X.D. (2005) Endoscopic Doppler optical coherence tomography in the human GI tract: initial experience. *Gastrointest. Endosc.*, **61**, 879–890.
72 Pitris, C., Goodman, A., Boppart, S.A., Libus, J.J., Fujimoto, J.G., and Brezinski, M.E. (1999) High-resolution imaging of gynecologic neoplasms using optical coherence tomography. *Obstet. Gynecol.*, **93**, 135–139.
73 Shakhova, N.M., Feldchtein, F.I., and Sergeev, A.M. (2002) Applications of optical coherence tomography in gynecology, in *Handbook of Optical Coherence Tomography* (eds. B.E. Bouma and G.J. Tearney), Marcel Dekker, New York, pp. 000–000.
74 Shakhova, N.M., Sapozhnikova, V.V., Kamensky, V.A., Kuranov, R.V., Loshenov, V.B. *et al.* (2003) Optical methods for diagnosis of neoplastic processes in the uterine cervix and vulva. *J. Appl. Res.*, **3**, 144–155.
75 Kuznetsova, I.A., Gladkova, N.D., Gelikonov, V.M., Belinson, J.L., Shakhova, N.M., and Feldchtein, F.I. (2008) OCT in gynecology, in *Optical Coherence Tomography* (eds. W. Drexler and J.G. Fujimoto), Springer, Berlin, pp. 1211–1240.
76 Zuluaga, A.F., Follen, M., Boiko, I., Malpica, A., and Richards-Kortum, R. (2005) Optical coherence tomography: a pilot study of a new imaging technique for noninvasive examination of cervical tissue. *Am. J. Obstet. Gynecol.*, **193**, 83–88.
77 Gladkova, N., Shakhova, N., and Sergeev, A. (eds.) (2007) *Handbook of Optical Coherence Tomography*, Fizmatlit, Moscow.
78 Brezinski, M.E. (2006) *Optical Coherence Tomography: Principles and Applications*, Academic Press/Elsevier, Burlington, MA
79 Park, B.H., Saxer, C., Srinivas, S.M., Nelson, J.S., and de Boer, J.F. (2001) *In vivo* burn depth determination by high-speed fiber-based polarization sensitive optical coherence tomography. *J. Biomed. Opt.*, **6**, 474–479.
80 Pierce, M.C., Sheridan, R.L., Park, B., Cense, B., and de Boer, J.F. (2004) Collagen denaturation can be quantified in burned human skin using polarization-sensitive optical coherence tomography. *Burns*, **30**, 511–517.
81 Strasswimmer, J., Pierce, M.C., Park, B.H., Neel, V., and de Boer, J.F. (2004) Polarization-sensitive optical coherence tomography of invasive basal cell carcinoma. *J. Biomed. Opt.*, **9**, 292–298.
82 Junqueira, L., Bignolas, G., and Brentani, R. (1979) Picrosirius staining plus polarization microscopy, a specific method for collagen detection. *J. Histochem.*, **11**, 447–455.
83 Borges, L., Gutierrez, P., Marana, H., and Taboga, S. (2007) Picrosirius-polarization

staining method as an efficient histopathological tool for collagenolysis detection in vesical prolapse lesions. *Micron*, **38**, 580–583.

84 Giattina, S.D., Courtney, B.K., Herz, P.R., Harman, M., Shortkroff, S. *et al.* (2006) Assessment of coronary plaque collagen with polarization sensitive optical coherence tomography (PS-OCT). *Int. J. Cardiol.*, **107**, 400–409.

85 de Boer, J.F., Srinivas, S.M., Nelson, J.S., Milner, T.E., and Ducros, M.G. (2002) Polarization-sensitive optical coherence tomography, in *Handbook of Optical Coherence Tomography* (eds. B.E. Bouma and G.J. Tearney), Marcel Dekker, New York, pp. 000–000.

86 de Boer, J.F., Srinivas, S.M., Park, B.H., Pham, T.H., Chen, Z.P. *et al.* (1999) Polarization effects in optical coherence tomography of various biological tissues. *IEEE J. Sel. Top. Quantum Electron.*, **5**, 1200–1204.

87 Herrmann, J.M., Pitris, C., Bouma, B.E., Boppart, S.A., Jesser, C.A. *et al.* (1999) High resolution imaging of normal and osteoarthritic cartilage with optical coherence tomography. *J. Rheumatol.*, **26**, 627–635.

88 Jiao, S., Yu, W., Stoica, G., and Wang, L.V. (2003) Contrast mechanisms in polarization-sensitive Mueller-matrix optical coherence tomography and application in burn imaging. *Appl. Opt.*, **42**, 5191–5197.

89 Terentéva, A.B., Shakhov, A.V., Maslennikova, A.V., Gladkova, N.D., Kamensky, V.A. *et al.* (2008) OCT in laryngology, in *Optical Coherence Tomography* (eds. W. Drexler and J.G. Fujimoto), Springer, Berlin, pp. 1123–1150.

90 Das, A., Sivak, M.V. Jr., Chak, A., Wong, R.C., Westphal, V. *et al.* (2001) High-resolution endoscopic imaging of the GI tract: a comparative study of optical coherence tomography versus high-frequency catheter probe EUS. *Gastrointest. Endosc.*, **54**, 219–224.

25
Intraoperative Optical Coherence Tomographic Monitoring
Alexandre Douplik

25.1
Optical coherence tomography as a Surgery-Assisting Tool

Optical coherence tomography (OCT) is a high-resolution imaging technology analogous to B-mode ultrasound except that it uses infrared light instead of sound. This technique, known as Michelson low-coherence interferometry, which measures the echo delay and intensity of back-reflected or backscattered infrared light from internal tissue microstructure, is used by comparing the reflected light beam with a beam that travels along a reference path. OCT can facilitate imaging close to histologic cross-sections at a high spatial resolution, up to 1–20 µm. Such a resolution is 1–2 orders of magnitude better than a conventional ultrasound. Imaging depths are limited to a few millimeters (2–3 mm) because of optical attenuation due to absorption and scattering; however, these imaging depths are sufficient for many diagnostic applications, including performing intraoperative monitoring [1]. OCT imaging is near real-time (but still lower than a standard PAL or NTSC video rate) and noncontact technique. With recent research advances, OCT imaging of optically scattering, nontransparent tissues has become possible using longer near-infrared (e.g., 1.3 µm) wavelengths where optical scattering is reduced, thus permitting a wide variety of biomedical applications [2]. Contrast in OCT images originates from differences in the reflectivity of different tissues, which are caused by their variations in (complex) refractive index n. Unfortunately, contrast is limited because for most tissues n ranges from 1.3 to 1.4. OCT imaging is sensitive to differences in terms of refraction index of light scattering between different tissues, whereas in histopathology, as in a microscopy, images are generated by differences in either optical reflection or transmission. Hence the inherent appearance of an OCT image is different from that in histopathology and, in general, further studies are required to develop a basis for interpreting OCT images in terms of clinically relevant pathology [3]. Examples of OCT images are provided in Figure 25.1a.

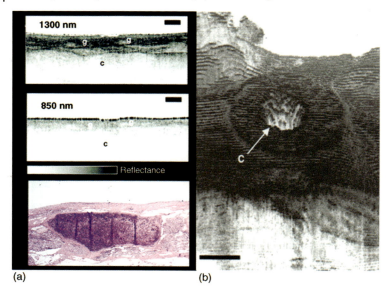

Figure 25.1 (a) Dependence of imaging depth on wavelength of the incident light. A human epiglottis was imaged at both 850 and 1300 nm. The underlying cartilage (c) could be visualized only using longer wavelengths, while a larger reflection at the surface occurred with the 850 nm light. Superficial secretory glands (g) were visualized at both wavelengths. Bars represent 500 μm. In this image, data are displayed as the logarithm of backscattering intensity versus depth (in gray scale) to emphasize deeper structures. Corresponding histology is included (lowest left) [26]. (b) Laser ablation crater. 3D projection illustrates central ablation crater (c) and concentric zones of thermal injury Numbers refer to depth below the surface. Bar represents 1 mm [1]. Courtesy of J. Fujimoto and S. Boppart.

As an intraoperative monitoring tool, OCT is exploited to differentiate normal tissues and recognize the normal and pathology to guide surgery. The intraoperative use of OCT can be invaluable in improving intraoperative evaluation, facilitating surgery, and supporting the surgeon's decision-making process on the fly, thus improving surgical outcomes. The OCT monitoring tools are required not to interfere with the surgery in intensive therapy procedures, either mechanically or using close light wavelengths such as during laser ablation or light interstitial thermotherapy. A 3D OCT image of ablation crater is shown in Figure 25.1b.

25.2
OCT Modalities and Implementations for Intraoperative Assistance

Since the introduction of OCT in 1991, it has been implemented as a time domain technique. The newest extension of OCT exploits frequency domain technology. The latter includes spatially encoded frequency domain OCT [spectral domain (SD) or Fourier domain OCT] and time-encoded frequency domain OCT (also swept-source

OCT) and provides much faster image acquisition relative to time domain technology. High-resolution and 3D OCT require longer data collection times than conventional OCT. Doppler optical coherence tomography (DOCT) is used to obtain information about blood perfusion spatial distribution and vessel reactions. Polarized sensitive (PS) or structural OCT, an extension of classical OCT, is sensitive to birefringence and consequently to anisotropies and stress within a material. The image acquisition in OCT is provided either with hand-held surgical probes, OCT fiber probes compatible with biopsy channels of laparoscopes, catheters, and endoscopes [4, 5] or an OCT system coupled with a surgical microscope (OCT camera) [6].

25.3
Advantages, Limitations, and Reproducibility of Intraoperative OCT Monitoring

Advantages of OCT are a high spatial resolution comparable to conventional histology, fast data collection building up an image nearly in real time, and the capability of acquiring images remotely. Unlike ultrasound, OCT does not require any conducting gel. The principal disadvantage of OCT, as with any optical imaging technique, is that light is highly scattered by most biotissues. In most tissues other than the eyes, optical scattering limits the image penetration depth to between 2 and 3 mm. Another serious limitation of OCT is a relatively small field of view (e.g., hundreds of microns to millimeters), demanding significant time for examination of the region of interest (ROI) and hence limiting the application of OCT for screening surveillance. The larger the area to be scanned, the greater is the probability of "smearing" the OCT image due to motion artifacts. Ho *et al.* reported that OCT imaging in ophthalmology is in general reproducible within 10–20% depending on the particular implementation [7]. High-resolution and 3D OCT require extensive scanning, making OCT imaging subject to motion artifacts when holding the OCT probe still relative to the object is not always possible in organs with high motility such as the gastrointestinal (GI) tract. Pressing the probe against the tissue to stabilize its position may lead to compression of the object, which can affect the measured OCT information [9], though this effect has even been exploited to enhance the OCT contrast [8].

25.4
Specificity and Sensitivity of OCT for Recognizing Normal and Pathologic Tissue

To evaluate the efficiency of the OCT method for the detection of different stages of malignization, three independent blind tests were performed using OCT images of the uterine cervix, urinary bladder, and larynx for diagnosing mucosal neoplasia. The studies showed that the OCT sensitivity was 82, 98, and 77% and specificity 78, 71, and 96%, respectively [9]. The diagnostic accuracy of OCT for recognizing normal and pathologic tissue depends on the particular clinical application and varies from

75 to 86% for sensitivity and 78 to 95% for specificity if H&E histology is used as a gold standard [10, 11]. More detailed analysis of OCT diagnostic accuracy including particular applications is provided in Chapter 2.1.2.4 Optical Coherence Tomography. As a result of providing detailed information on the architectural morphology of the retina at the level of individual retinal layers, OCT has been proposed for the detection of early pathologic changes, even before clinical signs or visual symptoms occur, and it has even been proposed as a new "gold standard" structural test for retinal abnormalities [12]. OCT was found to be a more accurate diagnostic technique than intravascular ultrasound [13].

25.5
Normal Tissue Differentiation with OCT

The high resolution of OCT permits the imaging of features such as tissue morphology and also cellular features. Tissue differentiation is important to avoid accidental destruction of critically important organs and tissues such as blood vessel walls and nerves [14], while accurate identification of pathologic or necrotic tissues assists delineation of the margins of resection. Attempts at localized measurement of the attenuation coefficient μ_t can provide additional information, and may increase the clinical potential of quantitative discrimination between different tissue types by OCT [15]. The ability to see beneath the surface of tissue in real time can guide surgery near sensitive structures such as vessels and nerves and also assist in microsurgical procedures [1, 16, 17]. OCT was integrated with surgical microscopes routinely used to magnify tissue to prevent iatrogenic injury and to guide delicate techniques in ear, nose, and throat (ENT) surgery [18]. Hand-held OCT surgical probes and laparoscopes have also been demonstrated [4]. Miyazawa et al. reported that PS-OCT was used for a tissue discrimination algorithm based on the optical properties of tissues. The 3D feature vector from the parameters intensity, extinction coefficient, and birefringence obtained by PS-OCT was calculated. The tissue type of each pixel was determined according to the position of the feature vector in the 3D feature space. The algorithm was applied for discriminating tissues of the human anterior eye segment. The conjunctiva, sclera, trabecular meshwork (TM), cornea, and uvea were well separated in the 3D feature space, and were observed with good contrast [19]. Liang et al. used structural OCT to investigate the longitudinal development of engineered tissues and cell dynamics such as migration, proliferation, detachment, and cell–material interactions [20]. Optical techniques that image functional parameters or integrate multiple imaging modalities to provide complementary contrast mechanisms have been developed, such as the integration of optical coherence microscopy with multiphoton microscopy to image structural and functional information from cells in engineered tissue, optical coherence elastography to generate images or maps of strain to reflect the spatially dependent biomechanical properties, and spectroscopic OCT to differentiate different cell types. Hauger et al. proposed a surgical system supported by an OCT surface scanner for tissue-differentiated tomography to identify a particular type of tissue [21].

25.6
OCT Biopsy Guidance

A technology capable of performing optical biopsy (defined as imaging tissue microstructure at or near the level of histopathology without the need for tissue excision) has a considerable impact on surgical intervention management such as in situations in which sampling errors severely restrict the effectiveness of excisional biopsy (e.g., the high failure rates associated with blind biopsies used to screen the premalignant conditions of ulcerative colitis or Barrett's esophagus) [22–24] or when conventional excisional biopsy is potentially hazardous in vulnerable regions (e.g., the retina, the central nervous system, the vascular system, the atherosclerotic plaque in the coronary arteries and articular cartilage) [25]. The ability to image at the cellular level could improve the effectiveness of many surgical and microsurgical procedures, including coronary atherectomy, transurethral prostatectomies, and microvascular repair [26]. Fibered OCT in conjunction with natural orifice transluminal endoscopic surgery (NOTES) could provide a facility for rapid, *in situ* pathologic diagnosis of intraperitoneal tissues in a truly minimally invasive fashion [27]. Iftimia *et al.* reported that SD-OCT could help in more precise needle placement during the fine needle aspiration (FNA) breast biopsy and therefore could substantially reduce the number of nondiagnostic aspirates and improve the sensitivity and specificity of the FNA procedures. Over 80% sensitivity and specificity in differentiating all tissue types and 93% accuracy in differentiating fatty tissue from fibrous or tumor tissue were obtained with this technology [28].

25.7
Identifying Malignancy

Intraoperative real-time detection of residual malignant tumor tissue remains an important challenge in surgery. Many malignancy-identifying OCT studies in the 1990s were focused on *in vitro* imaging and correlation with histopathology. *In vitro* studies have been performed to investigate OCT imaging in the GI tract [29], pancreatobiliary tissue [30], urinary system [31], respiratory system [32], and female reproductive tract [33] and in differentiating choroidal nevus from choroidal melanoma in the eye by 3D SD-OCT and OCT characteristics of other choroidal tumors [34]. During the last decade, a number of *in vivo* clinical studies have been performed in dermatology [35], GI tract [36], kidney [37, 38], human brain [39], breast [40], and ENT [41].

25.8
Identifying Atherosclerotic Plaque Composition

The majority of acute coronary events are precipitated by the rupture of a vulnerable atherosclerotic plaque in the coronary system, and subsequent thrombogenesis.

Thus, data on plaque composition and stability, complementing the image, may inform the decision on whether and how to treat a particular section of coronary artery. Van Soest et al. concluded from a survey of the complete OCT *in vivo* attenuation coefficient dataset that a high attenuation coefficient in OCT, $\mu_t > 10\,\text{mm}^{-1}$, is associated with two markers of plaque vulnerability: the presence of a necrotic core and macrophage infiltration. More stable forms of atherosclerotic tissue and healthy vessel wall have low attenuation coefficients: $\mu_t \approx 2\text{--}5\,\text{mm}^{-1}$ [42].

25.9
Identifying Necrotic Tissue

Van der Meer et al. used OCT to determine the optical properties of human fibroblasts in which necrosis or apoptosis had been induced *in vitro*. The OCT data included both the scattering properties of the medium and the axial point spread function of the OCT system. The optical attenuation coefficient in necrotic cells decreased from 2.2 ± 0.3 to $1.3 \pm 0.6\,\text{mm}^{-1}$, whereas, in the apoptotic cells an increase to $6.4 \pm 1.7\,\text{mm}^{-1}$ was observed. The results from cultured cells, as presented in this study, indicate the ability of OCT to detect and differentiate between viable, apoptotic, and necrotic cells, based on their attenuation coefficient [43]. Standish et al. exploited interstitial (IS) DOCT to predict photodynamic therapy (PDT)-induced tumor necrosis deep within solid rat prostate tumors, quantifying the PDT-induced microvascular shutdown rate *in vivo*. A strong relationship ($R^2 = 0.723$) was determined between the percentage tumor necrosis at 24 h post-treatment and the vascular shutdown rate: slower shutdown corresponded to higher treatment efficacy, that is, more necrosis [44].

25.10
Resection Delineation

OCT has already been used *in vitro* for the delineation of fine subtle structures of biotissue to help avoid damage to the normal structure, margins of the spots of the tissue damaged during surgical intervention, and margins of the malignant lesions. Results *in vivo* are expected soon. Thus, an OCT system has been utilized to display the multilayered structure of the airways delineating the subtle architectural differences in three separate anatomic locations, namely trachea, main bronchus, and tertiary bronchus, using fresh pig lung resections [45]. Spectral OCT sensitive to birefringes has been successfully applied for thermally damaged tissue margin delineation [46]. Tsai et al. applied swept-source (SS) OCT for the *ex vivo* delineation of an oral cancerous lesion margin sampling with the axial resolution of 8 μm in free space and a system sensitivity of 108 dB via two parameters [47]. One of the parameters was the decay constant in the exponential fitting of the SS-OCT signal intensity to depth. This decay constant decreases as the A-scan point moves laterally across the margin of a lesion. The other parameter was the standard deviation of the SS-OCT signal intensity fluctuation in an A-scan. This parameter increases significantly when the A-scan point is moved across the transition region between the normal and abnormal portions. Such parameters were used for determining the margins of oral cancer.

25.11
Particular Intraoperative OCT Applications

As follows from the snapshot of the distribution of OCT publications among different clinical applications presented in Figure 25.2, the greatest clinical impact of OCT is in ophthalmologic, cardiovascular, and dermatologic applications. The next impact group includes GI, urologic, osteologic, and neurologic applications. ENT, dental, gynecologic, and pulmonologic applications form the lowest impact group within the OCT research domain.

A similar distribution is expected for intraoperative OCT applications. Below, they are listed and reviewed in order of the priorities mentioned in Figure 25.2.

25.12
Intraoperative OCT Monitoring in Ophthalmology

During laser eye surgery, a surgeon uses a laser device to make permanent changes to the shape of the cornea. The laser used most often is the excimer laser, which produces a beam of ultraviolet light to vaporize tissue. Surgically altering the shape of the cornea can correct mild to moderate refractive errors in most people. However, laser eye surgery also poses certain risks. In general, refractive procedures enjoy very

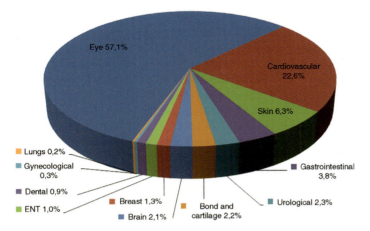

Figure 25.2 Snapshot of distribution of OCT publications among different clinical applications in 1991–2010 (based on ISI Web of Knowledge).

high success rates and are among the most commonly performed elective surgeries in medicine. However, all refractive surgeons must manage unsuccessful cases appropriately. Unsuccessful refractive surgery procedures may relate to each step of the refractive surgery process: preoperative screening, surgical planning, intraoperative events, and postoperative biomechanical or healing anomalies. OCT has demonstrated its ability to reduce these risks. Intraoperative OCT applied in ophthalmology is often coupled with a surgical microscope via adaptation to a slit lamp. Hüttmann et al. reported that through the surgical microscope, a volumetric imaging of epithelium, Bowman's and Descemet's membranes, limbus, iris, lens, conjunctiva, and sclera has been demonstrated with a working distance of several tens of centimeters [48]. Instruments and incisions in the cornea were visualized with 20 µm precision. SD-OCT permits wide-field noncontact real-time cross-sectional imaging of retinal structure, allowing for SD-OCT augmented intrasurgical microscopy for intraocular visualization. Real-time imaging and visualization of volumetric OCT data were also demonstrated. Bagayev et al. developed OCT for the determination of the removed corneal thickness profile in situ with laser ablation to improve the precision of refractive surgery [49]. The precision of measurements of the corneal thickness in preliminary in vivo experiments was found to be higher than in ex vivo experiments, which is promising for application of the method suggested here in refractive surgery. The ablation of corneal tissue with the excimer laser can be variable and can lead to miscorrections. To avoid that, the intraoperative ablation parameters during laser-assisted in situ keratomileusis (LASIK) were evaluated with online optical coherence pachymetry (OCP) [50]. The qualitative and quantitative evaluation of the corneal flap created by a femtosecond laser using anterior segment (AS) OCT before lifting the flap confirmed that the femtosecond laser is a practical intraoperative approach that offers safer surgery [51] as eyes with flat membranes showed strong adhesion to the retinal surface. Potential complications may lead to anterior chamber gas bubbles after LASIK flap formation with a femtosecond laser [52]. OCT has been used as an intraoperative tool for the management of unsuccessful LASIK procedures, focusing on significant advances related to wavefront-guided therapeutic ablations [53]. These studies showed that real-time OCT monitoring of the creation of thin flaps in LASIK using a femtosecond laser is possible, thus ensuring that the flap is created at the desired depth. AS-OCT has been applied to investigate the stromal demarcation line after corneal cross-linking and its influence on the short-term results of cross-linking in patients with progressive keratoconus [54]. A 1310 nm optical coherence tomograph coupled with a beamsplitter on the front lens of an operating microscope has been developed for intraoperative, noncontact visualization of anterior segment procedures by Geerling et al. [55]. This modification of OCT technology allowed intraoperative, high-resolution, cross-sectional imaging and pachymetry of the cornea and sclera during anterior segment surgery, which is particularly helpful for lamellae dissection techniques such as deep anterior lamellar keratoplasty and trabeculectomy.

For retinal OCT images, 830 nm wavelength OCT images have been evaluated for surgical indication, intraoperative management, and postoperative outcome of patients with idiopathic epiretinal macular membrane (IEMM) to reveal IEMM partially separated from the macula, thus guiding the surgeon intraoperatively in the peeling procedures [56]. OCT scans are a valuable adjunct for the pre- and postoperative analysis of staging of idiopathic macular holes [57]. Preoperative assessment of topographic features has the potential to improve surgical strategies in eyes with macular pucker for patients who have undergone full ophthalmologic evaluation including high-definition optical coherence tomography (Cirrus, Carl Zeiss Meditec) prior to a vitrectomy with peeling of the epiretinal membrane (ERM) [58].

Wykoff et al. described the use of the intraoperative OCT imaging in macular surgery facilitating by the use of hand-held SD-OCT and providing an efficient method for visualizing macular pathology. This technology may, in certain cases, help confirm or identify diseases that may be difficult to visualize during surgery [59]. Hand-held Fourier domain OCT can be used during Descemet stripping automated endothelial keratoplasty (DSAEK) to assess the donor–host interface in lamellar corneal transplantation surgery [60] or to find interface fluid that is clinically undetectable under the microscope. As such, all patients were able to leave the operating room with a fully attached graft [61].

OCT also has been applied to investigate the long-term mechanical biocompatibility of a polymer microtechnology that can be used to position electrodes in close proximity to the retina [polydimethylsiloxane (PDMS) array implants] in animal studies [62].

OCT has traditionally been used in the outpatient environment as an important diagnostic tool for retinal clinical decision-making. Recent advances in OCT technology have made the intraoperative use of OCT feasible [63].

25.13
Intraoperative OCT Monitoring in Cardiovascular Surgery

Coronary artery bypass grafting (CABG) is the most commonly performed major surgery in the United States. A critical determinant of its outcome has been postulated to be injury to the conduit vessel incurred during the harvesting procedure or pathology pre-existing in the harvested vessel. Brown et al. detected by OCT atherosclerotic lesions in the radial arteries radial artery (RA) and discerns plaque morphology as fibrous, fibrocalcific, or fibroatheromatous. OCT is also used to assess intimal trauma and residual thrombi related to endoscopic harvest and the quality of distal anastomosis. The feasibility of OCT imaging as an intraoperative tool to select conduit vessels for CABG has been demonstrated [64]. OCT has been found particularly useful in identifying so-called vulnerable or high-risk plaques which account for many acute coronary events arising from non-flow-limiting coronary lesions. OCT permits excellent resolution of coronary architecture and precise characterization of plaque architecture. Quantification of macrophages within the plaque is also possible. These capabilities allow precise identification of the most common type of vulnerable plaque, thin-cap fibroatheroma [65]. Acute coronary syndromes frequently arise consequent to rupture of an angiographically moderate plaque with occlusive thrombus formation [66]. Plaques prone to rupture have certain characteristics and are often referred to as vulnerable or high-risk plaques. Such plaques contain a large lipid pool, often covered by a thin fibrous cap, typically less than 65 μm thick [67]. Two-dimensional cross-sectional imaging of the exposed surface of the arterial segments has been performed *in vitro* with OCT by Brezinski *et al.* [26]. A 1300 nm wavelength super-luminescent diode light source was used that allows an axial spatial resolution of 20 μm. The signal-to-noise ratio was 109 dB. Images were displayed in gray scale or false color. Imaging was performed over 1.5 mm into heavily calcified tissue, and a high contrast was noted between lipid- and water-based constituents, making OCT attractive for intracoronary imaging. The 20 μm axial resolution of OCT allowed small structural details such as the width of intimal caps and the presence of fissures to be determined. The extent of lipid collections, which had a low backscattering intensity, was also well documented [26]. OCT, a catheter-based intravascular imaging modality, was used to measure the degree of radial artery spasm induced by means of harvest with electrocautery or a harmonic scalpel in patients undergoing CABG. OCT provides levels of speed and accuracy for quantifying endothelial injury and vasospasm that have not been described for any other modality, suggesting potential as an intraoperative quality assurance tool. Our OCT findings suggest that the harmonic scalpel induces less spasm and intimal injury than electrocautery [68]. Brazio *et al.* used OCT to evaluate the tendency of RAs to spasm when used as a bypass graft [68]. RA conduits otherwise considered acceptable for bypass grafting were often found by OCT imaging to have a substantial amount of lipid, which in turn strongly relates to the risk of postoperative spasm. Screening conduits based on characteristics of intimal quality may improve results following RA grafting [69]. Boppart *et al.* used a 3D-OCT system integrated with a surgical microscope in the assessment of the microsurgical anastomoses of vessels and nerves during microsurgery. 3D-OCT reconstructions depicted the structure within an arterial anastomosis and helped identify sites of luminal obstruction. The longitudinal spatial orientation of individual nerve fascicles was tracked in three dimensions to identify changes in position. *In vitro* human arteries and nerves embedded in highly scattering tissue and not visible on microscopy were located and imaged with OCT at eight frames per second [17]. OCT helped to reveal that intraluminal saphenous vein clot is frequently found after endoscopic vein harvest. In a study by Brown *et al.*, systemic heparinization before harvest or an open carbon dioxide endoscopic vein harvest system was recommended as a benign change in practice that can significantly ameliorate this complication [70].

25.14
Intraoperative OCT Monitoring in Dermatology

OCT has been shown to help clinicians to visualize and map the hidden spread of non-melanoma skin cancer (NMSC), such as basal cell carcinomas (BCCs), squamous cell carcinoma (SCC), and the premalignant actinic keratosis (AK) [71] and also for malignant melanomas (MMs) [72]. Mogensen *et al.* described a break-up of the characteristic layering of normal skin is in OCT images of both NMSC and MM lesions. Several other NMSC features in OCT images have also been described, the most important being focal changes including thickening of epidermis in AK lesions, and dark rounded areas, sometimes surrounded by a white structure in BCC lesions. SCC has mainly been studied on oral mucosal surfaces using OCT but changes similar to those with BCC have been described [73]. OCT has also been applied for monitoring the process of tissue healing of rat skin *in vivo* after laser irradiation [74] and collagen remodeling on opto-thermal response of photoaged skin irradiated with an Er:YAG laser [75].

25.15
Intraoperative OCT Monitoring in Gastroenterology

The intraoperative use of OCT in gastroenterology is mostly part of endoscopic or laparoscopic procedures and facilitated via fiber forward-looking [76] or side-looking fiber probes [77] at wavelength 1300 nm. A few studies assessed OCT guidance by autofluorescence imaging (AFI) in endoscopy [78] and laparoscopy [79]. The advantages of such guidance are a combination of large (AFI) and small (OCT) fields of view and also two different physical imaging modalities that may potentially improve the specificity and sensitivity of cancer diagnostics in the GI tract, particularly revealing cancer in Barrett's esophagus when OCT as a single approach facilitates only 71–85% sensitivity and 68% specificity [78]. OCT has been used as an optical biopsy tool for intraoperative *ex vivo* detection of transmural inflammation in Crohn disease [80].

25.16
Intraoperative OCT Monitoring in Urology

OCT allows imaging of the tissue structure of the bladder wall during cystoscopy with high resolution. OCT systems are currently being studied for potential intraoperative diagnostic use in laparoscopic and robotic nerve-sparing prostate cancer surgery for image-guided cancer removal, prostatectomy, and sperm retrieval.

According to Daniltchenko *et al.*, OCT of normal bladder mucosa clearly shows a differentiation between urothelium, lamina propria, and smooth muscle [81]. Cystitis and metaplasia are characterized by blurring of the laminated structure and thickening of the epithelial layer. In malignant areas, there is complete loss of the regular layered tissue structure. OCT improves the diagnosis of flat lesions of the urothelium and it has the potential for facilitating intraciperative staging of malignant areas in the bladder.

Positive surgical margins represent incomplete resection by the surgeon, and elimination of positive margins represents the only clinical feature during radical prostatectomy that can lead directly to improved cancer outcomes. Skarecky *et al.* reported the union of real-time OCT technology with the da Vinci robotic platform for identification of positive margin sites, and technical advances with wider excisions during surgery suggest promise for further reduction of surgical margins to zero [82].

Two clinical studies by Zagaynova et al. have reported 82% sensitivity and 85% specificity, recognizing flat cancer lesions in urinary bladder [78]. They also reported an improvement in OCT diagnostics combined with fluorescence imaging.

Rais-Bahrami et al. used OCT for imaging of the cavernous nerve (CN) and periprostatic tissues. The rates of nerve preservation and postoperative potency after radical prostatectomy might improve with better identification of the CN. Seven male Sprague–Dawley rats underwent surgery using a midline celiotomy to expose the bladder, prostate, and seminal vesicles. OCT was also used to image *ex vivo* human prostatectomy specimens differentiating the CN from the bladder, prostate, seminal vesicles, and periprostatic fat. OCT images of the CN and prostate correlated well with the histologic findings. OCT of *ex vivo* human prostatectomy specimens revealed findings similar to those in the rat experiments, with, however, less dramatic architecture being visualized, in part because of the thicker capsule and denser stroma of human prostates [83]. Chitchian et al. segmented 2D OCT images of the rat prostate to differentiate the cavernous nerves from the prostate gland. To detect these nerves, three image features were employed: Gabor filter, Daubechies wavelet, and Laws filter using a nearest-neighbor classifier. N-ary morphologic post-processing was used to remove small voids. The cavernous nerves were differentiated from the prostate gland with a segmentation error rate of only 0.058 ± 0.019 [84]. Aron et al. also found that identification of the neurovascular bundles (NVBs) required an experienced OCT operator to distinguish them from adipose tissue, small vessels, and lymphatics [85]. Parekattil et al. showed that OCT with its high-resolution images may improve sperm retrieval rates by better identifying isolated foci of spermatogenesis in men with nonobstructive azoospermia [86].

25.17
Intraoperative OCT Monitoring in Bone and Joint Surgery

Quantitative and nondestructive methods for clinical diagnosis and staging of articular cartilage degeneration are important to the evaluation of potential disease-modifying treatments in osteoarthritis (OA). OCT real-time imaging with 11 μm resolution at four frames per second was performed by Li et al. on six patients using a portable OCT system with a hand-held imaging probe during open knee surgery [87]. Structural changes including cartilage thinning, fissures, and fibrillations were observed at a resolution substantially higher than is achieved with any current clinical imaging technology. In addition to changes in architectural morphology, changes in the birefringent or polarization properties of the articular cartilage were observed with PS-OCT, suggesting collagen disorganization, an early indicator of OA supporting the hypothesis that PS-OCT may allow OA to be diagnosed before cartilage thinning.

Using Fourier domain common-path OCT, the characteristic cartilage zones and identity of various microstructured defects in an *ex vivo* chicken knee cartilage were differentiated, thus demonstrating that it could be used to conduct early arthritis diagnosis and intraoperative endomicroscopy [88].

In a study by Chu et al., the hypotheses were tested that OCT can be used clinically to identify early cartilage degeneration and that these OCT findings correlate with arthroscopy results [89]. It was found that both qualitative and quantitative OCT assessments of early articular cartilage degeneration correlated strongly with the arthroscopy results ($p = 0.004$ and $p = 0.0002$, respectively, by the Kruskal–Wallis test).

Meniscal tears are often associated with anterior cruciate ligament (ACL) injury and may lead to pain and discomfort in humans. Maximum preservation of meniscal tissue is highly desirable to mitigate the progression of OA. Images of microstructural changes in meniscus would potentially guide surgeons to manage the meniscal tears better, but the resolution of current diagnostic techniques is limited for this application. Ling et al. demonstrated the feasibility of OCT for the diagnosis of meniscal pathology [90]. OCT and scanning electron microscopy (SEM) images of torn menisci were compared and each sample was evaluated for gross and microstructural abnormalities and reduction or loss of birefringence from the OCT images. OCT holds promise in the nondestructive and rapid assessment of microstructural changes and tissue birefringence of meniscal tears. Future development of intraoperative OCT may help surgeons in decision-making in meniscal treatment.

OCT can eventually be developed for use through an arthroscope, having considerable potential for assessing early osteoarthritic cartilage and monitoring therapeutic effects for cartilage repair with resolution close to real time on a scale of micrometers. This technology for nondestructive quantitative assessment of human articular cartilage degeneration may facilitate the development of strategies to delay or prevent the onset of OA.

25.18
Intraoperative OCT Monitoring in Neurology

Intraoperative identification of brain tumors and tumor margins has been limited by either the resolution of the *in vivo* imaging technique or the time required to obtain histologic specimens.

OCT can effectively differentiate normal cortex from intracortical melanoma based on variations in optical backscatter. A portable hand-held OCT surgical imaging probe has been constructed by Boppart et al. for imaging within the surgical field. Two-dimensional images showed increased optical backscatter from regions of tumor, which was quantitatively used to determine the tumor margins [91]. The images correlated well with the histologic findings. Three-dimensional reconstructions revealed regions of tumor penetrating the normal cortex and could be resectioned at arbitrary planes. Subsurface cerebral vascular structures could be identified and were therefore avoided.

Detection of residual tumor during resection of glial brain tumors remains a challenge because of the low inherent contrast of adjacent edematous brain, the surrounding infiltration zone, and the solid tumor. Böhringer et al. reported the results of a study using a time-domain Sirius 713 tomograph with a central wavelength of 1310 nm and a coherence length of 15 µm equipped with a mono-mode fiber and a modified OCT adapter containing a lens system for imaging at a working distance of 2.5 cm [92]. A spectral-domain tomograph using 840 and 930 nm superluminescence diodes (SLDs) with a central wavelength of 900 nm was used as a second imaging modality in the same study. Both time-domain and spectral-domain coherence tomography delineated normal brain, the infiltration zone, and solid tumor in murine intracerebral gliomas. Histologic evaluation of H&E-stained sections parallel to the optical plain demonstrated that tumor areas of less than 1 mm could be detected and that not only solid tumor but also brain invaded by low-density single tumor cells which produced an OCT signal different from normal brain. SD-OCT demonstrated a significantly more detailed microstructure of tumor and normal brain up to a tissue depth of 1.5–2.0 mm, whereas the interpretation of time-domain OCT was difficult at tissue depths >1.0 mm.

An OCT integrated endoscope can image the endoventricular anatomy and other endoscopically accessible structures in a human brain specimen. Böhringer et al, mounted a Sirius 713 OCT device on a modified rigid endoscope for simultaneous OCT imaging and endoscopic video imaging of the visible spectrum using a graded index rod endoscope [93]. OCT imaging of a human brain specimen in water allowed an in-depth view into structures such as the walls of the ventricular system, the choroid plexus, and the thalamostriatal vein. OCT further allowed imaging of structures beyond tissue barriers or opaque media. In this fixed specimen, OCT allowed discrimination of vascular structures down to a diameter of 50 μm. In vessels larger than 100 μm, the lumen could be discriminated and within larger blood vessels a layered structure of the vascular wall and also endovascular plaques could be visualized. This *in vitro* pilot study demonstrated that OCT integrated into neuroendoscopes may add information that cannot be obtained by video imaging alone. Using an intracranial glioma model according to Böhringer *et al.* [93], it has been shown recently that the working distance of the OCT adapter and the A-scan acquisition rate conceptually allow integration of the OCT applicator into the optical path of the operating microscopes. This would allow a continuous analysis of the resection plane, providing optical tomography, thereby adding a third dimension to the microscopic view and information on the light attenuation characteristics of the tissue. Performing 3D data arrays for multiplanar analysis of the tumor to brain interface, time-domain OCT allows discrimination of normal brain, diffusely invaded brain tissue, and solid tumor [39].

In conjunction with the rapid image acquisition rates of OCT, this technology carries the potential for a novel intraoperative imaging tool for the detection of residual tumor and guidance of neurosurgical resections. OCT technology may provide an extra margin of safety by providing cross-sectional images of tissue barriers within optically opaque conditions.

25.19
Intraoperative OCT Monitoring in Breast Surgery

As breast cancer screening rates increase, smaller and more numerous lesions are being identified earlier, leading to more breast-conserving surgical procedures.

Nguyen et al. reported that achieving a clean surgical margin represents a technical challenge with important clinical implications where OCT has been introduced as an intraoperative high-resolution imaging technique that assesses surgical breast tumor margins by providing real-time microscopic images up to 2 mm beneath the tissue surface of the lumpectomy specimens [94]. Based on histologic findings, nine true positives, nine true negatives, two false positives, and no false negatives were found, yielding a sensitivity of 100% and a specificity of 82%.

Another study by the same group reported the use of OCT for the intraoperative *ex vivo* imaging and assessment of axillary lymph nodes from 17 human patients with breast cancer who were imaged intraoperatively with OCT [95]. These preliminary clinical studies identified scattering changes in the cortex, relative to the capsule, which can be used to differentiate normal from reactive and metastatic nodes. These optical scattering changes are correlated with inflammatory and immunologic changes observed in the follicles and germinal centers. The results show the potential of OCT as a real-time method for intraoperative margin assessment in breast-conserving surgeries without having to resect physically and process histologically specimens in order to visualize microscopic features.

25.20
Intraoperative OCT Monitoring in ENT

Shakhov et al. applied OCT for intraoperative control in laser surgery of laryngeal carcinoma operated with a surgical YAG:Nd laser at two switchable wavelengths of 1.44 and 1.32 µm by laryngofissure, direct microlaryngoscopy, and fibrolaryngoscopy [96]. Information on structural alterations in laryngeal mucosa to a depth of 2 mm, obtained by OCT, makes it possible to locate tumor borders precisely, thus giving an opportunity to control the surgical treatment of laryngeal carcinoma. Combination of the two wavelengths in a single laser unit and intraoperative OCT monitoring represent a new modality for minimally invasive larynx surgery. The use of biocompatible materials offers new possibilities in ossiculoplasty in middle ear surgery. The exact calculation concerning the length of the implant to be used, however, still poses considerable difficulties and is an additional cause of a remaining air conduction difference or the need for further surgical intervention.

Heermann et al. reported the use of an optical coherence interferometer coupled to an operating microscope in five stapedoplasties and five tympanoplasties type III in order to determine the length of the prosthesis to be used [97]. The measurement of middle ear structures has an accuracy of 30 µm. The postoperative audiologic results showed a good auditory performance. Wong et al. examined patients undergoing surgical head and neck endoscopy using a fiber-optic hand-held OCT imaging probe placed in near contact with the target site (1.3 µm broadband light source, FWHM = 80 nm, frame rate 1 Hz) in order to study and characterize microstructural anatomy and features of the larynx and benign laryngeal pathology *in vivo*, including information on the thickness of the epithelium, integrity of the basement membrane, and structure of the lamina propria, and also microstructural features such as glands, ducts, blood vessels, fluid collection/edema, and the transitions between pseudostratified columnar and stratified squamous epithelium [98].

Just *et al.* used a specially equipped operating microscope with an integrated SD-OCT apparatus for standard middle-ear surgical procedures [41]. Intraoperative OCT was evaluated for establishment of the cause of stapes fixation, assessment of the morphology of the stapes footplate in revision stapes surgery, and as an orientation guide in cochlear implantation in congenital anomalies. OCT displays the middle and inner ear structures precisely. Potential areas of application can be defined as a result of these studies: visualization of the oval window niche in revision stapesplasty and reconstructive middle ear surgery, and also during explorative tympanotomy for intraoperative assessment of perilymph fistula, and demonstration of structures of the exposed but not opened inner ear. OCT allowed *in vivo* visualization and documentation of the annular ligament, the different layers of the footplate, and the inner-ear structures, both in non-fixed and fixed stapes footplates. In cases of otosclerosis and tympanosclerosis, an inhomogeneous and irregularly thickened footplate was found, in contrast to the appearance of non-fixed footplates. In both fixed and non-fixed footplates, there was a lack of visualization of the border between the footplate and the otic capsule [99].

Using the Niris OCT imaging system (Imalux, Cleveland, OH, USA), images of benign and premalignant laryngeal disease in 33 patients undergoing surgical head and neck endoscopy were obtained and analyzed by Rubinstein *et al.* [100]. This imaging system has a spatial depth resolution of 10–20 mm and a depth scanning range of 2.2 mm, obtaining images of 200 × 200 pixels at a maximum frame rate of 0.7 Hz. The tip of the probe was inserted through a rigid laryngoscope, and still images of arytenoids, aryepiglottic folds, piriform sinus, epiglottis, and true and false vocal cords were acquired. In patients whose OCT images were taken from normal tissue, the normal microstructures were clearly identified, and also disruption of the latter in malignant pathologies.

OCT imaging potentially offers an efficient, quick, and reliable imaging modality in guiding surgical biopsies, intraoperative decision-making, and therapeutic options of various laryngeal pathologies and premalignant disease. OCT has the unique ability to image laryngeal tissue microstructure and can detail microanatomic changes in benign, premalignant, and malignant laryngeal pathologies.

25.21
Intraoperative OCT Monitoring in Gynecology

Previous studies have correlated cancer recurrence and progression with obtaining clear margins upon resection. The most common need to obtain clear margins with respect to conservative treatment in patients with cervical neoplasia occurs with women who wish to preserve fertility. However, current detection methods are limited and current treatments present additional fertility concerns. In order to provide the best care for patients wishing to retain fertility post-treatment for cervical dysplasia, a superior option for detecting tumor margins accurately at the microscopic scale must be further explored [101]. A novel prototype intraoperative system combining positron detection and OCT imaging was developed for early ovarian cancer detection by Gamelin et al. [102]. The probe employs eight plastic scintillating fiber tips for preferential detection of local positron activity surrounding a central scanning OCT fiber, providing volumetric imaging of tissue structure in regions of high radiotracer uptake. In conjunction with co-registered frequency-domain OCT measurements, the results demonstrated the potential for a miniaturized laparoscopic probe offering simultaneous functional localization and structural imaging for improved early cancer detection.

25.22
Prediction of Future Developments in Intraoperative OCT Monitoring

OCT is a powerful optical biopsy tool although with a small field of view, with nearly real-time imaging. OCT can be guided by a large field of view technique such as spectral or fluorescence imaging [78, 103]. As OCT technology improves, the ability to image cellular level features will increase the range of applicability of OCT imaging. Full use of OCT for guided surgical interventions can be achieved with rapid 3D imaging, thus reducing motion artifacts and improving reproducibility, since otherwise it is difficult to bring the image field in coincidence with the relevant but dynamically changing anatomic structures [48]. Pattern recognition and filter algorithms may improve the discrimination of pathology and consequences of the surgical intervention of the OCT methods [104]. OCT has the potential to improve the efficacy of surgical or therapeutic interventions and may permit new treatment approaches. For instance, further studies are needed to assess possible active control of the laser ablation through real-time OCT clinical feedback, which could improve current ablation efficacy. OCT holds the potential to guide surgical biopsies, direct therapy, and monitor disease, particularly when office-based systems are developed. ENT, dental, gynecologic, and pulmonologic applications have significant potential to be more widely adopted as areas of OCT clinical research.

Acknowledgments

This study was supported by Erlangen Graduate School in Advanced Optical Technologies (SAOT) by the German National Science Foundation (DFG) in the framework of the excellence initiative. The author thanks Ioulia Davydova for helping with the design of the charts.

References

1 Boppart, S.A., Herrmann, J.J., Pitris, C., Stamper, D.L., Brezinski, M.E., and Fujimoto, J.G. (1999) High-resolution optical coherence tomography-guided laser ablation of surgical tissue. *J. Surg. Res.*, **82**, 275–284.

2 Schmitt, J.M., Knuttel, A., Yadlowsky, M., and Eckhaus, M.A. (1994) Optical coherence tomography of a dense tissue – statistics of attenuation and backscattering. *Phys. Med. Biol.*, **38**, 1705–1720.

3 Fujimoto, J.G., Pitris, C., Boppart, S.A., and Brezinski, M.E. (2000) Optical coherence tomography: an emerging technology for biomedical imaging and optical biopsy. *Neoplasia*, **2** (1–2), 9–25.

4 Boppart, S.A., Bouma, B.E., Pitris, C., Tearney, G.J., Fujimoto, J.G., and Brezinski, M.E. (1997) Forward-imaging instruments for optical coherence tomography. *Opt. Lett.*, **22**, 1618–1620.

5 McLaughlin, R.A. and Sampson, D.D. (2010) Clinical applications of fiber-optic probes in optical coherence tomography. *Opt. Fiber Technol.*, **16** (6), 467–475.

6 Probst, J., Hillmann, D., Lankenau, E., Winter, C., Oelckers, S., Koch, P., and Hüttmann, G. (2010) Optical coherence tomography with online visualization of more than seven rendered volumes per second. *J. Biomed. Opt.*, **15** (2), 026014.

7 Ho, J., Sull, A.C., Vuong, L.N., Chen, Y., Liu, J., Fujimoto, J.G., Schuman, J.S., and Duker, J.S. (2009) Assessment of artifacts and reproducibility across spectral- and time-domain optical coherence tomography devices. *Ophthalmology*, **116** (10), 1960–1970.

8 Agrba, P., Kirillin, M., Abelevich, A., Zagaynova, E., and Kamensky, V. (2009) Compression as a method for increasing the informativity of optical coherence tomography of biotissues. *Opt. Spectrosc.*, **107** (6), 853–858.

9 Dolin, S., Feldstein, F., Gelikonov, G., Gelikonov, V., Gladkova, N., Iksanov, R., Kamensky, V., Kuranov, R., Sergeev, A., Shakhova, N., and Turchin, I. (2004) Fundamentals of OCT and clinical applications of endoscopic OCT, in *Handbook of Coherent Domain Optical Methods* (ed. V. Tuchin), Kluwer, Dordrecht, pp. 211–270.

10 Meissner, O.A., Rieber, J., Babaryka, G., Oswald, M., Reim, S., Siebert, U., Redel, T., Reiser, M., and Mueller-Lisse, U. (2006) Intravascular optical coherence tomography: comparison with histopathology in atherosclerotic peripheral artery specimens. *J. Vasc. Interv. Radiol.*, **17** (2 Pt 1), 343–349.

11 Hollo, G., Garas, A., and Kóthy, P. (2010) Comparison of diagnostic accuracy of Fourier-domain optical coherence tomography and scanning laser polarimetry to detect glaucoma. *Acta Ophthalmol.*, **88** (Suppl. s246), on line publication, doi: 10.5301/ejo.5000011.

12 Drexler, W. and Fujimoto, J.G. (2008) State-of-the-art retinal optical coherence tomography. *Prog. Retin. Eye Res.*, **27** (1), 45–88.

13 Suzuki, Y., Ikeno, F., Koizumi, T., Tio, F., Yeung, A.C., Yock, P.G., Fitzgerald, P.J., and Fearon, W.F. (2008) In vivo comparison between optical coherence tomography and intravascular ultrasound for detecting small degrees of in-stent neointima after stent implantation. *JACC Cardiovasc. Interv.*, **1** (2), 168–173.

14 Douplik, A., Morofke, D., Chiu, S., Bouchelev, V., Mao, L., Yang, V., and Vitkin, A. (2008) In vivo real time monitoring of vasoconstriction and vasodilation by a combined diffuse reflectance spectroscopy and Doppler optical coherence tomography approach. *Laser Surg. Med.*, **40** (5), 323–331.

15 Dirk, J.F., van der Meer, F.J., Aalders, M., and van Leeuwen, C.G. (2004) Quantitative measurement of attenuation coefficients of weakly scattering media using optical coherence tomography. *Opt. Express*, **12** (19), 4353–4365.

16 Brezinski, M.E., Tearney, G.J., Boppart, S.A., Swanson, E.A., Southern, J.F., and Fujimoto, J.G. (1997) Optical biopsy with optical coherence tomography, feasibility for surgical diagnostics. *J. Surg. Res.*, **71**, 32–40.

17 Boppart, S.A., Bouma, B.E., Pitris, C., Southern, J.F., Brezinski, M.E., and Fujimoto, J.G. (1998) Intraoperative assessment of microsurgery with three-dimensional optical coherence tomography. *Radiology*, **208**, 81–86.

18 Justa, T., Lankenaua, E., Hüttmann, G., and Paua, H.W. (2009) Intra-operative application of optical coherence

tomography with an operating microscope. *J. Laryngol. Otol.*, **123** (9), 1027–1030.

19 Miyazawa, A., Yamanari, M., Makita, S., Miura, M., Kawana, K., Iwaya, K., Goto, H., and Yoshiaki, Y. (2009) Tissue discrimination in anterior eye using three optical parameters obtained by polarization sensitive optical coherence tomography. *Opt. Express*, **17** (20), 17426.

20 Liang, X., Graf, B.W., and Boppart, S.A. (2009) Imaging engineered tissues using structural and functional optical coherence tomography. *J. Biophotonics*, **2** (11), 643–655.

21 Hauger, C., Luber, J., Kaschke, M., and Krause-Bonte, M. (2004) Surgical system supported by optical coherence tomography. U.S. Pat. US 6763259.

22 Sampliner, R.E. (2002) Updated guidelines for the diagnosis, surveillance and therapy of Barrett's esophagus. *Am. J. Gastroenterol.*, **97**, 1888–1895.

23 Tearney, G.I., Brezinski, M.E., Bouma, B.E., Boppart, S.A., Pitris, C., Southern, J.F., and Fujimoto, J.G. (1997) *In vivo* endoscopic optical biopsy with optical coherence. *Tomogr. Sci.*, **276** (5), 2037–2039.

24 Tearney, G.J., Brezinski, M.E. *et al.* (1997) Optical biopsy in human gastrointestinal tissue using optical coherence tomography. *Am. J. Gastroenterol.*, **92** (10), 1800–1804.

25 Brezinski, M.E., Tearney, G.J., Boppart, S.A., Swanson, E.A., Southern, J.F., and Fujimoto, J.G. (1997) Optical biopsy with optical coherence tomography: feasibility for surgical diagnostics. *J. Surg. Res.*, **71** (1), 32–40.

26 Brezinski, M.E., Tearney, G.J., Bouma, B.E., Izatt, J.A., Hee, M.R., Swanson, E.A., Southern, J.F., and Fujimoto, J.G. (1996) Optical coherence tomography for optical biopsy. Properties and demonstration of vascular pathology. *Circulation*, **93** (6), 1206–1213.

27 Cahill, R.A., Asakuma, M., Trunzo, J., Schomisch, S., Wiese, D., Saha, S., Dallemagne, B., Marks, J., and Marescaux, J. (2009) Intraperitoneal virtual biopsy by fibered optical coherence tomography (OCT) at natural orifice transluminal endoscopic surgery (NOTES). *J. Gastrointest. Surg.*, **14**, 732–738.

28 Iftimia, N.V., Mujat, M., Ustun, T., Ferguson, R.D., Danthu, V., and Hammer, D.X. (2009) Spectral-domain low coherence interferometry/optical coherence tomography system for fine needle breast biopsy guidance. *Rev. Sci. Instrum.*, **80** (2), 024302.

29 Izatt, J.A., Kulkarni, M.D., Hsing-Wen, W., Kobayashi, K., and Sivak, M.V. Jr. (1996) Optical coherence tomography and microscopy in gastrointestinal tissues. *IEEE J. Sel. Top. Quantum Electron.*, **2**, 1017–1028.

30 Tearney, G.J., Brezinski, M.E., Southern, J.F., Bouma, B.E., Boppart, S.A., and Fujimoto, J.G. (1998) Optical biopsy in human pancreatobiliary tissue using optical coherence tomography. *Dig. Dis. Sci.*, **43**, 1193–1199.

31 Tearney, G.J., Brezinski, M.E., Southern, J.F., Bouma, B.E., Boppart, S.A., and Fujimoto, J.G. (1997) Optical biopsy in human urologic tissue using optical coherence tomography. *J. Urol.*, **157**, 1915–1919.

32 Pitris, C., Brezinski, M.E., Bouma, B.E., Tearney, G.J., Southern, J.F., and Fujimoto, J.G. (1998) High resolution imaging of the upper respiratory tract with optical coherence tomography – A feasibility study. *Am. J. Respir. Crit. Care Med.*, **157**, 1640–1644.

33 Pitris, C., Goodman, A., Boppart, S.A., Libus, J.J., Fujimoto, J.G., and Brezinski, M.E. (1999) High-resolution imaging of gynecologic neoplasms using optical coherence tomography. *Obstet. Gynecol.*, **93**, 135–139.

34 Sayanagi, K., Pelayes, D.E., Kaiser, P.K., and Singh, A.D. (2010) 3D spectral domain optical coherence tomography findings in choroidal tumors. *Eur. J. Ophthalmol.*, **21** (3), 271–275.

35 Strasswimmer, J. (2010) Optical biopsy at the bedside. *Arch. Dermatol.*, **146** (8), 909–910.

36 Bouma, B.E., Tearney, G.J., Compton, C.C., and Nishioka, N.S. (2000) High-resolution imaging of the

human esophagus and stomach *in vivo* using optical coherence tomography. *Gastrointest. Endosc.*, **51**, 467–474.

37 Karl, A., Stepp, H., Willmann, E., Tilki, D., Zaak, D., Knüchel, R., and Stief, C. (2008) Optical coherence tomography (OCT): ready for the diagnosis of a nephrogenic adenoma of the urinary bladder? *J. Endourol.*, **22** (11), 2429–2432.

38 Linehan, J.A., Hariri, L.P., Sokoloff, M.H., Rice, P.S., Bracamonte, E.R., Barton, J.K., and Nguyen, M.M. (2010) Use of optical coherence tomography imaging to differentiate benign and malignant renal masses. *J. Urol.*, **183** (4, Suppl.), e835.

39 Böhringer, H.J., Lankenau, E., Stellmacher, F., Reusche, E., Hüttmann, G., and Giese, A. (2009) Imaging of human brain tumor tissue by near-infrared laser coherence tomography. *Acta Neurochir. (Wien)*, **151** (5), 507–517.

40 Mujat, M., Ferguson, R.D., Hammer, D.X., Gittins C., and Iftimia, N. (2009) Automated algorithm for breast tissue differentiation in optical coherence tomography. *J. Biomed. Opt.*, **14** (3), 034040.

41 Just, T., Lankenau, E., Hüttmann, G., and Pau, H.W. (2009) Optische Kohärenztomographie in der Mittelohrchirurgie. *HNO*, **57** (5), 421–427.

42 Van Soest, G., Goderie, T.P.M., Gonzalo, N., Koljenovic, S., van Leenders, G.L.J.H., Regar, E., Serruys, P.W., and van der Steen, A.F.W. (2009) Imaging atherosclerotic plaque composition with intracoronary optical coherence tomography. *Neth. Heart J.*, **17** (11), 448–450.

43 Van der Meer, F.J., Faber, D.J., Aalders, M.C.G., Poot, A.A., Vermes, I., and van Leeuwen, T.G. (2010) Apoptosis- and necrosis-induced changes in light attenuation measured by optical coherence tomography. *Lasers Med. Sci.*, **25** (2), 259–267.

44 Standish, B.A., Lee, K.K.C., Jin, X., Mariampillai, A., Wood, M.F.G., Wilson, B.C., Vitkin, I.A., Munce, N.R., and Yang, V.X.D. (2008) Interstitial Doppler optical coherence tomography as a local tumor necrosis predictor in photodynamic therapy of prostatic carcinoma: an *in vivo* study. *Cancer Res.*, **68**, 9987.

45 Yang, Y., Whiteman, S., Gey van Pittius, D., He, Y., Wang, R.K., and Spiteri, M.A. (2004) Use of optical coherence tomography in delineating airways microstructure: comparison of OCT images to histopathological sections. *Phys. Med. Biol.*, **49** (7), 1247–1255.

46 De Boer, J.F., Srinivas, S.M., Malekafzali, A., Chen, Z., and Nelson, J.S. (1998) Imaging thermally damaged tissue by polarization sensitive optical coherence tomography. *Opt. Express*, **3**, 212–218.

47 Tsai, M.T., Lee, H.C., Lu, C.W., Wang, Y.M., Lee, C.K., Yang, C.C., and Chiang, C.P. (2008) Delineation of an oral cancer lesion with swept-source optical coherence tomography. *J. Biomed. Opt.*, **13** (4), 044012.

48 Hüttmann, G., Lankenau, E., Schulz-Wackerbarth, C., Müller, M., Steven, P., and Birngruber, R. (2009) Übersicht der apparativen Entwicklungen in der optischen Kohärenztomografie: von der Darstellung der Retina zur Unterstützung therapeutischer Eingriffe. *Klin. Monatsbl. Augenheilkd.*, **226** (12), 958–964.

49 Bagayev, S.N., Gelikonov, V.M., Gelikonov, G.V., Kargapoltsev, E.S., Kuranov, R.V., Razhev, A.M., Turchin, I.V., and Zhupikov, A.A. (2002) Optical coherence tomography for *in situ* monitoring of laser corneal ablation. *J. Biomed. Opt.*, **7** (4), 633–642.

50 Wirbelauer, C., Aurich, H., and Pham, D.T. (2007) Online coherence pachymetry to evaluate intraoperative ablation parameters in LASIK. *Graefes Arch. Clin. Exp. Ophthalmol.*, **245** (6), 775–781.

51 Kucumen, R.B., Dinc, U.A., Yenerel, N.M., Gorgun, E., and Alimgil, M.L. (2009) Immediate evaluation of the flaps created by

femtosecond laser using anterior segment optical coherence tomography. *Ophthalmic Surg. Lasers Imaging*, **40** (3), 251–254.

52 Utine, C., Altunsoy, M., and Basar, D. (2008) Visante anterior segment OCT in a patient with gas bubbles in the anterior chamber after femtosecond laser corneal flap formation. *Int. Ophthalmol.*, **30** (1), 81–84.

53 Ambrósio, R., Jardim, D., Netto, M.V., and Wilson, S.E. (2007) Management of unsuccessful LASIK surgery. *Compr. Ophthalmol. Update*, **8** (3), 125–141; discussion 143–144.

54 Doors, M., Tahzib, N.G., Eggink, F.A., Berendschot, T.T.J.M., Webers, C.A.B., and Nuijts, R. (2009) Use of anterior segment optical coherence tomography to study corneal changes after collagen cross-linking. *Am. J. Ophthalmol.*, **148** (6), 844–851.

55 Geerling, G., Müller, M., Winter, C., Hoerauf, H., Oelckers, S., Laqua, H., and Birngruber, R. (2005) Intraoperative 2-dimensional optical coherence tomography as a new tool for anterior segment surgery. *Arch. Ophthalmol.*, **123** (2), 253–257.

56 Azzolini, C., Patelli, F., Codenotti, M., Pierro, L., and Brancato, R. (1999) Optical coherence tomography in idiopathic epiretinal macular membrane surgery. *Eur. J. Ophthalmol.*, **9** (3), 206–211.

57 Gobel, W., Schrader, W.F., Sehrenker, M., and Klink, T. (2000) Findings of optical coherence tomography (OCT) before and after macular hole surgery. *Ophthalmology*, **97** (4), 251–256.

58 Hattenbach, L.O., Höhn, F., Fulle, G., and Mirshahi, A. (2009) Präoperative Diagnostik topografischer Merkmale bei epiretinaler Gliose mittels hochauflösender optischer Kohärenztomografie. *Klin. Monatsbl. Augenheilkd.*, **226** (8), 649–653.

59 Wykoff, C.C., Berrocal, A.M., Schefler, A.C., Uhlhorn, S.R., Ruggeri, M., and Hess, D. (2010) Intraoperative OCT of a full-thickness macular hole before and after internal limiting membrane peeling. *Ophthalmic Surg. Lasers Imaging*, **41** (1), 7–11.

60 Knecht, P.B., Kaufmann, C., Menke, M.N., Watson, S.L., and Bosch, M.M. (2010) Use of intraoperative Fourier-domain anterior segment optical coherence tomography during Descemet stripping endothelial keratoplasty. *Am. J. Ophthalmol.*, **150** (3), 360–365.

61 Ide, T., Wang, J., Tao, A., Leng, T., Kymionis, G.D., O'Brien, T.P., and Yoo, S.H. (2010) Intraoperative use of three-dimensional spectral-domain optical coherence tomography. *Ophthalmic Surg. Lasers Imaging*, **41** (2), 250–254.

62 Güven, D., Weiland, J.D., Maghribi, M., Davidson, J.C., Mahadevappa, M., Roizenblatt, R., Qiu, G., Krulevitz, P., Wang, X., Labree, L., and Humayun, M.S. (2006) Implantation of an inactive epiretinal poly(dimethyl siloxane) electrode array in dogs. *Exp. Eye Res.*, **82** (1), 81–90.

63 Puliafito, C.A. (2010) Optical coherence tomography: a new tool for intraoperative decision making. *Ophthalmic Surg. Lasers Imaging*, **41** (1), 6.

64 Brown, E.N., Burris, N.S., Gu, J., Kon, Z.N., Laird, P., Kallam, S., Tang, C.M., Schmitt, J.M., and Poston, R.S. (2007) Thinking inside the graft: applications of optical coherence tomography in coronary artery bypass grafting. *J. Biomed. Opt.*, **12** (5), 051704.

65 Low, A.F., Tearney, G.J., Bouma, B.E., and Jang, I.-K. (2006) Technology Insight: optical coherence tomography – current status and future development. *Nat. Cardiovasc. Clin. Pract. Med.*, **3** (3), 154–162.

66 Libby, P. (2001) Current concepts of the pathogenesis of the acute coronary syndromes. *Circulation*, **104**, 365–372.

67 Burke, A.P., Farb, A., Malcom, G.T., Liang, Y.-h., Smialek, J., and Virmani, R. (1997) Coronary risk factors and plaque morphology in men with coronary disease who died suddenly. *N. Engl. J. Med.*, **336**, 1276–1282.

68 Brazio, P.S., Laird, P.C., Xu, C.Y., Gu, J.Y., Burris, N.S., Brown, E.N., Kon, Z.N., and Poston, R.S. (2008) Harmonic scalpel versus electrocautery for harvest of radial

artery conduits: reduced risk of spasm and intimal injury on optical coherence tomography. *J. Thorac. Cardiovasc. Surg.*, **136** (5), 1302–1308.
69 Brown, E.N., Burris, N.S., Kon, Z.N., Grant, M.C., Brazio, P.S., Xu, C., Laird, P., Gu, J., Kallam, S., Desai, P., and Poston, R.S. (2009) Intraoperative detection of intimal lipid in the radial artery predicts degree of postoperative spasm. *Arteriosclerosis*, **205** (2), 466–471.
70 Brown, E.N., Kon, Z.N., Tran, R., Burris, N.S., Gu, J., Laird, P., Brazio, P.S., Kallam, S., Schwartz, K., Bechtel, L., Joshi, A., Zhang, S., and Poston, R.S. (2007) Strategies to reduce intraluminal clot formation in endoscopically harvested saphenous veins. *J. Thorac. Cardiovasc. Surg.*, **134** (5), 1259–1265.
71 Steiner, R., Kunzi-Rapp, K., and Scharffetter-Kochanek, K. (2003) Optical coherence tomography: clinical applications in dermatology. *Med. Laser Appl.*, **18** (3), 249–259.
72 Gambichler, T., Regeniter, P., Bechara, F.G., Orlikov, A., Vasa, R., Moussa, G., Stücker, M., Altmeyer, P., and Hoffmann, K. (2007) Characterization of benign and malignant melanocytic skin lesions using optical coherence tomography in vivo. *J. Am. Acad. Dermatol.*, **57**, 629–637.
73 Mogensen, M., Lars, T., Jørgensen, T.M., Andersen, P.E., and Jemec, G.B.E. (2009) OCT imaging of skin cancer and other dermatological diseases. *J. Biophoton*, **2** (6–7), 442–451.
74 He, Y., Wu, S., Li, Z., Cai, S., and Li, H. (2010) Monitoring the process of tissue healing of rat skin *in vivo* after laser irradiation based on optical coherence tomography. *Proc. SPIE*, **7845**, 78452E.
75 Zhang, X., Wu, S., and Li, H. (2010) Monitoring collagen remodeling on opto-thermal response of photoaged skin irradiated by Er:YAG laser with optical coherence tomography. *Proc. SPIE*, **7845**, 78450F.
76 Shakhova, N.M., Gelikonov, V.M., Kamensky, V.A., Kuranov, R.V., and Turchin, I.V. (2002) Clinical aspects of the endoscopic optical coherence tomography and the ways for improving its diagnostic value. *Laser Phys.*, **12** (4), 617–626.
77 Yang, V.X.D., Tang, S., Gordon, M.L., Qi, B., Gardiner, G., Cirocco, M., Kortan, P., Haber, G.B., Kandel, G., Vitkin, I.A., Wilson, B.C., and Marcon, N.E. (2005) Endoscopic Doppler optical coherence tomography in the human GI tract: initial experience. *Gastrointest. Endosc.*, **61** (7), 879–890.
78 Zagaynova, E., Gladkova, N., Shakhova, N., Gelikonov, G., and Gelikonov, V. (2008) Endoscopic OCT with forward-looking probe: clinical studies in urology and gastroenterology. *J. Biophotonics*, **1** (2), 114–128.
79 Cahill, R.A. and Mortense, N.J. (2010) Intraoperative augmented reality for laparoscopic colorectal surgery by intraoperative near-infrared fluorescence imaging and optical coherence tomography. *Minerva Chir.*, **65** (4), 451–462.
80 Shen, B., Zuccaro, G., Gramlich, T., Gladkova, N., Delaney, C., Connor, J., Kareta, M., Lashner, B., Bevins, C., Feldchtein, F., Strong, S., Bambrick, M., Trolli, P., and Fazio, V. (2004) Intraoperative *ex vivo* histology-correlated optical coherence tomography in the detection of transmural inflammation in Crohn's disease. *Gastroenterology*, **126** (4), (Suppl. 2) A207–A208.
81 Daniltchenko, D., König, F., Lankenau, E., Sachs, M., Kristiansen, G., Huettmann, G., and Schnorr, D. (2006) Anwendung der optischen Kohärenztomographie (OCT) bei der Darstellung von Urothelerkrankungen der Harnblase. *Radiologie*, **46** (7), 584–589.
82 Skarecky, D.W., Brenner, M., Rajan, S., Rodriguez, E., Narula, N., Melgoza, F., and Ahlering, T.E. (2008) Zero positive surgical margins after radical prostatectomy: is the end in sight? *Expert Rev. Med. Dev.*, **5** (6), 709–717.
83 Rais-Bahrami, S., Levinson, A.W., Fried, N.M., Lagoda, G.A., Hristov, A., Chuang, Y., Burnett, A.L., and Su, L.M. (2008) Optical coherence tomography of cavernous nerves: a step toward real-time

intraoperative imaging during nerve-sparing radical prostatectomy. *Urology*, **72** (1), 198–204.

84 Chitchian, S, Weldon, T.P., and Fried, N.M. (2009) Segmentation of optical coherence tomography images for differentiation of the cavernous nerves from the prostate gland. *J. Biomed. Opt.*, **14** (4), 044033.

85 Aron, M., Kaouk, J.H., Hegarty, N.J., Colombo, J.R. Jr., Haber, G.P., Chung, B.I., Zhou, M., and Gill, I.S. (2007) Second prize: preliminary experience with the Niris optical coherence tomography system during laparoscopic and robotic prostatectomy. *J. Endourol.*, **21**, 814–818.

86 Parekattil, S., Yeung, L.L., and Su, L.-M. (2009) Intraoperative tissue characterization and imaging. *Urol. Clin. North Am.*, **36** (2), 213–221.

87 Li, X.D., Martin, S., Pitris, C., Ghanta, R., Stamper, D.L., Harman, M., Fujimoto, J.G., and Brezinski, M.E. (2005) High-resolution optical coherence tomographic imaging of osteoarthritic cartilage during open knee surgery. *Arthritis Res. Ther.*, **7** (2), R318–R323.

88 Han, J.H., Liu, X., Kang, J.U., and Song, C.G. (2010) High-resolution subsurface articular cartilage imaging based on Fourier-domain common-path optical coherence tomography. *Chin. Opt. Lett.*, **8** (2), 167–169.

89 Chu, C.R., Williams, A., Tolliver, D., Kwoh, K., Bruno, S., and Irrgang, J.J. (2010) Clinical optical coherence tomography of early articular cartilage degeneration in persons with degenerative meniscal tears. *Arthritis Care Res.*, **62** (5), 1412–1420.

90 Ling, C.H.Y., Pozzi, A., Thieman, K.M., Tonks, C.A., Guo, S.G., Xie, H.K., and Horodyski, M. (2010) The potential of optical coherence tomography for diagnosing meniscal pathology. *Meas. Sci. Technol.*, **21** (4), 045801.

91 Boppart, S.A., Brezinski, M.E., Pitris, C., and Fujimoto, J.G. (1998) Optical coherence tomography for neurosurgical imaging of human intracortical melanoma. *Neurosurgery*, **43** (4), 834–841.

92 Böhringer, H.J., Boller, D., Leppert, J., Knopp, U., Lankenau, E., Reusche, E., Hüttmann, G., and Giese, A. (2006) Time-domain and spectral-domain optical coherence tomography in the analysis of brain tumor tissue. *Lasers Surg. Med.*, **38** (6), 588–597.

93 Böhringer, H.J., Lankenau, E., Rohde, V., Hüttmann, G., and Giese, A. (2006) Optical coherence tomography for experimental neuroendoscopy. *Minim. Invasive Neurosurg.*, **49** (5), 269–275.

94 Nguyen, F.T., Zysk, A.M., Chaney, E.J., Kotynek, J.G., Oliphant, U.J., Bellafiore, F.J., Rowland, K.M., Johnson, P.A., and Boppart, S.A. (2009) Intraoperative evaluation of breast tumor margins with optical coherence tomography. *Cancer Res.*, **69** (22), 8790–8796.

95 Nguyen, F.T., Zysk, A.M., Chaney, E.J., Adie, S.G., Kotynek, J.G., Oliphant, U.J., Bellafiore, F.J., Rowland, K.M., Johnson, P.A., and Boppart, S.A. (2010) Optical coherence tomography: the intraoperative assessment of lymph nodes in breast cancer. *IEEE Eng. Med. Biol. Mag.*, **29** (2), 63–70.

96 Shakhov, A.V., Terentjeva, A.B., Kamensky, V.A., Snopova, L.B., Gelikonov, V.M., Feldchtein, F.I., and Sergeev, A.M. (2001) Optical coherence tomography monitoring for laser surgery of laryngeal carcinoma. *J. Surg. Oncol.*, **77** (4), 253–258.

97 Heermann, R., Hauger, C., Issing, P.R., and Lenarz, T. (2002) Erste Anwendungen der optischen Kohärenztomographie (OCT) in der Mittelohrchirurgie. *Laryngorhinootologie*, **81** (6), 400–405.

98 Wong, B.J., Jackson, R.P., Guo, S., Ridgway, J.M., Mahmood, U., Su, J., Shibuya, T.Y., Crumley, R.L., Gu, M., Armstrong, W.B., and Chen, Z. (2005) In vivo optical coherence tomography of the human larynx: normative and benign pathology in 82 patients. *Laryngoscope*, **115** (11), 1904–1911.

99 Just, T., Lankenau, E., Hüttmann, G., and Pau, H.W. (2009) Optical coherence tomography of the oval window niche. *J. Laryngol. Otol.*, **123** (6), 603–608.

100 Rubinstein, M., Fine, E.L., Sepehr, A., Armstrong, W.B., Crumley, R.L., Kim, J.H., Chen, Z., and Wong, B.J. (2010) Optical coherence tomography of the larynx using the Niris system. *J. Otolaryngol. Head Neck Surg.*, **39** (2), 150–156.

101 Bickford, L.R., Drezek, R.A., and Yu, T.K. (2007) Intraoperative techniques and tumor margin status-room for improvement for cervical cancer patients of childbearing age. *Gynecol. Oncol.*, **107** (1 Suppl 1), S180–S186.

102 Gamelin, J., Yang, Y., Biswal, N., Chen, Y., Yan, S., Zhang, X., Karemeddini, M., Brewer, M., and Zhu, Q. (2009) A prototype hybrid intraoperative probe for ovarian cancer detection. *Opt. Express*, **17** (9), 7245–7258.

103 Wang, Z.G., Durand, D.B., Schoenberg, M., and Pan, Y.T. (2005) Fluorescence guided optical coherence tomography for the diagnosis of early bladder cancer in a rat model. *J. Urol.*, **174** (6), 2376–2381.

104 Buranachai, C., Thavarungkul, P., Kanatharanaa, P., and Meglinski, I.V. (2009) Application of wavelet analysis in optical coherence tomography for obscured pattern recognition. *Laser Phys. Lett.*, **6** (12), 892–895.

Part 4
Cardiology, Angiology

26
Microvascular Blood Flow: Microcirculation Imaging
Martin J. Leahy

26.1
Introduction

Direct (optical) observation of skin and other tissues is of course the oldest method of biomedical imaging. We know that the ancient Greeks already understood that the pallor of the skin was a significant discriminator between health and disease. However, the decomposition of white light, the Doppler effect [1], and the identification of hemoglobin as the substance which changes its color and that of the skin when carrying oxygen provided a scientific basis for reliable measurements of the microcirculation developed over the past 150 years. Over the past 30 years in particular, tremendous progress has been made in single-point biophotonic measurements of microvascular state/activity and more recently imaging of the same.

Blood flow in the skin provides an efficient mechanism for thermoregulation, oxygen transport, mechanical integrity, blood pressure management, and the transport of repair materials. Little is known about the relative importance of these functions in the body's most extensive and accessible organ, which even in resting conditions it is estimated [2] to receive almost 9% of the total cardiac output.

The present understanding of microcirculation (e.g., [3]) assumes that there is little capacity for regulation of blood flow within individual capillaries, and that potential oversupply of oxygen at rest is prevented by alternate shutdown of capillaries. The open capillaries at any instant may be as much as 500 μm apart, contrasting with an actual capillary density of perhaps one per 50 μm. The mechanism for precapillary sphincter closure is under local control at a microregional level, and is triggered by a hyperoxic threshold. Additional regulation is provided by capillary pressure under both local (myogenic) and central (nervous) control mechanisms. These factors must be considered carefully when intending to measure capillary blood flow. If the chosen field of measurement is sufficiently small, the alternating flow/no-flow in individual capillaries may be observed. On the other hand, if the field is so large that a number of capillaries are always open, then the flow pattern will appear smoother, and an average perfusion for this area of tissue is recorded.

Visible and near-infrared light, particularly in the wavelength region 600–1200 nm, offer a window into human and animal tissues due to reduced light scattering and absorption. Nevertheless, the remaining scattering obscures images of the microcirculation everywhere but the eye and nailfold plexus. Although high-frequency ultrasound has been applied to imaging the microcirculation, it has significant limitations in imaging microvessels close to the surface. In this chapter we examine the main biophotonic methods applied to visualization and assessment of the microcirculation and document the progress made over the past 5 years in particular. Applications, particularly in human skin, are of special topical importance due to an improved knowledge of its role and its value as a surrogate for other organs in drug testing at a time when drug development is slowing under the weight of regulation.

Many techniques have been proposed for imaging the microcirculation, from X-rays to thermography, and still more new techniques are emerging, for example, optical coherence tomography (OCT), photoacoustic tomography (PAT), tissue viability (TiVi), and speckle imaging. Magnetic resonance imaging (MRI) and positron emission tomography (PET) are two excellent techniques, but their size and cost render them inaccessible to the majority of patients. Recently developed biophotonic technologies can satisfy the clinical need for simple to use, inexpensive technologies for bedside monitoring and to get to (research) patient groups who cannot attend the small number of locations which house elite equipment. With the exception of the nailfold plexus, light microscopy and direct visual inspection of the skin microcirculation are hampered by scattering, regular reflection from the surface, and absorption by superficial chromophores. Bloodless melanin-free skin provides little absorption but significant scattering across all colors of visible light. Since returning photons are scattered many times within the skin, they are diffuse and therefore the skin appears white.

The vital role of blood supply, and the oxygen it carries, in the health of the individual has ensured that many different techniques for assessing it have been investigated. The principle upon which the methods are based vary widely, as does the suitability, cost of the materials, and technology necessary. A small number of the methods are continuous and even fewer are truly noninvasive. Non-invasive *in vivo* techniques have obvious advantages for the user in providing information without disturbing the normal environment. In the same way, it poses considerable difficulties for developers through the need to make accurate measurements in complex environments subject to enormous (biological) variability.

The imaging technologies discussed here can be separated by physical resolution and sampling depth (Figure 26.1) with confocal microscopy and ultrasound included for completeness. Pixel resolution is improved by two orders of magnitude in commercially available units by going from laser Doppler perfusion imaging (LDPI) to laser speckle perfusion imaging (LSPI) and again from LSPI to TiVi imaging.

Figure 26.1 shows the resolution and sampling depth of the various techniques. The depth of interdisciplinary knowledge required to specify the appropriate technique often means that the wrong technique is used, or that the correct one is used

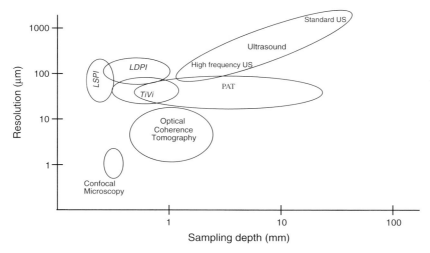

Figure 26.1 Microcirculation imaging domains, including laser speckle perfusion imaging (LSPI), laser Doppler perfusion imaging, tissue viability imaging (TiVi) and photoacoustic tomography (PAT).

inappropriately. This often leads to unfair criticism of techniques used in situations for which they were never designed. The techniques used in the clinical setting tend to have the common advantages of being mobile, inexpensive, or otherwise available as well as the required technical specification. The rush (fashion) for ever more sophisticated technologies, which suggests that expense is somehow an advantage, is not borne out by clinical use. As mentioned above, there is a clinical need for simple to use, inexpensive technologies for bedside monitoring and to get to (research) patient groups who cannot attend the small number of locations which house this elite equipment.

To understand why optical imaging of the skin and microcirculation (Figure 26.2) is not universally applied, it is instructive to consider how light interacts with tissue and which situations allow for direct visualization of the skin. Light microscopy and direct visual inspection are hampered by scattering, regular reflection from the surface, and absorption by superficial chromophores. Bloodless melanin-free skin provides little absorption but significant scattering across all colors of visible light. Since returning photons are scattered many times within the skin, they are diffuse therefore the skin appears white. Regular, mirror-like reflection accounts only for a small percentage of the incident photons. Ultrasound has many advantages due to its low scattering in this tissue matrix and can provide excellent images for larger, deeper vessels. However, it requires higher frequencies for smaller vessels, causing difficulties close to the surface. It is possible to view capillary vessels directly in the nailfold plexus and to view the retinal vessels directly due to the lack of significant absorption/scattering outside these vessels. However, absorption is the main contrast mechanism and backscattering is required to return photons to the detector/eye.

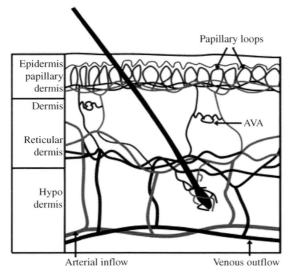

Figure 26.2 Microcirculation in the skin.

26.2
Nailfold Capillaroscopy

One of the most useful methods of assessing blood flow in a small number of blood vessels in the skin is by direct observation using capillary microscopy [4]. Capillaries are the smallest of the blood vessels and their purpose is to link the arterial and venous systems together, allowing the exchange of carbon and oxygen between the tissue and blood cells. Only in certain places in the body do the capillaries come close enough to the surface of the skin to be naturally visible, that is, without the use of specialized optical equipment or optical clearing agents. This is one of the reasons why the study of the skin overlapping the base of the fingernail and toenail is so important. In the nailfold, the capillaries come within 200 μm of the surface of the skin, meaning the methods of examining them are simple. Another reason is that the fingernail is easily fixed in position, free from any movement, due to arterial pulsations or respirations and the capillaries run parallel to the skin surface. Using a microscope with a magnification of between ×200 and ×600, it is possible to see clearly the structure of the blood vessels and, because each vessel is sufficiently transparent, the red blood cell (RBC) motion, in a single capillary, can be measured. If a video camera is attached to the microscope [5, 6], the motion can be examined in a frame-by-frame procedure yielding accurate velocity information for RBC flow in the capillary. The flow may be altered by the fixing procedure and/or the heating effect of the light source. Examining the capillary condition and density can aid the diagnosis of certain diseases. Capillary density and diameter are dependent on age, with younger children having a lower density and capillary thickness than adults [7]. The presence of abnormal vessels, in addition to these factors, has been attributed to

Figure 26.3 Nailfold capillary pattern of a healthy adult [9]. The fingernail is marked as N. A typical capillary density for a healthy adult is approximately 30–40 mm^{-2} [10]. Reproduced with permission from Wiley-Blackwell Publishing Ltd.

various diseases. Some conditions can be detected very early as capillary deformations can be observed before other symptoms occur [8].

Figure 26.3 is an image of a nailfold capillary arrangement of a healthy adult [9]. The capillaries are hairpin-like loops, arranged into regular rows. Each loop consists of two parallel limbs; a thinner arterial limb and a thicker venous limb. The walls of each blood vessel itself are transparent and the RBCs are seen passing through the capillaries [11].

26.2.1
Measurements

A disease-specific "pattern" can be identified by analyzing the geometry and density of the capillaries and the presence of abnormal or very large blood vessels [10]. Other relevant factors include avascular areas (absence of capillaries) [9] and the red blood cell velocity (RBV) through the capillaries [12]. A typical value of RBV would be ~ 1 mm s^{-1} [12]. Examples of avascular areas and giant and dilated capillaries can be seen in Figure 26.4.

The geometry of each capillary is defined by taking measurements at specific points. An example is shown in Figure 26.5.

The most difficult calculation in nailfold capillaroscopy is that of the RBV. There are many different methods of determining this. One such method is laser Doppler flowmetry (LDF). There are two main advantages of nailfold capillaroscopy over this method. The first is that LDF is not limited to a specific area, meaning the measurements are not taken on a specific RBC. As a result, the measurements are made less accurate by cells with various velocities [13]. The second drawback is that the signal produced is in proportion to microvascular perfusion, which is the product of both the velocity and concentration of the RBCs. It not currently possible to measure the absolute units of blood flow [14]. In nailfold capillaroscopy, because individual RBCs can be seen, direct measurement of the RBV is allowed. Several methods of assessing RBC perfusion exist.

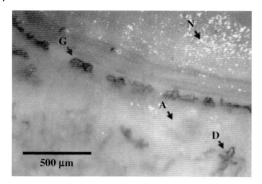

Figure 26.4 Nailfold capillaries in a patient with dermatomyositis. Examples of avascular areas (A), giant capillaries (G), and dilated loops (D) are shown [9]. Reproduced with permission from Wiley-Blackwell Publishing Ltd.

Measurements can be performed in a frame-by-frame analysis of specific patterns of RBC flow (RBCs or plasma gaps) [15]. By tracking the pattern over two consecutive frames, the velocity can be approximated. A drawback of this method is that these patterns are difficult to detect in larger vessels.

Other methods include cross-correlation methods, which can be further subdivided into two methods. The first works by measuring the average intensity at two independent regions of the capillary. Using a temporal cross-correlation and the distance between the regions, the RBV can be calculated [16]. The second method works by creating a series of spatial intensity distributions on a particular section. Using a temporal spatial correlation and the time separation between consecutive frames, the velocity can be calculated [17]. Sourice et al. made use of the spatiotemporal autocorrelation function to develop software to perform calculations of the RBV [18].

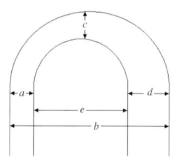

Figure 26.5 Measurements of dimensions of a capillary [10]. a, Arterial capillary diameter; b, venous capillary diameter; c, loop diameter; d, width of the capillary; e, distance between the limbs.

26.2.2
Experimental Considerations

A typical experimental setup would consist of a microscope, monitor, and video recording system [8, 11, 12, 18]. Fiber-optic illumination is often used [7, 10]. The finger can be immobilized easily by use of a clamp [12]. To increase the transparency of the skin, a clearing agent such as immersion oil is usually applied to the nailfold [7, 8, 11]. A fluorescent solution such as fluorescein can be used to distinguish the capillaries from the surrounding skin [18]. The software used for analyzing the images and the preparation of the subjects is dependent on what is being measured.

Ohtsuka et al. [8] showed evidence that for patients with primary Raynaud's phenomenon, the loop diameter, capillary width and capillary length (inner length + outer length/2) were greater in the patients who went on to develop undifferentiated connective tissue disease (UCTD) than those who did not. Bukhari et al. [10] studied the dimensions and density of capillaries and showed that there was a significant increase in all dimensions except the distance between the limbs in the patients with Raynaud's phenomenon and systemic sclerosis (SSc) compared to a control group. Also, capillary density is reduced in patients with SSc. Bhushan et al. [19] found that there was a decrease in capillary density in the nailfold of patients with psoriasis compared with a control group. Wong et al. [20] showed that a characteristic pattern of capillary abnormalities occurred in patients with scleroderma. Aguiar et al. [12] studied the RBV, functional capillary density, and capillary morphology differences in patients with primary Sjögren syndrome and a healthy control group. It was found that the affected patients had a higher functional capillary density and lower RBV than the control group.

26.2.3
Limitations and Improved Analysis

The measurements required for analyzing the capillary dimensions and patterns take much longer than the examination itself. Computerized systems have been developed [21–23] for analyzing images of capillary networks and improving their quality by excluding disturbances caused by hair, liquid, reflections, and so on. Combining videocapillaroscopy and various mathematical methods, the statistical properties of the capillaries can be analyzed automatically by extracting capillary count, position, density, size, and so on from the images. Sainthillier et al. [21] and Zhong et al. [22] used triangulation methods to allow statistical analysis of distances between nearest neighboring capillaries, which is useful for looking at areas that are susceptible to the development of necrosis. Hamar et al. [23] are currently developing a Markov chain-based detection algorithm for updating capillaroscopy images to reduce them to a grayscale image containing only the capillaries. These technologies are essential to the development of nailfold microscopy into a more sophisticated method of disease diagnosis.

26.2.4
Conclusion

Although nailfold capillaroscopy is important in the diagnosis of certain connective-tissue diseases, some of the methods described above cannot be applied to other areas of the microcirculation for the simple reason that the capillaries are not clearly visible. This necessitates other methods of measuring the condition and density of the capillaries and RBV.

26.3
Laser Doppler Perfusion Monitoring

Laser Doppler perfusion monitoring (LDPM) refers to the general class of techniques using the Doppler effect to measure changes in blood perfusion in the microcirculation noninvasively. Perfusion has previously been defined as the product of local velocity and concentration of blood cells [24]. The principles behind the technique were first developed by Riva et al. [25], who developed a technique to measure RBV in a glass flow tube model; however, the development of LDPM for assessing tissue microcirculation was first demonstrated in 1975 by Stern [26]. Measurement of the velocity of particles in solution by interpreting the Doppler frequency-shifted light backscattered there from had been described by Cummins and Swinney [27], only 4 years after the first working laser was demonstrated by Maiman (1960). However, all commercial devices now rely on the signal processor developed by Bonner and Nossal [28], which, in contrast to laser speckle perfusion monitoring, provides laser Doppler with a sound theoretical base.

The technique operates by using a coherent laser light source to irradiate the tissue surface. A fraction of this light propagates through the tissue in a random fashion and interacts with the different components within the tissue. The tissue consists of both static and moving components. The light scattered from the moving components will undergo a frequency shift which can be explained by the Doppler effect whereas the light scattered from the static components will not undergo any shift. It is almost impossible to resolve the frequency shifts directly. To overcome this, the backscattered light from the tissue is allowed to fall on the surface of a photodetector where a beat frequency is produced. The typical frequency of this beat detected using a wavelength of 780 nm ranges from 0 to 20 kHz. However, for a 633 nm source the range is smaller, typically 0–12 kHz [24]. The apparent contradiction with the theory that the frequency shift is inversely proportional to wavelength is caused by the deeper vessels probed by the longer wavelengths, which contain faster flowing blood. From this beat frequency, it is possible to determine the speed and concentration of the blood cells. The LDPM technique correlates well with other measurements of skin blood flow, such as venous occlusion plethysmography [33] and xenon washout [34].

LDPM offers substantial advantages over other methods, being very sensitive and able to follow easily changes in local blood perfusion over less than 0.05 s, making it useful for continuous perfusion monitoring. As the probe is not required to touch the surface of the tissue, the technique can be noninvasive, avoiding any impact on the microcirculation. Where invasive measurements are required, the probes can be as small as a single optical fiber, so that with multichannel devices measurements can be made from several depths or microregions within a tumor, for example. The technique registers tissue perfusion at a single point over time, but microcirculation is known to be heterogeneous, especially in human skin, so a small area of perfusion is not necessarily representative of perfusion in the region of interest [35–37]. Furthermore, the measurements are intrinsically of a relative nature, the measurements are proportional to blood flow. However, the factor of proportionality will be different for different tissues and tissue sites. On the other hand, "correct" values of blood perfusion (ml mg^{-1} s^{-1}) are not generally known, so it is difficult to imagine how absolute values would be used.

LDPM has been revised and improved to remove many of the preliminary issues and has been widely used both in research and more recently as a clinical tool. Guidelines have been published on how to perform measurements using the technique [36]. There are many different types of probes available, each suited for different applications such as fiber probe systems [39] and integrated probe systems [40].

26.3.1
Light Sources

The incident laser light beam has a depth of penetration of ~1 mm depending on the wavelength and configuration of the equipment used. This estimate originates from mathematical modeling of photon diffusion through "imaginary tissues" using Monte Carlo techniques. Therefore, all elements of the dermis may be included, from the superficial nutritional to deeper thermoregulatory vessels.

For a given wavelength of light, the absorption spectrum of components of the tissue determines the interaction that occurs [29]. The penetration depth of light is predetermined by the path the light takes through the skin and is limited by the absorption and scattering effects of the tissue. Another limiting factor is the level of pigmentation of the skin [30].

The absorption of hemoglobin and water is lower towards the red end of the visible spectrum. Because of commercial availability, HeNe lasers (632.8 nm) dominated in earlier LDPM; however, longer wavelengths produced by laser diodes (780, 810 nm) are now preferred owing to increased penetration depth and lower absorption by melanin in the near-infrared region, which shows less dependence on skin color [30]. Also, because the absorption of oxidized and deoxygenated hemoglobin is almost identical at this wavelength, any dependence on oxygen saturation is eliminated. However, systems exist that use dual wavelengths to obtain additional information about the tissue structure [31, 32].

26.3.2
Laser Doppler Perfusion Imaging (LDPI)

The need to study perfusion over larger tissue areas led to the construction LDPI [41, 42]. LDPI is a technique that scans a series of perfusion measurements over a section of tissue. This is converted into a color-coded image representing the distribution of perfusion over an area of interest.

This technique offers some advantages over the single-probe technique; the blood flow is measured over an area rather than at a single site, which removes some of the difficulties that arise in the LDPM technique such as movement artifacts and site-to-site perfusion variations due to the heterogeneous nature of tissue. The technique requires no contact with the tissue being examined, making it of particular advantage when assessing open, and often infected, wounds.

The technique has limitations; one issue is that several minutes are required to complete a full scan. This can introduce problems in trying to examine a rapid change in blood flow. However, more recent techniques such as line-scanning and full-field laser Doppler offer to improve this considerably. For example, the MK2 moorLDLS (Moor Instruments, Axminster, UK) is an FPGA (Field Programmable Gate Array)-based line-scanning LDPI instrument which acquires 64 pixels simultaneously and the best time for a 256×64 pixel image is 13 s. LDPI provides arbitrary perfusion measurements [35], meaning that the tool is valid for comparative data only and that all measurements must be performed under the same conditions. It is also necessary to consider temporal variations in blood flow over short periods [38] (seconds to minutes) and over extended periods due to seasonal changes [43] while attempting to monitor long-term changes in perfusion to a region.

Another issue with the technique is that due the scanner not being held in contact with the tissue, the distance between the imager and the tissue affects the measured level of perfusion. If this distance changes between subsequent measurements, it will affect the results. There are conflicting opinions on whether the measured perfusion increases [44] or decreases [45] as the distance from the tissue increases. It may be that the algorithms used for signal processing by different machines determine this [46]. In order to make comparable measurements between different systems, a standard calibration and system design is required [24].

Between 1997 and 2002, a standardization project was undertaken by several international research institutes, in which a perfusion simulator for calibration and standardization of LDPM and use of nomenclature was agreed. In 2002, a report on a project named HIRELADO (High Resolution Laser Doppler) from the standardization group of the European society of contact dermatitis published guidelines on the measuring of skin blood flow [47]. This project addressed mainly the technical aspects of how experiments should be performed and what precautions should be taken to ensure valid results. Several different techniques have been developed for producing LDPI data, each with different advantages and disadvantages.

26.3.3
Point Raster Technique

A raster scan laser Doppler image operates by recording a series of point perfusion measurements. This type was the first laser Doppler image system to be developed. The basic schematic is show in Figure 26.6. The imaging is achieved by using a moveable mirror to scan the laser beam over the area to be measured in a raster

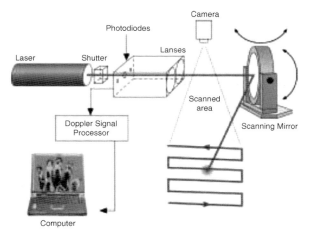

Figure 26.6 Schematic of raster scan laser Doppler imaging system. From moorLDI brochure, Moor Instruments Ltd.

fashion. At each site, an individual perfusion measurement is taken. When the scan is complete, the system generates a 2D map image from the single-point perfusion measurements.

The problem with the technique is the time taken to make measurements. Each individual measurement can be achieved in 4–50 ms; however, shortening the acquisition time results in increased noise in the signal [47]. For time considerations it is important to minimize the number of measured points. This is best achieved if the step length between measurements equals the diameter of the laser beam. For smaller step lengths, the same physiological information is partly collected by neighboring measurements, which increase the overall time without retrieving more information. The moorLDI system reports a capability of capturing a 64×64 pixel perfusion image in 20 s and has a spatial resolution of 0.2 mm at a distance of 20 cm. The system is capable of scanning a region up to 50×50 cm and as small as a single point measurement.

26.3.4
Line Scanning Technique

Laser Doppler line scanning imaging is a new development in LDI technology by Moor Instruments. This technique does not utilize the standard point raster approach but works by scanning a divergent laser line over the skin or other tissue surfaces while photodetection is performed in parallel by a photodiode array. This technique allows a series of perfusion measurements to be performed in parallel. This greatly reduces acquisition time and is bringing laser Doppler imaging one step closer to real-time imaging. A schematic of the system is shown in Figure 26.7. The moorLDLS system utilizes 64 diodes to obtain the perfusion measurements in parallel. The system is capable of measuring each line in times of 100–200 ms.

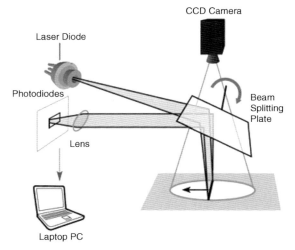

Figure 26.7 Schematic of line scanning laser Doppler system. From moorLDLS brochure, Moor Instruments Ltd.

Regions up to 20 × 15 cm can be imaged and a 64 × 64 pixel perfusion measurement can be produced in 6 s.

26.3.5
Full-Field Scanning Technique

High-speed laser Doppler perfusion imaging using an integrating CMOS (complementary metal–oxide–semiconductor) sensor for full field scanning has been developed by Serov et al. [48]. The basic schematic of the system is shown in Figure 26.8. A laser source is diverged to illuminate the area of the sample under investigation and the tissue surface is imaged through the objective lens on to a CMOS camera sensor. This new CMOS image sensor has several specific advantages: first, the imaging time is 3–4 times faster than the current commercial raster scan LDI systems. Second, the refresh rate of the perfusion images is approximately 90 s for a 256 × 256 pixel perfusion image. This time includes acquisition, signal processing, and data transfer to the display. For comparison, the specified scan speed of a commercial laser Doppler imaging system, the moorLDI, is ∼5 min for a 256 × 256 pixel resolution image. However, the scanning imager can only measure areas of up to ∼50 × 50 cm in size whereas the CMOS system at present does not allow imaging of areas larger than ∼50 × 50 mm [49, 50].

26.3.6
Applications

The laser Doppler technique has many medical applications; however, it still has not been fully integrated into clinical settings and is mainly used in research. LDPI

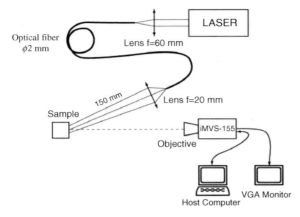

Figure 26.8 Schematic of full-field laser Doppler system [49].

systems have been reported in a wide range of applications from neuroscience [51] to the assessment of burns [52–54]. The use of LDPI is reported to outperform existing methods of assessment of burn wound depth and provides an objective, real-time method of evaluation. Accuracy of assessment of burn depth is reported to be up to 97% using LDPI, compared with 60–80% for established clinical methods [52]. High perfusion corresponds to superficial dermal burns, which heal with dressings and conservative management; burns with low perfusion require surgical management, that is, skin grafts. Correct assessment is of particular importance here as one in 50 grafts fail; if this failure is caught within 24 h, 50% of these cases can be saved [55].

The use of LDPI has also been investigated in different aspects of cancer research and treatment; it has been found that there are higher levels of perfusion in skin tumors which can be detected by the LDPI system [56]. The technique has also been applied to examine how a tumor is responding to different treatments such as photodynamic therapy [57]. Wang *et al.* applied the technique to the assessment of postoperative malignant skin tumors [58]. It has also been applied in many other fields, such as examining allergic reactions and inflammatory responses to different irritants [59–61]. Bjarnason and co-workers applied the technique in the assessment of patch tests [62, 63]. Ferrell *et al.* used the technique to study arthritis and inflammation of joints [64]. It has also been applied in investigating the healing response of ulcers [65] and for examining the blood flow pattern in patients at risk from pressure sores [66].

26.4
Laser Speckle Perfusion Imaging

As the angles in LDP are more or less random and a wide range of velocities are present in the microcirculation, a continuous Doppler frequency spectrum is produced. It is interesting that a similar result can be achieved by considering the scattering particle to be a moving reflector of light. The reflected light will interfere with the reference beam and the resultant intensity will vary with the difference in optical path between the two beams. The number of interference fringes to pass the detector in time is related to the speed of the moving cells. Each time the mirror (cell) moves through a distance equal to half the wavelength of the light, the detector "sees" an interference fringe pass. That is, a fluctuation in light intensity is recorded. Variations in speckle contrast will then be dependent on the velocity distribution of the scatterers. Therefore, this velocity can be determined using a measurement of the statistical behavior of the speckles over time. The integration time of the detector should be small in comparison with the correlation time of the intensity fluctuations to avoid "averaging out" of the signal [67]. This is similar to the reading given by a laser Doppler measurement, and in fact it has been demonstrated that Doppler and time-varying speckle imaging are two methods of arriving at the same result [68].

When recorded over a finite integration time, the moving scatterers result in a speckle pattern that appears "blurred" or decorrelated. The amount of decorrelation depends on the speed and volume of the RBCs in the tissue [69]. The time-varying component of the speckle can be quantified by comparing the intensity recorded at each pixel in successive scans. The average difference between the two measurements is dependent on flow – large in high-flow regions and small in low-flow areas. A two-dimensional map can be created by scanning the line across a region of interest. Experiments showed a decrease in the blood flow parameter during occlusion of the finger, and an increase was observed across a scar on the back of the hand [70].

26.4.1
Full-Field Measurements

The set-up for full-field laser speckle measurements is shown in Figure 26.9. The tissue is illuminated by an expanded laser beam and the resulting image speckle is recorded by a charge-coupled device (CCD) camera.

The source used is typically a low-power expanded helium–neon (632.8 nm) laser in the 3–30 mW range [71, 72] although the use of 660 [73], 514 [74], 780 [75, 76], and 808 nm [77] wavelengths have also been reported. Typically for studies of capillary perfusion, this is set between 5 and 20 ms [69, 71, 72, 75, 78]. Yuan et al. [79] found that the optimal integration time for rodent cerebral blood flow studies was 5 ms.

This technique, originally developed for retinal blood flow imaging [80], uses a short integration time, as mentioned previously, to photograph the illuminated area and produces a high-contrast speckle pattern. The contrast K, is calculated based on small areas of the pattern, typically a sliding 7×7 or 5×5 pixel window [67, 71, 72]. The higher the velocity of the RBCs, the higher is the frequency of the fluctuations. This means that for a given integration time, more averaging will occur and there will be a decrease in contrast. Conversely, an area where blood flow is low will appear as a high-contrast reading. These contrast measurements can therefore be used to build up a two-dimensional map of relative velocities.

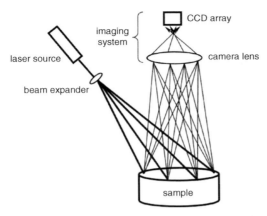

Figure 26.9 Full-field laser speckle contrast imaging. The tissue is illuminated by a laser that has been diverged by a lens. The speckle pattern is then recorded by the CCD camera and the speckle contrast, K, is computed.

26.4.2
Limitations and Improved Analysis

The effects of specular reflection can be eliminated with the addition of polarizing filters [69, 73, 81–83]. Multiple scattering can have the effect of blurring the image by influencing the apparent size of the object [83]. Also, the effects of scattering by static tissue must also be considered [73]. As the laser speckle signal is dependent on the fluctuating part of the pattern, static tissue has the effect of reducing the signal-to-noise ratio. Zakharov *et al.* developed a model taking this into account that includes a correction to the equation relating speckle contrast to correlation time [83]. Attempts to limit the effect of static scatterers have also been made in single-point measurements using fiber-optic probes by adjusting the source–detector separation [84]. This limiting effect may be one of the most significant reasons why the speckle technique is suited to more superficial measurements. Indeed, many of the applications reported, as described in Section 26.4.3, involve measurements on surgically exposed tissue.

The main difficulty with using imaging systems based the Laser Speckle Contrast Analysis (LASCA) Algorithm is that the relationship between speckle contrast and red blood cell velocity is nonlinear. Forrester *et al.* developed an improved version of Briers' LASCA system called laser speckle perfusion imaging (LSPI) [69]. A different approach to producing the contrast image is taken by first calculating a reference image by averaging the speckles spatially and temporally. The intensity fluctuations due to blood flow can then be examined by calculating the intensity difference between the corresponding pixels in the reference and newly acquired image. The result is then normalized by the reference in order to negate the effect of nonuniform illumination, or changes in laser output or tissue optical properties. The intensity difference is inversely proportional to the speed of the scatterers: in low-flow regions, a sharp speckle contrast is observed, and a large difference between the reference and speckle intensities will be recorded. In high-flow regions, a large amount of decorrelation occurs and the difference approaches zero.

An *in vitro* blood flow model was devised to test the response of the LSPI system [81] consisting of a 0.95 mm bore glass tube fixed in a gelatin-set tissue phantom 1 mm below the surface. Red cell concentrations of 0.1–5% were imaged at flow rates in the range 0–800 µl min^{-1} (corresponding to RBC velocities of 0–18.8 mm s^{-1}). These parameters cover the typical physiological range [85]. In comparison with a commercially available laser Doppler imager (moorLDI), it was found that at a constant flow rate, LSPI had a similar response to LDPI to changes in RBC concentration (correlation $r^2 = 0.93$). At a constant concentration (1%), LSPI also had a similar response to LDPI to changing red blood cell velocity, with its perfusion index increasing linearly over the range of flow rates ($r^2 = 0.99$).

Speckle imaging systems using the "averaging window" suffer from a loss of resolution (for example, a 7 × 7 pixel window will produce a speckle image 1/49th the resolution of the original). However, the temporal averaging technique used by the LSPI algorithm means that the full resolution of the CCD can be retained to produce high-detail images of the vascular structure. For higher speed acquisition, however, a spatial averaging mode may be employed [69].

The main advantage over LDPI is the speed at which the images may be acquired. Acquisition speeds at video rates [25 frames per second (fps)] can be attained, making it possible to measure the blood flow response to occlusion and hyperemia (typical LDPI scans are of the order of minutes). Also, the much higher resolution [limited only by the CCD, maximum LDI resolution is 256 × 256 (moorLDI)] permits more detailed imaging of tissue structures [69].

26.4.3
Microcirculation

The early use of laser speckle imaging of the microvasculature has been reported [67]. However, more recent work incorporating the improvements outlined above, and in particular the advent of video rate acquisition has led to some interesting applications.

Human retinal blood flow has been examined using the technique [77] in a similar method to Briers *et al.* and an *in vitro* calibration model was developed to relate speckle decorrelation to actual readings of velocity. *In vivo* experiments conducted in parallel with electrocardiogram measurements showed that the speckle contrast varied with the heart rate and red blood cell velocity reached a peak corresponding to that of the cardiac cycle. Yaoeda *et al.* [86] adapted the technique developed by Fujii *et al.* [70] to monitor blood flow to the optic nerve head, showing differences between the right and left eyes.

The LSPI system developed by Forrester *et al.* [69] was used to measure changes in flow in the knee joint capsule of a rabbit before and after femoral artery occlusion, recording a 56.3% decrease after the maneuver. The system was also incorporated into a hand-held endoscope for the simultaneous measurement of the same event. The response was validated as before with an *in vitro* model, showing the linear response over a large flow range (0–800 μl min^{-1}). Equipped with a motion-detection algorithm to limit motion artifact, the endoscopic set-up recorded a similar decrease (58.7%) [73]. Clinical data from human subjects has also been acquired using this endoscopic LSPI system (eLSPI), recording a decrease in blood flow in the medial compartment of the knee after application of a tourniquet. A dose-dependent response to the vasoconstrictor epinephrine was also recorded [87].

Choi et al. demonstrated the use of laser speckle imaging in a rodent skin fold model [72] and were able to provide quantitative measurements of relative blood flow in areas of tissue that were otherwise obscured by attached muscle and fat. Selected areas of the tissue were irradiated with a 585 nm laser pulse to stop blood flow locally, and this was seen in the LSI images. Similar animal models were investigated by Cheng et al. [71] while testing the effect of varying doses of phentolamine, a vasodilator, on intestinal tissue. In both cases, image processing was conducted offline.

Extensive studies of cerebral blood flow in rodents have been undertaken using various laser speckle imaging techniques to monitor the vascular response to stimuli and chemically induced changes. Cerebral blood flow was monitored in rats to demonstrate its heterogenic response to mild hypotension (induced by withdrawal of blood from the femoral artery). Regional variations in perfusion were recorded [88]. Blood perfusion measurements have also been made on the surgically exposed cerebrum of rodents that show a drop in signal over time after a photochemically induced infarction [76], while Royl et al. [89] recorded an increase in flow in the somatosensory cortex during electrical stimulation of the forepaw.

Laser speckle measurements of cerebral blood flow through the intact rat skull (minus the skin) have been presented Li et al. [75]. A temporal averaging algorithm was employed that used data from 40 consecutive speckle patterns and computed the difference between the intensity at a pixel of interest and the average intensity calculated for that pixel over the 40 frames. The resulting so-called laser speckle temporal contrast analysis image (LSTCA) was found to be more immune to the effects of static scatterers (such as the skull) than spatial techniques such as LASCA.

The effect of photodynamic therapy (PDT) and pulsed dye laser (PDL) irradiation on the microvasculature in a rodent dorsal skin-fold model was examined using laser speckle contrast analysis (LASCA) [90]. Images taken 18 h after treatment show vascular flow reduction in the exposed area (Figure 26.10).

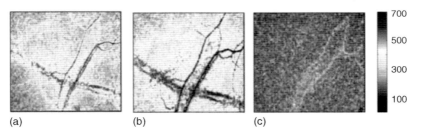

Figure 26.10 Laser speckle contrast maps of blood flow: (a) prior to PDT treatment, (b) immediately after and (c) 18 h after. Reproduced with permission from Wiley-Liss, Inc., a subsidiary of John Wiley & Sons, Inc. [90].

The versatility of the technique has been illustrated by conducting laser speckle analysis in parallel with PDT using a 514 nm argon ion PDT laser as the source for the speckle measurements [74]. The processing algorithm was the speckle reduction technique developed by Forrester *et al.* [69]. Other animal models studied include temperature-induced changes in blood flow in the rodent [78] and wound-healing in porcine subjects [82].

26.4.4
Agreement with Laser Doppler

While accurate absolute measurements of RBV cannot be calculated, all of the above LSI systems can successfully track relative changes in perfusion. Recent developments and *in vitro* experiments have shown that laser speckle images can respond in the same way as laser Doppler to changes in RBC concentration and velocity [81]. *In vivo* experiments have also shown a strong correlation between the techniques [89]. In fact, recent studies on burn scars show a correlation with an r^2 value of 0.86 between the techniques [91]. It should also be noted that the perfusion measurements correlated strongly with the clinical grading of the scar. As each imager used the same wavelength (632.8 nm), Stewart *et al.* stated that the penetration depth was, in theory, the same for each [91]. However, Moor Instruments report that the penetration depth of the "FLPI" system is less than that of the corresponding LDI system. This is due to two factors: first, unlike the laser Doppler signal, the laser speckle signal is not frequency weighted, meaning that photons arriving from the faster moving RBCs in deeper blood vessels cannot be isolated. Second, as the source in laser speckle systems is an expanded beam, the power density is much lower than in the corresponding Doppler system, hence the depth at which noise dominates is more superficial. The first commercial speckle imager, the moorFLPI, could acquire and process images at video rates (25 fps) at high resolution (576×768 pixels) [92].

Laser speckle, which is a confounding phenomenon in some applications such as laser Doppler [93], is proving to be a very useful tool in blood perfusion measurements. Full-field images can be recorded in a single shot containing information on RBC concentration and velocity. Whereas traditional laser Doppler images can take several minutes to acquire and are of limited resolution (256×256 pixels), LSI can show rapid changes in blood flow in real time.

The most popular processing algorithm reported is that based on speckle contrast maps (LASCA) [71, 72, 89] and it is this method that is used in the commercial device [92]. However, a newer algorithm has improved on this, responding linearly to increasing RBV, limiting the effect of static scatterers and specular reflection, and compensating for variations in tissue optical properties. This "reduced speckle" technique is promising as it follows laser Doppler measurements very closely but at a much higher resolution and speed [69, 73, 74, 81, 91]. The low-cost nature of the system, acquisition speed, resolution, and compatibility with existing laser Doppler measurements make LSPI a very attractive technology for studies of blood perfusion in the microvasculature.

26.5
Polarization Spectroscopy

The technique of polarization spectroscopy allows the gating of photons returning from different compartments of skin tissue under examination [94]. Figure 26.11 details the operation of basic polarization imaging, showing that by use of simple polarization filters, light from the superficial layers of the skin can be differentiated from light backscattered from the dermal tissue matrix. When monochromatic or white incoherent light is linearly polarized by a filter and is incident on the surface of the skin, a small fraction of the light (∼5%) is specularly reflected as surface glare (Fresnel reflection) from the skin surface due to refractive index mismatching between the two media. A further 2% of the original light is reflected from the superficial subsurface layers of the stratum corneum. These two fractions of light retain the original polarization state, determined by the orientation of the first filter. The remaining portion of light (∼93%) penetrates through the epidermis to be absorbed or backscattered by the epidermal or dermal matrix. Approximately 46% of this remaining light is absorbed by the tissue and not re-emitted, while 46% is diffusely backscattered in the dermal tissue. This backscattered portion is exponentially depolarized due to scattering centers in the tissue [95], and also by tissue birefringence due to collagen fibers [96].

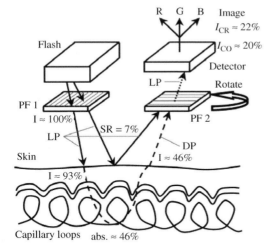

Figure 26.11 Fundamental operation of polarization spectroscopy. This is the foundation of TiVi. Remaining percentages of light intensity are represented by I, and 100% intensity is observed after the polarization filter. SR represents specular reflection, a combined effect of Fresnel reflection and light returning after few scattering events from the upper layers of the epidermis. LP and DP represent linear and depolarized light, respectively, and I_{CO} and I_{CR} represent the remaining intensity which contributes to CO or CR data, depending on the two possible states of polarization filter PF2; abs. represents the percentage of light absorbed in the tissue. PF2 can be arranged with pass direction parallel (CO) or perpendicular (CR) to PF1.

The depth of penetration of polarized light is heavily dependent on the optical properties of the medium at each wavelength present [97]. Upon re-emerging from the tissue structure as diffusely reflected light, the remaining fraction of light is almost completely depolarized, and consists of two 22% fractions of parallel and perpendicular polarizations, with respect to the direction of the original filter. This light contains information about the main chromophores in the epidermis (melanin) and dermis (hemoglobin), whereas the surface reflections contain information about the surface topography, such as texture and wrinkles. On detection by a CCD or similar light-collecting device, one can differentiate between detecting the surface reflections or a fraction of the diffusely backscattered light by placing another polarization filter over the detector parallel or perpendicular with respect to the direction of the filter over the light emitter. With both filters oriented parallel or perpendicular with respect to each other, co-polarized (CO) or cross-polarized (CR) data, respectively, are obtained. This allows the gating of photons, and the technique is based on the assumption that weakly scattered light (the surface reflections) retains its polarization state, whereas strongly scattered light will successively depolarize with each scattering event. It has been suggested that more than 10 scattering events are required to depolarize light sufficiently [98, 99].

26.5.1
Current Technology

Polarization gating has been employed in many different technologies, in order to investigate surface details of the skin structure by accepting light scattered from the superficial layers of the tissue. Examples of technologies that have applied polarization filtering include polarization-sensitive optical coherence tomography (PS-OCT), LDPI, Raman spectroscopy, and simple microscopy. The use of polarization filters has been shown to reduce tissue motion artifact in LDPI, while also reducing the overestimation in LDPI readings due to the amplification of specular reflection [100]. Detail of tissue birefringent axes from various layers via phase-sensitive light recording can be extracted by PS-OCT, and the technique can be used to generate 3D images of the polarization state of backscattered light from skin tissue *in vivo* [101, 102]. Raman spectroscopy of layered media will detect chemical signatures from multiple layers, and polarization gating has been successfully employed to generate Raman spectra of only the superficial skin layers [103]. The combination of polarization gating with multiphoton excitation microscopy allows the characterization of fibrous structures in the skin layers, thus verifying the mechanics of transdermal drug delivery and the significance of dermal structural dermal changes [104]. Gating has also been used with laser speckle imaging in the reduction of specular reflection for investigation of the blood vessels in the microcirculation [69].

Polarization cameras and probes employing the gating technique have been developed in order to investigate both the physiology and microcirculation of skin tissue [105, 106]. Orthogonal polarization spectroscopy (OPS) technology is a single-wavelength (548 nm – an isosbestic point of hemoglobin close to peak absorption) video microscopy technique operating at 30 fps, which increases contrast and detail by accepting only depolarized backscattered light from the tissue into the probe. Single blood vessels are imaged *in vivo* at a typical depth of 0.2 mm and magnification of ×10 between target and image, and information about vessel diameter and RBV are easily obtained. OPS in human studies is limited to only easily accessible surfaces, but can produce similar values for RBV and vessel diameter to those given by conventional capillary microscopy at the human nailfold plexus [107]. It has also been applied to study superficial and deep burns to the skin [108, 109], adult brain microcirculation, and preterm infant microcirculation [110].

Polarization gating images by sequential acquisition of superficial photons (CO) and photons undergoing multiple scattering (i.e., photons interacting with the reticular dermis – CR) have shown cancer, lesion, and burn scar boundaries. A fluorescence method of imaging skin tumors stained with fluorescent dyes rejects excitation light by way of sequential CO and CR image acquisition with a tunable monochromatic source, and details tumor boundaries similar to Mohs surgery, also known as chemosurgery. boundaries [111, 112]. A similar cost-effective method combining polarization imaging with spectroscopy has also been developed to visualize the water content in the superficial layers of the skin, and has shown good comparison with a well-recognized capacitance measurement technique [113]. Another method of polarized light imaging to examine surface skin pathologies in grayscale at 435×548 pixels is capable of distinguishing between freckles, nevii, and carcinomas [114, 115]. An extension of this preliminary technology produced a handheld cost-effective polarization imaging camera, which operates at 7 fps at 400×400 pixels (15 fps at 250×250 pixels) [116]. Parallel and perpendicular images are acquired simultaneously from two CCDs, and an image processing algorithm is applied which examines only the histologic aspects of the skin surface, with no investigation of the underlying microcirculation. The prototype technology has been successfully examined on melanoma margins.

26.5.2
Image and Data Processing

Image (or data) processing algorithms have generally employed an equation that requires acquisition of both CO and CR images in order to investigate the superficial layers of the tissue [104, 105, 112–114]. The equation represents a normalized difference between the CO and CR images:

$$P = \frac{CO - CR}{CO + CR} \qquad (26.1)$$

where P represents a composite image observing surface skin histology. The denominator in Eq. (26.1) represents the normalization of the image, which cancels any nonuniform illumination. P images have been shown to be comparable to histopathology examination, the invasive gold standard of skin examination.

Recent technology [116] analyzes only the RBC concentration in skin microcirculation by way of single-image acquisition or video acquisition by low-cost cameras utilizing orthogonally placed polarization filters. For single-image acquisition, the flash of a consumer-end RGB digital camera is used as a broadband light source, and the camera CCD acquires an instantaneous image in three 8-bit primary color planes. Color filtering is performed on-camera by three 100 nm bandwidth color filters, blue (∼400–500 nm), green (∼500–600 nm), and red (∼600–700 nm). This represents the first low-cost technique designed to image in real time tissue hematocrit.

Using polarization imaging theory and Kubelka–Munk theory, a spectroscopic algorithm was developed which is not dependent on incident light intensity, taking advantage of the physiological fact that green light is absorbed more by RBCs than red light. For each arriving image, the equation applied is

$$M_{out} = \frac{M_{red} - M_{green}}{M_{red}} \exp\left[-p\left(\frac{M_{red} - M_{green}}{M_{red}}\right)\right] \quad (26.2)$$

where M_{red} and M_{green} represent the red and green color planes of the image, respectively, and p is an empirical factor to produce the best linear fit between output variable (called TiVi index) and RBC concentration. M_{out} represents a matrix of maximum 3648×2736 TiVi index values for single-image acquisition.

Figure 26.12 shows an example of a TiVi image color coded in a method similar to laser Doppler images where high and low RBC concentrations are represented by blue and red, respectively. The technique has high spatial and temporal aspects, with lateral resolution estimated as 50 μm [116], and has been compared with LDPM and LSPI by O'Doherty et al. [117]. The sampling depth in Caucasian skin tissue is estimated approximately from Monte Carlo simulations of light transport in tissue [116]. Topical application of vasoconstricting and vasodilating agents demonstrate the capacity to document increases (erythema) and decreases (blanching) in RBC concentration [118].

Our group is currently developing a real-time system, whereby RBC concentration can be displayed online. Current technology can capture CR frames at 25 fps at a frame size of 400×400 pixels. Video studies will lead to increased temporal

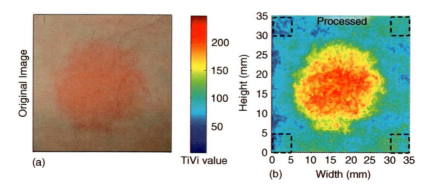

Figure 26.12 Examples of (a) a CR and (b) TiVi image. The CR image was taken of a UVB burn to the volar forearm of a young healthy volunteer. The image was taken 30 h after a 32 s exposure to an 8 mm wide UVB source. The TiVi image shows spatial heterogeneity in the blood distribution over the affected area.

26.5 Polarization Spectroscopy | 425

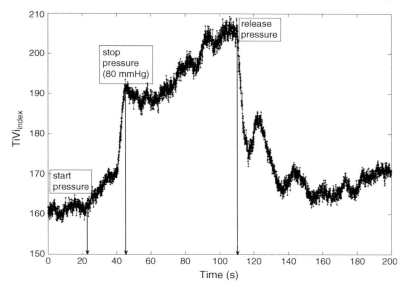

Figure 26.13 Time trace of a 50 × 50 pixel region of the volar forearm of a subject. A pressure cuff on the upper forearm causes an occlusion of the cephalic vein. This demonstration shows clear vasomotion during occlusion and release, and further investigation will focus on pulsatile activity.

resolution, with 0.04 s between each image. Thus, theoretically, the human pulse can be resolved from the time average (duplex) mode of the video imager, and also spatial data over time can be observed. Occlusion of the cephalic vein for the upper arm with 90 mmHg leads to higher RBC concentration due to the RBCs being trapped by the occlusion of the cephalic vein, as the brachial artery pumps RBCs into the area. An example of the duplex mode is shown in Figure 26.13, detailing the average TiVi index of a 50 × 50 pixel region of interest (ROI) on the volar forearm for an acquisition of 25 fps for cephalic vein occlusion. Clear vasomotion of the vessels can be observed at a rate of ∼5 cycles per minute during and post-occlusion and release of the area. Future work will investigate pulsations of the vascular bed, and pulse models have already shown promising results.

Limitations of this new technique include the lack of SI units in measurements, and inaccurate measurements on persons with dark skin due to high absorption of melanin. Light levels are also reduced due to the polarization filters, with only 22.5% of total light intensity reaching the detector. The effect of melanin in the epidermis causes an increase in TiVi index. Image subtraction algorithms have been developed so as to display only the reaction on the skin site, while subtracting out the static components in the CR images. This is also helpful in removing the unwanted effect of deep veins in the volar forearm, a site which is used commonly in routine skin tests. Owing to the availability of measuring tissue hematocrit only, low cost, ease of use, and portability, the TiVi imaging systems are an attractive alternative to expensive and cumbersome equipment such as LDPI and LSPI for the investigation of tissue RBC concentration.

26.6
Photoacoustic Tomography

In recent years, PAT has attracted considerable interest. This is because the technique seeks to combine the best features of biophotonics with those of ultrasound; it offers high optical contrast and high resolution.

The technique operates by irradiating a sample with a laser pulse. When the laser light is absorbed it will produce a pressure wave due to an increase in the temperature and volume. These pressure waves will emanate from the source in all directions. A high-frequency ultrasound receiver can then be used to detect these waves and build a 3D picture of the structure. This is made possible by the much lower speed of sound and lack of scattering or diffusion of ultrasound en route to the receiver. Diffuse, randomized scattering of the light before reaching the microvessels is not a disadvantage and is likely to be useful in providing relatively uniform illumination even to vessels that could not receive ballistic photons due to other vessels in the direct light path.

In PAT, light in the visible to near-infrared region of the spectrum is used. This is because hemoglobin in the blood dominates the optical absorption in most soft tissue. If a tissue sample is irradiated by light in this range, a map of the vessel network within the tissue can be produced. The pulse width is chosen such that it is much shorter than the thermal diffusion time of the system and in this way heat conduction can be neglected [119]. A schematic of the system is shown in Figure 26.14.

The data are processed by a computer and can be output in an image format. Image reconstruction is achieved by various reconstruction algorithms [121, 122].

The technique is capable of diagnosing different physiology conditions. This is because different physiological conditions can change the light absorption properties

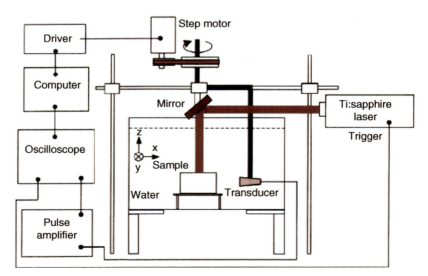

Figure 26.14 Schematic of photoacoustic tomography system [120]. Reproduced with permmission from The American Association of Physicists in Medicine.

of the tissue; for example, absorption contrast between breast tumors and normal breast tissue as high as 500% in the infrared range have been reported [123]. The technique is also capable of detecting hemoglobin levels in tissue with very high contrast. One group has used the technique to acquire 3D data of the hemoglobin level in rat brain [124]. This was achieved using a 532 nm laser source, as the absorption coefficient at this wavelength of whole blood is 100 cm^{-1}, which is much larger than the averaged absorption coefficient of the gray and white matter of the brain, 0.56 cm^{-1}.

The main advantage of combining ultrasound and light is that one can obtain much improved resolution (diffraction-limited) and can image vessels with 60 μm resolution [120]. The Wang laboratory at the University of Washington in St. Louis, MO, has built a system with 5 μm resolution and has already been able to produce exquisite 3D images of microvascular architecture in the rat.

One issue with the technique is the time it takes to produce large 3D scans of a region. Times of up to 16 h have been reported to image a region of radius 2.8 cm in the XY plane and a depth 4.1 cm in the Z plane [123]. However, the use of an ultrasonic transducer array would greatly improve imaging time.

26.7
Optical Microangiography (OMAG)

Optical microangiography (OMAG) has its origins in Fourier domain OCT and is similar to Doppler OCT. It is a recently developed high-resolution 3D method of imaging blood perfusion and direction at capillary level resolution within microcirculatory beds. A typical setup of an OMAG system is shown in Figure 26.15. The technique itself does not rely on the phase information of the OCT signals to extract the blood flow information. This in turn means that the technique is not sensitive to the environmental phase stability. This allows OMAG to compensate for the inevitable sample movement using the phase data, thus limiting noise through post-processing of the data [124].

An example of an OMAG image is shown in Figure 26.16. The acquisition time to obtain the 3D OMAG data shown in Figures 26.12 and 26.13 was about 50 s. However, the processing time required was much longer.

The OMAG technique appears to offer a much greater signal-to-noise ratio than standard Doppler OCT, allowing the system to visualize more features, as shown in Figure 26.17.

Figure 26.17 compares the two imaging modalities *in vivo* on an adult mouse brain. The images were both obtained on the same mouse with the skull intact. The images were then normalized by the maximum pixel value in each image and log compressed [126]. The improved contrast using this technique is clearly visible. OMAG allows for much smaller features to be visualized deeper into the tissue.

Another advantage of the OMAG technique is that it can determine the direction of the flow by applying a Hilbert transformation for each wavenumber along the positions separately. An example of this is shown in Figure 26.18.

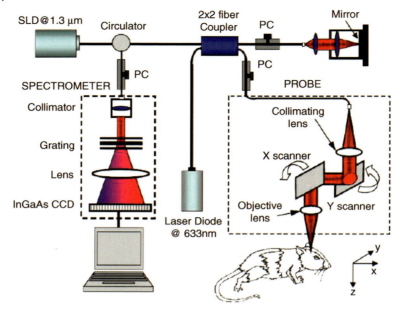

Figure 26.15 Schematic of a typical OMAG system [126]. Reproduced with permmission from The Institute of Physics.

In this experiment, the probe beam was shone from the top on to the sample. The mirror was then modulated using a triangle waveform. The images were generated for (a) the ascending and (b) the descending portion of the wave form. When the mirror moves toward the incident beam, blood perfusion that flows away from the beam direction is determined, and vice versa [127]. The figure shows that the flow direction is clearly determined using this technique. The minimum flow velocity that can be recorded using OMAG is reported to be $\sim 170\,\mu m\,s^{-1}$ and the maximum flow velocity that can be measured is $\sim \pm 2.1\,mm\,s^{-1}$ [124].

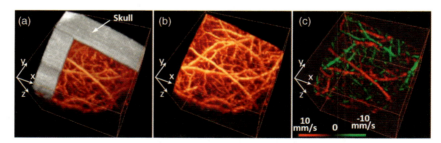

Figure 26.16 In vivo 3D OMAG imaging of the cortical brain of a mouse with the skull left intact. The volumetric visualization was rendered by (a) merging the 3D microstructural image with the 3D cerebral blood flow image, (b) the 3D signals of cerebral blood flow only, and (c) the corresponding DOMAG imaging of velocities within the 3D blood flow network in (b). The red color in (c) represents the blood moving towards the incident probe beam, and otherwise the green color. The physical image size was $2.5 \times 2.5 \times 2.0$ (x–y–z) mm^3 [126]. Reproduced with permmission from The Institute of Physics.

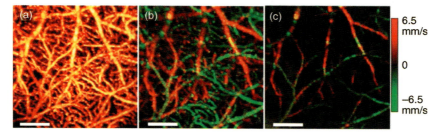

Figure 26.17 Comparison between 3D OMAG and PRDOCT imaging of the cortical brain in mice with the intact skull *in vivo*. Shown are the maximum x–y projection views of (a) OMAG cerebral blood flow image, (b) DOMAG flow velocity image, and (c) PRDOCT flow velocity image. The physical size of scanned tissue volume was $2.5 \times 2.5 \times 2.0\,\text{mm}^3$. White bar = 500 μm.

26.8
Comparison of Planar Microcirculation Imagers TiVi, LSPI, and LDPI

In order to validate the results of any microcirculation techniques, the opinion of a clinical expert is usually required. So far there is no modern technology that has supplanted the fundamental visual observation technique in the clinical environment, even though clinicians' interpretations of images can be bypassed entirely by a computerized analysis, such as neural networks. Limitations of skin assessment

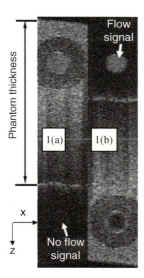
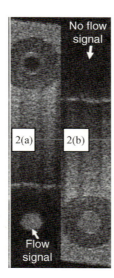

Figure 26.18 Determination of directionality of fluid flow using OMAG. (1) flow is oriented towards scanning beam (2) flow is reversed such that is it oriented away from scanning beam. (a) positive slope of reference mirror; (b) negative slope of reference mirror. The physical dimensions of the image are $1.3 \times 3.1\,\text{mm}$ [127]. Reproduced with permmission from osa.

Table 26.1 Comparison table outlining mechanical and operational principles of the different devices evaluated [117]. Reproduced with permmission from SPIE.

Property	LDLS	FLPI	TiVi
Principle	Doppler effect	Reduced speckle contrast	Polarization spectroscopy
Variable recorded	Perfusion (speed × concentration)	Perfusion (speed × concentration)	Relative concentration
Variable units	Perfusion units (PU)	Perfusion units (PU)	TiVi$_{index}$ (AU)
Measurement sites	16 384	431 680	Up to 12 million
Best resolution (μm)	~500	~100	~50
Measurement depth (μm)	~1000	~300	~500
Repetition rate (s)	6 (50 × 64 pixels)	4 (568 × 760 pixels)	5 (4000 × 3000 pixels)
Best capture time (s)	6	4	1/60
Zoom function	No	Yes	Yes
Laser driver	Required	Not required	Not required
Movement artifact sensitivity	Substantial	Dependent on temporal averaging	None
Video mode	No	Yes	Yes
Video rate	—	25 fps	25 fps
Video size (pixels)	—	113 × 152	Up to 720 × 576
Pixel resolution (cm^{-2})	100	40 000	600 000
View angle	Top only	Top only	User selectable
Imager dimensions (cm)	19 × 30 × 18	22 × 23 × 8	15 × 7 × 5
Weight (kg)	Scanner 3.2	2	1
Calibration	Motility standard	Motility standard	None
Ambient light sensitivity	Intermediate	Substantial	Negligible

visual scoring systems such as the skin blanching assay include the requirement for a trained observer to achieve comparable results, the need to make observations in standardized lighting conditions, application of steroids in random order, and assessment by an independent observer.

The following section represents a brief introduction to the instrumentation chosen to perform basic skin tests. A summary of principles and operation of each instrument is presented in Table 26.1. The actual instruments used are briefly described here.

26.8.1
Laser Doppler Imaging – Moor Instruments moorLDLS

The Moor Instruments moorLDLS is a line scanning laser Doppler imager with a 780 nm laser diode.

Owing to the capture of 64 points simultaneously, the temporal resolution is dramatically improved from conventional LDPI. The minimum image acquisition time is dependent on the number of data points per fast Fourier transform (FFT): at maximum resolution (256 × 64 pixels) the best image acquisition time is 13 s for 256 FFTs and 20 s for 1024 FFTs. The best image acquisition time for 64 × 64 pixels is 4 s. The maximum scan area is 20 × 15 cm and the scan distance to the tissue surface (given by the manufacturer) is between 10 and 20 cm. The device is also capable of single-point multichannel acquisition at a sample rate up to 40 Hz for the entire line of detectors simultaneously. The background threshold level was determined automatically by centering of the laser line on the background under laboratory conditions.

26.8.2
Laser Speckle Imaging – Moor Instruments moorFLPI

The Moor Instruments moorLPI system was used for comparison with laser speckle imaging. The laser source (780 nm) is expanded over an area, rather than a line as in LDLS. There is a high-resolution/low-speed temporal mode or a high-speed/low-resolution spatial mode. The integration time is 0.3 s, providing 25 fps at a size of 152 × 113 pixels for the spatial mode. The temporal mode allows an adjustable integration time, constant for temporal averaging (1 s per frame – 25 frames; 4 s per frame – 100 frames; 10 s per frame – 250 frames; 60 s per frame – 1500 frames) at 568 × 760 pixels for each image. The camera gain can be adjusted to increase the image acquisition area, but saturating the CCD must be avoided.

26.8.3
Polarization Spectroscopy – WheelsBridge TiVi Imager

The TiVi imager is a small and portable device for high-resolution imaging of the concentration of RBCs in superficial skin tissue. A white xenon flash lamp provides illumination over the 450–750 nm wavelength region, and a wide area CCD detects the backscattered light from the tissue. The available commercial system utilizes a high-resolution CCD offering instantaneous image acquisition (1/60th of a second) with a refresh rate of one image every 5 s at a resolution of 3648 × 2736 pixels. An enabled macro mode allows focusing at a closest distance of 4 cm from the tissue surface. A research prototype high-speed mode was also tested, allowing 8 fps at 400 × 400 pixels for real-time online viewing and 30 fps at 640 × 480 pixels for offline analysis. This prototype employed a constant broadband light source with the same CCD as used in the high-resolution version of TiVi. The spectra of the light sources are equivalent for both the high-resolution and high-speed TiVi systems.

Occlusion (occlusion of blood vessels using a sphygmomanometer) reactive hyperemia tests were performed on three nonsmoking, healthy subjects (average age 25 years), with no history of skin diseases, who had not used topical or systemic corticosteroid preparations in the previous 2 months. The subjects were acclimatized in a laboratory at 21 °C for 30 min and rested in a seated position with the volar forearm placed on a flat table at heart level. The instruments were positioned above the forearm for data acquisition.

A reactive hyperemia caused by brachial arterial occlusion using a sphygmomanometer on the upper arm of the test subjects was used to compare the systems in temporal mode (with the high-speed TiVi system used). A baseline level was taken for 1 min, then arterial occlusion was performed by inflating the cuff to 130 mmHg for 2 min. In previous studies with LDPI, upon complete release of the cuff, the resulting increase in blood perfusion is seen, and blood flux oscillations are regularly observed. This phase was observed for a further 3 min. In order to examine further the sensitivity of the TiVi system alone with pressure, a further experiment was conducted in which both the high-speed and high-resolution TiVi systems were arranged to image the volar forearm of a healthy individual at increasing and decreasing pressure steps of 10 mmHg.

Physical size prevented all three technologies being used at the same time on the same test area, meaning that tests with the FLPI had to be carried out separately from TiVi and LDLS studies. The white flash light source on the TiVi did not interfere with the LDLS system, as there are filters present to remove much of the extraneous light. Finally, the TiVi readings were not affected by the scanning beam of the LDLS system as it was outside the sensitive region of the detector.

Imaging comparisons can be carried out by inducing reactions via topical application of different drugs, including vasodilators (e.g., topical analgesic and methyl nicotinate) and vasoconstrictors (e.g., clobetasol-17-propionate). The active ingredient of the analgesic, menthol, induces vasodilation, and caused erythema in all three subjects. Due to percutaneous absorption of methyl salicylate, it was important to note that this analgesic should be used with care on persons sensitive to aspirin or suffering from asthma. None of the subjects were sensitive to aspirin or had asthma, and the analgesic was applied to an unbroken area of skin. The cream was applied to cover an area of $\sim 24\,cm^2$, in the shape of the letters "UL." A small amount of the cream was applied to the volar forearm and imaged 10 min later. The resulting erythema was imaged with the three devices, and can be seen in Figure 26.20. The images show the same affected area, and clear border discrimination can be seen with the TiVi and FLPI imaging systems. The LDLS however, does not show an increase in perfusion with the shape made by the analgesic cream, and it is proposed that the LDLS imaging system is probing much deeper vessels in the microcirculatory structure.

26.8.4
Results

In order to assess the time acquisition properties, it was decided to observe the commonly studied postocclusive reactive hyperemia at the nailfold of the second finger. Lower and upper pressure values of 90 and 130 mmHg were employed in order to arrest completely the venous and arterial circulation, respectively. Standard baseline was registered for 1 min, a pressure of 130 mmHg was applied using the sphygmomanometer, occluding the blood vessels, and maintained for 2 min.

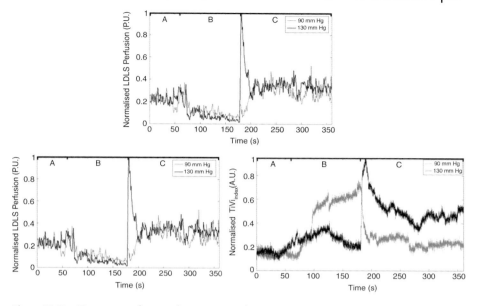

Figure 26.19 Time traces of postocclusive reactive hyperemia. The reaction was sectioned into three; A is the baseline measurement, B is the occlusion for 2 min, and C represents the reactive hyperemia upon removal of the occlusion. Pressure was 130 mmHg except for the TiVi curves, where 90 mmHg pressure was also used to show the TiVi system's (high resolution and high speed) sensitivity to pressure.

The resulting hyperemia upon removal of the occlusion was monitored for 3 min. The areas of the response representing the baseline, occlusion, and hyperemia are labeled A, B and C, respectively, in Figure 26.19. It should be noted that upon application of the sphygmomanometer and subsequent analysis for the TiVi, the curve increased for both 90 and 130 mmHg pressures. This is because the instrument is sensitive to the concentration of RBCs only, and not the speed. This new technology adds a new dimension to the study of postocclusive reactive hyperemia, with a nontraditional time trace sensitive to the concentration of RBCs only. Care has to be taken with the interpretation of these new data, and this is addressed further below.

26.8.5
Discussion

Even with the massive technological advancements of the past 30 years, current "gold standard" methods of examination of skin lesions, erythema and tumors, still involve simple microscope techniques, that is, a closer look at the affected area, carried out by a skilled clinician, be it a pathologist or dermatologist. Similarly, no modern imaging technique has been proven consistently to outperform the opinion of a clinical expert in relation to patch or prick testing [125]. Often, the capital cost of a technology

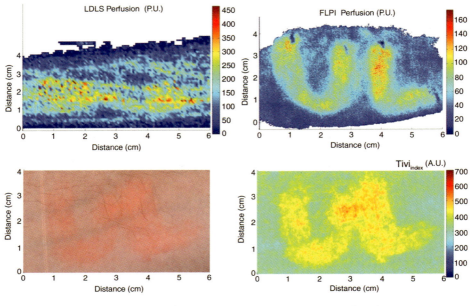

Figure 26.20 Comparison of a topical application of a common analgesic (Deep Heat) as assessed by the LDLS, FLPI, and TiVi imaging technologies. The color visual image is a cross-polarized image from the TiVi imager, which serves as a visual aid, and also the image for subsequent processing. Neither of the other two technologies investigated offers a standard color image [117]. Reproduced with permmission from SPIE.

without an extremely high success rate is not justified, even if the workload of the clinician is drastically reduced.

The ideal blood flow/concentration monitor should have the characteristics of observational measurements, be able to provide reproducible results, have little or no dependence on the tissue optical properties, instantaneous image acquisition (to help against patient motion), and provide a variable in standard SI units. The commercial devices examined in this chapter provide relative changes in flow/RBC concentration due to the random nature of the microvascular architecture. Since LDPI is the oldest technology used in this work (30 years), more research and commercial products exist on LDPI than the other technologies. A full-field laser Doppler imager has been developed and is in commercial production, with Aïmago, providing 256×256 points every 2–10 s [49].

The sensitivity of each device to pressure and the different dynamic reactions can be seen from Figure 26.19, where the concentration of RBCs in the nailfold of the second finger increased upon application of the sphygmomanometer. In the case where 130 mmHg pressure was applied, the arteries and veins were quickly closed, trapping the red blood cells in the tissue at a level not much higher than the baseline. Upon release of the occlusion, the concentration increased rapidly due to the backlog of blood rushing into the area to feed the nutrient-starved tissue. When only

90 mmHg was applied, the arterial inflow remained open while the venous outflow was in stasis. Thus, the concentration of RBCs increased gradually in an almost logarithmic function, and began to reach a plateau, either due to system saturation (the system is linearly calibrated to a physiological RBC concentration of 4% in tissue) or more likely due to a plateau reached by the amount of RBCs that can fit into the vasculature. This section of the curve showed the dynamics of the blood vessels filling up quickly with blood. After the occlusion had been released, the venous outflow was reopened and the RBC concentration decreased rapidly to empty the excess RBCs from the area, and showed signs of returning to a normal state 3 min after release of the occlusion.

The sensitivity of the TiVi imaging system to RBC concentration showed two distinct curves, depending on the pressure of the sphygmomanometer. Since the same CCD was used for both TiVi studies, it was assumed that there were no instrumentation differences. "Biological zero" is a well-known feature of laser Doppler studies and manifests as an offset from instrumental zero seen during occlusion. Total flow arrest would mean that concentration × velocity would be zero, but Brownian motion and occasionally opening and closing of local shunts, vasodilation, and vasoconstriction of local vessels can all contribute to the nonzero flow registered. A similar situation arises with FLPI as there is still some blurring of the speckle due to RBC movement. However, TiVi is sensitive to concentration alone and would require completely bloodless tissue to register a "zero" reading. The most important feature of the technologies presented is that both the LDLS and FLPI measure perfusion, albeit at different depths in dermal tissue, whereas TiVi technology measures relative concentration in the tissue. It is important to remember that the new TiVi imaging technology does not provide measurements of blood flow. Indeed, it is similar to processing of the unweighted moment of the laser Doppler power spectrum to provide measurement of concentration, except that it is valid only with a low number of scattering events at low concentrations, and is dependent on the total light intensity.

With the larger measurement depth of LDLS compared with the other techniques, it proves to be inefficient for very superficial measurements, such as the erythema produced by the topical analgesic. Acquisition on the temporal scale is also hindered with the LDLS due to rastering of the laser line. As the microcirculation is a highly dynamic environment, its perfusion values can change rapidly. Cardiac pulse (~1 Hz), breathing (~0.3 Hz), and vasomotion (~0.1 Hz) in highly perfused tissues are evidence of the rapid temporal nature of microvascular activity. Thus, in laser Doppler studies and the relatively long acquisition time of a single image, the temporal changes in perfusion may often be misinterpreted as spatial heterogeneity [24]. This occurs because it has to be assumed that the perfusion of the site under investigation is stationary and constant, hence the study of dynamic changes is severely compromised. It remains to be seen if the recent CMOS LDPI technology can bridge the gap in temporal resolution in LDPI.

The laser Doppler and speckle signals are a complex function of many parameters and are also dependent on source wavelength, coherence properties of the laser, tissue optical properties, and distance between the scanner and tissue surface. Additional processing is required to correct for nonlinear behavior at high RBC concentrations due to the presence of multiple scattering. It should be noted that all three techniques merely provide a ratiometric measurement to aid any clinical observation. The techniques are not to be used for diagnosis independently of clinical observation.

26.8.5.1 Depth of Measurement

Much variation is reported in the literature as to the depth of penetration of a light beam in tissue, which is due to light power and wavelength, and biological factors such as pigmentation level, age, skin transparency, and epidermal thickness. However, the important parameter is measurement depth rather than penetration or even sampling depth. Indeed, the laser Doppler imaging mean measurement depth in the forearm can be from 300 μm to even 600 μm. This measurement depth is highly wavelength dependent, and therefore depends on which device has been employed.

Physiologically, in the vessel compartments the total cross-sectional area (A) increases from the arterioles to the capillaries of the superficial dermal plexus and then decreases in the veins because the blood flow is constant in a closed system and the speed must be proportional to $1/A$. Thus, the speed is lower in the narrower vessels than in the larger (and deeper) vessels. It is for this reason that low and high frequencies are attributed to flux originating in the superficial vascular plexus and the deeper vessels, respectively. Indeed, ~15% of skin blood flow originates from the nutritional flow, while 85% results from the deeper thermoregulatory flow, especially in the presence of arteriovenous shunts, for example, in the palms of the hands.

Use of the laser wavelength for laser Doppler studies is also an issue; the frequency shift for HeNe laser light at 633 nm is 0–12 kHz, but for a 780 nm laser is 0–20 kHz. The difference in frequency shifts occurs due to the longer wavelength penetrating deeper into the microvasculature, thus probing faster flowing blood. However, both the LDLS and FLPI used for studies carried out in this work employed a 780 nm source. Speckle technology probes a shallower depth than laser Doppler for two main reasons. First, the speckle signal is not frequency weighted, and is looking at first-order statistics, whereas laser Doppler looks at second-order statistics. Also for speckle studies, the laser source is expanded over a much larger area than the point for laser Doppler studies, hence there is a reduction in the power density of the beam. This reduction means that the depth at which noise dominates will be more superficial so that the contribution from faster flowing blood in veins (which is enhanced in laser Doppler technology due to frequency weighting) is lost for speckle imaging. The observation that many speckle studies are carried out on surgically exposed tissue and the appearance of microvessels in the processed images can be explained by the choice of tissue and its preparation. In general, the reduced measurement depth of the FLPI system renders it unsuitable for monitoring nonsuperficial vessels. The FLPI imager was also used more frequently for imaging smaller areas more rapidly and at higher resolutions than are commonly used by laser Doppler systems and involve measurements on exposed tissues because the measurement conditions best suit the technique rather than because the technique best suits the measurement. This is an important factor to remember when optimizing the use of an expensive piece of equipment destined for use in a clinical setting. An interesting comparison should be carried out between the new full-field laser Doppler imager and current commercial laser Doppler technology to examine the depth selection of the signal.

The depth of measurement for TiVi has previously been estimated at 400–500 μm from Monte Carlo simulations of polarized light propagation into simulated dermal tissue [116]. Any future experimentally determined values should verify this model value. It should be noted that the technique is based on polarized light absorption spectroscopy, and does not bear any reference to the moving particles in the tissue. Although the system is not responsive to blood flow, it does return a signal from the deeper veins in the forearm due to the high concentration of blood. Because of the appearance of the blue veins, the TiVi returns an almost zero value in its variable, $TiVi_{index}$, even though a nearby skin site in the upper microcirculation can have a high $TiVi_{index}$ value. Again, it should be remembered that the TiVi imager is designed to investigate the upper microcirculation and that RBC concentration values from anywhere other than this region may not be accurate. The measurement volume (the volume from which the bulk of the signal is acquired) is well inside the reticular dermis of the microcirculation. For laser Doppler measurements, the measurement volume has been determined as ~ 1 mm^3 or smaller, and in laser speckle systems the volume is highly dependent on the type of system being used.

26.8.6
Conclusions

We have compared the operation of an established microcirculation imaging technique, laser Doppler (LDPI), with the more recently available laser speckle and TiVi in human skin tissue using the occlusion and reactive hyperemia response. However, TiVi differs from the other techniques in its response to occlusion of the brachial artery. Since this is a global occlusion of the forearm all instruments were sensitive to the effect. LDPI and LSPI showed a large decrease as they are both sensitive to RBC concentration and velocity. However, TiVi, being sensitive to concentration only, showed stasis when a full arterial occlusion (130 mmHg) was applied quickly and an increase in of RBC concentration when the arterial inflow remained open with the venous outflow occluded (90 mmHg). LSPI and TiVi are both welcome tools in the study of the microcirculation, but care must be taken in the interpretation of the images since blood flow and blood concentration in tissue are essentially different parameters. However, LDPI is the only one of these imaging techniques which has been approved by regulatory bodies, including the FDA. Laser Doppler imaging has been shown accurately to assess burn depth and predict wound outcome, with a large weight of evidence [128].

The work described here may motivate further studies of comparisons between noninvasive optical techniques for the investigation of tissue microcirculation.

We have discussed the *modus operandi* and applications of capillaroscopy, LDPI, LSPI, PAT, OCT, and TiVi. Each has had its own peak time and all retain niche applications. Proper application of each is dependent on user knowledge of light interaction with tissue and the basic workings of the device used. The choice of technology for *in vivo* imaging of the microcirculation is dependent on the sampling depth (nutritional or thermoregulatory vessels), resolution (physical and pixel), field of view, and whether structure or function is to be investigated.

Acknowledgments

The authors are grateful to HEA PRTLI4 (national development plan) funding of the National Biophotonics and Imaging Platform Ireland and IRCSET for supporting this research. They also thank Dr. Rodney Gush of Moor Instruments and Professor Gert Nilsson of WheelsBridge AB, Sweden.

References

1 Doppler, J.C. (1842) Über das farbige Licht der Doppelsterne und einiger anderer Gestirne des Himmels. [On the coloured light from double stars and some other Heavenly Bodies]. *Abhandl. Konigl. Bohmischen Ges. Wiss.*, 5th Ser., **2**, 465 [Treatises of the Royal Bohemian Society of Science].

2 Ostergren, J. and Fagrell, B. (1984) Videophotometric capillaroscopy for evaluating drug effects on skin microcirculation – a double-blind study with nifedipine. *Clin. Physiol.*, **4** (2), 169–176.

3 Schmidt-Nielsen, B. (1994) August Krogh and capillary physiology. *Int. J. Microcirc. Clin. Exp.*, **14** (1–2), 104–110.

4 Ryan, T.J. (1973) Measurement of blood flow and other properties of the vessels of the skin, in *The Physiology and Pathophysiology of the Skin*, vol. 1 (ed. A. Jarrett), Academic Press, London, pp. 653–679.

5 Butti, P., Intaglietta, M., Reimann, H., Holliger, C., Bollinger, A., and Anliker, M. (1975) Capillary red blood cell velocity measurements in human nailfold by video densitometric method. *Microvasc. Res.*, **10**, 220–227.

6 Fagrell, B., Fronek, A., and Intaglietta, M. (1977) A microscope television system for studying flow velocity in human skin capillaries. *Am. J. Physiol.*, **233**, 318–321.

7 Dolezalova, P., Young, S.P., Bacon, P.A., and Southwood, T.R. (2003) Nailfold capillary microscopy in healthy children and in childhood rheumatic diseases: a prospective single blind observational study. *Ann. Rheum. Dis.*, **62**, 444–449.

8 Ohtsuka, T., Tamura, T., Yamakage, A., and Yamazaki, S. (1998) The predictive value of quantitative nailfold capillary microscopy in patients with undifferentiated connective tissue disease. *Br. J. Dermatol.*, **139**, 622–629.

9 Nagy, Z. and Czirják, L. (2004) Nailfold digital capillaroscopy in 447 patients with connective tissue disease and Raynaud's disease. *J. Eur. Acad. Dermatol. Venereol.*, **18**, 62–68.

10 Bukhari, M., Herrick, A.L., Moore, T., Manning, J., and Jayson, M.I.V. (1996) Increased nailfold capillary dimensions in primary Raynaud's phenomenon and systemic sclerosis. *Br. J. Rheumatol.*, **35**, 1127–1131.

11 Allen, P.D., Hillier, V.F., Moore, T., Anderson, M.E., Taylor, C.J., Herrick, A.L., (2003). Computer Based System for Acquisition and Analysis of Nailfold Capillary Images. *Medical Image Understanding and Analysis*, eScholarID:1d32761.

12 Aguiar, T., Furtado, E., Dorigo, D., Bottino, D., and Bouskela, E. (2006) Nailfold videocapillaroscopy in primary Sjögren's syndrome. *Angiology*, **57**, 593–599.

13 Nilsson, G.E., Jakobsson, A., and Wårdell, K. (1992) Tissue perfusion monitoring and imaging by coherent light scattering. *Proc. SPIE*, **1524**, 90–109.

14 Stern, M.D. (1975) *In vivo* evaluation of microcirculation by coherent light scattering. *Nature*, **254**, 56–58.

15 Bollinger, A., Butti, P., Barras, J.P., Trachsler, H., and Siegenthaler, W. (1974) Red blood cell velocity in nailfold capillaries of man measured by a television microscopy technique. *Microvasc. Res.*, **7**, 62–72.

16 Mawson, D.M. and Shore, A.C. (1998) Comparison of capiflow and frame by frame analysis for the assessment of capillary red blood cell velocity. *J. Med. Eng. Technol.*, **22**, 53–63.

17 Rosen, B. and Paffhausen, W. (1993) On-line measurement of microvascular diameter and red blood cell velocity by a line-scan CCD image sensor. *Microvasc. Res.*, **45**, 107–121.

18 Sourice, A., Plantier, G., and Saumet, J.-L. (2005) Red blood cell velocity estimation in microvessels using the spatiotemporal autocorrelation. *Meas. Sci. Technol.*, **16**, 2229–2239.

19 Bhushan, M., Moore, T., Herrick, A.L., and Griffiths, C.E.M. (2000) Nailfold video capillaroscopy in psoriasis. *Br. J. Dermatol.*, **142**, 1171–1176.

20 Wong, M.L., Highton, J., and Palmer, D.G. (1988) Sequential nailfold capillary microscopy in scleroderma and related disorders. *Ann. Rheum. Dis.*, **47**, 53–61.

21 Sainthillier, J.M., Degouy, A., Gharbi, T., Pieralli, C., and Humbert, P. (2003) Geometrical capillary network analysis. *Skin Res. Technol.*, **9**, 312–320.

22 Zhong, J., Asker, C.L., and Salerud, G. (2000) Imaging, image processing and pattern analysis of skin capillary ensembles. *Skin Res. Technol.*, **6**, 45–57.

23 Hamar, G., Horváth, G., Tarján, Z., and Virág, T. (2007) Markov chain based edge detection algorithm for evaluation of capillary microscopic images. Proceedings of the 13th PhD Mini-Symposium, pp. 18–21.

24 Leahy, M.J., de Mul, F.F., Nilsson, G.E., and Maniewski, R. (1999) Principles and practice of the laser-Doppler perfusion technique. *Technol. Health Care*, **7**, 143–162.

25 Riva, C., Ross, B., and Benedek, G.B. (1972) Laser Doppler measurements of blood flow in capillary tubes and retinal arteries. *Invest. Ophthalmol.*, **11**, 936–944.

26 Stern, M.D. (1975) *In vivo* evaluation of microcirculation of coherent light scattering. *Nature*, **254**, 56–58.

27 Cummins, H.Z. and Swinney, H.L. (1970) Light beating spectroscopy. *Prog. Opt.*, **8**, 133–200.

28 Bonner, R. and Nossal, R. (1981) Model for laser Doppler measurements of blood flow in tissue. *Appl. Opt.*, **20**, 2097–2107.

29 Boulnois, J.L. (1985) Photophysical processes in recent medical laser developments: a review. *Lasers Med. Sci.*, **1**, 47–66.

30 Abbot, N., Ferrell, W., Lockhart, J., and Lowe, G. (1996) Laser Doppler perfusion imaging of skin blood flow using red and near-infrared sources. *J. Invest. Dermatol.*, **107**, 882–886.

31 Duteil, L., Bernengo, J.C., and Schalla, W. (1985) A double wavelength laser Doppler system to investigate skin microcirculation. *Trans. Biomed. Eng.*, **32**, 439–447.

32 Gush, R.J. and King, T.A. (1991) Discrimination of capillary and arterio-venular blood flow in skin by laser Doppler flowmetry. *Med. Biol. Eng. Comput.*, **29**, 387–392.

33 Johnson, J.M., Taylor, W.F., and Shepherd, A.P. (1984) Laser Doppler measurements of skin blood flow: comparison with plethysmography. *J. Appl. Physiol. Respir. Environ. Excercise Physiol.*, **56**, 796–803.

34 Holloway, G.A. and Watkins, D.W. (1977) Laser Doppler measurement of cutaneous blood flow. *J. Invest. Dermatol.*, **69**, 306–309.

35 Murray, A.K., Herrick, A.L., and King, T.A. (2004) Laser Doppler imaging: a developing technique for application in the rheumatic diseases. *Rheumatology*, **43**, 1210–1218.

36 Braverman, I.M., Keh, A., and Goldminz, D. (1990) Correlation of

laser Doppler wave patterns with underlying microvascular anatomy. *J. Invest. Dermatol.*, **95**, 283–286.

37 Tenland, T., Salerud, E.G., Nilsson, G.E., and Oberg, P.A. (1983) Spatial and temporal variations in human skin blood flow. *Int. J. Microcirc. Clin. Exp.*, **2**, 81–90.

38 Bircher, A., de boer, E.M., Agner, T., Wahlberg, J.E., and Serup, J. (1994) Guidelines for measurement of cutaneous blood flow by laser Doppler flowmetry. A report from the Standardization Group of the European Society of Contact Dermatitis. *Contact Dermatitis*, **30**, 65–72.

39 Gush, R.J. and King, T.A. (1987) Investigation and improved performance of optical fibers in laser Doppler blood flow measurements. *Med. Biol. Eng. Comput.*, **25**, 391–396.

40 de Mull, F.F.M., van der Plas, D., Greve, J., Aarnoudse, J.G., and van Spijker, J. (1984) Mini laser-Doppler (blood) flow monitor with diode laser source and detection integrated in the probe. *Appl. Opt.*, **23**, 2970–2973.

41 Wårdell, K. (1993) Laser Doppler perfusion imaging by dynamic light scattering. *IEEE Trans. Biomed. Eng.*, **40**, 309–316.

42 Essex, T.J. and Byrne, P. (1991) A laser Doppler scanner for imaging blood flow in skin. *J. Biomed. Eng.*, **13**, 189–194.

43 Gardner-Medwin, J.M., Mac Donald, I.A., Taylor, J.Y., Rikey, P.H., and Powell, R.J. (2001) Seasonal differences in finger skin temperature and microvascular blood flow in healthy men and women are exaggerated in women with primary Reynaud's phenomenon. *Br. J. Clin. Pharmacol.*, **52**, 17–23.

44 Kernick, D.P. and Shore, A.C. (2000) Characteristics of laser Doppler perfusion imaging *in vitro* and *in vivo*. *Physiol. Meas.*, **4**, 222–240.

45 Droog, E.J., Steenbergen, W., and Sjoberg, F. (2001) Measurement of depth of burns by laser Doppler perfusion imaging. *Burns*, **27**, 561–568.

46 Kan, F. and Newton, D. (2003) Laser Doppler imaging in the investigation of lower limb wounds. *Int. J. Lower Extrem. Wounds*, **2**, 74–86.

47 Fullerton, A., Stucker, M., Wilhelm, K.P., Wårdell, K., Anderson, C., Fisher, T., Nilsson, G.E., and Serup, J. (2002) Guidelines for visualization of cutaneous blood flow by laser Doppler perfusion imaging. *Contact Dermatitis*, **46**, 129–140.

48 Serov, A., Steenbergen, W., and de Mul, F. (2002) Laser Doppler perfusion imaging with a complementary metal oxide semiconductor image sensor. *Opt. Lett.*, **27**, 300–302.

49 Serov, A., Steinacher, B., and Lasser, T. (2005) Full-field laser Doppler perfusion imaging and monitoring with an intelligent CMOS camera. *Opt. Express*, **13**, 3681–3689.

50 Serov, A. (2005) High-speed laser Doppler perfusion imaging using an integrating CMOS image sensor. *Opt. Express*, **13**, 6416–6428.

51 Nielsen, A., Fabricius, M., and Lauritzen, M. (2000) Scanning laser-Doppler flowmetry of rat cerebral circulation during cortical spreading depression. *J. Vasc. Res.*, **37**, 513–522.

52 Pape, S., Skouras, C., and Byrne, P. (2001) An audit of the use of laser Doppler imaging (LDI) in the assessment of burns of intermediate depth. *Burns*, **27** (3), 233–239.

53 Brown, R., Rice, P., and Bennett, N. (1998) The use of laser Doppler imaging as an aid in clinical management decision making in the treatment of vesicant burns. *Burns*, **24**, 692–698.

54 Kloppenberg, F., Beerthuizen, G., and ten Duis, H. (2001) Perfusion of burn wounds assessed by laser Doppler imaging is related to burn depth and healing time. *Burns*, **27**, 359–363.

55 Repež, A., Oroszy, D., and Arnež, Z.M. (2007) Continuous postoperative monitoring of cutaneous free flaps using near infrared spectroscopy. *J. Plast. Reconstr. Aesthet. Surg.*, **61** (1), 71–77.

56 Stucker, M., Springer, C., Paech, V., Hermes, N., Hoffmann, M., and Altmeyer, P. (2006) Increased laser Doppler flow in skin tumors corresponds

to elevated vessel density and reactive hyperemia. *Skin Res. Technol*, **12**, 1–6.
57 Hassan, M., Little, R., Vogel, A., Aleman, K., Wyvill, K., Yarchoan, R., and Gandjbakhche, A. (2004) Quantitative assessment of tumor vasculature and response to therapy in Kaposi's sarcoma using functional noninvasive imaging. *Technol. Cancer Res. Treat.*, **3**, 451–457.
58 Wang, I., Andersson-Engels, S., Nilsson, G.E., Wårdell, G., and Svanberg, K. (1997) Superficial blood flow following photodynamic therapy of malignant non-melanoma skin tumours measured by laser Doppler perfusion imaging. *Br. J. Dermatol.*, **136**, 184–189.
59 Fullerton, A. and Serup, J. (1995) Laser Doppler image scanning for assessment of skin irritation, in *Irritant Dermatitis. New Clinical and Experimental Aspects. Current Problems in Dermatology*, Vol. 23 (ed P. Elsner and H.I. Maibach), Karger, Basel, pp. 159–168.
60 Fullerton, A., Rode, B., and Serup, J. (2002) Studies of cutaneous blood flow of normal forearm skin and irritated forearm skin based on high-resolution laser Doppler perfusion imaging (HR-LDPI). *Skin Res. Technol.*, **8**, 32–40.
61 Fullerton, A., Rode, B., and Serup, J. (2002) Skin irritation typing and grading based on laser Doppler perfusion imaging. *Skin Res. Technol.*, **8**, 23–31.
62 Bjarnason, B. and Fischer, T. (1998) Objective assessment of nickel sulfate patch test reactions with laser Doppler perfusion imaging. *Contact Dermatitis*, **39**, 112–1188.
63 Bjarnason, B., Flosadottir, E., and Fischer, T. (1999) Objective non-invasive assessment of patch tests with the laser Doppler perfusion scanning technique. *Contact Dermatitis*, **40**, 251–260.
64 Ferrell, W., Balint, P., and Sturrock, R. (2000) Novel use of laser Doppler imaging for investigating epicondylitis. *Rheumatology*, **39**, 1214–1217.
65 Newton, D., Khan, F., Belch, J., Mitchell, M., and Leese, G. (2002) Blood flow changes in diabetic foot ulcers treated with dermal replacement therapy. *J. Foot Ankle Surg.*, **41**, 233–237.
66 Nixon, J., Smye, S., Scott, J., and Bond, S. (1999) The diagnosis of early pressure sores: report of the pilot study. *J. Tissue Viability*, **9**, 62–66.
67 Briers, J.D. (2001) Laser Doppler, speckle and related techniques for blood perfusion mapping and imaging. *Physiol. Meas.*, **22**, R35–R36.
68 Briers, J.D. (1996) Laser Doppler and time-varying speckle: a reconciliation. *J. Opt. Soc. Am. A*, **13**, 345–350.
69 Forrester, K.R., Stewart, C., Tulip, J., Leonard, C., and Bray, R.C. (2002) Comparison of laser speckle and laser Doppler perfusion imaging: measurement in human skin and rabbit articular tissue. *Med. Biol. Eng. Comput.*, **40**, 687–697.
70 Fujii, H., Nohira, K., Yamamoto, Y., Ikawa, H., and Ohura, T. (1987) Evaluation of blood flow by laser speckle image sensing. *Appl. Opt.*, **26**, 5321–5325.
71 Cheng, H., Luo, Q., Liu, Q., Lu, Q., Gong, H., and Zeng, S. (2004) Laser speckle imaging of blood flow in microcirculation. *Phys. Med. Biol.*, **49**, 1347–1357.
72 Choi, B., Kang, N.M., and Nelson, J.S. (2004) Laser speckle imaging for monitoring blood flow dynamics in the *in vivo* rodent dorsal skin fold model. *Microvasc. Res.*, **68**, 143–146.
73 Forrester, K.R., Stewart, C., Leonard, C., Tulip, J., and Bray, R.C. (2003) Endoscopic laser imaging of tissue perfusion: new instrumentation and technique. *Lasers Surg. Med.*, **33**, 151–157.
74 Kruijt, B., de Bruijn, H.S., van der Ploeg-van den Heuvel, A., Sterenborg, H.J.C.M., and Robinson, D.J. (2006) Laser speckle imaging of dynamic changes in flow during photodynamic therapy. *Lasers Med. Sci.*, **21**, 208–212.
75 Li, P., Ni, S., Zhang, L., Zeng, S., and Luo, Q. (2006) Imaging cerebral blood flow through the intact rat skull with temporal laser speckle imaging. *Opt. Lett.*, **31**, 1824–1826.
76 Paul, J.S. (2006) Imaging the development of an ischemic core

following photochemically induced cortical infarction in rats using laser speckle analysis. *Neuroimage*, **29**, 38–45.

77 Nagahara, M., Tamaki, Y., Araie, M., and Fujii, H. (1999) Real-time blood velocity measurements in human retinal vein using the laser speckle phenomenon. *Jpn. J. Opthalmol.*, **43**, 186–195.

78 Zhu, D., Lu, W., Weng, Y., Cui, H., and Luo, Q. (2007) Monitoring thermal-induced changes in tumor blood flow and microvessels with laser speckle contrast imaging. *Appl. Opt.*, **46**, 1911–1917.

79 Yuan, S., Devor, A., Boas, D.A., and Dunn, A.K. (2005) Determination of optimal exposure time for imaging of blood flow changes with laser speckle contrast imaging. *Appl. Opt.*, **44**, 1823–1830.

80 Briers, J.D. and Fercher, A.F. (1982) Retinal blood-flow visualisation by means of laser speckle photography. *Invest. Ophthalmol. Vis. Sci.*, **22**, 255–259.

81 Forrester, K.R., Tulip, J., Leonard, C., Stewart, C., and Bray, R.C. (2004) A laser speckle imaging technique for measuring tissue perfusion. *IEEE Trans. Biomed. Eng.*, **51**, 2074–2084.

82 Stewart, C.J., Gallant-Behm, C.L., Forrester, K., Tulip, J., Hart, D.A., and Bray, R.C. (2006) Kinetics of blood flow during healing of excisional full-thickness skin wounds in pigs as monitored by laser speckle perfusion imaging. *Skin Res. Technol.*, **12**, 247–253.

83 Zakharov, P., Völker, A., Buck, A., Weber, A., and Scheffold, F. (2006) Quantitative modeling of laser speckle imaging. *Opt. Lett.*, **31**, 3465–3467.

84 Gonik, M.M., Mishin, A.B., and Zimnyakov, D.A. (2002) Visualization of blood microcirculation parameters in human tissues by time-integrated dynamic speckles analysis. *Ann. N. Y. Acad. Sci.*, **972**, 325–330.

85 Sugii, Y., Nishio, S., and Okamoto, K. (2002) *In vivo* PIV measurement of red blood cell velocity field in microvessels considering mesentery motion. *Physiol. Meas.*, **23**, 403–416.

86 Yaoeda, K., Shirakashi, M., Funaki, S., Funaki, H., Nakatsue, T., Fukushima, A., and Abe, H. (2000) Measurement of microcirculation in optic nerve head by laser speckle flowgraphy in normal volunteers. *Am. J. Opthalmol.*, **130** (5), 606–610.

87 Bray, R.C., Forrester, K.R., Reed, J., Leonard, C., and Tulip, J. (2006) Endoscopic laser speckle imaging of tissue blood flow: applications in the human knee. *J. Orthop. Res.*, **24** (8), 1650–1659.

88 Kharlamov, A., Brown, B.R., Easley, K.A., and Jones, S.C. (2004) Heterogeneous response of cerebral blood flow to hypotension demonstrated by laser speckle imaging flowmetry in rats. *Neurosci. Lett.*, **368**, 151–156.

89 Royl, G., Leithner, C., Sellien, H., Müller, J.P., Megow, D., Offenhauser, N., Steinbrink, J., Kohl-Bareis, M., Dirnagl, U., and Lindauer, U. (2006) Functional imaging with laser speckle contrast analysis: vascular compartment analysis and correlation with laser Doppler flowmetry and somatosensory evoked potentials. *Brain Res.*, **1121**, 95–103.

90 Smith, T.K., Choi, B., Ramirez-San-Juan, J.C., Nelson, J.S., Osann, K., and Kelly, K.M. (2006) Microvascular blood flow dynamics associated with photodynamic therapy, pulsed dye laser irradiation and combined regimens. *Laser Surg. Med.*, **38**, 532–539.

91 Stewart, C.J., Frank, R., Forrester, K.R., Tulip, J., Lindsay, R., and Bray, R.C. (2005) A comparison of two laser-based methods for determination of burn scar perfusion: laser Doppler versus laser speckle imaging. *Burns*, **31**, 744–752.

92 Tyrell, J. (2007) Video capability revives interest in laser method. *Opt. Laser Eur.*, **147**, 15–16.

93 Rajan, V., Varghese, B., van Leeuwen, T.G., and Steenbergen, W. (2006) Speckles in laser Doppler perfusion imaging. *Opt. Lett.*, **31**, 468–470.

94 Demos, S.G. and Alfano, R.R. (1997) Optical polarization imaging. *Appl. Opt.*, **36**, 150–155.

95 Ishimaru, A. (1989) Diffusion of light in turbid media. *Appl. Opt.*, **28**, 2210–2215.
96 da Silva, D.d.F.T., Vidal, B.d.C., Zezell, D.M., Zorn, T.M.T., Nunez, S.C., and Ribeiro, M.S. (2006) Collagen birefringence in skin repair to red-polarized laser therapy. *J. Biomed. Opt.*, **11**, 204002.
97 Liu, Y., Kim, Y.Y., Li, X., and Backman, V. (2005) Investigation of depth selectivity of polarization gating for tissue characterization. *Opt. Express*, **13**, 601–611.
98 Schmitt, J.M., Gandjbakhche, A.H., and Bonnar, R.F. (1992) Use of polarized light to discriminate short-pass phonons in a multiply scattering medium. *Appl. Opt.*, **31**, 6535–6539.
99 MacKintosh, F.C., Zhu, J.X., Pine, D.J., and Weitz, D.A. (1989) Polarization memory of multiply scattered light. *Phys. Rev. B*, **40**, 9342–9345.
100 Karlsson, M.G. and Wårdell, K. (2005) Polarized laser Doppler perfusion imaging – reduction of movement-induced artifacts. *J. Biomed. Opt.*, **10**, 064002.
101 Pircher, M., Goetzinger, E., Leitgeb, R., and Hitzenberger, C.K. (2004) Transversal phase resolved polarization sensitive optical coherence tomography. *Phys. Med. Biol.*, **49**, 1257–1263.
102 Goetzinger, E., Pircher, M., Sticker, M., Dejaco-Ruhswurm, I., Kaminski, S., Findl, O., Skorpik, C., Fercher, A.F., and Hitzenberger, C.K. (2003) Three dimensional polarization sensitive optical coherence tomography of normal and pathologic human cornea. *Proc. SPIE*, **5140**, 120–124.
103 Smith, Z.J. and Berger, A.J. (2005) Surface-sensitive polarized Raman spectroscopy of biological tissue. *Opt. Lett.*, **30**, 1363–1365.
104 Sun, Y., Su, J.W., Lo, W., Lin, S.J., Jee, S.H., and Dong, C.Y. (2003) Multiphoton imaging of the stratum corneum and the dermis in *ex-vivo* human skin. *Opt. Express*, **11**, 3377–3382.
105 Jacques, S.L., Ramella-Roman, J.C., and Lee, K. (2002) Imaging skin pathology with polarized light. *J. Biomed. Opt.*, **7**, 329–340.
106 Groner, W., Winkelman, J.W., Harris, A.G., Ince, C., Bouma, G.J., Messmer, K., and Nadeau, R.G. (1999) Orthogonal polarization spectral imaging: a new method for study of the microcirculation. *Nat. Med.*, **5**, 1209–1213.
107 Mathura, K.R., Vollebregt, K.C., Boer, K., DeGraaff, J.C., Ubbink, D.T., and Ince, C. (2001) Comparison of OPS imaging and conventional capillary microscopy to study the human microcirculation. *J. Appl. Physiol.*, **91**, 74–78.
108 Milner, S.M., Bhat, S., Gulati, S., Gherardini, G., Smith, C.E., and Bick, R.J. (2005) Observations on the microcirculation of the human burn wound using orthogonal polarization spectral imaging. *Burns*, **31**, 316–319.
109 Langer, S., Harris, A.G., Biberthaler, P., von Dobschuetz, E., and Messmer, K. (2001) Orthogonal polarization spectral imaging as a tool for the assessment of hepatic microcirculation: a validation study. *Transplantation*, **71**, 1249–1256.
110 Genzel-Boroviczeny, O., Strontgen, J., Harris, A.G., Messmer, K., and Christ, F. (2002) Orthogonal polarization spectral imaging (OPS): a novel method to measure the microcirculation in term and preterm infants transcutaneously. *Pediatr. Res.*, **51**, 386–391.
111 Yaroslavsky, A.N., Salomatina, E.V., Neel, V., Anderson, R., and Flotte, T. (2007) Fluorescence polarization of tetracycline derivatives as a technique for mapping nonmelanoma skin cancers. *J. Biomed. Opt.*, **12**, 014005.
112 Yaroslavsky, A.N., Neel, V., and Anderson, R.R. (2004) Fluorescence polarization imaging for delineating nonmelanoma skin cancers. *Opt. Lett.*, **29**, 2010–2012.
113 Arimoto, H. (2007) Estimation of water content distribution in the skin using dualband polarization imaging. *Skin Res. Technol.*, **13**, 49–54.
114 Jacques, S.L., Roman, J.R., and Lee, K. (2000) Imaging superficial tissues with polarized light. *Lasers Surg. Med.*, **26**, 119–129.

115 Ramella-Roman, J.C., Lee, K., Prahl, S.A., and Jacques, S.L. (2004) Design, testing, and clinical studies of a handheld polarized light camera. *J. Biomed. Opt.*, **9**, 1305–1310.

116 O'Doherty, J., Henricson, J., Anderson, C., Leahy, M.J., Nilsson, G.E., and Sjoberg, F. (2007) Sub-epidermal imaging using polarized light spectroscopy for assessment of skin microcirculation. *Skin Res. Technol.*, **13** (4), 472–484.

117 O'Doherty, J., McNamara, P., Clancy, N.T., Enfield, J.G., and Leahy, M.J. (2009) Comparison of instruments for investigation of microcirculatory blood flow and red blood cell concentration. *J. Biomed. Opt.*, **14** (3), 034025.

118 Leahy, M.J., O'Doherty, J., McNamara, P., Henricson, J., Nilsson, G.E., Anderson, C., and Sjoberg, F. (2006) Diffuse reflection imaging of sub-epidermal tissue haematocrit using a simple RGB camera. *Proc. SPIE*, **6535**, 653503.

119 Diebold, G.J., Sun, T., and Khan, M.I. (1992) In *Photoacoustic and Photothermal Phenomena III* (ed. D. Bicanic), Springer, Berlin, pp. 263–296.

120 Wang, X., Xu, Y., Xu, M., Yokoo, S., Fry, E.S., and Wang, L.V. (2002) Photoacoustic tomography of biological tissues with high cross-section resolution: reconstruction and experiment. *Med. Phys.*, **12**, 2799–2805.

121 Xu, M. and Wang, L.V. (2002) Time domain reconstruction of thermoacoustic tomography in a spherical geometry. *IEEE Trans. Med. Imaging*, **21**, 814–822.

122 Xu, M. and Wang, L.V. (2005) Universal back-projection algorithm for photoacoustic computed tomography. *Phys. Rev.*, **71** (1 Pt. 2), 016706.

123 Yang, D., Xing, D., Tan, Y., Gu, H., and Yang, S. (2006) Integrative prototype B-scan photoacoustic tomography system based on a novel hybridized scanning head. *Appl. Phys. Lett.*, **88** (17), 174101.

124 Wang, X., Pang, Y., Ku, G., Stoica, G., and Wang, L. (2003) Three-dimensional laser-induced photoacoustic tomography of mouse brain with the skin and skull intact. *Opt. Lett.*, **28**, 1739–1741.

125 Lamminen, H. and Voipio, V. (2008) Computer-aided skin prick test. *Exp. Dermatol.*, **17** (11), 975–976.

126 Wang, R.K. (2007) Three-dimensional optical micro-angiography maps directional blood perfusion deep within microcirculation tissue beds *in vivo*. *Phys. Med. Biol.*, **52**, N531–N537.

127 Wang, R.K., Jacques, S.L., Ma, Z., Hurst, S., Hanson, S.R., and Gruber, A. (2007) Three dimensional optical angiography. *Opt. Express*, **15**, 4083–4097.

128 Monstrey, S., Hoeksema, H., Verbelen, J., Pirayesh, A., and Blondeel, P. (2008) Assessment of burn depth and burn wound healing potential. *Burns*, **34** (6), 761–769.

27
Optical Oximetry
Stephen J. Matcher

27.1
Introduction

The most immediate requirement of most eukaryotic life forms is the continued availability of molecular oxygen. The physiology of all animals has evolved to ensure that an adequate supply of oxygen can always be maintained between the atmosphere and the mitochondria within the cells of the body. Oxygen reduces free protons to water in the terminal stage of the oxidative phosphorylation pathway, by which glucose is ultimately converted to CO_2 and water, with the resulting release of energy. Oxidative phosphorylation is the most effective pathway for the synthesis of adenosine triphosphate (ATP), and if this pathway is halted then anaerobic glycolysis alone must meet the body's demand for ATP. Only in a few rare cases (e.g., the cornea of the eye) can glycolysis meet this demand for any appreciable time, so in the absence of oxygen there is an initial decrease in tissue pH (acidosis) and accumulation of lactate, a halt in ATP synthesis followed later by a failure of cell membrane ion pumps, and ultimately hypoxic cell death. In the brain, irreversible damage occurs within just a few minutes. The ability to monitor oxygen delivery to the cells is therefore a major goal of physiological measurement science.

In mammals, oxygen is extracted from the air and transported to cells by a mixture of active and passive (diffusive) transport. The respiratory system extracts O_2 from the air and this is diffusively transported into the blood via close arrangement of alveoli and capillaries in the lungs. Blood is a suspension of erythrocytes in plasma and each erythrocyte (a biconcave anuclear cell about 8 μm in diameter and 2 μm thick) contains the 64 kDa metalloprotein hemoglobin. Each molecule of hemoglobin can physically bind four O_2 molecules. At normal physiological hematocrit, this results in an O_2 concentration nearly 100 times greater than the dissolved O_2 concentration in the plasma. The pulmonary and systemic circulations, powered by the heart, then actively transport this oxygenated blood through the systemic arterial circulation and into the capillary bed. The smallest capillaries are ~10 μm in diameter and all cells of the body that rely on oxidative phosphorylation are located within about 100 μm of a

---

Handbook of Biophotonics. Vol.2: Photonics for Health Care, First Edition. Edited by Jürgen Popp,
Valery V. Tuchin, Arthur Chiou, and Stefan Heinemann.
© 2012 Wiley-VCH Verlag GmbH & Co. KGaA. Published 2012 by Wiley-VCH Verlag GmbH & Co. KGaA.

capillary, so that passive diffusion, driven by the local O_2 partial pressure gradient, then delivers the oxygen through the cell membrane to the mitochondria.

From the previous discussion, we can see the importance of technologies that can monitor all stages of this O_2 transport process, especially in a critical care setting. The need to monitor the correct function of the heart has led to major developments in electrocardiography and lung function is routinely measured by electrical impedance techniques. Likewise, the corresponding systemic blood perfusion pressure can be measured noninvasively by techniques such as oscillometry and also invasively via catheter-based pressure transducers. The resulting volume flow rate of blood can be measured directly using clearance or dilution techniques (although this is not common in critical care medicine) or Doppler ultrasound, and this leaves one final parameter to be determined in order to characterize the supply rate of oxygen to the capillary bed: the overall fraction of blood hemoglobin molecules that are physically bound to oxygen. Denoting the overall concentration of oxygenated hemoglobin in blood as $[O_2Hb]$ and that of deoxygenated hemoglobin as $[HHb]$, then the hemoglobin saturation is defined as $[O_2Hb]/([O_2Hb] + [HHb])$. Given that the majority of the extraction of O_2 occurs in the capillary level, it is clear that this parameter will vary between the main vascular beds, with the arterial saturation exceeding the capillary saturation, which in turn will exceed the venous saturation.

27.2
Co-Oximetry

Oximetry is the name collectively given to techniques that estimate hemoglobin saturation using changes in the optical or near-infrared (NIR) absorption spectrum. It is well known that systemic arterial blood appears vivid crimson whereas systemic venous blood appears a duller brown (and vice versa in the pulmonary circulation). Hemoglobin physically binds oxygen via a conformational change that displaces the iron centers in the molecule, causing a shift in their spectral absorption peaks. Figure 27.1 shows the specific (i.e., molar) extinction coefficient spectra of the two forms of hemoglobin. The large peak in extinction at around 400 nm for both forms accounts for the red hue of blood. In the orange part of the spectrum (450–600 nm), subtle differences occur between the two forms and only at a few specific wavelengths (the "isosbestic points") are the extinction coefficients equal. The extinction coefficients are much lower than in the blue region, which makes it feasible to obtain transillumination spectra across thin samples of undiluted blood. This trend continues into the NIR region (600–1000 nm), where again distinct spectra arise for HHb and O_2Hb and a single isosbestic point occurs at around 800 nm. It is of great significance that the molar extinction coefficient is nearly 100 times lower than in the visible region. This makes it technically feasible to transilluminate intact biological tissue across several centimeters of pathlength and hence obtain spectroscopic information on blood in tissue noninvasively.

A very important application of oximetry is the co-oximeter. This instrument is a bench-top device that accepts blood samples drawn from a patient and automatically

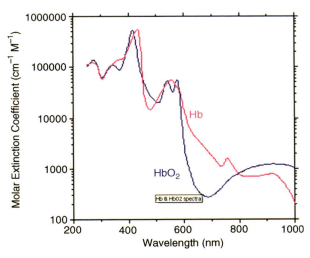

Figure 27.1 Molar extinction coefficient spectra of O_2Hb (HbO$_2$) and HHb (Hb). Reproduced from http://omlc.ogi.edu/spectra/hemoglobin/index.html (last accessed 3 May 2011).

obtains a visible transillumination spectrum at a number of wavelengths in the visible region. The Beer–Lambert law and multilinear regression are used to convert measurements of light transmission at distinct wavelengths into species concentrations by using the known molar extinction coefficient spectra of the species. Formally, the machine measures the incident and transmitted light intensities I_0 and I when transilluminating a standard pathlength l of blood. Expressing the attenuation in optical density (OD) units, then by the Beer–Lambert law:-

$$OD = \log_{10}\left(\frac{I_0}{I}\right) = \alpha l = (C_{HHb}\varepsilon_{HHb} + C_{O_2Hb}\varepsilon_{O_2Hb} + C_{COHb}\varepsilon_{COHb} + C_{MetHb}\varepsilon_{MetHb})l \tag{27.1}$$

where C_{HHb} is the concentration of HHb and ε_{HHb} is the molar extinction coefficient at the measurement wavelength, and similarly for the other species. Repeating the measurements at a series of wavelengths allows the above equation to be written in matrix form:

$$\mathbf{OD} = \varepsilon \cdot \mathbf{C} \tag{27.2}$$

where **OD** and **C** are column vectors of OD measurements and concentrations, respectively, and ε is a matrix of molar extinction coefficients, each row of which relates to a specific species and each column to a specific measurement wavelength. **C** is then obtained from **OD** by a suitable matrix inversion procedure. Evidently measurements are needed to at least two wavelengths if there are two species present in the sample. Although human blood is predominantly composed of either HHb or O_2Hb, in certain situations there can also be significant amounts of other forms, chiefly carboxyhemoglobin (COHb) and methemoglobin (MetHb). Elevated COHb is associated with carbon monoxide poisoning and leads directly to tissue hypoxia since CO has a binding affinity to hemoglobin that is over 200 times greater than that for O_2. Elevated methemoglobin (an abnormal form of hemoglobin in which the normally ferrous iron Fe^{2+} center in the heme group is replaced by ferric iron Fe^{3+}) indicates methemoglobinemia and again leads to hypoxia as MetHb is unable to bind reversibly to O_2. The co-oximeter therefore uses multiple measurement wavelengths to determine the concentrations of all these distinct hemoglobin species. A potential complication of the measurement is that the measured light attenuation is determined not just by absorption by the hemoglobin species but also by elastic light scattering caused by the red blood cell membranes. This is analogous to the light attenuation caused by fog, which is an aerosol formed by the dispersion of one nonabsorbing substance (water) in another (air). The co-oximeter avoids this problem by lysing the blood cells using detergents or sonication, thereby releasing the hemoglobin into nonscattering solution.

27.3
Pulse Oximetry

In 1942, Glenn Millikan in the United States developed the first practical *in vivo* oximeter. This device transilluminated the pinna of the ear using visible and infrared light and used mechanical pressure from a pneumatic bulb to exsanguinate blood from the tissue and hence collect a blood-free reference measurement. This aimed to estimate the contribution to light attenuation arising purely from tissue scattering. This device was used successfully in a number of physiological studies, especially in military aviation, where transient cerebral hypoxia was a major cause of blackouts experienced by fighter pilots engaged in high-G maneuvers. However, issues surrounding the reproducibility of the oxygenation values and also instrumental stability issues posed challenges to its widespread adoption in critical care medicine. In 1964, Hewlett-Packard commercialized an ear oximeter based on eight separate wavelengths which attempted to measure oxygenation *in vivo* by calibrating some of the scattering-induced background signals using measurements at these extra wavelengths. However, the bulk and expense of this machine meant that it has rarely been used outside specialist physiology laboratories. A major breakthrough occurred in 1975, when Nakajima *et al.* of Minolta Corporation demonstrated that most of the calibration and reproducibility problems generated by tissue scattering could be eliminated by making dynamic measurements of light attenuation between the systolic and diastolic phases of the cardiac cycle. A combination of the elastic compliance of arterial blood vessels and the large pulse pressure amplitudes in the arterial compartment means that the blood volume in this compartment shows measurable pulsatility at the cardiac frequency whereas the capillary and venous compartments do not. By measuring $\Delta OD = OD_{systolic} - OD_{diastolic}$, all contributions due to tissue scattering cancel, as do contributions due to capillary and venous hemoglobin absorption. This results in the determination of the relative proportions of O_2Hb and HHb in the arterial blood, that is, the arterial saturation SaO_2 can be derived. The great strengths of pulse oximetry are its simplicity, cost, and above all robustness and reliability in a clinical setting. Pulse oximetry is indicated in essentially all critical care and surgical situations and consequently is now ubiquitous in hospital-based medicine. To quote Severinghaus and Astrup: "Pulse oximetry is arguably the most significant technological advance ever made in [critical care] monitoring" [1].

27.4
Near-Infrared Spectroscopy and Imaging

Despite its enormous success, pulse oximetry does have some limitations:

- Only arterial saturation is measured, but venous saturation is also needed in order to estimate O_2 consumption.
- Scattering effects can still influence results indirectly via wavelength-dependent changes in the effective light pathlength and so empirical algorithms are frequently built in to pulse oximeters to obtain acceptable results.
- There is no direct spatial localization available, in contrast to imaging techniques for blood flow such as Doppler ultrasound.
- Only peripheral oxygenation is measured, common sites being the pinna of the ear or the finger-tip.

In 1977, Franz Jöbsis published a landmark paper in which he demonstrated that oxygen-dependent data could be obtained *in vivo* by transilluminating the intact skull of a cat using NIR light [2]. The original paper actually attempted to measure the oxidative phosphorylation pathway directly by detecting redox-dependent changes in the NIR absorption of cytochrome oxidase, a redox-active enzyme which forms part of the electron transport chain in the mitochondrial inner membrane. However, interest soon shifted to oximetry, in part because the inherent concentration of cytochrome oxidase in tissue is so low that hemoglobin absorption generally dominates over cytochrome oxidase absorption. In 1988, Cope and Delpy demonstrated the first clinically useable NIR spectrometer which used laser diodes and photomultipliers to provide deep tissue oxygenation data over 40 mm or more of transilluminated tissue [3]. This machine was soon commercialized by Hamamatsu Photonics as the NIRO-1000, and the field of clinical tissue near-infrared spectroscopy (NIRS) was born.

The NIRO-1000 was a "first-generation" NIRS system that did not provide an absolute value either for the total tissue hemoglobin concentration or the saturation. This was because the scattering offset was not removed by using pulsatile signals, as the aim of this device is to measure signals from all three compartments simultaneously. It is also impractical to obtain a blood-free reference using mechanical exsanguination on such large tissue volumes. Hence first-generation systems function as a trend monitor, that is, a baseline spectrum is recorded at a particular time and then the change in attenuation ΔOD relative to that baseline is recorded. The result of linear regression calculation is then the change in hemoglobin concentration in the tissue relative to the starting point. Unlike the pulse oximeter, the tissue NIRS system records the change in concentration of both species that is, it records $\Delta HHb(t)$ and $\Delta O_2Hb(t)$. It is therefore incapable of measuring the absolute level of hemoglobin saturation which is required to measure O_2 delivery. Worse still, in the absence of direct information on the photon pathlength l, it cannot quantify the magnitude of the change in absolute units such as mM; instead the units are unquantified mM cm. The physical distance between the light source and detector does not relate directly to the photon pathlength l through the tissue because intense scattering of photons causes them to travel in a quasi-random path from source to detector. The low volume fraction of blood in tissue (typically only a few percent) means that scattering is over 100 times more intense than absorption for NIR photons propagating through tissue. This is different to the situation in cuvette-based spectroscopy of nonscattering solutions. Fortunately, a number of technical solutions have been demonstrated that can provide "quantified trend" measurements by measuring the mean photon path length l. In 1988, Delpy *et al.* reported the first *in vivo* measurements of l (generally known as the differential pathlength) in human volunteers using a direct time-of-flight approach [4]. By illuminating the tissue with a picosecond pulse of light from a mode-locked titanium:sapphire laser and recording the transmitted light intensity as a function of time using a streak camera, the mean transit time of the pulse was determined. For a source–detector separation of 40 mm of tissue it was found that the photons had actually traveled over 200 mm, indicating a fivefold path lengthening. As explained, this arises because the photons do not travel directly from source to detector but follow a random path, with changes of direction occurring several times per millimeter. Shortly afterwards, Benaron *et al.* obtained similar measurements by using an intensity-modulated laser beam rather than a pulsed one [5]. By measuring the phase shift of the transmitted light relative to the incident light the mean transit time can again be derived. The phase method is generally cheaper and simpler to implement than the time-domain method. In 1994 Matcher, Cope and Delpy showed that pathlength could be obtained by measuring the apparent strength of the absorption band of water and comparing this with that of pure water [6]. Since the concentration of water in biological tissues is generally known, this provides information on the photon pathlength. At the time of writing, to the author's knowledge, none of these approaches has been implemented into a commercial NIRS system; however, reference values derived using these techniques in laboratory-based studies are incorporated into the software of most NIRS systems, so that quantified trend methods are routinely available.

Quantified trend monitors fall short of what clinicians ideally want in a critical care monitoring situation, namely a continuous absolute measurement of tissue saturation; however, quantitative hemodynamic measurements can be obtained indirectly. Blood flow can be inferred in several ways. In skeletal muscle, occlusion plethysmography can be performed using the NIRS-derived hemoglobin concentration as the volume measurement [7]. In the head, a modified clearance technique can be implemented in which O_2Hb itself is the tracer [8] and blood volume can be inferred using a similar idea: a dilution measurement is effectively performed using O_2Hb as the tracer [9]. Venous saturation can be inferred if a moderate venous occlusion is performed, again by using the plethysmographic principle and assuming that the resulting hemodynamic change arises solely through the pooling of venous blood [10]. A completely passive approach has been proposed in which respiratory fluctuations are used: changes in blood volume that occur at the respiratory frequency are assumed to arise solely from oscillations in venous blood volume because of the resulting fluctuations in venous pressure and the very high elasticity of the veins [11].

Despite these advances, the clinical uptake of NIRS has been impeded by the lack of a direct measurement of saturation. Responding to this challenge, both researchers and manufacturers have developed a number of approaches to the problem. Fundamentally, the difficulty arises because in a highly scattering medium the Beer–Lambert law no longer strictly applies, that is, the measurement (OD) is not a linear combination of the hemoglobin concentrations. Instead, two new parameters must be introduced, the tissue absorption and modified scattering coefficients μ_a and μ'_s. μ_a is linearly related to the hemoglobin concentrations; however, the measured OD is related in a complicated nonlinear way to both μ_a and μ'_s. In all cases, then, the approach to "absolute" oximetry involves attempting to determine directly the underlying absolute values of μ_a and μ'_s of the tissue given measurements of ODs. If μ_a is available at several wavelengths, then standard multilinear regression can be used to obtain the hemoglobin concentrations. Several distinct approaches have been introduced. Time-resolved spectroscopy uses the temporal distribution of transmitted light resulting from an incident picosecond light pulse to determine μ_a and μ'_s [12]. This distribution $I(t)$ is generally called the temporal point spread function (TPSF). Its mean value gives the differential path length already described. By comparing the TPSF with one predicted by a forward model of light transport in tissue, μ_a and μ'_s can be derived by least-squares fitting. This model is usually the diffusion equation. A variant of this idea is to use frequency-domain measurements. An intensity-modulated laser is used and the mean transit time and attenuation are measured for a variety of modulation frequencies [13]. One can show that this provides equivalent information to the time-domain approach. A third approach is to make measurements of light attenuation for a series of different source–detector spacings [14]. This approach is often referred to as spatially resolved spectroscopy (SRS) and has the major advantage of simplicity as only continuous-wave laser diodes and DC signal detectors are needed. Consequently, this approach has been commercialized in "second-generation" NIRS systems such as the Hamamatsu NIRO-100. A significant challenge for these machines is that they rely on a mathematical model of light transport through the tissue and this model will depend on the specific geometry of the tissue under study. In skeletal muscle studies, for example, the model must account for the thickness of overlying and weakly vascularized adipose layers whereas in the head the effects of extracranial tissues and the CSF layer are important. This points to a general area of research in tissue NIRS involving the need both to quantify and to localize the signals; the interested reader is directed elsewhere to a more extensive discussion of this field [15].

The need to combine NIRS with signal localization has spawned the related field of near-infrared imaging (NIRI). The use of a mathematical model to extract tissue absorption and scattering coefficients from measurements of light attenuation is taken to a greater level of sophistication by using measurements at a large number (tens to hundreds) of source–detector positions and then attempting to determine spatial maps of μ_a and hence oxygenation. This technology has evolved extensively over the last decade; however, routine clinical application has not yet been established. Spatially resolved trend monitors have proven to be valuable in functional studies of brain activation. This is a partial approach to imaging known as "optical topography," and seems to be capable of producing maps of cortical activation in response to a variety of cognitive and motor tasks similar to those obtainable by much more expensive and complicated techniques such as functional magnetic resonance imaging. The interested reader is directed to recent review articles [16].

27.5
New Approaches

No review of current approaches to noninvasive oximetry would be complete without a mention of emerging technologies such optical coherence tomography, hyperspectral imaging, and photoacoustic imaging. All of these techniques provide much higher spatial resolution than NIRS and NIRI at the expense of tissue penetration depth, and the emphasis is on physiological studies rather than clinical monitoring (although photoacoustic imaging has potential for clinical monitoring).

Optical coherence tomography (OCT) is an optical analog of ultrasound imaging with lower tissue penetration (<2 mm) but higher resolution (2–10 μm). Conventional OCT derives contrast from tissue boundaries, similarly to ultrasound; however, variants have been proposed to provide molecular contrast to hemoglobin and other species. The challenge is that the absorption coefficient of hemoglobin is very low in the NIR region, so the contrast generated over the very short pathlengths that characterize OCT is low. One potential solution to this is "pump–probe" OCT, in which transient changes in μ_a are induced by saturating the molecular ground state of the molecule using short pulses of pump radiation [17]. This pump light can lie in the visible range, whereas the imaging (i.e., "probe") OCT beam can be in the infrared region, thus retaining good depth penetration and resolution. This technology is interesting but has yet to demonstrate a clinical impact; the next few years could well clarify the situation.

Interest in the use of the visible spectrum has seen resurgence in recent years through the introduction of multispectral and hyperspectral cameras which can produce 3D data cubes whereby a 2D image gains a third dimension of wavelength information. 2D real-time images of hemoglobin saturation can thus be created with cellular level resolution. The main applications of this technology currently relate to basic physiological studies in animal models; in particular, the hemoglobin saturation in tumor vasculature can be determined [18]. Hypoxia is a characteristic of the tumor microenvironment because of a mismatch between metabolic demand from the hyperproliferating cells and blood supply. It is therefore important to monitor the effects of pharmacological treatments such as vasculature-targeting combretastatins on this parameter.

Photoacoustic imaging is a rapidly emerging new modality which effectively gives ultrasound imaging sensitivity to molecular absorption. The technique locates areas of high optical absorption, such as a blood vessels, by causing rapid and transient heating using a nanosecond pulse of light whose wavelength matches a peak in the absorption spectrum of compound. The rapidity of this heating and the resulting mechanical expansion of the surrounding water generate a pulse of ultrasound that can be detected on the tissue surface using a conventional ultrasound hydrophone. The amplitude of the ultrasound relates directly to the magnitude of the absorption coefficient and so, by tuning the pulsed light to various NIR wavelengths, the hemoglobin saturation can be determined [19]. A great advantage of this technique is that the arrival time of the ultrasound directly encodes the depth of the blood vessel and so an image can be formed as in conventional ultrasound. The major limitation of optical imaging of biological tissues is poor resolution and depth penetration, which arise because of intense light scattering. Photoacoustic imaging uses optical radiation only as a source of heating, image information is carried by the acoustic wave, and so resolution and depth penetration are fundamentally superior. This technology holds great promise to deliver clinically useful images of tissue oxygen delivery with a resolution matched to the dimensions of the human vasculature. The next few years will doubtless bring many interesting new developments in this area.

27.6
Conclusion

The routine clinical application of oximetry is now over 70 years old and still there is great scope for continued technological innovation and new clinical applications. The co-oximeter and pulse oximeter are standard equipment in the hematology laboratory and intensive care room, respectively, and the advent of second-generation NIRS systems that can measure absolute saturation on the brain and skeletal muscle is pushing this technology into the clinic. It should be noted, however, that debate still exists about whether NIRS technology is currently delivering clinically relevant information, owing to ongoing issues surrounding reproducibility and stability [20]. The combination of oximetry with spatially resolved mapping of the cerebral cortex can provide a viable alternative to functional magnetic resonance imaging (fMRI) in some cases, especially where the use of fMRI is impractical, such as in the neonate. New variants on old methods, especially hyperspectral imaging, are finding applications in nonclinical studies of tumor models and may also find clinical applications in, for example, studies of nonhealing wounds and burns in the skin.

Imaging technologies such as OCT and photoacoustic imaging can also be adapted to provide oximetry data. Photoacoustic imaging in particular could develop into a tool with a unique combination of spectral and spatial resolution combined with depth penetration.

References

1 Severinghaus, J.W. and Astrup, P.B. (1986) History of blood gas analysis. VI. Oximetry. *J. Clin. Monit.*, **2** (4), 270–288.

2 Jöbsis, F.F. (1977) Non-invasive infrared monitoring of cerebral and myocardial oxygen sufficiency and circulatory parameters. *Science*, **198** (4323), 1264–1267.

3 Cope, M. and Delpy, D.T. (1988) System for long-term measurement of cerebral blood and tissue oxygenation on newborn infants by near-infrared transillumination. *Med. Biol. Eng. Comput.*, **26**, 289–294.

4 Delpy, D.T., Cope, M., van der Zee, P., Arridge, S.R., Wray, S., and Wyatt, J. (1988) Estimation of optical pathlength through tissue from direct time of flight measurement. *Phys. Med. Biol.*, **33** (12), 1433–1442.

5 Benaron, D.A., Gwiazdowski, S., Kurth, C.D., Steven, J., Delivoria-Papadopoulos, M., and Chance, B. (1990) Optical pathlength of 754 nm and 816 nm light emitted into the head of infants. *Proc. IEEE Eng. Med. Biol.*, **12**, 1117–1119.

6 Matcher, S.J., Cope, M., and Delpy, D.T. (1994) Use of the water absorption spectrum to quantify tissue chromophore concentration changes in near-infrared spectroscopy. *Phys. Med. Biol.*, **39** (1), 177–196.

7 De Blasi, R.A., Ferrari, M., Natali, A., Conti, G., Mega, A., and Gasparetto, A. (1994) Noninvasive measurement of forearm blood flow and oxygen consumption by near-infrared spectroscopy. *J. Appl. Physiol.*, **76** (3), 1388–1393.

8 Edwards, A.D., Wyatt, J.S., Richardson, C., Delpy, D.T., Cope, M., and Reynolds, E.O. (1988) Cotside measurement of cerebral blood flow in ill newborn infants by near-infrared spectroscopy. *Lancet*, **ii** (8614), 770–771.

9 Wyatt, J.S., Cope, M., Delpy, D.T., Richardson, C., Edwards, A.D., Wray, S., and Reynolds, E.O. (1990) Quantitation of cerebral blood volume in human infants by near-infrared spectroscopy. *J. Appl. Physiol.*, **68** (3), 1086–1091.

10 Yoxall, C.W., Weindling, A.M., Dawani, N.H., and Peart, I. (1995) Measurement of cerebral venous oxyhemoglobin saturation in children by near-infrared spectroscopy and partial jugular venous occlusion. *Pediatr. Res.*, **38** (3), 319–323.

11 Franceschini, M.A., Zourabian, A., Moore, J.B., Arora, A., Fantini, S., and Boas, D.A. (2001) Local measurement of venous saturation in tissue with non-invasive near-infrared respiratory-oximetry. *Proc. SPIE*, **4250**, 164–170.

12 Patterson, M.S., Chance, B., and Wilson, B.C. (1989) Time resolved reflectance and transmittance for the noninvasive measurement of tissue optical properties. *Appl. Opt.*, **28**, 2331–2336.

13 Fantini, S., Franceschini, M.A., Maier, J.S., Walker, S.A., Barbieri, B., and Gratton, E. (1995) Frequency-domain multichannel optical detector for noninvasive tissue spectroscopy and oximetry. *Opt. Eng.*, **34**, 32–42.

14 Matcher, S., Kirkpatrick, P., Nahid, K., Cope, M., and Delpy, D.T. (1994) Absolute quantification methods in tissue near infrared spectroscopy. *Proc. SPIE*, **2389**, 486–495.

15 Matcher, S.J. (2002) Signal quantification and localization in tissue near-infrared spectroscopy, in *Handbook of Optical Biomedical Diagnostics* (ed. V. Tuchin), SPIE Press, Bellingham, WA, pp. 487–584.

16 Wolf, M., Ferrari, M., and Quaresima, V. (2007) Progress of near-infrared spectroscopy and topography for brain and muscle clinical applications. *J. Biomed. Opt.*, **12** (6), 062104.

17 Applegate, B.E. and Izatt, J.A. (2007) Molecular imaging of hemoglobin using ground state recovery pump–probe optical coherence tomography. *Proc. SPIE*, **6429**, 64291Y.

18 Sorg, B.S., Moeller, B.J., Donovan, O., Cao, Y., and Dewhirst, M.W. (2005) Hyperspectral imaging of hemoglobin saturation in tumor microvasculature and

tumor hypoxia development. *J. Biomed. Opt.*, **10** (4), 044004.

19 Laufer, J., Delpy, D.T., Elwell, C., and Beard, P. (2007) Quantitative spatially resolved measurement of tissue chromophore concentrations using photoacoustic spectroscopy: application to the measurement of blood oxygenation and haemoglobin concentration. *Phys. Med. Biol.*, **52**, 141–168.

20 Greisen, G. (2006) Is near-infrared spectroscopy living up to its promises? *Semin. Fetal Neonatal Med*, **11** (6), 498–502.

28
Photonic Therapies
Igor Peshko

28.1
Atrial Fibrillation

Atrial fibrillation (AF) is a heart-rhythm abnormality, which involves irregular and rapid heartbeats. During AF, irregular electrical impulses cause erratic and incomplete activation of the atria. Multiple stray currents (shown as round arrows in Figure 28.1a) result in chaotic short pulses that activate the heart muscle. The surgical maze procedure was the first treatment that offered a permanent solution for maintaining a normal sinus rhythm in patients with AF [1]. This procedure involves the creation of a maze-like series of incisions in the atrium, resulting in isolation of the pulmonary vein (Figure 28.1b). The scarring resulting from the incisions permanently blocks (isolates) the path of the erratic electrical impulses and prevents the arrhythmia. From an "electrical point of view", it does not matter which technology is used to make an incision. The incisions may be done mechanically (scalpel, ultrasound), by strong cooling (cryosurgery), or by strong heating (radio-frequency or microwave irradiation) (see review in [2]). Visible, near-infrared, or mid-infrared light can also be successfully used to initiate the scars. The optical power is easily delivered to any target, through an optical fiber, and can be focused to an area with a size spanning microns due to the small wavelength of the light. To provide an effective AF treatment, one needs a light source/catheter system that will create deep, homogeneous, electrophysiologically inactive scars with a controllable lesion size and a clear-cut lesion border. The fiber photocather (FPC) addresses all these requirements [3].

During the last decade, the most successful medical laser applications have been non- or less-invasive: laser trabeculoplasty, hair removal, photodynamic therapy, skin rejuvenation and cosmetology, laser treatments of psoriasis and acne, and laser acupuncture. Laser-based diagnostic tools, such as optical coherent tomography, nonlinear spectroscopy and microscopy, have become routine techniques. Any treatment that was provided with laser technologies can now be done with FPCs, with the added benefit of being used in parts of the human body that are difficult to access. In general, optical fiber technologies can be used for optical power delivery, to

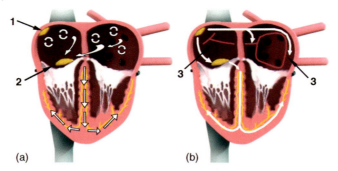

Figure 28.1 Simplified diagrams of the heart with: (a) atrial fibrillation and (b) after the treatment. 1, Sinus node; 2, atrioventricular node; 3, scars, blocking abnormal paths.

act as sensors, and as multifunctional devices controlling the laser's radiation and transferring the optical data.

28.1.1
Fiber Photocatheters

28.1.1.1 What is a Fiber Photocatheter?

A regular catheter is supposed to deliver liquid drugs, dyes, or some microinstrumentation that cannot be directly injected into the tissue. An FPC delivers light power, some optical signals, and/or contains micro-optical devices. Depending on the level of optical power, catheters may be roughly grouped into three main categories: (a) high power delivery (watts to tens of watts): scalpel-free surgery and light therapy; (b) middle (milliwatts to hundreds of milliwatts): precise microsurgery and "soft" light treatment; (c) low (microwatts to hundreds of microwatts): imaging, spectroscopy, microscopy, and so on.

The "photo" in FPC also refers to the fact that optical technologies are used for the fabrication of some parts of the catheter, such as long-period gratings, which can be photoprinted with point-by-point or interference-pattern technologies.

Historically, the first FPCs were applied in therapeutic treatments [4]. The uncoated fiber, which was mechanically scratched and/or chemically thinned, was covered with a special paste containing small particles that effectively scattered the radiation delivered by a fiber [5]. This type of FPC is still in use and is commercially available. However, this FPC is rigid, has a limited length of about 2–5 cm, and is fairly thick – 1–2 mm in diameter. An optically fabricated flexible FPS has been developed for use in photodynamic therapy [4] (Figure 28.2) and surgical treatment of atrial fibrillation with laser radiation [3, 6].

Photocatheters, being fabricated from inert materials such as quartz, Teflon, silicone rubber, and stainless steel, are ideally matched with medical requirements of sterility. Being fabricated from thin fibers (200–50 μm core), catheters are flexible enough to follow curved surfaces of typical tissue or blood-vessel "labyrinths" (inset in Figure 28.2). This makes it possible to act precisely in the intended area without

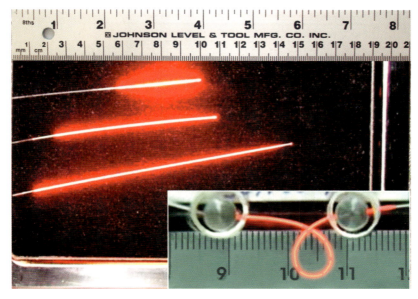

Figure 28.2 Photograph demonstrating FPCs of length 4, 9, and 14 cm. The inset shows a coiled FPC ~1.5 cm in diameter.

affecting surrounding tissue. Moreover, the FPC may be imprinted anywhere along the fiber, not only at the fiber end as a classical diffuser.

28.1.1.2 Fiber Photocatheter Design

A photocatheter with a flexible side-emitting optical fiber is based on specific long-period gratings (LPGs) [3, 4, 6] or Bragg gratings (BGs) [7] recorded in the core area of the fiber. These gratings produce a refractive index modulation at the fiber core–cladding interface that disrupts the total internal reflection conditions in the fiber core and enables the light traveling along the fiber to be scattered sideways. These LPGs or BGs operate as diffusers. The significant advantage of the proposed technology is that the output light intensity profile (along the catheter) is preliminarily designable [8]. It may be flat, dashed, decreasing, increasing, or any special shape. By varying the refractive index modulation value, density, and period of the gratings, it is possible to control the energy scattered along the entire length of the diffuser. For constant longitudinal radiant emission, the remaining power inside the fiber core decreases linearly with a corresponding increase in the strength of scattering as a function of the length of the grating. The local transmittance of each diffuser segment falls exponentially, reaching a minimum at the diffuser's end.

Figure 28.3 shows an example of the catheter design and demonstrates the photocatheter prototype [8]. The system includes: an up to 15 cm long, 50–400 μm core fiber diffuser (1); a 20–50 W laser diode; a flexible elliptical or cylindrical reflector (2); an end fiber cooler; a series of openings (3) for rapid self-attachment to the tissue (vacuum holder–gripper); a closed-loop irrigating chamber with circulating coolant

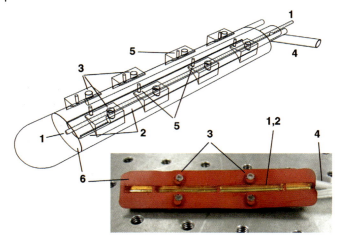

Figure 28.3 Design of the FPC for the atrial fibrillation surgical treatment: 1, Fiber with long period gratings; 2, reflector; 3, suction openings; 4, irrigation channel; 5, thermocouples. The photograph shows the FPC prototype.

(4) to cool the optical diffuser; and thermocouples for temperature control (5). Two or more light sources may be connected to the FPC with scattering of different wavelengths (colors) at different diffuser areas. The reflector is used to condense optical power on the irradiated tissue; alternatively, an LPG, recorded with a slope to reflect power to one side of the catheter, can be used. The fiber catheter body (6) was cast from high-temperature silicone rubber and was then gold coated to form a reflector.

28.2
All-in-One Endoscope Devices

An optical system can be installed at the end of a fiber, effectively transforming the fiber into a remote microscope, an optical tomograph, a spectrometer, or another endoscope instrument. A few examples of optical diagnostic catheters are mentioned below.

1) The feasibility of a time-resolved fluorescence spectroscopy technique (TRST) for intraluminal investigation of arterial vessel composition under intravascular ultrasound has been demonstrated [9]. A catheter combines a side-viewing optical fiber and an intravascular ultrasound catheter. The catheter can locate a fluorophore in the vessel wall, steer the fiber into place, perform blood flushing under flow conditions, and acquire high-quality TRFS data using ultraviolet wavelength excitation. This system has the potential for *in vivo* arterial plaque composition identification using TRFS.
2) A two-axis scanning catheter was developed for 3D endoscopic imaging with spectral domain optical coherence tomography (SD-OCT) [10]. The catheter

incorporates a micromirror scanner implemented with microelectromechanical systems (MEMS) technology. A micromirror is mounted on a two-axis gimbal comprised of folded flexure hinges and is actuated by a magnetic field. The catheter was incorporated with a multifunctional SD-OCT system for 3D endoscopic imaging.

3) An optical coherence tomography catheter–endoscope for micrometer-scale, cross-sectional imaging in internal organ systems has been demonstrated [11]. The catheter–endoscope uses single-mode fiber optics with a transverse scanning design. The distal end of the catheter–endoscope uses a gradient-index lens with a microprism to emit and collect a single spatial-mode optical beam with specific focusing characteristics. The beam is scanned in a circumferential pattern and can image transverse cross-sections through the structure into which it is inserted. An imaging of *in vitro* human venous morphology was demonstrated. A technique for performing an optical biopsy (high-resolution micrometer-scale cross-sectional optical imaging) of tissue architectural morphology, without the need to excise tissue specimens, would have a powerful effect on the diagnosis and clinical management of many diseases.

4) A one-dimensional optical fiber-based imaging catheter, specifically developed for the atherosclerotic plaque detection of emerging novel near-infrared fluorescence imaging agents, has been demonstrated [12]. It was shown that femtomole amounts of fluorochromes can be detected, especially in the presence of a blood-free medium. To detect the fluorescent molecular tags reliably, a dual-wavelength catheter-based system, using different fiber designs tuned to the near-infrared range, has been developed. The ability of the catheter to detect fluorescence signals in human carotid atherosclerotic plaque has also been demonstrated.

A multifunctional device may contain both active (optical surgical instruments) and passive (information channels) components. Devices that provide monitoring, diagnostics, and treatment by the same all-in-one device are currently being developed.

References

1 Cox, J.L., Schuessler, R.B., D'Agostino, H.J. Jr., Stone, C.M., Chang, B.C., Cain, M.E., Corr, P.B., and Boineau, J.P. (1991) The surgical treatment of atrial fibrillation, III. Development of a definitive surgical procedure. *J. Thorac. Cardiovasc. Surg.*, **101**, 569–583.

2 Song, H.K. and Puskas, J.D. Recent advances in surgery for atrial fibrillation, http://www.ctsnet.org/sections/newsandviews/specialreports/article-1.html (last accessed 20 April 2011).

3 Peshko, I., Rubtsov, V., Vesselov, L., Sigal, G., and Laks, H. (2007) Fiber photo-catheters for laser treatment of atrial fibrillation. *Opt. Laser Eng.*, **45**, 495–502.

4 Lilge, L., Vesselov, L.M., and Whittington, W. (2005) Thin cylindrical diffusers created in multimode Ge-doped silica fibers. *Lasers Surg Med.*, **36**, 245–251.

5 Fried, N.M., Lardo, A.C., Berger, R.D., Calkins, H., and Halperin, H.R. (2000) Linear lesions in myocardium created by Nd:YAG laser using diffusing optical fibers: *in vitro* and *in vivo* results. *Laser Surg. Med.*, **27**, 295–304.

6 Rubtsov, V., Laks, H., Levin, M., Akbarian, M., and Peshko, I. (2008) Fiber optic tissue ablation. U.S. Pat. Appl. US 2008/0082091 A1 3 April 2008.

7 Gu, X. and Tam, R. (2002) Optical fiber diffuser. U.S. Pat. US 6398778, filed 18 June 1999, issued 4 June 2002.

8 Peshko, I. and Rubtsov, V. (2007) Fiber photo-catheters with spatially modulated diffusers for laser treatment of atrial fibrillation. *Proc. SPIE*, **6424**, 642421.

9 Stephens, D.N., Park, J., Sun, Y., Papaioannou, T., and Marcu, L., (2009) Intraluminal fluorescence spectroscopy catheter with ultrasound guidance. *J. Biomed. Opt. Lett.*, **14** (3), 030505.

10 Kim, K.H., Park, B.H., Maguluri, G.N., Lee, T.W., Rogomentich, F.J., Bancu, M.G., Bouma, B.E., de Boer, J.F., and Bernstein., J.J. (2007) Two-axis magnetically-driven MEMS scanning catheter for endoscopic high-speed optical coherence tomography. *Opt. Express*, **15** (26), 18130–18140.

11 Tearney, G.J., Boppart, S.A., Bouma, B.E., Brezinski, M.E., Weissman, N.J., Southern, J.F., and Fujimoto., J.G. (1996) Scanning single-mode fiber optic catheter–endoscope for optical coherence tomography. *Opt. Lett.*, **21** (7), 543–545.

12 Zhu, B., Jaffer, F.A., Ntziachristos, V., and Weissleder., R. (2005) Development of a near infrared fluorescence catheter: operating characteristics and feasibility for atherosclerotic plaque detection. *J. Phys. D: Appl. Phys.*, **38**, 2701–2707.

Part 5
Pulmonology

29
Bronchoscopy

Lutz Freitag, Kaid Darwiche, and Dirk Hüttenberger

29.1
Diagnostics

The first report about a direct inspection of the airways of a living human being was published in 1897 [1]. Gustav Killian, an ENT surgeon had removed an aspirated chicken bone from the trachea of a farmer. For this first direct laryngoscopy he had used a rigid tube, inserted into the airways under local anesthesia. The next step was the development of angled rigid telescope lenses in combination with external light sources. This enabled physicians to detect abnormalities in lobar bronchi. Today, rigid bronchoscopy is performed mainly for therapeutic purposes. Interventional procedures, for example, tumor resection and stent insertion, are done faster and more safely with rigid bronchoscopes under general anesthesia and jet ventilation. In 1966, the Japanese thoracic surgeon Shigeto Ikeda built the first flexible fiber-optic bronchoscope. Every pulmonologist in the world knows his slogan, "more hope with the bronchoscope." In the following years, light sources became brighter, photo- and video-cameras were added, and the number of fibers and consequently the resolution were improved. Within the last 10 years, fiber bronchoscopes have been replaced by video-chip bronchoscopes. Figure 29.1 demonstrates the tremendous progress in image quality. The resolution of a routine instrument is so good that even the smallest surface abnormalities and also vessel distortions are visualized. Flexible bronchoscopes are easy to handle, and they have working channels that allow all kinds of optical probes, catheters, brushes, and biopsy forceps to pass. Therapeutic instruments such as laser fibers, electrocautery, and cryo-probes can be guided to targets down to subsegmental bronchi. The latest progress that has extended the use of bronchoscopy is the incorporation of an ultrasound sector scanner. With a modern endobronchial ultrasound (EBUS) bronchoscope, one can not only see the inner surface of the airways but also visualize adjacent anatomic structures. EBUS has revolutionized lung cancer staging. It has become common practice to puncture hilar and mediastinal lymph nodes under ultrasound guidance. Even small forceps biopsies from nodes can be taken (Figure 29.1c). Ideally, the aspirated cells are examined by the pathologist directly in the bronchoscopy suite. By combining flexible

Handbook of Biophotonics. Vol.2: Photonics for Health Care, First Edition. Edited by Jürgen Popp,
Valery V. Tuchin, Arthur Chiou, and Stefan Heinemann.
© 2012 Wiley-VCH Verlag GmbH & Co. KGaA. Published 2012 by Wiley-VCH Verlag GmbH & Co. KGaA.

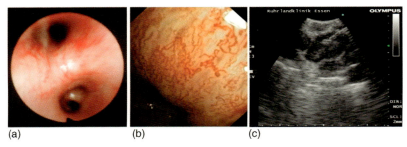

(a) (b) (c)

Figure 29.1 Progress of white light bronchoscopy. (a) Fiber-optic bronchoscopy image from 1995; (b) routine video-chip bronchoscope image from 2010; (c) lymph node puncture with endobronchial ultrasound (EBUS) bronchoscope.

bronchoscopy, EBUS, and on-site cytology endoscopic diagnosis, tumor type determination and lymph node staging are accomplished within less than 30 min.

29.1.1
Cancer Detection

Pulmonologists performing bronchoscopy are dealing with one of the most devastating diseases at all, lung cancer. It is estimated that 1.2 million people die every year from this cancer, which is mainly caused by cigarette smoking. Despite all medical progress, the prognosis remains poor because most patients are diagnosed at a stage when curative treatment is no longer possible. Screening and early detection programs are the only hope unless the tobacco industry decides to go out of business voluntarily. A typical squamous cell cancer in a lobar bronchus develops over a time span of more than 15 years. Without causing any symptoms during the first 12 years, it evolves from a tiny dysplastic lesion to a carcinoma *in situ* and then to a mass tumor that can occlude a bronchus and induce metastasis in other organs. Within the next few years the patient becomes incurable. Tumor-associated symptoms such as hemoptysis (coughing up blood) or shortness of breath that would indicate a possible tumor presence occur only in advanced stages. This long period of asymptomatic tumor growth leaves only a small time window for diagnosis. We do not know why many but not all dysplasias progress to squamous cell lung cancers. There are other premalignant changes that can result in deadly tumors. Adenocarcinomas develop from atypical adenomatous hyperplasias. They are usually located more peripherally than squamous cell cancers. Diffuse idiopathic pulmonary neuroendocrine cell hyperplasias (DIPNECH) advance into carcinoids. The technical challenge is to diagnose these tumors preferably in their earliest stages when curative treatment is still possible. Screening programs include computed tomography (CT), sputum cytology, use of molecular markers, and analysis of exhaled breath. If any of these methods indicates the presence of an early cancer, patients are administered bronchoscopy. If the tumor mass is still low, cancer formations are not seen on chest X-rays or CT scans. Bronchoscopy, mostly with a flexible endoscope under mild conscious sedation and local anesthesia, is used to find the tumor spot and take a biopsy for pathologic examination. Older studies aiming to find radio-occult cancers with white light bronchoscopy had sensitivities lower than 30%, and the detection rates of early invasive cancers were in the region of 60% [2, 3]. However, judging from the results, it is important to realize that these findings were made with fiber-optic scopes with limited resolutions, as shown in Figure 29.1a. It is reasonable to assume that detection rates of early lesions might be higher with current state-of-the-art video-chip endoscopes. Using white light bronchoscopy, early lesions do not differ much in color from normal mucosa and they are especially difficult to find in the presence of mild inflammation that is typical for smokers. There are two main approaches aimed at increasing the contrast between preneoplastic, neoplastic, and normal tissue: filtering parts of the remitted white light or visualization of fluorescence on excitation with non-white light.

29.1.2
Filter-Based Techniques for Tumor Detection

Early stages of tumor development are characterized by changes in cell composition, tissue layer thicknesses, and neo-vascularization. The tumor demands nutrition and it satisfies its need by stimulating angiogenesis. Blood vessels in the mucosa change shape, pattern, and density. This phenomenon can be used for diagnostic purposes with various optical techniques considering that blood has high absorption properties [4]. State-of-the-art light sources are xenon arc lamps. The bright white light is passed through an infrared light filter and transmitted to the bronchoscope via a fiber-optic bundle or a liquid light guide. Under normal conditions, the remitted light is picked up by a color chip camera and the image is displayed on a video screen as unaffected as possible. Vessels appear red on a pale mucosal background. If the remitted light is filtered properly, absorption differences can be emphasized.

To enhance the contrast of the mucosal microvasculature and possible alterations, only parts of the white light spectrum are used for illumination [5–8]. This is accomplished by placing a narrowband filter in front of the white light source. Basically, the absorption properties of blood are used in order to increase the diagnostic yield. One popular device is the EVIS EXERA narrowband imaging (NBI) system from Olympus (Tokyo, Japan). Figure 29.2 shows the enhanced contrast of vascular abnormalities if the system is switched from regular mode to NBI mode. The NBI technique is a well-known method in photography, especially in astrophysics, where the scattered light (for example, from street lights) is suppressed by narrow filters in the red, green, and blue. Emitted light from stars, with a characteristic spectrum, can be selected. For bronchoscopic use, the filter bands are selected to produce the highest contrast between vessels and surrounding tissue. In the aforementioned NBI system, filters are used in the blue and green spectral range. The different depths of penetration of light components lead to different signals. Blue light is absorbed by superficial capillaries and green light is absorbed by blood vessels beneath the mucosa. One main absorption peak of hemoglobin can be found at 415 nm. Structures with high blood content appear

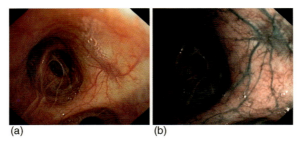

Figure 29.2 (a) White light bronchoscopic image and (b) narrowband image using the same endoscope (Olympus NBI). Mucosal and submucosal vessels and abnormalities become clearly visible.

darker than the surrounding mucosa that reflects the light. Deeper vessels are seen as green light at a secondary absorption maximum at 540 nm. In the digital camera system, the 415 nm signal is linked to the blue and green channels; the superficial vessels are displayed in brown. The red channel of the camera system monitors the reflected light of the green illumination and shows the deeper vessels in cyan. The filters have a bandwidth of about 30 nm. Vessel accumulation, pattern changes such as corkscrew distortions, abrupt stops, or caliber changes are indications of malignant transformations. A clinical study comparing autofluorescence bronchoscopy with narrowband imaging bronchoscopy in 62 patients showed a higher specificity for the filter-based technique [9].

A further step is combination with a magnifying endoscope. Truly microvascular changes are visualized and can be used for diagnostic purposes. In a recent study in 48 high-risk patients, an experimental setup of a high-magnification bronchovideoscope coupled to a narrowband imaging system was used. Angiogenic squamous dysplasia with capillary loops could be clearly differentiated from other premalignant lesions [10].

There have been further developments of band filter techniques. In a three-channel system, the blue channel is linked to 415 nm, the green channel to 445 nm and the red channel to 500 nm. Working with filters is not the only approach. In contrast to NBI, the FICE system (Fuji Intelligent Color Enhancement, Fujinon, Saitama, Japan) uses multi-band imaging (MBI). No optical filters are placed between the light source and the endoscope. A software-driven spectral estimation algorithm calculates and displays single wavelengths of a charge-coupled device (CCD) camera picture [11]. Up to 60 wavelength channels can be selected and varied to enhance the contrast. MBI creates a brighter image than NBI.

For practical purposes, all systems can be easily switched between white light and enhancing mode. In daily practice, the diagnostic yield depends greatly on the experience and concentration of the bronchoscopist. Good pattern recognition ability is required. Blood vessels that are typical for an early tumor are "different." They have a different pattern but they do not show up in easily recognizable color differences such as we see in competing techniques that use fluorescence abnormalities.

29.1.3
Fluorescence Techniques, Drug Induced Fluorescence

More or less accidentally, substances with fluorescent properties have been found that have higher affinities to neoplastic tissues. The enrichment of these substances in cancer can be due either to higher uptakes or to defects in cellular clearance mechanisms. Several hours after intravenous injection, such a fluorophore remains at a detectable concentration in tumor cells while it has already been cleared from normal tissue. Another possible mechanism is a metabolic difference of cancer cells, for example, an uncontrolled synthesis without competitive inhibition. Typical exogenous fluorophores for tumor detection are porphyrin based. Injected hematoporphyrins are retained in cancer tissue and this can be visualized using their fluorescence. Aminolevulinic acid (ALA) and its derivatives are non-fluorescent precursors of protoporphyrin IX in the heme biosynthetic pathway. After administration, for example as an aerosol, ALA is taken up by tumor cells and converted into a bright fluorescent porphyrin. These drugs have been used for the detection of lung cancer in early stages. Experimental systems in the 1970s required expensive and bulky krypton or helium–cadmium lasers to illuminate the bronchial mucosa with violet light. Bandpass filters with photodiodes and later spectroscopic devices were used to detect the typical fluorescent peak at 630 nm indicating the presence of porphyrins in neoplastic lesions [12–14]. When highly sensitive video cameras became available, images could be obtained and these devices were used for the detection of radio-occult tumors in clinical studies [15]. In order to facilitate interpretation, false-color techniques were used.

Eventually, the lasers could be replaced by mercury or xenon arc lamps and small spectrometers or CCD video cameras were attached to the bronchoscopes. Following Photofrin injection or ALA inhalation, flat tumors unrecognizable to the naked eye can be visualized. Figure 29.3 shows examples of such a system for early lung cancer detection that we used until the late 1990s. A major hurdle for daily practice is the skin sensitivity associated with the injection of the photoactive drugs (see later in this chapter). Attempts were made to lower the amount of the drugs in order to avoid these side effects while still obtaining sufficient tumor signals. The "noise" that limits the

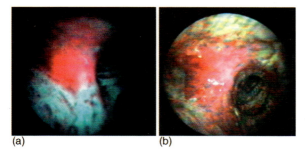

Figure 29.3 Drug induced fluorescence. The tumor lesion shows up bright red 30 min after inhalation of ALA (a) and in another patient 24 h after intravenous injection of Photofrin (b).

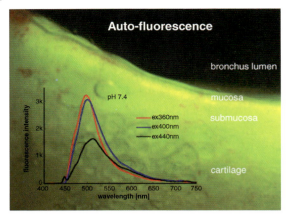

Figure 29.4 Autofluorescence of bronchial tissue. Blue light comes from below, perpendicular to bronchoscopy condition.

dose reduction is autofluorescence. It is well known that biological tissue shows a naturally occurring autofluorescence when excited by UV or visible blue light illumination (about 200–460 nm). Of course, only the visible and harmless range is suitable for medical applications. It was already known that malignant tissue yielded lower fluorescence than normal tissue when excited with UV light [16, 17]. On excitation with blue light (400–450 nm), bronchial tissue appears green when observed through a yellow filter, as shown in Figure 29.4. Like many other groups, we have studied the optical properties of normal and lung cancer cells and tissues with and without photosensitizers (Figure 29.5). Almost by serendipity it was found that there are significant differences in the autofluorescence intensities and compositions of normal bronchial tissue and early cancers on excitation with blue light (404–450 nm) used for porphyrin detection. Figure 29.6 shows the autofluorescence of a normal mucosa and the attenuated signal remitted from a tumor-affected area. The green signal drops considerably more than the red signal. The ratio can be used

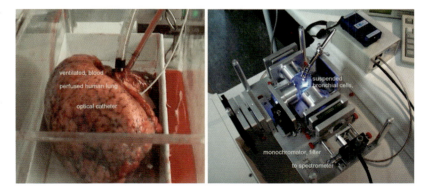

Figure 29.5 Laboratory tests for studying optical properties of bronchial and lung tissue.

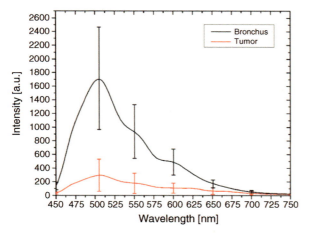

Figure 29.6 Autofluorescence of normal and tumor-affected bronchial mucosa.

for tumor detection. Suddenly there was no longer any need a for exogenous sensitizers with possible side effects. Lung cancer detection systems based on abnormalities of endogenous autofluorescence were developed.

29.1.4
Fluorescence Techniques, Auto-Fluorescence

Most fluorophores that are present in bronchial and lung tissue have been identified [18]. The strongest signal comes from flavins: flavin mononucleotide (FMN) and flavin adenine dinucleotide (FAD), with excitation at 530 nm. Elastin and collagen are architectural proteins with a yellow–green fluorescence giving the background fluorescence in most applications (excitation from 300 to 480 nm). Other optically active substances are the already mentioned porphyrins, which are important in the heme biosynthetic pathway (emission at 610–620 nm). The redox state of the tissue determines how much NADH is present: NADH (excitation at 340 nm/emission at 435 nm), NADPH (excitation at 360 nm/emission at 460 nm). In the deprotonated form (NAD/NADP), the molecules are not fluorescent. Absorption and emission spectra of human tissue are shown in Figures 29.7 and 29.8.

There are many autofluorescence-based systems available for lung cancer detection. The original LIFE system (Xillix Technologies, Richmond, BC, Canada) consisted of a helium–cadmium laser (442 nm) and two image-intensified BW cameras with green (520 nm) and red (>630 nm) filters hanging from the ceiling because they were too heavy to be carried by the physician. The clinical results were very encouraging. The detection rate for moderate and severe dysplasia increased from 38.5 and 40% with white light bronchoscopy to 73.1 and 91.4%, respectively, with the LIFE system [19]. Today, systems from all leading companies use modified xenon lamp white light sources and video-chip bronchoscopes. For the end user, these devices have become very convenient and most pulmonary departments have an

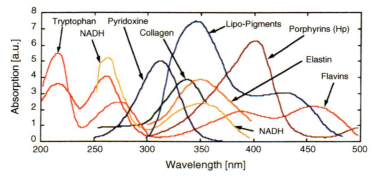

Figure 29.7 Absorption spectra of fluorophores in human tissue.

autofluorescence bronchoscope for clinical routine. However, despite the established application, some questions about the origin of autofluorescence remain unanswered. There are two main explanations for the decreased intensity in tumor tissue in comparison with normal tissue. On the one hand, the tumor has an increased epithelial thickness and therefore has a stronger absorption of excitation and fluorescence light. This architectural effect was first proposed by Qu et al. [20] and later confirmed by other groups (e.g., [21]). On the other hand, due to an abnormally high metabolism in tumor cells, there is a high enrichment of different kinds of molecules, as is known from the applications of photodynamic therapy (PDT) [22] or positron emission tomography (PET) techniques [23]. Biochemical effects (pH, metabolic specificities) have influences on the concentration and the spectra of certain fluorophores [24–27]. A third reasonable explanation for the attenuation of visible fluorescence would be increased absorption of the illuminating and/or the fluorescent light. As mentioned earlier, the main absorber in the visible range in human tissue is hemoglobin. Its influence is shown in Figure 29.9a and b. We compared fluorescence signals (405 nm excitation) in lungs that were perfused with blood and with saline solution [28]. The exsanguinated tissue yields a higher fluorescence with a "dromedary curve." The hemoglobin absorption curve [4] is

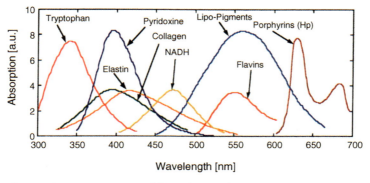

Figure 29.8 Emission spectra of fluorophores in human tissue.

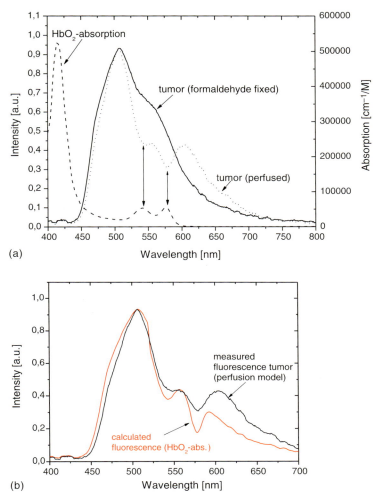

Figure 29.9 (a) Comparison of the fluorescence spectra of perfused tumor and formaldehyde-fixed tumor specimen. (b) Calculated and measured fluorescence signal of tumor tissue.

overlayed in the picture. The characteristic "camel curve" shape with two humps as usually seen in endobronchial fluorescence measurements results from this absorption. Changes in the red to green hump ratio indicate dysplasias or early cancers. We have used these curves in an optical catheter system [29] as shown in Figure 29.10. Excitation light from a mercury lamp or a laser diode (405 nm) is guided through a catheter. The remitted light is analyzed by a PC spectrometer after passing a chopper wheel (Ocean Optics, Duiven, The Netherlands). The continuous white light is replaced by stroboscopic light. Due to the latency of the video chip, the endoscopist sees a normal endoscopic image with a blue spot on the field of interest. During the long (unnoticed) dark phase, spectroscopy is carried out. On the video screen, the

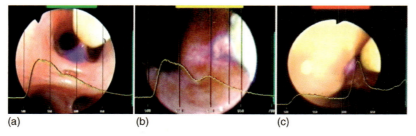

Figure 29.10 Optical catheter *in vivo* showing fluorescence signals of normal mucosa (a), early cancer (b), and drug-induced fluorescence of lung cancer (c).

fluorescence spectrum curve is overlayed on the endoscopic image in real time. The dromedary shape of scar tissue, the deepened dip characteristic of inflammation, and the flattened camel-shaped curves of a carcinoma *in situ* can be clearly differentiated. Most likely, changes in tissue texture, layer thickness, and intracellular metabolism and also changes in blood density contribute to the autofluorescence attenuation phenomenon.

29.1.5
Commercially Available Systems

All major companies offer autofluorescence systems for bronchoscopy; examples of screen images are shown in Figures 29.11–29.15.

The diagnostic autofluorescence endoscopy (DAFE) system of Wolf (Knittlingen, Germany) uses a 300 W xenon lamp as light source and a three-chip CCD camera. A flip-flop filter holder permits a change from white light to blue–violet emission. The bandpass filter allows transmission from 390 to 470 nm, giving the well-known green fluorescence picture behind the yellow blocking filter. Specific to the Wolf system is a variation in the filter behind the light source. A second low transmission (around 665 nm) results in red backscattered light that makes the final image brighter and enhances the contrast by a factor of 2.7 [30]. In the second generation of the DAFE system, the filters in front of the camera have also been modified. Small amounts of

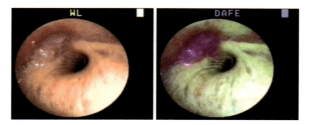

Figure 29.11 White light image and autofluorescence images of a small tumor. First generation of the Wolf DAFE system.

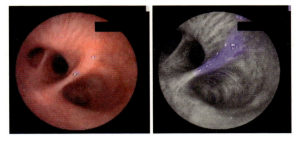

Figure 29.12 White light and auto-fluorescence images of a carcinoma *in situ*. Second generation of the Wolf DAFE system. Image courtesy of Dr. P. Ramon, Lille, France.

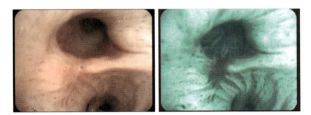

Figure 29.13 Twin mode images of a small carcinoma *in situ* using the Pentax SAFE system. Images courtesy of Dr. M. Wagner, Nürnberg, Germany.

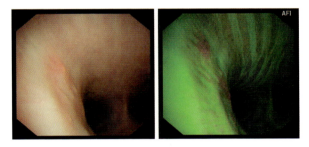

Figure 29.14 Carcinoma *in situ*, visualized by the Olympus LUCERA system.

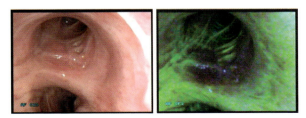

Figure 29.15 Carcinoma *in situ* visualized by the STORZ D-light system (software version currently not released for clinical use).

backscattered blue and white light in the fluorescence mode are detected. This allows better differentiation between premalignant lesions, inflammation, and blood [31].

Pentax (Tokyo, Japan) offers the SAFE system. The first generation (SAFE 1000) used a xenon light source. The imaging system consisted of a relatively bulky camera that could be mounted on any standard fiber bronchoscope. The second generation (SAFE 2000) used a video-chip bronchoscope. Brightness differences in the autofluorescence image could be manipulated electronically. Suspicious areas were rendered in pseudo-color yellow. The actually marketed SAFE 3000 system uses a 300 W xenon white light source and a blue diode laser (408 nm). Unique is the feature of displaying the fluorescence image and the white light image simultaneously on the video screen. This twin mode makes an examination very convenient. Regarding the detection of carcinoma *in situ* and dysplasia, improvements from 65% (white light only) to 90% (fluorescence mode) have been reported [32]. Finally, the brightness information of the autofluorescence image can be used to modify the chroma signal of the white light image. This Multiple Image Xposition (MIX) mode is the latest attempt to improve the specificity and distinguish between precancerous and inflammatory lesions [33].

Olympus Optical (Tokyo, Japan) introduced the EVIS LUCERA system in 2006. It uses a xenon white light source equipped with two switchable wheels containing appropriate filters and a bronchoscope with a triggered BW CCD camera chip at its tip. The RGB filter wheel is used to create high-resolution color images. The other filter wheel is used to excite an autofluorescence image. Similarly to other systems, parts of the reflected light are used to create a false color image with high contrasts. Normal tissue appears green, bronchitis purple, and cancer more reddish.

Xillix Technologies has modified the original LIFE system described above, which can be considered a milestone in fluorescence imaging bronchoscopy [34]. The new generation, called Onco-LIFE, also uses small amounts of red light (675–720 nm) in addition to the blue light (395–445 nm) from a mercury arc lamp for illumination. An area demarcated by a small target in the center of the displayed fluorescence–reflectance image is further processed and a value based on the ratio of reflectance to green fluorescence signal intensities in that area is calculated. The re-emitted diffuse light is used as the reference. In a multicenter, prospective study with 170 individuals, the relative sensitivity on a per lesion basis of white light plus fluorescence bronchoscopy versus white light bronchoscopy alone was determined as 1.5. For a quantified fluorescence reflectance response value of more than 0.4, a sensitivity and specificity of 51 and 80%, respectively, could be achieved for detection of moderate/severe dysplasia, carcinoma *in situ*, and microinvasive cancer [35].

Karl Storz (Tuttlingen, Germany) offers several versions of its D-Light system. A xenon light source with switchable filters is used in all generations. Most studies have been performed with CCD cameras attached to rigid telescope lenses or fiberscopes [21, 36]. Recently, a video-chip bronchoscope has become available for clinical studies. Early malignant lesions appear almost blue whereas inflamed areas look more brownish. The image quality in white light mode is outstanding, and no sacrifices regarding the resolution, dynamics, and brightness have to be made.

The system can be used in combination with exogenous drugs such as ALA, providing high-contrast images with bright red cancerous lesions. Currently, several image processing algorithms are being studied in order to improve the specificity. Results of clinical studies can be expected by the end of 2011.

29.1.6
Remaining Problems

The clinical problem with all systems is the relatively high number of false-positive biopsies. A typical statement from a clinician is that the systems are "too sensitive." The reported specificities for autofluorescence bronchoscopy range from 4 to 94% in various studies [37]. The challenge is to maintain the high sensitivity but improve the specificity. Most recently, attempts have been made to incorporate spectral measurements into imaging systems [38], use color fluorescence ratios in algorithms of fluorescence image analysis [39], or combine autofluorescence and narrowband imaging techniques [40]. The title of the Editorial [41] in the issue in which the last paper [40] appeared summarizes our current knowledge: "Optical diagnosis for preneoplasia, the search continues."

29.2
Therapy

Lung cancer causes symptoms from involvement of the trachea or major bronchi in up to 30% of all cases; 17% of all patients suffer of symptomatic airway obstruction [42]. Post-obstructive pneumonia is a severe consequence but even worse is the impairment of gas exchange. Suffocation is the most feared fate of cancer patients. All efforts are made to establish open airways. The armamentarium includes chemotherapy and radiation therapy and interventional bronchoscopic methods [43]. Compressed bronchi are kept open by placement of airway stents. In emergency cases, endoluminally growing tumors can be cored out mechanically. However, cutting techniques with coagulation options are preferred. The most established techniques for endobronchial tissue removal are electrocautery, cryotherapy, argon plasma coagulation, and laser photo-resection.

29.2.1
Lasers in Bronchology

Lasers are commonly used in pulmonary medicine. Figure 29.16 shows various lasers that are routinely used in the authors' department. Labeled 1 is a VISULAS diode laser emitting at 662 nm for PDT with chlorins. PDT with Foscan is performed with a Biolitec CERALAS (labeled 4) emitting at 652 nm and PDT with Photofrin with a Zeiss diode laser emitting 630 nm (labeled 6). Cutting instruments are the Biolitec Evolve laser (labeled 2), which is available with wavelengths of 980 or 1470 nm, and the Dornier Nd:YAG laser emitting 1064 nm (labeled 5). In addition to arc lamps with

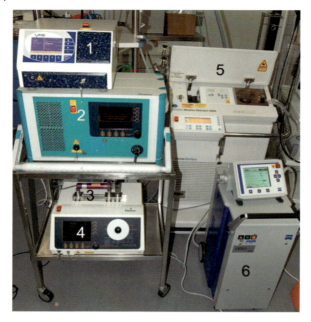

Figure 29.16 Lasers used in interventional bronchology. For detailed descriptions, see the text.

filters, we use small diode laser such as the OXIUS, which delivers monochromatic light at 405 nm for diagnostic and dosimetry excitation of photosensitizers.

29.2.2
Thermal Laser Applications

The classical workhorse in interventional bronchology is the Nd:YAG laser [44]. Its infrared light (1064 nm) is guided with air-cooled fibers through the working channel of the bronchoscope. Low power of 10–20 W is used for coagulation, as

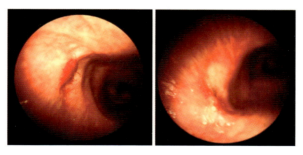

Figure 29.17 Endobronchial hemangioma (Osler disease). Photocoagulation using the Dornier Nd:YAG laser at 1064 nm.

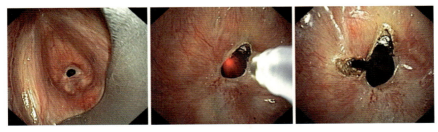

Figure 29.18 Scar tissue causing a subglotic web stenosis. Thermo-photoresection of fibrotic tissue using the Biotitec EVOLVE diode laser at 1470 nm.

demonstrated in Figure 29.17. While argon plasma coagulation has a depth of penetration in the region of 2 mm, YAG lasers are feasible for coagulating blood vessels 5–10 mm below the surface. Even highly vascularized tumors can be coagulated and removed safely. The technique has an excellent safety record, with thousands of applications worldwide [45]. The great advantage over other techniques such as cryotherapy and PDT is the immediate effect [46]. The main reason why other devices endanger the dominance of these thermal lasers is the high cost of the equipment. Recently, other thermal lasers have become available for endobronchial applications. Diode lasers are smaller, less noisy, and less expensive than Nd:YAG lasers. The air-cooled EVOLVE diode laser (Biolitec, Jena, Germany) delivers up to 40 W at 1470 nm. This wavelength matches an absorption maximum of water. The laser is suitable for precise cutting without undesired deep tissue necrosis, as demonstrated in Figure 29.18. Web-type airway stenosis can be treated with such a laser. However, surgery with end-to-end anastomosis remains the treatment of choice for benign tracheal stenosis. Any local treatment, including laser resection, can potentially induce further granulation tissue development. Coagulation in the laryngeal region may affect tissue nutrition, promote peribronchitis, and affect laryngeal nerve function with the risk of speech loss. For tissue removal at the level of the vocal cords where deeper tissue effects are risky, devices such as the CO_2 laser are safer.

29.2.3
Photodynamic Therapy

The method of PDT is based on the enrichment of exogenous chromophores in metaplastic tissue, which has been discussed in the diagnostics section. Most of the fluorophores used for photodynamic diagnostics are also potentially phototoxic. The excitation of those molecules with light leads to intermolecular interactions with molecular oxygen or surrounding molecules. Depending on rate constants, either oxygen radicals, surrounding molecules, or singlet oxygen, which is a highly reactive molecule, will be generated. As the sensitizer molecules are retained in cellular compartments, such as mitochondria or membranes, the reactive species created are able to destroy the cells. Endobronchial PDT is realized with laser systems that emit in the red spectral range because its depth of penetration makes it feasible for the destruction of mucosal tumors. Highly effective sensitizers need to have absorption spectra with maxima in the far-red region. Ideally, drugs for PDT should also have fluorescent properties as this facilitates tumor detection, therapy planning, and dosimetry. In addition to the penetration depth, which is wavelength dependent, another important parameter for therapy efficacy is the tumor to tissue ratio. A high enrichment in the target area and a low concentration in the surrounding tissues lead to the most specific tumor therapy without collateral tissue damage. A large variety of photosensitizers have been used in preclinical and clinical studies.

ALA and its esters are precursors of protoporphyrin IX as parts of the heme biosynthetic pathway. Administering excess ALA causes visible fluorescence of endobronchial tumors. As discussed above, enrichment in cancer cells is due to a higher metabolism in malignant cells and their lack of several enzymes for the decomposition of protoporphyrin IX. Application of ALA is orally or topically; the esters are synthesized for topical use. Most institutions use ALA for diagnosis only. A single study has been published on ALA PDT in lung cancer [47]. The comparative drug, an intravenously administered hematoporphyrin derivative (Photosan), proved to be superior.

Photofrin is the most commonly used representative of blood-based porphyrin formulations. Early studies had been made with Photofrin I. The currently marketed Photofrin II, which was developed by Dougherty *et al.* [48, 49], has a higher purity. The absorption maximum is in the typical Soret band at 400 nm [50]. Because of the higher light penetration in the red region, a smaller absorption band at 630 nm is used for therapy. A maximum of concentration is reached 48 h after systemic application. Bare fibers, micro-lens fibers, radial diffusers, or balloons are available to illuminate the tumor region. Superficial tumors or carcinoma *in situ* are illuminated with lens fibers; bulky tumors can be treated interstitially. The diffusing fiber can be pushed into such a tumor under direct bronchoscopic vision, as shown in Figure 29.19. The bulky and difficult to manage argon-pumped dye lasers that we had to use until the mid-1990s have been replaced by turn-key diode lasers. They can easily deliver up to 1000 mW, resulting in treatment times of less than 10 min; 100–200 J per centimeter of tumor length are applied in standard protocols. The first reaction to endobronchial PDT is the formation of fibrin plugs, debris, and necrosis

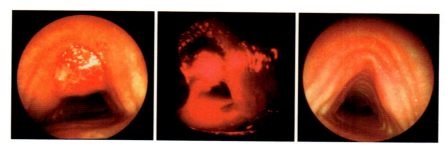

Figure 29.19 PDT of an endotracheal tumor using Photofrin. The 9 mm tumor is treated by interstitial application of red light (630 nm) applied by a cylindrical fiber of 10 mm length. Six months after therapy, no signs of residual tumor.

within 48 h after illumination. This may obstruct airways even further and could result in suffocation unless a clean-up bronchoscopy is performed. If we treat tracheal lesions, we perform these flexible bronchoscopies for removal of debris the next morning. PDT with Photofrin has been used for early [51], recurrent [52], and advanced lung cancers [53, 54].

There are numerous original articles and reviews stating the superior efficacy and cost effectiveness of PDT when compared with thermal YAG laser therapy [55–57]. The depth of penetration of red light at 630 nm is limited to a few millimeters. In small tumors less than 0.5 cm in diameter, cure rates of 95% have been accomplished. If tumors extended to 2 cm in length, the cure rate dropped to 43%, and if they were exophytic, they could be eradicated in only 14% of cases [51]. The response rate is determined by the tumor extension and the depth of infiltration [58]. In a routine clinical setting, early invasive lung cancers can only be cured by Photofrin PDT if they respect the bronchial wall. The method of choice to resolve this problem is endobronchial ultrasonography [58], which has become standard of care to check the depth of tumor infiltration with a balloon ultrasound probe before a decision is made as to whether PDT alone is sufficient to eradicate the tumor. If the tumor depth exceeds 3 mm or if the tumor is longer than 10 mm along any axis, we combine PDT with endobronchial brachytherapy (five times 4 Gy). Using this regime, eradication rates of >80% can be accomplished in these cases with Photofrin [59]. Another option for the treatment of these early invasive tumors is the use of other sensitizers.

As only Photofrin has been approved for treatment of lung cancer in Europe, the use of other drugs requires exceptional approvals from ethical committees. *m*-Tetrahydroxyphenylchlorin (mTHPC) (Foscan) is licensed for head and neck cancer PDT. It is a chlorin, not soluble in water. The drug is injected 96 h before the treatment. mTHPC has a higher quantum yield than porphyrins. Treatment times are shorter and the cytocidal effect is greater [61]. The most relevant clinical advantage is the excitation wavelength of 652 nm for maximum therapeutic effects. The depth of penetration is higher and additional brachytherapy can be spared. Used in clinical studies, we have seen very good tumor responses even in larger invasive cancers of the airways (Figure 29.20). Photofrin and Foscan accumulate in bronchogenic

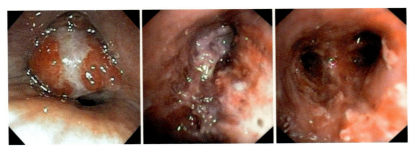

Figure 29.20 PDT of a squamous cell cancer occluding the middle lobe bronchus. PDT with Foscan. Two days after illumination with red light at 650 nm, tumor appears infarcted and necrotic. One week later, still inflammation but the bronchus is open without residual tumor.

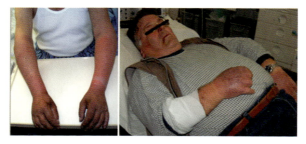

Figure 29.21 Severe skin toxicity after use of first-generation PDT drugs. Despite all warnings, patients went out into sunlight for cigarette smoking.

tumors and show visible fluorescence. Through a yellow blocking filter at the ocular of the bronchoscope, the cancer extension can be seen under blue light illumination which facilitates the treatment. The degree of drug retention can be estimated using the above-mentioned optical catheter and spectrometer (Figures 29.3 and 29.10).

A major disadvantage of PDT with the currently available drugs is the long-lasting photoxicity of the skin. The light sensitivity lasts for up to 6 weeks and certainly impairs the quality of life of patients as they are requested to avoid bright artificial light and especially sunlight. Figure 29.21 shows sunburn of patients who did not follow the recommendations. There is a need for new drugs with shorter retention times. Potential candidates are other chlorins, which are reduced, hydrophilic porphyrins with a strong absorption band in the 640–700 nm range. Chelation with metals can alter the absorption properties. A promising drug is chlorin e6. Derived from chlorophyll-*a* ("green porphyrin"), chlorin e6 is monomeric in solution, has a quantum yield of singlet oxygen of 0.7, and, like other sensitizers, has photo-oxidative effects on amino acids, enzymes, and lysosomes. It binds strongly to albumin [63, 64]. The highest accumulation in tumor tissue is reached 2–4 h after drug injection. The drug fluorescence is clearly visible under illumination with blue–violet light. A sharp peak at 670 nm over the tumor proves selectivity (Figure 29.22). The high absorption

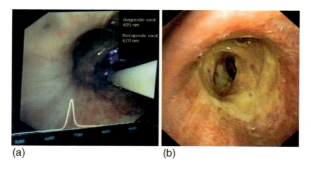

(a) (b)

Figure 29.22 Very large area of tumor, extending from mid-trachea to lower lobe bronchus. Experimental treatment with chlorin e6 (Fotolon) and 5 cm long cylindrical fibers. Sharp fluorescence peak at 670 nm during bronchoscopy (a). Unusual yellow fibrin covering of the treated area but histologically no residual tumor (b).

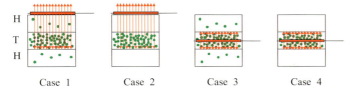

Figure 29.23 Cases of different enrichment and optimization of light delivery in tumor tissue.

at 670 nm allows deep penetration, especially in lung tissue. By 24 h after application the substance is cleared from mucosa and skin. Hence the phototoxic side effect is critical for only about 12 h, which makes PDT with this drug easy and practical. In preliminary studies, we have seen very convincing results with eradication of tumors exceeding several centimeters in length.

There are many other photosensitizers, such as 4-hydroxyphenylpyruvate dioxygenase (HPPD), phthalocyanines, NPe6 (talaporfin), Purlytin, Padoporfinin, lutrin, texaphryrin, and benzoporphynn, under investigation in clinical studies. A review was provided by Allison and Sibata [65]. For practical application, some critical parameters are important. It is mandatory for monitoring and dosimetry that the sensitizer has good visible fluorescence and little photobleaching. Nonfluorescent phototoxic drugs are difficult to use, and side effects can hardly be predicted or avoided [66].

Compared with thermal laser treatment, the advantage of PDT is a relative selectivity resulting from drug accumulation in the target. Submucosal tumors cannot be attacked with electrocautery or conventional laser treatment. At least theoretically, this could be accomplished with PDT. Ideally, there would be no sensitizer molecules in the healthy tissue. The therapeutic light would penetrate healthy tissue without doing any harm (Figure 29.23, case 2). This could be realized by injecting a sensitizer directly into the tumor. In the more realistic case 1 (depending on the threshold for cell destruction), damage to the sound tissue cannot be excluded. For a solid tumor embedded in healthy tissue, an additional reduction of the light can be realized by puncturing the tissue down into the tumor with an illumination fiber. The wavelength that activates the sensitizer will determine the maximum absorbance capacity and, thus, the wavelength of light to be used. The second-generation sensitizers allow the use of light with wavelengths between 600 and 800 nm, allowing an increase in penetration depths compared with first-generation porphyrins [68]. Figure 29.24 shows transmission spectra for therapeutic light of excised lung tissue with and without exsanguinations. The highest transmission is reached at around 670–700 nm. For endoscopic PDT of lung tumors, these wavelengths would be ideal. The knowledge about the transmission characteristics of human tissue gives us the opportunity for sufficiently precise dosimetry. The dose rate can be described by an extended Lambert's law. The dose D_{depth} in the tissue has to be above a threshold D_S^* for cell death, that is known [69]. D_{depth} is calculated by the photodynamic dose D^*, which results from the quality of light, its fluence rate,

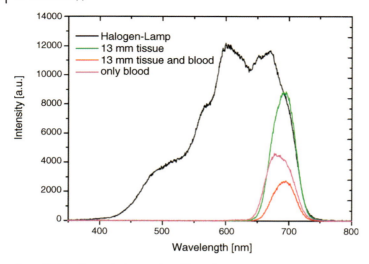

Figure 29.24 Transmission spectra of lung tissue.

the time of illumination, the extinction coefficient, and the density of the tissue. Table 29.1 shows the calculations for a first-generation drug excited at 630 nm and for a second-generation chlorin excited at 670 nm. As a clinical consequence, tumors penetrating the bronchial wall and even outside the airways could be treated effectively through a bronchoscope (Figure 29.25).

It is often argued that PDT has never made it into general pulmonary medical practice despite the fact that is has been available for almost 30 years. There is no reason for nihilism. Progress in pharmacology and technology makes it attractive today and even more in the foreseeable future.

Table 29.1 Calculation of dose rate.

$$D_{\text{depth}} = D^* e^{-\frac{z}{\delta}} \quad \text{with} \quad D^* = \varepsilon D \Phi t \frac{\lambda}{hc\varrho}$$

Parameter	Symbol	Photofrin	Chlorin e6
Extinction ($cm^{-1} M^{-1}$)	ε	10 000	40 000
Average tissue density (lung) ($g\,cm^{-3}$)	ϱ	0.3–0.48	
Concentration of sensitizer in tumor (M)	D	6×10^{-5}	6×10^{-5}
Illumination time	t		
Necrosis threshold [$p\,g^{-1}$ (photodynamic active units per gram)]	D_S^*	2×10^{19}	
Excitation wavelength (nm)	λ	630	670

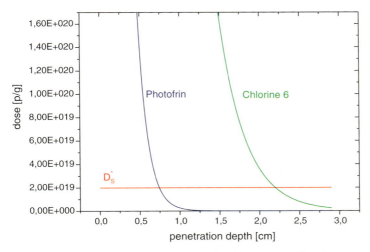

Figure 29.25 Depth of penetration and calculated therapeutic effect of PDT using excitation wavelengths of 630 vs. 670 nm.

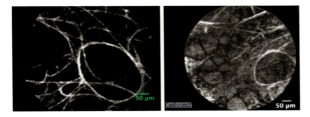

Figure 29.26 Probe-based confocal fluorescence endomicroscopy images using the CellVizio system (Mauna Kea Technologies, Newtown, PA, USA). Images courtesy of Professor Luc Thiberville, Rouen, France.

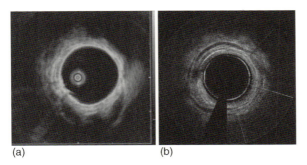

Figure 29.27 Endobronchial ultrasound image (a) and optical coherence tomography (b). OCT images using the OCTIS system (Tomophase, Burlington, MA, USA). Images courtesy of Professor A. Ernst, Boston, MA, USA.

29.3
New Techniques: Fluorescence Microimaging, Optical Coherence Tomography

There are several other promising new techniques that have been applied for bronchoscopic use. Fiber confocal fluorescence microscopy provides microimages from the smallest airways and even from the alveolar sacs (Figure 29.26). We obtain fascinating pictures from formerly inaccessible regions of the lung [70]. The lung parenchyma can be studied through the working channel of a bronchoscope. The combination with fluorescence measurements provides further information [71]. Combined with electromagnetic navigation techniques, optical biopsies from the smallest lung nodules becomes feasible.

An equally fascinating technique is optical coherence tomography (OCT). The already commercially available devices allow the study of airway wall structures with much better resolution than endobronchial ultrasound, as shown in Figure 29.27.

There is no question that optical methods will further boost the use of bronchoscopy. We are privileged to take part in a fascinating journey where the combination of techniques allows insights into bronchi and lungs and where we can help patients suffering from pulmonary diseases. Advances in molecular biology, optical cancer detection methods, and light-based methods provide us with very promising tools in the fight against lung cancer. There is more hope with NBI, PDD, PDT, OCT, and a high-resolution bronchoscope.

References

1 Kollofrath, O. (1897) Entfernung eines Knochenstücks aus dem rechten Bronchus auf natürlichem Wege und unter Anwendung der directen Laryngoscopie. *MMW*, **38**, 1038–1039.

2 Lam, S. and Becker, H.D. (1996) Future diagnostic procedures. *Chest Surg. Clin.*, **6**, 363–380.

3 Thiberville, L. (1999) Precancerous lesions of the bronchi: endoscopic detection. *Eur. Respir. J.*, **14S**, 2475.

4 Prahl, S. and Jacques, S. (1999) Optical absorption of hemoglobin. *Oregon Med. Laser Center Proc.*

5 Gono, K., Obi, T., Yamaguchi, M. et al. (2004) Appearance of enhanced tissue features in narrow-band endoscopic imaging. *J. Biomed. Opt.*, **9**, 568–577.

6 Yoshida, T., Inoue, H., Usui, S. et al. (2004) Narrow-band imaging system with magnifying endoscopy for superficial esophageal lesions. *Gastrointest. Endosc.*, **59**, 288–295.

7 Muto, M., Katada, C., Sano, Y. et al. (2005) Narrow band imaging: a new diagnostic approach to visualize angiogenesis in superficial neoplasia. *Clin. Gastroenterol. Hepatol.*, **3**, 16–20.

8 Kuznetsov, K., Lambert, R., and Rey, J.F. (2006) Narrow-band imaging: potential and limitations. *Endoscopy*, **38**, 76–81.

9 Herth, F.J., Eberhardt, R., Anantham, D. et al. (2009) Narrow-band imaging bronchoscopy increases the specificity of bronchoscopic early lung cancer detection. *J. Thorac. Oncol.*, **4** (9), 1060–1065.

10 Shibuya, K., Hoshino, H., Chiyo, M., Iyoda, A., Yoshida, S., Sekine, Y., Iizasa, T., Saitoh, Y., Baba, M., Hiroshima, K., Ohwada, H., and Fujisawa, T. (2003) High magnification bronchovideoscopy combined with narrow band imaging could detect capillary loops of angiogenic squamous dysplasia in heavy smokers at

high risk for lung cancer. *Thorax*, **58**, 989–995.
11 Miyake, Y., Kouzu, T., Takeuchi, S.,*et al.* (2005) Development of new electronic endoscopes using the spectral images of an internal organ. Proceedings of the IS&T/SID's Thirteen Color Imaging Conference, Scottsdale (AZ), pp. 261–269.
12 Profio, A.E., Doiron, D.R., and King, E.G. (1979) Laser fluorescence bronchoscope for localization of occult lung tumors. *Med. Phys.*, **6** (6), 523–525.
13 Doiron, D.R., Profio, E., Vincent, R.G., and Dougherty, T.J. (1979) Fluorescence bronchoscopy for detection of lung cancer. *Chest*, **76**, 27–32.
14 Kato, H. and Cortese, D.A. (1985) Early detection of lung cancer by means of hematoporphyrin derivative fluorescence and laser photoradiation. *Clin. Chest Med.*, **6**, 237–253.
15 Lam, S., Palcic, B., McLean, D., Hung, J., Korbelik, M., and Profio, A.E. (1990) Detection of early lung cancer using low dose Photofrin II. *Chest*, **97**, 333–337.
16 Sutro, C., and Burman, M. (1933) Examination of pathological tissue by filtered ultraviolet radiation. *Arch. Pathol.*, **16**, 346–349.
17 Herly, L. (1944) Studies in selective differentiation of tissues by means of filtered ultraviolet light. *Cancer Res.*, **4**, 227–231.
18 Wagnières, G., Star, W., and Wilson, B. (1998) In vivo fluorescence spectroscopy and imaging for oncological applications. *Photochem. Photobiol.*, **68** (5), 603–632.
19 Lam, S., McAulay, C., LeRiche, J.C., Ikeda, N., and Palcic, B. (1994) Early localization of bronchogenic carcinoma. *Diagn. Ther. Endosc.*, **1**, 75–78.
20 Qu, J.Y., MacAulay, C.E., Lam, S., and Palcic, B. (1995) Laser-induced fluorescence spectroscopy at endoscopy: tissue optics, Monte Carlo modeling and in vivo measurements. *Opt. Eng.*, **34** (11), 3334–3343.
21 Häussinger, K., Stanzel, F., Hubner, R.M., Pichler, J., and Stepp, H. (1999) Autofluorescence detection of bronchial tumors with the D-Light/AF. *Diagn. Ther. Endosc.*, **5**, 105–112.
22 Kennedy, J.C., Pottier, R.H., and Pross, D.C. (1990) Photodynamic therapy with endogenous Protoporphyrin IX: basic principles and present clinical experience. *J. Photochem. Photobiol. Biol.*, **6**, 143–148.
23 Yamamoto, Y.T., Thompson, C.J., Diksic, M., Meyer, E., and Feindel, W.H. (1984) Positron emission tomography. *Experientia Suppl., Neurosurg. Rev.*, **7** (4), 233–252.
24 Hung, J., Lam, S., LeRiche, J.C., and Palcic, B. (1991) Autofluorescence of normal and malignant bronchial tissue. *Lasers Surg. Med.*, **11**, 99–105.
25 Yang, Y., Ye, Y., Li, F., Li, Y., and Ma, P. (1987) Characteristic autofluorescence for cancer diagnosis and its origin. *Lasers Surg. Med.*, **7**, 528–532.
26 Harris, D.M., and Werkhaven, J. (1987) Endogenous porphyrin fluorescence in tumors. *Lasers Surg. Med.*, **7**, 467–472.
27 Wolfbeis, O. (1985) The fluorescence of organic natural products, in *Molecular Luminescence Spectroscopy: Methods and Applications – Part I* (ed. S.G. Schulman), John Wiley and Sons, Inc., New York, pp. 167–370.
28 Hüttenberger D., Gabrecht, T., Wagnières, G., Weber, B., Linder, A., Foth, H.J., and Freitag, L. (2008) Autofluorescence detection of tumors in the human lung – spectroscopical measurements in situ, in an in vivo model and in vitro. *Photodiagnosis Photodyn. Ther.*, **5** (2), 139–4729.
29 Freitag, L. and Hüttenberger, D. (2007) Autofluorescence bronchoscopy using the Hemer optical catheter and the Wolf DAFE system, in *Autofluorescence Bronchoscopy* (eds. M. Wagner and J.H. Ficker), Unimed Verlag, Bremen, pp. 79–89.
30 Gabrecht, F.T., Glanzmann, T., Freitag, L., Weber, B.C., van den Bergh, H., and Wagnières, G. (2007) Optimized autofluorescence using additional backscattered red light. *J. Biomed. Opt.*, **12** (6), 6401–6406.
31 Gabrecht, T., Radu, A., Grosjean, P., Weber, B., Reichle, G., Freitag, L., Monnier, P., van den Bergh, H., and Wagnières, G. (2008) Improvement of the

specificity of cancer detection by autofluorescence imaging in the tracheobronchial tree using backscattered violet light. *Photodiagnosis Photodyn Ther.*, **5**, 2–9.

32 Ikeda, N., Honda, H., Hayashi, A., Usuda, J., Kato, Y., Tsuboi, M., Ohira, T., Kato, H., Serizawa, H., and Aoki, Y. (2006) Early detection of bronchial lesions using newly developed videoendoscopy-based autofluorescence bronchoscopy. *Lung Cancer*, **52** (1), 21–27.

33 Wagner, M. and Ficker, J.H. (2007) Video-Autofluorescence SAFE-3000, in *Autofluorescence Bronchoscopy* (eds. M. Wagner and J.H. Ficker), Unimed Verlag, Bremen, pp. 19–58.

34 Lam, S., McAulay, C., Hung, J., LeRiche, J., Profio, A.E., and Palcic, B. (1993) Detection of dysplasia and carcinoma *in situ* using a lung imaging fluorescence endoscopy (LIFE) device. *J. Thorac. Cardiovasc. Surg.*, **105**, 1035–1040.

35 Edell, E., Lam, S., Pass, H., Miller, Y.E. et al. (2009) Detection and localization of intraepithelial neoplasia and invasive carcinoma using fluorescence–reflectance bronchoscopy: an international, multicenter clinical trial. *J. Thorac. Oncol.*, **4** (1), 49–54.

36 Stanzel, F. (2007) The Storz D-Light system, in *Autofluorescence Bronchoscopy* (eds. M. Wagner and J.H. Ficker), Unimed Verlag, Bremen, pp. 65–77.

37 Kennedy, T.C., McWilliams, A., and Edell, E. (2007) Bronchial intraepithelial neoplasia/early central airways lung cancer. *Chest*, **132**, 221S–233S.

38 Reinders, D., Snead, D., Dhillon, P., and Fawzy, Y. (2009) Endobronchial cancer detection using an integrated bronchoscope system for simultaneous imaging and noncontact spectral measurement. *J. Bronchol. Intervent. Pulmonol.*, **16**, 158–166.

39 Lee, P., van den Bergh, R.M., Lam, S., Gazdar, A.F., Grunberg, K., McWilliams, A., LeRiche, J., Postmus, P., and Sutedja, T. (2009) Color fluorescence ratio for the detection of bronchial dysplasia and carcinoma *in situ*. *Clin. Cancer Res.*, **15** (14), 4700–4705.

40 Nguyen, P.T., Salvado, O., Masters, I.B., Farah, C.S., and Fielding, D. (2010) Combining autofluorescence and narrow band imaging with image analysis in the evaluation of preneoplastic lesions in the bronchus and larynx. *J. Bronchol. Intervent. Pulmonol.*, **17** (2), 109–116.

41 Lee, P. (2010) Optical diagnosis for preneoplasia, the search continues. *J. Bronchol. Intervent. Pulmonol.*, **17** (2), 101–102.

42 Minna, J.D., Higgins, G.A., and Glatstein, E.J. (1985) Cancer of the lung, in *Cancer, Principles and Practice of Oncology*, 2nd edn (eds. V.T. DeVita, S. Hellman, and S.A. Rosenberg), J.P. Lippincott, Philadelphia, pp. 518–526.

43 Freitag, L., Macha, H.N., and Loddenkemper, R. (2001) Interventional bronchoscopic procedures. *Lung Cancer*, **6** (17), 272–304.

44 Dumon, J.F., Shapshay, S., Bourcereau, J., Cavaliere, S., Meric, B., Garbi, N., and Beamis, J. (1984) Principles for safety in application of neodymium–YAG laser in bronchology. *Chest*, **86** (2), 163–168.

45 Cavaliere, S., Foccoli, P., Toninelli, C., and Feijo, S. (1994) Nd:YAG laser therapy in lung cancer: an 11 year experience with 2253 applications in 1585 patients. *J. Bronchol.*, **1**, 105–111.

46 Bolliger, C.T., Sutedja, T.G., Strausz, J., and Freitag, L. (2006) Therapeutic bronchoscopy with immediate effect: laser, electrocautery, argon plasma coagulation and stents. *Eur. Respir. J.*, **27** (6), 1258–1271.

47 Maier, A., Tomaselli, F., Matzi, V., Woltsche, M., Anegg, U., Fell, B., Rehak, P., Pinter, H., and Smolle-Jüttner, F.M. (2002) Comparison of 5-aminolaevulinic acid and porphyrin photosensitization for photodynamic therapy of malignant bronchial stenosis: a clinical pilot study. *Lasers Surg. Med.*, **30** (1), 12–17.

48 Dougherty, T.J., Kaufman, J.E., Goldfarb, A., Weishaupt, K.R., Bouyle, D.G., and Mitleman, A. (1978) Photoradiation for the treatment of malignant tumors. *Cancer Res.*, **38**, 2628–2635.

49 Dougherty, T.J. (1989) Photodynamic therapy – new approaches. *Semin. Surg. Oncol.*, **5**, 151–152.

50 Pass, H.L. (1991) Photodynamic therapy for lung cancer. *Chest Surg. Clin. North Am.*, **1**, 135–151.

51 Furuse, K., Fukuoka, M., Kato, H., Horrai, T., Kubota, K., Kodama, N., Kusunoki, Y., Takifuji, N., Okunaka, T., Konaka, C., Wada, H., and Hayata, Y. (1993) A prospective phase II study on photodynamic therapy with Photofrin II for centrally located early-stage lung cancer. *J. Clin. Oncol.*, **11**, 1852–1857.

52 Freitag, L., Korupp, A., Itzigehl, I., Dankwart, F., Tekolf, E., Reichle, G., Kullmann, H.J., and Macha, H.N. (1996) Experience with fluorescence diagnosis and photodynamic therapy in the multimodality treatment concept of operated, recurrent bronchial carcinoma. *Pneumonologie*, **50** (10), 693–699.

53 Sutedja, T.G. (1999) Photodynamic therapy in advanced tracheobronchial cancers. *Diagn. Ther. Endosc.*, **5**, 245–251.

54 Moghissi, K., Dixon, K., Stringer, M., Freeman, T., Thorpe, A., and Brown, S. (1999) The place of bronchoscopic photodynamic therapy in advanced unresectable lung cancer: experience of 100 cases. *Eur. J. Cardiothorac Surg*, **15** (1), 1–6.

55 Diaz-Jimenez, J.P., Martinez-Ballarin, J.E., Llunell, A., Farrero, E., Rodriguez, A., and Castro, M.J. (1999) Efficacy and safety of photodynamic therapy versus Nd:YAG laser resection in NSCLC with airway obstruction. *Eur. Respir. J.*, **14** (4), 800–805.

56 Furukawa, K., Tetsuya, O., Yamamoto, H., Tsuchida, T., Usuda, J., Kumasaka, H., Ishida, J., Konaka, C., and Kato, H. (1999) Effectiveness of photodynamic therapy and Nd:YAG laser treatment for obstructed tracheobronchial malignancies. *Diagn. Ther. Endosc.*, **5**, 161–166.

57 Moghissi, K. and Dixon, K. (2003) Is bronchoscopic photodynamic therapy a therapeutic option in lung cancer? *Eur. Respir. J.*, **22**, 535–541.

58 Sutedja, T., Lam, S., and LeRiche, I. (1994) Response and pattern of failure after photodynamic therapy for intraluminal stage I lung cancer. *J. Bronchol.*, **1**, 259–289.

59 Miyazu, Y., Miyazu, T., Kurimoto, N., Iwamoto, Y., Kanoh, K., and Kohno, N. (2002) Endobronchial ultrasonography in the assessment of centrally located early-stage lung cancer before photodynamic therapy. *Am. J. Respir. Crit. Care Med.*, **165**, 832–837.

60 Freitag, L., Ernst, A., Thomas, M., Prenzel, R., Wahlers, B., and Macha, H.N. (2004) Sequential photodynamic therapy (PDT) and high dose brachytherapy for endobronchial tumour control in patients with limited bronchogenic carcinoma. *Thorax*, **59** (9), 790–793.

61 Ma, L., Moan, J., and Berg, K. (1994) Evaluation of a new photosensitizer, meso-tetra-hydroxyphenyl-chlorin, for use in photodynamic therapy: a comparison of its photobiological properties with those of two other photosensitizers. *Int. J. Cancer*, **57** (6), 883–888.

62 Spikes, J.D. (1990) Chlorins as photosensitizers in biology and medicine. *J. Photochem. Photobiol. B*, **6**, 259–274.

63 Osbroff, A.R., Ohuoha, D., Hasan, T. et al. (1986) Antibody-targeted photolysis: selective photodestruction of human T-cell leukemia cells using monoclonal antibody–chlorin e6 conjugates. *Proc. Natl. Acad. Sci. U. S. A.*, **83**, 8744–8788.

64 Kochubeev, G.A., Frolov, A.A., and Zenkevtch, E.L. (1988) Characteristics of complex formation of chlorin e6 with human and bovine serum albumins. *Mol. Biol.*, **22**, 968–975.

65 Chin, W.W.L., Heng, P.W.S., and Olivo, M. (2007) Chlorin e6 – polyvinylpyrrolidone mediated photosensitization is effective against human non-small cell lung carcinoma compared to small cell lung carcinoma xenografts. *BMC Pharmacol.*, **7**, 15.

66 Allison, R.R. and Sibata, C.H. (2010) Oncological photodynamic therapy photosensitizers: a clinical review. *Photodiagn. Photodyn. Ther.*, **7**, 61–75.

67 Tremblay, A., Leroy, S., Freitag, L., Copin, M.C., Brun, P.H., and Marquette, C.H. (2003) Endobronchial phototoxicity of WST 09 (Tookad), a new fast-acting photosensitizer for

photodynamic therapy: preclinical study in the pig. *Photochem. Photobiol.*, **78** (2), 124–130.

68 Gomer, C.J. (1991) Preclinical examination of first and second generation photosensitizers used in photodynamic therapy. *Photochem. Photobiol.*, **54**, 1093–1107.

69 Jacques, S. (1998) The mathematics of PDT dosimetry for cancer treatment. *Proc. Oregon Med. Laser Center.*

70 Thiberville, L., Moreno-Swirc, S., Vercauteren, T., Peltier, E., Cave, C., and Bourg-Heckly, G. (2007) *In vivo* imaging of the bronchial wall microstructure using fibered confocal fluorescence microscopy. *Am. J. Respir. Crit. Care Med.*, **175**, 22–31.

71 Thiberville, L., Salaun, M., Lachkar, S., Dominique, S., Moreno-Swirc, S., Vever-Bizet, C., and Bourg-Heckly, G. (2009) Human *in vivo* fluorescence microimaging of the alveolar ducts and sacs during bronchoscopy. *Eur. Respir. J.*, **33**, 974–985.

72 Michel, R.G., Kinasewitz, G.T., Fung, K.M., and Keddissi, J.I. (2010) Optical coherence tomography as an adjunct to flexible bronchoscopy in the diagnosis of lung cancer: a pilot study. *Chest*, **138** (4), 984–988.

30
Laser Therapy
Anant Mohan

Laser is an acronym for *Light Amplification by Stimulated Emission of Radiation*. In simpler language, lasers are devices that produce light that becomes transformed into heat upon interacting with living tissue. Three properties differentiate laser light from naturally occurring light [1]:

1) **Monochromaticity**: the laser light contains only one color or a narrow band of wavelengths, unlike natural light that is a blend of various wavelengths.
2) **Spatial coherence**: laser light hardly diverges and maintains its intensity as it travels forwards.
3) **Temporal coherence**: The packets of energy travel in uniform time with equal alignment.

Lasers have become an important diagnostic and therapeutic tool in pulmonary medicine.

Although Albert Einstein recognized the existence of stimulated emissions in 1917, the first description of stimulated optical emissions of monochromatic light occurred in 1960 with the advent of the ruby laser. The utility of lasers stems from their ability to transfer precisely high levels of energy to tissue (a property exploited for photoresection of endobronchial lesions), and also the unique properties of monochromatic (utilized for photodynamic therapy (PDT) and autofluorescence bronchoscopy (AFB).

The use of lasers in pulmonology may be both diagnostic and therapeutic. These are discussed below along with the current status:

30.1
Diagnostic Applications of Lasers

30.1.1
Autofluorescence Bronchoscopy (AFB)

In AFB, bronchial mucosa is illuminated by blue light. Normal mucosa emits autofluorescent light with a peak in the green range. In diseased mucosa (neoplastic or chronic inflammatory), the thickened epithelium reduces the fluorophore

Handbook of Biophotonics. Vol.2: Photonics for Health Care, First Edition. Edited by Jürgen Popp, Valery V. Tuchin, Arthur Chiou, and Stefan Heinemann.
© 2012 Wiley-VCH Verlag GmbH & Co. KGaA. Published 2012 by Wiley-VCH Verlag GmbH & Co. KGaA.

Table 30.1 Indications for the use of autofluorescence bronchoscopy [2].

Evidence-based indications

1. Evaluation of patients with high-grade sputum atypia and normal chest imaging studies
2. Bronchoscopic surveillance of moderate/severe dysplasia and carcinoma *in situ*
3. Guiding of curative endobronchial therapy of microinvasive lesions
4. Evaluation of patients with suspected, known, or previous lung cancer
   Preoperative:
   - Determination of tumor margins
   - Detection of synchronous lesions
   Follow-up:
   - Detection of recurrence
   - Detection of metachronous lesions

concentration and hyperemia, hence the intensity of autofluorescence (AF) emission is significantly reduced and the emission peak shifts from the green to the red range. Therefore, the diseased mucosa shows shadowing and discoloration.

In routine clinical practice, the primary indication for using AFB is for the early detection of intraepithelial malignant disease. The various clinical situations where AFB may be useful are listed in Table 30.1.

Of the several AFB systems available, two models using laser light are the LIFE and SAFE-3000. The LIFE Lung system (Lung Imaging Fluorescence Endoscope, Xillix Technologies, Richmond, BC, Canada) (Figure 30.1) was introduced in 1991 and has been commercially available since 1998. It uses a helium–cadmium laser to illuminate the bronchial mucosa with 442 nm light. The laser beam is conducted to the mucosa via a conventional fiber-optic bronchoscope. The reflected beams of the bronchial mucosa appear at two characteristic spectral bands in the green (480–520 nm) and red (4630 nm) range. Normal mucosa appears light green and neoplastic mucosa reddish brown (Figure 30.2).

The second laser-based system, the SAFE-3000 (Pentax Corporation, Tokyo, Japan) (Figure 30.3) uses a semiconductor laser diode that emits 408 nm light to induce AF and a xenon lamp for white light bronchoscopy (WLB) examination. A single high-sensitivity color charge-coupled device (CCD) sensor detects AF in the green and red range and also reflected blue light (430–700 nm), and generates a combined fluorescence–reflectance image. Both light sources and the video processor are integrated in one device. The white light and AF images can easily be switched using a button on the bronchoscope.

Several trials have compared the relative sensitivity of AFB and WLB for the detection of early cancerous changes in the bronchial mucosa [3–6] (Table 30.2). Although initial studies demonstrated a rather high relative sensitivity of up to 6.2, subsequent larger and better conducted multicenter studies showed a more realistic assessment of the potential of AFB, the relative sensitivity settling down to around 1.5. However, AFB suffers from inferior specificity to WLB due to false-positive results obtained in inflammation and scarring, leading to increases in costs and the number of biopsies.

30.1 Diagnostic Applications of Lasers | 493

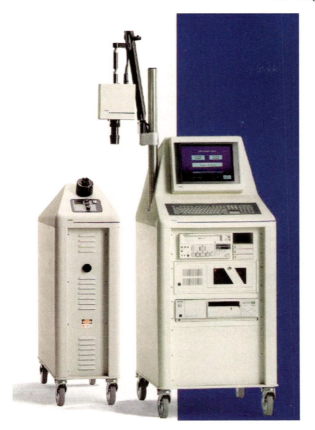

Figure 30.1 The LIFE Lung system (Lung Imaging Fluorescence Endoscope).

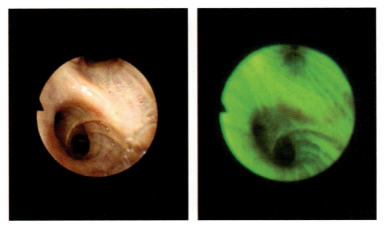

Figure 30.2 Simultaneous display of white light and autofluorescence images.

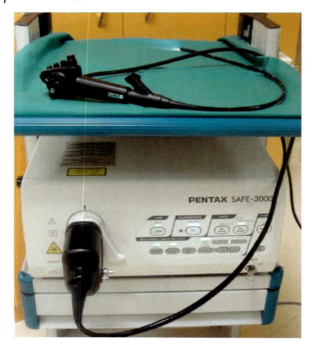

Figure 30.3 Pentax SAFE-3000 autofluorescence bronchoscopy system with flexible video-chip bronchoscope (Pentax Corporation, Tokyo, Japan).

Table 30.2 Comparative studies of white light and autofluorescence.

Study	No. of patients	Sensitivity (%)		Relative sensitivity	Specificity (%)	
		WLB	WLB + AFB		WLB	WLB + AFB
Lam et al. (1998) [3]	173	9	56	6.2	90	66
Khanavkar et al. (1998) [4]	165	32	86	2.7	75	31
Ikeda et al. (1999) [5]	158	58	92	1.6	62	66
Häussinger (2005) [6]	1173	58	82	1.4	62	58

30.2
Fibered Confocal Fluorescence Microscopy (FCFM)

Confocal fluorescence microscopy is a relatively new optical technology, which uses a laser as light source. It produces real-time dynamic images of living tissue, which due to their high resolution are referred to as "optical biopsies." The technique can be applied endoscopically to the lung via flexible fiber-optic miniprobes. The main endogenous fluorophore of the lung is elastin. In bronchial mucosa, fibered confocal fluorescence microscopy (FCFM) visualizes the network of elastic fibers in the subepithelial layer. Thiberville et al. [7] attempted to establish normal and abnormal patterns of FCFM imaging of the bronchial mucosa in a study of 29 subjects at high risk for lung cancer. They found that the absence of a regular fibered pattern seems to be sensitive for preinvasive lesions. Thus FCFM could be used as an adjunct to AFB, compensating for the low specificity of the latter by *in vivo* differentiation of AF-positive lesions in benign and suspicious samples. Although current data defining FCFM alveolar imaging of the normal lung are limited, the method appears promising because it is minimally invasive and can be repeated serially whenever indicated.

30.3
Therapeutic Uses of Laser

30.3.1
Photodynamic Therapy (PDT)

PDT utilizes the principle of a photosensitizer, which accumulates in tumor tissue, is activated by laser light of a specific wavelength, and in the presence of oxygen in the tissue environment produces reactive oxygen species. These cause cell death by inducing cellular damage, vascular ischemia, and inflammatory and immune responses. The primary clinical indications for using PDT are for curative therapy of early central lung cancer (ECLC) and palliative treatment of advanced lung cancer with bronchial obstruction. In addition, PDT has been tried as experimental therapy in early peripheral lung cancer and for the adjuvant treatment of mesothelioma. PDT is an alternative option in patients not eligible for surgery due to a high operative risk.

A recent review of the literature on PDT of ECLC, comprising 15 studies with 626 patients and 715 early cancers, revealed that almost all patients (99%) experienced at least partial remission while the rate of complete remission varied between 30 and 100% [8]. Overall, 5 year survival was 61%. The complication rate was low and acceptable and included photosensitivity skin reactions in 5–25%, respiratory complications in 13–18%, and nonfatal hemorrhage in 7.8%. Compared with other modalities used for treating central lung cancers, such as electrocautery, cryotherapy, brachytherapy, and photo-resection with the Nd:YAG laser, PDT was considered superior regarding the level of evidence, benefit, and grade of recommendation [9]. Moreover, PDT can be combined easily with other modalities of treatment such as Nd:YAG laser therapy, electrocautery, and brachytherapy. The role of PDT in advanced lung cancer has also been reviewed [10], comprising 12 studies with 636 patients, and it was concluded that PDT is safe and provides symptomatic benefit in patients with advanced lung cancer with a significant endobronchial component but good performance status and without extrathoracic metastasis. However, the cost–benefits of this procedure with respect to other modalities are not yet established. In addition, since it provides delayed response, PDT is not suitable for use in acute, life-threatening situations.

The recommended light dose for endobronchial PDT is 200 J cm^{-1} tumor length (150–300 J cm^{-1}). The light used for PDT can be generated by an argon laser and various types of dye lasers. More recently, modern diode lasers have served as light sources. The light is delivered through optical fibers, which can be introduced via the working channel of a flexible bronchoscope.

The use of PDT for treating early peripheral lung cancer is still experimental. However, after further development and evaluation of the technique, it could be a promising alternative for those patients who are unfit for surgery or radiotherapy. Similarly, PDT is an experimental tool for treating malignant mesothelioma, mainly due to the high rate of local recurrence.

30.3.2
Laser Therapy

Lasers are extensively used for the palliative treatment of symptomatic central lung cancer. The use of lasers in bronchoscopy started almost three decades ago when Strong and Jako in 1972 used a CO_2 laser to treat various laryngeal disorders [11]. Several types of lasers have been used in medical science over the past three decades (Table 30.3). Of these, the neodymium-doped yttrium aluminum garnet (Nd:YAG) laser is currently the most popular among the various lasers available. Although laser radiation gives the best results in lesions of short length located in the trachea and

Table 30.3 Commonly used medical lasers.

Laser	Wavelength (nm)	Characteristics	Advantages/disadvantages
Nd:YAG	1064	Invisible; absorption proportional to tissue pigmentation	Deep tissue penetration with excellent hemostasis; can be used with flexible scope
KTP	532	Green light; low absorption in water	Delivery via flexible scope possible; more precise cutting than Nd:YAG; useful in endoscopic sinus surgery
CO_2	10 600	Invisible; complicated delivery system required; poor hemostatic effect	Precise cutting but more bleeding; only possible via rigid scope
Argon	488–514	Blue–green; good affinity for hemoglobin	Useful for small cutaneous and retinal vessel coagulation; needs contact with tissue for resection
Xenon-chloride	400	Precise cutting without much scatter	Delivery via flexible scope possible; useful in corneal surgery, angioplasty

mainstem or proximal lower lobe bronchi, which are easily accessible to the rigid bronchoscope, the use of flexible bronchoscopes has expanded the reach to more distal lesions. Laser resection provides significant and rapid improvement in the patient's symptoms, and is at least as effective as other palliative methods such as cryotherapy, PDT, and stent placement, if not more so.

The effect of laser radiation on the tissue depends on several factors, such as the power settings and wavelength employed, distance of the laser tip from the target, duration of impact, and certain physical characteristics of the tissue, mainly its color, surface, and water content. The Nd:YAG laser can penetrate up to 5–10 mm from the focal point and coagulate blood vessels up to 5 mm in diameter, making it an ideal choice for large vascular tumors. However, laser energy has a tendency to scatter and damage the adjoining tissues. Reducing the duration of impact and giving short energy bursts of 0.5 s each instead of a continuous beam can minimize this complication.

It is now clear that relief of central airway obstruction by intraluminal growths, either malignant or benign, constitutes the primary indication for this procedure, but laser therapy also has several other uses (Table 30.4). In the emergency setting, lasers may be used to control acute pulmonary hemorrhage in children. Another common indication for laser usage is in the treatment of tracheal and bronchial stenosis, which may occur as a result of endotracheal intubation or tracheostomy or as a consequence of systemic disorders such as sarcoidosis, tuberculosis, and Wegener granulomatosis. The laser is applied prior to mechanical dilatation using a rigid

Table 30.4 Indications for the Nd:YAG laser.

1	Intrabronchial growth	Benign	Hamartoma
			Chondroma
			Lipoma
			Neurinoma
			Papillomatosis
		Malignant	Carcinoma *in situ* (CIS)
			Early lung cancer
			Typical carcinoid
			Mucoepidermoid carcinoma
			Adenoid–cystic carcinoma
			Metastatic lesion from thyroid, colon, kidney, esophagus, and melanoma
2	Tracheal or bronchial stenosis		After endotracheal intubation or tracheostomy
			Sarcoidosis
			Tuberculosis
			Wegener granulomatosis
3	Others		To cut metallic stents and foreign bodies prior to removal
			To close small proximal bronchopleural fistulas occurring after lung resection
			To control acute pulmonary hemorrhage in children

Table 30.5 Contraindications for the Nd:YAG laser.

Related to anatomic location of obstruction	Extrinsic lesions without endobronchial growth
	Lesions bordering or infiltrating adjoining vascular structures, esophagus or mediastinum
	Total obstruction with inadequate landmarks
Related to patient condition	Coagulation disorders
	Terminal metastatic disease

bronchoscope or balloon; however, distortion of the airways and scarring are frequent complications. Lasers are also used to cut metallic stents and foreign bodies prior to their removal and to close small proximal bronchopleural fistulas occurring after lung resection. Lasers have also been tried as an adjunct in tuberculosis, but a Cochrane review found insufficient evidence to recommend their use at present [12].

30.4
Contraindications

Table 30.5 lists the various common contraindications for laser procedures.

30.5
Laser Bronchoscopy Procedure

30.5.1
Anesthesia

The type of anesthesia used depends on whether a rigid, flexible, or combined bronchoscopic delivery is planned. Combustible gases, such as halothane, should be strictly avoided. Intravenous propofol supplemented with midazolam and fentanyl is a popular regimen. The inspired oxygen concentration (FiO_2) should be kept below 40% and oxygen saturation by pulse oximetry should be maintained above 90%.

30.5.2
Technique

This procedure requires a combined team effort of the bronchoscopist, anesthetist, laser operator, and nurses. Nd:YAG laser bronchoscopy using a fiber-optic bronchoscope can be performed on awake, spontaneously breathing patients. The instrument is passed transnasally or transorally and then the laser fiber is guided through it until the target lesion is reached. Keeping the tip of the laser fiber at least 1 cm away from the tip of the bronchoscope and 4–10 mm from the lesion to be treated, the laser is fired in pulses of 0.5–1 s at energy settings between 20 and 40 W. The beveled end of the rigid bronchoscope can sometimes be used to shear off a portion of protruding tissue. The bronchoscopist should be able to vaporize a small lesion completely or cause sufficient sloughing off of the tissue to be easily coughed out or removed using biopsy forceps.

Bronchoscopy should be repeated 2–4 days later to assess the results of treatment. These sessions should be repeated every 2–4 days until maximum benefit is achieved. A complete response is defined as complete clearing of the tumor whereas a partial response is a reduction in tumor size or improvement in the smallest tracheobronchial diameter. Symptomatic improvement of the patient's symptoms, improvement in pulmonary functions, resolution of atelectasis on chest radiograph, and improved performance status as noted by dyspnea indices are other parameters to gage the success of this procedure.

The following safety precautions should be strictly adhered to during laser resection:

- Use the lowest possible oxygen concentration during laser firing.
- Avoid laser firing towards the airway wall.
- Remove all flammable material from the vicinity of the equipment.
- Cover the patient's eyes with saline pads and aluminum foil.
- All personnel in the room should wear goggles.
- Perform frequent bronchial toilet to keep the airway patent.
- Avoid prolonged periods of apnea.
- Use the laser sparingly; rely more on mechanical resection.

30.6
Complications

These procedures are usually well tolerated in experienced hands, as is evident from several large series conducted over the last two decades where the complication rates ranged from 2.3 to 6.5%. The commonly encountered complications are listed in Table 30.6.

30.7
Outcome

The short-term results of endobronchial laser treatment are usually good (Figure 30.4). Immediate success depends more on the location and extent of the lesion in the tracheobronchial tree than on the histology. Personne *et al.* [13] reported the first large series in which they performed 2284 laser sessions on 1310 patients; 25% of their patients survived more than 1 year and more than half of all subjects had significant relief of airway obstruction, with an overall mortality of only 1.6%. In another large case series involving 1838 patients [14], successful de-obstruction after the first laser treatment was achieved with 93% of lesions located in the trachea, main bronchi, and bronchus intermedius, whereas the success rate dropped in lobar

Table 30.6 Complications of laser therapy..

	Complication	Management
Common	Hemorrhage	Continuous suctioning, lavage with saline and epinephrine and photocoagulation of the surrounding tissue
	Airway fire	Can be avoided by keeping the laser tip clean to minimize chances of self-ignition, removing all flammable objects from the laser path, avoiding combustible inhaled anesthetics, keeping the FiO_2 below 50%, and delivering laser pulses of 40 W or less
	Airway perforation	The most important preventive strategy against perforation is to use the minimum energy setting for short durations
	Pneumothorax	Avoid laser perforating the bronchial wall or due to a ruptured bleb in a hyperinflated lung
	Hypoxemia	As a result of bronchospasm or laryngospasm; should be treated before proceeding with the laser procedure. Maintaining a good pulmonary toilet and clear airways is the most important precaution against hypoxia.[27]
	Cardiac events	Rarely seen now with better monitoring of cardiac rhythm and oxygen saturation, and can be prevented to a great extent by following the safety measures for bronchoscopic laser therapy
Uncommon	Noncardiogenic pulmonary edema Focal pulmonary hyperinflation Airway scarring/stenosis Systemic/air embolism	

bronchi to between 77% in the lower lobes and 49% in the upper lobes, where more bending of the bronchoscope is required.

Hence the utility of lasers in pulmonary medicine is multifactorial and of both diagnostic and therapeutic importance. Further refinement of the technique and physician expertise is likely to expand its utility and improve patient outcomes following their use.

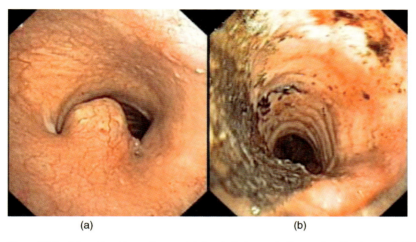

(a)   (b)

Figure 30.4 Adenoid-cystic carcinoma. (a) Exophytic tumor in the proximal trachea and (b) restored tracheal lumen immediately after laser resection with the Nd:YAG laser.

References

1. Van Der Spek, A.F., Spargo, P.M., and Norton, M.L. (1988) The physics of lasers and implications for their use during airway surgery. *Br. J. Anaesth.*, **60**, 709–729.
2. Kennedy, T.C., McWilliams, A., Edell, E., Sutedja, T., Downie, G., Yung, R., et al. (2007) Bronchial intraepithelial neoplasia/early central airways lung cancer: ACCP evidence-based clinical practice guidelines (2nd edition). *Chest*, **132** (3 Suppl.), 221S–233S.
3. Lam, S., Kennedy, T., Unger, M., Miller, Y.E., Gelmont, D., Rusch, V., et al. (1998) Localization of bronchial intraepithelial neoplastic lesions by fluorescence bronchoscopy. *Chest*, **113** (3), 696–702.
4. Khanavkar, B., Gnudi, F., Muti, A., Marek, W., Muller, K.M., Atay, Z., et al. (1998) Basic principles of LIFE-autofluorescence bronchoscopy. Results of 194 examinations in comparison with standard procedures for early detection of bronchial carcinoma – overview. *Pneumologie*, **52** (2), 71–76.
5. Ikeda, N., Honda, H., Katsumi, T., Okunaka, T., Furukawa, K., Tsuchida, T., et al. (1999) Early detection of bronchial lesions using lung imaging fluorescence endoscope. *Diagn. Ther. Endosc.*, **5** (2), 85–90.
6. Häussinger, K., Becker, H., Stanzel, F., Kreuzer, A., Schmidt, B., Strausz, J., et al. (2005) Autofluorescence bronchoscopy with white light bronchoscopy compared with white light bronchoscopy alone for the detection of precancerous lesions: a European randomized controlled multicentre trial. *Thorax*, **60** (6), 496–503.
7. Thiberville, L., Moreno-Swirc, S., Vercauteren, T., Peltier, E., Cave, C., and Bourg Heckly, G. (2007) *In vivo* imaging of the bronchial wall microstructure using fibered confocal fluorescence microscopy. *Am. J. Respir. Crit. Care Med.*, **175** (1), 22–31.
8. Moghissi, K. and Dixon, K. (2008) Update on the current indications, practice and results of photodynamic therapy (PDT) in early central lung cancer (ECLC). *Photodiagn. Photodyn. Ther.*, **5** (1), 10–18.

9 Mathur, P.N., Edell, E., Sutedja, T., and Vergnon, J.M. (2003) American College of Chest Physicians. Treatment of early stage non-small cell lung cancer. *Chest*, **123** (1 Suppl.), 176S–180S.

10 Moghissi, K. and Dixon, K. (2003) Is bronchoscopic photodynamic therapy a therapeutic option in lung cancer? *Eur. Respir. J.*, **22** (3), 535–541.

11 Strong, M.S. and Jako, J.G. (1972) Laser surgery in the larynx: early clinical experience with continuous CO_2 laser. *Ann. Otol. Rhinol. Laryngol.*, **81**, 791–798.

12 Vlassov, V.V. and Reze, A.G. (2006) Low level laser therapy for treating tuberculosis. *Cochrane Database Syst. Rev.*2 (Art. No.: CD003490), DOI: 10.1002/14651858.CD003490.pub2.

13 Personne, C., Colchen, A., Leroy, M., *et al.* (1986) Indications and technique for endoscopic laser resections in bronchology. *J. Thorac. Cardiovasc. Surg.*, **91**, 710–715.

14 Cavaliere, S., Venuta, F., Focolli, P. *et al.* (1996) Endoscopic treatment of malignant airway obstruction in 2008 patients. *Chest*, **110**, 1536–1542.

Part 6
Urology and Nephrology

31
Bladder Biopsy with Optical Coherence Tomography
Alexandre Douplik

The American Cancer Society estimates that in 2010 there will be 70 530 new cases and an estimated 14 680 deaths from bladder cancer in the United States [1]. The 5 year survival rate drops from 98% [when cancer is found at early (0) stage or carcinoma *in situ*] to 63% (when cancer is found at stage 3 or invasive tumor), and then to 15% for patients with stage 4 (metastasized) [2]. Hence it is critical to discover bladder cancer at an early stage. Another critical issue in cancer care of the bladder is to identify adequately the resection margins and maximally preserve the adjacent healthy tissue while eradicating cancer lesion(s). Downsizing of the bladder affects patients' quality of life as they may not be able to hold as much urine in the bladder as before surgery and may need to empty it more often [3].

People with bladder cancer sometimes have a similar tumor in the lining of another part of the urinary system. Therefore, when someone is found to have cancer in one part of their urinary system, the entire urinary tract needs to be checked for tumors. Particularly in the case of early cancerous changes, in most cases the lesion cannot be discovered, delineated, or checked for malignancy under conventional endoscopy or intraoperatively and the gold standard remains a biopsy histology examination that usually takes a few hours at least. The technique of optical coherence tomography (OCT) has been proposed to facilitate a biopsy directly during regular cystoscopic surveillance [4] or therapeutic intervention at an image resolution close to standard H&E histology nearly in real time [5]. As the OCT probe can be implemented in a small-diameter fiber probe (1–2 mm), it has been applied in both transurethral [6] and open bladder access [7].

The type of bladder cancer can affect the treatment options because different types respond to different treatments. Transitional cell carcinoma is by far the most common type of bladder cancer. It starts in the urothelial cells lining the bladder. These tumors are divided into grades based on how the cells look under the microscope. If the cells look more like normal cells or are well differentiated, the cancer is called a low-grade cancer. When the cells look very different from normal or are poorly differentiated, the cancer is high grade. Lower grade cancers tend to grow slower and to have a better outcome than higher grade cancers. The much less common other types of malignancy in the urinary system are squamous cell

Handbook of Biophotonics. Vol.2: Photonics for Health Care, First Edition. Edited by Jürgen Popp, Valery V. Tuchin, Arthur Chiou, and Stefan Heinemann.
© 2012 Wiley-VCH Verlag GmbH & Co. KGaA. Published 2012 by Wiley-VCH Verlag GmbH & Co. KGaA.

carcinoma, adenocarcinoma, and small cell cancer [8]. A recent study by Cauberg *et al.* showed *ex vivo* that based on OCT-measured optical attenuation (μ_t), the grade of bladder urothelial carcinoma could potentially be assessed in real time [9].

One of the limitations of OCT imaging is a relatively small field of view (of the order of 1–6 mm^2), making screening of the entire area of interest a considerably time-consuming procedure that is usually unaffordable for endoscopists and surgeons. Hence fluorescence imaging-navigated OCT imaging has been proposed as a small field of view approach to OCT guided by a large field of view technique (fluorescence imaging) [10]. Another advantage of such a combination stems from the different origins of the normal versus pathologic tissue imaging contrast and can improve the combined sensitivity and specificity. Schmidbauer *et al.* reported that using fluorescence cystoscopy in combination with OCT increases the specificity of fluorescence cystoscopy in targeting urothelial carcinoma by 19% at the lesion level and by 25% per patient [11].

OCT images of normal and cancerous bladder tissue demonstrate that well-organized layers can be seen in normal bladder tissue, whereas such clearly defined structures are not present in the invasive tumor [4, 12]. as can be seen in Figure 31.1. Karl *et al.* recently reported an OCT biopsy study of the bladder based on 166 scanned OCT images with 100% sensitivity and 65% specificity [13]. In a study by Zagayanova *et al.*, OCT images of the bladder were evaluated as being either malignant or benign based on the absence of a layering structure with a sensitivity of 98% and a specificity of 72% [14]. The relatively low specificity can be interpreted by the fact that OCT images of squamous metaplasia may be confused with carcinoma *in situ*. Squamous metaplasia is diagnosed when the urothelial transitional cells are replaced by benign squamous cells, with or without evidence of keratinization occurring, for instance, in response to chronic cystitis. The OCT image could have a layered structure; however, the urothelium is thickened and brighter than normal, frequently leading to poor or no contrast with the lamina propria [6].

In order to improve the diagnostics accuracy further, computational methods have been applied for bladder OCT image analysis, including texture analysis [15].

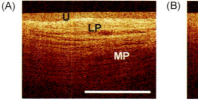 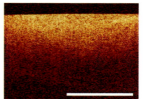

Figure 31.1 Images taken from a patient at the Cleveland Clinic after bladder OCT study using the Imalux Niris Imaging System. (a) The normal bladder (anterior wall) shows a thin urothelium (U) over a bright lamina propria (LP). A well-demarcated border with good contrast is presented between the urothelium and the lamina propria. The muscularis propria (MP) is below the lamina propria, and the border is also demarcated. (b) Image from invasive bladder cancer lesion. Courtesy of Imalux Corporation.

An overview of intraoperative OCT imaging modalities applied in the bladder, including benign tissues and inflammation, can be found in Chapter 25. Interoperative OCT monitoring (Douplik). The results described in this chapter justify the development of OCT biopsy in the bladder for further clinical studies.

31.1
Acknowledgments

This study was supported by Erlangen Graduate School in Advanced Optical Technologies (SAOT) and by the German National Science Foundation (DFG) in the framework of the excellence initiative. The author is grateful for support from Imalux Corporation (USA) in providing the OCT images.

References

1. American Cancer Society (2010) Cancer facts and figures, www.cancer.org (last accessed: Dec 15, 2010).
2. National Cancer Institute (2010) Annual report to the nation on the status of cancer 1975–2001, with a special feature on survival: questions and answers, http://www.cancer.gov/newscenter/qa/2004/2001-annual-report-survivalqa.
3. National Cancer Institute (2010) Bladder cancer surgery, http://www.cancer.gov/cancertopics/wyntk/bladder/page8#a.
4. Jesser, C.A., Boppart, S.A., Pitris, C., Stamper, D.L., Nielsen, G.P., Brezinski, M.E., and Fujimoto, J.G. (1999) High resolution imaging of transitional cell carcinoma with optical coherence tomography: feasibility for the evaluation of bladder pathology. *Br. J. Urol.*, **72**, 1170–1176.
5. Fujimoto, J.G., Pitris, C., Boppart, S.A., and Brezinski, M.E. (2000) Optical coherence tomography: an emerging technology for biomedical imaging and optical biopsy. *Neoplasia*, **2** (1–2), 9–25.
6. Zagaynova, E.V., Gladkova, N.D., Streltsova, O.S., Gelikonov, G.V., Tresser, N., and Feldchtein, F.I. (2008) Optical coherence tomography in urology, in *Optical Coherence Tomography: Technology and Applications (Biological and Medical Physics: Biomedical Engineering)* (eds W. Drexler and J.G. Fujimoto), Springer, Berlin, pp. 1241–1268.
7. Zagaynova, E., Gladkova, N., Shakhova, N., Gelikonov, G., and Gelikonov, V. (2008) Endoscopic OCT with forward-looking probe: clinical studies in urology and gastroenterology. *J. Biophotonics*, **1** (2), 114–128.
8. University of Virginia Health System. Urinary tract pathology tutorials, http://www.med-ed.virginia.edu/courses/path/urinary/uroth4.cfm (last accessed: Dec 15, 2010).
9. Cauberg, E.C.C., de Bruin, D.M., Faber, D.J., de Reijke, T.M., Visser, M., de la Rosette, J.J.M.C.H., and van Leeuwen, T.G. (2010) Quantitative measurement of attenuation coefficients of bladder biopsies using optical coherence tomography for grading urothelial carcinoma of the bladder. *J. Biomed. Opt.*, **15** (6), 066013.
10. Wang, Z.G., Durand, D.B., Schoenberg, M., and Pan, Y.T. (2005) Fluorescence guided optical coherence tomography for the diagnosis of early bladder cancer in a rat model. *J. Urol.*, **174** (6), 2376–2381.
11. Schmidbauer, J., Remzi, M., Klatte, T., Waldert, M., Mauermann, J., Susani, M., and Marberger, M. (2009) Fluorescence cystoscopy with high-resolution optical coherence tomography imaging as an

adjunct reduces false-positive findings in the diagnosis of urothelial carcinoma of the bladder. *Eur. Urol.*, **56**, 914–919.

12 Manyak, M.J., Gladkova, N.D., Makari, J.H., Shwartz, A., Zagaynova, E.V., Zolgahari, L., Zara, J.M., Iksanov, R., and Feldchtein, F.I. (2005) Evaluation of superficial bladder transitional-cell carcinoma by optical coherence tomography. *J. Endourol.*, **19** (5), 570–574.

13 Karl, A., Stepp, H., Willmann, E., Buchner, A., Hocaoglu, Y., Stief, C., and Tritschler, S. (2010) Optical coherence tomography for bladder cancer – ready as a surrogate for optical biopsy? Results of a prospective mono-centre study. *Eur. J. Med. Res.*, **15** (3), 131–134.

14 Zagayanova, E.V., Streltsova, O.S., Gladkova, N.D., Shakova, N.M., Feldchtein, F.I., Kamensky, V.A., Gelikonov, G.V., Snopova, L.B., and Donchenko, E.V. (2004) Optical coherence tomography in diagnostics of precancer and cancer of human bladder. *Proc. SPIE*, **5312**, 75–81.

15 Lingley-Papadopoulos, C.A., Loew, M.H., Manyak, M.J., and Zara, J.M. (2008) Computer recognition of cancer in the urinary bladder using optical coherence tomography and texture analysis. *J. Biomed. Opt.*, **13** (2), 024003.

32
Advanced Laser Endoscopy in Urology: Laser Prostate Vaporization, Laser Surgery, and Laser Removal of Renal Calculi

Ronald Sroka, Michael Seitz, and Markus Bader

32.1
Introduction

Clinical medical laser application using high-power laser light offers the opportunity to destroy malignant and benign soft and hard tissue. Innovative developments in clinical endoscopic techniques allow for the delivery of high-power laser light in continuous-wave and pulsed modes in the spectral wavelength region 500–2100 nm. Thus covering high-power lasers using second harmonic generation in the visible spectral region, high-power diode laser systems in the near-infrared (NIR) spectral region, and also lasers emitting in the 2 µm region such as solid-state lasers and fiber laser systems, all laser systems are today clinically available. The laser output power ranges from 250 W continuous wave to more than 2 J per pulse at repetition rates of 50 Hz in pulsed mode which can be applied by single fibers. For the treatment of benign prostate hyperplasia (BPH), high-power laser vaporization [1–9] and pulsed techniques for enucleation procedures [1, 2, 10–15] are clinically proven to be reliable, safe, and effective treatment procedures. Recently, a new chapter in BPH laser treatment began with investigations of the laser–tissue interaction of high-power diode laser systems either for vaporization or for enucleation purposes. Although this era is still open and a matter of debate, the basics of preclinical and clinical investigations can be presented. Further, the potential of laser light application in the upper urinary tract is outlined. Urologic stone fragmentation can be conducted with pulsed infrared (IR) laser application. The suitable combination of a laser wavelength and laser power provides the basis for laser–tissue interaction, thus achieving successful treatment.

32.2
High-Power Diode Laser for BPH-Treatment

After the successful clinical introduction of prostate laser vaporization and laser enucleation techniques, diode laser systems have been investigated for the treatment

Handbook of Biophotonics. Vol.2: Photonics for Health Care, First Edition. Edited by Jürgen Popp,
Valery V. Tuchin, Arthur Chiou, and Stefan Heinemann.
© 2012 Wiley-VCH Verlag GmbH & Co. KGaA. Published 2012 by Wiley-VCH Verlag GmbH & Co. KGaA.

of benign prostatic enlargement. Using light in the NIR spectral range (800–1100 nm), these devices offer high simultaneous absorption in water and hemoglobin; with lasers emitting in the spectral range 1320–2100 nm, the absorption is dominated by the tissue water. Diode lasers for the treatment of BPH are available at different wavelengths. Whereas low-power diode lasers at wavelengths between 805 and 850 nm have been used for interstitial laser coagulation [16–21], diode lasers emitting light at 940, 980, 1320, and 1470 nm are currently being tested for vaporization of the prostate [22–27].

32.2.1
Preclinical Investigations

To evaluate the coagulation and vaporization properties of these laser wavelengths at high powers, investigations of their effects in different kinds of tissue models have been performed under reproducible conditions. In preclinical studies using an *ex vivo* autologous blood-perfused porcine kidney model, diode lasers at output power $P > 50$ W emitting light at 940, 980, and 1470 nm showed major amounts of tissue ablation in comparison with a 532 nm laser at 80 W, but similar hemostasis. In contrast, the coagulation zones at 50 W differ substantially. Whereas the 532 nm laser gave a coagulation depth of about 1.5 mm, diode lasers emitting light at 940 and 980 nm gave a coagulation depth between 8.4 and 9.6 mm in the kidney model, indicating deeper optical tissue penetration and greater tissue damage [24, 26, 27]. With a coagulation depth of 3.4 mm, the diode laser at 1470 nm seems to be more comparable to the 532 nm laser. Although it is questionable whether the *ex vivo* perfused kidney model mimics the condition in human prostates in every respect, the specific heat capacities in kidney (3.89 kJ kg^{-1} K^{-1}) and prostate (3.80 kJ kg^{-1} K^{-1}) are almost identical. Therefore, the isolated, blood-perfused, porcine kidney seems to be a valuable model, especially for the investigation of laser procedures, in which the heat sink is of the utmost importance [28, 29].

Another set of preclinical experiments on human cadaver prostates gave similar results to those in the porcine kidney model. In this model, the macroscopic destruction capacity of the diode laser at 980 nm at 100–200 W is 2–3 times larger than the vaporization zone of the 532 nm laser at 80 W and the diode laser at 1470 nm at 50 W. In this human cadaver prostate model also, the diode laser at 980 nm demonstrated more than twofold deeper coagulation than the 532 nm laser at 80 W and the diode laser at 1470 nm at 50 W [24]. In this respect, the diode laser is able to combine very high tissue ablation properties with the benefit of excellent hemostasis due to deep coagulation. However, owing to the laser's deep coagulation zones in the 940 and 980 nm generators, there may be a risk of violating underlying structures such as the neurovascular bundles and the external sphincter, which may cause erectile dysfunction and incontinence [30].

Unfortunately, the value of the human cadaver prostates is limited, since there is no blood perfusion, which is one of the main absorbers in the wavelength range 500–1000 nm. Therefore, *in vivo* studies on canine prostates were carried out to evaluate the diode laser emitting at 940 nm using a generator output level of 200 W. The mean extension of the coagulation zone was measured as 4.3 mm and the ablation capacity was estimated to be up to 1500 mm^3 min^{-1} [26]. In comparison, the diode laser at 1470 nm produced a coagulation rim of 2.3 mm and the tissue removal ability reached about 400 mm^3 min^{-1} [27]. These findings indicate that diode laser vaporization is feasible and might be very effective for acutely relieving bladder outlet obstruction in an *in vivo* setting. Due to the different amounts of coagulation and ablation while showing similar overall destruction capabilities, in the clinical setting it is essential to protect underlying structures such as the neurovascular bundles and the external sphincter from thermal exposure, which may result in erectile dysfunction and incontinence.

32.2.2
Clinical Laser Application for BPH

Currently, only a limited number of studies have investigated the clinical applications of diode laser prostatectomy with a maximum follow-up of 1 year. Although all studies showed high intraoperative safety and excellent hemostatic properties, reports on the long-term durability and safety are inconsistent. In a preliminary clinical study on 10 patients, the diode laser at 1470 nm demonstrated an improvement in the international prostate symptom score (IPSS) after 12 months from 16.3 to 5.0, and the quality of life score (QoL) fell from 3.3 to 0.9. Also, the maximum flow rate (Q_{max}) increased from 8.9 ml s^{-1} preoperatively to 22.4 ml s^{-1} after the laser intervention. Unfortunately, reoperation rates were 20% due to laser failure in the treatment of a prominent midlobe [23]. In a single-center prospective study comparing a 200 W high-power diode laser emitting at 980 nm with a 120 W laser emitting at 532 nm, 117 patients were treated. Both laser devices provided rapid tissue ablation; the high-power diode laser at 980 nm was more favorable in terms of hemostasis during and after surgery. The higher rates of urinary retention (20%), urgency (24%), urge incontinence (7%), bladder neck strictures (15%), stress incontinence (9%), and tissue necrosis after diode laser ablation of the prostate are indications of deep tissue damage. The authors found high-power diode laser vaporization of the prostate at 980 nm to be in a premature state for treating lower urinary tract symptoms (LUTS) secondary to BPH [30]. Later, others investigated the diode laser at 980 nm at 100 W with a 0.1 s pulse duration and 0.01 s pulse interval in 52 patients and found significant durable improvements in Q_{max}, post-voiding residual urine (PVR), IPSS, and QoL score. In this single-center and single-surgeon series, no severe complications were reported, including any cases of urinary incontinence, significant irritative symptoms, retrograde ejaculation, or worsening of erectile function [30]. A subsequent investigation on diode laser vaporization at 980 nm in 47 patients by a Turkish group revealed a distinct improvement in IPSS and QoL score. There was no deterioration in erectile function. The most common postoperative complications were retrograde ejaculation (31.7%) and irritative symptoms (23.4%) [31].

32.2.3
Summary

In summary, diode laser prostatectomy is feasible. Nevertheless, despite its favorable intraoperative safety, long-term follow-up and large-scale trials will be needed to evaluate the technique finally in the future. From the clinical point of view, diode laser vaporization with generators emitting light between 900 and 1000 nm should be treated with caution owing to the large coagulation depth, which might lead to damage of underlying structures such as the neurovascular bundle, rectum, or external sphincter. On the other hand, diode lasers at 1470 nm have shown not only a good vaporization ability but also a reasonable and safe tissue coagulation capability. This might offer not only a routine operation technique but also a HoLEP (holmium laser enucleation of the prostate)-like procedure with a high safety profile.

32.3
Laser Therapy for the Upper Urinary Tract Tumors

In the treatment of upper urinary tract transitional cell carcinoma (UUTT), laser-guided techniques are mainly used through flexible and semi-rigid ureteroscopes. Nephroureterectomy has been considered as the gold standard for treating UUTT. Ureteroscopically guided laser ablation has been used successfully, resulting in recurrence rates ranging from 31% to 65% and disease-free rates of 35–86%, depending on stage and grade at diagnosis. To obtain the highest treatment success, the initial staging and grading of the tumor are crucial. Owing to the lack of comparable randomized, multicenter trials, the indications for an endoscopic laser treatment option have to be defined based on the patient's individual situation.

The endoscopic treatment of UUTT with a laser requires a light-energy transport via flexible quartz fibers introduced through the working channel of the endoscope. Depending on the laser source and the laser parameters (e.g., output power, fluence, fluence rate), the induced effects on soft tissue vary among coagulation, ablation, vaporization, and incision. Although the application of Nd:YAG and Ho:YAG lasers in urology, including in the treatment of UUTT [32–36], are widespread, innovative thulium laser developments have shown promising initial results with minimal collateral damage under experimental conditions. The roles of the thulium laser emitting in the 2 µm region and other newly evolved diode laser devices have yet to be defined in clinical studies and to produce long-term data.

Ureteral access sheaths facilitate easy introduction and re-introduction of flexible ureterorenoscopes for laser treatment throughout the whole collecting system. The new generation of scopes combine excellent intraoperative visualization and maneuverability together with low-pressure, continuous-flow irrigation and working channels of up to 3.6 Fr [36]. The degree of active and passive deflection and the amount of irrigation flow are least affected when utilizing the smallest fibers available. Attention must always be paid to prevent the laser harming the endoscope. The laser fiber tip should always be visualized a few millimeters beyond the tip of the scope [33, 36].

32.3.1
Laser Interventions

Ho:YAG laser energy is highly absorbed by water and water-containing tissues. As the heat is rapidly dispersed, minimal thermal damage to the surrounding tissue occurs. To achieve tumor ablation, the Ho:YAG laser fiber must be used in full-contact manner. For debulking of papillary urothelial tumors, the Ho:YAG laser parameters should be set in the range 0.6–1.0 J per pulse at repetition rates between 5 and 10 Hz [37, 38]. When tissue adherence to the fiber tip during ablation occurs, interruption of laser intervention and cleaning of the fiber tip are mandatory.

Although the Ho:YAG laser allows tissue ablation under direct vision, it is not possible to coagulate larger tumor vessels [32, 34, 39], and therefore circumferential treatment in the surrounding tissue is indicated [38, 40, 41]. Incomplete resection may be a problem in the case of large tumors (diameter >1.5 cm). Patients with bulky, low-grade, upper-tract tumors frequently require staged ureteroscopic therapy, meaning that the tumor is ablated to the highest amount possible; subsequently, a ureteral stent is placed, and the patient returns after an interval of healing for an additional intervention [36, 37, 39]. Staged therapy is also employed when extensive or multifocal tumor resection is required and when visualization is a constraint [37, 38, 42]. Post-laser operation, most patients require ureteral stents for 1–6 weeks [36, 37, 39]. In the case of vascular and bulky tumors, Nd:YAG laser energy should be used for coagulation because of its greater optical depth of penetration, which provides deeper coagulation and an ablative effect. Nd:YAG laser parameter settings of 20–30 W for 2–3 s using fiber diameters in the range 200–600 µm are suggested. Laser application should be performed in noncontact mode until the tumor coagulation occurs, identified by tissue bleaching and whitening [33, 34, 37–39]. Using the Nd:YAG laser, the risk of subsequent ureteral stricture is significantly higher compared with the Ho:YAG laser due to its deep tissue penetration [34]. Whenever diode laser devices with suitable fibers become clinical available, the first Nd:YAG laser treatment procedures could be replaced.

Ureteroscopic laser treatment has been reported in a relatively small but increasing number of published studies. The rate of disease-specific mortality for low-grade tumors is zero in the majority of the series [37]. Patients with grade 1 and 2 tumors treated ureteroscopically did not progress in grade or stage during close endoscopic follow-up [38]. Tumors larger than 1.5 cm requiring staged treatment showed significantly higher rates of local recurrence [43]. The localization of the tumor is less important [44]. Indications for ureteroscopic therapy initially included a solitary kidney, bilateral disease, renal insufficiency, and the presence of significant medical co-morbidities. The level of acceptance of ureteroscopically guided laser treatment of grade 1 superficial tumors of the renal pelvis is high. A less invasive approach with endoscopic laser ablation and surveillance is favored [45].

32.3.2
Complications

The rate of complication associated with retrograde endoscopic therapy has decreased with innovative developments in instrumentation and technique. Perforation while using the Ho:YAG or Nd:YAG laser can occur if the power settings are too high and the laser light is directed on to the ureteral or pelvic wall rather than the tumor. As the optical penetration depth of the Ho:YAG laser is 0.5 mm and that of the Nd:YAG laser is 5–6 mm, their coagulation effects in issue depth differ significantly. The laser fiber tip should always be visualized a few millimeters beyond the tip of the ureteroscope and directed on to the tumor under direct vision control [33, 37–39]. Owing to the induced thermal effects, there is a risk of scarring and of subsequent stricture formation after laser ablation in the ureter [33, 34, 37]. The reported stricture rate in large series ranged from 5 to 14% [33].

32.4
Laser-Assisted Lithotripsy

The surgical management of nephrolithiasis has undergone tremendous design and development changes over the past 25 years, beginning with the introduction of shockwave lithotripsy (SWL) and percutaneous management of intrarenal calculi [percutaneous nephrolithotomy (PCNL)] in the early 1980s, followed by the further development of semi-rigid ureteroscopes and the advent of actively deflecting endoscopes. Together with the miniaturization of endurologic equipment, the field of ureteroscopy expanded from the distal ureter beyond the proximal ureter to the intrarenal collecting system. With emerging new-generation flexible ureteroscopes with greater angles of maximum active tip deflection and improved durability, the complete collecting system was made accessible for ureteroscopic laser lithotripsy.

32.4.1
Experience from Preclinical Investigations

Clinically available pulsed laser systems emitting in either the infrared (IR) or visible (VIS) spectral region could be compared in reproducible standardized experiments with respect to their impact on phantom stones in underwater laboratory setups. So far, there are three pulsed laser systems emitting light either in the IR ($\lambda = 2100$ nm, Ho:YAG laser) or VIS ($\lambda = 532$ nm/1064 nm, FREDDY laser; 598 nm, FLPD laser) spectral range available. The fragmentation rates could be determined in relation to the fluence (depending on pulse energy and fiber diameter) using artificial stones. The threshold value of the laser pulse energy to induce an ablation of artificial stones induced by the different laser systems showed that even the lowest laser settings induced significant ablation without regard to the repetition rate and fiber diameter. The VIS lasers showed higher fragmentation rates than the IR lasers. However, VIS lasers are useful solely for laser-induced shockwave lithotripsy of soft stones, whereas IR lasers are also able to fragment the hardest stone composites.

32.4.2
Laser Lithotripsy from the Clinical Point of View

The combination of the newest innovative endoscopic equipment and the Ho:YAG laser, with its precise and powerful thermal decomposition mechanism, has the ability to treat large urinary calculi irrespective of size or chemical consistency [46–50].

The Ho:YAG laser emits light of wavelength 2100 nm with a pulse duration of $250\,\mu s < t_p < 300\,\mu s$ at repetition rates between 3 and 50 Hz. The mechanism of stone fragmentation is based on a plasma-induced shock wave that is generated following the creation of a water channel between the fiber tip and the stone. The laser energy is guided directly on to the stone surface through this channel ("Moses effect"). A primary photothermal mechanism causing photochemical decomposition of the calculus is the dominant mechanism behind Ho:YAG laser lithotripsy. The pressure waves generated at cavitation bubble collapse make little contribution to the fragmentation process. Chemical decomposition as a consequence of a prevalent photothermal mechanism plays the main role in the Ho:YAG fragmentation process, whereas photomechanical forces make little contribution to fragmentation. Close contact between the fiber tip and the stone surface is therefore a precondition for effective Ho:YAG laser lithotripsy that is capable of breaking stones irrespective of their composition.

In addition to the improved accessibility of the collection system due to small-diameter fibers, Ho:YAG laser lithotripsy has another major advantage: the stone debris generated is finer and therefore more likely to pass spontaneously than that from any other lithotripsy device [51, 52]. In addition, soft tissue injury and bleeding are less likely owing to the small depth of optical penetration of the Ho:YAG laser.

Even with a large stone burden (e.g., partial staghorn calculi), Ho:YAG laser lithotripsy has become the procedure of choice for patients who are poor candidates (e.g., bleeding diathesis, severe cardiopulmonary disease, certain anatomic factors, and body mass index >30), for standard percutaneous therapy. In these selected patients, ureteroscopy combined with Ho:YAG laser lithotripsy may allow a decrease in morbidity and hospital stay with a stone-free rate touching that with the PCNL approach.

When treating large upper urinary tract calculi, the prevention of inadvertent laser firing, patience, and clear visibility are essential. Decreased visibility during fragmentation leads to a prolonged operative time and the potential risk of injuring the surrounding tissue or the ureteroscope, making a staged approach ultimately necessary in several cases.

Major complications secondary to ureteroscopy are less common and ureteroscopy has proved to be safe in patients where shock wave lithotripsy is likely to fail or in whom PCNL may be contraindicated (e.g., pregnancy, obese patients, patients with coagulopathy or scoliosis or other body deformities) [53–55]. Due to the decrease in ureteroscope size, significant complications such as ureteral avulsion and stricture formation are exceedingly rare and have been reported to be as low as 1.5% [56].

The combination of the retrograde ureteroscopic approach and the holmium laser as the lithotrite of choice appears to be an adequate tool to disintegrate all kinds of calculi, independent of stone size or location, throughout the urinary tract. On the downside, generating small fragments or fine debris is a time-consuming procedure resulting in long operative times, making staged therapy necessary especially for large stone burdens in some cases.

32.5
Conclusion

During recent years, the application of thermal lasers in urology has increased. Both benign and malignant operations can be managed effectively. The primary indication for thermal laser treatment is symptomatic prostate enlargement due to BPH which can be treated either by laser enucleation techniques or by laser vaporization. Both procedures are equally effective as standard treatment [57]. Although diode laser applications have shown encouraging results and further lasers systems such as the thulium laser [58–61] are clinically available, the applicability of these laser treatment procedures needs to be confirmed in the future. Latest investigations in clinical laser application in urology report on successfull laparascopic laser assisted partial nephrectomy without need of ischemia [62]. With respect to treating patients suffering from clinical stone diseases, the combination of the retrograde ureteroscopic approach and the Ho:YAG laser as the lithotrite of choice appears to be an adequate tool to disintegrate all kinds of calculi, independent of stone size or location, throughout the urinary tract. Overall, laser assisted endoscopic treatments in urology are widespread and show promising advantages over conventional methods.

References

1 Kuntz, R.M. (2006) Current role of lasers in the treatment of benign prostatic hyperplasia (BPH). *Eur. Urol.*, **49** (6), 961–969.
2 Naspro, R., Bachmann, A., Gilling, P., Kuntz, R., Madersbacher, S., Montorsi, F., et al. (2009) A review of the recent evidence (2006–2008) for 532-nm photoselective laser vaporisation and holmium laser enucleation of the prostate. *Eur. Urol.*, **55** (6), 1345–1357.
3 McAllister, W.J. and Gilling, P.J. (2004) Vaporization of the prostate. *Curr. Opin. Urol.*, **14** (1), 31–34.
4 Te, A.E. (2004) The development of laser prostatectomy. *BJU Int.*, **93** (3), 262–265.
5 Hai, M.A. and Malek, R.S. (2003) Photoselective vaporization of the prostate: initial experience with a new 80 W KTP laser for the treatment of benign prostatic hyperplasia. *J. Endourol.*, **17** (2), 93–96.
6 Bachmann, A., Ruszat, R., Wyler, S., Reich, O., Seifert, H.H., Müller, A., et al. (2005) Photoselective vaporization of the prostate: the Basel experience after 108 procedures. *Eur. Urol.*, **47** (6), 798–804.

7 Sarica, K., Alkan, E., Lüleci, H., and Taşci, A.I. (2005) Photoselective vaporization of the enlarged prostate with KTP laser: long-term results in 240 patients. *J. Endourol.*, **19** (10), 1199–1202.

8 Ruszat, R., Seitz, M., Wyler, S.F., Abe, C., Rieken, M., Reich, O., *et al.* (2008) Green light laser vaporization of the prostate: single-center experience and long-term results after 500 procedures. *Eur. Urol.*, **54** (4), 893–901.

9 Hamann, M.F., Naumann, C.M., Seif, C., van der Horst, C., Jünemann, K.P., and Braun, P.M. (2008) Functional outcome following photoselective vaporisation of the prostate (PVP): urodynamic findings within 12 months follow-up. *Eur. Urol.*, **54** (4), 902–907.

10 Kuntz, R.M., Lehrich, K., and Ahyai, S.A. (2008) Holmium laser enucleation of the prostate versus open prostatectomy for prostates greater than 100 grams: 5-year follow-up results of a randomised clinical trial. *Eur. Urol.*, **53** (1), 160–166.

11 Naspro, R., Suardi, N., Salonia, A., Scattoni, V., Guazzoni, G., Colombo, R., *et al.* (2006) Holmium laser enucleation of the prostate versus open prostatectomy for prostates >70 g: 24-month follow-up. *Eur. Urol.*, **50** (3), 563–568.

12 Wilson, L.C., Gilling, P.J., Williams, A., Kennett, K.M., Frampton, C.M., Westenberg, A.M. *et al.* (2006) A randomised trial comparing holmium laser enucleation versus transurethral resection in the treatment of prostates larger than 40 grams: results at 2 years. *Eur. Urol.*, **50** (3), 569–573.

13 Ahyai, S.A., Lehrich, K., and Kuntz, R.M. (2007) Holmium laser enucleation versus transurethral resection of the prostate: 3-year follow-up results of a randomized clinical trial. *Eur. Urol.*, **52** (5), 1456–1463.

14 Gilling, P.J., Aho, T.F., Frampton, C.M., King, C.J., and Fraundorfer, M.R. (2008) Holmium laser enucleation of the prostate: results at 6 years. *Eur. Urol.*, **53** (4), 744–749.

15 Tan, A., Liao, C., Mo, Z., and Cao, Y. (2007) Meta-analysis of holmium laser enucleation versus transurethral resection of the prostate for symptomatic prostatic obstruction. *Br. J. Surg.*, **94** (10), 1201–1208.

16 Pow-Sang, M., Orihuela, E., Motamedi, M., Pow-Sang, J.E., Cowan, D.F., Dyer, R., and Warren, M.M. (1995) Thermocoagulation effect of diode laser radiation in the human prostate: acute and chronic study. *Urology*, **45**, 790–794.

17 Prapavat, V., Roggan, A., Walter, J., Beuthan, J., Klingbeil, U., and Muller, G. (1996) *In vitro* studies and computer simulations to assess the use of a diode laser (850 nm) for laser-induced thermotherapy (LITT). *Lasers Surg. Med.*, **18**, 22–33.

18 Cromeens, D.M., Johnson, D.E., Stephens, L.C., and Gray, K.N. (1996) Visual laser ablation of the canine prostate with a diffusing fiber and an 805-nanometer diode laser. *Lasers Surg. Med.*, **19**, 135–142.

19 Martenson, A.C. and de la Rosette, J.J. (1999) Interstitial laser coagulation in the treatment of benign prostatic hyperplasia using a diode laser system: results of an evolving technology. *Prostate Cancer Prostatic Dis.*, **2**, 148–154.

20 Daehlin, L. and Hedlund, H. (1999) Interstitial laser coagulation in patients with lower urinary tract symptoms from benign prostatic obstruction: treatment under sedoanalgesia with pressure-flow evaluation. *BJU Int.*, **84**, 628–636.

21 de la Rosette, J.J., Muschter, R., Lopez, M.A., and Gillatt, D. (1997) Interstitial laser coagulation in the treatment of benign prostatic hyperplasia using a diode-laser system with temperature feedback. *Br. J. Urol.*, **80**, 433–438.

22 Sroka, R., Ackermann, A., Tilki, D., Reich, O., Steinbrecher, V., Hofstetter, A., Stief, C.G., and Seitz, M. (2007) *In-vitro* comparison of the tissue vaporisation capabilities of different lasers. *Med. Laser Appl.*, **22**, 227–231.

23 Seitz, M., Sroka, R., Gratzke, C., Schlenker, B., Steinbrecher, V., Khoder, W., Tilki, D., Bachmann, A., Stief, C., and Reich, O. (2007) The diode laser: a novel side-firing approach for laser vaporisation of the human prostate – immediate efficacy

and 1-year follow-up. *Eur. Urol.*, **52**, 1717–1722.

24 Seitz, M., Reich, O., Gratzke, C., Schlenker, B., Karl, A., Bader, M., Khoder, W., Fischer, F., Stief, C., and Sroka, R. (2009) High-power diode laser at 980 nm for the treatment of benign prostatic hyperplasia: *ex vivo* investigations on porcine kidneys and human cadaver prostates. *Lasers Med. Sci.*, **24**, 172–178.

25 Seitz, M., Ackermann, A., Gratzke, C., Schlenker, B., Ruszat, R., Bachmann, A., Stief, C., Reich, O., and Sroka, R. (2007) Diode laser: *ex vivo* studies on vaporization and coagulation characteristics. *Urologe A*, **46**, 1242–1247.

26 Seitz, M., Bayer, T., Ruszat, R., Tilki, D., Bachmann, A., Gratzke, C., Schlenker, B., Stief, C., Sroka, R., and Reich, O. (2009) Preliminary evaluation of a novel side-fire diode laser emitting light at 940 nm, for the potential treatment of benign prostatic hyperplasia: *ex-vivo* and *in-vivo* investigations. *BJU Int.*, **103**, 770–775.

27 Seitz, M., Ruszat, R., Bayer, T., Tilki, D., Bachmann, A., Stief, C., Sroka, R., and Reich, O. (2009) *Ex vivo* and *in vivo* investigations of the novel 1470 nm diode laser for potential treatment of benign prostatic enlargement. *Lasers Med. Sci.*, **24**, 419–424.

28 Cooper, T.E. and Trezek, G.J. (1972) A probe technique for determining the thermal conductivity of tissue. *J. Heat Transfer*, **94**, 133.

29 Reich, O., Bachmann, A., Schneede, P., Zaak, D., Sulser, T., and Hofstetter, A. (2004) Experimental comparison of high power (80 W) potassium titanyl phosphate laser vaporization and transurethral resection of the prostate. *J. Urol.*, **171**, 2502–2504.

30 Ruszat, R., Seitz, M., Wyler, S.F., Muller, G., Rieken, M., Bonkat, G., Gasser, T.C., Reich, O., and Bachmann, A. (2009) Prospective single-centre comparison of 120-W diode-pumped solid-state high-intensity system laser vaporization of the prostate and 200-W high-intensive diode-laser ablation of the prostate for treating benign prostatic hyperplasia. *BJU Int.*, **104**, 820–825.

31 Erol, A., Cam, K., Tekin, A., Memik, O., Coban, S., and Ozer, Y. (2009) High power diode laser vaporization of the prostate: preliminary results for benign prostatic hyperplasia. *J. Urol.*, **182**, 1078–1082.

32 Schilling, A., Bowering, R., and Keiditsch, E. (1986) Use of the neodymium–YAG laser in the treatment of ureteral tumors and urethral condylomata acuminata: clinical experience. *Eur. Urol.*, **12** (Suppl 1), 30–33.

33 Chen, G.L. and Bagley, D.H. (2001) Ureteroscopic surgery for upper tract transitional-cell carcinoma: complications and management. *J. Endourol.*, **15**, 399–404.

34 Schmeller, N.T. and Hofstetter, A.G. (1989) Laser treatment of ureteral tumors. *J. Urol.*, **141**, 840–843.

35 Marks, A.J. and Teichmann, J.M.H. (2007) Laser in clinical urology: state of the art and new horizons. *World J. Urol.*, **25**, 227–233.

36 Lam, J.S. and Gupta, M. (2004) Ureteroscopic management of upper tract transitional cell carcinoma. *Urol. Clin. North Am.*, **31**, 115–128.

37 Johnson, G.B. and Grasso, M. (2005) Ureteroscopic management of upper urinary tract transitional cell carcinoma. *Curr. Opin. Urol.*, **15**, 89–93.

38 Grasso, M., Fraiman, M., and Levine, M. (1999) Ureteropyeloscopic diagnosis and treatment of upper urinary tract urothelial malignancies. *Urology*, **54**, 240–249.

39 Grasso, M. (2000) Ureteroscopic management of upper urinary tract urothelial malignancies. *Rev. Urol.*, **2**, 116–121.

40 Elliott, D.S., Blute, M.L., and Patterson, D.E. (1996) Long-term follow-up of endoscopically treated upper urinary tract transitional cell carcinoma. *Urology*, **47**, 819–825.

41 Razvi, H.A., Shun, S.S., and Denstedt, J.D. (1995) Soft-tissue applications of the holmium:YAG laser in urology. *J. Endourol.*, **9**, 387–390.

42 Gerber, G.S. and Lyon, E.S. (1993) Endourological management of

upper tract urothelial tumors. *J. Urol.*, **150**, 2–7.
43 Keeley, F.X., Jr., Bibbo, M., and Bagley, D.H. (1997) Ureteroscopic treatment and surveillance of upper urinary tract transitional cell carcinoma. *J. Urol.*, **157**, 1560–1565.
44 Tawfiek, E.R. and Bagley, D.H. (1999) Upper-tract transitional cell carcinoma. *Urology*, **50**, 321–329.
45 Razdan, S., Johannes, J., Cox, M., and Bagley, D.H. (2005) Current practice patterns in urologic management of upper-tract transitional cell carcinoma. *J. Endourol.*, **19**, 366–371.
46 Grasso, M., Conlin, M., and Bagley, D. (1998) Retrograde ureteropyeloscopic treatment of 2 cm or greater upper urinary tract and minor staghorn calculi. *J. Urol.*, **160**, 346–351.
47 Busby, J.E. and Low, R.K. (2004) Ureteroscopic treatment of renal calculi. *Urol. Clin North Am.*, **31**, 89–98.
48 Mariani, A.J. (2007) Combined electrohydraulic and holmium: YAG laser ureteroscopic nephrolithotripsy of large (greater than 4 cm) renal calculi. *J. Urol.*, **177**, 168–173; discussion 73.
49 Johnson, G.B., Portela, D., and Grasso, M. (2006) Advanced ureteroscopy: wireless and sheathless. *J. Endourol.*, **20**, 552–555.
50 Vassar, G.J., Chan, K.F., Teichman, J.M. et al. (1999) Holmium:YAG lithotripsy: photothermal mechanism. *J. Endourol.*, **13**, 181–190.
51 Mariani, A.J. (2007) Combined electrohydraulic and holmium:YAG laser ureteroscopic nephrolithotripsy of large (greater than 4 cm) renal calculi. *J. Urol.*, **177**, 168–173; discussion 73.
52 Teichman, J.M., Vassar, G.J., Bishoff, J.T., and Bellman, G.C. (1998) Holmium:YAG lithotripsy yields smaller fragments than lithoclast, pulsed dye laser or electrohydraulic lithotripsy. *J. Urol.*, **159**, 17–23.
53 Soucy, F., Ko, R., Duvdevani, M. et al. (2009) Percutaneous nephrolithotomy for staghorn calculi: a single center's experience over 15 years. *J. Endourol.*, **23** (10), 1669–1673.
54 Busby, J.E. and Low, R.K. (2004) Ureteroscopic treatment of renal calculi. *Urol. Clin North Am.*, **31**, 89–98.
55 Miller, N.L. and Lingeman, J.E. (2007) Management of kidney stones. *BMJ*, **334**, 468–472.
56 Harmon, W.J., Sershon, P.D., Blute, M.L., Patterson, D.E., and Segura, J.W. (1997) Ureteroscopy: current practice and long-term complications. *J. Urol.*, **157**, 28–32.
57 Rieken, M. and Bachmann, A. (2010) Thermal lasers in urology. *Med. Laser Appl.*, **25** (1), 20–26.
58 Bach, T., Wendt-Nordahl, G., Michel, M.S., Herrmann, T.R., and Gross, A.J. (2009) Feasibility and efficacy of thulium:YAG laser enucleation (VapoEnucleation) of the prostate. *World J. Urol.*, **27** (4), 541–545.
59 Bach, T., Herrmann, T.R., Haecker, A., Michel, M.S., and Gross, A. (2009) Thulium:yttrium-aluminium-garnet laser prostatectomy in men with refractory urinary retention. *BJU Int.*, **104** (3), 361–364.
60 Bach, T., Herrmann, T.R., Ganzer, R., Burchardt, M., and Gross, A.J. (2007) RevoLix vaporesection of the prostate: initial results of 54 patients with a 1-year follow-up. *World J. Urol.*, **25** (3), 257–262.
61 Xia, S.J., Zhuo, J., Sun, X.W., Han, B.M., Shao, Y., and Zhang, Y.N. (2008) Thulium laser versus standard transurethral resection of the prostate: a randomized prospective trial. *Eur. Urol.*, **53** (2), 382–389.
62 Khoder, W.Y., Sroka, R., Hennig, G., Seitz, M., Siegert, S., Zillinberg, K., Gratzke, C., Stief, C.G., Becker, A.j. (2011) The 1318-nm diode laser supported partial nephrectomy in larascopic and open surgery: prelimnary results of a prospective feasability study. *Lasers Med. Sci.* doi 10.1007/s10103-011-0897-y.

**Part 7
Gastroenterology**

33
Barrett's Esophagus and Gastroesophageal Reflux Disease – Diagnosis and Therapy

Jason M. Dunn, Steven G. Bown, and Laurence B. Lovat

33.1
Introduction

The rationale for endoscopic surveillance in patients with Barrett's esophagus (BE) is to detect progression to cancer and to allow early intervention when the disease is curable. Dysplasia arising in BE increases cancer risk, although this risk is variable according to the grade of dysplasia, and is often occult. The four-quadrant biopsy protocol, one biopsy in every quadrant of the Barrett's segment every 1–2 cm, has been shown to detect high-grade dysplasia (HGD) [1] and is recommended for Barrett's surveillance by several professional bodies, including the American Gastroenterology Association (AGA) and the British Society of Gastroenterology (BSG) [2, 3]. Even using these rigorous biopsy protocols, only about 1% of a patient's BE is examined histologically and up to one-third of cases with HGD or early cancer may be missed [4–7]. Furthermore, dysplasia itself is an imperfect marker of cancer risk, with progression rates to cancer varying between 13 and 59% after 5 years with HGD to between 1 and 8% after 5 years with low-grade dysplasia (LGD). The diagnosis of dysplasia is difficult to assess correctly and subject to disagreement between expert gastrointestinal pathologists, 15% of the time for HGD and 80% of the time for LGD [8, 9].

There is a need for more accurate assessment of the Barrett's columnar lined mucosa. Technological advances in gastrointestinal endoscopy since the introduction of conventional white light endoscopes have led to a plethora of potential tools to aid the endoscopist. In scope technologies such as narrow band imaging, magnification endoscopy by optical zoom, and high-definition charge-coupled device (CCD) chips (which produce pixel densities up to 1.25 million) are now widely used. Enhanced identification of intestinal metaplasia has been demonstrated by simple spray techniques, either by altering mucosal topography using acetic acid or chromoendoscopy with methylene blue [10, 11]. Confocal laser endomicroscopy (CLE) and optical coherence tomography (OCT) are promising research tools which inform on tissue characteristics at a cellular level. Despite these technical advances, a significant diagnostic advantage over four-quadrant biopsy has yet to be achieved. The main disadvantage of these examiner-dependent modalities is end-user interpretation,

Handbook of Biophotonics. Vol.2: Photonics for Health Care, First Edition. Edited by Jürgen Popp,
Valery V. Tuchin, Arthur Chiou, and Stefan Heinemann.
© 2012 Wiley-VCH Verlag GmbH & Co. KGaA. Published 2012 by Wiley-VCH Verlag GmbH & Co. KGaA.

with inter-observer variability [12] and a learning curve for the procedure [13] contributing significant bias to studies. Spectroscopic techniques also inform on tissue at a nuclear level, and are advantageous as end-user interpretation is removed. The first section of this chapter reviews the evidence for these novel optical techniques for the assessment of BE, with focus on translational research from bench to bedside.

The second section explores recent advances of minimally invasive treatment for dysplasia and early cancer arising in BE. The principle of endoscopic therapy for BE is that the mucosa of columnar lined esophagus may be effectively removed and is replaced, when healing is in an acid-controlled environment, by squamous epithelium (squamous regeneration). The use of biophotonics for this purpose was first demonstrated in 1992 with Nd:YAG laser therapy [14]. Since then, there has been much interest in the use of biophotonic techniques to provide endosopcially delivered therapy for dysplastic BE. These may be broadly categorized into those that utilize thermal energy (argon plasma coagulation and, more recently, radiofrequency ablation) and photodynamic therapy (PDT).

33.2
Diagnosis

33.2.1
Spectroscopy

All spectroscopic modalities analyze light–tissue interactions to yield diagnostic information about the structural (scattering) and molecular (Raman, absorption, fluorescence) composition of tissue. When light of a certain wavelength is used to stimulate tissue, distinct optical signals are produced. These "spectra" can then be correlated with histopathology and analyzed to create algorithms for tissue classification and grading. Spectroscopy offers the potential for real-time endoscopic differentiation of neoplastic and normal mucosa. and has the additional advantage of dispensing with end-user interpretation.

33.2.1.1 Scattering (Reflectance) Spectroscopy

Elastic and Light Scattering Spectroscopy When tissue is interrogated by white light, typically delivered through a fiber, *elastic* scattering events occur that redirect incident photons without a change in wavelength (or energy). Elastic scattering (also known as diffuse reflectance) spectra, which may be detected by either the same fiber or a second collection fiber, are a measure of the behaviour of white light emerging from the tissue surface following these multiple elastic scattering events. These spectra may contain both multiply scattered photons (primarily from the stroma) and photons that have undergone single or few scattering events following their interaction with the epithelium. Incident photons are also sensitive to tissue absorbers, such as hemoglobin, that reduce the intensity of the diffusely reflected light at particular wavelengths. Elastic scattering spectroscopy (ESS) thus informs on the tissue morphology (predominantly scatter) and tissue biochemistry (predominantly absorption by hemoglobin). The interpretation of these spectra, and how they correlate with the physical properties of preneoplastic and neoplastic tissue, have attracted much interest.

Physical Correlation ESS spectra relate to the wavelength dependence and angular probability of scattering efficiency of tissue micro-components. The size, structure, and refractive indices of the denser subcellular components (e.g., the nucleus, nucleolus, and mitochondria) are known to change upon transformation to premalignant or malignant conditions. Indeed, histopathologists use these nuclear changes to grade dysplasia, that is, the sizes and shapes of nuclei and organelles, the ratio of nuclear to cellular volume (nuclear:cytoplasmic ratio) and clustering patterns (nuclear crowding).

Mourant *et al.* examined the influence of scatterer size in rat tumor cell lines and compared them with normal cells [15]. An optical geometry of 400 nm optical fiber and a 200 nm collection fiber spaced 550 nm away (center-to-center) was used. ESS measurements demonstrated a small but significant difference in optical properties between the two cell lines, with this difference in the scattering rather than the absorption properties of the cells. The authors postulated either that this was due to the average size of a scattering particle being larger in cancer cell lines (e.g., size of nuclei or mitochondria increase), or that there could be a change in the number and distribution of scattering particles (e.g., number of mitochondria increase).

The same group went on to hypothesize that spectral signatures will be altered if the refractive index of the nucleus changes, due to an increase in the amount of deoxyribonucleic acid (DNA) content [16]. By measuring DNA content by flow cytometry, the DNA indices DNA index (DI) were calculated for two different cell lines, and plotted against high-angle scatter (between $110°$ and $140°$). Cell suspensions with higher DI (analogous to abnormal DNA content or aneuploidy) scattered more light than cell suspensions with smaller DI (analogous to normal DNA content or diploid).

Backman *et al.* reported on the correlation of tissue morphology with reflectance spectra using polarized light, which allows the measurement of single scattering events, called light-scattering spectroscopy (LSS) [17]. LSS was evaluated on colonic cell lines and *ex vivo* colonic polyps, where the spectrum of the single backscattering events provided quantitative information about the epithelial cell nuclei such as nuclear size, degree of pleomorphism, degree of hyperchromasia, and amount of chromatin [18].

Clinical Studies of ESS/LSS The largest clinical *in vivo* study of scattering spectroscopy to date evaluated the value of ESS in discriminating dysplastic and nondysplastic tissue in BE [9]. A total of 181 matched biopsy sites from 81 patients, where histopathologic consensus was reached, were analyzed. There was good pathologist agreement in differentiating HGD and cancer from other pathology ($\kappa = 0.72$). Spectral data was analyzed by linear discriminant analysis (LDA) + principle component analysis (PCA) to form a model which was then tested by leave-one-out cross-validation. Elastic scattering spectroscopy detected HGD or cancer with 92% sensitivity and 60% specificity. If used to target biopsies during endoscopy, the number of low-risk biopsies taken would decrease by 60% with minimal loss of accuracy. ESS had a negative predictive value of 99.5% for HGD or cancer. This group went on to publish, in abstract form, on the accuracy of detection of DNA content abnormalities (aneuploidy/tetraploidy) *in vivo* [19]. A total of 258 biopsies without HGD were collected and analyzed by image cytometric DNA analysis from 60 patients. The sensitivity and specificity for the diagnosis of DNA content abnormalities was 83 and 78% respectively.

There have been two clinical studies of LSS. The first, on 13 BE patients (76 biopsies, four HGDs), found a sensitivity and specificity of 90% [20]. The second study evaluated LSS with ESS and autofluorescence spectroscopy (AFS) as part of a putative "tri-modal spectroscopy" system. Sixteen patients with BE (40 biopsies, seven HGDs) were studied and the sensitivity and specificity for differentiating HGD from non-HGD BE was evaluated. LSS gave a sensitivity and specificity of 100 and 91%, respectively, ESS 86 and 100%, respectively and AFS 100 and 97%, respectively [21]. When all three modalities were combined, the sensitivity and specificity were both reported as 100%, although as only seven sites contained dysplasia, and multiple spectral parameters were evaluated, the possibility of statistical error by overfitting is high.

Angle-resolved low-coherence interferometry (a/LCI) is a recent advance, a probe-based optical biopsy technique that combines OCT with LSS. Data from an *ex vivo* pilot study suggested that a/LCI may inform on nuclear density and the nuclear diameter [22]. Although these data are encouraging, the sample size was small (18 sites) and the device is yet to be tested *in vivo*.

The Field Effect – 4DELF and LEBS Backman's group went on to develop two new approaches that are extensions of LSS, called four-dimensional elastic light-scattering fingerprinting (4D-ELF) and low-coherence enhanced backscattering spectroscopy (LEBS). Both of these techniques may detect field changes in carcinogenesis – the proposition that the genetic/environmental milieu that results in neoplasia in one mucosal region should be detectable at a distant mucosal site [23]. In a landmark clinical study, both LEBS and 4D-ELF were able to detect pancreatic adenocarcinoma from optical measurements taken from the histologically normal duodenal periampullary mucosa [24].

Studies of the azoxymethane-treated rat carcinogenesis model and colonic adenomas in humans have also demonstrated changes in LEBS and 4D-ELF, with the authors suggesting these may be used as novel optical biomarkers for cancer [25, 26]. In addition to changes in tissue organization, 4D-ELF also measures subtle changes in microvascular blood supply [so-called early increase in blood supply (EIBS)] that may inform on field carcinogenesis elsewhere [27]. A recent study using a simplified 4D-ELF probe supports this theory in colon cancer [28]. No studies have been undertaken on patients with BE or early gastric cancer to date. Further large-scale clinical studies of these promising techniques are awaited.

33.2.1.2 Raman Spectroscopy

Raman spectroscopy provides molecular-specific information about tissue based on the detection of *inelastically* scattered light. Following tissue excitation by a laser, a small fraction of the scattered light undergoes "wavelength shifts" relative to the excitation wavelength due to energy transfer between incident photons and tissue molecules. The emitted signal is very weak because only a very small proportion of light interacting with tissue results in the inelastic scattering (compared with elastic scattering and autofluorescence, which are up to 10^6 times and 10^3 times stronger, respectively). Recent developments in instrumentation have permitted its application in tissue to allow the objective identification of molecular markers associated with neoplastic progression [29–31]. Despite promising results *ex vivo*, however, *in vivo* studies in BE are limited because readings take a comparatively long time to collect (around 5 s) and the esophagus is a dynamic organ. In a single-center *in vivo* study of 65 patients, 26 out of 192 biopsy sites showed HGD or early cancer and Raman spectra correctly identified these with a sensitivity of 88% and a specificity of 89% [32].

33.2.1.3 Absorption Spectroscopy

Fourier Transform Infrared Spectroscopy Fourier transform infrared (FTIR) spectroscopy probes the vibrational properties of amino acids and cofactors which are sensitive to minute structural changes [33]. FTIR spectroscopy is advantageous over Raman spectroscopy as it produces large signals that are insensitive to tissue autofluorescence. In the first study of FTIR spectroscopy in BE, Wang *et al.* showed that DNA, protein, glycogen, and glycoprotein comprise the principal sources of infrared absorption [34]. Using LDA with leave-one-out cross-validation on 98 *ex vivo* specimens from 32 patients, squamous mucosa could be distinguished from columnar mucosa (BE and gastric mucosa) with a sensitivity and specificity of 100 and 97% respectively. Moreover, using FTIR to detect dysplastic mucosa resulted in a sensitivity and specificity of 92 and 80%, respectively. Dysplastic Barrett's mucosa was found to have a higher mean DNA content and greater average glycoprotein concentration than nondysplastic BE. This study was limited by its use of partially dehydrated tissue samples. Martin *et al.* commented that "hydrated or *in vivo* samples are a problem for FTIR spectroscopy because the water peak obscures much of the useful information in the biomolecular range" [35].

33.2.1.4 Fluorescence Spectroscopy

Autofluorescence results when a tissue surface is illuminated by short wavelengths, either ultraviolet (400 nm) or short visible light (400–550 nm). The resulting fluorescence is of a longer wavelength and emanates from tissue biomolecules ("fluorophores") that are distributed in varying concentrations in the different layers of the esophagus. The dominant fluorescent layer is the submucosa, which contains highly green-fluorescing connective tissues (collagen and elastin), while a weaker mucosal component arises from epithelium and lamina propria. Different excitation wavelengths activate different groups of fluorophores, each of which emits at different wavelengths. HGD is characterized by low collagen fluorescence and high nicotinamide adenine dinucleotide phosphate [NAD(P)H] fluorescence, which manifests as an increase in red fluorescence. Although first described in BE using spectroscopic point measurements (AFS), the technology has since been adapted for incorporation into video endoscopes [autofluorescence imaging (AFI)] [36–39].

Autofluorescence Spectroscopy Panjehpour et al. were the first to report the in vivo use of AFS in the esophagus [40]. Thirty-six patients were studied using a laser-induced fluorescence (LIF) system with an excitation wavelength of 410 nm, which corresponds to the excitation of flavins. Normal esophagus could be distinguished from esophageal cancer with a sensitivity of 100% and a specificity of 98%. The specificity, however, dropped to 28% for the 56 biopsies that demonstrated focal HGD.

Two studies assessed the accuracy of diagnosis of dysplasia when the excitation wavelengths are in the near-ultraviolet range. With 330 nm excitation light, the major endogenous fluorophores contributing to tissue emission are collagen (absorption maximum 335 nm), NADH (absorption maximum 340 nm) and, to a much lesser extent, flavins (absorption maximum 450 nm). Hence, as several fluorophores are involved, more spectral information may be collected.

Bourg-Heckly et al. used various excitation ratios of different wavelengths to assess 18 BE, 15 nondysplastic, and three HGD/intramucosal cancer patients [41]. An excitation ratio of 390/550 nm gave the best results, with a sensitivity of 86% and a specificity of 95%.

Another study reported disappointing results using 337 nm excitation, with a sensitivity of 74% and a specificity of 67% for differentiating dysplastic and non-dysplastic Barrett's mucosa [42]. The authors of this study also evaluated time-resolved fluorescence, with the hypothesis that using time- or frequency-domain techniques, changes in tissue constituents or factors such as pH or oxygenation may be evaluated. A previous study had successfully applied this technique in the colon for the detection of adenomatous polyps, with a subsequent increase in diagnostic accuracy [43]. In the study on BE, however, no such improvement was noted.

Drug-Induced Fluorescence Spectroscopy Fluorescence can be enhanced by the addition of exogenous prodrugs that induce endogenous photosensitization. Aminolevulinic acid (ALA) is perhaps the best studied. Administration causes an increase in porphobilinogen deaminase levels and decreased ferrochelatase levels in the tissue, which leads to greater production and retention of protoporphyrin IX (PpIX) in neoplastic cells. Excitation by ultraviolet light from a xenon lamp (375–440 nm) causes emission of fluorescence signals which are stronger than AFS.

Clinical studies of drug-induced fluorescence spectroscopy (DFS) with oral and topical ALA have yet to demonstrate a significant diagnostic advantage over AFS, however [44], and due to the problems of variable drug pharmacokinetics, adverse effects, costs, and regulatory approval, its use is unlikely to gain wider application in Barrett's surveillance.

33.2.1.5 Summary

The understanding of the physical correlation of spectroscopic data and tissue morphology continues to evolve. A correlation with genomic instability both at the site of biopsy and from a field effect has now been shown. In addition, the recent introduction of both CLE and OCT allows real-time *in vivo* assessment at the cellular level. These advances in optical diagnostics may pave the way for a new strategy in Barrett's surveillance, whereby a multimodal approach would allow the endoscopist to quantify accurately an individual's risk of progression to cancer *in vivo*. Furthermore, as endoscopic therapy becomes more commonly employed as a first-line treatment for dysplasia and early cancer in BE, so too will there be a need for accurate assessment of the mucosa, both prior to and after therapy. Indeed, it may be possible for real-time management decisions to be made at endoscopy, based on information from these modalities. Ultimately, the adoption of these advanced optical techniques as standard practice will be based on ease of use, diagnostic accuracy, and cost.

33.3 Therapy

33.3.1 Argon Plasma Coagulation

Argon plasma coagulation (APC) is a noncontact technique in which monopolar energy is delivered to tissue using ionized argon gas, via a probe held a few millimeters from the target surface. There are many studies that report short-term success rates for complete reversal of dysplasia and metaplasia of between 60 and 100%, with efficacy apparently dependent on the power used and total energy delivered to the tissue [45–50]. The long-term efficacy of APC is questionable, however, with recurrence rates as high as 66% in one study [51]. This may be a consequence of the relatively superficial injury inflicted by APC, and also the inherent difficulty in applying the energy uniformly over the area of abnormal mucosa. Serious adverse events reported include esophageal perforations, pneumomediastinum, strictures (4–9%), and significant gastrointestinal bleeding following therapy at higher powers (60–90 W) in 4–9% [48, 52]. These complication rates, and the introduction of newer ablation techniques, have rendered APC largely obsolete for the eradication of BE, although APC could still be used to eliminate small islands of Barrett's metaplasia [53].

33.3.2 Photodynamic Therapy

PDT is a field ablation technique that is potentially an effective lower risk alternative to esophagectomy for the treatment of HGD in BE. It is a two step, nonthermal, photochemical process, which requires the interaction of a photosensitizer, light, and oxygen.

The first step is the administration of the chemical photosensitizer, which becomes concentrated in the abnormal target tissue. The majority of PDT photosensitizers possess a heterocyclic ring structure similar to that of either chlorophyll (chlorine-based photosensitizers) or, more commonly, heme (porphyrin-based photosensitizers). The second step is the application of visible light of appropriate wavelength and energy, which photoactivates the porphyrin molecule. The resulting production of cytotoxic singlet oxygen, or superoxide, contributes to the destruction of the abnormal cells where it is concentrated [54]. The most commonly used PDT light sources are lasers as they produce high-power monochromatic light of a specific wavelength with a narrow bandwidth that can be matched to the absorption peak for specific photosensitizers.

Many parameters may be varied to obtain optimal conditions for PDT, including drug dose, light power, total energy delivered, wavelength of activating light, and the time interval between photosensitizer administration and activation (drug–light interval). Hence, although there are many photosensitizers under development, only the first-generation hematoporphyrin derivative Photofrin (porfimer sodium; Axcan Pharma Inc.), has received Food and Drug Administration (FDA) and National Institute for Health and Clinical Excellence (NICE) approval for the treatment of BE. Both Levulan (5-aminolevulinic acid HCl; DUSA Pharmaceuticals, Inc.) and Foscan (temoporfin, m-tetrahydroxyphenylchlorin, mTHPC; Biolitec AG) have been investigated and the results of clinical trials are summarized below.

33.3.2.1 Photofrin

Photofrin is a hematoporphyrin derivative (HpD), a mixture of different porphyrins, and dihematoporphyrin ester (DHE). NICE approved Photofrin PDT for treating HGD in BE in 2006. The drug is administered by intravenous injection, 3 days prior to therapy, at a dose of $2\,\mathrm{mg\,kg^{-1}}$ body weight. Photofrin is activated by 630 nm laser light, typically by using a balloon-based cylindrical fiber-optic diffuser (e.g., Wizard X-cell balloon) that is placed endoscopically in the esophageal lumen, and inflated so as to hold position and flatten the esophageal folds [55]. Photosensitivity lasts up to 3 months after drug delivery.

Clinical Studies Early studies of cohort series demonstrated encouraging results for ablation of BE, with reversal of HGD in 88–95% [56–58]. A multicenter, international randomized controlled trial of 200 patients with high-grade dysplasia in BE confirmed the effectiveness of Photofrin PDT [59, 60]. Patients were randomized in a 2:1 ratio to either Photofrin PDT and a proton pump inhibitor (PPI), or PPI alone. Complete reversal of high-grade dysplasia (CR-HGD) at 1 year was achieved with Photofrin PDT in 71% versus 30% in the control group ($p < 0.001$). In the follow-up evaluation at 5 years, CR-HGD was 59% compared with 5% in the control arm [60]. A significant decreased incidence of esophageal adenocarcinoma (13% with Photofrin PDT, 28% with PPI alone) was reported as a secondary end point.

A retrospective study of 199 patients with HGD in BE compared outcomes of Photofrin PDT versus surgery [61]. There was no significant difference in long-term survival, with overall mortality 9% (11/129) for PDT and 8.5% (6/70) in the esophagectomy group, over a median 5 years of follow-up. None of the patients treated either with surgery or PDT died from esophageal cancer. A more recent study from the same group, comparing outcomes for early intramucosal cancer with PDT plus endoscopic mucosal resection versus surgery, came to a similar conclusion [62].

Several studies have evaluated the cost-effectiveness of PDT, and all have found that Photofrin PDT for HGD is cost-effective in terms of quality-adjusted life-years compared with esophagectomy [63–65].

Adverse Events Two common side effects of PDT using Photofrin are skin photosensitivity reactions and esophageal strictures. Photosensitivity can occur up to 3 months after treatment and is reported in more than 60% of patients. This can lead to severe burns requiring hospital admission and patients are advised to undertake strict light precautions after therapy. In a randomized controlled trial severity was graded as mild in 66%, moderate in 24%, and severe in 1% [59].

As Photofrin is taken up by both the esophageal mucosa and the submucosa, the associated esophageal stricture rate is high: 22% with one treatment and rising to 50% with multiple treatments [56, 66]. Several studies have investigated ways to lower this stricture rate. In a nonblinded dosimetry study, there was a threefold decrease in severe stricture occurrence when light doses below $115\,\mathrm{J\,cm^{-1}}$ (105–$85\,\mathrm{J\,cm^{-1}}$) were used [67]. This occurred, however, at the expense of a reduction in efficacy for ablation of HGD and intramucosal cancer. The same group evaluated the effect of oral steroids after PDT, but this intervention did not reduce the incidence of stricture formation [57].

Other adverse events include pleural effusions, atrial fibrillation [68], nausea and vomiting, and odynophagia. These complications significantly affect the acceptability of Photofrin PDT.

33.3.2.2 Levulan

Levulan is the chemically stable hydrochloride salt of 5-ALA, a naturally occurring five-carbon amino ketone carboxylic acid. It is a recent drug that has come into prominence over the last decade for its potential use in the PDT of superficial, benign, and malignant skin disorders. ALA is not a new chemical entity, however. It is present in virtually all human cells as the first committed intermediate of the biochemical pathway resulting in heme synthesis in humans and chlorophyll in plants. It differs from other types of PDT in that it is not a preformed photosensitizer, but instead a metabolic precursor of the endogenously formed photosensitizer PpIX. The human body is estimated to synthesize 350 mg of ALA per day to support endogenous heme production. The synthesis of ALA is normally tightly controlled by feedback inhibition of ALA synthetase, presumably by intracellular heme levels. ALA, when provided to the cell, bypasses this control point and enters the heme synthesis pathway, which results in the accumulation of PpIX. As in the chemical chain from ALA to heme, the rate-limiting step is the final conversion of PpIX to heme [69]. Free PpIX, unlike heme, does not appear to have intrinsic biological activity. However, PpIX is a potent natural photosensitizer, and irradiation of tissues that have been photosensitized, at sufficient dose rates and doses of light, can lead to significant photodynamic effects on cells, subcellular elements, and macromolecules via production of singlet oxygen [70].

Clinical Studies The first study of ALA PDT for HGD in BE treated five patients with a median segment length of 5 cm [71]. Oral ALA 60 mg kg^{-1} was administered, followed by 630 nm laser light via a 3 cm cylinder diffuser-tipped fiber optic at a power density of 150 mW cm^{-2} and an energy fluence of 90–150 J cm^{-2}. All healed with squamous regeneration. Two patients were found to have BE buried beneath squamous islands (buried glands), although no remaining HGD was identified.

Gossner *et al.* treated 32 patients with ALA PDT using 60 mg kg^{-1} ALA given orally and 635 nm laser light delivered 4–6 h after drug dosing through a 2 cm cylinder diffuser-tipped fiber optic at a power density of 100 mW cm^{-2} to a light dose of 150 J cm^{-2} [72]. As no balloon size was published, however, it is not possible to establish dosimetry. Ten patients had HGD and the remainder mucosal cancer. Patients were maintained on omeprazole for the duration of the study. Dysplasia was eradicated in all patients with HGD, and there was complete remission of the cancers in 17/22 patients (77%), giving an overall success rate of 27/32 (84%). This remission was maintained during follow-up of 1–30 months (mean 9.9 months). Squamous regeneration was seen in 68%, although the presence of nondysplastic buried glands was noted in some patients. A mean of 1.7 treatment sessions was required for the eradication of mucosal tumors. Of note, no tumor more than 2 mm in depth was included in the study.

In the only double-blind, randomized, placebo-controlled study reported for BE with LGD, 36 patients were randomized to receive oral ALA (30 mg kg^{-1}) or placebo [73]. This was followed 4 hours later by endoscopy and treatment of up to 6 cm of BE using green (514 nm) light. Green light is potentially advantageous as it is more strongly absorbed by the mucosa, although the depth of effect is less. A response was seen in 16/18 (89%) of patients in the ALA group, with a median area decrease of BE of 30% (range 0–60%). In the placebo group, a median area decrease of 0% was seen (range 0–10%). Of those who responded in the ALA PDT group, all maintained regression to normal squamous epithelium over the entire 24 month follow-up period, and no dysplasia was observed in the treated areas of BE which remained. In the placebo group, persistent LGD was found in 12/18 (67%) of patients followed for 24 months ($p < 0.001$). Although this study is encouraging, there are significant problems in performing studies in patients with LGD as there is a low intra- and inter-observer agreement on the diagnosis between pathologists [9, 74]. In this study, although the presence of LGD was confirmed by two independent pathologists, it was only at a single time point prior to PDT.

The optimum parameters for ALA PDT were demonstrated in a series of studies at our center. We recognized that when using a diffuser fiber to deliver light to a length of BE, the total energy delivered was best expressed per centimeter length of BE treated rather than in J cm^{-2}, as the latter measure varied depending on the size of the balloon used, whereas the total number of cells treated did not. When using 60 mg kg^{-1} ALA with red laser light, a light dose of less than 1000 J cm^{-1} was less effective than this highest light dose [75]. In a randomized controlled trial of 27 patients, we went on to demonstrate that red laser light (635 nm) was more effective than green laser light (512 nm) when using ALA 60 mg kg^{-1} [76]. Additionally, patients receiving 30 mg kg^{-1} of ALA relapsed to HGD significantly more frequently than those receiving 60 mg kg^{-1}. Kaplan–Meier analysis of the 21 patients who were subsequently treated with this optimal regimen demonstrated an eradication rate of 89% for HGD and a cancer-free proportion of 96% at 36 months' follow-up.

The first long-term results of ALA PDT in patients with HGD were described by Pech et al. [77]. This was a 6 year observational study that evaluated both efficacy and survival rates of 35 patients with a total of 44 ALA PDT procedures. All patients received 60 mg kg^{-1} ALA activated by red laser light. Complete remission was achieved in 34 of 35 (97%) of patients with HGD. Only 6% of patients with HGD had progression to cancer during the median follow-up period of 37 months. Six patients (18%) developed a metachronous lesion, and five of these underwent successful repeat treatment. The calculated 5 year survival was 97%. A further 31 patients with intramucosal cancer were studied with a total of 38 PDT treatments carried out; 29% had recurrence of metachronous lesions during the median follow-up period of 37 months. The calculated 5 year survival in this group was 80%.

Adverse Events Esophageal strictures following ALA occur less commonly than with Photofrin, as the drug is mostly converted to PpIX in the mucosa [78]. Photosensitivity reactions are rare and often mild as patients are sensitive to light for only 36 h. More common adverse reactions from oral ALA are nausea and vomiting and transient increases in serum aspartate aminotransferase (AST) [79]. Other serious adverse reactions reported to date include hypotension and unmasking of angina pectoris, although these appear to be self-limited and manageable in an inpatient setting.

There has been one case of acute neuropathy reported after ALA [80]. The authors surmised that the patient likely had a silent porphyria prior to treatment. On the basis of available evidence, current doses used in PDT are highly unlikely to result in neurologic or other systemic toxicologic findings.

Two early deaths have been reported following ALA PDT in two separate studies [81, 82]. One death was reported as occurring 24 h post-PDT in a patient who was treated in an ambulatory setting. The patient was randomized to receive 13-*cis*-retinoic acid 14 days prior to treatment with ALA PDT and died of acute cardiopulmonary failure; autopsy revealed aspiration pneumonitis [81]. The cause of death in the other study was not identified but also occurred the day after ALA PDT which had been done in an ambulatory setting. In our view, ALA PDT should be limited to inpatients.

33.3.2.3 Foscan

Foscan (*m*-tetrahydroxyphenylchlorin or mTHPC) is a powerful photosensitizer approved in Europe for the palliative treatment of advanced squamous cell carcinoma of the head and neck. Foscan is administered by intravenous injection after a drug–light interval of 4 days at a wavelength of 652 nm (Summary of Product Characteristics, biolitec Pharma).

Clinical Studies Javaid et al. in 2002, successfully treated five out of seven patients with HGD or intramucosal cancer at a median follow-up of 1 year [83]. Another study reported successfully treating 12/12 patients with HGD or intramucosal cancer using green light to activate Foscan for PDT with a median follow-up of 18 months. Just one patient required further PDT and this was successful [84].

Our center has reported a nonrandomized study of 19 patients treated with Foscan PDT which compared red and green light activation via a diffuser balloon and a bare-tipped fiber activated by red light [85]. Seven had HGD and 12 had intramucosal cancer; all were unfit for surgery. The red light diffuser group achieved CR-HGD of 70% (4/6 patients with cancer and 3/4 with HGD); 0/3 patients achieved remission in the green light group. When using the bare-tipped fiber, there was one procedure-related death as described below and only 1/5 patients with cancers were successfully treated. Two others were downgraded to HGD.

Adverse Events There have been two esophageal perforations reported with Foscan PDT, from taking multiple biopsy specimens too soon after therapy [85]. Additional contributory factors were that one patient received a very high light dose to a single area, and the other patient had received Nd:YAG laser irradiation prior to treatment. Skin photosensitivity is comparable to that with Photofrin, up to 1 month after treatment. Stricture formation occurred in 13% (5/38) of patients presented in the three studies above. Foscan is mainly distributed in the submucosa, although more drug is found in the mucosa than with Photofrin [86].

In summary, although isolated studies showed promising initial results of Foscan PDT for HGD in BE, long-term efficacy and safety data are lacking. The data on the use of Foscan for intramucosal cancers are inferior to those with newer techniques for these lesions (Endoscope Mucosel Resection) and serious adverse events are more common. For these reasons, Foscan PDT is rarely used for the treatment of BE.

33.3.3
Radiofrequency Ablation

Radiofrequency ablation (RFA) using the HALO System (Barrx Medical, Sunnyvale, CA, USA) is a new technique for field ablation in the esophagus. The HALO System uses ultra-short pulse radiofrequency energy that delivers a constant power density of $40 \, W \, cm^{-2}$ and energy density $10–15 \, J \, cm^{-2}$. A uniform ablation depth of 0.5–1 mm is achieved, thereby affecting the mucosa while preserving the submucosa. There are two devices, a balloon-mounted electrode for field ablation (HALO^{360}) and a smaller paddle device for focal ablation (HALO^{90}). All catheters are single-use disposable devices. Clinical trials in the United States and Europe have suggested that it is safe and effective for treating nondysplastic BE, LGD, and HGD. The HALO Ablation System was most recently reviewed by the FDA in 2006, although the device has been cleared for human use since 2001. These devices also have a Conformité Européene

(CE) mark for use in Europe and have recently been approved in the United Kingdom by NICE for the treatment of HGD arising in BE. The technical setup is simpler than in PDT, with no need for drug administration or hospital admission.

33.3.3.1 Clinical Studies

A randomized controlled trial of 127 patients (63 with HGD and 64 with LGD) treated by RFA or a sham procedure reported complete eradication of BE in 77% (65/84) and 2% (1/43) of patients, respectively, at 12 month follow-up ($p < 0.001$) [87]. The rate of progression from HGD to cancer was significantly lower in those patients treated by RFA than those treated by sham procedure at 12 month follow-up [2% (1/42) and 19% (4/21), respectively] ($p = 0.04$). No LGD patients progressed to cancer in either group. A United States Registry has reported efficacy data on 92 patients with HGD treated by RFA with at least one follow-up endoscopy [88]. Among those patients, CR-HGD was 90% (83/92) at a median 1 year follow-up; 80% (74/92) had no dysplasia and 54% (50/92) had complete reversal of BE with no intestinal metaplasia. Other cases series have reported CR-HGD rates between 79 and 100% at 12 months [89, 90]. Long-term follow-up data, however, are currently lacking.

33.3.3.2 Adverse Events

Perhaps the largest study on safety experience with HALO RFA was of 508 procedures in the United States, with no serious adverse events [91]. Symptom diaries showed mild and transient symptoms after ablation which generally resolved by day 4. Within the randomized controlled trial, adverse events included esophageal stricture formation (6% per patient), gastrointestinal bleeding (1% per patient), and chest pain requiring hospitalization (2% per patient) [87].

33.3.4
Subsquamous specialized intestinal metaplasia SSIM

A concern following all ablative therapies is the partial treatment of Barrett's epithelium with healing and squamous regeneration over the top of the Barrett's mucosa, which is hidden from the endoscopist at follow-up, so-called subsquamous specialized intestinal metaplasia (SSIM). The presence of SSIM is reported following all ablative therapies, with a frequency of 0–40% after PDT and 0–60% after APC. There have been several case reports of adenocarcinoma arising unnoticed underneath normal squamous tissue, at a rate of 0–3.7% [56, 66, 72]. One case series of 52 patients undergoing Photofrin PDT found the rate to be as high as 35.1% [92]. A more recent study, the largest of SSIM to date, evaluated 33 000 esophageal biopsies in patients undergoing PDT within a randomized trial [93]. The rate of SSIM was 5.8–30% of patients treated, but a similar frequency was found in the group treated with PPI alone (2.9–33%). In a recent randomized controlled trial of a new ablative therapy, radiofrequency ablation, the frequency of SSIM was reduced post-therapy (from 25 to 5%) and increased in the PPI alone group (from 25 to 40%) [87]. Other studies have shown that the neoplastic potential of SSIM post-PDT, as measured by Ki67 (a proliferative marker) and DNA ploidy, may be lower than that of pre-PDT SSIM [94]. The significance of SSIM is yet to be determined.

33.4
Conclusion

The emergence of endoscopic therapy for the treatment of HGD in BE has been paralleled by studies into the use of PDT and APC. These techniques have been shown to be effective in reversing dysplasia and present a viable alternative to esophagectomy. More recent techniques for ablation of Barrett's epithelium are RFA and cryospray ablation with liquid nitrogen, which has shown promising results in single-center studies [95]. These techniques are potentially advantageous over PDT, with no need for drug administration, hospital admission, or general anesthetic, meaning that patients can have the procedure as a day case. Initial efficacy and safety data are promising, although long-term follow-up data are currently unavailable.

Dysplasia is, however, an imperfect marker of cancer risk and convincing evidence of long-term cancer prevention is lacking. Questions also remain regarding the long-term durability of the ablated Barrett's mucosa, and there is much interest in the use of biomarkers as surrogate prognostic biomarkers. Genetic abnormalities, including DNA ploidy, p16, and proliferation markers, have been shown to persist post-PDT and predict recurrent dysplasia [96–98]. Prospective studies on the use of biomarkers to guide further treatment and surveillance intervals are awaited.

In conclusion, endoscopically delivered minimally invasive therapy presents a viable alternative to surgery for patients with HGD in BE. Which therapy is chosen will depend on many factors, including regional availability, institutional expertise, cost, and ultimately patient acceptability.

References

1 Reid, B.J., Weinstein, W.M., Lewin, K.J., Haggitt, R.C., VanDeventer, G., DenBesten, L., and Rubin, C.E. (1988) Endoscopic biopsy can detect high-grade dysplasia or early adenocarcinoma in Barrett's esophagus without grossly recognizable neoplastic lesions. *Gastroenterology*, **94** (1), 81–90.

2 Playford, R.J. (2006) New British Society of Gastroenterology (BSG) guidelines for the diagnosis and management of Barrett's oesophagus. *Gut*, **55** (4), 442.

3 Wang, K.K. and Sampliner, R.E. (2008) Updated Guidelines 2008 for the Diagnosis, Surveillance and Therapy of Barrett's Esophagus. *Am. J. Gastroenterol.*, **103** (3), 788–797.

4 Kara, M.A., Peters, F.P., Rosmolen, W.D., Krishnadath, K.K., ten Kate, F.J., Fockens, P., and Bergman, J.J. (2005) High-resolution endoscopy plus chromoendoscopy or narrow-band imaging in Barrett's esophagus: a prospective randomized crossover study. *Endoscopy*, **37** (10), 929–936.

5 Kara, M.A., Smits, M.E., Rosmolen, W.D., Bultje, A.C., ten Kate, F.J., Fockens, P., Tytgat, G.N., and Bergman, J.J. (2005) A randomized crossover study comparing light-induced fluorescence endoscopy with standard videoendoscopy for the

detection of early neoplasia in Barrett's esophagus. *Gastrointest. Endosc.*, **61** (6), 671–678.

6 Reid, B.J., Blount, P.L., Feng, Z., and Levine, D.S. (2000) Optimizing endoscopic biopsy detection of early cancers in Barrett's high-grade dysplasia. *Am. J. Gastroenterol.*, **95** (11), 3089–3096.

7 Falk, G.W., Rice, T.W., Goldblum, J.R., and Richter, J.E. (1999) Jumbo biopsy forceps protocol still misses unsuspected cancer in Barrett's esophagus with high-grade dysplasia. *Gastrointest. Endosc.*, **49** (2), 170–176.

8 Reid, B.J., Haggitt, R.C., Rubin, C.E., Roth, G., Surawicz, C.M., Van Belle, G., Lewin, K., Weinstein, W.M., Antonioli, D.A., and Goldman, H. (1988) Observer variation in the diagnosis of dysplasia in Barrett's esophagus. *Hum. Pathol.*, **19** (2), 166–178.

9 Lovat, L.B., Johnson, K., Mackenzie, G.D., Clark, B.R., Novelli, M.R., Davies, S., O'Donovan, M., Selvasekar, C., Thorpe, S.M., Pickard, D., Fitzgerald, R.C., Fearn, T., Bigio, I., and Bown, S.G. (2006) Elastic scattering spectroscopy accurately detects high grade dysplasia and cancer in Barrett's oesophagus. *Gut*, **55** (8), 1078–1083.

10 Guelrud, M., Herrera, I., Essenfeld, H., and Castro, J. (2001) Enhanced magnification endoscopy: a new technique to identify specialized intestinal metaplasia in Barrett's esophagus. *Gastrointest. Endosc.*, **53** (6), 559–565.

11 Ngamruengphong, S., Sharma, V.K., and Das, A. (2009) Diagnostic yield of methylene blue chromoendoscopy for detecting specialized intestinal metaplasia and dysplasia in Barrett's esophagus: a meta-analysis. *Gastrointest. Endosc.*, **69** (6), 1021–1028.

12 Meining, A., Rösch, T., Kiesslich, R., Muders, M., Sax, F., and Heldwein, W. (2004) Inter- and intra-observer variability of magnification chromoendoscopy for detecting specialized intestinal metaplasia at the gastroesophageal junction. *Endoscopy*, **36** (02), 160–164.

13 Adler, A., Pohl, H., Papanikolaou, I.S., Abou-Rebyeh, H., Schachschal, G., Veltzke-Schlieker, W., Khalifa, A.C., Setka, E., Koch, M., Wiedenmann, B., and Rösch, T. (2008.) A prospective randomised study on narrow-band imaging versus conventional colonoscopy for adenoma detection: does narrow-band imaging induce a learning effect? *Gut*, **57** (1), 59–64.

14 Brandt, L.J. and Kauvar, D.R. (1992) Laser-induced transient regression of Barrett's epithelium. *Gastrointest. Endosc.*, **38** (5), 619–622.

15 Mourant, J.R., Hielscher, A.H., Eick, A.A., Johnson, T.M., and Freyer, J.P. (1998) Evidence of intrinsic differences in the light scattering properties of tumorigenic and nontumorigenic cells. *Cancer Cytopathol.*, **84** (6), 366–374.

16 Mourant, J.R., Canpolat, M., Brocker, C., Esponda-Ramos, O., Johnson, T.M., Matanock, A., Stetter, K., and Freyer, J.P. (2000) Light scattering from cells: the contribution of the nucleus and the effects of proliferative status. *J. Biomed. Opt.*, **5** (2), 131–137.

17 Backman, V., Wallace, M.B., Perelman, L.T., Arendt, J.T., Gurjar, R., Muller, M.G., Zhang, Q., Zonios, G., Kline, E., McGilligan, J.A., Shapshay, S., Valdez, T., Badizadegan, K., Crawford, J.M., Fitzmaurice, M., Kabani, S., Levin, H.S., Seiler, M., Dasari, R.R., Itzkan, I., Van Dam, J., and Feld, M.S. (2000) Detection of preinvasive cancer cells. *Nature*, **406** (6791), 35–36.

18 Gurjar, R.S., Backman, V., Perelman, L.T., Georgakoudi, I., Badizadegan, K., Itzkan, I., Dasari, R.R., and Feld, M.S. (2001) Imaging human epithelial properties with polarized light-scattering spectroscopy. *Nat. Med.*, **7** (11), 1245–1248.

19 Mackenzie, G.D., Oukrif, D., Green, S., Novelli, M.R., Bown, S.G., and Lovat, L.B. (2007) Elastic scattering spectroscopy for the detection of aneuploidy in Barrett's oesophagus. *Gut*, **56**, A71.

20 Wallace, M.B., Perelman, L.T., Backman, V., Crawford, J.M., Fitzmaurice, M., Seiler, M., Badizadegan, K., Shields, S.J., Itzkan, I., Dasari, R.R., Van Dam, J., and Feld, M.S.

(2000) Endoscopic detection of dysplasia in patients with Barrett's esophagus using light-scattering spectroscopy. *Gastroenterology*, **119** (3), 677–682.

21 Georgakoudi, I., Jacobson, B.C., Van Dam, J., Backman, V., Wallace, M.B., Muller, M.G., Zhang, Q., Badizadegan, K., Sun, D., Thomas, G.A., Perelman, L.T., and Feld, M.S. (2001) Fluorescence, reflectance, and light-scattering spectroscopy for evaluating dysplasia in patients with Barrett's esophagus. *Gastroenterology*, **120** (7), 1620–1629.

22 Pyhtila, J.W., Chalut, K.J., Boyer, J.D., Keener, J., D'Amico, T., Gottfried, M., Gress, F., and Wax, A. (2007) In situ detection of nuclear atypia in Barrett's esophagus by using angle-resolved low-coherence interferometry. *Gastrointest. Endosc.*, **65** (3), 487–491.

23 Kopelovich, L., Henson, D.E., Gazdar, A.F., Dunn, B., Srivastava, S., Kelloff, G.J., and Greenwald, P. (1999) Surrogate anatomic/functional sites for evaluating cancer risk: an extension of the field effect. *Clin. Cancer Res.*, **5** (12), 3899–3905.

24 Liu, Y., Brand, R.E., Turzhitsky, V., Kim, Y.L., Roy, H.K., Hasabou, N., Sturgis, C., Shah, D., Hall, C., and Backman, V. (2007) Optical markers in duodenal mucosa predict the presence of pancreatic cancer. *Clin. Cancer Res.*, **13** (15), 4392–4399.

25 Roy, H.K., Kim, Y.L., Liu, Y., Wali, R.K., Goldberg, M.J., Turzhitsky, V., Horwitz, J., and Backman, V. (2006) Risk stratification of colon carcinogenesis through enhanced backscattering spectroscopy analysis of the uninvolved colonic mucosa. *Clin. Cancer Res.*, **12** (3 Pt 1), 961–968.

26 Roy, H.K., Turzhitsky, V., Kim, Y., Goldberg, M.J., Watson, P., Rogers, J.D., Gomes, A.J., Kromine, A., Brand, R.E., Jameel, M., Bogovejic, A., Pradhan, P., and Backman, V. (2009.) Association between rectal optical signatures and colonic neoplasia: potential applications for screening. *Cancer Res.*, **69** (10), 4476–4483.

27 Wali, R.K., Roy, H.K., Kim, Y.L., Liu, Y., Koetsier, J.L., Kunte, D.P., Goldberg, M.J., Turzhitsky, V., and Backman, V. (2005) Increased microvascular blood content is an early event in colon carcinogenesis. *Gut*, **54** (5), 654–660.

28 Gomes, A.J., Roy, H.K., Turzhitsky, V., Kim, Y., Rogers, J.D., Ruderman, S., Stoyneva, V., Goldberg, M.J., Bianchi, L.K., Yen, E., Kromine, A., Jameel, M., and Backman, V. (2009.) Rectal mucosal microvascular blood supply increase is associated with colonic neoplasia. *Clin. Cancer Res.*, **15** (9), 3110–3117.

29 Stone, N., Kendall, C., Smith, J., Crow, P., and Barr, H. (2004) Raman spectroscopy for identification of epithelial cancers. *Faraday Discuss.*, **126**, 141–157.

30 Shetty, G., Kendall, C., Shepherd, N., Stone, N., and Barr, H. (2006) Raman spectroscopy: elucidation of biochemical changes in carcinogenesis of oesophagus. *Br. J Cancer*, **94** (10), 1460–1464.

31 Kendall, C., Stone, N., Shepherd, N., Geboes, K., Warren, B., Bennett, R., and Barr, H. (2003) Raman spectroscopy, a potential tool for the objective identification and classification of neoplasia in Barrett's oesophagus. *J. Pathol.*, **200** (5), 602–609.

32 Song, L.M.W.K., Molckovsky, A., Wang, K., Burgart, L., Buttar, N., Papenfuss, S., Lutzke, L., Wilson, B., Dolenko, B., and Somorjai, R. (2005.) Diagnostic performance of near-infrared Raman spectroscopy in Barrett's esophagus. *Gastroenterology*, **128** (4), A51.

33 Berthomieu, C. and Hienerwadel, R. (2009) Fourier transform infrared (FTIR) spectroscopy. *Photosynth. Res.*, **101** (2), 157–170.

34 Wang, T.D., Triadafilopoulos, G., Crawford, J.M., Dixon, L.R., Bhandari, T., Sahbaie, P., Friedland, S., Soetikno, R., and Contag, C.H. (2007) Detection of endogenous biomolecules in Barrett's esophagus by Fourier transform infrared spectroscopy. *Proc. Natl. Acad. Sci. U. S. A.*, **104** (40), 15864–15869.

35 Martin, F.L. and Fullwood, N.J. (2007) Raman vs. Fourier transform spectroscopy in diagnostic medicine. *Proc. Natl. Acad. Sci. U. S. A.*, **104** (51), E1.

36 Haringsma, J., Tytgat, G.N., Yano, H., Iishi, H., Tatsuta, M., Ogihara, T., Watanabe, H., Sato, N., Marcon, N., Wilson, B.C., and Cline, R.W. (2001) Autofluorescence endoscopy: feasibility of detection of GI neoplasms unapparent to white light endoscopy with an evolving technology. *Gastrointest. Endosc.*, **53** (6), 642–650.

37 Ortner, M.A., Ebert, B., Hein, E., Zumbusch, K., Nolte, D., Sukowski, U., Weber-Eibel, J., Fleige, B., Dietel, M., Stolte, M., Oberhuber, G., Porschen, R., Klump, B., Hortnagl, H., Lochs, H., and Rinneberg, H. (2003.) Time gated fluorescence spectroscopy in Barrett's oesophagus. *Gut*, **52** (1), 28–33.

38 Niepsuj, K., Niepsuj, G., Cebula, W., Zieleznik, W., Adamek, M., Sielanczyk, A., Adamczyk, J., Kurek, J., and Sieron, A. (2003) Autofluorescence endoscopy for detection of high-grade dysplasia in short-segment Barrett's esophagus. *Gastrointest. Endosc.*, **58** (5), 715–719.

39 Kara, M.A., Peters, F.P., ten Kate, F.J., Van Deventer, S.J., Fockens, P., and Bergman, J.J. (2005) Endoscopic video autofluorescence imaging may improve the detection of early neoplasia in patients with Barrett's esophagus. *Gastrointest. Endosc.*, **61** (6), 679–685.

40 Panjehpour, M., Overholt, B.F., Vo-Dinh, T., Haggitt, R.C., Edwards, D.H., and BuckleyIII, F.P. (1996) Endoscopic fluorescence detection of high-grade dysplasia in Barrett's esophagus. *Gastroenterology*, **111** (1), 93–101.

41 Bourg-Heckly, G., Blais, J., Padilla, J.J., Bourdon, O., Etienne, J., Guillemin, F., and Lafay, L. (2000) Endoscopic ultraviolet-induced autofluorescence spectroscopy of the esophagus: tissue characterization and potential for early cancer diagnosis. *Endoscopy*, **32** (10), 756–765.

42 Pfefer, T.J., Paithankar, D.Y., Poneros, J.M., Schomacker, K.T., and Nishioka, N.S. (2003) Temporally and spectrally resolved fluorescence spectroscopy for the detection of high grade dysplasia in Barrett's esophagus. *Lasers Surg. Med.*, **32** (1), 10–16.

43 Mycek, M.A., Schomacker, K.T., and Nishioka, N.S. (1998) Colonic polyp differentiation using time-resolved autofluorescence spectroscopy. *Gastrointest. Endosc.*, **48** (4), 390–394.

44 Brand, S., Wang, T.D., Schomacker, K.T., Poneros, J.M., Lauwers, G.Y., Compton, C.C., Pedrosa, M.C., and Nishioka, N.S. (2002) Detection of high-grade dysplasia in Barrett's esophagus by spectroscopy measurement of 5-aminolevulinic acid-induced protoporphyrin IX fluorescence. *Gastrointest. Endosc.*, **56** (4), 479–487.

45 Madisch, A., Miehlke, S., Bayerdorffer, E., Wiedemann, B., Antos, D., Sievert, A., Vieth, M., Stolte, M., and Schulz, H. (2005) Long-term follow-up after complete ablation of Barrett's esophagus with argon plasma coagulation. *World J. Gastroenterol.*, **11** (8), 1182–1186.

46 Attwood, S.E., Lewis, C.J., Caplin, S., Hemming, K., and Armstrong, G. (2003) Argon beam plasma coagulation as therapy for high-grade dysplasia in Barrett's esophagus. *Clin. Gastroenterol. Hepatol.*, **1** (4), 258–263.

47 Basu, K.K., Pick, B., Bale, R., West, K.P., and de Caestecker, J.S. (2002) Efficacy and one year follow up of argon plasma coagulation therapy for ablation of Barrett's oesophagus: factors determining persistence and recurrence of Barrett's epithelium. *Gut*, **51** (6), 776–780.

48 Pereira-Lima, J.C., Busnello, J.V., Saul, C., Toneloto, E.B., Lopes, C.V., Rynkowski, C.B., and Blaya, C. (2000) High power setting argon plasma coagulation for the eradication of Barrett's esophagus. *Am. J. Gastroenterol.*, **95** (7), 1661–1668.

49 Schulz, H., Miehlke, S., Antos, D., Schentke, K.U., Vieth, M., Stolte, M., and Bayerdorffer, E. (2000) Ablation of Barrett's epithelium by endoscopic argon plasma coagulation in combination with high-dose omeprazole. *Gastrointest. Endosc.*, **51** (6), 659–663.

50 Morris, C.D., Byrne, J.P., Armstrong, G.R., and Attwood, S.E. (2001) Prevention

of the neoplastic progression of Barrett's oesophagus by endoscopic argon beam plasma ablation. *Br. J. Surg.*, **88** (10), 1357–1362.

51 Mork, H., Al Taie, O., Berlin, F., Kraus, M.R., and Scheurlen, M. (2007) High recurrence rate of Barrett's epithelium during long-term follow-up after argon plasma coagulation. *Scand. J. Gastroenterol.*, **42** (1), 23–27.

52 Ragunath, K., Krasner, N., Raman, V.S., Haqqani, M.T., Phillips, C.J., and Cheung, I. (2005) Endoscopic ablation of dysplastic Barrett's oesophagus comparing argon plasma coagulation and photodynamic therapy: a randomized prospective trial assessing efficacy and cost-effectiveness. *Scand. J. Gastroenterol.*, **40** (7), 750–758.

53 Spechler, S.J., Fitzgerald, R.C., Prasad, G.A., and Wang, K.K. (2010) History, molecular mechanisms, and endoscopic treatment of Barrett's esophagus. *Gastroenterology*, **138** (3), 854–869

54 Bown, S.G. and Lovat, L.B. (2000) The biology of photodynamic therapy in the gastrointestinal tract. *Gastrointest. Endosc. Clin North Am.*, **10** (3), 533–550.

55 Panjehpour, M., Overholt, B.F., and Haydek, J.M. (2000) Light sources and delivery devices for photodynamic therapy in the gastrointestinal tract. *Gastrointest. Endosc. Clin North Am.*, **10** (3), 513–532.

56 Overholt, B.F., Panjehpour, M., and Haydek, J.M. (1999) Photodynamic therapy for Barrett's esophagus: follow-up in 100 patients. *Gastrointest. Endosc.*, **49** (1), 1–7.

57 Panjehpour, M., Overholt, B.F., Haydek, J.M., and Lee, S.G. (2000) Results of photodynamic therapy for ablation of dysplasia and early cancer in Barrett's esophagus and effect of oral steroids on stricture formation. *Am. J. Gastroenterol.*, **95** (9), 2177–2184.

58 Wang, K.K. (2000) Photodynamic therapy of Barrett's esophagus. *Gastrointest. Endosc. Clin North Am.*, **10** (3), 409–419.

59 Overholt, B.F., Lightdale, C.J., Wang, K.K., Canto, M.I., Burdick, S., Haggitt, R.C., Bronner, M.P., Taylor, S.L., Grace, M.G., and Depot, M. (2005) Photodynamic therapy with porfimer sodium for ablation of high-grade dysplasia in Barrett's esophagus: international, partially blinded, randomized phase III trial. *Gastrointest. Endosc.*, **62** (4), 488–498.

60 Overholt, B.F., Wang, K.K., Burdick, J.S., Lightdale, C.J., Kimmey, M., Nava, H.R., Sivak J Jr., M.V., Nishioka, N., Barr, H., Marcon, N., Pedrosa, M., Bronner, M.P., Grace, M., and Depot, M. (2007.) Five-year efficacy and safety of photodynamic therapy with Photofrin in Barrett's high-grade dysplasia. *Gastrointest. Endosc.*, **66** (3), 460–468.

61 Prasad, GA., Wang, KK., Buttar, NS., Wongkeesong, L., Krishnadath, K.K., Nichols J III, F.C., Lutzke, L.S., and Borkenhagen, L.S. (2007) Long-term survival following endoscopic and surgical treatment of high-grade dysplasia in Barrett's esophagus. *Gastroenterology*, **132** (4), 1226–1233.

62 Prasad, G.A., Wu, T.T., Wigle, D.A., Buttar, N.S., Wongkeesong, L.M., Dunagan, K.T., Lutzke, L.S., Borkenhagen, L.S., and Wang, K.K. (2009) Endoscopic and surgical treatment of mucosal (T1a) esophageal adenocarcinoma in Barrett's esophagus. *Gastroenterology*, **137** (3), 815–823.

63 Shaheen, N.J., Inadomi, J.M., Overholt, B.F., and Sharma, P. (2004) What is the best management strategy for high grade dysplasia in Barrett's oesophagus? A cost effectiveness analysis. *Gut*, **53** (12), 1736–1744.

64 Vij, R., Triadafilopoulos, G., Owens, D.K., Kunz, P., and Sanders, G.D. (2004) Cost-effectiveness of photodynamic therapy for high-grade dysplasia in Barrett's esophagus. *Gastrointest. Endosc.*, **60** (5), 739–756.

65 Comay, D., Blackhouse, G., Goeree, R., Armstrong, D., and Marshall, J.K. (2007) Photodynamic therapy for Barrett's esophagus with high-grade dysplasia: a

66 Overholt, B.F., Panjehpour, M., and Halberg, D.L. (2003) Photodynamic therapy for Barrett's esophagus with dysplasia and/or early stage carcinoma: long-term results. *Gastrointest. Endosc.*, **58** (2), 183–188.
67 Panjehpour, M., Overholt, B.F., Phan, M.N., and Haydek, J.M. (2005) Optimization of light dosimetry for photodynamic therapy of Barrett's esophagus: efficacy vs. incidence of stricture after treatment. *Gastrointest. Endosc.*, **61** (1), 13–18.
68 Overholt, B.F., Panjehpour, M., and Ayres, M. (1997) Photodynamic therapy for Barrett's esophagus: cardiac effects. *Lasers Surg. Med.*, **21** (4), 317–320.
69 Marcus, S.L., Sobel, R.S., Golub, A.L., Carroll, R.L., Lundahl, S., and Shulman, D.G. (1996) Photodynamic therapy (PDT) and photodiagnosis (PD) using endogenous photosensitization induced by 5-aminolevulinic acid (ALA): current clinical and development status. *J. Clin. Laser Med. Surg.*, **14** (2), 59–66.
70 Poh-Fitzpatrick, M.B. (1986) Molecular and cellular mechanisms of porphyrin photosensitization. *Photodermatology*, **3** (3), 148–157.
71 Barr, H., Shepherd, N.A., Dix, A., Roberts, D.J., Tan, W.C., and Krasner, N. (1996) Eradication of high-grade dysplasia in columnar-lined (Barrett's) oesophagus by photodynamic therapy with endogenously generated protoporphyrin IX. *Lancet*, **348** (9027), 584–585.
72 Gossner, L., Stolte, M., Sroka, R., Rick, K., May, A., Hahn, E.G., and Ell, C. (1998) Photodynamic ablation of high-grade dysplasia and early cancer in Barrett's esophagus by means of 5-aminolevulinic acid. *Gastroenterology*, **114** (3), 448–455.
73 Ackroyd, R., Brown, N.J., Davis, M.F., Stephenson, T.J., Marcus, S.L., Stoddard, C.J., Johnson, A.G., and Reed, M.W. (2000) Photodynamic therapy for dysplastic Barrett's oesophagus: a prospective, double blind, randomised, placebo controlled trial. *Gut*, **47** (5), 612–617.

cost-effectiveness analysis. *Can. J. Gastroenterol.*, **21** (4), 217–222.

74 Kerkhof, M., van Dekken, H., Steyerberg, E.W., Meijer, G.A., Mulder, A.H., de Bruine, A., Driessen, A., ten Kate, F.J., Kusters, J.G., Kuipers, E.J., and Siersema, P.D. (2007) Grading of dysplasia in Barrett's oesophagus: substantial interobserver variation between general and gastrointestinal pathologists. *Histopathology*, **50** (7), 920–927.
75 Mackenzie, G.D., Jamieson, N.F., Novelli, M.R., Mosse, C.A., Clark, B.R., Thorpe, S.M., Bown, S.G., and Lovat, L.B. (2007) How light dosimetry influences the efficacy of photodynamic therapy with 5-aminolaevulinic acid for ablation of high-grade dysplasia in Barrett's esophagus. *Lasers Med. Sci.*, **23** (2), 203–210.
76 Mackenzie, G.D., Dunn, J.M., Selvasekar, C.R., Mosse, C.A., Thorpe, S.M., Novelli, M.R., Bown, S.G., and Lovat, L.B. (2009) Optimal conditions for successful ablation of high-grade dysplasia in Barrett's oesophagus using aminolaevulinic acid photodynamic therapy. *Lasers Med. Sci.*, **24** (5), 729–734.
77 Pech, O., Gossner, L., May, A., Rabenstein, T., Vieth, M., Stolte, M., Berres, M., and Ell, C. (2005) Long-term results of photodynamic therapy with 5-aminolevulinic acid for superficial Barrett's cancer and high-grade intraepithelial neoplasia. *Gastrointest. Endosc.*, **62** (1), 24–30.
78 Mackenzie, G.D., Dunn, J.M., Novelli, M.R., Mosse, S., Thorpe, S.M., Bown, S.G., and Lovat, L.B. (2008) Preliminary results of a randomised controlled trial into the safety and efficacy of ALA versus Photofrin photodynamic therapy for high grade dysplasia in Barrett's oesophagus 1. *Gut*, **57**, A14.
79 Regula, J., MacRobert, A.J., Gorchein, A., Buonaccorsi, G.A., Thorpe, S.M., Spencer, G.M., Hatfield, A.R., and Bown, S.G. (1995) Photosensitisation and photodynamic therapy of oesophageal, duodenal, and colorectal tumours using 5-aminolaevulinic acid induced protoporphyrin IX – a pilot study. *Gut*, **36** (1), 67–75.

80 Sylantiev, C., Schoenfeld, N., Mamet, R., Groozman, G.B., and Drory, V.E. (2005) Acute neuropathy mimicking porphyria induced by aminolevulinic acid during photodynamic therapy. *Muscle Nerve*, **31** (3), 390–393.

81 Forcione, D.G., Hasan, T., Ortel, B.J., and Nishioka, N.S. (2004) Optimization of aminolevulinic acid-based photodynamic therapy of Barrett's esophagus with high grade dysplasia. *Gastrointest. Endosc.*, **59** (5), 251.

82 Haringsma, J., Siersema, P.D., and Kuipers, E.J. (2004) Endoscopic ablation of Barrett's neoplasia. Rotterdam results. *Gastrointest. Endosc.*, **59** (5), AB252.

83 Javaid, B., Watt, P., and Krasner, N. (2002) Photodynamic therapy (PDT) for oesophageal dysplasia and early carcinoma with mTHPC (*m*-tetrahydroxyphenyl chlorin): a preliminary study. *Lasers Med. Sci.*, **17** (1), 51–56.

84 Etienne, J., Dorme, N., Bourg-Heckly, G., Raimbert, P., and Flejou, J.F. (2004) Photodynamic therapy with green light and *m*-tetrahydroxyphenyl chlorin for intramucosal adenocarcinoma and high-grade dysplasia in Barrett's esophagus. *Gastrointest. Endosc.*, **59** (7), 880–889.

85 Lovat, L.B., Jamieson, N.F., Novelli, M.R., Mosse, C.A., Selvasekar, C., Mackenzie, G.D., Thorpe, S.M., and Bown, S.G. (2005) Photodynamic therapy with *m*-tetrahydroxyphenyl chlorin for high-grade dysplasia and early cancer in Barrett's columnar lined esophagus. *Gastrointest. Endosc.*, **62** (4), 617–623.

86 Mlkvy, P., Messmann, H., Regula, J., Conio, M., Pauer, M., Millson, C.E., MacRobert, A.J., and Bown, S.G. (1998) Photodynamic therapy for gastrointestinal tumors using three photosensitizers – ALA induced PPIX, Photofrin(R) and MTHPC. A pilot study. *Neoplasma*, **45** (3), 157–161.

87 Shaheen, N.J., Sharma, P., Overholt, B.F., Wolfsen, H.C., Sampliner, R.E., Wang, K.K., Galanko, J.A., Bronner, M.P., Goldblum, J.R., Bennett, A.E., Jobe, B.A., Eisen, G.M., Fennerty, M.B., Hunter, J.G., Fleischer, D.E., Sharma, V.K., Hawes, R.H., Hoffman, B.J., Rothstein, R.I., Gordon, S.R., Mashimo, H., Chang, K.J., Muthusamy, V.R., Edmundowicz, S.A., Spechler, S.J., Siddiqui, A.A., Souza, R.F., Infantolino, A., Falk, G.W., Kimmey, M.B., Madanick, R.D., Chak, A., and Lightdale, C.J. (2009) Radiofrequency ablation in Barrett's esophagus with dysplasia. *N. Engl. J. Med.*, **360** (22), 2277–2288.

88 Ganz, R.A., Overholt, B.F., Sharma, V.K., Fleischer, D.E., Shaheen, N.J., Lightdale, C.J., Freeman, S.R., Pruitt, R.E., Urayama, S.M., Gress, F., Pavey, D.A., Branch, M.S., Savides, T.J., Chang, K.J., Muthusamy, V.R., Bohorfoush, A.G., Pace, S.C., DeMeester, S.R., Eysselein, V.E., Panjehpour, M., and Triadafilopoulos, G. (2008) Circumferential ablation of Barrett's esophagus that contains high-grade dysplasia: a U.S. Multicenter Registry. *Gastrointest. Endosc.*, **68** (1), 35–40.

89 Sharma, V.K., Jae, K.H., Das, A., Wells, C.D., Nguyen, C.C., and Fleischer, D.E. (2009) Circumferential and focal ablation of Barrett's esophagus containing dysplasia. *Am. J. Gastroenterol.*, **104** (2), 310–317.

90 Gondrie, J.J., Pouw, R.E., Sondermeijer, C.M., Peters, F.P., Curvers, W.L., Rosmolen, W.D., Krishnadath, K.K., ten Kate, F., Fockens, P., and Bergman, J.J. (2008) Stepwise circumferential and focal ablation of Barrett's esophagus with high-grade dysplasia: results of the first prospective series of 11 patients. *Endoscopy*, **40** (5), 359–369.

91 Rothstein, R.I., Chang, K., Overholt, B.F., Bergman, J.J., and Shaheen, N.J. (2007) Focal ablation for treatment of dysplastic and non-dysplastic Barrett esophagus: safety profile and initial experience with the Halo90 device in 508 cases. *Gastrointest. Endosc.*, **65** (5), AB147.

92 Mino-Kenudson, M., Ban, S., Ohana, M., Puricelli, W., Deshpande, V., Shimizu, M., Nishioka, N.S., and Lauwers, G.Y. (2007) Buried dysplasia and early adenocarcinoma arising in Barrett esophagus after

porfimer-photodynamic therapy. *Am. J. Surg. Pathol.*, **31** (3), 403–409.

93 Bronner, M.P., Overholt, B.F., Taylor, S.L., Haggitt, R.C., Wang, K.K., Burdick, J.S., Lightdale, C.J., Kimmey, M., Nava, H.R., Sivak, M.V., Nishioka, N., Barr, H., Canto, M.I., Marcon, N., Pedrosa, M., Grace, M., and Depot, M. (2009) Squamous overgrowth is not a safety concern for photodynamic therapy for Barrett's esophagus with high-grade dysplasia. *Gastroenterology*, **136**1), 56–64.

94 Hornick, J.L., Mino-Kenudson, M., Lauwers, G.Y., Liu, W., Goyal, R., and Odze, R.D. (2008) Buried Barrett's epithelium following photodynamic therapy shows reduced crypt proliferation and absence of DNA content abnormalities. *Am. J. Gastroenterol.*, **103** (1), 38–47.

95 Johnston, M.H., Eastone, J.A., and Horwhat, J.D. (2005) Cryoablation of Barrett's esophagus: a pilot study. *Gastrointest. Endosc.*, **62**, 842–848.

96 Dunn, J.M., Mackenzie, G.D., Oukrif, D., Mosse, C.A., Banks, M.R., Thorpe, S., Sasieni, P., Bown, S.G., Novelli, M.R., Rabinovitch, P.S., and Lovat, L.B. (2010) Image cytometry accurately detects DNA ploidy abnormalities and predicts late relapse to high-grade dysplasia and adenocarcinoma in Barrett's oesophagus following photodynamic therapy. *Br. J. Cancer*, **102** (11), 1608–1617.

97 Krishnadath, K.K., Wang, K.K., Taniguchi, K., Sebo, T.J., Buttar, N.S., Anderson, M.A., Lutzke, L.S., and Liu, W. (2000) Persistent genetic abnormalities in Barrett's esophagus after photodynamic therapy. *Gastroenterology*, **119** (3), 624–630.

98 Prasad, G.A., Wang, K.K., Halling, K.C., Buttar, N.S., Wongkeesong, L.M., Zinsmeister, A.R., Brankley, S.M., Westra, W.M., Lutzke, L.S., Borkenhagen, L.S., and Dunagan, K. (2008) Correlation of histology with biomarker status after photodynamic therapy in Barrett esophagus. *Cancer*, **113** (3), 470–476.

Part 8
Rheumatology

34
Noninvasive Fluorescence Imaging in Rheumatoid Arthritis – Animal Studies and Clinical Applications
Andreas Wunder and Bernd Ebert

34.1
Introduction

In rheumatoid arthritis (RA), most of the noninvasive imaging techniques in clinical use visualize anatomic, physiologic, or metabolic heterogeneity rather than identifying specific cellular or molecular events that underlie the disease. Most imaging agents used are unspecific. In consequence, the methods used in the clinic report relatively late in the course of the disease, assessment of disease activity and early monitoring of treatment response are challenging, and little or no information about cellular and molecular mechanisms is provided [1–4]. More specific imaging methods are urgently needed. Such techniques are highly relevant not only clinically, but also as tools for arthritis research in animal disease models [5–7]. It has been shown that new imaging agents that bind specifically to molecules crucially involved in the pathophysiology of the disease permit the visualization, characterization, and quantification of molecular and cellular processes noninvasively within intact living organisms. Such compounds have been developed and are under development for different imaging techniques, including positron emission tomography (PET), single photon emission computed tomography (SPECT), magnetic resonance tomography (MRT), ultrasound (US), and optical imaging techniques [8–10]. In this chapter, we describe the use of fluorescent imaging agents and noninvasive fluorescence imaging techniques for the visualization of arthritis in animal disease models and we discuss possible clinical applications of fluorescence imaging in RA.

34.2
Non-Invasive Fluorescence Imaging Techniques and Fluorescent Imaging Agents

Noninvasive fluorescence imaging techniques are increasingly important tools in biomedical research. Thus, fluorescence imaging systems are most frequently found in research facilities. Several fluorescence imaging systems for small animals with different characteristics are commercially available. These systems are comparatively simple. They consist of one ore more light sources for excitation of fluorochromes and one or more fluorescence detectors, mostly a highly sensitive charge-coupled device (CCD) camera [11]. The major advantage of fluorescence imaging methods is their high sensitivity. Very small amounts (in the range of nanomoles to femtomoles or less) of an injected fluorochrome located inside the body of laboratory animals can be detected. Furthermore, fluorescence imaging techniques are fast, easy to use, and inexpensive. The major disadvantages of fluorescence imaging techniques are a comparatively low spatial resolution (in the range of millimeters) and depth limit (in the range of centimeters if near-infrared light is used), attributed to absorption and scattering of light in biological tissue. Moreover, absolute quantitation of the imaging signal is challenging [11–13].

Numerous fluorescent imaging agents have been developed, permitting the visualization of a series of biological processes such as increased vessel permeability or tissue perfusion, blood flow, hypoxia, proliferation, cell death, enzyme activity, receptor expression, cell migration, and many more. In most of these studies, the fluorescent imaging agents used absorb and emit light in the near-infrared fluorescence (NIRF) range, because in this spectral range the penetration depth of light is comparatively high and the tissue autofluorescence is comparatively low [14, 15].

Fluorescent imaging agents have been successfully used to study physiology and pathophysiology in different animal disease models, including animal models of cancer [16], vascular diseases [17] such as atherosclerosis [18] and stroke [19], as well as arthritis [20]. Currently, two fluorescent dyes are clinically approved: indocyanine green (ICG), an NIRF dye, and the red fluorescent protoporphyrin IX, which generates from the precursor δ-aminolevulinic acid (ALA). These compounds are relatively small molecules of low molecular weight (775 and 168 Da, respectively). ICG is used clinically in, for instance, angiography [21, 22]. ALA can be applied for the fluorescence diagnosis or for fluorescence-guided resection of both premalignant and nonmalignant diseases [23]. Recently, a pilot study using fluorescence imaging after systemic administration of the fluorescent dye ICG in patients with RA was reported [24]. Further, a fluorescence camera device, Xiralite (mivenion, Berlin, Germany), designed for ICG imaging of patients has been developed. In the following, we first describe the use of noninvasive fluorescence imaging and fluorescent imaging agents in arthritis models. We then describe clinical approaches, followed by a critical overview of fluorescence imaging in arthritis.

34.3
Experimental Studies in Arthritis Models

Imaging agents can be classified into three different groups: nonspecific, targeted, and activatable probes. All three classes of probes have been evaluated in animal models of arthritis. Nonspecific probes do not have a specificity for a certain molecular target. They can be used to assess, for instance, perfusion, blood flow, or the extravasation through blood–tissue barriers. Targeted probes are characterized by specific binding to a certain molecular target, for example, a receptor. With these probes, the presence of a specific molecular target can be visualized. In contrast to nonspecific and targeted probes, activatable or "smart" sensor probes do not give a signal in the injected state: only after interaction with a certain molecular target is the signal activated. Examples are enzyme-activatable probes, which are activated upon cleavage by the target enzyme. Owing to the low background, this strategy results in excellent signal-to-noise ratios [5].

34.3.1
Nonspecific Fluorescent Imaging Agents in Arthritis Models

In several approaches, nonspecific fluorescent imaging agents were administered to animals with arthritis and thereafter noninvasive fluorescence imaging was performed. The idea behind using nonspecific imaging agents in arthritis is that inflamed joints show a an increased perfusion, a higher rate of extravasation, and a large number of cells with increased metabolism and phagocytosis, which increases the retention of injected compounds in tissues. The interplay of these parameters leads to contrast generation in inflamed joints compared with normal joints [25–27].

Hansch *et al.* injected the NIRF dye Cy5.5 intravenously into mice with antigen-induced arthritis (AIA). Noninvasive NIRF imaging was performed at different time points after administration of the fluorochrome. Fluorescence intensities of arthritic and normal knee joints by region of interest (ROI) analysis were estimated [25]. Figure 34.1a shows that significantly higher fluorescence intensities were found in inflamed knee joints compared with normal knee joints between 2 and 72 h after injection of the nonspecific dye. The ratios between inflamed and normal knee joints that could be achieved in this setting were never higher than 2, decreasing after 24 h. The fluorochrome used in this approach is highly water soluble and has a comparably short half-life in blood (in the range of a few hours). The dye binds *in vivo* to serum proteins, mainly to albumin, and is therefore protected from an even more rapid excretion by the kidneys. By covalent binding of fluorochromes to macromolecular carriers, the contrast between inflamed and normal joints can be increased significantly. Moreover, the contrast decreases less rapidly. In mice with collagen-induced arthritis (CIA), we could demonstrate strong fluorescence in inflamed toes and paws after injection of albumin labeled with aminofluorescein [26]. The high fluorescence intensity in inflamed toes and paws could easily be documented by taking pictures with a conventional camera and a slide film, as shown in Figure 34.1b. The pharmacokinetic data after radiolabeling of the protein showed a 6–7-fold higher uptake in inflamed paws compared with normal paws even 48 h after injection. Fischer *et al.* evaluated two nonspecific NIRF dyes, indocyanine green (ICG) and a hydrophilic carbocyanine derivative, SIDAG [1,1′-bis(4-sulfobutyl)indotricarbocyanine-5,5′-dicarboxylic acid diglucamide monosodium salt], in a mouse model of *Borrelia*-induced Lyme arthritis [27]. ICG is a lipophilic fluorochrome, which is rapidly cleared by the liver [21, 22]. The dye is characterized by high plasma protein binding (98%), whereas the more hydrophilic fluorochrome SIDAG also used in this

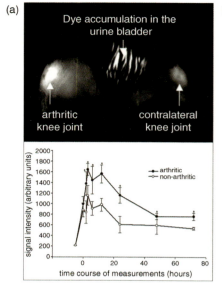 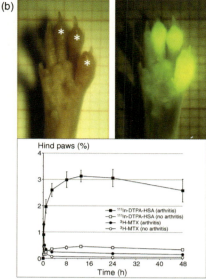

Figure 34.1 Noninvasive visualization of arthritis in animal models using nonspecific fluorescent imaging probes (Cy5.5 and fluorescently labeled albumin). (a) Noninvasive NIRF image of a mouse with AIA affecting the right knee joint taken 2 h after i.v. injection of the nonspecific NIRF dye Cy5.5. The knee joint affected by arthritis shows a higher fluorescence intensity than the normal knee joint. The time-dependent fluorescence signal intensities in arthritic and contralateral joints are depicted in the graph. Significantly higher fluorescence intensities were found in inflamed knee joints compared with normal knee joints between 2 and 72 h after injection. Ratios between inflamed knee joints compared with normal knee joints were about 2 or less, decreasing rapidly. Adapted with permission from [25]. (b) Noninvasive white light (left) and fluorescence image (right) of a paw of a mouse with CIA affecting three toes (indicated by asterisks) 3 h after injection of human serum albumin (HSA) labeled with aminofluorescein, a green fluorescent dye. Only toes affected by arthritis showed strong fluorescence demonstrating the high rate of albumin accumulation in inflamed toes. The uptake kinetics of radiolabeled HSA ([^{111}In]-DTPA-HSA) and radiolabeled methotrexat ([^{3}H]-MTX) in hind paws of mice with and without CIA after i.v. injection is shown in the graph (percentage of the initially administered radioactivity). Significant amounts of albumin accumulated in arthritic hind paws, exceeding the uptake in noninflamed hind paws by 6–7-fold. In contrast, uptake of radiolabeled MTX in arthritic hind paws was found to be significantly less, decreasing rapidly over time. Adapted with permission from [26].

study shows plasma protein binding of only 10% [27]. As shown in Figure 34.2, the pharmacokinetic behavior of the two dyes differs significantly. After injection of ICG, the fluorescence intensity in inflamed and normal joints increased rapidly within the first few seconds after injection and started to decrease about 60 s after injection. The washout of ICG from inflamed joints compared with normal joints was surprisingly even faster. After injection of ICG, significantly lower fluorescence intensities were found in the joints compared with joints of mice injected with SIDAG. After injection of SIDAG, fluorescence intensities in inflamed and normal joints increased within

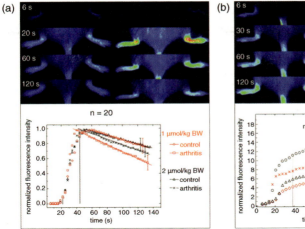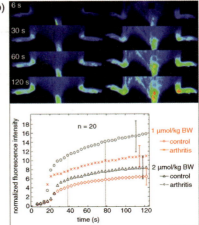

Figure 34.2 Noninvasive visualization of arthritis in animal models using nonspecific fluorescent imaging probes (ICG and a hydrophilic carbocyanine derivative called SIDAG). Color-coded noninvasive NIRF images of mice with *Borrelia*-induced Lyme arthritis after intravenous injection of ICG (a) or SIDAG (b). The graphs show the time course of the measured fluorescence intensities over normal and arthritic ankle joints after injection of the two compounds in two different doses. Mice injected with ICG showed a rapid increase in fluorescence intensity over both inflamed and normal joints, decreasing about 60 s after injection. In contrast, in mice injected with SIDAG, significantly higher fluorescence intensities were detected in inflamed joints compared with normal joints. The ratio between inflamed and normal joints increased over time. Adapted with permission from [27].

the observation time of 120 s, with a stronger increase in inflamed joints. The ratios between inflamed and normal joints were about 2. The highest ratios were measure at 24 h after injection. In most animals with Lyme arthritis, no cartilage destruction at the time of imaging was observed, indicating that early stages can be imaged with this technique.

Taking the results of these three studies together, it has been demonstrated that in arthritic mice acceptable ratios between inflamed and normal joints can be achieved by injection of nonspecific imaging agents. This might help in the diagnosis of arthritis. The feasibility of diagnosing early stages of arthritis with such nonspecific imaging agents has to be further supported. The results of these three studies tell one more important story: all approaches with specific compounds need a careful evaluation of the extent to which unspecific distribution contributes to the signal. This is especially true under inflammatory conditions, where this effect can be substantial.

34.3.2
Targeted Fluorescent Imaging Agents in Arthritis Models

Numerous fluorescently labeled ligands that bind specifically to target molecules involved in the pathophysiology of diseases have been developed and evaluated in different animal disease models, including cancer models, models for vascular diseases, and arthritis models [16–20]. Some examples of the use of targeted fluorescence imaging probes in arthritis models are described in the following:

In patients with RA, it has been shown that the nonepithelial isoform beta of the folate receptor is expressed on activated but not on resting synovial macrophages. Against this background, Chen et al. [28] evaluated NIRF-labeled folate receptor targeted imaging probe followed by noninvasive NIRF imaging in two different arthritis models: arthritis induced by intra-articular injection of lipopolysaccharide (LPS) and the KRN serum transfer model. After intravenous injection of the targeted imaging agent and noninvasive NIRF imaging, the authors found about twofold higher fluorescence intensities in the inflamed joints compared with normal joints. As a control, the free, uncoupled NIRF dye was used (see Figure 34.3a). The authors concluded that this folate receptor targeted imaging method might facilitate improved arthritis diagnosis, especially at early stages, by providing an *in vivo* characterization of active macrophage status in inflammatory joint diseases. Moreover, the presented technique might also be helpful in the evaluation of anti-inflammatory treatment [28].

Sensitive imaging strategies for monitoring treatment response would indeed be valuable for facilitating appropriate therapy and dosing, evaluating clinical outcome, and developing more effective drugs in RA. In mice with CIA, we evaluated NIRF-labeled annexin V as a marker for treatment response to methotrexate (MTX), one of the most commonly used drugs in the treatment of RA [29]. Annexin V binds tightly and specifically to phosphatidylserine (PS). In normal cells, PS is almost exclusively restricted to the inner leaflet of the cell membrane. When cells die by apoptosis or necrosis, exposure of PS occurs. Therefore, annexin V is a suitable marker for specific imaging of cell death, which has been shown in several animal disease models and in patients with different diseases using different imaging modalities (for a review, see [30]). Three hours after injection of NIRF-labeled annexin V and noninvasive NIRF imaging, arthritic paws of MTX-treated mice showed a sevenfold higher fluorescence intensity than arthritic paws of untreated mice and a fourfold higher fluorescence intensity than non-arthritic paws of MTX-treated mice. With this approach, we could demonstrate that fluorescent annexin V and NIRF imaging can be used successfully to monitor response to treatment in a mouse model of RA [29].

Vollmer et al. investigated a fluorescently labeled single-chain antibody fragment covalently linked to the NIRF dye tetrasulfocyanine (TSC), which binds with high specificity to the extracellular domain B (ED-B) of the extracellular matrix protein fibronectin, as a targeted imaging probe in rats with CIA [31]. ED-B is a highly conserved domain that occurs after alternative splicing, for instance during vascular remodeling processes such as angiogenesis. Since the hallmark of RA is synovitis leading to both angiogenesis in the synovium and the promotion of cartilage and bone disruption, the antibody fragment might be a suitable marker in the clinical diagnostics and in the evaluation of treatment response in RA. The authors labeled the marker with an NIRF fluorochrome, injected the labeled compound intravenously into rats with and without CIA, and performed noninvasive NIRF imaging. As a control, ovalbumin labeled with the same NIRF dye was used. Fluorescence

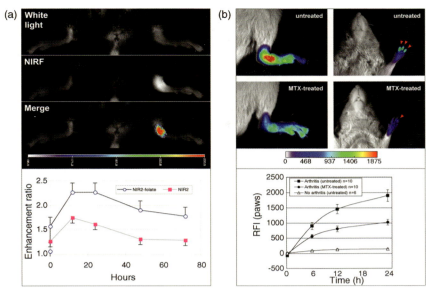

Figure 34.3 Noninvasive visualization of arthritis in animal models using targeted (folate receptor) and activatable (matrix degrading enzymes) fluorescent imaging probes. (a) Noninvasive white light, NIRF image, and color-coded merged image of a mouse that received an intra-articular injection of LPS into the left ankle joint. Images were taken 24 h after i.v. injection of a folate receptor-targeted NIRF imaging probe. In the graph, the time course of the enhancement ratios (fluorescence intensity over the inflamed joint divided by the fluorescence intensity over the opposite joint) after injection of the targeted imaging probe (highest ratio: 2.3:1) and free, uncoupled NIRF dye as a control is depicted. Adapted with permission from [28]. (b) Color-coded noninvasive NIRF images superimposed on white light images of untreated (upper row) and MTX-treated (bottom row) mice with collagen-induced arthritis (CIA) affecting the right hind paw (left) or toes of the left fore paw (right; red arrowheads) 24 h after i.v. injection of a "smart" fluorescence imaging probe that is activated by proteases such as cathepsin B. The time course of the fluorescence intensities in arthritic paws of untreated and MTX-treated mice and nonarthritic paws of untreated mice after probe injection is shown in the graph. Inflamed paws of untreated arthritic mice showed a sevenfold higher fluorescence intensity than paws of untreated healthy mice at 24 h. MTX treatment significantly reduced the fluorescence intensities from arthritic paws compared with untreated mice at 6, 12, and 24 hours. Adapted with permission from [34].

intensity ratios between inflamed joints and the nose of the animals were about 4–6-fold for the antibody fragment and about twofold for ovalbumin.

34.3.3
Activatable Fluorescent Imaging Agents in Arthritis Models

Activatable imaging agents based on different principles have been developed for different imaging modalities, including magnetic resonance imaging (MRI) and optical imaging, targeting different processes, including enzyme activity [32, 33]. Matrix-degrading enzymes such as matrix metalloproteinases (MMPs) and cathepsins have a substantial role in arthritic joint destruction. Using an enzyme-cleavable NIRF imaging agent that can be activated by cathepsins, we were able to visualize the presence of these enzymes in inflamed joints of mice with CIA noninvasively. Fluorescence intensities were 5–7-fold higher in inflamed compared with normal paws or toes 24 h after injection. With this approach, we could also show that the probe can be used for early monitoring of treatment response to MTX (see Figure 34.3b). The design of this imaging agent nicely illustrates how activatable probes work in principle: multiple Cy5.5 fluorochromes are bound to a graft copolymer backbone of polylysine, to which methoxypolyethylene glycol (MPEG) side chains are attached. Owing to interactions between the fluorochromes, fluorescence quenching occurs. Enzymatic cleavage of the backbone by cathepsins releases Cy5.5 and results in activation of fluorescence [34].

RA is also characterized by high expression levels of reactive oxygen species (ROS) and hyaluronidase, inducing progressive degradation of joint hyaluronic acid (HA) in the extracellular matrix. Lee *et al.* [35] immobilized NIRF-labeled HA on the surface of gold nanoparticles, which resulted in quenching of the fluorochromes. The presence of ROS and hyaluronidase leads to activation of the fluorescence. After intravenous injection of the particles into mice with CIA, about fivefold higher fluorescence intensities were found in inflamed compared with normal limbs, decreasing between 6 and 24 h after injection [35]. Similarly to our approach described above, this activatable imaging agent might also be useful as a marker in early detection of arthritis and for evaluation of treatment efficacy in arthritis.

34.3.4
Monitoring Cell Migration in Arthritis Models

Noninvasive monitoring of the fate of injected cells, also called cell tracking, is highly relevant in biomedical research and in the clinic. Tracking the fate of implanted cells, for example, in stem cell approaches or visualization of leukocyte trafficking, are just two out of numerous examples (for a review, see [36]). Simon *et al.* [37] labeled allogenic leukocytes with a fluorochrome called DID, a lipophilic NIRF dye, and injected 10^7 fluorescent cells intravenously into rats with AIA followed by noninvasive NIRF imaging. At 4 and 24 h post-injection, the fluorescence intensities of arthritic knees were significantly higher than those of normal knees, and of arthritic knees and knees of cortisone treated rats, decreasing between 4 and 24 h. Ratios between arthritic and normal rat knees were around 2:1 or less. This study shows that the infiltration of leukocytes into inflamed joints of rats can be visualized with noninvasive fluorescence imaging. The authors concluded that inflammation imaging with labeled leukocytes might help in the monitoring of new drugs that specific interact with T cells and monocytes in the synovium [37].

Sutton *et al.* [38] labeled human mesenchymal stem cells (hMSCs) with DID, the same dye as mentioned above. Fluorescent hMSCs (from 10^4 up to 10^7 cells) were administered into the peritoneum of athymic rats with an immune-mediated polyarthritis induced by intraperitoneal injection of peptidoglycan polysaccharide. Evaluating the noninvasive NIRF images, the authors reported that the fluorescence intensity of arthritic ankle joints was significantly higher 24 and 48 h after injection of the cells compared with normal ankle joints, decreasing at 72 h (see Figure 34.4). The data presented led the authors to the conclusion that the described technique might be useful for monitoring the factors that influence the survival of implanted stem cells and therapeutic progress [38].

34.4
Clinical Applications

Optical methods have also been applied to patients with RA. For this purpose, sophisticated imaging systems were designed. Two basically different approaches are described, both using near-infrared light, lasers, and highly sensitive detectors. One strategy is the measurement of intrinsic tissue parameters such as absorption and scattering of light in joints, which are altered under inflammatory conditions. For detailed descriptions of techniques based on intrinsic parameters, the interested reader is referred to [39–43]. The second approach is noninvasive NIRF imaging after injection of a fluorescent compound. Our recent pilot study demonstrates that noninvasive fluorescence imaging of human finger joints after systemic administration of the nonspecific dye ICG is feasible [24]. ICG was used as a fluorochrome because it is the only dye in the NIRF spectral range that is clinically approved. Five patients with clinically active RA and five healthy volunteers were injected with ICG as an intravenous bolus injection at a dose of $0.1\,\mathrm{mg\,kg}^{-1}$, which is lower than that used for other diagnostic investigations. After injection of the compound, fluorescence images of one hand were acquired over a period of 15 min. All participants underwent contrast-enhanced MRI (Gd-DTPA) 1 day before fluorescence imaging. Figure 34.5 shows fluorescence images of RA patients and a healthy volunteer, and also the time course of the detected fluorescence intensities in different regions of interest. The fluorescence imaging data and the MRI data showed a good correlation [24]. The results of an ongoing study with a higher number of subjects are eagerly awaited.

34.5
Critical Overview of Fluorescence Imaging in Arthritis

Noninvasive fluorescence imaging is a powerful tool for studying physiology and pathophysiology in animal models of arthritis. It is easy to use, fast, and inexpensive. If the target tissue is located superficially, highly sensitive imaging can be performed with an acceptable spatial resolution. Rapidly increasing numbers of fluorescent imaging agents are available, targeting different biological processes and offering multiple opportunities in arthritis research. However, absolute quantitation in

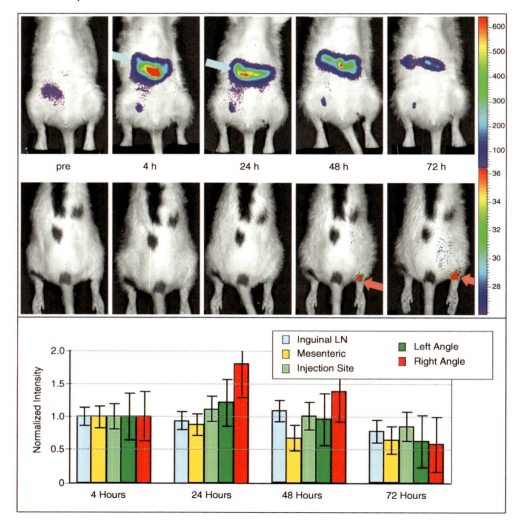

Figure 34.4 Monitoring cell migration in an arthritis model with noninvasive fluorescence imaging. Color-coded noninvasive fluorescence images superimposed on white light images of an athymic rat with an immune-mediated polyarthritis induced by intraperitoneal injection of peptidoglycan polysaccharide before and after i.p. injection of fluorescently labeled hMSCs show fluorescence of the abdomen on early post-injection images (indicated by blue arrow) and a progressively enhanced fluorescence of the right ankle (red arrow). The graphs show the normalized intensity of the fluorescence image at 4 h post-injection within the ROI in the inguinal LN, mesenteric, injection site, and right and left ankle joints after background subtraction (obtained from preinjection image) at different time points post-injection. The fluorescence intensity of arthritic ankle joints was significantly higher 24 and 48 h after injection of the cells compared with normal ankle joints, decreasing at 72 h. Reproduced with permission from [38].

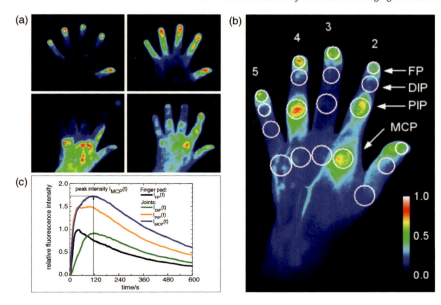

Figure 34.5 Noninvasive visualization of arthritis in patients using a nonspecific fluorescent imaging probe (ICG). (a) Fluorescence images of a hand of a healthy volunteer (upper row) and of a patient with rheumatoid arthritis (lower row) after application of ICG (0.1 mg kg^{-1}). Images were taken 40 s (left) and 65 s (right) after injection of ICG. (b) Location of investigated areas, that is, ROIs indicated as white circles on the dorsal side of the hand. The hand is shown in a normalized false color presentation, corresponding to the color bar at the right side. FP, finger pad; DIP, distal interphalangeal joint; PIP, proximal interphalangeal joint; MCP, metacarpal phalangeal joint. (c) Temporal behavior of 90th percentiles of the fluorescence intensity in selected ROIs of the left forefinger of a patient [see (b)] after injection of ICG. Only the finger pad shows a clear arrival and washout of the ICG bolus. Adapted with permission from [24].

fluorescence imaging is still challenging, which is a clear disadvantage compared with nuclear imaging methods such as PET and SPECT, especially when absolute quantitation of the imaging signal is mandatory. Our first proof-of-concept study with RA patients published in 2010 gives hope that in the near future noninvasive fluorescence imaging after injection of fluorescent reporters can be performed in the clinical routine. Especially the use of specific fluorescent reporter agents for imaging physiology and pathophysiology at the cellular and molecular level will, in our opinion, improve the diagnostic power, and therefore the management, of the disease considerably. Generally, these novel tools will be helpful in evaluating disease processes, facilitating diagnosis, and monitoring therapeutic regimens, to permit a reliable prognosis and to support the development of new therapies. Furthermore, the techniques promise to be safe, simple, cost-effective, and rapid. However, the penetration depth of near-infrared light limits the applicability of fluorescence imaging to finger and foot joints.

References

1 Fouque-Aubert, A. and Chapurlat, R.D. (2010) A comparative review of the different techniques to assess hand bone damage in rheumatoid arthritis. *Joint Bone Spine*, **77** (3), 212–217.
2 Yazici, Y., Sokka, T., and Pincus, T. (2009) Radiographic measures to assess patients with rheumatoid arthritis: advantages and limitations. *Rheum. Dis. Clin. North Am.*, **35** (4), 723–729.
3 Brown, A.K. (2009) Using ultrasonography to facilitate best practice in diagnosis and management of RA. *Nat. Rev. Rheumatol.*, **5** (12), 698–706.
4 Boesen, M., Østergaard, M., Cimmino, M.A., Kubassova, O., Jensen, K.E., and Bliddal, H. (2009) MRI quantification of rheumatoid arthritis: current knowledge and future perspectives. *Eur. J. Radiol.*, **71** (2), 189–196.
5 Wunder, A., Straub, R.H., Gay, S., Funk, J., and Müller-Ladner, U. (2005) Molecular imaging: novel tools in visualizing rheumatoid arthritis. *Rheumatology (Oxford)*, **44** (11), 1341–1349.
6 McBride, H.J. (2010) Nuclear imaging of autoimmunity: focus on IBD and RA. *Autoimmunity*, **43** (7), 539–549.
7 Malviya, G., Conti, F., Chianelli, M., Scopinaro, F., Dierckx, R.A., and Signore, A. (2010) Molecular imaging of rheumatoid arthritis by radiolabelled monoclonal antibodies: new imaging strategies to guide molecular therapies. *Eur. J. Nucl. Med. Mol. Imaging*, **37** (2), 386–398.
8 Jaffer, F.A. and Weissleder, R. (2005) Molecular imaging in the clinical arena. *JAMA*, **293** (7), 855–862.
9 Massoud, T.F. and Gambhir, S.S. (2007) Integrating noninvasive molecular imaging into molecular medicine: an evolving paradigm. *Trends Mol. Med.*, **13** (5), 183–191.
10 Contag, C.H. (2007) *In vivo* pathology: seeing with molecular specificity and cellular resolution in the living body. *Annu. Rev. Pathol.*, **2**, 277–305.
11 Leblond, F., Davis, S.C., Valdés, P.A., and Pogue, B.W. (2010) Pre-clinical whole-body fluorescence imaging: review of instruments, methods and applications. *J. Photochem. Photobiol. B*, **98** (1), 77–94.
12 Ntziachristos, V. (2006) Fluorescence molecular imaging. *Annu. Rev. Biomed. Eng.*, **8**, 1–33.
13 Hilderbrand, SA. and Weissleder, R. (2010) Near-infrared fluorescence: application to *in vivo* molecular imaging. *Curr. Opin. Chem. Biol.*, **14** (1), 71–79.
14 Licha, K., Schirner, M., and Henry, G. (2008) Optical agents. *Handb. Exp. Pharmacol.*, **185** (1), 203–222.
15 Escobedo, J.O., Rusin, O., Lim, S., and Strongin, R.M. (2010) NIR dyes for bioimaging applications. *Curr. Opin. Chem. Biol.*, **14** (1), 64–70.
16 Pierce, M.C., Javier, D.J., and Richards-Kortum, R. (2008) Optical contrast agents and imaging systems for detection and diagnosis of cancer. *Int. J. Cancer*, **123** (9), 1979–1990.
17 Wunder, A. and Klohs, J. (2008) Optical imaging of vascular pathophysiology. *Basic Res. Cardiol.*, **103** (2), 182–190.
18 Jaffer, F.A., Libby, P., and Weissleder, R. (2009) Optical and multimodality molecular imaging: insights into atherosclerosis. *Arterioscler. Thromb. Vasc. Biol.*, **29** (7), 1017–1024.
19 Wunder, A. and Klohs, J. (2010) Noninvasive optical imaging in small animal models of stroke, in *Rodent Models of Stroke*, Neuromethods, vol. **47**, 1st edn (ed. U. Dirnagl), Springer, Berlin, pp. 167–177.
20 Gompels, L.L., Lim, N.H., Vincent, T., and Paleolog, E.M. (2010) *In vivo* optical imaging in arthritis – an enlightening future? *Rheumatology (Oxford)*, **49** (8), 1436–1446.
21 Dzurinko, V.L., Gurwood, A.S., and Price, J.R. (2004) Intravenous and indocyanine green angiography. *Optometry*, **75** (12), 743–755.
22 Rodrigues, E.B., Costa, E.F., Penha, F.M., Melo, G.B., Bottós, J., Dib, E., Furlani, B., Lima, V.C., Maia, M., Meyer, C.H., Höfling-Lima, A.L., and Farah, M.E.

(2009) The use of vital dyes in ocular surgery. *Surv. Ophthalmol.*, **54** (5), 576–617.

23 Krammer, B. and Plaetzer, K. (2008) ALA and its clinical impact, from bench to bedside. *Photochem. Photobiol. Sci.*, **7** (3), 283–289.

24 Fischer, T., Ebert, B., Voigt, J., Macdonald, R., Schneider, U., Thomas, A., Hamm, B., and Hermann, K.G. (2010) Detection of rheumatoid arthritis using non-specific contrast enhanced fluorescence imaging. *Acad. Radiol.* **17** (3), 375–381.

25 Hansch, A., Frey, O., Hilger, I., Sauner, D., Haas, M., Schmidt, D., Kurrat, C., Gajda, M., Malich, A., Bräuer, R., and Kaiser, W.A. (2004) Diagnosis of arthritis using near-infrared fluorochrome Cy5.5. *Invest. Radiol.*, **39** (10), 626–632.

26 Wunder, A., Müller-Ladner, U., Stelzer, E.H., Funk, J., Neumann, E., Stehle, G., Pap, T., Sinn, H., Gay, S., and Fiehn, C. (2003) Albumin-based drug delivery as novel therapeutic approach for rheumatoid arthritis. *J. Immunol.*, **170** (9), 4793–4801.

27 Fischer, T., Gemeinhardt, I., Wagner, S., Stieglitz, D.V., Schnorr, J., Hermann, K.G., Ebert, B., Petzelt, D., Macdonald, R., Licha, K., Schirner, M., Krenn, V., Kamradt, T., and Taupitz, M. (2006) Assessment of unspecific near-infrared dyes in laser-induced fluorescence imaging of experimental arthritis. *Acad. Radiol.*, **13** (1), 4–13.

28 Chen, W.T., Mahmood, U., Weissleder, R., and Tung, C.H. (2005) Arthritis imaging using a near-infrared fluorescence folate-targeted probe. *Arthritis Res. Ther.*, **7** (2), R310–R317.

29 Wunder, A., Schellenberger, E., Mahmood, U., Bogdanov, A. Jr., Müller-Ladner, U., Weissleder, R., and Josephson, L. (2005) Methotrexate-induced accumulation of fluorescent annexin V in collagen-induced arthritis. *Mol. Imaging*, **4** (1), 1–6.

30 Schutters, K. and Reutelingsperger, C. (2010) Phosphatidylserine targeting for diagnosis and treatment of human diseases. *Apoptosis*, **15** (9), 1072–1082.

31 Vollmer, S., Vater, A., Licha, K., Gemeinhardt, I., Gemeinhardt, O., Voigt, J., Ebert, B., Schnorr, J., Taupitz, M., Macdonald, R., and Schirner, M. (2009) Extra domain B fibronectin as a target for near-infrared fluorescence imaging of rheumatoid arthritis affected joints *in vivo*. *Mol. Imaging*, **8** (6), 330–340.

32 Blum, G. (2008) Use of fluorescent imaging to investigate pathological protease activity. *Curr. Opin. Drug Discov. Dev.*, **11** (5), 708–716.

33 Schellenberger, E., Rudloff, F., Warmuth, C., Taupitz, M., Hamm, B., and Schnorr, J. (2008) Protease-specific nanosensors for magnetic resonance imaging. *Bioconjug. Chem.*, **19** (12), 2440–2445.

34 Wunder, A., Tung, C.H., Müller-Ladner, U., Weissleder, R., and Mahmood, U. (2004) In vivo imaging of protease activity in arthritis: a novel approach for monitoring treatment response. *Arthritis Rheum.*, **50** (8), 2459–2465.

35 Lee, H., Lee, K., Kim, I.K., and Park, T.G. (2008) Synthesis, characterization, and *in vivo* diagnostic applications of hyaluronic acid immobilized gold nanoprobes. *Biomaterials*, **29**, 4709–4718.

36 Arbab, A.S., Janic, B., Haller, J., Pawelczyk, E., Liu, W., and Frank, J.A. (2009) In vivo cellular imaging for translational medical research. *Curr. Med. Imaging Rev.*, **5** (1), 19–38.

37 Simon, G.H., Daldrup-Link, H.E., Kau, J., Metz, S., Schlegel, J., Piontek, G., Saborowski, O., Demos, S., Duyster, J., and Pichler, B.J. (2006) Optical imaging of experimental arthritis using allogeneic leukocytes labeled with a near-infrared fluorescent probe. *Eur. J. Nucl. Med. Mol. Imaging*, **33** (9), 998–1006.

38 Sutton, E.J., Boddington, S.E., Nedopil, A.J., Henning, T.D., Demos, S.G., Baehner, R., Sennino, B., Lu, Y., and Daldrup-Link, H.E. (2009) An optical imaging method to monitor stem cell migration in a model of immune-mediated arthritis. *Opt. Express*, **17** (26), 24403–24413.

39 Scheel, A.K., Krause, A., Mesecke-von Rheinbaben, I., Metzger, G., Rost, H., Tresp, V., Mayer, P., Reuss-Borst, M., and Müller, G.A. (2002) Assessment of

proximal finger joint inflammation in patients with rheumatoid arthritis, using a novel laser-based imaging technique. *Arthritis Rheum.*, **46** (5), 1177–1184.

40 Scheel, A.K., Backhaus, M., Klose, A.D., Moa-Anderson, B., Netz, U.J., Hermann, K.-G.A., Beuthan, J., Müller, G.A., Burmester, G.R., and Hielscher, A.H. (2005) First clinical evaluation of sagittal laser optical tomography for detection of synovitis in arthritic finger joints. *Ann. Rheum. Dis.*, **64** (2), 239–245.

41 Hielscher, A.H., Bluestone, A.Y., Abdoulaev, G.S., Klose, A.D., Lasker, J., Stewart, M., Netz, U., and Beuthan, J. (2002) Near-infrared diffuse optical tomography. *Dis. Markers*, **18** (5–6), 313–337.

42 Hielscher, A.H., Klose, A.D., Scheel, A.K., Moa-Anderson, B., Backhaus, M., Netz, U., and Beuthan, J. (2004) Sagittal laser optical tomography for imaging of rheumatoid finger joints. *Phys. Med. Biol.*, **49** (7), 1147–1163.

43 Klose, C.D., Klose, A.D., Netz, U., Beuthan, J., and Hielscher, A.H. (2008) Multiparameter classifications of optical tomography images. *J. Biomed. Opt.*, **13** (5), 050503.

35
Diffuse Optical Tomography of Osteoarthritis
Zhen Yuan and Huabei Jiang

35.1
Motivation

Osteoarthritis (OA) is the most common arthritic condition worldwide and is estimated to affect nearly 60 million Americans. Classically, OA is most often found in the large weight-bearing joints of the lower extremities, particularly the knees and hips. However, there is also a subset of individuals with a predilection for developing OA of the hands and a more generalized form of OA. To diagnose cartilage abnormalities and alterations in the composition of synovial fluid in joints affected by OA, a variety of imaging methods have been developed and tested, such as X-ray methods, ultrasound (US), computed tomography (CT), and magnetic resonance imaging (MRI). While X-ray methods are able to visualize joint space narrowing and osteophyte formation, it is insensitive to changes in cartilage and fluid and therefore incapable of capturing the primary features of the early stage of OA. MRI, another commonly used modality in clinical practice, can reliably detect early-stage OA when high-contrast agents are used. However, it is costly and time consuming. CT has also been employed in the diagnosis of OA. However, it is expensive and provides only qualitative structural information in severe OA. Musculoskeletal US has been the subject of much recent interest in evaluating rheumatoid arthritis and regional musculoskeletal pathology. US is, however, a strongly operator-dependent modality and sensitive only to changes in the boundary layer, limiting its utility in the evaluation of the early stages of OA. Although present therapy is symptomatic, remarkable advances have been made in our understanding of the pathophysiology of OA. Much of the degradation of cartilage is mediated through matrix metalloproteinases, and the development of small-molecule inhibitors has been an area of active research interest. In anticipation of the development of new products with the potential to alter the natural history of OA, it will be crucial to have noninvasive technologies that can detect early-stage OA and monitor the efficacy of therapy [1–7].

35.2
Diffuse Optical Tomography Imaging of Osteoarthritis in the Hands

Owing to its numerous advantages of low cost, portability, and use of non-ionizing radiation, near-infrared (NIR) diffuse optical tomography (DOT) is emerging as a potential tool for imaging bones and joint tissues [1–7]. DOT imaging methods are able to provide a variety of functional and structural information with high sensitivity and specificity compared with other imaging modalities. This is especially true for the joints of the fingers, where the small dimensions and much higher transmitted light intensities should result in better signal-to-noise ratios and greatly improved spatial resolution. In a recent pilot case study, we have shown that the optical contrast between OA and normal joints is high, suggesting that DOT indeed has the potential for detecting OA joints in the hands and for assessing its treatment [1]. Although DOT appears to be especially suited for imaging of the finger joints because of the high signal-to-noise ratio associated with the small volume, the spatial resolution is still relatively low due to light scattering. In addition, X-ray imaging is currently the gold standard imaging method for the detection and quantification of joint destruction in patients with OA with high resolution. To take advantage of the complementary information from the optical and X-ray imaging modalities, an optimized approach that combines X-ray and DOT imaging has been developed for *in vivo* imaging of OA in the finger joints [2, 3]. The basic idea of this multi-modality imaging approach is to incorporate the high-resolution structural X-ray images into the DOT reconstruction so that both the resolution and quantitative accuracy of optical image reconstruction are enhanced.

35.2.1
DOT and Hybrid X-Ray–DOT Systems

The DOT system (Figure 35.1) consists of laser modules, a hybrid light delivery subsystem, a fiber optics–tissue interface, a data acquisition module, and light

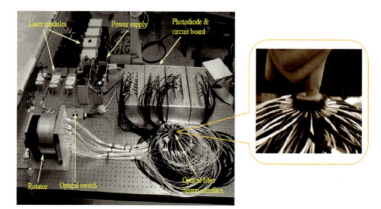

Figure 35.1 Photograph of the DOT imaging system. The inset is a close-up photograph of the finger–coupling medium–fiber optics interface.

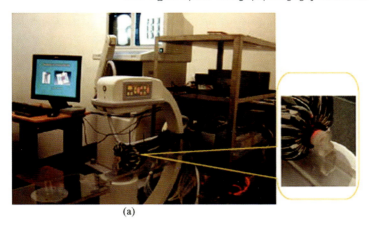

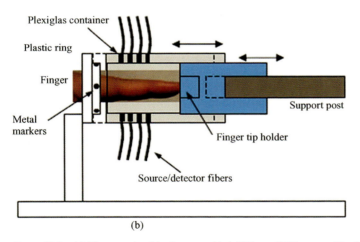

Figure 35.2 (a) Photograph of the integrated hybrid X-ray–DOT system. The inset is a close-up photograph of the finger–fiber optics–X-ray interface. (b) Schematic of the interface.

detection modules [1]. The cylindrical fiber optics–tissue interface is composed of 64 source and 64 detector fiber bundles that are positioned in four layers along the surface of a Plexiglas container and cover a volume of 15 × 30 mm. In each layer, 16 source and 16 detector fiber bundles are arranged alternately. The space between the finger and the wall of the Plexiglas container is filled with tissue-like phantom materials as coupling media consisting of distilled water, agar powder, Indian ink and Intralipid, giving an absorption coefficient of 0.014 mm^{-1} and a reduced scattering coefficient of 1.0 mm^{-1}. A total of eight laser modules in the NIR region are available. The hybrid X-ray–DOT imaging system (Figure 35.2) integrates a modified mini C-arm X-ray system (MiniView 6800, GE-OEC, Salt Lake City, UT, USA) with the 64 × 64-channel photodiode-based DOT system [2]. In the hybrid imaging of joint tissues, the X-ray imaging was performed immediately after the DOT data acquisition. To eliminate the artifacts in the X-ray projections possibly caused by the optical interface [2], we used a co-axial post to support the optical interface so that the interface can be translated along the post (see the inset in Figure 35.2a and the schematic of the interface shown in Figure 35.2b).

35.2.2
DOT Reconstruction Methods

When the data acquisition was finished, the next step was to generate the optical images using a robust 3D reconstruction algorithm. Our existing reconstruction algorithm is based on the following diffusion equation and type III boundary conditions [1]:

$$\nabla \cdot D(r)\nabla\Phi(r) - \mu_a(r)\Phi(r) = -S(r), \quad -D\nabla\Phi \cdot n = \alpha\Phi \qquad (35.1)$$

where $\Phi(r)$ is the photon intensity, $D(r)$ the diffusion coefficient, α is a coefficient related to the boundary, $\mu_a(r)$ is the absorption coefficient, and $S(r)$ is the source term. The discretized finite element form of Eq. (35.1) is written as $[A]\{\Phi\} = \{b\}$. The inverse solution is obtained through the following equations:

$$[A]\{\partial\Phi/\partial\chi\} = \{\partial b/\partial\chi\} - [\partial A/\partial\chi]\{\Phi\}, \quad (\Im^T\Im + \lambda I)\Delta\chi = \Im^T(\Phi^o - \Phi^c) \qquad (35.2)$$

where χ expresses D and μ_a, and $\Im$ is the Jacobian matrix formed by $\partial\Phi/\partial\chi$ at the boundary measurement sites. λ is a scalar, I is the identity matrix and $\Delta\chi$ is the updating vector for the optical properties. $\Phi^o = (\Phi_1^o, \Phi_2^o, \ldots, \Phi_M^o)^T$ and $\Phi^c = (\Phi_1^c, \Phi_2^c, \ldots, \Phi_M^c)^T$, where Φ_i^o and Φ_i^c are observed and computed photon intensity for $i = 1, 2, \ldots, M$ boundary locations. For X-ray guided DOT reconstruction, the following updating equation is used:

$$\Delta\chi = (\Im^T\Im + \Im^T\Im + \lambda I + L^T L)^{-1}[\Im^T(\Psi^o - \Psi^c)] \qquad (35.3)$$

where the X-ray structural *a priori* information is incorporated into the iterative process using a spatially variant filter matrix L [2].

35.2.3
Reconstructed Results

From the absorption and scattering images shown in Figures 35.3 and 35.4, we find that the bones are clearly identified. Importantly, compared with the optical parameters of the bones, we observe a significant decrease in the strength of absorption and scattering properties of the healthy joint tissues in the joint space/cavity. However, we see only a small drop for the OA joint tissues. Interestingly, the difference in joint tissues between the OA patients and healthy controls seems apparent from the ratio of the optical properties of joint soft tissues to that of bone [2, 3]. We note that these ratios for the diseased joints are significantly larger than those from the healthy joints [2].

35.2 Diffuse Optical Tomography Imaging of Osteoarthritis in the Hands

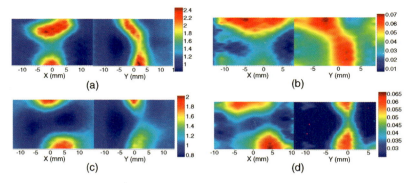

Figure 35.3 Reconstructed low-resolution optical images at 805 nm with selected dorsal/coronal planes: scattering (a) and absorption slices (b) for an OA joint; scattering (c) and absorption slices (d) for a healthy joint. Compared with the optical parameters of the bones, a significant decrease in the values of absorption and scattering properties of the healthy joint tissues surrounding the joint space (c, d) can be observed, whereas the drop for the OA joint is not so significant (a, b).

As shown in Figure 35.4, when a subset of the prior X-ray information on the joint structure was used in the DOT reconstruction, distinct boundaries separating different tissues were clearly recovered, indicating a significant improvement of DOT resolution because of the incorporation of prior X-ray structural information. These optical images show accurate delineation of the joint space and bone geometry, consistent with the X-ray findings. Both the absorption and scattering images reconstructed without X-ray guidance (Figure 35.3) show significantly overestimated thickness of the joint tissues and also increased boundary artifacts. The image quality obtained from the hybrid system is significantly improved over that from DOT alone. We see that the ratios for the diseased joints are significantly larger than those from normal joints. The differences in the ratio between the OA and normal joints

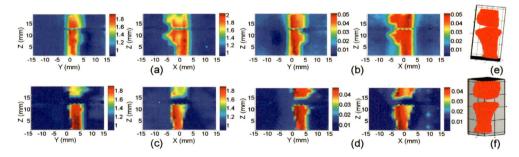

Figure 35.4 Reconstructed scattering (a) and absorption slices (b) for the OA joint with X-ray guidance; scattering (c) and absorption slices (d) for the healthy joint with X-ray guidance; tomographic X-ray image for the OA joint (e) and healthy joint (f). Joint space narrowing is observed for the OA joint (a, b). Importantly, compared with the optical parameters of the bones a significant decrease in the values of optical properties of the healthy joint tissues (c, d) is identified.

estimated from the X-ray-guided DOT reconstruction are notably increased relative to that without X-ray guidance [2].

We have seen that the OA-induced joint changes are quantitatively revealed by DOT. It is noted from the results that the optical properties and also the joint spacing between OA and healthy joints are different, and both can be used as an OA indicator. Our statistical analysis revealed that sensitivity and specificity up to 92 and 100%, respectively, can be achieved when the optical properties of joint tissues are used as classifiers [3].

35.2.4
Diffuse Model Versus Transport Model

Owing to computational complexity, the model used for describing photon migration in biological tissues has usually been limited to the diffusion approximation (DA) to the radiative transport equation (RTE). However, the DA-based method is challenging in imaging small volumes including arthritis in the finger joints. In such cases, the DA is not able to describe the photon migration accurately owing to the small optical distance between sources and detectors. To overcome this limitation, the discrete ordinate approximated RTE has been utilized to recover the optical parameters of small volume tissues. However, this type of reconstruction algorithm is time consuming. We have developed a 3D reconstruction method based on simplified spherical harmonics approximated RTE which is only 1.2 times slower than the DA-based method [4].

As shown in Figure 35.5, the difference in recovered quantitative optical properties between the two models can be as large as 15% [4]. This demonstrates that the RTE

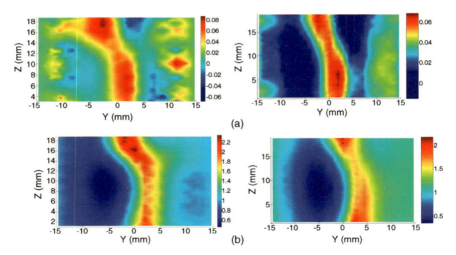

Figure 35.5 Reconstructed absorption (a) and scattering (b) images at selected coronal planes for the OA joint using the DA (left column) and the RTE model (right column). Importantly, boundary effect is significantly reduced for the optical image recovered using the RTE without X-ray guidance (right column).

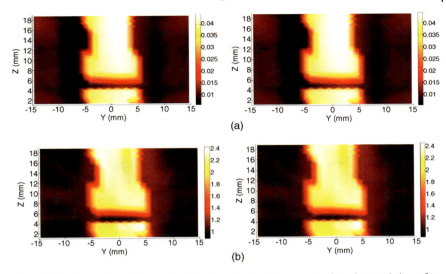

Figure 35.6 Reconstructed absorption (a) and scattering (b) images at selected coronal planes for the healthy joint using DA (left column) and RTE model (right column) with X-ray guidance. The image quality using the RTE is not enhanced when X-ray guidance is available.

model can provide significantly improved reconstruction accuracy for joint imaging. More interestingly, model errors appear to lead to smaller errors in reconstruction when the X-ray *a priori* structural information is incorporated into DOT reconstruction, as shown in Figure 35.6. This is because more accurate modeling of photon migration in tissue can be achieved using the DA when anatomic *a priori* information becomes available. The use of prior structural information eliminates the need to look for spatial/anatomy information in the optical inversion, which ensures that optical property profiles are the only parameter(s) that need to be recovered.

35.3
Image-Guided Optical Spectroscopy in Diagnosis of OA

There is increasing evidence that OA is a disease involving a metabolic dysfunction of bone [3]. It is likely that this metabolic dysfunction of bone, often associated with high metabolism of subchondral bone and connected joint soft tissues, will cause changes in tissue oxygen saturation (S_TO_2), deoxyhemoglobin, oxyhemoglobin, and water content (H_2O). When multiple wavelengths are used, the absorption spectra determine the tissue concentration of S_TO_2, Hb, HbO_2, and H_2O, the dominant NIR molecular absorbers in joint tissues. These quantitative physiological parameters of joint tissues may possibly allow us to detect metabolic changes associated with OA joints. In the spectroscopic experiments conducted, six wavelengths from the laser sources were used to irradiate the finger joint systems (633, 670, 723, 805, 853, and 896 nm) to ensure the reconstruction accuracy and minimize the parameter cross-talk.

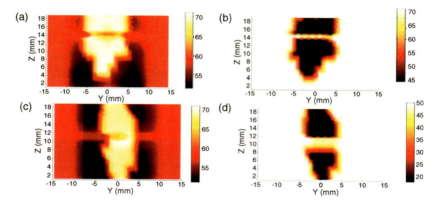

Figure 35.7 Reconstructed images at selected coronal planes: the oxygen saturation (a) and water content slice (b) for an OA finger joint; oxygen saturation (c) and water content slice (d) for a healthy joint. The axes (left and bottom) indicate the spatial scale in millimeters, whereas the color scale gives the oxygen saturation (%) or water content (%). Compared with the S_TO_2 values of bones, a significant decrease in the magnitude of the S_TO_2 value of soft tissues for the OA joints is observed (a) whereas only a small decrease for the healthy one is identified (c); the H_2O content of soft tissues for the OA joints (b) overall is significantly higher than that for the healthy joints (d).

We found that both S_TO_2 and H_2O of joint soft tissues are able to distinguish well between OA and normal joints in the majority of subjects [3], as observed from Figure 35.7. We noted that in most OA cases, the mean S_TO_2 values of joint soft tissues are smaller than those for bones. Further, compared with the S_TO_2 values of bones, we observed a significant decrease in the magnitude of the S_TO_2 value of soft tissues for the OA joints, whereas we saw only a small decrease in this parameter of soft tissues for the healthy joints. In some healthy cases, the S_TO_2 values of soft tissues are comparable to those of the bones. We can also see from Figure 35.7 that the H_2O content of soft tissues for the OA joints is overall significantly higher than that for the healthy joints. The high-resolution metabolic/functional images of joint tissues achievable by X-ray–multispectral DOT may shed light on the very early detection of OA in the finger joints.

35.4
Future Directions

Based on joint morphology and the optical and physiological/metabolic properties recovered, we observed statistically significant differences between healthy and OA finger joints [1–7]. This suggests that these imaging parameters are potential indicators for diagnosing OA and monitoring its progression.

Further large-scale clinical studies are necessary to evaluate prospectively the potential of DOT for finger joint imaging. In addition, there is a need to explore the possibility of DOT imaging in large joints such as the knee. Finally, it is definitely necessary to study whether this technique can distinguish well between OA and other types of arthritis such as rheumatoid arthritis and psoriatic arthritis.

References

1 Yuan, Z., Zhang, Q., Sobel, E., and Jiang, H.B. (2007) *J. Biomed. Opt.*, **12**, 034001.
2 Yuan, Z., Zhang, Q., Sobel, E., and Jiang, H.B. (2008) *J. Biomed. Opt.*, **13**, 044006.
3 Yuan, Z., Zhang, Q., Sobel, E., and Jiang, H.B. (2009) *Proc. SPIE*, **7174**, 71740K.
4 Yuan, Z., Zhang, Q., Sobel, E., and Jiang, H.B. (2009) *J. Biomed. Opt.*, **14**, 054013.
5 Hielscher, A.H., Klose, A.D., and Scheel, A.K. (2004) *Phys. Med. Biol.*, **49**, 1147.
6 Schwaighofer, A., Tresp, V., Mayer, P., Krause, A., *et al.* (2003) *IEEE Trans. Biol. Eng.*, **50**, 375.
7 Scheel, A.K., Krause, A., Rheinbaben Mesecke-von, I., *et al.* (2002) *Arthritis Rheum.*, **46**, 1177.

36
Laser Doppler Imaging

Jamie Turner, Bernat Galarraga, and Faisel Khan

In patients with rheumatoid arthritis (RA), the redness and heat associated with articular inflammation are directly related to increased blood flow to the area (hyperemia). It is therefore desirable to measure synovial blood flow *in vivo* since abnormalities in this are indicative of active arthritis and may give information about the extent and development of the disease. One of the more widely used noninvasive techniques for measuring microcirculatory blood flow is laser Doppler flowmetry (LDF). Until relatively recently, its use in rheumatology had been largely confined to research purposes, especially so with the traditional instruments that record perfusion continuously from a small area of tissue. With the advent of scanning laser Doppler imaging, which allows blood flow to be measured over much larger areas, its use has been extended to provide useful clinical information regarding disease severity in various rheumatologic conditions. In addition, laser Doppler imaging can also be used to measure microvascular endothelial function, a key marker of early atherosclerotic disease. This is potentially an extremely useful tool in a condition associated with an increased cardiovascular (CV) morbidity and mortality such as RA.

36.1
Development and Use of Laser Doppler Flowmetry

The development of LDF owes much to the study of the Doppler effect, whereby light waves that have been scattered by, for example, moving red blood cells in tissue undergo a shift in frequency. This phenomenon is utilized in LDF by directing a low-power monochromatic laser beam at the skin surface, which, depending on skin pigmentation, penetrates up to 2 mm into tissue. Light that is reflected off stationary tissue undergoes no shift whereas light that is reflected off moving blood cells with velocity undergoes the aforementioned Doppler shift in proportion to the velocity of the red blood cells. An analyzer/recorder detects this, giving an output of red blood cell flux (number of red blood cells times their velocity), which determines circulation [1].

In vivo measurements of blood flow with this technique were first attempted in animal models using single vessels such as the retinal artery [2] and femoral vein (via a fiber-optic catheter) of the rabbit [3]. Stern [4] first adapted the technique to look at tissue perfusion in humans, specifically in fingertip skin, and showed that the amplitude of the Doppler signal in the 10–12 kHz band varied predictably with challenges in the microcirculation, such as arterial occlusion, and after a vasodilator drug.

Correlation between laser Doppler and other measurements of blood flow, such as venous occlusion plethysmography, has been demonstrated, [5, 6]. However, the laser Doppler instrument cannot be calibrated directly against these other techniques because the anatomic variation of the microcirculation between different sites and different tissues would mean that the calibrations are valid only for that particular tissue and measurement site [7]. An LDF-derived flux value is therefore generally expressed in arbitrary units.

36.2
Laser Doppler Imaging

A major limitation of LDF is that it measures blood flow at only a single point, and with respect to the cutaneous microcirculation there is considerable spatial heterogeneity of blood flow, meaning that tissue perfusion can vary considerably over short distances [8]. This reduces the reproducibility of the measurements, making it difficult to monitor changes in perfusion over time, particularly within an individual where even subtle changes in positioning of the laser probe can lead to significant variations in perfusion.

A recent development to counteract this has been the ability to scan the laser with a motor-driven mirror in a raster fashion over a defined area of tissue to produce a detailed color-coded perfusion map. Scanning LDF or laser Doppler imaging (LDI) therefore allows investigators to compensate for spatial heterogeneity and also allows for the study of large areas of tissue, for example, the ability to obtain noncontact images of perfusion in inflamed joints in RA patients.

Standard laser Doppler imagers employ a helium–neon laser (red, wavelength 632.8 nm), which provides measures of perfusion from the relatively superficial dermis (1–2 mm depth) (Figure 36.1). However, perfusion from deeper levels can be obtained with the use of near-infrared light (wavelength 780 nm) [9]. The latter has been used both in animal models of inflamed joints and in patients with RA, giving an understanding of the development and treatment of joint injury and arthritis and the study of the mechanisms and mediators that may contribute to joint inflammation.

Forrester *et al.* used standard (633 nm) and near-infrared (780 nm) light to measure blood flow changes in the connective tissue of surgically exposed rabbit medial ligaments [10]. LDI has also been used to examine blood perfusion in the medial aspect of an exposed rat knee joint capsule after acute inflammation had been induced via carrageenan injection. This work showed an enhancement of vasodilator

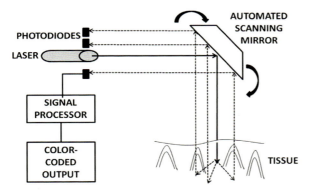

Figure 36.1 Schematic diagram of laser Doppler imager.

responses to substance P and calcitonin gene-related peptide, suggesting that these neuropeptides contained in the articular sensory C-fibers are important in mediating neurogenic joint inflammation [11]. This LDI method for measuring perfusion in inflamed and arthritic rat joints has been utilized in subsequent studies [12, 13].

LDI has also been used to show increased perfusion over the proximal interphalangeal and metacarpophalangeal joint of patients with RA, compared with control subjects [14, 15]. These differences were only seen with the more penetrating near-infrared laser light, confirming that these changes were indeed related to perfusion in the joint rather than the skin microcirculation (Figure 36.2). This was confirmed in a later study in which a red laser source failed to detect any hyperemic areas associated

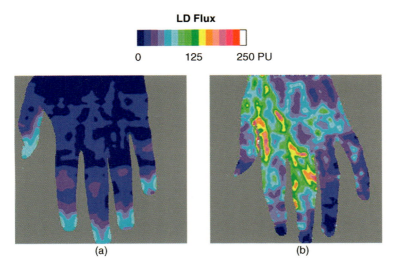

Figure 36.2 Near-infrared laser Doppler images from the hand of: (a) a normal subject, and (b) a patient with rheumatoid arthritis showing increased perfusion over the inflamed joints.

with joint inflammation in patients with osteoarthritis [16]. In future, this near-infrared laser Doppler scanning could provide a potential method for quantification of joint inflammation in patients with RA and permit noninvasive monitoring of the effects of drug therapy targeted at reducing inflammation, especially in longitudinal clinical trials.

36.3
Endothelial Dysfunction

Patients with RA have an increased mortality compared with the general population [17–19], with life expectancy being reduced by 10–15 years [20]. This increase in mortality is largely attributable to CV disease, with CV morbidity also increased in comparison with the general population [21, 22]. Factors that contribute to this excess CV risk include traditional risk factors [e.g., dyslipidemia, diabetes mellitus, hypertension, higher body mass index (BMI), impaired physical fitness], along with manifestations of the disease itself [23]. However, traditional CV risk factors cannot completely explain the increased morbidity and mortality observed in RA.

RA is characterized by inflammation, which also is a key component in the development of atherosclerosis [24, 25]. Inflammation leads to the activation of endothelial cells, which, through an increase in the expression of leukocyte adhesion molecules, promotes a pro-atherosclerotic environment. Pro-inflammatory markers such as C-reactive protein (CRP) levels and tumor necrosis factor alpha (TNFα) have an important role in atherosclerosis and CV mortality in RA [26].

The vascular endothelium is the key site through which inflammation exerts its deleterious effects on atherogenesis and its subsequent progression are intimately related to endothelial dysfunction. The vascular endothelium is the innermost layer of the blood vessels and has an essential role in maintaining the health of blood vessels through the maintenance of vascular tone and through the release of a variety of vasoactive substances and mediators of inflammation and coagulation. One the most important of these substances is nitric oxide (NO); it maintains blood flow by causing vasodilatation, inhibiting platelet aggregation and leukocyte adhesion, and preventing smooth-muscle proliferation.

In endothelial dysfunction, an imbalance between NO and other substances produced in the endothelium creates an environment that promotes vasoconstriction, inflammation, and coagulation, which can lead to both thrombosis and atherosclerotic disease. Endothelial dysfunction is therefore an early preclinical marker of atherosclerosis, and is commonly found in patients with RA.

36.4
LDI and Iontophoresis

Given the importance of endothelial function on the CV health of RA patients, numerous techniques to measure endothelial function in patients have been developed. A major use of LDI is to measure cutaneous perfusion accompanied by iontophoresis of vascular test drugs as a measure of endothelial function. Iontophoresis involves delivery of ions of soluble salts across the skin (typically in the inner forearm) under the influence of a relatively weak electrical current. Increasing either the current or the time of delivery can increase the total drug administered [27].

Using this technique, drugs such as acetylcholine (ACh) and sodium nitroprusside (SNP) have been tested on the skin and have given some insight into the involvement of endothelial function and NO activity in rheumatologic conditions [28]. ACh is an endothelium-dependent vasodilator that causes normal vasodilatation in the presence of an intact and fully functional endothelium. ACh mediates vasodilatation through the production of NO by NO synthase, with an accessory role for vasodilator prostanoids (e.g., prostacyclin) and endothelium-derived hyperpolarizing factor (EDHF). In contrast, SNP is an NO donor which allows examination of the response of the vasculature to exogenous NO, independently of the endothelium, and is therefore an established test drug for measuring vasculature smooth muscle cell function. Once the iontophoresed ACh/SNP reaches the microvessels of the skin, subsequent vasodilatation is recorded by LDI, which provides a relative measure of skin perfusion.

There is increasing evidence linking microvascular dysfunction to CV outcomes in a range of conditions [29, 30], hence measuring the response of the microcirculation to pharmacologic stimuli could improve our understanding of vascular dysfunction and disease progression. The technique has good reproducibility [31], and appears to be reflective of global endothelial function and CV health. Abnormalities in the microvascular bed have been shown to correlate with CV risk factors [32] and with established coronary artery disease [33]. The association reported between microvascular function and coronary flow reserve in healthy individuals [34] provides further evidence that assessments made in the skin do indeed provide a reliable global measure of microvascular function.

36.5
Studies Using LDI in RA Patients

Given the strong correlation between microvascular function and the presence of CV risk factors, assessment of the microvascular bed could provide a sensitive measure of the early stages of atherosclerosis in RA patients. Studies have shown that endothelial dysfunction in RA is closely associated with inflammation, and therapeutic reduction of inflammation leads to improvements in endothelial function. As such, assessments of microvascular endothelial function could prove to be useful in the identification and monitoring of CV risk in patients with RA. However, to date, not many studies have focused on microvascular function in RA (Figure 36.3). One small-scale study reported microvascular dysfunction in female patients with RA [35], and a pilot study showed improvements in microvascular function in eight patients with RA following anti-inflammatory treatment [36]. The most convincing evidence to date connecting systemic inflammation and microvascular endothelial function in RA was provided by a large, cross-sectional study of 128 patients with RA [37]. In this study, we reported

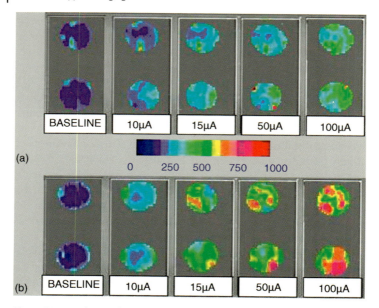

Figure 36.3 Colour-coded LDI output showing skin perfusion response to concurrent iontophoresis of ACh (top) and SNP (bottom) at increasing currents (10-100 uA). There is a clearly reduced response to both drugs in the RA patient (A) compared to the healthy control (B).

that microvascular endothelial dysfunction is directly correlated with CRP level, a key marker of systemic inflammation, independent of other conventional vascular risk factors [37]. We have also reported that vascular function improves in patients who respond to anti-TNF therapy or DMARDs (disease-modifying antirheumatic drugs), suggesting that, in the microcirculation at least, reduction in inflammation, rather than choice of treatment *per se*, can be beneficial [38].

36.6
Conclusion

Laser Doppler flowmetry has great potential in rheumatology in clinical practice, especially with the development of laser Doppler imagers that can measure the progression of the disease in synovial tissue. The noninvasive nature of the instruments and the fact that measurements can be made without contact make them highly acceptable to patients. Many of the inherent difficulties with traditional single-point techniques, such as variability of measurements due to spatial heterogeneity of skin blood flow, can be overcome with the use of imagers.

To date, LDI has been used mostly as a research tool, where it has provided a better understanding of the abnormalities of microvascular endothelial function in RA patients. Increased inflammation and subsequent endothelial dysfunction contribute to the many factors that lead to increased CV morbidity in RA patients. Hence LDI can become a vital tool for measuring CV risk in future studies of RA patient populations.

References

1 Nilsson, G.E., Tenland, T., and Oberg, P.A. (1980) Evaluation of a laser Doppler flowmeter for measurement of tissue blood flow. *IEEE Trans. Biomed. Eng.*, **27** (10), 597–604.
2 Riva, C., Ross, B., and Benedek, G.B. (1972) Laser Doppler measurements of blood flow in capillary tubes and retinal arteries. *Invest. Ophthalmol.*, **11** (11), 936–944.
3 Tanaka, T. and Benedek, G.B. (1975) Measurement of the velocity of blood flow (*in vivo*) using a fiber optic catheter and optical mixing spectroscopy. *Appl. Opt.*, **14** (1), 189–196.
4 Stern, M.D. (1975) *In vivo* evaluation of microcirculation by coherent light scattering. *Nature*, **254** (5495), 56–58.
5 Saumet, J.L., Dittmar, A., and Leftheriotis, G. (1986) Non-invasive measurement of skin blood flow: comparison between plethysmography, laser-Doppler flowmeter and heat thermal clearance method. *Int. J. Microcirc. Clin. Exp.*, **5** (1), 73–83.
6 Tooke, J.E., Ostergren, J., and Fagrell, B. (1983) Synchronous assessment of human skin microcirculation by laser Doppler flowmetry and dynamic capillaroscopy. *Int. J. Microcirc. Clin. Exp.*, **2** (4), 277–284.
7 Oberg, P.A. (1990) Laser-Doppler flowmetry. *Crit. Rev. Biomed. Eng.*, **18** (2), 125–161.
8 Wardell, K., Braverman, I.M., Silverman, D.G., and Nilsson, G.E. (1994) Spatial heterogeneity in normal skin perfusion recorded with laser Doppler imaging and flowmetry. *Microvasc. Res.*, **48** (1), 26–38.
9 Abbot, N.C., Ferrell, W.R., Lockhart, J.C., and Lowe, J.G. (1996) Laser Doppler perfusion imaging of skin blood flow using red and near-infrared sources. *J. Invest. Dermatol.*, **107** (6), 882–886.
10 Forrester, K., Doschak, M., and Bray, R. (1997) *In vivo* comparison of scanning technique and wavelength in laser Doppler perfusion imaging: measurement in knee ligaments of adult rabbits. *Med. Biol. Eng. Comput.*, **35** (6), 581–586.
11 Lam, F.Y. and Ferrell, W.R. (1993) Acute inflammation in the rat knee joint attenuates sympathetic vasoconstriction but enhances neuropeptide-mediated vasodilatation assessed by laser Doppler perfusion imaging. *Neuroscience*, **52** (2), 443–449.
12 Egan, C.G., Lockhart, J.C., and Ferrell, W.R. (2004) Pathophysiology of vascular dysfunction in a rat model of chronic joint inflammation. *J. Physiol. (Lond).*, **557** (2), 635–643.
13 McDougall, J.J. (2001) Abrogation of α-adrenergic vasoactivity in chronically inflamed rat knee joints. *Am. J. Physiol. Regul. Integr. Comp. Physiol.*, **281**, R821–R827.
14 Ferrell, W.R., Sturrock, R.D., Mallik, A.K., Abbot, N.C., Lockhart, J.C., and Edmondson, W.D. (1996) Laser Doppler perfusion imaging of proximal interphalangeal joints in patients with rheumatoid arthritis. *Clin. Exp. Rheumatol.*, **14** (6), 649–652.
15 Ferrell, W.R., Balint, P.V., Egan, C.G., Lockhart, J.C., and Sturrock, R.D. (2001) Metacarpophalangeal joints in rheumatoid arthritis: laser Doppler imaging – initial experience. *Radiology*, **220** (1), 257–262.
16 Ng, E.Y.K. and How, T.J. (2003) Laser-Doppler imaging of osteoarthritis in proximal interphalangeal joints. *Microvasc. Res.*, **65** (1), 65–68.
17 Myllykangas-Luosujarvi, R.A., Aho, K., and Isomaki, H.A. (1995) Mortality in rheumatoid arthritis. *Semin. Arthritis Rheum.*, **25** (3), 193–202.

18 Solomon, D.H., Goodson, N.J., Katz, J.N., Weinblatt, M.E., Avorn, J., Setoguchi, S., Canning, C., and Schneeweiss, S. (2006) Patterns of cardiovascular risk in rheumatoid arthritis. *Ann. Rheum. Dis.*, **65** (12), 1608–1612.

19 Van Doornum, S., McColl, G., and Wicks, I.P. (2002) Accelerated atherosclerosis: an extraarticular feature of rheumatoid arthritis? *Arthritis Rheum.*, **46** (4), 862–873.

20 Van Doornum, S., Jennings, G.L.R., and Wicks, I.P. (2006) Reducing the cardiovascular disease burden in rheumatoid arthritis. *Med. J. Aust.*, **184** (6), 287–290.

21 Shoenfeld, Y., Gerli, R., Doria, A., Matsuura, E., Cerinic, M.M., Ronda, N., Jara, L.J., Abu-Shakra, M., Meroni, P.L., and Sherer, Y. (2005) Accelerated atherosclerosis in autoimmune rheumatic diseases. *Circulation*, **112** (21), 3337–3347.

22 Sitia, S., Atzeni, F., Sarzi-Puttini, P., Di Bello, V., Tomasoni, L., Delfino, L., Antonini-Canterin, F., Di Salvo, G., De Gennaro Colonna, V., La Carrubba, S., Carerj, S., and Turiel, M. (2009) Cardiovascular involvement in systemic autoimmune diseases. *Autoimmun. Rev.*, **8** (4), 281–286.

23 Nurmohamed, M.T. (2009) Cardiovascular risk in rheumatoid arthritis. *Autoimmun. Rev.*, **8** (8), 663–667.

24 Ku, I.A., Imboden, J.B., Hsue, P.Y., and Ganz, P. (2009) Rheumatoid arthritis – a model of systemic inflammation driving atherosclerosis. *Circ. J.*, **73** (6), 977–985.

25 Ridker, P.M., Hennekens, C.H., Buring, J.E., and Rifai, N. (2000) C-reactive protein and other markers of inflammation in the prediction of cardiovascular disease in women. *N. Engl. J. Med.*, **342** (12), 836–843.

26 Dixon, W.G. and Symmons, D.P.M. (2007) What effects might anti-TNFα treatment be expected to have on cardiovascular morbidity and mortality in rheumatoid arthritis? A review of the role of TNFα in cardiovascular pathophysiology. *Ann. Rheum. Dis.*, **66** (9), 1132–1136.

27 Khan, F., Davidson, N.C., Littleford, R.C., Litchfield, S.J., Struthers, A.D., and Belch, J.J.F. (1997) Cutaneous vascular responses to acetylcholine are mediated by a prostanoid-dependent mechanism in man. *Vasc. Med.*, **2** (2), 82–86.

28 Cerinic, M.M., Generini, S., and Pignone, A. (1997) New approaches to the treatment of Raynaud's phenomenon. *Curr. Opin. Rheumatol.*, **9** (6), 544–556.

29 Turner, J., Belch, J.J.F., and Khan, F. (2008) Current concepts in assessment of microvascular endothelial function using laser Doppler imaging and iontophoresis. *Trends Cardiovasc. Med.*, **18** (4), 109–116.

30 Huang, A.L., Silver, A.E., Shvenke, E., Schopfer, D.W., Jahangir, E., Titas, M.A., Shpilman, A., Menzoian, J.O., Watkins, M.T., Raffetto, J.D., Gibbons, G., Woodson, J., Shaw, P.M., Dhadly, M., Eberhardt, R.T., Keaney, J.F. Jr., Gokce, N., and Vita, J.A. (2007) Predictive value of reactive hyperemia for cardiovascular events in patients with peripheral arterial disease undergoing vascular surgery. *Arterioscler. Thromb. Vasc. Biol.*, **27** (10), 2113–2119.

31 Newton, D.J., Khan, F., and Belch, J.J.F. (2001) Assessment of microvascular endothelial function in human skin. *Clin. Sci.*, **101** (6), 567–572.

32 Khan, F., Elhadd, T.A., Greene, S.A., and Belch, J.J.F. (2000) Impaired skin microvascular function in children, adolescents, and young adults with type I diabetes. *Diabetes Care*, **23** (2), 215–220.

33 Ijzerman, R.G., De Jongh, R.T., Beijk, M.A.M., Van Weissenbruch, M.M., Delemarre-van De Waal, H.A., Serne, E.H., and Stehouwer, C.D.A. (2003) Individuals at increased coronary heart disease risk are characterized by an impaired microvascular function in skin. *Eur. J. Clin. Invest.*, **33** (7), 536–542.

34 Khan, F., Patterson, D., Belch, J.J.F., Hirata, K., and Lang, C.C. (2008)

Relationship between peripheral and coronary function using laser Doppler imaging and transthoracic echocardiography. *Clin. Sci.*, **115** (9), 295–300.

35 Arosio, E., De Marchi, S., Rigoni, A., Prior, M., Delva, P., and Lechi, A. (2007) Forearm haemodynamics, arterial stiffness and microcirculatory reactivity in rheumatoid arthritis. *J. Hypertens.*, **25** (6), 1273–1278.

36 Datta, D., Ferrell, W.R., Sturrock, R.D., Jadhav, S.T., and Sattar, N. (2007) Inflammatory suppression rapidly attenuates microvascular dysfunction in rheumatoid arthritis. *Atherosclerosis*, **192** (2), 391–395.

37 Galarraga, B., Khan, F., Kumar, P., Pullar, T., and Belch, J.J.F. (2008) C-reactive protein: the underlying cause of microvascular dysfunction in rheumatoid arthritis. *Rheumatology (Oxford)*, **47** (12), 1780–1784.

38 Galarraga, B., Belch, J.J.F., Pullar, T., Ogston, S., and Khan, F. (2010) Clinical improvement in rheumatoid arthritis is associated with healthier microvascular function in patients who respond to antirheumatic therapy. *J. Rheumatol.*, **37** (3), 521–528.

Part 9
Ophthalmology and Optometry

37
Ocular Diagnostics and Imaging
Michael Larsen

37.1
Ocular Diagnosis, Imaging and Therapy

The eye is an optical imaging system with transparent refractive components that permit the projection of an image on its sensory component, the retina, where a neuronal signal is produced an relayed to the brain via the optic nerve (Figure 37.1). The same refractive media can be used to observe the interior of the eye and to apply therapeutic optical radiation to the eye. This chapter presents an overview of the current state of ophthalmic optical technology and the diagnostic use of optics in the study of the eye and its diseases.

37.2
Outline of Anatomy and Physiology

The optical interface between the eye and the surrounding world is the surface of the tear film, a delicate, multilayered film composed of an outer layer of lipid, a fluid layer of electrolytes dissolved in water in the middle, and a bottom of seaweed-like polysaccharide chains that anchor the film to the cells that cover the surface of the cornea [1]. Without a tear film, there is no glimpse (specular reflex) in an eye, there is intense discomfort and blurred vision, and the surface cells of the cornea will soon be lost, as will eventually the transparency of the cornea (Figure 37.2). Two-thirds of the refractive power of the eye is found on the interface between air and tear film.

After the tear film comes the cornea, which is a living tissue supported internally by an ordered network of collagen fibers analogous to the component of skin that is used to make leather. Behind it, the anterior chamber of the healthy eye is filled with clear, watery fluid. The iris forms the pupil, an adjustable aperture located close to the front nodal plane of the eye. The iris covers the anterior surface of the lens to a variable extent. The lens (Figure 37.3) differs from conventional artificial lenses in having the properties of a gradient refractive index (GRIN) lens. It is responsible for one-third of the refractive power of the eye. Behind the lens, a gelatinous structure with a few fibrils, the vitreous body (or simply *the vitreous* in clinical jargon), provides

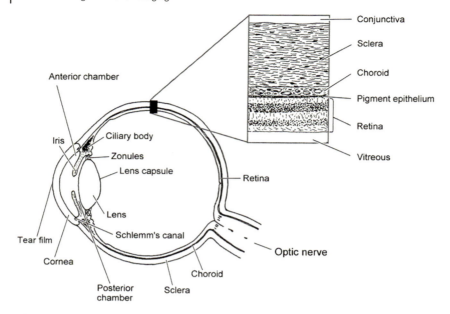

Figure 37.1 Cross-section of the right eye of a human seen from above. The refractive media include the tear film, the cornea, the aqueous humor in the anterior and posterior chambers, the lens, the vitreous body, and the retina. The photoreceptors are located in the outer retina, next the to retinal pigment epithelium.

supplementary mechanical support for the lens. It has no apparent optical role other than being clear and its delicate structure is a rudiment of fetal vessels that is responsible for the phenomenon of seeing "floaters." The posterior inside of the eye is covered by its actual sensory tissue, the retina, the arrayed photoreceptors of which

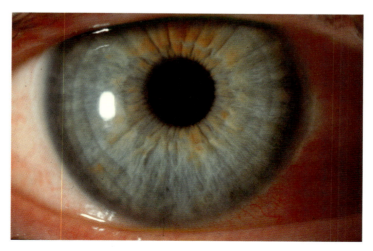

Figure 37.2 The smooth corneal reflex is produced by a contiguous tear film covering the entire surface of the eye, thus forming the primary and most powerfully refracting optical surface of the eye.

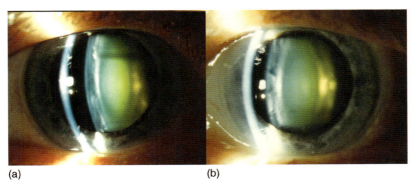

Figure 37.3 The anterior segment of the eye seen in diffuse illumination combined with a narrow vertical beam of intense white light from a slit-lamp illuminating the eye from the left at a 30° angle from the line of sight. The left eye (b) is nearly normal and shows some backscatter of light as the beam passes the cornea (narrow and curved, to the left) and the lens (biconvex and broader, at the center of the pupil). The right eye (a) has a cortical lens opacity that casts a shadow into the upper part of the lens and a highly scattering lens nuclear cataract.

convert incoming photons to neuronal signals that are processed and compressed in the inner layers of the retina before being conveyed to the brain via the optic nerve. Behind the retinal pigment epithelium, which is the outermost layer of the retina, the highly perfused network of blood vessels called the choroid delivers nutrition that diffuses into the retina to supplement the relatively sparse network of retinal vessels. A small depression in the retina near the posterior pole of the eye marks the fovea and its center, the foveola, where the daylight-optimized cone photoreceptors are found at maximum density and where fixation, reading, face recognition, and other high angular resolution tasks are subserved.

Anatomic terminology refers the corneal direction as being anterior and the direction towards the fovea as being posterior, or it may refer to the eye as a globe, the direction toward its center being inside or internal and the direction away from its center being outside or external. In some contexts, the posterior part of the eye may be referred to as though the patient were lying down on an operating table, the choroid being below the retinal pigment epithelium, and so on.

The optical components of the eye are not perfectly aligned on a linear optical axis and the axis of viewing defined by the location of the foveola and the center of the pupil deviates from the geometric axis of the eye by a few degrees. While such theoretical imperfections can be pointed out, there are indications that they help compensate for a number of classical problems in the manufacture of artificial optics. An elegant example is the bowl-shaped sensor, that is, the photoreceptor layer of the retina, the shape of which is a near-perfect match to the curved image plane of the optical apparatus in the anterior segment of the eye. This example shows that what would be considered imperfections on an optical assembly line – a grotesquely curved image plane and a sensor that is not a perfectly flat – combine to form a magnificently compact wide-angle biological camera.

37.3
Basic Optical Instrumentation for Examination of the Eye

The slit-lamp biomicroscope was designed for stereoscopic observation of the anterior segment of the eye at high magnification and with adjustable illumination that can be focused to a narrow slit-shaped beam at the focus of the microscope (Figure 37.4). The light source can be rotated about this focus so that the various tissues can be seen in the optical cross-section provided by light scattered from the semi-transparent tissues (cornea, lens, vitreous, retina) and the aqueous, which is normally free from visible scatter.

Observation of the interior of the eye requires nothing but coaxial illumination. The unaided eye can see all the way to the posterior inside of the eye, if assisted by a suitable source of coaxial illumination. This is the principle of direct ophthalmoscopy (Figure 37.5). Limitations of direct ophthalmoscopy include lack of stereoscopic viewing and a narrow field of view, often less than 5° as seen from outside the eye.

Stereoscopic viewing and a wide field are provided by indirect ophthalmoscopy, where a real image of the fundus is formed between the observer and a lens held close to the eye (Figures 37.4 and 37.6).

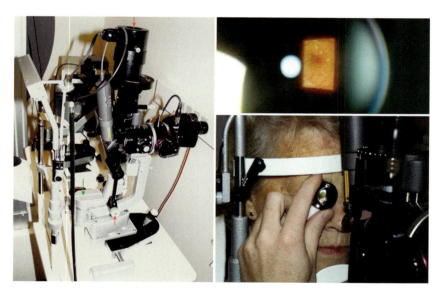

Figure 37.4 Binocular microscope for examination of a seated subject, with his or her head rested on a fixed support, the examiner being seated on the other side of an adjustable table. The light source and the microscope can be moved on the table with the aid of a joystick and can be rotated independently about a common vertical axis (green arrow). The lamp house (upper red arrow) and its optical shutters can be rotated and adjusted by knobs at the lower end of this unit (lower red arrow).

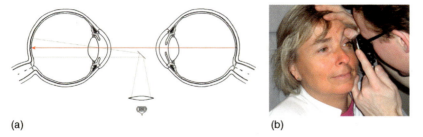

Figure 37.5 Direct ophthalmoscopy. The left eye of the observer views the fovea of the eye of the patient with the aid of a quasi-coaxial light source. The modern hand-held direct ophthalmoscope includes a rotating disk with lenses of various dioptric powers that allow correction for spherical refractive errors in the examiner and the patient. The sketch in (a) shows the eye of the patient to the left, with a pharmaceutically dilated pupil and the observer to the right, both seen from above.

37.4
Fundus Photography

The principle of conventional fundus photography (Figure 37.7) is the same as that of indirect ophthalmoscopy, except that the eye of the observer is replaced by an artificial refractive system and the retina with a film. The source of illumination is built into the camera and the lens in front of the eye is fixed to the camera housing. Alignment of the camera is accomplished by moving the entire camera housing while the patient is seated and fixating a target light with his or her other eye.

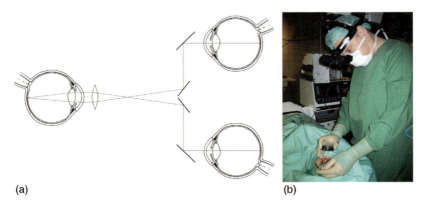

Figure 37.6 Indirect ophthalmoscopy. The fundus of the eye of the patient, shown to the left in (a), is observed with both eyes of the observer at a viewing angle of nearly 7° angular separation inside the eye. This is made possible by using mirrors to narrow the effective distance between the observer's pupils to a few mm. The illumination (not shown on the sketch) is mounted on the axis of symmetry, slightly above the plane of the viewing optics. The two lines of sight are crossing between the physician (b) and the hand-held lens, at a conjugate plane of the fundus where a real image is formed.

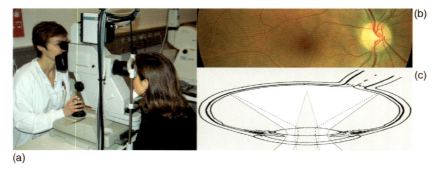

Figure 37.7 (a) Fundus camera with photographer (left) and patient (right). (c) Schematic diagram of the pathways of optical illumination (white) and imaging (stippled outline), which are separated as they pass through the back-scattering media of the cornea and the lens. The layout of these pathways determines how small a pupil the camera can work with. Optical cross-talk between the incoming and outgoing pathways increases with increasing scatter in the lens as it ages or suffers from opacification from other causes.

High-quality fundus photography requires pharmaceutical pupil dilation. Maximum dilation is achieved by combining two mydriatic eyedrops, one of which is an anticholinergic and the other an adrenergic agent, thus paralyzing the circular pupil sphincter muscle while stimulating the radial dilator muscle. Pupil dilation requires about 30 min of preparation (instillation, waiting for the effect to set in, and observing for side effects) and vision will be blurred for 1–3 h after the examination. To provide faster and more convenient examinations, fundus cameras for smaller pupils have been designed. These non-mydriatic cameras provide less lateral resolution, because of the smaller numerical aperture, and less contrast than mydriatic fundus cameras because of the poorer separation between illumination and imaging pathways. Systematic use of non-mydriatic cameras in settings such as diabetic retinopathy screening clinics is problematic because ~25% of patients will be found to need pharmaceutical pupil dilation and the extra work involved tends to offset the time gained in the remainder of the population.

Conventional fundus cameras cover an angle of 30–60 degrees in a single image. Wider coverage can be obtained by pointing the camera in multiple directions. Systematic fundus photography protocols divide the fundus into a number of fields. Systematic screening for diabetic retinopathy is often restricted to a fovea-centered and a disk-centered image (Figure 37.8).

Fundus photographs can be made in stereoscopic pairs by parallel displacement of the camera horizontally away from the centerline of the pupil, as much as can be done without sacrificing the illumination of the fundus (Figure 37.9). Before the development of optical coherence tomography (OCT), stereoscopic biomicroscopy and fundus photography were the only means whereby information about retinal thickening could be obtained.

Digital fundus cameras are made with three-channel image sensors that capture the red, green, and blue components or layers of an object illuminated by white light. The effective illumination varies considerably with the spectral characteristics of the

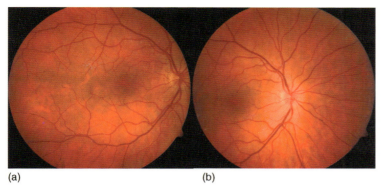

Figure 37.8 Color fundus photographs, each 60° horizontal subtense seen from outside the eye. The image in (a) is centered on the temporal side of the fovea and that in (b) is centered on the optic disk.

crystalline lens. With increasing age comes increasing absorption from the blue end of the spectrum (Figure 37.10). It would be desirable to have fundus cameras that interactively regulate spectral illumination so that it could be tuned to suit the individual eye.

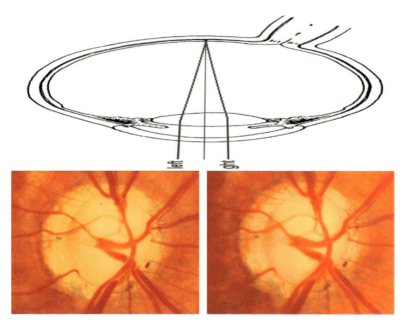

Figure 37.9 Schematic presentation of parallel displacement of fundus camera to record stereoscopic fundus photographs. The pair below have been swapped to allow stereoscopic viewing by crossing one's eyes. Successful viewing is accomplished when one is looking at the middle of three images and perceiving a profound cupping of the optic nerve head.

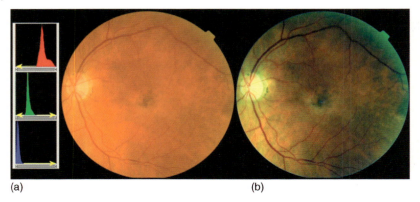

Figure 37.10 Raw color fundus photograph from an elderly person made using a nonmydriatic fundus camera (a) with curves to the left showing the distribution of pixel intensities over the 256-bit dynamic range of the image sensor. The red channel is well exposed, bordering on overexposure, the green channel is moderately underexposed, and the blue channel is severely underexposed. By spreading the signal over the full dynamic range of the display (histogram stretching), as indicated by yellow arrows, a color-selective contrast-enhanced image can be produced (b).

Gray-scale fundus photographs provide superior contrast, the balance between the various features of the normal and diseased fundus varying with the color of the illumination. Briefly, blue light provides optimal viewing of the retinal nerve fiber layer, green light shows blood in high contrast, and red light shows pigmented lesions to advantage. Quantitative spectral analysis of the fundus image has few applications, but retinal oximetry is currently being studied as a means of investigating retinal metabolism. This method of estimating oxygen saturation in the blood is based on a characteristic shift in the absorption spectrum of hemoglobin when it changes between being oxygenated and deoxygenated (Figure 37.11).

Fundus photography can be performed in the infrared region of the spectrum. There is poor contrast, except when densely pigmented structures such as choroidal nevi are present. Infrared fundus video photography is used to monitor aiming and focusing of nonmydriatic fundus cameras because the invisible radiation does not stimulate pupil contraction. Because the infrared image is useless in itself, a flash of white light must be fired to obtain a regular fundus image. Digital nonmydriatic cameras work with low flash settings, thanks to the high sensitivity of digital image sensors, but it is obvious that in many subjects some degree of pupil constriction persists for several minutes after the first image has been shot.

Fundus image analysis has been shown experimentally to be able to separate fundus images with diabetic retinopathy from images without retinopathy with an accuracy that is comparable to that of trained observers [2] (Figure 37.12). The method has been introduced in selected settings as a means of prescreening images to reduce the number of screening images that need to be assessed by graders. Additional uses of digital image analysis include measurement of retinal

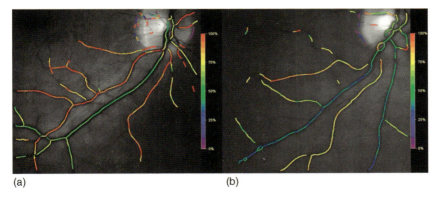

Figure 37.11 Oxygen saturation in retinal vessels of a healthy subject (a) and a patient with severe systemic hypoxia because of a congenital cardiac septum defect (b). A pair of fundus images obtained with 605 nm and 586 nm illumination, respectively, were subjected to computerized analysis of the variation in optical density of the retinal vessels, which varies with oxygen saturation at 605 nm but not at the reference wavelength of 586 nm. The ratio of the optical densities is approximately linearly related to the oxygen saturation of hemoglobin.

vessel diameters, which has shown epidemiological associations with systemic cardiovascular disease but which has yet to find an application in routine clinical practice.

The virtue of fundus photography in blue light as a means of imaging the retinal nerve fiber layer is evident in young subjects (Figure 37.13), whereas in elderly people the aging of the lens makes the method less useful. A range of alternative methods for quantitative characterization of the retinal nerve fiber and ganglion cell layers are available, as will be discussed in the following.

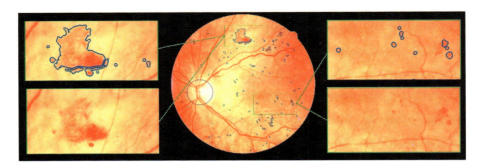

Figure 37.12 Fundus photograph analyzed by a digital algorithm that subtracts the optic disk and the retinal blood vessels, after which it identifies circumscribed dark lesions in the green channel and marks the border of each lesion, so that an observer can be alerted to the presence of abnormalities in the image.

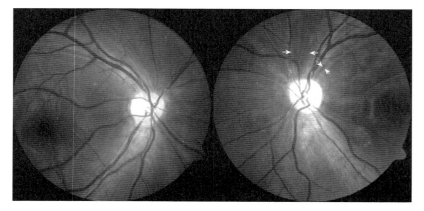

Figure 37.13 Fundus photographs recorded in red-free illumination (white incandescent light source behind red-blocking filter) using a panchromatic digital image sensor. The images are from a 21-year-old man with retinal nerve fiber damage in his left eye. Wedge-shaped defects of the retinal nerve fiber layer (arrows) are seen more readily in the green and blue light. Pure blue illumination is ideal for nerve fiber imaging but in elderly people with glaucoma, where such defects are most common, high absorption and scatter in the aged lens and other age-related changes in the eye make the method less reliable.

37.5
Fluorescence Imaging

Many molecules can be made to fluoresce if struck by photons of sufficient energy to excite an electron to a higher energy level, yet the energy should be small enough not to break covalent bonds, in which case damage to the molecule would occur. Within or near the visible range of electromagnetic radiation, molecules with conjugated double bonds, especially aromatic systems, are likely to have the critical property of being able to absorb a photon, store the absorbed energy in an excited electron with little risk of dissipation of the energy, and then re-emit most of the energy as a single new photon. Because some of the energy stored in the excited state tends to be lost to molecular vibration, the emitted fluorescent photon is generally of a longer wavelength than the absorbed photon. The shift in energy/wavelength is called the Stokes shift. Fluorescence permits imaging of various properties of tissue in an attractive manner. In the case of the fundus of the eye, blue light can be used to excite fluorescence from the retinal pigment epithelium. The intensity of the fundus autofluorescence is so weak that it is overwhelmed by reflected light in most image sensing systems, but colored optical filters can block the exciting light and allow only the fluorescent light to reach an observer or a camera aimed at the fundus. Visual observation was indeed used to make the first observations of the retinal circulation of blood stained with fluorescein (peak excitation at 494 nm, peak fluorescence emission at 521 nm). The intrinsic fluorescence or autofluorescence of the fundus is much weaker and cannot be seen by the observer unless the illumination reaches toxic levels. This happens during photocoagulation treatment (Figure 37.14).

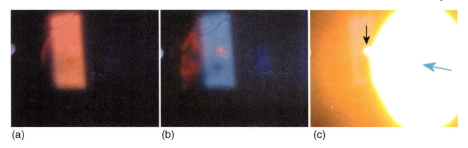

Figure 37.14 View of the fundus of the eye during 532 nm laser photocoagulation treatment. The treating physician is observing the posterior inside of the eye in white light (a) or blue–green light that provides better contrast. (b) The direction of incoming light is from the right. A red aiming spot shows where the laser is aimed. Four bright spots (b) indicate where coagulation has already been applied. The dark area below the aiming beam is the fovea. A barrier filter blocks all 532 nm light from reaching the physician's eye and the monitoring camera. When the physician activates the 532 nm laser by pressing a footswitch, a bright yellow flash of fluorescence is seen from the fundus (black arrow) and from the lens (blue arrow, the direction of which indicates the direction of the incoming light) (c).

At illumination intensities that are tolerable for the eye, the weak fluorescence from the intrinsic fluorophores that are naturally present in the fundus is best imaged by summation of multiple frames recorded by a confocal scanning laser ophthalmoscope (Figure 37.15).

If a conventional fundus camera is used, the powerful fluorescence of the lens, although it is out of focus, will add a veil of background fluorescence that obscures much of the detail in the fundus image (Figure 37.16). This effect is also very noticeable on indocyanine green angiograms.

Fluorescence can be elicited from multiple natural components of the eye, in health and in disease. Aged proteins of the lens are the most prominent source. In the fundus, N-retinylidene-N-retinylethanolamine (A2E) is an important fluorophore. A2E is a byproduct of the biochemical cycle commonly known as the visual cycle that generates photopigment. The photopigment drives vision by capturing photons in the photoreceptors. The most important artificial fluorophores are indocyanine green (peak excitation at 780 nm, peak fluorescence at 830 nm) and fluorescein, both of which can be given intravenously. This is usually done as a rapid bolus injection into a vein in the forearm. A fundus camera with an appropriate set of excitation and barrier filters, or a light source with an appropriate wavelength such as a 488 nm laser for excitation of fluorescein, is used to capture the image of stained blood flowing through the blood vessels of the fundus of the eye (Figures 37.17 and 37.18).

37.6
Photocoagulation Therapy

Photocoagulation treatment of the fundus of the eye is a crude and rather brutal treatment, the direct effect of which is to amputate photoreceptors and

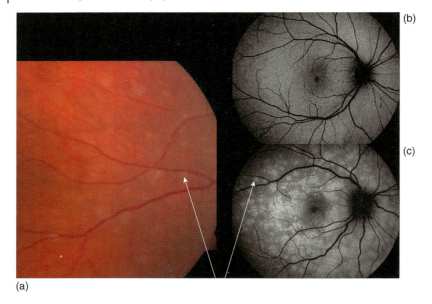

Figure 37.15 Fundus autofluorescence image from a healthy subject (b) and from a patient with the multiple evanescent white dot syndrome (c). The patient's fundus was also photographed in color (a). Arrows indicate white dots that are difficult to discern in color but are very distinct on the fluorescence image.

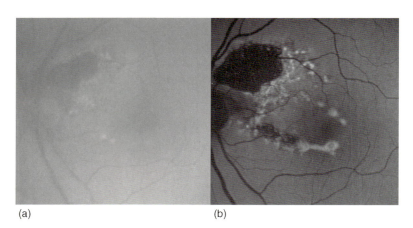

Figure 37.16 Fundus autofluorescence images from a 50-year old phakic man (phakic meaning with an intact lens) with sequels of central serous chorioretinopathy. With a conventional camera (a), the diffuse overlay of lens fluorescence adds a diffuse veil of background fluorescence that severely reduces contrast. An image recorded with a confocal scanning laser ophthalmoscope (b) has better contrast because fluorescence from sources outside the focal plane is effectively rejected.

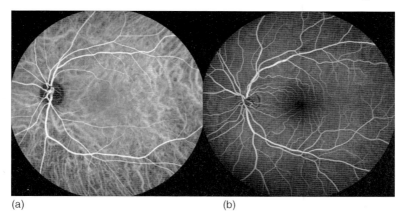

Figure 37.17 Indocyanine green (a) and fluorescein fundus angiography (b) images from healthy eyes. The infrared light used to excite indocyanine green penetrates the retinal pigment epithelium, thus allowing visualization of the dense network of large choroidal vessels behind the retina. Most of the blue light used to excite fluorescein is absorbed by the pigment epithelium, thus producing a more selective visualization of the vessels of the retina.

pigment epithelial cells [3]. The virtue of the treatment is that it is noninvasive, selective for the outer layers of the retina, and applicable in small, selected locations in the fundus. It is also inexpensive and highly effective in controlling conditions such as proliferative diabetic retinopathy. The lesions are often called laser burns, although no combustion is involved; hence photocoagulation scar is a more precise term. The desired temperature to be reached at the target site during the application of the treatment pulse is 70–90 °C, which leads to coagulation of protein in and around the retinal pigment epithelium, the melanin pigment of which is responsible for the bulk of light absorption in the retina (Figure 37.19). The reason why coagulation can be confined to the outer retina is that the inner retina is transparent, except for its blood vessels. The treating physician therefore aims between, not on, the major blood vessels of the retina.

The photocoagulation lesions are swollen and white like boiled egg-white immediately after the laser application, if sufficient energy to achieve coagulation has been

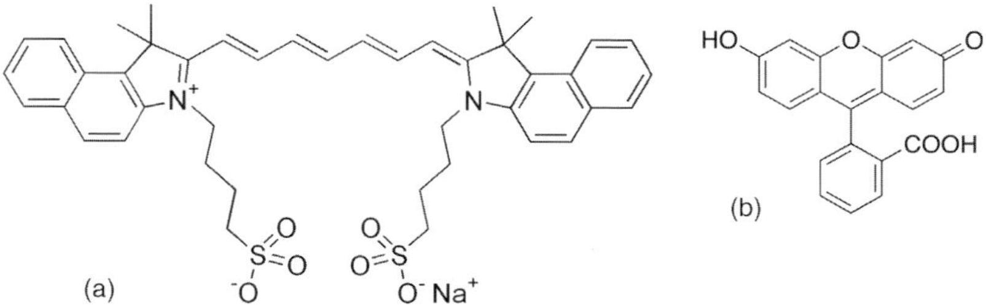

Figure 37.18 Indocyanine green (a) and fluorescein (b), two fluorescent dyes used in clinical practice to visualize the flow of blood through the vessels of the fundus of the eye.

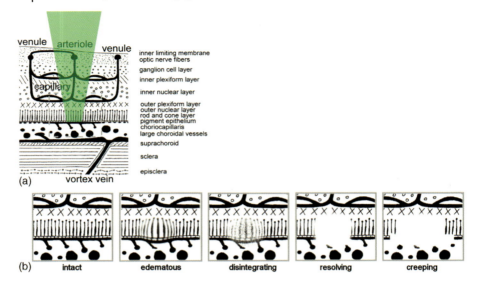

Figure 37.19 (a) Absorption of green light during photocoagulation treatment occurs deep in the retina in the pigment epithelium. Melanin absorbs light throughout the visible and near-ultraviolet range, its absorption wearing off towards the red and infrared regions. The tissue response that follows immediately and over months to years after photocoagulation is outlined in (b).

used (Fig. 37. 19(b), edematous phase). Coagulation leads to immediate cell death, but the dead tissue disintegrates only slowly, over weeks to months, when a scar characterized by loss of tissue and occasionally with hyperpigmentation appears. The resolution of the lesion involves invasion of the retina by macrophages, cells that eat and digest the components of the disintegrating tissue. Subsequently, various degrees of creeping atrophy (secondary withering of cells that were not killed by coagulation) may occur. This process is believed to be the result of a deficit of biochemical stimuli from the cells that were lost to photocoagulation and not a result of sublethal injury during photocoagulation (Figure 37.20).

The complex evolution of the photocoagulation lesions and their considerable variation between individuals demonstrate the dynamic interaction between treatment and tissue response. Photocoagulation is currently made with laser light sources because they are compact, rugged, energy-efficient, and easy to couple to an optical delivery system, but the treatment has not changed fundamentally since the pioneering work was done by using sunlight captured and led to the retina by a system of mirrors.

Photocoagulation of the anterior chamber angle between the cornea and the iris is used to lower intraocular pressure in glaucoma patients. Argon laser trabeculoplasty is applied using a contact lens (Figure 37.21) that allows focused delivery under direct observation of thermal effects to the trabecular meshwork, a filter through which the aqueous leaves the eye to be drained away by veins on the outside of the eye. The trabecular meshwork is involved in controlling the intraocular pressure and can be

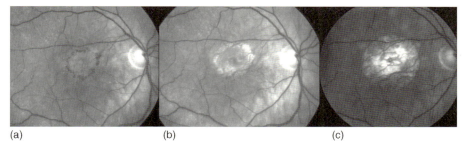

(a) (b) (c)

Figure 37.20 Subfoveal choroidal neovascularization (a) in a patient with neovascular (wet) age-related macular degeneration (AMD). The round mat of new vessels is demarcated by a rim of small dark hemorrhages (bleedings) that are visible through the clear neurosensory retina. The hemorrhages and new vessels are found in front of the retinal pigment epithelium and are an example of what is called a classic neovascularization. The new vessels are connected to the choroidal vessels by a vascular stalk that penetrates Bruch's membrane and the retinal pigment epithelium. Because the new vessels were actively growing and loss of residual central vision was imminent, confluent photocoagulation treatment was applied to limit the damage (514.5 nm argon laser, 300 mW, 0.2 s, 200 μm laser spot diameter). Three months later (b), the treated area had lost all traces of pigment epithelium and photoreceptors. Two years later (c), the scar had expanded to a 40% larger diameter because of creeping outer retinal atrophy at the rim of the scar.

manipulated to increase drainage. The laser energy is titrated so that "champagne bubbles" are produced when the 488/514.5 nm light is delivered in pulses of millisecond duration. Selective laser trabeculoplasty is a term that denotes pulsed 532 nm treatment using a laser exposure time (3 ns) that is shorter than the thermal reaction time of melanin. In other words, the exposure is stopped before significant diffusion of heat has occurred beyond the melanin-containing cells that are the intended target of treatment. The treatment is performed using a frequency-doubled 532 nm Q-switched Nd:YAG (neodymium-doped yttrium aluminum garnet) laser with a spot size of 400 mm.

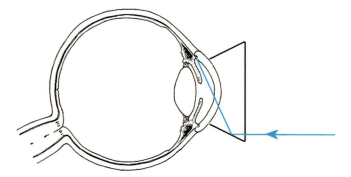

Figure 37.21 Thermal laser treatment of the trabecular meshwork in the anterior chamber angle promotes drainage of the intraocular fluid (aqueous humor or simply aqueous) and lowering of the intraocular pressure, which is useful in selected patients with glaucoma.

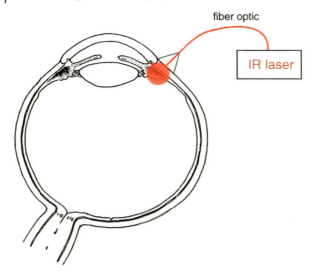

Figure 37.22 Infrared laser cyclodestruction using fiber-optic delivery to the surface of the eye, the infrared light penetrating the unpigmented sclera and exerting its heating effect in the pigmented ciliary body.

A minor application of photocoagulation is in the trans-scleral delivery of infrared light from a fiber-optic system. By placing the tip of the fiber on the outside of the eye over the ciliary body (Figure 37.22), sufficient heat can be generated inside the eye to coagulate the pigmented tissue of the ciliary body inside the eye, thus reducing the production of intraocular fluid. This treatment, cyclodestruction, is often the last resort in the fight against intraocular pressure elevation. It is mainly used in eyes with neovascular glaucoma.

Slow infrared heating of the fundus, called transpupillary thermotherapy (TTT), has been tested as a means of arresting the progression of neovascular age-related macular degeneration. The method failed to demonstrate a significant effect in a well-planned clinical trial. Although the method is believed to be effective, in principle, there is a fundamental problem in that there is no clinically applicable method of monitoring the rise in temperature in the posterior pole of the eye during the treatment. The use of standard parameters across subjects with pronounced variations in fundus pigmentation may not be a viable solution to this problem.

37.7
Photodisruption

Photodisruption is an elegant noninvasive technique based on the unique properties of ultra-short laser pulses that can produce microscopic explosions at precisely targeted locations deep inside the eye. The most common procedure is laser capsulotomy, the splitting of a brittle membranous part of the lens that is left behind after cataract surgery to hold the artificial implant lens. Lens epithelial cells left behind in the peripheral fold of the lens capsule proliferate and migrate over the posterior aspect of what is left of the lens capsule, leading to opacification of the capsule a year or more after cataract surgery. The alternative to laser capsulotomy is to insert a hooked needle behind the lens and use it to split the capsule [4]. Cataract surgery, as we know it today, would never have reached its current popularity had a convenient, noninvasive technology for capsulotomy not been invented.

The explosive effect of the ultrafast pulsed laser is not based on absorption of light in a chromophore, as happens in photocoagulation or fluorescence. The interaction is more subtle and much faster; it is the interaction that occurs between light and any refractive medium during the propagation of light in matter. It is this interaction that is responsible for the speed of light in matter being slower than in vacuum. The interaction it is extremely weak and over the distances covered in the eye it has no measurable effect at normal light intensities. The typical light source used for photodisruption in the eye, however, generates 1 mJ of photonic energy in the form of infrared photons within 1 ns. This is equivalent to the electricity consumption of a major city during the same time interval. When this radiation is made to converge on a small focus, the flux of photons reaches such a density that the sum of weak interactions becomes large enough to tear electrons from their nuclei and create plasma (photodisruption). The extreme heat causes expansion of the gas-like plasma, which cools very rapidly under the emission of white light (bremsstrahlung) and with the collapse of the plasma bubble. This cavitation event is accompanied by an audible click and often with the release of gas from decomposed molecules and by dissolution of gasses dissolved in the biological fluids. The phenomenon has features in common with a flash of lightning.

The photodisruption laser is housed in a delivery system based on the slit-lamp biomicroscope design. The commonly used light source is an Nd:YAG laser. The operator focuses a red helium—neon aiming beam to the smallest possible spot on the capsule, so that the entire flow of photons is delivered to as small a target area as possible. The beam should be focused precisely on the posterior lens capsule or slightly behind it, to avoid damaging the intraocular implant lens. If no effect is produced with a low initial energy setting, the pulse energy is increased and so on. An abrupt transition will occur, from producing no effect to producing plasma and cavitation (Figure 37.23). This is an example of a nonlinear optical phenomenon where nothing happens below a critical threshold.

If the focus of the Nd:YAG laser is fixed while the pulse energy is increased, it will be observed that with increasing intensity the focus where plasma is formed moves closer to the instrument. This is because the higher photonic flux means that the threshold flux is reached before the light has fully converged at the focus of the instrument.

Nd:YAG lasers are perfectly capable of producing cavitation in air. Because the free particles of the plasma absorb photons of any wavelength, the plasma will form a shield that stops any further advance of the light. This means that unlike photocoagulation systems, pulsed cavitation laser energy cannot be delivered reliably to the posterior segment of the eye using optics that relay the radiation into the eye from a

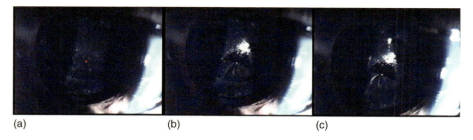

Figure 37.23 Sequences of photographs recorded during photodisruptive Nd:YAG laser treatment of secondary cataract in an eye with a dilated pupil and an artificial implant lens somewhat smaller than the dilated pupil. First, a red aiming beam is focused on the posterior part of the lens capsule that remains after cataract surgery (a). The pulsed infrared laser is fired and 80 ms later (b) the capsule is seen to have been split by the explosive formation of a plasma that also led to the release of a bubble of gas, which is seen 400 ms later (c) to have rise–behind the posterior capsule.

primary focus between the instrument and a lens such as a 90-diopter precorneal lens. If a strand of fibrous tissue has to be transected in front of the retina, or a hemorrhage is to be produced to facilitate development of a chorioretinal anastomosis in an eye with a central retinal vein occlusion, it will have to be done using a corneal contact lens such as the classical planoconvex Goldmann corneal contact lens.

The latest use of photodisruption in ophthalmology is in refractive cornea surgery where infrared femtosecond fiber-optic lasers are being used to produce very rapid sequences of tightly spaced cavitation effects inside the cornea, so that it can be cut in quasi-contiguous planes in preparation for stromal ablation. After the femtosecond procedure, a blunt probe is used to lift the corneal flap while gently breaking residual strands of corneal stroma connecting the two layers that have been prepared. This principle is now replacing the microkeratomes that were originally used in the LASIK procedure. Experimental studies are being conducted where such femtosecond laser instruments are used to cut the cornea in two planes nearly parallel to its surface, thus producing an intrastromal lenticule that can be extracted, leaving the residual corneal tissue with the desired refractive characteristics.

37.8
Photodynamic Therapy

Certain organic molecules, notably those containing a porphyrin ring, a structural motif found also in hemoglobin, can absorb light and expend the energy in the production of free oxygen radicals. New vessels formed under the fovea in neovascular age-related macular degeneration can be treated using verteporfin (Figure 37.24), which accumulates in the endothelial cells of newly formed blood vessels. Here, photochemical activation at 693 nm reaction leads to production of extremely reactive singlet oxygen and other reactive oxygen radicals that are toxic to

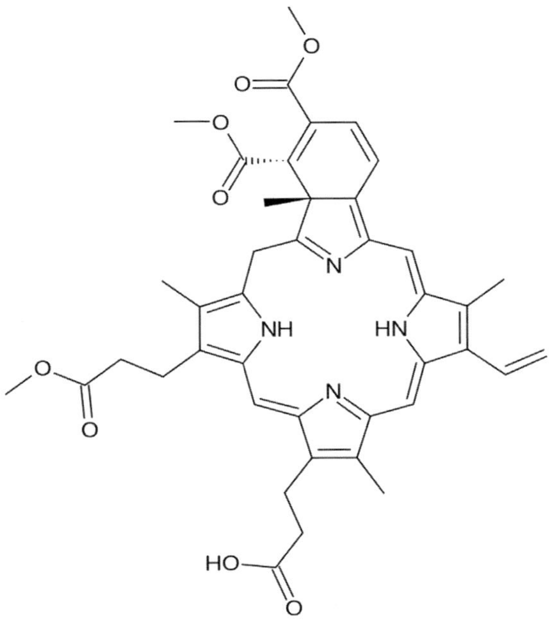

Figure 37.24 Verteporfin, a porphyrin derivative used in photodynamic therapy of diseases of the retina and choroid. Originally developed for the treatment of neovascular age-related macular degeneration, verteporfin is now used mainly in polypoidal choroidal vasculopathy, choroidal hemangioma, and central serous chorioretinopathy.

the endothelial cells and leads to occlusion of the new vessels. Photodynamic therapy includes psoralens for psoriasis and δ-aminolevulinic acid, a porphyrin precursor, for skin cancer. Porphyria cutanea tarda, a hereditary deficiency of the conversion of uroporphyrinogen, a precursor of hemoglobin, leads to accumulation of this intermediate and to phototoxic production of skin blisters.

37.9
Optical Coherence Tomography

Optical coherence tomography (OCT) allows selective registration of reflections from a given optical pathlength inside the eye [5]. The method is based on the principle that light from a light source that is phase-coherent only near the emmitting surface of the light source will be coherent at any conjugate image of the light source, be it inside the eye or on the surface of the reference mirror. By recombining light from the reference mirror with light returned from the eye, constructive interference can be obtained corresponding to the target inside the eye, whereas stray light that can be orders of magnitude more intense than the specular reflection is lost to destructive interference (Figure 37.25).

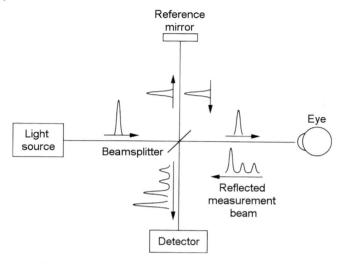

Figure 37.25 Principle of time-domain optical coherence tomography.

The focus of the OCT instrument can be scanned in various patterns over the retina or any other optically accessible part of the eye after which a tomographic reconstruction of the anatomy of the retina can be made (time-domain OCT). State-of-the art systems have progressed to spectral-domain OCT, which is faster and more robust in terms of axial alignment. Furthermore, scans can be aligned with respect to a simultaneously recorded fundus image and summation and averaging of multiple scans can be used to reduce speckle noise (Figure 37.26).

OCT has become an important addition to and to some extent a replacement for biomicroscopy and fluorescein angiography (Figure 37.19). The first applications were mainly in conditions where the retina is thickened or detached, such as diabetic macular edema, neovascular age-related macular degeneration, and central serous chorioretinopathy (Figure 37.27). With increasing resolution OCT has become a valuable tool for the objective characterization of tissue loss in the atrophic conditions that dominate the hereditary retinal degenerations.

The most common atrophic condition in the retina and optic nerve is glaucoma, which is characterized by loss of retinal ganglion cells and nerve fibers (Figure 37.28). OCT is one of the techniques that compete for a place in glaucoma diagnostics and monitoring. The modern definition of glaucoma is that it is a condition where loss of nerve fibers in the inner retina and on the optic disk is accompanied by cupping of the disk of no identifiable cause other than, in some cases, intraocular hypertension. The diagnosis needs to be backed up by the demonstration of subnormal visual field sensitivity, that is, loss of function. Assessment of intraocular pressure is relevant, not only for making the diagnosis of glaucoma, but also to assess the effect of therapy. Temporal arteritis is another cause of nerve fiber loss with cupping of the disk. Nerve fiber loss without cupping is seen in intracranial hypertension, Leber hereditary optic

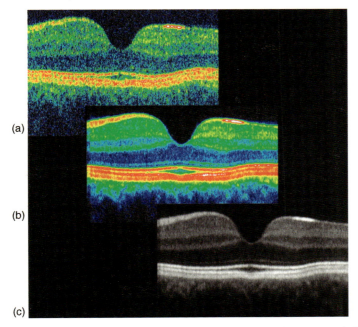

Figure 37.26 Time-domain optical coherence tomography single scan (a) and after summation and averaging of 12 scans, shown in false-color code (b) and in gray-scale reflection intensity (c).

neuropathy, uveitis, optic neuritis, and other diseases with an apparent focus of injury away from the optic disk. The concomitant loss of nerve fibers and the supporting stroma on the disk is one of several indications that the focus of injury is on the disk, another being the occasional observation of hemorrhages among the nerve fibers on the disk. Other conditions are characterized, to variable extents, by stationary nerve fiber deficits, for example, optic nerve hypoplasia. There is no loss or deficit of nerve fibers on the disk, however, without a deficit of retinal ganglion cells in the retina. When ophthalmoscopy and conventional fundus photography were the only means of retinal imaging, however, there was no method of assessing ganglion cell loss in the retina. This has now changed and noninvasive optical methods for nerve fiber and ganglion cell assessment are becoming a diagnostic mainstay in optic nerve disease.

Ophthalmoscopic examination of the optic disk, especially when stereoscopic, provides an impression of the surface contour of the disk, the fullness and vascularization of its nerve fiber rim, and the occasional splinter hemorrhage. Inspection of the peripapillary area gives a view of the delicate, silky gray bundles of nerve fibers as they arch toward the disk. Nerve fiber defects in patients with glaucoma and other optic neuropathies can be immediately obvious to the trained clinician. In these cases, objective methods of nerve fiber mapping are relevant only as a secondary means of measuring disease progression, the primary one being visual

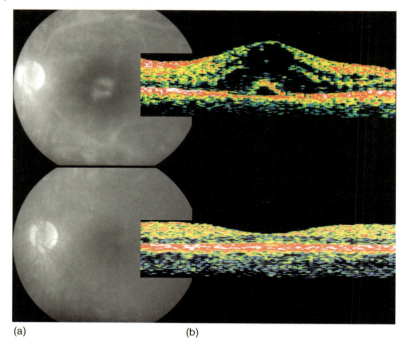

(a) (b)

Figure 37.27 Fundus fluorescein angiography (a) and transfoveal optical coherence tomography (b) (first-generation technology) during an attack of retinal vasculitis with exudation and accumulation of fluid in the fovea (above) and after treatment with systemic glucocorticosteroid with subsequent resolution of the edema and normalization of the foveal contour (below). The case illustrates the ability of OCT, a noninvasive method, to replace fluorescein angiography, an invasive method based on intravenous injection of a synthetic dye.

field assessment by computerized perimetry. Objective optical methods have the advantage, in principle, of allowing structural mapping independently of confounding factors such as inability to cooperate with computerized perimetry, a change in optical media quality as a result of cataract surgery, and, to some extent, poor optical quality of the refractive media of the eye.

The rational use of objective biometric measurement in clinical practice requires a base of evidence from normative studies and, preferably, from prospective studies to show that the instrument can help provide a better functional outcome for patients over time. New diagnostic medical equipment is generally authorized for sale with minimal documentation of its diagnostic utility, physicians being expected to know what to do with it based on their biomedical insight. This is good for innovation but leaves many questions about an instrument's practical use unanswered, and years may pass before the necessary research has been completed. Glaucoma diagnostics has been supplemented with a succession of optical instruments that are described in the following.

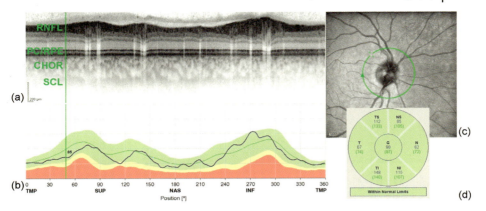

Figure 37.28 Spectral domain optical coherence tomography scan around the optic nerve head (optic disk) (c) in the right eye of a healthy subject (c). The unfolded scan (a) shows the variation in retinal layer thickness, with shadows cast by the major blood vessels. By convention, the scan is shown at a height:width aspect ratio of 2:1. Computerized segmentation of the retinal nerve fiber layer (RNFL) is overlaid on a reference curve collected in a population of healthy subjects. PC/RPE, photoreceptor/retinal pigment epithelium layer; CHOR, choroid; SCL, sclera.

Scanning laser ophthalmoscopy is an imaging technique that mimics essential features of conventional fundus photography but has better spatial selectivity for the plane of interest in the fundus. Because it picks up information only from a small volume in three-dimensional space at any given time and then reconstructs an image from a three-dimensional scanning array of such points, scatter and fluorescence from the anterior segment of the eye is very effectively rejected. Built on the flying-spot scanner principle, it only illuminates a spot with a diameter of a few micrometers at the fundus and then sweeps the spot across the horizontal and vertical directions of an image plane inside the eye it either accepts all reflected light from the fundus at all times or only from a small spot that is confocal with the illumination spot. The confocal method rejects reflected or fluorescent light outside the small sampling volume. By scanning the image plane along the optical axis of the eye, images from various depths can be obtained, provided that there is a transparent optical medium in front of it. Because the neurosensory retina is transparent, sectional images from various depths can be reconstructed, providing information about reflection, back-scatter, and fluorescence.

Additional techniques for objective characterization of optic nerve disease include three-dimensional contour mapping of the disk using confocal scanning laser ophthalmoscopy and assessment of nerve fiber layer thickness by measurement of the optical anisotropy induced by the birefringence of the nerve fibers and their uniform direction. For all technologies, follow-up studies are more informative than single examinations. Normative studies generally lag years to decades behind the marketing of new instruments, meaning that the evidence base for rational application in clinical practice is often weak when the product is launched.

Figure 37.29 Adaptive optics fundus photography of a section of the fundus 2° superior of the fovea showing a normal cone photoreceptor matrix. Faint shadows are cast by the blood vessels anterior of the photoreceptors. The photoreceptors vary in brightness depending on their spectral absorption characteristics and the color of the illumination.

37.10
Adaptive Optics Fundus Photography

Lateral and axial resolution in fundus photography are limited by the optical quality of the anterior segment of the eye. In practice, there are higher orders of aberrations than spherical errors and astigmatism. By focusing on a small area of the fundus and using wavefront analysis to interactively control the properties of a multipoint-deformable mirror, these aberrations can be overcome and sufficient resolution can be achieved to image single photoreceptors [6, 7] (Figure 37.29).

37.11
Refractive Surgery

The prerequisite for correction of refractive errors, be it with spectacles, contact lenses, or refractive surgery, is a precise preoperative characterization of the refractive properties of the eye [8]. The common clinical refractioning procedure consists of a simple titration with spherical and cylindrical trial lenses using tests of foveal angular resolution for high-contrast, single-symbol optotypes (letters, numbers, simple objects). These techniques require no information about how the refractive components of the eye combine to yield the actual refraction of the individual eye, except that contact lenses must fit the surface contour of the eye (Figure 37.30). To achieve this a keratometric measurement of corneal radii of curvature is made, based on the specular reflex of the tear film (Figure 37.31). Subsequently, the actual fit of a selected contact lens can be tested on the eye (Figure 37.32).

Refractive surgery requires additional information because one of the components of the eye is going to be altered and the effect of, say, flattening the corneal

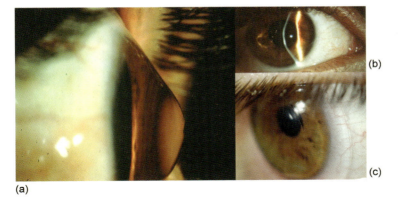

Figure 37.30 Irregular corneal contour in an eye with keratoconus, a condition of progressive attenuation of the corneal stroma where the degree of thinning increases with the distance from the limbus corneae (the transition between the sclera and the cornea) (a). The side view of a slit-beam transecting a section of the cornea (b) can be computer analyzed and used to estimate corneal curvature. A contact lens (c) can be used to bridge the irregularities of the corneal surface and substitute the role of the cornea by forming a regularly shaped interface with the atmosphere.

radius of curvature by 0.5 mm will depend on the axial length of the eye and other parameters of variable importance. The most common refractive surgery procedure is cataract surgery with implantation of an artificial intraocular lens [9]. The two most important parameters used to calculate the power of an implant lens that will produce a desired postoperative refraction is the corneal curvature and the axial length of the eye. The latter is defined as the distance from the apex of the cornea to

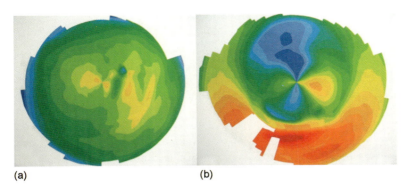

Figure 37.31 Corneal topography maps in color code indicating the local radius of curvature. A healthy eye with normal visual acuity (a) shows little irregularity whereas a cornea that underwent unsuccessful refractive surgery (b) has a highly irregular surface contour.

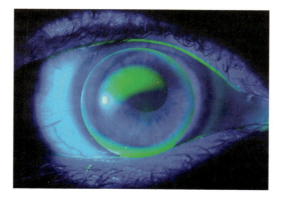

Figure 37.32 The fit between a hard contact lens and the surface of the cornea can be tested after the tear film has been stained using fluorescein. In this patient with keratoconus, the contour is irregular, causing accumulation of fluid in a crescent-shaped pocket under the contact lens right above the apex of the cornea.

the photoreceptor layer of the retina. The techniques for corneal topography can be divided into those that analyze the image formed by the specular reflex of the cornea, for instance the image of a series of concentric rings, and those techniques that observe the corneal contour from the side while a selected section the cornea is intensely illuminated by a slit-beam (Figure 37.30). The Scheimpflug projection (Figure 37.33) is ideal for the latter and is employed in a range of topography instruments.

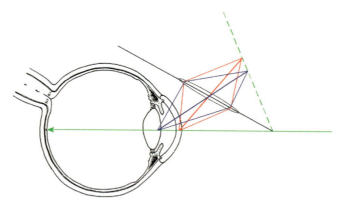

Figure 37.33 Principle of Scheimpflug projection enabling the full depth of an optical section of the anterior segment of the eye (green arrow representing incoming slit-beam of illuminating light) to be in focus on an image sensor placed at the plane indicated by the dashed green line. The lens placed across the black line is projecting the image of the cornea and the lens on the image sensor.

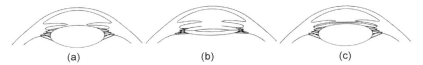

Figure 37.34 Schematic diagram of the normal eye (a), an eye with an implant lens in the bag created out of the lens capsule (b), and a posterior chamber phakic implant lens for high myopia (phakic meaning that the natural lens or phakos remains intact) (c). Lenses made out of polymers with high refractive indices achieve the necessary refractive power with much less substance than the natural lens. in daylight the pupil is rarely dilated beyond the diameter of the optical part of the implant lens. Implant lens dimensions are kept small to avoid having to make large incisions for implantation.

37.12
Intraocular Lens Implantation

Implantation of an intraocular lens in the capsular bag that remains after removal of the contents of the lens through a circular opening of its anterior surface is the standard solution for optical rehabilitation after cataract surgery (Figure 37.34) [10, 11]. The refractive power of the lens to be implanted is calculated based on preoperative measurements of corneal surface radii of curvature and axial length. The latter was previously made using ultrasonography but optical coherence line scanning has better precision and is now the standard of care.

Bifocal (multifocal) implant lenses are available in various designs made to compensate for the presbyopia that follows from the rigidity of implant lenses. All of these lenses sacrifice contrast for depth of focus [12]. Patient satisfaction is highly variable and dependent upon subjective preferences. An alternative solution to presbyopia is to make one eye emmetropic (focused at infinity) while the other eye is optimized for reading (focused at 33 cm). This again is a compromise, now of binocular vision, and again there are patients who are perfectly pleased with this solution whereas others are deeply dissatisfied. This compromise, called monovision, can be tested in a reversible manner using contact lenses.

37.13
Photoablation

Excited dimer or excimer lasers can produce powerful pulses of ultraviolet radiation that disrupt the molecules in a thin surface layer of biological material, thus spending the energy without heating the nonablated substance [13, 14]. Excimer lasers used in ophthalmology include argon fluoride (ArF) lasers operating at 193 nm. Typical pulse durations are 10 ns and repetition rates are from 100 Hz and higher. Because of the short wavelength, mirrors rather than lenses are used to control the beam. Various delivery systems have been designed, beginning with simple circular apertures with a controllable diameter to smaller moving spots. Ablation patterns have developed

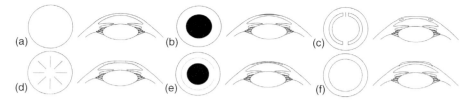

Figure 37.35 Types of refractive cornea surgery: (a) normal anatomy; (b) apical photoablation; (c) intracorneal stromal ring implant; (d) radial keratotomy (2/3 incision to relax pericentral cornea); (e) intrastromal ablation after creation of a corneal flap that is folded away during ablation (LASIK); (f) epikeratophakia (sutured corneal transplant shaped from donor tissue that is eventually covered by the recipient's corneal epithelium).

from one of simple correction for spherical errors to custom ablation in complex patterns calculated on the basis of a preoperative wavefront analysis that includes higher orders of aberration (Figure 37.35) [15].

37.14
Thermal Keratoplasty

Holmium lasers emit at 2.08 μm, a wavelength that is very effectively absorbed in water. Holmium laser radiation has been used experimentally to induce thermal shrinkage of corneal collagen in a limbus-concentric zone at a distance from the limbus. When successful, the treatment produces a steeper curvature of the apical prepupillary cornea. This effect can help correct hypermetropia (far-sightedness). Holmium lasers have also been used to create an artificial passage from the anterior chamber through the sclera at or near the anterior chamber, through the sclera to an opening under the conjunctiva. The aqueous is supposed to seep out at a rate that allows a reduction of the intraocular pressure without inducing hypotony (too low a pressure, for practical purposes below 5 mmHg). The radiation is delivered through a 0.5 mm diameter fiber-optic probe inserted through a surgical opening in the cornea. The probe is advanced toward the angle opposite the point of entry. The pulse energy is 60–150 mJ and the repetition rate 5 Hz. Energy levels of 1.35–6.6 J are sufficient to produce full-thickness tunnels through the sclera. All such types of filtration surgery, which ever way the opening is made, suffer from the consequences of the tissue reaction to the surgical injury. Consequently, long-term success depends not only on the surgical procedure but also very much on postoperative care and pharmaceutical modulation of the healing and scarring processes.

37.15
Blood Flow Measurement

The rates of blood flow through the retina and the choroid are of fundamental interest for the study of the eye. Noninvasive flowmetry in the eye is problematic.

Ultrasonography is very useful in other organs, but in relation to the eye only the vessels on the outside of the sclera in the posterior pole are sufficiently large to permit measurements that are reproducible and reliable. Optical methods include laser speckle flowmetry, which shows movement of particulate matter through the vessels of the fundus but fails to identify direction, volume, or sum in a manner that can be converted to volumetric or linear flow rates. Laser Doppler velocimetry has been studied in great detail and its virtues and limitations are well characterized. A laser pointed at a selected major retinal vessel and a detector that measures the frequency shift of the light reflected by erythrocytes and leukocytes in motion allow determination of peak velocity. According to Poiseuille, volumetric flow can be estimated if the vessel diameter is also known and if the fluid follows the characteristics of isotropic fluids. Laser Doppler flowmetry has found no application outside experimental physiology.

Stroboscopic subtraction imaging of retinal blood flow is a recent invention. By firing apid sequences of flashes and recording a sequence of fundus images without the eye having moved significantly, frame-by-frame subtraction can yield an image in which the faint trace of formed elements of the blood moving through the retinal capillaries stand out against an empty background. The method permits measurement of retinal capillary blood flow is a manner that is comparable to the blue-field entoptic technique a technique, that relies on the patients themselves to observe the rate of flow and compare it with a simulated reference shown to their fellow eye on a video screen.

37.16
Perspectives

Ophthalmic biophotonics is under constant development and several experimental techniques are expected to mature within the next decade, including femtosecond laser treatment of the lens [16], adaptive optics imaging of the fundus, and multimodal imaging and spectral analysis of the various tissues of the eye. The application of sophisticated optical devices outside university hospitals and specialized clinics will be spurred by increasing compactness and affordability.

References

1 Westheimer, G. (2009) Visual acuity: information theory, retinal image structure and resolution thresholds. *Prog. Retin. Eye Res.*, **28** (3), 178–186.

2 Cheung, N., Mitchell, P., and Wong, T.Y. (2010) Diabetic retinopathy. *Lancet*, **376** (9735), 124–136. Epub 2010 Jun 26.

3 Wong, T.Y. and Scott, I.U. (2010) Clinical practice. Retinal-vein occlusion. *N. Engl. J. Med.*, **363** (22), 2135–2144.

4 Buehl, W. and Findl, O. (2008) Effect of intraocular lens design on posterior capsule opacification. *J. Cataract Refract. Surg.*, **34** (11), 1976–1985.

5 Drexler, W. and Fujimoto, J.G. (2008) State-of-the-art retinal optical coherence

tomography. *Prog. Retin. Eye Res.*, **27** (1), 45–88.
6. Godara, P., Dubis, A.M., Roorda, A., Duncan, J.L., and Carroll, J., (2010) Adaptive optics retinal imaging: emerging clinical applications. *Optom. Vis. Sci.*, **87** (12), 930–941.
7. Lombardo, M. and Lombardo, G. (2009) New methods and techniques for sensing the wave aberrations of human eyes. *Clin. Exp. Optom.*, **92** (3), 176–186.
8. Ehlers, N., Hjortdal, J., and Nielsen, K. (2009) Corneal grafting and banking. *Dev. Ophthalmol.*, **43**, 1–14.
9. Gwiazda, J. (2009) Treatment options for myopia. *Optom. Vis. Sci.*, **86** (6), 624–628.
10. Norrby, S., Piers, P., Campbell, C., and van der Mooren, M. (2007) Model eyes for evaluation of intraocular lenses. *Appl. Opt.*, **46** (26), 6595–6605.
11. Olsen, T. (2007) Calculation of intraocular lens power: a review. *Acta Ophthalmol. Scand.*, **85** (5), 472–485.
12. Glasser, A., (2008) Restoration of accommodation: surgical options for correction of presbyopia. *Clin. Exp. Optom.*, **91** (3), 279–295.
13. Krueger, R.R., Rabinowitz, Y.S., and Binder, P.S. (2010) The 25th anniversary of excimer lasers in refractive surgery: historical review. *J. Refract. Surg.*, **26** (10), 749–760.
14. Krueger, R.R., Thornton, I.L., Xu, M., Bor, Z., and van den Berg, T.J. (2008) Rainbow glare as an optical side effect of IntraLASIK. *Ophthalmology*, **115** (7), 1187–1195.
15. Myrowitz, E.H. and Chuck, R.S. (2009) A comparison of wavefront-optimized and wavefront-guided ablations. *Curr. Opin. Ophthalmol.*, **20** (4), 247–250.
16. Binder, P.S. (2010) Femtosecond applications for anterior segment surgery. *Eye Contact Lens*, **36** (5), 282–285.

38
Glaucoma Diagnostics
Marc Töteberg-Harms, Cornelia Hirn, and Jens Funk

38.1
Introduction

Glaucoma is a widespread disease leading to progressive loss of visual function. It is still one of the leading causes of blindness around the world [1, 2] and the number one reason for blindness in industrialized countries [3–5]. Worldwide there are more cases of primary open angle glaucoma than of angle closure glaucoma [3]. Glaucoma is a chronic and progressive neurodegenerative disorder causing loss of retinal ganglion cells and their axons [6]. Characteristically, cupping of the optic disc is seen with a possible loss of visual field. In most cases, the intraocular pressure (IOP) is elevated above the normal [7] range.

IOP is the major risk factor for glaucoma. In addition, some other risk factors are well known, for example, age, family history, and race (e.g., African or Caucasian descent) [8].

Aqueous humor is formed by the ciliary processes [9–14] and drained mainly through the trabecular meshwork into Schlemm's canal [15]. Schlemm's canal communicates with the episcleral veins. This is called the trabecular outflow (83–96% of the aqueous humor outflow) [16–19]. In addition, 5–15% of the aqueous humor is drained via the uveoscleral pathway [17–23]. Usually an increased drain resistance is the reason for elevated IOP [17–25], whereas aqueous humor production is nearly constant [26–28]. The main location of outflow resistance could probably be found in the juxtacanalicular tissue of the trabecular meshwork [29, 30].

In managing glaucoma patients, lowering the IOP is the only treatment, with excellent quality of evidence [31–36]. Medical reduction of IOP is the first-line therapy in most cases [31, 37, 38]. If medical treatment fails, there are several well-established surgical procedures to reduce IOP.

Trabeculectomy (TE) as it is performed today was introduced in 1968 by Cairns [39], and it is still the gold standard in glaucoma surgery. The aqueous flows via a scleral flap from the anterior chamber into the subconjunctival space [40]. TE is very effective in lowering the IOP for long periods [41, 42]. The use of antimetabolites during

Figure 38.1 Typical OCT test setting with patient and examiner.

surgery provides even better long-term IOP control [43–49]. Other surgical procedures include laser surgery of the trabecular meshwork {argon laser trabeculoplasty (ALT) [50–57] and selective laser trabeculoplasty [50, 51, 54, 58–65]} and cyclophotocoagulation (CPC) [66–73] for the ciliary body or shunt implants (e.g., Molteno [74–76], Ahmed [77, 78], and Baerveldt tube [79, 80]).

Managing the glaucoma patient over time by measuring the IOP and testing the visual field is very important. By the time the breakdown of retinal ganglion cells is clinically detected, extended and irreversible damage has already occurred [81, 82]. Since effective therapy can slow the progress of glaucoma, early diagnosis of worsening is one of the main goals in treatment of this disease. It is strongly believed that the thinning of the retinal nerve fiber layer (RNFL) correlates highly with or even precedes visual field loss in glaucoma [83–89]. Well established in detecting early RNFL thinning are optical coherence tomography (OCT) (Figure 38.1), scanning laser ophthalmoscopy with the Heidelberg retina tomograph (HRT) (Heidelberg Engineering, Heidelberg, Germany), and laser scanning polarimetry (GDx) (Carl Zeiss Meditec, Oberkochen, Germany).

38.2
Optical Coherence Tomography

Using OCT, Huang *et al.* were the first to present a noncontact, noninvasive method of using low-coherence interferometry to determine the echo time delay and magnitude of backscattered light reflected off different layers of a structured tissue sample [90]. The unique optic free pathway through the eye made OCT highly applicable to the visualization of retina layers. In 1995, time domain optical coherence tomography (TD-OCT) was introduced as an imaging technique for glaucoma diagnosis [87]. In spectral domain optical coherence tomography (SD-OCT) (or Fourier domain OCT), a moving reference mirror as used in TD-OCT is no longer needed. SD-OCT provides higher resolution at faster scanning speeds [91, 92].

In TD-OCT, a prism splits an 820 nm infrared light source into two beams. One spreads through the layers of the eye while a reference beam is deflected to a moving reference mirror. For every measurement point detected along an axial depth scan (A-scan), the moving reference mirror has to be readjusted. If the reference mirror is positioned at virtually the same distance from the light source as the backscattering tissue, the reunited light beams will interfere and be detected by an interferometer. Processing multiple A-scans along a scanning line (B-scan) allows for the visualization of retinal layers. TD-OCT provides axial resolutions of up to 10 µm [87, 90, 93, 94].

A major revolution in OCT imaging was the introduction SD-OCT. Instead of time delay as used in TD-OCT, SD-OCT applies Fourier analysis to light wavelengths to determine the spatial location of reflected light [92, 95]. The first ophthalmic SD-OCT *in vivo* scans were introduced by Wojtkowski *et al.* in 2002 [91]. With SD-OCT, all axial data information along one A-scan can be gathered within a single scan. In 2006, the first high-speed, high-resolution OCTs became commercially available. Today, the latest generation of SD-OCT devices provide resolutions that are five times higher and scanning speeds 100 times faster compared to TD-OCT [95, 96]. Owing to the faster scanning speed, more data (A-scans) can be acquired in a given time, allowing SD-OCT devices to generate 3D reconstructions of retina areas [97].

Another development available in recent OCT devices is the implementation of specific algorithms and software to enhance the scanning resolution further and decrease motion artifacts. In 2006, the Spectralis SD-OCT system (Heidelberg Engineering, Heidelberg, Germany) was introduced for retinal imaging. This instrument features two different options to enhance reproducibility. An online eye tracking device (eye tracker) compensates for involuntary eye movements during the scanning process, and a retest function assures that follow-up measurements are taken from the same area of the retina as the baseline examination. The Heidelberg Spectralis is the only commercially available SD-OCT system with an eye tracking function. Both options can be switched off separately. To measure the RNFL circle, scans could be performed around the optic disc with a scanning angle of 12°, which equates to a retinal diameter of 3.5 mm when assuming a standard corneal curvature of 7.7 mm. The SD-OCT Super uses a luminescent diode with a center wavelength of 840 mm. The Spectralis SD-OCT obtains up to 40 000 A-scans per second with a depth resolution of 7 µm in tissue. Transversal resolution depends on the tightness of the A-scans and is up to 5 µm.

With a dual beam, a corresponding scanning laser ophthalmoscope (SLO) fundus image can be captured at the same time as the OCT measurement, enabling the system to link every OCT scan (Figure 38.2a) to its corresponding position on the SLO fundus image (Figure 38.2b). Processing of SLO data and identification of specific patterns in retinal structures such as blood vessels allow scans to be marked as a reference and baseline. In follow-up examinations, the system recognizes the former scanning area on the retina and automatically positions the retest scan on the same location. During the measurement, online eye tracking provides a

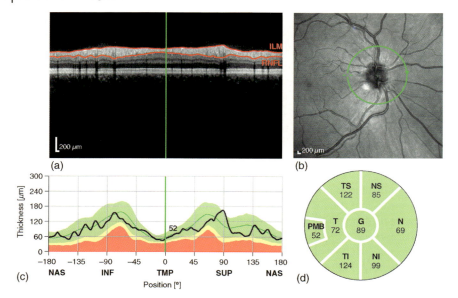

Figure 38.2 Circular peripapillary optical coherence tomogram of a normal right eye (a) with its corresponding fundus image (b) obtained with the dual-beam scanning laser ophthalmoscope. Red lines in the OCT B-scan (a) indicate the inner and outer border of the RNFL found by the algorithm. In (c), measured RNFL thickness is plotted on the thickness values measured in healthy subjects of the same age. The mean RNFL thickness of sectors, the peripapillary bundle, and the global mean RNFL thickness are shown in (d).

real-time adjustment of the OCT scanner on the simultaneously gathered SLO image in order to decrease motion artifacts, and the high scanning speed reduces the time for distracting eye movements to occur. To increase image quality, the Spectralis SD-OCT includes an ART (automatic real time) function. With ART activated, multiple frames (B-scans) are gathered during the scanning process and images are averaged for noise reduction [98, 99]. The number of frames can be adjusted from one to 100.

The SD-OCT devices provide an algorithm to determine the inner and outer borders of the RNFL (Figure 38.2a). OCT data were analyzed by this algorithm to detect the RNFL thickness along the circular scan in micrometers. The median thickness was plotted on a pie chart diagram representing different sectors of the optic disc (Figure 38.2d). The calculated thickness profile (Figure 38.2c) refers to a set of normative RNFL thickness data in healthy subjects [100]. Green areas represent the 95% normal range found in healthy subjects of the same age whereas values outside the 99% confidence interval of the normal distribution ($0.01 < p < 0.05$) are indicated in red. Yellow areas represent values outside the 95% confidence interval but within the 99% confidence interval of the normal distribution. Figure 38.3 shows a typical printout of an OCT RFNL scan of a glaucoma patient.

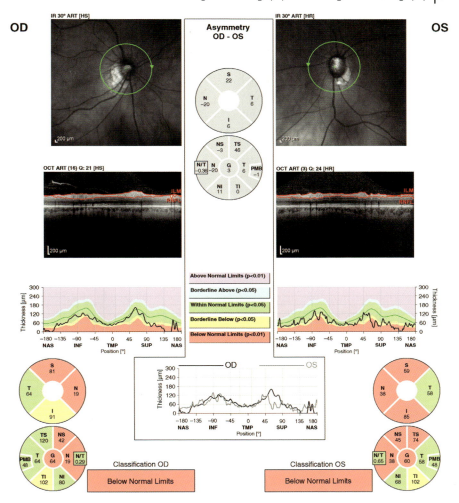

Figure 38.3 Typical printout of OCT RFNL scan of a glaucoma patient.

38.3
Scanning Laser Tomography – Heidelberg Retina Tomography

The aim of optic nerve head (ONH) imaging is to assist in the classification of normal and glaucomatous optic discs on the one hand and to detect glaucoma progression on the other.

The latest version of the HRT system (HRT3) is a confocal scanning laser system for acquisition and analysis of three-dimensional images of the ONH and RNFL.

It allows quantitative assessment of the topography of the ONH and also the follow-up of topographic changes.

Pupil dilation is not required for the examination.

A diode laser with a wavelength of 670 nm is focused and deflected periodically for point-by-point two-dimensional scanning of the posterior pole perpendicular to the optical axis. The intensity of the light reflected from the object is detected separately from the incident laser beam. In confocal scanning laser systems, a diaphragm with a pinhole is placed before the detector. Only light reflected from the focal plane is focused on the pinhole, passes the diaphragm, and is detected; reflections from layers above or below the focal plane are suppressed. This technique is called scanning laser tomography.

The location of the light-reflecting surface allows the determination of the height of the scanned structure. HRT permits image acquisition at different depth locations, providing a series of optical section images forming a layer-by-layer three-dimensional image of the ONH and the peripapillary RNFL. The size of the field of view is set at $15° \times 15°$; the images are digitized in frames of 384×384 pixels. The distance between the focal planes is set at 62.5 µm.

Sets of three images are acquired in one measurement. By translating each specific height location into a specific color, a color-coded mean topography image in addition to a mean reflectance image is computed.

For stereometric measurements and analysis of the ONH, a contour line has to be drawn around the disc margin by the operator; the structure enclosed is analyzed three-dimensionally. Based on the contour line, a reference plane is automatically defined. This reference plane is set parallel to the peripapillary retinal surface and is located 50 µm posterior to the retinal surface of the papillomacular bundle. All structures located below the reference plane are considered to be cup, and all structures located above the reference plane and within the contour line are considered to be rim.

The result of the analysis is a set of stereometric parameters, which are printed in a table and shown in color-coded images and graphs. The most important are disc area, cup and rim area, cup and rim volume, cup to disc ratio, mean and maximum cup depth, and thickness of the RNFL along the contour line. The only parameters independent of the reference plane are the three-dimensional cup shape measure and the height variation of the contour.

Moorfields regression analysis (MRA) is a tool to determine normal or glaucomatous discs by a regression analysis between disc size and rim area [101]. The results of the MRA are superimposed on the reflectance image with a subdivision into six sectors, each marked according to the classification as normal, borderline, or outside normal limits, with red crosses for outside normal, yellow exclamation marks for borderline, and green ticks for within normal limits.

Additional graphs are horizontal and vertical cross-sections of the topography image with a marked line for the reference plane and the optic disc margin, a graph of the height variation of the RNFL following the contour line, representing the RNFL thickness at the disc margin, and the results of the MRA displayed as seven bars – one for each of the six sectors of the ONH and a seventh for the overall optic disc assessment.

When interpreting the results of an HRT examination, classification as normal or glaucomatous based on individual stereometric parameters is rather difficult.

So far, the major drawback in interpretation of the results is that the stereometric parameters are dependent on the reference plane. In addition to an inter-observer variability of the drawing of the contour line [102, 103], a high physiological variation of ONH stereometric parameters adds to changes in the height of the RNFL during the course of the disease and also a physiological reduction in RNFL thickness with age [104]. Therefore, the reference plane is moving posteriorly, leading to underestimation of the cupping and possible progression [105].

Additionally, the IOP has a direct influence on cup volume, cup depth, rim area, and cup to disc area ratio [106, 107].

The coefficients of variation for area and volume parameters are 2–6 and 6–14%, respectively [102, 108].

To improve the diagnostic power, in HRT3 the normative database has been expanded and now offers 215 healthy African-American eyes and 733 healthy eyes of white individuals, and a further database of $\sim$100 Indian (south Asian) eyes is available.

A new tool is the OU report, which compares the subject's stereometric parameters against the normative database and which features an image quality assessment for both eyes, based on the mean pixel height standard deviation, scaled from excellent to very poor quality, with acceptable image quality with standard deviation (SD) less than 35 µm, and good quality images categorized as images with SD less than 25 µm.

A second new tool is the glaucoma probability score (GPS), which aims to discriminate between normal and glaucomatous eyes using a mathematical model of ONH shape [109]. Included in the analysis are horizontal and vertical curvature of the peripapillary retina, cup depth, cup width, and cup wall slope. The advantage of this analysis is that it is independent of the contour line and the reference plane position and therefore not operator dependent.

A graphical representation is used similar to the MRA, with red crosses for outside normal, yellow exclamation marks for borderline, and green ticks for within normal limits.

The GPS shows a sensitivity and specificity of 77.1 and 90.3%, compared with 71.4 and 91.9% for MRA [110].

Data indicate that an abnormal MRA classification is more useful to confirm that a disc is glaucomatous, and MRA better discriminates in more severe glaucoma, whereas a within normal limits GPS classification is more likely to confirm that a disc is normal [111].

Although the MRA is based on the relation between disc size and rim area, the diagnostic precision seems to be reduced in large discs [112, 113].

For progression analysis, there are currently two algorithms: trend analysis examines the change of stereometric parameters over time, and topographic change analysis (TCA) estimates the probability that a change in surface height between baseline and follow up occurred by chance alone [114]. The change probability map combines clusters of 4×4 adjacent height measurements or pixels to give so-called superpixels. The variability of baseline measurements is compared with the combined variability of the baseline and follow-up measurements. The resulting probability maps are displayed color-coded [107, 114, 115].

In summary, the HRT provides moderate discrimination between normal and glaucomatous ONH, but is promising when it comes to identifying and quantifying progression.

38.4
Scanning Laser Polarimetry – GDx

Thinning of the peripapillary RNFL has been shown to be one of the earlier signs of glaucomatous damage [81, 84, 116], and scanning laser polarimetry (SLP) is a technique developed to evaluate the peripapillary RNFL thickness.

SLP is based on the birefringence of the RNFL, which is due to the parallel alignment of the microtubuli in the nerve fibers. This birefringence induces a change in retardation of polarized light passing through the RNFL.

The GDx uses a diode laser with a wavelength of 780 nm; the scanning area is $15° \times 15°$. The retardation of the reflected laser light is detected with a scanning raster of 256×256 pixels. Pupil dilation is not necessary. Based on experimental conversion factors, the RNFL thickness is estimated [117, 118].

In addition to RNFL, in the anterior segment the cornea and the lens also are birefringent and affect the measurement. To compensate for this effect, a variable corneal compensation (VCC) was developed. In the GDx VCC, initially a macular scan is performed. As there are no retinal nerve fibers in the fovea centralis, all changes in retardation should be due to corneal birefringence. Hence the polarization axis and magnitude of the cornea can be calculated [119]. For subsequent scans of the peripapillary RNFL, those values are applied for correction of the measurement.

For the analysis of the scans, after marking of the ONH margin, a ring (the so-called ellipse) is placed around the optic disc with a diameter of 1.75 mm. Along that ring, the modulation and thickness of the RNFL are calculated. On the printout, there are an ONH image and a color-coded RNFL thickness map and also a deviation map and the TSNIT graph. The TSNIT graph shows the peripapillary height modulation of the RNFL.

Additionally, a number of modulation and thickness parameters are presented, and also a global parameter called the nerve fiber indicator (NFI), which is a support-vector machine-derived parameter for discrimination between healthy and glaucomatous eyes [120, 121]. The coefficient of variation for the GDx VCC is 3–6%.

One drawback with the VCC technique is that corneal compensation is dependent on an intact fovea. In patients with macular diseases, VCC can be severely hindered. In patients with peripapillary atrophy, the ellipse has to be enlarged to spare the areas of atrophy to avoid artifacts.

Typical retardation patterns display larger amounts of retardation superior and inferior to the optic disc, and decreasing amounts of retardation with increasing distance from the optic disc.

Atypical patterns occur in ~10–25% of healthy and 15–51% of glaucomatous eyes [121–124]. They seem to be more frequent in patients with high myopia and lightly pigmented fundi. Therefore, a new method was introduced a few years ago: enhanced corneal compensation (ECC). With this method, the signal-to-noise ratio is improved in cases of atypical retardation patterns.

In contrast to VCC, where the measurement was adjusted to neutralize the corneal birefringence, in ECC an additional amount of birefringence is added to the measurements, thus shifting the measurement into a more sensitive detection range [125, 126]. Hence images showing atypical retardation patterns are reduced compared with the VCC technique at similar levels of reproducibility [127]. The ability to discriminate between healthy and glaucomatous eyes seems to be improved in ECC [124].

Progression analysis is shown in a serial printout with differences between measurements flagged based on arbitrarily chosen cut-off marks.

References

1. Thylefors, B., et al. (1995) Global data on blindness. *Bull. World Health Organ.*, **73** (1), 115–121.
2. Resnikoff, S.,et al. (2004) Global data on visual impairment in the year 2002. *Bull. World Health Organ.*, **82**, 844–851.
3. Quigley, H.A. and Broman, A.T. (2006) The number of people with glaucoma worldwide in 2010 and 2020. *Br. J. Ophthalmol.*, **90** (3), 262–267.
4. Quigley, H.A. (1996) Number of people with glaucoma worldwide. *Br. J. Ophthalmol.*, **80** (5), 389–393.
5. Thylefors, B. and Negrel, A.D. (1994) The global impact of glaucoma. *Bull. World Health Organ.*, **72** (3), 323–326.
6. Quigley, H.A.,et al. (1981) Optic nerve damage in human glaucoma. II. The site of injury and susceptibility to damage. *Arch. Ophthalmol.*, **99** (4), 635–649.
7. Cellini, M.,et al. (2008) Matrix metalloproteinases and their tissue inhibitors after selective laser trabeculoplasty in pseudoexfoliative secondary glaucoma. *BMC Ophthalmol.*, **8**, 20.
8. Boland, M.V. and Quigley, H.A. (2007) Risk factors and open-angle glaucoma: classification and application. *J. Glaucoma*, **16** (4), 406–418.
9. Green, K. and Pederson, J.E. (1972) Contribution of secretion and filtration to aqueous humor formation. *Am. J. Physiol.*, **222** (5), 1218–1226.
10. Weinbaum, S.,et al. (1972) The role of secretion and pressure-dependent flow in aqueous humor formation. *Exp. Eye Res.*, **13** (3), 266–277.
11. Green, K. and Pederson, J.E. (1973) Aqueous humor formation. *Exp. Eye Res.*, **16** (4), 273–286.
12. Pederson, J.E and Green, K. (1973) Aqueous humor dynamics: experimental studies. *Exp. Eye Res.*, **15** (3), 277–297.
13. Pederson, J.E. and Green, K. (1973) Aqueous humor dynamics: a mathematical approach to measurement of facility, pseudofacility, capillary pressure, active secretion and X c. *Exp. Eye Res.*, **15** (3), 265–276.
14. Bill, A. (1975) Blood circulation and fluid dynamics in the eye. *Physiol. Rev.*, **55** (3), 383–417.
15. Grant, W.M. (1963) Experimental aqueous perfusion in enucleated human eyes. *Arch. Ophthalmol.*, **69**, 783–801.
16. Toris, C.B.,et al. (2002) Aqueous humor dynamics in ocular hypertensive patients. *J. Glaucoma*, **11** (3), 253–258.
17. Toris, C.B.,et al. (1999) Aqueous humor dynamics in the aging human eye. *Am. J. Ophthalmol.*, **127** (4), 407–412.
18. Jocson, V.L. and Sears, M.L. (1971) Experimental aqueous perfusion in

enucleated human eyes. Results after obstruction of Schlemm's canal. *Arch. Ophthalmol.*, **86** (1), 65–71.
19. Bill, A. and Phillips, C.I. (1971) Uveoscleral drainage of aqueous humour in human eyes. *Exp. Eye Res.*, **12** (3), 275–281.
20. Inomata, H. and Bill, A. (1977) Exit sites of uveoscleral flow of aqueous humor in cynomolgus monkey eyes. *Exp. Eye Res.*, **25** (2), 113–118.
21. Sherman, S.H., Green, K., and Laties, A.M. (1978) The fate of anterior chamber fluorescein in the monkey eye. 1. The anterior chamber outflow pathways. *Exp. Eye Res.*, **27** (2), 159–173.
22. Townsend, D.J. and Brubaker, R.F. (1980) Immediate effect of epinephrine on aqueous formation in the normal human eye as measured by fluorophotometry. *Invest. Ophthalmol. Vis. Sci.*, **19** (3), 256–266.
23. Pederson, J.E., Gaasterland, D.E., and MacLellan, H.M. (1977) Uveoscleral aqueous outflow in the rhesus monkey: importance of uveal reabsorption. *Invest. Ophthalmol. Vis. Sci.*, **16** (11), 1008–1017.
24. Peterson, W.S., Jocson, V.L., and Sears, M.L. (1971) Resistance to aqueous outflow in the rhesus monkey eye. *Am. J. Ophthalmol.*, **72** (2), 445–451.
25. Grant, W.M. (1958) Further studies on facility of flow through the trabecular meshwork. *AMA Arch. Ophthalmol.*, **60** (4 Part 1), 523–533.
26. Becker, B. (1958) The decline in aqueous secretion and outflow facility with age. *Am. J. Ophthalmol.*, **46** (5 Part 1), 731–736.
27. Brubaker, R.F. (1991) Flow of aqueous humor in humans [The Friedenwald Lecture]. *Invest. Ophthalmol. Vis. Sci.*, **32** (13), 3145–3166.
28. Brubaker, R.F.,*et al.* (1981) The effect of age on aqueous humor formation in man. *Ophthalmology*, **88** (3), 283–288.
29. Bill, A. and Svedbergh, B. (1972) Scanning electron microscopic studies of the trabecular meshwork and the canal of Schlemm – an attempt to localize the main resistance to outflow of aqueous humor in man. *Acta Ophthalmol. (Copenh.)*, **50** (3), 295–320.
30. Maepea, O. and Bill, A. (1992) Pressures in the juxtacanalicular tissue and Schlemm's canal in monkeys. *Exp. Eye Res.*, **54** (6), 879–883.
31. Heijl, A.,*et al.* (2002) Reduction of intraocular pressure and glaucoma progression: results from the Early Manifest Glaucoma Trial. *Arch. Ophthalmol.*, **120** (10), 1268–1279.
32. Grant, W.M. and Burke, J.F. Jr. (1982) Why do some people go blind from glaucoma? *Ophthalmology*, **89** (9), 991–998.
33. Collaborative Normal-Tension Glaucoma Study Group (1998) The effectiveness of intraocular pressure reduction in the treatment of normal-tension glaucoma. *Am. J. Ophthalmol.*, **126** (4), 498–505.
34. Collaborative Normal-Tension Glaucoma Study Group (1998) Comparison of glaucomatous progression between untreated patients with normal-tension glaucoma and patients with therapeutically reduced intraocular pressures. *Am. J. Ophthalmol.*, **126** (4), 487–497.
35. The AGIS Investigators (2000) The Advanced Glaucoma Intervention Study (AGIS): 7. The relationship between control of intraocular pressure and visual field deterioration. *Am. J. Ophthalmol.*, **130** (4), 429–440.
36. Palmberg, P. (2001) Risk factors for glaucoma progression: where does intraocular pressure fit in? *Arch. Ophthalmol.*, **119** (6), 897–898.
37. Wilensky, J.T. (1999) The role of medical therapy in the rank order of glaucoma treatment. *Curr. Opin. Ophthalmol.*, **10** (2), 109–111.
38. Feiner, L. and Piltz-Seymour, J.R. (2003) Collaborative Initial Glaucoma Treatment Study: a summary of results to date. *Curr. Opin. Ophthalmol.*, **14** (2), 106–111.
39. Cairns, J.E. (1968) Trabeculectomy. Preliminary report of a new method. *Am. J. Ophthalmol.*, **66** (4), 673–679.
40. Spencer, W.H. (1972) Symposium: microsurgery of the outflow channels.

Histologic evaluation of microsurgical glaucoma techniques. *Trans. Am. Acad. Ophthalmol. Otolaryngol.*, **76** (2), 389–397.
41 Migdal, C., Gregory, W., and Hitchings, R. (1994) Long-term functional outcome after early surgery compared with laser and medicine in open-angle glaucoma. *Ophthalmology*, **101** (10), 1651–1656; discussion 1657.
42 Jay, J.L. and Allan, D. (1989) The benefit of early trabeculectomy versus conventional management in primary open angle glaucoma relative to severity of disease. *Eye (Lond.)*, **3** (Pt 5), 528–535.
43 Yoon, P.S. and Singh, K. (2004) Update on antifibrotic use in glaucoma surgery, including use in trabeculectomy and glaucoma drainage implants and combined cataract and glaucoma surgery. *Curr. Opin. Ophthalmol.*, **15** (2), 141–146.
44 WuDunn, D.,*et al.* (2002) A prospective randomized trial comparing intraoperative 5-fluorouracil vs mitomycin C in primary trabeculectomy. *Am. J. Ophthalmol.*, **134** (4), 521–528.
45 Khaw, P.T.,*et al.* (2001) Modulation of wound healing after glaucoma surgery. *Curr. Opin. Ophthalmol.*, **12** (2), 143–148.
46 Katz, G.J.,*et al.* (1995) Mitomycin C versus 5-fluorouracil in high-risk glaucoma filtering surgery. Extended follow-up. *Ophthalmology*, **102** (9), 1263–1269.
47 Towler, H.M.,*et al.* (2000) Long-term follow-up of trabeculectomy with intraoperative 5-fluorouracil for uveitis-related glaucoma. *Ophthalmology*, **107** (10), 1822–1828.
48 Suzuki, R.,*et al.* (2002) Long-term follow-up of initially successful trabeculectomy with 5-fluorouracil injections. *Ophthalmology*, **109** (10), 1921–1924.
49 Lama, P.J. and Fechtner, R.D. (2003) Antifibrotics and wound healing in glaucoma surgery. *Surv. Ophthalmol.*, **48** (3), 314–346.
50 Damji, K.F.,*et al.* (2006) Selective laser trabeculoplasty versus argon laser trabeculoplasty: results from a 1-year randomised clinical trial. *Br. J. Ophthalmol.*, **90** (12), 1490–1494.
51 Cioffi, G.A., Latina, M.A., and Schwartz, G.F. (2004) Argon versus selective laser trabeculoplasty. *J. Glaucoma*, **13** (2), 174–177.
52 Rouhiainen, H., Terasvirta, M., and Tuovinen, E. (1987) Low power argon laser trabeculoplasty. *Acta Ophthalmol. (Copenh.)*, **65** (1), 67–70.
53 Blondeau, P., Roberge, J.F., and Asselin, Y. (1987) Long-term results of low power, long duration laser trabeculoplasty. *Am. J. Ophthalmol.*, **104** (4), 339–342.
54 Kramer, T.R. and Noecker, R.J. (2001) Comparison of the morphologic changes after selective laser trabeculoplasty and argon laser trabeculoplasty in human eye bank eyes. *Ophthalmology*, **108** (4), 773–779.
55 Grinich, N.P., Van Buskirk, E.M., and Samples, J.R. (1987) Three-year efficacy of argon laser trabeculoplasty. *Ophthalmology*, **94** (7), 858–861.
56 Bergea, B. (1986) Repeated argon laser trabeculoplasty. *Acta Ophthalmol. (Copenh.)*, **64** (3), 246–250.
57 Bergea, B. (1986) Intraocular pressure reduction after argon laser trabeculoplasty in open-angle glaucoma. A two-year follow-up. *Acta Ophthalmol. (Copenh.)*, **64** (4), 401–406.
58 Babighian, S.,*et al.* (2010) Excimer laser trabeculotomy vs 180 degrees selective laser trabeculoplasty in primary open-angle glaucoma. A 2-year randomized, controlled trial. *Eye (Lond.)*, **24** (4), 632–638.
59 Latina, M.A. and Gulati, V. (2004) Selective laser trabeculoplasty: stimulating the meshwork to mend its ways. *Int. Ophthalmol. Clin.*, **44** (1), 93–103.
60 Latina, M.A.,*et al.* (1998) Q-switched 532-nm Nd:YAG laser trabeculoplasty (selective laser trabeculoplasty): a multicenter, pilot, clinical study. *Ophthalmology*, **105** (11), 2082–2088; discussion 2089–2090.
61 Nagar, M.,*et al.* (2005) A randomised, prospective study comparing selective laser trabeculoplasty with latanoprost for the control of intraocular pressure in ocular hypertension and open angle glaucoma. *Br. J. Ophthalmol.*, **89** (11), 1413–1417.

62 Chen, E., Golchin, S., and Blomdahl, S. (2004) A comparison between 90 degrees and 180 degrees selective laser trabeculoplasty. *J. Glaucoma*, **13** (1), 62–65.

63 Song, J.,*et al.* (2005) High failure rate associated with 180 degrees selective laser trabeculoplasty. *J. Glaucoma*, **14** (5), 400–408.

64 Harasymowycz, P.J.,*et al.* (2005) Selective laser trabeculoplasty (SLT) complicated by intraocular pressure elevation in eyes with heavily pigmented trabecular meshworks. *Am. J. Ophthalmol.*, **139** (6), 1110–1113.

65 Latina, M.A. (1995) A technique to evaluate fluid flow in glaucoma shunts. *Ophthalmic Surg. Lasers*, **26** (6), 576–578.

66 Schuman, J.S.,*et al.* (1990) Contact transscleral continuous wave neodymium:YAG laser cyclophotocoagulation. *Ophthalmology*, **97** (5), 571–580.

67 Flamm, C. and Wiegand, W. (2004) Intraocular pressure after cyclophotocoagulation with the diode laser. *Ophthalmologe*, **101** (3), 263–267 (in German).

68 Pucci, V.,*et al.* (2003) Long-term follow-up after transscleral diode laser photocoagulation in refractory glaucoma. *Ophthalmologica*, **217** (4), 279–283.

69 Beckman, H. and Sugar, H.S. (1973) Neodymium laser cyclocoagulation. *Arch. Ophthalmol.*, **90** (1), 27–28.

70 Hennis, H.L. and Stewart, W.C. (1992) Semiconductor diode laser transscleral cyclophotocoagulation in patients with glaucoma. *Am. J. Ophthalmol.*, **113** (1), 81–85.

71 Wilensky, J.T., Welch, D., and Mirolovich, M. (1985) Transscleral cyclocoagulation using a neodymium: YAG laser. *Ophthalmic Surg.*, **16** (2), 95–98.

72 Fankhauser, F.,*et al.* (1986) Transscleral cyclophotocoagulation using a neodymium YAG laser. *Ophthalmic Surg.*, **17** (2), 94–100.

73 Beckman, H.,*et al.* (1972) Transscleral ruby laser irradiation of the ciliary body in the treatment of intractable glaucoma. *Trans. Am. Acad. Ophthalmol. Otolaryngol.*, **76** (2), 423–436.

74 Molteno, A.C. (1969) New implant for drainage in glaucoma. Clinical trial. *Br. J. Ophthalmol.*, **53** (9), 606–615.

75 Molteno, A.C. (1983) The use of draining implants in resistant cases of glaucoma. Late results of 110 operations. *Trans. Ophthalmol. Soc. N. Z.*, **35**, 94–97.

76 Molteno, A.C. and Haddad, P.J. (1985) The visual outcome in cases of neovascular glaucoma. *Aust. N. Z. J. Ophthalmol.*, **13** (4), 329–335.

77 Prata, J.A. Jr.,*et al.* (1995) *In vitro* and *in vivo* flow characteristics of glaucoma drainage implants. *Ophthalmology*, **102** (6), 894–904.

78 Albis-donando, O. (2006) The Ahmed valve, in *Atlas of Glaucoma Surgery* (eds T. Shaarawy and A. Meermoud), Anshan, Tunbridge Wells, pp. 58–76.

79 Lloyd, M.A.,*et al.* (1994) Intermediate-term results of a randomized clinical trial of the 350- versus the 500-mm^2 Baerveldt implant. *Ophthalmology*, **101** (8), 1456–1463; discussion 1463–1464.

80 Britt, M.T.,*et al.* (1999) Randomized clinical trial of the 350-mm^2 versus the 500-mm^2 Baerveldt implant: longer term results: is bigger better? *Ophthalmology*, **106** (12), 2312–2318.

81 Sommer, A.,*et al.* (1991) Clinically detectable nerve fiber atrophy precedes the onset of glaucomatous field loss. *Arch. Ophthalmol.*, **109** (1), 77–83.

82 Mikelberg, F.S., Yidegiligne, H.M., and Schulzer, M. (1995) Optic nerve axon count and axon diameter in patients with ocular hypertension and normal visual fields. *Ophthalmology*, **102** (2), 342–348.

83 Horn, F.K.,*et al.* (2009) Correlation between local glaucomatous visual field defects and loss of nerve fiber layer thickness measured with polarimetry and spectral domain OCT. *Invest. Ophthalmol. Vis. Sci.*, **50** (5), 1971–1977.

84 Tuulonen, A., Lehtola, J., and Airaksinen, P.J. (1993) Nerve fiber layer defects with

normal visual fields. Do normal optic disc and normal visual field indicate absence of glaucomatous abnormality? *Ophthalmology*, **100** (5), 587–597; discussion 597–598.
85 Kerrigan-Baumrind, L.A.,*et al.* (2000) Number of ganglion cells in glaucoma eyes compared with threshold visual field tests in the same persons. *Invest. Ophthalmol. Vis. Sci.*, **41** (3), 741–748.
86 Wollstein, G.,*et al.* (2004) Optical coherence tomography (OCT) macular and peripapillary retinal nerve fiber layer measurements and automated visual fields. *Am. J. Ophthalmol.*, **138** (2), 218–225.
87 Schuman, J.S.,*et al.* (1995) Optical coherence tomography: a new tool for glaucoma diagnosis. *Curr. Opin. Ophthalmol.*, **6** (2), 89–95.
88 Ajtony, C.,*et al.* (2007) Relationship between visual field sensitivity and retinal nerve fiber layer thickness as measured by optical coherence tomography. *Invest. Ophthalmol. Vis. Sci.*, **48** (1), 258–263.
89 Quigley, H.A., Dunkelberger, G.R., and Green, W.R. (1989) Retinal ganglion cell atrophy correlated with automated perimetry in human eyes with glaucoma. *Am. J. Ophthalmol.*, **107** (5), 453–464.
90 Huang, D.,*et al.* (1991) Optical coherence tomography. *Science*, **254** (5035), 1178–1181.
91 Wojtkowski, M.,*et al.* (2002) *In vivo* human retinal imaging by Fourier domain optical coherence tomography. *J. Biomed. Opt.*, **7** (3), 457–463.
92 Drexler, W.,*et al.* (2003) Enhanced visualization of macular pathology with the use of ultrahigh-resolution optical coherence tomography. *Arch. Ophthalmol.*, **121** (5), 695–706.
93 Fercher, A.F.,*et al.* (1993) *In vivo* optical coherence tomography. *Am. J. Ophthalmol.*, **116** (1), 113–114.
94 Hee, M.R.,*et al.* (1995) Optical coherence tomography of the human retina. *Arch. Ophthalmol.*, **113** (3), 325–332.
95 Wollstein, G.,*et al.* (2005) Ultrahigh-resolution optical coherence tomography in glaucoma. *Ophthalmology*, **112** (2), 229–237.
96 Kiernan, D.F., Mieler, W.F., and Hariprasad, S.M. (2010) Spectral-domain optical coherence tomography: a comparison of modern high-resolution retinal imaging systems. *Am. J. Ophthalmol.*, **149** (1), 18–31.
97 Wojtkowski, M.,*et al.* (2005) Three-dimensional retinal imaging with high-speed ultrahigh-resolution optical coherence tomography. *Ophthalmology*, **112** (10), 1734–1746.
98 Sakamoto, A., Hangai, M., and Yoshimura, N. (2008) Spectral-domain optical coherence tomography with multiple B-scan averaging for enhanced imaging of retinal diseases. *Ophthalmology*, **115** (6), 1071–1078 e7.
99 Sander, B.,*et al.* (2005) Enhanced optical coherence tomography imaging by multiple scan averaging. *Br. J. Ophthalmol.*, **89** (2), 207–212.
100 Bendschneider, D.,*et al.* (2010) Retinal nerve fiber layer thickness in normals measured by spectral domain OCT. *J. Glaucoma*, **19** (7), 475–482.
101 Wollstein, G., Garway-Heath, D.F., and Hitchings, R.A. (1998) Identification of early glaucoma cases with the scanning laser ophthalmoscope. *Ophthalmology*, **105** (8), 1557–1563.
102 Miglior, S.,*et al.* (2002) Intraobserver and interobserver reproducibility in the evaluation of optic disc stereometric parameters by Heidelberg retina tomograph. *Ophthalmology*, **109** (6), 1072–1077.
103 Tan, J.C.,*et al.* (2003) Reasons for rim area variability in scanning laser tomography. *Invest. Ophthalmol. Vis. Sci.*, **44** (3), 1126–1131.
104 Garway-Heath, D.F., Wollstein, G., and Hitchings, R.A. (1997) Aging changes of the optic nerve head in relation to open angle glaucoma. *Br. J. Ophthalmol.*, **81** (10), 840–845.
105 Chen, E., Gedda, U., and Landau, I. (2001) Thinning of the papillomacular bundle in the glaucomatous eye and its influence on the reference plane of the Heidelberg

retinal tomography. *J. Glaucoma*, **10** (5), 386–389.
106 Bowd, C.,*et al.* (2000) Optic disk topography after medical treatment to reduce intraocular pressure. *Am. J. Ophthalmol.*, **130** (3), 280–286.
107 Harju, M. and Vesti, E. (2001) Scanning laser ophthalmoscopy of the optic nerve head in exfoliation glaucoma and ocular hypertension with exfoliation syndrome. *Br. J. Ophthalmol.*, **85** (3), 297–303.
108 Sihota, R.,*et al.* (2002) Variables affecting test–retest variability of Heidelberg Retina Tomograph II stereometric parameters. *J. Glaucoma*, **11** (4), 321–328.
109 Swindale, N.V.,*et al.* (2000) Automated analysis of normal and glaucomatous optic nerve head topography images. *Invest. Ophthalmol. Vis. Sci.*, **41** (7), 1730–1742.
110 Harizman, N.,*et al.* (2006) Detection of glaucoma using operator-dependent versus operator-independent classification in the Heidelberg Retina Tomograph-III. *Br. J. Ophthalmol.*, **90** (11), 1390–1392.
111 Zangwill, L.M.,*et al.* (2007) The effect of disc size and severity of disease on the diagnostic accuracy of the Heidelberg retina tomograph glaucoma probability score. *Invest. Ophthalmol. Vis. Sci.*, **48** (6), 2653–2660.
112 De Leon-Ortega, J.E.,*et al.* (2007) Comparison of diagnostic accuracy of Heidelberg Retina Tomograph II and Heidelberg Retina Tomograph 3 to discriminate glaucomatous and nonglaucomatous eyes. *Am. J. Ophthalmol.*, **144** (4), 525–532.
113 Ferreras, A.,*et al.* (2007) Diagnostic ability of Heidelberg Retina Tomograph 3 classifications: glaucoma probability score versus Moorfields regression analysis. *Ophthalmology*, **114** (11), 1981–1987.
114 Chauhan, B.C.,*et al.* (2000) Technique for detecting serial topographic changes in the optic disc and peripapillary retina using scanning laser tomography. *Invest. Ophthalmol. Vis. Sci.*, **41** (3), 775–782.
115 Burk, R.O. and Rendon, R. (2001) Clinical detection of optic nerve damage: measuring changes in cup steepness with use of a new image alignment algorithm. *Surv. Ophthalmol.*, **45** (Suppl 3), S297–S303; discussion S332–334.
116 Quigley, H.A.,*et al.* (1992) An evaluation of optic disc and nerve fiber layer examinations in monitoring progression of early glaucoma damage. *Ophthalmology*, **99** (1), 19–28.
117 Huang, X.R. and Knighton, R.W. (2002) Linear birefringence of the retinal nerve fiber layer measured *in vitro* with a multispectral imaging micropolarimeter. *J. Biomed. Opt.*, **7** (2), 199–204.
118 Weinreb, R.N.,*et al.* (1990) Histopathologic validation of Fourier-ellipsometry measurements of retinal nerve fiber layer thickness. *Arch. Ophthalmol.*, **108** (4), 557–560.
119 Knighton, R.W., Huang, X.R., and Greenfield, D.S. (2002) Analytical model of scanning laser polarimetry for retinal nerve fiber layer assessment. *Invest. Ophthalmol. Vis. Sci.*, **43** (2), 383–392.
120 Reus, N.J. and Lemij, H.G. (2004) Diagnostic accuracy of the GDx VCC for glaucoma. *Ophthalmology*, **111** (10), 1860–1865.
121 Medeiros, F.A.,*et al.* (2004) Comparison of the GDx VCC scanning laser polarimeter, HRT II confocal scanning laser ophthalmoscope, and stratus OCT optical coherence tomograph for the detection of glaucoma. *Arch. Ophthalmol.*, **122** (6), 827–837.
122 Toth, M. and Hollo, G. (2005) Enhanced corneal compensation for scanning laser polarimetry on eyes with atypical polarisation pattern. *Br. J. Ophthalmol.*, **89** (9), 1139–1142.
123 Bagga, H., Greenfield, D.S., and Feuer, W.J. (2005) Quantitative assessment of atypical birefringence images using scanning laser polarimetry with variable corneal compensation. *Am. J. Ophthalmol.*, **139** (3), 437–446.
124 Mai, T.A., Reus, N.J., and Lemij, H.G. (2007) Diagnostic accuracy of scanning laser polarimetry with enhanced versus variable corneal compensation. *Ophthalmology*, **114** (11), 1988–1993.
125 Reus, N.J., Zhou, Q., and Lemij, H.G. (2006) Enhanced imaging algorithm for scanning laser polarimetry with variable

corneal compensation. *Invest. Ophthalmol. Vis. Sci.*, **47** (9), 3870–3877.
126 Zhou, Q. (2006) Retinal scanning laser polarimetry and methods to compensate for corneal birefringence. *Bull. Soc. Belge Ophtalmol.*, (302), 89–106.
127 Sehi, M., Guaqueta, D.C., and Greenfield, D.S., (2006) An enhancement module to improve the atypical birefringence pattern using scanning laser polarimetry with variable corneal compensation. *Br. J. Ophthalmol.*, **90** (6), 749–753.

39
Early Detection of Cataracts
Rafat R. Ansari and Manuel B. Datiles III

Cataracts, the clouding of the eye lens (Figure 39.1), accounts for half of all blindness worldwide [1–3]. Although there is very active research into finding the cause of cataracts, there is no medical (nonsurgical) cure. Cataract surgery is one of the most successful operations in the world today, but requires highly skilled surgeons and complex and expensive equipment and facilities. Up to 7 million cataract operations are performed worldwide each year at considerable cost. In the United States, cataract accounts for 45% of visits to eye doctors and cataract surgery is the most frequently performed surgical procedure among Medicare beneficiaries [4].

In medicine, early detection of disease leads to early treatment while the disease is still reversible. Noncataractous (before the appearance of haze and cloudiness) lens changes occur with normal aging [5, 6] and the clinical difference between early cataract and age-related change is often not clear-cut in transparent lenses. This makes early detection of cataract very difficult, especially with conventional photographic techniques in use today. Patient follow-up examinations help differentiate nonprogressive aging changes from slowly progressive cataractous changes.

This chapter summarizes currently available methods for the evaluation of cataracts and describes briefly promising methods for the early detection of cataracts. Table 39.1 gives a list of these methods. However, light scattering in various forms remains the main mode of operation in these methods. Simply, more light is scattered as a cataract is formed.

39.1
Visual Acuity Testing

Visual acuity testing (for example, the Snellen test originally introduced in 1862) and survey-type questions are very subjective and dependent upon patients' response. These tests (shown as A and B in the table) do not adequately describe the ability to see large but low-contrast patterns such as faces or nearby objects. A cataract may affect the results of the Snellen acuity test minimally, and yet a patient may already

Handbook of Biophotonics. Vol.2: Photonics for Health Care, First Edition. Edited by Jürgen Popp, Valery V. Tuchin, Arthur Chiou, and Stefan Heinemann.
© 2012 Wiley-VCH Verlag GmbH & Co. KGaA. Published 2012 by Wiley-VCH Verlag GmbH & Co. KGaA.

Figure 39.1 Photograph of an eye showing a mature or advanced cataract seen through a dilated pupil.

experience difficulties in daily activities such as driving or walking, especially in bright sunlight or at night, or have difficulty in reading fine print accurately.

39.2
Ophthalmologic Clinical Examinations

39.2.1
Hand-Held Light Examination

A hand-held light in conjunction with a magnifying loupe is the basic tool used for gross examination such as in-field examination of the lens.

39.2.2
Ophthalmoscopy

The direct ophthalmoscope's built-in +10 lens allows the gross detection of opacities in the lens. The indirect ophthalmoscope may also be useful in making a gross assessment of the clarity of the media as one observes the fundus (back of the eye).

39.2.3
Slit-Lamp Examination

The standard way of examining the lens clinically is using the slit-lamp biomicroscope through a widely dilated pupil. This instrument provides a three-dimensional view of the lens. One can focus on specific areas of the lens from different angles, and at the same time vary the location, direction, and intensity of the illuminating beam independently. The following techniques can be used: (1) direct focal illumination using either a wide or narrow beam; (2) retroillumination; and (3) others, including specular reflection, indirect illumination, diffuse illumination, and use of the light reflected from the iris and posterior capsule.

Table 39.1 Standard methods used in cataract evaluation.

A. Visual acuity/function tests	B. Functional impairment/quality-of-life tests	C. Clinical examination and documentation of physical lens changes
1. Snellen/ETDRS acuity charts or projectors [7]	1. NEI VFQ-25 [13]	1. Clinical examination with hand-held light, slit-lamp biomicroscopy and ophthalmoscopy; and accessory devices [20]
2. Glare and contrast sensitivity tests [8]	2. VF-14 [14, 15]	2. Standardized clinical grading and photographic systems (comparing a patient's cataract with standard photographs)
3. Potential acuity tests	3. Others: Short Form 36 (SF-36) activities of daily vision (ADV), sickness impact profile [16–19]	a. Lens opacities classification system (LOCS), clinical and photographic grading system [21]
a. Pinhole aperture [9]		b. Wisconsin clinical and photographic cataract grading system [22]
b. Entoptic phenomenon [10]		c. Wilmer clinical and photographic cataract grading system [23]
c. Macular function tests: potential acuity meter (PAM), clinical interferometers [11]		d. Oxford clinical cataract grading system [24]
4. Tests for refractive distortions in the lens		e. Age-Related Eye Disease Study (AREDS), clinical and photographic cataract grading system [25, 26]
a. Resolution test target projection ophthalmoscope (acuityscope, Oqual) [12]		f. World Health Organization cataract grading system [27]
b. Wavefront aberrometer, double-pass optical quality analysis system (OQAS II) [18]		3. Modified slit-lamp photography/imaging
		a. Scheimpflug slit-lamp imaging and densitometry systems (Pentacam, Symphony) [28, 29, 32]
		b. Others: laser slit-lamp, sequential color imaging and analysis [33, 34]

| New methods currently used in research | Quasi-elastic or dynamic light scattering (QELS or DLS) [45–52] | Optical coherence tomography [33, 34, 54–56] | MRI and NMR spectroscopy [42–44] | Wavefront technology [40] | Raman spectroscopy [35–39] | Autofluorescence [39, 57] |

39.2.4
Slit-Lamp Photography of the Lens and Grading of Cataracts

Slit-lamp digital photography documents abnormalities and opacities in the lens. Variables to consider in its use in the lens include the limited depth of field, the variabilities in light intensities with the slit beam, limits of magnification with corresponding limits on the area that can be photographed, limits in the angle of the slit beam used, and limits imposed by pupil size. Recently, cataract classification systems have been developed that use carefully selected slit-lamp photographs of cataracts as standards for comparison with the patient's cataracts. These are enumerated in Table 39.1. These systems provide lens photographs showing various severities or grades of cortical, nuclear, and posterior subcapsular cataracts to be used as standards, which a clinician can then compare with the patient's cataract as seen directly on the slit lamp. For nuclear cataracts, slit photographs of the lens are used (Figure 39.2), and for cortical and posterior subcapsular cataracts, retroillumination photographs are used in which the light is reflected back from the retina (Figure 39.3).

39.2.5
Modified Slit-Lamp Photography

Several instruments have been developed to convert the cataract photographic image into numbers. In nuclear cataracts, densitometric analysis can be performed and clouding of the lens nucleus can be expressed in terms of optical density similarly to turbidity measurements. In cortical and posterior subcapsular cataracts, the area occupied by the cataract inside the pupillary space that is blocking light can be measured. These objective measurements can then be analyzed statistically for research studies.

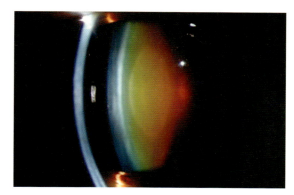

Figure 39.2 Photograph of a saggital section of a nuclear cataract using the slit beam of a slit-lamp camera (note the brownish color or brunesence of nuclear or central core area of the lens).

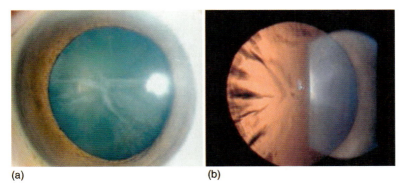

Figure 39.3 Photograph of a cortical cataract seen using direct illumination (a) and using retroillumination (reflection of light from the retina) of a slit-lamp camera (b). Note spoke like opacities.

39.2.6
Scheimpflug Cameras

Slit-lamps modified along the Scheimpflug principle [28, 29] can capture lens images with sufficient depth of focus that the entire anterior chamber from the cornea to the posterior capsule of the lens is in sharp detail. Figure 39.4 shows Scheimpflug images of the front of the eye. Densitometric analysis can be also performed in an area of interest, including the cornea and the lens nucleus. Longitudinal studies to follow the progression of nuclear cataracts can thus be conducted in an objective and masked fashion [29]. A recent review outlined the development of the technique and its introduction into ophthalmology [30].

39.2.7
Retroillumination

Retroillumination cameras [31, 32] can capture images of cortical and posterior subcapsular cataracts using light reflected from the retina as shown in Figure 39.5 and allow the calculation of the size of opacities.

39.3
New Methods Under Development

New technologies are being applied in the early detection of cataracts and show great promise for providing new insights into the cataract problem. In this section, the following devices are discussed: optical coherence tomography (OCT), Raman spectroscopy, autofluorescence, magnetic resonance imaging (MRI), nuclear magnetic resonance (NMR) spectroscopy, wavefront technology, and dynamic light scattering (DLS) [also known as quasi-elastic light scattering (QELS) and photon

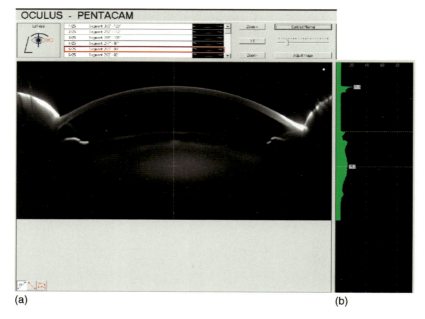

Figure 39.4 (a) Scheimpflug image of the lens with a nuclear cataract showing the automated densitometric measurement through the center [the graph in (b) shows the density across the lens in pixel units]. Currently, the Pentacam Scheimpflug imaging device performs a 360° sweep of the anterior eye segment through the pupillary axis to create a three-dimensional image. It is used routinely to map the cornea for LASIK surgery measurements, but is used here to portray the lens.

correlation spectroscopy (PCS)]. These new noninvasive techniques add to the armamentarium available to cataract researchers.

39.3.1
Optical Coherence Tomography

OCT is a near-infrared optical ranging imaging technique [33]. High-resolution cross-sectional images of the cornea, anterior lens, and retina are obtained by measuring the echo time delay of reflected infrared light using a technique known as low coherence interferometry. The images obtained by OCT are of much higher resolution ($\sim$1–15 µm) than images obtained by low-frequency ultrasound pulse-echo imaging ($\sim$100 µm). OCT provides more detailed structural information than any other noninvasive ophthalmic imaging technique at present. OCT of the lens and cataracts is limited by the depth of scanning, hence only the anterior half can usually be imaged. It has not yet been used extensively in this area but holds promise in the study of cortical cataracts. Recently, OCT nucleus density measurements correlated well with LOCS III for nuclear opalescence and color [34].

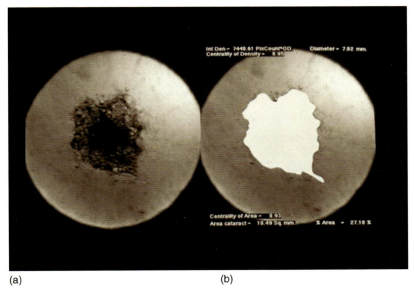

Figure 39.5 Digitized retroillumination image of the lens with a posterior subcapsular cataract (a) and the area of cataract highlighted for measurement (b) using an automated system using thresholding to determine what is an opaque versus nonopaque area.

39.3.2
Raman Spectroscopy for the Detection of Cataract

Raman scattering is the result of inelastic collisions in which the scattered photons exchange energy with the vibrational energy modes of an atom. This frequency shift (or the difference in frequency of an incident photon and the scattered photon) points to specific structural information about a constituent molecule analogous to a certain specific fingerprint that can identify any species present in the system being investigated. However, the Raman signal is very weak. Of 10^6–10^{10} incident photons, only one scattered photon exhibits a Raman shift. Raman spectroscopy has furthered our knowledge of normal aging and pathologic processes in the lens [35], which would not have been possible with other currently available methods. The structural information it provided includes SH, S–S, H_2O, Trp, Try, Phe, and protein secondary structure. Studies can be carried out in the intact living lens. Raman spectroscopy has been used to demonstrate regional swelling of the lens in diabetes. In mildly diabetic rats, the overall increase in lens hydration is hardly detectable. However, regional swelling was demonstrated by Mizuno *et al.* [36]. This method permitted the determination of water content from the periphery of the lens to the center. The advantage of this type of noninvasive technique, similar to that of NMR spectroscopy and QELS, is that it permits the analysis of discrete areas of the lens. Hence these methods may be helpful in determining the changes that occur in certain regions of the lens during cataract formation. The Raman spectra of animal and human lenses have been discussed by Ozaki in a review article [37]. Raman spectroscopy is very useful for cataract research; it gives unique information about changes in water content and structural alterations in the lens proteins with cataract development. Clinical *in vivo* use of this technology is limited by the need to use high laser power, and comprehensive spectral data libraries must first be generated and established. It can then be used to give searchable fingerprints (indices) for ocular and other diseases.

39.3.3
Autofluoresence for the Detection of Cataract

Ocular tissues exhibit natural or autofluorescence (AF) and it has been found to increase with age in healthy individuals [38]. Accumulation of fluorescent proteins in ocular tissue can result from long-term exposure to the UV or UVA radiation in sunlight. This accumulation of fluorophores may also be responsible for lens opacification and can be considered a risk factor for cataract formation. These fluorophores can be found during cataract formation. In the initial stages these can be characterized by exhibiting fluorescence in the near-ultraviolet and violet regions of the spectrum (340 and 411 nm). However, in advanced stages of cataract development, an increase in the intensity of the long-wave fluorescence of the lipids in the blue–green region (430/480 nm) occurs [39]. AF from transparent (noncataractous) lenses exhibits a strong correlation with age. The increased level of fluorescence from the lens can be attributed to oxidative stress or absorbance of UV light as a function of age. Because the cornea does not absorb UV light, its AF level remains constant. However, both diabetic lens and cornea show significantly increased AF levels.

39.3.4
Wavefront in Cataract Detection

Wavefront aberrometry was introduced in ophthalmologic practice to evaluate visual performance objectively. A wavefront analyzer is used to evaluate the quality of an optical system. The wavefront is the pack of light beams or bundle of rays reflected from the retina and coming out of the eye, which is then detected by a sensor called a Shack–Hartmann wavefront sensor (originally developed for high-energy laser and astronomy applications). The light used in the measurement is in the near-infrared region of the spectrum where the eye is nearly insensitive. The sensor has lenslets which divide the wavefront into parts.

Image degradation in the macula because of cataract may be caused not only by light scattering but also by optical aberrations. As discussed in Section 39.1, several devices such as the Oqual and the resolution test target test have been devised as simple ways to test for the effect of optical degradation in the retina.

A new technology using wavefront analysis to study optical aberrations of the eye and in particular the cornea to enhance the results of refractive surgery in patients has also been used on the lens. Kuroda *et al.* [40], using the Hartmann–Shack aberrometer (Topcon, Tokyo, Japan), found that the ocular total higher order optical aberration in eyes with a cortical or nuclear cataract was significantly higher than in normal subjects. Corneal total high-order optical aberration in eyes with mild cortical or nuclear cataracts did not differ from that in normal subjects. This suggests that high-order optical aberration increases in eyes with cataract because of the local refractive change in the lens. Another finding was that the polarity of spherical aberration was different between nuclear and cortical cataracts. In nuclear cataracts, the polarity is always negative, suggesting that a delay of the light wavefront occurs when the ray travels inside the hard nucleus with increased refractive index. In contrast, in cortical cataract, the polarity was always positive.

These findings suggest that in mild cortical and nuclear cataracts, not only light scattering but also optical aberrations in the lens contribute to loss of visual function as measured by loss of contrast sensitivity [40]. Hence this new technique may be useful in studying the total effect of early cataracts on visual function, and explain some patients' complaints such as monocular diplopia in the presence of mild lens changes.

39.3.5
Magnetic Resonance Imaging and Nuclear Magnetic Resonance Spectroscopy of the Lens

As any other human tissue, the lens contains carbon and hydrogen atoms in which protons spin around their nuclei in random directions. On application of a magnetic field, these "microscopic magnets" are aligned in a particular (north–south) direction (higher energy state). On turning off the magnetic field, the microscopic magnets return to their original random state (lower energy state). The frequency of rotation is equal to the energy of a photon (normally a known radiofrequency) that would cause the nuclei to flip between these two energy levels. This provides measurements of relaxation rates between different energy states of the nuclei in relation to the applied excitation photon field. Because they are dependent on the hydrogen nuclei densities in the tissue, the relaxation rate information can be translated into images.

MRI provides the ability to probe the chemical and metabolic status of the lens noninvasively. Thus, in response to normal and pathophysiologic conditions, aspects such as lens metabolism, ion concentrations, the state (bound versus free) of lens water, and metabolite and macromolecular motional dynamics may be investigated. Valuable biochemical and biophysical information pertinent to the factors that govern lens transparency, and conversely the medical condition of the cataract, can thus be studied.

MRI has been used to image the eye but problems have been encountered, including poor resolution; limited access to the surface coil, and poor resultant magnetic signal (because of the location of the eye within the bony structure of the orbit); motion artifacts (caused by microsaccadic eye movements, breathing and heartbeat); and the presence of high susceptibility gradients around the eye. Lizak *et al.* [41] used a special technique, magnetization transfer constant enhancement (MTCE), to enhance the lens image successfully and study diabetic and galactosemic animal models of cataract, and applied it to clinical use. MTCE takes advantage of the magnetic interactions between water and macromolecule hydrogen atoms.

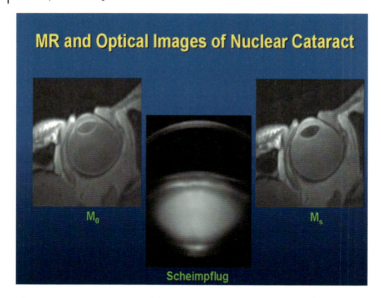

Figure 39.6 An MRI image of the lens compared to a Scheimpflug image of the same lens with nuclear cataract. M_0, magnetic resonance image of a patient's eye with a nuclear cataract; Scheimpflug, image of the same nuclear cataract (LOCS II nuclear opalescence grade 2) taken using a Zeiss Scheimpflug slit-lamp camera (optical/digital); M_s, magnetic resonance image of the same eye with magnetic transfer contrast enhancement [41].

Preliminary clinical studies suggest that cortical lens changes can be better observed with unenhanced magnetic resonance images, whereas nuclear lens changes are better observed by the addition of the MTCE preparation pulse (Figure 39.6). MRI therefore promises to be an imaging method independent of optical imaging that will allow clinicians to monitor metabolic processes in the lens [41]. In a recent study, a new ultra-high resolution MRI instrument with a magnetic field of 7.1 T was used to study the anterior segment of the eye including the crystalline lens of postmortem animal and human eyes and in *in vivo* rabbit experiments [42]. ^{13}C NMR spectroscopy of the intact lens, on the other hand, has provided information about the production, turnover, and inhibition of sorbitol by aldose reductase inhibitors. Proton NMR spectroscopy of ^{13}C-labeled metabolites offers the ability to monitor the reactivation and dynamics of the hexose monophosphate shunt (HMPS), a pathway important for the maintenance of the lens redox state, in real time and noninvasively [43]. ^{31}P NMR spectroscopy allows the monitoring of phosphorus-containing metabolites, thereby permitting the real-time assessment of lens tissue metabolic response to pathophysiologic conditions. Important metabolites, such as adenosine triphosphate, phosphomonoesters, and phosphodiesters, may be monitored. Furthermore, intralenticular pH may be measured. However, no clear correlation between phosphorus metabolite levels and lens clarity has been established to date, despite numerous NMR and classic biochemical studies. This lack of correlation suggests the importance of biophysical investigations

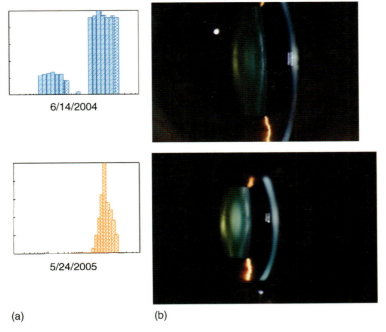

Figure 39.7 A sequential study of an eye which developed a nuclear cataract in a year (b). The dynamic light scattering particle size distribution graphs are shown in (a). The first peak represent the α-crystallin content of the lens, which disappeared by the time the nuclear cataract developed (seen in the photograph). (Reproduced from Datiles et al. [50]).

aimed at the interaction behavior and organization of the constituent lens proteins in the cytoplasm, the macromolecular entities responsible for light scattering associated with cataract. NMR spectroscopy may be viewed as an important adjunct to the better established laser light scattering studies of the lens, and has remained mainly a laboratory, rather than a clinical, method of studying the human lens [44].

39.3.6
Dynamic Light Scattering (DLS)

DLS is an established laboratory technique to measure the average size or size distribution of microscopic particles (3 nm to 3 μm) suspended in a fluid medium in which they undergo random Brownian (or thermal) motion. The intensity of light scattered by the particles from a laser beam passing through such dispersion will fluctuate in proportion to the Brownian motion of the particles. Because the size of the particles influences their Brownian motion, analysis of the fluctuations in scattered light intensity yields a distribution of the diffusion coefficient(s) of the suspended particles from which average particle size or particle size distribution can be extracted [45, 46]. Clinically, DLS can be used to study cataracts noninvasively at the molecular level [47, 48]. *In vivo* DLS measurements at the molecular level correlate well with established laboratory analytical methods such as high-performance liquid chromatography [49, 50]. DLS is safe and fast to use in early cataract evaluation because of the very low laser power (50–100 μW) and short data acquisition time (5 s). In a cold-induced cataract model experiment in which the cataract was simultaneously monitored with both the DLS device and Scheimpflug camera, the DLS picked up subtle changes in the lens faster (2–3 orders of magnitude earlier) than the Scheimpflug camera [51]. The DLS measures the Brownian motion of the crystallins and aggregated proteins inside the lens. The α-crystallins have recently been found to prevent the aggregation of other crystallins and also other proteins such as membranes that have been damaged by oxidative stress. For clinical use, the DLS probe for early cataract detection was successfully integrated into a Keratron device with a three-dimensional aiming system. This aiming arrangement permits consistent remeasurement of changes in the same location in the lens over time. Data obtained from patients in clinical studies have shown good reproducibility [51, 52]. Using this DLS device, it was shown that DLS can measure α-crystallins clinically and noninvasively and that loss of α-crystallins correlated highly with the development of cataract. Hence DLS can measure the reserve of α-crystallins in a given lens and thus assess the risk for cataract development (Figure 39.7) in a patient [53].

39.4
Conclusion

New developments in biophotonics have led to compact and efficient devices showing great potential in aiding researchers to elucidate cataract formation processes *in vivo* and noninvasively. Further improvements and testing on patients will be helpful in moving this field forward and one day help find new ways to prevent or treat cataracts without the need for surgery.

Disclaimer: The views and opinions expressed in this talk are those of the authors and not those of NASA, NIH, and/or the Government of the United States of America.

References

1 Javitt, J. and Wang, F. (1996) Blindness due to cataract: epidemiology and prevention. *Annu. Rev. Public Health*, **17**, 159.

2 Foster, A. (1999) Cataract – a global perspective: output, outcome and outlay. *Eye*, **3**, 449.

3 Kupfer, C. (1985) Bowman Lecture. The conquest of cataract: a global challenge. *Trans. Ophthalmol. Soc. UK*, **104**, 1.

4 National Eye Institute, National Institutes of Health. Vision Research: a National Plan: 1999–2003, Report of the Cataract Panel. US Department of Health and Human Services, NIH Publication No. 98-4120, pp. 59–75.

5 Weekers, R., Delmarcelle, Y., Luyckx-Bacus, J., *et al.* (1973) Morphological changes of the lens with age and cataract, in *The Human Lens – in Relation to*

Cataract (eds. K. Elliott and D. Fitzsimons), Elsevier, Amsterdam, p. 25.
6. Weale F R.A. (1981) Physical changes due to age and cataract, in *Mechanism of Cataract Formation in the Human Lens* (ed. G Duncan), Academic Press, New York, p. 47.
7. Ferris, F.L., Kassoff, A., Bresnick, G.H., et al. (1982) New visual acuity charts. *Am. J. Ophthalmol.*, **4**, 91.
8. American Academy of Ophthalmology Committee on Ophthalmic Procedures (1990) Contrast sensitivity and glare testing in the evaluation of anterior segment disease: ophthalmic procedures assessment. *Ophthalmology*, **97**, 1233.
9. Melki, S., Safar, A., Martin, J., et al. (1999) Potential acuity pinhole. *Ophthalmology*, **106**, 1262.
10. Sinclair, S.H., Loebl, M., and Riva, C. (1979) Blue field entoptic phenomenon in cataract patient. *Arch. Ophthalmol.*, **97**, 1092.
11. Lasa, M.S.M., Datiles, M.B. III, and Friedlin, V. (1995) Potential vision tests in patients with cataracts. *Ophthalmology*, **102**, 1007.
12. Lobo, R.F. and Weale, R.A. (1993) The objective clinical assessment of the quality of the anterior segment by means of the Oqual. *Doc. Ophthalmol.*, **83**, 71.
13. Mangione, C.M., Lee, P.P., Gutierrez, P.R., et al. (2001) Development of the 25-item National Eye Institute Visual Function Questionnaire. *Arch. Ophthalmol.*, **119**, 1050.
14. Steinberg, E.P., Tielsch, J.M., Schein, O.D., et al. (1994) The VF-14. An index of functional impairment in patients with cataract. *Arch. Ophthalmol.*, **112**, 630.
15. Cassard, S.D., Patrick, D.L., Damiano, A.M., et al. (1995) Reproducibility and responsiveness of the VGF-14. An index of visual impairment in patients with cataracts. *Arch. Ophthalmol.*, **113**, 1508.
16. Lee, P., Spritzer, K., and Hays, R. (1997) The impact of blurred vision on functioning and well being. *Ophthalmology*, **104**, 390.
17. Ware, J.E. and Shelbourne, C.D. (1992) The MOS 36 Item Short Form Health Survey (SF-36): I. Conceptual framework and item selection. *Med. Care*, **30**, 473.
18. Mangione, C.M., Phillips, R.S., Seddon, J., et al. (1992) Development of the Activities of Daily Vision Scale. A measure of visual function status. *Med. Care*, **30**, 1111.
19. Berner, M., Bobbitt, R.A., Carter, W.B., et al. (1981) The Sickness Impact Profile: development and final revision of a health status measure. *Med. Care*, **19**, 787.
20. Berliner, L. (1949) *Biomicroscopy of the Eye*, vol. II, Paul B. Hoeber, New York, p. 948.
21. Chylack, L., Leske, C., Sperduto, R., et al. (1988) Lens Opacities Classification System. *Arch. Ophthalmol.*, **106**, 330.
22. Klein, B.E.K., Klein, R., Linto, K.L.P., et al. (1990) Assessment of cataracts from photography in the Beaver Dam Eye Study. *Ophthalmology*, **97**, 1428.
23. Taylor, H.R. and West, S.K. (1989) The clinical grading of lens opacities. *Aust. N. Z. J. Ophthalmol.*, **17**, 81.
24. Sparrow, J., Bron, A., Brown, N., et al. (1986) The Oxford clinical cataract classification and grading system. *Int. Ophthalmol.*, **9**, 207.
25. Kassoff, A., Kassoff, J., Mehu, M., et al. (2001) The Age Related Eye Disease Study (AREDS) system for classifying cataracts from photographs: AREDS Report No. 4. *Am. J. Ophthalmol.*, **131**, 167–175.
26. Braccio, L., Campirini, M., Graziosi, P., et al. (1998) An independent evaluation of the Age Related Eye Disease Study (AREDS) cataract grading system. *Curr. Eye Res.*, **17**, 53.
27. Thyleflors, B., Chylack, L., Konyama, K., et al. (2002) A simplified cataract grading system. *Ophthalmic Epidemiol.*, **9**, 83.
28. Dragomirescu, V., Hockwin, H., Koch, H.R., et al. (1978) Development of a new equipment of rotating slit image photography according to Scheimpflug's principle. *Interdiscipl. Top. Gerontol.*, **13**, 118.
29. Datiles, M., Magno, B., and Friedlin, V. (1995) Study of nuclear cataract progression using the NEI Scheimpflug system. *Br. J. Ophthalmol.*, **79**, 527.

30 Wegener, A. and Laser-Junga, H. (2009) Photography of the anterior eye segment according to Scheimpflug's principle: options and limitations. *Clin. Exp. Ophthalmol.*, **37**, 144–154.

31 Kawara, T. and Obazawa, H. (1980) A new method of retroillumination photography of cataractous lens opacities. *Am. J. Ophthalmol.*, **90**, 186.

32 Brown, N., Bron, A., Ayliffe, W., et al. (1987) The objective assessment of cataract. *Eye*, **1**, 234.

33 Drexler, W., Morgner, V., Ghanta, R., Kastner, F., et al. (2001) Ultrahigh resolution OCT. *Nat. Med.*, **7**, 502–507.

34 Wong, A.L., Leung, C.K., Weinreb, R.N., et al. (2009) Quantitative assessment of lens opacities with anterior segment optical coherence tomography. *Br. J. Ophthalmol.*, **93** (1), 61–65.

35 McCreery, R.L. (1996) Analytical Raman spectroscopy: an emerging technology for practical applications. *Am. Lab.*, **28**, 34.

36 Mizuno, A., Nozawa, H., Taginuma, T., et al. (1987) Diabetic cataracts. *Exp. Eye Res.*, **45**, 185.

37 Ozaki, Y. (1988) Medical application of Raman spectroscopy. *Appl. Spectrosc. Rev.*, **24**, 259.

38 Bursel, S.E. and Yu, N.T. (1990) Fluorescence and Raman spectroscopy of the crystalline lens, in *Noninvasive Techniques in Ophthalmology* (ed. B.R. Masters), Springer, New York. 342–365.

39 Babizhayev, M.A. (1989) Lipid fluorophores of the human crystalline lens with cataract. *Graefes. Arch. Clin. Exp. Ophthalmol.*, **227**, 384.

40 Kuroda, T., Fujikado, T., Maeda, N., et al. (2002) Wavefront analysis in eyes with nuclear or cortical cataract. *Am. J. Ophthalmol.*, **134**, 1.

41 Lizak, M., Datiles, M.B., Aletras, A., et al. (2000) MRI of the human eye using magnetization contrast enhancement. *Invest. Ophthalmol. Vis. Sci.*, **41**, 3878.

42 Langner, S., Martin, H., Terwee, T., et al. (2010) 7.1 T MRI to assess the anterior segment of the eye. *Invest. Ophthalmol. Vis. Sci.*, **51**, 6575.

43 Gonzales, R.G., Willis, J., Aguayo, J., et al. (1982) C-13 nuclear magnetic resonance studies of sugar cataractogenesis in the single intact rabbit lens. *Invest. Ophthalmol. Vis. Sci.*, **22**, 808.

44 Williams, W.F., Austin, C.D., Farnsworth, P.N., et al. (1988) Phosphorus and proton magnetic resonance spectroscopic studies on the relationship between transparency and glucose metabolism in the rabbit lens. *Exp. Eye Res.*, **47**, 97.

45 Chu, B. (1991) *Laser Light Scattering: Basic Principles*, Academic Press, New York.

46 Bern, B. and Pecora, A. (1976) *Dynamic Light Scattering*, John Wiley and Sons, Inc., New York.

47 Benedek, G.B., Clark, J.I., Serrallad, E.N., et al. (1979) Light scattering and reversible cataracts in the calf and human lens. *Philos. Trans. R. Soc. Lond. Ser. A Math. Phys. Sci.*, **293**, 329.

48 Ansari, R.R. (2004) Ocular static and dynamic light scattering: a non-invasive diagnostic tool for eye research and clinical practice. *J. Biomed. Opt.*, **9** (1), 22–37.

49 Simpanya, M.F., Ansari, R.R., Leverenz, V., and Giblin, F.J. (2008) Measurement of lens protein aggregation *in vivo* using dynamic light scattering in a guinea pig/UVA model for nuclear cataract. *J. Photochem. Photobiol.*, **84**, 1589–1595.

50 Datiles, M.B. III, Ansari, R.R., Suh, K.I., Vitale, S., Reed, G.F., Zigler, J.S. Jr., and Ferris, F.L. III (2008) Clinical detection of precataractous lens protein changes using dynamic light scattering. *Arch. Ophthalmol*, **126** (12), 1687–1693.

51 Datiles, M.B., Ansari, R.R., and Reed, G.F. (2002) A clinical study on the human lens with a dynamic light scattering device. *Exp. Eye Res.*, **74**, 93.

52 Thurston, G., Hayden, D.L., Burrows, P., et al. (1997) Quasielastic light scattering study of the living human lens as a function of age. *Curr. Eye Res.*, **16**, 197.

53 Steinberg, E.P., Tielsch, J.M., Schein, O.D., et al. (1994) The VF-14. An index of functional impairment in

patients with cataract. *Arch. Ophthalmol.*, **112**, 630.
54 Fercher, A.F. (1996) Optical coherence tomography. *J. Biomed. Opt.*, **1**, 15.
55 Izatt, J.A., Hee, M.R., Swanson, E.A., *et al.* (1994) Micrometer-scale resolution imaging of the anterior eye *in vivo* with optical coherence tomography. *Arch. Ophthalmol.*, **112**, 1584.
56 DiCarlo, C.D., Roach, W.P., Gagliano, D.A., *et al.* (1999) Comparison of optical coherence tomography (OCT) imaging of cataracts with histopathlogy. *J. Biomed. Opt.*, **4**, 450.
57 Dochio, F. and Van Best, J.A. (1998) Simple, low-cost, portable corneal fluorometer for detection of the level of diabetic retinopathy. *Appl. Opt*, **37** 4303.

40
Diabetic Retinopathy

Adzura Salam, Sebastian Wolf, and Carsten Framme

40.1
Introduction

Diabetic retinopathy (DR) is the commonest microvascular complication occurring in diabetes mellitus and remains one of the leading causes of vision loss and blindness among adults aged 40 years and older. According to the latest report on "Prevalence of Diabetic Retinopathy in the United States, 2005–2008," the prevalence of diabetic retinopathy and vision-threatening diabetic retinopathy was 28.5 and 4.4%, respectively, among American adults with diabetes [1]. The prevalence of DR among Americans aged 40 years or older with vision-threatening diabetic retinopathy (VTDR) is predicted to triple by 2050, from 5.5 million in 2005 to 16.0 million for DR and from 1.2 million in 2005 to 3.4 million for VTDR [2].

Vision loss occurs in DR due to the development of maculopathy, especially diabetic macular edema (DME), and due to proliferative diabetic retinopathy (PDR). The Wisconsin Epidemiologic Study of Diabetic Retinopathy (WESDR) reported that the prevalence of PDR was 23% in the WESDR younger-onset group, 10% in the WESDR older-onset group who take insulin, and 3% in the group who do not take insulin [3]. The same study reported that hyperglycemia, longer duration of diabetes, and more severe retinopathy at baseline were associated with an increased 4 year risk of developing PDR.

In 2009, another study by the same group, The Wisconsin Epidemiologic Study of Diabetic Retinopathy XXIII, reported that the 25 year cumulative incidence was 29% for DME and 17% for clinically significant macula edema (CSME) [4]. They also concluded that the relatively high 25 year cumulative rate of incidence of DME was related to glycemia and blood pressure.

Both DR and DME share common risk factors, namely the duration of diabetes and severity of hyperglycemia. Others risk factors for progression of DR include coexisting hypertension, hyperlipidemia, renal disease, and pregnancy [5–8]. The WESDR also reported that the incidence of DME was associated with more severe DR, higher glycosylated hemoglobin, proteinuria, higher systolic and diastolic blood pressure, and chronic smoking [4].

40.2
Diagnosis of DR and DME

Traditionally, DR is diagnosed by using slit-lamp biomicroscopy and stereo fundus photography [9, 10]. The gold standard for grading the severity of DR is stereoscopic fundus photography through dilated pupils, using seven standard fields [11–13], and grading guidelines for these photographs were established by the Early Treatment Diabetic Retinopathy Study (ETDRS) group [14]. Clinically, DR can be classified into two stages: nonproliferative diabetic retinopathy (NPDR) and proliferative diabetic retinopathy (PDR) [15] (Table 40.1). The stereo fundus photograph helps in documentation of the clinical findings in DR. NPDR appear as dot-blot hemorrhage, cotton-wool spot, hard exudates, and venous beading depending on the severity (Figure 40.1). The presence of new vessels elsewhere (NVE), new vessels on the disc (NVD), and enhanced epiretinal and intravitreal hemorrhage are the characteristic features of PDR (Figure 40.2).

DME is a common microvascular complication which may appear at any stage of DR. DME is further classified into focal, diffuse, and ischemic maculopathy [16] (Table 40.2). In developed countries, stereo fundus photography is being used in DR screening programs. Such screening programs effectively help to reduce the incidence of blindness secondary to DR [11, 13]. However, in developing countries, the utility of fundus photography as a large-scale screening procedure is limited because of its cost and the requirements for special equipment and trained personnel.

In a further diagnostic step, fluorescein angiography (FA) is frequently used, especially for baseline examination. It is a procedure in which sodium fluorescein is administered intravenously followed by rapid sequence photography of the retina to evaluate its circulation. FA is usually used to evaluate the extent and origin of fluid

Table 40.1 International clinical diabetic retinopathy disease severity scale.

Proposed disease severity level	Findings observable upon dilated ophthalmoscopy
No apparent retinopathy	No abnormalities
Mild nonproliferative diabetic retinopathy	Microaneurysms only
Moderate nonproliferative diabetic retinopathy	More than microaneurysm but less than severe NPDR
Severe nonproliferative diabetic retinopathy	Any of the following: 1. More than 20 intraretinal hemorrhages in each of four quadrants 2. Definite venous beading in two or more quadrants, prominent intraretinal microvascular abnormalities (IRMA) in one or more quadrants, *and* no signs of proliferative retinopathy
Proliferative diabetic retinopathy	One or both of the following: 1. Neovascularization 2. Vitreous/preretinal hemorrhage

Source: Wilkinson *et al.* [16].

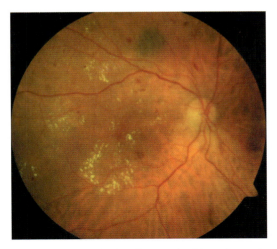

Figure 40.1 Color fundus photograph of the right eye showing severe nonproliferative diabetic retinopathy with diffuse clinically significant macular edema. Note the presence of dot-blot hemorrhage, hard exudates in "succinate ring," cotton-wool spots, and venous beading.

leakage and also the extent of capillary ischemia in DME [17]. The flow and permeability of retinal vessels can be correlated with anatomic changes by using FA (Figure 40.3). FA findings in DME can be categorized into three different types of leakage: (1) focal leakage, (2) diffused leakage, and (3) diffused cystoid leakage. The most important information gained from FA is whether there is presence of macular ischemia, a condition which responds poorly to most conventional DME treatments [17]. However, FA has its limitations: an invasive technique, time constraints, expensive equipment

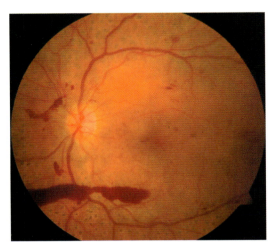

Figure 40.2 Color fundus photograph of the left eye showing proliferative diabetic retinopathy with presence of new vessels at the disc, new vessels elsewhere, and subhyaloid hemorrhage. Note old panretinal laser photocoagulation scars outside the vessel arcades.

Table 40.2 International clinical diabetic macular edema disease severity scale.

Proposed disease severity level	Findings observable upon dilated ophthalmoscopy[a]
Diabetic macular edema apparently absent	No apparent retinal thickening or hard exudates in posterior pole
Diabetic macular edema apparently present	Some apparent retinal thickening or hard exudates in posterior pole
If diabetic macular edema is present, it can be categorized as follows:	
Diabetic macular edema present	Mild diabetic macular edema: some retinal thickening or hard exudates in posterior pole but distant from the center of the macula
	Moderate diabetic macular edema: retinal thickening or hard exudates approaching the center of the macula but not involving the center
	Severe diabetic macular edema: retinal thickening or hard exudates involving the center of the macula

a) Hard exudates are a sign of current or previous macular edema. Diabetic macular edema is defined as retinal thickening; this requires a three-dimensional assessment that is best performed by dilated examination using slit-lamp biomicroscopy and/or stereo fundus photography.

required, and possible allergic-type reactions to sodium fluorescein. Hence FA should be performed only under strong indication or for treatment planning.

The most recent step in diagnostics is provided by optical coherence tomography (OCT), which was first introduced in 1991 by Huang *et al.*, offering a noninvasive, noncontact imaging technology that can image retinal structures *in vivo* [18]. Time-domain OCT uses low-coherence interferometry to provide absolute measurements of retinal thickness and achieves a high axial resolution of $\sim 10\,\mu m$ (Figure 40.4). OCT is used in clinical routine to assess the posterior pole pathology in various retinal diseases and specifically to analyze DME in DR [19].

40.3
Spectral-domain OCT in DME

The technology of OCT has undergone rapid evolution over the past decade. The most recent advance, spectral-domain optical coherence tomography (SD-OCT), made both high resolution and fast scanning speeds possible, thus improving the quality of images. Various SD-OCT instruments are available: Cirrus HD-OCT (Carl Zeiss Meditec), RTVue-Fourier Domain OCT (Optovue), Copernicus OCT (Reichert/Optopol Technology), Spectral OCT/SLO (Opko/OTI), Spectralis HRA +OCT (Heidelberg Engineering), Topcon 3D OCT-1000 (Topcon), and RS-3000 Retiscan (Nidek). All instruments provide high-quality OCT images and produce three-dimensional images.

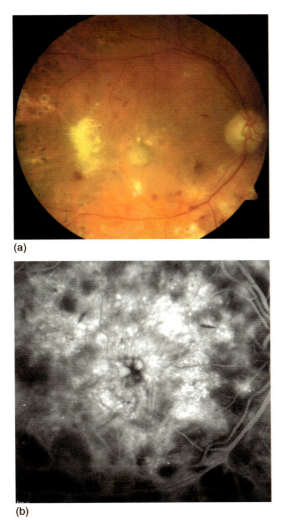

Figure 40.3 Color fundus photograph showing CSMO (a) and hyperfluorescein leakage during fluorescein angiography (b).

SD-OCT reveals various pathologic findings at qualitative and quantitative levels, and also abnormal morphology of retinal layers. The qualitative interpretation includes hyperreflective structures (hard exudates and cotton-wool spots), hyporeflective structures (intraretinal edema, exudative retinal detachment, and cystoid macular edema), and also shadow effects (hemorrhage, exudates, and retinal vessels) [20]. In the past, OCT has been used primarily to analyze macular thickness in DME (Figure 40.5). Various studies have documented changes in retinal thickness, central foveal thickness, and the macular volume among diabetes patient as compared with nondiabetes patients [21, 22]. In addition, some researchers have reported

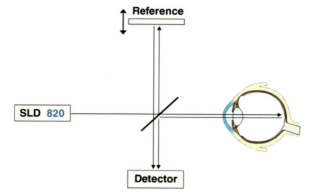

Figure 40.4 Diagram showing the principle of OCT. OCT uses low-coherence interferometry technology to produce a two-dimensional images. A low-coherence near-infrared light beam (820 nm) is directed towards the target tissue. The magnitude and relative location of backscattered light from the internal tissue microstructures are interpreted by the OCT to generate an image.

a modest correlation between macular thickness and best-corrected visual acuity (BCVA) in diabetic eyes [23, 24]. These findings may be useful for early detection of macular thickening and may be indicators for closer follow-up of patients with diabetes. Strøm *et al.* reported close agreement between subjective and objective assessments of retinal thickness; the same study suggested that DME can be accurately and prospectively measured with OCT [25].

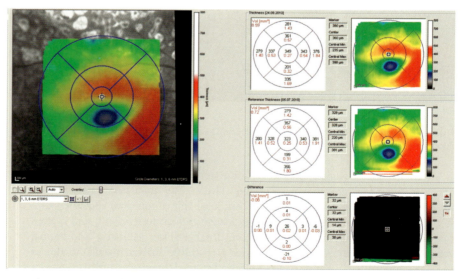

Figure 40.5 Spectralis HRA+OCT retinal thickness scan. Note that the central macular thickness and volume can be compared before and after a treatment.

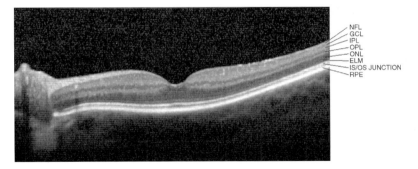

Figure 40.6 Normal retina as imaged by the Spectralis HRA+OCT: ganglion cell layer (GCL), inner plexiform layer (IPL), inner nuclear layer (INL), outer plexiform layer (OPL), outer nuclear layer (ONL), external limiting membrane (ELM), photoreceptor inner segments (IS), outer segments (OS), and retinal pigment epithelium (RPE).

The Spectralis HRA+OCT, used as a standard in our clinic, combines high-resolution SD-OCT with an scanning laser ophthalmoscopy (SLO). The system allows for simultaneous OCT scans with high-resolution scanning laser retinal imaging. The instrument uses broadband 870 nm SLD for the OCT channel. The retina is scanned at 40 000 A-scans per second, creating highly detailed images of the structure of the retina. The OCT optical depth resolution is 7 μm and the digital depth resolution is 3.5 μm. The combination of high-resolution scanning laser retinal images and SD-OCT allows for real-time tracking of eye movements and real-time averaging of scanning laser images and OCT scans, reducing speckle noise of the OCT images [26].

Recent advances in the SD-OCT technique made both high resolution and fast scanning speeds possible. High resolution allows for differentiation of as many as 11 structural characteristics within the retina (Figure 40.6). Other additional information available using SD-OCT includes the structural changes in DME, which can include epiretinal membrane (ERM), retinal swelling, cystoid macular edema (CME), subretinal fluid (SRF) accumulation, and intraretinal fluid (IRF) accumulation (Figure 40.7). SD-OCT imaging allows visualization of the integrity of the outer

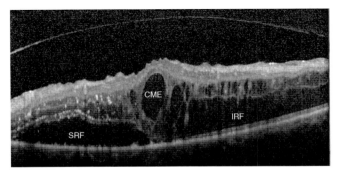

Figure 40.7 The Spectralis HRA+OCT shows cystoid macular edema (CME), intraretinal fluid (IRF), and subretinal fluid (SRF) with subretinal detachment.

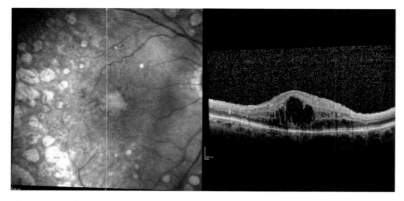

Figure 40.8 The Spectralis HRA+OCT shows disturbed outer and inner segment junction of photoreceptor and external limiting membrane.

retinal layers in DME, which are most important for vision. These include the external limiting membrane (ELM), the photoreceptor inner segment (IS), the outer segment (OS), the retinal pigment epithelium (RPE), and Bruch's membrane (Figure 40.8).

Otani *et al.* were among the earliest researchers to observe three basic structural changes in DME by OCT, namely sponge-like retinal swelling, edema with cystic spaces, and edema with serous retinal detachment [27]. The same study reported a correlation between BCVA and retinal thickness, regardless of the different tomographic features [27]. Panozzo *et al.* introduced vitreo-macular traction (VMT) into the OCT-derived classification of DME and this led to more research into DME [28]. The role of VMT is particularly important in considering possible surgical intervention for DME.

OCT is less invasive than FA for diagnosis of DME in DR patients. Several groups demonstrated that OCT provides an objective documentation of foveal structural changes in eyes with DR [29–31]. Some of the diabetic structural changes such as foveal traction, serous foveal detachment, and CME can be detected by OCT, which may not be evident using ophthalmoscopy and FA. Interestingly, serous macular detachment (SMD) and VMT in the fovea are observable in OCT but not in FA and also not by funduscopy [32].

Koleva-Georgieva *et al.* reported that SD-OCT is useful in diagnosing subclinical SMD in DME and the presence of VMT [33]. VMT appeared as a relative hyperreflective line in the nonreflective space of the vitreous body (Figure 40.9). However, the major drawback of OCT over FA is its incapability to detect macular ischemia, denoted by nonperfusion of the retinal capillaries. Hence FA is still the best diagnostic tool in ischemic DME, as OCT will not be able to give much information regarding this particular condition. Therefore, FA is the most important diagnostic tool in baseline examination in DME to rule out significant macular ischemia, obviating the need for further macular treatments such as laser or intravitreal drug treatment.

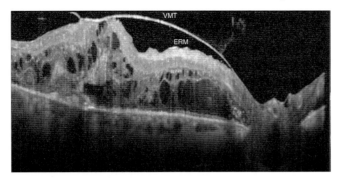

Figure 40.9 The Spectralis HRA+OCT shows vitreo-macular traction (VMT) and epiretinal membrane (ERM) in diabetic macular edema.

40.4
How OCT Changes Our View in Management of DR

OCT technology has changed our perspective in the management of DR, particularly DME. OCT helps the retina physician in accurately diagnosing the different types of DME, especially in the early stages when structural changes may occur that would not be evident with slit-lamp biomicroscopy or FA. In addition, OCT helps in deciding the treatment protocols, either surgical or medical, in DME. Finally, OCT also aids as a noninvasive tool in monitoring the disease progress and the treatment outcome in both DR and DME.

Based on the ETDRS report, the macular photocoagulation laser (MPL) used to be the gold standard in treating DME [34]. Over the last decade, this trend has changed where in cases of refractory DME diagnosed by OCT, which is obviously not be responding to MPL treatment, medical treatment has become the alternative solution. The Intravitreal Triamcinolone for Clinically Significant Macular Edema That Persists After Laser Treatment (TDMO) study reported a reduction in central macular thickness observed by OCT and significant improvement in visual acuity, especially during the initial treatment [35]. In another study, Avitabile *et al.* reported reduced central macular thickness and improved BCVA more in a group of patient treated with intravitreal triamcinolone compared with macular laser grid photocoagulation [36]. It is also important to note that the risk of glaucoma due to steroid response has to be considered in patients who are to be treated with triamcinolone.

In the era of antivascular endothelial growth factors (anti-VEGFs), new drugs such as bevacizumab and ranibizumab have shown promising results in the treatment of DME. The Pan-American Collaborative Retina Study Group (PACORES) reported that primary intravitreal bevacizumab for DME seems to provide stability or improvement in BCVA, OCT, and FA in diffused DME at 12 months [37]. Similarly, results of The RESOLVE Study also showed a promising outcome regarding the safety and efficacy of ranibizumab treatment in patients with DME at 12 months [38].

Framme *et al.* demonstrated postoperative proliferation of the retinal pigment epithelium (RPE), RPE atrophy, and neurosensory retina alteration seen with

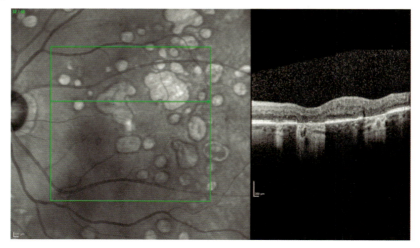

Figure 40.10 The Spectralis HRA+OCT demonstrates postoperative RPE atrophy and neurosensory retina alteration following macular laser photocoagulation.

SD-OCT following MPL [39] (Figure 40.10). Therefore, in the future, treatment of DME will probably be focused more on medical therapy or mild laser treatment, such as selective laser therapy using Nd:YLF laser methods, to prevent such structural damage.

Clinical and anatomic evidence indicates that VMT has been established as one of the known factors in the pathogenesis of CME. A few researchers have reported improvement of BCVA and a significant reduction in macular thickness following vitrectomy for diffused DME combined with VMT [40, 41]. Recchia *et al.* suggested that pars plana vitrectomy with peeling of the inner limitans membrane (ILM) may provide anatomical and visual benefit in DME [42]. In contrast, Shah *et al.* [43] and Dhingra *et al.* [44] reported that vitrectomy and ILM peeling for refractory DME in the absence of VMT failed to improve visual acuity. Hence pars plana vitrectomy with ILM peeling should be reserved for selected cases.

40.5
SD-OCT Scans as a Prognostic Features for DME

The use of SD-OCT is becoming more valuable in the management of DME. SD-OCT imaging enables us to assess the status of the photoreceptor layers. A few researchers have reported a close relationship between this photoreceptor layer status and visual function in various macular diseases such as branch retinal vein occlusion [45], central serous chorioretinopathy [46], age-related macular degeneration [47], retinitis pigmentosa [48], and retinal detachment [49]. Interestingly, an association has been detected between VA and abnormal OCT findings such as those pertaining to the external limiting membrane and the junction between the photoreceptor inner and outer segments (IS/OS) [45–48].

Einbock *et al.* were the first to report an improvement in VA after intravitreal anti-VEGF therapy in DME patients with normal appearance of the external limiting membrane, photoreceptor inner and outer segments, and the RPE [50]. In contrast, patients with disturbed outer retinal layers on SD-OCT showed only a reduction in retinal thickness, without much visual improvement after intravitreal anti-VEGF therapy. Another important finding from the same study was that patients with discontinuity of the inner retinal layer and disturbed outer retinal layers failed to achieve anatomic or visual improvement after intravitreal anti-VEGF in DME (Figures 40.11–40.13).

Forooghian *et al.* observed a stronger correlation of photoreceptor outer segment (PROS) length with VA in DME and suggested that the PROS measures may be more directly related to visual function [51]. In another study, Otani *et al.* reported a

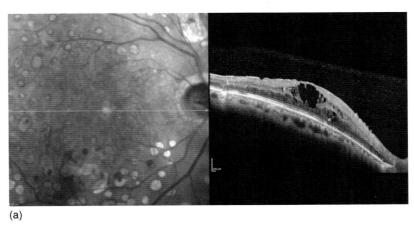

(a)

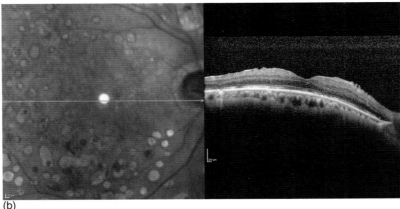

(b)

Figure 40.11 Patient with cystoid macular edema before (a) and 3 months after repeated treatment (b) with intravitreal anti-VFGF therapy (ranibizumab). Cystoid edema disappeared and VA improved from 20/60 to 20/30. Note the normal outer retinal layers at baseline. At the inner surface of the retina, an epiretinal membrane was observable.

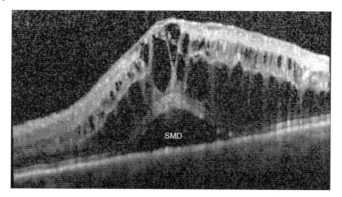

Figure 40.12 Patient with DME and severe cystoid changes. Note that the outer retinal layers such as the external limiting membrane and photoreceptor layers are severely disturbed with serous macular detachment (SMD). VA was unchanged in this patient after anti-VEGF therapy.

significant correlation between the integrity of the external limiting membrane and inner and outer segments of the photoreceptors with best-corrected VA when compared with central subfield thickness in diabetic macular edema [52]. In 2009, Sakamoto et al. reported a significant association between foveal photoreceptor status and VA after resolution of diabetic macular edema by pars plana vitrectomy [53]. These findings suggested that integrity of the photoreceptor outer segment plays a significant role in prediction of visual outcome after DME treatment. Such an enormous increase in medical knowledge was only possible due to this OCT technique, which has become one of the most important noninvasive diagnostic tools in ophthalmology.

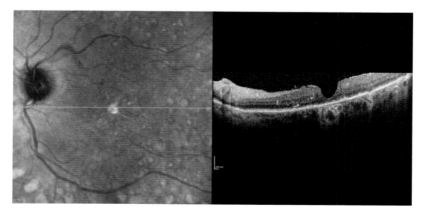

Figure 40.13 Chronic diabetic macular edema patient with discontinuity of the inner retinal layer and disturbed outer retinal layers. This patient failed to achieve anatomic or visual improvement after intravitreal anti-VEGF.

40.6
Summary for the Clinician

- The development of advanced diagnostic instrumental techniques such as SD-OCT has changed our understanding of this disease's pathophysiology, and improved the ability of clinicians to individualize the approach to treatment.
- The integrity of retinal layers can be analyzed with high-resolution OCT scans.
- The presence and integrity of the external limiting membrane, the photoreceptor inner and outer segments, and the retinal pigment epithelium play a significant role in the prediction of visual outcome after DME treatment.

References

1 Zhang, X., Saaddine, J.B., Chou, C.-F., Cotch, M.F., Cheng, Y.J., Geiss, L.J., Gregg, E.W., Albright, A.L., Klein, B.E.K., and Klein, R. (2010) Prevalence of diabetic retinopathy in the United States, 2005–2008. *JAMA*, **304** (6), 649–656.

2 Saaddine, J.B., Honeycutt, A.A., Narayan, K.M.V., Zhang, X., Klein, R., and Boyle, J.P. (2008) Projection of diabetic retinopathy and other major eye diseases among people with diabetes mellitus United States, 2005–2050. *Arch. Ophthalmol.*, **126** (12), 1740–1747.

3 Klein, B.E. and Moss, S.E. (1992) Epidemiology of proliferative diabetic retinopathy. *Diabetes Care*, **15** (12), 1875–1891.

4 Klein, R., Knudtson, M.D., Lee, K.E., Gangnon, R., and Klein, B.E.K. (2009) The Wisconsin Epidemiologic Study of Diabetic Retinopathy XXIII: the twenty-five-year incidence of macular edema in persons with type 1 diabetes. *Ophthalmology*, **116**, 497–503.

5 Klein, R., Klein, B.E.K., Moss, S.E., Davis, M.D., and DeMets, D.L. (1988) Glycosylated hemoglobin predicts the incidence and progression of diabetic retinopathy. *JAMA*, **260**, 2864–2871.

6 Vitale, S., Maguire, M.G., Murphy, R.P., Hiner, C.J., Rourke, L., Sackett, C., and Patz, A. (1995) Clinically significant macular edema in type I diabetes. *Ophthalmology*, **102**, 1170–1176.

7 Chew, E.Y., Klein, M.L., Ferris, F.L., Remaley, N.A., Murphy, R.P., Chantry, K., Hoogwerf, B.J., and Miller, D. (1996) ETDRS Research Group: association of elevated serum lipid levels with retinal hard exudate in diabetic retinopathy: Early Treatment of Diabetic Retinopathy Study (ETDRS) Report No. 22. *Arch. Ophthalmol.*, **114**, 1079–1084.

8 Klein, R., Klein, B.E.K., Moss, S.E., and Cruickshanks, K. (1998) The Wisconsin Epidemiologic Study of Diabetic Retinopathy, XVII: the 14-year incidence and progression of diabetic retinopathy and associated risk factors in type I diabetes. *Ophthalmology*, **105**, 1801–1815.

9 Early Treatment Diabetic Retinopathy Study Research Group (1991) Fundus photographic risk factors for progression of diabetic retinopathy. ETDRS Report Number 12. *Ophthalmology Suppl.*, **98**, 823–833.

10 Early Treatment Diabetic Retinopathy Study Research Group (1991) Grading diabetic retinopathy from stereoscopic color fundus photographs – an extension of the modified Airlie House classification. ETDRS Report Number 10. *Ophthalmology Suppl.*, **98**, 786–805.

11 Hutchinson, A., McIntosh, A., Peters, J., O'Keeffe, C., Khunti, K., Baker, R., and Booth, A. (2000) Effectiveness of screening and monitoring tests for diabetic retinopathy: a systematic review. *Diabet. Med.*, **17**, 495–506.

12 American Diabetes Association (2000) Diabetic retinopathy. *Diabetes Care*, **23** (Suppl. 1), S73–S76.

13 Singer, D.E., Nathan, D.M., Fogel, H.A., and Schachat, A.P. (1992) Screening for diabetic retinopathy. *Ann. Intern. Med.*, **116**, 660–671.

14 Early Treatment Diabetic Retinopathy Study Research Group (1991) Grading diabetic retinopathy from stereoscopic color fundus photographs: an extension of the modified Airlie House classification: ETDRS Report No 10. *Ophthalmology*, **98** (Suppl.), 786–806.

15 American Academy of Ophthalmology. The Eye M.D. Association. Year 2011 (last review 5.7.2011) http://one.aao.org/CE/PracticeGuidelines/ClinicalStatements_Content.aspx? cid=5e96758f-c10d-41aa-a9b2-fc4d4faac6b9

16 Wilkinson, C.P., Ferris, F.L., Klein, R.E., Lee, P.P., Agardh, C.D., Davis, M., Dills, D., Kampik, A., Pararajasegaram, R., Verdaguer, J.T., and Global Diabetic Retinopathy Project Group (2003) Proposed international clinical diabetic retinopathy and diabetic macular edema disease severity scales. *Ophthalmology*, **110** (9), 1677–1682.

17 Cunha-Vaz, J.G. (2000) Diabetic retinopathy: surrogate outcomes for drug development for diabetic retinopathy. *Ophthalmologica*, **214**, 377–380.

18 Huang, D., Swanson, E.A., Lin, C.P., Schuman, J.S., Stinson, W.G., Chang, W., Hee, M.R., Flotte, T., Gregory, K., and Puliafito, C.A. (1991) Optical coherence tomography. *Science*, **254** (5035), 1178–1181.

19 Wolf, S. and Wolf-Schnurrbusch, U. (2010) Spectral-domain optical coherence tomography use in macular diseases: a review. *Ophthalmologica*, **224** (6), 333–340.

20 Lang, G.E. (2007) Optical coherence tomography findings in diabetic retinopathy. *Dev. Ophthalmol.*, **39**, 31–47.

21 Sánchez-Tocino, H., Alvarez-Vidal, A., Maldonado, M.J., Moreno-Montañés, J., and García-Layana, A. (2002) Retinal thickness study with optical coherence tomography in patients with diabetes. *Invest. Ophthalmol. Vis. Sci.*, **43**, 1588–1594.

22 Massin, P., Erginay, A., Haouchine, B., Mehidi, B., and Gaudric, A. (2002) Retinal thickness in healthy and diabetic subjects measured using optical coherence tomography mapping software. *Eur. J. Ophthalmol.*, **12**, 102–108.

23 Nunes, S., Pereira, I., Santos, A., Bernardes, R., and Cunha-Vaz, J. (2010) Central retinal thickness measured with HD-OCT shows a weak correlation with visual acuity in eyes with CSME. *Br. J. Ophthalmol.*, **94** (9), 1201–1204.

24 Diabetic Retinopathy Clinical Research Network (2007) The relationship between OCT-measured central retinal thickness and visual acuity in diabetic macular edema. *Ophthalmology*, **114** (3), 525–536.

25 Strøm, C., Sander, B., Larsen, N., Larsen, M., and Lund-Andersen, H. (2002) Diabetic macular edema assessed with optical coherence tomography and stereo fundus photography. *Invest. Ophthalmol. Vis. Sci.*, **43**, 241–245.

26 Wolf-Schnurrbusch, U.E., Ceklic, L., Brinkmann, C.K., Iliev, M., Frey, M., Rothenbuehler, S.P., Enzmann, V., and Wolf, S. (2009) Macular thickness measurements in healthy eyes using six different optical coherence tomography instruments. *Invest. Ophthalmol. Vis. Sci.*, **50** (7), 3432–3437.

27 Otani, T., Kishi, S., and Maruyama, Y. (1999) Patterns of diabetic macular edema with optical coherence tomography. *Am. J. Ophthalmol.*, **127**, 688–693.

28 Panozzo, G., Gusson, E., Parolini, B., and Mercanti, A. (2003) Role of optical coherence tomography in the diagnosis and follow up of diabetic macular edema. *Semin. Ophthalmol.*, **18**, 74–81.

29 Özdek, S.C., Erdinç, M.A., Grelik, G., Aydin, B., Bahçeci, U., and Hasanreisoǧlu, B. (2005) Optical coherence tomography assessment of diabetic macular edema: comparison with fluorescein angiographic and clinical findings. *Ophthalmologica*, **219**, 86–92.

30 Soliman, W., Sander, B., Hasler, P.W., and Larsen, M. (2008) Correlation between intraretinal changes in diabetic macular oedema seen in fluorescein angiographic and optical coherence tomography. *Acta Ophthalmol.*, **86**, 34–39.

31 Kang, S.W., Park, C.Y., and Ham, D.I. (2004) The correlation between fluorescein angiographic and optical coherence tomography feature in clinically significant diabetic macular edema. *Am. J. Ophthalmol.*, **137**, 313–322.

32 Ozdemir, H., Karacorlu, M., and Karacorlu, S. (2005) Serous macular detachment in

diabetic cystoid macular edema. *Acta Ophthalmol. Scand.*, **83**, 63–66.

33 Koleva-Georgieva, D. and Sivkova, N. (2009) Assessment of serous macular detachment in eyes with diabetic macula edema by use of spectral-domain optical coherence tomography. *Graefes Arch. Clin. Exp. Ophthalmol.*, **247**, 1461–1469.

34 The Diabetes Control and Complications Trial Research Group (1993) The effect of intensive treatment of diabetes on the development and progression of long-term complications in insulin-dependent diabetes mellitus. *N. Engl. J. Med.*, **329**, 977–986.

35 Gillies, M.C., Sutter, F.K., Simpson, J.M., Larsson, J., Ali, H., and Zhu, M. (2006) Intravitreal triamcinolone for refractory diabetic macular edema: two-year results of a double-masked, placebo-controlled, randomized clinical trial. *Ophthalmology*, **113**, 1533–1538.

36 Avitabile, T., Lungo, A., and Reibaldi, A. (2005) Intravitreal triamcinolone compared with macular laser grid photocoagulation for the treatment of cystoid macular edema. *Am. J. Ophthalmol.*, **140**, 695–702.

37 Arevalo, J.F., Sanchez, J.G., Fromow-Guerra, J., Wu, L., Berrocal, M.H., Farah, M.E., Cardillo, J., Rodríguez, F.J., and Pan-American Collaborative Retina Study Group (PACORES) (2009) Comparison of two doses of primary intravitreal bevacizumab (Avastin) for diffused diabetic macular oedema: result from the Pan American Collaborative Retina Study Group (PACORES) at 12 months follow up. *Graefes Arch. Clin. Exp. Ophthalmol.*, **247** (6), 735–743.

38 Massin, P., Bandello, F., Garweg, J.G., Hansen, L.L., Harding, S.P., Larsen, M., Mitchell, P., Sharp, D., Wolf-Schnurrbusch, U.E.K., Gekkieva, M., Weichselberger, A., and Wolf, S. (2010) Safety and efficacy of ranibizumab in diabetic macular edema (RESOLVE Study). A 12-month, randomized, controlled, double-masked, multicenter phase II study. *Diabetes Care*, **33**, 2399–2405.

39 Framme, C., Alt, C., Theisen-Kunde, D., and Brinkmann, R. (2009) Structural changes of the retina after conventional laser photocoagulation and selective retinal treatment (SRT) in spectral-domain optical coherence tomography. *Curr. Eye Res.*, **34**, 568–579.

40 Massin, P., Duguid, G., Erginay, A., Haouchine, B., and Gaudric, A. (2003) Optical coherence tomography for evaluating diabetic macular edema before and after vitrectomy. *Am. J. Ophthalmol.*, **135**, 169–177.

41 Shimonagono, Y., Makiuchi, R., Miyazaki, M., Doi, N., Uemura, A., and Sakamoto, T. (2007) Result of visual acuity and foveal thickness in diabetic macular edema after vitrectomy. *Jpn. J. Ophthalmol.*, **51**, 204–209.

42 Recchia, F.M., Ruby, A.J., and Carvalho Recchia, C.A. (2005) Pars plana vitrectomy with removal of internal limiting membrane in the treatment of persistent diabetic macular edema. *Am. J. Ophthalmol.*, **139**, 447–454.

43 Shah, S.P., Patel, M., Thomas, D., Aldington, S., and Laidlaw, D.A.H. (2006) Factors predicting outcome of vitrectomy for diabetic macular oedema: results of a prospective study. *Br. J. Ophthalmol.*, **90**, 33–36.

44 Dhingra, N., Sahni, J., Shipley, J., Harding, S.P., Groenewald, C.I., Pearce, A., Stanga, P.E., and Wong, D. (2005) Vitrectomy and internal limiting membrane (ILM) removal for diabetic macular edema in eyes with absent vitreo–macular traction fails to improve visual acuity: results of a 12 months prospective randomized controlled clinical trial. Presented at the Annual Meeting of the Association for Research in Vision and Ophthalmology, 1–5 May 2005, Fort Lauderdale, FL.

45 Ota, M., Tsujikawa, A., Murakami, T., Kita, M., Miyamoto, K., Sakamoto, A., Yamaike, N., and Yoshimura, N. (2007) Association between integrity of foveal photoreceptor layer and visual acuity in branch retinal vein occlusion. *Br. J. Ophthalmol.*, **91**, 1644–1649.

46 Piccolino, F.C., Longrais, R.R.D.L., Ravera, G., Eandi, C.M., Ventre, L., Abdollahi, A., and Manea, M. (2005) The foveal photoreceptor layer and visual acuity loss in central serous chorioretinopathy. *Am. J. Ophthalmol.*, **139**, 87–99.

47 Hayashi, H., Yamashiro, K., Tsujikawa, A., Ota, M., Otani, A., and Yoshimura, N. (2009) Association between foveal

photoreceptor integrity and visual outcome in neovascular age-related macular degeneration. *Am. J. Ophthalmol.*, **148** (1), 83–89.

48 Oishi, A., Otani, A., Sasahara, M., Kojima, H., Nakamura, H., Kurimoto, M., and Yoshimura, N. (2009) Photoreceptor integrity and visual acuity in cystoids macular oedema associated with retinitis pigmentosa. *Eye*, **23**, 1411–1416.

49 Wakabayashi, T., Oshima, Y., Fujimoto, H., Murakami, Y., Sakaguchi, H., Kusaka, S., and Tano, Y. (2009) Foveal microstructure and visual acuity after retinal detachment repair. Imaging analysis by Fourier-domain optical coherence tomography. *Ophthalmology*, **116**, 519–528.

50 Einbock, W., Berger, L., Wolf-Schnurrbusch, U., Fleischhauer, J., and Wolf, S. (2008) Predictive factors for visual acuity of patients with diabetic macular edema interpreted from the Spectralis HRA+OCT (Heidelberg Engineering). *Invest. Ophthalmol. Vis. Sci.*, **49**, 3432–3473.

51 Forooghian, F., Stetson, P., Scott, M., Chew, E.Y., Wong, W.T., Chukras, C., Meyerle, C.B., and Frederick, L. (2010) Relationship between photoreceptor outer segment length and visual acuity in diabetic macular edema. *Retina*, **30** (1), 63–70.

52 Otani, T., Yamaguchi, Y., and Kishi, S. (2010) Correlation between visual acuity and foveal microstructural changes in diabetic macular edema. *Retina*, **30** (5), 774–780.

53 Sakamoto, A., Nishijima, K., Kita, M., Oh, H., Tsujikawa, A., and Yoshimura, N. (2009) Association between foveal photoreceptor status and visual acuity after resolution of diabetic macular edema by pars plana vitrectomy. *Graefes Arch. Clin. Exp. Ophthalmol.*, **247**, 1325–1330.

41
Glaucoma Laser Therapy
Marc Töteberg-Harms, Peter P. Ciechanowski, and Jens Funk

41.1
Introduction

There are multiple possibilities for lowering intraocular pressure (IOP). Medical therapy is the first-line treatment (http://www.worldglaucoma.org; http://www.eugs.org, http://www.seagig.org). When medical therapy is no longer sufficient, there are many surgical ways to control IOP. Trabeculectomy is still the gold standard. The European Glaucoma Society recommends laser surgery of the trabecular meshwork (TM) before or in addition to medical therapy (see guidelines of the European Glaucoma Society, http://www.eugs.org).

In this chapter, laser trabeculoplasty (LTP), argon laser trabeculoplasty (ALT), and selective laser trabeculoplasty (SLT), and also excimer laser trabeculotomy (ELT) and cyclophotocoagulation (CPC), are described.

For about 35 years, laser therapy has been used to lower IOP [1, 2]. Different laser types could be used, such as the continuous-wave, frequency-doubled neodymium: yttrium aluminum garnet (Nd:YAG) laser (wavelength 1064 and 532 nm), argon laser (488 and 514.5 nm), diode laser (810 nm), excimer laser (308 nm), and krypton laser (647 or 568 nm) [3–6]. The most important parameter of these lasers with regard to lowering the IOP is the energy density delivered to specific tissues.

By decreasing or even replacing the need for topical medications, especially in patients with primary open-angle glaucoma (POAG), laser therapy can reduce systemic [7] and local [8] side effects, or can even avoid the need for surgery.

41.2
Laser Trabeculoplasty (LTP) and Selective Laser Trabeculoplasty (SLT)

In 1973, Krasnov [1] reported the first temporary lowering of IOP via a Q-switched ruby laser. The laser was meant to increase the outflow of aqueous humor within the TM [2, 9].

Different types of lasers are used for LTP: argon laser (488 or 514.5 nm), solid state laser (532 nm) or diode laser (810 nm).

There are several theories about the mechanism of LTP (including ALT and SLT), but it is not yet definitely known. From a biochemical point of view, LTP and SLT could increase the replication of cells which are involved in the outflow of the TM [10–13]. Another theory focuses on the mechanical distortion of the TM.

With ALT, a series of 50 μm laser beam spots (about 20–25 spots within 90° of the TM) with an exposure time of 0.1 s and a power of up to 1000 mW are placed 180–360° around the TM to increase the aqueous humor outflow. If only 180° is used, it is advisable to re-evaluate the patient after 4–6 weeks. This intervention causes thermal coagulative damage and can decrease the IOP by up to 9–10 mmHg [14]. However, filtering surgery should be considered first, if a greater decrease in IOP is needed.

The SLT device in clinical use is a frequency-doubled, Q-switched Nd:YAG laser (532 nm) [15]. Usually 180° or 360° of the TM are treated with 50–100 spots. The spot size is 400 μm, the pulse duration is 3 ns and the power is 0.6–1.0 mJ per pulse on average.

In contrast to ALT, SLT does not cause significant coagulative damage. Despite this, because of a wider beam and a very short exposure time, each SLT pulse delivers less than 0.1% energy in comparison with ALT.

In both laser procedures (ALT and SLT), an IOP spike within the first few hours after treatment can occur. Local medication or systemic carboanhydrase inhibitors can be used to prevent this peak.

For the laser treatment, a gonioscopic lens (e.g., Goldmann three-mirror lens, Latina lens or other gonioscopic lenses) is used under topical anesthesia to focus the laser beam to the TM.

After LTP, most patients require continuing medication; about 30–50% can reduce their level of treatment within the following months.

ALT causes scars in the TM, hence repeated coagulations are not recommended, whereas SLT does not and can be repeated [16].

Both ALT and SLT are favorable ways of lowering the eye pressure in patients who need their IOP to be reduced by around 5–10 mmHg.

41.3
Excimer Laser Trabeculotomy (ELT)

In addition to the methods mentioned above, ELT is mostly used in POAG and also in pseudoexfoliative glaucoma. The excimer laser is short pulsed (80 ns) and uses xenon chloride at a wavelength of 308 nm. ELT delivers photoablative energy and decreases the outflow obstruction at the inner wall of Schlemm's canal and the juxtacanalicular TM [17, 18].

This intervention can be performed with topical anesthesia either in combination with cataract surgery (Phako-ELT) or as monotherapy. A paracentesis is needed to insert a fiber-optic probe and an endoscope or a goniolens is used for observation. Ten pores are usually created in Schlemm's canal.

ELT is an elegant way of reducing the IOP, in particular in combination with cataract surgery. Usually the IOP can be reduced by up to 35%, depending on the level of initial IOP [19–21].

41.4
Cyclophotocoagulation (CPC)

For CPC, different lasers are used: trans-sclerally a 1064 nm wavelength Nd:YAG laser or an 810 nm diode laser [22].

This procedure destroys the ciliary processes and reduces the production of aqueous humor. Because other laser therapies have fewer complications, CPC is often considered just if other attempts at IOP reduction have failed.

For the intervention, short anesthesia or retro-bulbar block is usually required. It is recommended to treat at 270° with 16–18 applications each with an exposure time of 2500 ms and a power of 0.5–2.75 J. The tip is placed at a distance of 0.5–1.5 mm from the limbus and held as perpendicular to the sclera as possible.

In special cases, it is also possible to perform CPC endoscopically (endo-cyclophotocoagulation).

Cyclophotocoagulation can be repeated several times, but because of its possible complications, including inflammation, pain, uveitis, reduction of visual acuity, hypotony, and phthisis bulbi or hemorrhage, the indication has to be carefully considered.

References

1 Krasnov, M.M. (1973) Laseropuncture of anterior chamber angle in glaucoma. *Am. J. Ophthalmol.*, **75** (4), 674–678.
2 Wise, J.B. and Witter, S.L. (1979) Argon laser therapy for open-angle glaucoma. A pilot study. *Arch. Ophthalmol.*, **97** (2), 319–322.
3 Chung, P.Y., et al. (1998) Five-year results of a randomized, prospective, clinical trial of diode vs argon laser trabeculoplasty for open-angle glaucoma. *Am. J. Ophthalmol.*, **126** (2), 185–190.
4 Englert, J.A., et al. (1997) Argon vs. diode laser trabeculoplasty. *Am. J. Ophthalmol.*, **124** (5), 627–631.
5 Guzey, M., et al. (1999) Effects of frequency-doubled Nd:YAG laser trabeculoplasty on diurnal intraocular pressure variations in primary open-angle glaucoma. *Ophthalmologica*, **213** (4), 214–218.
6 Spurny, R.C. and Lederer, C.M. Jr. (1984) Krypton laser trabeculoplasty. A clinical report. *Arch. Ophthalmol.*, **102** (11), 1626–1628.
7 Nelson, W.L., et al. (1986) Adverse respiratory and cardiovascular events attributed to timolol ophthalmic solution, 1978–1985. *Am. J. Ophthalmol.*, **102** (5), 606–611.
8 Osborne, S.A., et al. (2005) Alphagan allergy may increase the propensity for multiple eye-drop allergy. *Eye (Lond.)*, **19** (2), 129–137.
9 Wickham, M.G. and Worthen, D.M. (1979) Argon laser trabeculotomy: long-term follow-up. *Ophthalmology*, **86** (3), 495–503.
10 Alvarado, J.A., et al. (2005) A new insight into the cellular regulation of aqueous outflow: how trabecular meshwork endothelial cells drive a mechanism that

11. Ticho, U., et al. (1978) Argon laser trabeculotomies in primates: evaluation by histological and perfusion studies. *Invest. Ophthalmol. Vis. Sci.*, **17** (7), 667–674.
12. Van Buskirk, E.M., et al. (1984) Argon laser trabeculoplasty. Studies of mechanism of action. *Ophthalmology*, **91** (9), 1005–1010.
13. Wickham, M.G., Worthen, D.M., and Binder, P.S. (1977) Physiological effects of laser trabeculotomy in rhesus monkey eyes. *Invest. Ophthalmol. Vis. Sci.*, **16** (7), 624–628.
14. Starita, R.J., et al. (1984) The effect of repeating full-circumference argon laser trabeculoplasty. *Ophthalmic Surg.*, **15** (1), 41–43.
15. Latina, M.A., et al. (1998) Q-switched 532-nm Nd:YAG laser trabeculoplasty (selective laser trabeculoplasty): a multicenter, pilot, clinical study. *Ophthalmology*, **105** (11), 2082–2088; discussion 2089–90.
16. Damji, K.F., et al. (1999) Selective laser trabeculoplasty v argon laser trabeculoplasty: a prospective randomised clinical trial. *Br. J. Ophthalmol.*, **83** (6), 718–722.
17. Walker, R. and Specht, H. (2002) Theoretical and physical aspects of excimer laser trabeculotomy (ELT) *ab interno* with the AIDA laser with a wavelength of 308 nm. *Biomed. Tech. (Berl.)*, **47** (5), 106–110 (in German).
18. Babighian, S., Rapizzi, E., and Galan, A. (2006) Efficacy and safety of *ab interno* excimer laser trabeculotomy in primary open-angle glaucoma: two years of follow-up. *Ophthalmologica*, **220** (5), 285–290.
19. Pache, M., Wilmsmeyer, S., and Funk, J. (2006) Laser surgery for glaucoma: excimer-laser trabeculotomy. *Klin. Monbl. Augenheilkd.*, **223** (4), 303–307 (in German).
20. Wilmsmeyer, S., Philippin, H., and Funk, J. (2006) Excimer laser trabeculotomy: a new, minimally invasive procedure for patients with glaucoma. *Graefes Arch. Clin. Exp. Ophthalmol.*, **244** (6), 670–676.
21. Töteberg-Harms, M., et al. (2011) One-year results after combined cataract surgery and excimer laser trabeculotomy for elevated intraocular pressure. *Ophthalmologe*, 2011 Feb 27. [Epub ahead of print] PMID: 21359550 (in German).
22. Beckman, H. and Sugar, H.S. (1973) Neodymium laser cyclocoagulation. *Arch. Ophthalmol.*, **90** (1), 27–28.

42
Ocular Blood Flow Measurement Methodologies and Clinical Application

Yochai Z. Shoshani, Alon Harris, and Brent A. Siesky

42.1
Background and Significance

Open-angle glaucoma (OAG) is a chronic progressive optic neuropathy representing one of the leading causes of blindness in the United States and worldwide [1]. Elevated intraocular pressure (IOP) is the only current modifiable risk factor [2]; however, many patients with OAG continue to experience progression of the disease despite lowered IOP. This has led to the identification of other risk factors, including low ocular perfusion pressure, vascular dysregulation, and ischemia. Assessment of blood flow and its current clinical implications are discussed in this chapter.

42.2
Blood flow Methodologies

42.2.1
Color Doppler Imaging

Color Doppler imaging (CDI) measures blood flow in the retrobulbar blood vessels supplying ocular tissues. CDI combines two-dimensional structural ultrasound images with velocity measurements derived from the Doppler shift of sound waves reflected from erythrocytes as they travel through blood vessels. The peak systolic velocity (PSV) and end diastolic velocity (EDV) are measured and used to calculate Pourcelot's index of resistivity (RI), a marker of downstream resistance, as RI = (PSV − EDV)/PSV [3].

42.2.2
Laser Doppler Flowmetry and Scanning Laser Flowmetry

Laser Doppler flowmeter measures retinal capillary blood flow noninvasively by utilizing a modified fundus camera combined with a computer system. The

Heidelberg retinal flowmeter (HRF) is one commercially available system that combines laser Doppler flowmetry with scanning laser tomography, providing a two-dimensional map of blood flow to the optic nerve and surrounding retina. This technique is most sensitive to microvascular blood flow changes in the superficial layers of the optic nerve head.

42.2.3
Retinal Vessel Analyzer

The retinal vessel analyzer (RVA), which is composed of a fundus camera, a video camera, a monitor, and a computer with specialized software, permits continuous monitoring of vessels in real time with a maximum frequency of 50 Hz. Each vessel has a specific transmittance profile that is based on the absorbing properties of hemoglobin. The diameter of a vessel is determined using an algorithm converting this profile.

42.2.4
Ocular Pulse Amplitude and Pulsatile Ocular Blood Flow

Dynamic contour tonometry (DCT) is a noninvasive and direct method of continuously measuring IOP over time [4]. The difference between the highest and lowest IOP measurements defines the ocular pulse amplitude (OPA), which is believed to be a result of the changing blood volume of the eye. Similarly to DCT, the sinusoidal pattern of pulsatile ocular blood flow (POBF) can be estimated by quantifying the changes in ocular volume and pressure during the cardiac cycle using a variety of devices.

42.2.5
Laser Speckle Method (Laser Speckle Flowgraphy)

The laser speckle method is a technique based on the interference phenomenon, observed when coherent light sources are scattered by a diffusing surface. The scattered laser light forms a "speckle" pattern, which is imaged and then statistically characterized. The variation in the structure of the pattern changes with the velocity of erythrocytes in the retina.

42.2.6
Digital Scanning Laser Ophthalmoscope Angiography

Retinal blood flow can be directly visualized using sodium fluorescein dye, and choroidal flow using indocyanine green dye, during digital scanning laser ophthalmoscope (SLO) angiography. The fluorescent compound is injected into a vein and observed as it fills the ocular vasculature. A time-based stream of measured intensities is then used to construct a video signal [5]. The amount of time between the first appearance of dye in a retinal artery and the associated vein is called the arteriovenous passage time (AVP), which has been shown to be very sensitive to small changes in blood flow through the retinal vascular bed, and measurements can be localized to specific quadrants of the retina.

42.2.7
Doppler Optical Coherence Tomography

Doppler optical coherence tomography (DOCT) uses the Doppler frequency shift of backscattered light reflecting off red blood cells in the vasculature to measure blood flow in the retinal arteries in real time. Using Fourier domain OCT, it is possible to combine the high-resolution cross-sectional imaging of OCT with laser Doppler imaging to capture information from the retinal blood vessels in three dimensions in a time period that is only a fraction of the cardiac cycle [6]. DOCT is used to measure both the velocity and volumetric flow rate in the retinal branch vessels (Figure 42.1).

42.2.8
Retinal Oximetry

Retinal oximetry is the noninvasive measurement of hemoglobin oxygen saturation in the retinal vasculature. A linear relationship exists between the oxygen saturation and the ratio of optical densities measured at the two wavelengths. This allows for direct and quantitative mapping of retinal biochemistry [7]. Retinal oximetry is a

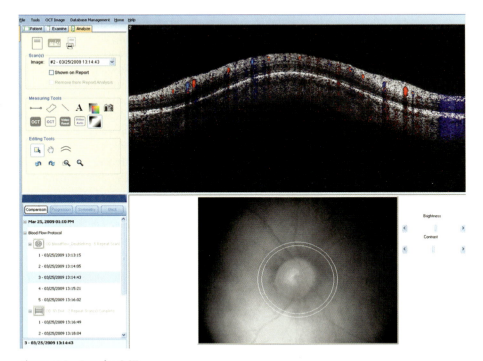

Figure 42.1 Doppler OCT.

novel technique that may help investigators understand the metabolic changes that may contribute to the pathology of glaucoma.

42.2.9
Newly Developed Techniques

In addition to these established techniques, recent developments in high-resolution magnetic resonance imaging (MRI) and its application to image anatomy, physiology, and function in the retina may provide future avenues for assessing the ocular circulation [8]. MRI offers unique advantages over existing retinal imaging techniques, including the ability to image multiple layers without depth limitation and to provide multiple clinically relevant data in a single setting. Functional laser Doppler flowmetry (FLDF), which is the coupling between visually evoked neural activity and vascular activity within the neural tissue of the optic nerve (neurovascular coupling), may also be useful [9]. The application of FLDF in patients with ocular hypertension and early glaucoma demonstrates that the visually evoked hyperemic responses are significantly depressed even when neural retinal activity may still be relatively preserved [9].

42.3
Relationship Between Blood Flow and Visual Function

The retrobulbar circulation of the orbit has been the most studied vascular bed in glaucoma, with dozens of research investigations demonstrating altered blood flow parameters in patients with glaucoma. A correlation between glaucomatous progression and decreased blood flow velocities in the A correlation between glaucomatous progression and decreased blood flow velocities in the short posterior ciliary arteries (SPCAs), the small retrobulbar vessels that supply the optic nerve head, decreased EDV in the central retinal artery (CRA), higher RI in the CRA and SPCA, and a high RI in the ophthalmic artery (OA) have been reported [10, 11]. Both blood flow and velocity in the retinal microcirculation were also demonstrated to be significantly decreased in eyes with worse damage compared with their fellow eyes with less damage [12]. Blood flow at the neuroretinal rim has been shown to correspond to regional VF defects in patients with normal-tension glaucoma (NTG) [13]. Circadian fluctuations of OPP were also found to correlate significantly with VF damage in a group of 132 patients with NTG [14]. Although a causal relationship has yet to be established, ocular blood flow likely plays an integral role in both the pathophysiology and progression of glaucomatous vision loss, and therefore warrants further investigation.

42.4
Relationship Between Blood Flow and Optic Nerve Structure

Several studies have investigated hemodynamic alterations in the SPCA and other retrobulbar blood vessels and also the retina and choroid in relation to ocular structure. Loss of capillaries was found in the neural tissue of the optic disc in glaucomatous optic atrophy [15]. Characteristic glaucomatous excavation of the optic disc has also been correlated with alterations in the retinal and retrobulbar circulations [16]. Neuroretinal rim damage was found to be associated with a local reduction in blood flow in glaucoma patients [17]. Ocular blood flow was found to be associated with thinning in the retinal nerve fiber layer (RNFL) [18]. A reduction in blood flow with thinning of the RNFL would be expected because capillary loss accompanies loss of neural tissue, which likely has decreased metabolic demands. However, several of the above studies found an increase in local retinal blood flow with thinning of the RNFL. This divergence may be explained by an autoregulatory compensatory response. Initially, blood flow may increase to a region of RNFL thinning; as glaucomatous damage progresses, blood flow may gradually decrease. Unfortunately, no study has examined this relationship in advanced glaucoma patients.

It has been suggested that glaucomatous optic neuropathy may be linked to unstable blood flow and reperfusion injury from either perfusion pressure fluctuations or dysfunctional autoregulation [19]. Larger fluctuations in perfusion and arterial pressure were correlated with reduced RNFL thickness. This suggests that faulty autoregulation during fluctuations in ocular perfusion pressure may cause daily ischemic damage, succeeded by reperfusion injury. Recent evidence from the Thessaloniki Eye Study [20] found that patients on antihypertensive therapy with diastolic blood pressure (DBP) less than 90 mmHg was positively correlated with cup area and cup-to-disc ratio when compared with both patients with high DBP and untreated patients with normal DBP. Low perfusion pressure was also positively associated with cup area and cup-to-disc ratio. The results remained consistent after adjusting for age, IOP, cardiovascular disease, diabetes, and duration of antihypertensive treatment. The findings suggest that blood pressure status is an independent risk factor for glaucomatous damage [20]. Systemic hypertension has also been linked as a risk factor for development of glaucoma. The association of systemic hypertension as a risk factor may be related to its association with elevated IOP. Because the Thessaloniki study adjusted for IOP, its results may more appropriately portray the relationship between blood pressure and cupping [20]. The translamina cribrosa pressure difference, the difference between the IOP and the pressure of the cerebrospinal fluid (CSF) around the optic nerve, may be an important consideration as CSF pressure is likely correlated with blood pressure and was found to be lowered in OAG and NTG patients [21].

The 2009 report of the World Glaucoma Association stressed that only larger, longitudinal multi-center investigations can confirm the relationship between ocular blood flow and visual function and optic nerve structure hinted at in the available pilot research. It is therefore vital that we determine at what levels of (and fluctuations in) IOP and ocular perfusion pressure there is an increase in the probability of glaucoma progression and how altered ocular blood flow may play a role in the development and progression of glaucoma [22].

References

1 Quigley, H.A. (1996) Number of people with glaucoma worldwide. *Br. J. Ophthalmol.*, **80** (5), 389–393.
2 Weinreb, R.N. (2001) A rationale for lowering intraocular pressure in glaucoma. *Surv. Ophthalmol.*, **45** (S4), s335–s336.
3 Pourcelot, L. (1974) Applications of cliniques de l'examinen Doppler transcutane. *INSERM*, **34**, 213–240.
4 Punjabi, O.S., Kniestedt, C., Stamper, R.L., and Lin, S.C. (2006) Dynamic contour tonometry: principle and use. *Clin. Exp. Ophthalmol.*, **34** (9), 837–840.
5 Wolf, S., Arend, O., Toonen, H., et al. (1991) Retinal capillary blood flow measurement with a scanning laser ophthalmoscope. Preliminary results. *Ophthalmology*, **98**, 996–1000.
6 Wehbe, H.M., Ruggeri, M., Jiao, S., Gregori, G., Puliafito, C.A., and Zhao, W. (2007) Automatic retinal blood flow calculation using spectral domain optical coherence tomography. *Opt. Express*, **15**, 15193–15206.
7 Alabboud, I., Muyo, G., Gorman, A., et al. (2007) New spectral imaging techniques for blood oximetry in the retina. *Proc. SPIE*, **6631**, 6310.
8 Duong, T.Q., Pardue, M.T., Thulé, P.M., et al. (2008) Layer-specific anatomical, physiological and functional MRI of the retina. *NMR Biomed.*, **21** (9), 978–996.
9 Riva, C.E. and Falsini, B. (2008) Functional laser Doppler flowmetry of the optic nerve: physiological aspects and clinical applications. *Prog. Brain Res.*, **173**, 149–163.
10 Zeitz, O., Galambos, P., Wagenfeld, L., et al. (2006) Glaucoma progression is associated with decreased blood flow velocities in the short posterior ciliary artery. *Br. J. Ophthalmol.*, **90** (10), 1245–1248.
11 Yamazaki, Y. and Drance, S. (1997) The relationship between progression of visual field defects and retrobulbar circulation in patients with glaucoma. *Am. J. Ophthalmol.*, **124**, 287–295.
12 Grunwald, J.E., Piltz, J.E., Hariprasad, S.M., and DuPont, J. (1998) Optic nerve and choroidal circulation in glaucoma. *Invest. Ophthalmol. Vis. Sci.*, **39** (12), 2329–2336.
13 Sato, E.A., Ohtake, Y., Shinoda, K., Mashima, Y., and Kimura, I. (2006) Decreased blood flow at neuroretinal rim of optic nerve head corresponds with visual field deficit in eyes with normal tension glaucoma. *Graefes Arch. Clin. Exp. Ophthalmol.*, **244** (7), 795–801.
14 Choi, J., Kim, K.H., Jeong, J., Cho, H.S., Lee, C.H., and Kook, M.S. (2007) Circadian fluctuation of mean ocular perfusion pressure is a consistent risk factor for normal-tension glaucoma. *Invest. Ophthalmol. Vis. Sci.*, **48** (1), 104–111.
15 Gottanka, J., Kuhlmann, A., Scholz, M., Johnson, D.H., and Lütjen-Drecoll, E. (2005) Pathophysiologic changes in the optic nerves of eyes with primary open angle and pseudoexfoliation glaucoma. *Invest. Ophthal. Vis. Sci.*, **46**, 4170–4181.
16 Hafez, A.S., Bizzarro, R.L.G., and Lesk, M.R. (2003) Evaluation of optic nerve head and peripapillary retinal blood flow in glaucoma patients, ocular hypertensives, and normal subjects. *Am. J. Ophthalmol.*, **136**, 1022–1031.
17 Nicolela, M.T., Drance, S.M., Rankin, S.J.A., Buckley, A.R., and Walman, B.E. (1996) Color Doppler imaging in patients with asymmetric glaucoma and unilateral visual field loss. *Am. J. Ophthalmol.*, **121**, 502–510.
18 Grieshaber, M.C., Mozaffarieh, M., and Flammer, J. (2007) What is the link between vascular dysregulation and glaucoma? *Surv. Ophthalmol.*, **52**, S144–S154.
19 Vulsteke, C., Stalmans, I., Fieuws, S., and Zeyen, T. (2008) Correlation between ocular pulse amplitude measured by dynamic contour tonometer and visual field defects. *Graefes Arch. Clin. Exp. Ophthalmol.*, **246** (4), 559–565.
20 Topouzis, F., Coleman, A.L., Harris, A., et al. (2006) Association of blood pressure status with the optic disk structure in non-glaucoma subjects: the Thessaloniki Eye Study. *Am. J. Ophthalmol.*, **142**, 60–67.
21 Berdahl, J.P., Allingham, R.R., and Johnson, D.H. (2008) Cerebrospinal fluid pressure is

decreased in primary open-angle glaucoma. *Ophthalmology*, **115**, 763–768.

22 Weinreb, R.N. and Harris, A. (eds.) (2009) Section V. What do we still need to know?, in *Blood Flow in Glaucoma: the Sixth Consensus Report of the World Glaucoma Association*, Kugler Publications, Amsterdam, pp. 161–163.

43
Photodynamic Modulation of Wound Healing in Glaucoma Filtration Surgery

Salvatore Grisanti

43.1
Introduction

Glaucoma is one of the major causes of blindness in the world. Different factors contribute to the progression of the disease and visual field defects (Figure 43.1a and b). An elevated intraocular pressure plays a pivotal role in most cases. The outflow of aqueous humor through its natural pathway at the trabecular meshwork is somehow reduced (Figure 43.1c). Trabeculectomy is the most frequently applied surgical method to reduce intraocular pressure in these cases (Figure 43.1d and e). The wound healing process and the developing fibrosis at the site of the filtration area (Figure 43.1f), however, lead to failure in about 30% of the cases within 6–8 weeks after surgery [1]. Characteristically, fibroblasts from Tenon's capsule and episclera lead to a fibro-proliferative response involving and closing the created fistula [2–12]. To increase the success rates of filtration surgery, agents such as mitomycin-C and 5-fluorouracil [4, 16–34] have been introduced and are used perioperatively as an antifibrotic therapy. Despite their positive effect on filtration surgery and success rates in cases with poor surgical prognosis, diffusion into adjacent ocular tissues cause toxic effects [35, 36]. New surgical complications, such as hypotony maculopathy, blebitis, endophthalmitis, and an increased incidence of already known postoperative complications, have hampered clinical use and stimulated a search for clinically less harmful alternatives [37, 38].

43.2
Photodynamic Therapy

Photodynamic therapy (PDT) is an alternative method for the treatment of localized pathologies, such as skin cancer [42]. Specific activation of pertinent drugs at the targeted area should avoid side effects. PDT has also been evaluated for some distinct

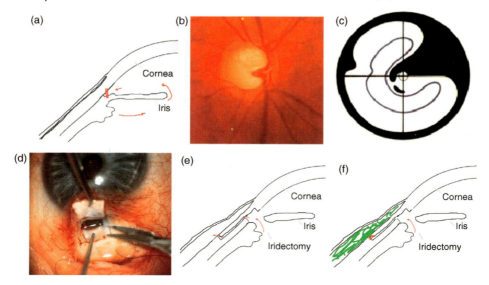

Figure 43.1 Filtration surgery to reduce IOP in glaucoma. (a) A reduction of aqueous humor outflow leads to elevated IOP. Chronic elevation of IOP leads to glaucomatous changes of the optic nerve (b) and visual field defects (c). Filtration surgery (d) allows outflow of aqueous humor into the subconjunctival space (e). (f) Postoperative wound healing and fibrosis (green).

ophthalmic diseases, such as ocular tumors, choroidal and corneal neovascularization, proliferative vitreoretinal disorders, and postoperative fibrosis in glaucoma surgery [41, 43–47]. Hill *et al.* investigated the feasibility of PDT in a rabbit model of filtration surgery. Using ethyletiopurin, a photosensitizer traditionally delivered by intravenous injection, they showed that subconjunctival delivery could have an impact on filtering bleb survival [41].

At about the same time, we had already started to evaluate BCECF-AM [2′,7′-bis (2-carboxyethyl)-5-(and -6)-carboxyfluorescein acetoxymethyl ester] in *in vitro* [39] and *in vivo* [40] studies in order to induce an antiproliferative photodynamic effect.

43.3
Photodynamic Modulation of Wound Healing in Glaucoma Filtration Surgery

43.3.1
Preclinical Studies

BCECF-AM is a cell membrane-permeable compound (Figure 43.2) rendered membrane impermeable and fluorescent upon cleavage by intracellular

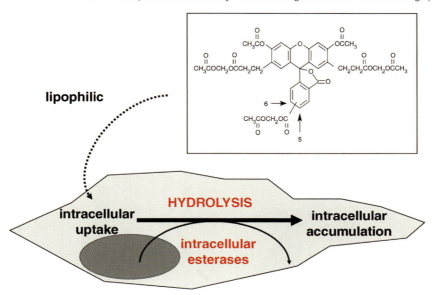

Figure 43.2 The fluorophore BCECF-AM [2′,7′-bis(2-carboxyethyl)-5-(and -6)-carboxyfluorescein acetoxymethyl ester] is lipophilic and therefore cell membrane permeable. After intracellular uptake, esterases and hydrolysis are responsible for the conversion into a lipophobic molecule and intracellular accumulation.

esterases [48–50]. Exposure of cells having incorporated BCECF-AM to light at an appropriate wavelength leads to cellular photoablation (Figure 43.3). PDT is a selective and localized treatment based on the photosensitized oxidation of biological matter [39–42]. A photosensitizer can be used as a mediator of controlled light-induced cell toxicity. The mechanism of action is reported to be due to photooxidative reactions of type I and II. Oxygen is transformed into singlet oxygen, which oxidizes amino acids, nucleic acids, and unsaturated fatty acids. Within hours, the cytotoxic effect is seen at cell membranes (blebs), plasma membranes, mitochondria, lysosomes, and nuclei. Selective activation of the photosensitizer by application of light at an appropriate wavelength limits the drug effect to a selected area [40, 41].

BCECF-AM activated at the appropriate wavelength has been shown to induce cellular damage (Figure 43.4) and inhibit the proliferation of human Tenon's fibroblasts *in vitro* after treatment with 10 µg and irradiation for 10 min [39] (Figure 43.5). In experimental eyes of pigmented rabbits, it was shown to inhibit scarring effectively after filtration surgery and demonstrated a dose–response curve from 40 to 100 µg after irradiation for 10 min [40]. BCECF-AM was injected preoperatively into the subconjunctival space (Figure 43.6a). Thereafter, similarly to surgery in patients, the conjunctiva was opened and the sclera and Tenon's capsule were exposed to illumination to induce the photoreactive effect

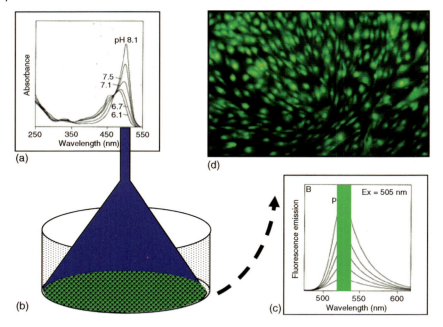

Figure 43.3 Photoactivation of intracellular BCECF-AM with diffuse blue light. (a) BCECF optimal absorbance ranges at a wavelength of 490 nm. (b) Cultured cells previously incubated with BCECF-AM illuminated with diffuse blue light (490 nm) have a peak emission (c) at a wavelength of 505 nm and (d) show a green fluorescence.

(Figure 43.6b). The postoperative phase did not show toxic effects or inappropriate postoperative inflammation (Figure 43.6c and d). The intraocular pressure was significantly reduced for a prolonged time compared with surgery without cellular photoablation. Macroscopic and histological analysis of the surgical site revealed less fibrosis (Figure 43.7).

43.3.2
Clinical Studies

The feasibility, efficacy, and safety shown in the animal experiments led to the first clinical application. An open-label phase II study was initiated to investigate the impact of cellular photoablation mediated by BCECF-AM on the postoperative fibrosis after filtration surgery in human glaucomatous eyes with poor prognosis. All patients had uncontrolled glaucoma due to high intraocular pressure (IOP) despite maximum tolerable medical therapy. Only eyes with advanced glaucomatous damage of the optic nerve head were included after approval by the institutional ethics review board. Each patient gave written informed consent after the purpose of the study and the method of surgery had been explained in detail.

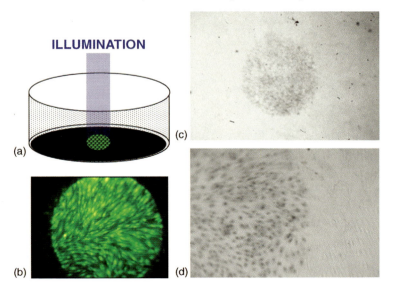

Figure 43.4 Photodynamic effect of intracellular BCECF-AM. (a) Spotlight illumination (diffuse blue light) of a cultured fibroblast monolayer previously incubated (5 min) with BCECF-AM (10 μg). (b) BCECF-AM is activated within the spot of illuminated cells. (c) Trypan blue exclusion test reveals the damaged cells within the round spot that had been previously illuminated. (d) Magnification of the spot edge reveals the demarcation line between damaged (stained) and unaffected (surrounding) cells.

The fluorescent probe BCECF-AM was provided by the pharmacy of the University of Lübeck in tuberculin syringes (Figure 43.8a). A dose of 80 μg was diluted in 300 μl of balanced salt solution and stored at $-70\,°C$. Fifteen minutes prior to surgery, the eye received a single subconjunctival injection of 80 μg of BCECF-AM solution in one superior quadrant where the trabeculectomy was to be performed (Figure 43.8b). Following a limbus-based conjunctival flap, the episcleral Tenon's capsule and the subconjunctival tissue were irradiated. Illumination with diffuse blue light (450–490 nm; $\sim 51.9 \times 10^3\,\text{cd}\,\text{m}^{-2}$) was performed for 8 min (Figure 43.8c). Photoactivating light was delivered by a portable Zeiss lamp equipped with a blue filter fixed at a distance of about 2 cm from the episclera. Light was focused to an area of about 20 mm in diameter, comprising the subconjunctival and episcleral tissue and sparing the cornea, which in addition was covered with tape. The illuminated area encompassed the consequently produced scleral flap of 3.5×3.5 mm (Figure 43.8d). A standard trabeculectomy and a full thickness iridectomy followed.

The application of BCECF-AM and photoactivation in this group of patients resulted in successful and prolonged reduction of IOP. Slit-lamp microscopy revealed no detectable damage of the conjunctiva or cornea (Figure 43.9). As BCECF-AM is a lipophilic drug, it may diffuse into other tissues adjacent the area of subconjunctival injection. However, subconjunctival injection of BCECF-AM

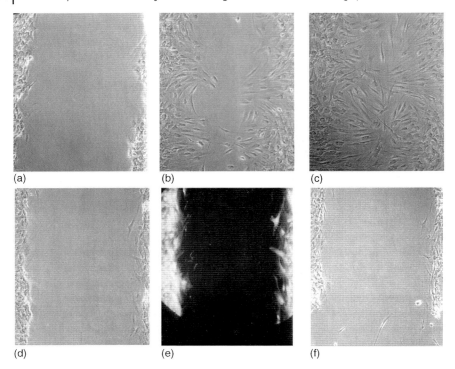

Figure 43.5 Photodynamic effect on an *in vitro* wound model. Photomicrographs showing a monolayer of fibroblasts at time 0 after creating a "wound" by scratching (a, d). The same site as in (a) shown after 12 h (b) and 24 h (c) when remaining untreated. (c) After 24 h the previous "wound" is covered by proliferating cells. (e) Illumination of the same site as in (d) at time 0. (f) Same site as in (d) and (e) after 96 h. The "wound" was not closed by cells.

followed by illumination and surgery did not cause a conjunctival defect. Owing to the promising results after more than 1 year of follow-up, additional patients were included. A subsequent study comprising 42 human glaucomatous eyes confirmed the previous results, showing the effectiveness of the approach. The clinical safety and tolerability were represented by a functioning filtering bleb with prolonged filtration, no signs of local toxicity or intraocular inflammation, and the lack of any discomfort or adverse effects for the patient.

Cellular photoablation seems to be an effective therapeutic approach to control postoperative fibrosis in human glaucomatous eyes with poor surgical prognosis. Various parameters such as light dose, light application, wavelength, irradiation area, total dose of the dye and multiple dosing may be altered in the future to improve the antifibrotic effect of PDT during glaucoma surgery. The safety and efficacy however need to be tested in randomized controlled clinical studies.

The complex issue of modulating the wound healing process after glaucoma filtration surgery has been the target of many experimental and clinical studies. So

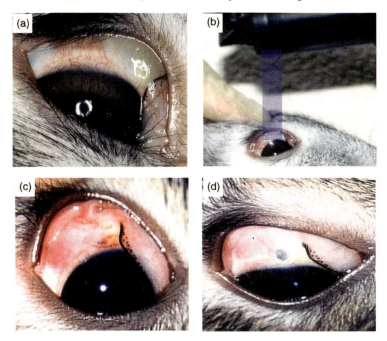

Figure 43.6 PDT in an *in vivo* glaucoma filtration surgery model. (a) The anterior segment of a rabbit eye after subconjunctival injection of BCECF-AM solution displaying a subconjunctival bleb filled with the yellowish solution. (b) Photographs reveal the intraoperative illumination (450–490 nm; 51.9×10^3 cd m^{-2}; 10 min). The same eye is shown (c) 1 day and (d) 2 weeks after surgery and PDT.

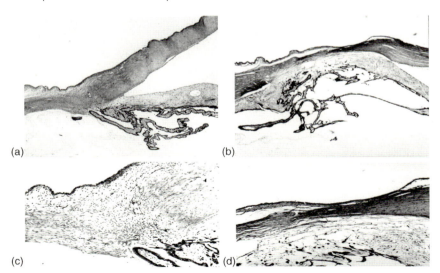

Figure 43.7 PDT in an *in vivo* glaucoma filtration surgery model. Photomicrographs of the surgical site 10 days after filtering intervention without (a, c) adjuvans and combined with photodynamic wound modulation (b, d).

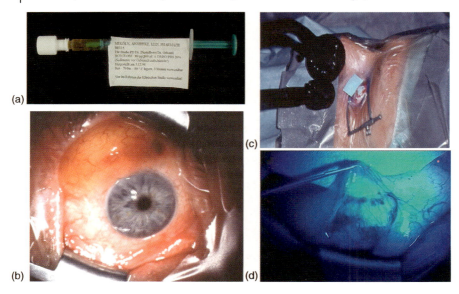

Figure 43.8 PDT in clinical glaucoma filtration surgery. (a) BCECF-AM in a syringe (80 µg diluted in 300 µl of BSS solution) as provided by the pharmacy. (b) Intraoperative site after a single subconjunctival injection of BCECF-AM 15 min. preoperatively. Intraoperative situation depicting the illumination with diffuse blue light (450–490 nm; 51.9×10^3 cd m^{-2}; 8 min).

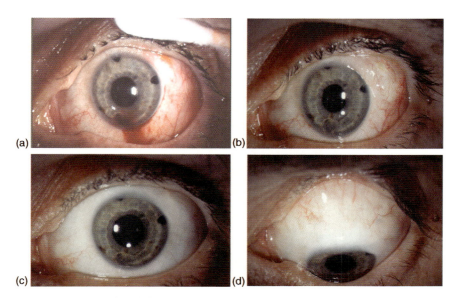

Figure 43.9 PDT in clinical glaucoma filtration surgery. Photographs of the anterior segment 1 (a), 19 (b), and 139 days (c, d) after surgery.

far, no completely suitable method has been established to achieve satisfactory long-term postoperative surgical results with only minimal or no side-effects for the patient, but PDT with BCECF-AM has proved to be a promising approach.

References

1. Diestelhorst, M., Khalili, M.A., and Krieglstein, G.K. (1999) Trabeculectomy: a retrospective follow-up of 700 eyes. *Int. Ophthalmol.*, **22**, 211–220.
2. Maumenee, A.E. (1960) External filtering operations for glaucoma: the mechanisms of function and failure. *Trans. Am. Ophthalmol. Soc.*, **58**, 319–328.
3. Addicks, E.M., Quigley, H.A., Green, W.R., and Robin, A.L. (1983) Histologic characteristics of filtering blebs in glaucomatous eyes. *Arch. Ophthalmol.*, **101**, 795–798.
4. Hitchings, R.A. and Grierson, I. (1983) Clinicopathological correlation in eyes with failed fistulizing surgery. *Trans. Ophthalmol. Soc. U. K.*, **103**, 84–88.
5. Teng, C.C., Chi, H.H., and Katzin, H.M. (1959) Histology and mechanism of filtering operations. *Am. J. Ophthalmol.*, **47**, 16–34.
6. Jampel, H.D., McGuigan, L.J.B., Dunkelberger, G.R., L'Hernault, N.L., and Quigley, H.A. (1988) Cellular proliferation after experimental glaucoma filtration surgery. *Arch. Ophthalmol.*, **106**, 89–94.
7. Costa, V.P., Spaeth, G.L., Eiferman, R.A., and Orengo-Nania, S. (1993) Wound healing modulation in glaucoma filtration surgery. *Ophthalmic Surg.*, **24**, 152–170.
8. Heuer, D.K., Parrish, R.K. II, Gresel, M.G., Hodapp, E., Palmberg, F.F., and Anderson, D.R. (1984) 5-Fluorouracil and glaucoma filtration surgery. II. A pilot study. *Ophthalmology*, **91**, 384–394.
9. Joseph, J.P., Miller, M.H., and Hitchings, R.A. (1988) Wound healing as a barrier to successul filtration surgery. *Eye*, **2**, S113–S123.
10. Skuta, G.L. and Parrish, R.K. II (1987) Wound healing in glaucoma filtering surgery. *Surv. Ophthalmol.*, **32**, 149–170.
11. Tahery, M.M. and Lee, D.A. (1989) Review: pharmacologic control of wound healing in glaucoma filtering surgery. *J. Ocular Pharmacol.*, **5**, 155–179.
12. Weinreb, R.N. (1994) Wound healing in filtration surgery. *Curr. Opin. Ophthalmol.*, **13**, 8–25.
13. Starita, R.J., Fellmann, R.L., Spaeth, G.L., Poryzees, E.M., Greenidge, K.C., and Traverso, C.E. (1985) Short- and long-term effects of postoperative corticosteroids on trabeculectomy. *Ophthalmology*, **92**, 938–946.
14. Sugar, H.S. (1965) Clinical effect of corticosteroids on conjuctival filtering blebs: a case report. *Am. J. Ophthalmol.*, **59**, 854–860.
15. Roth, S.M., Spaeth, G.L., Starita, R.J., Birbillis, E.M., and Steinmann, W.C. (1991) The effect of postoperative corticosteroids on trabeculectomy and the clinical course of glaucoma: five-year follow-up study. *Ophthalmic Surg.*, **22**, 724–729.
16. Skuta, G.L., Beeson, C.C., Higginbotham, E.J., Lichter, P.R., Musch, D.C., Bergstrom, T.J., Klein, T.B., and Falck, F.Y. (1992) Intraoperative mitomycin versus postoperative 5-fluorouracil in high risk glaucoma filtering surgery. *Ophthalmology*, **99**, 438–444.
17. Costa, V.P., Wilson, R.P., Moster, M.R., Schmidt, C.M., and Gandham, S. (1993) Hypotony maculopathy following the use of topical mitomycin C in glaucoma filtration surgery. *Ophthalmic Surg.*, **24**, 389–394.
18. Shields, M.B., Scroggs, M.W., Sloop, C.M., and Simmons, R.B. (1993) Clinical and histopathologic observations concerning hypotony after trabeculectomy with adjunctive mitomycin C. *Am. J. Ophthalmol.*, **116**, 673–683.
19. Xu, Y., Yang, G.H., Gin, W.M., Chen, K.Q., and Song, X.H. (1993) Effect of subconjunctival daunorubicin on glaucoma surgery in rabbits. *Ophthalmic Surg.*, **24**, 382–388.

20. Lee, D.A., Hersh, P., Kersten, D., and Melamed, S. (1987) Complications of subconjuctival 5-fluorouracil following glaucoma filtering surgery. *Ophthalmic Surg.*, **18**, 187–190.
21. Lee, D.A., Lee, T.C., Cortes, A.E., and Kitada, S. (1990) The effect of mithramycin, mitomycin, daunorubicin, and bleomycin on human subconjunctival fibroblast attachment and proliferation. *Ophthalmol. Vis. Sci.*, **31**, 2136–2144.
22. Lee, D.A., Shapourifar-Tehrani, S., and Kitada, S. (1990) The effect of 5-fluorouracil and cytarabine on human fibroblasts from Tenon's capsule. *Invest. Ophthalmol. Vis. Sci.*, **31**, 1848–1855.
23. Bansal, R.K. and Gupta, A. (1992) 5-Fluorouracil in trabeculectomy for patients under the age of 40 years. *Ophthalmic Surg.*, **23**, 278–280.
24. Heuer, D.K., Parrish, R.K. II, Gressel, M.G., Hodapp, E., Desjardins, D.C., Skuta, G.L., Palmberg, P.F., Nevarez, J.A., and Rockwood, E.J. (1986) 5-Fluorouracil and glaucoma filtering surgery. Intermediate follow-up of a pilot-study. *Ophthalmology*, **93**, 1537–1546.
25. Jampel, H.D., Jabs, D.A., and Quigley, H.A. (1990) Trabeculectomy with 5-fluorouracil for adult inflammatory glaucoma. *Am. J. Ophthalmol.*, **109**, 168–173.
26. Patitsas, C.J., Rockwood, E.J., Meisler, D.M., and Lowder, C.Y. (1992) Glaucoma filtering surgery with post-operative 5-fluorouracil in patients with intraocular inflammatory disease. *Ophthalmology*, **99**, 594–599.
27. Rockwood, E.J., Parrish, R.K. II, Heuer, D.K., Skuta, G.L., Hodapp, E., Palmberg, P.F., Gressel, M.G., and Feuer, W. (1987) Glaucoma filtering surgery with 5-fluorouracil. *Ophthalmology*, **94**, 1071–1078.
28. The Fluorouracil Filtering Surgery Study Group (1993) Three-year follow-up of the fluorouracil filtering surgery study. *Am. J. Ophthalmol.*, **115**, 82–92.
29. Bergstrom, T.J., Wilkinson, S., Skuta, G.L., Watnick, R.L., and Elner, V.M. (1991) The effect of subconjuctival mitomycin C on glaucoma filtration surgery in rabbits. *Arch. Ophthalmol.*, **109**, 1725–1730.
30. Khaw, P.T., Doyle, J.W., Sherwood, M.B., Smith, M.F., and McGorray, S. (1993) Effects of intraoperative 5-fluorouracil or mitomycin C on glaucoma filtration surgery in the rabbit. *Ophthalmology*, **100**, 367–372.
31. Liang, L.L. and Epstein, D.L. (1992) Comparison of mitomycin C and 5-fluorouracil on filtration surgery success in rabbit eyes. *J. Glaucoma*, **1**, 87–93.
32. Pasquale, L.R., Thibault, D., Dorman-Pease, M.E., Quigley, H.A., and Jampel, H.D. (1992) Effect of topical mitomycin C on glaucoma filtering surgery in monkeys. *Ophthalmology*, **99**, 14–18.
33. Wilson, M.R., Lee, D.A., Baker, R.S., Goodwin, L.T., and Wooten, F. (1991) The effects of topical mitomycin on glaucoma filtration surgery in rabbits. *J. Ocular Pharmacol.*, **7**, 1–8.
34. Chen, C.W., Huang, H.T., Bair, J.S., and Lee, C.C. (1990) Trabeculectomy with simultaneus topical application of mitomycin C in refractory glaucoma. *J. Ocular Pharmacol.*, **6**, 175–182.
35. Mietz, H., Arnold, G., Kirchhof, B., Diestelhorst, M., and Krieglstein, G.K. (1996) Histopathology of episcleral fibrosis after trabeculectomy with and without mitomycin C. *Graefes Arch. Clin. Exp. Ophthalmol.*, **234**, 364–368.
36. Ustundag, C. and Diestelhorst, M. (1998) Effect of mitomycin C on aqueous humor flow, flare and IOP in eyes with glaucoma 2 years after trabeculectomy. *Graefes Arch. Clin. Exp. Ophthalmol.*, **236**, 734–738.
37. Mietz, H., Addicks, K., Diestelhorst, M., and Krieglstein, G.K. (1995) Intraocular toxicity to ciliary nerves after extraocular application of mitomycin C in rabbits. *Int. Ophthalmol.*, **19**, 89–93.
38. Schraermeyer, U., Diestelhorst, M., Bieker, A., Theison, M., Mietz, H., Ustundag, C., Joseph, G., and Krieglstein, G.K. (1999) Morphologic proof of the toxicity of mitomycin C on the ciliary body in relation to different application methods. *Graefe's Arch. Clin. Exp. Ophthalmol.*, **237**, 593–600.
39. Grisanti, S., Gralla, A., Maurer, P., Diestelhorst, M., Krieglstein, G., and Heimann, K. (2000) Cellular photoablation to control postoperative

fibrosis in filtration surgery: *in vitro* studies. *Exp. Eye Res.*, **70**, 145–152.

40 Grisanti, S., Diestelhorst, M., Heimann, K., and Krieglstein, G. (2000) Cellular photoablation to control postoperative fibrosis in a rabbit model of filtration surgery. *Br. J. Ophthalmol.*, **83**, 1353–1359.

41 Hill, R.A., Crean, D.H., Doiron, D.R., McDonald, T.J., Liaw, L.H., Ghosheh, F., Hamilton, A., and Berns, M.W. (1997) Photodynamic therapy for antifibrosis in a rabbit model of filtration surgery. *Ophthalmic Surg. Lasers*, **28**, 574–581.

42 Bissonnette, R. and Lui, H. (1997) Current status of photodynamic therapy in dermatology. *Dermatol. Clin.*, **15** (3), 507–519.

43 Weissgold, D.J., Hu, L.K., Gragoudas, E.S., and Young, L.H.Y. (1996) Photodynamic therapy (PDT) of pigmented choroidal melanoma via a trans-scleral approach. *Invest. Ophthalmol. Vis. Sci.* **37** (Suppl), S123.

44 Moshfeghi, D.M., Peyman, G.A., Khoobehi, B., Moshfeghi, A., and Crean, D.H. (1995) Photodynamic occlusion of retinal vessels using tin ethyl etiopurin (SnET2): an efficacy study. *Invest. Ophthalmol. Vis. Sci.*, **36** (Suppl), S115.

45 Cox, K.W., Shepperd, J.D., Lattanzio, F.A., and Williams, P.B. (1997) Photodynamic therapy of corneal neovascularization using topical dihematoporphyrin ester. *Invest. Ophthalmol. Vis. Sci.*, **38** (Suppl), S512.

46 Smyth, R.J., Nguyen, K., Ahn, S.S., Panek, W.C., and Lee, D.A. (1993) The effects of photophrin on human Tenon's capsule fibroblasts *in vitro*. *J. Ocular Pharmacol.*, **9**, 171–178.

47 Dacheux, R. and Guidry, C. (1995) Cellular photoablation as a therapy to control proliferative vitreoretinopathy: *in vitro* studies. *Invest. Ophthalmol. Vis. Sci.*, **36** (Suppl), S751.

48 Grimes, P.A., Stone, R.A., Laties, A.M., and Li, M. (1982) Carboxyfluorescein. A probe of the blood–ocular barriers with lower membrane permeability than fluorescein. *Arch. Ophthalmol.*, **100**, 635–639.

49 Hofmann, J. and Sernetz, M. (1983) A kinetic study on the enzymatic hydrolysis of fluorescein diacetate and fluorescein di-beta-D-galactopyranoside. *Anal. Biochem.*, **131**, 180–186.

50 Rotman, B. and Papermaster, B. (1966) Membrane properties of living mamalian cells as studied by enzymatic hydrolysis of fluorogenic esters. *Proc. Natl. Acad. Sci. U. S. A.*, **55**, 134–141.

44
Correction of Wavefront Aberrations of the Eye
Dan Reinstein and Manfred Dick

44.1
Introduction

A plane optical wavefront with no aberrations is in general considered to give the best performance of optical systems. However, especially in the field of refractive corrections of the human eye, only the correction of low-order optical aberrations such as tilt, sphere, and cylinder was possible in the past. With the advent of wavefront aberrometers (e.g., on the basis of Shack–Hartmann sensors) around 2000, monochromatic higher order aberrations such as spherical aberration, trefoil, and coma became measurable and their influence on visual performance could also be studied. The correction of these higher order aberrations, fully or to some extent, was investigated for all vision correction modalities. The aim was to improve visual acuity significantly up to supervision, which is defined as a visual acuity of 20/10 or 20/8 [1].

As the eye, due to its accommodating power, is a dynamic system which is age dependent because of presbyopia, an additional challenge concerning the correction of aberrations arises. Mostly a higher order aberration correction for distance vision and large pupils is performed to achieve the best visual acuity, but it is still limited by retinal resolution and diffraction. Other concepts, especially with LASIK and intraocular lenses, divide the pupil into radial optical zones with different corrections of aberrations, for example, for near and distance vision.

With regard to glasses, the correction of higher order aberrations can be realized for one viewing direction and a fixed position of the pair of glasses, but for all viewing directions and tolerances in position only partial optimization is possible. With the help of the i.Scription wavefront-based refraction glasses, patients had a considerable improvement in night vision [2, 3].

Contact lenses are better suited for correction of higher order aberrations as only one viewing direction is predetermined. The problem with contact lenses is the lateral shift behavior, which could worsen the performance of optical correction. Our own experiments have shown that a decentration of 0.3 mm leads to a 20% reduction in the Strehl intensity ratio.

Handbook of Biophotonics. Vol.2: Photonics for Health Care, First Edition. Edited by Jürgen Popp, Valery V. Tuchin, Arthur Chiou, and Stefan Heinemann.
© 2012 Wiley-VCH Verlag GmbH & Co. KGaA. Published 2012 by Wiley-VCH Verlag GmbH & Co. KGaA.

Intraocular lenses have great potential for the correction of higher order aberrations as biometric diagnosis and also lens calculation algorithms become more and more accurate, in addition to improved surgical techniques [4–7].

As laser refractive surgery of the cornea represents the most stable target for the correction of higher order aberrations, we will concentrate here on excimer laser-based customized refractive surgery [8, 9].

44.2
Excimer Laser-Based Refractive Surgery for Correction of Wavefront Aberrations

All excimer laser platforms now allow the surgeon to include a patient's wavefront data to try to reduce higher order aberrations at the same time as treating the refraction. The micronic accuracy of excimer laser systems means that irregular ablation profiles based on wavefront data can be delivered with confidence. For example, Carl Zeiss Meditec performed laboratory closed-loop experiments to validate the use of the MEL80 for the correction of higher order aberrations. The MEL80 was used to create transparent polymer phase plates with a defined profile that would represent a specific Zernike coefficient of determined amplitude on the polymer surface. The resulting ablated phase plate surface shape was verified independently using a profilometer, and was then used to aberrate a collimated planar wavefront beam and generate a wavefront distortion to be detected by the WASCA. The WASCA and profilometry measurements were used as measures of accuracy in the ability to produce a given wavefront aberration on a phase plate. The WASCA measurement was then imported into a CRS-Master workstation to generate an ablation profile designed to reverse the specific aberration induced. Figure 44.1 depicts a profilometer-based topography measurement and horizontal wavefront section measured after a collimated beam has passed through a phase plate ablated to produce a Z(3,–1) Zernike shape. The agreement between the measured horizontal section and the intended Zernike polynomial can be seen to be very high.

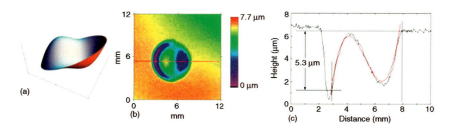

Figure 44.1 Laboratory validation of WASCA and the MEL80 excimer laser as components of a wavefront-guided ablation system: 3D plot of a planned Z(3,–1) Zernike polynomial to be ablated into PMMA (a), 3D profilometry of the surface of the PMMA plate after ablation (b), and a horizontal section (red line) showing the actual profile achieved (c). The attempted shape (red curve) is seen to be very close to the achieved shape (blue curve).

The efficacy of wavefront customized ablation profiles will therefore depend on the accuracy of the higher order aberration measurements (i.e., the aberrometer used), the centration of the ablation profile, and the biomechanical response to the ablation.

Given the fact that all current ablation profiles induce higher order aberrations, particularly spherical aberration, it is interesting to investigate whether simply including the wavefront for primary myopic treatments will be of benefit. It is likely that including the wavefront for primary myopic treatments would be more beneficial for eyes presenting with significant higher order aberrations, whereas, the wavefront ablation profile will only have minor differences from the base ablation profile for a patient presenting with negligible aberrations. Any possible benefit should be weighed against the small increase in ablation depth when including wavefront data.

The known induction of spherical aberration should also be taken into account when planning a wavefront customized ablation for a primary procedure. For example, myopic ablation induces positive spherical aberration (OSA notation), so if the patient's wavefront has negative spherical aberration and is used for treatment, this will increase the amount of induced positive spherical aberration, which would be detrimental to the outcome of the treatment. To account for this, some laser systems, such as the MEL80, allow the surgeon to include specific types of aberration while excluding others so that the ablation profile can be tailored to offer the optimal treatment.

Using the MEL80, we set out to study whether the incorporation of higher order aberrations into the standard aspheric profile would be of further benefit. In this study, carried out between July 2003 and May 2004, 24 consecutive patients recruited for study were given a MEL80 standard aspheric profile in the dominant eye, while receiving a CRS-Master derived wavefront-guided profile in the nondominant eye, which included the nascent higher order aberrations of the patient [10]. Preoperatively, the right and left eyes of each patient were within ± 0.75 D spherical equivalent refraction and cylinder. At a minimum follow-up of 6 months, there were no statistically significant differences between eyes in efficacy (UDVA), accuracy, or safety (changes in CDVA). However, positive improvements were noted in the induction of higher order aberrations for the eyes treated with a wavefront-guided profile. There was 22% less induction in spherical aberration (measured at a 6 mm pupil) in the eyes treated with a wavefront guided profile, which was a statistically significant difference ($p = 0.005$). Contrast sensitivity was equal in right and left eyes preoperatively, but statistically significantly higher in the eyes treated with a wavefront-guided profile at 3 and 6 cpd (cycles per degree) ($p < 0.015$). This study demonstrated that wavefront-guided treatments may indeed offer a small benefit over standard aspheric treatments. However, given the amount of aberrations induced by the base ablation profiles compared with the relatively small amount of higher order aberrations in the majority of untreated eyes, the final solution of actually being able to correct nascent higher order aberrations will have to wait until aberration-free ablation profiles are developed. Many groups are working on this problem.

Wavefront-guided treatments are likely to be most useful for treating highly aberrated eyes, such as are encountered in corneas that have undergone refractive surgery previously with complaints of night vision disturbances. In these cases, there is often little refraction and so there should not be much induction of higher order aberrations due to any further surgery. Therefore, it is reasonable to treat the higher order aberrations with a wavefront-guided treatment to help alleviate visual difficulties.

We performed a study to investigate the amount of aberrations in post-refractive surgery eyes complaining of night vision disturbances compared with those not complaining after primary myopic LASIK using the standard aspheric MEL80 profile [11]. The repair eye group were found to have 42% more spherical aberration than the post-LASIK control group. This supports the theory that higher order aberrations, especially spherical aberration, are the root cause of night vision disturbances. The repair eye group were also found to have significantly lower contrast sensitivity at all frequency levels; the normalized contrast sensitivity ratio [12] was about 0.80 for the repair eyes (below normal) compared with 1.02 for the post-LASIK control eyes (normal). After wavefront-guided repair, the repair eye group were found to have only 17% more spherical aberration than the post-LASIK control group. Furthermore, the normalized contrast sensitivity ratio was increased to 1.02 (Figure 44.2). The MEL80 wavefront-guided treatment managed to reduce the aberrations to a similar level to that which is commonly found after myopic LASIK with the MEL80. It was found that this reduction was sufficient to restore the contrast sensitivity to the normal range in the majority of cases; 80% of repair eyes had contrast sensitivity in the normal range.

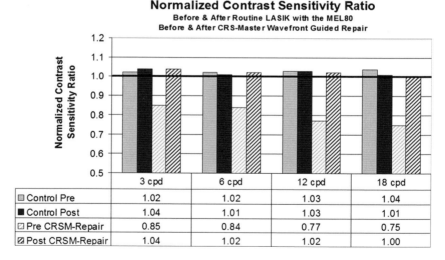

Figure 44.2 Bar chart of the average normalized contrast ratio before and after routine LASIK with the MEL80 and before and after MEL80 wavefront-guided repair. A ratio of 1 is considered normal and is marked on the graph as a dark solid line. The graph shows that the contrast sensitivity remained the same following routine LASIK with the MEL80. It also shows that the contrast sensitivity was restored to the normal range following MEL80 wavefront-guided repair.

This result showed that although the laser ablation was delivering the measured wavefront exactly (as proved by profilometry and described earlier), and the profile included only a small refractive component so there should have been few induced aberrations, the postoperative wavefront was a long way from being flat. The reasons for the undercorrection of the higher order aberrations could be due to a number of factors: (1) the aberrometers may not have high enough accuracy and repeatability, although the resolution and repeatability of instruments such as the WASCA [10, 13] appear to be good enough (Figure 44.3 shows the repeatability of each Zernike coefficient in 70 consecutive measurements of a highly aberrated eye after LASIK in a corneal transplant); (2) there may be diurnal changes in wavefront; (3) the wavefront changes as the pupil diameter changes; and (4) the biomechanical response of the cornea serves partially to reverse some of the intended change (it is possible that a nomogram is required for each Zernike value).

However, the result that the contrast sensitivity was restored to the normal level despite reducing the spherical aberration by only 27% suggests that the goal of wavefront treatments might not have to be to *flatten* the wavefront, but only to reduce the aberrations to a tolerable level. Aberrations are induced in every LASIK patient, but the majority of patients do not complain of night vision disturbances or a decrease in the quality of vision. Therefore, we can conclude that the brain has a tolerance level to aberrations.

Indeed, we can actually take advantage of the ability of the brain to filter a certain amount of aberrations. It has been shown that an increase in spherical aberration is associated with an increase in depth of field [14], and we have employed this increase

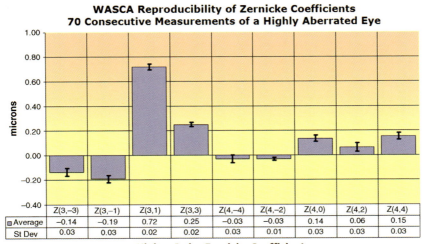

Figure 44.3 Zernike polynomial coefficient mean repeatability (standard deviation) for a consecutive series of 70 measurements for a highly aberrated eye. The x-axis shows the Zernike ordinal number. The value of the bar is the mean for each Zernike (Malacara notation) and the repeatability is shown as the error bars.

in depth of field to develop a procedure for the treatment of presbyopic patients using a new nonlinear aspheric ablation profile, a procedure known as laser blended vision [15, 16]. The initial goal was to increase the depth of field sufficiently to provide vision from a distance through intermediate to near; creating an eye that would see 20/20 at a distance, read a computer screen, but also read J1. However, as we learned, visual quality and contrast sensitivity can be compromised by excessive changes in aberrations of the cornea. We discovered that it was possible to safely increase the depth of field of the cornea to 1.50 D for any starting refractive error. However, given a 1.50 D depth of field, it would not be possible to achieve full distance and full near vision monocularly; therefore, we borrowed from the time-tested concept of monovision and set up the nondominant eye to be slightly myopic, so that the depth of field of the predominantly distance (dominant) eye was able to see from a distance down to intermediate, while the predominantly near (nondominant) eye was able to see in the intermediate and near range. In the intermediate region, both eyes have similar acuity, which draws on our knowledge of binocular fusion processing: the horopter, a volume centered on the fixation point that contains all points in space that yield single vision. Monovision, or in this case, micro-monovision, draws on the inherent cortical processes of neuronal gating and blur suppression (the ability for conscious attention to be directed to the part of the visual field with the best image quality).

In all, laser blended vision draws on six mechanisms for its success as a procedure: (1) depth of field increase by a specific controlled increase in corneal aberrations; (2) pupil constriction during accommodation affording a further depth of field increase on the retinal image; (3) retinal and cortical processing for increasing contrast of the retinal image changes produced by spherical aberration; (4) micro-monovision to allow continuous distance to intermediate to near vision between the two eyes; (5) relying on central cortical processing including neuronal gating and blur suppression; and (6) making use of the residual accommodation of the eye.

Customized ablation profiles can also be generated using front surface topography data and some laser systems offer a topography-guided ablation profile. Topography-guided ablation algorithms are designed to calculate an ablation profile that would alter an irregular topography to become a smooth, regular, aspheric surface. An irregular front surface topography will cause significantly higher order aberrations; therefore, the consequence of regularizing the front surface corneal topography will be to reduce higher order aberrations.

Using the MEL80 system, topography-guided ablations are actually more effective in reducing aberrations, particularly spherical aberration, than wavefront-guided ablations. In our study using the MEL80 topography-guided system for post-refractive surgery complications, the spherical aberration was reduced by 41% on average [17]. Topography-guided ablations were found to be particularly successful for the correction of decentrations and for optical zone enlargement. One of the major advantages of topography-guided ablation over wavefront-guided ablation is the centration. By definition, topographic data are centered about the corneal vertex, which approximates the visual axis. On the other hand, wavefront data are currently calculated centered on the entrance pupil center, as decided by the OSA Committee, since wavefront data can only be obtained within the pupil. Therefore, in an eye with a large angle kappa, the wavefront will not represent the patient's vision since the patient is not looking through the center of the pupil (where the wavefront is centered). This means that using a wavefront custom ablation in eyes with a large angle kappa could result in worsening of the visual symptoms. For this reason, a topography-guided ablation is a much more reliable and effective treatment option for decentration.

However, before a wavefront- or topography-guided ablation is performed, it is important first to have made a confident diagnosis and in particular to have considered the effect of epithelial thickness changes. The epithelium will always remodel itself to compensate for underlying stromal surface irregularities; the epithelium overlying bumps in the stromal surface becomes progressively thinner and the epithelium overlying troughs in the stromal surface becomes progressively thicker [18–25]. The epithelium effectively acts as a low-pass filter over the stoma, regularizing high-frequency noise. The epithelium is regulated by the blinking motion of the eyelid [25, 26] and is therefore designed to improve regularly irregular astigmatism about the visual axis. Unfortunately, this evolutionary power of the epithelium to remodel itself to a smooth surface poses a problem when trying to diagnose subsurface causes of irregularities in corneal front surface shape. We believe that epithelial thickness mapping has the potential to improve greatly our diagnostic capabilities, and also improve treatment planning in the correction of irregular astigmatism. The presence of epithelial thickness profile changes also means that neither wavefront- nor topography-guided ablation will ever be a complete solution because they do not account for epithelial thickness changes.

References

1 Krueger, R.R., Applegate, R.A., and MacRae, S.M. (2004) *Wavefront Customized Visual Corrections: the Quest for Super Vision II*, Slack, Thorofare, NJ.

2 Cabeza, J.M., Mendel, L., and Kratzer, T.C. (2003) The performance of wavefront-based refraction can be proven. *Dtsch Opt Ztg*, (DOZ 4), 68–71.

3 Cabeza, J.M. (2010) Optimized low order prescription by incorporating wavefront data of the eye. *MAFO Ophthalmic Labs Ind.*, **7** (MAFO 4-10), 10–16.

4 Fiala, W. (2009) Remarks on wavefront designed aberraton correcting intraocular lenses. *Optom. Vis. Sci.*, **86** (5), 529–536.

5 Pieh, S., Fiala, W. Malz, A., and Stork, W. (2009) *In vitro* Strehl ratios with spherical, aberration free, average, and customized spherical aberration-correcting intraocular lenses. *Invest. Ophthalmol. Vis. Sci.*, **50** (3), 1264–1269.

6 Nochez, Y., Favard, A. Majzoub, S., and Pisella, P.-J. (2010) Measurement of corneal aberrations for customistion of intraocular lens asphericity: impact on quality of vision after midro-incision cataract surgery. *Br. J. Ophthalmol.*, **94**, 440–444.

7 Preussner, P.-R. and Wahl, J. (2003) Simplified mathematics for customized refractive surgery. *J. Cataract Refract. Surg.*, **29**, 462–470.

8 Mrochen, M. and Seiler, T. (2001) Grundlagen der wellenfront-geführten refraktiven Hornhautchirurgie. *Ophthalmologe*, **98**, 703–714.

9 Mrochen, M. (2001) Aberrationsmessung und wellenfront-geführte refraktive Chirurgie. *Ophthalmochirurg*, **13**, 45–51.

10 Reinstein, D.Z., Neal, D.R., Vogelsang, H., Schroeder, E., Nagy, Z.Z., Bergt, M. *et al.* (2007) Optimized and wavefront guided corneal refractive surgery using the Carl Zeiss Meditec platform:

11. Reinstein, D.Z., Archer, T.J., Couch, D., Schroeder, E., and Wottke, M. (2005) A new night vision disturbances parameter and contrast sensitivity as indicators of success in wavefront-guided enhancement. *J. Refract. Surg.*, **21** (5), S535–S540.
12. Wachler, B.S. and Krueger, R.R. (1998) Normalized contrast sensitivity values. *J. Refract. Surg.*, **14** (4), 463–466.
13. Reinstein, D.Z., Archer, T.J., and Couch, D. (2006) Accuracy of the WASCA aberrometer refraction compared to manifest refraction in myopia. *J. Refract. Surg.*, **22** (3), 268–274.
14. Marcos, S., Moreno, E., and Navarro, R. (1999) The depth-of-field of the human eye from objective and subjective measurements. *Vision Res.*, **39** (12), 2039–2049.
15. Reinstein, D.Z., Couch, D.G., and Archer, T.J. (2009) LASIK for hyperopic astigmatism and presbyopia using micro-monovision with the Carl Zeiss Meditec MEL80. *J. Refract. Surg.*, **25** (1), 37–58.
16. Reinstein, D.Z., Archer, T.J., and Gobbe, M. (2011) LASIK for myopic astigmatism and presbyopia using non-linear aspheric micro-monovision with the Carl Zeiss Meditec MEL 80 platform. *J. Refract. Surg.*, **27** (1), 23–37.
17. Reinstein, D.Z., Archer, T.J., and Gobbe, M. (2009) Combined corneal topography and corneal wavefront data in the treatment of corneal irregularity and refractive error in LASIK or PRK using the Carl Zeiss Meditec MEL80 and CRS master. *J. Refract. Surg.*, **25** (6), 503–515.
18. Reinstein, D.Z. and Archer, T. (2006) Combined Artemis very high-frequency digital ultrasound-assisted transepithelial phototherapeutic keratectomy and wavefront-guided treatment following multiple corneal refractive procedures. *J. Cataract Refract. Surg.*, **32** (11), 1870–1876.
19. Holland, S.P., Srivannaboon, S., and Reinstein, D.Z. (2000) Avoiding serious corneal complications of laser assisted *in situ* keratomileusis and photorefractive keratectomy. *Ophthalmology*, **107** (4), 640–652.
20. Reinstein, D.Z., Silverman, R.H., Sutton, H.F., and Coleman, D.J. (1999) Very high-frequency ultrasound corneal analysis identifies anatomic correlates of optical complications of lamellar refractive surgery: anatomic diagnosis in lamellar surgery. *Ophthalmology*, **106** (3), 474–482.
21. Reinstein, D.Z., Rothman, R.C., Couch, D.G., and Archer, T.J. (2006) Artemis very high-frequency digital ultrasound-guided repositioning of a free cap after laser *in situ* keratomileusis. *J. Cataract Refract. Surg.*, **32** (11), 1877–1882.
22. Reinstein, D.Z. and Archer, T.J. (2007) Evaluation of irregular astigmatism with Artemis very high-frequency digital ultrasound scanning, in *Irregular Astigmatism: Diagnosis and Treatment* (ed. M. Wang), Slack, Thorofare, NJ, pp. 29–42.
23. Reinstein, D.Z., Archer, T.J., and Gobbe, M. (2009) Corneal epithelial thickness profile in the diagnosis of keratoconus. *J. Refract. Surg.*, **25** (7), 604–610.
24. Reinstein, D.Z., Archer, T.J., Gobbe, M., Silverman, R.H., and Coleman, D.J. (2010) Epithelial thickness after hyperopic LASIK: three-dimensional display with Artemis very high-frequency digital ultrasound. *J. Refract. Surg*, **26** (8), 555–564.
25. Reinstein, D.Z., Archer, T.J., Gobbe, M., Silverman, R.H., and Coleman, D.J. (2010) Epithelial, stromal and corneal thickness in the keratoconic cornea: three-dimensional display with Artemis very high-frequency digital ultrasound. *J. Refract. Surg.*, **26** (4), 259–271.
26. Reinstein, D.Z., Archer, T.J., Gobbe, M., Silverman, R.H., and Coleman, D.J. (2008) Epithelial thickness in the normal cornea: three-dimensional display with Artemis very high-frequency digital ultrasound. *J. Refract. Surg.*, **24** (6), 571–581.

Part 10
Otolaryngology (ENT)

45
Optical Coherence Tomography of the Human Larynx
Marc Rubinstein, Davin Chark, and Brian Wong

45.1
Anatomy

The larynx is a complex organ composed of cartilage, connective tissue, muscle, and mucosal epithelium that acts in a coordinated fashion to allow sphincteric control of the airway. The larynx has respiratory and digestive functions; it protects the lungs from aspiration of food and water, and it helps in the passage of air, phonation, and in the clearance of secretions. The larynx is divided anatomically into the supraglottic, glottic, and subglottic regions, illustrated in Figure 45.1. The supraglottic region extends from the tip epiglottis superiorly to the false vocal folds inferiorly, the glottic region encompasses the true vocal folds, and the subglottic region extends from the inferior border of the true vocal folds superiorly to the cricoid cartilage inferiorly. The vocal folds (i.e., the vocal "cords") consist of a multi-layered structure of nonkeratinized stratified squamous epithelium covering three layers of dense connective tissue; the epithelial layer is separated from the connective tissue by the basement membrane. The basement membrane is a critical tissue layer as cancer cells that spread through this layer gain access to blood vessels and lymphatics. Deep to these layers is the vocal ligament of the thyroarytenoid muscle, illustrated in Figure 45.2 [1–5].

45.2
Clinical Problems

45.2.1
Laryngeal Cancer

Laryngeal cancer is the second most common cancer in the upper aerodigestive tract, with an estimated 12 290 new cases per year in the United States and roughly 3660 annual deaths [6]. Approximately 95% of laryngeal cancers are squamous cell carcinoma (SCC), and the majority occur in patients between the ages of 60 and 70 years [7]. The 5 year overall survival rate for laryngeal SCC is around 64% [8]. The

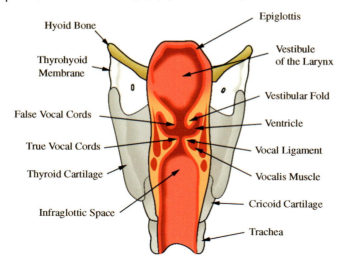

Figure 45.1 Diagram showing the human larynx with its corresponding structures.

risk of developing laryngeal SCC has been closely related to the use of tobacco and alcohol and it is proportional to the amount and duration of use, and when combined they act synergistically [9].

45.2.2
Importance of Laryngeal Optical Coherence Tomography

The cardinal symptom of early laryngeal cancer is hoarseness, but since this is a very innocuous and nonspecific symptom, laryngeal cancer often is not diagnosed until

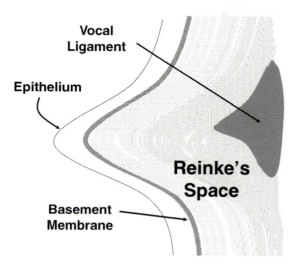

Figure 45.2 Diagram showing the microstructure of the vocal fold.

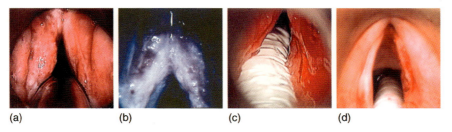

Figure 45.3 Disorders that can mimic laryngeal cancer: (a) chronic inflammation; (b) hyperkeratosis; (c) severe dysplasia; (d) invasive cancer (the white plastic object in the center is the endotracheal tube).

advanced stages. Clinical evaluation relies on using a flexible fiber-optic laryngoscope or a laryngeal mirror, which provides only a two-dimensional view of the laryngeal structures, which in the case of early cancers is generally inadequate to distinguish early malignancy from benign diseases and processes of the larynx. Early diagnosis of laryngeal cancer is key for optimal treatment. The "gold standard" for accurate diagnosis is a biopsy of the suspicious lesion. In general, this has to be performed under general anesthesia. Vocal fold biopsy is not without risk, and injury to the delicate epithelium of the folds may occur. Permanent changes can occur such as scarring, which may alter fold vibration and voice quality.

The spread of cancer cells through the basement membrane into the deeper layers of the vocal fold is the most important feature needed to establish a diagnosis of invasive cancer as opposed to premalignant entities such as dysplasia and carcinoma *in situ* and benign conditions such as chronic laryngitis (Figure 45.3).

Because of the risks associated with biopsy, and the considerable decision an ENT surgeon must make to take a patient to the operating room for a biopsy of the vocal folds under general anesthesia, noninvasive methods of imaging vocal fold microstructure have received significant attention. In particular, optical coherence tomography (OCT) may be used to evaluate the structural integrity of the basement membrane in patients at risk for cancer, and also provide surgeons with a means to guide biopsies towards regions with a greater diagnostic yield. OCT may also potentially be used to monitor the progression of the disease, providing a three-dimensional archive of vocal fold micro-architecture, reducing the reliance upon biopsies for disease surveillance.

45.3
Clinical *In Vivo* Applications

45.3.1
OCT During Microsurgical Endoscopy

In laryngology, the first clinical studies were performed during surgical endoscopy (i.e., under general anesthesia), where the OCT probe was passed through the working channel of a flexible endoscope to obtain images of the laryngeal mucosa [10–12]. Observing that in OCT images taken from the regions with cancer no structures were identified, as compared with OCT images taken from normal mucosa, it was concluded that OCT has the capability to determine the border of the tumor [12]. Later studies expanded the applications, increased the patient numbers, and broadened the disease classes imaged using OCT in the operating room [13–20]. Most studies used slender rigid or semi-rigid probes. Wong et al. [14] imaged normative anatomy and benign lesions in 82 patients. OCT images compared favorably with results obtained using conventional histology. Epithelial thickness measurements were also recorded for specific laryngeal subsites, although these measurements were not directly compared with histology. Armstrong et al. [15] imaged 22 patients with laryngeal cancer, where integrity of the basement membrane was identified; disruption of the basement membrane was seen in those patients with histologic diagnosis of cancer, and also the transition zone was identify at the cancer margin. Klein et al. [16] used both OCT and polarization-sensitive optical coherence tomography (PS-OCT) to perform preliminary studies in 13 patients in the operating room and in the outpatient clinic, and examined the collagen content in the subepithelial tissue. A landmark study by Kraft et al. [18] imaged normal, benign, and malignant laryngeal diseases in 193 patients using a commercial OCT system [20]. In all cases, OCT images were compared with histopathology. They concluded that the use of OCT during microsurgical endoscopy yielded a correct diagnosis in 89% of cases compared with a rate of only 80% during microsurgical endoscopy alone. They also observed a higher sensitivity for predicting invasive tumor growth and epithelial dysplasia when using microlaryngoscopy with OCT compared with microlaryngoscopy alone.

Other clinical studies have focused on more specific applications. Shakhov et al. [13] used OCT as an aid for intraoperative surgical guidance and targeting of resection margins in laryngeal cancer surgery. Twenty-six patients were imaged during surgery and OCT was used to identify the transition zone between normal tissue and frank invasive cancer, thus increasing the accuracy of the tumor resection. Kaiser et al. [19] used OCT to measure the *in vivo* epithelial thickness of six different laryngeal subsites and correlated these measurements with those obtained on fixed anatomic laryngectomy specimens. Ridgway et al. [21] used OCT to evaluate the *in vivo* normal anatomy of the upper aerodigestive tract in 15 children. The normal microanatomy of different regions including the pediatric larynx were identified, and also various benign pathologic conditions. In neonates, Ridgway et al. [22] used OCT to image the larynx and trachea of 12 newborns who were admitted to the neonatal intensive care unite and required endotracheal intubation. Their study demonstrated that OCT may be capable of functioning as a noninvasive imaging technique to study and monitor changes in the newborn airway during intubation, and possibly reduce the incidence of acquired disorders such as subglottic stenosis. An example of laryngeal OCT is illustrated in Figures 45.4 and 45.5.

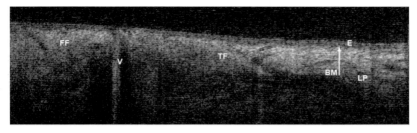

Figure 45.4 OCT image of a newborn's larynx. FF, false vocal folds; V, ventricle; TF, true vocal folds; BM, basement membrane locus; E, epithelium; LP, lamina propria.

45.3.2
Office-Based OCT

With the immense body of data supporting the use of OCT to image the larynx under general anesthesia, the focus has shifted towards developing systems for use in the office with awake, fully conscious patients using only topical anesthetics. An office-based OCT system would allow ENT surgeons to identify suspicious lesions better in the office and thus select patients better to undergo surgical biopsy. An office-based system also provides a means to archive three-dimensional cross-sectional images of suspicious lesions over time, and hence monitor disease progression. Although there are several reports of the use of OCT in the office, a major challenge that clinicians must contend with is motion of the patient's body, the patient's larynx, and the clinicians hand, and the sway of the fiber-optic endoscope or rigid telescope [16, 23]. There are two basic device platforms that are used to image the larynx in the office: flexible fiber-optic laryngoscopes, which are inserted through the nares and passed into the pharynx, and rigid laryngeal endoscopes, which are inserted into the oral cavity to view the larynx from the oropharynx.

Klein et al. [16] first reported using OCT in the office, in three patients, although the study did not focus on the feasibility of this technique or the methods used to obtain results. Sepehr et al. [23] studied the feasibility of using OCT in the office-based setting in 17 patients using a flexible OCT probe in tandem with a flexible fiber-optic laryngoscope. The instruments were inserted through the nose, and the OCT fiber tip was placed in direct contact with the larynx. They concluded that using this system in the office setting is feasible and could aid in the diagnosis of laryngeal lesions.

Figure 45.5 OCT image of a pediatric larynx. FF, false vocal folds; V, ventricle; TF, true vocal folds; S, subglottis; C, cricoid cartilage.

An alternative approach is to couple an OCT imaging system with a rigid Hopkins rod endoscope or similar rigid optical device. Using this approach allows the clinician simultaneously to obtain real-time OCT images while examining the patient's larynx [24–27]. This approach has several limitations. First, because the distance from any point in the oropharynx to the vocal folds varies greatly from patient to patient, the working distance also varies and this has to be manually adjusted during imaging, making it a challenge to acquire good images. The image quality is also affected by light dispersion as related to the image working distance. Also, the motion of the clinician's hand or the patients head, neck, or larynx affects the ability to identify the image plane. Finally, some patients may have a strong gag reflex and cannot tolerate imaging with this system configuration. Yu *et al.* [27] recently reported obtaining real-time *in vivo* OCT images at the same time as measuring the vocal fold vibrations. These vocal fold vibrations were recorded in both male and female patients and were successfully correlated with the vocal fundamental frequencies. Different systems have been used to image the human larynx, as summarized in Table 45.1.

45.3.3
OCT Integrated with a Surgical Microscope

The hand-held OCT probes used during laryngeal microsurgery are useful, but are limited in that the surgeon cannot use both hands (and two surgical instruments) during surgery. This also lengthens the surgical time, since the surgeon has to remove the OCT probe in order to introduce a second surgical instrument, which limits real-time imaging surgery *per se*. The probes also occupy space in what is already a narrow working port within a surgical laryngoscope. One approach to overcome these limitations is to integrate an OCT system with a surgical microscope [28, 29]. Vokes *et al.* [28] first reported using this type of integrated system to perform noncontact OCT imaging in 10 patients. The device consisted of an OCT scanner mounted on the objective of a surgical microscope much like conventional laser surgical micromanipulators. Although they reported identifying the vocal fold epithelium and lamina propria, limitations included reduced image resolution and operational challenges such as aligning the OCT beam and adjusting the focal length. Just *et al.* [29], using a different optical design incorporating the native optics of a surgical microscope, reported imaging patients during laryngeal microsurgery. This system effectively addressed some of the limitations previously reported by Vokes *et al.* [28], although with limitations of reduced lateral and axial resolution, which are due to the limited numerical aperture of the system. The potential of these combined systems is promising, although additional improvements need to be made before they can be widely used during laryngeal surgery.

45.4
Other Preclinical Applications of OCT in the Larynx

There are several *ex vivo* imaging studies using OCT in both the human [30, 31] and animal larynx [32–37]. Bibas *et al.* [30] imaged 10 human *ex vivo* laryngeal specimens

Table 45.1 Comparison of different OCT systems used to image the laryngeal tissue in humans.

Authors	Year	Central wavelength (nm)	System[a]	Speed (kHz)	Axial resolution (μm)	Lateral resolution (μm)	Commercially available	In vivo	Type of study
Sergeev et al. [10]	1997	830	TD	1	15	17	No	Yes	Direct contact
Feldchtein et al. [11]	1998								
Shakhov et al. [12, 13]	1999, 2001								
Bibas et al. [30]	2004	830	TD	0.7	13	5	No	No	Noncontact
Wong et al. [14]	2005	1310	TD	0.5	10	10	No	Yes	Direct contact
Armstrong et al. [15]	2006								
Ridgway et al. [21, 22]	2007, 2008								
Sepehr et al. [23]	2008								
Kaiser et al. [19]	2009								
Burns [31]	2005	1310	FD	10	12	25	No	No	Noncontact
Klein et al. [16]	2006	1310	FD	10	12	30	No	Yes	Noncontact
Lueerssen et al. [26]	2007	1278	TD	0.1	20	30–55	No	Yes	Noncontact
Kraft et al. [18]	2008	1300	TD	0.5	15	25	Yes	Yes	Direct contact
Just et al. [29]	2009	840	SD	5	12	24	No	Yes	Noncontact
Yu et al. [27]	2009	1310	SS	20	8	25	No	Yes	Noncontact
Rubinstein et al. [20]	2010	1300	TD	0.75	10–20	25	Yes	Yes	Direct contact

a) TD, time domain; SS, swept source; SD, spectral domain.

using OCT. After the images were reconstructed in 3D, they were compared with histopathology obtained after processing of the same specimens. Burns et al. [31] used a combination of OCT and PS-OCT to image three human cadaver larynges; PS-OCT showed an increase in the tissue birefringence in the outer surface of the vocal ligament, which delineates it from the superficial lamina propria, suggesting a potential use for PS-OCT for detecting scar areas within the lamina propria and also to aid OCT in diagnosing subepithelial benign lesions. Several *ex vivo* animal model studies have focused on measuring and characterizing the lesions in the vocal folds [32, 34, 35]. Karamzadeh et al. [32] used rabbit models to mimic three different vocal folds lesions: scar formation, edema, and trauma. Other studies have focused on evaluating the injury caused by using a CO_2 laser, demonstrating that OCT can potentially be used in real time to assist surgeons during laryngeal laser surgery [34, 35]. In a more recent study, Burns et al. [37] used OCT to track real-time injection of hydrogel into to the vocal folds, and thus provide feedback for phonosurgical operations.

Although OCT may have tremendous potential in helping clinicians to diagnose laryngeal pathology, monitor disease progression and even aid in guiding surgical biopsies, the technology still needs to be improved before its use can spread beyond academic centers. Such limitations are signal penetration, the acquisition speed, and the resolution. Implementing better office-based OCT systems will also increase the usefulness and widespread use of this imaging technique.

References

1 Lorenz, R.R., Netterville, J.L., and Burkey, B.B. (2008) Head and neck, in *Sabiston Textbook of Surgery* (eds C.M. Townsend, R.D. Beauchamp, B.M. Evers, and K.L. Mattox), Saunders Elsevier, Philadelphia, pp. 813–847.

2 Woodson, G.E. (2005) Laryngeal and pharyngeal function, in *Cummings Otolaryngology: Head and Neck Surgery* (eds C.W. Cummings, P.W. Flint, B.H. Haughey, K.T. Robbins et al.), Elsevier Mosby, Philadelphia, pp. 1963–1975.

3 Noordzij, J.P. and Ossoff, R.H. (2006) Anatomy and physiology of the larynx. *Otolaryngol. Clin. North Am.*, 39, 1–10.

4 Merati, A.L. and Rieder, A.A. (2003) Normal endoscopic anatomy of the pharynx and larynx. *Am. J. Med.*, 115 (Suppl 3A), 10S–14S.

5 Armstrong, W.B. and Netterville, J.L. (1995) Anatomy of the larynx, trachea, and bronchi. *Otolaryngol. Clin. North Am.*, 28, 685–699.

6 Jemal, A., Siegel, R., Ward, E., Hao, Y. et al. (2009) Cancer statistics, 2009. *CA. Cancer J. Clin.*, 59, 225–249.

7 Barnes, L., Tse, L.L.Y., Hunt, J.L., Brandwein-Gensler, M. et al. (2005) Tumours of the hypopharynx, larynx and trachea: Introduction, in *World Health Organization Classification of Tumours. Pathology and Genetics of Head and Neck Tumours* (eds L. Barnes, J.W. Eveson, P. Reichart, and D. Sidransky), IARC Press, Lyon, pp. 111–117.

8 Hoffman, H.T., Porter, K., Karnell, L.H., Cooper, J.S. et al. (2006) Laryngeal cancer in the United States: changes in demographics, patterns of care, and survival. *Laryngoscope*, 116, 1–13.

9 Pelucchi, C., Gallus, S., Garavello, W., Bosetti, C., and La Vecchia, C. (2008) Alcohol and tobacco use, and cancer risk for upper aerodigestive tract and liver. *Eur. J. Cancer Prev.*, 17, 340–344.

10 Sergeev, A., Gelikonov, V., Gelikonov, G., Feldchtein, F. et al. (1997) In vivo endoscopic OCT imaging of precancer and cancer states of human mucosa. *Opt. Express*, 1, 432–440.

11 Feldchtein, F.I., Gelikonov, G.V., Gelikonov, V.M., and Kuranov, R.V. et al.

(1998) Endoscopic applications of optical coherence tomography. *Opt. Express*, **3**, 257–270.

12 Shakhov, A., Terentjeva, A., Gladkova, N.D., Snopova, L. *et al.* (1999) Capabilities of optical coherence tomography in laryngology. *Proc. SPIE*, **3590**, 250–260.

13 Shakhov, A.V., Terentjeva, A.B., Kamensky, V.A., Snopova, L.B. *et al.* (2001) Optical coherence tomography monitoring for laser surgery of laryngeal carcinoma. *J. Surg. Oncol.*, **77**, 253–258.

14 Wong, B.J., Jackson, R.P., Guo, S., Ridgway, J.M. *et al.* (2005) *In vivo* optical coherence tomography of the human larynx: normative and benign pathology in 82 patients. *Laryngoscope*, **115**, 1904–1911.

15 Armstrong, W.B., Ridgway, J.M., Vokes, D.E., Guo, S. *et al.* (2006) Optical coherence tomography of laryngeal cancer. *Laryngoscope*, **116**, 1107–1113.

16 Klein, A.M., Pierce, M.C., Zeitels, S.M., Anderson, R.R. *et al.* (2006) Imaging the human vocal folds *in vivo* with optical coherence tomography: a preliminary experience. *Ann. Otol. Rhinol. Laryngol.*, **115**, 277–284.

17 Kraft, M., Luerssen, K., Lubatschowski, H., Glanz, H., and Arens, C. (2007) Technique of optical coherence tomography of the larynx during microlaryngoscopy. *Laryngoscope*, **117**, 950–952.

18 Kraft, M., Glanz, H., von Gerlach, S., Wisweh, H. *et al.* (2008) Clinical value of optical coherence tomography in laryngology. *Head Neck*, **30**, 1628–1635.

19 Kaiser, M.L., Rubinstein, M., Vokes, D.E., Ridgway, J.M. *et al.* (2009) Laryngeal epithelial thickness: a comparison between optical coherence tomography and histology. *Clin. Otolaryngol.*, **34**, 460–466.

20 Rubinstein, M., Fine, E.L., Sepehr, A., Armstrong, W.B. *et al.* (2010) Optical coherence tomography of the larynx using the Niris system. *J. Otolaryngol. Head Neck Surg.*, **39**, 150–156.

21 Ridgway, J.M., Ahuja, G., Guo, S., Su, J. *et al.* (2007) Imaging of the pediatric airway using optical coherence tomography. *Laryngoscope*, **117**, 2206–2212.

22 Ridgway, J.M., Su, J., Wright, R., Guo, S. *et al.* (2008) Optical coherence tomography of the newborn airway. *Ann. Otol. Rhinol. Laryngol.*, **117**, 327–334.

23 Sepehr, A., Armstrong, W.B., Guo, S., Su, J. *et al.* (2008) Optical coherence tomography of the larynx in the awake patient. *Otolaryngol. Head Neck Surg.*, **138**, 425–429.

24 Guo, S., Hutchison, R., Jackson, R.P., Kohli, A. *et al.* (2006) Office-based optical coherence tomographic imaging of human vocal cords. *J. Biomed. Opt.*, **11**, 30501.

25 Guo, S., Yu, L., Sepehr, A., Perez, J. *et al.* (2009) Gradient-index lens rod based probe for office-based optical coherence tomography of the human larynx. *J. Biomed. Opt.*, **14**, 014017.

26 Lueerssen, K., Wisweh, H., Ptok, M., and Lubatschowski, H. (2007) Optical characterization of vocal folds by OCT-based laryngoscopy. *Proc. SPIE*, **6424**, 64241O.

27 Yu, L., Liu, G., Rubinstein, M., Saidi, A. *et al.* (2009) Office-based dynamic imaging of vocal cords in awake patients with swept-source optical coherence tomography. *J. Biomed. Opt.*, **14**, 064020.

28 Vokes, D.E., Jackson, R., Guo, S., Perez, J.A. *et al.* (2008) Optical coherence tomography-enhanced microlaryngoscopy: preliminary report of a noncontact optical coherence tomography system integrated with a surgical microscope. *Ann. Otol. Rhinol. Laryngol.*, **117**, 538–547.

29 Just, T., Lankenau, E., Huttmann, G., and Pau, H.W. (2009) Intra-operative application of optical coherence tomography with an operating microscope. *J. Laryngol. Otol.*, **123**, 1027–1030.

30 Bibas, A.G., Podoleanu, A.G., Cucu, R.G., Bonmarin, M. *et al.* (2004) 3-D optical coherence tomography of the laryngeal mucosa. *Clin. Otolaryngol. Allied Sci.*, **29**, 713–720.

31 Burns, J.A., Zeitels, S.M., Anderson, R.R., Kobler, J.B. *et al.* (2005) Imaging the mucosa of the human vocal fold with optical coherence tomography. *Ann. Otol. Rhinol. Laryngol.*, **114**, 671–676.

32 Karamzadeh, A.M., Jackson, R., Guo, S., Ridgway, J.M. *et al.* (2005) Characterization of submucosal lesions using optical coherence tomography in the rabbit

subglottis. *Arch. Otolaryngol. Head Neck Surg.*, **131**, 499–504.
33 Luerssen, K., Lubatschowski, H., Gasse, H., Koch, R., and Ptok, M. (2005) Optical characterization of vocal folds with optical coherence tomography. *Proc. SPIE*, **5686**, 328–332.
34 Nassif, N.A., Armstrong, W.B., de Boer, J.F., and Wong, B.J. (2005) Measurement of morphologic changes induced by trauma with the use of coherence tomography in porcine vocal cords. *Otolaryngol. Head Neck Surg.*, **133**, 845–850.
35 Torkian, B.A., Guo, S., Jahng, A.W., Liaw, L.H. *et al.* (2006) Noninvasive measurement of ablation crater size and thermal injury after CO_2 laser in the vocal cord with optical coherence tomography. *Otolaryngol. Head Neck Surg.*, **134**, 86–91.
36 Wisweh, H., Merkel, U., Huller, A.-K., Luerssen, K., and Lubatschowski, H. (2007) Optical coherence tomography monitoring of vocal fold femtosecond laser microsurgery. *Proc. SPIE*, **6632**, 663207.
37 Burns, J.A., Kim, K.H., Kobler, J.B., deBoer, J.F. *et al.* (2009) Real-time tracking of vocal fold injections with optical coherence tomography. *Laryngoscope*, **119**, 2182–2186.

46
Optical Coherence Tomography of the Human Oral Cavity and Oropharynx

Marc Rubinstein, Davin Chark, and Brian Wong

46.1
Anatomy

The oral cavity (OC) is a complex organ, composed of salivary glands, teeth, muscles, and sensory receptors. It is bounded by the palate above, the floor of the mouth below and the cheek on each side. The OC has several functions, including sense of taste, mastication, deglutition, vocalization, and respiration. The OC can also be divided by subsites, which include the lip, buccal mucosa, floor of mouth, hard palate, oral tongue, alveolar ridges, and retromolar trigone, illustrated in Figure 46.1 [1–3].

The area where the soft palate joins the hard palate is the transition between the OC and the oropharynx (OP). The OP is bounded anteriorly by the soft palate, uvula, and tonsillar pillars, posteriorly it goes from the soft palate to the epiglottis, and laterally by the palatine tonsils. The OP has both respiratory and digestive functions; it is involved in the deglutition, respiration, and phonation process. The OP subsites include the base of the tongue, soft palate, palatine tonsils and posterior pharyngeal wall, illustrated in Figure 46.2 [1–3].

46.2
Progression of Disease

The American Cancer Society estimated that 35 720 new cases of OC and OP cancer was diagnosed in 2009 and accounted for 7600 deaths [4]. In the United States, these malignant neoplasms comprise 3.1% of all cancers in men and 2.5% in women [5]. Approximately 80% of the cancer in the OC and OP is squamous cell carcinoma (SCC) [6]. Premalignant lesions are usually precursors of this type of cancer. Lesions can present as a white lesion (leukoplakia) or red lesion (erythroplakia); although these lesions are not normal, they are not necessarily neoplastic. Currently the only way to differentiate if the lesion is benign, premalignant, or malignant is with a biopsy. Premalignant lesions are called dysplasia, atypical cells

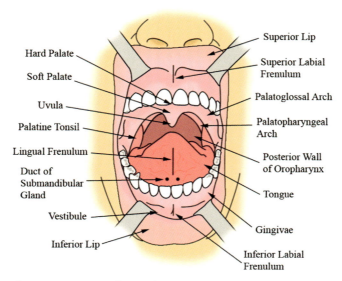

Figure 46.1 Diagram showing the human oral cavity.

that have not invaded the underlying tissue. Although dysplasia can progress to cancer, it can also regress. Erythroplakia is more likely to progress to cancer (51%) than leukoplakia (5%) [7]. The most important factor in long-term survival is early lesion detection and adequate treatment.

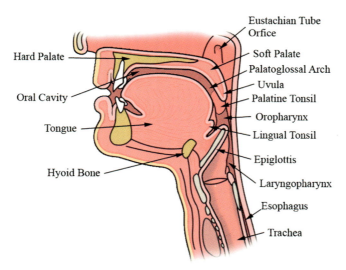

Figure 46.2 Diagram showing the human oropharynx.

46.3
Usefulness of OCT

Although there are many diagnostic tools and techniques for oral cancer management, none at present have shown significantly greater efficacy than conventional clinical examination [8], although this is an evolving situation. Recent human *in vivo* clinical studies have shown the efficacy of diagnosing premalignant and malignant diseases in the OC using optical coherence tomography (OCT) [9–11]. Tsai *et al.* [9] imaged 32 patients using OCT and diagnosed and distinguished epithelial hyperplasia, mild dysplasia, and SCC, differentiating between these three distinct pathologic entities and also from normal mucosa using A-line imaging-derived features. Likewise, Wilder-Smith *et al.* [10] imaged a total of 50 patients with normal mucosa, dysplasia, or malignant lesions. They found that OCT images compared favorably with histology, with high sensitivity and specificity for detecting these different types of lesions. Fomina *et al.* [11] used OCT to detect OC neoplasia *in vivo* in 35 patients, comparing OCT images with histology in benign or dysplastic/malignant lesions. They concluded that OCT has clinical application for diagnosis and treatment guidance and also for cancer screening (sensitivity 83% and specificity 98%).

OCT may also be used to guide surgical biopsies toward regions of higher diagnostic yield. Jerjes *et al.* [12] examined 34 *ex vivo* biopsies of 27 patients, and examined changes in the keratin layer, epithelial layer, and the lamina propria, and for the presence or absence of an intact basement membrane. Based on the architectural changes identified using OCT, biopsies were then performed in the sites of interest. One of the areas where OCT has its greatest clinical use in the OC and OP is in the management of patients with "field cancerization." Field cancerization is the presence of histologically abnormal tissue surrounding primary cancerous lesions of squamous cell carcinomas [13]. Tsai *et al.* [14] reported using an *ex vivo* model to delineate the cancer from normal tissue. Using two parameters, decay constant and standard deviation A-scan, they observed that while the decay constant decreased as the A-scan moved laterally away from the abnormal portion, the standard deviation A-scan increased when it was performed in the abnormal portion of the lesion. Also, having the ability to identify the transition zone between normal epithelium and malignant epithelium makes OCT a useful clinical tool (Figures 46.3 and 46.4).

Another clinical application of OCT is to monitor disease progression and change in a noninvasive manner. Tsai *et al.* [15] evaluated OC lesions in different oncologic stages and were able to differentiate between mild and moderate dysplasia.

OCT has also been used to evaluate and diagnose different pathologic conditions throughout the OC and OP [16–21]. One of the most common side effects of radiotherapy and/or chemotherapy is mucositis. OCT has the capability of detecting, in patients receiving radiotherapy and/or chemotherapy, early microstructural changes in the OC before the clinical manifestations appear, which may provide an effective and noninvasive method to evaluate early mucosal changes, and thus lead to early treatment of mucositis [16–18]. Ridgway *et al.* [21] studied normal and pathologic conditions in the OC and OP in 41 patients, and compared OCT images with histopathology, confirming the feasibility of using OCT to identify differences in

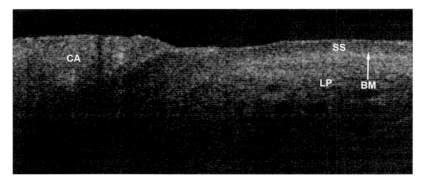

Figure 46.3 OCT image showing the transition zone in the buccal mucosa; note the loss of the basement membrane in the area of the cancer. LP, lamina propria; SS, stratified squamous epithelium; BM, basement membrane locus; CA, cancer.

the mucosal and submucosal tissue of the OC and OP. Other studies imaging the OC have focused on specific disease processes, such as oral submucous fibrosis [19] and vascular anomalies [20].

Several studies focused on imaging only normal healthy mucosa of the OC [22, 23]. Recently, Prestin *et al.* [24] used OCT to measure the epithelial thickness in different areas of the OC; this is an important study as thickness is one factor distinguishing neoplastic from normal tissues. Multiple OCT systems have been used to image the OC and OP; Table 46.1 presents a summary of some of these systems. There have also been several animal studies using OCT to evaluate various benign and malignant processes in the OC [25–31].

Although OCT has potential to aid clinicians in diagnosing OC and OP pathologies, there are limitations with this technology, such as a limited depth of penetration, the acquisition speed and the resolution. Improvements in theses areas are important to obtain better images at a faster rate and thus aid in the diagnosis and treatment of neoplasia.

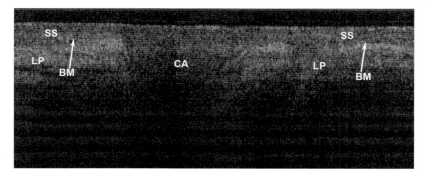

Figure 46.4 OCT image showing transition zones in the buccal mucosa; note the loss of the basement membrane in the area of the cancer. LP, lamina propria; SS, stratified squamous epithelium; BM, basement membrane locus; CA, cancer.

Table 46.1 Comparison of different OCT systems used to image the OC and OP in humans.

Authors	Year	Central wavelength (nm)	System[a]	Speed (kHz)	Axial resolution (μm)	Lateral resolution (μm)	Commercially available
Feldchtein et al. [22]	1998	830	TD	1	13	17	No
		1280			17	22	
Fomina et al. [11]	2004	1270	TD	0.75	15–20		Yes
Gladkova et al. [16, 17]	2005, 2006						
Ridgway et al. [21]	2006	1310	TD	0.5	10	10	No
Kawakami-Wong et al. [18]	2007						
Tsai et al. [9, 14, 15, 19]	2008, 2009	1310	SS	20	8	15	Yes
Lee et al. [19]	2009						
Ozawa et al. [20, 23]	2009	1260–1360	SS	20	11 (air) 8 (tissue)	37	Yes
Jerjes et al. [12]	2009	1310	SS	<10	<10	<7.5	Yes
Wilder-Smith et al. [10]	2009	1300	TD	0.75	10–20	25	Yes
Prestin et al. [24]	2010						

a) TD, time domain; SS, swept-source.

References

1 Travers, J.B. and Travers, S.P. (2005) Oral cavity, pharynx, esophagus, in *Cummings Otolaryngology: Head and Neck Surgery* (eds C.W. Cummings, P.W. Flint, B.H. Haughey, K.T. Robbins, *et al.*), Elsevier Mosby, Philadelphia, pp. 1409–1437.

2 Dhillon, N. (2007) Anatomy, in *Current Diagnosis and Treatment in Otolaryngology—Head and Neck Surgery* (ed. A. Lalwani), McGraw-Hill Medical, New York, pp. 1–30.

3 Lorenz, R.R., Netterville, J.L., and Burkey, B.B. (2008) Head and neck, in *Sabiston Textbook of Surgery* (eds C.M. Townsend, R.D. Beauchamp, B.M. Evers, and K.L. Mattox), Saunders Elsevier, Philadelphia, pp. 813–847.

4 American Cancer Society (2009) *Cancer Facts and Figures*, American Cancer Society, Atlanta, GA.

5 Jemal, A., Siegel, R., Ward, E., Hao, Y. *et al.* (2009) Cancer statistics. *CA. Cancer J. Clin.*, **59**, 225–249.

6 Horner, MJ., Ries, LAG., Krapcho, M., Neyman, N. *et al.* (2009) SEER Cancer Statistics Review, 1975–2006, National Cancer Institute. Bethesda, MD, http://seer.cancer.gov/csr/1975_2006/, based on November 2008 SEER data submission, posted to the SEER web site (last accessed 15 May 2011).

7 Lee, N. and Chan, K. (2007) Benign and malignant lesions of the oral cavity, oropharynx, and nasopharynx, *Current Diagnosis and Treatment in Otolaryngology—Head and Neck Surgery* (ed. A. Lalwani), McGraw-Hill Medical, New York, pp. 356–366.

8 DeCoro, M. and Wilder-Smith, P. (2010) Potential of optical coherence tomography for early diagnosis of oral malignancies. *Expert Rev. Anticancer Ther.*, **10**, 321–329.

9 Tsai, M.T., Lee, H.C., Lee, C.K., Yu, C.H. *et al.* (2008) Effective indicators for diagnosis of oral cancer using optical coherence tomography. *Opt. Express*, **16**, 15847–15862.

10 Wilder-Smith, P., Lee, K., Guo, S., Zhang, J. *et al.* (2009) *In vivo* diagnosis of oral dysplasia and malignancy using optical coherence tomography: preliminary studies in 50 patients. *Lasers Surg. Med.*, **41**, 353–357.

11 Fomina, J.V., Gladkova, N.D., Snopova, L. B., Shakhova, N.M. *et al.* (2004) *In vivo* study of neoplastic alterations of the oral cavity mucosa. *Proc. SPIE*, **5316**, 41–47.

12 Jerjes, W., Upile, T., Conn, B., Hamdoon, Z. *et al.* (2009) *In vitro* examination of suspicious oral lesions using optical coherence tomography. *Br. J. Oral Maxillofac. Surg.*, **48**, 18–25.

13 Slaughter, D.P., Southwick, H.W., and Smejkal, W. (1953) Field cancerization in oral stratified squamous epithelium; clinical implications of multicentric origin. *Cancer*, **6**, 963–968.

14 Tsai, M.T., Lee, H.C., Lu, C.W., Wang, Y.M. *et al.* (2008) Delineation of an oral cancer lesion with swept-source optical coherence tomography. *J. Biomed. Opt.*, **13**, 044012.

15 Tsai, M.T., Lee, C.K., Lee, H.C., Chen, H.M. *et al.* (2009) Differentiating oral lesions in different carcinogenesis stages with optical coherence tomography. *J. Biomed. Opt.*, **14**, 044028.

16 Gladkova, N., Maslennikova, A., Balalaeva, I., Vyseltseva, Y. *et al.* (2006) OCT visualization of mucosal radiation damage in patients with head and neck cancer : pilot study. *Proc. SPIE*, **6078**, 60781J–60788.

17 Gladkova, N., Maslennikova, A., Terentieva, A., Fomina, Y. *et al.* (2005) *Proc. SPIE*, 58610T–58618T.

18 Kawakami-Wong, H., Gu, S., Hammer-Wilson, M.J., Epstein, J.B. *et al.* (2007) *In vivo* optical coherence tomography-based scoring of oral mucositis in human subjects: a pilot study. *J. Biomed. Opt.*, **12**, 051702.

19 Lee, C.K., Tsai, M.T., Lee, H.C., Chen, H.M. *et al.* (2009) Diagnosis of oral submucous fibrosis with optical coherence tomography. *J. Biomed. Opt.*, **14**, 054008.

20 Ozawa, N., Sumi, Y., Chong, C., and Kurabayashi, T. (2009) Evaluation of oral vascular anomalies using optical coherence tomography. *Br. J. Oral Maxillofac. Surg.*, **47**, 622–626.

21 Ridgway, J.M., Armstrong, W.B., Guo, S., Mahmood, U. et al. (2006) *In vivo* optical coherence tomography of the human oral cavity and oropharynx. *Arch. Otolaryngol. Head Neck Surg.*, **132**, 1074–1081.

22 Feldchtein, F., Gelikonov, V., Iksanov, R., Gelikonov, G. et al. (1998) *In vivo* OCT imaging of hard and soft tissue of the oral cavity. *Opt. Express*, **3**, 239–250.

23 Ozawa, N., Sumi, Y., Shimozato, K., Chong, C., and Kurabayashi, T. (2009) *In vivo* imaging of human labial glands using advanced optical coherence tomography. *Oral Surg. Oral Med. Oral Pathol. Oral Radiol. Endod.*, **108**, 425–429.

24 Prestin, S., Betz, C., and Kraft, M. (2010) Measurement of epithelial thickness within the oral cavity using optical coherence tomography. Proc. SPIE, 7548, 75482F–75489F.

25 Kim, C.S., Wilder-Smith, P., Ahn, Y.C., Liaw, L.H. et al. (2009) Enhanced detection of early-stage oral cancer *in vivo* by optical coherence tomography using multimodal delivery of gold nanoparticles. *J. Biomed. Opt.*, **14**, 034008.

26 Hanna, N.M., Waite, W., Taylor, K., Jung, W.G. et al. (2006) Feasibility of three-dimensional optical coherence tomography and optical Doppler tomography of malignancy in hamster cheek pouches. *Photomed. Laser Surg.*, **24**, 402–409.

27 Wilder-Smith, P., Hammer-Wilson, M.J., Zhang, J., Wang, Q. et al. (2007) *In vivo* imaging of oral mucositis in an animal model using optical coherence tomography and optical Doppler tomography. *Clin. Cancer Res.*, **13**, 2449–2454.

28 Wilder-Smith, P., Krasieva, T., Jung, W.G., Zhang, J. et al. (2005) Noninvasive imaging of oral premalignancy and malignancy. *J. Biomed. Opt.*, **10**, 051601.

29 Matheny, E.S., Hanna, N.M., Jung, W.G., Chen, Z. et al. (2004) Optical coherence tomography of malignancy in hamster cheek pouches. *J. Biomed. Opt.*, **9**, 978–981.

30 Jung, W.G., Zhang, J., Chung, J.R., Wilder-Smith, P. et al. (2005) Advances in oral cancer detection using optical coherence tomography. *IEEE J. Sel. Top. Quantum Electron.*, **11**, 811–817.

31 Wilder-Smith, P., Jung, W.G., Brenner, M., Osann, K. et al. (2004) *In vivo* optical coherence tomography for the diagnosis of oral malignancy. *Lasers Surg. Med.*, **35**, 269–275.

**Part 11
Neurology**

47
In Vivo Brain Imaging and Diagnosis
Christoph Krafft and Matthias Kirsch

47.1
Introduction

In general, biophotonic imaging methods give information regarding intrinsic optical properties of brain tissue and the presence or absence of endogenous or exogenous chromophores. The benefits of using light in imaging living tissue include that light can provide high sensitivity to functional changes and can reveal the dynamics of cells in the nervous system, via either absorption of light, emission of light (fluorescence and phosphorescence), and elastic or inelastic scattering. Further advantages are diffraction-limited spatial resolution in the micrometer and even submicrometer range, nondestructive sampling, use of nonionizing radiation, fast data collection, relatively inexpensive instrumentation and reduced infrastructure requirements compared with other clinical imaging methods. A disadvantage of all optical techniques is their limited penetration depth in brain tissue, which is dependent on its absorption and scattering properties. High absorption exists in both the visible wavelength range (350–700 nm), mainly due to hemoglobin, and in the wavelength range above 900 nm, due to lipids, proteins, and water. However, in the range 700–900 nm, the absorption of all biomolecules is weak; hence imaging using light in this interval maximizes tissue penetration while simultaneously minimizing autofluorescence from nontarget tissue. The most widely measured intrinsic chromophores are oxy- and deoxyhemoglobin, alongside cytochromes and metabolites which have distinctive absorption and fluorescence properties. Cytochrome oxidase is a mitochondrial enzyme that plays an important role in oxygen metabolism in the brain. The concentration of cytochrome oxidase in the brain is lower than that of hemoglobin (2–3 µM compared with 40–60 µM). Extrinsic chromophores such as absorbing and fluorescent dyes in addition to transgenic methods can also provide specific optical contrast and are often able actively to report functional parameters. Optical imaging modalities which probe molecular vibrations as an inherent feature of all molecules do not depend on chromophores. Molecular vibrations can be excited by absorption of mid-infrared (MIR) radiation and inelastic scattering of monochromatic light.

Handbook of Biophotonics. Vol.2: Photonics for Health Care, First Edition. Edited by Jürgen Popp,
Valery V. Tuchin, Arthur Chiou, and Stefan Heinemann.
© 2012 Wiley-VCH Verlag GmbH & Co. KGaA. Published 2012 by Wiley-VCH Verlag GmbH & Co. KGaA.

Among the first applications of optical brain imaging was a paper by Jöbsis in 1977 [1]. He measured blood and tissue oxygenation changes in the brain of a cat using near-infrared (NIR) light. Since then, the field has grown, encompassing both basic science and diagnostic applications that exploit the interactions between light and tissue to image the brain. Optical *in vivo* imaging of the diseased nervous system [2], optical brain imaging *in vivo* [3], and vibrational spectroscopic imaging in the context of neuro-oncology [4] have recently been summarized. Currently, numerous parts of the nervous system can be visualized by optical *in vivo* imaging, including the cortex, cerebellum, olfactory bulb, retina, spinal cord, peripheral nerves, and autonomic ganglia. Transgenic labels have been developed for tracing many structures in the nervous system, including neurons, axon tracts, most glial populations, oligodendrocytes, brain vasculature, and invading tumor cells. This has allowed a growing number of models to be studied in mice.

Magnetic resonance imaging (MRI), X-ray computed tomography (CT) and positron emission tomography (PET) are valuable and established tools for brain imaging. MRI and X-ray CT require shielded rooms for protection, and PET requires synchrotrons to generate the short-lived radioactive isotopes of carbon, nitrogen, fluorine, or oxygen. X-ray CTs and structural MRI have excellent spatial resolution and allow the detection of morphologic changes in the brain, for example, to delineate tumors. In particular, functional brain imaging has seen significant growth since the availability of functional magnetic resonance imaging (fMRI) in 1990. Applications cover neurosurgical planning, the investigation of the physiologic basis of neurologic diseases such as epilepsy, Alzheimer's disease, and stroke, the development of diagnostic methods, drugs, treatments, and interventions, and the study of cognitive and perceptual responses and developmental changes, to name just a few [5]. Owing to their broad range of contrast mechanisms and their ability to detect and image a wide range of functional parameters, optical imaging technologies can complement the established clinical modalities.

For example, to visualize regions of the brain exhibiting functional changes in response to stimuli, fMRI typically exploits the blood oxygen level-dependent (BOLD) signal, widely thought to correlate inversely with deoxygenated hemoglobin (HbR) concentration. However, a BOLD increase (an HbR decrease) may correspond to an increase in oxygenation, or a decrease in blood volume. Newer fMRI methods such as arterial spin labeling can provide measures of flow, and intravenous contrast agents such as monocrystalline iron oxide nanocompounds can allow cerebral blood volume changes to be observed. These measurements cannot easily be acquired simultaneously and only the combination of the methods provides insight into the true hemodynamic responses and oxygenation dynamics in the brain. PET is performed using contrast agents such as 2-fluoro-2-deoxy-D-glucose containing radioactive isotopes. Subsequently, glucose utilization in the brain can be imaged via the localization of these isotopes. However, resolution is typically poor, data acquisition is slow, and hence dynamics cannot be readily evaluated. Further disadvantages are that contrast agents must be manufactured locally prior to each scan, and radioactive dose limits its use in certain patient populations, especially infants. The major drawback of both fMRI and PET imaging is that they cannot be used during a surgical procedure or for extensive amounts of time. In addition, especially fMRI is not as reliable as morphologic analysis [6, 7].

A major challenge in optical imaging for noninvasive clinical applications is to overcome the effects of elastic light scattering, which limits the penetration depth and achievable imaging resolution. Therefore, optical imaging encompasses a wide range of measurement techniques with the aim of overcoming the effects of scatter. These range from laser scanning microscopy of surface accessible regions with submicron resolution to diffuse optical tomography of deeper and larger volumes of tissue. According to Abbe's law of diffraction, the lateral resolution of conventional light microscopy is limited to $\delta_{lat} = 0.61\lambda/NA$ and the axial resolution to $\delta_{ax} = 2\lambda n/NA^2$ (where λ = wavelength; n = refractive index; NA = numerical aperture). Although ultraviolet and visible light offer higher resolutions due to smaller wavelengths, clinical optical brain imaging generally uses NIR light to obtain improved penetration through the scalp, skull, and brain due to minimal absorption of intrinsic chromophores.

Optical brain imaging is finding widespread applications as a research tool for both clinical and animal studies of brain functions, diseases, and treatments of pathologies. The structures of the human brain and principles to assess brain tissue composition and brain function using optical brain imaging are displayed in Figure 47.1. However, little is known about the way in which the normal brain functions, in part due to the difficulties of measuring such a complex organ without disturbing or damaging its *in vivo* function. Optical imaging allows the living brain to be closely observed, and many functional interactions and changes to be investigated over various length and time scales. Imaging of animals (commonly mouse, rat, cat, or primate) rather than humans provides more flexibility since preparations can be much more controlled and diseases and treatments can be systematically compared. Furthermore, extrinsic dyes and cross-validation techniques can be utilized and developed without the need for clinical regulatory approval. Since the adverse effects

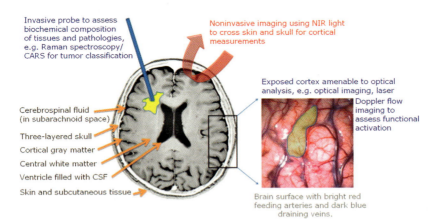

Figure 47.1 Structures of the human brain and principal technologies to assess brain tissue composition and brain function using biophotonic technologies.

of light scattering worsen as the size of the tissue being interrogated increases, imaging of smaller animals therefore also offers significant technical advantages, allowing higher resolution imaging and improved sensitivity and quantitation. For very high-resolution imaging, the cortex can be surgically exposed, allowing direct optical imaging of the brain's surface with only minimal disturbance to normal brain activity. For long-term experiments, the cortex is covered by implanting a permanent cranial window. A small cover-slip is placed over the opening of the skull and is sealed using acrylic cement. This procedure effectively reseals the skull, allows intracranial pressure to re-equalize, prevents contamination or dissolution of ambient gas, and controls brain movement which would otherwise be substantially affected by heart rate and breathing. Exposed cortex optical imaging is highly versatile, and most commonly performed in animals, although it has also been performed on the human brain, for example, during epilepsy surgery. Optical imaging of the exposed cortex is currently the only technique that can provide sufficient resolution to reveal detailed information of functional responses for sensory and cognitive processing research.

In this overview, we describe various optical imaging approaches to brain imaging in neurology and neurosurgery as summarized in Table 47.1. We introduce the techniques concisely in the context of brain imaging. All techniques are described

Table 47.1 Summary of optical imaging techniques.

Technique	Attributes
Optical imaging	Techniques including microscopy and biophotonic methods that use light to visualize the distribution of cells and molecules.
Wide-field microscopy	The entire field of view is illuminated simultaneously to produce an image.
Scanning microscopy	A focused beam of light (usually a laser) is moved (scanned) across the field of view to detect the signal of each spot sequentially (includes confocal microscopy and multiphoton microscopy).
Confocal microscopy	A technique that allows optical sectioning by placing a pinhole in front of the detector to exclude out-of-focus light and generate a sharp image of structures in the focal plane.
Tomography	Imaging approaches in which the object is imaged from multiple angles and mathematical algorithms are used to reconstruct the three-dimensional structure of the signal source.
Intrinsic optical imaging	A technique that detects changes in the optical properties of brain tissue such as absorption or scattering.
Fluorescence microscopy	Detection of fluorescence emission of intrinsic or extrinsic chromophores after absorption of light.
Calcium imaging	Calcium indicator dyes (that is, calcium-dependent fluorophores) in combination with microscopy are used to measure changes in intracellular calcium concentration.
Multiphoton microscopy	Scanning microscopy technique in which a pulsed near-infrared laser is used to achieve optical sectioning by limiting (multiphoton) excitation to the perifocal plane (includes two-photon microscopy and second-harmonic generation).

Table 47.1 (Continued)

Technique	Attributes
Second-harmonic generation microscopy	Near-infrared lasers generate frequency-doubled signals from the scattering of light as it interacts with specific macromolecules. Structures such as cytoskeleton and collagen can be visualized.
Two-photon microscopy	Two photons of a pulsed near-infrared laser are absorbed and excite a fluorophore which emits a higher energy photon.
Raman microscopy	Inelastic scattering of monochromatic light probes molecular vibrations (includes nonlinear variant coherent anti-Stokes Raman scattering using high intensity pulsed lasers).
Optical coherence tomography	A technique that uses interferometry of reflected or backscattered light from biological tissues.

more comprehensively in Part I of this Handbook. We then focus on examples of *in vivo* brain imaging: Section 47.3 details the use in neuro-oncology and imaging of functional areas, Section 47.4 covers vascularity and Section 47.5 regeneration.

47.2
Basic Principles

47.2.1
Two-Dimensional Camera-Based Imaging: Absorption

2D camera–based imaging of the exposed cortex is the simplest optical imaging technique and is often referred to as "optical imaging" in the neuroscience literature. The term "optical imaging of intrinsic signals" (OIS) is also known because intrinsic optical properties such as absorption and scattering are detected which can be measured directly at the brain surface. This brain imaging technique can visualize brain compartments with micrometer and millisecond resolution. The most significant absorbers in the brain at visible and NIR wavelengths are oxyhemoglobin (HbO_2) and deoxyhemoglobin (HbR). Their distinctive absorption spectra yield quantitative chemical information about their concentration, blood flow, blood volume, and oxygenation. Blood delivers oxygen to tissue by locally converting HbO_2 to HbR, so changes in the oxygenation state of blood correspond to changes in the relative concentrations of HbO_2 and HbR. Consequently, variations in the amount and oxygenation state of hemoglobin modulate the absorption properties of the brain. By simply shining light into the exposed cortex and taking pictures with a camera, the hemodynamic properties can be readily observed. Furthermore, since HbR and HbO_2 have different absorption spectra, measurements with different wavelengths of light can produce images that are preferentially sensitive to changes in the concentration of either HbO_2 or HbR. At isosbestic points where the HbR and HbO_2 absorptions are approximately the same (e.g., 500, 530, 570, and 797 nm), changes in total hemoglobin (HbT) concentration (HbT = HbR + HbO_2) can be measured independently of changes in blood oxygenation. To calculate the actual changes in the concentrations of HbO_2, HbR, and HbT, the absorption of the brain must be measured at two or more wavelengths. This is expressed by the term hyperspectral imaging. OIS is usually described as an invasive imaging technique because it requires surgery to make the cortex visible. However, as intraoperative OIS is performed in the operating room by attaching a charge-coupled device (CCD) camera and optical filters to the operating microscope, it is noninvasive compared with the current gold standard intraoperative electrophysiologic techniques, which require placing electrodes directly on to the brain. Intraoperative IOS has been demonstrated to be a potentially useful neurosurgical tool for both functional brain mapping and lesion delineation (see overview by Prakash *et al.* [8]).

47.2.2
Two-Dimensional Camera Based Imaging: Fluorescence

The fluorescence of intrinsic and extrinsic chromophores can be imaged by a two-dimensional camera-based setup which consists of an excitation filter of the appropriate wavelength and a long-pass filter in front of the camera to block excitation light and isolate the fluorescence emission. In some cases, a dichroic mirror is used such that the illumination light is reflected on to the brain, and emitted fluorescence passes through the dichroic mirror to the camera.

Extrinsic voltage-sensitive dyes (VSDs) provide a way to detect optically the neuronal activity in the *in vivo* exposed cortex. VSDs are molecules that generally bind across a neuron's membrane. Changes in membrane potential cause measurable changes in the fluorescence of the dye. VSDs can only be used in animals at present, and are usually topically applied directly to the exposed cortex for 1–2 h before imaging commences. Some VSDs have been shown to photobleach quickly or have phototoxic effects, and so illumination times are typically kept to a minimum. Calcium-sensitive dyes (CaSDs) are compounds that increase their fluorescence in response to increases in calcium ion concentration. When cells such as neurons are loaded with CaSDs, it becomes possible to monitor intracellular calcium. Since calcium influx occurs when a neuron fires an action potential, the calcium dye's fluorescence can optically report this, providing a second tool with which to image neuronal activity optically. The challenge with CaSDs is to introduce them into the neurons successfully. Concentration changes in intrinsic flavoproteins (FADs) in the brain can be measured from their intrinsic fluorescence *in vivo*. FADs provide a measure of metabolic changes in cells (neurons and possibly glia), and have a distinct time course compared with shorter neuronal and longer hemodynamic responses.

47.2.3
Two-Dimensional Camera-Based Imaging: Laser Speckle

Laser speckle flow imaging is capable of imaging the blood flow dynamics in the superficial cortex. Flow is a very important parameter if the rate of oxygen delivery and consumption are to be calculated. Laser Doppler imaging has been widely used to provide point measurements of blood flow in the brain. However, laser speckle flow imaging of the exposed cortex can image the spatial distribution of the flow throughout the vascular network during functional activation. In this case, the light source is replaced with a divergent laser diode. The camera then simply acquires rapid and high-resolution images of the exposed cortex, which appear as low-contrast images of the speckle pattern from the laser illumination. The speckle pattern is caused by the coherent laser light scattering within the brain. If red blood cells are moving within the image, they cause the speckle pattern to vary over time. The rate at which the speckles change relates to how fast the red blood cells are moving.

47.2.4
Two-Dimensional Fiber Bundle-Based Imaging

Although most measurements use anesthetized animals, some studies of native behavior require that the animal be both awake and able to move around freely. Recently, a technique was developed in which the exposed cortex of a mouse could be imaged via a flexible fiber bundle with high temporal and spatial resolution [9]. The principle is that excitation light propagates in one direction through an optical fiber bundle while emitted fluorescence returns via the same path, is reflected by a dichroic mirror, and is imaged by a camera. Microlenses project and enlarge small fluorescent brain structures on to a fiber bundle containing well-ordered arrays of glass cores for flexible image transfer from the mouse to the camera. These fiber bundles typically contain thousands of fiber cores whose centers are spaced by around $10\,\mu m$ apart. Using a fast CCD camera, the authors achieved image acquisition rates up to 100 Hz over imaging fields of 0.25–0.33 mm.

47.2.5
Three-Dimensional Imaging: Optical Coherence Tomography

Optical coherence tomography (OCT) is analogous to ultrasound imaging except that reflections of NIR radiation, rather than sound, are detected. In ultrasonography, the time delay of reflected sound waves is used to generate an image of tissue structures. Because of the high velocity of light, a time delay of reflected light cannot be measured directly. Therefore, light is split into probe and reference light. The reference light and the probe light are combined by a beamsplitter and registered by a detector. Interference of low-coherence light occurs only when the optical pathlengths of the reference and probe are matched within the coherence length of the light source, which means that the difference between the optical path of the reference and the probe light determines the depth in the sample at which the magnitude of reflection is registered. Interferometers with fiber-optic couplers and beam scanning systems have been implemented to perform OCT measurements through microscopes, fiber-optic catheters, endoscopes, and hand-held probes. The contrast in OCT is generally due to refractive index mismatches. OCT is typically performed using NIR wavelengths $>1\,\mu m$. OCT synthesizes cross-sectional images from a series of laterally adjacent depth scans. OCT allows noninvasive real-time *in vivo* imaging of brain tissues up to a 3 mm penetration depth with a spatial resolution of 10–15 μm and in ultrahigh-resolution mode even better (0.9 μm axial, 2 μm lateral) [10, 11].

47.2.6
Confocal Microscopy

Confocal microscopy relies on the rejection of scattered light by isolating signals originating from the focus of the scanning beam. Confocal excitation wavelengths are typically in the visible spectrum, where tissue scatter and absorption are high. Therefore, when using confocal microscopy to image depths beyond 200–300 μm, the laser beam can no longer focus, and images become blurry and lack sensitivity. Confocal microscopy can also be combined with fluorescence emission. While it is assumed that the detected fluorescence light originates from the focus of the scanning beam, some of the detected signal is also generated by excitation and emission light that is scattered or absorbed within the tissue above and below the focus. This light not only contributes to image blurring, but can also result in cumulative photodamage to areas of the tissue that are not actively being imaged.

47.2.7
Two-Photon Microscopy

Since two-photon microscopy was first demonstrated in 1990 [12], it has become an invaluable tool for imaging intact biological specimens at very high resolution, to depths of up to 600 μm. The advantages have been summarized and discussed in depth [13]. Two-photon microscopy has also been applied to *in vivo* imaging of the brain, which has previously been reviewed in the context of brain function monitoring [14]. Although fluorescent contrast is required, two-photon microscopy provides an unprecedented view of *in vivo* brain activity on a cellular and microvascular level. In particular, sophisticated transgenic technology allows the labeling of single cells in the nervous system and permits their function to be monitored. Similarly to confocal microscopy, two-photon microscopy requires a focused beam of laser light to be steered within the tissue, sensing the properties of each location and using them to form a 2D or 3D image. This technique overcomes many of the disadvantages of confocal microscopy. Instead of fluorescence excitation at visible wavelengths, two-photon microscopy utilizes laser light of twice the excitation wavelength. When sufficient photon flux is achieved, it is usually possible for a fluorophore to be excited by two of these lower energy photons arriving in quick succession. Once excited, the fluorophore will emit a photon at or near its usual emission wavelength. This high photon flux is achievable if a pulsed laser is used which delivers high-energy, rapid pulses. Typically, a Ti:sapphire laser is used, which is tunable between 700 and 1000 nm, with 80 MHz pulse repetition, <500 fs pulse width, mean power 1 W, and peak power 25 kW. Even with such a high peak power, the nonlinear effect of two-photon excitation will only occur at the tight focus of the laser beam. Consequently, two-photon microscopy can use NIR light to image the same fluorophores as confocal microscopy. However, since scattering and absorption of NIR light in tissue are much lower than for visible light, the focus of the laser beam can be maintained at depths of >600 µm, depending on the tissue type. The fact that the fluorophore is excited only at the very focus of the beam means that surrounding tissue does not experience significant photodamage. As NIR excitation is generally spectrally well separated from the emission wavelength, excitation light is rejected much more simply than for confocal microscopy. Further, since any light emerging from the tissue is expected at the fluorescent emission wavelength, it can only have originated from the focus of the scanning beam and it is not necessary to reject scattered light as required in confocal microscopy. Sensitive detectors can be placed as close to the tissue as possible to collect all of the emerging emission light. This signal, when reformed according to the beam scanning pattern, produces a high-resolution image of the fluorescent structures within tissue. Whereas traditional microscopy required tissue to be stained and sectioned, two-photon microscopy allows high-resolution 3D imaging of intact, functioning tissue.

47.2.8
Second-Harmonic Generation

Although endogenous multiphoton fluorescence imaging has found a number of applications in brain imaging, the small number of naturally fluorescent compounds limits its general use. Only small modifications are required to equip a laser-scanning two-photon microscope for second-harmonic generation (SHG) imaging microscopy. The nonlinear optical effect SHG is also commonly called frequency doubling. The phenomenon requires intense laser light passing through a highly polarizable material with a noncentrosymmetric molecular organization. Collagen within connective tissue, muscle thick filaments, and microtubule arrays within interphase and mitotic cells belong to the biological material that can produce SHG signals. The second-harmonic light emerging from the material is at precisely half the wavelength of the light entering the material. Hence the SHG process within the material changes two NIR incident photons into one emerging visible photon at exactly twice the energy (and half the wavelength). Like that of two-photon microscopy, the amplitude of SHG is proportional to the square of the incident light intensity. Therefore, SHG microscopy has the same intrinsic optical sectioning characteristics. Hence SHG microscopy offers a contrast mechanism that does not require excitation of fluorescent molecules. The fact that only selected proteins reveal contrast in SHG imaging is a limiting factor. Applications for visualizing biomolecular arrays in cells, tissues, and organisms have been summarized [15].

47.2.9
Inelastic Light Scattering: Raman Spectroscopy

Raman spectroscopy is based on inelastic scattering of monochromatic light. This process can excite molecular vibrations. As numerous vibrations of biomolecules are probed simultaneously, more bands are observed in vibrational spectra of cells and tissue than in other optical spectra, giving a fingerprint-like signature. These bands provide information about the biochemistry, structure, and composition of the underlying sample. Among the main advantages is that this information is obtained without labels. Potential autofluorescence is reduced and penetration of incident and scattered light is enhanced by excitation with NIR lasers. The relatively weak signals require intense laser excitation and highly sensitive instrumentation. Continuous-wave lasers are operated at powers of a few hundred milliwatts to prevent tissue degradation. Optical components for detection are optimized for high throughput and low light losses. Furthermore, signal amplification methods (resonance Raman scattering, surface-enhanced scattering, coherent anti-Stokes Raman scattering, stimulated Raman scattering) have been developed to increase sensitivity. Raman images which provide molecular contrast are usually registered in the sequential point mapping mode, which means that the laser focus or the sample is moved from one spot to the other.

47.2.10
Noninvasive Optical Brain Imaging: Topography, Tomography

For routine optical brain imaging in humans, it is clearly necessary to develop noninvasive techniques. They must overcome the effects of light scattering, while maintaining the benefits of optical contrast to study functional activation and to investigate pathologies. Since light penetration and scattering are significant obstacles for noninvasive imaging, measurements on children and babies are easier to achieve than on adults. It is often difficult to image these infants and small children using fMRI and PET, so optical imaging provides a unique opportunity to study many aspects of functional brain development. Optical imaging of the adult brain can also provide valuable functional information that complements modalities such as fMRI. As a particular advantage of its ability to image HbO_2, HbR, and HbT simultaneously, optical imaging is widely adopted for studies of the cortical hemodynamic response to a wide range of stimuli.

Direct topography is the most widespread approach to clinical optical brain imaging and is often referred to as near-infrared spectroscopy (NIRS). Optical fibers are generally used to transmit light to and from the head. Instrumentation may be continuous wave (cw), frequency domain (FD), or time domain (TD) using two or more wavelengths. In its simplest form, NIRS uses light source and detector pairs positioned on the head. Light that passes though the skull scatters within the cortex of the brain. Despite significant attenuation, variations in absorption due to changes in local HbT, HbO_2, and HbR concentrations result in detectable modulations in the intensity of the emerging light in the cortical surface of the brain where the majority of functional processing occurs. If an array of sources and detectors is used, the changes measured between a source–detector pair can be approximately mapped or interpolated to the underlying cortex and a low-resolution (>5 mm) 2D topographical map of cortical activation can be produced. Data processing is an important issue to consider systematic errors due to breathing or hearth rate. A further difficulty is reliable placement of sources and detectors on the head. Hair on the head is a major challenge as it can significantly attenuate light entering and leaving the head, and can cause measurement instability over time if the probes shift. Finally, light propagation in the human head has to be modeled using either Monte Carlo algorithms or the diffusion approximation to the radiative transfer equation.

In addition to absorption, photon emission can be detected, preferentially using NIR fluorescence or red-shifted bioluminescence reporters that take advantage of the low absorption of biological tissues in this spectral window. However, one limitation is the low spatial and temporal resolution of this technique. Although photons from deep within the brain can reach the surface, they are scattered along the way so that their origin is obscured. Another limitation is that biophotonic signals are weakened by absorption, which restricts temporal resolution by long acquisition times. In general, detection of light through >7 cm of tissue will cause around 10^{-7} attenuation. New fluorophores that might improve sensitivity are becoming available, such as quantum dots (that is, semiconductor nanocrystals that have unique fluorescence features).

The most complex approach to optical brain imaging involves performing a full 3D reconstruction of the entire head using an approach similar to X-ray CT. As for tomographic imaging, a series of measurements or projections through the tissue need to be imaged and light has to be transmitted across the whole head; this approach has only been demonstrated on small infants. However, the benefits are that in addition to functional imaging, this approach can potentially permit imaging of pathologies. For reconstructed topography, measurements between sources and detectors with different separations allow some depth-resolved information to be deduced. In tomography, the same basic image reconstruction approach is used, but a larger number of projections through the head at different angles are required to calculate the 3D structure deep within the brain. Fluorescence or bioluminescence imaging requires a 3D reconstruction of the location of the cells that emit the signal. A comprehensive introduction to biophotonic imaging techniques and their current capabilities and limitation has been given elsewhere [16].

47.3
Applications in Neuro-Oncology and Functional Imaging

Surgical removal of brain tumors is the most common initial treatment received by brain tumor patients because it relieves the mass effect of the tumor on neurologic tissue and allows histopathologic diagnosis of the tumor, which directly affects the direction of follow-up therapeutic strategy. Aggressive surgical resection results in prolonged survival of patients with a malignant glioma, may lower the risk of anaplastic progression in a low-grade glial tumor, and may thereby increase the survival length and quality of life in these patients. To achieve this goal with minimal neurologic deficits, an accurate identification of brain tumor margins during surgery is required. However, the intraoperative detection of residual tumor tissue, especially in a low-grade tumor, is difficult because of the low optical contrast between tumor and adjacent brain. Moreover, a glioma lacks a true tumor to brain interface; instead, single cells migrate along preformed anatomic structures into the adjacent brain beyond the highly cellular margin of the tumor. Currently, neurosurgeons determine brain tumor margins intraoperatively by visual inspection and information provided by surgical navigation systems that are based on X-ray CT/MRI and/or intraoperative ultrasound (IOUS). However, limitations of these techniques include the following:

- The true infiltrating margins may not be visible on X-ray CT/MRI images.
- Registration errors and intraoperative brain shift can degrade the spatial accuracy of the surgical navigation systems.
- Primary brain tumors, unlike most metastatic tumors, do not typically possess clear boundaries, that is, the margins appear blurred in the IOUS images.
- It is often difficult, even for experienced neurosurgeons, to differentiate visually low-grade gliomas and associated tumor margins from normal brain tissue.

Because of their complexity, IOUS and intraoperative MRI cannot provide a continuous online imaging of the resection cavity, and integration of these technologies into microsurgical instruments or operating microscopes is limited. Applications of IOUS and MRI require a pause in the course of the brain tumor resection for intermittent analyses, which is realistic only for a limited number of investigations. Furthermore, the spatial resolution of IOUS in brain tissue is typically no higher than 150–250 μm, and with the advancement of the surgical procedure, artifacts diminish the use of IOUS. In particular, the creation of new surfaces, alteration of the acoustic properties of the tumor–tissue interfaces by surgical interactions, and the introduction of foreign material (e.g., for hemostasis or therapeutic interactions) generate new structures that limit comparability of IOUS images within the same procedure. The spatial resolution of MRI is between 0.5 and 1 mm which is insufficient to resolve tissue microstructure and to discriminate between tissue alterations due to tumor infiltrations and those resulting from surgery. However, the main advantages of both IOUS and MRI are the penetration depth and the routine volumetric sampling. This eases interpretation and co-registration to other imaging modalities.

Biophotonic imaging techniques are attractive candidates for analyzing brain tissue nondestructively in real time and for providing intraoperative *in vivo* diagnosis. It is of clinical interest to obtain data on both the morphology and the composition of the tissues. A combination of biophotonic imaging techniques with conventional imaging should allow for rapid introduction into clinical routine.

47.3.1
Absorption-Based Imaging

Determination of tumor oxygenation at the microvascular level will provide important insight into tumor growth, angiogenesis, necrosis, and therapeutic response. NIRS has the potential to differentiate lesions and hemoglobin dynamics. The vascular status and the pathophysiologic changes that occur during tumor vascularization were studied in an orthotopic brain tumor model [17]. A noninvasive multimodal approach based on NIRS along with MRI was applied for monitoring the concentrations of HbO_2, HbR, and water within the tumor region. The concentrations of these compounds were determined using 15 discrete wavelengths in a spectral window of 670–780 nm. A direct correlation between tumor size, intratumoral microvessel density, and tumor oxygenation was found.

To clarify the characteristics of the evoked cerebral blood oxygenation changes occurring in stroke and brain tumors, noninvasive NIRS and BOLD fMRI were compared during functional brain activation [18]. NIRS enabled functional brain stimulation and activation to be imaged through measurement of changes in hemoglobin concentrations. The approach was applied to monitor 12 glioma patients during recovery after neurosurgery [19]. The sensitivity and specificity of NIRS for hemoglobin were used to image functional brain stimulation and activation through measurement of changes in hemoglobin concentrations. Functional activation of specific areas of the brain can be elicited by, for example, sensory stimulation of the arm, hand, or leg (somato-sensory evoked potentials). A corresponding region of activation can be correlated with the stimulated area, and the image can be overlaid into an operating microscope and used to guide an intracerebral approach.

47.3.2
Fluorescence-Based Imaging

Autofluorescence with 337 nm excitation can discriminate solid tumors from normal brain tissues based on their reduced fluorescence emission at 460 nm compared with normal tissue. False-positive rates constitute a general problem because not all regions showing reduced fluorescence are tumors [20]. A hand-held optical spectroscopic probe was applied in clinical trials that combined autofluorescence with diffuse reflection spectroscopy [21, 22]. Spectra of brain tumors and infiltrating tumor margins were separated from spectra of normal brain tissues *in vivo* using empirical algorithms with high sensitivity and specificity. Blood contamination was found to be a major obstacle.

Multiphoton-excited autofluorescence using NIR femtosecond laser pulses allows the reconstruction of 3D microanatomic images of native tissues at a subcellular level without the need for contrast-enhancing markers or histologic stains. Multiphoton autofluorescence microscopy could visualize solid tumor, the tumor–brain interface, and single invasive tumor cells in gliomas of mouse brains and of human biopsies [23, 24]. Furthermore, fluorescence lifetime imaging of endogenous fluorophores provides an additional parameter in so-called 4D microscopy, since the different decay times of the fluorescent signal could differentiate tumor and normal brain tissue [16, 25].

47.3.2.1 5-ALA
In order to improve selectivity, fluorescent markers have been applied to malignant brain tissue. These markers include indocyanine green [26], fluorescein [27], and fluorescein–albumin [28]. Fluorescent porphyrins accumulate in malignant brain tissue after administration of the metabolic precursor 5-aminolevulinic acid (ALA) [29]. These fluorophores are taken up by individual glioma cells, but not in

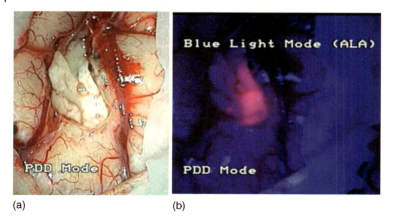

Figure 47.2 Intraoperative macroscopic (∼5× magnification) view of subcortical ALA + glioma tissue. (a) Normal white light image of the operative field. The adjacent normal cortical structures and vessels are easily delineated. (b) The ALA-derived fluorescence depicts viable pink fluorescent tumor within the opened cortex. Although being revolutionary to neurosurgery, the images also reveal the limitations: macroscopic resolution, no morphologic information, single pathology classification.

normal brain tissue [29]. Single glioma cells can be detected. However, given the limited magnification usable for a surgical procedure, the ratio of infiltrative cells to normal parenchyma is seen as a shift from bright orange–pink to a soft pinkish tint of the tumor margin (Figure 47.2).

The technique can be integrated in an operating microscope and the fluorescence images permitted intraoperative detection of the brain tumors and a safer tumor resection (Figure 47.2). Results from a human clinical trial showed that ALA-derived fluorescence with peaks at 635 and 704 nm could predict the presence of tumor tissue with a specificity of 100% and a sensitivity of 85% with respect to histopathology [30]. A multicenter clinical phase III trial of 322 patients revealed that this approach led to improved progression-free survival of patients with malignant gliomas [31]. However, identification of low-grade tumors and tumor margins could be problematic. It remains unclear whether the fluorophore requires an opening of the blood–brain barrier which is disturbed within solid glioma tissue, since the current recommendations suggest a combined application of ALA and dexamethasone, a drug that seals the blood–brain barrier and has been shown to improve fluorescence. The technique needs planning and is not available any time or on demand during surgery, since a long metabolic phase has to be taken into account. The substance has to be applied several hours prior to the operation. Second, a certain degree of photobleaching of the fluorophore occurs with fluorescence decay within 25 min for excitation light and ∼90 min for white light [29]. Third, ALA-derived fluorescence is only positive in malignant gliomas as a pathognomic sign. Many other brain tumor types are only positive in a small percentage of cases, such as pituitary adenomas, meningiomas, and the frequent cerebral metastases. These problems have diminished the capability of induced fluorescence for brain tumor demarcation so far. Normal intraoperative magnification of up to 20-fold does not allow for individual cell detection. Moreover, infiltrating glioma cannot be discerned from normal tissue. This exemplifies the need for better imaging tools to delineate fine morphologic structures and their classification with regard to tissue types. ALA-derived imaging is revolutionary to neurosurgery, but is monocentric with regard to its possible applications and does not allow morphologic information to be obtained. It is not available to all patients since contraindications exist with regard to hepatic co-morbidities. Since ALA is given systemically, the patient's skin is also sensitive to bright light exposure. This light sensitivity is mild and temporary as it is attenuated to a normal level within 24 h. However, patients should be exposed for 1 day to ambient light only.

47.3.2.2 Nanoparticles

Nanoparticle-based-platforms have attracted considerable attention for their potential effect in oncology and other biomedical fields. A fluorescent nanoprobe was presented that is able to cross the blood–brain barrier and specifically target brain tumors in a genetically engineered mouse model [32]. Animals were investigated *in vivo* using magnetic resonance and biophotonic imaging, and histologic and biodistribution analysis. The nanoprobe consists of an iron oxide nanoparticle coated with biocompatible poly(ethylene glycol)-grafted chitosan copolymer, to which the tumor-targeting agent chlorotoxin and an NIR fluorophore are conjugated. After the nanoparticles have been injected, they accumulate and remained in brain tumors for up to 5 days. The study demonstrated that the nanoparticles can improve contrast in both MRI and optical imaging during surgery. The authors called their technique brain tumor illumination or brain tumor painting.

47.3.2.3 Quantum Dots

Quantum dots (QDs) are optical semiconductor nanocrystals that offer prospects as labels due to their stable, bright fluorescence. The intravenous injection of QDs is accompanied by macrophage sequestration. Macrophages infiltrate brain tumors and phagocytize intravenously injected QDs, optically labeling the tumors. Macrophage-mediated delivery of QDs to brain tumors may represent a novel technique to label tumors preoperatively [33, 34]. The surface of QDs can be coupled with antibodies specific to one or several tumor epitopes, leading to selective enhancement of tumor cells rather than indirect labeling by macrophage sequestration [35]. Proven technologies for antibody coupling using streptavidin-conjugated QDs with biotinylated antibodies may soon be employed [36]. This technology is particularly interesting for neurosurgery since many different fluorochromes can be used and different epitopes can be labeled. This extends the potential usability to other pathologies such as Parkinson's disease and other degenerative central nervous system (CNS) disorders. QDs might serve for both detection and delivery of therapeutic substances. QDs within tumors may be detected within 30 min after application [37] employing optical imaging and optical spectroscopy, providing the surgeon with real-time optical feedback during the resection and biopsy of brain tumors.

47.3.3
Bioluminescence Imaging

Bioluminescence imaging (BLI) has emerged as a sensitive imaging technique for small animals. BLI exploits the emission of photons based on energy-dependent reactions catalyzed by luciferases with a maximum depth of 2–3 cm. Luciferases comprise a family of photoproteins that emit photons in the presence of oxygen and adenosine triphosphate (ATP) during metabolism of substrates such as luciferin into oxyluciferin. Luciferins are injected immediately before data acquisition. The light from these enzyme reactions typically has very broad emission spectra that frequently extend beyond 600 nm, with the red components of the emission spectra being the most useful for imaging by virtue of easy transmission through tissues [38]. In the context of neuro-oncology, BLI was used to monitor the formation of grafted tumors *in vivo*, to measure cell numbers during tumor progression and response to therapy [39], and to monitor the proliferative activity of glioma cells and cell cycle in a genetically engineered mouse model of glioma *in vivo* [40]. The ability to image two or more biological processes in a single animal can greatly increase the utility of luciferase imaging by offering the opportunity to distinguish the expression of two reporters biochemically. Dual BLI was used to monitor gene delivery via a therapeutic vector and to follow the effects of the therapeutic protein TRAIL (tumor necrosis factor-related apoptosis-inducing ligand) in gliomas [41]. TRAIL has been shown to induce apoptosis in neoplastic cells and may offer new prospects for tumor treatment.

47.3.4
OCT-Based Imaging

OCT-based noncontact imaging of brain tissue during neurosurgical procedures is challenging because after opening of the dura, the target volume follows the respiratory and arterial cycle, resulting in movements several millimeters in amplitude. Slow scan times will result in distortion of the tissue surface contour or may result in the area of interest moving out of the measurement window. Therefore, short scan acquisition times are crucial. In particular, spectral-domain (SD)-OCT allows rapid scanning times of three images per second, which would be sufficient to suppress motion artifacts of the relatively slow movements of the brain exposed during operations. Both time domain (TD)- and SD-OCT of experimental gliomas in mice and human brain tumor specimens delineated normal brain, infiltration zone, and solid tumor based on the tissue microstructure and signal characteristics [10, 42]. The same authors also presented the first feasibility study of intraoperative OCT [43]. Post-image acquisition processing for noninvasive imaging of the brain and brain tumor was used to compensate for image distortion caused by pulse- and respiration-induced movements of the target volume. Based on different microstructure and characteristic signal attenuation profile, OCT imaging discriminated normal brain, areas of tumor-infiltrated brain, solid tumor, and necrosis. Ultrahigh-resolution OCT discriminated between healthy and pathologic human brain tissue biopsies by visualizing and identifying microcalcifications, enlarged cell nuclei, small cysts, and blood vessels [44]. The working distance of the OCT adapter recently allowed integration of the OCT applicator into the optical path of operating microscopes. This permits a continuous analysis of the resection plain in neurosurgery, the identification of vocal cord tumors, and *in situ* measurements of the auditory ossicles.

OCT imaging, which is based on reflection of radiation, has its strength in clarifying the architectural tissue morphology. However, for many diseases, including cancer, diagnosis based on such architectural features is not sufficient since the most important diagnostic indicators of neoplastic changes are features such as accelerated rate of growth, mass growth, local invasion, lack of differentiation, anaplasia, and metastasis, which mainly occur on the subcellular and molecular level. An approach to overcome this limitation is the combination of complementary techniques. Optical intrinsic signal imaging (OISI) provides two-dimensional, depth-integrated activation maps of brain activity. OCT provides depth-resolved, cross-sectional images of functional brain activation. Co-registered OCT and OISI imaging was performed simultaneously on the rat somatosensory cortex through a thinned skull during forepaw electrical stimulation, which correlated with OISI relating to the functional activation patterns, indicating retrograde vessel dilation [45].

47.3.5
Raman-Based Techniques

One of the potential advantages of Raman spectroscopy in neuro-oncology is that brain tissue can be characterized label free at the microstructural and/or molecular level with spatial resolution in the single cell range. This permits accurate delineation of tumor margins, and sensitive and specific identification of tumor remnants upon preservation of normal tissue. Raman images of pristine human brain tissue specimens were collected *ex vivo* by a Raman spectrometer coupled to a microscope with a low magnification objective ($10 \times / NA\ 0.25$) [46]. Regions of 2×2 mm of 2 mm tissue sections were covered by a window to prevent drying during data acquisition. Raman images were acquired using a step size of 100 μm. The laser intensity of 60 mW was focused to a spot ~60 μm in diameter. No tissue degradation was observed during signal collection of 20 s per spectrum. Raman spectra of white matter were characterized by maximum spectral contributions of lipids and cholesterol. In the Raman spectra of gray matter, the bands of proteins and water were more prominent whereas bands of lipids were less intense. In Raman spectra of gliomas, spectral contributions of lipids further decreased and spectral contributions of hemoglobin increased, which was consistent with hemorrhage or higher blood perfusion. An interesting feature was the accumulation of phosphatidylcholine, which was indicated by a Raman marker band near $717\,\text{cm}^{-1}$ and which was confirmed by NMR spectroscopy and chromatography.

Raman systems can be coupled to fiber-optic probes. This approach has been applied to detect metastasis in mouse brains [47]. Injection of tumor cells in the carotid artery induces tumor development preferentially in one brain hemisphere. The unaffected brain hemisphere can be used as a control for normal murine brain tissue. Owing to the improved sensitivity of the fiber-optic probe compared with the microscope, Raman images were acquired with a shorter exposure time of 8 s per spectrum. Raman spectra of brain metastases from malignant melanomas showed additional spectral contributions of the pigment melanin. The band intensities were resonance enhanced with 785 nm excitation because the pigment showed electronic absorption of the excitation wavelength. The presence of melanin in brain metastasis of malignant melanomas demonstrated that secondary tumors possess molecular properties of the primary tumors. As Raman spectroscopy probes molecular fingerprint, Raman spectra can be applied to identify the origin the brain metastases. The related vibrational spectroscopy technique of Fourier transform infrared (FTIR) imaging classified the primary tumor of the four most frequent brain metastases [48, 49]. As modern FTIR imaging spectrometers combine the Fourier transform and the multi-channel advantage of IR-sensitive focal plane array detectors, FTIR image acquisition of extended tissue sections took just minutes. However, owing to the high water content of tissue and cells and the strong absorption of MIR radiation by water, the penetration depth is just a few micrometers and application under *in vivo* conditions is limited.

The acquisition time can be further decreased using coherent anti-Stokes Raman scattering (CARS) microscopy, which is a nonlinear variant of Raman spectroscopy using picosecond pulsed lasers. It has been applied to image structure and pathology in fresh unfixed and unstained *ex vivo* brain tissue and tumor [50]. CARS was tuned to specific vibrational bands to provide image contrast with subcellular resolution and near real-time temporal resolution. However, only single bands can be probed in short exposure times. Most frequently, bands near $2845\,\mathrm{cm}^{-1}$ are probed that are particularly intense in lipid-rich structures.

47.4
Applications in Neurovascular Pathologies

Diseases that affect the neurovascular unit of the CNS, such as ischemic stroke, are particularly suited to optical *in vivo* studies. Ischemia can be easily induced in animals, and vessels can be outlined by intravenous injection of fluorescent dextran conjugates [51]. The resultant negative contrast of erythrocytes which exclude the dextran can be used to assess blood flow. Earlier, flow measurements were taken in the pia mater and in the upper cortical layers using wide-field or confocal microscopy. Later, measuring changes in cortical blood flow in response to sensory stimulation was one of the first applications of multiphoton microscopy which allowed the visualization of capillary flow down to cortical layer IV in rats (see overview by Misgeld and Kerschensteiner [2]). Transcranial analysis of the vascular architecture was used to monitor tumor vascularization and also to monitor antivascular anti-angiogenic therapeutic effects in tumors [52–55] and to monitor therapeutic pro-angiogenic responses of ischemic lesions.

The intravasal fluorescent dye indocyanine green (ICG), which labels serum proteins in the blood, has been introduced into clinical routine in neurovascular

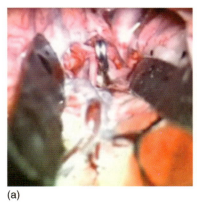 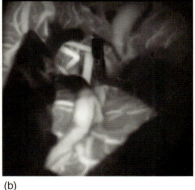

Figure 47.3 Intraoperative vascular fluorescence imaging using ICG. Perfused vessels reveal NIR-fluorescence (b) that can be superimposed on the white light picture (a). The fluorescent bolus of labeled blood can be analyzed using video microscopy and allows one to assess direction of flow, diameter of vessels, and type of vessel since arteries are perfused before and veins after the parenchymal phase.

procedures (Figure 47.3). Operative microscope-integrated ICG video angiography, as an intraoperative method for detecting vascular flow, was found to be rapid, reliable, and cost-effective and possibly a substitute or adjunct for Doppler ultrasonography or intraoperative conventional digital subtraction angiography.

The simplicity of the method, the speed with which the investigation can be performed, and the quality of the images help to improve manipulation of cerebral vessels and to control the surgical manipulations. Using video microscopy, the temporal changes in perfusion of arteries, then tissue perfusion followed by venous drainage, allow quantification of the type of vessel, direction of flow, and the perfusion of the supplied tissue. Perfusion was imaged using ICG in acute stroke patients undergoing decompressive craniotomy to visualize the extent of nonperfused and penumbral tissue [56]. The delineation of the nonperfused tissue can facilitate the placement of parenchymal probes for pO_2/pCO_2 and pH measurements *in situ*.

47.4.1
Absorption-Based Imaging

The growth and expansion of gliomas are highly dependent on vascular neogenesis. An association of microvascular density and tumor energy metabolism is assumed in most human gliomas. Intraoperative NIRS was applied to 13 glioma patients in order to elucidate the relationship between microvascular blood volume, oxygen saturation, histology, and patient survival [57]. High intratumoral blood volume and high oxygen saturation in malignant brain tumors of the glioblastoma multiforme type correlated with a low median survival time. Limitations of oxygen saturation data acquired using NIR technology are that (i) only tumors with appearance on the pial surface are accessible for noninvasive NIR spectroscopy probes and (ii) the total blood volume of the whole tumor usually cannot be assessed since intraoperative NIR measurements analyze the tumor surface up to a depth of only 4 mm.

47.4.2
Thermography

The local analysis of spontaneous emission of infrared radiation, and therefore of thermal radiation, is termed thermography. Thermography is based on infrared radiation of wavelength 1200–2500 nm. In experimental neurosurgical applications, small differences in temperature have been used to delineate surgically exposed arterial from venous vessels, especially in complex cerebral and spinal vascular lesions [58, 59], and to detect alterations of cortical blood flow [60] as early as 1970. In particular, the control of tissue perfusion during vascular procedures might be feasible in the exposed brain [61, 62]. More complex analysis of thermographic images might help to verify functional areas on the cortical surface during neurosurgical procedures using similar physiologic effects as for intrinsic optical imaging: an increase in blow flow due to local increase in metabolism should result in thermographic differences that correlate with active functional regions [63, 64].

47.5
Applications to Regeneration, Transplantation, and Stem Cell Monitoring

In vivo optical imaging is a powerful tool to study how axons in the peripheral nervous system behave during degeneration and regeneration. Axonal transaction (axotomy) is a common form of neurologic damage. In the peripheral nervous system, axotomy is followed by regeneration and recovery. This is not the case in the CNS, where regeneration fails with the dire consequences that are known from brain and spinal injuries. Axotomy was the first pathologic alteration of the nervous system to be imaged *in vivo* using intrinsic contrast [65, 66]. Further examples are summarized in a review by Misgeld and Kerschensteiner [2].

47.5.1
Identification of Cellular Fractions and Monitoring of Implants in the CNS

The identification of neural stem and progenitor cells (NPCs) using *in vivo* brain imaging could have important implications for diagnostic, prognostic, and therapeutic purposes. In several rodent models using various subtypes of stem cells for targeting lesions of the CNS, the cells were labeled prior to implantation. Labeling with, for example, magnetic iron particles led to improved identification of these cells after systemic or focal injection [67–71]. It was possible to visualize whether the stem cells would migrate to and remain within the lesions. It was not possible to monitor their local differentiation and whether they participated in cellular regeneration. Using a cranial window technique and multiphoton microscopy, cellular movements and vascular growth can be visualized [52, 72–74].

Recently, magnetic resonance spectroscopy defined a fingerprint for neural stem cells in humans using a biomarker in which NPCs are enriched and demonstrated its use as a reference for monitoring neurogenesis [75]. *In vivo* magnetic resonance spectroscopy in rodents and humans helped to identify regions with high expression of the signature indicating higher concentrations of neural stem cells. However, this notion has been challenged since then [76]. Similar purposes can be met using biophotonic techniques, for example, to assess cellular composition for targeting of specific brain regions. An example is the deposition of implants in the degenerated substantia nigra or areas within the basal ganglia. A combination of established electrophysiologic analyses along with newer spectroscopic techniques to verify the biochemical composition of the target area might improve the biological accuracy of the procedure. A second example is based on the current inability of any method to monitor transplants and differentiation of transplanted cells except for biopsies. A biophotonic approach seems feasible to overcome these limitations.

47.6 Conclusions

The potential applications of biophotonic imaging techniques in the field of basic and of clinical neurosciences are numerous. The possibility of probing for morphologic, that is, structural information and also for biochemical information holds great promise for the different modalities. Neurosurgery in particular might benefit from these evolving technologies. They might offer, alone or in combination with existing technologies and routine applications in brain surgery, enhancement of the visual details, and analysis of tissue for both pathologic changes – for example, by Raman spectroscopy or CARS – or to obtain online, label-free information on the functional status of an exposed cortical region (Figure 47.3). From a neurosurgical perspective, intraoperative optical imaging technologies should comply with the following requirements:

- integration into existing intraoperative setups, especially the operating microscope
- wide-field data generation, usually several cm^2
- depth information, generation of volume data sets
- image fusion into volumetric conventional morphologic MRI data sets
- lateral resolution: depiction of morphology
- chemometric resolution: depiction of biochemical composition = classification of tissue types
- temporal resolution: depiction of function
- rapid (online), continuous, or repeatable, reproducible, and robust
- low affection due to frequent intraoperative artifacts, such as contamination by CSF, blood or irrigation
- no time interval for preparation of tissue/patient before/after imaging
- no toxicity from excitation light source.

In spite of numerous successful applications, *in vivo* optical imaging is not without pitfalls. There are two fundamentally different modes of *in vivo* imaging and each has its own problems. On the one hand, real-time observations visualize a process as it happens. On the other hand, intermittent imaging collects multiple observations. Although real-time imaging is more direct, intermittent imaging alleviates many concerns about toxicity. However, finding and orienting cells for each session pose a new set of challenges. Artifacts can be induced by anesthesia, labeling, and surgery. Such invasive procedures can change the normal behavior of cells and the brain might be susceptible to such stress factors. In the future, novel miniature microscopes and fiber optics might permit improved experiments.

Too much light is toxic to cells. Toxic effects of light on cells limit most *in vivo* microscopy experiments. They are caused at least in part by bleaching and associated photochemical generation of free radicals. Phototoxicity itself is unacceptable unless used for therapeutic means, for example, in photodynamic therapy (PDT). Phototoxic damage which is induced in brain structures can be misinterpreted as pathology-related changes. Potential solutions include the following

- using as little excitation light as possible and the most sensitive detector possible
- filtering out irrelevant wavelengths from the excitation light
- considering multiphoton microscopy
- controlling temperature, oxygenation, and pH to reduce cellular stress.

The living brain is in constant motion due to motor activity, respiration, and heartbeat that induce movement and orientation artifacts. Repeatedly identifying and orienting the same cells under the microscope can be challenging during intermittent imaging protocols. Well-engineered equipment (such as stereotactic holders) helps to immobilize the target. Pressure gradients can be reduced by, for example, closing a craniotomy in small animals with agarose and a cover-slip. Image acquisition or illumination can be triggered from the electrocardiogram. Ventilation may be interrupted for short periods of time (<1 min) while collecting images. Fast imaging allows images to be taken that are little affected by pulse-related blurring. Subsequent x–y alignment and averaging can improve the signal-to-noise ratio.

Finally, instruments for *in vivo* imaging, especially multiphoton microscopy, can be expensive. To reduce costs, many parts of the nervous system can be imaged with a relatively affordable wide-field microscope. Furthermore, detailed descriptions have been published about how to build one's own multiphoton microscope. These instruments can be tailored to specific needs, and have often outperformed those available commercially [13].

In summary, an optimistic picture is emerging of what optical *in vivo* imaging could offer to basic and clinical neuroscientists. Further progress will depend on technological innovations and integration with accepted standards in clinical care and clinical technology. Among the most exciting perspectives is a new field, dubbed nanoscopy, which might overcome the ultimate frontier of light microscopy – that is, the diffraction limit of resolution – and which aims to expand light microscopy into the subcellular or even the molecular realm. The fundamental principle is to create excitation volumes that are much smaller than those that can be generated in conventional microscopy. Current nanoscopic approaches use complicated optical tools. Newly identified proteins that have light-inducible fluorescent properties could permit nanoscopic imaging with simpler optics that is applicable to living multicellular organisms. Therefore, rather than depending on physicists alone, the future development of optical *in vivo* imaging might be a multidisciplinary task with contributions made by chemists and molecular biologists.

References

1 Jöbsis, F.F. (1977) Noninvasive, infrared monitoring of cerebral and myocardial oxygen sufficiency and circulatory parameters. *Science*, **198**, 1264–1267.
2 Misgeld, T. and Kerschensteiner, M. (2006) In vivo imaging of the diseased nervous system. *Nat. Rev. Neurosci.*, **7**, 449–463.
3 Hillman, E.M. (2007) Optical brain imaging *in vivo*: techniques and applications from animal to man. *J. Biomed. Opt.*, **12**, 051402.
4 Krafft, C. and Salzer, R. (2008) Neuro-oncological applications of infrared and Raman spectroscopy, in *Vibrational Spectroscopy for Medical Diagnosis* (eds M. Diem, J.M. Chalmers, and P.R. Griffiths), John Wiley&Sons, Ltd, Chichester, pp. 231–259.
5 Matthews, P.M., Honey, G.D., and Bullmore, E.T. (2006) Applications of fMRI in translational medicine and clinical practice. *Nat. Rev. Neurosci.*, **7**, 732–744.
6 Havel, P., Braun, B., Rau, S., Tonn, J.C., Fesl, G., Bruckmann, H., and Ilmberger, J. (2006) Reproducibility of activation in four motor paradigms. An fMRI study. *J. Neurol.*, **253**, 471–476.
7 Rau, S., Fesl, G., Bruhns, P., Havel, P., Braun, B., Tonn, J.C., and Ilmberger, J. (2007) Reproducibility of activations in Broca area with two language tasks: a functional MR imaging study. *AJNR Am. J. Neuroradiol.*, **28**, 1346–1353.
8 Prakash, N., Uhlemann, F., Sheth, S.A., Bookheimer, S., Martin, N., and Toga, A.W. (2009) Current trends in intraoperative optical imaging for functional brain mapping and delineation of lesions of language cortex. *Neuroimage*, **47** (Suppl 2), T116–T126.
9 Flusberg, B.A., Nimmerjahn, A., Cocker, E.D., Mukamel, E.A., Barretto, R.P., Ko, T.H., Burns, L.D., Jung, J.C., and Schnitzer, M.J. (2008) High-speed, miniaturized fluorescence microscopy in freely moving mice. *Nat. Methods*, **5**, 935–938.
10 Böhringer, H.J., Lankenau, E., Rohde, V., Huttmann, G., and Giese, A. (2006) Optical coherence tomography for experimental neuroendoscopy. *Minim. Invasive Neurosurg.*, **49**, 269–275.
11 Giese, A., Kantelhardt, S.R., Oelckers, S., Lankenau, E., Huttmann, G., and Rohde, V. (2008) Certified prototype of an OCT–integrated operating microscope for intraoperative optical tissue analysis. Presented at 59. Jahrestagung der Deutschen Gesellschaft für Neurochirurgie, 3. Joint Meeting mit der Italienischen Gesellschaft für Neurochirurgie, Würzburg, 1–4 June 2008
12 Denk, W., Strickler, J.H., and Webb, W.W. (1990) Two-photon laser scanning fluorescence microscopy. *Science*, **248**, 73–76.
13 Zipfel, W.R., Williams, R.M., and Webb, W.W. (2003) Nonlinear magic: multiphoton microscopy in the biosciences. *Nat. Biotechnol.*, **21**, 1369–1377.
14 Garaschuk, O., Milos, R.I., Grienberger, C., Marandi, N., Adelsberger, H., and Konnerth, A. (2006) Optical monitoring of brain function *in vivo*: from neurons to networks. *Pflugers Arch.*, **453**, 385–396.
15 Campagnola, P.J. and Loew, L.M. (2003) Second-harmonic imaging microscopy for visualizing biomolecular arrays in cells, tissues and organisms. *Nat. Biotechnol.*, **21**, 1356–1360.

16 Ntziachristos, V., Ripoll, J., Wang, L.V., and Weissleder, R. (2005) Looking and listening to light: the evolution of whole-body photonic imaging. *Nat. Biotechnol.*, **23**, 313–320.

17 Saxena, V., Gonzalez-Gomez, I., and Laug, W.E. (2007) A noninvasive multimodal technique to monitor brain tumor vascularization. *Phys. Med. Biol.*, **52**, 5295–5308.

18 Sakatani, K., Murata, Y., Fujiwara, N., Hoshino, T., Nakamura, S., Kano, T., and Katayama, Y. (2007) Comparison of blood-oxygen-level-dependent functional magnetic resonance imaging and near-infrared spectroscopy recording during functional brain activation in patients with stroke and brain tumors. *J. Biomed. Opt.*, **12**, 062110.

19 Fujiwara, N., Sakatani, K., Katayama, Y., Murata, Y., Hoshino, T., Fukaya, C., and Yamamoto, T. (2004) Evoked-cerebral blood oxygenation changes in false-negative activations in BOLD contrast functional MRI of patients with brain tumors. *Neuroimage*, **21**, 1464–1471.

20 Goujon, D., Zellweger, M., Radu, A., Grosjean, P., Weber, B.C., van den Bergh, H., Monnier, P., and Wagnieres, G. (2003) In vivo autofluorescence imaging of early cancers in the human tracheobronchial tree with a spectrally optimized system. *J. Biomed. Opt.*, **8**, 17–25.

21 Lin, W.C., Toms, S.A., Johnson, M., Jansen, E.D., and Mahadevan-Jansen, A. (2001) In vivo brain tumor demarcation using optical spectroscopy. *Photochem. Photobiol.*, **73**, 396–402.

22 Toms, S.A., Lin, W.C., Weil, R.J., Johnson, M.D., Jansen, E.D., and Mahadevan-Jansen, A. (2005) Intraoperative optical spectroscopy identifies infiltrating glioma margins with high sensitivity. *Neurosurgery*, **57**, 382–391.

23 Kantelhardt, S.R., Leppert, J., Krajewski, J., Petkus, N., Reusche, E., Tronnier, V.M., Huttmann, G., and Giese, A. (2007) Imaging of brain and brain tumor specimens by time-resolved multiphoton excitation microscopy *ex vivo*. *Neuro-Oncology*, **9**, 103–112.

24 Kantelhardt, S.R., Diddens, H., Leppert, J., Rohde, V., Huttmann, G., and Giese, A. (2008) Multiphoton excitation fluorescence microscopy of 5-aminolevulinic acid induced fluorescence in experimental gliomas. *Lasers Surg. Med.*, **40**, 273–281.

25 Leppert, J., Krajewski, J., Kantelhardt, S.R., Schlaffer, S., Petkus, N., Reusche, E., Huttmann, G., and Giese, A. (2006) Multiphoton excitation of autofluorescence for microscopy of glioma tissue. *Neurosurgery*, **58**, 759–767.

26 Haglund, M.M., Berger, M.S., and Hochman, D.W. (1996) Enhanced optical imaging of human gliomas and tumor margins. *Neurosurgery*, **38**, 308–317.

27 Kabuto, M., Kubota, T., Kobayashi, H., Nakagawa, T., Ishii, H., Takeuchi, H., Kitai, R., and Kodera, T. (1997) Experimental and clinical study of detection of glioma at surgery using fluorescent imaging by a surgical microscope after fluorescein administration. *Neurol. Res.*, **19**, 9–16.

28 Kremer, P., Wunder, A., Sinn, H., Haase, T., Rheinwald, M., Zillmann, U., Albert, F.K., and Kunze, S. (2000) Laser-induced fluorescence detection of malignant gliomas using fluorescein-labeled serum albumin: experimental and preliminary clinical results. *Neurol. Res.*, **22**, 481–489.

29 Stummer, W., Stocker, S., Novotny, A., Heimann, A., Sauer, O., Kempski, O., Plesnila, N., Wietzorrek, J., and Reulen, H.J. (1998) In vitro and in vivo porphyrin accumulation by C6 glioma cells after exposure to 5-aminolevulinic acid. *J. Photochem. Photobiol. B*, **45**, 160–169.

30 Stummer, W., Stocker, S., Wagner, S., Stepp, H., Fritsch, C., Goetz, C., Goetz, A.E., Kiefmann, R., and Reulen, H.J. (1998) Intraoperative detection of malignant gliomas by 5-aminolevulinic acid-induced porphyrin fluorescence. *Neurosurgery*, **42**, 518–525.

31 Stummer, W., Pichlmeier, U., Meinel, T., Wiestler, O.D., Zanella, F., and Reulen, H.J. (2006) Fluorescence-guided surgery with 5-aminolevulinic acid for resection of malignant glioma: a randomised controlled multicentre phase III trial. *Lancet Oncol.*, **7**, 392–401.

32 Veiseh, O., Sun, C., Fang, C., Bhattarai, N., Gunn, J., Kievit, F., Du, K., Pullar, B., Lee, D., Ellenbogen, R.G., Olson, J., and Zhang, M. (2009) Specific targeting of

brain tumors with an optical/magnetic resonance imaging nanoprobe across the blood–brain barrier. *Cancer Res.*, **69**, 6200–6207.
33 Jackson, H., Muhammad, O., Daneshvar, H., Nelms, J., Popescu, A., Vogelbaum, M.A., Bruchez, M., and Toms, S.A. (2007) Quantum dots are phagocytized by macrophages and colocalize with experimental gliomas. *Neurosurgery*, **60**, 524–529.
34 Muhammad, O., Popescu, A., and Toms, S.A. (2007) Macrophage-mediated colocalization of quantum dots in experimental glioma. *Methods Mol. Biol.*, **374**, 161–171.
35 Arndt-Jovin, D.J., Kantelhardt, S.R., Caarls, W., de Vries, A.H., Giese, A., and Jovin Ast, T.M. (2009) Tumor-targeted quantum dots can help surgeons find tumor boundaries. *IEEE Trans. Nanobiosci.*, **8**, 65–71.
36 Sabharwal, N., Holland, E.C., and Vazquez, M. (2009) Live cell labeling of glial progenitor cells using targeted quantum dots. *Ann. Biomed. Eng.*, **37** (10), 1967–1973.
37 Wang, J., Yong, W.H., Sun, Y., Vernier, P.T., Koeffler, H.P., Gundersen, M.A., and Marcu, L. (2007) Receptor-targeted quantum dots: fluorescent probes for brain tumor diagnosis. *J. Biomed. Opt.*, **12**, 044021.
38 Shah, K. and Weissleder, R. (2005) Molecular optical imaging: applications leading to the development of present day therapeutics. *NeuroRx*, **2**, 215–225.
39 Burgos, J.S., Rosol, M., Moats, R.A., Khankaldyyan, V., Kohn, D.B., Nelson M.D. Jr., and Laug, W.E. (2003) Time course of bioluminescent signal in orthotopic and heterotopic brain tumors in nude mice. *Biotechniques*, **34**, 1184–1188.
40 Uhrbom, L., Nerio, E., and Holland, E.C. (2004) Dissecting tumor maintenance requirements using bioluminescence imaging of cell proliferation in a mouse glioma model. *Nat. Med.*, **10**, 1257–1260.
41 Shah, K., Tang, Y., Breakefield, X., and Weissleder, R. (2003) Real-time imaging of TRAIL-induced apoptosis of glioma tumors *in vivo*. *Oncogene*, **22**, 6865–6872.
42 Bohringer, H.J., Boller, D., Leppert, J., Knopp, U., Lankenau, E., Reusche, E., Huttmann, G., and Giese, A. (2006) Time-domain and spectral-domain optical coherence tomography in the analysis of brain tumor tissue. *Lasers Surg. Med.*, **38**, 588–597.
43 Bohringer, H.J., Lankenau, E., Stellmacher, F., Reusche, E., Huttmann, G., and Giese, A. (2009) Imaging of human brain tumor tissue by near-infrared laser coherence tomography. *Acta Neurochir. (Wien)*, **151**, 507–517.
44 Bizheva, K., Unterhuber, A., Hermann, B., Povazay, B., Sattmann, H., Fercher, A.F., Drexler, W., Preusser, M., Budka, H., Stingl, A., and Le, T. (2005) Imaging *ex vivo* healthy and pathological human brain tissue with ultra-high-resolution optical coherence tomography. *J. Biomed. Opt.*, **10**, 11006.
45 Chen, Y., Aguirre, A.D., Ruvinskaya, L., Devor, A., Boas, D.A., and Fujimoto, J.G. (2009) Optical coherence tomography (OCT) reveals depth-resolved dynamics during functional brain activation. *J. Neurosci. Methods*, **178**, 162–173.
46 Krafft, C., Sobottka, S.B., Schackert, G., and Salzer, R. (2005) Near infrared Raman spectroscopic mapping of native brain tissue and intracranial tumors. Analyst, 130, 1070–1077.
47 Krafft, C., Kirsch, M., Beleites, C., Schackert, G., and Salzer, R. (2007) Methodology for fiber-optic Raman mapping and FTIR imaging of metastases in mouse brains. *Anal. Bioanal. Chem.*, **389**, 1133–1142.
48 Krafft, C., Shapoval, L., Sobottka, S.B., Geiger, K.D., Schackert, G., and Salzer, R. (2006) Identification of primary tumors of brain metastases by SIMCA classification of IR spectroscopic images. *Biochim. Biophys. Acta*, **1758**, 883–891.
49 Krafft, C., Shapoval, L., Sobottka, S.B., Schackert, G., and Salzer, R. (2006) Identification of primary tumors of brain metastases by infrared spectroscopic imaging and linear discriminant analysis. *Technol. Cancer Res. Treat.*, **5**, 291–298.
50 Evans, C.L., Xu, X., Kesari, S., Xie, X.S., Wong, S.T., and Young, G.S. (2007) Chemically-selective imaging of brain

structures with CARS microscopy. *Opt. Express*, **15**, 12076–12087.

51 Nishimura, N., Schaffer, C.B., Friedman, B., Tsai, P.S., Lyden, P.D., and Kleinfeld, D. (2006) Targeted insult to subsurface cortical blood vessels using ultrashort laser pulses: three models of stroke. *Nat. Methods*, **3**, 99–108.

52 Read, T.A., Farhadi, M., Bjerkvig, R., Olsen, B.R., Rokstad, A.M., Huszthy, P.C., and Vajkoczy, P. (2001) Intravital microscopy reveals novel antivascular and antitumor effects of endostatin delivered locally by alginate-encapsulated cells. *Cancer Res.*, **61**, 6830–6837.

53 Fukumura, D., Xavier, R., Sugiura, T., Chen, Y., Park, E.C., Lu, N., Selig, M., Nielsen, G., Taksir, T., Jain, R.K., and Seed, B. (1998) Tumor induction of VEGF promoter activity in stromal cells. *Cell*, **94**, 715–725.

54 Brown, E., McKee, T., DiTomaso, E., Pluen, A., Seed, B., Boucher, Y., and Jain, R.K. (2003) Dynamic imaging of collagen and its modulation in tumors *in vivo* using second-harmonic generation. *Nat. Med.*, **9**, 796–801.

55 Jain, R.K., Munn, L.L., and Fukumura, D. (2002) Dissecting tumour pathophysiology using intravital microscopy. *Nat. Rev. Cancer*, **2**, 266–276.

56 Woitzik, J., Pena-Tapia, P.G., Schneider, U.C., Vajkoczy, P., and Thome, C. (2006) Cortical perfusion measurement by indocyanine-green videoangiography in patients undergoing hemicraniectomy for malignant stroke. *Stroke*, **37**, 1549–1551.

57 Asgari, S., Rohrborn, H.J., Engelhorn, T., and Stolke, D. (2003) Intra-operative characterization of gliomas by near-infrared spectroscopy: possible association with prognosis. *Acta Neurochir. (Wien)*, **145**, 453–459.

58 Nakagawa, A., Hirano, T., Uenohara, H., Utsunomiya, H., Suzuki, S., Takayama, K., Shirane, R., and Tominaga, T. (2004) Use of intraoperative dynamic infrared imaging with detection wavelength of 7–14 microm in the surgical obliteration of spinal arteriovenous fistula: case report and technical considerations. *Minim. Invasive Neurosurg.*, **47**, 136–139.

59 Okudera, H., Kobayashi, S., and Toriyama, T. (1994) Intraoperative regional and functional thermography during resection of cerebral arteriovenous malformation. *Neurosurgery*, **34**, 1065–1067.

60 Anderson, R.E., Waltz, A.G., Yamaguchi, T., and Ostrom, R.D. (1970) Assessment of cerebral circulation (cortical blood flow) with an infrared microscope. *Stroke*, **1**, 100–103.

61 Nakagawa, A., Hirano, T., Uenohara, H., Sato, M., Kusaka, Y., Shirane, R., Takayama, K., and Yoshimoto, T. (2003) Intraoperative thermal artery imaging of an EC–IC bypass in beagles with infrared camera with detectable wave-length band of 7–14 microm: possibilities as novel blood flow monitoring system. *Minim. Invasive Neurosurg.*, **46**, 231–234.

62 Okada, Y., Kawamata, T., Kawashima, A., and Hori, T. (2007) Intraoperative application of thermography in extracranial–intracranial bypass surgery. *Neurosurgery*, **60**, 362–365.

63 George, J.S., Lewine, J.D., Goggin, A.S., Dyer, R.B., and Flynn, E.R. (1993) IR thermal imaging of a monkey's head: local temperature changes in response to somatosensory stimulation. *Adv. Exp. Med. Biol.*, **333**, 125–136.

64 Gorbach, A.M., Heiss, J., Kufta, C., Sato, S., Fedio, P., Kammerer, W.A., Solomon, J., and Oldfield, E.H. (2003) Intraoperative infrared functional imaging of human brain. *Ann. Neurol.*, **54**, 297–309.

65 Hollander, H. and Mehraein, P. (1966) On the mechanics of myelin sphere formation in Wallerian degeneration. Intravital microscopic studies of single degenerating motor fibers of the frog. *Z. Zellforsch. Mikrosk. Anat.*, **72**, 276–280.

66 Williams, P.L. and Hall, S.M. (1971) Prolonged *in vivo* observations of normal peripheral nerve fibres and their acute reactions to crush and deliberate trauma. *J. Anat.*, **108**, 397–408.

67 Zhu, W.Z., Li, X., Qi, J.P., Tang, Z.P., Wang, W., Wei, L., and Lei, H. (2008) Experimental study of cell migration and functional differentiation of transplanted neural stem cells co-labeled with superparamagnetic iron oxide and BrdU in an ischemic rat model. *Biomed. Environ. Sci.*, **21**, 420–424.

68 Ruiz-Cabello, J., Walczak, P., Kedziorek, D.A., Chacko, V.P., Schmieder, A.H.,

Wickline, S.A., Lanza, G.M., and Bulte, J.W. (2008) In vivo "hot spot" MR imaging of neural stem cells using fluorinated nanoparticles. *Magn. Reson. Med.*, **60**, 1506–1511.

69 Walczak, P., Zhang, J., Gilad, A.A., Kedziorek, D.A., Ruiz-Cabello, J., Young, R.G., Pittenger, M.F., van Zijl, P.C., Huang, J., and Bulte, J.W. (2008) Dual-modality monitoring of targeted intraarterial delivery of mesenchymal stem cells after transient ischemia. *Stroke*, **39**, 1569–1574.

70 Zhu, W., Li, X., Tang, Z., Zhu, S., Qi, J., Wei, L., and Lei, H. (2007) Superparamagnetic iron oxide labeling of neural stem cells and 4.7T MRI tracking *in vivo* and *in vitro*. *J. Huazhong Univ. Sci. Technol. Med. Sci.*, **27**, 107–110.

71 Curtis, M.A., Kam, M., Nannmark, U., Anderson, M.F., Axell, M.Z., Wikkelso, C., Holtas, S., van Roon-Mom, W.M., Bjork-Eriksson, T., Nordborg, C., Frisen, J., Dragunow, M., Faull, R.L., and Eriksson, P.S. (2007) Human neuroblasts migrate to the olfactory bulb via a lateral ventricular extension. *Science*, **315**, 1243–1249.

72 Schaffer, C.B., Friedman, B., Nishimura, N., Schroeder, L.F., Tsai, P.S., Ebner, F.F., Lyden, P.D., and Kleinfeld, D. (2006) Two-photon imaging of cortical surface microvessels reveals a robust redistribution in blood flow after vascular occlusion. *PLoS Biol.*, **4**, e22.

73 Spires, T.L., Meyer-Luehmann, M., Stern, E.A., McLean, P.J., Skoch, J., Nguyen, P.T., Bacskai, B.J., and Hyman, B.T. (2005) Dendritic spine abnormalities in amyloid precursor protein transgenic mice demonstrated by gene transfer and intravital multiphoton microscopy. *J. Neurosci.*, **25**, 7278–7287.

74 Yuan, F., Dellian, M., Fukumura, D., Leunig, M., Berk, D.A., Torchilin, V.P., and Jain, R.K. (1995) Vascular permeability in a human tumor xenograft: molecular size dependence and cutoff size. *Cancer Res.*, **55**, 3752–3756.

75 Manganas, L.N., Zhang, X., Li, Y., Hazel, R.D., Smith, S.D., Wagshul, M.E., Henn, F., Benveniste, H., Djuric, P.M., Enikolopov, G., and Maletic-Savatic, M. (2007) Magnetic resonance spectroscopy identifies neural progenitor cells in the live human brain. *Science*, **318**, 980–985.

76 Loewerbrück, K.F., Fuchs, B., Hermann, A., Brandt, M., Werner, A., Kirsch, M., Schwarz, J., Schiller, J., and Storch, A. (2011) Proton MR spectroscopy of neural stem cells: does te proton MNR peak at 1.28 ppm function as a biomarker for cell type or state? Rejuvenation Res. DOI: 10.1089/rej.2010.1102.

48
Assessment of Infant Brain Development
Nadege Roche-Labarbe, P. Ellen Grant, and Maria Angela Franceschini

48.1
Introduction

Studying the infant's brain has been one of the major applications of *in vivo* optical imaging since the mid-1980s [1–4], for reasons both scientific and technical.

There are very few tools available for noninvasive bedside study of the neonatal brain, and they do not assess local cerebral metabolism and hemodynamics. Electroencephalography (EEG) and amplitude-integrated electroencephalography (aEEG) measure electrical activity, head ultrasound (US) gives structural images, and transcranial Doppler (TCD) ultrasonography estimates blood flow but only in the major arteries. Our current understanding of metabolic and vascular regional brain development in infants is derived from positron emission tomography (PET) [5–8], single photon emission computed tomography (SPECT) [9], and functional magnetic resonance imaging (fMRI) [10–12] studies. However, these techniques are performed rarely, and primarily in clinical populations, because they cannot be used at the bedside and may require sedation. In addition PET and SPECT come with an increased risk from radiation exposure. In contrast, near-infrared spectroscopy (NIRS) and diffuse correlation spectroscopy (DCS) are portable, noninvasive, and safe optical imaging techniques that use nonionizing radiation (near-infrared light) and emit less energy than a head US [13, 14]. For this reason, NIRS and DCS are particularly well suited to neonates and infants.

From a technical point of view, NIRS is particularly optimal for neonatal brain studies because, in addition to being safe and noninvasive, it takes advantage of the thinner scalp and skull layer, smaller head diameter, and smaller amount of hair compared with older children and adults. These characteristics of newborn heads (especially premature newborns) allow for deeper penetration of light into the brain, and a better signal-to-noise ratio [15]. These characteristics also

Handbook of Biophotonics. Vol.2: Photonics for Health Care, First Edition. Edited by Jürgen Popp,
Valery V. Tuchin, Arthur Chiou, and Stefan Heinemann.
© 2012 Wiley-VCH Verlag GmbH & Co. KGaA. Published 2012 by Wiley-VCH Verlag GmbH & Co. KGaA.

necessitate shorter source–detector distances, approximately 2 cm as opposed to 4 cm in adults [16].

Optical imaging in neonates and infants uses various NIRS and DCS techniques described below. Applications can be divided into functional and baseline studies of normal and abnormal brain development.

48.2
Techniques

48.2.1
Measurement of Local Oxy- and Deoxyhemoglobin Concentrations

NIRS is the most widely used optical imaging technique in studies of the infant brain. It uses the specific absorption properties of oxyhemoglobin (HbO_2) and deoxyhemoglobin (HbR) in the near-infrared range (around 650–850 nm) to measure the concentration of these molecules in tissues [17].

The first NIRS systems were simple continuous wave (CW) systems introduced in the late 1980s and 1990s, which were used to quantify hemoglobin concentration from tissue absorption. With CW-NIRS systems, light is emitted continuously with a constant or low-frequency (kHz) amplitude, and light attenuation through tissues is measured. Because biological tissues are highly scattering, the Beer–Lambert law, an empirical relationship between light attenuation and chromophore concentration [18], cannot be directly applied. Delpy et al. developed a modified Beer–Lambert law [Eq. (48.1)], which takes into account the photon's longer path distribution due to the multiple scattering events [19].

$$OD(\lambda) = \{\varepsilon_{HbR}(\lambda)[HbR] + \varepsilon_{HbO_2}(\lambda)[HbO_2]\} L\, DPF(\lambda) + G(\lambda) \qquad (48.1)^{1)}$$

where OD is the optical density, λ is the wavelength used, ε is the extinction coefficient of each chromophore, L is the source–detector separation, G is the geometry factor, and DPF is the differential pathlength factor, the term that takes into account the longer photon path in a multiple scattering regime.

The DPF is either assumed or measured with more complex NIRS instruments (see below). The geometry factor is not known but is eliminated if we consider relative changes of hemoglobin instead of quantifying absolute concentrations. By repeating the same measurements at two or more wavelengths, the changes in concentration of HbO_2 and HbR can then be derived. CW-NIRS systems are the simplest, cheapest, and most widely used type of NIRS system. They are optimal for following changes in hemoglobin at a high temporal resolution, and are therefore often used in functional experiments [20–24]. Because the DPF cannot be directly measured with CW

---

1) In biological tissues, we also need to consider water absorption. Typically for brain applications, water concentration is assumed to be 75–85%. To measure water concentration, wavelengths higher than 900 nm are needed, and such wavelengths are not commonly available in commercial NIRS systems.

systems, CW-NIRS is not optimal for quantifying hemoglobin concentration and total oxygenation index (TOI)[2], but it is still widely used for this purpose.

Under physiologically induced changes in blood volume [4, 26, 27], CW-NIRS can quantify total hemoglobin concentration (HbT = HbR + HbO$_2$), and from that derive cerebral blood volume (CBV):

$$CBV = \frac{HbT \times MW_{Hb}}{HGB \times D_{bt}} \tag{48.2}$$

where MW_{Hb} = molecular weight of hemoglobin = 64 500 g mol^{-1}, D_{bt} = brain tissue density = 1.05 g ml^{-1}, and HGB = blood hemoglobin measured in g/dl^{-1}.

In the early 1990s a more rigorous theory than the Beer–Lambert law derived from the radiative transport theory [28], the diffusion approximation to the Boltzmann transport equation [29–31], allowed separation of the absorption and scattering contributions and quantification of absolute hemoglobin concentration in highly scattering media. Time-domain (TD) and frequency-domain (FD) NIRS instruments are able to quantify absorption (μ_a) and scattering (μ_s) coefficients if the light temporal profile is acquired in TD systems [32] or if the light attenuation is acquired at multiple frequencies [33] and multiple source–detector distances [34] in FD systems. From the absorption coefficient at two or more wavelengths, HbO$_2$ and HbR concentrations can be quantified, and HbT, CBV, and total oxygenation (StO$_2$ = HbO$_2$/HbT) can be derived.

With TD systems, light is emitted as a pulse and its temporal profile is measured. TD is the most sophisticated and sensitive type of NIRS system [1, 35–38]. With FD systems, light is sinusoidally modulated in the radiofrequency range, and light amplitude attenuation and phase delay are measured. FD-NIRS provides a good compromise between performance (absolute quantification) and simplicity, and is especially well suited for clinical and physiological applications when quantification of baseline hemoglobin is needed [34, 39–42].

48.2.2
Local Cerebral Blood Flow Measurement

There have been attempts to measure optically regional cerebral blood flow (CBF) in neonates using NIRS, using either an oxygen bolus obtained by changing the inspired oxygen [43–46], a method applicable only in ventilated infants, or an

---

[2] In the 20 + years of NIRS, hemoglobin concentration and oxygenation have been abbreviated in many different ways by many different groups. Unfortunately, there is still no unified nomenclature. Moreover, especially with CW-NIRS, where several assumptions need to be made, the algorithms used to calculate oxygenation differ between commercial systems [25]. Here, we use TOI for total oxygenation index, as is done with the tissue oxygenation index measured with the NIRO systems (Hamamatsu Photonics, Hamamatsu City, Japan). We prefer to define TOI as total oxygenation index instead of tissue oxygenation index because it better reflects that it measures the sum of arterial and venous hemoglobin oxygenation, and not tissue oxygenation. The regional cerebral oxygen saturation (rScO$_2$) measured with the INVOS (Somanetics Corp., Troy, MI, USA) is calculated with a different algorithm but represents the same quantity.

indocyanine green bolus [47, 48], obtained with the injection of the dye that cannot be performed in healthy infants for ethical reasons. These methods give correct estimates of CBF but are not generally applicable in all infants.

DCS is a more recent technique, providing a measure of tissue perfusion based on the movement of scatterers (i.e., blood cells) inside the tissue [49, 50]. As NIRS, DCS can directly measure tissue perfusion without having to rely on a bolus or oxygen manipulations. DCS was introduced in the early 1990s and has been validated in a number of animal and human studies [50–54] for measuring regional and temporal variations in blood flow. We have recently proven the ability of DCS to quantify an absolute parameter, blood flow index (BF_i), proportional to CBF in the neonatal brain [55].

48.2.3
Estimates of Cerebral Metabolic Rate of Oxygen

To determine the balance between oxygen delivery and oxygen consumption, CW systems can be used to derive an index of the regional oxygen extraction fraction, the relative fractional tissue oxygen extraction (FTOE) [56]:

$$FTOE = (SaO_2 - TOI)/SaO_2 \qquad (48.3)$$

More importantly, following the PET and fMRI literature, NIRS can be used to derive the regional metabolic rate of oxygen ($CMRO_2$) of the infant brain. $CMRO_2$ describes oxygen consumption in relation to oxygen delivery and is a very important parameter in assessing brain metabolism during normal development and disease [5].

Using the Fick principle, estimation of $CMRO_2$ requires a measurement of CBF, HGB, and hemoglobin oxygenation in the arterial (SaO_2) and venous (SvO_2) compartments, that is,

$$CMRO_2 = \frac{HGB \times CBF \times (SaO_2 - SvO_2)}{MW_{hb}} \qquad (48.4)$$

SaO_2 is provided by standard pulse oximetry, but SvO_2 is unknown. With NIRS, as defined above, we measure total oxygenation, a weighted sum of arterial and venous oxygenation:

$$StO_2 = aSaO_2 + bSvO_2 \qquad (48.5)$$

with $a + b = 1$ [57]. By assuming a and b constant with age and disease, from Eq. (48.4) we can quantify relative $CMRO_2$ without having to measure SvO_2:

$$rCMRO_2 = \frac{CMRO_2}{CMRO_{2o}} = \frac{HGB}{HGB_o} \times \frac{CBF}{CBF_o} \times \frac{SaO_2 - StO_2}{SaO_{2o} - StO_{2o}} \qquad (48.6)$$

with the subscript o indicating the reference measure.

Using DCS, CBF can be replaced by BF_i. In CW systems, StO_2 is approximated with TOI, and CBF can be quantified with a bolus perturbation [27, 58, 59]. With NIRS alone, by assuming a constant power law relation between changes in blood flow and blood volume with age:

$$\frac{\text{CBF}}{\text{CBF}_o} = \left(\frac{\text{CBV}}{\text{CBV}_o}\right)^\beta \tag{48.7}$$

with $2 \leq \beta \leq 5$ [60, 61], one can recover rCMRO$_2$ [40, 62]. While the flow–volume relationship has been validated in healthy adult human subjects with MRI functional studies [63], the relationship may not hold in infants or disease states [55].

48.3
Functional and Cognitive Studies

When a cortical area becomes active, there is a local increase in perfusion to respond to the increased metabolic demand driven by neuronal activity [64]. Increases in oxygen consumption are significantly lower than increases in cerebral blood flow and blood volume caused by the release of vasoactive neurotransmitters [65, 66]. As a result, we see a net increase in the amount of oxygen in the blood and tissue [67–69]. The typical vascular response observed with NIRS following neural activity is an increase in HbT and HbO$_2$ but a decrease in HbR. Although the decrease in HbR is equivalent to a positive blood oxygen level-dependent (BOLD) signal in fMRI, fMRI is unable to detect changes in HbO$_2$ or HbT.

For more than a decade, CW-NIRS has been used in functional studies in infants and has been shown to detect vascular changes in response to visual stimulation of the primary visual cortex in newborns [70–72] and infants [21, 23, 72–75], auditory stimulation of the temporal and frontal cortices in newborns [24, 76–79] and infants [73, 80], passive stimulation of the somatosensory cortex in premature [81, 82] and term newborns [83], noxious stimulation of the somatosensory cortex in premature and term newborns [81, 84], and olfactory stimulation of the orbital gyri in newborns [20]. NIRS has also been used to monitor neuronal activation due to spontaneous arousal [85] or spontaneous neuronal activity bursts during slow sleep [86].

In several cases, there have been reports of a "negative" vascular response, that is, an increase in HbR and a decrease in HbO$_2$ concentration [24, 74, 75] consistent with negative BOLD responses in fMRI studies [10, 87–91]. The cause of these inverted responses to stimulation remains unclear. It might be due to differences in the subjects' vigilance level across studies, or to differences in neurovascular coupling maturation [10, 87–91]. Because of the inconsistency in the signs of HbO$_2$ and HbR in newborns and infants, and the lack of a plausible explanation for this, the current literature may appear confusing and the results discouraging. Understanding these inverted responses is of paramount importance, and further systematic studies with large samples and multiple modalities are needed.

In the past several years, an increasing number of successful NIRS cognitive studies have been performed. Valuable information was gathered in older infants from the study of facial and social stimuli processing [92–97], object processing [98–101], language processing [102–105], and habituation/novelty auditory paradigms [106].

For a detailed review of language processing NIRS studies in infants, see [107]; for a comprehensive review of all functional/cognitive NIRS studies in infants, and also infant probes and specific issues, see [108] and [109].

48.4
Baseline Hemodynamic Assessment and Clinical Applications

Baseline hemoglobin concentration and blood flow in the cerebral cortex can be related to neurovascular development and brain health. For more than 20 years, researchers have attempted to establish NIRS as a bedside cerebral monitoring tool in neonatal intensive care units. We can only mention here some of the many CW-NIRS baseline applications in clinical settings.

Autoregulation has been studied in response to peripheral saturation [110], blood pressure [111, 112], pCO_2 [113], and posture changes [114]. The effects of respiratory support on cerebral StO_2 have been studied extensively, including the effect of surfactants [115–117], endotracheal suctioning [118, 119], ventilation methods [46, 120–124], and extracorporeal membrane oxygenation [125, 126]. NIRS has been used as a monitoring tool during surgery [127–130]. There is also a vast literature on the use of NIRS in newborns and infants with congenital heart disease [131–133].

Wardle et al. [134] studied the effect of hypotension and anemia on the cerebral fraction of oxygen extraction in premature neonates, Lemmers et al. [135] showed that StO_2 is more variable in newborns with respiratory distress syndrome, and several groups have characterized the effect of hypoxia on the newborn brain [2, 136–139]. Initial findings in the study of perinatal asphyxia and subsequent hypoxic–ischemic injury were sometimes inconsistent [4, 140–142], but NIRS measures seemed to correlate with outcome [56, 143]. Despite their prevalence, intraventricular hemorrhage and periventricular leukomalacia studies are still under-represented in the literature. However, CBF [144] and StO_2 [145] at birth have been found to correlate with the occurrence and severity of intraventricular hemorrhage. Roche-Labarbe et al. also described different hemodynamic responses to spontaneous bursts of neuronal activity between healthy newborns and patients with intraventricular hemorrhage [86].

For a synthetic review of CW-NIRS baseline hemodynamic studies in newborns and infants before 2003, see [146]. For a comprehensive review of functional and baseline studies of NIRS in neonates, see [109].

Although the efforts to implement NIRS in the NICU generated a large number of publications, there has not been a clear breakthrough leading to the use of NIRS technology as a routine diagnostic tool. Past studies have been very important in opening up the field [25, 56, 147–149]; however there are limitations that CW-NIRS and perturbation approaches cannot overcome [146, 150]. We believe that a new generation of TD, FD, and DCS systems will allow developers to establish NIRS successfully as the technology of choice for bedside monitoring of brain health neonatal intensive care units.

As we stressed in the techniques section, TD, FD, and DCS instruments are able to quantify baseline hemodynamic parameters and metabolism with fewer assumptions. This offers several advantages with respect to the traditional CW systems:

- It allows for absolute measurements, and as a result optical probes do not need to be kept in place for long periods of time, but can be frequently repositioned on the subject's head without losing the reference value. The variability is about 5–10% for FD measurements and 10–15% for DCS measurements [40, 55, 151], and differences in the recovered parameters are due more to tissue heterogeneity than to system calibration issues.
- Blood parameters can be directly quantified without external perturbations, or bolus techniques, that can alter cerebral vascular physiology (thus biasing the results), are not applicable in every infant, require additional expertise, and cannot be performed continuously or frequently.
- As opposed to typical commercial NIRS systems that acquire the minimum number of parameters necessary to resolve oxygenated and deoxygenated hemoglobin, some newer systems acquire additional distances and/or wavelengths and use redundancy in the measurements to assess data quality. This enables preset rejection criteria to be used to discard bad measurements based on statistical results [40, 55].
- The majority of previous work using CW systems focused on determining TOI differences between the healthy and diseased brain. Although TOI and SaO_2 are very sensitive to changes during hypoxic episodes, recent studies suggest that oxygenation saturation is not the best biomarker for detecting brain injury after the insult, because the brain tends to maintain a constant oxygenation saturation level by adjusting perfusion. Across ages [40] and diseases [62], total oxygenation is remarkably constant. On the other hand, CBV and $CMRO_2$ acquired using FD-NIRS seem to discriminate much more effectively between brain-injured and healthy newborns [62]. The same issues are encountered with the oxygen extraction fraction (OEF): PET studies showed that regional OEF is constant across ages [152] and does not predict brain injury as well as CBF or $CMRO_2$ [153].
- Finally, we recently showed that in the neonatal population, the flow–volume relationship [Eq. (48.7)] is not valid, because the term β does not take into account the large decrease in blood viscosity during the first few months of life due to the transition from fetal to adult hemoglobin [154]. Therefore, we cannot replace the CBF ratio with the CBV ratio to calculate $CMRO_2$ using Eq. (48.6). Rather, we need to measure CBF directly, which is made possible by DCS [55].

48.5
Conclusion and Perspectives

NIRS is especially well suited to the study of newborn and infant brain development, and recent technical improvements such as the ability to measure absolute values with TD- and FD-NIRS, and a reliable and safe CBF index with DCS, have made it suitable for large-scale clinical applications. In the future, multimodal approaches coupling optical imaging with other imaging techniques should allow a more extensive understanding of the developing brain and the incorporation of NIRS into the clinical routine. Although there have been promising attempts at optical tomography, that is, the three-dimensional imaging of newborns' brains with NIRS [35, 37, 38], to date the technique is too complex and time consuming for routine clinical application. The trend is towards more flexible and patient-friendly setups allowing repeated measurements.

With respect to functional and cognitive studies, we are looking forward to more complex experimental designs allowing us to study the details of cognitive development during the first year of life.

References

1 Benaron, D.A. et al. (1992) Noninvasive methods for estimating in vivo oxygenation. Clin. Pediatr. (Phila.), 31, 258–273.

2 Brazy, J.E., Lewis, D.V., and Mitnick, M.H. (1985) Noninvasive monitoring of cerebral oxygenation in preterm infants: preliminary observations. Pediatrics, 75, 217–225.

3 Rea, P.A. et al. (1985) Non-invasive optical methods for the study of cerebral metabolism in the human newborn: a technique for the future? J. Med. Eng. Technol., 9 (4), 160–166.

4 Wyatt, J.S. et al. (1986) Quantification of cerebral oxygenation and haemodynamics in sick newborn infants by near infrared spectrophotometry. Lancet, ii, 1063–1066.

5 Altman, D.I. et al. (1993) Cerebral oxygen metabolism in newborns. Pediatrics, 92 (1), 99–104.

6 Chugani, H.T. (1998) A critical period of brain development: studies of cerebral glucose utilization with PET. Prev. Med., 27, 184–188.

7 Chugani, H.T. and Phelps, M.E. (1986) Maturational changes in cerebral function in infants determined by 18FDG positron emission tomography. Science, 231, 840–843.

8 Shi, Y. et al. (2009) Brain positron emission tomography in preterm and term newborn infants. Early Hum. Dev., 85 (7), 429–432.

9 Tokumaru, A.M. et al. (1999) The evolution of cerebral blood flow in the developing brain: evaluation with iodine-123 iodoamphetamine SPECT and correlation with MR imaging. AJNR Am. J. Neuroradiol., 20, 845–852.

10 Arichi, T. et al. (2010) Somatosensory cortical activation identified by functional MRI in preterm and term infants. Neuroimage, 49 (3), 2063–2071.

11 Fransson, P. et al. (2009) Spontaneous brain activity in the newborn brain during natural sleep – an fMRI study in infants born at full term. Pediatr. Res., 66 (3), 301–305.

12 Sander, K., Frome, Y., and Scheich, H. (2007) FMRI activations of amygdala, cingulate cortex, and auditory cortex by infant laughing and crying. Hum. Brain Mapp., 28 (10), 1007–1022.

13 Bozkurt, A. and Onaral, B. (2004) Safety assessment of near infrared light emitting diodes for diffuse optical measurements. BioMed. Eng. OnLine, 3 (9), 1–10.

14 Ito, Y. et al. (2000) Assessment of heating effects in skin during continuous wave near infrared spectroscopy. J. Biomed. Opt., 5 (4), 383–391.

15 Faris, F. et al. (1991) Non-invasive in vivo near-infrared optical measurement of the penetration depth in the neonatal head. Clin. Phys. Physiol. Meas., 12 (4), 353–358.

16 Taga, G., Homae, F., and Watanabe, H. (2007) Effects of source-detector distance

of near infrared spectroscopy on the measurement of the cortical hemodynamic response in infants. *Neuroimage*, **38** (3), 452–460.

17 Jöbsis, F.F. (1977) Noninvasive, infrared monitoring of cerebral and myocardial oxygen sufficiency and circulatory parameters. *Science*, **198**, 1264–1267.

18 Beer, A. (1853) Einleitung in die höhere Optik. Published by F. Vieweg und Sohn, Braunschweig, Germany.

19 Delpy, D. *et al.* (1988) Estimation of optical path length through tissue from direct time of flight measurements. *Phys. Med. Biol.*, **33**, 1433–1442.

20 Bartocci, M. *et al.* (2000) Activation of olfactory cortex in newborn infants after odor stimulation: a functional near-infrared spectroscopy study. *Pediatr. Res.*, **48** (1), 18–23.

21 Meek, J.H. *et al.* (1998) Regional hemodynamic responses to visual stimulation in awake infants. *Pediatr. Res.*, **43** (6), 840–843.

22 Taga, G. *et al.* (2000) Spontaneous oscillation of oxy- and deoxy-hemoglobin changes with a phase difference throughout the occipital cortex of newborn infants observed using non-invasive optical topography. *Neurosci. Lett.*, **282**, 101–104.

23 Watanabe, H. *et al.* (2008) Functional activation in diverse regions of the developing brain of human infants. *NeuroImage*, **43** (2), 346–357.

24 Zaramella, P. *et al.* (2001) Brain auditory activation measured by near-infrared spectroscopy (NIRS) in neonates. *Pediatr. Res.*, **49**, 213–219.

25 Toet, M. and Lemmers, P.M. (2009) Brain monitoring in neonates. *Early Hum. Dev.*, **85**, 77–84.

26 Skov, L. *et al.* (1993) Estimation of cerebral venous saturation in newborn infants by near infrared spectroscopy. *Pediatr. Res.*, **33** (1), 52–55.

27 Yoxall, C.W., Weindling, A.M., and Yoxall, C.W. (1998) Measurement of cerebral oxygen consumption in the human neonate using near infrared spectroscopy: cerebral oxygen consumption increases with advancing gestational age. *Pediatr. Res.*, **44** (3), 283–290.

28 Ishimaru, A. (1978) *Wave Propagation and Scattering in Random Media*, Academic Press, New York.

29 Fantini, S. *et al.* (1994) Quantitative determination of the absorption spectra of chromophores in strongly scattering media: a light-emitting-diode based technique. *Appl. Opt.*, **33** (22), 5204–5213.

30 Farrell, T., Patterson, M., and Wilson, B. (1992) A diffusion theory model of spatially resolved, steady-state diffuse reflectance for the noninvasive determination of tissue optical properties *in vivo*. *Med. Phys.*, **19** (4), 879–888.

31 Fishkin, J.B., Gratton, E., and Fishkin, J.B. (1993) Propagation of photon-density waves in strongly scattering media containing an absorbing semi-infinite plane bounded by a straight edge. *J. Opt. Soc. Am. A*, **10** (1), 127–140.

32 Patterson, M.S. *et al.* (1995) Absorption spectroscopy in tissue-simulating materials: a theoretical and experimental study of photon paths. *Appl. Opt.*, **34**, 22–30.

33 Pham, T.H. *et al.* (2000) Broad bandwidth frequency domain instrument for quantitative tissue optical spectroscopy. *Rev. Sci. Instrum.*, **71** (6), 2500–2513.

34 Fantini, S. *et al.* (1995) Frequency-domain multichannel optical detector for non-invasive tissue spectroscopy and oximetry. *Opt. Eng.*, **34**, 34–42.

35 Austin, T. *et al.* (2006) Three dimensional optical imaging of blood volume and oxygenation in the neonatal brain. *Neuroimage*, **31** (4), 1426–1433.

36 Gibson, A.P. *et al.* (2006) Three-dimensional whole-head optical tomography of passive motor evoked responses in the neonate. *NeuroImage*, **30**, 521–528.

37 Hebden, J.C. *et al.* (2004) Imaging changes in blood volume and oxygenation in the newborn infant brain using three-dimensional optical tomography. *Phys. Med. Biol.*, **49** (7), 1117–1130.

38 Hebden, J.C. *et al.* (2002) Three-dimensional optical tomography of

the premature infant brain. *Phys. Med. Biol.*, **47**, 4155–4166.

39 Cerussi, A. *et al.* (2005) Noninvasive monitoring of red blood cell transfusion in very low birthweight infants using diffuse optical spectroscopy. *J. Biomed. Opt.*, **10** (5), 051401.

40 Franceschini, M.A. *et al.* (2007) Assessment of infant brain development with frequency-domain near-infrared spectroscopy. *Pediatr. Res.*, **61** (5), 546–551.

41 Spichtig, S. *et al.* (2009) Multifrequency frequency-domain spectrometer for tissue analysis. *Rev. Sci. Instrum.*, **80** (2), 024301.

42 Zhao, J. *et al.* (2005) *In vivo* determination of the optical properties of infant brain using frequency-domain near-infrared spectroscopy. *J. Biomed. Opt.*, **10** (2), 024028.

43 Edwards, A.D. *et al.* (1988) Cotside measurement of cerebral blood flow in ill infants by near-infrared spectroscopy. *Lancet*, **332** (8614), 770–771.

44 Elwell, C.E. *et al.* (2005) Measurement of $CMRO_2$ in neonates undergoing intensive care using near infrared spectroscopy. *Adv. Exp. Med. Biol.*, **566**, 263–268.

45 Meek, J.H. *et al.* (1998) Cerebral blood flow increases over the first three days of life in extremely preterm neonates. *Arch. Dis. Child. Fetal Neonatal Ed.*, **78**, F33–F37.

46 Noone, M.A. *et al.* (2003) Postnatal adaptation of cerebral blood flow using near infrared spectroscopy in extremely preterm infants undergoing high-frequency oscillatory ventilation. *Acta Paediatr.*, **92** (9), 1079–1084.

47 Patel, J. *et al.* (1998) Measurement of cerebral blood flow in newborn infants using near infrared spectroscopy with indocyanine green. *Pediatr. Res.*, **43** (1), 34–39.

48 Kusaka, T. *et al.* (2001) Estimation of regional cerebral blood flow distribution in infants by near-infrared topography using indocyanine green. *NeuroImage*, **13**, 944–952.

49 Boas, D.A. and Yodh, A.G. (1997) Spatially varying dynamical properties of turbid media probed with diffusing temporal light correlation. *J. Opt. Soc. Am.*, **14** (1), 192–215.

50 Cheung, C. *et al.* (2001) In vivo cerebrovascular measurement combining diffuse near-infrared absorption and correlation spectroscopies. *Phys. Med. Biol.*, **46**, 2053–2065.

51 Buckley, E.M. *et al.* (2009) Cerebral hemodynamics in preterm infants during positional intervention measured with diffuse correlation spectroscopy and transcranial Doppler ultrasound. *Opt. Express*, **17** (15), 12571–12581.

52 Culver, J.P. *et al.* (2003) Diffuse optical measurement of hemoglobin and cerebral blood flow in rat brain during hypercapnia, hypoxia and cardiac arrest. *Adv. Exp. Med. Biol.*, **510**, 293–297.

53 Durduran, T. *et al.* (2004) Diffuse optical measurement of blood flow, blood oxygenation, and metabolism in a human brain during sensorimotor cortex activation. *Opt. Lett.*, **29** (5), 1766–1768.

54 Li, J. *et al.* (2005) Noninvasive detection of functional brain activity with near-infrared diffusing-wave spectroscopy. *J. Biomed. Opt.*, **10** (4), 044002.

55 Roche-Labarbe, N. *et al.* (2010) Noninvasive optical measures of CBV, StO_2, CBF index, and $rCMRO_2$ in human premature neonates' brains in the first six weeks of life. *Hum. Brain Mapp.*, **31**, 341–352.

56 Toet, M.C. *et al.* (2006) Cerebral oxygenation and electrical activity after birth asphyxia: their relation to outcome. *Pediatrics*, **117** (2), 333–339.

57 Watzman, H.M. *et al.* (2000) Arterial and venous contributions to near-infrared cerebral oximetry. *Anesthesiology*, **93**, 947–953.

58 Brown, D.W. *et al.* (2003) Near-infrared spectroscopy measurement of oxygen extraction fraction and cerebral metabolic rate of oxygen in newborn piglets. *Pediatr. Res.*, **54** (6), 861–867.

59 Tichauer, K.M. *et al.* (2006) Measurement of cerebral oxidative metabolism with near-infrared spectroscopy: a validation

study. *J. Cereb. Blood Flow Metab.*, **26** (5), 722–730.

60 Boas, D.A. *et al.* (2003) Can the cerebral metabolic rate of oxygen be estimated with near-infrared spectroscopy? *Phys. Med. Biol.*, **48** (15), 2405–2418.

61 Grubb, R.L.J. *et al.* (1974) The effects of changes in $PaCO_2$ on cerebral blood volume, blood flow, and vascular mean transit time. *Stroke*, **5** (5), 630–639.

62 Grant, P.E. *et al.* (2009) Increased cerebral blood volume and oxygen consumption in neonatal brain injury. *J. Cereb. Blood Flow Metab.*, **29** (10), 1704–1713.

63 Chen, J.J., Pike, G.B., and Chen, J.J. (2009) BOLD-specific cerebral blood volume and blood flow changes during neuronal activation in humans. *NMR Biomed.*, **22** (10), 1054–1062.

64 Sherrington, C. (1933) *The Brain and Its Mechanism*, Macmillan, New York.

65 Fox, P.T. and Raichle, M.E. (1986) Focal physiological uncoupling of cerebral blood flow and oxidative metabolism during somatosensory stimulation in human subjects. *Proc. Natl. Acad. Sci. U. S. A.*, **83** (4), 1140–1144.

66 Fox, P.T. *et al.* (1988) Nonoxidative glucose consumption during focal physiologic neural activity. *Science*, **241** (4), 462–464.

67 Malonek, D. and Grinvald, A. (1996) Interactions between electrical activity and cortical microcirculation revealed by imaging spectroscopy: implications for functional brain mapping. *Science*, **272**, 551–554.

68 Meek, J.H. *et al.* (1995) Regional changes in cerebral haemodynamics as a result of a visual stimulus measured by near infrared spectroscopy. *Proc. Biol. Sci.*, **261** (1362), 351–356.

69 Obrig, H. *et al.* (1996) Cerebral oxygenation changes in response to motor stimulation. *J. Appl. Physiol.*, **81** (3), 1174–1183.

70 Hoshi, Y. *et al.* (2000) Hemodynamic responses to photic stimulation in neonates. *Pediatr. Neurol.*, **23** (4), 323–327.

71 Karen, T. *et al.* (2008) Hemodynamic response to visual stimulation in newborn infants using functional near-infrared spectroscopy. *Hum. Brain Mapp.*, **29**, 453–460.

72 Taga, G. *et al.* (2003) Hemodynamic responses to visual stimulation in occipital and frontal cortex of newborn infants: a near-infrared optical topography study. *Early Hum. Dev.*, **75** (Suppl.), S203–S210.

73 Bortfeld, H., Wruck, E., and Boas, D.A. (2007) Assessing infants' cortical response to speech using near-infrared spectroscopy. *NeuroImage*, **34**, 407–415.

74 Csibra, G. *et al.* (2004) Near infrared spectroscopy reveals neural activation during face perception in infants and adults. *J. Pediatr. Neurol.*, **2** (2), 85–89.

75 Kusaka, T. *et al.* (2004) Noninvasive optical imaging in the visual cortex in young infants. *Hum. Brain Mapp.*, **22** (2), 122–132.

76 Kotilahti, K. *et al.* (2005) Bilateral hemodynamic responses to auditory stimulation in newborn infants. *NeuroReport*, **16** (12), 1373–1377.

77 Pena, M. *et al.* (2003) Sounds and silence: an optical topography study of language recognition at birth. *Proc. Natl. Acad. Sci. U.S.A.*, **100** (20), 11702–11705.

78 Saito, Y. *et al.* (2007) The function of the frontal lobe in neonates for response to a prosodic voice. *Early Hum. Dev.*, **83** (4), 225–230.

79 Sakatani, K. *et al.* (1999) Cerebral blood oxygenation changes induced by auditory stimulation in newborn infants measured by near infrared spectroscopy. *Early Hum. Dev.*, **55** (3), 229–236.

80 Taga, G. and Asakawa, K. (2007) Selectivity and localization of cortical response to auditory and visual stimulation in awake infants aged 2 to 4 months. *Neuroimage*, **36** (4), 1246–1252.

81 Bartocci, M. *et al.* (2006) Pain activates cortical areas in the preterm newborn brain. *Pain*, **122** (1–2), 109–117.

82 Hintz, S.R. *et al.* (2001) Bedside functional imaging of the premature infant brain during passive motor activation. *J. Perinat. Med.*, **29** (4), 335–343.

83 Isobe, K. *et al.* (2001) Functional imaging of the brain in sedated newborn infants using near infrared topography during

passive knee movement. *Neurosci. Lett.*, **299** (3), 221–224.
84 Slater, R. et al. (2006) Cortical pain responses in human infants. *J. Neurosci.*, **26** (14), 3662–3666.
85 Zotter, H. et al. (2007) Cerebral hemodynamics during arousals in preterm infants. *Early Hum. Dev.*, **83** (4), 239–246.
86 Roche-Labarbe, N. et al. (2007) Coupled oxygenation oscillation measured by NIRS and intermittent cerebral activation on EEG in premature infants. *Neuroimage*, **36** (3), 718–727.
87 Anderson, A.W. et al. (2001) Neonatal auditory activation detected by functional magnetic resonance imaging. *Magn. Reson. Imaging*, **19** (1), 1–5.
88 Heep, A. et al. (2009) Functional magnetic resonance imaging of the sensorimotor system in preterm infants. *Pediatrics*, **123** (1), 294–300.
89 Morita, T. et al. (2000) Difference in the metabolic response to photic stimulation of the lateral geniculate nucleus and the primary visual cortex of infants: a fMRI study. *Neurosci. Res.*, **38**, 63–70.
90 Yamada, H. et al. (1997) A rapid brain metabolic change in infants detected by fMRI. *NeuroReport*, **8**, 3775–3778.
91 Yamada, H. et al. (2000) A milestone for normal development of the infantile brain detected by functional MRI. *Neurology*, **55** (2), 218–223.
92 Blasi, A. et al. (2007) Investigation of depth dependent changes in cerebral haemodynamics during face perception in infants. *Phys. Med. Biol.*, **52** (23), 6849–6864.
93 Carlsson, J. et al. (2008) Activation of the right fronto-temporal cortex during maternal facial recognition in young infants. *Acta Paediatr.*, **97** (9), 1221–1225.
94 Lloyd-Fox, S. et al. (2009) Social perception in infancy: a near infrared spectroscopy study. *Child Dev.*, **80** (4), 986–999.
95 Nakato, E. et al. (2009) When do infants differentiate profile face from frontal face? A near-infrared spectroscopic study. *Hum. Brain Mapp.*, **30** (2), 462–472.
96 Otsuka, Y. et al. (2007) Neural activation to upright and inverted faces in infants measured by near infrared spectroscopy. *Neuroimage*, **34** (1), 399–406.
97 Shimada, S. and Hiraki, K. (2006) Infant's brain responses to live and televised action. *Neuroimage*, **32** (2), 930–939.
98 Baird, A.A. et al. (2002) Frontal lobe activation during object permanence: data from near-infrared spectroscopy. *NeuroImage*, **16** (4), 1120–1126.
99 Wilcox, T. et al. (2009) Hemodynamic changes in the infant cortex during the processing of featural and spatiotemporal information. *Neuropsychologia*, **47** (3), 657–662.
100 Wilcox, T. et al. (2005) Using near-infrared spectroscopy to assess neural activation during object processing in infants. *J. Biomed. Opt.*, **10** (1), 11010.
101 Wilcox, T. et al. (2008) Hemodynamic response to featural changes in the occipital and inferior temporal cortex in infants: a preliminary methodological exploration. *Dev. Sci.*, **11** (3), 361–370.
102 Bortfeld, H., Fava, E., and Boas, D.A. (2009) Identifying cortical lateralization of speech processing in infants using near-infrared spectroscopy. *Dev. Neuropsychol.*, **34** (1), 52–65.
103 Homae, F. et al. (2006) The right hemisphere of sleeping infant perceives sentential prosody. *Neurosci. Res.*, **54** (4), 276–280.
104 Homae, F. et al. (2007) Prosodic processing in the developing brain. *Neurosci. Res.*, **59** (1), 29–39.
105 Minagawa-Kawai, Y. et al. (2007) Neural attunement processes in infants during the acquisition of a language-specific phonemic contrast. *J. Neurosci.*, **27** (2), 315–321.
106 Nakano, T. et al. (2009) Prefrontal cortical involvement in young infants' analysis of novelty. *Cereb. Cortex*, **19** (2), 455–463.
107 Minagawa-Kawai, Y. et al. (2008) Optical imaging of infants' neurocognitive development: recent advances and perspectives. *Dev. Neurobiol.*, **68** (6), 712–728.
108 Lloyd-Fox, S., Blasi, A., and Elwell, C.E. (2010) Illuminating the developing brain: the past, present and future of functional

near infrared spectroscopy. *Neurosci. Biobehav. Rev.*, **34** (3), 269–284.
109 Wolf, M. and Greisen, G. (2009) Advances in near-infrared spectroscopy to study the brain of the preterm and term neonate. *Clin. Perinatol.*, **36** (4), 807–834.
110 Wolf, M. *et al.* (2000) Tissue oxygen saturation measured by near infrared spectrophotometry correlates with arterial oxygen saturation during induced oxygenation changes in neonates. *Physiol. Meas.*, **21** (4), 481–491.
111 De Smet, D. *et al.* (2009) New measurements for assessment of impaired cerebral autoregulation using near-infrared spectroscopy. *Adv. Exp. Med. Biol.*, **645**, 273–278.
112 Wong, F.Y. *et al.* (2008) Impaired autoregulation in preterm infants identified by using spatially resolved spectroscopy. *Pediatrics*, **121** (3), e604–e611.
113 Vanderhaegen, J. *et al.* (2009) The effect of changes in tPCO$_2$ on the fractional tissue oxygen extraction – as measured by near-infrared spectroscopy – in neonates during the first days of life. *Eur. J. Paediatr. Neurol.*, **13** (2), 128–134.
114 Ancora, G. *et al.* (2009) Effect of posture on brain hemodynamics in preterm newborns not mechanically ventilated. *Neonatology*, **97** (3), 212–217.
115 Dorrepaal, C.A. *et al.* (1993) Cerebral hemodynamics and oxygenation in preterm infants after low- vs. high-dose surfactant replacement therapy. *Biol. Neonate*, **64** (4), 193–200.
116 Edwards, A.D. *et al.* (1992) Cerebral hemodynamic effects of treatment with modified natural surfactant investigated by near infrared spectroscopy. *Pediatr. Res.*, **32** (5), 532–536.
117 Roll, C. *et al.* (2000) Effect of surfactant administration on cerebral haemodynamics and oxygenation in premature infants – a near infrared spectroscopy study. *Neuropediatrics*, **31** (1), 16–23.
118 Kohlhauser, C. *et al.* (2000) Effects of endotracheal suctioning in high-frequency oscillatory and conventionally ventilated low birth weight neonates on cerebral hemodynamics observed by near infrared spectroscopy (NIRS). *Pediatr. Pulmonol.*, **29** (4), 270–275.
119 Mosca, F.A. *et al.* (1997) Closed versus open endotracheal suctioning in preterm infants: effects on cerebral oxygenation and blood volume. *Biol. Neonate*, **72** (1), 9–14.
120 Dani, C. *et al.* (2007) Brain haemodynamic effects of nasal continuous airway pressure in preterm infants of less than 30 weeks' gestation. *Acta Paediatr.*, **96** (10), 1421–1425.
121 Milan, A. *et al.* (2009) Influence of ventilation mode on neonatal cerebral blood flow and volume. *Early Hum. Dev.*, **85** (7), 415–419.
122 Palmer, K. *et al.* (1994) Negative extrathoracic pressure ventilation – evaluation of the neck seal. *Early Hum. Dev.*, **37** (1), 67–72.
123 Palmer, K.S. *et al.* (1995) Effects of positive and negative pressure ventilation on cerebral blood volume of newborn infants. *Acta Paediatr.*, **84** (2), 132–139.
124 van de Berg, E. *et al.* (2010) The effect of the "InSurE" procedure on cerebral oxygenation and electrical brain activity of the preterm infant. *Arch. Dis. Child. Fetal Neonatal Ed.*, **95** (1), F53–F58.
125 Ejike, J.C. *et al.* (2006) Cerebral oxygenation in neonatal and pediatric patients during veno-arterial extracorporeal life support. *Pediatr. Crit. Care Med.*, **7** (2), 154–158.
126 Rais-Bahrami, K., Rivera, O., and Short, B.L. (2006) Validation of a noninvasive neonatal optical cerebral oximeter in veno-venous ECMO patients with a cephalad catheter. *J. Perinatol.*, **26** (10), 628–635.
127 Dotta, A. *et al.* (2005) Effects of surgical repair of congenital diaphragmatic hernia on cerebral hemodynamics evaluated by near-infrared spectroscopy. *J. Pediatr. Surg.*, **40** (11), 1748–1752.
128 Toet, M.C. *et al.* (2005) Cerebral oxygen saturation and electrical brain activity before, during, and up to 36 hours after arterial switch procedure in neonates without pre-existing brain damage: its relationship to neurodevelopmental outcome. *Exp. Brain Res.*, **165**, 343–350.

129 Vanderhaegen, J. et al. (2008) Surgical closure of the patent ductus arteriosus and its effect on the cerebral tissue oxygenation. *Acta Paediatr.*, **97** (12), 1640–1644.

130 Zaramella, P. et al. (2006) Surgical closure of patent ductus arteriosus reduces the cerebral tissue oxygenation index in preterm infants: a near-infrared spectroscopy and Doppler study. *Pediatr. Int.*, **48** (3), 305–312.

131 Maher, K.O., Phelps, H.M., and Kirshbom, P.M. (2009) Near infrared spectroscopy changes with pericardial tamponade. *Pediatr. Crit. Care Med.*, **10** (1), e13–e15.

132 Mascio, C.E. et al. (2009) Near-infrared spectroscopy as a guide for an intermittent cerebral perfusion strategy during neonatal circulatory arrest. *ASAIO J.*, **55** (3), 287–290.

133 Tobias, J.D., Russo, P., and Russo, J. (2009) Changes in near infrared spectroscopy during deep hypothermic circulatory arrest. *Ann. Card Anaesth.*, **12** (1), 17–21.

134 Wardle, S.P., Yoxall, C.W., and Weindling, A.M. (2000) Determinants of cerebral fractional oxygen extraction using near infrared spectroscopy in preterm neonates. *J. Cereb. Blood Flow Metab.*, **20** (2), 272–279.

135 Lemmers, P.M. et al. (2006) Cerebral oxygenation and cerebral oxygen extraction in the preterm infant: the impact of respiratory distress syndrome. *Exp. Brain Res.*, **173** (3), 458–467.

136 Jenni, O.G. et al. (1996) Impact of central, obstructive and mixed apnea on cerebral hemodynamics in preterm infants. *Biol. Neonate*, **70** (2), 91–100.

137 Livera, L.N. et al. (1991) Effects of hypoxaemia and bradycardia on neonatal cerebral haemodynamics. *Arch. Dis. Child*, **66** (4 Spec. No.), 376–380.

138 Petrova, A. and Mehta, R. (2006) Near-infrared spectroscopy in the detection of regional tissue oxygenation during hypoxic events in preterm infants undergoing critical care. *Pediatr. Crit. Care Med.*, **7** (5), 449–454.

139 Urlesberger, B. et al. (2000) Changes in cerebral blood volume and cerebral oxygenation during periodic breathing in term infants. *Neuropediatrics*, **31** (2), 75–81.

140 Huang, L. et al. (2004) Assessment of the hypoxic–ischemic encephalopathy in neonates using non-invasive near-infrared spectroscopy. *Physiol. Meas*, **25** (3), 749–761.

141 Naulaers, G. et al. (2004) Continuous measurement of cerebral blood volume and oxygenation during rewarming of neonates. *Acta Paediatr.*, **93** (11), 1540–1542.

142 van Bel, F. et al. (1993) Changes in cerebral hemodynamics and oxygenation in the first 24 hours after birth asphyxia. *Pediatrics*, **92** (3), 365–372.

143 Meek, J.H. et al. (1999) Abnormal cerebral haemodynamics in perinatally asphyxiated neonates related to outcome. *Arch. Dis. Child. Fetal Neonatal Ed.*, **81** (2), F110–F115.

144 Meek, J.H. et al. (1999) Low cerebral blood flow is a risk factor for severe intraventricular haemorrhage. *Arch. Dis. Child. Fetal Neonatal. Ed.*, **81** (1), F15–F18.

145 Sorensen, L.C. et al. (2008) Neonatal cerebral oxygenation is not linked to foetal vasculitis and predicts intraventricular haemorrhage in preterm infants. *Acta Paediatr.*, **97** (11), 1529–1534.

146 Nicklin, S.E. et al. (2003) The light still shines, but not that brightly? The current status of perinatal near infrared spectroscopy. *Arch. Dis. Child. Fetal Neonatal Ed.*, **88** (4), 263–268.

147 Kane, J.M., Steinhorn, D.M., and Kane, J.M. (2009) Lack of irrefutable validation does not negate clinical utility of near-infrared spectroscopy monitoring: learning to trust new technology. *J. Crit. Care*, **24** (3), 472–477.

148 Tina, L.G. et al. (2010) S100B protein and near infrared spectroscopy in preterm and term newborns. *Front. Biosci. (Elite Ed.)*, **2**, 159–164.

149 Wolf, M. et al. (2007) Progress of near-infrared spectroscopy and topography for brain and muscle clinical applications. *J. Biomed. Opt.*, **12** (6), 062104.

150 Greisen, G. (2006) Is near-infrared spectroscopy living up to its promises?

Semin. Fetal Neonatal Med., **11** (6), 498–502.

151 D'Arceuil, H.E. *et al.* (2005) Near-infrared frequency-domain optical spectroscopy and magnetic resonance imaging: a combined approach to studying cerebral maturation in neonatal rabbits. *J. Biomed. Opt.*, **10**, 011011.

152 Takahashi, T. *et al.* (1999) Developmental changes of cerebral blood flow and oxygen metabolism in children. *Am. J. Neuroradiol.*, **20** (5), 917–922.

153 Gupta, A.K. *et al.* (2002) Measurement of brain tissue oxygenation performed using positron emission tomography scanning to validate a novel monitoring method. *J. Neurosurg.*, **96** (2), 263–268.

154 Oski, F.A. and Naiman, J.L. (1982) *Hematologic Problems in the Newborn*, 3rd edn., Major Problems in Clinical Pediatrics, vol. **4**, Saunders, Philadelphia, pp. 1–360.

49
Revealing the Roles of Prefrontal Cortex in Memory
Hui Gong and Qingming Luo

49.1
Introduction

The major breakthroughs in neuroscience in the past two decades have generated the opportunity to achieve an integrated understanding of the brain's structure and functions [1]. The brain is highly specialized, differentiated, and organized so that different regions of the neocortex simultaneously carry out computations on separate features of the external world. Memory is the process of retaining informations in storage, which can possibly be drawn later on. It is localized in the brain as physical changes produced by experience through years of research [2]. Information is accumulating about how memory is organized and what structures and connections are involved. The prefrontal cortex (PFC) is one of the latest cortices in the human developing process, having attained maximum relative growth in the human brain. The PFC function relies closely on its connections with a vast array of other cerebral structures [3].

There are several non- or minimally invasive neuromonitoring techniques available to examine functional brain activity for psychiatric researcher and clinicians. These methods can be categorized as direct or indirect based on the information acquisition principles. Direct methods include magnetoencephalography (MEG), electroencephalography (EEG), and event-related potentials (ERPs). Among these, EEG and ERPs record the electrical fields generated by neuronal activity, whereas MEG records the magnetic fields induced by such activity. The direct methods have a high speed response to the brain activity but a limited spatial resolution. By monitoring hemodynamic changes consequent to brain electrical activity, positron emission tomography (PET), single positron emission computed tomography (SPECT) and functional magnetic resonance imaging (fMRI) are regarded as indirect methods. PET and SPECT operate by monitoring the decay of blood-borne radioactive isotopes as they pass through the brain. fMRI, in contrast, detects changes in the local concentration of deoxyhemoglobin via its effect on imposed magnetic fields. The indirect methods can only detect neuronal activity after it has been filtered by a complex and poorly understood neurovascular coupling [4]. Human neuropsychology and neuroimaging studies have begun to provide a broad view of the task conditions in which it is engaged. However, an understanding of the mechanisms by which PFC controls has remained elusive [5].

As a new approach in indirect methods, near-infrared spectroscopy (NIRS) is a noninvasive optical method that utilizes the sensitivity of near-infrared light to hemoglobin oxygenation-state shifts [6]. Since Chance et al. [7] proved that hemodynamic signals in the brain could be quantified by NIRS, many studies have demonstrated the validity of this method [8–10]. In comparison with fMRI, NIRS can measure relative changes in blood volume, oxyhemoglobin (Δ[oxy-Hb]), deoxyhemoglobin (Δ[deoxy-Hb]), and cytochrome oxidase redox state [11]. In most cases, Δ[oxy-Hb], Δ[deoxy-Hb], and total hemoglobin (Δ[total-Hb], the sum of Δ[oxy-Hb] and Δ[deoxy-Hb]), act as NIRS parameters that vary with brain cognitive activity [12].

In this chapter, we review the optical approaches to human PFC activation during cognition function for working memory and long-term memory.

49.2
Working Memory Modality

The concept of working memory proposes that a dedicated system maintains and stores information in the short term, and that this system underlies human thought processes [13]. Recently, applications of functional optical methods have begun to shed light on the contributions of specific regions to working memory in humans.

The forward digit span (DF) and backward digit span (DB) tasks are widely used to assess short-term memory in neuropsychologic research and clinical evaluations. Hoshi et al. [14] used a 64-channel time-resolved optical tomographic imaging system to demonstrate that the DB task activated the dorsolateral prefrontal cortex (DLPFC) of each hemisphere more than the DF task in healthy adult volunteers. They found that higher performance of the DB task was closely related to the activation of the right DLPFC.

The N-back task combines maintenance and manipulation, which requires the monitoring of a continuous sequence of stimuli. A positive response occurs whenever the current stimulus matches the stimulus n positions back in the sequence. The value of n is often viewed as proportional to the "working memory load" – the total demand placed on the maintenance and/or manipulation process [15]. We have reported a functional NIRS study on the PFC activation caused by a WM task, a verbal n-back task [16]. A portable multichannel continuous-wave (cw) NIRS instrument was developed for monitoring the PFC activity by measuring the local relative hemodynamic changes [17]. The cognition task is a verbal parametric "n-back" task in which working memory load increases with n from 0 to 3 and a sequence of 45 letters as test presentation. We found that, in 2- and 3-back conditions, the behavior performances are negatively correlated with cerebral activation. Lower accuracy is accompanied by longer response time and stronger PFC activation, which suggests that a subject with difficulty in solving a problem will lead to more significant hemodynamic changes than one without difficulty in solving the same problem. Increasing memory load

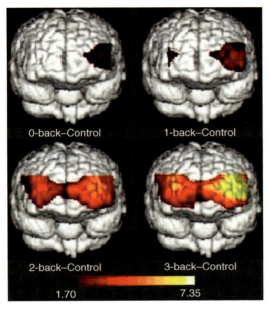

Figure 49.1 Activation distribution in the prefrontal lobe averaged across all the 23 subjects in 0–3-back memory tasks [37].

causes strengthening activation in bilateral polar and DLPFC and the left ventrolateral prefrontal cortex (VLPFC). Figure 49.1 shows the significant activation distribution in the prefrontal lobe averaged across all the 23 subjects in 0–3-back memory tasks. Using the same task, we compared hemodynamic parameters between males and females [18]. Δ[oxy-Hb] and Δ[tot-Hb] exhibited obvious gender differences, but Δ[deoxy-Hb] did not. Males showed bilateral activation with slight left-side dominance, whereas females showed left activation. The activation in males was more widespread and stronger than in females. Furthermore, females required a lower hemodynamic supply than males to obtain comparable performance. Our results suggest that females possess more efficient hemodynamics in the PFC during working memory.

Hoshi used a multichannel cw-NIRS system to monitoring regional cerebral blood flow (rCBF) on the forehead with random digits as a 2-back test presentation [19]. There were no significant changes during the 1-back task, whereas increases in oxy-Hb and tot-Hb with a decrease in deoxy-Hb were observed during the 2-back task. The activated areas correspond to the left DLPFC and no activated areas were seen in the right side. In contrast, in another study, Hoshi *et al.* employed a multichannel cw-NIRS system and a single-channel time-resolved spectroscopy (TRS) instrument to examine changes in rCBF during the performance of an *n*-back task [20]. A sequence of single random digits was used as *n*-back test presentation. It was found that activation from the bilateral VLPFC was prominent with the *n*-back task, but hemispheric dominance was not clearly observed. This study demonstrated that there are overlaps in brain activations across the tasks in each participant.

Schreppel *et al.* used sequences of task-relevant and task-irrelevant faces as the *n*-back test presentation [21]. The analysis of bilateral enhancements of oxygenated hemoglobin showed that the lateral PFC was for remembering relevant faces and the right lateral PFC was for ignoring irrelevant faces. However, oxygenation analysis showed no significant differences between relevant and irrelevant faces, indicating that memory processes and interference resolution were interdependent functions, which were subserved by comparable prefrontal regions. This study supports the notion that the prefrontal activity during working memory tasks reflects not only maintenance processes but also attentional monitoring and selection processes.

Conflict processing, which includes conflict monitoring and solving, is necessary to accomplish a specific task in a complex situation. This solving process typically uses brain function of working memory. The Stroop task is a classical method of studying conflict processing, which was developed in 1935 by John Ridley Stroop, who discovered a cognitive delay when the meaning of a printed word was inconsistent with the color of the word. This phenomenon was dubbed the Stroop effect. Because the Stroop effect reflects conflicts between different pieces of information, it has been widely used in both clinical neuropsychology and experimental psychology. By combining cw-NIRS and event-related potentials (ERPs), we used Chinese character color–word as the Stroop task presentation. Hemodynamic and electrophysiologic signals in the PFC were monitored simultaneously by cw-NIRS and ERPs. We found that P600 signals correlated significantly with the hemodynamic parameters, suggesting that the PFC executes conflict-solving function. Additionally, we observed that the change in deoxy-Hb concentration showed higher sensitivity in response to the Stroop task than other hemodynamic signals [22].

Leon-Carrion *et al.* employ a multichannel cw-NIRS system and used a modified Stroop paradigm to establish the relationship between the hemodynamic response of the PFC and individual differences in cognitive control [23]. Mean concentration levels of oxy-Hb were correlated with behavioral performance in the conflict task. Those with shorter reaction times had higher levels of oxy-Hb concentration in the superior dorsolateral PFC. Negoro *et al.* assessed PFC dysfunction in attention-deficit/hyperactivity disorder (ADHD) children during the Stroop color–word task, using a 24-channel NIRS instrument [24]. During the Stroop color–word task, they found that the oxy-Hb changes in the control group were significantly larger than those in the ADHD group in the inferior prefrontal cortex, especially in the inferior lateral prefrontal cortex bilaterally.

It is well known that the PFC play a central role in working memory, but it remains unknown whether the lateral area of prefrontal cortex (LPFC) of children of preschool age is responsible for working memory. To address this issue, Tsujimoto *et al.* applied optical topography to compare activity in the LPFC between adult subjects and preschool children [25]. The cognitive task is an item recognition test, which requires working memory, under different memory-load conditions. Areas and properties of such activity were similar in adults and preschool children. Data suggest that the LPFC of 5- and 6-year-old children is active during working memory processes; the LPFC has already developed processing of this important cognitive function.

Tsujii *et al.* also used a spatial item recognition working memory task to compare frontal cortical activation in children between 5 and 7 years old using an optical topography system [26]. Behavioral data confirmed that older children performed the working memory task more precisely and more rapidly than younger children. NIRS data showed that older children characterized right hemisphere dominance, but younger children had no significant hemispheric differences. Children with strengthened lateralization showed improved performance from 5 to 7 years old. This study suggests that the right lateralization for spatial working memory may start in children between 5 and 7 years old.

Schroeter *et al.* measured brain activation in the LPFC of children and adults with functional NIRS during an event-related, color–word matching Stroop task. In children, the Stroop task elicited significant brain activation in the left LPFC comparable to adults. The results reveal that brain activation due to Stroop interference increased with age in the DLPFC in correlation with an improvement in behavioral performance and the hemodynamic response occurred later in children than adults [27].

To compare the activity in the left LPFC, Zhang *et al.* employed multichannel cw-NIRS to monitor the concentration changes of rCBF in normal and dyslexic Chinese children, during phonological processing testing. The results show that the degree and the areas of the PFC activity are different between these two groups. During the phonological awareness section, the Δ[deoxy-Hb] in dyslexia-reading children was significantly higher than that in normal-reading children in the left VLPFC. In the phonological decoding section, both normal-reading and dyslexic children had more activity in the left VLPFC, but only normal-reading children had activity in the left mid-dorsal prefrontal cortex [28].

Herrmann *et al.* used multi-channel NIRS to compare the PFC activation between young and older subjects during a verbal fluency task [29]. The results clearly show that the memory test activated the left and right DLPFC with increases in HbO_2 and more localized decreases in deoxygenated hemoglobin (HHb), with an obvious left-hemispheric dominance. The elderly subjects generally exhibited less activation and no left hemispheric lateralization effect.

To study the neural activation regions associated with separable cognitive components of task, Cutini *et al.* used multichannel NIRS to measure brain cortical activity in a task-switching paradigm [30]. The results suggest that factors associated with load and maintenance of distinct stimulus–response mapping rules in working memory are likely contributors to the activation of the LPFC, whereas only activity in the left superior frontal gyrus can be linked unequivocally to switching between distinct cognitive tasks.

49.3
Long-Term Memory Modality

The application of functional neuroimaging techniques to human long-term memory (LTM) has created growing interest in the contribution of frontal lobe function to these processes [31]. Compared with the working memory of maintaining information temporarily over periods of seconds, the long-term memory has the ability to retain information for much longer periods. Data analysis suggests that PFC activations during LTM tasks reflect control processes that aid and optimize memory encoding and retrieval, rather than more automatic storage processes. Most neuroimaging experiments for LTM study include two phases. One is the study phase, in which multiple stimuli are presented, with or without explicit instruction to remember the stimuli. The other is the test phase, during which these stimuli must be recalled, or recognized from among other stimuli. These studies allow a clear distinction between two types of LTM: semantic memory and episodic memory [15].

To examine the hemodynamic response of the PFC, Matsui *et al.* used a portable NIRS system and verbal learning in a words memory learning task [32]. The results show that [oxy-Hb] increased and [deoxy-Hb] decreased during the memory task. The relative change in [oxy-Hb] during encoding from the first to the second condition was significantly larger than that during retrieval. This study suggests that memory organization is facilitated during encoding of the first condition, and that the retrieval period through two conditions still involves more activation in the prefrontal area than the encoding period. For comparison of verbal and spatial memory effects in PFC, Li *et al.* used a multichannel cw-NIRS system to examine the hemodynamic response during stem recognition [33]. The experimental protocol comprised four test blocks, including verbal control, verbal memory, spatial control, and spatial memory. The results show that the left PFC has more significant hemodynamic changes of verbal memory than that of spatial memory. The oxygenation of spatial memory mainly increased in the middle and right PFC. Kubota *et al.* used a two-channel NIRS system and monitoring of PFC hemodynamics in real time, while subjects studied word lists and subsequently recognized unstudied items (false recognition) [34]. The results show that bilateral increases in the [oxy-Hb] were observed during false recognition compared with true recognition, and a left PFC dominant increase in [oxy-Hb] was observed during encoding phases where subjects later claimed that they recognized unstudied words. Reflected primarily in the left PFC activity, traces of semantic processing could eventually predict whether subjects falsely recognize nonexperienced events.

Yang *et al.* explored the role of the prefrontal cortex in semantic encoding of unrelated word pairs by using a cw-NIRS system [35]. The cognitive tasks of unrelated pairs of Chinese words under both nonsemantic and semantic encoding conditions were presented. Under the nonsemantic conditions, subjects judged whether the two words had similar orthographic structures; under the semantic condition, they generated a sentence involving the presented word pairs. The regions that corresponded to the PFC showed greater activation under semantic than nonsemantic conditions in both the left and right hemispheres, although the extent of the activation was larger in the left than right prefrontal regions. This result was consistent with other neuroimaging studies on unrelated word pairs processing, but did not conform to the strict interpretations of the hemispheric encoding/retrieval asymmetry model. This study suggests that material specificity is one of the important factors to influence hemispheric asymmetry in memory encoding. Sanefuji *et al.* examined the correlation between old/new effects and task performance during the recognition judgments about old (studied) or new (unstudied) meaningless shapes, by using event-related NIRS [36]. The data show that lateralization of the VLPFC old/new effects may also depend on the type of stimulus and the VLPFC could be associated with retrieval success.

In summary, we have reviewed the NIRS studies for PFC activation of both working memory and long-term memory with different stimulus protocols. As one of the indirect approaches, NIRS has the advantages of simple monitoring of the PFC and easily coupling with other measurements, either direct or indirect methods.

References

1. Pechura, C.M. and Martin, J.B. (eds.) (1991) *Mapping the Brain and Its Functions: Integrating Enabling Technologies into Neuroscience Research*, National Academies Press, Washington, DC.
2. Wilson, R.A. and Keil, F.C. (eds.) (1999) *The MIT Encyclopedia of the Cognitive Sciences*, MIT Press, Cambridge, MA.
3. Fuster, J.M. (2001) The prefrontal cortex – an update: time is of the essence. *Neuron*, **30**, 319–333.
4. Strangman, G., Boas, D.A., and Sutton, J.P. (2002) Non-invasive neuroimaging using near-infrared light. *Biol. Psychiatry*, **52**, 679–693.
5. Miller, E.K. and Cohen, J.D. (2001) An integrative theory of prefrontal cortex function. *Annu. Rev. Neurosci.*, **24**, 167–202.
6. Jöbsis, F.F. (1977) Noninvasive infrared monitoring of cerebral and myocardial oxygen sufficiency and circulatory parameters. *Science*, **198**, 1264–1267.
7. Chance, B., Leigh, J.S., Miyake, H. *et al.* (1988) Comparison of time-resolved and -unresolved measurements of deoxyhemoglobin in brain. *Proc. Natl. Acad. Sci. U. S. A.*, **88**, 4971–4975.
8. Hoshi, Y. (2003) Functional near-infrared optical imaging: utility and limitations in human brain mapping. *Psychophysiology*, **40**, 511–520.
9. Wolf, M., Ferrari, M., and Quaresima, V. (2007) Progress of near-infrared spectroscopy and topography for brain and muscle clinical applications. *J. Biomed. Opt.*, **12** (6), 062104–062114.
10. Izzetoglu, M., Izzetoglu, K., Bunce, S. *et al.* (2005) Functional near-infrared neuroimaging. *IEEE Trans. Neural Syst. Rehabil. Eng.*, **13** (2), 153–159.
11. Chance, B., Luo, Q., Nioka, S. *et al.* (1997) Optical investigations of physiology: a study of intrinsic and extrinsic biomedical contrast. *Philos. Trans. R. Soc. Lond. B Biol. Sci.*, **352**, 707–716.
12. Villringer, A. and Chance, B. (1997) Non-invasive optical spectroscopy and imaging of human brain function. *Trends Neurosci.*, **20** (10), 435–442.
13. Baddeley, A. (2003) Working memory: looking back and looking forward. *Nat. Rev. Neurosci.*, **4**, 829–839.
14. Hoshi, Y., Oda, I., Wada, Y. *et al.* (2000) Visuospatial imagery is a fruitful strategy for the digit span backward task: a study with near-infrared optical tomography. *Brain Res. Cogn. Brain Res.*, **9**, 339–342.
15. Fletcher, P.C. and Henson, R.N.A. (2001) Frontal lobes and human memory: insights from functional meuroimaging. *Brain*, **124**, 849–881.
16. Li, C., Gong, H., Zeng, S. *et al.* (2005) Verbal working memory load affects prefrontal cortices activation: evidence from a functional NIRS study in humans. *Proc. SPIE*, **5696**, 33–40.
17. Lv, X., Zheng, Y., Li, T. *et al.* (2008) A portable functional imaging instrument for psychology research based on near-infrared spectroscopy. *Frontiers Optoelectron. China*, **1** (3–4), 279–284.

18 Li, T., Luo, Q., and Gong, H. (2010) Gender-specific hemodynamics in prefrontal cortex during a verbal working memory task by near-infrared spectroscopy. *Behav. Brain Res.*, **209**, 148–153.

19 Hoshi, Y. (2003) Functional near-infrared optical imaging: utility and limitations in human brain mapping. *Psychophysiology*, **40**, 511–520.

20 Hoshi, Y., Tsou, B.H., Billock, V.A. *et al.* (2003) Spatiotemporal characteristics of hemodynamic changes in the human lateral prefrontal cortex during working memory tasks. *Neuroimage*, **20**, 1493–1504.

21 Schreppel, T., Egetemeir, J., Schecklmann, M. *et al.* (2008) Activation of the prefrontal cortex in working memory and interference resolution processes assessed with near-infrared spectroscopy. *Neuropsychobiology*, **57** (4), 188–193.

22 Zhai, J., Li, T., Zhang, Z., and Gong, H. (2009) Hemodynamic and electrophysiological signals of conflict processing in the Chinese-character Stroop task: a simultaneous near-infrared spectroscopy and event-related potential study. *J. Biomed. Opt.*, **14** (5), 054022.

23 Leon-Carrion, J., Damas-Lopez, J., and Martin-Rodriguez, J.F. (2008) The hemodynamics of cognitive control: the level of concentration of oxygenated hemoglobin in the superior prefrontal cortex varies as a function of performance in a modified Stroop task. *Behav. Brain Res.*, **193** (2), 248–256.

24 Negoro, H., Sawada, M., Iida, J. *et al.* (2010) Prefrontal dysfunction in attention-deficit/hyperactivity disorder as measured by near-infrared spectroscopy. *Child Psychiatry Hum. Dev.*, **41** (2), 193–203.

25 Tsujimoto, S., Yamamoto, T., Kawaguchi, H. *et al.* (2004) Prefrontal cortical activation associated with working memory in adults and preschool children: an event-related optical topography study. *Cereb. Cortex*, **14** (7), 703–712.

26 Tsujii, T., Yamamoto, E., Masuda, S. *et al.* (2009) Longitudinal study of spatial working memory development in young children. *Neuroreport*, **20** (8), 759–763.

27 Schroeter, M.L., Zysset, S., Wahl, M. *et al.* (2004) Prefrontal activation due to Stroop interference increases during development: an event-related fNIRS study. *Neuroimage*, **23** (4), 1317–1325.

28 Zhang, Z., Li, T., Zheng, Y. *et al.* (2006) Study the left prefrontal cortex activity of Chinese children with dyslexia in phonological processing by NIRS. *Proc. SPIE*, **6078**, 607833.

29 Herrmann, M.J., Walter, A., Ehlis, A.C. *et al.* (2006) Cerebral oxygenation changes in the prefrontal cortex: effects of age and gender. *Neurobiol. Aging*, **27**, 888–894.

30 Cutini, S., Scatturin, P., Menon, E. *et al.* (2008) Selective activation of the superior frontal gyrus in task-switching: an event-related fNIRS study. *Neuroimage*, **42** (2), 945–955.

31 Frackowiak, R.S.J., Friston, K.J., Frith, C. *et al.* (eds.) (2004) *Human Brain Function*, Elsevier, Amsterdam.

32 Matsui, M., Tanaka, K., Yonezawa, M. *et al.* (2007) Activation of the prefrontal cortex during memory learning: near-infrared spectroscopy study. *Psychiatry Clin. Neurosci.*, **61**, 31–38.

33 Li, T., Li, L., Luo, Q. *et al.* (2009) Assessing working memory in real-life situations with functional near-infrared spectroscopy. *J. Innov. Opt. Health Sci.*, **2** (4), 423–430.

34 Kubota, Y., Toichi, M., Shimizu, M. *et al.* (2006) Prefrontal hemodynamic activity predicts false memory: a near-infrared spectroscopy study. *Neuroimage*, **31**, 1783–1789.

35 Yang, J., Zeng, S., Luo, Q. *et al.* (2005) Hemispheric asymmetry for encoding unrelated word pairs? A functional near-infrared spectroscopy study. *Space Med. Med. Eng.*, **18** (5), 318–323.

36 Sanefuji, M., Nakashima, T., Kira, R. *et al.* (2007) The relationship between retrieval success and task performance during the recognition of meaningless shapes: an event-related near-infrared spectroscopy study. *Neurosci. Res.*, **59**, 191–198.

37 Gong, H., Li, C., Li, T. *et al.* (2007) Near infrared Optical imaging of prefrontal cortex for working memory. *Sci China Ser G*, **37** (51), 110–117.

50
Cerebral Blood Flow and Oxygenation
Andrew K. Dunn

50.1
Introduction

The hemodynamic state of the brain, including blood flow and oxygenation, is critical for maintaining normal brain function. In diseases such as stroke, blood flow is disrupted, leading to inadequate oxygen supply which results in a cascade of harmful events [1]. Many stroke therapies currently under development target the cerebral vasculature. Therefore, techniques for monitoring blood flow and oxygen levels in the brain with good spatial resolution are essential for evaluating the efficacy of these new treatments. Optical methods are particularly well suited for imaging hemodynamic parameters *in vivo* because exogenous contrast agents are not required for most techniques. Optical methods have been applied to the brain at multiple spatial scales ranging from optical microscopy at the micron scale to noninvasive diffuse optical tomography at the centimeter length scale [2]. In this chapter, we focus specifically on three different optical methods for quantifying hemodynamics on the surface of the cortex in preclinical animal models: laser speckle contrast imaging of blood flow, multispectral reflectance imaging of hemoglobin oxygenation and volume, and phosphorescence quenching for pO_2 mapping using a digital micromirror device.

50.2
Laser Speckle Contrast Imaging of Cerebral Blood Flow

Laser Doppler flowmetry has been used extensively for *in vivo* monitoring of cerebral blood flow (CBF) at a single spatial location in animals under a variety of physiologic conditions. Although laser Doppler flowmetry provides high temporal resolution (millisecond) measurement of relative CBF, the lack of any spatial information is a significant limitation. Scanning laser Doppler can be used to provide some spatial information [3], but is limited in both its spatial and temporal resolutions, since the spatial information is typically obtained at the expense of temporal resolution due to the need for scanning [4–6].

Handbook of Biophotonics. Vol.2: Photonics for Health Care, First Edition. Edited by Jürgen Popp,
Valery V. Tuchin, Arthur Chiou, and Stefan Heinemann.
© 2012 Wiley-VCH Verlag GmbH & Co. KGaA. Published 2012 by Wiley-VCH Verlag GmbH & Co. KGaA.

Over the past several years, laser speckle contrast imaging has emerged as a powerful tool for imaging CBF changes with high spatial resolution [7–9]. Speckle imaging is a full-field imaging technique that utilizes a charge-coupled device (CCD) camera, does not require any scanning components or tissue contact, and provides millisecond temporal and tens of microns spatial resolution [9–12]. Therefore, speckle contrast imaging is a relatively simple technique for obtaining the detailed spatiotemporal dynamics of CBF changes.

The basic concept of speckle contrast imaging is straightforward. When the cortex is illuminated with coherent laser light and the reflected light is imaged on to a camera, a dynamic speckle pattern (Figure 50.1) is produced. Because some of the scattering particles are in motion (i.e., blood cells), the speckle pattern fluctuates in time. When the exposure time of the camera (5–10 ms) is longer than the time scale of the speckle intensity fluctuations (typically <1 ms), the camera integrates the intensity variations resulting in blurring of the speckle pattern. In areas of increased motion there is more blurring of the speckles during the camera exposure, resulting in a lower spatial contrast of the speckles in these areas.

Since the motion of the scattering particles (i.e., blood flow) is encoded in the dynamics of the speckle pattern, a measure of blood flow can be obtained by quantifying the spatial blurring of the speckle pattern. This is accomplished by calculating the local speckle contrast (K), defined as the ratio of the standard deviation, σ_s, to the mean intensity of pixel values, $\langle I \rangle$, in a small region of the image [11]:

$$K = \frac{\sigma_s}{\langle I \rangle} \tag{50.1}$$

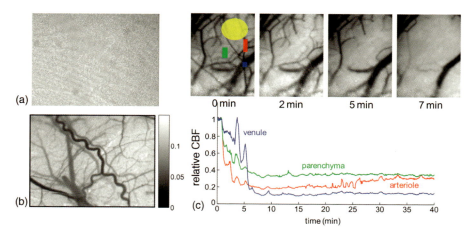

Figure 50.1 Example of raw speckle image (a) and the corresponding speckle contrast image (b) calculated with Eq. (50.1). The gray levels in the speckle contrast image are inversely related to blood flow. The field of view is ~4 mm. (c) Images show changes in speckle contrast following local activation of photothrombosis (green shaded region). The time course of the CBF changes in three regions reveals a significant drop in CBF as vessels are occluded.

A typical example of a raw speckle image of the rat cortex, taken through a thinned skull, and the computed speckle contrast are shown in Figure 50.1a and b. The raw speckle image illustrates the grainy appearance of the speckle pattern. The speckle contrast image, computed directly from the raw speckle image using the Eq. (50.1), represents a two-dimensional map of blood flow. Areas of higher baseline flow, such as large vessels, have lower speckle contrast values and appear darker in the speckle contrast images.

Laser speckle contrast imaging has been widely used to investigate the CBF changes during ischemia in animal models. Figure 50.1c illustrates the speckle contrast and CBF changes in the cortex during photothrombotic stroke in a rat. The series of speckle contrast images illustrates that as the clot is formed, the speckle contrast values decrease and the vessels seem to disappear from the images. This apparent disappearance occurs because the flow in those vessels is greatly reduced and therefore the speckle contrast values increase. The plots show the time course of the actual CBF changes in three particular regions of interest and also highlight the importance of spatially resolved CBF measurements, something that is not possible with traditional laser Doppler flowmetry.

50.3
Multispectral Reflectance Imaging of Hemoglobin Oxygenation

In addition to imaging of CBF changes, optical techniques can be used to determine the changes in hemoglobin oxygenation. Optical imaging of intrinsic signals (OIS) is based on the fact that perfusion related changes lead to changes in the reflectance of the cortical tissue due to hemoglobin absorption. In most OIS measurements, the cortical surface is exposed and illuminated with narrowband (~10 nm) visible light. OIS has provided numerous insights into the functional organization of the cortex in a number of animal models [13–15].

Despite the success of OIS in revealing detailed changes in the cortex, single-wavelength OIS techniques are unable to quantify the changes in hemoglobin oxygenation and volume. Recently, there has been growing interest in the quantitative determination of the hemodynamic changes within the cortex, such as changes in concentrations of HbO, HbR, and total hemoglobin (HbT). However, full-field imaging of these parameters has been challenging owing to the need to obtain images at a minimum of two wavelengths. Acquisition of this spectroscopic information has been achieved by sacrificing spatial information [16, 17]. These approaches have enabled the temporal dynamics of hemoglobin volume and oxygenation to be studied during functional activation in animal models [18, 19]. However, a major limitation of these techniques is the need to sacrifice spatial information in order to obtain spectroscopic information. To overcome this limitation, we have developed a spectroscopic imaging method that permits full-field imaging of reflectance changes from the cortex at multiple wavelengths by rapid switching of the illumination wavelength using a continuously moving filter wheel [20, 21]. This technique allows quantitative imaging of the concentration

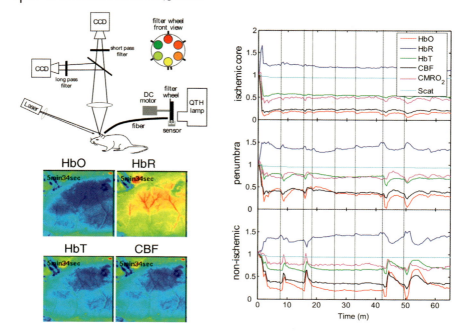

Figure 50.2 Combined laser speckle imaging of CBF and multispectral imaging of hemoglobin oxygenation and saturation. The images demonstrate the spatial changes in each parameter 5 min after stroke induction in a mouse model, and the time courses reveal a complex set of hemodynamic changes that vary depending on the location within the ischemic territory [22].

changes in HbO, HbR, and HbT with the same spatial and temporal resolution as traditional OIS.

Figure 50.2 illustrates how multispectral reflectance imaging (MSRI) can be combined with speckle imaging of CBF to permit simultaneous imaging of CBF, HbO, HbR, and HbT. The multispectral imaging consists of sequential acquisition of reflectance images at illumination wavelengths in the range 560–630 nm. Each set of spectral images is converted to maps of changes in HbO, HbR, and HbT using an appropriate model [20–22]. The images and plots in Figure 50.2 show the spatial and temporal dynamic of the CBF and hemoglobin changes in a mouse model of focal ischemia [22].

50.4
Imaging of pO$_2$ with Phosphorescence Quenching

In vivo measurement of dissolved oxygen (pO$_2$) has been accomplished using several different methods, including oxygen-sensitive electrodes, oxygen-dependent phosphorescence quenching, and electron paramagnetic resonance [23]. Oxygen-sensitive electrodes can provide pO$_2$ measurements from small tissue volumes (a few μm^3) due to the small size of many electrodes, but suffer from the fact that they are single-point measurements and they are invasive and can lead to tissue damage. Determination of pO$_2$ using phosphorescence quenching has been shown to be highly effective for *in vivo* measurements due to recent advances in imaging technology and probe design [24–28]. This method is based on the fact that dissolved oxygen quenches the phosphorescence or fluorescence of certain compounds such as porphyrin dyes and ruthenium complexes, resulting in changes in excited-state lifetimes [29]. As a result, the pO$_2$ can be quantified from the measured lifetime [30] using the Stern–Volmer relationship:

$$\frac{\tau_0}{\tau} = 1 + k_q \tau_0 [\text{pO}_2] \tag{50.2}$$

where τ is the measured lifetime, τ_0 is the unquenched lifetime, and k_q is a quenching constant that depends on the properties of the dye and the surrounding environment. Therefore, absolute oxygenation can be quantified from the measured decay time provided that the quenching constant and unquenched lifetime are known. In addition, lifetime-based measurements are not dependent on absolute intensities [29], which eliminates the need to correct for the presence of other tissue chromophores or tissue scattering.

Despite their widespread use, phosphorescence quenching methods are almost always limited to a single spatial location and only a few reports have demonstrated pO$_2$ imaging or spatial mapping [31–34]. This limitation arises from the need to resolve the phosphorescence lifetime, which makes most cameras extremely difficult to use. Single-point detectors such as avalanche photodiodes and photomultiplier tubes are significantly more sensitive and have sufficient temporal resolution to resolve the phosphorescence decay times.

In order to obtain spatially resolved pO$_2$ measurements, we have utilized a digital micromirror device (DMD) combined with a sensitive single-element detector such as a photomultiplier or avalanche photodiode [35]. This approach retains the high sensitivity of single-element detectors while still permitting spatially resolved phosphorescence decay measurements (Figure 50.3). A speckle contrast image is first obtained, and several regions of interest are selected from the image of the surface vasculature. Each region is transformed to the DMD coordinates and the appropriate mirrors of the DMD are turned on to restrict the pulsed phosphorescence excitation light to this region. The entire phosphorescence emission from that region is collected by the detector and the decay time is determined and converted to a pO$_2$ value. The entire system is controlled by custom software and the DMD switches illumination regions at approximately 10 regions per second.

An example of the use of this system in a rodent stroke model is illustrated in Figure 50.3d–f. In this example, eight spatial regions were selected for pO$_2$ measurement. The CBF (Figure 50.3e) and pO$_2$ (Figure 50.3f) time courses over the eight regions demonstrate that the changes in these two parameters during the course of ischemia differ, particularly in branches of the arteriole. In particular, the CBF drop is larger than the corresponding decrease in pO$_2$. This example illustrates the great

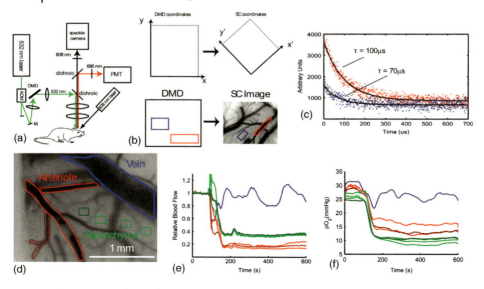

Figure 50.3 (a, b) Combined instrument for simultaneous imaging of CBF and spatial mapping of pO$_2$. Spatially resolved pO$_2$ is obtained by spatially restricting phosphorescence excitation light and utilizing a single-element detector to record time-resolved phosphorescence decays (c). (d–f) Measured temporal dynamics of CBF and pO$_2$ changes in eight different spatial regions following ischemia induction (at $t = 100$ s).

potential of combined optical imaging instruments for revealing the subtleties of hemodynamic changes in the cortex during cerebrovascular pathologies.

References

1. Lo, E.H., Dalkara, T., and Moskowitz, M.A. (2003) Mechanisms, challenges and opportunities in stroke. *Nat. Rev. Neurosci.*, **4** (5), 399–415.
2. Hillman, E.M.C. (2007) Optical brain imaging *in vivo*: techniques and applications from animal to man. *J. Biomed. Opt.*, **12** (5), 051402.
3. Wardell, K., Jakobsson, A., and Nilsson, G.E. (1993) Laser Doppler perfusion imaging by dynamic light scattering. *IEEE Trans. Biomed. Eng.*, **40** (4), 309–316.
4. Ances, B.M., Greenberg, J.H., and Detre, J.A. (1999) Laser Doppler imaging of activation-flow coupling in the rat somatosensory cortex. *Neuroimage*, **10** (6), 716–723.
5. Nielsen, A.N., Fabricius, M., and Lauritzen, M. (2000) Scanning laser-Doppler flowmetry of rat cerebral circulation during cortical spreading depression. *J. Vasc. Res.*, **37** (6), 513–522.
6. Serov, A., Steinacher, B., and Lasser, T. (2005) Full-field laser Doppler perfusion imaging and monitoring with an intelligent CMOS camera. *Opt. Express*, **13**, 3681–3689.
7. Dunn, A.K., Bolay, H., Moskowitz, M.A., and Boas, D.A. (2001) Dynamic imaging of cerebral blood flow using laser speckle. *J. Cereb. Blood Flow Metab.*, **21** (3), 195–201.
8. Boas, D.A. and Dunn, A.K. (2010) Laser speckle contrast imaging in biomedical optics. *J. Biomed. Opt.*, **15** (1), 011109.

9 Briers, J.D. (2001) Laser Doppler, speckle and related techniques for blood perfusion mapping and imaging. *Physiol. Meas.*, **22** (4), R35–R36.

10 Briers, J.D., Richards, G., and He, X. (1999) Capillary blood flow monitoring using laser speckle contrast analysis. *J. Biomed. Opt.*, **4**, 164–175.

11 Briers, J.D. and Webster, S. (1996) Laser Speckle contrast analysis (LASCA): a nonscanning, full-field technique for monitoring capillary blood flow. *J. Biomed. Opt.*, **1**, 174–179.

12 Fercher, A. and Briers, J. (1981) Flow visualization by means of single-exposure speckle photography. *Opt. Commun.*, **37**, 326–329.

13 Grinvald, A., Lieke, E., Frostig, R., Gilbert, C., and Wiesel, T. (1986) Functional architecture of cortex revealed by optical imaging of intrinsic signals. *Nature*, **324**, 361–364.

14 Grinvald, A., Frostig, R., Siegel, R., and Bartfeld, E. (1991) High resolution optical imaging of neuronal activity in the awake mokey. *Proc. Natl. Acad. Sci. U. S. A.*, **88**, 11559–11563.

15 Masino, S. and Frostig, R. (1996) Quantitative long-term imaging of the functional representation of a whisker in rat barrel cortex. *Proc. Natl. Acad. Sci. U. S. A.*, **93**, 4942–4947.

16 Malonek, D. and Grinvald, A. (1996) Interactions between electrical activity and cortical microcirculation revealed by imaging spectroscopy: implications for functional brain mapping. *Science*, **272** (5261), 551–554.

17 Kohl, M., Lindauer, U., Royl, G., Kuhl, M., Gold, L., Villringer, A., and Dirnagl, U. (2000) Physical model for the spectroscopic analysis of cortical intrinsic optical signals. *Phys. Med. Biol.*, **45** (12), 3749–3764.

18 Mayhew, J., Johnston, D., Martindale, J., Jones, M., Berwick, J., and Zheng, Y. (2001) Increased oxygen consumption following activation of brain: theoretical footnotes using spectroscopic data from barrel cortex. *Neuroimage*, **13** (6 Pt 1), 975–987.

19 Jones, M., Berwick, J., Johnston, D., and Mayhew, J. (2001) Concurrent optical imaging spectroscopy and laser-Doppler flowmetry: the relationship between blood flow, oxygenation, and volume in rodent barrel cortex. *Neuroimage*, **13** (6 Pt 1), 1002–1015.

20 Dunn, A.K., Devor, A., Bolay, H., Andermann, M.L., Moskowitz, M.A., Dale, A.M., and Boas, D. (2003) Simultaneous imaging of total cerebral hemoglobin concentration, oxygenation, and blood flow during functional activation. *Opt. Lett.*, **28**, 28–30.

21 Dunn, A.K., Devor, A., Dale, A.M., and Boas, D.A. (2005) Spatial extent of oxygen metabolism and hemodynamic changes during functional activation of the rat somatosensory cortex. *Neuroimage*, **27** (2), 279–290.

22 Jones, P.B., Shin, H.K., Boas, D.A., Hyman, B.T., Moskowitz, M.A., Ayata, C., and Dunn, A.K. (2008) Simultaneous multispectral reflectance imaging and laser speckle flowmetry of cerebral blood flow and oxygen metabolism in focal cerebral ischemia. *J. Biomed. Opt.*, **13** (4), 44007–44011.

23 Tsai, A., Johnson, P.C., and Intaglietta, M. (2003) Oxygen gradients in the microcirculation. *Physiol. Rev.*, **83** (3), 933–963.

24 Wilson, D.F., Rumsey, W.L., Green, T.J., and Vanderkooi, J.M. (1988) The oxygen dependence of mitochondrial oxidative phosphorylation measured by a new optical method for measuring oxygen concentration. *J. Biol. Chem.*, **263** (6), 2712–2718.

25 Shonat, R.D., Wilson, D.F., Riva, C.E., and Pawlowski, M. (1992) Oxygen distribution in the retinal and choroidal vessels of the cat as measured by a new phosphorescence imaging method. *Appl. Opt.*, **31** (19), 3711–3718.

26 Shonat, R.D., Richmond, K.N., and Johnson, P.C. (1995) Phosphorescence quenching and the microcirculation: an automated, multipoint oxygen tension measuring instrument. *Rev. Sci. Instrum.*, **66**, 5075–5084.

27 Vinogradov, S.A., Fernandez-Searra, M., Dugan, B.W., and Wilson, D.F. (2001) Frequency domain instrument for measuring phosphorescence lifetime

28. Dunphy, I., Vinogradov, S.A., and Wilson, D.F. (2002) Oxyphor R2 and G2: phosphors for measuring oxygen by oxygen-dependent quenching of phosphorescence. *Anal. Biochem.*, **310** (2), 191–198.

29. Lakowicz, J.R. (2006) *Principles of Fluorescence Spectroscopy*, Springer, New York.

30. Vanderkooi, J.M., Maniara, G., Green, T.J., and Wilson, D.F. (1987) An optical method for measurement of dioxygen concentration based upon quenching of phosphorescence. *J. Biol. Chem.*, **262** (12), 5476–5482.

31. Shonat, R.D. and Kight, A.C. (2003) Oxygen tension imaging in the mouse retina. *Ann. Biomed. Eng.*, **31** (9), 1084–1096.

32. Shonat, R.D., Wachman, E.S., Niu, W., Koretsky, A.P., and Farkas, D.L. (1997) Near-simultaneous hemoglobin saturation and oxygen tension maps in mouse brain using an AOTF microscope. *Biophys. J.*, **73** (3), 1223–1231.

33. Helmlinger, G., Yuan, F., Dellian, M., and Jain, R.K. (1997) Interstitial pH and pO_2 gradients in solid tumors *in vivo*: high-resolution measurements reveal a lack of correlation. *Nat. Med.*, **3** (2), 177–182.

34. Sakadzić, S., Yuan, S., Dilekoz, E., Ruvinskaya, S., Vinogradov, S.A., Ayata, C., and Boas, D.A. (2009) Simultaneous imaging of cerebral partial pressure of oxygen and blood flow during functional activation and cortical spreading depression. *Appl. Opt.*, **48** (10), D169–D177.

35. Ponticorvo, A. and Dunn, A.K. (2010) Simultaneous imaging of oxygen tension and blood flow in animals using a digital micromirror device. *Opt. Express*, **18** (8), 8160–8170.

(continued from previous) distributions in heterogeneous samples. *Rev. Sci. Instrum.*, **7** (8), 3396–3406.

Part 12
Dermatology

51
Skin Diagnostics

Martin Kaatz, Susaane Lange-Asschenfeldt, Joachim Fluhr, and Johannes Martin Köhler

51.1
Introduction

With an overall size of approximately 2 m^2 human skin is considered the largest organ of the human body and is exposed to the environment. Due to this unique position, the skin and its appendages are susceptible to a great variety of physical, biological, and chemical influences. Owing to its specific features, however, it is able to protect us effectively from many of these influences and to ensure the organism's integrity.

Nevertheless, a considerable number of diseases are manifest on the skin, which may be limited to the skin but may also affect other organ systems. On the other hand, a number of visceral diseases present with cutaneous symptoms, often providing characteristic hints at their true origin. Also, stress factors or emotional challenges may trigger the onset of a skin disease or contribute to its worsening, making the skin a potential reflector of emotional imbalance.

Human skin is easily accessible to clinical evaluation, and also to analysis by invasive and noninvasive examination methods. In recent years, a number of noninvasive *in vivo* diagnostic devices have been developed for the evaluation of skin lesions. This is partly due to the fact that the incidence of many dermatoses, skin cancer in particular, is increasing considerably, thus prompting the need for adjunct diagnostic devices to ensure early diagnosis and treatment. By means of biophotonics, light may be used for diagnosis and therapy of skin diseases. This chapter provides an overview of possible applications of biophotonics in clinical and investigational dermatology.

51.2
Dermoscopy

51.2.1
Technical Principle and Development

Dermoscopy or epiluminescence microscopy is a noninvasive technique for the *in vivo* evaluation and differential diagnosis of skin lesions. It is mainly used as a hand-held device with light-emitting diodes (LEDs) as internal light source. The objective requires optical coupling with the skin surface by oil, water, or gel in order to achieve translucency of the stratum corneum. Recently developed models use polarized light that permit its use without direct contact with the skin. The combination with digital cameras and computer systems allows not only efficient image analysis and documentation but also software-assisted evaluation of the skin lesions. Furthermore, resolution and magnification can be increased (10–60-fold), thus improving the visualization of dermoscopic criteria.

51.2.2
Applications

51.2.2.1 Pigmented Skin Lesions

Dermoscopy is an important examination method for differentiation between melanocytic (Figure 51.1 and 51.2) and nonmelanocytic skin lesions [1]. Also, the status of many tumors can be evaluated more specifically and more sensitively than by visual inspection alone [2]. The application of digital image acquisition systems allows the storage of additional information of examined lesions for re-evaluation and comparison of follow-up pictures. This facilitates longitudinal monitoring for patients at high risk for skin cancer and allows the early identification of malignant melanomas that do not display sure signs of dysplasia upon initial evaluation.

Chapter 53 provides detailed information on the dermoscopic features.

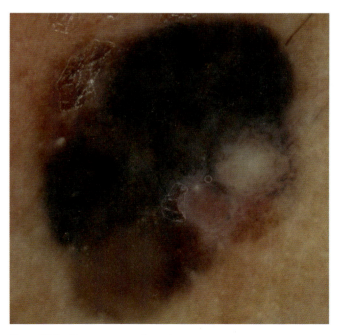

Figure 51.1 Dermatoscopy of a melanoma.

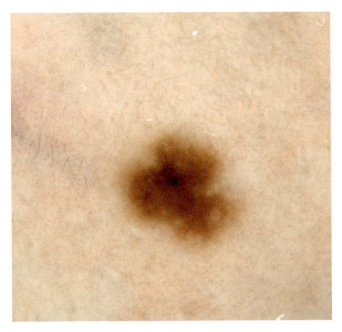

Figure 51.2 Dermatoscopy of a melanocytic nevus.

In addition, dermoscopic criteria that need to be considered in differential diagnosis have been described for other pigmented skin lesions.

Among them, *seborrheic keratoses* are characterized by multiple milia-like cysts, comedo-like openings, light-brown fingerprint-like structures, and/or a cerebriform pattern [1].

Pigmented basal cell carcinomas are diagnosed by characteristic features such as arborizing vessels, leaf-like structures, large blue–gray ovoid nests, multiple blue–gray globules, spoke-wheel areas, and ulceration [3].

The essential dermoscopic features of *pigmented actinic keratoses*, which may be difficult to distinguish from lentigo maligna, are an annular–granular pattern, rhomboidal structures, pseudo-network, black globules, slate-gray globules, black dots, asymmetric pigmented follicular openings, a hyperpigmented rim of follicular openings, and slate-gray areas and streaks.

A clinical discrimination of solar Lentigo from Lentigo maligna is often impossible. The consideration of dermoscopic criteria, however, may help with the differentiation. Important diagnostic features are comedo-like openings, diffuse opaque-brown pigmentation, light-brown fingerprint-like structures, and milia-like cysts [4].

Red–blue lacunae or homogeneous areas are indicators for *benign vascular lesions* such as *cherry angiomas* or *venous malformations*.

51.2.2.2 Non-Pigmented Skin Lesions

Dermoscopy is also able to improve diagnostic accuracy in the examination of non-melanoma skin cancer and nontumorous, particularly vascular, lesions.

For nonpigmented skin disease, the vascular structures are of primary importance. A three-step algorithm has recently been proposed [5]. The first step comprises the assessment of the morphology of vascular patterns, described as linear curved (comma), dotted (globules), linear irregular, linear looped (hairpin), linear coiled (glomerular), or linear serpentine (arborizing), and three specific global patterns: crown, strawberry, and clods. In the second step, the vascular architecture is categorized. The vessel arrangement may be regular, in strings, in clusters, radial, branched, or irregular. The third step consists in the assessment of additional criteria such as duct openings, surface scales, ulcerations, and others. The diagnosis is made based on the combination of the features assessed in the three-step algorithm. For a variety of nonpigmented skin tumors, evidence-based diagnosis and management recommendations have been proposed [5].

Dermoscopic criteria for a broad range of inflammatory and nonpigmented neoplastic skin lesions have been summarized previously [3]. However, in most cases the diagnostic value has not been investigated in larger studies, but only in case reports and smaller case series. However, a number of specific dermoscopic criteria have been defined for some inflammatory diseases, for example, lichen planus, which shows whitish lines in a reticular distribution (corresponding to Wickham striae). Mature lesions also display additional radial capillaries intermingled with the reticular striae. In regressive lesions, the vascular component disappears and gray–blue dots with or without reticular striae become visible.

Dermoscopy may also aid in the diagnosis of contagious skin diseases, including scabies, larva migrans, and tungiasis [3].

A special indication for dermoscopy is the microscopic evaluation of the nailfold capillaries. Here, dermoscopy is used as a capillaroscopic instrument mainly in autoimmune diseases, where characteristic vascular changes may already be detected in early stages [6].

Chapter 52 provides supplementary information on the dermoscopy of non-melanoma skin cancer.

51.2.3
Conclusion

The advantages of dermoscopy in clinical practice are its time-effective and cost-effective use, an improvement in diagnostic accuracy compared with clinical examination alone, and a moderate training effort. Disadvantages – compared with the (invasive) gold standard histology – are the low magnification and the restriction to mainly superficial skin layers.

51.3
Optical Coherence Tomography (OCT)

51.3.1
Technical Principle and Development

Optical coherence tomography (OCT) as a diagnostic tool was first presented by Huang *et al.* [7]. It is based on the principle of Michelson interferometry using light of short coherence length. In this process, a light beam is directed into the piece of tissue under examination. The optical pathlength distribution of the sample beam measured by the interference modulation of the axial OCT scan can be interpreted as the depth-resolved reflection signal of the sample. Interference occurs only when the propagation distance of both beams match within the coherence length of the light source. This coherence length also determines the vertical resolution of the system [7].

51.3.2
Applications

In ophthalmology, OCT has been implemented in clinical routine diagnostics for the evaluation of cornea, lens, and corpus vitreum. As early as in the 1990s, however, first studies on cutaneous tissue were performed whereby hyperkeratosis, epidermis, and dermis could be differentiated [8]. In the following years, the value of OCT has been investigated for the diagnosis of a broad range of skin disorders.

51.3.2.1 Skin Tumors
The continuously improving resolution of OCT allowed the examination of both pigmented and nonpigmented skin tumors. At the same time, the penetration depth was measured and compared with histology and high-resolution sonography. Detailed results are given in Chapter 52 (Figures 51.1 and 51.2) Chapter 53 (Figures 53.3 and 53.4).

51.3.2.2 Inflammatory Skin Diseases
In a first study, patients with irritant contact dermatitis ($n = 15$) and psoriasis vulgaris ($n = 16$) were examined by OCT. The findings were compared with the results obtained by biophysical methods. Whereas intact control skin presented with a

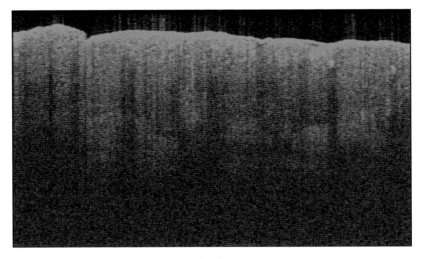

Figure 51.3 OCT of a melanoma at costal arch.

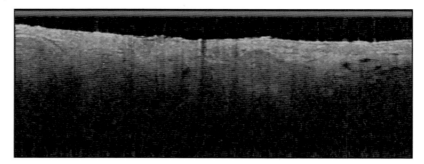

Figure 51.4 OCT of basal cell carcinoma at forehead.

small entrance echo and a clearly defined epidermis, irritated skin caused an increased entrance echo and a thickening of the epidermis [8]. Psoriatic skin presented similarly, with a broadened entrance signal and acanthosis. Intracorneal pustules were depicted as hypo-dense spaces. Longitudinal evaluations following therapy demonstrated a marked regression of these changes [8]. The results were recently confirmed in a separate study including 23 patients. Epidermal thickness of psoriatic and control skin was measured before and after therapy and OCT results correlated well with histologic findings, including a significant decrease in epidermal thickness under therapy [9].

OCT was also applied in the evaluation of contact patch test reactions. Compared with an untreated control area, there was a thickened and/or disrupted entrance signal, pronounced skin folds, and significant acanthosis. Moreover, clearly demarcated signal-free cavities within the epidermis and a considerable reduction in dermal reflectivity are demonstrated by OCT. This result clearly correlated with the clinical characteristics of the patch test result [10].

51.3.2.3 Wound Healing

The different phases of wound healing have also been studied by OCT, although to date only in animal models. Individual wound healing phases could be identified after artificial biopsy in mice. The method allowed the description of wound size over time, early re-epithelialization, dermoepidermal junction formation, and differences in wound tissue composition [11]. In another study, performed on rabbits, polarization-sensitive optical coherence tomography (PS-OCT) was employed for monitoring wound healing processes *in vivo* and *in vitro*. Also, several drugs either fostering (sphingosylphosphorylcholines) or hampering (tetraacetylphytosphingosines) wound healing were administered. The evaluation was performed in comparison with an untreated control in correlation with a routine histology. The variations of phase retardation values caused by the collagen morphology changes on wound sites are quantified for all cases. Collagen fiber regeneration could be visualized by means of PS-OCT, and the effects of various drugs on the process of wound healing process were quantified. Different phases of wound healing such as inflammation, re-epithelialization, and collagen remodeling could be identified in the examination of infected and uninfected wounds. Infected wounds were marked by prominent edema. In the comparison of infected and uninfected wounds, a significantly delayed re-epithelization and collagen remodeling phases were demonstrated [12].

In vivo studies on the longitudinal evaluation of wound healing processes are yet to be performed in order to prove the usefulness of OCT in this area of research.

51.3.2.4 Bullous Skin Disorders

OCT allows for morphologic discrimination between different types of bullae. Single patients with bullous pemphigoid, pemphigus vulgaris, Darier disease, subcorneal pustulosis, and burn blisters were examined by OCT [13].

The tense blisters of bullous pemphigoid were seen in OCT images as dark, round, or ovoid, mostly well-differentiated lobules or clefts. Also, in some cases, gray-appearing portions could be detected that corresponded histologically to the intrabullous eosinophilic infiltrate. Slack blisters (as in pemphigus vulgaris) rather appear gray in OCT images. On the other hand, thermal burns show both intra- and subepidermal blisters. The characteristic skin lesions of subcorneal pustulosis filled by neutrophils are difficult to distinguish from the surrounding area in OCT. They appear as faint gray lobules in the OCT image. The OCT image of a papular lesion such as in Darier disease presents first a disruption of well-defined layering. Also, the epidermis cannot be discriminated from the dermis. With OCT, the depth of bullae development could be clearly defined. Hence OCT can be considered a promising adjunct tool for the classification of bullous dermatoses [13].

51.3.2.5 Other

Another possible clinical application is the longitudinal evaluation of topical therapy of benign inflammatory dermatoses. In one study, an unspecific decrease in surface reflexivity was demonstrated by OCT, which was associated with a lowered A scan entrance peak. The application of glycerin resulted in an increase in skin "transparency" by convergence of the refractive indices. At the same time, an increased thickness of the horny layer was observed, likely due to swelling and occlusion effects. Steroid-induced epidermal atrophy was also investigated by OCT [14]. However, further studies to prove the applicability of OCT for morphometric measurement of normal or atrophic epidermis are necessary.

51.3.3
Conclusion

Currently, OCT is applicable as an adjunct to high-resolution sonography in the noninvasive imaging of structural alterations of the skin *in vivo* and for the evaluation of epidermal and tumor thickness [10]. The resolution achieved by this method is constantly improving, enabling also microscopic structures to be imaged. OCT allows for the imaging of the skin from the stratum corneum down to the reticular dermis, thus allowing the real-time evaluation of numerous dermatoses, skin tumors and therapeutic effects. In order to judge the diagnostic value of this novel method, further studies have to be performed.

51.4
Laser Doppler Imaging

51.4.1
Technical Principle and Development

In principle, the method is based on the Doppler effect. Frequency changes of monochromatic light reflected by erythrocytes are measured. A distinction is made between laser Doppler imaging (LDI) which allows an extensive perfusion measurement, and the use of laser Doppler flowmetry (LDF) for a selective definition of the blood flow.

In LDI, the area under examination is scanned in a meandering pattern. The result is a two-dimensional image of perfusion differences in the area examined. Owing to technological progress and the continually improving lateral resolution, ($<100\,\mu m$), the method allows for precise perfusion mapping in the area examined.

51.4.2
Applications

51.4.2.1 Skin Tumors
The application of LDI in diagnostics of skin tumors is described in more detail in the respective chapters (Chapter 52 and 53).

51.4.2.2 Inflammatory Skin Diseases
Being an inflammatory dermatosis, psoriasis was supposed to be accompanied by abnormal skin perfusion. Speight *et al.* [15] searched for differences in perfusion within the psoriatic plaques compared with the surrounding skin [15]. A fourfold increase in perfusion was detected within the plaque area compared with surrounding tissue. Even an increased blood flow that, according to clinical evaluation, had not been expected to occur, was detected in the close proximity to the plaques within a diameter of 2–4 mm [15]. Davison *et al.* [16] also were able to prove this fringe of increased perfusion outside the clinical edges of the psoriasis plaques [16]. Krogstad *et al.* [17] defined the histamine release during local anesthesia by microdialysis, and could thus eliminate the impact of histamine as a cause of this increased perfusion [17].

Yosipovitch *et al.* [18] examined the blood flow following a barrier dysfunction by tape stripping in both psoriatic plaques and in normal skin of psoriasis patients and also in healthy test persons [18]. An increased perfusion within the plaques was accompanied by a decreased heat pain threshold, whereas the healthy skin displayed an increased heat pain threshold.

Also in psoriasis, LDI was applied for the evaluation of therapeutic efficacy. Stücker *et al.* employed LDI in order to examine the inflammatory response of psoriasis patients to topical retinoids [19]. The initial irritation of the skin, which is often observed, can be reduced by adaptation of the therapy.

51.4.2.3 Microcirculation of Chronic Wounds
The possible fields of application for the laser Doppler method are broad. Above all they comprise diseases that are accompanied by an impaired microcirculation,

mainly focusing on the chronic venous and arterial wounds. The method has been employed not only for examinations on pathogenesis, but also for reviewing the therapeutic success of other methods of treatment.

In a study performed by Mayrovitz et al., perfusion was measured in the area surrounding the ulcers of patients with venous ulcerations [20]. The perfusion measured was highly dependent on the place measurement and its characteristics. Given good granulation, a high perfusion was measured right inside the ulcer. In cases with absence of granulation, perfusion was shown to decrease significantly, an observation that was also made in the area of normal skin surrounding the ulcer. This study proved not only an increased basal periulcerous flow, but also its causes, being a combination of both increased blood flow velocity and circulating blood volume [20].

Ambrozy et al. defined the microcirculation in mixed arterial–venous ulcers and the surrounding skin [21]. In this study, perfusion was found to be highest in the granulation tissue area, whereas the granulation-free parts of the ulcer showed a weak flux only. A high perfusion was also detected in scars. In all areas examined, however, capillary density measured by capillaroscopy was lower than in normal skin [21].

The laser Doppler method is also employed for the evaluation of therapeutic interventions. Jünger et al. examined the impact of a low-frequency pulsed current on the microcirculation of venous ulcers [22]. Arora et al. [23] showed a significantly improved microcirculation in neuropathic diabetic feet after a successful lower extremity revascularization [23]. Also, its effectiveness was examined with regard to the application of wound dressings to chronic wounds, and to comparison with a vasodilating therapy of peripheral artery occlusire disease [23].

Moreover, the method was also employed for the evaluation of the extent of skin burns. Findings showed a significantly improved predictability of the outcome of a burn wound compared with a merely clinical evaluation [24]. In 1993, a study by Niazi et al. showed 100% accord of LDI with histology, whereas the agreement in clinical assessment was only 65% [25]. These findings were confirmed in numerous studies.

51.4.2.4 Collagenosis and Rheumatic Diseases

In addition to progressive sclerosis of the skin and viscera, rarefication especially of the terminal vessels is considered one of the most important pathogenetic mechanisms in systemic scleroderma. This rarefication can lead to necroses of the distal phalanges of the finger, being a sure visible sign of the advancement of the disease. At the same time, vascular regulation is faulty, which soon leads to a secondary Raynaud's phenomenon in patients with scleroderma.

In 1994, Seifalian et al. [26] examined LDI, LDF, and thermography on both healthy patients and those with systemic scleroderma [26]. Whereas a good correlation was found between LDI and LDF, this did not apply to thermography. Perfusion of the finger in patients with systemic scleroderma was lower than in healthy test persons [26].

Clark et al. [27] performed a study of 40 patients with Raynauds's phenomenon (33 of whom suffered from systemic scleroderma), and compared them with 17 healthy volunteers [27]. Hands and fingers of the persons were examined by both LDI and thermography, and the difference between them was measured. The study's focus was on the correlation of these two methods. Only a poor correlation was shown in this examination. LDI is more sensitive to blood flow changes and therefore more likely to show inhomogeneities than the highly damped temperature response. Therefore, one method cannot substitute for the other [27].

In a study of 44 patients suffering from systemic scleroderma, Correa et al. [28] demonstrated a significantly decreased blood flow in the fingers of the patients compared with healthy test persons [28]. The findings could be confirmed at any point in time after cryostimulation. Both in patients and in healthy test persons, blood flow decreased significantly after cryostimulation. In the healthy test persons, however, blood flow recovered after an average of 27 min, whereas this recovery did not occur in the patients with scleroderma. These findings, however, did not correlate with the morphologic microvascular abnormalities detected by nailfold capillaroscopy [28].

Further studies aimed to evaluate the impact of therapies on systemic scleroderma. Sadik et al. [29] could not prove an improvement in perfusion by the application of atorvastatin [29]. Rosato et al., however, showed that patients suffering from systemic scleroderma who were administered bosentan because of their pulmonal hypertonia also presented a higher perfusion of the hands [30].

Moreover, LDI has also been applied in inflammatory joint disease. Ferrell et al. examined 11 patients with rheumatoid arthritis of the interphalangeal joints [31]. Here, a significantly increased perfusion of the affected joints was detected. The method was also employed in order to prove the impact of locally applied gels or intra-articular steroid injections on the hyperemia [31]. In a later study, Ferrell et al. compared LDI with ultrasound and pain score [32]. The results proved particularly the advantages of LDI in the examination of inflammatory joint disease in its early stages [32].

51.4.3
Conclusion

LDI is particularly well suited for the evaluation of blood perfusion, and in that regard for the examination of cutaneous microcirculation. Therefore, LDI has a role in the evaluation of diseases with impaired microcirculation such as chronic wounds, or as an aid in the evaluation of blood flow changes in autoimmune skin disorders including collagenoses. In these diseases, LDI may be useful both for a diagnostic evaluation and for monitoring of therapeutic efficacy. Whether this method will play a role in the routine diagnostic work-up of other diseases remains to be evaluated in future studies and also depends on the continuing development of this methodology.

51.5
Confocal Laser Scanning Microscopy (CLSM)

51.5.1
Technical Principle and Development

The principle of confocal laser scanning microscopy (CLSM) was first described in the 1950s and patented by Minsky in 1961, long before the invention of lasers [33, 34].

The first commercially available system was realized by Rajadhyaksha et al. at Harvard Medical School in 1995, based on an 830 nm diode laser that is used as a point light source in a scanning mode providing horizontal tomographic images [35]. The visualization of tissues is based on the different refractive indices of the distinct tissue chromophores and cell structures. This effect is called reflectancy, which is caused by the reflection and scattering of light from the piece of tissue under examination [36]. Owing to the very small aperture, most of the reflected light from outside the focal plane is eliminated. This permits a lateral resolution of 0.5–1 μm and an axial resolution (section thickness) of 3–5 μm. In contrast to routine histology, CLSM imaging is based on optical horizontal sectioning.

For morphologic evaluation, CLSM is generally applied in the reflectance mode without exogenously applied fluorophores. This facilitates its use for imaging of human skin *in vivo*. On the other hand, the application of fluorescent dyes may be useful for the evaluation of cutaneous metabolism, cell to cell contacts, or cell migration, thereby allowing the identification of targets by imaging in the fluorescent mode. Fluorescein sodium is routinely used as a fluorophore (application intradermal or to the skin surface). This fluorophore is safe for human use *in vivo*, whereas many other dyes are permitted only for use *ex vivo* [37].

51.5.2
Applications

51.5.2.1 Topography of Normal Skin
Using CLSM, the selective imaging of single cells and single layers of the epidermis is possible. Due to the horizontal sectioning, the first layer to be evaluated is the stratum corneum, which is characterized by anucleated corneocytes arranged in highly refractive cell clusters. The stratum granulosum consists of 2–4 cell layers, the single cells being between 20 and 25 μm in size. The cell nuclei are displayed in the center as dark, oval, or round structures that are surrounded by a narrow ring of bright, granulous-appearing cytoplasm. The stratum spinosum displays polygonal, rather small cells of size 15–20 μm. These are arranged in a typical honeycomb pattern that, at the papillary tips, shows first signs of pigmented basal cells. The cells of the basal layer are between 10 and 12 μm in size. The basal cell layer itself consists of more or less refractive cells that vary with respective Fitzpatrick skin phototypes, according to their different melanin contents. Here, the content of melanin correlates with the respective reflectancy and, thus, with the image brightness.

At the level of the dermoepidermal junction, the basal cell layer surrounds centrally positioned dark papillae. Often, blood flow can be visualized within the papillae due to the high reflectance of erythrocyte hemoglobin [38].

51.5.2.2 Skin Tumors
The value of CLSM in diagnostics of epithelial and pigmented skin tumors has already been surveyed in numerous clinical studies. Thereby, CLSM may be useful not only for the diagnostic classification of proliferative skin lesions but also for the definition of tumor margins prior to Mohs' surgery or the detection of subclinical

disease. Chapter 52 and 53 provides a detailed outline of the applicability of CLSM for examination of melanoma and non-melanoma skin cancer.

51.5.2.3 Vascular Skin Lesions

Reflectance confocal microscopy allows for the examination of small vessels and their anatomic and flow characteristics *in vivo* and in real time. Hence the method is applicable to the evaluation of vascular lesions. Some key features are listed below.

CLSM examination of *spider angioma* produces images of whorled canalicular structures in the mid- and upper dermis. Vessels measure 40–50 µm in diameter and show a high flow [39].

CLSM examination of *cherry angioma* shows a large number of dilated and tortuous dermal capillaries which are up to 50–100 µm wide and of medium flow. The structure is lobular and shows fine fibrous septa [39].

CLSM evaluation of *pyogenic granuloma* shows an increased number of elongated dermal capillaries only 5–10 µm in size. Further characteristics are the detection of lymphocyte rolling and the marked luminescence of canalicular structures. The epidermis additionally shows spongiosis and exocytosis [39].

Kaposi's sarcoma displays inflammatory single cells or aggregates in epidermis and dermis, and also numerous spindle cells. Apart from erythrocyte extravasates and hemosiderin deposits, an increased number of dilated and anastomosed cells can be detected in the stroma [40].

Owing to the small number of cases, the ultimate role of CLSM in the diagnosis of vascular alterations remains to be determined in larger clinical and correlative studies.

51.5.2.4 Inflammatory Skin Diseases

Pilot studies have examined a large number of inflammatory skin lesions by CLSM. For most entities, however, randomized multicenter studies are still lacking.

Typical histologic features for acute contact dermatitis could be visualized by CLSM: cell nuclei in the stratum corneum (parakeratosis), intraepidermal vesiculation, inflammatory cells within the epidermis, and an increase in papillary vessels [41].

In search of allergen-specific results of CLSM compared with the clinical evaluation of the patch test (PT), varying images of the reactions to 2% cobalt chloride and 5% nickel sulfate could be proved [41]. The examination of positive patch test reactions showed an increased suprabasal epidermal thickness on days 2 (D2), and 3 (D3) for cobalt chloride. Fairly often, vesicle formation and an overall increased suprabasal epidermal thickness for nickel sulfate appeared in patients with a positive reaction to nickel. In two out of three doubtful positive PTs to cobalt chloride, CLSM images presented characteristics of irritant reactions and one characteristic of a positive allergic reaction [41]. The dynamic evolution of allergic contact dermatitis (ACD) and irritant contact dermatitis (ICD) was also investigated [42]. Here, significant differences in both the time course and the severity of the single features were observed. In this study, structural changes of the stratum corneum were identified as a discriminative feature of ICD. Similarly, the initial presence of superficial disruption is highly indicative of irritant reactions [42]. In ACD, this pattern may only be observed later in the course, or in very extensive reactions.

The degree of epidermal spongiosis, vesicle formation, and exocytosis, however, did not show significant differences between ICD and ACD.

CLSM criteria for the diagnosis of psoriasis correspond well with those established by histology and include hyperparakeratosis, loss of the stratum granulosum, elongation of the rete ridges, and dilatation of the papillary vessels [43]. In a study of 75 patients with psoriasis, a sensitivity of 89% and a specificity of 95% were shown [44]. Further entities examined in this study include mycosis fungoides and lupus erythematodes. In both diseases, the specificity was high (>92%), but the sensitivity was low (63%) [44].

51.5.2.5 Infections

Infections of the skin are the cause of many dermatoses. At the same time, they are able to trigger existing dermatoses, especially as a super antigen. Bacterial, mycotic, or viral infections have been studied by CLSM in both reflectance and fluorescence modes.

CLSM was applied to the diagnostics of secondary syphilis. The method allowed for the proof of treponema in the stratum spinosum, but again, these findings are awaiting more extensive evaluations [45].

Staphylococcus and *Candida* strains have also been examined. The yeast and hyphal forms of *C. albicans* could be proved by CLSM within a biofilm using the fluorescence mode (FITC-labeled). These findings allow future studies on the effectiveness of antimicrobial agents in order to eliminate the biofilms, a review of prophylactic application of antimycotics, and examination of biomaterials [46].

51.5.2.6 Drug Delivery

CLSM has been applied in numerous surveys in order to study the extent of substance deposition in the skin and their routes of penetration. Owing to improvements of this method by the supplementary implementation of a reflectance mode, examinations can be performed without administration of any additional fluorophores. This permits examinations *in vivo* in humans, and thus a more precise evaluation of locally applied drugs.

In most cases, fluorescent samples were used in order to permit of quantitative measurements. Examination of liposomes and the impact of chemical penetration enhancers were of priority here. Most of the studies, however, were performed on skin models, on skin *ex vivo*, or in animal experiments in order to avoid damage caused by the potential toxicity of the fluorophores applied.

Far-reaching investigations were also performed on matter transportation prior to photodynamic therapy, which plays an important role in the therapy of non-melanoma skin cancer. In these studies, the influence of dimethyl sulfoxide on the penetration of 5-aminolevulinic acid *in vitro* and the resulting protoporphyrin IX accumulation *in vivo* were analyzed by confocal microscopy [47]. Also, the impact of physical methods such as iontophoresis and ultrasound on the penetration were examined [48].

51.5.3
Conclusion

Owing to its high resolution and a relatively large imaging area, CLSM allows for the noninvasive evaluation of inflammatory and neoplastic alterations of the skin.

In contrast to routine histology, CLSM permits the imaging of dynamic processes, which allows for evaluation of disease evolution and therapeutic effects. Also, research has shown that CLSM permits the detection of subclinical disease manifestations.

51.6
Multiphoton Laser Imaging

51.6.1
Technical Principle and Development

Maria Goeppert-Mayer described the principle of fluorescence excitation of molecules by multiple photons in 1931, but could not prove the process experimentally [90]. The experimental confirmation of the theoretical prediction was provided by Kaiser and Garrett in 1961 [50], shortly after the development of the first laser systems. On the basis of these findings, the first microscope-based on multiphoton process was developed in 1990 [51]. The effect of the second-harmonic generation (SHG) was already detected at the beginning of the 1960s. In 1974, a microscope was developed based on the SHG effect, which was later refined into a scanning microscope.

The principle of multiphoton imaging is the simultaneous absorption of two or more photons by a fluorophore and emission of one photon at a shorter wavelength, that is, with higher energy. The process requires high photon flux densities that are obtained either by a consistently high laser power or by very short pulses with high peak power [49]. The advantage of using ultra-short pulses is the low average total energy in the tissue, thus avoiding tissue damage.

Owing to the complex structure of the skin, there is a variety of fluorescent biomolecules. Among them, the most important include the nicotinic acid derivatives NADH and NADPH in their reduced state, which contribute significantly to tissue fluorescence [52–54]. Oxidized (NAD and NADP) nicotinic acid derivatives have, however, no fluorescent properties. The highest concentration of NADH and NADPH is found in the mitochondria, but they are also present in the cytoplasm. As an electron transporters and cofactors of numerous redox reactions, they are important indicators of cell metabolism.

With regard to endogenous pigments in cutaneous tissue, melanin and lipofuscin have fluorescent properties. Porphyrins are found as organic dyes in the hemoglobin of red blood cells and show a high absorption in the near-infrared spectral range. Another important and prominent epidermal fluorophore is keratin.

While most of the fluorophores are located intracellularly, elastin and collagen as extracellular matrix proteins also have fluorescent properties. A characteristic phenomenon of fibrillar collagen is its ability for SHG. In sum, the photons emitted from the dermis consist of both autofluorescence and SHG effects [55].

51.6.2 Applications

51.6.2.1 Topography of Normal Skin

The high lateral and axial resolution, each of only a few microns, allows the *in vivo* visualization of microstructures of cutaneous tissue, whereby single cells and even subcellular structures may be described (Figure 51.4). To date, both penetration depth and the field of view are restricted to about 200 μm in *in vivo* multiphoton laser tomography (MLT). Nevertheless, its use in dermatologic science has been proved in several studies. By depth-resolved semiquantitative assessment of autofluorescence (AF) and SHG, it was shown that MLT provides a novel *in vivo* imaging tool for the measurement of epidermal morphometric parameters such as epidermal thickness, which may on the other hand be useful for the observation over time of dynamic changes in skin disorders and therapeutic side effects or in cosmetic research and development [56].

51.6.2.2 Skin Tumors

Multiphoton laser tomography or microscopy has been applied both in the diagnostics of malignant melanomas and in the classification of nonmelanocytic skin tumors. The specific diagnostic criteria and the preciseness that can be achieved by the method are described in Chapter 52 and 53 (Figure 51.5).

51.6.2.3 Skin Aging

Skin aging is essentially characterized by alterations of the elastic and the collagenous fiber network. In the course of the aging process of the skin, so-called solar elastosis

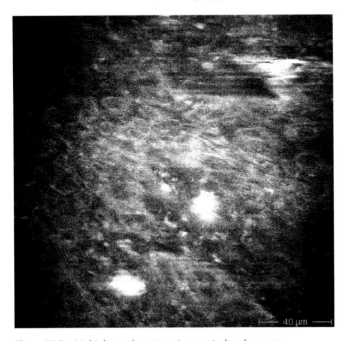

Figure 51.5 Multiphoton laser imaging atypical melanocytes.

appears, which is an accumulation of inferior elastic fibers in the skin. At the same time, the number of mature collagenous fibers decreases. Both fiber qualities can be recorded by MLT. Elastic fibers can be quantified by measuring the autofluorescence caused by multiphoton processes. Collagenous fibers, however, can be recorded via the SHG effect. Since the excitation maximum of both the SHG effect and the AF of elastic fibers lies within the blue spectral range, fibers of both qualities are best excited at 820 nm. In an *ex vivo* study of a few skin samples, Lin *et al.* [57] were able to show for the first time a negative correlation of test persons' age with an index derived from both these values (SAAID). SAAID is defined as: $SAAID = (SHG - AF)/(SHG + AF)$, SHG and AF being a dimension of corresponding numbers of photons measured in a rectangle [57]. These results were reviewed *in vivo*. In this survey, which covered 18 test persons aged between 21 and 84 years, both the AF of the elastic fibers and the SHG effect were measured. SAAID was calculated. The score proved a statistically verified negative correlation with the age of the test persons [58].

In another study, young healthy volunteers using indoor tanning facilities and older people were compared with appropriate controls by semiquantitative evaluation of the dermal matrix composition with a DermaInspect multiphoton laser tomograph. Highly significant differences between the sun-protected volar forearm and the dorsal forearm and also between young and old test persons were found [59].

Another study focused on age-dependent morphologic changes of the dermal fiber network. Using a filter system, the fibers of collagen and elastin were optically separated. Several parameters were defined to describe the fiber network and the parameters were found to be different on comparing young and old volunteers [59] (Figures 51.6–51.9).

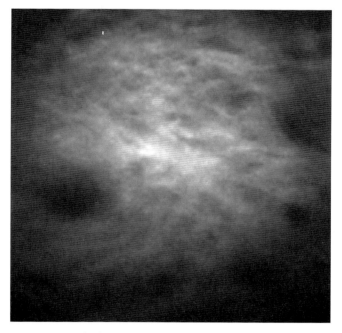

Figure 51.6 Multiphoton laser imaging collagen in young skin.

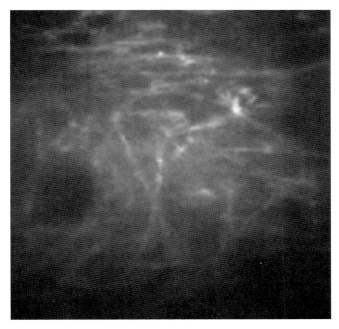

Figure 51.7 Multiphoton laser imaging elastin in young skin.

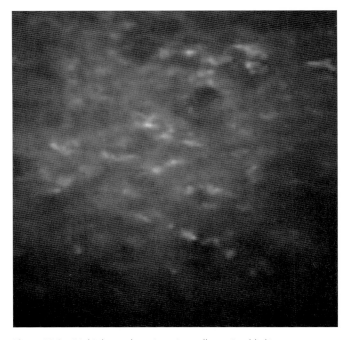

Figure 51.8 Multiphoton laser imaging collagen in old skin.

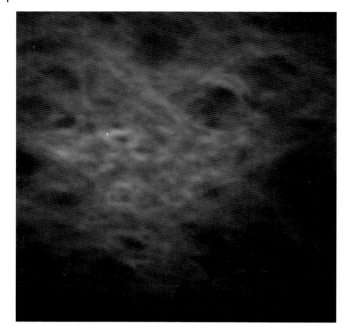

Figure 51.9 Multiphoton laser imaging elastin in old skin.

51.6.2.4 Drug Delivery

As a noninvasive imaging tool, MLT is suitable for the measurement of the effectiveness and the kinetics of percutaneous substance transport [60]. After application of different NaCl solutions, swelling and shrinking of corneocytes were observed depending on the salt concentration [61]. Application of penetration enhancers resulted in visible corneocyte disruption, disturbance of the intracellular keratin network, and pore formation [62]. The penetration of fluorescing nanoparticles into hair follicles and wrinkles could be proved by MLT [63]. In a recent study, Gratieri et al. examined the penetration characteristics of quantum dot particles (4 nm). Even for those small particles, a penetration into intact skin could not be proved [64]. However, penetration was proved only 10 min after rubbing the particles into damaged skin. Hence, for the time being, nanoparticles are not considered suitable carriers in substance transport [64].

Moreover, MPT results obtained by Lee et al. [61] showed the disruption of the intracellular keratin networks and intercellular cohesion of corneocytes by the penetration enhancer of a depilatory cream [61]. Prior treatment of the stratum corneum with a depilatory cream can detach corneocytes and lead to the development of intracellular pores in the remaining corneocytes by disrupting the intracellular keratin. Due to these processes, shunts for drug penetration are created [54].

Furthermore, Yu and co-workers applied various model fluorescent drugs *ex vivo*, the AF characteristics of which differ from those of keratin, thus enabling the dynamics of drug penetration in the stratum corneum to be elucidated [63, 65].

The penetration dynamics of the model substance oleic acid have also been visualized and quantified by MLT [66].

51.6.3
Conclusion

To date, multiphoton imaging is not a diagnostic tool in routine clinical practice, but it has been applied in study settings for the diagnosis of various skin diseases, including skin tumors (malignant melanoma [67], basal cell carcinoma [68], squamous cell carcinoma [69], inflammatory skin disorders (pemphigus vulgaris [70], psoriasis [71], photoaging [72], and intrinsic aging [73]).

The advantages of MLT as a noninvasive method are its high resolution, the specific effect of SHG, and the possibility of fluorescence lifetime detection.

51.7
Raman Spectroscopy

51.7.1
Technical Principle and Development

Raman spectroscopy (RS) helps in gathering information on the molecular architecture, the structure, and the interactions within a sample. In the process, light radiation leads to changes in oscillation and rotational energy levels of the molecules. RS is a method that measures Raman dispersion radiation, that is, nonelastic photon scattering.

Different laser types are employed in Raman spectrometers, including continuously operating gas lasers, Nd:YAG lasers which are mainly employed in the form of Fourier transform spectrometers, and semiconductor junction lasers. The excitation radiation requires monochromatic light in the ultraviolet–visible range and the infrared light of a highly intense radiation beam.

Depending on different wavelengths, RS allows the examination of the stratum corneum, the entire epidermis, and even the chemical architecture down to the dermis. Owing to their maximum penetration depth of 200 µm, shorter wavelengths below 830 nm are only suitable for examinations of the epidermis. Excitation wavelengths of more than 1000 nm, however, also permit examinations of the dermis [74].

When combined with confocal microscopy, RS allows the capture of both the chemical structure and morphologic aspects of the tissue of interest [75]. Thus, the chemical composition of a tissue can be assigned to a specific morphologic structure. At the same time, the sensitivity of substance transport and substance penetration are improved [75].

51.7.2
Applications

51.7.2.1 Molecular Structure of Normal Skin
Gniadecka et al. were able to prove significant differences between the Raman spectra of healthy skin, hairs, and nails [76]. The major spectral differences in the confor-

mational behavior of lipids and proteins were of causal nature. Although the examined proteins of all three samples were of the same secondary α-helix structure, there were differences with regard to the positioning of the S-S stretching bands and to the degree of folding. The analysis of the lipid peaks showed a highly ordered, lamellar crystalline structure with an increased lipid fluidity of the entire skin compared with nails and hairs [76].

Detailed analyses performed in subsequent studies focused on the stratum corneum as the primary component of the epidermal barrier function. The thickness of the stratum corneum (axial resolution 2 µm) could be exactly defined by means of confocal RS. Also, the semiquantitative analysis of important components such as stratum corneum lipids or natural moisturizer factor (NMF) components such as lactate, urea, and urocanic acid could be accomplished [77]. The composition and arrangement of lipids in the epidermis, especially in the stratum corneum, are of particular importance for the barrier function. Thus, a subtyping of neutral lipids, (cholesterol, cholesterol esters, cholesterol sulfate, triglycerides, free fatty acids, squalene, and alkanes), polar lipids (phosphatidylcholine, phosphatidylserine, phosphatidylethanolamine, spingomyelin), and sphingolipids (cermamides I–VII, monohexosylceramides) was analyzed based on Raman spectra and supplemented by X-ray diffraction, differential calorimetry, and fluorescence or infrared spectroscopy [78]. Confocal RS also allowed the localization and analysis of substance mixtures of complex structure, such as NMF (main components: 2-pyrrolidone-5-carboxylic acid and urocanic acid). These factors, which are formed by a filaggrin protein breakdown, are useful biomarkers for the filaggrin genotype [79]. RS was also suitable for comparing different methods for distance measurement of the stratum corneum.

In the course of the further development of laser technology, longer excitation wavelengths were applied, allowing the examination of the epidermis, and also the qualitative analysis of the upper dermis. By using a wavelength of 1064 nm, Naito *et al.* were able to examine lipids and proteins of the dermis, but also the cutaneous appendages (sebaceous glands and hair follicles) [74].

Nakagawa *et al.* employed RS in order to analyze the water content of the dermis [80]. By adjusting the exposure time and depth increment, the signal-to-noise ratio of the Raman spectra could be significantly improved. A comparison of different age groups showed a significantly higher water content in the skin of older volunteers. Furthermore, Lademann *et al.* succeeded in defining the contents of β-carotene and lycopene (carotenoids) in the skin [81].

51.7.2.2 Inflammatory Skin Diseases

Psoriasis Vulgaris
Psoriasis is a chronic inflammatory disease with a pronounced disorder of the epidermal metabolism and subsequent alterations in the molecular composition of the stratum corneum.

Wohlrab *et al.* examined the stratum corneum of healthy test persons compared with patients suffering from atopic dermatitis and psoriasis vulgaris, applying RS to unaffected skin areas [82]. The *in vitro* Fourier transformed Raman spectra showed first of all differences with regard to lipids (spectral range 1112–1142 nm). After statistical evaluation, this approach allowed the differentiation of the skin of healthy test persons and both the inflammatory dermatoses even in unaffected skin areas.

The analysis of psoriatic scaling from different areas of the body and of normal skin of healthy test persons showed no differences between the spectra with regard to the anatomic locations. The Raman spectra did not show any significant differences in lipid band positions whereas the crystalline lipid structure was disrupted in psoriatic scales. Nevertheless, there were great spectral differences with regard to the molecular structure of the proteins. Compared with normal skin, the peak position of the amide I band was shifted towards longer wavelengths, suggesting an unfolding of proteins [82].

Atopic Dermatitis
Being a very heterogeneous inflammatory skin disease, atopic dermatitis (AD) is modified and triggered by numerous immunologic, genetic, and environmental factors. The filaggrin gene (FLG), which codes for the protein filaggrin, is of great importance in pathogenesis. The classification of Raman profiles indirectly allows for an evaluation of different filaggrin genotypes which strongly influence the composition of the NMF. O'Regan et al. examined 132 well-characterized patients with medium to severe atopic dermatitis [83]. The study was performed using confocal RS, and the results were correlated with the clinical findings and also with the FLG genotype. Based on the examination results, a clear distinction could be made between FLG-related and an unrelated AD. Furthermore, the analysis of the NMF composition allowed the definition of a subset of FLG-related AD, and also differentiation between one or two mutations [83].

Gonzáles et al. examined newborns over the course of 1 year, studying the protein filaggrin. Owing to significant differences between the respective Raman spectra, a statistical analysis allowed for prognoses on the development of a filaggrin-related atopic dermatitis [84].

Infections and Wound Healing
In different studies, especially in animal models, RS has been employed for the evaluation of wound healing processes. The definition of specific peaks that correspond to elastin or keratin permits a differentiation between normally healing wounds and those undergoing a retarded healing process. Andrew Chan et al. evaluated all phases of acute wound healing (inflammation, proliferation including re-epithelialization and remodeling) after acute traumatization [85]. The protein profiles of the proteins involved in the wound healing process, and also their subclasses, were analyzed. Along with acute wounds, mainly chronic wounds are the target of noninvasive diagnostic methods. This is due to the great impairment of affected patients, and to the limited therapeutic success. Among others, the interaction of oleic acid with wound dressings applied for proteinase inhibition was tested by means of RS. Proteases are of critical importance for wound

chronification. A profile definition may effectively support the choice of suitable local therapy methods.

Not only in wound healing but also in numerous dermatoses, infections are highly relevant. Willemse-Erix *et al.* examined various skin isolates of multiresistant coagulase-negative strains of *Staphylococcus epidermidis* [86]. Here, RS allowed a clear differentiation between the specific isolates. Additionally, the high throughput allows for a similar analysis of multiple colonies in a patient.

Drug Delivery

RS, especially combined with CLSM, is a method suitable for examination of substance transport and substance penetration within and through the skin.

As carotenoids are antioxidant substances, various studies have been performed to analyze their content and distribution in the skin. Owing to their capacity as radical scavenger, carotenoids significantly influence skin aging processes. It has been shown recently that skin with a higher carotenoid content displays less aging signs than skin with a lower carotenoid content [87]. Specific carotenoid profiles could be determined by means of RS. These Raman spectra also permit the semiquantitative definition of carotenoid (e.g., lycopene) and retinol (vitamin A_1) contents. For example, Hesterberg *et al.* showed that a special diet with ecologic eggs led to a significantly increased content of carotenoids in the skin. Also, protective antioxidant effects of carotenoids after exposure of the skin to infrared light has been shown [88].

Further studies were performed on the effectiveness of penetration enhancers, on tolubuterol, and on the quantification of metronidazole in the skin [89].

51.7.3
Conclusion

RS, originally developed for *in vitro* studies, is gaining rapid importance in clinical studies. Especially superficial parts of the skin, that is, stratum corneum and in some cases down to the dermis, can be studied. The composition of the skin and subsequently specific skin diseases have been studied by RS. Recently, alterations at the molecular level have been addressed by RS in atopic dermatitis, psoriasis, infections, and wound healing, and drug delivery studies have also been performed. However many questions are still remain to be answered in this field, for example, standards on the way in which clinic studies can be performed. RS is a powerful tool allowing clinical and basic research to gain rapid *in vivo* insight into molecular processes in the skin.

Acknowledgement

For technical support: Nancy Schmidt, Christine Weidl; for scientific support: Prof. Karsten König, Rainer Bückle, Martin Weinigel, Marcel Kellner-Höfer, Susanne Metz, Enrico Dimitrow.

Figures: OCT (Chapter 51–53): Prof. Julia Welzel, Augsburg; CLSM: Dr. Susanne Lange–Asschenfeldt, Berlin.

References

1. Malvehy, J. et al. (2007) Dermoscopy report: proposal for standardization. Results of a consensus meeting of the International Dermoscopy Society. *J. Am. Acad. Dermatol.*, **57**, 84–95.
2. Argenziano, G. et al. (2003) Dermoscopy of pigmented skin lesions: results of a consensus meeting via the Internet. *J. Am. Acad. Dermatol.*, **48**, 679–693.
3. Zalaudek, I. et al. (2006) Dermoscopy in general dermatology. *Dermatology.*, **212**, 7–18.
4. Tanaka, M. et al. (2011) Key points in dermoscopic differentiation between lentigo maligna and solar lentigo. *J. Dermatol.*, **38**, 53–58.
5. Zalaudek, I. et al. (2010) How to diagnose nonpigmented skin tumors: a review of vascular structures seen with dermoscopy. Part I. Melanocytic skin tumors. *J. Am. Acad. Dermatol.*, **63**, 361–374, quiz 375–376.
6. Beltrán, E. et al. (2007) Assessment of nailfold capillaroscopy by ×30 digital epiluminescence (dermoscopy) in patients with Raynaud phenomenon. *Br. J. Dermatol.*, **156**, 892–898.
7. Huang, D. et al. (1991) Optical coherence tomography. *Science*, **254**, 1178–1181.
8. Welzel, J. et al. (2003) Optical coherence tomography in contact dermatitis and psoriasis. *Arch. Dermatol. Res.*, **295**, 50–55.
9. Morsy, H. et al. (2010) Optical coherence tomography imaging of psoriasis vulgaris: correlation with histology and disease severity. *Arch. Dermatol. Res.*, **302**, 105–111.
10. Gambichler, T. et al. (2007) Characterization of benign and malignant melanocytic skin lesions using optical coherence tomography in vivo. *J. Am. Acad. Dermatol.*, **57**, 629–637.
11. Cobb, M.J. et al. (2006) Noninvasive assessment of cutaneous wound healing using ultrahigh-resolution optical coherence tomography. *J. Biomed. Opt.*, **11**, 064002.
12. Sahu, K. et al. (2010) Non-invasive assessment of healing of bacteria infected and uninfected wounds using optical coherence tomography. *Skin Res. Technol.*, **16**, 428–437.
13. Mogensen, M. et al. (2008) Optical coherence tomography imaging of bullous diseases. *J. Eur. Acad. Dermatol. Venereol.*, **22**, 1458–1464.
14. Cossmann, M. and Welzel, J. (2006) Evaluation of the atrophogenic potential of different glucocorticoids using optical coherence tomography, 20-MHz ultrasound and profilometry; a double-blind, placebo-controlled trial. *Br. J. Dermatol.*, **155**, 700–706.
15. Speight, E.L. et al. (1993) The study of plaques of psoriasis using a scanning laser-Doppler velocimeter. *Br. J. Dermatol.*, **128**, 519–524.
16. Davison, S.C. et al. (2001) Early migration of cutaneous lymphocyte-associated antigen (CLA) positive T cells into evolving psoriatic plaques. *Exp. Dermatol.*, **10**, 280–285.
17. Krogstad, A.L. et al. (1999) Capsaicin treatment induces histamine release and perfusion changes in psoriatic skin. *Br. J. Dermatol.*, **141**, 87–93.
18. Yosipovitch, G. et al. (2003) Thermosensory abnormalities and blood flow dysfunction in psoriatic skin. *Br. J. Dermatol.*, **149**, 492–497.
19. Stücker, M. et al. (2002) Instrumental evaluation of retinoid-induced skin irritation. *Skin Res. Technol.*, **8**, 133–140.
20. Mayrovitz, H.N. and Larsen, P.B. (1994) Periwound skin microcirculation of venous leg ulcers. *Microvasc. Res.*, **48**, 114–123.
21. Ambrozy, E. et al. (2009) Microcirculation in mixed arterial/venous ulcers and the surrounding skin: clinical study using a laser Doppler perfusion imager and capillary microscopy. *Wound Repair Regen.*, **17** (1), 19–24.
22. Jünger, M. et al. (1997) Treatment of venous ulcers with low frequency pulsed current (Dermapulse): effects on cutaneous microcirculation. *Hautarzt*, **48**, 897–903.

23. Arora, S. et al. (2002) Cutaneous microcirculation in the neuropathic diabetic foot improves significantly but not completely after successful lower extremity revascularization. *J. Vasc. Surg.*, **35** (3), 501–505.
24. La Hei, E.R. et al. (2006) Laser Doppler imaging of paediatric burns: burn wound outcome can be predicted independent of clinical examination. *Burns*, **32**, 550–553.
25. Niazi, Z.B. et al. (1993) New laser Doppler scanner, a valuable adjunct in burn depth assessment. *Burns*, **19**, 485–489.
26. Seifalian, A.M. et al. (1994) Comparison of laser Doppler perfusion imaging, laser Doppler flowmetry, and thermographic imaging for assessment of blood flow in human skin. *Eur. J. Vasc. Surg.*, **8**, 65–69.
27. Clark, S. et al. (2003) Comparison of thermography and laser Doppler imaging in the assessment of Raynaud's phenomenon. *Microvasc. Res.*, **66**, 73–76.
28. Correa, M.J. et al. (2010) Comparison of laser Doppler imaging, fingertip lacticemy test, and nailfold capillaroscopy for assessment of digital microcirculation in systemic sclerosis. *Arthritis Res. Ther.*, **12**, R157.
29. Sadik, H.Y. et al. (2010) Lack of effect of 8 weeks atorvastatin on microvascular endothelial function in patients with systemic sclerosis. *Rheumatology (Oxford).*, **49**, 990–996.
30. Rosato, E. et al. (2010) Bosentan improves skin perfusion of hands in patients with systemic sclerosis with pulmonary arterial hypertension. *J. Rheumatol.*, **37**, 2531–2539.
31. Ferrell, W.R. et al. (1996) Laser Doppler perfusion imaging of proximal interphalangeal joints in patients with rheumatoid arthritis. *Clin. Exp. Rheumatol.*, **14**, 649–652.
32. Ferrell, W.R. et al. (2001) Metacarpophalangeal joints in rheumatoid arthritis: laser Doppler imaging – initial experience. *Radiology*, **220**, 257–262.
33. Minsky, M. Microscopy apparatus. U.S. Pat. US 3013467; http://www.freepatentsonline.com/3013467.html (last accessed 16 May 2011).
34. Minsky, M. Brief Academic Biography of Marvin Minsky, http://web.media.mit.edu/~minsky/minskybiog.html (last accessed 16 May 2011).
35. Rajadhyaksha, M. et al. (1995) In vivo confocal scanning laser microscopy of human skin: melanin provides strong contrast. *J. Invest. Dermatol.*, **104**, 946–952.
36. Rajadhyaksha, M. et al. (1999) Video-rate confocal scanning laser microscope for imaging human tissues in vivo. *Appl. Opt.*, **38**, 2105–2115.
37. Suihko, C. et al. (2005) Fluorescence fibre-optic confocal microscopy of skin in vivo: microscope and fluorophores. *Skin Res. Technol.*, **11**, 254–267.
38. Sauermann, K. et al. (2002) Histometric data obtained by in vivo confocal laser scanning microscopy in patients with systemic sclerosis. *BMC Dermatol.*, **2**, 8.
39. Astner, S. et al. (2010) Preliminary evaluation of benign vascular lesions using in vivo reflectance confocal microscopy. *Dermatol. Surg.*, **36**, 1099–1110.
40. Grazziotin, T.C. et al. (2010) Preliminary evaluation of in vivo reflectance confocal microscopy features of Kaposi's sarcoma. *Dermatology*, **220**, 346–354.
41. Astner, S. et al. (2005) Pilot study on the sensitivity and specificity of in vivo reflectance confocal microscopy in the diagnosis of allergic contact dermatitis. *J. Am. Acad. Dermatol.*, **53**, 986–992.
42. Astner, S. et al. (2006) Noninvasive evaluation of allergic and irritant contact dermatitis by in vivo reflectance confocal microscopy. *Dermatitis*, **17**, 182–191.
43. Ardigo, M. et al. (2009) Concordance between in vivo reflectance confocal microscopy and histology in the evaluation of plaque psoriasis. *J. Eur. Acad. Dermatol. Venereol.*, **23**, 660–667.

44 Koller, S. et al. (2009) *In vivo* reflectance confocal microscopy of erythematosquamous skin diseases. *Exp. Dermatol.*, **18**, 536–540.

45 Venturini, M. et al. (2009) Reflectance confocal microscopy for the *in vivo* detection of Treponema pallidum in skin lesions of secondary syphilis. *J. Am. Acad. Dermatol.*, **60**, 639–642.

46 Peters, BM. et al. (2010) Microbial interactions and differential protein expression in *Staphylococcus aureus–Candida albicans* dual-species biofilms. *FEMS Immunol. Med. Microbiol.*, **59**, 493–503.

47 De Rosa, FS. et al. (2000) A vehicle for photodynamic therapy of skin cancer: influence of dimethylsulphoxide on 5-aminolevulinic acid *in vitro* cutaneous permeation and *in vivo* protoporphyrin IX accumulation determined by confocal microscopy. *J. Control. Release*, **65**, 359–366.

48 Alvarez-Román, R. et al. (2004) Visualization of skin penetration using confocal laser scanning microscopy. *Eur. J. Pharm. Biopharm.*, **58**, 301–316.

49 Kaiser, W. et al. (1961) Two-photon excitation in $CaF_2:Eu^{2+}$. *Phys. Rev. Lett.*, **7**, 229.

50 Denk, W. et al. (1990) Two-photon laser scanning fluorescence microscopy. *Science*, **248**, 73–76.

51 König, K. (2000) Robert Feulgen Prize Lecture. Laser tweezers and multiphoton microscopes in life sciences. *Histochem. Cell Biol.*, **114**, 79–92.

52 Masters, B.R. et al. (1997) Multiphoton excitation fluorescence microscopy and spectroscopy of *in vivo* human skin. *Biophys. J.*, **72**, 2405–2412.

53 Masters, B.R. et al. (1998) Multiphoton excitation microscopy of *in vivo* human skin. Functional and morphological optical biopsy based on three-dimensional imaging, lifetime measurements and fluorescence spectroscopy. *Ann. N. Y. Acad. Sci.*, **838**, 58–67.

54 Zoumi, A. et al. (2001) Imaging cells and extracellular matrix *in vivo* by using second-harmonic generation and two-photon excited fluorescence. *Proc. Natl. Acad. Sci. U. S. A.*, **99**, 11014–11019.

55 Koehler, M.J. et al. (2010) *In vivo* measurement of the human epidermal thickness in different localizations by multiphoton laser tomography. *Skin Res. Technol.*, **16**, 259–264.

56 Dimitrow, E. et al. (2009) Sensitivity and specificity of multiphoton laser tomography for *in vivo* and *ex vivo* diagnosis of malignant melanoma. *J. Invest. Dermatol.*, **129**, 1752–1758.

57 Lin, S.-J. et al. (2005) Evaluating cutaneous photoaging by use of multiphoton fluorescence and second-harmonic generation microscopy. *Opt Lett.*, **30**, 2275–2277.

58 Lin, SJ. et al. (2006) Prediction of heat induced collagen shrinkage by use of second harmonic generation microscopy. *J. Biomed. Opt.* (2006) **11**, 34020.

59 Koehler, M.J. et al. (2008) Morphological skin ageing criteria by multiphoton laser scanning tomography: non-invasive *in vivo* scoring of the dermal fibre network. *Exp. Dermatol.*, **17**, 519–523.

60 Konig, K. et al. (2006) *In vivo* drug screening in human skin using femtosecond laser multiphoton tomography. *Skin Pharmacol. Physiol.*, **19**, 78–88.

61 Lee, J. et al. (2008) The effects of depilatory agents as penetration enhancers on human stratum corneum structures. *J. Invest. Dermatol.*, **128**, 2240–2247.

62 Luengo, J. et al. (2006) Influence of nanoencapsulation on human skin transport of flufenamic acid. *Skin Pharmacol. Physiol.*, **19**, 190–197.

63 Yu, B. et al. (2001) *In vitro* visualization and quantification of oleic acid induced changes in transdermal transport using two-photon fluorescence microscopy. *J. Invest. Dermatol.*, **117**, 16–25.

64 Gratieri, T. et al. (2010) Penetration of quantum dot particles through human skin. *J. Biomed. Nanotechnol.*, **6** (5), 586–595.

65 Yu, B. et al. (2003) Visualization of oleic acid induced transdermal diffusion pathways using two-photon fluorescence microscopy. *J. Invest. Dermatol.*, **120**, 448–455.

66 Lin, S. et al. (2006) Discrimination of basal cell carcinoma from normal dermal stroma by quantitative multiphoton imaging. *Opt. Lett.*, **31**, 2756–2758.

67 Dimitrow, E. et al. (2009) Spectral fluorescence lifetime detection and selective melanin imaging by multiphoton laser tomography for melanoma diagnosis. *Exp. Dermatol.*, **18**, 509–515.

68 Paoli, J. et al. (2008) Multiphoton laser scanning microscopy on non-melanoma skin cancer: morphologic features for future non-invasive diagnostics. *J. Invest. Dermatol.*, **128**, 1248–1255.

69 König, K. et al. (2009) Clinical optical coherence tomography combined with multiphoton tomography of patients with skin diseases. *J. Biophotonics*, **2**, 389–397.

70 König, K. et al. (2010) Clinical application of multiphoton tomography in combination with high-frequency ultrasound for evaluation of skin diseases. *J. Biophotonics* **3**, 759–773

71 Kaatz, M. et al. (2010) Depth-resolved measurement of the dermal matrix composition by multiphoton laser tomography. *Skin Res. Technol.*, **16**, 131–136.

72 Koehler, M.J. et al. (2009) Intrinsic, solar and sunbed-induced skin aging measured *in vivo* by multiphoton laser tomography and biophysical methods. *Skin Res. Technol.*, **15**, 357–363.

73 Koehler, M.J. et al. (2006) *In vivo* assessment of human skin aging by multiphoton laser scanning tomography. *Opt. Lett.*, **31**, 2879–2881.

74 Naito, S. et al. (2008) *In vivo* measurement of human dermis by 1064 nm-excited fiber Raman spectroscopy. *Skin Res. Technol.*, **14**, 18–25.

75 Puppels, G.J. et al. (1990) Studying single living cells and chromosomes by confocal Raman microspectroscopy. *Nature*, **347**, 301–303.

76 Gniadecka, M. et al. (1998) Structure of water, proteins, and lipids in intact human skin, hair, and nail. *J. Invest. Dermatol.*, **110**, 393–398.

77 Wu, J.Q. and Kilpatrick-Liverman, L. (2011) Characterizing the composition of underarm and forearm skin using confocal Raman spectroscopy. *Int. J. Cosmet. Sci.*, **33**, 257–262.

78 Martini, M.C. (2003) Biochemical analysis of epidermal lipids. *Pathol. Biol. (Paris)*, **51**, 267–270.

79 Kezic, S. (2008) Methods for measuring *in-vivo* percutaneous absorption in humans. *Hum. Exp. Toxicol.*, **27**, 289–295.

80 Nakagawa, N. et al. (2010) In vivo measurement of the water content in the dermis by confocal Raman spectroscopy. *Skin Res. Technol.*, **16**, 137–141.

81 Lademann, J. et al. (2011) Carotenoids in human skin. *Exp. Dermatol.*, **20**, 377–382.

82 Wohlrab, J. et al. (2001) Noninvasive characterization of human stratum corneum of undiseased skin of patients with atopic dermatitis and psoriasis as studied by Fourier transform Raman spectroscopy. *Biopolymers*, **62**, 141–146.

83 O'Regan, G.M. et al. (2010) Raman profiles of the stratum corneum define 3 filaggrin genotype-determined atopic dermatitis endophenotypes. *J. Allergy Clin. Immunol.*, **126**, 574–580. e1.

84 González, F.J., Alda, J., Moreno-Cruz, B., et al. (2011) Use of Raman spectroscopy for the early detection of filaggrin-related atopic dermatitis. *Skin Res. Technol.*, **17**, 45–50.

85 Andrew Chan et al. (2008) A coordinated approach to cutaneous wound healing: vibrational microscopy and molecular biology. *J. Cell Mol. Med.*, **12** (5B), 2145–2154.

86 Willemse-Erix, D.F. (2010) Towards Raman-based epidemiological typing of *Pseudomonas aeruginosa*. *J. Biophotonics*, **3**, 506–511.

87 Lademann, J. et al. (2011) Interaction between carotenoids and free radicals in human skin. *Skin Pharmacol. Physiol.*, **24**, 238–244.

88 Hesterberg, K. et al. (2009) Raman spectroscopic analysis of the increase of the carotenoid antioxidant concentration in human skin after a 1-week diet with

ecological eggs. *J. Biomed. Opt.*, **14**, 024039.

89 Mélot, M. *et al.* (2009) Studying the effectiveness of penetration enhancers to deliver retinol through the stratum cornum by *in vivo* confocal Raman spectroscopy. *J. Control. Release*, **138**, 32–39.

90 Goeppert-Mayer, M. (1931) Über elementarakte mit zwei quantensprüngen. *Ann Phys*, **9** (3), 273–295.

52
Non-Melanoma Skin Cancer

Martin Kaatz, Susanne Lange-Asschenfeldt, Martin Johannes Koehler, and Uta Christina Hipler

52.1
Introduction

Basal cell carcinoma (BCC) is the most common skin tumor, and overall one of the most frequently occurring tumors. Over the last three decades, incidence rates have increased steadily and about one-third of all Caucasians must be expected to develop a BCC at some point in their lives. Although the tumor is metastasizing only under special circumstances, it may nonetheless pose a considerable therapeutic problem, depending on its size and localization.

Squamous cell carcinoma (SCC) of the skin often develops out of precursors – mainly actinic keratoses – which have histopathologically been classified as *in situ* SCCs. Although early actinic keratoses may undergo a regression process, the probability of developing an invasive SCC increases depending on the thickness of neoplastic alterations, and host factors such as genetic or iatrogenic immunosuppression.

Exposure to UV light poses the most important etiological factor in skin tumorigenesis, hence the preferred development of SCC in sun-exposed parts of the skin, such as the face hands, and forearms. Other possible causes of SCC development include chronic wounds, burn injuries, and exposure to radiation.

Numerous issues arise with regard to the application of noninvasive imaging. These issues cover all areas of tumorigenesis, prevention, diagnostics, and therapy, and also the diagnostic classification of tumors and the assessment of their lateral expansion and penetration depth. This applies particularly to the evaluation of tumor margins, part of which, depending on the tumor type, are clinically not easily determined. Other possibilities are the assessment of preclinical lesions and the monitoring of therapeutic success. Efficacy and tolerance of protective methods also play a role in matters of prevention. Additionally, noninvasive imaging can help clarify scientific issues with regard to tumorigenesis.

52.2
Dermoscopy

52.2.1
Basal Cell Carcinoma

The diagnosis of (pigmented) basal cell carcinomas is based on features such as arborizing vessels, leaf-like structures, large blue–gray ovoid nests, multiple blue–gray globules, spoke-wheel areas, and ulceration [1–3].

52.3
Optical Coherence Tomography (OCT)

52.3.1
Basal Cell Carcinoma

Epithelial skin tumors and basal cell carcinomas are more easily visualized in optical coherence tomography (OCT) than pigmented lesions due to the usually low melanin content of these tumors. This makes it easier to define clearly the dermoepidermic junction zone [4, 5]. In 2004, Bechara *et al.* employed OCT in order to perform initial studies on the specific features of basal cell carcinomas [6]. The analysis of three superficial BCCs proved subepidermally an echo-rich band that corresponded to the tumor aggregates verified by histology. These aggregates were separated from the surrounding stroma by a narrow, echo-poor band. In a follow-up examination of 43 BCCS, among which there were 23 nodular and 10 multifocal-superficial, OCT allowed the definition of distinctions between different types of BCC. The most important OCT criteria for a BCC were the plug-like signal-intensive structures corresponding to the histology of the dense tumor cell aggregates, honeycomb-like signal-poor structures correlating to the adjacent tumor lobules, and signal-free cavities in the imaging of a supervascularization. The frequency of occurrence of these criteria was found to be different in the single BCC types. The last two of these criteria (70 and 80%, respectively) were found particularly in infiltrative BCC, whereas the plug-like signal-intense structures were detected mainly in the superficial (80%) and, slightly less frequently, in the nodular BCCs (52%). The comparison of penetration depth measurement methods for BCCs <2 mm in size proved an overestimated thickness for both high-resolution ultrasound and OCT. The OCT results, however, were significantly closer to those achieved by histology than the penetration depths measured by high-frequency ultrasound (HFUS). In a blind study, OCT images of 64 BCCs and 39 actinic keratoses were evaluated by dermatologists and pathologists. It was found that dermal tumor cell clusters can be distinguished from the surrounding stroma with sensitivities and specificities of 79–96%. OCT also allowed for the differentiation of solid versus sclerodermatous BCCs. However, a differentiation between BCC and actinic keratosis had an error rate of 50 versus 52% and was thus proved impossible (Figure 52.1).

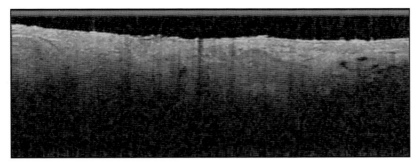

Figure 52.1 OCT of BCC at forehead.

52.3.2
Actinic Keratosis and Squamous Cell Carcinoma

Actinic keratoses predominantly emerge in parts of the body that are exposed to UV light, and that in many cases have already been obviously damaged by sunlight. In several studies, OCT could prove a differentiation between sun-damaged skin and surrounding unaffected areas by detection of a significant thickening of the viable epidermis and/or stratum corneum. Images of sun-damaged skin were characterized by an increased signal in the epidermis and rapid attenuation of light. The imaging of actinic keratoses under examination was characterized by a great variability and heterogeneity, with dark bands in the epidermis possibly corresponding to hyperkeratosis.

The compacted keratin is weakly scattering, and a thickened keratin layer was seen in corresponding histology. In OCT, the epidermis of actinic keratoses often displays distinct horizontal reflections due to flaking within the keratinized region (Figure 52.2). In cases of a significantly widened stratum corneum, a distinct

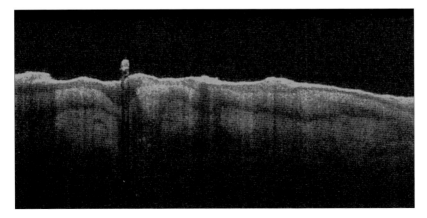

Figure 52.2 OCT of actinic keratosis.

boundary was observed between the keratinized region and underlying epidermis. On comparing intralesional skin with perilesional skin, both OCT and histology could prove an acanthosis. Unlike undiseased skin, the epidermis in actinic keratosis images could appear relatively hyperintense, possibly because of an increased backscatter from the dysplastic cells or parakeratosis. Also, a distinct boundary at a depth consistent with the epidermal–dermal junction was regularly observed. Vertical shadowing from skin flakes was common. Attenuation in the dermis, however, was often strong, similar to severely sun-damaged sites. There were, however, great differences in the comparison of the single lesions. The differentiation of actinic keratoses from the surrounding skin resulted in a sensitivity of 73% and a specificity of 65%. These could even be increased for single OCT criteria (presence of a dark band in the epidermis) to 86% sensitivity and 83% specificity. The clinical–dermatologic assessment achieved a sensitivity of 98% and a specificity of 62%.

The preciseness of thickness measurement in OCT compared with HFUS was also surveyed on actinic keratoses. Results for the lesion thickness were more exact in OCT, even though in actinic keratoses also there was an apparent overestimation [7].

In summary, it can be stated that OCT could prove qualitative and statistically significant quantitative differences in OCT image features of skin with varying degrees of sun damage, and between undiseased skin and actinic keratosis.

Nevertheless, the differentiation between actinic keratoses and BCCs was limited, with a high error rate of ~50% [6, 8].

52.4
Confocal Laser Scanning Microscopy (CLSM)

52.4.1
Basal Cell Carcinoma

In the first study on the noninvasive diagnostics of BCCs by confocal laser scanning microscopy (CLSM), 12 patients with histologically diagnosed BCC were examined. Typical features included an increased number and bore of the vessels, loss of normal architecture, and a strongly reflectant stroma contrasting with the dark tumor parenchyma [9].

Specific criteria were established in 2004 in a multicenter study on 152 lesions, 83 of which were diagnosed as BCC. The criteria comprised elongated, monomorphous cell nuclei, a polarization of these cell nuclei along the same axis (palisading), an inflammatory cell infiltrate, an increase and dilatation of dermal vessels, and a loss of the epidermal honeycomb structure (Figure 52.3). With the presence of four out of five criteria, sensitivity and specificity were 82.9 and 95.7%, respectively. At the same time, CLSM allows for a distinction of BCC subtypes, especially of superficial and solid BCCs. Also, pigmented BCC may be identified by detection of strong contrast due to the high melanin content, whereas the differentiation of the morpheaform

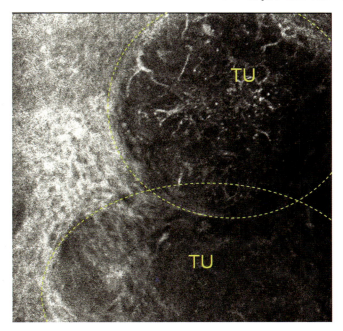

Figure 52.3 CLSM of BCC. Morpheaform variant, tumor lobules (dashed circle, TU), pigmented cells of dendritic shape.

BCC appears to be more difficult due to the few strayed tumor nests extending into the dermis [10].

By means of horizontal mapping, the architecture of larger tumor lobules may be evaluated, allowing the use of CLSM for margin-controlled surgery. One study on 45 excisions reported sensitivity and specificity values of 97 and 89%, respectively, which were confirmed in another study. Since the duration of the CLSM examination was only about 7.5 min, the use of this method promises an acceleration of those surgical procedures [11, 12].

52.4.2
Actinic Keratoses and Squamous Cell Carcinoma

The value of CLSM in diagnostics of epithelial and pigmented skin tumors has already been surveyed in numerous clinical studies. Epithelial *in situ* carcinomas of the actinic keratosis type were examined in 2000 [13]. In that pilot study, hyperkeratosis, pleomorphism, and architectural disarray were described as criteria for actinic keratoses. In subsequent studies, both a high sensitivity (80%) and specificity (96%) of the method compared with routine histology could be proved. In a study on 46 actinic keratoses, sensitivities and specificities of 80–99% were reached [14]. In another study on 30 actinic keratoses, values of 89–94% were obtained. The main features were parakeratosis and/or hyperkeratosis, stratum corneum disruption,

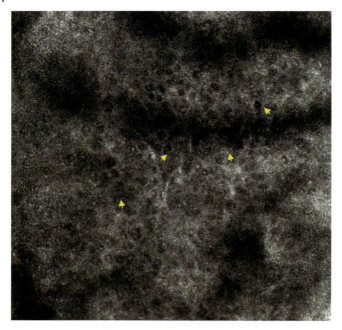

Figure 52.4 CLSM of an actinic keratosis. Clinical II epidermal disarray and keratinocyte pleomorphism (arrowheads) with different shapes and sizes of dark appearing nuclei.

individual corneocytes, pleomorphism, and architectural disruption in stratum granulosum and stratum spinosum [15] (Figure 52.4).

However, in this study, hyperkeratotic actinic keratoses were excluded due to the limited penetration depth allowed by the method. The assessment of the lesion's horizontal extension beyond the clinically visible lesion (subclinical changes) turned out to be excellent. Hence, the method also qualifies for monitoring. Invasive SCC displays similar, if more severe, criteria in CLSM. Vertical extension, however, cannot be recorded owing to the limited penetration depth of the method [13, 16].

52.5
Multiphoton Tomography (MPT)

52.5.1
Basal Cell Carcinoma

Especially superficial BCCs were suitable for examination by multiphoton tomography (MPT), which was mainly due to the limited penetration depth of this method. For nearly all of the lesions examined, a significantly increased epidermal thickness compared with perilesional skin could be proved. Also, a large part of the BCCs examined showed a pronounced hyperkeratosis that was at least 60 μm thick and highly fluorescent, which hindered the imaging of the underlying epidermis. In these cases, no specific characteristics could be defined in the epidermis. In a large part of the lesions that showed a less pronounced hyperkeratosis, the epidermis was twice as thick as in the surrounding areas of unaffected skin. Therefore, no dermal papillae could be proved even at a depth of 100–130 μm. The tumor cells produced a monomorphous image. Arrangement was, as far as could be evaluated, palisade-like. Also, although rarely, elongated nuclei and cytoplasms could be proved in the lower third of the epidermis, which were oriented in the same direction in the x–y plane, a feature described as nuclei polarization. Speckled perinuclear fluorescence was observed in the subcorneal epidermis of all examined lesions but was also present in the corresponding normal perilesional skin in most cases.

In nodular BCC, in which the important histological alterations of the dermis can be proved, evaluation was limited. Comparing intralesional conditions with those in the surrounding areas, MPT could not detect any differences in the epidermis. Only one-third of the tumors examined presented typical nodular aggregates of tumor cells within the dermis. The tumor nests lacked contiguity with the overlying normal epidermis, and the basal cells within these nests presented large, oval nuclei, little cytoplasm, and peripheral palisading. The autofluorescence of these cells was comparable to that of the cells found in the superficial BCC.

In a further study, additional characteristics of the BCC were defined by multiphoton imaging. In the multiphoton images, BCC was characterized by clumps of autofluorescent cells in the dermis. The tumor cells showed a relatively large nucleus and a higher nucleus-to-cytoplasm ratio. Another prominent feature revealed by multiphoton imaging is the alteration of extracellular matrix in the BCC stroma. In the stroma within and surrounding the cancer clumps, second-harmonic generation (SHG) diminishes whereas autofluorescence signals increase. The decrease in SHG indicates that the collagen molecule packing has been disrupted and can reflect an up-regulated collagenase activity in cancer tissue [17, 18].

52.5.2
Actinic Keratoses and Squamous Cell Carcinomas

In most of the actinic keratoses examined, the stratum corneum proved to be abnormally thick, reaching values up to 30–40 μm. Inside the thickened stratum corneum, fluorescent nuclei compartments within the corneocytes were observed. This morphologic and fluorescent feature has earlier been described to occur in the presence of hyperkeratosis. In two hyperkeratotic lesions, large, rounded bundles of keratin were observed within the stratum corneum, corresponding to so-called keratin pearls. The lower layers of the epidermis displayed an architectural disarray with polymorphous keratinocytes (Figure 52.5). The stratum basale, however, could be submitted to only a limited evaluation, which was due to the limited penetration

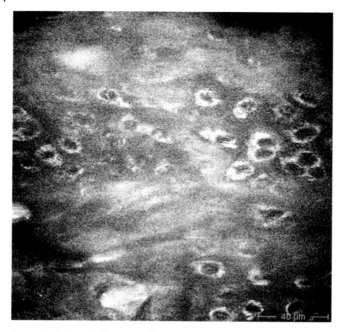

Figure 52.5 MPT of an actinic keratosis. Architectural disarray stratum spinosum.

depth of the method. The reasons for this limitation were the hyperkeratosis and the epidermis widened due to acanthosis [19].

52.6
Laser Doppler Imaging

52.6.1
Basal Cell Carcinoma

Histology shows that expanding tumors, and BCCs in particular, are accompanied by increasing vascularization. Among others, laser Doppler imaging is one way of imaging this vascularization *in vivo*. Stücker *et al.* [20] performed a study on 16 BCCs with this method, providing a two-dimensional image of the blood flow. In the tumor regions, blood flow was always higher than in the surrounding areas, although not as strongly increased as in patients with melanocytic nevi or malignant melanomas. In BCCs, however, a fairly homogeneous increase in perfusion all over the tumor was found. Nevertheless, the peak flow reached lower values than those reached especially in malignant melanomas. Owing to the wide variability of results, a reliable status differentiation by means of laser Doppler imaging is considered impossible [20].

Well-structured studies on the perfusion of actinic keratoses and squamous cell carcinomas are as yet lacking.

52.7
Raman Spectroscopy

52.7.1
Basal Cell Carcinoma

The capabilities of Raman spectroscopy in diagnostics of BCCs have been reviewed in numerous *ex vivo* surveys.

Raman spectral intensities of carotenoids in human skin have been found to be increased in BCC, compared with the surrounding skin. At the same time, differences in lipid and protein structures between BCC and normal skin biopsies were proved. With supplementary employment of the respective Raman intensities, a complete separation between these tissue types could be achieved [21]. Simple analysis of confocal Raman spectra obtained from various skin depths showed a 95% separation between normal skin and BCC [22], Confocal Raman maps of BCC sections have also been shown to identify tumor margins accurately, with 100% sensitivity and 93% specificity. When applying high-wavenumber Raman bands (2800–3125 cm^{-1}), BCC could be discriminated from perilesional tissue.

These results formed the basis for depth-resolved *in vivo* diagnostics using confocal Raman spectroscopy in the form of a hand-held device. The *in vivo* studies achieved a specificity of 100% and a sensitivity of 91%.

Morphometric measurements of the depth of the proximal tumor margin from the skin surface was possible in 66% of cases (six out of the nine BCCs). The measured depths of the tumor margins were 49, 69, 89, 169, 223, 234, 593, 888, and 961 μm from the surface.

Several of these differences reveal that the pathologic spectra can largely be separated by protein- and lipid-related Raman activity.

In addition, compared with conventional Raman spectroscopy, polarized Raman microspectroscopy provides additional relevant information on the molecular ordering of molecular biomolecules. Also, it offers a refined discrimination between the tumor tissue and the normal epidermis.

The depolarization ratios of the bands corresponding to phenylalanine (623, 1339 cm^{-1}), tyrosine (644 cm^{-1}), polysaccharides (1128 cm^{-1}), and *trans* hydrocarbon chains (1128 cm^{-1}) in spectra of the tumor were lower than those recorded in the normal epidermis, representing potential markers to discriminate the tumor tissue and the epidermis. The depolarization ratios of the bands of the peritumoral stroma corresponding mainly to lipids (1065 cm^{-1}), collagens, and various other proteins (920, 940, 1065, and 1128 cm^{-1}) were different from those of the healthy dermis.

The differences in spectra between the peritumoral and the normal dermis were lower than the differences between the tumor tissue and the normal epidermis.

The use of polarized Raman microspectroscopy shows that depolarization ratios of selected vibrations could be potential diagnostic markers to distinguish between the tumor and normal epidermis, and between peritumoral and normal dermis [23].

52.7.2
Actinic Keratoses and Squamous Cell Carcinomas

The spectral intensities of carotenoids were also increased in actinic keratoses, compared with the surrounding skin. At the same time, there were obvious variations in Raman spectra between normal skin and hyperkeratotic lesion samples, which are considered to be related to lipid concentration. In subsequent depth-resolved *in vivo* studies, a correct allocation of abnormal spectra could be achieved in 75% of the actinic keratoses or SCCs. Other than in BCCs, the spectra of SCC tissues generally show a moderate probability for normal tissue [24].

References

1 Zalaudek, I. *et al.* (2006) Dermoscopy in general dermatology. *Dermatology*, **212**, 7–18.
2 Braun, R.P. *et al.* (2005) Dermoscopy of pigmented skin lesions. *J. Am. Acad. Dermatol.*, **52**, 109–121.
3 Campos-do-Carmo, G. and Ramos-e-Silva, M. (2008) Dermoscopy: basic concepts. *Int. J. Dermatol.*, **47**, 712–719.
4 Mogensen, M. *et al.* (2009) Assessment of optical coherence tomography imaging in the diagnosis of non-melanoma skin cancer and benign lesions versus normal skin: observer-blinded evaluation by dermatologists and pathologists. *Dermatol. Surg.*, **35**, 965–972.
5 Jørgensen, T.M. *et al.* (2008) Machine-learning classification of non-melanoma skin cancers from image features obtained by optical coherence tomography. *Skin Res. Technol.*, **14**, 364–369.
6 Bechara, F.G. *et al.* (2004) Histomorphologic correlation with routine histology and optical coherence tomography. *Skin Res. Technol.*, **10**, 169–173.
7 Korde, V.R. *et al.* (2007) Using optical coherence tomography to evaluate skin sun Damage and precancer. *Lasers Surg. Med.*, **39**, 687–695.
8 Mogensen, M. *et al.* (2009) *In vivo* thickness measurement of basal cell carcinoma and actinic keratosis with optical coherence tomography and 20-MHz ultrasound. *Br. J. Dermatol.*, **160**, 1026–1033, Epub 2009 Jan 12.
9 Sauermann, K. *et al.* (2002) Investigation of basal cell carcinoma [correction of carcinoma] by confocal laser scanning microscopy *in vivo*. *Skin Res. Technol.*, **8**, 141–147.
10 Nori, S. *et al.* (2004) Sensitivity and specificity of reflectance-mode confocal microscopy for *in vivo* diagnosis of basal cell carcinoma: a multicenter study. *J. Am. Acad. Dermatol*, **51**, 923–930.
11 Karen, J.K. *et al.* (2009) Detection of basal cell carcinomas in Mohs excisions with fluorescence confocal mosaicing microscopy. *Br. J. Dermatol.*, **160**, 1242–1250.
12 Ziefle, S. *et al.* (2010) Confocal laser scanning microscopy vs 3-dimensional histologic imaging in basal cell carcinoma. *Arch. Dermatol.*, **146**, 843–847.
13 Aghassi, D. *et al.* (2000) Confocal laser microscopic imaging of actinic keratoses *in vivo*: a preliminary report. *J. Am. Acad. Dermatol.*, **43**, 42–48.
14 Ulrich, M. *et al.* (2008) Clinical applicability of *in vivo* reflectance confocal microscopy for the diagnosis of actinic keratoses. *Dermatol. Surg.*, **34**, 610–619.
15 Horn, M. *et al.* (2008) Discrimination of actinic keratoses from normal skin with reflectance mode confocal microscopy. *Dermatol. Surg.*, **34**, 620–625.
16 Ulrich, M. *et al.* (2007) Noninvasive diagnostic tools for nonmelanoma skin cancer. *Br. J. Dermatol.*, **157**, 56–58.

17. Gonzalez, S. and Tannous, Z. (2002) Real-time, *in vivo* confocal reflectance microscopy of basal cell carcinoma. *J. Am. Acad. Dermatol.*, **47**, 869–874.
18. Nori, S. *et al.* (2004) Sensitivity and specificity of reflectance-mode confocal microscopy for *in vivo* diagnosis of basal cell carcinoma: a multicenter study. *J. Am. Acad. Dermatol.*, **51**, 923–930.
19. König, K. and Riemann, I. (2003) High-resolution multiphoton tomography of human skin with subcellular spatial resolution and picosecond time resolution. *J. Biomed. Opt.*, **8**, 432–439.
20. Stücker, M. *et al.* (2002) *In vivo* differentiation of pigmented skin tumors with laser Doppler perfusion imaging. *Hautarzt*, **53**, 244–249.
21. Gniadecka, M. *et al.* (1997) Distinctive molecular abnormalities in benign and malignant skin lesions: studies by Raman spectroscopy. *Photochem. Photobiol.*, **66**, 418–423.
22. Choi, J. *et al.* (2005) Direct observation of spectral differences between normal and basal cell carcinoma (BCC) tissues using confocal Raman microscopy. *Biopolymers*, **77**, 264–272.
23. Lieber, C.A. *et al.* (2008) *In vivo* nonmelanoma skin cancer diagnosis using Raman microspectroscopy. *Lasers Surg. Med.*, **40**, 461–467.
24. Hata, T.R. *et al.* (2000) Non-invasive raman spectroscopic detection of carotenoids in human skin. *J. Invest. Dermatol.*, **115**, 441–448.

53
Pigmented Skin Lesions

Martin Kaatz, Susanne Lange-Asschenfeldt, Peter-Elsner, and Sindy Zimmermann

53.1
Introduction

Cutaneous melanoma is a malignant tumor that arises from melanocytic cells and is potentially the most dangerous form of skin tumor. Simultaneously, cutaneous malignant melanoma is the most rapidly increasing cancer in white populations, increasingly also affecting adolescents and young adults [1, 2]. The frequency of its occurrence is closely associated with the constitutive color of the skin and depends on the geographic zone. The incidence of malignant melanoma in Europe varies from 6–10 per 100 000 per year in Mediterranean countries to 12–20 per 100 000 per year in Nordic countries. Individuals with large numbers of common nevi, congenital nevi and with atypical nevi (dysplastic nevi) are at greater risk [1].

The prognosis for patients with high-risk or advanced metastatic melanoma remains poor, despite advances in this field. Patient outcome and curability depend on timely recognition and excision at early stages of tumor progression. This situation encourages extensive research on new imaging technologies for early melanoma detection and differentiation from benign melanocytic nevi or other non-malignant pigmented lesions.

53.2
Dermoscopy

Dermoscopy (epiluminescence microscopy) allows for visualization of characteristic features of selected skin lesions that would otherwise be invisible to the naked eye [3–6]. This method has been studied extensively for the evaluation of pigmented skin lesions, where by predictions with respect to the status of the lesion may be made.

Two extensive meta-analyses in which studies from different clinical and experimental settings were assessed showed an increase in the diagnostic accuracy in the detection of melanomas when employing dermoscopy [7–9]. The diagnostic accuracy expressed as the relative diagnostic odds ratio was 15.6 (95% CI 2.9–83.7) times higher for dermoscopy than for mere visual inspection [10]. It must be kept in mind, however, that the user's experience is crucial to diagnostic accuracy, allowing a decreased excision rate in melanocytic lesions.

According to an expert consensus, after documentation of clinical information on the patient and the lesion, the first dermoscopic step is the distinction between melanocytic and non-melanocytic lesions [6, 11]. In this context, dermoscopic criteria for melanocytic lesions are taken into consideration, including the presence of a pigment network, aggregated globules, streaks, homogeneous blue pigmentation and parallel pattern (Figure 53.1). The second step comprises the description of dermoscopic structures and patterns by standardized terms and optionally the use of a diagnostic algorithm in order to differentiate between benign and malignant tumors. Finally, a specific final diagnosis or differential diagnoses should be provided. With respect to pigmented lesions, the most critical step is the differentiation of benign melanocytic nevi from malignant melanoma. In a study on inter-observer diagnostic agreement, pattern analysis showed the best diagnostic performance among the second-step procedures with an inter-observer agreement of $\kappa = 0.55$ and an intra-observer agreement of $\kappa = 0.85$. Alternative algorithms including the ABCD rule [12], the Menzies score [13], and the seven-point list [14] had inferior diagnostic values. Pattern analysis includes the standardized description of the global pattern, determination of absence or presence and description (typical vs. atypical or regular vs. irregular) of a pigment network, dots/globules, streaks, a

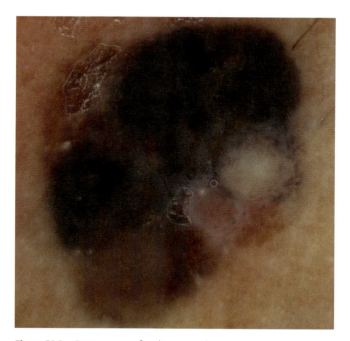

Figure 53.1 Dermoscopy of malignant melanoma.

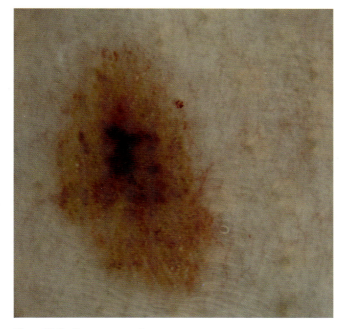

Figure 53.2 Dermoscopy of nevus.

blue–whitish veil, blotchy hypopigmentation, regression structures, and vascular structures [11]. Among these criteria, the most powerful predictors for a melanoma were an atypical pigment network, irregular dots and globules, irregular blotches, and the presence of regression structures, whereas global patterns described as globular, cobblestone, homogeneous, or starburst were most indicative for nevi (Figures 53.1 and 53.2).

Recent advances in computer-assisted dermoscopy may be applied for the follow-up of suspicious melanocytic skin lesions. Sequential digital dermoscopy (SDD) comprises the digital acquisition of dermoscopic images and their storage, allowing for a longitudinal observation of the dynamic changes of a lesion. In several studies, an improvement in the diagnostic power compared with routine dermoscopy was shown [15–17]. These elaborated methods were applied especially for the follow-up of high-risk patients and for the evaluation of the dynamic change of lesions that do not fulfill dermoscopic criteria [8, 18, 19].

53.3
Optical Coherence Tomography (OCT)

Optical coherence tomography (OCT), which is based on the principle of Michelson interferometry using light of short coherence length, has been employed for a variety of dermatologic applications in the past decade. However, few studies evaluated the

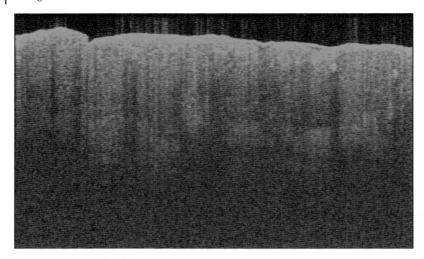

Figure 53.3 OCT of malignant melanoma.

use of this method for the differentiation of melanocytic skin lesions. In a study with 75 patients, criteria for the differentiation of benign and malignant melanocytic lesions by OCT were established by Gambichler et al. [20]. Criteria that were more frequently associated with malignant melanoma compared with melanocytic nevi include marked architectural disarray and a blurred dermo-epidermal junction [20] (Figures 53.3 and 53.4). Controlled studies on sensitivity and specificity of the method are lacking, however.

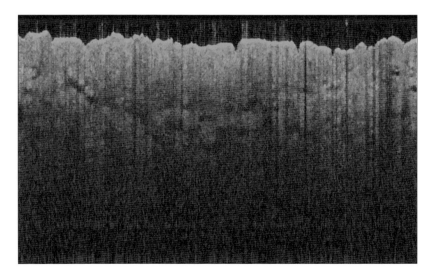

Figure 53.4 OCT of nevus.

53.4
Confocal Laser Scanning Microscopy (CLSM)

Confocal laser scanning microscopy (CLSM) is based on an 830 nm diode laser that is used as a point light source in a scanning mode providing horizontal tomographic images. The visualization of tissues is based on the different refractive indices of the distinct tissue chromophores and cell structures.

There have been numerous studies on CLSM for the evaluation of pigmented skin lesions. Initial investigations have shown that CLSM is particularly suitable for the evaluation of melanocytic lesions, since melanin provides a very strong contrast. In an investigation on 117 melanocytic lesions including 27 malignant melanomas, the sensitivities and specificities of five observers were reported to be 59–96 and 94–100%, respectively [21]. In another study of 125 melanocytic lesions, CLSM was compared with dermoscopy, resulting in similar specificities of 83 and 84%, respectively. In contrast, a much higher sensitivity was obtained by CLSM (97 vs. 89%) [22].

Typical features described for the malignant melanoma include the "non-edged papillae," irregular nests of atypical melanocytes, a pronounced dissolution of the normal epidermal structure, the existence of large, highly refractive cells with a prominent nucleus in the epidermis, and large dendritic cells with long, irregular branches, and also atypical nests of cerebriform morphology [23].

In contrast, benign melanocytic nevi are characterized by small, monomorphic, round to oval cells at the dermo-epidermal junction (junctional nevi), or at the junction and in the superficial dermis (compound nevi). Papillae are particularly well defined (so-called "edged papillae"). Furthermore, both the visualization of regular, homogeneously structured nests of melanocytes and a regular structure of the epidermis are considered to be decisive criteria for the classification of a lesion as benign nevus. Here, the structure of the epidermis may follow either a honeycomb or a cobblestone pattern. Dysplastic nevi display irregularly structured papillae with modified shapes, and also the existence of atypical melanocytes at the junction zone [21, 24] (Figure 53.5).

A number of further skin lesions with pigment disturbance were examined. *Seborrheic keratoses* are characterized by a cerebriform architecture of the epidermis, well-defined, round, black areas filled with whorled refractive material (horn cysts), and enlarged papillary rings lined by brightly refractive cells (pigmented keratinocytes). Plump bright cells (melanophages) may be present in the dermis. Vitiligo presents a complete loss of dermal papillary rings at the dermo-epidermal junction and no bright cells in the epidermis. Dendritic melanocytes may be found at the basal layer in repigmented skin. The features of *lentigo* include hyperrefractive dermal papillary rings with ovoid to annular or polycyclic contours at the dermo-epidermal junction. Plump bright cells (melanophages) may be present in the dermis. *Melasma* is characterized by increased cobblestoning and the presence of dendritic melanocytes at the basal cell layer [25].

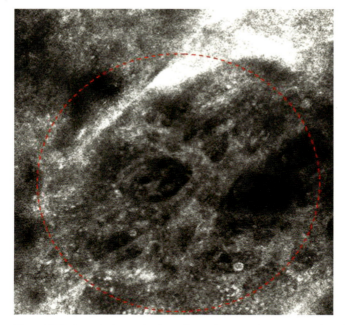

Figure 53.5 Irregular/non-edged papillae with sparse clusters/some large bright cells (circle).

53.5
Multiphoton Laser Tomography (MPT)

The principle of multiphoton imaging is the simultaneous absorption of two or more photons by a fluorophore and emission of one photon at a shorter wavelength, that is, with higher energy. While most of the fluorophores are located intracellularly, elastin and collagen as extracellular matrix proteins also have fluorescent properties. A characteristic phenomenon of fibrillar collagen is its ability for second-harmonic generation (SHG). In sum, the photons emitted from the dermis consist of both autofluorescence and SHG effects. Due to high resolution and the properties of melanin and its components for autofluorescence, a recent study examined the significance of MPT in the diagnostics of malignant melanomas. The investigation included 83 melanocytic skin lesions, including 26 melanomas. In this study, distinct morphologic differences in melanoma compared with melanocytic nevi were identified by multiphoton laser tomography (MLT). Sensitivities and specificities values up to 95 and 97%, respectively, were achieved for diagnostic classification.

The most significant diagnostic criteria include architectural disarray of the epidermis, poorly defined keratinocyte cell borders and the presence of pleomorphic or dendritic cells [26] (Figures 53.6 and 53.7). Remarkable differences

53.5 Multiphoton Laser Tomography (MPT) | 823

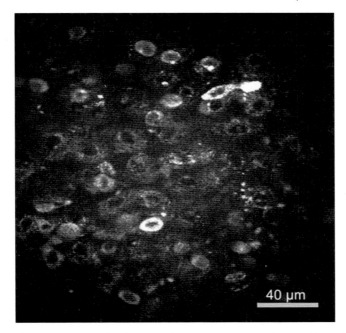

Figure 53.6 MPT of malignant melanoma: stratum granulosum with atypical melanocytes.

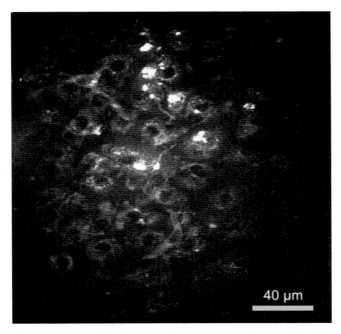

Figure 53.7 MPT of malignant melanoma: stratum spinosum, dendritic cells.

in real-time behavior of keratinocytes in contrast to melanocytes were detected. The fluorescence lifetime distribution was found to correlate with the intracellular amount of melanin [27]. Spectral analysis of melanoma revealed a main fluorescente peak around 470 nm in combination with an additional peak close to 550 nm throughout all epiclermal layers [27].

53.6
Raman Spectroscopy

Raman spectroscopy is an optical technique that probes the vibrational activity of chemical bonds, and each molecule has a spectral signature characteristic of its modes of vibration. These spectral signatures can be used to identify unknown substances in a sample, or to differentiate samples according to their chemical constituency.

Cartaxo *et al.* conducted a survey of 10 excision biopsy samples of malignant melanoma, nine benign melanocytic nevi, and 10 healthy skin samples *ex vivo* [28]. In this pilot study, a high correlation of the Fourier transform Raman spectra was observed between the elements of the same group. A great variation was found in the pigmented nevi and in the malignant melanoma group, whereas no significant variation of the spectra was observed for normal control skin sites. Measured in the vibrational modes for polysaccharides, tyrosine, and amide-I, the discriminatory analysis showed an efficiency of 75.3% in the differentiation of the three groups. In further examinations of small groups of patients, lymph node and brain metastases of malignant melanoma could also be differentiated [29, 30].

53.7
Laser Doppler Imaging

Laser Doppler perfusion imaging (LDPI) is a sophisticated technique for measurement of cutaneous blood flow that possesses broad applications within dermatology. The major advantages of LDPI include measurement of blood flow over an area rather than at a single site and measurement without contacting the skin, which could potentially influence blood flow.

Vascularization is an important element especially of tumor growth. Therefore, for the noninvasive differentiation of skin tumors, the vascular pattern is taken into account, in most examination methods. Stücker *et al.* measured blood flow patterns in basal cell carcinomas, benign nevi, and malignant melanomas with a laser Doppler imager [31]. The results showed the highest perfusion in junctional nevi and malignant melanomas. This technique, however, is preferably employed in combination with other examination methods owing to its low sensitivity and specificity.

References

1. Garbe, C. *et al.* (2010) Diagnosis and treatment of melanoma: European consensus-based interdisciplinary guideline. *Eur. J. Cancer*, **46**, 270–283.
2. Bleyer, A. *et al.* (2006) Cancer in 15- to 29-year-olds by primary site. *Oncologist*, **11**, 590–601.
3. Menzies, S.W. *et al.* (2005) The performance of SolarScan: an automated dermoscopy image analysis instrument for the diagnosis of primary melanoma. *Arch. Dermatol.*, **141**, 1388–1396.
4. Menzies, S.W. and Zalaudek, I. (2006) Why perform dermoscopy? The evidence for its role in the routine management of pigmented skin lesions. *Arch. Dermatol.*, **142**, 1211–1212.
5. Bowling, J. *et al.* (2007) Dermoscopy key points: recommendations from the International Dermoscopy Society. *Dermatology*, **214**, 3–5.
6. Malvehy, J. *et al.* (2007) Dermoscopy report: proposal for standardization. Results of a consensus meeting of the International Dermoscopy Society. *J. Am. Acad. Dermatol.*, **57**, 84–95.
7. Kittler, H. and Binder, M. (2002) Follow-up of melanozytic skin lesions with digital dermoscopy: risks and benefits. *Arch. Dermatol.*, **138**, 1379.
8. Kittler, H. *et al.* (2006) Identification of clinically featureless incipient melanoma using sequential dermoscopy imaging. *Arch. Dermatol.*, **142**, 1113–1119.
9. Bafounta, M.L. *et al.* (2001) Is dermoscopy (epiluminescence microscopy) useful for the diagnosis of melanoma? Results of a meta-analysis using techniques adapted to the evaluation of diagnostic tests. *Arch. Dermatol.*, **137**, 1343–1350.
10. Vestergaard, M.E. *et al.* (2008) Dermoscopy compared with naked eye examination for the diagnosis of primary melanoma: a meta-analysis of studies performed in a clinical setting. *Br. J. Dermatol.*, **159**, 669–676.
11. Argenziano, G. *et al.* (2003) Dermoscopy of pigmented skin lesions: results of a consensus meeting via the Internet. *J. Am. Acad. Dermatol.*, **48**, 679–693.
12. Nachbar, F. *et al.* (1994) The ABCD rule of dermatoscopy. High prospective value in the diagnosis of doubtful melanocytic skin lesions. *J. Am. Acad. Dermatol.*, **30**, 551–559.
13. Menzies, S.W. *et al.* (1996) A sensitivity and specificity analysis of the surface microscopy features of invasive melanoma. *Melanoma Res.*, **6**, 55–62.
14. Argenziano, G. *et al.* (1998) Epiluminescence microscopy for the diagnosis of doubtful melanocytic skin lesions. Comparison of the ABCD rule of dermatoscopy and a new 7-point checklist based on pattern analysis. *Arch. Dermatol.*, **134**, 1563–1570.
15. Haenssle, H.A. *et al.* (2009) Large speckled lentiginous naevus superimposed with Spitz naevi: sequential digital dermoscopy may lead to unnecessary excisions triggered by dynamic changes. *Clin. Exp. Dermatol.*, **34**, 212–215.
16. Rademaker, M. and Oakley, A. (2010) Digital monitoring by whole body photography and sequential digital dermoscopy detects thinner melanomas. *J. Prim. Health Care*, **2**, 268–272.
17. Altamura, D. *et al.* (2008) Assessment of the optimal interval for and sensitivity of short-term sequential digital dermoscopy monitoring for the diagnosis of melanoma. *Arch. Dermatol.*, **144**, 502–506.
18. Menzies, S.W. *et al.* (2009) Impact of dermoscopy and short-term sequential digital dermoscopy imaging for the management of pigmented lesions in primary care: a sequential intervention trial. *Br. J. Dermatol.*, **161**, 1270–1277.
19. Robinson, J.K. and Nickoloff, B.J. (2004) Digital epiluminescence microscopy monitoring of high-risk patients. *Arch. Dermatol.*, **140**, 49–56.
20. Gambichler, T. *et al.* (2007) Characterization of benign and malignant melanocytic skin lesions using optical coherence tomography *in vivo*. *J. Am. Acad. Dermatol.*, **57**, 629–637.
21. Gerger, A. *et al.* (2005) Diagnostic applicability of *in vivo* confocal laser scanning microscopy in melanocytic skin tumors. *J. Invest. Dermatol.*, **124**, 493–498.

22 Langley, R.G.B. *et al.* (2007) The diagnostic accuracy of *in vivo* confocal scanning laser microscopy compared to dermoscopy of benign and malignant melanocytic lesions: a prospective study. *Dermatology (Basel)*, **215**, 365–372.

23 Pellacani, G. *et al.* (2007) The impact of *in vivo* reflectance confocal microscopy for the diagnostic accuracy of melanoma and equivocal melanocytic lesions. *J. Invest. Dermatol.*, **127**, 2759–2765.

24 Gonzalez, S. (2008) *Reflectance Confocal Microscopy of Cutaneous Tumors. An Atlas with Clinical, Dermoscopic and Histological Correlations*, Informa Healthcare, London.

25 Kang, H.Y., *et al.* (2010) Reflectance confocal microscopy for pigmentary disorders. *Exp. Dermatol.*, **19**, 233–239.

26 Dimitrow, E. *et al.* (2009) Sensitivity and specificity of multiphoton laser tomography for *in vivo* and *ex vivo* diagnosis of malignant melanoma. *J. Invest. Dermatol.*, **129**, 1752–1758.

27 Dimitrow, E. *et al.* (2009) Spectral fluorescence lifetime detection and selective melanin imaging by multiphoton laser tomography for melanoma diagnosis. *Exp. Dermatol.*, **18**, 509–515.

28 Cartaxo, S.B. *et al.* (2010) FT-Raman spectroscopy for the differentiation between cutaneous melanoma and pigmented nevus. *Acta Cir. Bras.*, **25**, 351–356.

29 Kirsch, M. *et al.* (2010) Raman spectroscopic imaging for *in vivo* detection of cerebral brain metastases. *Anal. Bioanal. Chem.*, **398**, 1707–1713.

30 de Oliveira, A.F. (2010) Differential diagnosis in primary and metastatic cutaneous melanoma by FT-Raman spectroscopy. *Acta Cir. Bras.*, **25**, 434–439.

31 Stücker, M. *et al.* (1999) Blood flow compared in benign melanozytic naevi, malignant melanoma and basal cell carcinomas. *Clin. Exp. Dermatol.*, **24**, 107–111.

54
Monitoring of Blood Flow and Hemoglobin Oxygenation

Sean J. Kirkpatrick, Donald D. Duncan, and Jessica Ramella-Roman

54.1
Introduction

The primary means of supplying nutrition and oxygen to the skin is the microvasculature. This same network of vessels also plays a key role in wound repair and thermoregulation. Because of this, reliable measures of cutaneous blood flow are necessary for a variety of pharmacological and physiological studies [1], and also for a number of direct clinical applications, including evaluating skin blood flow reserve in diabetic patients [2], assessing cutaneous flow in patients suffering from hypertriglyceridemia [3], determining vascular resistance for the differential diagnosis of vascular disease [4], investigating the effects of locally applied pressure on the development of pressure ulcers [5], studying the changes in blood flow following burns [6], and assessing the changes in the microvascular blood flow dynamics associated with photodynamic therapy [7]. From a historical perspective, there have been numerous approaches for measuring blood flow and inferring perfusion (the product of the scalar flow velocity and the concentration of red blood cells in the probed area). In general, these approaches can be divided into two categories: laser Doppler velocimetry techniques (both point and scanning approaches) and laser speckle techniques. The distinction between the two categories may be somewhat artificial; however, for ease of discussion we will maintain this classification.

The assessment of oxygen saturation of skin has been used in the monitoring of edema and erythema [8], in the demarcation of certain skin lesions and tumors [9, 10], including the evolution of Kaposi's sarcoma (KS) [11], and for monitoring the progress of skin ulcers [12, 13]. Near-infrared imaging spectroscopy [14] was used to obtain maps of oxygen saturation of thermal burns and to separate superficial from deep burns, while other groups showed that SO_2 can be used to quantify skin inflammation due to an irritant [15]. Finally, SO_2 was shown to be a metric of interest when assessing the onset of autonomic dysreflexia [16]. The experimental methodologies mentioned utilize the time averages of skin SO_2, thus implying that the arterial, venous, and capillary SO_2, can be lumped into one unique number (reported SO_2 values vary

Handbook of Biophotonics. Vol.2: Photonics for Health Care, First Edition. Edited by Jürgen Popp,
Valery V. Tuchin, Arthur Chiou, and Stefan Heinemann.
© 2012 Wiley-VCH Verlag GmbH & Co. KGaA. Published 2012 by Wiley-VCH Verlag GmbH & Co. KGaA.

between 0.5 and 0.7). When arterial flow is isolated, such as in pulse oximetry, normal values of SO_2 exceed 0.98. Only time-averaged techniques are considered here.

54.2
Laser Doppler Approaches to Quantifying Cutaneous Blood Flow

For cutaneous blood flow, one can generally divide laser Doppler measurements into the categories of point-wise and regional. While both categories rely on the same physical phenomenon, the frequency shift observed when light of a given wavelength is scattered by a moving object, the hardware and software implementations are different. Moreover, the character of the motion estimates is also different. Point-wise measurements rely on a small measurement volume, typically $\sim 0.1\,mm^3$ and less, that is defined by two or more laser beams that cross at an oblique angle. Many commercial systems exist (e.g., TSI, Dantec Dynamics, Measurement Science Enterprise) that are capable of providing (without calibration) absolute velocities in one, two, or three dimensions. Such systems can provide vectorial velocity measurements within vessels of sizes down to the region defined by the beam crossing volume. These systems are typically referred to as heterodyne because one of the laser beams, which is typically frequency shifted, provides an interferometric reference for the other. Because of this frequency shift, the measurement provides both direction and velocity.

Regional measurement is provided by laser Doppler perfusion systems [17]. In this category are the optical fiber-based probe systems [laser Doppler perfusion monitoring (−LDPM] and systems that employ a scanned laser or area illumination and CCD or CMOS detectors [laser Doppler perfusion imaging (−LDPI] (Moor Instruments). All of the LDPx systems rely on a measurement of Doppler shift as well, but have a limiting resolution that is large compared with the cutaneous vasculature structure. The Doppler shift that they perceive is due to motion in a variety of directions, thus resulting in a broadened Doppler spectrum. Multiple scattering within the interrogation volume further broadens this Doppler spectrum [18]. A measure of motion is derived by inspecting the width of this Doppler spectrum. As a result, such measurement concepts provide qualitative measurements of regional flow, that is, perfusion. Such systems are often termed homodyne because the reference is provided by scatter from supposedly stationary tissues outside the vasculature. They have been useful in a variety of research [19] and clinical applications [20] where the objective is to measure perfusion *changes* resulting from some stimulus. Any attempt to provide quantitative assessment of volumetric flow using such systems, however, is fraught with difficulties. An example is the difficulty in establishing a "biological zero" [21], that is, a zero instrument reading when biological motion has supposedly been halted. Such a condition is typically established using a pressure cuff. Motion persists, nonetheless, due simply to Brownian motion. A standardization of sorts for these LDPx measurements can be provided by using an *in vitro* calibration with an aqueous suspension of microspheres. The subsequent *in vivo* measurements, however, still represent a regional, qualitative estimate of motion for the reasons outlined above.

One reoccurring issue with Doppler measurements of whatever kind is that velocity does not equal flow. Although this is true, they are in fact highly correlated. For example, for flow of a Newtonian fluid within a vessel, the velocity profile is parabolic (Poiseuille flow). As a result, the volume flow can be inferred from a measurement of the centerline velocity. Blood, however is non-Newtonian; at low shear rates it displays an elevated viscosity, attributed to the agglomeration of cells, whereas at high shear rates the lowered viscosity is attributed to deformation of cells [22]. Nevertheless, a number of studies [23–25] suggest that for vessels between 155 and 2000 µm the effect can be ignored. As a result, a measurement of centerline velocity together with knowledge of the configuration of the vessel, can be used to infer volume flow rate.

54.3
Laser Speckle Methods for Blood Flow Estimations

The concept of estimating flow based on the contrast of a time-integrated laser speckle pattern was introduced 30 years ago by Fercher and Briers [26]. As originally proposed, the concept was to illuminate a vascularized tissue with a coherent light source and take a photograph with a particular exposure time. The concept relies on the fact that an image formed in coherent illumination has a high-contrast, speckled appearance. Further, if the illuminated object moves, the speckles translate and scintillate. If the exposure time is long compared with the speckle fluctuation time, the imaged speckles are blurred, that is, the contrast is reduced. Contrast, K, is simply defined here as $K \equiv \sigma_I/\mu_I$, where σ_I and μ_I are the standard deviation and the mean of the measured intensity, respectively, typically assessed over a small window of say 5×5 or 7×7 pixels. Therefore, in principle, the idea is to infer a temporal correlation time constant from the observed speckle contrast and subsequently relate this time constant to the flow velocity. Since that time, numerous variations on the same theme have been proposed and implemented [27–30]. Regardless of the specifics, the general concept has remained the same, namely, the idea is to infer a temporal correlation time constant from the observed speckle contrast and subsequently relate this time constant to the flow velocity.

Laser speckle contrast imaging (LASCI), laser speckle contrast analysis (LASCA), and all of the numerous other titles given to this same general approach have strong appeal, namely, the hardware requirements are minimal, requiring only a laser source and a camera, and the technique is full field, thus holding promise for full field velocity estimates. In addition, the approach always yields results (i.e., a moving speckle pattern) and it can be relatively straightforward to demonstrate a correlation between velocity and speckle contrast [27, 31, 32]. The challenge is to provide a *quantitative* relationship between these parameters, and this has yet to be demonstrated in a convincing fashion. A number of the issues that have impeded the progress of LASCI from a correlative technique to a truly quantitative technique were reviewed by Kirkpatrick and co-workers [33, 34]. Recently, Boas and Dunn [35] reviewed the history of and developments in laser speckle contrast imaging, using the formalisms of quasi-elastic light scattering and diffusing wave spectroscopy as a theoretical starting point. These related techniques have very similar underlying statistical theory; however, their data collection and data analysis requirements differ substantially from those of LSCI, and care must be taken by the user when viewing LSCI from this vantage point.

LSCI has a further advantage, that is, the simplicity at which the spatial resolution can be altered from the size of small arteries and veins up to larger areas. Dunn et al. [36] and Zhou et al. [37], for example, have exploited the temporal effects in LSCI to investigate changes in regional blood perfusion.

Laser speckle contrast imaging is a promising tool for assessing flow in the cutaneous microvasculature. At present, its greatest potential lies perhaps in assessing relative changes in regional blood perfusion. Challenges such as speckle movement not associated with blood flow (e.g., rigid body motion associated with respiration), multiple scattering, and the presence of a spectrum of decorrelation times in the observed speckle patterns still need to be rigorously investigated and dealt with in order to make LSCI a truly quantitative tool for blood flow velocity measurement.

54.4
Oxygenation

Experimental systems for the assessment of SO_2 in the skin are varied and application dependent. Nevertheless, two main categories can be established – fiber-based systems and imaging systems. When using fiber-based systems, only very localized information can be gathered as fiber probes tend to be only a few millimeters in size. These systems have several advantages, first and foremost being that diffusion-based models of light propagation in the skin may be used (depending on fiber geometry [38]). Moreover, since fibers generally are interfaced with spectrometers, hundreds of wavelengths can be measured in relatively short time. The redundancy in the data can be used to quantify different metrics of interest apart from SO_2, such as melanin, water, or biliriubin content. Imaging systems generate spatial maps of SO_2 and are preferable when monitoring extended lesions. These systems are noncontact and can be useful in clinical scenarios where infection is an issue.

The measurement of skin oxygen saturation starts generally with the assessment of the skin absorption μ_a, scattering coefficient μ_s, or reduced scattering coefficient μ_s'. Once these global parameters have been calculated, the specific chromophores and scatterers can be inferred [11, 39]. The assessment of the skin optical properties from skin reflectance spectra or wavelength-dependent images depends largely on the instrumentation used. When fibers are used, diffusion theory may apply provided that the source–detector separation is longer than five mean free paths, $MFP = 1/(\mu_a + \mu_s')$ [10]. Several authors have used this approach [16, 38, 39] using probes arranged in a configuration with one source fiber and several detector fibers at increasing distance from the source. The detectors are connected either to multiple spectrometers [16] or to optical switches and a single spectrometer [9, 40]. This type of probe can be used to measure oxygen saturation at different depths, where the longest source–detector separation corresponds to the deepest penetration in the tissue. Some uncertainties exist when using diffusion-based models due to boundary conditions and the fact that shallow penetration is desired. A diffusive probe can be used in these cases, [41] where a thin Spectralon slab is inserted between the collector fiber tips and the skin; the slabs act as a diffuser. Yet another approach is the use of look-up table (LUT) inverse models [42], where an LUT is generated from diffuse reflection data obtained with tissue-simulating phantoms. The phantom optical properties at several wavelengths need to be known and are measured with benchtop techniques, such as integrating spheres and inverse adding doubling (IAD) models [40]. An iterative least-squares fit is ultimately used to formulate the inverse model.

Reflectance-based imaging systems are constructed with a black and white imager (CCD or CMOS) combined with either a filter wheel and several narrowband filters or a liquid crystal tunable filter (LCTF). Spectroscopic snapshot systems have also been used based either on division of aperture [13] or the addition of a custom mosaic filter [40]. Often cross-polarizers are used on the source and detector to eliminate specular reflection. Imaging spectroscopy for the assessment of SO_2 relies heavily on empirical parametric models, as even in diffusive media such as skin, diffusion theory cannot be used. Exceptions are structured illumination systems, where the light source is spatially modulated, and contains DC and AC components that can be separated with a phase-shifting technique [43]. From these two components, the bulk absorption and scattering coefficient of skin may be calculated and ultimately oxygen saturation obtained [39].

References

1 Pershing, L.K., Huether, S., Connklin, R.L., and Kreuger, G.G. (1989) Cutaneous blood flow and percutaneous absorption: a quantitative analysis using a laser Doppler velocimeter and a blood flow meter. *J. Invest. Dermatol.*, **92**, 355–359.

2 Rendell, M., Saxena, S., and Shah, D. (2004) Cutaneous blood flow and peripheral resistance in type II diabetes as compared to intermittent claudication patients. *Int. J. Angiol.*, **12** (3), 166–171.

3 Tur, E., Politi, Y., and Rubinstein, A. (1994) Cutaneous blood flow abnormalities in hypertriglyceridemia. *J. Invest. Dermatol.*, **103**, 597–600.

4 Nitzan, M., Goss, D.E., Chagne, D., and Roberts, V.C. (1988) Assessment of regional blood flow and specific microvascular resistance in the foot by means of the transient thermal clearance method. *Clin. Phys. Physiol. Meas.*, **9**, 347–352.

5 Fromy, B., Abraham, P., Bouvet, C., Bouhanick, B., Fressinaud, P., and Saumet, J.L. (2002) Early decrease of skin blood flow in response to locally applied pressure in diabetic subjects. *Diabetes*, **51**, 1214–1217.

6 Jaskille, A.D., Ramella-Roman, J.C., Shupp, J.W., Jordan, M.H., and Jeng, J.C. (2010) Critical review of burn depth assessment techniques: Part II. Review of laser Doppler technology. *J. Burn Care Res.*, **31**, 151–157.

7 Smith, T.K., Choi, B., Ramirez-San-Juan, J.C., Nelson, J.S., Osann, K., and Kelly, K.M. (2006) Microvascular blood flow dynamics associated with photodynamic therapy, pulsed dye laser irradiation and combined regimens. *Lasers Surg. Med.*, **38**, 532–539.

8 Stamatas, G.N., Southall, M., and Kollias, N. (2006) In vivo monitoring of cutaneous edema using spectral imaging in the visible and near infrared. *J. Invest. Dermatol.*, **126**, 1753–1760.

9 Yu, C.-C., Lau, C., O'Donoghue, G., Mirkovic, J., McGee, S., Galindo, L., Elackattu, A., Stier, E., Grillone, G., Badizadegan, K., Dasari, R.R., and Feld, M.S. (2008) Quantitative spectroscopic imaging for non-invasive early cancer detection. *Opt. Express*, **16**, 16227–16239.

10 Zonios, G., Perelman, L.T., Backman, V., Manoharan, R., Fitzmaurice, M., Van Dam, J., and Feld, M.S. (1999) Diffuse reflectance spectroscopy of human adenomatous colon polyps in vivo. *Appl. Opt.*, **38**, 6628–6637.

11 Vogel, A., Chernomordik, V.V., Riley, J.D., Hassan, M., Amyot, F., Dasgeb, B., Demos, S.G., Pursley, R., Little, R.F., Yarchoan, R., Tao, Y., and Gandjbakhche, A.H. (2004) Using noninvasive multispectral imaging to quantitatively assess tissue vasculature. *J. Biomed. Opt.*, **12**, 051604.1–051604.12.

12 Noordmans, H.J., De Roode, R., Staring, M., and Verdaasdonk, R. (2006) Registration and analysis of in-vivo multispectral images for correction of motion and comparison in time. *Proc. SPIE*, **7**, 608106.1–608109.

13 Basiri, A., Nabili, M., Mathews, S., Libin, A., Groah, S., Noordmans, H.J., and Ramella-Roman, J.C. (2010) Use of a multi-spectral camera in the characterization of skin wounds. *Opt. Express*, **18**, 3244–3257.

14 Cross, K.M., Leonardi, L., Payette, J.R., Gomez, M., Levasseur, M.A., Schattka, B.J., Sowa, M.G., and Fish, J.S. (2007) Clinical utilization of near-infrared spectroscopy devices for burn depth assessment. *Wound Repair Regen.*, **15**, 332–340.

15 Kollias, N., Gillies, R., Muccini, J.A., Uyeyama, R.K., Phillips, S.B., and Drake, L.A. (1995) A single parameter, oxygenated hemoglobin, can be used to quantify experimental irritant-induced inflammation. *J. Invest. Dermatol.*, **104**, 421–424.

16 Ramella-Roman, J.C. and Hidler, J.M. (2008) The impact of autonomic dysreflexia on blood flow and skin response in individuals with spinal cord injury. Hindawi Publishing Corporation Advances in Optical Technologies Volume 2008, Article ID 797214, 7 pages doi:10.1155/2008/797214.

17 Leahy, M.J., Enfield, J.G., Clancy, N.T., O'Doherty, J., McNamara, P., and Nilsson, G.E. (2007) Biophotonic methods in microcirculation imaging. *Med. Laser Appl.*, **22**, 105–126.

18 Bonner, R. and Nossai, R. (1981) Model for laser Doppler measurements of blood flow in tissue. *Appl. Opt.*, **20** (12), 2097–2107.

19 Jaskille, A.D., Ramella-Roman, J.C., Shupp, J.W., Jordan, M.H., and Jeng, J.C. (2010) Critical review of burn depth assessment techniques: Part II. Review of laser Doppler technology. *J. Burn Care Res.*, **31** (1), 151–157.

20 Ng, E.Y., Fok, S.C., and Goh, C.T. (2003) Case studies of laser Doppler imaging system for clinical diagnosis applications and management. *J. Med. Eng. Technol.*, **27** (5), 200–206.

21 Kernick, D.P., Tooke, J.E., and Shore, A.C. (1999) The biological zero signal in laser doppler fluximetry – origins and practical implications. *Eur. J. Physiol.*, **437**, 624–631.

22 Vennemann, P., Lindken, R., and Westerweel, J. (2007) In vivo whole-field blood velocity measurement techniques. *Exp. Fluids*, **42**, 495–511.

23 Charm, S.E. and Kurland, G.S. (1966) On the significance of the Reynolds number in blood flow. *Rheology*, **3** (3), 163–164.

24 Charm, S.E., Kurland, G.S., and Brown, S.L. (1968) The influence of radial distribution and marginal plasma layer on the flow of red cell suspensions. *Biorheology*, **5** (1), 15–43.

25 Charm, S.E. and Kurland, G.S. (1974) *Blood Flow and Microcirculation*, John Wiley & Sons, Inc., New York.

26 Fercher, A.R. and Briers, J.D. (1981) Flow visualization by means of single exposure photography. *Opt. Commun.*, **37**, 326–330.

27 Aizu, Y. and Asakura, T. (1999) Coherent optical techniques for diagnostics of retinal blood flow. *J. Biomed. Opt.*, **4**, 61–75.

28 Ramirez-San-Juan, J.C., Ramos-Garcia, R., Guizar-Iturbide, I., Martinez-Niconoff, G., and Choi, B. (2008) Impact of velocity

distribution assumption on simplified laser speckle imaging equation. *Opt. Express*, **16**, 3197–3203.

29 Li, P., Ni, S., Zhang, L., Zeng, S., and Luo, Q. (2006) Imaging cerebral blood flow through the intact rat skull with temporal laser speckle imaging. *Opt. Lett.*, **31**, 1824–1826.

30 Kirkpatrick, S.J., Duncan, D.D., Wang, R.K., and Hinds, M.K. (2007) Quantitative temporal contrast imaging for tissue mechanics. *J. Opt. Soc. Am. A*, **24**, 3728–3734.

31 Isono, H., Kishi, S., Kimura, Y., Hagiwara, N., Konishi, N., and Fujii, H. (2003) Observation of choroidal circulation using index of erythrocyte activity. *Arch. Ophthalmol.*, **121**, 225–231.

32 Durduran, T., Burnett, M.G., Yu, G., Zhou, C., Furuya, D., Yodh, A.G., Detre, J.A., and Greenburg, J.H. (2004) Spatiotemporal quantification of cerebral blood flow during functional activation in rat somatosensory cortex using laser-speckle flowmetry. *J. Cereb. Blood Flow Metab.*, **24**, 518–525.

33 Duncan, D.D. and Kirkpatrick, S.J. (2008) Can laser speckle flowmetry be made a quantitative tool? *J. Opt. Soc. Am. A*, **25**, 2088–2094.

34 Kirkpatrick, S.J., Duncan, D.D., and Wells-Gray, E.M. (2008) Detrimental effects of speckle-pixel size matching in laser speckle contrast imaging. *Opt. Lett.*, **33**, 2886–2888.

35 Boas, D.A. and Dunn, A.K. (2010) Laser speckle contrast imaging in biomedical optics. *J. Biomed. Opt.*, **15**, 011109.

36 Dunn, A.K., Bolay, H., Moskowitz, M.A., and Boas, D.A. (2001) Dynamic imaging of cerebral blood flow using laser speckle. *J. Cereb. Blood Flow Metab.*, **21**, 195–201.

37 Zhou, C., Yu, G., Furuya, D., Greenburg, J.H., Yodh, A., and Durduran, T. (2006) Diffuse optical correlation tomography of cerebral blood flow during cortical spreading depression in rat brain. *Opt. Express*, **14**, 1125–1144.

38 Farrell, T.J., Patterson, M.S., and Wilson, B. (1992) A diffusion theory model of spatially resolved, steady-state diffuse reflectance for the noninvasive determination of tissue optical properties in vivo. *Med. Phys.*, **19**, 879–888.

39 Jacques, S.L. (2005) Spectroscopic determination of tissue optical properties using optical fiber spectrometer, http://omlc.ogi.edu/news/feb06/skinspect/index.htm, Oregon Health and Science University, Portland, OR (last accessed 18 May 2011).

40 Pickering, J.W., Prahl, S.A., van Wieringen, N., Beek, J.F., Sterenborg, H.J.C.M., and van Gemert, M.J.C. (1993) Double-integrating-sphere system for measuring the optical properties of tissue. *Appl. Opt.*, **32**, 399–410.

41 Tseng, S.-H., Bargo, P., Durkin, A., and Kollias, N. (2009) Chromophore concentrations, absorption and scattering properties of human skin in-vivo. *Opt. Express*, **17**, 14599–14617.

42 Rajaram, N., Aramil, T., Lee, K., Reichenberg, J.S., Nguyen, T.H., and Tunnell, J.W. (2010) Design and validation of a clinical instrument for spectral diagnosis of cutaneous malignancy. *Appl. Opt.*, **49**, 142–152.

43 Kong, L., Yi, D., Sprigle, S., Wang, F., Wang, C., Liu, F., Adibi, A., and Tummala, R. (2010) Single sensor that outputs narrowband multispectral images. *J. Biomed. Opt.*, **15** (1), 010502.

55
Sensing Glucose and Other Metabolites in Skin
Elina A. Genina, Kirill V. Larin, Alexey N. Bashkatov, and Valery V. Tuchin

55.1
Introduction

Recent technological advances in the photonics industry have led to a resurgence of interest in optical imaging technologies and real progress towards the development of noninvasive clinical functional imaging systems. Application of optical methods to physiological condition monitoring and cancer diagnostics, and also for treatment, is a growing field owing to the simplicity, low cost, and low risk of these methods. The development of noninvasive measurement techniques for monitoring of endogenous (metabolic) agents in human tissues is very important for the diagnosis and therapy of various human diseases and might play a key role in the proper management of many devastating conditions.

Glucose is a monosaccharide sugar with chemical formula $C_6H_{12}O_6$. It is one of the most important carbohydrate nutrient sources and is fundamental to almost all biological processes as it required for the production of ATP and other essential cellular components. The normal range of glucose in human blood is 70–160 mg dl^{-1} (3.9–8.9 mM; 1 mM = 18.0 mg dl^{-1}) depending on the time of the last meal, the extent of physical tolerance, and other factors [1]. Freely circulating glucose molecules in the bloodstream stimulate the release of insulin from the pancreas. Insulin (a large peptide hormone composed of two proteins bound together) helps glucose molecules to penetrate the cell wall by binding to specific receptors in cell membranes, which are normally impermeable to glucose. Diabetes is a disorder caused by decreased production of insulin, or by decreased ability to utilize insulin in transport of the glucose across cell membrane. As a result, a high and potentially dangerous concentration of glucose can be accumulated in the blood (hyperglycemia) during the disease [2]. Therefore, it is of great importance to maintain blood glucose concentration within the normal range in order to prevent possibly severe complications.

A significant role for physiological glucose monitoring is in the diagnosis and management of several metabolic diseases, such as diabetes mellitus (or simply diabetes). A number of invasive and noninvasive techniques have been investigated

Handbook of Biophotonics. Vol.2: Photonics for Health Care, First Edition. Edited by Jürgen Popp,
Valery V. Tuchin, Arthur Chiou, and Stefan Heinemann.
© 2012 Wiley-VCH Verlag GmbH & Co. KGaA. Published 2012 by Wiley-VCH Verlag GmbH & Co. KGaA.

for glucose monitoring [3]; however, the problem of noninvasive glucose monitoring in a clinically acceptable form has not yet been solved.

While the standard analysis of blood currently involves puncturing a finger and subsequent chemical analysis of collected blood samples, in recent decades noninvasive blood glucose monitoring has become an increasingly important topic of investigation in the realm of biomedical engineering. In particular, the introduction of optical approaches has brought exciting advances to this field [4].

Tissue metabolites play an important role in the state of the human organism. For example, bilirubin-IXα, a metabolite of normal heme degradation, occurs in blood as a water-soluble complex with human serum albumin. It can cause life-threatening neurotoxic effects in babies with hyperbilirubinemia if its level becomes too high and it partitions into the brain [5]. Therefore, noninvasive monitoring of the bilirubin level in serum is very important in some clinical cases.

The defense mechanism of the skin is based on the action of antioxidant substances such as carotenoids, vitamins, and enzymes. β-Carotene and lycopene represent more than 70% of the carotenoids in the human organism. The topical or systemic application of β-carotene and lycopene is the general strategy for improving the defense system of the human body. For the evaluation and optimization of this treatment, it is necessary to measure the β-carotene and lycopene concentrations in the tissue, especially in the human skin as the barrier to the environment [6].

Many optical techniques for sensing different tissue metabolites and glucose in living tissue have been developed over the last 50 years. Methods based on fluorescence [7–17], near-infrared (NIR) and mid-infrared (mid-IR) spectroscopic [18–26], Raman spectroscopic [22, 27–29], photoacoustic [30, 31], optical coherence tomography (OCT) [32–36], and other techniques are discussed in this chapter.

55.2
Light–Tissue Interaction

Optical sensing techniques are defined by light–tissue interaction. This interaction depends on the optical properties of tissue components and structures. The absorption depends on the type of predominant absorption centers and water content in tissues. Absorption for most tissues in the visible region is insignificant except for the absorption bands of blood hemoglobin and some other chromophores [37]. The absorption bands of protein molecules are mainly in the near-ultraviolet (UV) region. Absorption in the IR region is essentially defined by the water absorption spectrum and its concentration in tissues. In the NIR spectral range, the main light absorbers are water and lipids, which are present in different tissues in various quantities. In this spectral range, the absorption bands of water in skin with maxima at 930, 970 [38], 1197, 1430, and 1925 nm [39] and lipids with maxima at 1212, 1710, and 1780 nm [40, 41] are observed. The absorption bands of oxyhemoglobin with maxima at about 415, 540, and 575 nm are observed in the visible spectral range [42, 43]. Absorption of water in this spectral range is negligible.

Fluorescence arises upon light absorption and is related to an electronic transition from the excited state to the ground state of a molecule. Fluorescence spectra often give detailed information on fluorescent molecules, their conformation, binding sites, and interaction within cells and tissues or other molecules. On excitation of biological objects by UV light ($\lambda \leq 370$ nm), autofluorescence of proteins and also of nucleic acids can be observed. Autofluorescence of proteins is related to the amino acids tryptophan, tyrosine, and phenylalanine with absorption maxima at 280, 275, and 257 nm, respectively, and emission maxima between 280 nm (phenylalanine) and 350 nm (tryptophan) [44–47]. Fluorescence from collagen or elastin is excited between 400 and 600 nm with maxima around 400, 430, and 460 nm. Fluorescence of collagen and elastin can be used to distinguish various types of tissues, for example, epithelial and connective tissues [46–49].

The reduced form of coenzyme nicotinamide adenine dinucleotide (NADH) is excited selectively in the wavelength range 330–370 nm. NADH is most concentrated within mitochondria, where it is oxidized within the respiratory chain located within the inner mitochondrial membrane, and its fluorescence is an appropriate parameter for detection of ischemic or neoplastic tissues [46, 49]. Flavin mononucleotide (FMN) and flavin adenine dinucleotide (FAD) with excitation maxima around 380 and 450 nm have also been reported to contribute to intrinsic cellular fluorescence [46].

55.3
Fluorescence Measurements

Fluorescent kinetic probes have been used to obtain information on the antioxidant activity of the skin and eye pigment melanin and its biogenetic precursors 5,6-dihydroxyindole and 5,6-dihydroxyindole-2-carboxylic acid [50]. These constitute the major monomeric building blocks of eumelanin, the brown and dark melanin type. These precursors occur in relatively high local abundance in the melanocytes and keratinocytes, and their enhanced blood or urine levels provide biochemical markers for malignant melanoma and a diagnostic tool for its therapy [50].

Porphyrin molecules, for example, protoporphyrin, coproporphyrin, uroporphyrin, and hematoporphyrin, occur within the biosynthetic pathway of hemoglobin, myoglobin, and cytochromes. Abnormalities in heme synthesis, occurring in cases of porphyries and some hemolytic diseases, may enhance considerably the porphyrin level within tissues [44]. Several bacteria, for example, *Propionibacterium acnes*, and bacteria within dental plaque (biofilm), such as *Porphyromonas gingivalis*, *Prevotella intermedia*, and *Prevotella nigrescens*, accumulate considerable amounts of protoporphyrin [51, 52]. Therefore, acne and oral and tooth lesion detection based on measurements of intrinsic fluorescence specific to porphyrin molecules appears to be a promising method of disease diagnostics. For example, time-gated fluorescence spectroscopy was successfully applied to the detection of the tumor within tissues based on the strong autofluorescence signal coming from accumulated porphyrins [53].

A number of fluorescence-based techniques have been developed and applied in attempts to sense blood glucose concentration [4, 7–10]. The techniques can be subdivided into two general categories: glucose oxidase (GOx)-based and affinity-binding sensors. Sensors in the first category use the electroenzymatic oxidation of glucose by GOx in order to generate an optically detectable glucose-dependent signal [7]. In the late 1990s, research towards a reagentless glucose sensor, using the deactivated apo-GOx enzyme [8], was started, which used the enzyme as a receptor rather than a catalyst. This work showed that apo-GOx retained its high specificity and in many ways paved the way for advances in the development of a new biosensor genre [9, 10]. Several methods for optical detection of the products of this reaction and, hence, the glucose concentration driving the reaction have been presented [11–13]. A simple approach for creating a sensor based on this reaction is to incorporate an oxygen-sensitive dye such as a ruthenium complex [7, 13]. Ruthenium(II) complexes exhibit fluorescence quenching in the presence of oxygen; therefore, a decrease in local oxygen concentration can be detected as an increase in fluorescence of a ruthenium-based dye. Incorporating GOx and a ruthenium dye together results in a sensor whose fluorescence increases with increase in glucose concentration, since increased glucose concentration will lead to an increased consumption of O_2. A disadvantage of this simple approach is that a decrease in local oxygen content may not be distinguished from a rise in glucose concentration.

Fluorescent affinity-binding sensors utilize competitive binding between glucose and a suitable labeled fluorescent compound to a common receptor site. Several methods are based on Förster resonance energy transfer (FRET) and on competition between glucose and dextran for concanavalin-A (ConA) binding sites [7]. The assay components are ConA labeled with an energy acceptor, or an energy donor, and dextran labeled with the complementary molecule for FRET [14, 15]. Glucose displaces the dextran labeled with the fluorophore at the ConA binding sites and consequently FRET-based quenching occurs as its fluorescence it critically dependent on the distance between donor molecule (e.g., ConA) and acceptor molecule (e.g., dextran). The use of ConA as the membrane-bound molecule affords a very sensitive technique because of the strong affinity of ConA to glucose. Using this approach, glucose concentrations can be measured by a decrease in amplitude of the fluorescence peak with increasing glucose concentration. In either approach, the fluorescence techniques appears to be very specific to glucose and sensitive to glucose concentration, without interference from other constituents frequently found in blood and tissues.

Although no fluorescence-based sensing devices are yet commercially available, it appears that a number of the techniques have been sufficiently developed to expect that clinically-viable devices are just over the horizon, and more activity in trials will be observed in the next few years. Advances in nanomaterials such as quantum dots will likely allow improvements in optical stability and choice of excitation/emission wavelengths for various transduction methods [16, 17].

55.4
IR Spectroscopy

Measurements of the absorption and scattering of NIR light traveling through tissue may provide a useful means to quantify functional parameters of tissues, such as total hemoglobin concentration and blood oxygen saturation and glucose [54]. The optical absorption methods for glucose measurements are based on the concentration-dependent absorption of specific wavelengths of light by glucose or other metabolites. In theory, a beam of radiation may be directed through a blood-containing portion of the body and the exiting light analyzed to determine the content of glucose by assessing the absorption spectrum.

For example, the mid-IR spectral bands of glucose and other carbohydrates have been assigned and are dominated by C–C, C–H, O–H, O–C–H, C–O–H and C–C–H stretching and bending vibrations [22]. The 800–1200 cm^{-1} fingerprint region of the IR spectrum of glucose has bands at 836, 911, 1011, 1047, 1076, and 1250 cm^{-1}, which have been assigned to C–H bending vibrations [22]. A 1026 cm^{-1} band corresponds to a C–O–H bending vibration and the 1033 cm^{-1} band can be associated with the ν(C–O–H) vibration [22] or the ν(C–O–C) vibration [55]. Therefore, it might be possible to utilize these spectrum "fingerprints" to quantify the concentration of carbohydrates in the sample with good specificity.

Despite the specificity offered by IR absorption spectroscopy, its application to quantitative blood glucose measurement is limited. A strong background absorption by water and other components of blood and tissues severely limits the pathlength that may be used for transmission spectroscopy to roughly 100 μm or less. Further, the magnitude of the absorption peaks and the dynamic range required to record them make quantification based on these sharp peaks difficult. Nonetheless, attempts have been made to quantify blood glucose using IR absorption spectroscopy *in vitro* and *in vivo* [8, 23–26].

A new concept for *in vivo* glucose detection in the mid-IR spectral range is based on the use of two quantum cascade lasers emitting at wavelengths 9.26 and 9.38 μm to produce photoacoustic signals in the forearm skin [56]. One of the wavelengths correlates with glucose absorption, whereas the other does not. Determination of glucose concentration in the extracellular fluid of the stratum spinosum permits the deduction of the glucose concentration in blood, because the two factors correlate closely with each other. This method allows one to improve glucose specificity and to remove the effect of other blood substances.

In contrast to the mid-IR range, the incident radiation in the NIR spectral range passes relatively easily through water and body tissues, allowing moderate pathlengths to be used for the measurements. Therefore, a large amount of effort has been devoted to the development of NIR spectroscopic techniques for the noninvasive measurement of blood glucose [18]. The NIR region is ideal for the noninvasive measurement of human body compositions because biological tissue is relatively transparent to light in this region, the so-called therapeutic window. The molecular formula of glucose is $C_6H_{12}O_6$ and several hydroxyl and methyl groups are present.

They are mainly hydrogen functional groups whose absorption occurs in the NIR region. For glucose, the second overtone absorption is in the spectral region 1100–1300 nm and the first overtone absorption is in the region 1500–1800 nm [19]. In the range 1400–1500 nm, there is a peak that corresponds to the absorption peak of water. This information provides the theoretical basis for the measurement of blood glucose using NIR spectroscopy [19, 20].

The concept of noninvasive blood glucose sensing using the scattering properties of blood is an alternative to the spectral absorption method. The method of NIR frequency-domain reflectance is based on changes in glucose concentration, which affect the refractive index mismatch between the interstitial fluid and tissue fibers, and hence reduced scattering coefficient μ'_s [57, 58]. The refractive index n of the interstitial fluid modified by glucose is defined by the equation [44]

$$n_{glw} = n_w + 1.515 \times 10^{-6} \times C_{gl} \tag{55.1}$$

where C_{gl} is glucose concentration in mg dl^{-1} and n_w is the refractive index of water [58]. As the subject's blood glucose rise, μ'_s decreases (see Figure 55.1). Key factors for the success of this approach are the precision of the measurements of the reduced scattering coefficient and the separation of the scattering changes from

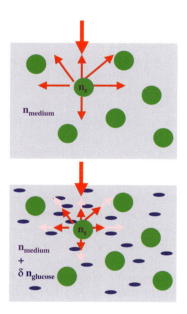

Figure 55.1 Mechanism of glucose-induced changes in the scattering coefficient of a medium: the increase in glucose concentration in the medium (shown as blue ellipses) changes the refractive index of the medium by $+\delta n_{glucose}$, decreases the refractive index mismatch between scattering centers (shown as green circles) and the medium, and, hence, reduces the scattering coefficient (the scattering of light is shown as thin red arrows and incident light is shown as thick red arrows).

absorption changes, as obtained with the NIR frequency-domain spectrometer [57]. Evidently, other physiological effects related to glucose concentration could account for the observed variations of μ'_s and, as it was mentioned earlier, the effect of glucose on the blood flow in the tissue may be one of the sources of the errors in μ'_s measurements.

One critical difficulty associated with *in vivo* blood glucose assay is the extremely low signal-to noise ratio of the glucose peak in the NIR spectrum of human skin tissue. Therefore, the main problem in NIR glucose monitoring is the construction of calibration models [19, 20, 59, 60].

The calibration models typically developed from controlled experiments such as oral glucose tolerance tests might have a better opportunity to provide an acceptable correlation between blood glucose level and the reduced scattering coefficient. At present, the utilization of several experiment data sets for quantitative modeling is most useful to avoid the chance of poor temporal correlation. In one study [59], partial least-squares regression was carried out for the NIR data (total 250 paired data set) and calibration models were built for each subject individually. The selection of informative regions in NIR spectra for analysis can significantly refine the performance of these full-spectrum calibration techniques [60].

A floating-reference method of calibration has been described [20, 21]. The key factor and precondition of the method are the existence of a reference position where diffuse reflectance light is not sensitive to the variations in glucose concentration. Using the signal at the reference position as the internal reference for human body measurement can improve the specification for extraction of glucose information [20].

55.5
Photoacoustic Technique

The photoacoustic technique is based on sensitive time-resolved detection and analysis of laser-induced pressure waves using an ultrasound transducer. Both *in vitro* and *in vivo* studies have been carried out in the spectral range of the transparent "tissue window" around the 1 μm, to assess the feasibility of photoacoustic spectroscopy (PAS) for noninvasive glucose detection [30]. The photoacoustic (PA) signal is obtained by probing the sample with monochromatic radiation, which is modulated or pulsed. Absorption of probe radiation by the sample results in localized short-duration heating. Thermal expansion then gives rise to a pressure wave, which can be detected with a suitable acoustic transducer. An absorption spectrum for the sample can be obtained by recording the amplitude of generated pressure waves as a function of probe beam wavelength [30]. The pulsed PA signal is related to the properties of a turbid medium by the equation [30]

$$PA = k(\beta v^n / C_p) E_0 \mu_{\text{eff}} \qquad (55.2)$$

where PA is the signal amplitude, k is a proportionality constant, E_0 is the incident pulse energy, β is the coefficient of volumetric thermal expansion, v is the speed of sound in the medium, C_p is the specific heat capacity, n is a constant between one and two, depending on the particular experimental conditions, and $\mu_{eff} = \sqrt{3\mu_a(\mu_a + \mu'_s)}$. In the PAS technique, the effect of glucose can be analyzed by detecting changes in the peak-to-peak value of laser-induced pressure waves due to changes in the absorption or scattering as a function of the glucose concentration [31, 61].

Investigations have demonstrated the applicability of PAS to the measurement of glucose concentration [31, 32]. The greatest percentage change in the PA response was observed in region of the C—H second overtone at 1126 nm, with a further peak in the region of the second O—H overtone at 939 nm [30]. In addition, the generated pulsed PA time profile can be analyzed to detect the effect of glucose on tissue scattering, which is reduced by an increase in glucose concentration [62].

Monitoring of the photoisomerization of bilirubin–human serum albumin (BR–HSA) complex is very important in the phototherapy of jaundiced newborns. This complex represents a chromophore–protein interaction. Isomerization reactions of BR–HSA have revealed a detectable structural volume change when studied by photothermal techniques such as laser-induced optoacoustic spectroscopy. The PA signal from solutions of BR–HSA together with the signal amplitude from the calorimetric reference was used to calculate the structural volume change within detection limits [63].

55.6
Raman Spectroscopy

Raman spectroscopy is a powerful technique for the analysis and identification of molecules in a sample. Raman spectroscopy measures inelastically scattered light that results from oscillations and rotations of atoms and atomic groups in molecules (see Figure 55.2). It can provide potentially precise and accurate analysis of metabolite concentrations and biochemical composition. Raman spectra can be utilized to identify molecules because these spectra are characteristic of variations in the molecular polarizability and dipole momentum. Enejder et al. [28] accurately measured glucose concentrations in 17 nondiabetic volunteers following an oral glucose tolerance protocol.

β-Carotene and lycopene have different absorption values at 488 and 514.5 nm and, consequently, the Raman lines for β-carotene and lycopene have different scattering efficiencies with 488 and 514.5 nm excitations. These differences were used for the determination of the concentrations of β-carotene and lycopene. The Raman signals are characterized by two prominent Stokes lines at 1160 and 1525 cm^{-1}, which have nearly identical relative intensities [6]. Thus, resonance Raman spectroscopy was used as a fast and noninvasive optical method of measuring the absolute concentrations of β-carotene, lycopene, and other carotenoids in human skin and mucosal tissue [6, 64] and for the determination of the influence of IR radiation on their degradation [65].

In contrast to IR and NIR spectroscopy, Raman spectroscopy has a spectral signature that is less affected by water, which is very important for tissue studies. In addition, Raman spectral bands are considerably narrower (typically

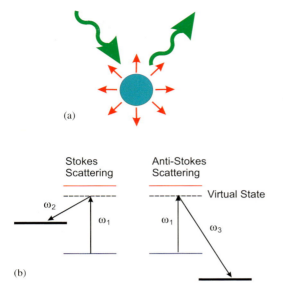

Figure 55.2 Basic principle of Raman spectroscopy: (a) a small portion of photons incident on a target molecule scatters with a shift in photon energy due to molecular vibrations; (b) energy diagram representing the process of Stokes and anti-Stokes scattering (note: there is no transfer of electron population to the "virtual state").

10–20 cm^{-1} [29]) than those produced in NIR spectroscopy. Raman spectroscopy also has the ability for the simultaneous estimation of multiple metabolites, requires minimum sample preparation, and would allow for direct sample analysis [7]. Like IR absorption spectra, Raman spectra exhibit highly specific bands that are dependent on concentration. As a rule, for Raman analysis of tissue, the spectral region between 400 and 2000 cm^{-1}, commonly referred to as the "fingerprint region," is employed.

Different molecular vibrations lead to Raman scattering in this part of the spectrum. In many cases, bands can be assigned to specific molecular vibrations or molecular species, aiding the interpretation of the spectra in terms of biochemical composition of the tissue [7].

Because of the reduction of elastic light scattering on tissue optical clearing, more effective interaction of a probing laser beam with the target molecules is expected. The optical clearing agents increase the signal-to-noise ratio, reduce the systematic error incurred as a result of incompletely resolved surface and subsurface spectra, and significantly improve the Raman signal [66].

55.7
Occlusion Spectroscopy

An approach known as occlusion spectroscopy (OS) was established as a noninvasive method to evoke blood-specific variations of spectral characteristics [67]. The OS technique is based on the observation that the cessation of blood flow triggers a change in time of the optical characteristics of blood *in vivo*. Because a state of temporary blood flow cessation at the measurement site is created, the average size of aggregates, which are the main scattering centers, begins to grow. Consequently, the mean free path, the light scattering pattern, and the mean absorption coefficients start to evolve with a growth in average size of the scattering particles [68].

The key element of OS is the fact that light scattering changes, originated solely by the blood, drive the optical response of the media. Light scattering fluctuations are associated with both the red blood cells aggregation process and the mismatch of refractive indices of the aggregates and their surrounding media. Glucose present in blood increases the refractive index of plasma and thus decreases the scattering coefficient of blood. Based on a method of parameterization, increased sensitivity to the changes in glucose concentration in blood plasma is achieved [68]. There are still many obstacles in the modeling of aggregation assistance signals in the blood since red blood cells aggregation kinetics vary strongly from subject to subject.

55.8
Reflectance Spectroscopy

The measurement of bilirubin levels in the serum of newborn infants is a very important problem. Hyperbilirubinemia is a common symptom of neonates due to the accumulation of bilirubin in the serum because the liver has not yet developed the enzymes for oxidizing bilirubin for excretion by the kidneys. An optical fiber-based spectrometer has been used for noninvasive monitoring of cutaneous bilirubin (which correlates with serum levels of bilirubin) [69]. It is possible to deduce the bilirubin levels in the skin directly regardless of variations in background scattering, cutaneous melanin, and cutaneous blood content using first principles of optical transport.

The bilirubin absorption adds to the optical density in the region around 480 nm. The bilirubin concentration, C (mg cm^{-3}), is related to the absorption coefficient, $\mu_{a,bili}$ (cm^{-1}), by the extinction coefficient, ε [cm^{-1} (mg cm^{-3})$^{-1}$]:

$$\mu_{a,bili} = \varepsilon C \tag{55.3}$$

Hence the absorption coefficient $\mu_{a,bili}$ must be deduced in order to specify the amount of bilirubin in the skin. The optical reduced scattering coefficient, μ'_s, can be determined by measurement of skin specimens using an integrating sphere spectrophotometer. μ_a and μ'_s allow the prediction of the reflectance of the skin R and the collection efficiency f [69]. Also, the factor $f_{std}R_{std}$ for the Teflon standard has to be determined. The predicted measurement M^* is calculated as $M^* = f^*R$. Then the predicted and measurement values, M^* and M, are compared using a simplex routine and the value of μ_a is updated. Iterative cycling of this algorithm converges on a stable deduced value of μ_a using the appropriate f^*. This algorithm is applied to each wavelength being considered for analysis. The result is a true absorption spectrum, $\mu_a(\lambda)$, which is then analyzed for the blood and bilirubin content.

This method can be useful to deduce directly the skin bilirubin levels optically without resorting to calibration by a training set of optical versus laboratory serum measurements on some test newborn population [69].

55.9
Polarimetric Technique

The polarimetric technique for blood glucose monitoring utilizes a physical effect such as optical rotation of the polarization plane by glucose molecules. This feature has been used extensively in industry to detect glucose in aqueous solutions. The basic principle of polarimetry is the following: linearly polarized light passing through a medium containing chiral molecules (such as glucose) experiences rotation of a plane of polarization (see Figure 55.3). The degree of this rotation, ϕ, depends on the optical pathlength, L, the concentration of chiral molecules, C, and an intrinsic property of chiral molecules to rotate polarized light, called the specific rotation constant, α:

$$[\phi]_{\lambda,\text{pH}}^{T} = \frac{\alpha}{LC} \qquad (55.4)$$

where the subscripts λ and pH and the superscript T indicate dependence of the specific rotation constant on wavelength, pH, and temperature, respectively.

Several groups have tried to utilize this idea to measure glucose concentration in the aqueous humor of the eye [70–73]. A major disadvantage of this method is that glucose-induced changes in the signal are very small and difficult to measure: the angle of rotation for a 1 cm optical pathlength is less than $0.00004°$ per $1\,\text{mg}\,\text{dl}^{-1}$.

With the exception of transparent ocular tissues, the human body is highly absorbing and scattering in the UV–IR range, and the validity of Eq. (55.4) is questionable. Specifically, (i) light is highly depolarized upon tissue multiple scattering, so even the initial detection of a polarization-preserved signal from which to attempt glucose concentration extraction is a formidable challenge; (ii) the optical pathlength in turbid media is a difficult quantity to define; (iii) other optically active chiral species are present in tissue, thus contributing to the observed optical rotation and hiding/confounding the specific glucose contribution; and (iv) several optical polarization effects occur in tissue simultaneously (e.g., optical rotation, birefringence, absorption, depolarization), contributing to the resultant polarization signals in a complex interrelated way and hindering their unique interpretation [74, 75].

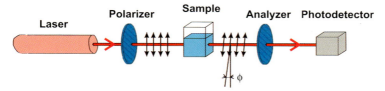

Figure 55.3 Basic principle of the polarimetric approach for glucose monitoring: linearly polarized light passing through a medium containing molecules of glucose experiences rotation of a plane of polarization; ϕ is the degree of this rotation.

Despite these difficulties, it was shown recently that even in the presence of severe depolarization, measurable polarization signals can be obtained reliably from highly scattering media such as biological tissue. Vitkin and co-workers [75] demonstrated surviving linear and circular polarization fractions of light scattered from optically thick turbid media, and measured the resulting optical rotations of the linearly polarized light with the help of a comprehensive turbid polarimetry platform, comprising a highly sensitive experimental system, an accurate forward model that can handle all the complex simultaneous polarization effects manifested by biological tissues, and an inverse signal analysis strategy that can be applied to complex tissue polarimetry data to isolate specific quantities of interest (such as small optical rotation that can be linked to glucose concentration). Additionally, application of this method to relatively transparent tissues (such as the cornea of the human eye) might improve the detection sensitivity of the polarimetric methods [76].

55.10
OCT Technique

An OCT-based method for glucose assessment in tissues has been proposed and its capability for highly sensitive noninvasive monitoring of blood glucose concentration was demonstrated in number of animal and clinical studies [32–35, 77]. The basic principle of this method is to detect and analyze backscattered photons from a tissue layer of interest within the coherence length of a laser source with a two-beam interferometer in the NIR spectral range. OCT technology originates from low-coherence interferometry (a nonscanning/imaging version of OCT) and typically employs a Michelson interferometer, with a broadband laser in the source arm, a beamsplitter, and a photodetector in the detector arm (see Figure 55.4). Light backscattered from the sample is combined with light returned from a mirror in the reference arm of the interferometer, and a photodiode detects the resulting interferometric signal. Interferometric signals can be formed only when the optical pathlength in the sample arm matches the reference arm length within the coherence length of the source. Therefore, by changing the optical pathlength in the reference arm, one-dimensional in-depth profiles of light scattering (A-scans) could be reconstructed in real time with a resolution of about a few microns at depths up to several millimeters.

As mentioned earlier, diffusion of glucose into the interstitial space leads to an increase in the refractive index of the interstitial fluid (ISF), and thus to refractive index matching of collagen fiber (and other tissue components – scatterers) and ISF [64]. Since the scattering coefficient of tissues depends on the refractive index, n, mismatch between ISF and tissue components (fibers, cell components), an increase in tissue glucose concentration will increase the refractive index of the ISF, which will decrease the scattering coefficient of the tissue as a whole. Since the OCT technique measures the in-depth light distribution with high resolution, changes in the in-depth distribution of the tissue scattering coefficient and/or refractive index are reflected in changes in the OCT signal that could be analyzed [32–35, 77].

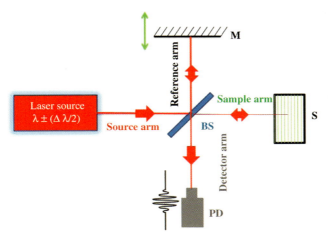

Figure 55.4 General schematic of an OCT system. λ is the central wavelength of the laser source, $\Delta\lambda$ is the bandwidth of the laser source, M is the mirror, S is the sample, BS is the beamsplitter, and PD is the photodetector.

The OCT studies performed in animals (New Zealand white rabbits and Yucatan micropigs) and normal human subjects (during oral glucose tolerance tests) demonstrated (1) capability of the OCT technique to detect changes in scattering coefficient with an accuracy of about 1.5%; (2) a sharp and linear decrease in the OCT signal slope in the dermis with increase in blood glucose concentration; and (3) the accuracy of glucose concentration monitoring may be substantially improved if optimal dimensions of the probed skin area are used [32–35, 77]. The results suggest that the high-resolution OCT technique has a potential for noninvasive, accurate, and continuous glucose monitoring with high sensitivity. However, an OCT method based on analysis of the in-depth amplitude distribution of low-coherent light in skin might have insufficient sensitivity and repeatability and lack of a proper calibration method. Therefore, the development of novel algorithms (e.g., phase-sensitive measurements) and/or combination with other noninvasive sensing modalities are currently being actively investigated by several groups.

55.11
Conclusion

This chapter demonstrates the variety of optical techniques for sensing glucose and other metabolites in skin. Significant efforts have been made by numerous groups and companies in the past two decades to develop a biosensor for noninvasive blood glucose analysis. These approaches include NIR absorption and scattering, polarimetry, Raman spectroscopy, photoacoustics and OCT. The ability of these techniques to assess blood glucose and other metabolites concentration is summarized in Table 55.1.

Table 55.1 Optical techniques for sensing glucose and other metabolites.

Method	Ability	Ref.
Fluorescent glucose-oxidase based sensors	Use the electroenzymatic oxidation of glucose alone or in combination with oxygen-sensitive dye. Fluorescence of the sensor increases with increase in glucose concentration. Sensitivity to decrease in O_2	[7–13]
Fluorescent affinity-binding sensors	Use binding between agent and a labeled fluorescent compound. Fluorescence of the sensor increases with increase in glucose concentration. Specificity to glucose	[14, 15]
NIR absorption spectroscopy	Uses absorption spectral fingerprints of glucose in "therapeutic window." Low sensitivity of agent concentration measurement in physiologic range	[19, 20]
NIR scattering spectroscopy	Use of the change in reduced scattering coefficient with blood glucose concentration due to matching effect. Sensitivity to other physiological effects related to glucose increase	[57, 58]
Mid-IR spectroscopy	Uses spectrum "fingerprints" to quantify the concentration of carbohydrates in tissue with good specificity	[22–26]
Raman spectroscopy	Uses specific spectral bands that are less affected by water. Provides potentially precise and accurate analysis of metabolite concentration. However, it is difficult to isolate glucose from the background spectrum caused by other tissue components	[6, 28, 64, 65]
Occlusion spectroscopy	Uses light scattering fluctuations, originated by the blood. Sensitivity to the changes in glucose concentration in blood plasma	[67, 68]
Reflectance spectroscopy	Uses optical measurements of body composition. Allows noninvasive estimation of cutaneous bilirubin concentration	[69]
Photoacoustic technique	Uses absorption of probe radiation by the sample that results in localized short-duration heating and then gives rise to a pressure wave, which can be detected with a transducer. Sensitivity to the changes in tissue glucose concentration and bilirubin structural volume	[30–32, 63]
Polarimetry technique	Uses optical rotation of the polarization plane by glucose molecules. Measurable polarization signals can be obtained from highly scattering media such as biological tissue. A major disadvantage is that glucose-induced changes in the signal are very small and difficult to measure: the angle of rotation for a 1 cm optical pathlength is less than $0.00004°$ per $1\,mg\,dl^{-1}$	[70–75]
OCT technique	Uses backscattered photons from a tissue of interest within the coherence length of the laser source with a two-beam interferometer in the NIR spectral range. Sensitivity to the changes in blood glucose concentration. Requires proper calibration	[32–35, 77]

Despite significant efforts, the described techniques for noninvasive monitoring of glucose concentration have faced limitations associated with low sensitivity and accuracy and insufficient specificity of glucose concentration measurement in the physiologic range. Therefore, further development of the methods is important for noninvasive diagnosis of various human diseases and blood glucose monitoring and is the subject of current research efforts by several groups.

Acknowledgments

This work was supported in part by the following: 224 014 Photonics4life-FP7-ICT-2007-2; RUB1-2932-SR-08 CRDF; Russian Federation Ministry of Science and Education 2.1.1/4989 and 2.2.1.1/2950, Project 1.4.09 of Federal Agency of Education of the Russian Federation; RFBR-08-02-92224-NNSF_a (Russian Federation–China); RFBR-10-02-90039-Bel_a; and Russian Federation governmental contracts 02.740.11.0770 and 02.740.11.0879.

References

1 Goldstein, D.E., Little, R.R., Lorenz, R.A., Malone, J.I., Nathan, D., Peterson, C.M., and Sacks, D.B. (2004) Tests of glycemia in diabetes. *Diabetes Care*, **27**, 1761–1773.

2 Diabetes Control and Complications Trial Research Group (1997) Hypoglycemia in the diabetes control and complications trial. *Diabetes*, **46**, 271–286.

3 Tuchin, V.V. (ed.) (2009) *Handbook of Optical Sensing of Glucose in Biological Fluids and Tissues*, Taylor & Francis/CRC Press, Boca Raton, FL.

4 Klonoff, D.C. (2005) Continuous glucose monitoring. *Diabetes Care*, **28**, 1231–1239.

5 Williams, R.M., McDonagh, A.F., and Braslavsky, S.E. (1998) Structural volume changes upon photoisomerization of the bilirubin–albumin complex: a laser-induced optoacoustic study. *Photochem. Photobiol.*, **68** (4), 433–437.

6 Darvin, M.E., Gersonde, I., Meinke, M., Sterry, W., and Lademann, J. (2005) Non-invasive *in vivo* determination of the carotenoids beta-carotene and lycopene concentrations in the human skin using the Raman spectroscopic method. *J. Phys. D Appl. Phys.*, **38**, 2696–2700.

7 McNichols, R.J. and Cote, G.L. (2000) Optical glucose sensing in biological fluids: an overview. *J. Biomed. Opt.*, **5**, 5–16.

8 D'Auria, S., Herman, P., Rossi, M., and Lakowicz, J.R. (1999) The fluorescence emission of the apo-glucose oxidase from *Aspergillus niger* as probe to estimate glucose concentrations. *Biochem. Biophys. Res. Commun.*, **263**, 550–553.

9 Mack, A.C., Mao, J., and McShane, M.J. (2005) Transduction of pH and glucose-sensitive hydrogel swelling through fluorescence resonance energy transfer, in *Sensors, 2005 IEEE Conference*, IEEE, (ed. A. Shkel), University of California, Irvine, CA, USA, pp. 912–916.

10 Chinnayelka, S. and McShane, M.J. (2005) Microcapsule biosensors using competitive binding resonance energy transfer assays based on apoenzymes. *Anal. Chem.*, **77**, 5501–5511.

11 Moreno-Bondi, M.C., Wolfbeis, O.S., Leiner, M.J., and Schaffar, B.P.H. (1990) Oxygen optrode for use in a fiber-optic glucose biosensor. *Anal. Chem.*, **62**, 2377–2380.

12 Rosenzweig, Z. and Kopelman, R. (1996) Analytical properties and sensor size effects of a micrometer-sized optical fiber glucose biosensor. *Anal. Chem.*, **68**, 1408–1413.

13 Li, L. and Walt, D.R. (1995) Dual-analyte fiber-optic sensor for the simultaneous and continuous measurement of glucose and oxygen. *Anal. Chem.*, **67**, 3746–3752.
14 McCartney, L.J., Pickup, J.C., Rolinski, O.J., and Birch, D.J.S. (2001) Near-infrared fluorescence lifetime assay for serum glucose based on allophycocyanin-labeled concanavalin A. *Anal. Biochem.*, **292**, 216–221.
15 Rolinski, O.J., Birch, D.J.S., McCartney, L.J., and Pickup, J.C. (2001) Fluorescence nanotomography using resonance energy transfer: demonstration with a protein–sugar complex. *Phys. Med. Biol.*, **46**, N221–N226.
16 Coté, G.L. and McShane, M. (2009) Fluorescence-based glucose biosensors, in *Handbook of Optical Sensing of Glucose in Biological Fluids and Tissues* (ed. V.V. Tuchin), Taylor & Francis/CRC Press, Boca Raton, FL, pp. 331–364.
17 Cordes, D.B., Gamsey, S., and Singaram, B. (2006) Fluorescent quantum dots with boronic acid substituted viologens to sense glucose in aqueous solution. *Angew. Chem. Int. Ed.*, **45**, 3829–3832.
18 Marbach, R., Koschinsky, Th., Gries, F.A., and Heise, H.M. (1993) Noninvasive blood glucose assay by near-infrared diffuse reflectance spectroscopy of the human inner lip. *Appl. Spectrosc.*, **47**, 875–881.
19 Heise, H.M., Lampen, P., and Marbach, R. (2009) Near-infrared reflection spectroscopy for non-invasive monitoring of glucose – established and novel strategies for multivariate calibration, in *Handbook of Optical Sensing of Glucose in Biological Fluids and Tissues* (ed. V.V. Tuchin), Taylor & Francis/CRC Press, Boca Raton, FL, pp. 115–156.
20 Xu, K. and Wang, R.K. (2009) Challenges and countermeasures in NIR non-invasive blood glucose monitoring, in *Handbook of Optical Sensing of Glucose in Biological Fluids and Tissues* (ed. V.V. Tuchin), Taylor & Francis/CRC Press, Boca Raton, FL, pp. 281–316.
21 Luo, Y., An, L., and Xu, K. (2006) Discussion on floating-reference method for noninvasive measurement of blood glucose with near-infrared spectroscopy. *Proc. SPIE*, **6094**, 60940K.
22 Vasko, P.D., Blackwell, J., and Koenig, J.L. (1971) Infrared and Raman spectroscopy of carbohydrates. I: Identification of O–H and C–H vibrational modes for D-glucose, malose, cellobiose, and dextran by deuterium substitution methods. *Carbohydr. Res.*, **19**, 297–310.
23 Hahn, S. and Yoon, G. (2006) Identification of pure component spectra by independent component analysis in glucose prediction based on mid-infrared spectroscopy. *Appl. Opt.*, **45**, 8374–8380.
24 Kim, Y.J., Hahn, S., and Yoon, G. (2003) Determination of glucose in whole blood samples by mid-infrared spectroscopy. *Appl. Opt.*, **42**, 745–749.
25 Martin, W.B., Mirov, W.B., and Venugopalan, R. (2002) Using two discrete frequencies within the middle infrared to quantitatively determine glucose in serum. *J. Biomed. Opt.*, **7**, 613–617.
26 Shen, Y.C., Davies, A.G., Linfield, E.H., Elsey, T.S., Taday, P.F., and Arnone, D.D. (2003) The use of Fourier-transform infrared spectroscopy for the quantitative determination of glucose concentration in whole blood. *Phys. Med. Biol.*, **48**, 2023–2032.
27 Yaroslavskaya, A.N., Yaroslavsky, I.V., Otto, C., Puppels, G.J., Guindam, H., Vrensen, G.F.J.M., Greve, J., and Tuchin, V.V. (1998) Water exchange in human eye lens monitored by confocal Raman microspectroscopy. *Biophysics*, **43**, 109–114.
28 Enejder, A.M.K., Scecina, T.G., Oh, J., Hunter, M., Shih, W.-C., Sasic, S., Horowitz, G.L., and Feld, M.S. (2005) Raman spectroscopy for noninvasive glucose measurements. *J. Biomed. Opt.*, **10**, 031114.
29 Hanlon, E.B., Manoharan, R., Koo, T.W., Shafer, K.E., Motz, J.T., Fitzmaurice, M., Kramer, J.R., Itzkan, I., Dasari, R.R., and Feld, M.S. (2000) Prospects for *in vivo* Raman spectroscopy. *Phys. Med. Biol.*, **45**, R1–R59.
30 MacKenzie, H.A., Ashton, H.S., Spiers, S., Shen, Y., Freeborn, S.S., Hannigan, J., Lindberg, J., and Rae, P. (1999) Advances in photoacoustic noninvasive glucose testing. *Clin. Chem.*, **45**, 1587–1595.
31 Kinnunen, M. and Myllyla, R. (2005) Effect of glucose on photoacoustic signals at the

wavelengths of 1064 and 532 nm in pig blood and intralipid. *J. Phys. D Appl. Phys.*, **38**, 2654–2661.
32 Kinnunen, M., Myllylä, R., Jokela, T., and Vainio, S. (2006) In vitro studies toward noninvasive glucose monitoring with optical coherence tomography. *Appl. Opt.*, **45**, 2251–2260.
33 Larin, K.V., Eledrisi, M.S., Motamedi, M., and Esenaliev, R.O. (2002) Noninvasive blood glucose monitoring with optical coherence tomography. *Diabetes Care*, **25**, 2263–2267.
34 Esenaliev, R.O., Larin, K.V., Larina, I.V., and Motamedi, M. (2001) Noninvasive monitoring of glucose concentration with optical coherence tomography. *Opt. Lett.*, **26**, 992–994.
35 Larin, K.V., Akkin, T., Esenaliev, R.O., Motamedi, M., and Milner, T.E. (2004) Phase-sensitive optical low-coherence reflectometry for the detection of analyte concentrations. *Appl. Opt.*, **43**, 3408–3414.
36 Esenaliev, R.O. and Prough, D.S. (2009) Noninvasive monitoring of glucose concentration with optical coherence tomography, in *Handbook of Optical Sensing of Glucose in Biological Fluids and Tissues* (ed. V.V. Tuchin), Taylor & Francis/CRC Press, Boca Raton, FL, pp. 563–586.
37 Young, A.R. (1997) Chromophores in human skin. *Phys. Med. Biol.*, **42**, 789–802.
38 McBride, T.O., Pogue, B.W., Poplack, S., Soho, S., Wells, W.A., Jiang, S., Osterberg, U.L., and Paulsen, K.D. (2002) Multispectral near-infrared tomography: a case study in compensating for water and lipid content in hemoglobin imaging of the breast. *J. Biomed. Opt.*, **7**, 72–79.
39 Palmer, K.F. and Williams, D. (1974) Optical properties of water in the near infrared. *J. Opt. Soc. Am.*, **64**, 1107–1110.
40 Martin, K.A. (1993) Direct measurement of moisture in skin by NIR spectroscopy. *J. Soc. Cosmet. Chem.*, **44**, 249–261.
41 Lauridsen, R.K., Everland, H., Nielsen, L.F., Engelsen, S.B., and Norgaard, L. (2003) Exploratory multivariate spectroscopic study on human skin. *Skin Res. Technol.*, **9**, 137–146.
42 Bashkatov, A.N., Genina, E.A., Kochubey, V.I., and Tuchin, V.V. (2005) Optical properties of human skin, subcutaneous and mucous tissues in the wavelength range from 400 to 2000 nm. *J. Phys. D Appl. Phys.*, **38**, 2543–2555.
43 Prahl, S.A. Optical Absorption of Hemoglobin, http://omlc.ogi.edu/spectra/hemoglobin/index.html, Oregon Medical Laser Center, Portland, OR (last accessed 20 December 2010).
44 Tuchin, V.V. (2007) *Tissue Optics: Light Scattering Methods and Instruments for Medical Diagnosis*, SPIE Press, Bellingham, WA.
45 Lakowicz, J.R. (1999) *Principles of Fluorescence Spectroscopy*, 2nd edn, Kluwer Academic/Plenum, New York.
46 Schneckenburger, H., Steiner, R., Strauss, W., Stock, K., and Sailer, R. (2002) Fluorescence technologies in biomedical diagnostics, in *Optical Biomedical Diagnostics* (ed. V.V. Tuchin), vol. **PM107**, SPIE Press, Bellingham, WA, pp. 827–874.
47 Sinichkin, Yu.P., Kollias, N., Zonios, G., Utz, S.R., and Tuchin, V.V. (2002) Reflectance and fluorescence spectroscopy of human skin in vivo, in *Optical Biomedical Diagnostics* (ed. V.V. Tuchin), vol. **PM107**, SPIE Press, Bellingham, WA, pp. 725–785.
48 Zhadin, N.N. and Alfano, R.R. (1998) Correction of the internal absorption effect in fluorescence emission and excitation spectra from absorbing and highly scattering media: theory and experiment. *J. Biomed. Opt.*, **3** (2), 171–186.
49 Drezek, R., Sokolov, K., Utzinger, U., Boiko, I., Malpica, A., Follen, M., and Richards-Kortum, R. (2001) Understanding the contributions of NADH and collagen to cervical tissue fluorescence spectra: modeling, measurements, and implications. *J. Biomed. Opt.*, **6** (4), 385–396.
50 Zhang, X., Erb, C., Flammer, J., and Nau, W.M. (2000) Absolute rate constants for the quenching of reactive excited states by melanin and related 5,6-dihydroxyindole metabolites: implications for their antioxidant activity. *Photochem. Photobiol.*, **71** (5), 524–533.
51 Lucchina, L.C., Kollias, N., Gillies, R., Phillips, S.B., Muccini, J.A., Stiller, M.J., Trancik, R.J., and Drake, L.A. (1996) Fluorescence photography in the

evaluation of acne. *J. Am. Acad. Dermatol.*, **35**, 58–63.

52 Soukos, N.S., Som, S., Abernethy, A.D., Ruggiero, K., Dunham, J., Lee, C., Doukas, A.G., and Goodson, J.M. (2005) Phototargeting oral black-pigmented bacteria. *Antimicrob. Agents Chemother.*, **49**, 1391–1396.

53 Schneckenburger, H., Steiner, R., Strauss, W., Stock, K., and Sailer, R. (2002) Fluorescence technologies in biomedical diagnostics, in *Optical Biomedical Diagnostics* (ed. V.V. Tuchin), SPIE Press, Bellingham, WA, pp. 825–874.

54 Cheng F X., Mao, J.-M., Bush, R., Kopans, D.B., Moore, R.H., and Chorlton, M. (2003) Breast cancer detection by mapping hemoglobin concentration and oxygen saturation. *Appl. Opt.*, **42** (31), 6412–6421.

55 Petibois, C., Rigalleau, V., Melin, A.-M., Perromat, A., Cazorla, G., Gin, H., and Deleris, G. (1999) Determination of glucose in dried serum samples by Fourier-transform infrared spectroscopy. *Clin. Chem.*, **45**, 1530–1535.

56 Von Lilienfeld-Toal, H., Weidenmuller, M., Xhelaj, A., and Mantele, W. (2005) A novel approach to non-invasive glucose measurement by mid-infrared spectroscopy: the combination of quantum cascade lasers (QCL) and photoacoustic detection. *Vibr. Spectrosc.*, **38**, 209–215.

57 Maier, J.S., Walker, S.A., Fantini, S., Franceschini, M.A., and Gratton, E. (1994) Possible correlation between blood glucose concentration and the reduced scattering coefficient of tissues in the near infrared. *Opt. Lett.*, **19**, 2062–2064.

58 Kohl, M., Esseupreis, M., and Cope, M. (1995) The influence of glucose concentration upon the transport of light in tissue-simulating phantoms. *Phys. Med. Biol.*, **40**, 1267–1287.

59 Maruo, K., Tsurugi, M., Tamura, M., and Ozaki, Y. (2003) *In vivo* nondestructive measurement of blood glucose by near-infrared diffuse-reflectance spectroscopy. *Appl. Spectrosc.*, **57**, 1236–1244.

60 Jiang, J.H., Berry, R.J., Siesler, H.W., and Ozaki, Y. (2002) Wavelength interval selection in multicomponent spectral analysis by moving window partial least-squared regression with applications to mid-infrared and near-infrared spectroscopic data. *Anal. Chem.*, **74**, 3555–3565.

61 Larin, K.V. and Oraevsky, A.A. (1999) Optoacoustic signal profiles for monitoring glucose concentration in turbid media. *Proc. SPIE*, **3726**, 576–583.

62 Tuchin, V.V. (2006) *Optical Clearing of Tissues and Blood*, SPIE Press, vol. **PM154**, Bellingham, WA,.

63 Williams, R.M., McDonagh, A.F., and Braslavsky, S.E. (1998) Structural volume changes upon photoisomerization of the bilirubin-albumin complex: a laser-induced optoacoustic study. *Photochem. Photobiol.*, **68** (4), 433–437.

64 Ermakov, I.V., Sharifzadeh, M., Ermakova, M., and Gellermann, W. (2005) Resonance Raman detection of carotenoid antioxidants in living human tissue. *J. Biomed. Opt.*, **10** (6), 064028.

65 Darvin, M.E., Gersonde, I., Albrecht, H., Zastrow, L., Sterry, W., and Lademann, J. (2007) In vivo Raman spectroscopic analysis of the influence of IR radiation on the carotenoid antioxidant substances beta-carotene and lycopene in the human skin. Formation of free radicals. *Laser Phys. Lett.*, **4** (4), 318–321.

66 Schulmerich, M.V., Cole, J.H., Dooley, K.A., Morris, M.D., Kreider, J.M., and Goldstein, S.A. (2008) Optical clearing in transcutaneous Raman spectroscopy of murine cortical bone tissue. *J. Biomed. Opt.*, **13** (2), 021108.

67 Fine, I. (2002) Non-invasive method and system of optical measurements for determining the concentration of a substance in blood. U.S. Patent US 6400972.

68 Fine, I. (2009) Glucose correlation with light scattering patterns, in *Handbook of Optical Sensing of Glucose in Biological Fluids and Tissues* (ed. V.V. Tuchin), Taylor & Francis/CRC Press, Boca Raton, FL, pp. 250–292.

69 Jacques, S.L., Saidi, I., Ladner, A., and Oelberg, D. (1997) Developing an optical fiber reflectance spectrometer to monitor bilirubinemia in neonates. *Proc. SPIE*, **2975**, 115–124.

70 Cameron, B.D., Gorde, H.W., Satheesan, B., and Coté, G.L. (1999) The use of polarized laser light through the eye for

noninvasive glucose monitoring. *Diabetes Technol. Ther.*, **1**, 135–143.
71 Cameron, B.D. and Anumula, H. (2006) Development of a real-time corneal birefringence compensated glucose sensing polarimeter. *Diabetes Technol. Ther.*, **8**, 156–164.
72 Pu, C., Zhu, Z., and Lo, Y.H. (2000) A surface-micromachined optical self-homodyne polarimetric sensor for noninvasive glucose monitoring. *IEEE Photon. Technol. Lett.*, **12**, 190–192.
73 Ansari, R.R., Boeckle, S., and Rovati, L. (2004) New optical scheme for a polarimetric-based glucose sensor. *J. Biomed. Opt.*, **9**, 103–115.
74 Wood, M.F.G., Guo, X., and Vitkin, I.A. (2007) Polarized light propagation in multiply scattering media exhibiting both linear birefringence and optical activity: Monte Carlo model and experimental methodology. *J. Biomed. Opt.*, **12**, 014029.
75 Wood, M.F.G., Ghosh, N., Guo, X., and Vitkin, I.A. (2009) Towards noninvasive glucose sensing using polarization analysis of multiply scattered light, in *Handbook of Optical Sensing of Glucose in Biological Fluids and Tissues* (ed. V.V. Tuchin), Taylor & Francis/CRC Press, Boca Raton, FL, pp. 536–574.
76 Malik, B.H. and Coté, G.L. (2010) Real-time, closed-loop dual-wavelength optical polarimetry for glucose monitoring. *J. Biomed. Opt.*, **15** (1), 017002.
77 Larin, K.V., Motamedi, M., Ashitkov, T.V., and Esenaliev, R.O. (2003) Specificity of noninvasive blood glucose sensing using optical coherence tomography technique: a pilot study. *Phys. Med. Biol.*, **48**, 1371–1390.

56
Psoriasis and Acne
Bernhard Ortel and Piergiacomo Calzavara-Pinton

56.1
Psoriasis

Psoriasis is a common skin disease, where areas of affected skin become red, thickened, and scaly, and may be pruritic. Large body areas can be affected. In a susceptible genetic background, the pathogenetic process involves an immune response involving T cells, dendritic cells, polymorphonuclear cells, multiple cytokines, angiogenesis, epidermal hyperplasia, and altered keratinocyte differentiation. Many successful therapies target T lymphocytes that drive the inflammatory process [1].

56.1.1
General Principles of Phototherapies

Photon-based treatments of psoriasis consist of repeated controlled phototoxic exposures of involved skin. Solar phototherapy (heliotherapy) is based on empirical improvement of psoriasis with sun exposure [2]. Artificial radiation sources have made therapeutic applications more available, controllable, and versatile. The availability of ultraviolet (UV)-emitting fluorescent tubes permits the widespread use of phototherapies in modern dermatology. Specific modalities include phototherapy, where UV radiation alone is used, whereas photosensitizers are administered in photochemotherapy and photodynamic therapy. The advantage of the photon-based therapies over other treatment modalities is that photons only reach superficial layers of the skin organ, limiting potential adverse effects.

Treatments are commonly done by whole-body exposure (Figure 56.1), but localized irradiation may be preferable with limited disease. The response of unaffected skin is dose limiting with total body irradiations, as skin affected by psoriasis is more resistant to phototoxic exposures. Initial irradiations are determined by phototesting (quantified erythema response threshold in normal skin) [3] or by estimation of individual sensitivity depending on the skin phototype. Subsequent doses are gradually increased depending on the skin response. Increments are

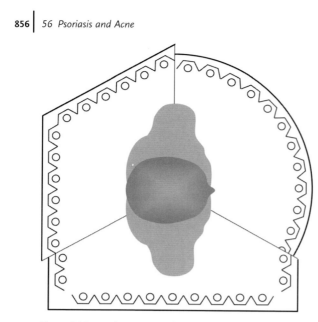

Figure 56.1 Bird's-eye view of a patient in an irradiation device lined by vertical fluorescent tubes with reflectors. The segments of the scheme represent different booth geometries.

either by the same amount or by a percentage of the previous exposure [4]. Linear increases result in slower response and higher cumulative dose. Percentage increases usually yield faster response and lower cumulative exposure, but have a higher frequency of unwanted phototoxicity. Certain protocols do not increase irradiations beyond a fixed target dose. Treatment frequency is two to five times per week for ultraviolet B (UVB) phototherapy and two to four times weekly for photochemotherapy. Some regimens include maintenance therapy, a continued therapy after at least 95% improvement or complete response, with stable radiation dose and progressively reduced frequency [5]. No standard regimens exist for photodynamic therapy. Multiple photosensitizers and different radiation sources make phototherapies very versatile (Figure 56.2).

56.1.2
Ultraviolet B Phototherapy

UVB radiation is delivered by fluorescent tubes, metal halide lamps, or excimer sources. Broadband (280–320 nm), narrowband (311 nm), and monochromatic UVB (308 nm) are available. Most therapeutic radiation sources also emit wavelengths outside the therapeutic spectrum, such as ultraviolet A (UVA), visible, and infrared radiation. Because of the latter, cooling by ventilation may be beneficial to avoid patient discomfort from heat. Moderate doses of unsensitized UVA radiation (320–400 nm) are of limited benefit in psoriasis by itself or if added to UVB. The action spectrum of antipsoriatic efficiency peaks in the wavelength range around

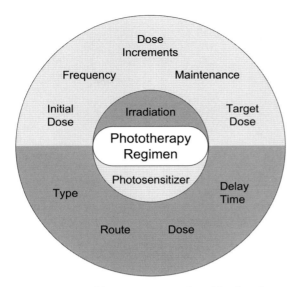

Figure 56.2 Variable parameters are selected for phototherapies of psoriasis, and their multitude illustrates the versatility of photon-based regimens in this common dermatosis.

310–315 nm (Figure 56.3). Several technical developments have applied this finding and removed the shorter, more erythemogenic while therapeutically less effective UVB wavelengths from the treatment spectrum.

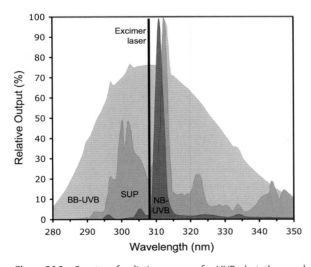

Figure 56.3 Spectra of radiation sources for UVB phototherapy demonstrate the progressive narrowing of the therapeutic spectrum to the wavelength range around 310 nm, where antipsoriatic efficiency is best.

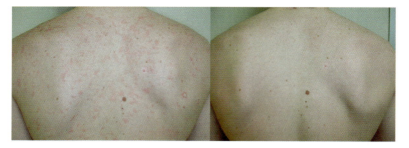

Figure 56.4 A patient with psoriasis before and at the end of a course of phototherapy using NB-UVB fluorescent tubes.

Selective ultraviolet phototherapy (SUP) applied this principle first in a commercial setting and was more efficient than broadband UVB. SUP did not work as well as psoralen with ultraviolet A (PUVA) photochemotherapy [6] and was therefore not very widely used. The more recent availability of narrowband ultraviolet B (NB-UVB, 311 nm) fluorescent tubes coincided with increasing concerns about adverse effects of PUVA. NB-UVB compared favorably with both broadband UVB and PUVA [7, 8]. Therefore NB-UVB is at present the first choice in most phototherapy centers (Figure 56.4). Additional radiation sources include xenon chloride excimer lasers that deliver high-intensity pulse trains of 308 nm radiation, thus requiring only short exposure times, but only of small areas [9]. Excimer lamps are useful for treating larger areas, but their output spectrum may include shorter wavelengths.

56.1.3
Photochemotherapy

PUVA photochemotherapy uses psoralens that are delivered topically or orally to photosensitize skin. The subsequent UVA (320–400 nm) exposure excites the psoralen molecule resulting in photochemical processes that cause a phototoxic reaction. Repeated controlled PUVA exposures lead to remission of psoriasis (Figure 56.5) [10]. Three psoralens are being used for PUVA of psoriasis (8-methoxypsoralen, 5-methoxypsoralen, trimethylpsoralen). Topical delivery may use aqueous solutions of psoralens for soaks of small areas, such as hands or feet, or for whole-body immersion (bath PUVA). Other topical preparations of psoralens are available, such as lotions, gels, and creams. 8-Methoxypsoralen (8-MOP) and 5-methoxypsoralen (5-MOP) are available for oral delivery for treatment of psoriasis. Total body irradiations are done in units as shown in Figure 56.1 that are equipped with UVA fluorescent tubes. Small irradiation units are available for localized exposure, for example, as needed in palmoplantar psoriasis. Metal halide lamps generally have a higher output and therefore require shorter exposure times; however, they require careful filtering of the UVB portion [11]. The spectral sensitivity of psoriasis treated with 8-MOP and UVA appears to be higher in the UVA-2 range (320–340 nm) than in the UVA-1 range

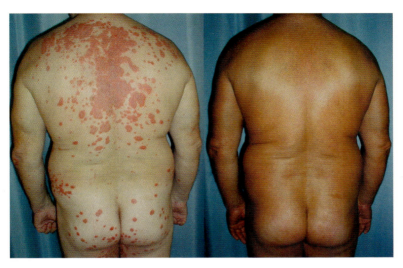

Figure 56.5 A patient before and after a course of oral photochemotherapy using 5-MOP. Note that PUVA induces an intense tan in addition to clearing psoriasis.

(340–400 nm). The fluorescent UVA lamps used most often for PUVA have an emission continuum peaking at 352 nm. Sources peaking around 325 nm appear to be much more efficient but have not been used widely for PUVA [12]. Differences between PUVA and UVB phototherapy include a delayed erythema response after PUVA. Maximum erythema is observed after 72–96 h with oral 8-MOP and at 120 h after bathwater sensitization, while the skin response to UVB peaks within 24 h. Therefore, PUVA exposures are given on two consecutive days only (not more than four per week) to avoid cumulative excessive phototoxicity [13].

56.1.4
Photodynamic Therapy

Photodynamic therapy (PDT) is a modality in which photochemical generation of singlet oxygen plays a central role. Originally devised as cancer treatment, a wide range of applications have been developed over recent decades. The treatment of psoriasis with PDT has been reported with several different sensitizers. Hematoporphyrin derivative [14], tin protoporphyrin [15], and benzoporphyrin derivative [16] have been used for systemic sensitization. Persistent remission in a treated area after a single exposure has been reported, but most regimens require repeated exposures for consistent improvement of psoriatic skin. Topical sensitization has been evaluated with a number of exogenous photosensitizers, but no standard PDT regimen is available at present [17, 18].

Most recently, a specific type of PDT, aminolevulinic acid (ALA)-based PDT (ALA-PDT) has been explored for the treatment of psoriasis. This modality is discussed detail in Chapter 52 for the treatment of skin cancers. In brief, exogenous ALA [or its methyl ester, methyl aminolevulinate (MAL)] induces the formation of excess protoporphyrin IX in target cells. Psoriatic skin has constitutively elevated levels of endogenous porphyrins, but not enough to be clinically photosensitive [19]. Additional porphyrin accumulation due to ALA exposure sensitizes psoriatic skin sufficiently for PDT [20]. At present, ALA-induced porphyrins are prevalent as photosensitizers. ALA-PDT treatment is complex, requiring topical ALA application to the psoriatic skin, a waiting period for porphyrins to accumulate in the lesions, and subsequent irradiation with visible light [21, 22]. In addition to the need for repeated, time-consuming treatments, ALA-PDT of psoriasis is often intensely painful during irradiation, severely compromising its acceptance by patients [23]. It is safe to say that no convenient or broadly applicable clinical regimen for ALA-PDT of psoriasis is available at present [24].

56.1.5
Laser Therapy

Visible radiation from the pulsed dye laser (PDL, 585–595 nm) is being widely used for the treatment of telangiectasia. Widened capillary loops in the papillary dermis of psoriatic skin have been similarly targeted with good success (Figure 56.6) [25, 26]. The treatment is based on the principle of selective photothermolysis. There is no standard regimen; a range from three to six treatments every 2–6 weeks has been

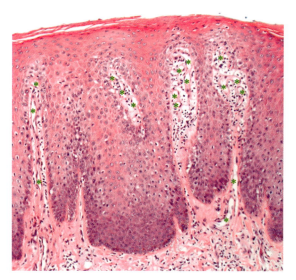

Figure 56.6 Micrograph of psoriatic skin. The capillary loops in the papillary dermis (asterisks) are obliterated by PDL treatments. Other phototherapies primarily target the inflammatory infiltrate cells in the dermis around the vessels.

reported. Although fairly effective, the cost and the small field of laser light delivery make this approach useful only in patients with limited disease.

56.1.6
Mechanisms of Action

Each phototherapeutic modality has its own set of molecular targets and responses. UVB exposure leads to DNA photoproduct formation, predominantly pyrimidine dimers, which result in DNA repair, alterations of transcriptional regulation, cytokine release, and – in those cells that are not reparable – apoptotic cell death. Apoptosis of infiltrating lymphocytes has been associated with therapeutic response to NB-UVB [27]. With UVA exposure, psoralens form cross-links between complementary strands of the DNA helix. Although singlet oxygen production and formation of other psoralen adducts, such as with RNA and proteins, have been demonstrated, the DNA cross-links appear critical for the therapeutic effect. Cross-link formation results in reduced epidermal keratinocyte proliferation. In addition to effects on keratinocytes, the critical effect of both UVB and PUVA in psoriasis appears to be the modulation of cutaneous immunological activities and the induction of apoptosis of lesional lymphocytes [28]. ALA-PDT is mediated predominantly through the action of singlet oxygen and other reactive species. Multiple cellular–molecular effects on cytokine expression and lymphocyte apoptosis are in concordance with the effects of other phototherapeutic modalities [29], but details are yet to be explored. ALA-PDT and other forms of PDT may also target the vasculature in psoriatic skin. In addition, low-dose whole-body exposures to verteporfin have demonstrated immunomodulatory effects that may be beneficial in psoriatic disease [30].

Pulsed laser treatments decimate papillary neovasculature that plays a pathogenetic role in psoriasis. Reduced blood flow and decreased T cell numbers in responsive plaques have been demonstrated [31].

56.1.7
Adverse Effects

The most common short-term side effects of phototherapies are symptoms of excessive phototoxicity, ranging from pruritus over sunburn to painful blistering. Careful dosimetry and patient monitoring help prevent these. If they occur, they are managed by symptomatic treatment and subsequent dose adjustment. Increased pigmentation is an effect that – besides esthetic considerations (that may be favorable) – is not desirable for phototherapy, as it increases the dose requirements; consequently, patients who tan more easily will require longer exposures and higher cumulative irradiations. Tanning is most relevant with PUVA (Figure 56.5) and UVB. Oral 8-MOP can cause nausea and vomiting. 5-MOP is tolerated much better and can be substituted for 8-MOP in patients with such symptoms where available [32]. Uncontrolled environmental UVA exposure has to be carefully avoided after psoralen ingestion to avoid added skin phototoxicity; eye protection must also be worn after taking psoralens. Other photosensitizers (ALA-derived porphyrins, exogenous tetrapyrroles) also induce various, substance-specific periods of persistent photosensitivity and require protection from accidental light exposure for that time. With ALA-PDT, pain during light exposure is a very common feature and may be ultimately limiting to this application [24]. Oral intake of ALA may cause nausea and hypotension, but this route is not commonly used for psoriasis treatment. The most important unwanted long-term effect of phototherapies is carcinogenesis. UVB is a known complete carcinogen, but the carcinogenic effect of therapeutic exposures to broadband UVB is small compared with PUVA. Extended use and high cumulative exposures to PUVA are associated with increased risks of non-melanoma skin cancer and possibly also melanoma [33]. Interestingly, bath PUVA with trimethylpsoralen (TMP) does not appear to be associated with increased skin cancer rates [34]. NB-UVB therapy does not significantly increase the risk of skin cancers according to recent analysis [35]. PUVA and UVB are immunosuppressive and this effect may contribute to both therapeutic efficacy and carcinogenesis. PDT has not been associated with increased cancer rates, but the observation period and patient numbers are too limited for a reliable assessment. Pulsed visible light is not associated with carcinogenesis.

Additional long-term effects of chronic UVA and UVB exposure include accelerated photoaging with elastosis, wrinkling, dryness, and lentigines.

56.1.8
Summary

Photon-based treatments psoriasis are versatile, cost-effective, and among the most powerful tools in the therapeutic armamentarium for psoriasis. Well-developed designs of the regimens, careful documentation of adverse effects, and a track record of safety and efficacy make phototherapies an important part in everyday management of psoriatic skin disease [4].

56.2
Acne

Acne is a very common dermatosis affecting most teenagers and people at other ages less frequently. It is a disease of the pilosebaceous unit (hair follicle plus sebaceous gland) with a multifactorial pathogenic process involving hormonal sensitivity, bacterial growth, inflammation, and altered epidermal differentiation [36]. Patients affected by acne may present with comedones (open and closed), inflammatory papules, pustules, and deep-seated cysts. Therapy includes daily topical and sometimes oral administration of antibiotics, anti-inflammatory agents, antiseptics, retinoids, and other agents. Many patients desire alternative treatments, and several light-based methods have been promoted.

Figure 56.7 Fluorescence image of the cheek of a 13-year-old girl with untreated acne. The comedones show faint red fluorescence of coproporphyrin in the resident *P. acnes*.

56.2.1
General Principles of Phototherapies

Many acne patients experience transitory improvement with sun exposure but it is not certain what part of the solar spectrum contributes most to this effect. Comedones show red fluorescence, which has been correlated with constitutive coproporphyrin accumulation in *Propionibacterium acnes* that populates pilosebaceous units (Figure 56.7). *P. acnes* has been considered an important pathogenic factor. Light therapy aims at destroying *P. acnes* through photosensitization by endogenous porphyrins. In addition, prominent damage to sebaceous glands was reported in mouse skin after PDT with systemic ALA [37]. Also, topical indocyanine green (ICG) appears to penetrate the pilosebaceous unit [38]. These findings indicated the possibility of relatively selective targeting of sebaceous glands in acne patients using photon-based regimens. Conceptually, infrared radiation thermally targets inflammatory infiltrate cells and sebaceous glands.

56.2.2
Ultraviolet Phototherapy and PUVA

UV radiation was considered a useful therapeutic option at a time when few specific agents were available for the management of acne. Critical evaluation showed that PUVA is useless in acne management and unsensitized UV radiation is of very limited therapeutic benefit at best [39]. At present, UV radiation-based treatments are discouraged as acne is often managed using multiple agents, some of which are UV photosensitizers. In addition, UV radiation is considered comedogenic. Long-term adverse effects of UV exposure in young patients are an additional concern.

56.2.3
Visible Light Phototherapy

Light therapy of acne is based on the concept of photosensitizing *P. acnes* by its endogenous porphyrins [40]. Multiple porphyrin absorption peaks allow photoactivation by a number of visible light sources, which differ in relative tissue penetration depth and dose requirements. Fluorescent lamps and light-emitting diodes (LEDs) in the blue and red range, the potassium titanyl phosphate (KTP) laser (532 nm), the PDL (585 nm), and intense pulsed light (IPL) (530–750 nm) have been used with variable success in mild and moderate acne. Regimens vary largely in terms of dosimetry and number and frequency of exposures, ranging from single PDL treatments to twice-daily broadband red light exposures for 8 weeks. One study compared blue and blue plus red light exposures with topical 5% benzoyl peroxide and found the combined blue and red light therapy to be most effective [41]. Infrared-emitting lasers, such as the Nd:YAG laser (1320 nm) and a diode laser (1450 nm), have also been explored. Interestingly, whereas the Nd:YAG laser improved the comedonal components [42], the diode laser persistently reduced inflammatory lesions [43]. Only a few studies included placebo control groups. Because of the large variation of protocols and the multitude of light sources, a champion among acne phototherapies has not evolved. Neither has an action spectrum been established. At present, it appears that phototherapies are definitely not useful for severe and cystic acne. Recent evidence-based review articles evaluated in detail relevant clinical trials of acne phototherapy [44, 45]. A careful comparison showed recently that red light phototherapy was no less effective than PDT using MAL and red light [46].

56.2.4
Photodynamic Therapy

The first study using ALA-PDT in acne showed efficacy of a single and four consecutive weekly exposures [47]. Subsequent studies compared photosensitized acne areas with areas treated with light alone. ALA, MAL, and ICG have been evaluated and several lasers and broadband radiation sources have been used [48]. The difficulty in evaluating PDT regimens lies in the heterogeneity of the protocols. In general, multiple treatments with weekly to monthly intervals work better than single exposures, and the photosensitized areas do better than those treated with light alone. In one study, in which all patients received topical 0.1% adapelene gel, only the control group showed significant improvement, whereas with IPL treatments and PDT using MAL and IPL significance was not attained [49]. Only a few trials have compared photon-based therapies with conventional treatments such as topical 1% clindamycin or 5% benzoyl peroxide, and none compared light-based treatments with standard multi-agent therapies. Finally, side effects may be significant with ALA- and MAL-PDT. Recent reviews are available that compare and critically evaluate various approaches [44, 45]. Although sebaceous gland targeting was an important concept initially, inflammatory acne lesions generally respond better than comedones.

56.2.5
Mechanisms and Adverse Effects

Very little is known about the mechanisms of phototherapy in acne, although the lethal photosensitization of *P. acnes* was considered an important component [47]. However, several authors refute the relevance of this mechanism [46, 50]. A number of protocols have been shown to reduce sebum excretion, including ALA-PDT and infrared laser treatment. In addition, shedding of infundibular keratinocytes that obstruct the pilosebaceous opening and direct damage to the sebocytes may hypothetically contribute to the therapeutic effect [47], although no data exist to support these claims. It is more likely that sebaceous glands are modified in their function rather than sustaining direct damage [50]. With laser-based therapies, surface cooling reduces unwanted epidermal effects while allowing impact on subepidermal structures. Unsensitized continuous-wave light treatments are not associated with adverse effects. Laser treatments and IPL may cause pain and epidermal damage and topical anesthesia and surface cooling are advisable. Adverse effects during ALA-PDT treatment include pain, burning, and itching. After the treatment, an exacerbation (a folliculitis-like reaction) is common, and epidermal hyperpigmentation may last for months [47]. A conceptual concern about long-term compromise of sebaceous glands is currently not supported by any evidence of their persistent involution after light-based therapies.

56.2.6
Summary

Because of the great efficacy of pharmacotherapies in acne, it is difficult to equal or surpass them with photonic regimens. Phototherapy is an option in patients who cannot use other medications, such as pregnant women. The advantage of light-based approaches lies in the high patient acceptance and the reduced frequency of applications. On the other hand, adverse effects, such as pain, exacerbation, and pigmentation, which accompany the more efficient PDT regimens, compromise their appeal. The use of photon-based therapies in combination regimens deserves further evaluation.

Acknowledgments

We thank Dr. Justin Wasserman, University of Chicago, for providing Figure 56.7 and Saalmann GmbH (Herford, Germany) for the emission spectra of the SUP and NB-UVB lamps.

References

1 Krueger, J.G. (2002) The immunologic basis for the treatment of psoriasis with new biologic agents. *J. Am. Acad. Dermatol.*, **46** (1), 1–23; quiz 23–26.
2 Nilsen, L.T., Soyland, E., and Krogstad, A.L. (2009) Estimated ultraviolet doses to psoriasis patients during climate therapy. *Photodermatol. Photoimmunol. Photomed.*, **25** (4), 202–208.
3 Palmer, R.A., Aquilina, S., Milligan, P.J., Walker, S.L., Hawk, J.L., and Young, A.R. (2006) Photoadaptation during narrowband ultraviolet-B therapy is independent of skin type: a study of 352 patients. *J. Invest. Dermatol.*, **126** (6), 1256–1263.
4 Menter, A., Korman, N.J., Elmets, C.A., Feldman, S.R., Gelfand, J.M., Gordon, K.B., Gottlieb, A., Koo, J.Y., Lebwohl, M., Lim, H.W., Van Voorhees, A.S., Beutner, K.R., and Bhushan, R. (2010) Guidelines of care for the management of psoriasis and psoriatic arthritis: Section 5. Guidelines of care for the treatment of psoriasis with phototherapy and photochemotherapy. *J. Am. Acad. Dermatol.*, **62** (1), 114–135.
5 Stern, R.S., Armstrong, R.B., Anderson, T.F., Bickers, D.R., Lowe, N.J., Harber, L., Voorhees, J., and Parrish, J.A. (1986) Effect of continued ultraviolet B phototherapy on the duration of remission of psoriasis: a randomized study. *J. Am. Acad. Dermatol.*, **15** (3), 546–552.
6 Hönigsmann, H., Fritsch, P., and Jaschke, E. (1977) UV-therapy of psoriasis. Half-side comparison between oral photochemotherapy (PUVA) and selective UV-phototherapy (SUP). *Z. Hautkr.*, **52** (21), 1078–1082.
7 Coven, T.R., Burack, L.H., Gilleaudeau, R., Keogh, M., Ozawa, M., and Krueger, J.G. (1997) Narrowband UV-B produces superior clinical and histopathological resolution of moderate-to-severe psoriasis in patients compared with broadband UV-B. *Arch. Dermatol.*, **133** (12), 1514–1522.
8 Tanew, A., Ortel, B., and Hönigsmann, H. (1999) Half-side comparison of erythemogenic versus suberythemogenic UVA doses in oral photochemotherapy of psoriasis. *J. Am. Acad. Dermatol.*, **41** (3 Pt 1), 408–413.
9 Asawanonda, P., Anderson, R.R., Chang, Y., and Taylor, C.R. (2000) 308-nm excimer laser for the treatment of psoriasis: a dose-response study. *Arch. Dermatol.*, **136** (5), 619–624.
10 Hönigsmann, H. (2001) Phototherapy for psoriasis. *Clin. Exp. Dermatol.*, **26** (4), 343–350.
11 Calzavara-Pinton, P.G. (1997) Efficacy and safety of stand-up irradiation cubicles with UVA metal-halide lamps (and a new filter) or UVA fluorescent lamps for photochemotherapy of psoriasis. *Dermatology*, **195** (3), 243–247.
12 Farr, P.M., Diffey, B.L., Higgins, E.M., and Matthews, J.N. (1991) The action spectrum between 320 and 400 nm for clearance of psoriasis by psoralen photochemotherapy. *Br. J. Dermatol.*, **124** (5), 443–448.
13 Wolff, K.W., Fitzpatrick, T.B., Parrish, J.A., Gschnait, F., Gilchrest, B., Honigsmann, H., Pathak, M.A., and Tannenbaum, L. (1976) Photochemotherapy for psoriasis with orally administered methoxsalen. *Arch. Dermatol.*, **112** (7), 943–950.
14 Diezel, W., Meffert, H., and Sonnichsen, N. (1980) Therapy of psoriasis by means of hematoporphyrin derivative and light. *Dermatol. Monatsschr.*, **166** (12), 793–797.
15 Emtestam, L., Berglund, L., Angelin, B., Drummond, G.S., and Kappas, A. (1989) Tin-protoporphyrin and long wave length ultraviolet light in treatment of psoriasis. *Lancet*, **333** (8649), 1231–1233.
16 Lui, H. (1994) Photodynamic therapy in dermatology with porfimer sodium and benzoporphyrin derivative: an update. *Semin. Oncol.*, **21** (6 Suppl 15), 11–14.
17 McCullough, J.L., Weinstein, G.D., Lemus, L.L., Rampone, W., and Jenkins, J.J. (1983) Development of a topical hematoporphyrin derivative formulation: characterization of photosensitizing effects *in vivo*. *J. Invest. Dermatol.*, **81** (6), 528–532.

18 Takahashi, H., Itoh, Y., Nakajima, S., Sakata, I., and Iizuka, H. (2004) A novel ATX-S10(Na) photodynamic therapy for human skin tumors and benign hyperproliferative skin. *Photodermatol. Photoimmunol. Photomed.*, **20** (5), 257–265.

19 Maari, C., Viau, G., and Bissonnette, R. (2003) Repeated exposure to blue light does not improve psoriasis. *J. Am. Acad. Dermatol.*, **49** (1), 55–58.

20 Boehncke, W.H., Sterry, W., and Kaufmann, R. (1994) Treatment of psoriasis by topical photodynamic therapy with polychromatic light. *Lancet*, **343** (8900), 801.

21 Radakovic-Fijan, S., Blecha-Thalhammer, U., Schleyer, V., Szeimies, R.M., Zwingers, T., Honigsmann, H., and Tanew, A. (2005) Topical aminolaevulinic acid-based photodynamic therapy as a treatment option for psoriasis? Results of a randomized, observer-blinded study. *Br. J. Dermatol.*, **152** (2), 279–283.

22 Collins, P., Robinson, D.J., Stringer, M.R., Stables, G.I., and Sheehan-Dare, R.A. (1997) The variable response of plaque psoriasis after a single treatment with topical 5-aminolaevulinic acid photodynamic therapy. *Br. J. Dermatol.*, **137** (5), 743–749.

23 Warren, C.B., Karai, L.J., Vidimos, A., and Maytin, E.V. (2009) Pain associated with aminolevulinic acid-photodynamic therapy of skin disease. *J. Am. Acad. Dermatol.*, **61** (6), 1033–1043.

24 Schleyer, V., Radakovic-Fijan, S., Karrer, S., Zwingers, T., Tanew, A., Landthaler, M., and Szeimies, R.M. (2006) Disappointing results and low tolerability of photodynamic therapy with topical 5-aminolaevulinic acid in psoriasis. A randomized, double-blind phase I/II study. *J. Eur. Acad. Dermatol Venereol*, **20** (7), 823–828.

25 Katugampola, G.A., Rees, A.M., and Lanigan, S.W. (1995) Laser treatment of psoriasis. *Br. J. Dermatol.*, **133** (6), 909–913.

26 Zelickson, B.D., Mehregan, D.A., Wendelschfer-Crabb, G., Ruppman, D., Cook, A., O'Connell, P., and Kennedy, W.R. (1996) Clinical and histologic evaluation of psoriatic plaques treated with a flashlamp pulsed dye laser. *J. Am. Acad. Dermatol.*, **35** (1), 64–68.

27 Ozawa, M., Ferenczi, K., Kikuchi, T., Cardinale, I., Austin, L.M., Coven, T.R., Burack, L.H., and Krueger, J.G. (1999) 312-nanometer ultraviolet B light (narrow-band UVB) induces apoptosis of T cells within psoriatic lesions. *J. Exp. Med.*, **189** (4), 711–718.

28 Vallat, V.P., Gilleaudeau, P., Battat, L., Wolfe, J., Nabeya, R., Heftler, N., Hodak, E., Gottlieb, A.B., and Krueger, J.G. (1994) PUVA bath therapy strongly suppresses immunological and epidermal activation in psoriasis: a possible cellular basis for remittive therapy. *J. Exp. Med.*, **180** (1), 283–296.

29 Bissonnette, R., Tremblay, J.F., Juzenas, P., Boushira, M., and Lui, H. (2002) Systemic photodynamic therapy with aminolevulinic acid induces apoptosis in lesional T lymphocytes of psoriatic plaques. *J. Invest. Dermatol.*, **119** (1), 77–83.

30 Ratkay, L.G., Waterfield, J.D., and Hunt, D.W. (2000) Photodynamic therapy in immune (non-oncological) disorders: focus on benzoporphyrin derivatives. *BioDrugs*, **14** (2), 127–135.

31 Hern, S., Allen, M.H., Sousa, A.R., Harland, C.C., Barker, J.N., Levick, J.R., and Mortimer, P.S. (2001) Immunohistochemical evaluation of psoriatic plaques following selective photothermolysis of the superficial capillaries. *Br. J. Dermatol.*, **145** (1), 45–53.

32 Calzavara-Pinton, P., Ortel, B., Carlino, A., Honigsmann, H., and De Panfilis, G. (1992) A reappraisal of the use of 5-methoxypsoralen in the therapy of psoriasis. *Exp. Dermatol.*, **1** (1), 46–51.

33 Patel, R.V., Clark, L.N., Lebwohl, M., and Weinberg, J.M. (2009) Treatments for psoriasis and the risk of malignancy. *J. Am. Acad. Dermatol.*, **60** (6), 1001–1017.

34 Hannuksela-Svahn, A., Sigurgeirsson, B., Pukkala, E., Lindelof, B., Berne, B., Hannuksela, M., Poikolainen, K., and Karvonen, J. (1999) Trioxsalen bath PUVA did not increase the risk of squamous cell skin carcinoma and cutaneous malignant melanoma in a joint analysis of 944 Swedish and Finnish patients with

psoriasis. *Br. J. Dermatol.*, **141** (3), 497–501.

35 Hearn, R.M., Kerr, A.C., Rahim, K.F., Ferguson, J., and Dawe, R.S. (2008) Incidence of skin cancers in 3867 patients treated with narrow-band ultraviolet B phototherapy. *Br. J. Dermatol.*, **159** (4), 931–935.

36 Zouboulis, C.C., Eady, A., Philpott, M., Goldsmith, L.A., Orfanos, C., Cunliffe, W.C., and Rosenfield, R. (2005) What is the pathogenesis of acne? *Exp. Dermatol.*, **14** (2), 143–152.

37 Divaris, D.X., Kennedy, J.C., and Pottier, R.H. (1990) Phototoxic damage to sebaceous glands and hair follicles of mice after systemic administration of 5-aminolevulinic acid correlates with localized protoporphyrin IX fluorescence. *Am. J. Pathol.*, **136** (4), 891–897.

38 Lloyd, J.R. and Mirkov, M. (2002) Selective photothermolysis of the sebaceous glands for acne treatment. *Lasers Surg. Med.*, **31** (2), 115–120.

39 Mills, O.H. and Kligman, A.M. (1978) Ultraviolet phototherapy and photochemotherapy of acne vulgaris. *Arch. Dermatol.*, **114** (2), 221–223.

40 Ashkenazi, H., Malik, Z., Harth, Y., and Nitzan, Y. (2003) Eradication of *Propionibacterium acnes* by its endogenic porphyrins after illumination with high intensity blue light. *FEMS Immunol. Med. Microbiol.*, **35** (1), 17–24.

41 Papageorgiou, P., Katsambas, A., and Chu, A. (2000) Phototherapy with blue (415 nm) and red (660 nm) light in the treatment of acne vulgaris. *Br. J. Dermatol.*, **142** (5), 973–978.

42 Orringer, J.S., Kang, S., Maier, L., Johnson, T.M., Sachs, D.L., Karimipour, D.J., Helfrich, Y.R., Hamilton, T., and Voorhees, J.J. (2007) A randomized, controlled, split-face clinical trial of 1320-nm Nd:YAG laser therapy in the treatment of acne vulgaris. *J. Am. Acad. Dermatol.*, **56** (3), 432–438.

43 Jih, M.H., Friedman, P.M., Goldberg, L.H., Robles, M., Glaich, A.S., and Kimyai-Asadi, A. (2006) The 1450-nm diode laser for facial inflammatory acne vulgaris: dose–response and 12-month follow-up study. *J. Am. Acad. Dermatol.*, **55** (1), 80–87.

44 Haedersdal, M., Togsverd-Bo, K., and Wulf, H.C. (2008) Evidence-based review of lasers, light sources and photodynamic therapy in the treatment of acne vulgaris. *J. Eur. Acad. Dermatol. Venereol.*, **22** (3), 267–278.

45 Hamilton, F.L., Car, J., Lyons, C., Car, M., Layton, A., and Majeed, A. (2009) Laser and other light therapies for the treatment of acne vulgaris: systematic review. *Br. J. Dermatol.*, **160** (6), 1273–1285.

46 Hörfelt, C., Stenquist, B., Halldin, C.B., Ericson, M.B., and Wennberg, A.M. (2009) Single low-dose red light is as efficacious as methyl-aminolevulinate – photodynamic therapy for treatment of acne: clinical assessment and fluorescence monitoring. *Acta Derm. Venereol.*, **89** (4), 372–378.

47 Hongcharu, W., Taylor, C.R., Chang, Y., Aghassi, D., Suthamjariya, K., and Anderson, R.R. (2000) Topical ALA-photodynamic therapy for the treatment of acne vulgaris. *J. Invest. Dermatol.*, **115** (2), 183–192.

48 Tuchin, V.V., Genina, E.A., Bashkatov, A.N., Simonenko, G.V., Odoevskaya, O.D., and Altshuler, G.B. (2003) A pilot study of ICG laser therapy of acne vulgaris: photodynamic and photothermolysis treatment. *Lasers Surg. Med.*, **13** (5), 296–310.

49 Yeung, C.K., Shek, S.Y., Bjerring, P., Yu, C.S., Kono, T., and Chan, H.H. (2007) A comparative study of intense pulsed light alone and its combination with photodynamic therapy for the treatment of facial acne in Asian skin. *Lasers Surg. Med.*, **39** (1), 1–6.

50 Pollock, B., Turner, D., Stringer, M.R., Bojar, R.A., Goulden, V., Stables, G.I., and Cunliffe, W.J. (2004) Topical aminolaevulinic acid-photodynamic therapy for the treatment of acne vulgaris: a study of clinical efficacy and mechanism of action. *Br. J. Dermatol.*, **151** (3), 616–622.

57
Wound Healing
Bernard Choi

57.1
Introduction

Wound repair clearly is a critically important function for the human body. Several factors affect the wound-healing process, including age and gender. The manner in which the injury occurs (e.g., electrical burn wound versus scald burn wound) can also modulate the resultant wound repair dynamics. Specific diseases, such as diabetes, can impede the wound-healing process.

To accelerate wound repair, a large variety of methods have been proposed. Wound occlusion with a bandage is one common approach, to maintain a hydrated wound bed and also to restrict access to the wound site by pathogens. A more advanced method of considerable interest is the use of light to modulate cellular function and stimulate the wound-healing process.

The objectives of this chapter are twofold. First, it focuses on the wound-healing response which ensues from selective optical injury to the microvasculature, with application of a method known as selective photothermolysis [1]. Second, a summary is presented of recent applications of optical-based methods for both therapeutic and diagnostic applications in wound healing. The list of applications is not comprehensive; instead, emphasis is placed to give the reader a sense of the diversity of optical-based methods available.

57.2
Wound-Healing Response to Selective Optical Injury to the Microvasculature

Histological and functional analysis of normal and abnormal vasculature irradiated with high-power, pulsed laser irradiation demonstrates that selective photothermal injury can be achieved with minimal perceivable structural alteration of perivascular tissue. This precise outcome is based on the principles of selective photothermolysis originally proposed by Anderson and Parrish [1]. A clinical application of selective photothermolysis is targeted therapy of subsurface microvasculature, such as port-wine stain (PWS) birthmarks [2]. Despite the immediate stoppage of blood flow which routinely occurs, complete clearance of PWS birthmarks rarely occurs, even after multiple (5–20) treatment sessions.

Hypotheses which describe the reason for the resistance of PWS birthmarks to selective laser therapy are difficult to test, in large part because an animal model with PWS currently does not exist. Nevertheless, one *in vivo* model which has served as an important translational platform for novel treatment discovery is the rodent dorsal window chamber [3]. With this animal model, Babilas *et al.* [4] compared the 24 h microvascular response to selective laser injury with the predictions of photodamage derived from computational modeling data using the finite-element method. They observed that pulsed-dye laser therapy can reliably photocoagulate vessels with diameters >20 μm, but this therapy is ineffective at photocoagulating smaller vessels; these experimental findings were in agreement with results of the finite-element modeling. According to the authors, these findings may also offer a hypothesis describing the basis for healing and restoration of the injured vasculature.

Heger *et al.* [5] described, in a comprehensive manner, the biological response to coagulum formation in vessels. Based on published literature describing the impact of specific growth factors and other molecules on thrombus organization, neovascularization, and angiogenesis, they extrapolated these past findings to the biological response of vasculature to photocoagulation. Similarly to Babilas *et al.* [4], they postulated that incomplete photocoagulation, and hence partial thrombus formation, are the primary cause of suboptimal outcomes after laser therapy of PWS birthmarks. They suggested that proposed hypothesis may help to define the role of a combined optical–pharmacologic treatment approach, which could be used to aid phototherapy of PWS birthmarks. This approach would involve photocoagulation accompanied by local release of prothrombotic agents, to enhance the degree of vascular shutdown in damaged vasculature.

Because some PWS birthmarks in patients are known to be resistant to selective laser injury, monitoring the wound-healing response in a longitudinal fashion may offer clues to the underlying reasons. With the rodent dorsal window chamber model of the microvasculature, Choi *et al.* [6] studied the response of the microvascular network to selective optical injury. Results up to 24 h after laser injury support previous work [4, 7] showing acute shutdown of the irradiated blood vessels, presumably due to photocoagulation. The results also suggest that, thereafter, a dramatic remodeling process occurs which involves blood-flow redistribution, vascular repair, and ultimate restoration of blood flow to the injured site (Figure 57.1). The longitudinal investigation of the microvascular hemodynamics in response to selective optical injury may offer a method for future studies to help researchers to understand better the biological factors responsible for the repair process and also to identify novel therapeutic approaches to achieve persistent vascular shutdown. One promising method, identified with the methods described by Choi *et al.*, is the use of combined photodynamic and photothermal therapies [8].

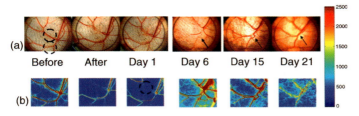

Figure 57.1 Vascular remodeling and blood flow dynamics were evident during the 21 day monitoring period, with the Day 0 "Before" and Day 21 structural images having similar appearances. Methods: a time sequence of wide-field color reflectance images (a) and corresponding speckle flow index images (b) was acquired over a 21 day monitoring period after pulsed laser irradiation of selected sites. Two arteriole–venule pairs (dashed circles in "Before" image) were irradiated with simultaneous 532 and 1064 nm laser pulses (upper circle, five 1 ms laser pulses at 27 Hz repetition rate, 2 J cm^{-2} at 532 nm, 3.6 J cm^{-2} at 1064 nm; lower circle, single 1 ms laser pulse, 4 J cm^{-2} at 532 nm, 7.2 J cm^{-2} at 1064 nm). Note the diffuse appearance of the speckle flow index image at Day 21, due to the presence of a scattering, overlying fascial layer. Color reflectance image dimensions (H × V), 13 × 10 mm; speckle flow index image dimensions, 9 × 7 mm. Adapted with permission from Ref. [4].

Previous research has demonstrated that rapamycin can inhibit angiogenesis. Phung et al. [9] and Jia et al. [10] reported that topically applied rapamycin can inhibit revascularization of sites irradiated with pulsed laser irradiation. The results suggest that a combined light–drug-based treatment protocol may be appropriate for enhanced therapy of PWS birthmarks and other cutaneous vascular abnormalities.

With microvascular photocoagulation, a purported initial biological response is secretion of angiogenic growth factors by perivascular stromal cells. These growth factors are expected to stimulate angiogenesis and revascularize the injured site. In addition to stromal cells, pluripotent stem cells in the dermis may also modulate the microvascular repair process. Loewe et al. [11] used immunohistochemistry to develop a biological model describing the role of stem cells in the overall wound-healing response to selective laser injury. They demonstrated that, following pulsed laser irradiation of *in vivo* human skin bearing PWS vasculature, the expression of Ki-67 (a marker of cell proliferation) and nestin (a stem cell marker) is strongly upregulated. In contrast, markers for circulating endothelial stem cells and mesenchymal stem cells were absent in the histologic sections, suggesting that stem cells are not recruited during the wound-healing process following selective laser injury. From these data, it was postulated that (a) pulsed laser irradiation induces endothelial cell proliferation, hence leading to revascularization of the injured site; and (b) terminally differentiated cells may become activated or "dedifferentiated" to modulate the wound-healing process. Loewe et al. also reported that the use of topical rapamycin in concert with pulsed laser irradiation reduces the expression of Ki-67 and nestin and enhances the overall therapeutic efficacy. The results are in agreement with previous preclinical findings [9, 10] and collectively support a framework for a combined light–drug-based treatment protocol for PWS birthmarks.

57.3
Recent Applications of Optical-Based Methods in Wound Healing

57.3.1
Low-Level Laser Therapy (LLLT) – Wound Healing

An excellent, concise overview of low-level laser therapy (LLLT) was provided by Pinheiro [12]. Here we focus only on recent findings.

LLLT of biological tissue can affect tissue repair processes *in vitro*. However, several groups have reported that *in vivo* application of LLLT does not improve wound repair. A possible explanation is that LLLT can stimulate bacterial infection of the treated wound site. Nussbaum et al. [13] studied bacterial growth at incisional wound sites created on *in vivo* rat dorsal skin and irradiated with either 635 or 808 nm laser light. When compared with bacterial growth at sites 3 days after wound creation, irradiation with 808 nm light resulted in a decrease in normal skin flora and an increase in *Staphylococcus aureus* growth. The research suggests that disruption of normal skin flora enables pathogens such as *S. aureus* to colonize the wound.

Previous research has demonstrated that LLLT can modulate tissue repair processes. However, papers have reported negative findings in terms of the efficacy of LLLT in human subjects. Fulop et al. [14] performed a meta-analysis of 23 papers published between 2000 and 2007 to calculate a composite metric of LLLT efficacy in preclinical animal studies and clinical studies. They reported that a significantly positive effect of LLLT was observed in both animal and clinical studies. The calculated efficacy was stronger in animals than in human subjects; it was postulated that this trend is due to the higher degree of experimental control that is present in animal studies. According to the authors, the results demonstrate that LLLT, overall, does improve tissue repair after injury.

Mitomycin C is a chemotherapeutic drug with antineoplastic antibiotic properties, but its effects on wound healing remain to be ascertained. Silva Santos et al. [15] studied the effects of LLLT on skin wounds treated with mitomycin C. They postulated that LLLT, which is known to stimulate fibroblast proliferation and collagen production, can modulate the inhibitory effects that mitomycin C on collagen synthesis. They reported that LLLT reduced the inflammatory response and increased both collagen deposition and fibroblast proliferation in scalpel wounds created in the dorsa of a rodent model and subsequently treated with mitomycin C.

57.3.2
LLLT to Aid Cerebral Nervous System Repair Processes

Anders [16] provided an excellent, brief overview of the current state of LLLT to enhance repair of traumatic injury to the central nervous system. Here, we focus on reviewing several key studies published recently.

Tissue plasminogen activator (tPA) is the only treatment option for patients suffering from ischemic stroke. Successful treatment relies on tPA administration within 3 h after stroke onset. As an alternate treatment approach, LLLT purportedly stimulates mitochondrial function and mitigates the risk of apoptosis in cerebral regions affected by stroke. Zivin et al. [17] demonstrated a favorable outcome of ischemic stroke with the use of LLLT. They observed that 36% of patients treated with LLLT achieved a favorable outcome, compared with 31% of patients receiving a sham therapy (i.e., no irradiation with laser light). It is important to note that the difference between patient treatment response was not significant. LLLT can be performed at any time after ischemic stroke, making it an attractive alternative in the clinical management of stroke victims.

Knowledge of the precise mechanisms by which LLLT works could help in the design of robust LLLT-based treatment protocols. One proposed mechanism by which LLLT enhances wound healing is stimulation of mitochondrial activity via absorption of the incident light by cytochrome oxidase c. Lapchak and de Taboada [18] employed a luminescence assay to estimate ATP content in excised parietal/occipital cortex from rabbits, after embolization. They performed experiments on three groups of rabbits: (1) negative control (no embolization, no LLLT), (2) embolization without LLLT, and (3) embolization with LLLT. It was found that embolization without LLLT decreases ATP content, and embolization with LLLT increases ATP content by as much as 77% above the content in the negative control group. It was suggested that the potential mechanism by which LLLT improves the outcome of ischemic stroke is enhanced mitochondrial function.

57.3.3
Wound Healing after Laser Cartilage Reshaping

Selective photothermal heating of cartilage can achieve a persistent shape change in the irradiated sample. Recent studies have focused on the effects of laser irradiation on chondrocytes, the cells in cartilage responsible for development of the extracellular matrix in cartilage. Holden et al. [19] demonstrated that only collagen type II gene expression, and not that of collagen type I, is observed following laser irradiation of cartilage. They used an Nd:YAG laser to irradiate nasal septum freshly extracted from rabbits and assessed the gene expression of collagen types I and II using reverse transcription polymerase chain reaction (RT-PCR). They determined that collagen type II was absent from the irradiated site but expressed in regions surrounding the site. Collagen type I gene expression was not observed, suggesting that photothermal modulation of cartilage does not induce a traditional wound-healing response. Instead, based on the findings, it was suggested that laser irradiation of cartilage results in the formation of hyaline cartilage, which consists primarily of collagen type II and thus indicates an optimal wound-healing response.

57.3.4
Novel Diagnostic Applications of Biophotonics to Study Wound Healing

Current methods to monitor dermal wound repair are limited. Gross clinical assessment of the wound is highly dependent on the prior experience and training of the clinician. Histologic analysis requires the use of destructive biopsies.

Zhuo et al. [20] found that nonlinear optical microscopy (NLOM) is capable of chronic, nondestructive evaluation of scald wound repair in an *in vivo* rodent model. They specifically analyzed second harmonic generation and two-photon excited fluorescence signals. When they transplanted fluorescently labeled bone marrow-derived mesenchymal stem cells (MSCs) to specific wound sites, they were able to monitor the location of the MSCs and the dynamic collagen architecture using NLOM. With the development of clinic-friendly NLOM instruments, clinicians may soon be able to monitor, in a routine manner, the progress of cutaneous wound repair.

57.4
Summary

This chapter has presented a summary of our current knowledge on the biological response of the microvasculature to selective optical injury, followed by an overview of recently published studies describing the application of LLLT to enhance wound healing and optical imaging methods to study noninvasively the wound-healing response. With the current shift towards interdisciplinary preclinical and clinical research, researchers are now well poised to conduct integrated studies at the interface of the cutting edge of both optical technology and biology, to address key fundamental questions on the biological processes which underlie the dynamics of wound healing.

References

1 Anderson, R.R. and Parrish, J.A. (1983) Selective photothermolysis – precise microsurgery by selective absorption of pulsed radiation. *Science*, **220** (4596), 524–527.

2 Kelly, K.M., Choi, B., McFarlane, S., Motosue, A., Jung, B.J., Khan, M.H., Ramirez-San-Juan, J.C., and Nelson, J.S. (2005) Description and analysis of treatments for port-wine stain birthmarks. *Arch. Facial Plast. Surg.*, **7** (5), 287–294.

3 Papenfuss, H.D., Gross, J.F., Intaglietta, M., and Treese, F.A. (1979) Transparent access chamber for the rat dorsal skin fold. *Microvasc. Res.*, **18** (3), 311–318.

4 Babilas, P., Shafirstein, G., Baumler, W., Baier, J., Landthaler, M., Szeimies, R.M., and Abels, C. (2005) Selective photothermolysis of blood vessels following flashlamp-pumped pulsed dye laser irradiation: *in vivo* results and mathematical modelling are in agreement. *J. Invest. Dermatol.*, **125** (2), 343–352.

5 Heger, M., Beek, J.F., Moldovan, N.I., van der Horst, C.M.A.M., and van Gemert, M.J.C. (2005) Towards optimization of selective photothermolysis: prothrombotic pharmaceutical agents as potential adjuvants in laser treatment of port wine stains – a theoretical study. *Thromb. Haemost.*, **93** (2), 242–256.

6 Choi, B., Jia, W.C., Channual, J., Kelly, K.M., and Lotfi, J. (2008) The importance of long-term monitoring to evaluate the microvascular response to light-based therapies. *J. Invest. Dermatol.*, **128** (2), 485–488.

7 Barton, J.K., Hammer, D.X., Pfefer, T.J., Lund, D.J., Stuck, B.E., and Welch, A.J. (1999) Simultaneous irradiation and imaging of blood vessels during pulsed laser delivery. *Lasers Surg. Med.*, **24** (3), 236–243.

8 Channual, J., Choi, B., Pattanachinda, D., Lotfi, J., and Kelly, K.M. (2008) Long-term vascular effects of photodynamic and pulse dye laser therapy protocols. *Lasers Surg. Med.*, **40**, 644–650.

9 Phung, T.L., Oble, D.A., Jia, W., Benjamin, L.E., Mihm, M.C., and Nelson, J.S. (2008) Can the wound healing response of human skin be modulated after laser treatment and the effects of exposure extended? Implications on the combined use of the pulsed dye laser and a topical angiogenesis inhibitor for treatment of port wine stain birthmarks. *Lasers Surg. Med.*, **40** (1), 1–5.

10 Jia, W.C., Sun, V., Tran, N., Choi, B., Liu, S.W., Mihm, M.C.J., Phung, T.L., and Nelson, J.S. (2010) Long-term blood vessel removal with combined laser and topical rapamycin antiangiogenic therapy: implications for effective port wine stain treatment. *Lasers Surg. Med.*, **42** (2), 105–112.

11 Loewe, R., Oble, D.A., Valero, T., Zukerberg, L., Mihm, M.C., and Nelson, J.S. (2010) Stem cell marker upregulation in normal cutaneous vessels following pulsed-dye laser exposure and its abrogation by concurrent rapamycin administration: implications for treatment of port-wine stain birthmarks. *J. Cutan. Pathol.*, **37**, 76–82.

12 Pinheiro, A.L.B. (2009) Advances and perspectives on tissue repair and healing. *Photomed. Laser Surg.*, **27** (6), 833–836.

13 Nussbaum, E.L., Mazzulli, T., Pritzker, K.P.H., Heras, F.L., Jing, F., and Lilge, L. (2009) Effects of low intensity laser irradiation during healing of skin lesions in the rat. *Lasers Surg. Med.*, **41** (5), 372–381.

14 Fulop, A.M., Dhimmer, S., Deluca, J.R., Johanson, D.D., Lenz, R.V., Patel, K.B., Douris, P.C., and Enwemeka, C.S. (2009) A meta-analysis of the efficacy of phototherapy in tissue repair. *Photomed. Laser Surg.*, **27** (5), 695–702.

15 Silva Santos, N.R., dos Santos, J.N., Macedo Sobrinho, J.B., Ramalho, L.M.P., Carvalho, C.M., Soares, L.G.P., and Pinheiro, A.L.B. (2010) Effects of laser photobiomodulation on cutaneous wounds treated with mitomycin C: a histomorphometric and histological study in a rodent model. *Photomed. Laser Surg.*, **28** (1), 81–90.

16 Anders, J.J. (2009) The potential of light therapy for central nervous system injury and disease. *Photomed. Laser Surg.*, **27** (3), 379–380.

17 Zivin, J.A., Albers, G.W., Bornstein, N., Chippendale, T., Dahlof, B., Devlin, T., Fisher, M., Hacke, W., Holt, W., Ilic, S., Kasner, S., Lew, R., Nash, M., Perez, J., Rymer, M., Schellinger, P., Schneider, D., Schwab, S., Veltkamp, R., Walker, M., Streeter, J., and NeuroThera Effectiveness and Safety Trial-2 Investigators (2009) Effectiveness and safety of transcranial laser therapy for acute ischemic stroke. *Stroke*, **40** (4), 1359–1364.

18 Lapchak, P.A. and De Taboada, L. (2010) Transcranial near infrared laser treatment (NILT) increases cortical adenosine-5′-triphosphate (ATP) content following embolic strokes in rabbits. *Brain Res.*, **1306**, 100–105.

19 Holden, P.K., Li, C., Da Costa, V., Sun, C.H., Bryant, S.V., Gardiner, D.M., and Wong, B.J.F. (2009) The effects of laser irradiation of cartilage on chondrocyte gene expression and the collagen matrix. *Lasers Surg. Med.*, **41** (7), 487–491.

20 Zhuo, S.M., Chen, JX., Xie, S.S., Fan, L., Zheng, L.Q., Zhu, X.Q., and Jiang, X.S. (2010) Monitoring dermal wound healing after mesenchymal stem cell transplantation using nonlinear optical microscopy. *Tissue Eng. Part C Methods*, **16** (5), 1107–1110.

Part 13
Gynecology and Obstetrics

58
Diagnosis of Neoplastic Processes in the Uterine Cervix

Natalia Shakhova, Irina Kuznetsova, Ekaterina Yunusova, and Elena Kiseleva

58.1
Introduction

Malignant neoplasms of the female reproductive organs, including cervical cancer, are among the principal causes of death: in 2005, about 500 000 cases of cervical cancer and 260 000 related deaths were reported worldwide [1]. A significant increase in the incidence of cervical cancer is observed for women in the early reproductive age group (by 2% yearly on average) [2]. Detection and treatment of precancerous abnormalities and early-stage cervical cancer can prevent up to 80% of invasive diseases [1]. This situation requires accurate diagnostic techniques providing effective management of women with abnormal screening tests.

The paradigm of colposcopic examination after primary screening (Pap and HPV tests) is still considered the standard for the evaluation of cervical neoplasia. The aim of colposcopy is to identify disease, to obtain representative specimens for histological verification, and to direct patient management. Unfortunately, recent research shows that colposcopic grade and histologic lesion grade are not closely correlated [3]. The sensitivity of colposcopy-guided biopsy for precancerous abnormalities [cervical intraepithelial neoplasia grade 2+ (CIN2+)] was estimated as only 54%, and for individual physicians this statistical value varies between 28.6 and 81.5% [4, 5]. This is connected with significant errors in colposcopic assessment, subjectivism, and a considerable dependence on the experience of clinicians. However, even when performed by trained and experienced personnel, colposcopic biopsy misses 26–42% of prevalent CIN2+ [6, 7]. The only compensating strategy for timely and accurate diagnosis of early cervical cancer and high-grade premalignant lesion is considered to be morphologic examination of specimens obtained by a large excisional procedure or multiple random biopsies [7, 8]. Such an approach can induce complications due to invasiveness and increase costs and time. In our opinion, complementary application of noninvasive technologies [e.g., fluorescence imaging, high-frequency ultrasound, optical coherence tomography (OCT)] in colposcopy is appropriate for adequate management of Pap and HPV positive women. It allows one to ensure adequate diagnosis and, at the same time, to avoid over-treatment and over-expense.

The goal of this chapter is to demonstrate the efficacy of OCT in the management of cervical neoplasia (early cervical cancer and premalignant abnormalities).

58.2
Materials and Methods

This research was conducted in the Nizhny Novgorod Regional Hospital (Russia). The study was approved by the Ethical Committee for scientific studies with human subjects; written informed consent was obtained from all patients. A total of more than 500 female patients with different cervical conditions have been examined with OCT. The inclusion criterion was indication for colposcopy and the exclusion criterion was age under 18 years.

The standard equipment for colposcopy and methodology with 5% acetic acid local application (traditional colposcopic test) were used. OCT examinations were conducted with a time-domain OCT device (Institute of Applied Physics, Russian Academy of Sciences, Russia). The system has the following technical characteristics: central wavelength, 1300 nm; radiation power, not exceeding 6.0 mW; spatial resolution, 15–20 µm; in-depth scanning range, up to 2 mm; lateral scanning range, 1–2 mm; and acquisition time for two-dimensional 200 × 200 pixel image, 1.5–2 s. The OCT system for cervical inspection performs using a detachable forward-looking OCT probe with an outer diameter of 2.7 mm. The probe is applied to the cervical area of interest under colposcope control. To reduce artifacts related to unintentional movements of the probe and the studied object, the distal end of the probe has to be in contact with the tissue. The images obtained are interpreted directly during the patient examination.

In the previous stages, the initial approval of OCT for the detection of cervical pathology and adaptation of OCT for the basic colposcopic procedure were carried out. Criteria for OCT detection of cervical neoplasia were elaborated and estimated in a blind recognition test. The sensitivity was shown to be 82%, specificity 78%, and error 19.2%, and the kappa value equals 0.65 [9].

At the present stage of clinical study, 135 female patients of reproductive age (mean age 32 ± 1.3 years) have been examined after primary cervical screening using OCT–colposcopy. Analysis of results was based on morphologic verification (a group of 100 patients) and 3 years of follow-up (a group of 35 patients).

58.3
Results

The suggested approach for increasing the efficacy of cervical cancer and precancer diagnosis is based on the complementary use of colposcopy and OCT, and the method includes the following steps:

- colposcopy with acetic acid test;
- determination of atypical colposcopic findings (ACFs);

- OCT examination of zones with ACFs;
- determination of OCT indications for biopsy;
- biopsy based on OCT indications;
- management of patient depending on histological results.

In previous studies, we distinguished three classes of OCT images: "benign," "suspicious," and "malignant."

We identify three types of images as being in the "benign" class (Figure 58.1). One is two-layer structure with different optical properties of the layers (a moderately scattering upper layer and a strongly scattering lower layer) and a highly contrasted boundary between layers (Figure 58.1a). The second type represents images with a disturbed structure of layers or without layer stratification identified, but with low-scattering areas with high-contrast boundaries (Figure 58.1b). Keeping the imaging depth at about 1.5 mm is an important indicator of "benign"-type images. The third type includes non-structured images but demonstrating a high signal to a depth of 1.5 mm (Figure 58.1c).

The "malignant" class includes two types of images (Figure 58.2). The most typical indicator of "malignant" class is a non-structured image with a large decrease in signal resulting in a low imaging depth (usually 0.5–1 mm) (Figure 58.2a). The other type, which is observed much more rarely than the first type, is a two-layered structure where each layer demonstrates high scattering and a substantial decrease in the signal (Figure 58.2b). The images with a disturbed structure which cannot be clearly identified as "benign" or "malignant" are classified as "suspicious."

Our method of using OCT in cervical diagnostics required revision of the approaches for image interpretation. At present we distinguish two groups of images: "biopsy indication" and "no biopsy indication." In our opinion, the biopsy must be performed for cervical sites characterized by "malignant" or "suspicious" OCT image types.

Approval of this approach was based on 100 clinical cases (randomized group), which showed a high efficiency of OCT–colposcopy. All OCT images obtained in this approval process were verified morphologically by biopsy performed based on colposcopic data according to standard indications. It was shown that the colposcopic data gave an indication for biopsy in 83% of cases whereas cervical neoplasia (CIN2 + and cervical microcarcinoma) was diagnosed in only 37% of cases. OCT images with "biopsy indication" were obtained in only 36% of cases. According to our study, the

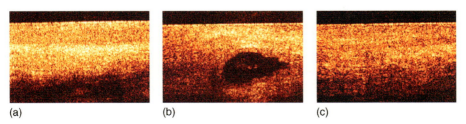

(a) (b) (c)

Figure 58.1 Examples of "benign" types of OCT images. All data have been morphologically verified.

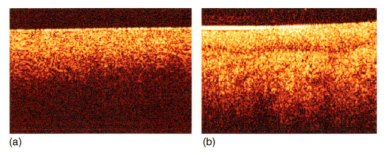

Figure 58.2 Examples of "malignant" types of OCT images. All data have been morphologically verified.

complementary performance of OCT and colposcopy allows the specificity of diagnostics of early-stage cervical cancer to be increased significantly. Concerning the sensitivity of the method, OCT demonstrated "benign" types of images in 1% of cases corresponding morphologically to an in-between state (CIN2) which did not allow us to replace totally the biopsy with OCT inspection in the final protocol. In our opinion, refusal of a biopsy procedure should be accompanied by an OCT follow-up.

The final protocol for early-stage neoplasia (OCT–colposcopy and OCT–follow-up) was performed with the main group of 35 patients of reproductive age. The total follow-up duration was 3 years. The OCT–colposcopy procedures were performed every 6 months. The control group included 20 patients. It was found that in the control group the biopsy was performed for 13 out of 20 patients; in seven cases repeated biopsy was required, and finally cervical neoplasia was diagnosed in two patients. In the main group, indications for biopsy were found for 11 patients out of 35, and indications for repeated biopsy were absent; cervical cancer was diagnosed in six patients. Hence over-biopsy in the control group was performed in 25 cases compared with five cases in the main group. It is worth mentioning that during the 3 year follow-up, no cases of neoplastic changes in the cervix were diagnosed in patients with OCT-based biopsy refusal.

58.4
Discussion and Conclusion

Colposcopy-directed biopsy remains the "gold standard" for cervical cancer diagnosis. However, the shortcomings of colposcopy require the development of new techniques, and OCT is a promising candidate. Improvement of the specificity, but not the sensitivity, by adding OCT to colposcopy was demonstrated in our study. Our results are in agreement with those of other researchers [10].

In spite of some limitations, OCT–colposcopy can be successfully applied in cervical cancer diagnosis. This technique is very helpful in female patients of reproductive age, particularly in cases of questionable interpretation of colposcopic data and limitations of or contraindications to biopsy. For example, 1% of cervical cancers are detected during pregnancy and there are an increasing number of premalignant lesions in these cases for which diagnosis and evaluation are not easy. Colposcopy in pregnancy does not differ from that in the non-pregnant state, but as the pregnancy advances, interpretation of colposcopy becomes progressively more difficult [10]. OCT could be the method of choice in pregnancy. Another reasonable category for adding OCT to colposcopy is aged patients, as they have a high risk for biopsy complications.

Unfortunately, OCT in some way "suffers" from subjectivism like colposcopy. To minimize this factor, the development of mathematical algorithms for OCT image interpretation is suggested and epithelial brightness has been shown to be a statistically significant distinguishing feature of cervical OCT images [11]. The sensitivity of OCT–colposcopy could be improved by increasing the scanning rate, using a newly designed probe with a larger diameter, and so on [12].

In conclusion:

- OCT is an attractive technique for the detection of cervical neoplasia.
- It should be applied according to relevant indications and in an appropriate scenario.
- Further improvements of the method are required.

Acknowledgments

The Russian Academy of Sciences (Fundamental Sciences for Medicine), the Russian Foundation for Basic Research (RFBR) (07-08-00803, 08-02-99049), and the Science and Innovations Federal Russian Agency (2007-2-2.2-04-01-012) are thanked for financial support.

References

1 World Health Organization (2009) Human papillomavirus vaccines – WHO position paper. *WHO Weekly Epidemiol. Rec.*, **84** (15), 117–132.

2 Ronco, G. and Rossi, P. (2008) New paradigms in cervical cancer prevention: opportunities and risks. *BMC Women's Health*, **8**, 23.

3 Ferris, D.G., Litaker, M.S., and ALTS Group (2006) Prediction of cervical histologic results using an abbreviated reid colposcopic index during ALTS. *Am. J. Obstet. Gynecol.*, **194**, 704–710.

4 The ASCUS-LSIL Triage Study (ALTS) Group (2003) Results of a randomized trial on the management of cytology interpretations of atypical squamous cells of undetermined significance. *Am. J. Obstet. Gynecol.*, **188**, 1383–1392.

5 Pretorius, R.G., Belinson, J.L., and Qiao, Y.L. (2010) Regardless of colposcopic skill, performing more biopsies increases the yield of CIN 3 or cancer. Presented at the Eurogin 2010 Congress. Cervical Cancer Prevention: 20 Years of Progress and a Path to the Future. February 17–20, Monte Carlo, Monaco.

6 Massad, L.S. and Collins, Y.C. (2003) Strength of correlations between colposcopic impression and biopsy histology. *Gynecol. Oncol.*, **89**, 424–428.

7 Stoler, M. (2010) Accuracy and limitations of colposcopic performance. Presented at the Eurogin 2010 Congress. Cervical Cancer Prevention: 20 Years of Progress and a Path to the Future. February 17–20, Monte Carlo, Monaco.

8 Gage, J.C., Hanson, V.W., Abbey, K., Dippery, S. *et al.* (2006) Number of cervical biopsies and sensitivity of colposcopy. *Obstet. Gynecol.*, **108** (2), 264–272.

9 Zagaynova, E.V., Gladkova, N.D., Shakhova, N.M., Streltsova, O.S., *et al.* (2011) Interoperative OCT monitoring, in *Handbook of Biophotonics,* vol. **1** (ed. J. Popp), Wiley-VCH Verlag GmbH, Weinheim.

10 Diakomanolis, S.E. (2010) HPV associated diseases in pregnancy. Presented at the Eurogin 2010 Congress. Cervical Cancer Prevention: 20 Years of Progress and a Path to the Future. February 17–20, Monte Carlo, Monaco.

11 Belinson, S., Belinson, J., Ledford, K., Na, W., *et al.* (2010) Cervical epithelial brightness by optical coherence tomography (OCT) distinguishes grades of dysplasia with statistical significance. Presented at the Eurogin 2010 Congress. Cervical Cancer Prevention: 20 Years of Progress and a Path to the Future. February 17–20, Monte Carlo, Monaco.

12 Liu, Z., Belinson, S.E., Li, J., Yang, B. *et al.* (2010) Diagnostic efficacy of real-time optical coherence tomography in the management of preinvasive and invasive neoplasia of the uterine cervix. *Int. J. Gynecol. Cancer,* **20** (2), 283–287.

Part 14
Reproductive Medicine

59
The Use of Single-Point and Ring-Shaped Laser Traps to Study Sperm Motility and Energetics

Linda Z. Shi, Bing Shao, Jaclyn Nascimento, and Michael W. Berns

59.1
Introduction

Single-point laser trapping is a noninvasive biophysical tool which has been widely applied to study physiologic and biomechanical properties of cells [1, 2]. Over the past three decades, researchers have used laser trapping (tweezers) to quantify sperm motility by measuring swimming forces [3–7]. These studies determined that the minimum amount of laser power needed to hold the sperm in the trap (or the threshold escape power) is directly proportional to the sperm's swimming force according to the equation $F = QP/c$, where F is the swimming force, P is the laser power, c is the speed of light in the medium, and Q is the geometrically determined trapping efficiency parameter [3]. Sperm swimming force measurements have been used to evaluate sperm viability after cryopreservation [5], as a way to evaluate the motility of epididymal versus ejaculated sperm [4], and to examine infertility in human patients [5]. In addition, a relationship between sperm velocity and swimming force (escape power from trap) was found for human and dog sperm [7, 8].

59.2
Single-Point Laser Trap

The more common single-point gradient trap that is used for sperm analysis is shown in Figure 59.1. An Nd:YVO$_4$ laser operating at 1064 nm is coupled to a Zeiss Axiovert S100 microscope equipped with a phase III, 40×, NA 1.3, oil immersion objective. The laser power in the specimen plane is attenuated by rotating an optical polarizer mounted in a stepper-motor-controlled rotating mount. Two dual video adapters are used to bring the laser into the microscope and simultaneously image the sperm with phase contrast and fluorescence microscopy. The laser beam enters the side port of the first dual video adapter and is transmitted to the microscope. A filter is used to prevent back-reflections of infrared laser light from exiting the top port of the adapter and allow reflected visible light coming from the specimen

Handbook of Biophotonics. Vol.2: Photonics for Health Care, First Edition. Edited by Jürgen Popp, Valery V. Tuchin, Arthur Chiou, and Stefan Heinemann.
© 2012 Wiley-VCH Verlag GmbH & Co. KGaA. Published 2012 by Wiley-VCH Verlag GmbH & Co. KGaA.

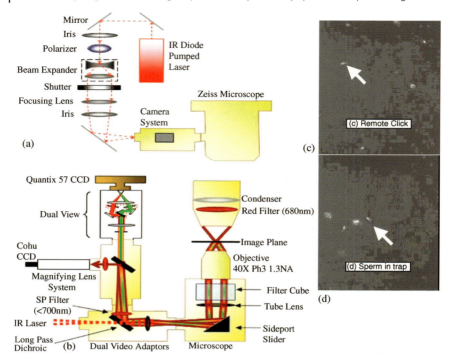

Figure 59.1 (a) Optical schematic showing the components used to generate and control the single-point laser tweezers; (b) imaging setup showing illumination sources, filters, and cameras used to image the sperm in both phase contrast and fluorescence; (c) a collaborator from Australia clicked on a selected swimming dog sperm on the image screen in Australia, which then activated RATTS in the California laboratory, which tracked the sperm for 5 s, and then held it in the trap for 10 s (d).

to pass to the second video adapter. The specimen is viewed in phase contrast using red light from the halogen lamp and viewed in fluorescence mode using the arc lamp as the excitation source. The fluorescence filter cube contains an HQ 500/20 nm excitation filter and a dichroic beam splitter with a 505 nm cut-off wavelength. The second dual video adapter attached to the top port of the first video adapter uses a filter cube to separate the phase information (reflects >670 nm) from the fluorescence (transmits 500–670 nm). The phase contrast images are filtered through a filter and acquired by a charge-coupled device (CCD) camera (operating at 40 frames per second) coupled to a variable zoom lens system (0.33–1.6×) to increase the field of view. For the fluorescent images, a dual-view system splits the red and green specimen fluorescence emitted resulting in an image in each color. Fluorescent emission filters are placed in this emission-splitting system (green fluorescence emitter, HQ 535/40 nm M; red fluorescence emitter, HQ 605/50 nm M). The dual-view system is coupled to a digital camera that captures the images [9].

An integrated system has been developed to analyze individual sperm motility and energetics automatically. A two-level real-time automated tracking and trapping system (RATTS) has been developed to quantify the motility and energetics of sperm using real-time tracking (done by the upper-level system) and fluorescent ratio imaging (done by the lower-level system). The communication between these two systems is achieved by a gigabit network. The RATTS system can be operated remotely through the Internet using the logmein website. Individual sperm can be automatically trapped using laser tweezers to measure sperm swimming forces during real-time tracking. Once the sperm is stably trapped, the custom-built program can gradually reduce the laser power until the sperm is capable of escaping the trap. This escape power can be converted to the actual amount of swimming force needed to escape from the trap. By taking the ratio of the wavelengths of a mitochondrial membrane dye, mitochondrial energetics can be monitored and correlated with both sperm escape power (converted to swimming force in piconewtons by $F = QP/c$) and sperm swimming speed. The latter two parameters are linearly correlated. This two-level system to study individual sperm motility and energetics has not only increased experimental throughput over single-level systems by an order of magnitude, but also allows monitoring of sperm energetics prior to and after exposure to the laser trap [10].

59.3 Ring-Shaped Laser Traps

The experimental setup for a ring-shaped laser trap system is presented in Figure 59.2. The light beam from a continuous-wave ytterbium fiber laser (1070 nm wavelength) is collimated and expanded via the 3× beam expander. For better performance of the trap, a refractive beam shaper is used to convert a Gaussian laser beam into a collimated flat-top beam. A telescope lens pair shrinks the shaped beam so that the thickness of the light cone input to the objective is equal to the diameter

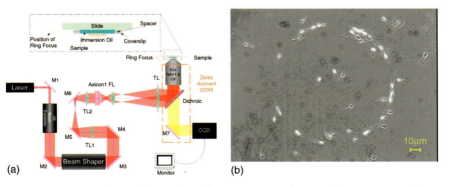

Figure 59.2 (a) Optical setup for ring-shaped laser trap system. (b) Tens of sperm are stopped or moving along the ring focus.

of the back aperture, hence the numerical aperture of the trapping beam is maximized. The laser beam is divided and bent towards the optical axis at an angle upon the Axicon lens, $\beta = \arcsin(n \sin \gamma) - \gamma$, where γ is the base angle and n is the refractive index of the Axicon. At the back focal plane of the focusing lens (FL), a ring image is formed that is conjugate to the ring focus at the specimen plane. After the tube lens (TL), the laser is sent into the microscope and to the objective by the dichroic mirror.

A continuous 3D ring-shaped laser trap can be used for multi-level and high-throughput (tens to hundreds of sperm) sperm sorting and analysis. One potential advantage of the ring trap is that it acts as a force shield for protecting the sperm that are being measured from interference of other sperm. Additionally, the ring-shaped laser trap could potentially be used for parallel sperm sorting based on motility that may be stimulated by chemical agents (chemotaxis). This is a critical feature of sperm in response to the diffusion gradient of chemicals released by the egg and surrounding cells of the cumulus oophorus, which may lead to a better understanding of infertility and provide new approaches for contraception [11]. Differentiating it from the single-spot laser trap, which focuses hundreds of milliwatts on a sperm to achieve trapping, the ring trap utilizes only tens of milliwatts, which allow the sperm to swim along the ring without stopping. As a result, the effect of the optical force on sperm swimming patterns and physiology could be investigated in more detail [2, 12].

59.4
Biological Studies

59.4.1
Sperm Competition

An important and controversial question in evolution is sperm competition. This can be studied in primates using a laser trap in combination with custom-designed computer tracking and trapping algorithms. Semen samples from chimpanzee, rhesus macaque, human, and gorilla were analyzed. The mating systems for both the chimpanzee and the rhesus macaque are multi-male, multi-female (females of these species mate with more than one male within a short period of time). Gorillas are polygynous, defined here as one male, multiple females – that is, the single dominant male mates with the females in the "harem" unit [13] and therefore, from the female point of view, the gorilla is strictly monogamous. Human mating patterns are variable, differing across cultures, but can be considered to be predominantly polygynous (83% of societies), more rarely monogamous (16%), and only very occasionally polyandrous (<1% [13]). Therefore, the sperm analyzed in this study came from primates that represent a variety of mating patterns, ranging from strictly polygynous (gorilla) to multi-male, multi-female (chimpanzee and rhesus macaque).

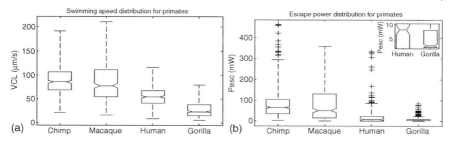

Figure 59.3 Box plots of the distributions of (a) swimming speed (VCL) and (b) escape force (Pesc) for all four primates. Inset in (b) shows an expanded view of human and gorilla distributions to emphasize the difference in median values.

The box plots for distributions of sperm swimming speed (VCL, Figure 59.3a) and escape power (Pesc, indicative of force, Figure 59.3b) for the four primates are shown. Pesc is measured by reducing the laser power after the sperm is trapped until the sperm is capable of escaping the trap. Each species' swimming speed and escape power distributions are statistically different ($p < 0.05$) using the Wilcoxon rank sum test for equal medians. The medians of both measurements, VCL and Pesc, show that rhesus and chimpanzee sperm swim with the fastest speeds and strongest forces, whereas gorilla sperm swim with the slowest speeds and weakest forces. Human sperm swimming speeds and forces lie between these two extremes [14].

59.4.2
Sperm Energetics

The combination of single-point laser tweezers with custom computer tracking software and robotics can be used to analyze motility (speed and force) and energetics [mitochondrial membrane potential (MMP)] of individual sperm. Domestic dog sperm are labeled with a cationic fluorescent probe, $DiOC_2(3)$, that reports the MMP across the inner membranes of the mitochondria located in the sperm's midpiece. Individual sperm are tracked to calculate VCL. The MMP is measured every second over a 5 s interval during the tracking phase (sperm are swimming freely) and continuously during the trapping phase. Figure 59.4 shows the ratio value prior to trapping, during trapping, and after trapping plotted over time for two different sperm. For the sperm in Figure 59.4a, there was an overall decline in ratio value over time as the sperm was held in the trap. Once released from the optical trap, the ratio value increased. However, within 5 s, it did not fully recover to the original value that it had prior to being trapped. Similarly, the sperm swimming speed, VCL, did not recover to its pretrapping value. For the sperm in Figure 59.4b, there was a slight decrease in ratio value while the sperm was in the trap. Again, neither swimming speed nor ratio value fully recovered [15].

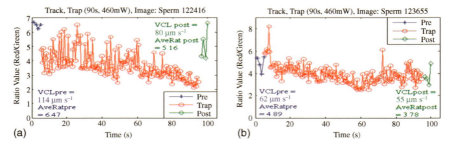

Figure 59.4 Track, trap, and fluorescence of two randomly selected sperm (a) and (b). The ratio value of red/green is plotted against time for the three different phases: prior to, during, and after trapping. Both the average ratio and VCL prior and after trapping are inset in the figures.

59.5 Conclusion

We have described two methods of optical trapping, the more common single-point gradient trap and the ring-shaped laser trap. The single-point gradient trap has found wide application in cell biology, in particular for studies on sperm motility, energetics, evolution (sperm competition), and human infertility. The ring-shaped laser trap is a more recent development, and shows promise for sperm sorting and analytics because of the potential for greater throughput than trapping of single sperm in a point trap.

Acknowledgment

This work was supported by a grant from the Lucille and David Packard Foundation endowment gift to the Beckman Laser Institute Inc. Foundation.

References

1. Berns, M.W. (1998) Laser scissors and tweezers. *Sci. Am. (Int. Ed.)*, **278**, 52–57.
2. Shao, B., Zlatanovic, S., and Esener, S.C. (2004) Microscope-integrated micromanipulation based on multiple VCSEL traps. *Proc. SPIE*, **5514**, 62–72.
3. Konig, K., Svaasand, L., Liu, Y., Sonek, G., Patrizio, P., Tadir, Y., Berns, M.W., and Tromberg, B.J. (1996) Determination of motility forces of human spermatozoa using an 800 nm optical trap. *Cell. Mol. Biol. (Noisy-le-Grand)*, **42**, 501–509.
4. Araujo E. Jr., Tadir, Y., Patrizio, P., Ord, T., Silber, S., Berns, M.W., and Asch, R.H. (1994) Relative force of human epididymal sperm. *Fertil. Steril.*, **62**, 585–590.
5. Dantas, Z.N., Araujo E. Jr., Tadir, Y., Berns, M.W., Schell, M.J., and Stone, S.C. (1995) Evect of freezing on the relative escape force of sperm as measured by a laser optical trap. *Fertil. Steril.*, **63**, 185–188.
6. Patrizio, P., Liu, Y., Sonek, G.J., Berns, M.W., and Tadir, Y. (2000) Evect of pentoxifylline on the intrinsic swimming forces of human sperm assessed by optical tweezers. *J. Androl.*, **21**, 753–756.

7 Tadir, Y., Wright, W.H., Vafa, O., Ord, T., Asch, R.H., and Berns, M.W. (1990) Force generated by human sperm correlated to velocity and determined using a laser generated optical trap. *Fertil. Steril.*, **53**, 944–947.

8 Nascimento, J., Botvinick, E., Shi, L., Durrant, B., and Berns, M.W. (2006) Analysis of sperm motility using optical tweezers. *J. Biomed. Opt.*, **11** (4), 044001.

9 Nascimento, J.M., Shi, L.Z., Tam, J., Chandsawangbhuwana, C., Durrant, B., Botvinick, E.L., and Berns, M.W. (2008) Comparison of glycolysis and oxidative phosphorylation as energy sources for mammalian sperm motility, using the combination of fluorescence imaging, laser tweezers, and real-time automated tracking and trapping. *J. Cell. Physiol.*, **217**, 745–751.

10 Shi, L., Nascimento, J., Chandsawangbhuwana, C., Botvinic, E.L., and Berns, M.W. (2008) An automatic system to study sperm motility and energetics. *Biomed. Microdevices*, **10** (4), 573–583.

11 Eisenbach, M. and Tur-Kaspa, I. (1999) Do human eggs attract spermatozoa? *BioEssays*, **21**, 203–210.

12 Shi, L., Shao, B., Chen, T., and Berns, M.W. (2009) Automatic annular laser trapping: a system for high-throughput sperm analysis and sorting. *J. Biophotonics*, **2** (3), 167–177.

13 Dixson, A.F. (1998) *Primate Sexuality: Comparative Studies of the Prosimians, Monkeys, Apes, and Human Beings*, Oxford University Press, Oxford.

14 Nascimento, J., Shi, L., Meyers, S., Gagneux, P., Loskutoff, N., Botvinick, E.L., and Berns, M.W. (2008) The use of optical tweezers to study sperm competition and motility in primates. *J. R. Soc. Interface*, **5** (20), 297–302.

15 Nascimento, J., Shi, L., Chandsawangbhuwana, C., Tam, J., Botvinic, E.L., and Berns, M.W. (2008) Use of laser tweezers to analyze sperm motility and mitochondrial membrane potential. *J. Biomed. Opt.*, **13** (1), 014002.

60
Laser Thinning of the Zona Pellucida
Hanna Balakier

60.1
The Zona Pellucida

The zona pellucida (ZP) is a specialized extracellular matrix that surrounds mammalian oocytes and early-stage embryos [1]. It is composed of a network of cross-linked filaments usually consisting of three or four highly conserved, zona-specific glycoproteins. The ZP is multifunctional and it plays a crucial role in oogenesis, fertilization, and preimplantation development. The ZP is produced during the early development of the ovarian follicles and it increases in thickness as oocytes increase in diameter [1, 2]. At the blastocyst stage, the ZP is no longer essential and the embryo must hatch out of its coat in order to implant in the uterine endometrium [3, 4]. Any disturbance of this process may cause implantation failure, leading to infertility.

Increased ZP thickness, deficiency in the production of embryonic enzymatic lysins which thin the ZP during embryo development, and the phenomenon of zona hardening are all possible causes of unsuccessful hatching and reduced implantation rates in human IVF (*in vitro* fertilization [3–5]) cycles. It seems that zona hardening may especially influence the implantation potential of the embryo, since although it occurs naturally at fertilization or in aging oocytes, it is increased further by suboptimal *in vitro* culture conditions and oocyte/embryo cryopreservation.

Several experimental techniques have been used to assist embryo hatching in order to improve implantation and clinical pregnancy rates after IVF treatment. Using mechanical and chemical methods, piezo-vibrators, or lasers, ZP manipulations can be performed either by opening the ZP by creating a hole or by zona pellucida thinning (ZPT), reducing zona thickness without breaching it [3, 4]. Regardless of the assisted hatching (AH) technique used, the type of ZP handling (intact or fully-breached) may have important implications for the developmental fate of the embryo [6].

60.2
Laser Assisted Hatching

The introduction of laser techniques in the field of human-assisted reproduction has led to a variety of applications facilitating the efficient manipulation of the ZP and spermatozoa [7, 8]. These include procedures such as laser AH, biopsy of embryonic cells, laser-assisted intracytoplasmic sperm injection (ICSI), sperm viability testing, and the production of hemizone. The use of several types of lasers has been validated in animal and human clinical studies, and currently a noncontact 1.48 μm diode laser system is the simplest, most reproducible, and safest method available for use in IVF laboratories [9].

Laser AH is on course to supersede previous mechanical and chemical methods for opening or thinning the ZP. Of the two methods, laser ZPT appears to be less invasive and associated with higher clinical outcomes than ZP opening [4, 10]. This suggests that breaching of the ZP could be detrimental to the implantation potential of the embryo due to the loss of cells and blastocyst "herniation" through the hole, and it could deprive the embryo of protection by the ZP against infectious or immunologic attacks. Therefore, laser ZPT is more frequently utilized for AH of fresh or frozen embryos before their transfer to the uterus [7, 8].

In theory, the AH by artificial laser ZPT should increase both the hatching potential of the embryo and IVF clinical outcomes. However, the number of studies, especially randomized ones, is limited [6], and the existing data are controversial, showing decreased [11], similar [12], or increased [13–15] implantation/pregnancy rates in the laser ZPT group versus control group without AH. These conflicting results might arise from the study design, patient characteristics, sample size, and variability in the laser technique used, such as the length/depth of ZP thinning and laser pulse duration.

Although clinical indications and the impact of laser ZPT are not yet fully defined, this technique could be particularly beneficial for patients with repeated, unexplained implantation failures [14]. Some embryo implantation problems in these patients may be explained by the inability of the embryos to hatch, but it is unknown whether underlying defects are related to the structure or function of the ZP, the impairment of the production of embryonic/uterine lysins, or other factors.

As recently shown, laser ZPT can also be successfully applied to frozen–thawed human embryos by the slow freezing or vitrification method [13, 15]. It has also become apparent that increasing the area of laser thinning from a small ablation (30–40 μm) to one-quarter or half of the zona may considerably improve IVF outcomes [13–15]. Evaluation of the impact of the size of ZPT area, 25 versus 50%, on vitrified–warmed cleavage-stage embryos, clearly demonstrates that thinning half of the ZP results in much higher implantation/pregnancy rates (32/46.7%) compared with one-quarter of ZPT (16/25.0%) [15]. Interestingly, retrospective analysis of laser AH by the creation of small (40 μm) and large (50% of ZP) holes in the ZP at the blastocyst stage in frozen–thaw cycles suggests that the pregnancy, implantation, and delivery rates are better in the 50% opening group (74, 52, 65%, respectively) compared with the 40 μm opening group (43, 27, 38%) [15].

It is unclear why the ZPT facilitates embryo implantation, and why AH of half rather that one-quarter of the ZP generates better results. Is this simply a matter of the different patterns of embryo hatching, that it is more "natural" and easier to hatch via a wider area versus a smaller "8"-shaped neck in which the embryo can become trapped? Other questions are which ZP structural changes are important, and whether they are related to the immune recognition of the embryo in ways that might enhance implantation? Are such changes induced by the laser ZPT? It is currently hypothesized that the ZP can act as an intrinsic source of signals activating the maternal immune recognition of the developing mammalian embryo [16]. Moreover, blastocyst hatching in different species, including humans, appears to be regulated by trophectodermal projections, which most likely are involved in the delivery of embryo-derived zona lysins to the ZP, causing its lysis [5]. Does AH affect the secretion of embryonic lysins which thin the ZP? A study on mouse embryos has shown that neither ZP thinning nor opening by acid Tyrode's solution changes the secretion of trypsin-like proteases involved in blastocyst hatching [17]. Whether similar mechanisms operate in human embryos awaits further investigations. In summary, the information on the cellular/molecular mechanisms of blastocyst hatching and ZP lysis is scarce and these areas require more systematic and thorough study.

60.3
Laser-Assisted ICSI

Laser ZPT has been proposed as a method to improve the efficiency of routine ICSI procedures, simultaneously providing both a reduction of mechanical stress on the oocytes and AH. This approach may increase oocyte survival rate, embryo hatching, and possibly IVF clinical outcomes, but recent results are preliminary and need to be confirmed by a large randomized study. The size of the thinned area of the ZP prior to ICSI should also be carefully evaluated with regard to safe embryo hatching and avoidance of additional AH at later stages that might create multiple ZP herniations which are known to contribute to monozygotic twinning (MZT; embryo splits into genetically identical embryos). Similar laser ZPT has been adopted recently in animal experiments [7, 8]. This ZP treatment appears to be extremely beneficial, especially in mice, where regular ICSI results in very low survival rates. There is also a potential for future applications of ZPT in domestic and endangered species.

60.4
Safety Aspects of ZPT

The technical and clinical safety aspects of laser ZPT require further investigation. For example, ZPT over a larger area (25 and 50%) of the embryo involves a relatively high amount of energy, which may lead to penetration beyond the ZP, harming the embryo. To avoid adverse effects, proper laser calibration, short pulse duration, and safe working distance are recommended [18]. Evidently, safety also depends on the properties of the laser. Improvements in noncontact laser technology have achieved better accuracy and safety in ZPT [9]. The earliest laser used pulses as long as 15 ms, later 2–5 ms, and currently it is possible to apply pulses of only 300 μs [15]. It remains to be seen whether a superior laser system can be developed which will be more precise and generate minimal heat.

The recent promising results of laser ZPT need to be confirmed on a larger scale in randomized clinical trials. The other challenge will be to determine the impact of ZPT on several important clinical outcomes, including live birth rate, multiple pregnancies, MZT, congenital malformations, and chromosomal aberrations in children born from this technique [6]. So far, there are only three published follow-up reports on children born after laser ZP opening, and there has been no evidence of increased incidence of chromosomal aberrations or major congenital malformations from the laser treatment [4]. Similar data for laser ZPT are not available. MZT is a particularly important unsolved problem in IVF. The incidence of MZT in assisted conceptions is 2.25 times higher than in natural conceptions, and some techniques such as blastocyst transfer and ICSI seem to carry a higher risk of MZT than the others. The etiology and exact mechanisms leading to this phenomenon are controversial, and the literature provides multiple theories [19]. A combination of factors is likely to be involved, including ovarian stimulation, ICSI, AH, blastocyst culture, and ZP structural changes. Larger scale clinical studies on MZT from single embryo transfers and confirmation of zygosity with DNA analysis are warranted before definitive conclusions can be drawn. Further multidisciplinary studies are also required to define the impact of laser ZPT for optimizing IVF outcomes.

References

1 Wassarman, P.M. and Litscher, E.S. (2008) Mammalian fertilization: the egg's multifunctional zona pellucida. *Int. J. Dev. Biol.*, **52** (5–6), 665–676.

2 Gook, D.A., Edgar, D.H., Borg, J., and Martic, M. (2008) Detection of zona pellucida proteins during human folliculogenesis. *Hum. Reprod.*, **23** (2), 349–402.

3 De Vos, A. and Van Steirteghem, A. (2000) Zona hardening, zona drilling and assisted hatching: new achievements in assisted reproduction. *Cell Tissues Organs*, **166** (2), 220–227.

4 Al-Nuaim, L.A. and Jenkins, J.M. (2002) Assisted hatching in assisted reproduction. *Br J. Obstet. Gynecol.*, **109** (9), 856–862.

5 Seshagiri, P.B., Roy, S.S., Sireesha, G., and Rao, R.P. (2009) Cellular and molecular regulation of mammalian blastocyst hatching. *J. Reprod. Immunol.*, **83** (1–2), 79–84.

6 Das, S., Blake, D., Farquhar, C., and Seif, M.M.W. (2009) *Assisted Hatching on Assisted Conception (IVF and ICSI) Review. The Cochrane Collaboration*, John Wiley & Sons, Ltd., Chichester.

7 Ebner, T., Moser, M., and Tews, G. (2005) Possible application of a non-contact 1.48 μm wavelength diode laser in assisted reproduction technologies. *Hum. Reprod. Update*, **11** (4), 425–435.

8 Montag, M.H.M., Klose, R., Koster, M., Rosing, B., Van der Ven, K., Rink, K., and Van der Ven, H. (2009) Application of non-contact laser technology in assisted

reproduction. *Med. Laser Appl.*, **24** (1), 57–64.

9. Tadir, Y. and Douglas-Hamilton, D.H. (2007) *Laser Effects in the Manipulation of Human Eggs and Embryos for In Vitro Fertilization, Methods in Cell Biology*, vol. 82 (eds M.W. Berns and K.O. Greulich,), Elsevier, Amsterdam, pp. 409–431.

10. Ghobara, T. (2006) Effects of assisted hatching method and age on implantation rates of IVF and ICSI. *RMB Online*, **13** (2), 261–267.

11. Valojerdi, M.R., Eftekhari-Yazdi, P., Karimian, L., Hassani, F., and Movaghar, B. (2010) Effect of laser zona thinning on vitrified–warmed embryo transfer at the cleavage stage: a prospective, randomized study. *RMB Online*, **20** (2), 234–242.

12. Ciray, H.N., Bener, F., Karagene, L., Ulug, U., and Bahceci, M. (2005) Impact of assisted hatching on ART outcome in women with endometriosis. *Hum. Reprod.*, **30** (9), 2546–2549.

13. Ge, H. (2008) Impact of assisted hatching on fresh and frozen–thawed embryo transfer cycles: a prospective, randomized study. *RMB Online*, **16** (4), 689–596.

14. Debrock, S., Spiessens, C., Peeraer, K., and De Loecker, P. (2009) Higher implantation rate using modified quarter laser-assisted zona thinning in repeated implantation failure. *Gynecol. Obstet. Invest.*, **67** (2), 127–133.

15. Hiraoka, K., Hiraoka, K., Horiuchi, T., Kusuda, T., Okano, S., Kinutani, M., and Kinutani, K. (2009) Impact of the size of zona pellucida thinning area on vitrified-warmed cleavage- stage embryo transfers: a prospective, randomized study. *J. Assist. Reprod. Genet.*, **26** (9–10), 515–521.

16. Fujiwara, H., Ariaki, Y., and Toshimori, K. (2009) Is the zona pellucida an intrinsic source of signals activating maternal recognition of the developing mammalian embryo? *J. Reprod. Immunol.*, **81** (1), 1–8.

17. Hwang, S.S., Lee, E.Y., Chung, Y.C., Yoon, B.K., Lee, J.H., and Choi, D.S. (2000) Intactness of zona pellucida does not affect the secretion of a trypsin-like protease from mouse blastocyst. *J. Korean Med. Sci.*, **15** (5), 529–532.

18. Chatzimeletiou, K. (2005) Comparison of effects of zona drilling by non-contact infrared laser or acid Tyrode's on the development of human biopsied embryos as revealed by blastomere viability, cytoskeletal analysis and molecular cytogenetics. *RMB Online*, **11** (6), 697–710.

19. Vitthala, S., Gelbaya, T.A., Brison, D.R., Fitzgerald, C.T., and Nardo, L.G. (2009) The risk of monozygotic twins after assisted reproductive technology: a systematic review and meta-analysis. *Hum. Reprod. Update*, **15** (1), 45–55.

Part 15
Genetics

61
Transfection of Cardiac Cells by Means of Laser-Assisted Optoporation

Alena V. Nikolskaya, Vladimir P. Nikolski, and Igor R. Efimov

61.1
Introduction

The first optoporating lasers developed were in the ultraviolet (UV) range because of strong absorption by the membrane constituents in this spectral region [1, 2]. However, the use of UV light could irreversibly damage cells [1, 3, 4]. Alternatively, the use of near-infrared (800 nm) femtosecond and infrared (1064 nm) nanosecond lasers for efficient injection of impermeable dyes and gene transfection has been suggested [5, 6].

The use of visible laser irradiation for optoporation at 488 nm has also been demonstrated. It exploited the fact that the dye Phenol Red, a usual component of cell culture media, has strong absorption at this wavelength. It was hypothesized that laser absorption induced a local rise in temperature, which led to changes in membrane permeability that were considered reversible and relatively harmless to the irradiated cells [4, 7].

However, several cellular components have significant absorption at 488 nm and the possibility of deleterious effects still exists, such as generation of toxic reactive oxygen species (ROS) and irreversible mitochondrial permeability transition pore (mPTP) opening [8–10].

We therefore aimed to improve the laser-assisted optoporation techniques with several impermeable fluorochromes, which could facilitate deposition of energy at the cell membrane and limit damage to intracellular compartments.

61.2
Methods

61.2.1
Primary Cultures of Neonatal Rat Cardiac Cells

Primary neonatal cardiac cells were obtained from ventricles of 3–4-day-old Wistar rat pups (Harlan) using a modification of the method described by Fast and Cheek [11].

The small pieces of ventricle were minced with scissors and dissociated in Hank's balanced salt solution (HBSS, without Ca^{2+} and Mg^{2+}) (Invitrogen) containing trypsin (0.1%) (Boehringer) and pancreatin (60 µg ml^{-1}) (Sigma). The solution of dispersed cells was supplemented with neonatal calf serum (10%) (Boehringer) to stop enzyme activity, centrifuged, and resuspended in UltraCulture medium (BioWhittaker) supplemented with vitamin B_{12} (20 µg ml^{-1}) (Sigma), L-glutamine with penicillin–streptomycin (100 µg ml^{-1}) (Sigma), and bromodeoxyuridine (0.1 mM) (Sigma). The cell suspension was preplated in large culture flasks and incubated for 2 h to isolate the fibroblast-enriched fraction. The myocytes remaining in suspension were collected and plated at a density of 2.5×10^5 cells cm^{-2} on to 300 mm diameter glass-bottomed Petri dishes (EMS) precoated with a collagen, laminin, and fibronectin mixture and grown for 2–7 days before experimentation. The culture medium was exchanged the day after preparation and every second day thereafter. All animal protocols were approved by the Washington University Animal Studies Committee.

61.3
Experimental Design

For all experiments on laser-assisted optoporation, a standard Nikon C1 confocal microscope with a three laser launcher integrated with a fluorescent upright Nikon i80 microscope was used. Transfection and optoporation tests were performed at 21 °C on the microscope stage.

To estimate the size of the effected targeted spot, a glass slide was painted with a marker pen, allowed to dry, and placed in the microscope light path with a 0.45 NA 10× lens. Following focusing, the blue laser pulse was applied, causing the formation of a hole in the paint. The effective size of the hole was estimated to be less than 4 µm, which corresponds to 4× lateral optical resolution ($r = 1.22\lambda/2NA$). The selected areas of cells were located in premarked regions. Cells were selected for transfection using a custom graphic software interface, which allowed us to define the set of targeted positions with individual exposure times on the imaged cell area. This procedure allows desirable patterns to be drawn for beam exposure, making targeted transfection possible. To minimize cell damage while obtaining the initial navigation image and focusing on the area of interest, an He–Ne laser (633 nm, continuous wave, 6.75 mW nominal output power) attenuated 32 times was used in the DIC transmitted light mode.

The blue beam of the argon laser (488 nm, continuous wave, 17 mW nominal output power) was focused at the cell membrane. Laser wounding was done by aiming the beam at a selected position for 2–10 s, depending on the enhancing dye concentration. The total procedure lasted 2–5 min for 20–30 targeted cells. At the site of beam incidence, the cell membrane had modified permeability and allowed impermeable dye or plasmid vector present in the culture media to penetrate inside the cell.

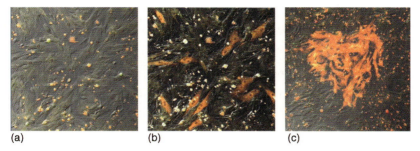

Figure 61.1 Laser-assisted injection of FM 1-43 into cultured cardiomyocytes. (a) Image of the area of interest with cells selected for optoporation in the presence of FM 1-43 in culture media; (b) the same area after optoporation; (c) drawing of specially shaped area.

61.4
Results

We investigated the use of 488 nm laser irradiation for dye-assisted microinjection of membrane-impermeable fluorochromes, such as propidium iodide (PI), and exogenous membrane dyes, such as FM 1-43, and also green fluorescent protein (GFP) encoding plasmid into neonatal cardiac cells. We also investigated the implementation of the same membrane dye, FM 1-43, for further absorption enhancement and perfecting focusing on the membrane surface with additional pharmaceutical treatments for apoptotic and necrotic damage prevention to improve transfection efficiency and ensure cell survival.

The exposure times for reversible membrane permeabilization were established by monitoring accumulation and retention of FM 1-43 (5 µg ml^{-1}) (Invitrogen), PI (20 µg ml^{-1}) (Sigma), and calcein-AM (10 µM) (Invitrogen). The short-term viability of exposed cells was evaluated 15–20 min after irradiation (Figures 61.1 and 61.2). PI membrane-impermeable cationic dye reveals strong red fluorescence after binding to nucleic acids. PI served two purposes: as an indicator of membrane permeabilization when added to the medium before irradiation, and as a vital stain when it was

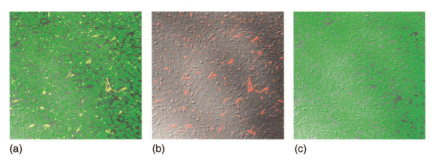

Figure 61.2 Cells optoporated in the presence of FM 1-43 and PI; calcein-AM was added after optoporation. (a) Dual-channel registration (PI, FM 1-43, and calcein) after 5 min of dye accumulation; (b) red channel (PI and FM 1-43) after 5 min; (c) green channel (calcein) after 10 min.

added to the medium 15–20 min after irradiation it distinguished dead from living cells. FM 1-43–styryl dye that has been widely used to demonstrate membrane disruption/resealing [12–14] was used to test membrane injury. This dye is impermeable to intact cell membranes, and specifically fluoresces only when incorporated into the lipid environment such as a cell membrane. The fluorescence of an optoporated cell increases several-fold (with changed emission spectra) owing to the presence of endo-membranes that become stained with the dye. Resealing of the membrane halts this increase in fluorescence. Calcein acetomethyl ester (calcein-AM) is widely used for cell survival assays. The viability of cells subjected to optoporation was assessed by incubating them with calcein-AM for 10 min. Accumulation and retention of this dye in irradiated cells indicate cell viability. For facilitating laser light absorption, in addition to Phenol Red (contained in media), cells were stained with various membrane dyes as further absorption enhancers. We selected FM 1-43 as a nontoxic dye the main absorption band of which corresponds to the laser wavelength. Membrane staining of cultured cells increased the optical density at the wavelength corresponding to the dye absorption spectrum. Moreover, as a membrane fluorescent stain, this dye allowed precise focusing on the membrane itself by means of z-stack positioning with maximum projection fluorescence intensity.

After a series of pilot experiments, exposure times of 10 s for 15 mg l^{-1} and 2 s for 40 mg l^{-1} of Phenol Red were selected for the control transfection procedures. These conditions correspond to the longest irradiation times that allowed transient optoporation of the cells but still did not produce detectable short-term injury to the cells.

The efficiency of optotransfection was assessed by monitoring expression of GFP in cardiac myocytes after laser-assisted optoporation in the presence of 5 µg ml^{-1} of plasmid containing the reporter gene for GFP from *Aequorea victoria* jellyfish.

In order to protect cells against long-term apoptotic damage that may be caused by mPTP irreversible opening or ROS accumulation generated through photodynamic treatment [15], cells were preincubated with cyclosporin A (0.2 µM) (Sigma) for 30 min before laser exposure and supplemented with the ROS scavengers, NO donors SIN-1 (125 µM) (Sigma), and SNAP (100 mM) (Invitrogen).

Before the experiment, regular culture medium was aspirated and substituted with fresh OPTI-MEM (Invitrogen) or M199 (Invitrogen) medium with HEPES buffer. OPTI-MEM was initially phenol free. The nominal concentration of the Phenol Red in M199 medium was 15 mg l^{-1}. In some experiments, culture media was additionally supplemented with Phenol Red solution (Sigma) to obtain the different concentrations (15, 20, 40 mg l^{-1}). In experiments with membrane-stained cells, the cells were incubated with FM 1-43 (5 µg ml^{-1}) (Invitrogen) for 10 min and unbound dye was then removed. GFP plasmids (5 µg ml^{-1}) were added to the culture dishes before irradiation. Assigned areas of individual cells were exposed to the incident laser beam, such that the beam was focused on their membrane.

Following optoporation, cells were incubated for 30 min with the GFP plasmids. After laser exposure, 1000× antioxidant supplement (1 µl ml^{-1}) (Sigma), CaCl$_2$ (2 mM) (Sigma), and surfactant Pluronic-F68 (1 mM) (Sigma) were added to the medium to facilitate membrane resealing. Poloxamer 188 (P188, Pluronic F68) was

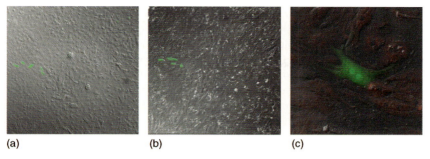

Figure 61.3 Cells optotransfected within selected areas: (a) fibroblast-enriched culture; (b) myocyte-enriched culture; (c) enlarged optotransfected myocyte.

shown to seal cells after electroporation [16–19], heat shock [20, 21], mechanical and barometric traumas [14, 22, 23], and high-dose ionizing radiation [24, 25]. Successful transfection was confirmed after 48–72 h through pronounced green fluorescence, which was at least five times higher than the intrinsic fluorescence of the cells and was distributed over the cytoplasm (Figure 61.3).

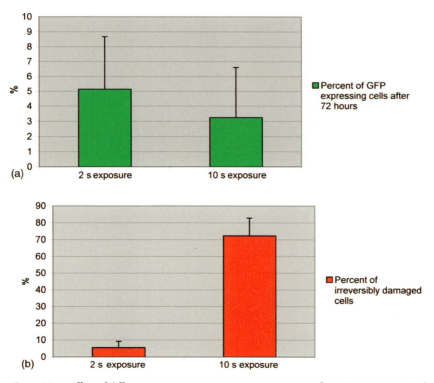

Figure 61.4 Effect of different optoporation parameters on optotransfection outcome: 2 s with 40 mg l^{-1} versus 10 s with 15 mg l^{-1} of Phenol Red. (a) Percentage of optotransfected cells expressing GFP; (b) percentage of irreversibly damaged cells.

The percentage of transfected cells was determined for each area of illuminated cells on the basis of counting individual cells expressing GFP; mean values and standard deviations were calculated from all areas of irradiated cells (Figure 61.4).

The percentages of cells expressing GFP 72 h after optotransfection are shown in Figure 61.4a for the two types of laser exposure: 2 s with 40 mg l^{-1} and 10 s with 15 mg l^{-1} of Phenol Red. From 24 separate experiments, the transfection efficiency for 2 s exposure was estimated as $5.1 \pm 3.5\%$, and from 20 experiments, the transfection efficiency for 10 s exposure was estimated as $3.25 \pm 3.4\%$. The percentage of irreversibly damaged cells for the same exposure conditions was determined through 12 short-term viability assays (PI accumulation and calcein retention) and was $5.6 \pm 3.6\%$ for 2 s exposure and $72.3 \pm 10.5\%$ for 10 s exposure (Figure 61.4b).

When cells were irradiated without light-absorbing dye in OPTI-MEM phenol-free media, the optoporation efficacy was almost negligible, and no short-term membrane damage was observed after several minutes of exposure.

For a comparison of transfection rates, a standard method of lipofection was used. The lipofectamine transfection rate was 10–15%. However, as expected, it was spatially indiscriminate.

61.5
Conclusion

Laser-assisted optoporation appears to be a resourceful tool for cultured, attached cardiac cell transfection. It is possible to perform this method with a standard confocal microscope. It can be applied to different cell types, including those which are not easy to transfect with conventional methods.

Decreasing exposure times sufficient for optoporation resulted in increasing transfection outcome, which could be due to reducing the risk of triggering apoptosis. Results of laser-assisted optoporation were similar to findings reported in the literature for noncardiac cells with the same concentration of Phenol Red. Additional enhancement of light absorbance with the membrane dye FM 1-43 allowed not only an improved focusing technique, but also a further decrease in exposure time sufficient for optoporation, and highlighted the opportunity to use this dye for cell viability and permeability tests.

Acknowledgment

This work was supported by NIH grants HL074283 and HL085369.

References

1 Shirahata, Y., Ohkohchi, N., Itagak, H., and Satomi, S. (2001) New technique for gene transfection using laser irradiation. *J. Investig. Med.*, **49**, 184–190.

2 Tao, W., Wilkinson, J., Stanbridge, E.J., and Berns, M.W. (1987) Direct gene transfer into human cultured cells facilitated by laser micropuncture of

the cell membrane. *Proc. Natl. Acad. Sci. U. S. A.*, **84**, 4180–4184.

3 Lebedeva, L.I., Akhmamet'eva, E.M., Razhev, A.M., Kochubei, S.A., and Rydannykh, O.V. (1990) Cytogenetic effects of UV laser radiation with wavelengths of 248, 223 and 193 nm on mammalian cells. *Radiobiologiia*, **30**, 821–826.

4 Palumbo, G., Caruso, M., Crescenzi, E., Tecce, M.F., Roberti, G., and Colasanti, A. (1996) Targeted gene transfer in eucaryotic cells by dye-assisted laser optoporation. *J. Photochem. Photobiol. B*, **36**, 41–46.

5 Mohanty, S.K., Sharma, M., and Gupta, P.K. (2003) Laser-assisted microinjection into targeted animal cells. *Biotechnol. Lett.*, **25**, 895–899.

6 Tirlapur, U.K. and Konig, K. (2002) Targeted transfection by femtosecond laser. *Nature*, **418**, 290–291.

7 Schneckenburger, H., Hendinger, A., Sailer, R., Strauss, W.S., and Schmitt, M. (2002) Laser-assisted optoporation of single cells. *J. Biomed. Opt.*, **7**, 410–416.

8 Halestrap, A.P., Clarke, S.J., and Javadov, S.A. (2004) Mitochondrial permeability transition pore opening during myocardial reperfusion – a target for cardioprotection. *Cardiovasc. Res.*, **61**, 372–385.

9 Nakajima, M., Fukuda, M., Kuroki, T., and Atsumi, K. (1983) Cytogenetic effects of argon laser irradiation on Chinese hamster cells. *Radiat. Res.*, **93**, 598–608.

10 Peng, T.I. and Jou, M.J. (2004) Mitochondrial swelling and generation of reactive oxygen species induced by photoirradiation are heterogeneously distributed. *Ann. N. Y. Acad. Sci.*, **1011**, 112–122.

11 Fast, V.G. and Cheek, E.R. (2002) Optical mapping of arrhythmias induced by strong electrical shocks in myocyte cultures. *Circ. Res.*, **90**, 664–670.

12 Bi, G.Q., Alderton, J.M., and Steinhardt, R.A. (1995) Calcium-regulated exocytosis is required for cell membrane resealing. *J. Cell Biol.*, **131**, 1747–1758.

13 Togo, T., Alderton, J.M., and Steinhardt, R.A. (2000) The mechanism of cell membrane repair. *Zygote*, **8** (Suppl 1), S31–S32.

14 Yasuda, S., Townsend, D., Michele, D.E., Favre, E.G., Day, S.M., and Metzger, J.M. (2005) Dystrophic heart failure blocked by membrane sealant poloxamer. *Nature*, **436**, 1025–1029.

15 Shanmuganathan, S., Hausenloy, D.J., Duchen, M.R., and Yellon, D.M. (2005) Mitochondrial permeability transition pore as a target for cardioprotection in the human heart. *Am. J. Physiol. Heart Circ. Physiol.*, **289**, H237–H242.

16 Lee, R.C. (2002) Cytoprotection by stabilization of cell membranes. *Ann. N. Y. Acad. Sci.*, **961**, 271–275.

17 Lee, R.C., Hannig, J., Matthews, K.L., Myerov, A., and Chen, C.T. (1999) Pharmaceutical therapies for sealing of permeabilized cell membranes in electrical injuries. *Ann. N. Y. Acad. Sci.*, **888**, 266–273.

18 Lee, R.C., River, L.P., Pan, F.S., Ji, L., and Wollmann, R.L. (1992) Surfactant-induced sealing of electropermeabilized skeletal muscle membranes *in vivo*. *Proc. Natl. Acad. Sci. U. S. A.*, **89**, 4524–4528.

19 Tung, L., Troiano, G.C., Sharma, V., Raphael, R.M., and Stebe, K.J. (1999) Changes in electroporation thresholds of lipid membranes by surfactants and peptides. *Ann. N. Y. Acad. Sci.*, **888**, 249–265.

20 Merchant, F.A., Holmes, W.H., Capelli-Schellpfeffer, M., Lee, R.C., and Toner, M. (1998) Poloxamer 188 enhances functional recovery of lethally heat-shocked fibroblasts. *J. Surg. Res.*, **74**, 131–140.

21 Padanilam, J.T., Bischof, J.C., Lee, R.C., Cravalho, E.G., Tompkins, R.G., Yarmush, M.L., and Toner, M. (1994) Effectiveness of poloxamer 188 in arresting calcein leakage from thermally damaged isolated skeletal muscle cells. *Ann. N. Y. Acad. Sci.*, **720**, 111–123.

22 Fischer, T.A., McNeil, P.L., Khakee, R., Finn, P., Kelly, R.A., Pfeffer, M.A., and Pfeffer, J.M. (1997) Cardiac myocyte membrane wounding in the abruptly pressure-overloaded rat heart under high wall stress. *Hypertension*, **30**, 1041–1046.

23 McNeil, P.L. and Steinhardt, R.A. (1997) Loss, restoration, and maintenance of plasma membrane integrity. *J. Cell Biol.*, **137**, 1–4.

24 Greenebaum, B., Blossfield, K., Hannig, J., Carrillo, C.S., Beckett, M.A., Weichselbaum, R.R., and Lee, R.C. (2004) Poloxamer 188 prevents acute necrosis of adult skeletal muscle cells following high-dose irradiation. *Burns*, **30**, 539–547.

25 Hannig, J., Zhang, D., Canaday, D.J., Beckett, M.A., Astumian, R.D., Weichselbaum, R.R., and Lee, R.C. (2000) Surfactant sealing of membranes permeabilized by ionizing radiation. *Radiat. Res.*, **154**, 171–177.

Part 16
Laser Surgery

62
Laser Surgery: an Overview
Roberto Pini, Francesca Rossi, and Fulvio Ratto

62.1
Introduction

Nowadays lasers are in routine clinical use, and the continual development of technological solutions is opening up new frontiers in micro- and nano-surgery, thus enabling the implementation of new surgical applications.

With the use of lasers and delivery systems that may be developed with the current technology on offer, new *minimally invasive* surgical procedures can be conceived. This expression denotes a wide variety of laser procedures, aimed at minimizing the surgical damage to the patient, with the advantages of safer and more effective operations, shorter healing times, and lesser risk of postoperative complications. Examples of these laser surgeries include endoscopic laser surgery, laparoscopic laser surgery, laser-induced suturing, and precise tissue manipulation and cutting.

Endoscopic laser surgery is usually performed with the help of small endoscopic cameras, several thin rigid instruments, and a laser delivery system, which may be either rigid or flexible depending on the particular application. The surgical tools are inserted towards the surgical target through natural body cavities and small artificial incisions (*keyhole surgery*). In comparison with the usual open surgery, there are several advantages for the patient, such as less pain, less strain on the organism, faster recovery, and smaller injuries and scar formation (esthetics). An example of endoscopic laser surgery is *laser lithotripsy*, in which kidney stones are photodisrupted noninvasively by delivering pulsed laser light with optical fibers through the urinary duct. Endoscopic procedures in the human abdomen are typically termed laparoscopy. Originally developed for minimally invasive treatment of the uterus and ovaries, the use of laparoscopic surgery has today expanded to a wide variety of conditions seen by the general surgeon, such as appendectomy and gall bladder removal. Its benefits to the patient over open surgery include faster healing with smaller scars and fewer postoperative infections. These advances, together with the advent of durable delivery fibers, have made laparoscopic laser surgery a cost-effective and convenient solution for patients and physicians alike.

Precise laser microcutting and microsuturing can be included among these minimally invasive laser procedures. An example is intra-tissue cutting by femtosecond lasers, which is currently applied for clinical use in corneal surgery for the resection of accurate corneal flaps during corneal reshaping in order to correct visual defects, and also in transplant of the cornea. In the latter case, femtosecond laser cutting can be associated with another minimally invasive technique, laser welding, which is used to replace conventional suturing, providing lesser inflammation reactions and faster healing times.

In general, laser surgery can be considered minimally invasive provided that most of the side effects are carefully evaluated and controlled, which means that the laser action must be confined to a well-defined volume of tissue, where the surgery is intended to occur. Possible side effects include the following:

1) Uncontrolled heat damage to surrounding healthy tissues, which is mainly due to two factors: (a) incorrect evaluation of the laser penetration depth into the tissue, owing to, for example, underestimation of the light scattering and nonlinear optical absorption processes; (b) diffusion of the heat generated by photothermal conversion well beyond the treatment volume, which may be due to laser pulse durations exceeding the conditions of heat confinement (see Section 62.2.1.3).
2) Mechanical damage to adjacent tissue structures, caused by the development of laser-induced photoacoustic effects, such as thermoelastic stresses within the tissue, recoil pressure pulses generated by material ejection from the tissue surface, cavitation bubble formation in soft tissues, and spallation of superficial tissue layers, up to dramatic photofragmentation processes induced by uncontrolled plasma formation at the laser focus, followed by shock waves expanding at supersonic speed.

In this introduction, we briefly overview the principles of the interactions between laser light and biotissues in order to provide basic and simple arguments which may help to make the application of lasers in surgical procedures safer, more precise, and better controlled.

62.2
Laser–Tissue Interactions

The interaction of light radiation with biological matter depends upon the irradiation parameters (such as wavelength, power/energy, pulse duration, and spot area), and also on the optical properties of the tissue. Despite the structural complexity of and the morphologic differences between organic tissues, in a first approximation the propagation of light radiation (at low intensities) can be described by fundamental optical properties, such as reflection, absorption, scattering, and transmission.

When laser light impinges on the surface of the tissue, a small fraction, typically 4–5%, is diffusively reflected in the backward direction due to the refractive index mismatch between the external environment and biotissue. The fraction of laser radiation remaining after reflection propagates inside the tissue and is subjected to the processes of absorption and multiple scattering.

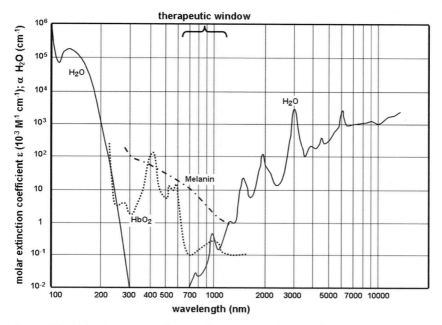

Figure 62.1 Light absorption coefficients of the main absorbers in a biological tissue.

The main absorption features of biotissues are summarized in Figure 62.1, which displays the spectral absorption of water and other biological components. A fundamental role is played by the water content of the tissue, which is the primary absorber in the infrared (IR) spectral region. Moreover, it can be noted that a low absorption of light occurs in the spectral region 700–1300 nm, the so called *therapeutic window* of biological tissues, which is exploited to perform laser treatments or for diagnostic purposes in depth inside the human body.

The process of light absorption in the tissue along the coordinate z, assumed parallel to the optical axis of the light beam and perpendicular to the tissue surface, can be described to a first approximation by the Lambert–Beer law:

$$I(z) = I_0 \exp(-\mu_a z) \tag{62.1}$$

where I_0 is the incident intensity and μ_a is the absorption coefficient, which may be expressed in terms of the *light penetration depth* (also called *extinction length*):

$$L_{ext} = 1/\mu_a \tag{62.2}$$

which represents the distance after which the intensity undergoes a decrease by a factor $1/e$ ($e \approx 2.7$) due to absorption.

In the visible and IR optical bands, large diffusion of light occurs mainly due to the abundance of cellular structures with size comparable to the wavelength of the propagating light. Schematically, three propagation types can be classified, depending on the light wavelength:

1) *Predominant absorption effects*, occurring at ultraviolet (UV) wavelengths in the range 190–300 nm, for example, those of excimer lasers, and also at IR wavelengths of 2–10 μm, for example, those of Ho:YAG, Er:YAG, and CO_2 lasers. In these cases, the light penetration depth is very short, typically in the range 1–20 μm.
2) *Comparable absorption and scattering effects*, occurring at visible wavelengths in the range 450–590 nm. Here the light penetration depth is about 0.5–2.5 mm. Schematically, the light beam propagating through the tissue may be represented as a collimated component surrounded by a region of multiple scattering phenomena. The backward component of the scattered light constitutes the major fraction of the total reflectance, which can be 15–40% of that of the incident beam.
3) *Predominant scattering effects*, occurring at wavelengths in the range 590 nm–1.5 μm, which exhibit light penetration depths of 2.0–8.0 mm. When the light beam enters the tissue, the collimated component is almost fully converted into diffused light. The backward scattering intensity increases significantly, reaching values as high as 35–70% of that of the incident beam.

According to this simplified picture, most biological tissues can be regarded as turbid media, in which the combination of absorption and scattering produces an effective reduction of the light penetration depth. One may account for this reduction by means of an *effective* absorption coefficient $\mu_{eff} = \mu_a + \mu_s$, which includes both absorption and scattering contributions. The *effective* counterpart of the light penetration depth is

$$L_{eff} = \frac{1}{\mu_{eff}} = \frac{1}{\mu_a + \mu_s} \qquad (62.3)$$

This description is strictly true as far as low laser intensities are concerned. Conversely, optical absorption and optical scattering of biotissues may vary significantly under intense laser irradiation, giving rise to nonlinear processes and peculiar spatial and temporal dynamics [1].

The use of lasers as surgical tools requires the emission of high doses of laser power or energy, with emission characteristics which are the key factors to induce particular therapeutic effects in biotissues, which may be otherwise impossible to perform. Lasers are used, for example, to cut, coagulate, weld, ablate, or simply heat the tissue, thus permitting minimally invasive treatments in many surgical fields.

When laser light hits the tissue, several interaction effects are induced, which may be classified schematically into three major types: *photochemical*, *photothermal*, and *photomechanical* effects [2, 3], as sketched in Figure 62.2.

Photochemical effects occur when the absorption of light is able to induce chemical reactions, such as in photodynamic therapies [4]. Photothermal effects are observed when the fraction of laser energy that is absorbed in the tissue

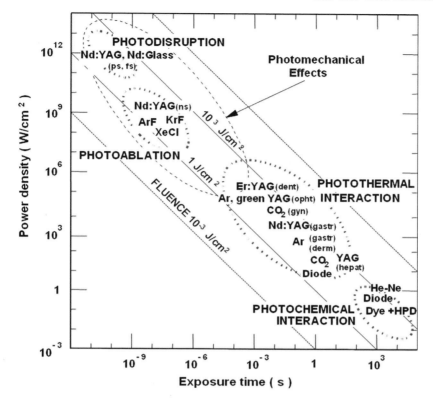

Figure 62.2 Map of laser–tissue interactions: the various types of laser–tissue interactions can be sketched like regions in a logarithmic map, where the abscissa represents the exposure time in seconds, the ordinate expresses the applied power density (W cm^{-2}), and the diagonals show constant energy fluences (J cm^{-2}).

becomes converted into heat; this phenomenon is used, for example, for retinal photocoagulation [5] and for the treatment of superficial angiomas [6]. Photomechanical effects instead are generated by the stress associated with the rapid heating and expansion of the illuminated tissue volume, occurring when short laser pulses (pulse durations from microseconds to femtoseconds) are used, thus inducing photoablation or photodisruption of tissues and cells. In most cases, these three phenomena may occur simultaneously during the laser–tissue interaction in different sites of the irradiated tissue volume, which experience different light and heat distributions. However, in the literature, the description of laser–tissue interactions is typically given through the effect which dominates under the irradiation conditions used for the specific treatment. The photochemical effects are beyond the scope of this chapter. In the following, we focus mainly on photothermal and photomechanical effects, which are particularly relevant for the most diffused laser surgical techniques.

62.2.1
Photothermal Effects

Most of the surgical applications of lasers rely on the conversion of radiation energy into thermal energy. At the microscopic level, the photothermal process consists of a two-step reaction: a molecule in the target tissue is brought to an excited state, as a consequence of the absorption of a photon delivered from the laser light. Then the excited molecule decays by collision with the surrounding molecules, which increase their kinetic energy. This nonradiative decay induces rapid heating (in 1–100 ps) around the illuminated volume.

Many observations suggest a schematic classification of the biological effects induced by photothermal processes according to the temperature range realized in the tissue [2, 3]:

- low-temperature effects (43–100 °C), causing different degrees of damage in the tissue as a function of the exposure time (damage accumulation processes) [2, 7];
- medium-temperature effects ($\geq$100 °C), dominated by water vaporization, with confinement and release of heated water vapor from tissues;
- high-temperature ablation or thermoablation ($\sim$300 °C) by tissue vaporization, combustion, molecular dissociation, and/or plasma formation.

These changes may occur simultaneously at different sites within the irradiated volume as a consequence of a nonhomogeneous distribution of the temperature, originated by the differential light penetration through the biological tissue, as depicted in Figure 62.3. The phenomenological description of biological changes as a function of temperature typically applies whenever continuous-wave (cw) laser emission is used. In the case of pulsed irradiation, in addition to the transient temperature rise, the total exposure time still plays a role in the heat-induced effects (see Figure 62.4).

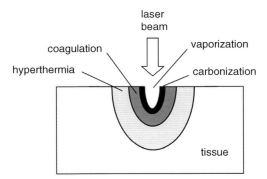

Figure 62.3 Distribution of photothermal effects that may possibly occur in the proximity of the irradiated volume.

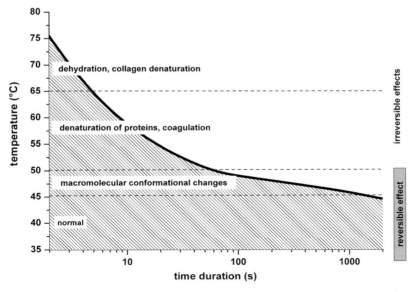

Figure 62.4 Schematic diagram of the low-temperature photothermal effects in biological tissues, depending on the duration of the treatment.

62.2.1.1 Low-Temperature Effects (43–100 °C)

In the case of low-temperature thermal interactions, individual cells and various tissues heated above ∼43–45 °C (*hyperthermic range*) can undergo reversible injury which becomes irreversible (cells death) after exposure times ranging from 25 min to hours, depending on the type of tissue and treatment conditions.

The first hyperthermic effects at about 45 °C are modifications of the macromolecular conformations, bond breakage, and membrane alterations, in addition to the onset of the denaturation of biomolecules and their aggregates (collagen, hemoglobin, other proteins, lipids, etc.), such as the rupture of intramolecular hydrogen bonds. In the case of cancerous cells, their viability experiences a significant reduction above these temperatures. This effect is at the basis of the interstitial thermotherapy of tumoral masses performed with cw Nd:YAG or CO_2 lasers.

The denaturation of proteins such as collagen and hemoglobin typically occurs in the range 50–70 °C. In this so-called *coagulative range*, coagulative necrosis of cells and vacuolization in tissues may be observed. The thermal coagulation of tissues is defined as a thermally induced, irreversible alteration of the proteins and other biological molecules, organelles, membranes, cells, and extracellular components observable with the naked eye and microscopic techniques. When inducing temperatures in this range in fibrous collagen (such as type I), rupture of the intermolecular bonds may occur, resulting in a spatial disorganization of its regular structure, macroscopic contraction and swelling of the tissue, and the loss of its opto-physical properties (such as transparency in corneal stroma and birefringence in muscular fibers). The contraction of the intercellular proteins induces a possible collapse of the

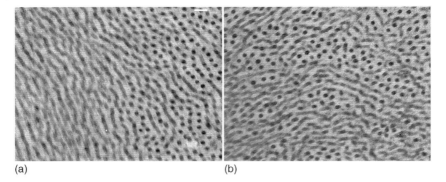

Figure 62.5 Low-temperature effects (at 55–65 °C) in a laser-welded corneal stroma. The transmission electron microscope images (scale bar 90 nm) show the spatial organization of fibrillar collagen in (a) native and (b) laser-welded porcine corneas. In (a), the fibers are organized parallel to each other and grouped in orthogonal lamellar planes; in (b), the individual collagen fibers are preserved, but their spatial distribution appears modified, indicating thermal denaturation of the ground materials which maintain the correct fiber organization (proteoglycan chains).

cytoskeletons, thus producing shrinkage of the coagulated cells. These photocoagulative processes are used, for example, in ophthalmic surgery for reducing retinal detachment and in dermatology for the treatment of vascular or pigmented lesions. The shrinkage of collagen (induced at temperatures higher than 70 °C) is at the basis of some surgical thermal treatments, such as laser capsulorrhaphy [8] and laser thermokeratoplasty [7], while other milder thermal modifications in the spatial distribution of the collagen fibers (see Figure 62.5) are at the basis of the low-power laser welding of corneal tissue, induced at temperatures in the range 55–65 °C) [9].

62.2.1.2 Medium- and High-Temperature Effects (≥ 100 °C)

Thermal effects in this temperature range are mainly dominated by the vaporization of the water component, which may be as high as ∼80% in soft tissues. As the tissue temperature approaches the threshold temperature of water vaporization, photothermal effects induced by the laser–tissue interaction are characterized by (1) energy absorption in the liquid–vapor phase transition, (2) tissue drying, and (3) formation of vapor vacuoles inside the tissue. Macroscopically, steam bubbles are formed in the hottest zones below the irradiated tissue surface. When a critical pressure is reached, the thin walls of the tissues including the vacuoles become broken and many vacuoles may aggregate. Under prolonged irradiation, the largest bubbles rupture explosively, causing the so-called *popcorn effect*. Histological analysis of the popcorn effect clearly evidences that the damage induced in the tissue as holes through the surface does not correspond to an actual loss of tissue mass. The hot vapor ejection causes immediate tissue cooling. The water loss diminishes the local heat conductivity and limits heat conduction to surrounding areas.

When the water content of the tissue has completely evaporated, the tissue temperature rises rapidly to >300 °C, which causes high-temperature ablation of the remaining tissue components by combustion, molecular dissociation, solid material vaporization, and carbonization.

62.2.1.3 Controlling Heating Effects

The control of heat effects is a key factor for the effectiveness of the surgical procedure. For this purpose, objective observations of the degree of thermal damage based on the surgeon's experience are not often accurate enough. This control should require in principle the use of noncontact monitoring systems, with high spatial accuracy (in the order of 0.1 mm) and a fast acquisition rate, since the temperature dynamics may develop over the range of milliseconds or shorter times. The available technology does not include any compact, low-cost system which may be effectively used for this purpose during surgery. The thermal analysis is usually performed by the use of an IR thermocamera, which gives a direct measurement of the temperature rise on the external surface of the tissue (see Figure 62.6). In the case of deep penetration of the light into the tissue, in addition to transient heating phenomena with durations shorter than the time acquisition rate of the camera, the temperature dynamics and the temperature distribution inside the tissue under the specific irradiation conditions may become accessible by the development of a mathematical model based on the *bio-heat equation* (see the example in Figure 62.7) [10].

When considering heating and its effects on biological tissues, it is important to control not only the maximum temperature value induced, but also the confinement of the heating to the region where the laser therapy or surgery is required. The results of the laser–tissue interactions may be significantly different whether or not the heated region is restricted to the volume under direct irradiation. Heat confinement

Figure 62.6 IR thermography as a tool to control the surface temperature dynamics during live laser surgery, while performing diode laser welding of the cornea. This image was recorded with a compact IR camera, equipped with software for visualization and analysis. The bright spot corresponds to the temperature rise on the surface of the cornea during irradiation by means of a 300 μm optical fiber operated in noncontact mode.

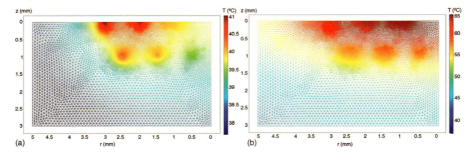

Figure 62.7 Example of a mathematical simulation developed to study spatial and temporal dynamics of laser-induced thermal effects in the upper layers of the skin (including melaninless dermis with blood capillaries of diameters varying in the range 10–200 μm). In this picture, the results of the heating effects induced by cw 300 mW, 405 nm (a) and 470 nm (b), light sources in a 3 mm section of skin are sketched. The model required the solution of the light diffusion and bio-heat equations with the Finite Element Method by the use of commercial software (Comsol Multiphysics).

can be important, for example, to limit the portion of tissue affected by heat damage, especially when precise coagulation has to be realized in the proximity of delicate structures. This result may be achieved by setting the time duration of the laser–tissue interaction process shorter than the time during which heat propagates from the directly irradiated volume to the adjacent tissue.

To provide a quantitative although simplified description of heat confinement during laser irradiation of biotissue, in addition to the laser pulse duration τ_{laser} one has to introduce the *thermal relaxation time* τ_{th}, that is, the time necessary for the propagation of heat over a distance equal to the optical penetration length of the laser light in that particular tissue; τ_{th} thus depends both on the optical absorption coefficient and on the thermal conductivity of the tissue itself. In order to define this parameter, let us introduce the thermal diffusion length L_{th}, which is the distance over which heat diffuses in the time t:

$$L_{th}^2 = 4Kt \tag{62.4}$$

where K is the thermal diffusivity of the tissue, which depends on its thermal conductivity, specific heat capacity, and density. For example, in water $K = 1.43 \times 10^{-3}$ cm^2 s^{-1}, hence heat diffuses over a length of 0.8 mm in 1 s. Similarly, considering for simplicity the thermal diffusivity of blood to be the same as that of water, it takes ~170 μs for heat to propagate through a 10 μm diameter blood capillary, and ~17 ms for a 100 μm diameter blood capillary.

If we impose L_{th} to be equal to the optical penetration length L_{ext} defined by Eq. (62.2) (or, in case of absorption and scattering, to the effective penetration length L_{eff}), the thermal relaxation time is given by

$$\tau_{th} = \frac{L_{ext}^2}{4K} \tag{62.5}$$

Table 62.1 Typical values of thermal relaxation time for different biological targets of selective photothermolysis.

Target	Thermal relaxation time
Hair follicle, 200–300 µm	40–100 ms
Capillary in port-wine stain, 100 µm	5–100 ms
Epidermis, 20–50 µm	0.2–1 ms
Erythrocytes, 7 µm	20 µs
Melanosome, 1 µm	1 µs
Tattoo cell, 0.1 µm	10 ns

which represents the time necessary for heat to propagate over the optical penetration length. If the laser pulse duration is shorter then the thermal relaxation time, that is, $\tau_{laser} < \tau_{th}$, laser heating is confined in a volume $V = A \cdot L_{ext}$, where A is the area of the irradiated tissue surface. Conversely, when $\tau_{laser} > \tau_{th}$, heat diffuses through the tissue over lengths longer than the optical penetration length, extending the heat-affected zone and eventually the thermal damage to the surrounding tissue volumes.

In a case of practical concern, let us consider the irradiation of soft tissues (assuming for K the value of water) by CO_2 laser emission, which penetrates a length $L_{th} \approx 50$ µm. From Eq. (62.5), one can calculate $\tau_{th} \approx 44$ µs. Therefore, if the CO_2 laser pulse duration is <40 µs, then the effect of irradiation will be a highly localized heating to a depth of 50 µm, which may result in vaporization of a very thin layer of tissue, with minimal heat damage to the deeper layers. By adjusting the energy delivered per pulse, the pulse duration, and the repetition rate, the heating effects may thus be controlled, avoiding irreversible thermal damage to sites adjacent to the treatment volume.

The control of heat confinement is at the basis of a specific laser treatment, called *selective photothermolysis*, which employs a green laser, highly absorbed by hemoglobin, which allows, by setting a proper pulse duration, selective coagulation of dilated blood vessels of, for example, port-wine stains, preserving at the same time the small capillaries, and also the overlying skin [11]. Typical values of thermal relaxation time for different biological targets of selective photothermolysis are given in Table 62.1.

62.2.2
Photomechanical Effects

Photomechanical effects become evident when short laser pulses are used to treat biological targets. As the laser pulse duration becomes shorter than a few microseconds, in addition to the purely thermal effects there may also appear significant photomechanical effects, such as pressure pulses propagating both in the air above the irradiated surface and inside the tissue. Depending on the type of interaction, the pressure pulse can be an acoustic pulse, that is, a low-pressure perturbation propagating at the speed of sound, or a shock wave, characterized by a high instantaneous pressure peak, propagating at ultrasound speed.

If tissue irradiation is performed under heat confinement conditions, very high temperatures may develop in a short time within a limited target volume, thus inducing significant photomechanical effects, which may be regarded as a drawback of the laser treatment or otherwise employed for therapeutic purposes, such as to induce fragmentation of hard tissues such as kidney stones, or to permit highly precise cutting of tissues in a controlled way as in the case of femtosecond laser ablation.

When laser irradiation takes place in heat confinement conditions, but at moderate intensities, the induced temperatures and pressure waves may not determine substantial and irreversible structural modifications in the tissue. This is the case in the *thermoelastic* regime, in which only acoustic waves are originated from the irradiated volume.

At higher levels of laser irradiation and for pulse durations in the nanosecond range, photomechanical effects become more evident and can play a fundamental role in the processes of laser ablation of biotissues (*photoablation*), thus increasing the effectiveness of tissue removal during processes of rapid vaporization.

At shorter pulse durations in the pico- to femtosecond regimes (or even in the nanosecond regime, but for very high laser intensities), a process called *photodisruption* can take place, mediated by nonlinear phenomena, such as avalanche ionization of the target materials with the formation of a plasma plume (a cloud of free electrons and ionized atoms) with subsequent microexplosive effects, which may induce the formation of cavitation bubbles inside the tissue. When suitably sized, these bubbles can be used, for example, to perform precise intra-tissue dissections for use in corneal surgery, as described in the following.

In the cases of both *photoablation* and *photodisruption*, one has to bear in mind that tissue removal is accompanied by mechanical stresses inside the tissue, control of which is essential to avoid significant damage to adjacent biological structures. For instance, the retina may by subjected to secondary damage when photorefractive surgery is performed [12–14].

62.2.2.1 Photoablation with Nanosecond Pulses

The term *photoablation* means the removal of material (biotissue) by laser irradiation. The photoablation modalities may be very different, depending on the tissue properties (optical, thermal, mechanical, and chemical characteristics of the target tissue), and also on the laser emission parameters and irradiation conditions. In the nanosecond pulse range, in homogeneous or quasi-homogeneous tissues, laser ablation is characterized by a so-called *ablation threshold*, F_{th}, which is the minimum level of laser fluence (i.e., laser energy per unit surface area, typically expressed in $J\,cm^{-2}$) above which material removal starts to occur. The processes taking place around the ablation threshold are displayed schematically in Figure 62.8.

Photoablation behaviors are usually described by the *ablation rate curves* (sketched in Figure 62.9), representing the etching rate (weight of the removed tissue or etching depth per pulse) as a function of the laser fluence. The typical behavior of an ablation rate curve is a quasi-linear dependence of the etching depth on fluence, when a threshold value F_{th} is exceeded. In general, below this threshold fluence, material

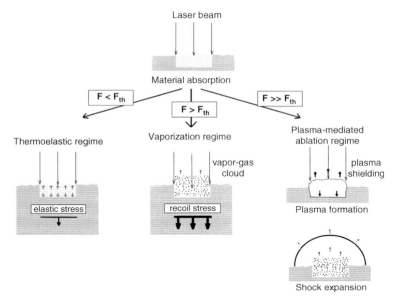

Figure 62.8 Scheme of the physical processes induced by pulsed laser irradiation around the ablation threshold of the target material.

removal does not take place (except for the two cases described in the following), even if photoacoustic effects can still be generated under pulsed irradiation by the thermoelastic effect. On the other hand, above the saturation value F_{sat}, the ablation rate is slowed by nonlinear absorption effects, which may reduce the optical penetration depth or even shield the laser light impinging on the tissue, as in the case of dense plasma formation. The ablation rate curves thus exhibit typical S-shaped behavior, which, for the same type of material, display different threshold values with the laser wavelength (shorter wavelengths exhibit lower threshold values).

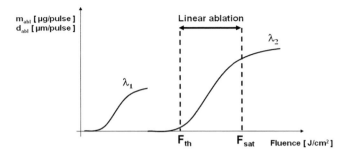

Figure 62.9 Curves of ablation rates for different laser wavelengths.

As the photoablation process is strongly related to the water content of tissues (the energy to vaporize water is about 2500 J cm^{-3} under static conditions), it may be useful to characterize two kinds of photoablative regimes, occurring below and above the vaporization threshold.

Photoablation Below the Vaporization Threshold To describe the processes involved better, let us introduce the *stress relaxation time*, τ_{st}, which is the time for a pressure wave to travel a distance equal to the optical penetration length. Below the vaporization threshold of water, material removal under laser irradiation (in the nanosecond regime) may be induced by:

- Pressure peaks due to *inertial confinement*. This is a regime which may occur when the laser pulse duration (typically <10 ns) is so short that the irradiated and heated tissue volume cannot undergo expansion during the pulse, that is, when $\tau_{laser} < \tau_{st}$. Such heat deposition at constant volume results in a rapid increase in the internal pressure, which may give rise to an explosive expansion, eventually causing material ejection.
- *Spallation*. This process consists of the detachment of a superficial layer, rather than a real ejection of vapor or particulate (see Figure 62.10). Spallation is a photoablation condition which is reached when the laser pulse duration is shorter than the thermal relaxation time (heat confinement), but longer than the stress relaxation time: $\tau_{st} < \tau_{laser} < \tau_{th}$. This process originates from thermoelastic expansion inside the tissue, which produces pressure waves propagating in all directions. Let us consider the component of this thermoelastic pulse propagating towards the tissue surface. At the air–tissue interface, due to the mismatch of acoustic impedance, this compressive (positive) pressure peak will be reflected, thus experiencing a phase change π and becoming a rarefaction (negative) pulse peak. The traction exerted by this negative pressure peak may exceed the mechanical resistance of the tissue and cause the detachment of a superficial layer.

Photoablation Above the Vaporization Threshold Tissue removal may be induced by a very rapid vaporization process involving mainly the water content of the tissue, with

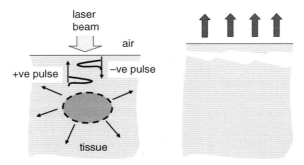

Figure 62.10 Scheme of the laser-induced spallation into biotissue.

explosive characteristics at the microscopic scale. The process is typically induced by laser pulses with durations of 1–100 ns and fluences in the range 0.1–5 J cm^{-2}, depending on the type of tissue. This kind of explosive phase change is one of the most important laser ablation mechanisms in surgical applications whenever controlled removal of soft tissue layers is required with minimal collateral heat effects. Photoablation by fast vaporization is actually the regime in which the description of tissue removal can be given through the ablation rate curves described above. In this respect, control of the laser fluence is important to operate safely in the linear part of the rate curve. Under these conditions, each laser pulse provides fast delivery of energy to the tissue, so as to induce its vaporization in a very short time, thus inducing rapid material ejection in the form of vapor and fine particulate from the surface.

Surgical applications of photoablation processes driven by fast vaporization include: photorefractive keratectomy, in which ablation and reshaping of the cornea are produced by ArF (190 nm) excimer laser pulses for the correction of visual defects, and excimer laser angioplasty, which employs XeCl (308 nm) laser pulses delivered through intraluminal fiber-optic catheters, in order to perform the recanalization of arterial stenoses and obstructions. In the fast vaporization regime, thermal damage to surrounding tissues can be negligible, as most of the delivered energy is used to vaporize the irradiated volume, and because heat confinement conditions are satisfied. In contrast, photomechanical damage may be not negligible, as for example when the rapid material ejection to the air induces a strong recoil and thus a pressure wave propagating in the opposite direction back into the tissue.

62.2.2.2 Photodisruption

At very high laser fluences, exceeding the saturation threshold of the typical ablation regime described above, plasma formation can occur, which may become so dense as to absorb the incoming laser light and give rise to a shock wave, which in turn may lead to macroscopically explosive effects, including tissue fragmentation. On the other hand, this disruptive process induced by highly intense nanosecond pulses may find beneficial applications, for example, in the photofragmentation of hard tissues such as kidney stones (laser lithotripsy).

Photodisruption processes in laser–tissue interactions are mediated by the dynamics of a plasma cloud, which is produced by ionization of the atoms of the irradiated material or of the surrounding gas medium [15]. The plasma is a macroscopically neutral gas phase of matter comprising a partially ionized gas, in which a certain proportion of electrons are free. The plasma is ignited when very high laser irradiances exceed the threshold to induce breakdown conditions (*optical breakdown*). The detailed mechanisms behind breakdown with optical pulses depend on the pulse duration. Particularly for femtosecond pulses, multiphoton ionization can efficiently generate free carriers in the initial phase of the pulse, followed by strong absorption by the generated plasma, which leads to further heating and ionization. For long pulses, multiphoton ionization is less important, and the breakdown starts primarily from the few carriers which are already present before the pulse. The random occurrence of such carriers makes optical breakdown less deterministic in the regime of longer pulses, whereas the breakdown threshold can be very well defined for femtosecond pulses. The conditions for laser breakdown may be reached either at the external surface of a soft or hard target, or in the bulk of a transparent material.

In the case of nontransparent media and pulses in the nanosecond regime or longer, the laser light is absorbed at the air–tissue interface, which induces extraction of free electrons near the surface of the target and then acceleration of these electrons by the electromagnetic field of the laser radiation, which is still present during the evolution of the process. This results in a laser-sustained *avalanche ionization*, which induces a rapid increase in the free electron density, overcoming the losses due to recombination and diffusion. The avalanche ionization thus originates the condition for plasma formation. Very high temperatures ($\sim 10^4\,^\circ$C) and high electron densities ($>10^{20}\,\text{cm}^{-3}$) are thus reached. In these conditions, the plasma becomes optically opaque, exhibiting a very high absorption coefficient and very high reflectivity. The consequence is a continuous growth of the plasma electron density and energy, and the formation of a plasma plume towards the external ambient (air), thus shielding further irradiation of the tissue target. The subsequent plasma expansion gives rise to a shock wave that may induce fragmentation and localized ruptures of the tissue.

For optically transparent media, the process is similar to that described above, although in this case the expansion of shock waves developing within the tissue drives the formation of cavitation bubbles with the particular dynamics of fast growth and successive collapse, which thus transmits significant mechanical stresses to the surrounding tissues. The disruptive effects may be enhanced by photochemical effects, with weakening of intra- and intermolecular bonds.

Plasma-mediated processes are at the basis of various laser techniques employed in the surgery of both hard and soft tissues. The former case, typically induced by highly intense nanosecond pulses, has an application in the already mentioned laser-induced fragmentation of kidney stones with Ho:YAG and Er:YAG pulsed lasers. Indeed, in this case it may be preferable to break the stone into small fragments by a few laser shots, instead of slowly ablating layer by layer.

The latter, realized with femtosecond lasers to produce controlled ablation confined to a very thin volume comparable to the focal region of the laser, is described in detail in the following section.

62.2.2.3 Photoablation with Femtosecond Pulses

Femtosecond laser (fs-laser)-induced microablation has recently been proposed in cellular and micro-surgery, and in some cases has even reached the phase of clinical applications, for example, for cutting transparent tissues such as the cornea [15–21]. Other intriguing biomedical applications are under study, conventionally indicated as fs-laser *nano-surgery*, which involve intra-cell cutting procedures, with the possibility of manipulating items inside a cell, thus opening up new frontiers in fields such as gene therapy.

The potential of such an fs-laser surgical regime rests on significant properties: when using femtosecond pulses, the confinement of multiphoton absorption in the focal spot and the very low energies required to induce optical breakdown can provide high precision in cutting biological tissues and, at the same time, a very limited extension of the collateral damage. For example, the optical breakdown threshold can be reduced ~100-fold on passing from 3 ns to 100 fs pulse durations, which results in tissue ablation with less transformation of light energy into destructive mechanical energy. Moreover, in some particular cases, the cutting precision can even exceed the optical diffraction limit by a factor of 2–3, which theoretically determines the minimum size of a focal spot. This can be realized considering that fs-laser photodisruption effects originate from the free-electron distribution. At the focus, the nonlinear absorption process can thus make the electron density distribution narrower than the laser intensity distribution, which results in a better spatial resolution of the fs-laser surgery compared with that obtained in a linear absorption regime (i.e., by cw lasers), where the temperature distribution corresponds to the laser intensity distribution.

There are two main parameter regimes used for micro- and nano-surgery:

1) Long series of pulses emitted from femtosecond oscillators at repetition rates of 80 MHz and pulse energies well below the optical breakdown threshold, which induce accumulative effects; in these conditions, heat-induced damage may occur.
2) Amplified series of femtosecond pulses, (e.g., by means of a regenerative amplifier injected by the femtosecond oscillator) with repetition rates of 1–150 kHz and pulse energies slightly above the threshold for bubble formation; in this case, the effect of each individual pulse can be regarded as independent of each other, and so heat accumulation is negligible.

Let us briefly describe the latter femtosecond regime of those listed above, which is much more within the scope of this chapter. The micrometric photodisruptive effect may be used in notable applications, such as the intra-tissue cutting of ocular structures, and in particular for the cutting and *sculpturing* of the cornea, which is nowadays routinely used in clinics to prepare corneal flaps in the LASIK procedure (laser *in situ* keratomileusis, a type of refractive surgery for correcting myopia, hyperopia, and astigmatism), and also in new procedures of corneal transplant.

Femtosecond laser pulses are used to ablate tissue layers with predefined geometric characteristics, such as thickness, diameter, and lateral profile [16]. Commercially available fs-laser systems may be of various types [22], but in general with emission in the near-infrared (NIR) region, such as Nd:YAG or Nd:glass lasers. This means that laser radiation is well transmitted through transparent tissues such as the cornea. The laser pulses typically have energies in the range 0.5–2.5 µJ, durations of hundreds of femtoseconds and repetition rates in the range 50–150 kHz. Each individual laser pulse is focused to a precise location inside the cornea. This can be achieved, for example, by applanation of the corneal surface by means of a quartz flat connected with the laser beam focusing system. The high laser intensity at the focus creates a microplasma which vaporizes a volume of tissue of a few microns thickness. Then the microplasma drives the creation of a cavitation bubble, which separates the corneal lamellas. The size of this microbubble is governed both by the size of the focal spot and by the pulse energy. The present fs-laser technology allows for an optimized choice of the emission parameters which permits the controlled production of microbubbles smaller than 10–15 μm. A scanning system moving the focal spot with micrometric precision allows for the delivery of thousands of laser pulses connected together in a raster pattern in order to define a resection plane, which may be horizontal, vertical, or tilted at any arbitrary angle with respect to the optical axis of the laser beam. The distance between the centers of contiguous microbubbles can be adjusted at will: a shorter distance increases the treatment time, but requires lower pulse energies, which increases the precision of the cleavage and minimizes the collateral damage.

The unique capabilities of fs-laser intra-tissue cutting procedures give micrometric precision and repeatability of the same cut pattern in different eyes, thus allowing the sculpturing of identical flaps in both the donor and recipient eyes when performing a lamellar or full thickness transplant of the cornea [20, 23–25].

62.3
Notes on Laser Systems Suitable for Surgical Applications

In this section, we briefly introduce the most common surgical lasers, as summarized in Table 62.2. In the UV region, the most widely used surgical lasers are the excimer lasers, such as the ArF (194 nm) and the XeCl (308 nm) lasers, which permit the so-called *cold ablation* of tissues, which corresponds to the process of fast vaporization described above. This technique is used whenever minimization of heat effects is critical, such as in the reshaping of the cornea (photorefractive keratectomy) and in the recanalization of atherosclerotic arteries. In the visible region, ion lasers such as the argon laser (several emission lines, the main ones at 488 and 514 nm) and krypton laser (several emission lines in the yellow–red region, the main one at 647 nm) were very common in the past, especially for retinal photocoagulation, but are now becoming obsolete owing to their poor efficiency. They are now being replaced by frequency-doubled Nd:YAG lasers emitting in the green (532 nm), which also allow for cw emission, and by diode lasers (630–960 nm). Dye lasers also emit lines in the visible (tunable, e.g., around 550–600 nm), which are mainly used for the treatment of superficial vascular lesions and for laser lithotripsy. In the NIR region, Nd:YAG lasers (1064 nm), operated in the cw, free-running (ms), Q-switching (ns), and mode-locking (ps–fs) regimes, remain the most diffused surgical lasers, used for photocoagulation and photoablation. In the IR region, the emissions of Ho:YAG (2100 nm), Er:YAG (2940 nm), and CO_2 (10 600 nm) lasers are used to excise and vaporize soft tissues, thanks to their high absorption by the high water content of such tissues. This list is, of course, not complete, as many more types of lasers are employed in experimental and preclinical surgery and in very specific medical applications. Moreover, this situation is rapidly changing with new developments in the laser technology, with a trend to replace the old lasers types characterized by high maintenance costs, high electric power consumption, and delicate components, with new compact and cost-effective devices, mainly based on

Table 62.2 Summary of relevant characteristics and surgical applications of the most commonly used lasers (not exhaustive).

Spectral region	Laser	Wavelength (nm)	Operation mode	Energy/power output	Repetition rate (Hz)	Delivery system	Surgical applications
UV	ArF	194	Pulsed: 10 ns	100–500 mJ	10–100	Optical channels with mirrors; no fiber-optic transmission	Photorefractive keratectomy Cardiovascular surgery
	XeCl	308	Pulsed: 10–100 ns	100–500 mJ	10–100	Fiber-optic transmission at moderate energies	Cardiovascular surgery Laser ablation of bones Dermatology: atopical dermatitis, vitiligo, alopecia areata, psoriasis
Visible and NIR	Argon	488–514	cw, pulsed	10 mW–10 W	—	Fiber-optic transmission	Retinal photocoagulation Laser trabeculoplasty Theleangectasia Tattoo removal
	Frequency-doubled Nd:YAG	532	cw and pulsed (ns)	500 mJ	30	Fiber-optic transmission: cw, yes; pulsed, no	Retinal photocoagulation Laser trabeculoplasty Treatment of vascular lesions (e.g. port-wine stains)

(Continued)

Table 62.2 (Continued)

Spectral region	Laser	Wavelength (nm)	Operation mode	Energy/power output	Repetition rate (Hz)	Delivery system	Surgical applications
	Dye	585–600	Pulsed	1–3 J²	—	Fiber-optic transmission	Cutaneous vascular lesions
	Diode	630–960	cw Pulsed: pulse width in ms range	1–30 W	—	Optimal fiber-optic transmission	Laser welding Dentistry Facial telangectasias
	Alexandrite	720–800	cw or pulsed	1–10 J	1–100	Fiber-optic transmission	Hair removal Tattoo removal Nevi
	Nd:YAG	1064	Q-switching (short pulse): 10 ns	100–500 mJ	10–100	Fiber-optic transmission at low energies	Laser capsulotomy Deep pigmented lesions Tattoo removal
			Free-running (long pulse): 1 ms	1–2 J	10–100	Optimal fiber-optic transmission	Endoscopic neurosurgery Laparoscopic surgery Laser hair removal Laser lipolysis and treatment of cellulite Treatment of vascular lesions
	Nd:glass	1060	500–800 fs	1–10 μJ	10–150 kHz	Optical channels with mirrors, no fiber-optic transmission	Intra-tissue cutting (e.g., cornea) Tissue flap creation

	Laser	Wavelength (nm)	Pulse duration	Energy	Rate	Transmission	Applications
	Ti:sapphire	700–1000	10–100 fs	1–10 nJ	80 MHz	Optical channels with mirrors, no fiber-optic transmission	Cellular and intracellular surgery
IR	Ho:YAG	2100	0.3 ms	1 J	10	Fiber-optic transmission	Endoscopic surgery, Urology, Lithotripsy, Orthopedics
	Er:YAG	2940	0.2 ms	10–50 J	10	Fiber-optic transmission with special fibers or hollow waveguides	Skin rejuvenation, Esthetics, Leg veins, Acne scars, Hair removal, Hemangiomas, Superficial pigmented lesions
	CO_2	10600	cw, pulsed (100 ms) and super-pulsed (100 μs)	10–100 W	5–100	Transmission through articulated arms or special fibers at low power	Noncontact neurosurgery, Ear, nose and throat surgery, Gynecology, Skin resurfacing, Dermatology, Tattoo removal, Laser welding

solid-state technology. Worth noting also is that high-power lamps and LEDs (light-emitting diodes) are penetrating many surgical application fields that were formerly occupied by lasers.

Table 62.2 also reports the types of delivery systems usable with the various laser sources. The delivery system is a crucial part of the whole laser system, which may either enable or hinder, for example, endoscopic surgical applications, easy access to the treatment site, precise operations under the surgical microscope, or delivery of high laser doses in contact or noncontact modes.

The available delivery systems provided with the most common surgical laser are mainly based on articulated arms, optical fibers, and the direct transmission of laser light through the lens of a surgical microscope. The choice of a particular system clearly depends on the specific surgical application, but is also determined in most practical cases by the capabilities of the delivery medium, which may hinder the transmission of certain wavelengths (e.g., for standard fused-silica optics and optical fibers in the UV and IR regions), and also of high power densities (which may exceed the damage threshold of the light guide).

Articulated arms [26] are based on the reflection of light by multiple mirrors. They are composed of at least two rotating and tilting mirrors, enclosed in linear tubing arrangements and pivot joints, all of which may be arranged in very different setups to satisfy several different surgical needs. A good beam-handling system for surgical applications requires at least five degrees of freedom (three for spatial positioning and two for angular displacement) and terminates in a handpiece operated manually by the surgeon. With the use of these mechanical delivery systems, it is usually not possible to perform minimally invasive surgery inside the human body because of their large size and limited movability, although they may represent the only choice for several UV and IR laser sources and for very high peak power laser pulses. Moreover, they offer good beam quality, preserving the characteristics of the light source, such as its polarization and coherence. A variant of the articulated arms are the motorized optical channels used in clinical fs-laser systems, in which the laser path is based on mirror reflections as above, but controlled by means of a joystick.

Optical fibers [27, 28] are the backbone of minimally invasive surgery. The development of the technology enabling better performance and applicability of this kind of beam handling system is at the basis of modern endoscopy. A fiber-optic is composed of three coaxial cylindrical layers, from the inner part of the fiber, the core, the cladding, and the jacket. The optical properties of the core and the cladding are responsible for the fiber transmission, and the jacket is a protective layer. The principle behind fiber-optic systems may be understood by simple geometric optics arguments, which apply when the size of the fiber core is larger than the wavelength of the light. The core and the cladding are composed of different materials, with refractive index n_{co} and n_{cl}, respectively. If $n_{co} > n_{cl}$, the light ray entering the fiber core with a small enough angle $\alpha < \alpha_{max} = \arccos(n_{cl}/n_{co})$ undergoes total internal reflection at the interface with the cladding. This condition is satisfied at every ray reflection, thus permitting propagation with low losses over long distances (see Figure 62.11).

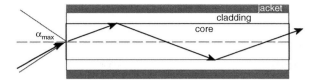

Figure 62.11 Visual representation of light propagation through an optical fiber by means of geometric optics. A light ray enters the fiber core and experiences total internal reflection at the core–cladding interface; α_{max} is the fiber acceptance angle, which defines the maximum angle that the light ray can form with the optical axis to become coupled into the fiber guide.

In surgical applications, because of the requirement to transmit high laser intensities, the core diameters of the fibers employed have to be in the range 100–400 μm. In a description of light propagation by means of electromagnetic fields, these relatively large cores correspond to a large number of electromagnetically guided modes, which mix together during the propagation and provide a homogeneous *top hat* intensity distribution at the fiber output. This intensity profile may be very useful to control the laser–tissue interaction.

Thanks to their flexibility and reduced size, optical fibers are the optimal tools to deliver light radiation inside the human body, thus permitting a wide variety of minimally invasive surgical procedures. The main limitations to their surgical use concern the range of transmitted wavelengths, which depends on the core material (e.g., fused silica gives a good transmission only in the range 300–2000 nm) and on the damage threshold of the fiber itself, which may prevent the transmission of the high laser intensities required to perform specific surgeries, as in the case of laser pulses in the nanosecond range or shorter. In practical medical use, fibers may be either disposable or reusable after sterilization. In some cases, they can be inserted in handpieces to allow easy handling of the fiber tip by the surgeon, especially when used under a microscope.

References

1 Vogel, A. and Venugopalan, V. (2003) Mechanisms of pulsed laser ablation of biological tissues. *Chem. Rev.*, **103**, 577–644.

2 Niemz, M.H. (1996) *Laser–Tissue Interactions*, Springer, Berlin.

3 Thomsen, S. (1991) Pathological analysis of photothermal and photomechanical effects of laser–tissue interactions. *Photochem. Photobiol.*, **53** (6), 825–835.

4 Mitton, D. and Ackroyd, R. (2008) A brief overview of photodynamic therapy in Europe. *Photodiagnosis Photodyn. Ther.*, **5** (2), 103–111.

5 Maguen, E., Chu, T.C., and Boyer, D. (2003) Lasers in ophthalmology, in *Biomedical Photonics Handbook* (ed. T. Vo-Dinh), CRC Press, Boca Raton, FL, pp. 43-9–43-10.

6 Babilas, P., Shafirstein, G., Baumler, W., Baier, J., Landthaler, M., Szeimies, R.M., and Abels, C. (2005) Selective photothermolysis of blood vessels following flashlamp-pumped pulsed dye laser irradiation: *in vivo* results and mathematical modelling are in

agreement. *J. Invest. Dermatol.*, **125** (2), 343–352.

7 Brinkmann, R., Radt, B., Flamm, C., Kampmeier, J., Koop, N., and Birngruber, R. (2000) Influence of temperature and time on thermally induced forces in corneal collagen and the effect on laser thermokeratoplasty. *J. Cataract Refract. Surg.*, **26**, 744–754.

8 Noonan, T.J., Tokish, J.M., Briggs, K.K., and Hawkins, R.J. (2003) Laser-assisted thermal capsulorrhaphy. *Arthroscopy J. Arthrosc. Relat. Surg.*, **19** (8), 815–819.

9 Pini, R., Rossi, F., Matteini, P., and Ratto, F. (2007) Laser tissue welding in minimally invasive surgery and microsurgery, in *Biophotonics* (eds. L. Pavesi and P.M. Fauchet), Springer Verlag Series in Medical Physics, Springer, Heidelberg, pp. 269–296.

10 Rossi, F., Pini, R., and Menabuoni, L. (2007) Experimental and model analysis on the temperature dynamics during diode laser welding of the cornea. *J. Biomed. Opt.*, **12** (1), 014031.

11 Anderson, RR., and Parrish, JA. (1983) Selective photothermolysis: precise microsurgery by selective absorption of pulsed radiation. *Science*, **220** (4596), 524–527.

12 Maatz, G., Heisterkamp, A., Lubatschowski, H., Barcikowski, S., Fallnich, C., Welling, H., and Ertmer, W. (2000) Chemical and physical side effects at application of ultrashort laser pulses for intrastromal refractive surgery. *J. Opt. A Pure Appl. Opt.*, **2** (1), 59–64.

13 Siano, S., Pini, R., Rossi, F., Salimbeni, R., and Gobbi, P.G. (1998) Acoustic focusing associated with excimer laser ablation of the cornea. *Appl. Phys. Lett.*, **72** (6), 647–649.

14 Schumacher, S., Sander, M., Stolte, A., Doepke, C., Baumgaertner, W., and Lubatschowski, H. (2006) Investigation of possible fs-LASIK induced retinal damage. *Proc. SPIE*, **6138**, 61381I-1–61381I-9.

15 Vogel, A., Noack, J., Hüttman, G., and Paltauf, G. (2005) Mechanisms of femtosecond laser nanosurgery of cells and tissues. *Appl. Phys. B*, **81**, 1015–1047.

16 Soong, H.K. and Malta, J.B. (2009) Femtosecond lasers in ophthalmology. *Am. J. Ophthalmol.*, **147**, 189–197.

17 Mian, S.H. and Shtein, R.M. (2007) Femtosecond laser-assisted corneal surgery. *Curr. Opin. Ophthalmol.*, **18**, 295–299.

18 König, K., Krauss, O., and Riemann, I. (2002) Intratissue surgery with 80 MHz nanojoule femtosecond laser pulses in the near infrared. *Opt. Express*, **10** (3), 171–176.

19 Ripken, T., Oberheide, U., Ziltz, C., Ertmer, W., Gerten, G., and Lubatschowski, H. (2005) Fs-laser induced elasticity changes to improve presbyopic lens accommodation. *Proc. SPIE*, **5688**, 278–287.

20 Menabuoni, L., Lenzetti, I., Cortesini, L., Susini, M., Rossi, F., Pini, R., and Parel, J.-M. (2008) Technical improvements in DSAEK performed with the combined use of femtosecond and diode lasers. *Invest. Ophthalmol. Vis. Sci.*, **49**, E-Abstract 2319.

21 Nishimura, N., Schaffer, C.B., and Kleinfeld, D. (2006) *In vivo* manipulation of biological systems with femtosecond laser pulses. *Proc. SPIE*, **6261**, 62611J.

22 Lubatschowski, H. (2008) Overview of commercially available femtosecond lasers in refractive surgery. *J. Refract. Surg.*, **24** (1), S102–S107.

23 Menabuoni, L., Lenzetti, I., Cortesini, L., Pini, R., and Rossi, F. (2009) Technical improvements in PK performed with the combined use of femtosecond and diode lasers. *Invest. Ophthalmol. Vis. Sci.*, **50**, E-Abstract 2220.

24 Menabuoni, L., Lenzetti, I., Rutili, T., Rossi, F., Pini, R., Yoo, S.H., and Parel, J.-M. (2007) Combining femtosecond and diode lasers to improve endothelial keratoplasty outcome. A preliminary study. *Invest. Ophthalmol. Vis. Sci.*, **48**, E-Abstract 4711.

25 Menabuoni, L., Pini, R., Fantozzi, M., Susini, M., Lenzetti, I., and Yoo, S.H. (2006) "All-laser" sutureless lamellar keratoplasty (ALSL–LK): a first case report. *Invest. Ophthalmol. Vis. Sci.*, **47**, E-Abstract 2356.

26 Greve, P. (2003) Beam handling systems, in *Applied Laser Medicine* (eds. H.-P. Berlien and G.J. Müller), Springer, Heidelberg, 137–140.

27 Gannot, I., and Ben-David, M. (2003) Optical fibers and waveguides for medical applications, in *Biomedical Photonics Handbook* (ed. T. Vo-Dinh), CRC Press, Boca Raton, FL, pp. 7-1–7-10.

28 Katzir, A. (1993) *Lasers and Optical Fibers in Medicine*, Academic Press, San Diego, pp. 107–125.

63
Endovenous Laser Therapy – Application Studies, Clinical Update, and Innovative Developments

Ronald Sroka and Claus-Georg Schmedt

63.1
Introduction

The first reports on clinical results of endovenous laser therapy (ELT) were published by Boné [1] and Navarro *et al.* [2] about 10 years ago. By means of ELT, the hemodynamic elimination of incompetent truncal veins should be feasible, which is already possible with other endovenous thermal procedures (radiofrequency ablation) and endovenous chemical procedures (foam sclerotherapy). Using these techniques, the endothermal damage to the vein wall results in consecutive occlusion of the treated vein. The clinical outcome looks very promising. Meta-analysis gives evidence that the innovative endoluminal techniques result in similar clinical outcomes to conventional surgical stripping. From the technical point of view, developments are still continuing. One characteristic of ELT is the broad spectrum of different treatment protocols using a variety of laser systems and forms of endovenous application. In recent years, systematic experimental investigations and the analysis of clinical results have increased our understanding of the connection between particular details of endovenous laser application and clinical results. During laser light application, bare and flat cut fibers are mainly used, causing inhomogeneous circumferential light application to the vessels wall, resulting in non-uniform tissue alteration and perforation. Different kinds of newly developed radial emitting laser fibers are tested in combination with a laser emitting light at 980 and 1470 nm in *ex vivo* model tissue using different treatment parameters. It could be demonstrated that the induced tissue alterations are circumferential uniform without any perforation. Those investigations led to the continuous development and optimization of ELT. This chapter outlines the principles of ELT and experimental developments for optimizing this minimally invasive procedure.

63.2
Clinical Application of ELT

In endovenous application, laser light is introduced into the vein lumen by means of a flexible optical fiber via a sheath system. Depending on the laser wavelength used and on the absorption characteristics of the primary target molecules in the tissue (e.g., water or hemoglobin), laser energy is absorbed and transformed into heat energy. Heating of the tissue leads to cell destruction and collagen shrinkage. Currently, laser systems emitting light of wavelengths in the near-infrared region (810, 940, 980, 1064, 1320, and 1470 nm) are used in clinical practice. In most systems, the light energy is transported via a bare fiber (core diameter of up to 600 μm) which is inserted into the vein via a catheter system.

In principle, for ELT the same indications and contraindications apply as for conventional operative treatment of incompetent truncal veins using crossectomy (high ligation) and stripping. Around 60% of patients suffering from an epifascial vein system were found to be suitable for an endovenous therapy procedure [3]. Contraindications for endovenous thermal application were found to be acute ascending thrombophlebitis, postphlebitic stenosis, segmental obliteration, and aneurysmatic widening of the vein. In the case of marked meandering path of the vein, insertion of the fiber is often impossible. Furthermore, large vein diameters (>20 mm) can also be regarded as a relative contraindication since this makes adequate circular thermal alteration of the vein wall more difficult.

The principle operative procedure of the vena saphena magna is outlined in the following. During the course of preoperative ultrasound diagnosis, the vein diameter should be measured for the precalculation of the appropriate energy density. In principle, ELT can be carried out as an outpatient procedure under local anesthesia (tumescence anesthesia). The great saphenous vein can be treated with ELT in its entire length from the sapheno-femoral junction to the distal lower leg. Particularly in the area of the lower leg, care has to be taken to ensure adequate tumescence and light dosimetry in order to avoid postoperative paresthesia due to the proximity of the saphenous nerve. The use of sonography during operation is an integral part of ELT treatment.

First, the vein is punctured in the area of the distal insufficiency point and a guide wire is introduced. Before introducing the sheath system (Seldinger technique), the bare fiber is inserted into the sheath so that the fiber tip extends ~1–2 cm beyond the end of the sheath. This position should be marked on the fiber. This is important since the fiber tip can be heated to several hundred degrees Celsius during the treatment and if there is insufficient distance from the sheath, this can lead to the melting of the sheath material. Thereafter the sheath system, without fiber, is pushed forward to the sapheno-femoral junction with the aid of the guide wire. After the removal of the guide wire, the bare fiber is inserted with respect to the mark. Together and under sonographic monitoring, the sheath and fiber are then positioned so that the tip of the fiber comes to rest ~1–2 cm distally from the sapheno-femoral junction. The pilot beam can be used to make a rough check of the location of the bare fiber tip as it is visible through the skin. The perivenous tumescence anesthesia is then carried out under ultrasound supervision. The fluid is to protect the perivenous tissue from thermal damage and reduce the lumen of the truncal vein by compression and spasm.

Two modes of light application are clinically in use. In stepwise pull-back, the fiber with sheath is pulled back by 1–5 mm between each irradiation pulses activated. Using continuous pull-back, continuous irradiation is applied while withdrawing the fiber with sheath at a velocity of 1–3 mm s^{-1}. After pull-back is finished at the distal point of vein incompetence, the vein is ligated after removing the sheath.

Since the damage caused to tissue is proportional to the laser energy applied, the parameter linear endovenous energy density (LEED) is defined, which is proportional to the appropriate level of energy to the length and diameter of the vein segment to be treated in joules per centimeter of vein [4]. Using LEED, the diameter of the vein or the irradiated inner surface of the vein is not taken into consideration. The laser energy applied in relation to the irradiated surface is described as endovenous fluence equivalent (EFE) in J cm^{-2} [4] and uses a cylindrical model of the vein segment, calculating the vein inner surface according to $F = 2r\pi l$ where r is the vein radius and l the length of the treated vein segment. Therefore, the following parameters should be documented for every treatment: vein diameter d (cm), length of the treated vein segment l (cm), laser wavelength λ (nm), laser power P (W), irradiation time t (s) or pull-back speed v (mm s^{-1}) for continuous pull-back. Only in this way LEED and EFE can be calculated for every treatment, thereby making it possible to compare clinical results with different treatment protocols. Clinical analyses suggest that failures in therapy with nonocclusion or recanalization after ELT treatment are observed less when sufficiently high energy density is used – compared with lower LEED or EFE [5–12]. On the other hand, energy densities above a certain level can lead to side effects such as transmural ablation, perforations, and alteration of perivenous tissue [13, 14]. So far it has not been possible to establish generally valid recommendations for appropriate energy density. Table 63.1 gives an overview of recommendations formulated so far by a number of authors. A recent trend of recommendations for higher energy density has been observed.

Table 63.1 Recommended energy density for effective ELT with varying wavelengths[a].

Study	Wavelength (nm)	LEED (J cm^{-1})	EFE (J cm^{-2})
Proebstle et al. [5]	940	23.6	13.0
Timperman et al. [6]	810/940	>80	n.d.
Proebstle et al. [7]	940	n.d.	20.0
Kim et al. [8]	980	32.7	9.82
Desmyttère et al. [9]	980	50–120	29–41
Kontothanassis et al. [10]	980	45–60	n.d.
Vuylsteke et al. [11]	980	n.d.	52.0
Elmore and Lackey. [12]	810	109.6	46.6

a) LEED, endovenous energy density; EFE, endovenous fluence equivalent; n.d., not defined.

Clinical and duplex ultrasound examinations should be repeated regularly to ensure the success of the treatment and to exclude possible postoperative thrombosis or thrombus propagation. Also, persistent or recurrent reflux should be ruled out. Directly after the operation, the treated truncal vein can be seen to be somewhat thickened under sonography. The lumen is usually reduced and filled with strongly echoing thrombus material. In the remainder of the vein's path, as a result of repair and scar formation, a stringy, fibrous change may appear or the vein disappears [12].

63.3
Clinical Studies on ELT

After the first clinical reports on ELT in 1999, more recently clinical series with small numbers of cases and relatively short periods of postoperative follow-up have been published. There have occasionally been reports, however, on series and multi-center studies with more than 500 treatments and postoperative observation periods of more than 60 months [9, 12, 15–18]. Today, there are data from more than 60 publications with altogether more than 15 000 ELT treatments available, which have been summarized and evaluated in overviews [19].

With the appropriate selection of patients, ELT treatment can technically be carried out in more than 99% of patients [12] where effective occlusion and hemodynamic elimination can be assumed. The cosmetic results are on the whole very good. However, lack of effectiveness and undesired side effects are also observed [20], for example, 2.3% nonocclusion or early recanalization, 4.5% expected recanalization. The follow-up in the studies analyzed seldom exceeded 6–12 months and the studies showed widely differing results. Recanalization was observed to have occurred with an incidence of up to 24% [3]. Furthermore, the connection between rate of occlusion and endovenously applied laser energy (LEED, EFE) became obvious. Postoperative pain is rated as slight to moderate by most patients (81.5%). However, in a series of 113 patients, 18.5% described the pain as severe to very severe. Postoperative induration of the truncal vein was observed in 78.1% of patients, and perivenous ecchymosis and hematoma in an average of 52% of patients [20]. In a number of studies, however, the incidence of postoperative ecchymosis and hematoma was reported in 60–80% of patients [3, 4]. Phlebitis of the truncal vein was observed in 2.8% of patients, but certain studies reported a significantly higher incidence of up to 17.2% [21]. Further complications such as persistent dysesthesia after nerve lesions (0.8%), burns on the skin (0.5%), thrombus propagation in the deep vein system (0.2%), and pulmonary thrombosis (0.02%) were seldom observed [19].

In Table 63.2, a number of randomized controlled studies comparing ELT with the open surgery techniques of crossectomy and stripping are listed. As the number of these studies is small and the period of postoperative observation is short, a final evaluation is not possible at present. In sum, these studies document, at least initially, effective elimination of the epifascial reflux through ELT accompanied by reduced side effects. Whether refraining from crossectomy in fact leads to an increased incidence of inguinal neo-reflux or even to a reduction in neo-vascularization at the

Table 63.2 Randomized controlled studies comparing ELT with crossectomy and stripping.

Study	Country	No. of procedures	Comparison
de Medeiros and Luccas [40]	Brazil	20	ELT vs. crossectomy/stripping
Rasmussen et al. [41]	Denmark	137	ELT vs. crossectomy/stripping in local anesthesia
Darwood et al. [42]	United Kingdom	103	ELT vs. crossectomy/stripping
Kalteis et al. [43]	Austria	95	ELT vs. crossectomy/stripping
Disselhoff et al. [44, 45]	The Netherlands	120	ELT vs. crossectomy/cryostripping

sapheno-femoral junction will have to be established in further controlled trials and, in particular, in further postoperative examinations of already randomized patients.

63.4
Analysis of Comparison Studies

Looking at meta-analyses and reviews, it becomes obvious there are some meta-studies available on the one hand evaluating the efficiency and the safety of endoluminal techniques in comparison with conventional surgical stripping and on the other hand comparing the different endoluminal techniques from the clinical point of view.

Although all investigations are of great interest, such studies must be assessed critically because they represent just the short history of about 10 years of the evaluation of innovative laser application techniques which have shown promising successful improvements in finding a more effective treatment protocol and dosimetric concept, including the bare fiber technique that has been improved by the application of radial firing energy. Unfortunately, such developments are so innovative that up to now only reports of single-center clinical studies are available.

In a clinical review, the effects on insufficient veins traditionally treated with high ligation and stripping were compared with minimally invasive techniques including endovenous laser surgery, radiofrequency ablation, and foam sclerotherapy. The authors discussed the mechanism of action of each modality, as well as the techniques for intervention, outcomes, and complications. Minimally invasive treatment of venous insufficiency has replaced open surgical stripping in most cases, with less morbidity than high ligation and stripping, allowing for successful treatment and improved patient satisfaction [22]. A clinical update and clinical hints on how to use ELT clinically was presented [23]. In a histological investigation of ELT using either 940 or 1319 nm light and the bare fiber technique, changes including acute effects such as loss of intima, vacuolization in the subintima layer and effects at 1 month post-ELT, inflammatory changes, vein wall thickening, intraluminal thrombus, and increased amounts of fibroblasts and collagen could be observed [24].

Meta-analytic studies were performed, comparing the overall outcome of endovenous treatment concepts including ELT with conventional methods. In a meta-analytic survey which included 59 studies of which seven directly compared ELT and surgery (involving ligation and stripping) with respect to safety and efficacy, it was reported that serious side effects are rare in both, whereas minor side effects are more common in ELT. A short-term comparison favored ELT over surgery, showing comparable safety aspects. However, ELT offers short-term benefits and appears to be as clinically effective as surgery up to 1 year after treatment [25]. In another meta-analysis, evidence from the collected data suggests that ELT and radiofrequency techniques are as safe and effective as surgery, particularly in the treatment of saphenous veins. However, most importantly, the type of varicose vein should govern the intervention of choice [26]. A further meta-analysis concluded that endovenous therapies have not yet been compared with surgical ligation and stripping in large randomized clinical trials, especially with long-term follow-up of more than 3 years. In the absence of large, comparative randomized clinical trials, the minimally invasive techniques appear to be at least as effective as surgery in the treatment of lower extremity varicose veins. The results from 64/119 eligible studies with an average follow-up of about 32 months showed that foam therapy and radiofrequency ablation were as effective as surgical stripping. Endovenous laser therapy using bare fiber techniques was significantly more effective than stripping, foam therapy, and radiofrequency ablation. The results of this meta-analysis support the increasing use of minimally invasive interventions in the treatment of lower extremity varicosities. However, large, long-term comparative randomized controlled trials that include patient-reported outcomes, cost-effectiveness analyses, and safety assessment are needed to achieve the highest level of evidence for these novel therapies [27].

ELT using 810 nm light and the bare fiber technique showed thermal damage of the intima and the internal part of the media. Whereas at lower LEEDs of 110–200 J cm^{-1} vascular perforations were observed in less than 10% of cases, at higher LEEDs of 40–880 J cm^{-1} nearly all cases were perforated [28]. As neovascularization is a major cause of recurrent varicosities following surgery, a prospective cohort study was performed comparing recurrence rates and the occurrence of neovascularization after surgery or ELT (810 nm, LEED <50 J cm^{-1}, bare fiber) for great saphenous vein reflux. Although the frequencies of recurrent varicosities 2 years after surgery and ELT were similar, neovascularization, a predictor of future recurrence, was less common following ELT. It was supposed that delivering larger LEEDs could reduce recanalization rates and recurrence [29].

The data from a study comparing radiofrequency ablation with 980 nm wavelength ELT (bare fiber technique) at comparable energy densities clearly showed that patients who received radiofrequency ablation recovered more quickly than those who received 980 nm endovenous laser treatment for the first 2 weeks, and then the recovery parameters began to equalize between 2 and 4 weeks after the procedure and were equal at 4 weeks. The authors agreed that the data may differ if different laser wavelengths are compared with radiofrequency ablation, such as the newer 1319, 1320, and 1470 nm wavelengths [30, 31]. Further comparison studies of radiofrequency ablation versus 980 nm ELT showed that radiofrequency ablation was significantly superior to ELT as measured by post-procedural recovery and quality of life parameters in a randomized prospective comparison between these two thermal ablation modalities for closure of the great saphenous vein within 2 weeks post-treatment [30].

The newer wavelength of 1470 nm was clinically tested in a randomized study using the bare fiber technique. It could be shown that the power of the laser (15 W versus 25 W) did not influence the occlusion rate when an appropriate high LEED (110–130 J cm^{-1}) was used. Side effects such as ecchymoses and pain were reduced compared with similar studies using wavelengths of 810 and 980 nm [32]. Furthermore, a prospective follow-up study using the bare fiber technique with a 1470 nm diode laser showed promising clinical results with this minimally invasive, safe, and efficient therapy. It appears that using LEED >100 J cm^{-1}, paresthesia is increased compared with LEED <100 J cm^{-1} at the same occlusion rate of 100% at 1 year follow-up [33].

63.5
Investigations to Improve Light Application

As mentioned earlier, during laser light application, bare and flat cut fibers are mainly used, causing inhomogeneous circumferential light application to the vessels wall, resulting in nonuniform tissue alteration and perforation. The endovenous laser treatment relies on the transformation of luminous energy into heat due to absorption of the photon energy. This process depends on the laser wavelength and on the optical properties of the irradiated tissue. Thus the endoluminal application of laser energy implies a variety of parameters, overall influencing the alterations of the vein wall. Variations in the laser wavelength and power settings result in different temperature levels and thermal alterations. In addition, the light application technique, more precisely the light distribution within the vessel, could also influence the type of response; for example, vessel wall perforation could be observed in the bare fiber technique in combination with special treatment protocols [14, 34, 35]. With respect to this, the blood as the primary medium around the laser fiber tip also influences the mechanism and the process. The purpose of the following study was to investigate in a standardized manner the tissue effects induced by an innovative 360° radial emitting laser fiber which is based upon an interstitial laser light application system for prostate tissue (patent DE 19703208.7).

63.5.1
Experiments

The *ex vivo* model consists of a cow's foot from freshly slaughtered cows (18–24 months of age, weight 550–650 kg, right hind leg) [14, 35]. The subcutaneous veins (V. saphena lateralis and V. digitalis dorsalis communis III, 20.0–25.0 cm in length,

5–8 mm in diameter) are large enough to allow for introduction of the endoluminal application systems and are available for acute *ex vivo* experiments. Leaving these veins *in situ*, appropriate sheath and catheter systems permit perfusion of fluid (e.g., blood, saline) and intraluminal pressure measurement during treatment. After surgical preparation, treatment was performed with endoluminal laser irradiation from the central end (joint) to the peripheral end (hoof).

Laser therapy was performed using a diode laser (Ceralas D15, Biolitec, Bonn, Germany) emitting light of wavelength 1470 nm which is transported via a 600 μm fiber. The 360° radial fiber tip (Biolitec, Bonn, Germany) (Figure 63.1) was polished to emit the light in air radial to the optical axis of the fiber. Retaining this condition, a transparent dome covers the fiber tip and is sealed to prevent blood inflow when used inside the vessel. From the technical point of view, the irradiation profile of the 360° radial fiber was measured using a goniometer setup. With respect to this, the fluence rate delivered through the dome could be calculated and compared with the bare fiber parameters. Experiments on the robustness of the new 360° radial fiber was performed with safety precautions. The fiber robustness under extreme crash conditions was tested by intravenous positioning of the 360° radial fiber without any pull velocity but with blood irrigation and application of an energy of up to 300 J.

For tissue experiments, the 360° radial fiber was introduced into the vein via an introduction sheath. The vein was irrigated using heparinized blood. The pull-back velocity was set to 1 mm s^{-1} for a 3–4 cm vein length. The laser power settings were 3, 6, 10, and 15 W. Thus the applied LEED could be calculated to be 30, 60, 100, and 150 J cm^{-1}, respectively. Each experiment was performed 5–7 times. After treatment, application devices and treated veins were examined macroscopically with standardized *in situ* and *ex situ* preparation and documentation, then the harvested tissues were processed for histology.

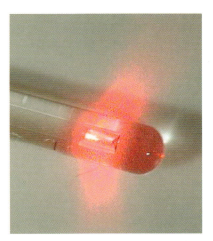

Figure 63.1 360° radial emitting fiber for endoluminal uniform circumferential irradiation of the vein wall.

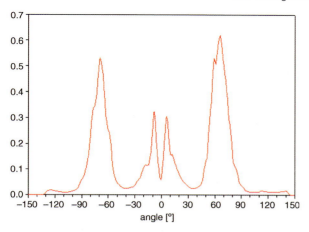

Figure 63.2 Irradiation pattern of the 360° radial emitting fiber tip showing the radial direction of 70° to the optical axis of the fiber and an irradiation divergency of 25–30°.

63.5.2
Experimental Results

Goniometric measurements showed a radial irradiation pattern of the 360° radial fiber with the main deflection axes in the 70° direction with respect to the optical axis of the fiber. As shown in Figure 63.2, the divergence angle of the emitting beam (full width at half-maximum) could be determined as 25–30°. The fiber robustness experiments showed that the fiber remained intact without any damage when a power of 10 W was applied for 30 s. Fiber damage could be observed when using a power of 15 W for more than 20 s. Regarding safety aspects, it should be mentioned that such a situation and such a high dose of light energy should never be reached during endovenous laser treatment. Such a crash situation is far beyond the intended clinical situation.

As shown in Table 63.3, calculations according to the geometric conditions as sketched in Figure 63.3 were performed for a set of vein lumen diameter.

Table 63.3 Calculations of the irradiation area A and the applied fluence rate FR due to geometric considerations when 10 W laser output power is delivered to 100% of the tissue surface[a].

Vein diameter, d (cm)	Distance from fiber tip or dome to tissue surface, a (cm)	A_{radial} (cm^2)	A_{fiber} (cm^2)	FR_{radial} (W cm^{-2})	FR_{fiber} (W cm^{-2})
0.18	0.00	0.027	0.003	370	3333
0.40	0.10	0.135	0.019	74	526
0.70	0.25	0.413	0.069	24	145
1.00	0.40	0.842	0.149	12	67

a) $A_{radial} = d^2 \pi \tan(\alpha_{radial}/2)$; $A_{fiber} = \pi \cdot [(a+x)\tan(\alpha_{fiber}/2)]^2$ with $x = 0.67$ mm with respect to the apex of the divergency.

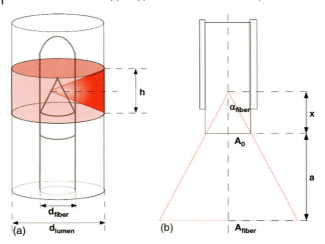

Figure 63.3 Sketch of the geometric conditions for the radial emitting fiber (a) and a bare fiber (b) used for calculations in Table 63.3.

The distance of the fiber, either the tip of the bare fiber (core diameter 0.6 mm) or the dome surface (dome diameter 1.8 mm), to the tissue varied from touching to up to 4 mm, thus affecting the size of the irradiation area (A) due to the beam divergency [α_{fiber}(bare fiber) = 50°; α_{radial}(radial-dome fiber) = 30°]. As a result, the irradiation area in touch of the 360° radial fiber (A_{radial}) differs by a factor of 9 compared with the bare fiber tip (A_{fiber}). Assuming 100% power transmission through the fiber and that no absorption by the blood occurred, it can be calculated that the fluence rate (FR) applied to such an area at a laser output power of 10 W resulted in an FR_{fiber}/FR_{radial} ratio of about 5–9 depending on the vein diameter.

In Table 63.4, the macroscopic observations during the 360° radial application of 1470 nm laser light are summarized, with differentiation between vessel diameters of $d > 7$ mm (large vessel) and $d < 5$ mm (small vessel). The observations were

Table 63.4 Macroscopic evaluation of the tissue alterations due to 360° radial laser light application at $\lambda = 1470$ nm.

Vein diameter (mm)	Power (W)	Rigidity, thickening (class 0–4)	Observations
>7	3	0–1	Boiling noise
<5	3	1–2	Low shrinkage
>7	6	2	Noise, hindered pulling, shrinkage
<5	6	3	Fast shrinkage, hindered pulling
>7	10	2–3	Boiling, shrinkage
<5	10	3	Hard pulling, fast shrinkage
>7	15	3	Shrinkage, sometimes sticking
<5	15	3	Fast shrinkage, sometimes sticking

Table 63.5 Therapeutic options for truncal varicose veins.

Open procedures	Endovenous procedures		
	Endothermal		Chemical
	Radiofrequency ablation	Laser therapy (ELT)	
High ligation Barrier techniques *Stripping* Invagination Cryostripping Electrical stripping Endoscopic stripping	VNUS Closure *Plus* VNUS Closure *Fast* Celon RFITT	810–1470 nm Bare fiber Cylindrical diffuser Radial fiber	Liquid sclerotherapy Foam sclerotherapy Catheter-based application

classified into five groups labeled from 0 to 4, where 0 is defined as normal unaffected tissue, 4 represents burned and carbonized sensations, and classes 1–3 represent different degrees of rigidity concomitant with wall thickening and change of color. The evaluation was performed by three independent investigators. The data in Table 63.4 indicate that sufficient shrinkage with alteration in rigidity and wall thickening could be obtained at applied laser powers of 6 and 10 W, for both vein lumens. Shrinkage was accompanied by increased effort for pulling, especially in small lumen veins. Boiling noise could be experienced in nearly every experiment due to cooking of the fluid inside the vein lumen. Sticking to the tissue occurred especially when moving the fiber through the vein valves.

63.6 Discussion

Endovenous laser therapy has become established in the last few years among the range of therapy options for the treatment of truncal vein insufficiency as listed in Table 63.5. In endovenous radiofrequency ablation (VNUS Closure), reproducible, defined thermal damage to tissue can already be produced through the implementation of feedback mechanisms with automatic generator-steered regulation of energy density and standardized treatment protocols. By contrast, the effects to ELT using the bare fiber technique on tissue, and therefore on the clinical results, are clearly very varied [18, 34]. Because of the use of different wavelengths, different pullback protocols, and different energy densities, ELT can hardly be standardized. Feedback mechanisms through which accurate energy density in ELT can be controlled are lacking. This makes the appropriate dosage of light for the sufficient thermal alteration of tissue more difficult to achieve. In addition to transmural damage, wall perforation is also observed [14], which has the potential for clinically relevant consequences such as ecchymosis and postoperative pain. On the other hand, a locally focused, not completely circular thermal damage of the vein wall leaves only slightly locally damaged wall areas, which can lead to a higher incidence of nonocclusion or recanalization.

It becomes obvious that a certain level of thermal dose is required to occlude the vein efficiently in a circumferential homogeneous manner. The thermal dose delivered at a specific vein segment is related to both treatment time and temperature. It appears that denaturation of almost all the collagen fibers within the vein wall and consecutive wound healing processes guarantee complete and durable occlusion of the vein [36, 37]. Investigations on mathematical modeling based on optical and thermal parameters of the vessel walls and the perivascular tissue were performed and could serve a useful tools to simulate the mechanisms of action of endovenous laser treatment [38].

Applying a high fluence rate to the tissue ($FR > 1000\,\mathrm{W\,cm^{-2}}$) and exceeding the ablation threshold of the tissue, local perforations could occur [14, 34, 35]. Such tissue effects could be observed during bare fiber light application, especially in cases where the fiber tip is in contact with the tissue. Using the bare fiber tip technique, the mechanism of heat induction and heat transfer to the vessel target has still not been investigated in detail. Hence, with reducing the side effects and clarification of the underlying mechanism in mind, improved and reproducible application of light energy to the vessels wall is required [14, 34, 35].

With respect to this, the innovative 360° radial fiber tip showed promising results in the *ex vivo* experiments. The fluence rate is dramatically reduced below the ablation threshold, so no perforation was observed during experiments. The induced tissue alterations were circumferential and homogeneous. Shrinkage of the vessel due to thickening, loss of flexibility, and change in color as a sign of effective treatment [36] could be observed throughout the complete vessels. Furthermore, shrinkage could be recognized additionally by feeling the improved power needed for pulling the 360° radial fiber through the lumen. As the absorption coefficient of water and blood at a wavelength 1470 nm is superior to that at all other laser wavelengths (810, 940, 980, 1064, 1320 nm) used in this field of laser treatment, an immediate tissue response could be observed already at laser output powers between 6 and 10 W. Hence a continuous pull-back velocity of about $1\,\mathrm{mm\,s^{-1}}$ seems appropriate for reproducible and efficient effects. Based on these experiments, it seems promising that a secure treatment protocol could be developed, depending on the vein lumen diameter, pull-back velocity, and light energy applied. Initial clinical results of ELT treatments with this radial light applicator using a 1470 nm laser generator indicate minimal side effects coupled with a high rate of closure [39]. Long-term studies using the innovative 360° radial fiber should be performed to evaluate the treatment procedure with respect to recanalization, side effects, and other aspects.

63.7
Conclusion

Laser light application in endovenous treatment procedures suffered from inhomogeneous tissue alterations after treatment. Laser treatment parameters showed, in comparison with radiofrequency ablation, a greater variety of parameters due to wavelength, power, pull-back velocity, compression by tumescence, bare fiber versus cylindrical emitting fibers, and so on. Therefore, a certain treatment protocol was not available. Meta-analysis support the increasing use of minimally invasive interventions such as ELT in the treatment of varicosities which are efficient and with reduced side effects. However, large, long-term comparative randomized controlled trials that include patient-reported outcomes, cost-effectiveness analyses, and safety assessment are needed to achieve the highest level of evidence for these novel therapies.

The latest innovative technologies, such as the 360° radial fiber, seem promising to develop secure, reliable, and reproducible tissue alteration treatments in the acute situation. By means of these optimizations, ELT treatment is coming closer to achieving the goal of standardizing an effective method for the treatment of varicose veins. Further controlled studies are required to compare the results of optimized ELT with those of other endothermal modes of treatment and conventional open surgery.

Acknowledgments

The authors would like to thank the following for supporting these investigations and studies: T. Beck, R. Blagova, C. Burgmeier, V. Hecht, N. Karimi-Poor, R. Meier, M. Sadeghi-Azandaryani, S. Scheibe, I. Sroka, B. Steckmeier, S. Steckmeier, A. von Conta, S. Winter. K. Weick, and B. Steckmeier.

References

1 Boné, C. (1999) *Rev. Patol. Vasc.*, **5**, 35–46.
2 Navarro, L., Min, R.J., and Boné, C. (2001) *Dermatol. Surg.*, **27** (2), 117–122.
3 Sharif, M.A., Soong, C.V., Lau, L.L., Corvan, R., Lee, B., and Hannon, R.J. (2006) *Br. J. Surg.*, **93** (7), 831–835.
4 Proebstle, T.M., Moehler, T., Gül, D., and Herdemann, S. (2005) *Dermatol. Surg.*, **31** (12), 1678–1683; discussion 1683–1684.
5 Proebstle, T.M., Krummenauer, F., Gül, D., and Knop, J. (2004) *Dermatol. Surg.*, **30** (2 Pt 1), 174–178.
6 Timperman, P.E., Sichlau, M., and Ryu, R.K. (2004) *J. Vasc. Interv. Radiol.*, **15** (10), 1061–1063.
7 Proebstle, T.M., Moehler, T., and Herdemann., S. (2006) *J. Vasc. Surg.*, **44** (4), 834–839.
8 Kim, H.S., Nwankwo, I.J., Hong, K., and McElgunn, P.S. (2006) *Cardiovasc. Interv. Radiol.*, **29** (1), 64–69.
9 Desmyttère, J., Grard, C., Wassmer, B., and Mordon, S. (2007) *J. Vasc. Surg.*, **46** (6), 1242–1247.
10 Kontothanassis, D., Di Mitri, R., Ferrari Ruffino, S., Ugliola, M., and Labropoulos, N. (2007) *Int. Angiol.*, **26** (2), 183–188.
11 Vuylsteke, M., Liekens, K., Moons, P., and Mordon, S. (2008) *Vasc. Endovasc. Surg.*, **42** (2), 141–149.
12 Elmore, F.A. and Lackey, D. (2008) *Phlebology*, **23** (1), 21–31.
13 Chang, C.J. and Chua, J.J. (2002) *Lasers Surg. Med.*, **31** (4), 257–262.
14 Schmedt, C.G., Sroka, R., Steckmeier, S., Meissner, O.A., Babaryka, G., Hunger, K., Ruppert, V., Sadeghi-Azandaryani, M.,

and Steckmeier, B.M. (2006) *Eur. J. Vasc. Endovasc. Surg.*, **32** (3), 318–325.
15. Agus, G.B., Mancini, S., and Magi, G. (2006) *Int. Angiol.*, **25** (2), 209–215.
16. Bush, R.G. (2006) *J. Vasc. Surg.*, **43** (3), 642; author reply 642–643.
17. Min, R.J. and Khilnani, N.M. (2005) *J. Cardiovasc Surg. (Torino)*, **46** (4), 395–405.
18. Ravi, R., Rodriguez-Lopez, J.A., Trayler, E.A., Barrett, D.A., Ramaiah, V., and Diethrich, E.B. (2006) *J. Endovasc. Ther.*, **13** (2), 244–248.
19. Schmedt, C.G. and Steckmeier, B.M. (2006) Endoluminale radiofrequenz- und lasertherapie zur therapie der stammveneninsuffizienz, in *Handbuch der Angiologie* (eds. M. Marshall and F.X. Breu), Ecomed, Landsberg, pp. 1–28.
20. Luebke, T. and Brunkwall, J. (2008) *J. Cardiovasc. Surg. (Torino)*, **49** (2), 213–233.
21. Disselhoff, B.C., der Kinderen, D.J., and Moll, F.L. (2005) *Endovasc. Ther.*, **12** (6), 731–738.
22. Brown, K. and Moore, C.J. (2009) Update on the treatment of saphenous reflux: laser, RFA or foam? *Perspect. Vasc. Surg. Endovasc. Ther.*, **21**, 226–231.
23. Schmedt, C.G., Blagova, R., Karimi-Poor, N., Burgmeier, C., Steckmeier, S., Беck, T., Hecht, V., Meier, R., Sadeghi-Azandaryanic, M., Steckmeier, B., and Sroka, R. (2010) Update of endovenous laser therapy and the latest application studies. *Med. Laser Appl.*, **25**, 34–43.
24. Bush, R.G., Shamma, H.N., and Hammond., K. (2008) Histological changes occurring after endoluminal ablation with two diode lasers (940 and 1319 nm) from acute changes to 4 months. *Lasers Surg. Med.*, **40**, 676–679.
25. Hoggan, B.L., Cameron, A.L., and Maddern., G.J. (2009) Systemic review of endovenous laser therapy versus surgery for the treatment of saphenous varicose veins. *Ann. Vasc. Surg.*, **23**, 277–287.
26. Leopardi, D., Hoggan, B.L., Fitridge, R.A., Woodruff, P.W.H., and Maddern., G.J. (2009) Systemic review of treatments for varicose veins. *Ann. Vasc. Surg.*, **23**, 264–276.
27. van den Bos, R., Arends, L., Kockaert, M., Neumann, M., and Nijsten, T. (2009) Endovenous therapies of lower extremity varicosities: a meta-analysis. *J. Vasc. Surg.*, **49**, 230–239.
28. der Kinderen, D.J., Disselhoff, C.V.M., Koten, J.W., de Bruin, P.C., Seldenrijk, C.A., and Moll, F.L. (2009) Histpathologic studies of the below-the-knee great saphenous vein after endovenous laser ablation. *Dermatol. Surg.*, **35**, 1985–1988.
29. Theivacumar, N.S., Darwood, R., and Gough, M.J. (2009) Neovascularisation and recurrence 2 years after varicose vein treatment for sapheno-femoral and great saphenous vein reflux: a comparison of surgery and endovenous laser ablation. *Eur. J. Vasc. Endovasc. Surg.*, **38**, 203–207.
30. Almeida, J.I., Kaufman, J., Göckeritz, O., Chopra, P., Evans, M.T., Hoheim, D.F., Makhoul, R.G., Richards, T., Wenzel, C., and Raines., J.K. (2009) Radiofrequency endovenous ClosureFAST versus laser ablation for the treatment of great saphenous reflux: a multicenter, single-blinded, randomized study (RECOVERY Study). *J. Vasc. Interv. Radiol.*, **20**, 752–759.
31. Lewis, B.D. (2010) Re: radiofrequency endovenous ClosureFAST versus laser ablation for the treatment of great saphenous reflux: a multicenter, single-blinded, randomized study (RECOVERY Study). *J. Vasc. Interv. Radiol.*, **21**, 302; author reply 302–303.
32. Maurins, U., Rabe, E., and Pannier, F. (2009) Does laser power influence the results of endovenous ablation (EVLA) of incompetent saphenous veins with the 1470 nm diode laser? A prospective randomized study comparing 15 and 25 W. *Int. Angiol.*, **28**, 32–37.
33. Pannier, F., Rabe, E., and Maurins., U. (2009) First results with a new 1470-nm diode laser for endovenous ablation of incompetent saphenous veins. *Phlebology*, **24** (1), 26–30.
34. Schmedt, C.G., Meissner, O.A., Hunger, K., Babaryka, G., Ruppert, V., Sadeghi-Azandaryani, M.,

Steckmeier, B.M., and Sroka, R. (2007) *J. Vasc. Surg.*, **45** (5), 1047–1058.

35 Sroka, R., Schmedt, C.G., Steckmeier, S., Meissner, O.A., Beyer, W., Babaryka, G., and Steckmeier, B. (2006) *Med. Laser Appl.*, **21** (1), 15–22.

36 Beck, T.J., Burgmeier, C., Blagova, R., Steckmeier, B., Hecht, V., Schmedt, C.G., and Sroka, R. (2007) *Med. Laser Appl.*, **22** (4), 238–241.

37 Blagova, R., Burgmeier, C., Steckmeier, S., Steckmeier, B., Babaryka, G., Beck, T.J., and Sroka, R. (2007) *Med. Laser Appl.*, **22** (4), 242–247.

38 Mordon, S.R., Wassmer, B., and Zemmouri, J. (2007) *Lasers Surg. Med.*, **39** (3), 256–265.

39 Maurins., U. (2008) Presentation No. 50 at Conference of Deutschen Gesellschaft für Phlebologie, Bochum, 17 October 2008.

40 de Medeiros, C.A. and Luccas, G.C. (2005) *Dermatol. Surg.*, **31** (12), 1685–1694, discussion 1694.

41 Rasmussen, L.H., Bjoern, L., Lawaetz, M., Blemings, A., Lawaetz, B., and Eklof, B. (2007) *J. Vasc. Surg.*, **46** (2), 308–315.

42 Darwood, R.J., Theivacumar, N., Dellagrammaticas, D., Mavor, A.I., and Gough, M.J. (2008) *Br. J. Surg.*, **95** (3), 294–301.

43 Kalteis, M., Berger, I., Messie-Werndl, S., Pistrich, R., Schimetta, W., Pölz, W., and Hieller, F. (2008) *J. Vasc. Surg.*, **47** (4), 822–829; discussion 829.

44 Disselhoff, B.C., der Kinderen, D.J., and Moll, F.L. (2008) *Phlebology*, **23** (1), 10–14.

45 Disselhoff, B.C., der Kinderen, D.J., Kelder, J.C., and Moll, F.L. (2008) *Br. J. Surg.*, **95** (10), 1232–1238.

64
Laser Treatment of Cerebral Ischemia

Ying-Ying Huang, Vida J. Bil De Arce, Luis De Taboada, Thomas McCarthy, and Michael R. Hamblin

64.1
Introduction

Low-level laser (light) therapy (LLLT) has been clinically applied for many indications in medicine that require the following processes: protection from cell and tissue death, stimulation of healing and repair of injuries, and reduction of pain, swelling, and inflammation. The notable lack of any effective drug-based therapies for stroke has rapidly increased researchers' interest in the use of light as a viable approach to mitigating stroke. The fact that near-infrared (NIR) light can penetrate into the brain and spinal cord allows noninvasive treatment to be carried out with a low likelihood of treatment-related adverse events. Although in the past it was generally accepted that the central nervous system could not repair itself, recent discoveries in the field of neuronal stem cells have brought this dogma into question. LLLT may have beneficial effects in the acute treatment of brain damage after stroke or traumatic brain injury (TBI).

64.2
The Problem of Cerebral Ischemia

Stroke is the third leading cause of death in the United States, after heart disease and cancer [1]. Strokes can be classified into two major categories, ischemic and hemorrhagic. Ischemic strokes account for over 80% of all strokes. The most common cause of ischemic stroke is the blockage of an artery in the brain by thrombosis, embolism, or stenosis. Cerebral ischemia is a condition in which there is insufficient blood flow to the brain to meet metabolic demand. This leads to poor oxygen supply or cerebral hypoxia and thus to the death of brain tissue or cerebral infarction/ischemic stroke [2].

64.3
Mechanisms of Brain Injury After Cerebral Ischemia

When a cerebral ischemia occurs, the blood supply to the brain is interrupted, and brain cells are deprived of the glucose and oxygen that they need to function. Normally, the brain requires a large amount of oxygen to generate sufficient adenosine triphosphate (ATP) by oxidative phosphorylation to maintain and restore ionic gradients [3]. The ischemic neuron becomes depolarized as ATP is depleted and membrane ion-transport systems fail. The resulting influx of calcium leads to the release of a number of neurotransmitters, including large quantities of glutamate, which, in turn, activates other excitatory receptors on other neurons. These neurons then become depolarized, causing further calcium influx, further glutamate release, and local amplification of the initial ischemic insult. This massive calcium influx also activates various degradative enzymes, leading to the destruction of the cell membrane and other essential neuronal structures [4]. Reactive oxygen species (ROS), nitric oxide (NO) and arachidonic acid are generated by this process, which lead to further neuronal damage. Within hours to days after a stroke, specific genes are activated, leading to the formation of cytokines and other factors that, in turn, cause further inflammation and microcirculatory compromise [4].

64.4
Current Treatment for Cerebral Ischemia

Thrombolytic therapy is the only intervention of proven and substantial benefit for select patients with acute cerebral ischemia [5]. The main agent that has been employed is recombinant tissue plasminogen activator (t-PA) [6], which works by converting the proenzyme plasminogen to the activated enzyme plasmin, which dissolves fibrin clots into low molecular weight fibrin degradation products. Other thrombolytics that have been used include streptokinase [7, 8]. Time lost is brain lost in acute cerebral ischemia. A pooled analysis of all 2775 patients enrolled in the first six intravenous tPA trials provided clear and convincing evidence of a time-dependent benefit of thrombolytic therapy [9].

64.5
Investigational Neuroprotectants and Pharmacologic Intervention

Neuroprotection is defined as any strategy, or combination of strategies, that antagonizes, interrupts, or slows the sequence of injurious biochemical and molecular events that, if left unchecked, would eventuate in irreversible ischemic injury to the brain. An enormous variety of agents and strategies have received clinical scrutiny, each justified by a pathophysiologic rationale. In all, nearly 165 ongoing or completed clinical trials have been published [10]. There has been an almost universal outcome of failure in all these trials, with exceptions of some small hint of efficacy in only a few cases.

64.6 Mechanism of Low-Level Laser Therapy

The most understood mechanism of action involves the relationship between lasers and cytochrome c oxidase. The first law of photobiology states that biological responses of living cells to photon irradiation are initiated by photon absorption in intracellular chromophores or photoacceptors [11]. The absorbed photon's energy excites the photoacceptor molecule into a more energetic (higher) electronic state, resulting in a physical and/or chemical molecular change, ultimately leading to the cell's biological response. It was suggested in 1989 by Karu that the mechanism of LLLT at the cellular level was based on the absorption of monochromatic visible and NIR light by cytochrome c oxidase [12] (see Figure 64.1). Cytochrome c oxidase is located in the inner mitochondrial membrane and plays a central role in eukaryotic cells. The cytochrome c oxidase enzyme complex contains two copper centers, the first of which contains a broad-wavelength absorption peak when oxidized. As a terminal enzyme in the respiratory chain, it delivers protons across the inner mitochondrial membrane and permits the formation of ATP by oxidative phosphorylation [4]. In cerebral ischemia, the neurons become seriously deprived of oxygen and their ATP levels fall precipitously. Laser stimulation of respiration not only increases ATP but also sets in motion several signaling pathways that lead to activation of genes involved in neuroprotection, anti-apoptosis, anti-inflammation, and activation of brain repair processes mediated by neural progenitor cells.

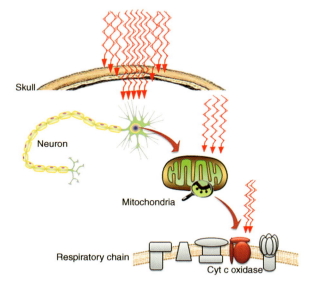

Figure 64.1 Mechanisms of action of transcranial laser therapy for stroke. NIR light penetrates through the scalp and skull and is absorbed by cytochrome c oxidase in the mitochondria of the damaged cortical neurons. Signaling pathways are activated that prevent neuronal death and stimulate brain repair.

64.7
LLLT on Neuronal Cells

In cultured human neuronal cells, LLLT resulted in doubling of the ATP content in normal human neural progenitor cells (NHNPs) [13]. LLLT also increased heat shock proteins and preserved mitochondrial function [4]. Ignatov et al. [14] showed that He–Ne laser irradiation of a pond snail neuron at a dose of 0.07 mJ increased the amplitude of the potential-dependent slow potassium current. More recently, Ying et al. reported that pretreatment with NIR light via a light-emitting diode (LED) significantly suppressed rotenone or 1-methyl-4-phenylpyridinium (MPP)-induced apoptosis in both striatal and visual cortical neurons from newborn rats [15]. These in vitro findings suggest that LLLT could have beneficial effects in animal models of neurologic disorders and could be a treatment for stroke.

64.8
LLLT for Stroke in Animal Models

Animals that have been used in cerebral ischemia models for various specific applications include rats, mice, gerbils, cats, rabbits, dogs, pigs, and nonhuman primates. Although there is no one animal model that identically mimics stroke in humans, the use of animal models is essential for the development of therapeutic interventions for stroke. Ischemia is typically induced by occluding the middle cerebral artery (MCA) in the animal. The MCA is most often used to simulate human stroke as most human strokes are due to occlusion of this vessel or one of its branches [16]. Laser treatment for stroke should penetrate the skull bones and scalp. Transmittance of all tissues increases progressively with wavelength from 600 to 814 nm. Tissue thickness, optical absorption, and scattering are major influencing factors [17]. Lychagov et al. presented results of measurements of transmittance of high power laser irradiation through skull bones and scalp [18].

Transcranial laser therapy (TLT) at 808–810 nm can penetrate the brain and was shown to lead to enhanced production of ATP in the rat cerebral cortex [4]. Findings of increased neurogenesis in the subventricular zone (SVZ) were reported in an ischemic stroke animal model treated with TLT [19]. Based on these findings, it is thought that TLT may have multiple mechanisms of action and could be beneficial in acute ischemic stroke [20]. Figure 64.2 shows non-invasive delivery of 810 nm laser light into the brain of a rat that had been subjected to stroke induction.

Lapchak and De Taboada [21] were the first to demonstrate that TLT substantially increased cortical ATP production in an embolic stroke model. They hypothesized that this amount should be sufficient to improve or maintain mitochondrial function, improve neuronal survival, and produce behavioral improvement. Given the evidence regarding the anti-apoptotic effect of NIR irradiation and the apparent

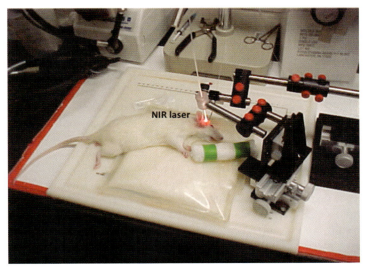

Figure 64.2 Rat model of transcranial laser therapy for stroke. NIR laser fiber allows 810 nm light to penetrate into the stroked rat brain.

extended time window of efficacy, during which apoptotic cell death could still be occurring in the penumbra, it is possible that TLT acts in a neuroprotective fashion. It is also likely that TLT enhances recovery of function because there is evidence that lesion volume is not altered, but neurologic function is restored [19].

For several years, enhanced mitochondrial function with subsequent preservation of the ischemic penumbra and improved outcomes has been the hypothesized mechanism. Only relatively recently has this been demonstrated using an *in vivo* model. In 2004, Lapchak *et al.* [22] used the rabbit small clot embolic stroke model and applied a gallium–arsenic diode NIR laser within 6 h of symptom onset. Significant and sustained improvement was demonstrated in behavioral function measured 24 h after treatment and again at 21 days from stroke onset [23]. Similar studies confirmed this beneficial effect despite using a rodent and a different model of inducing strokes (rat middle cerebral artery occlusion) [2]. There were differences in the treatment times for which TLT was effective, but both models resulted in statistically significant behavioral improvement that warranted human trials.

A recently completed trial using rabbits and an embolic model tested TLT applied after administration of intravenous tPA. The results showed that this combination did not adversely affect hemorrhage rate, volume, or survival compared with tPA alone [24]. Based on this study, and theoretically the different mechanisms of action of TLT and tPA could synergistically treat strokes, a clinical trial testing TLT plus tPA was recommended and is on the horizon.

64.9
TLT in Clinical Trials for Stroke

The NeuroThera effectiveness and safety trial-1 (NEST-1) evaluated the safety and preliminary effectiveness of the NeuroThera Laser System in the ability to improve 90-day outcomes in ischemic stroke patients treated within 24 h from stroke onset [25]. The NEST-1 study indicated that infrared laser therapy had shown initial safety and effectiveness for the treatment of ischemic stroke in humans when initiated within 24 h of stroke onset and a second larger trial was warranted.

Recently, the second clinical trial of the effectiveness and safety trial of NIR laser treatment within 24 h from stroke onset (NEST-2) was completed [26]. This study was a prospective, double-blind, randomized, sham controlled, parallel group, multi-center study that included sites in Sweden, Germany, Peru, and the United States, and enrolled 660 subjects. TLT was applied within 24 h from stroke onset. Comparable results were seen for the other outcome measures. Although no prespecified test achieved significance, a *post hoc* analysis of patients with a baseline National Institutes of Health Stroke Scale (NIHSS) score of <16 showed a favorable outcome at 90 days on the primary end point ($p < 0.044$). Mortality rates and serious adverse events did not differ between groups with 17.5 and 17.4% mortality and 37.8 and 41.8% serious adverse events for TLT and sham control patients, respectively.

64.10
Conclusions and Future Outlook

It is now generally accepted by the LLLT community that NIR (810 nm) laser light applied transcranially can beneficially impact ischemic stroke. It may take somewhat longer to be accepted by the stroke community. Many questions still remain to be answered: for instance, what is the best time after stroke to apply the light; does the use of multiple sessions of treatment (e.g., every day) present any increased benefit over a single treatment session; is pulsed light better that continuous-wave light and if so what are the best pulse parameters; are there any wavelengths that are better than 810 nm? We can envisage the time when all hospital emergency rooms will have the ability to deliver TLT to stroke patients and possibly also TBI patients.

Acknowledgments

This work was supported by Photothera Inc. and grants from the US Air Force MFEL Program (FA9550-04-1-0079), Center for Integration of Medicine and Innovative Technology (DAMD17-02-2-0006), and CDMRP Program in TBI (W81XWH-09-1-0514).

References

1 Wolf, P.A. (1990) An overview of the epidemiology of stroke. *Stroke*, **21**, II4–II6.

2 Ginsberg, M.D. (2009) Current status of neuroprotection for cerebral ischemia: synoptic overview. *Stroke*, **40**, S111–S114.

3 Pasantes-Morales, H. and Tuz, K. (2006) Volume changes in neurons: hyperexcitability and neuronal death. *Contrib. Nephrol.*, **152**, 221–240.

4 Streeter, J., De Taboada, L., and Oron, U. (2004) Mechanisms of action of light therapy for stroke and acute myocardial infarction. *Mitochondrion*, **4**, 569–576.

5 Adams, H., Adams, R., Del Zoppo, G., and Goldstein, L.B. (2005) Guidelines for the early management of patients with ischemic stroke – 2005 guidelines update – a scientific statement from the Stroke Council of the American Heart Association/American Stroke Association. *Stroke*, **36**, 916–923.

6 Barushka, O., Yaakobi, T., and Oron, U. (1995) Effect of low-energy laser (He–Ne) irradiation on the process of bone repair in the rat tibia. *Bone*, **16**, 47–55.

7 Donnan, G.A., Davis, S.M., Chambers, B.R., Gates, P.C., Hankey, G.J., McNeil, J.J., Rosen, D., Stewart-Wynne, E.G., and Tuck, R.R. (1995) Trials of streptokinase in severe acute ischaemic stroke. *Lancet*, **345**, 578–579.

8 Furlan, A., Higashida, R., Wechsler, L., Gent, M., Rowley, H., Kase, C., Pessin, M., Ahuja, A., Callahan, F., Clark, W.M., Silver, F., and Rivera, F. (1999) Intra-arterial prourokinase for acute ischemic stroke. The PROACT II study: a randomized controlled trial. Prolyse in acute cerebral thromboembolism. *JAMA*, **282**, 2003–2011.

9 Hacke, W., Donnan, G., Fieschi, C., Kaste, M., von Kummer, R., Broderick, J.P., Brott, T., Frankel, M., Grotta, J.C., Haley, E.C. Jr., Kwiatkowski, T., Levine, S.R., Lewandowski, C., Lu, M., Lyden, P., Marler, J.R., Patel, S., Tilley, B.C., Albers, G., Bluhmki, E., Wilhelm, M., and Hamilton, S. (2004) Association of outcome with early stroke treatment: pooled analysis of ATLANTIS, ECASS, and NINDS rt-PA stroke trials. *Lancet*, **363** 768–774.

10 Ginsberg, M.D. (2008) Neuroprotection for ischemic stroke: past, present and future. *Neuropharmacology*, **55**, 363–389.

11 Sutherland, J.C. (2002) Biological effects of polychromatic light. *Photochem. Photobiol.*, **76**, 164–170.

12 Karu, T.I. (1989) Laser biostimulation: a photobiological phenomenon. *J. Photochem. Photobiol. B*, **3**, 638–640.

13 Oron, U., Ilic, S., De Taboada, L., and Streeter, J. (2007) Ga–As (808 nm) laser irradiation enhances ATP production in human neuronal cells in culture. *Photomed. Laser Surg.*, **25**, 180–182.

14 Ignatov, Y.D., Vislobokov, A.I., Vlasov, T.D., Kolpakova, M.E., Mel'nikov, K.N., and Petrishchev, I.N. (2005) Effects of helium–neon laser irradiation and local anesthetics on potassium channels in pond snail neurons. *Neurosci. Behav. Physiol.*, **35**, 871–875.

15 Ying, R., Liang, H.L., Whelan, H.T., Eells, J.T., and Wong-Riley, M.T. (2008) Pretreatment with near-infrared light via light-emitting diode provides added benefit against rotenone- and MPP+-induced neurotoxicity. *Brain Res.*, **1243**, 167–173.

16 Philip, M., Benatar, M., Fisher, M., and Savitz, S.I. (2009) Methodological quality of animal studies of neuroprotective agents currently in phase II/III acute ischemic stroke trials. *Stroke*, **40**, 577–581.

17 Salomatina, E., Jiang, B., Novak, J., and Yaroslavsky, A.N. (2006) Optical properties of normal and cancerous human skin in the visible and near-infrared spectral range. *J. Biomed. Opt.*, **11**, 064026.

18 Lychagov, V.V., Tuchin, V.V., Vilensky, M.A., Reznik, B.N., Ichim, T., and De Taboada, L. (2006) Experimental study of NIR transmittance of the human skull. *Proc. SPIE*, **6085**, 60850T

19 Oron, A., Oron, U., Chen, J., Eilam, A., Zhang, C., Sadeh, M., Lampl, Y., Streeter, J., DeTaboada, L., and Chopp, M. (2006) Low-level laser therapy applied transcranially to rats after induction of stroke significantly reduces long-term

neurological deficits. *Stroke*, **37**, 2620–2624.
20 Lampl, Y. (2007) Laser treatment for stroke. *Expert Rev. Neurother.*, **7**, 961–965.
21 Lapchak, P.A. and De Taboada, L. (2010) Transcranial near infrared laser treatment (NILT) increases cortical adenosine-5′-triphosphate (ATP) content following embolic strokes in rabbits. *Brain Res.*, **1306**, 100–105.
22 Lapchak, P.A., Wei, J., and Zivin, J.A. (2004) Transcranial infrared laser therapy improves clinical rating scores after embolic strokes in rabbits. *Stroke*, **35**, 1985–1988.
23 Lapchak, P.A., Salgado, K.F., Chao, C.H., and Zivin., J.A. (2007) Transcranial near-infrared light therapy improves motor function following embolic strokes in rabbits: an extended therapeutic window study using continuous and pulse frequency delivery modes. *Neuroscience*, **148**, 907–914.
24 Lapchak, P.A., Han, M.K., Salgado, K.F., Streeter, J., and Zivin, J.A. (2008) Safety profile of transcranial near-infrared laser therapy administered in combination with thrombolytic therapy to embolized rabbits. *Stroke*, **39**, 3073–3078.
25 Lampl, Y., Zivin, J.A., Fisher, M., Lew, R., Welin, L., Dahlof, B., Borenstein, P., Andersson, B., Perez, J., Caparo, C., Ilic, S., and Oron, U. (2007) Infrared laser therapy for ischemic stroke: a new treatment strategy: results of the NeuroThera Effectiveness and Safety Trial-1 (NEST-1). *Stroke*, **38**, 1843–1849.
26 Zivin, J.A., Albers, G.W., Bornstein, N., Chippendale, T., Dahlof, B., Devlin, T., Fisher, M., Hacke, W., Do, W.H., Ilic, S., Kasner, S., Lew, R., Nash, M., Perez, J., Rymer, M., Schellinger, P., Schneider, D., Schwab, S., Veltkamp, R., Walker, M., Streeter, J., and NeuroThera Effectiveness and Safety Trial-2 Investigators (2009) Effectiveness and safety of transcranial laser therapy for acute ischemic stroke. *Stroke*, **40** (4), 1359–1364.

65
Laser Vision Correction
Diogo L. Caldas and Renato Ambrósio Jr.

65.1
Introduction

Refractive surgery evolved significantly during the last two decades, emerging as a separate ophthalmic sub-specialty when refractive surgical procedures became the most commonly performed elective surgery in medicine [1]. Refractive surgery is classified into three broad categories: refractive keratoplasty (keratoplasty means "molding the cornea"), refractive intraocular lenses (IOLs), and refractive scleral surgery [2]. This classification then divides into various specific surgical techniques, such as refractive keratotomy (RK) or keratomileusis. The advent of the excimer (shortened from excited dimer) laser, a rare gas and halogen argon fluoride (ArF) laser that emits a high-energy ultraviolet (UV) laser beam that is ultimately used to photoablate corneal stroma (i.e., reshaping corneal stroma by microscopically removing portions of the cornea in a precise, accurate, and safe fashion), was a great step forward in refractive surgery and, more specifically, refractive keratoplasty. Laser energy can be delivered either to the Bowman's layer on the corneal stromal surface in surface ablation (SA) procedures or deeper into the corneal stroma by the means of lamellar surgery in which a flap is created and then an excimer laser ablation is performed on the stromal bed in laser-assisted *in situ* keratomileusis (LASIK) [3]. The LASIK flap is created using a mechanical or laser (i.e., using a solid-state femtosecond laser) microkeratome.

Various SA and lamellar techniques are used for correcting myopia, hyperopia, and astigmatism, with high rates of success [95.4% patient satisfaction rate (range: 87.2–100%)] and predictability with few untoward negative consequences [4]. Like any other surgery, refractive surgery has indications and contraindications. In general, myopia up to $-10\,D$, hyperopia up to $+6\,D$, and astigmatism up to -6 are the current upper limits for keratorefractive laser vision correction. Each case, however, should be evaluated individually because these limits may change, depending on each individual patient's situation. In fact, currently one in four potential refractive candidates is deemed not a good candidate for refractive surgery. A stable refraction is critical, as also are other criteria (see http://www.fda.gov/

MedicalDevices/ProductsandMedicalProcedures/SurgeryandLifeSupport/LASIK/ucm061325.htm). Additionally, the patient must understand the risks, benefits, limitations including possible complications, and alternatives for these procedures.

65.2
Current Lasers Used in Refractive Surgery

65.2.1
The Excimer Laser

The 193 nm UV wavelength-emitting ArF excimer laser was introduced by Trokel *et al.* [5] in 1983 and was first used on human subjects by McDonald *et al.* in 1987 [6]. Since that time, there has been a tremendous evolution for refining its use in refractive keratoplasty procedures and these improvements in technology (hardware and software) have paralleled the evolution of surgical techniques and our basic scientific understanding of this type of surgery [3].

Early, first-generation excimer laser models used a broad beam with an expanding diaphragm to create simple spherical or spherical–cylindrical ablation profiles [7]. More sophisticated second- and third-generation laser models emerged using large scanning systems or slit beams to create more complicated ablation profiles without the need for masking agents. Further improvements in lasers occurred with the development of smaller beam delivery systems associated with eye trackers, resulting in fourth-generation laser models with more sophisticated algorithms to create smoother, more aspheric, and even custom ablation profiles. For example, the top two most popular lasers currently used by refractive surgeons in world are the VISX Star S4 and Allegretto Eye Q systems. The VISX Star S4 laser (Figure 65.1a) has variable spot scanning with beam diameters from 6.5 to 0.65 mm [3] with a variable repetition rate up to 20 Hz and the Allegretto Eye Q has a fixed flying spot with 0.8 mm and a frequency of 400 Hz (Figure 65.1b).

65.2.2
The Femtosecond Laser

The femtosecond laser produces a short electromagnetic pulse with a time duration on the order of femtoseconds (10^{-15} s) [8]. It therefore belongs in the category of ultrafast lasers and the generation of ultrashort pulses is nearly always achieved with the technique of passive mode locking. These ultrashort pulses are too brief to transfer heat or shock to the material being cut, which means that cutting, drilling, and machining occur with virtually no damage to surrounding material. Furthermore, this revolutionary laser can cut with extreme precision, making hairline cuts in thick materials along a computer-generated path. In the cornea, its automated computer application causes planar dissection possible due to the formation of cavitation microbubbles of CO_2 gas. The laser allows the surgeon to focus the laser energy at a particular depth in the transparent or even the mildly opaque cornea and

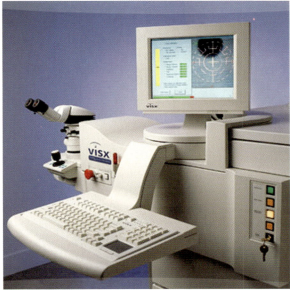

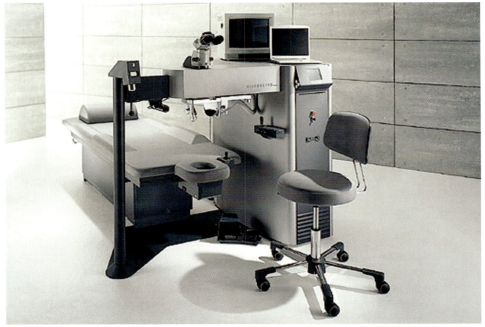

Figure 65.1 Excimer laser platforms: (a) VISX-AMO; (b) Allegretto Eye Q.

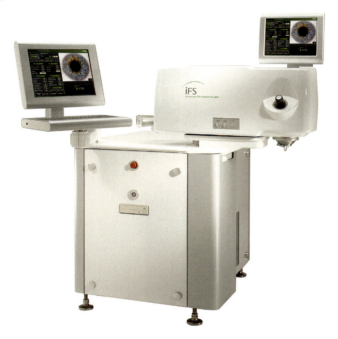

Figure 65.2 Femtosecond laser platform: Intralase iFs, 150 kHz (AMO).

then rapidly cut the tissue at that depth. Automated computer systems controlling the femtosecond laser applications allow the surgeon to pattern and size these cuts into customized shapes, creating a highly precise incision resulting in a perfect match of the donor and host tissue and a stronger junction (Figure 65.2).

While currently being used primarily for cutting LASIK flaps, femtosecond lasers are proving useful in other corneal applications also, including variously shaped and/or sized femtosecond laser-assisted keratoplasty procedures (e.g., geometric-shaped and varying sized penetrating grafts, varying depth and sized lamellar grafts), intracorneal ring segment implantation, astigmatic keratoplasty, and intrastromal inlays [8].

65.3
Refractive Surgery Techniques

The different laser refractive keratoplasty techniques are basically classified into two types: SA and lamellar surgery. Each has certain advantages and disadvantages compared with the other. As the cornea has five layers (epithelium, Bowman's layer, stroma, Descement's membrane, and endothelium) (Figure 65.3), SA techniques involve the removal of the epithelium and ablation of the anterior-most stromal surface (Bowman's layer and then the anterior stroma proper). Lamellar procedures

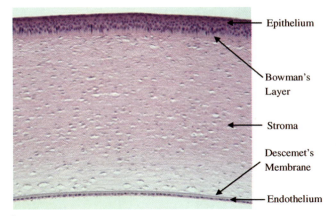

Figure 65.3 Histological section of the human cornea.

involve the creation of a flap containing the epithelium and a thin or thick layer of stroma and subsequent stromal ablation of the bed after the flap is reflected out of the way of the laser beam. At the conclusion of the procedure, the flap is repositioned back into its original place.

65.3.1
Surface Ablation

With SA, various techniques are used to remove the corneal epithelium before the excimer laser is applied to the surface of the corneal stroma. For example, in photorefractive keratectomy (PRK), the epithelium is either scraped or photoablated away. The latter is sometimes referred to as transepithelial PRK. Thus, after PRK or transepithelial PRK, a bare excimer ablated corneal stroma surface is left behind. In comparison, with laser-assisted subepithelial keratectomy (LASEK), instead of removing or excising the epithelium completely as in PRK, the surgeon cuts a round semi-circular area of epithelium with a fine-bladed instrument called a trephine. A dilute alcohol solution is then applied to loosen the cut epithelium inside the semi-circle where it attaches to Bowman's layer, and then the surgeon gently lifts and folds back the hinged loosened epithelial flap, exposing the underlying Bowman's layer and corneal stroma for the excimer laser ablation before repositioning the epithelial flap back into its original location. Finally, with epikeratome laser-assisted *in situ* keratomileusis (EpiLASIK), a unique epikeratome (a microkeratome with a blunt, oscillating blade) is used to separate the epithelium mechanically from the Bowman's layer and stroma, while suction is applied, to make a hinged epithelial flap, similar to a traditional LASIK flap. Unlike LASIK, no sharp blades or intrastromal cuts are made. It is very similar to LASEK, but no alcohol is required. These four techniques have the important advantage of having a high level of safety and security, especially in patients

Table 65.1 Cases where surface ablation may be the best option.

- Patients preference
- Predisposition for contact injury (e.g., martial arts practitioners)
- Anterior basement membrane (Cogan's) dystrophy
- Epithelial sloughing during LASIK in the contralateral eye
- Thin corneas in which the stromal residual bed would be less than 250–300 μm in LASIK
- Deep orbits or tight eyelid fissure causing poor exposure for the microkeratome base
- Flat corneas (<41 D) or steep corneas (>48 D)
- Previous surgery involving the conjunctiva: bleb associated with filtering procedure; scleral buckle for the treatment of retinal detachment
- Moderate dry eye prior to surgery

with thinner corneas or slight changes in its curvature. During wound healing of the corneal surface after the various SA procedures, transient pain or discomfort and blurred vision occur. Usually, the pain or discomfort is relieved by the third or fourth postoperative day and the blurred vision gradually improves during the first week until it stabilizes in at about 1–3 months postoperation. SA procedures typically do not reach the best or uncorrected visual potential of LASIK until the third month postoperatively, but it then tends to surpass LASIK surgery in the longer term in a statistically insignificant manner (Table 65.1). In any of these four techniques, if the subsequent excimer laser ablative step to Bowman's layer and stroma is refractively neutral, this technique is referred to as phototherapeutic keratectomy (PTK).

65.3.1.1 PRK

This technique has been used since the early era of excimer lasers, for nearly 20 years. It consists of mechanical removal of corneal epithelium using a scalpel blade, then ablating the anterior stroma with the excimer laser (Figure 65.4).

PRK has been eclipsed by LASIK, mainly because of the faster visual rehabilitation and less discomfort associated the latter during the early postoperative period. Despite this, PRK remains as an excellent option, particularly for mild to moderate corrections [6].

During early phases of corneal wound healing, haze formation associated with myopic treatments over 6 D was the major drawback and limitation of PRK. With the advent of single intraoperative application of topical mitomycin C (MMC), this SA-related wound healing complication has been substantially reduced and has broadened the treatment of SA procedures to higher refractive errors. Cases associated with thin corneas, recurrent erosions, or with a predisposition for eye trauma (martial arts, military, etc.) were among the most important indications for SA procedures since they retain the most biomechanical properties [9].

65.3.1.2 Transepithelial PRK

The excimer laser is used to remove the epithelium in transepithelial PRK [9].

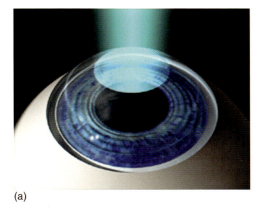

(a)

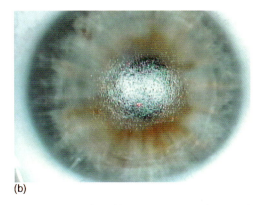

(b)

Figure 65.4 Surface ablation: (a) cartoon diagram; (b) perioperative image after laser ablation.

65.3.1.3 **LASEK**

In this technique, an epithelial flap is detached after application of a dilute alcohol solution (typically 18–25%). After laser ablation, the epithelial sheet is repositioned, like the stromal flap is in the LASIK procedure [3]. Most studies suggest that patients with LASEK will experience less pain, faster visual recovery, and less haze compared with PRK or transepithelial PRK [10–18]. The mechanism through which the epithelial flap could provide protection and improve clinical outcomes is not completely understood, but most likely is due to retained epithelial basement membrane, which acts as via bidirectional intrinsic binding and storage properties of wound healing-related cytokines and growth factors.

65.3.1.4 **EpiLASIK**

Here the dissection of the epithelium is carried out with an apparatus similar to the microkeratome, but it is blunt and does not penetrate the Bowman's layer or corneal stroma. It also avoids the use of chemical agents and, at least theoretically, increases the viability of the epithelial cells, which if intact and retain their superficial squamous layer zonula occludens-type junctional properties, prevent or abrogate exposure to tear film-derived cytokines and growth factors. Some surgeons deviate from conventional EpiLASIK and discard the epithelial flap during the procedure, thus making this latter variant another type of PRK technique since bare ablated stroma remains at the conclusion of the procedure [9].

65.3.2
LASIK

In LASIK, a flap including the epithelium and a portion of the anterior stroma with a total thickness of 120–180 μm is created with the mechanical microkeratome or femtosecond laser and then the stromal bed is accessible for excimer laser ablation [9] (Figure 65.5). The main advantage of LASIK over PRK is related to maintaining an intact, undamaged central corneal epithelium, which increases comfort during the early postoperative period, allows for rapid visual recovery,

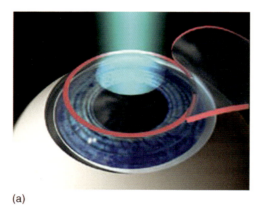

(a)

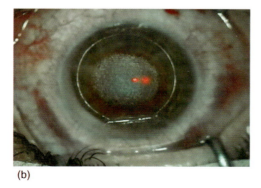

(b)

Figure 65.5 LASIK: (a) cartoon diagram; (b) perioperative image after laser ablation.

Table 65.2 Significant advantages of LASIK.

- Faster visual rehabilitation with earlier postoperative stabilization of visual acuity
- Less postoperative patient discomfort
- Attenuated wound healing and less stromal haze formation
- Possibly improved predictability, stability, and corneal clarity in higher correction groups
- Shorter duration of postoperative medications use
- Easier enhancement procedure

and attenuates wound healing response (Table 65.2) [9]. Microkeratome technology has undergone important developments and improvements over the last decade that have improved the overall safety of the LASIK procedure. This further increased the popularity of LASIK, making it the dominant procedure in refractive surgery today [3].

However, it should be mentioned that there are also complications with this procedure (see the FDA website cited in the Introduction), mainly related to long-term biomechanical instability, such as ectasia development or traumatic flap detachments, and infection [19].

65.3.2.1 Mechanical microkeratome LASIK

There are several models of mechanical microkeratomes, manual and automated. Automated microkeratomes most commonly have electric motors that drive blade oscillation, translation of the microkeratome head, or both. Hinge location, flap diameter, average flap thickness, suction pressure, and loss of suction cut-off are just a few of the features that vary from model to model. Average flap diameter and flap thickness will be important determinants of the dioptric range and maximum pupil sizes of patients who can be treated with a particular mechanical microkeratome. The chance of complications associated with creating the LASIK flap using the mechanical microkeratome is about one in 500 cases.

65.3.2.2 Femtosecond laser LASIK

The laser pulses are focused within the corneal stroma by a computerized system to produce hundreds of spots or cavitation bubbles placed close together in a spiral or rastered pattern to produce a stromal incision below the surface of the cornea. Pulses are stacked along the periphery of the lamellar incision until a cut is made in the corneal surface. One area of the periphery is spared to produce a hinge. Once the flap has been formed with the femtosecond laser, the surgeon peels back the flap and applies the ablation to the stromal bed with the excimer laser.

There are a number of advantages to the femtosecond laser compared with mechanical microkeratomes, including better control over flap thickness, diameter, and hinge position, the manufacture of oval flaps, and the creation of an inverted insertion angle, which increases the biomechanical strength of its adhesion.

65.4
Custom Corneal Ablations

Laser vision correction is intended primarily to fix the spherical and/or cylindrical refractive errors of the eye, which are called lower order aberrations. However, some patients experience unpleasant visual distortions after surgery, such as blurring, ghosting, and halos, especially associated with night vision [20].

New technologies available today allow the study of refractive optics and its application in refractive surgery through the use of computerized corneal topography, tomography, and wavefront analysis. This allows refractive surgeries to be programmed to correct high-order aberrations and sometimes complex corneal irregularities and, thus, improves the quality of vision of patients.

References

1 Ambrósio, R. Jr., Jardim, D., Netto, M.V., and Wilson, S.E. (2007) Management of unsuccessful lasik surgery. *Compr. Ophthalmol. Update*, **8** (3), 125–141; discussion 143–144.

2 Waring, G.O. (2009) Classification and terminology of refractive surgery, in *Corneal Surgery. Theory, Technique and Tissue*, 4th edn (eds F.S. Brightbill, P.J. McDonnell, C.N. McGhee, A.A. Farjo, and O.N. Serdarevic), Mosby Elsevier, St. Louis, MO, pp. 697–709.

3 Ambrósio, R. Jr. and Wilson, S.E. (2003) LASIK vs. LASEK vs. PRK: advantages and indications. *Semin. Ophthalmol.*, **18** (1), 2–10.

4 Solomon, K.D., Fernández de Castro, L.E., Sandoval, H.P., Biber, J.M., Groat, B., Neff, K.D., Ying, M.S., French, J.W., Donnenfeld, E.D., and Lindstrom, R.L. (2009) Joint LASIK Study Task Force. LASIK world literature review: quality of life and patient satisfaction. *Ophthalmology*, **116**, 691–701.

5 Trokel, S.L., Srinivasan, R., and Braren, B. (1983) Excimer laser surgery of the cornea. *Am. J. Ophthalmol.*, **96**, 710–715.

6 McDonald, M.B., Liu, J.C., Byrd, T.J., Abdelmegeed, M., Andrade, H.A., Klyce, S.D., Varnell, R., Munnerlyn, C.R., Clapham, T.N., and Kaufman, H.E. (1991) Central photorefractive keratectomy for myopia. Partially sighted and normally sighted eyes. *Ophthalmology*, **98**, 1327–1337.

7 Dawson, D.G., Manns, F., and Lee, Y. (2009) Excimer laser ablations principles and ablation profiles, in *Corneal Surgery. Theory, Technique and Tissue*, 4th edn (eds F.S. Brightbill, P.J. McDonnell, C.N. McGhee, A.A. Farjo, and O.N. Serdarevic), Mosby Elsevier, St. Louis, MO, pp. 777–787.

8 Farjo, Q.A. and Farjo, A.A. (2009) Femtosecond laser-assisted corneal surgery, in *Corneal Surgery. Theory, Technique and Tissue*, 4th edn (eds F.S. Brightbill, P.J. McDonnell, C.N. McGhee, A.A. Farjo, and O.N. Serdarevic), Mosby Elsevier, St. Louis, MO, pp. 861–870.

9 Ambrósio, R. Jr. and Guerra, F., Refractive laser ablations on the cornea: Understanding new terminology in refractive surgery. *Pearls in Ophthalmology*, http://www.medrounds.org/ophthalmology-pearls/2009/06/refractive-laser-ablations-on-cornea.html (last accessed 28 April 2010).

10 Azar, D.T., Ang, R.T., Lee, J.B., Kato, T., Chen, C.C., Jain, S., Gabison, E., and Abad, J.C. (2001) Laser subepithelial keratomileusis: electron microscopy and visual outcomes of flap photorefractive keratectomy. *Curr. Opin. Ophthalmol.*, **12**, 323–328.

11 Shah, S., Sebai Sarhan, A.R., Doyle, S.J., Pillai, C.T., and Dua, H.S. (2001) The epithelial flap for photorefractive keratectomy. *Br. J. Ophthalmol.*, **85**, 393–396.

12. Lee, J.B., Seong, G.J., Lee, J.H., Seo, K.Y. Lee, Y.G., and Kim F E.K. (2001) Comparison of laser epithelial keratomileusis and photorefractive keratectomy for low to moderate myopia. *J. Cataract Refract. Surg.*, **27**, 565–570.
13. Claringbold, T.V. II (2002) Laser-assisted subepithelial keratectomy for the correction of myopia. *J. Cataract Refract. Surg.*, **28**, 18–22.
14. Kornilovsky, I.M. (2001) Clinical results after subepithelial photorefractive keratectomy (LASEK). *J. Refract. Surg.*, **17**, S222–S223.
15. Scerrati, E. (2001) Laser *in situ* keratomileusis vs. laser epithelial keratomileusis (LASIK vs. LASEK). *J. Refract. Surg.*, **17**, S219–S221.
16. Lee, J.B., Choe, C.M., Seong, G.J., Gong, H.Y., and Kim, E.K. (2002) Laser subepithelial keratomileusis for low to moderate myopia: 6-month follow-up. *Jpn. J. Ophthalmol.*, **46**, 299–304.
17. Vinciguerra, P. and Camesasca, FI. (2002) Butterfly laser epithelial keratomileusis for myopia. *J. Refract. Surg.*, **18**, S371–S373.
18. Rouweyha, R.M., Chuang, A.Z., Mitra, S., Phillips, C.B., and Yee, R.W. (2002) Laser epithelial keratomileusis for myopia with the autonomous laser. *J. Refract. Surg.*, **18**, 217–224.
19. Ambrósio, R. Jr. and Wilson, S.E. (2001) Complications of laser in situ keratomileusis: etiology, prevention, and treatment. *J Refract Surg.*, **17**, 350–379.
20. Netto, M.V., Dupps, W. Jr., and Wilson, S.E. (2006) Wavefront-guided ablation: evidence for efficacy compared to traditional ablation. *Am. J. Ophthalmol.*, **141**, 360–368.

66
Laser Trabeculoplasty
Stephan Eckert

66.1
Laser Systems

Application of lasers to the trabecular meshwork has been used to lower the intraocular pressure (IOP) since the early 1970s [1]. Worthen and Wickham [2] first described an argon laser to perform trabeculoplasty in 1973 and Krasnov used a Q-switched ruby laser to perform goniopuncture or laseropuncture [3]. Wise and Witter described a modified technique which subsequently gained acceptance as an option for the treatment of primary open-angle glaucoma [4]. This procedure was originally, and most often in the past, performed using an argon laser at 488–514 nm. Several other lasers, including diode (810 nm) [5], continuous-wave (1064 nm), and frequency-doubled Nd:YAG (532 nm) [6] lasers, have been described to yield a reduction in IOP similar to that of argon laser trabeculoplasty (ALT) and are often used today (Tables 66.1 and 66.2). Latina and Park conducted a study designed to develop a procedure to target pigmented trabecular meshwork cells selectively while sparing adjacent cells and tissues from collateral thermal damage and to maintain the trabecular meshwork architecture [7]. This study formed the basis of the currently available selective laser trabeculoplasty (SLT) system (532 nm frequency-doubled Q-switched Nd:YAG laser) providing an only 3 ns pulse with a 400 µm beam diameter. Selective absorption of a very short laser pulse generates and spatially confines heat to pigmented targets. The thermal relaxation time of a chromophore is the time required to convert absorbed electromagnetic energy to heat energy. A short pulse duration is critical to prevent collateral tissue damage. If the energy deposition time is short, as in the case of a Q-switched laser, minimal heat transfer takes place. The 1 µs thermal relaxation time of melanin and a 3 ns SLT pulse essentially prevent thermal dissipation to surrounding tissue [8].

66.2
Mechanisms of Action

The exact mechanisms of action of laser trabeculoplasty have not yet been determined. The mechanical theory of IOP reduction proposes that a thermal burn

Table 66.1 Wavelengths of different laser systems for trabeculoplasty.

Laser	Wavelength (nm)
Argon laser (ALT)	514
Nd:YAG	1064, continuous wave
	532, frequency-doubled (SLT)
Diode laser	810

Table 66.2 Comparison of different laser systems for trabeculoplasty.

Parameter	ALT	SLT	DLT
No. of spots	50	50	50
Spot size (μm)	50	400	100
Pulse duration	100 ms	3 ns	100–200 ms
Power/energy	600–1200 mW	0.2–2.0 mJ	800–1200 mW

contracts tissue and stretches open adjacent, untreated regions of the trabecular meshwork to increase outflow [9]. However, histologic studies show that coagulative damage and tissue disruption to the trabecular meshwork are associated with ALT [10]. This injury might limit the effectiveness of future retreatment with ALT and theoretically also of other treatment modalities. It is also likely that there are contributing factors at the cellular level such as an increase in the division of trabecular cells and remodeling of the juxtacanalicular extracellular matrix representing a biological theory [11, 12]. Bradley et al. showed that laser trabeculoplasty induces the expression and secretion of both interleukin-1β and TNFα within the first 8 h after treatment. These cytokines then mediate increased trabecular stromelysin expression and initiate remodeling of the juxtacanalicular extracellular matrix [13]. After SLT, there was a total absence of coagulative damage with intact beams of the trabecular meshwork [14], supporting the view that the biological theory is the main mechanism for the IOP-lowering effect of SLT.

66.3
Methods of Treatment

Before laser trabeculoplasty, miotics (e.g., 1% pilocarpine) should be applied to ensure optimum access to the trabecular meshwork. With the patient seated in front of the laser slit-lamp system, a Goldmann three-mirror goniolens is coupled to the eye with 1% methylcellulose after application of topical anesthetic. In ALT, the smallest attainable laser spot (typically 50 μm) is placed in the anterior part of the trabecular

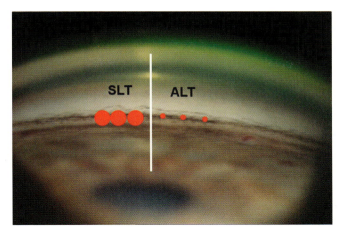

Figure 66.1 Application of laser spots in SLT compared with ALT.

meshwork (Figure 66.1) using 600–1500 mW and a 0.1 s pulse duration. The distance between the laser spots should be 2–3 spot diameters. Standard therapy is to deliver about 50 spots over 180°. Visible endpoints of the treatment are tissue blanching or a large vaporization bubble.

Diode laser trabeculoplasty (DLT) is a variation of ALT. Despite the use of a higher laser power, tissue blanching is less intense than with ALT. Vaporization bubbles are not observed.

In contrast, SLT uses a relatively large laser spot size of 400 μm. This is sufficient to cover the entire width of the trabecular meshwork, making accurate aiming less critical (Figures 66.1 and 66.2). Treatment of Sampaolesi's line should be avoided because it could cause corneal endotheliitis. Standard therapy is to deliver about 50 adjacent but nonoverlapping spots over 180°, or alternatively 100 spots over 360°, of the angle circumference. The energy is initially set at 0.7 mJ per pulse and can be adjusted to the desired treatment endpoint from 0.2 to 2.0 mJ. More heavily pigmented meshwork requires lower power. In contrast to ALT, blanching or large vaporization bubbles within the trabecular meshwork are not seen as an endpoint in SLT; however, tiny "champagne" bubble formation is used as an endpoint for setting the SLT pulse energy. They should be seen at least 50% of the time. Some authors recommend decreasing the energy level by 0.1 mJ when bubble formation is observed [8]. Pretreatment antiglaucoma medications should be resumed after laser trabeculoplasty. Usually the full effect of laser trabeculoplasty is reached after 4–6 weeks. IOP reduction after SLT is often seen even 1 day after treatment, but the period of time can be highly variable. If laser therapy is successful and the IOP reduction is sufficient, antiglaucomatous medication can be reduced.

It is important to check the IOP and exclude increased inflammation on the first postoperative day. Only if increased inflammation is present anti-inflammatory regimens are required (e.g., 1% prednisolone five times per day for 1 week).

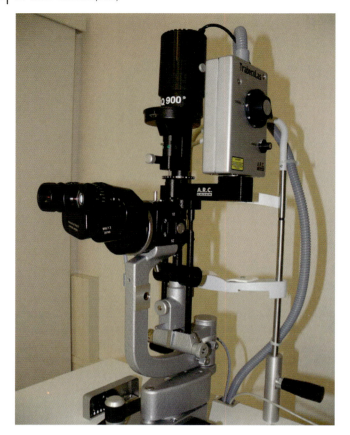

Figure 66.2 SLT laser (ARC Laser, Nürnberg, Germany).

66.4
Indications for Laser Trabeculoplasty

Laser trabeculoplasty is a method to reduce IOP in patients with primary open-angle glaucoma and also pseudoexfoliation glaucoma and pigmentary glaucoma. The low risk of adverse effects may make it a good alternative to long-term medical therapy, even in ocular hypertension patients without established glaucomatous optic nerve damage [8].

As an advantage, laser trabeculoplasty can lower IOP without relying on patient compliance with medications, which is a common problem in glaucoma therapy. It can replace or decrease the need for topical medications and therefore reduce systemic and local side effects such as allergy or chronic inflammation.

The IOP before laser trabeculoplasty should not exceed 25 mmHg and the target pressure should not be 25–30% below the preoperative pressure. A larger decrease in IOP cannot be achieved with laser treatment alone. Those patients who have a baseline IOP >21 mmHg are more likely to have a greater absolute value of IOP reduction and vice versa [15]. This could be the reason why patients with normal tension glaucoma are less successfully treated with laser trabeculoplasty.

Other indications can be old age and patients who refuse operation, have too high risk for operation, or have problems with application of their eyedrops (palsy, tremor).

66.5
Contraindications for Laser Trabeculoplasty

Contraindications for laser trabeculoplasty are neovascular glaucoma, goniosynechiae, post-traumatic glaucoma, and congenital glaucoma.

66.6
Effectiveness of Laser Trabeculoplasty

Clinical studies suggest that laser trabeculoplasty is efficacious in lowering IOP. Average reductions in IOP from 2.7 to more than 6 mmHg or 12–40% depending on the follow up-time have been reported. Response rates within the first postoperative year vary from 59 to 96% according to different definitions (Table 66.3) and usually decline with time. ALT and SLT seem to be equivalent in lowering IOP [8, 16]. Damji et al. conducted a prospective randomized trial comparing the effectiveness of a 180° ALT and SLT in lowering IOP in 176 eyes (87 cases of ALT and 89 cases of SLT). Eyes with ALT showed a significant decrease in IOP of 6.04 mmHg and eyes with SLT of 5.86 mmHg after 12 months. Failure rates were 32 and 36% for SLT and ALT, respectively [17]. The Glaucoma Laser Trial showed that in newly diagnosed patients with open-angle glaucoma, ALT was at least as effective as initial medical treatment with 0.5% timolol maleate even after 7 years [14]. In another study, selective laser trabeculoplasty was compared with 0,005% latanoprost. Differences in success rates between latanoprost and 360° SLT did not reach statistical significance [18]. The same study showed that 90° SLT is not effective.

Table 66.3 Studies on laser trabeculoplasty.

Study	Laser	No. of Patients	Follow-up (months)	IOP reduction (mmHg)	IOP reduction (%)
Latina et al. (1998) [19]	SLT	30	6	5.8	23.5
Cvenkel (2004) [20]	SLT	44	12	4.4	17.1
Best et al. (2005) [21]	SLT	269	24	2.7	12.1
Gracner et al. (2006) [22]	SLT	90	72	5.4	22.8
Juzych et al. (2004) [23]	ALT	154	60		23.5
Damji et al. (2006) [17]	ALT	87	12	6.04	25.8

Because of its nondestructive nature, multiple treatments with SLT are theoretically possible. If IOP reduction is inadequate after 6 weeks, treatment of the untreated 180° may be considered. The fibrosis and scarring that result after ALT do not occur after SLT because of the lack of structural damage to the trabecular meshwork. Hence repeated therapy may result in an additional reduction in IOP [8].

66.7
Complications After Laser Trabeculoplasty

Laser trabeculoplasty, especially SLT, is a very safe procedure, having only minor, transient, self-limiting, or easily controlled side effects [16]. For SLT, a complication rate of 4.5% was observed, much lower than the complication rate in ALT, which may reach up to 34% [8, 14]. Side effects of laser trabeculoplasty are postprocedure IOP spikes (6%), iritis [19], and peripheral anterior synechiae [24].

Early postoperative elevation of IOP in some patients has been observed in all published clinical studies. In all of those cases it resolved quickly with observation or additional antihypertensive treatment [5, 17, 19]. Highly significant IOP elevations have been reported after SLT in some cases of pigmentary glaucoma or glaucoma with heavy angle pigmentation [25]. Therefore, it is recommended to use lower energy or fewer applications, or to treat a smaller proportion of the angle (90°) in such patients. Close IOP monitoring and prophylactic oral carbonic anhydrase inhibitors may be helpful. Postoperative suppression of the inflammatory response may be relatively contraindicated because macrophage response may contribute to IOP lowering. Anterior synechiae appear more often after ALT, but may be prevented when ALT spots are placed in the anterior part of the trabecular meshwork [26].

Other side effects, namely nonspecific conjunctivitis (5%), eye pain (5%), blurred vision (<1%), corneal edema (<1%), and corneal lesion (<1%), have been described as transient and without sequelae [17, 19].

66.8
Conclusion

Laser trabeculoplasty has been shown to be safe, well tolerated, and effective in lowering IOP in patients with open-angle glaucoma. The preservation of the trabecular meshwork makes SLT a safe alternative to ALT that is potentially repeatable.

References

1 van der Zypen, E. and Frankhauser, F. (1982) Laser in the treatment of chronic simple glaucoma. *Trans. Ophthalmol. Soc. UK*, **102**, 147–153.

2. Worthen, D.M. and Wickham, M.G. (1974) Argon laser trabeculotomy. *Trans. Am. Acad. Ophthalmol. Otolaryngol.*, **78**, 371–375.

3. Krasnov, M.M. (1973) Laseropuncture of anterior chamber angle in glaucoma. *Am. J. Ophthalmol.*, **75**, 674–678.

4. Wise, J.B. and Witter, S.L. (1979) Argon laser therapy for open angle glaucoma: a pilot study. *Arch. Ophthalmol.*, **97**, 319–322.

5. Chung, P.Y., Schuman, J.S., and Netland, P.A. (1998) Five-year results of a randomized, prospective, clinical trial of diode vs argon laser trabeculoplasty for open-angle glaucoma. *Am. J. Ophthalmol.*, **126**, 185–190.

6. Guzey, M., Arslan, O., Tamcelik, N., and Satici, A. (1999) Effects of frequency-doubled Nd:YAG laser trabeculoplasty on diurnal intraocular pressure variations in primary open-angle glaucoma. *Ophthalmologica*, **213**, 214–218.

7. Latina, M.A. and Park, C. (1995) Selective targeting of trabecular meshwork cells: in vitro studies of pulsed and CW laser interactions. *Exp. Eye Res.*, **60** (4), 359–371.

8. Latina, M.A. and de Leon, J.M. (2005) Selective laser trabeculoplasty. *Ophthalmol. Clin. North Am.*, **18**, 409–419.

9. Van Buskirk, E.M., Pond, V., Rosenquist, R.C. et al. (1984) Argon laser trabeculoplasty, studies of mechanisms of action. *Ophthalmology*, **91**, 1005–1010.

10. Kramer, T.R. and Noecker, R.J. (2001) Comparison of the morphologic changes after selective laser trabeculoplasty and argon laser trabeculoplasty in human eye bank eyes. *Ophthalmology*, **108**, 773–779.

11. Parshley, D.E., Bradley, J.M., Fisk, A. et al. (1996) Laser trabeculoplasty induces stromelysin expression by trabecular juxtacanalicular cells. *Invest. Ophthalmol. Vis. Sci.*, **37**, 795–804.

12. Bylsma, S.S., Samples, J.R., Acott, T.S., and Van Buskirk, E.M. (1988) Trabecular cell division after argon laser trabeculoplasty. *Arch. Ophthalmol.*, **106**, 544–547.

13. Bradley, J.M.B., Anderssohn, A.M., Colvis, C.M. et al. (2000) Mediation of laser trabeculoplasty-induced matrix metalloproteinase expression by IL-1β and TNFα. *Invest. Ophthalmol. Vis. Sci.*, **41**, 422–430.

14. Glaucoma Laser Trial Research Group (1995) The Glaucoma Laser Trial (GLT) and glaucoma laser trial follow-up study, seven-year results. *Am. J. Ophthalmol.*, **120**, 718–731.

15. Johnson, P.B., Katz, L.J., and Rhee, D.J. (2006) Selective laser trabeculoplasty: predictive value of early intraocular pressure measurements for success at 3 months. *Br. J. Ophthalmol.*, **90** (6), 741–743.

16. Barkana, Y. and Belkin, M. (2007) Diagnostic and surgical techniques. *Surv. Ophthalmol.*, **52**, 634–654.

17. Damji, K.F., Bovell, A.M., Hodge, W.G. et al. (2006) Selective laser trabeculoplasty versus argon laser trabeculoplasty: results from a 1-year randomized clinical trial. *Br. J. Ophthalmol.*, **90**, 1490–1494.

18. Nagar, M., Ogunyomade, A., O'Brart, D.P.S., Howes, F., and Marshall, J. (2005) A randomized, prospective study comparing selective laser trabeculplasty with latanoprost for the control of intraocular pressure in ocular hypertension and open angle glaucoma. *Br. J. Ophthalmol.*, **89**, 1413–1417.

19. Latina, M.A. et al. (1998) Q-switched 532 nm Nd:YAG laser trabeculoplasty (selective laser trabeuloplasty): a multi-center, pilot, clinical study. *Ophthalmology*, **105**, 2082–2088.

20. Cvenkel, B. (2004) One-year follow-up of selective laser trabeculoplasty in open-angle glaucoma. *Ophthalmolojica*, **218**(1), 20–25.

21. Best, U.P., Domack, H., Schmidt, V. (2005) Long-term results after selective laser trabeculoplasty - a clinical study on 269 eyes. *Klin Monbl Augenheilkd*, **222**(4), 326–31.

22. Gracner, T., Falez, M., Gracner, B., Pahor, D. (2006) Long-term follow-up of selective laser trabeculoplasty in primary open-angle glaucoma. *Klin Monbl Augenheilkd*, **223**(9), 743–7.

23. Juzych, M.S., Chopra, V., Banitt, M.R., Hughes, B.A., Kim, C., Goulas, M.T., Shin, D.H. (2004) Comparison of long-term outcomes of selective laser trabeculoplasty versus argon laser

trabeculoplasty in open-angle glaucoma. *Ophthalmology*, **111**(10), 1853–9.
24 Reiss, G.R. *et al.* (1991) Laser trabeculoplasty. *Surv. Ophthalmol.*, **35**, 407–428.
25 Harasymowycz, P.J., Papamatheakis, D.G., Latina, M.A. *et al.* (2005) Selective laser trabeculoplasty complicated by intraocular elevation in eyes with heavily pigmented trabecular meshworks. *Am. J. Ophthalmol.*, **139**, 1110–1113.
26 Traverso, C.E., Greenridge, K.C., and Spaeth, G.L. (1984) Formation of peripheral anterior synechiae following argon laser trabeculoplasty. *Arch. Ophthalmol.*, **102**, 861–863.

67
Laser Photocoagulation
Bettina Fuisting and Gisbert Richard

67.1
Introduction

The penetration depth of laser light and the thermal expansion in ocular tissue depend on the wavelength of the laser light [1, 2]. The amount of laser energy required has to be chosen according to the pigmentation of the fundus and also the transmission of the optic media. Both factors influence the light absorption and temperature rise at the retina. Light absorption is mainly due to melanin. Hence the temperature rise caused by the light energy is higher in highly pigmented fundi. Laser photocoagulation in ophthalmology is used for the treatment of retinal and choroidal diseases. For safety reasons, at the beginning of a laser treatment the laser should be operated with low energy. The laser energy then should be raised carefully to achieve a desired laser effect.

Parameters for laser coagulation are laser spot size $= S$, duration (time) $= T$, and laser energy level $= E$ [2, 3].

In ophthalmology, several lasers with different wavelengths are used: argon green (532 nm), dye laser (510–630 nm), double-pulsed Nd:YAG laser (532 nm), and diode laser (532–810 nm) [2, 3].

The laser is fixed either to a slit lamp, (Figure 67.1) to a binocular head ophthalmoscope (indirect ophthalmoscopic laser delivery system), or to an endosonde during vitrectomy [4].

For laser treatment using a slit lamp, special laser contact lenses have to be used (Figure 67.2).

Handbook of Biophotonics. Vol.2: Photonics for Health Care, First Edition. Edited by Jürgen Popp, Valery V. Tuchin, Arthur Chiou, and Stefan Heinemann.
© 2012 Wiley-VCH Verlag GmbH & Co. KGaA. Published 2012 by Wiley-VCH Verlag GmbH & Co. KGaA.

Figure 67.1 Patient and ophthalmologist sitting at a slit lamp during laser treatment.

Figure 67.2 Laser contact lenses: Mainster wide-field lens and Goldmann three-mirror lens.

67.2
Retinal Degeneration

67.2.1
Retinal Tears, Retinoschisis, Retinal Detachment

Purpose: For retinal tears or detachments, a laser is used to create retinochoroidal adherence. Retinal tears are caused by vitreous traction and are often horseshoe-shaped [4] (Figure 67.3).

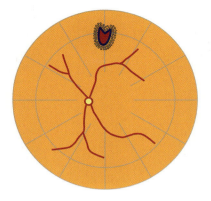 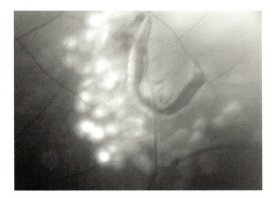

Figure 67.3 Retinal tear with laser treatment.

Laser treatment is applied in several rows (normally 3–4) circular to the retinal tear to avoid retinal detachment. Laser treatment for retinoschisis or small retinal detachments is applied centrally to the retinal pathology. If a vessel rupture due to vitreous traction occurs, so-called "feeder–vessel–treatment" is necessary to obliterate retinal vessels in order to avoid vitreous hemorrhage.

Laser parameters: $S = 200\text{–}500\,\mu\text{m}$, $T = 0.2\text{–}0.5\,\text{s}$, $E = 150\text{–}500\,\text{mW}$.

Visible effect: White to grayish effect.

Complications: Small laser spots with a high energy level can lead to ruptures of Bruchs' membrane. This may cause choroidal neovascularization, retinal holes with retinal hemorrhage, or retinal detachment. An extensive laser treatment may lead to epiretinal membranes, which might need vitrectomy with membrane peeling [5].

67.3 Diabetic Retinopathy (DRP)

67.3.1 Diabetic Macular Edema (DME)

Laser treatment in diabetic retinopathy (DRP) is applied to minimize hypoxic retinal areas and to reduce retinal oxygen demand. Thus the release of vascular endothelial growth factor (VEGF) could be avoided [5, 6].

Indications for focal laser treatment: Diabetic macular edema (DME), non-proliferative DRP without macular edema, combination of DME and DRP. In DRP, microaneurysms should be coagulated directly. Intraretinal hemorrhages and exudates should be surrounded by laser burns.

In DME, *grid laser* coagulation (see Figure 67.4) should be performed to reduce macular edema.

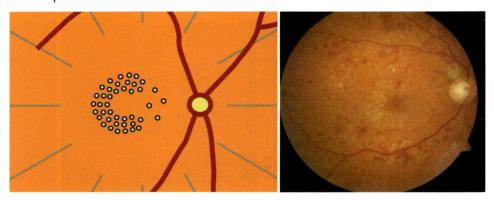

Figure 67.4 Grid laser coagulation.

Laser parameters: $S = 100–150\,\mu m$, $T = 0.1\,s$, $E = 80–150\,mW$.

The distance to the foveola is 300–500 μm. Burns should be barely visible and have a light grayish effect (off-whiteburn). In *ischemic* DRP, laser treatment is *not* indicated. If DRP occurs in combination with DME, DME has to be treated first to avoid an increase in DME.

67.3.2
Proliferative Diabetic Retinopathy

Proliferative retinopathy is the most severe ocular complication of diabetes.

Indications for *pan-retinal* laser treatment are retinal neovascularization or proliferative diabetic vitreoretinopathy (PDVP). Pan-coagulation should be performed between the retinal arcades and equator, placing about 2000 laser spots with a distance of 1–2 laser spot sizes between each other. During one laser session, no more than 500 laser spots should be applied. Laser treatment can be extended up to the outer periphery (see Figure 67.5).

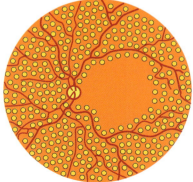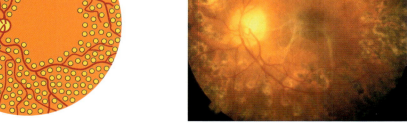

Figure 67.5 Pan-retinal coagulation.

Laser parameters: $S =$ up to 500 μm, $T =$ up to 0.5 s, $E =$ up to 500 mW.

Parameters depend on fundus pigmentation and transparency of optical media.

Visible effect: Grayish effect at the fundus.

Risks: Permanent visual field defects, chorioidal neovascularization, epiretinal membranes, retinal hemorrhage, retinal or choroidal detachment [5].

67.4
Retinal Vein Occlusions

67.4.1
Central Retinal Vein Occlusion (CRVO)

Retinal vein occlusions are common retinal vascular diseases [7, 8]. Central retinal vein occlusion (CRVO) is one of the most common reasons why an eye has to be removed completely by surgery (see Figure 67.6).

Indication: Ischemic central retinal vein occlusion.

Purpose: To avoid retinal neovascularization, macular edema, and rubeosis iridis caused by retinal ischemia (non-perfusion areas).

Laser parameters: $S = 200-500$ μm, $T = 0.2-0.5$ s, $E =$ up to 500 mW.

Visible effect: Gray–white burns; pan-retinal laser treatment (scatter treatment).

The distance between the laser spots has to be 0.5–2 laser spot size diameters. Laser spots (up to 2500 burns) should be placed between the hemorrhages.

If macular edema is present, grid laser coagulation (see above) may be indicated [7].

67.4.2
Branch Retinal Vein Occlusion (BRVO)

Branch retinal vein occlusion (BRVO) most often involves the superior or inferior temporal vein. BRVO occurs in older patients showing risk factors such as hyper-

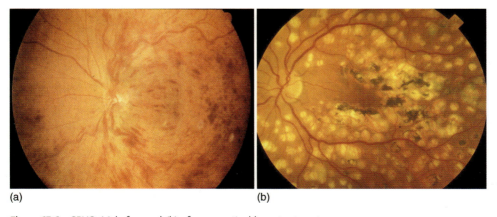

(a)　　　　　　　　　　　　　　　　　　(b)

Figure 67.6 CRVO (a) before and (b) after pan-retinal laser treatment.

tension, arteriosclerosis, diabetes, and heart disease. BRVO may occur when retinal arteries and veins share a common adventitia sheath; glaucoma is also a risk factor for BRVO.

Laser parameters: These are similar to the parameters in CRVO: mild scatter treatment in the affected hemisphere.

67.5
Macular Diseases and Choroidal Neovascularization

67.5.1
Age-Related Macular Degeneration (ARMD)

In patients older than 60 years, age-related macular degeneration (ARMD) is the leading cause of legal blindness in industrial countries (see Figure 67.7).

Purpose of laser treatment: Obliteration of abnormal vessels due to parafoveolar choroidal neovascularization (CNV).

Laser parameters: $S = 80–150\,\mu m$, $T = 0.1–0.3\,s$, $E = 100–250\,mW$.

Visible effect: Soft white and confluent.

67.5.2
Transpupillary Thermotherapy (TTT) for the Treatment of CNV

CNV can also be treated by transpupillary thermotherapy (TTT). For TTT, a diode laser with a wavelength of 810 nm is used (see Figure 67.8). After TTT treatment, *no* marked laser effect should be visible at the fundus. The spot size has to overlap the CNV completely [9].

Laser parameters: $S = 2000–3000\,\mu m$, $T = \leq 60\,s$, $E = 400–800\,mW$.

Risks and complications after TTT: Subretinal hemorrhage, iatrogen retinal burns followed by a retinal scar, central scotoma [5, 9].

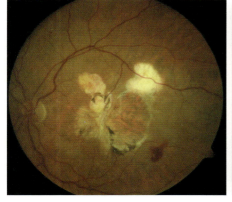
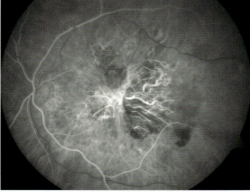

Figure 67.7 Age-related macular degeneration (ARMD).

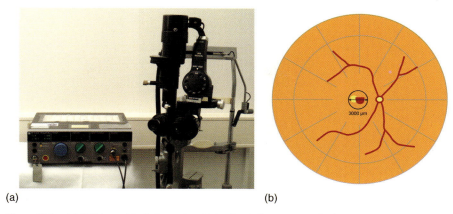

Figure 67.8 (a) TTT laser (diode laser: Ocu Light SLx, 810 nm; Iris Medical, Mountain View, CA, USA) and (b) treatment of ARMD with TTT laser.

67.5.3
Photodynamic Therapy (PDT)

For the sake of completeness, photodynamic therapy (PDT) should be mentioned for the treatment of ARMD, even though PDT is not a coagulative laser treatment. For PDT, a non-thermic laser with a wavelength of 689 nm is used [10].

Purpose: Treatment of subfoveal choroidal neovascularization.

About 15 min before laser treatment, a photosensitive dye (verteporphin, Visudyne®) is injected into a peripheral vein. The dye is activated by a diode laser.

Parameters: $S =$ variable, overlapping the CNV completely, $T = 83$ s, $E = 600 \, \text{W cm}^{-2}$.

PDT causes a photothrombosis of the CNV. In most patients, PDT has to be performed 2–3 times or even more often, every 3 months. PDT is used for so-called "classical" CNV and CNV due to pathologic high myopia.

67.5.4
Central Serous Chorioretinopathy (CSC)

Clinic: Idiopathic central serous neurosensory retinal detachment.

The subretinal fluid originates from the choroid [4]. Central serous chorioretinopathy (CSC) often affects younger men aged 20–50 years. It may also occur during pregnancy and in patients who have been treated with corticosteroids. Symptoms of CSC are blurred vision, metamorphopsia, and contrast sensitivity loss. In most cases only one eye is affected. Visual acuity ranges from 20/20 to 20/200 and may be improved by hyperopic correction [4]. The most important examination to evaluate CSC is fluorescein angiography (FLA). Fluorescein leaks through a focal defect in the retinal pigment epithelium, causing a "smokestack" configuration [4] (see Figure 67.9). Most patients have spontaneous resolution after 3–6 months. If leakage is still present after this time, laser coagulation has to be considered.

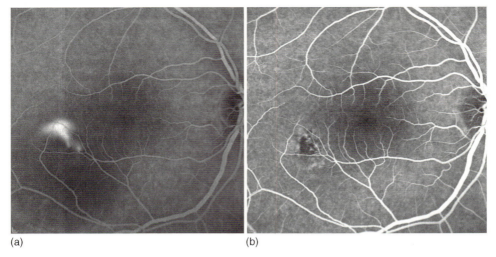

Figure 67.9 CSC (a) before laser treatment (note the typical "smokestack" configuration) and (b) after laser treatment.

Laser parameters: $S = 100\text{--}200\,\mu m$, $T = 0.1\text{--}0.2\,s$, $E = 80\text{--}200\,mW$.
Visible effect: Light gray burns, confluent laser treatment.

In some patients there may be a recurrence of the CSC, in which case additional laser treatment is indicated.

67.6
Choroidal Melanoma

Small choroidal melanoma $\leq 3.0\,mm$ can be treated either by laser or even better by Transpupillary Thermotherapy [11] (see Figure 67.10). The laser energy (E) is ≥ 800

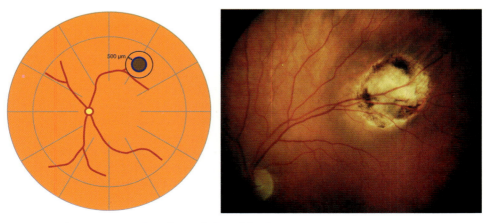

Figure 67.10 TTT laser for small choroidal melanoma.

Table 67.1 Differences between laser techniques.

	Classic laser coagulation	PDT	TTT
Wavelength	Green, yellow, red, and infrared laser (514–810 nm)	Red diode laser (689 nm)	Infrared diode laser (810 nm)
Laser duration	µs	83 s	60 s
Laser–tissue interaction	Photothermal	Photochemical	Photothermal
Retinal irradiation	80 W cm^{-2} (514 nm, 50 mW, with 200 µm spot size)	0.6 mW cm^{-2} (65 mW, with 3.0 mm spot size)	7.5 W cm^{-2} (800 mW, with 3.0 mm spot size)
Assumed maximum rise in temperature on the retina	42 °C (514 nm, 50 mW, 0.1 s, with 200 µm spot size)	2 °C (65 mW, 83 s, with 3.0 mm spot size)	10 °C (800 mW, 60 s, with 3.0 mm spot size)
Invasiveness	Not invasive	Invasive (vein puncture)	Not invasive

Source: modified from Mainster and Reichel [12].

mW for choroidal melanoma. After TTT treatment, a marked laser effect should be visible on the lesion (grayish coagulation effect). The spot size has to overlap the choroidal melanoma by at least 500 µm. Tumors with a larger diameter that cannot be overlapped completely with a single laser spot must be treated with several overlapping laser spots. In TTT for tumor patients, higher energy levels are necessary. This may lead to severe pain, and TTT therefore has to be performed under retrobulbar anesthesia.

Laser coagulation is also used for the treatment of the following rare diseases: retinal teleagiectasia, vascular tumors such as retinal hemangioma and choroidal hemangioma, peripheral retinal neovascularization such as sickle cell retinopathy, retinopathy of prematurity, and fundus alterations due to uveitis, Coats disease, Eales disease, and others [4].

Differences between laser techniques are summarized in Table 67.1. Further information can be found in ophthalmology laser handbooks.

References

1 Rol, P., Fankhauser, F., Giger, H., Dürr, U., and Kwasniewska, S. (2000) Transpupillar laser phototherapy for retinal and choroidal tumors: a rational approach. *Graefe's Arch. Clin. Exp. Ophthalmol.*, **238**, 249–272.

2 Framme, C., Roider, J., Brinkmann, R., Birngruber, R., and Gabel, V.-P. (2008) Grundlagen und klinische Anwendung der Lasertherapie an der Netzhaut. *Klin. Monatsbl. Augenheilkd.*, **225**, 259–268.

3 Augustin, A.J. (2007) *Augenheilkunde*, 3rd edn, chap. **28**, Springer, Berlin, pp. 811–820.

4 Folk, J.C. and Pulido, J.S. (1997) *Laser Photocoagulation of the Retina and Choroid*, American Academy of Ophthalmology Monograph Series, vol. **11**, American Academy of Ophthalmology, San Francisco.

5 Burk, A. and Burk, R. (2005) *Checkliste Augenheilkunde, 3. Auflage*, Thieme Verlag, pp. 456–460.

6 Early Treatment Diabetic Retinopathy Study Research Group (1987) Treatment techniques and clinical guidelines for photocoagulation of diabetic macular edema. ETDRS Report Number 2. *Ophthalmology*, **94**, 761–774.

7 Central Vein Occlusion Study Group (1995) Evaluation of grid pattern photocoagulation for macular edema in central vein occlusion. *Ophthalmology*, **102**, 1425–1433.

8 Central Vein Occlusion Study Group (1995) Randomized clinical trial of early panretinal photocoagulation for ischemic central vein occlusion. *Ophthalmology* **102**, 1434–1444.

9 Feucht, M., Fuisting, B., and Richard, G. (2006) Stand der TTT bei der Therapie der CNV. *Klin. Monatsbl. Augenheilkd.*, **223**, 802–807.

10 Treatment of Age-Related Macular Degeneration with Photodynamic Therapy (TAP) Study Group (2001) Photodynamic therapy of subfoveal choroidal neovascularization in age-related macular degeneration with verteporfin: two-year results of 2 randomized clinical trials – TAP Report 2. *Arch. Ophthalmol.*, **119**, 198–207.

11 Journee-de Korver, J.G., Oosterhuis, J.A., Kakebeeke-Kemme, H.M., and de Wolff-Rouendaal, D. (1992) Transpupillary thermotherapy (TTT) by infrared irradiation of choroidal melanoma. *Doc. Ophthalmol.*, **82**, 185–191.

12 Mainster, M.A. and Reichel, E. (2000) Transpupillary thermotherapy for age-related macular degeneration: long-pulse photocoagulation, apoptosis and heat shock proteins. *Ophthalmic Surg. Lasers*, **31**, 359–373.

Part 17
Dentistry

68
Diagnostic Imaging and Spectroscopy
Daniel Fried

68.1
Caries Detection

New diagnostic imaging tools are needed for the detection and characterization of caries lesions (dental decay) in the early stages of development [1]. Conventional methods, visual/tactile and radiographic, have numerous shortcomings and are inadequate for the detection of the early stages of the caries process [2–4]. Radiographic methods do not have sufficient sensitivity for early lesions, particularly occlusal lesions, and by the time the occlusal lesions are radiolucent they have often progressed well into the dentin, at which point surgical intervention is necessary [4–6]. At that stage in the decay process, it is too late for preventive and conservative intervention and a large portion of carious and healthy tissue will need to be removed, often compromising the mechanical integrity of the tooth. If left untreated, the decay will eventually infect the pulp, leading to loss of tooth vitality and possible extraction. The caries process is potentially preventable and curable. If carious lesions are detected in the enamel early enough, it is likely that they can be arrested/reversed by nonsurgical means through fluoride therapy, antibacterial therapy, dietary changes, or low-intensity laser irradiation [1, 7]. Therefore, one cannot overstate the importance of detecting the decay in the early stage of development, at which point noninvasive preventive measures can be taken to halt further decay.

Accurate determination of the degree of lesion activity and severity is of paramount importance for the effective employment of the treatment strategies mentioned above. Since optical diagnostic tools exploit changes in light scattering in the lesion, they have great potential for the diagnosis of the "activity" of the lesion, that is, whether or not the caries lesion is active and expanding or whether the lesion has been arrested and is undergoing remineralization. Such data are invaluable for caries management by risk assessment in the patient and for determining the appropriate form of intervention. A nondestructive, quantitative method of monitoring demineralization *in vivo* with high sensitivity would also be invaluable for use in short-term clinical trials for various anti-caries agents such as new or modified fluoride dentifrices and antimicrobials. In particular, methods that could be employed in the relevant high-risk areas of the tooth such as the occlusal pits and fissures, would be universally useful. The purpose of this chapter is to describe several recent optically based methods for the detection and imaging of tooth decay.

68.2
Optical Properties of Dental Hard Tissue

68.2.1
Optical Properties of Dental Hard Tissue in the Visible and Near-Infrared Regions

A fundamental understanding of how light propagates through sound and carious dental hard tissues is essential for the development of clinically useful optical diagnostic systems, since image contrast is based on changes in the optical properties of these tissues upon demineralization. Dental hard tissue optics are inherently complex owing to the nonhomogeneous and anisotropic nature of these biological materials. The scattering distributions depend on tissue orientation relative to the irradiating light source [8–11] in addition to the polarization of the incident light [12, 13]. The optical properties of biological tissue can be completely and quantitatively described by defining the optical constants, the absorption (μ_a) and scattering coefficients (μ_s), which represent the probability of the incident photons being absorbed or scattered, and the scattering phase function [$\Phi(\cos\theta)$], which is a mathematical function that describes the directional nature of scattering [14–17]. With knowledge of these parameters, light transport in dental hard tissue can be completely characterized and modeled. Accurate description of light transport in dental hard tissue relies on knowledge of the exact form of the phase function $\Phi(\cos\theta)$ for each tissue scatterer at each wavelength [17].

68.2.2
Optical Properties of Sound Enamel and Dentin

Light absorption by enamel is very weak in the visible and near-infrared (NIR) region ($\mu_a < 1 \text{ cm}^{-1}$, $\lambda = 400\text{–}1300 \text{ nm}$) and increases in the ultraviolet (UV) region ($\mu_a > 10 \text{ cm}^{-1}$, $\lambda < 240 \text{ nm}$) [18]. The absorption coefficient of dentin is essentially wavelength independent with a value of $\mu_a \approx 4 \text{ cm}^{-1}$ [19] above 400 nm. The optical behavior of enamel and dentin is dominated by scattering in the visible and NIR region [18]. The scattering coefficient of enamel, μ_s, decreases with increasing wavelength from 400 cm^{-1} in the near-UV region [18] to as low as $2\text{–}3 \text{ cm}^{-1}$ at 1310 and 1550 nm [10, 20] (see Figure 68.1). At longer wavelengths, water absorption increases and dominates the attenuation [21]. The scattering of dentin is variable with position in the tooth and exceeds $100\text{–}200 \text{ cm}^{-1}$ across the visible and NIR region [19]. Measurement of the enamel attenuation coefficients in the NIR region is a particular challenge since the scattering coefficient is almost two orders of magnitude lower than in the visible range, $\sim 2\text{–}3 \text{ cm}^{-1}$ at 1310 nm versus 105 cm^{-1} at 543 nm, and

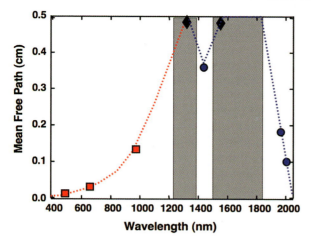

Figure 68.1 Plot of the mean free path of photons in sound dental enamel at visible and NIR wavelengths. The regions of highest transparency are shaded. The red squares are data from ref. [10], the black diamonds are from ref. [20], and the blue circles represent 10% absorption from water taken from ref. [21].

surface scattering can completely mask measurement of the bulk scattering properties [10].

The Monte Carlo (MC) method [22, 23] was used to determine the scattering coefficient and scattering phase function from angular-resolved scattering measurements for dentin and enamel at 543, 632, and 1053 nm. Although the MC method is somewhat labor intensive, it has been shown consistently to yield the most reliable values [14, 17]. The accuracy of the values computed was confirmed by successfully modeling the angular resolved scattering distributions measured for sections of various thickness from 100 μm to 2 mm in the multiple scattering regime. The measured angular resolved scattering distributions could not be represented by a single scattering phase function $\Phi(\cos\theta)$, and required a linear combination of a highly forward-peaked phase function, a Henyey–Greenstein (HG) function, and an isotropic phase function represented by the following equation [10]:

$$\Phi(\cos\theta) = f_d + (1-f_d)\left[\frac{1-g^2}{(1+g^2-2g\cos\theta)^{\frac{3}{2}}}\right] \qquad (68.1)$$

The parameter f_d for "fraction diffuse" is defined as the fraction of isotropic scatterers. The average value of the cosine of the scattering angle (θ) is called the scattering anisotropy (g), and g = 0.96 and 0.93 for enamel and dentin, respectively, at 1053 nm [10]. The fraction of isotropic scatterers (f_d) was measured to be 36% for enamel and less than 2% for dentin at 1053 nm. Note that Zijp and ten Bosch [8, 11] reported values of g = 0.4 for dentin and 0.68 for enamel, calculated by taking the ratio of the forward and backward scattered light. This latter approach is prone to error due to the contribution of surface scattering [10]. Most scattering biological tissues that can be represented by an HG function have measured g values >0.8 [14]; moreover, the g value should be determined within the context of an appropriate phase function based on the nature of the scatterers in the tissue [24], and subsequently validated through comparison of simulated scattering distributions with measured distributions of various thickness [10].

Light scattering in enamel and dentin is anisotropic, particularly for dentin due to the cylindrical-shaped enamel prisms and dentinal tubules. Altshuler and Grisimov suggested that these structures can act as waveguides to guide visible light [25, 26]. Kienle and co-workers extensively investigated the anisotropy of light scattering in dentin and showed how light is magnified through it [27–31].

68.2.3
Optical Property Changes of Demineralized Enamel

Increased backscattering from the demineralized region of early caries lesions is the basis for the visual appearance of white spot lesions [32, 33]. Attempts at measuring the optical properties of dental caries have been limited to measurements of backscattered light from optically thick, multilayered sections of simulated caries lesions [34–36]. The microstructure of enamel and dentin caries lesions is complex, consisting of various turbid and transparent zones [13, 37]. Highly porous demineralized areas of coronal caries appear whiter and are more opaque. During remineralization, pores and tubules are filled with mineral and those areas are typically more transparent. There have been some attempts to measure and simulate light scattering in artificial and natural lesions [32, 36, 38]. There has not been a major effort to measure and quantify the optical properties of carious lesions in terms of changes in the fundamental optical constants, μ_s and μ_a, and the scattering phase function. This is due in part to the difficulty in acquiring large uniform areas for measurement by conventional means and the highly variable nature of caries lesions. One challenge in quantifying the optical properties of carious lesion areas versus sound tissue areas is the method of index matching. Lesion areas are highly porous, allowing index matching fluids to imbibe into the lesion, effectively eliminating the lesion from an optical standpoint, that is, the lesion becomes more transparent than the sound tissue. One must assume that carious lesions will be filled with fluid in their natural state and that measurement in water replicates the conditions that will be encountered clinically. Darling et al. measured the optical properties of lesion areas imbided in water at 1310 nm and compared the increase in the magnitude of the scattering coefficient with mineral loss measured using microradiography [39]. The attenuation in the lesion exceeds $100\,\text{cm}^{-1}$, a factor of 50 times higher than that of sound enamel, $2–3\,\text{cm}^{-1}$, and is of a similar magnitude to that of the dentin. This difference can be seen in Figure 68.2a, which contains an image of a 200 μm thick tooth section with an interproximal lesion taken with 1310 nm light. The areas of the lesion and dentin are opaque (white) whereas the enamel is highly transparent (dark). A high-resolution digital microradiograph of the demineralized region of the section is shown in Figure 68.2b. The volume-% mineral is represented by the

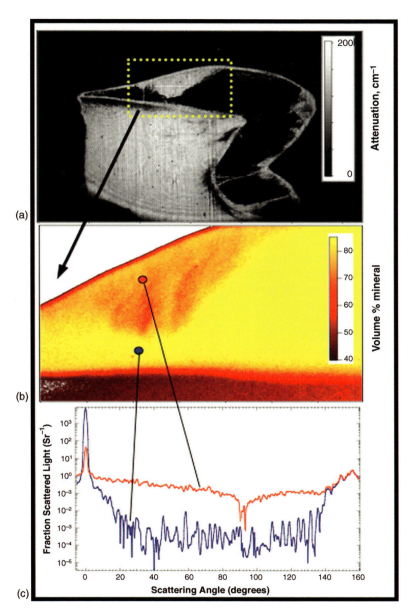

Figure 68.2 (a) NIR transillumination image of 200 μm thick tooth section with demineralization indicated in the yellow box. (b) A high-resolution digital microradiograph of the lesion area shows the volume percent mineral versus position in the lesion area. (c) The fraction of NIR 1310 nm light scattered versus angle in the sound blue and carious black areas of the sample measured with the scattering goniometer. From ref. [39].

yellow–red color table with the areas of lower mineral content, caries lesion and dentin, demarcated in red and the high-mineral areas, namely sound enamel, in yellow. Optical scattering measurements were taken in sound and carious regions of the same section as indicated by the two points, red and blue, in the microradiograph using a 1310 nm diode laser focused to a spot size of 100 μm. Angularly resolved scattering measurements taken at those two points are plotted in Figure 68.2c. The ballistic light is reduced by two orders of magnitude and the intensity of the light scattered at angles >10° is 2–3 orders of magnitude higher for the carious tissue over the sound enamel.

NIR optical property measurements during the longitudinal development of early artificial demineralization suggest a rapid exponential increase in optical scattering during initial lesion development followed by a more gradual increase in scattering as the lesion severity increases [39]. Angularly-resolved scattering measurements from artificially demineralized enamel over a period of 5 days show a rapid increase in attenuation after just 1 day followed by an increase in another order of magnitude after the next 4 days. The scattering anisotropy of demineralized enamel also appears to be highly forward directed. This is not entirely unexpected since demineralization proceeds along the core of the enamel rods, hence the scattering centers may be fairly large in the micron range and highly anisotropic.

68.2.4
Optical Properties of Dental Hard Tissue in the IR region

Infrared (IR) wavelengths can be used to monitor changes in the chemical composition and mineral phase crystalline orientation as a result of demineralization and remineralization, and after thermal treatments by lasers to modify the susceptibility to acid dissolution. In the mid-IR region, scattering is negligible and knowledge of the absorption coefficient and the reflectivity are sufficient to describe the optical behavior. The magnitude of the absorption coefficient is high in the mid-IR region due to resonant absorption by molecular groups in water, protein, and mineral. The reflectivity can exceed 50% at wavelengths coincident with mineral absorption bands due to the large increase in the imaginary component of the refractive index. The fraction of incident laser light reflected at the surface of the tooth is described by the Fresnel reflection equation:

$$R = \frac{(n_r - 1)^2 + k^2}{(n_r + 1)^2 + k^2} \tag{68.2}$$

where n_r is the real component of the refractive index and k is the attenuation index. The absorption coefficient (μ_a), k, and the wavelength (λ) are related by the expression $\mu_a = 4\pi k/\lambda$. Hence the reflectance of materials can increase markedly and approach 100% in regions of strong absorption; for example, some metals have absorption coefficients $>10^6$ cm^{-1} and reflectance >99% in the visible and IR region. The reflectance of dentin and enamel was measured at $\lambda = 2.79, 2.94, 10.6, 10.3,$

68.2 Optical Properties of Dental Hard Tissue

9.6, and 9.3 μm using a gold-coated integrating sphere [40]. The reflectance of enamel is substantially higher at $\lambda = 9.6$ and 9.3 μm than at $\lambda = 10.6$ μm, near 50%, and must be accounted for when calculating ablation efficiencies, ablation thresholds, and the heat deposition in the tooth when investigating laser interactions with these tissues.

Absorption coefficients corresponding to CO_2 laser lines that could be obtained using conventional transmission measurements were calculated to be $1168 \pm 49\,\text{cm}^{-1}$ at 10.3 μm and $819 \pm 62\,\text{cm}^{-1}$ at 10.6 μm for enamel and $1198 \pm 104\,\text{cm}^{-1}$ at 10.3 μm and $813 \pm 63\,\text{cm}^{-1}$ at 10.6 μm for dentin. Enamel absorption coefficients corresponding to Er:YAG (2.94 μm) and Er:YSGG (2.79 μm) were calculated to be 768 ± 27 and $451 \pm 29\,\text{cm}^{-1}$, respectively. The absorption coefficient of dentin at 2.79 μm was calculated to be $988 \pm 111\,\text{cm}^{-1}$. Those values are shown in Figure 68.3 along with $1/e$ absorption depth, thermal relaxation time, and percentage reflectance. Conventional transmission measurements are not possible for the determination of the optical properties of enamel and dentin at 9.3 and 9.6 μm, therefore an alternative method must be used such as angular-resolved reflection measurements of polarized light and time-resolved radiometric measurements. Previous studies have used angular resolved reflection measurements to estimate IR absorption coefficients in dental enamel. Duplain et al. [41] reported absorption coefficients of 6500 and $5200\,\text{cm}^{-1}$ at 10.3 and 10.6 μm, respectively. These coefficients were much higher than the dental enamel absorption coefficients determined using direct transmission measurements (1168 and $819\,\text{cm}^{-1}$ for 10.3 and 10.6 μm, respectively). Those previously reported values are also not consistent with observed surface modification thresholds based on the known melting range of dental enamel (800–1200 °C). More recent measurements employed time-resolved radiometry measurements coupled with numerical simulations of thermal relaxation in the tissue to measure 8000 and $5250\,\text{cm}^{-1}$ at 9.6 and 9.3 μm, respectively [42]. Those values are consistent with time-resolved temperature measurements during laser irradiation and surface melting thresholds [43].

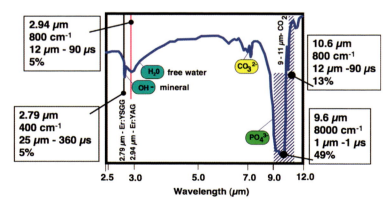

Figure 68.3 Optical properties of human dental enamel and dentin at relevant laser wavelengths. The wavelength, absorption coefficient ($1/e$), absorption depth, thermal relaxation time, and percentage reflectance are listed for each laser wavelength. From refs. [168] and [169].

68.3
Reflectance Spectroscopy

Since the magnitude of light scattering increases markedly in dental hard tissues upon demineralization producing visible white spots on tooth surfaces due to the increased porosity, reflectance spectroscopy is an obvious method for measuring and quantifying lesion severity. In 1984, ten Bosch *et al.* [33] introduced an optical monitor that used optical fibers for reflectance measurements on tooth surfaces. The reflectivity increased from the lesion area with increasing mineral loss [32]. Using the Kubelka–Munk equations, Ko *et al.* [36] showed that the optical scattering power correlated with mineral loss and yielded improved results over reflectance measurements. Benson *et al.* [44] demonstrated that the visibility of scattering structures on highly reflective surfaces such as teeth can be enhanced by the use of crossed polarizers to remove the glare from the surface due to the strong specular reflection from the enamel surface [44, 45]. The contrast between sound and demineralized enamel can be further enhanced by depolarization of the scattered light in the area of demineralized enamel [46, 47]. Blodgett and Webb [48] measured the optical bidirectional reflectance distribution functions (BRDF) and bidirectional scattering distribution functions (BSDF) from the surfaces of human incisors at 632 nm, 1054 nm, and 3.39-μm. Analoui *et al.* [49] showed that multi-spectral La*b* color coordinates measured using a diode-array spectrometer in the range 380–780 nm did not correlate well with the depth of artificial lesions.

The contrast between sound and demineralized enamel is greatest in the NIR region due to the minimal scattering of sound enamel (see Figure 68.1) and this can be exploited for reflectance imaging of early demineralization [39]. Wu and Fried [50] reported the high-contrast polarized reflectance images of early demineralization on buccal and occlusal tooth surfaces measured at $\lambda = 1310$ nm and found that the contrast between early demineralization was significantly higher at 1310 nm than in the visible range. Figure 68.4a and b show cross-polarization images of enamel demineralization in the visible and NIR region, showing the marked difference in contrast between the sound and demineralized enamel in these two spectral regions.

68.4
Optical Transillumination

Optical transillumination was used extensively before the discovery of X-rays for the detection of dental caries. Over the past two decades, there has been continued interest in this method, especially with the availability of high-intensity, fiber optic-based illumination systems for the detection of interproximal lesions [51–56]. During fiber-optic transillumination (FOTI), a carious lesion appears dark upon transillumination because of decreased transmission due to increased scattering and absorption by the lesion. A digital fiber-optic transillumination (DIFOTI) system, that utilizes visible light for the detection of caries lesions, has recently been developed by Electro-Optical Sciences (Irvington, NY, USA) [57]. Several studies have been carried

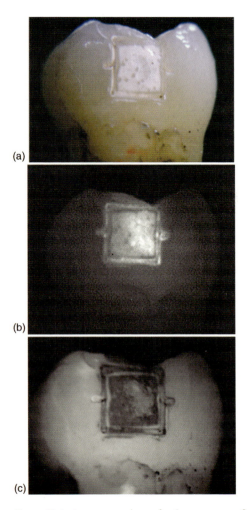

Figure 68.4 Images are shown for the buccal surface of a tooth with a 2×2 mm^2 window of demineralization. (a) Visible reflectance with crossed polarizers, (b) NIR reflectance with crossed polarizers, and (c) laser induced fluorescence image ($\lambda = 473$ nm excitation, emission $\lambda > 500$ nm). From ref. [50].

out using visible light transillumination either as an adjunct to bitewing radiography or as a competing method for the detection of interproximal caries lesions [57–62]. However, since FOTI and DIFOTI operate in the visible range, the photon mean free path is limited (see Figure 68.1) and the results are mixed.

Near-IR light can penetrate much further through tooth enamel due to the markedly longer mean free path of the photons [63], that is, enamel is virtually transparent in the NIR region with optical attenuation 1–2 orders of magnitude less than in the visible range (see Figure 68.1). Transmission measurements through demineralized tooth sections at 1310 nm show that the demineralized enamel

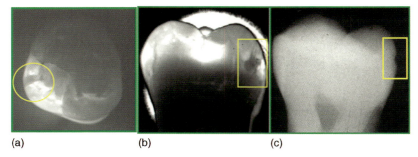

Figure 68.5 Multiple NIR views of a tooth with an interproximal lesion. The radiograph is shown in (c). From ref. [66].

attenuates NIR light by a factor of 20–50 times greater than the sound enamel [64, 65], hence high contrast can be achieved at 1310 nm between demineralized and sound tissues. High-contrast images have been acquired of simulated and natural interproximal lesions on extracted human teeth [66, 67]. Since the tooth manifests such high transparency in the 1310 and 1550 nm NIR regions, multiple imaging configurations can be used to image the decay. Interproximal lesions can be viewed using transillumination with the light source and detector placed on opposite sides of the tooth or NIR light can be delivered near or below the gum-line and both interproximal and occlusal lesions can be imaged from directly above the occlusal surface, as demonstrated in Figure 68.5 for a natural interproximal lesion. Shorter NIR wavelengths have also been investigated in the region that is accessible to conventional silicon-based charge-coupled device (CCD) cameras with the NIR filter removed, namely at 830 nm [68]. Figure 68.6 compares the image contrast of tooth sections of increasing thickness with simulated lesions in the visible region at 830 and 1310 nm [68]. Simulated lesions were produced by drilling small 1 mm diameter cavities on the mesial or distal aspect of tooth sections of varying thickness or whole teeth. The imaging system operating near 830 nm, utilizing a low-cost silicon CCD optimized for the NIR region, is capable of significantly higher performance than a visible system, but the contrast is significantly lower then that attainable at 1310 nm [68]. Owing to the high transparency of the enamel in the NIR region, NIR imaging also has great potential to examine defects in tooth structure and cracks in enamel can be clearly resolved with this method.

68.5
Optical Coherence Tomography

Optical coherence tomography (OCT) is a noninvasive technique for creating cross-sectional images of internal biological structure [69]. The intensity of backscattered light is measured as a function of its axial position in the tissue. Low-coherence interferometry is used to selectively remove or gate out the component of backscattered signal that has undergone multiple scattering events, resulting in

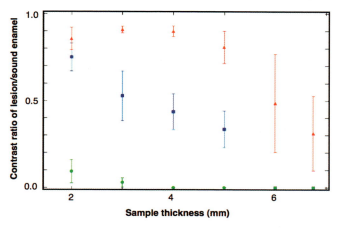

Figure 68.6 Contrast ratio between sound enamel and the simulated lesion for the three imaging systems for varying section thickness: triangles, 1310 nm; squares, 830 nm; circles, visible standard CCD, no filters. From ref. [68].

very high-resolution images (<15 μm). Lateral scanning of the probe beam across the biological tissue is then used to generate a two-dimensional intensity plot, similar to ultrasound images, called a b-scan. The one-dimensional analog of OCT, optical coherence domain reflectometry (OCDR), was first developed as a high-resolution ranging technique for the characterization of optical components [70, 71]. Huang *et al.* [72] combined transverse scanning with a fiber-optic OCDR system to produce the first OCT cross-sectional images of biological microstructure. Excellent books by Bouma and Tearney [69] and Brezinski [73] explain the mechanics of OCT and polarization-sensitive optical coherence tomography (PS-OCT). The first images of the soft and hard tissue structures of the oral cavity were acquired by Colston and co-workers [74–76]. Feldchtein *et al.* [77] presented the first high-resolution dual-wavelength 830 and 1280 nm images of dental hard tissues, enamel, and dentin caries and restorations *in vivo*. OCT has also been used to look at different restorative materials and identify pit and fissure sealants [77, 78]. High-speed Fourier domain (FD)-OCT systems have also been investigated for imaging dental caries [79–82]. OCT images of sound, natural, and artificial demineralized enamel demonstrate that one can image up to 3 mm deep in sound enamel. Although the image depth in enamel is limited to 1–2 mm, highly scattering structures such as the dentin at the dentin–enamel junction (see Figure 68.7) and carious dentin can be detected to a depth of 2–3 mm beneath sound enamel, so that "hidden" dentinal caries can be detected. Figure 68.7 shows an *in vivo* PS-OCT image of the buccal surface of a premolar scanned from the crown to the root showing the excellent optical penetration through the enamel. Identification of caries lesions in OCT images is challenging because the complex signal generation and interpretation of images requires considerable training and experience with the system being used. Since enamel is a weak scatterer, the reflectivity decreases exponentially with depth in sound enamel and subsurface lesions and the reflectivity of a stronger scattering

Figure 68.7 *In vivo* PS-OCT images of a sound premolar tooth. The dentinal enamel junction is visible along the entire length from the crown to the root. Banding is caused by tooth birefringence and developmental growth lines.

underlying layer such as dentin or a subsurface lesion can show up even though the signal from sound enamel is no longer visible at that depth. Strong scattering from lesions or developmental defects at the enamel surface can mask the signal from the underlying sound tissues. Moreover, in many cases the reflectivity from the lesion areas can be resolved at greater depths than in sound tissue even though the mean free path of the photons in the sound enamel is two to three orders of magnitude higher than for the caries tissue. Amaechi and co-workers [83, 84] demonstrated that the loss of reflectivity with depth in OCT images correlated with quantitative light fluorescence (QLF) measurements of smooth surface artificial caries. However, as we pointed out above, for some OCT imaging systems the reflectivity actually increases with depth in caries lesions so this approach is not likely to be feasible for natural lesions; however, the rapid loss of reflectivity with depth can be valuable in identifying highly attenuating lesion areas. Enamel and dentin are birefringent and this birefringence can cause bands in OCT images that can be misinterpreted as subsurface caries lesions (see Figure 68.7). Bands are regularly spaced for differentiation from caries lesions. An OCT system with polarization sensitivity can aid in differentiating such structures attributed to birefringence (see below). The incremental growth lines in teeth or striae of Retzius actually scatter light within the tooth and also cause subsurface bands in the images as can be seen in Figure 68.7.

Polarization sensitivity is particularly valuable for imaging caries lesions due to the enhanced contrast of caries lesions caused by depolarization of the incident light by the lesion. Another important advantage is that the confounding influence of the strong surface reflectance of the tooth surface is reduced in the image of the orthogonal polarization state to the incident polarized light. Baumgartner and co-workers [85–87] presented the first polarization-resolved images of dental caries. PS-OCT images are typically processed in the form of phase and intensity images [45, 88]; such images best show variations in the birefringence of the tissues. Areas of demineralization rapidly depolarize incident polarized light and the image of the orthogonal polarization relative to that of the incident polarization can provide improved contrast of caries lesions [46]. One approach to quantifying the severity of caries lesion is by directly integrating the reflectivity of the orthogonal axis or perpendicular polarization ($\perp$) [46]. There are two mechanisms in which intensity can arise in the perpendicular axis. The native birefringence of the tooth enamel can rotate the phase angle of the incident light beam between the two orthogonal axes (similar to a wave-plate) as the light propagates through the enamel without changing the degree of polarization. The other mechanism is depolarization from scattering in which the degree of polarization is reduced. It is this latter mechanism that can be exploited to measure the severity of demineralization. Complete depolarization of the incident linearly polarized light leads to equal distribution of the intensity in both orthogonal axes. Demineralization of the enamel due to dental decay causes an increase in the scattering coefficient by 1–2 orders of magnitude, thus demineralized enamel induces a very large increase in the reflectivity along with depolarization. This in turn causes a large rise in the perpendicular polarization channel or axis. This can be seen in the PS-OCT images of a premolar with shallow demineralization around the fissure presented in Figure 68.8. The parallel ($\parallel$)-axis image represents the linearly polarized light reflected in the original polarization incident on the tooth and the $\perp$-axis image represents the light in the orthogonal polarization to the

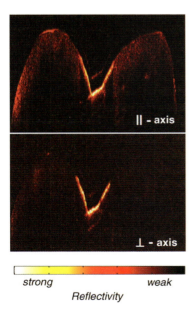

Figure 68.8 PS-OCT images of an area of demineralization in the occlusal pit and fissure of a premolar. From ref. [89].

incident light. The surface of the tooth is less visible in the ⊥-axis image and the lesion appears with higher contrast to the surrounding sound enamel.

This approach also has the added advantage of reducing the intensity of the strong reflection from the tooth surface. The lesion surface zone is very important since this zone can potentially provide information about the lesion activity and remineralization. A conventional OCT system cannot differentiate the strong reflectance from the tooth surface from increased reflectivity from the lesion itself. This facilitates direct integration of the lesion reflectivity to quantify the lesion severity, regardless of the tooth topography and the difficult task of having to deconvolute the strong surface reflection from the lesion surface can be circumvented. The integrated reflectivity of the ⊥-PS-OCT images of natural lesions correlate well with the integrated mineral loss and can thus be used to measure lesion severity directly [46, 89, 90]. Figure 68.9 shows ⊥-axis PS-OCT images, polarized light microscopy (PLM) images, and transverse microradiographs (TMR) of a pigmented interproximal lesion. Five PS-OCT line profiles were taken at different positions in the lesion and each was integrated to yield a total integrated reflectivity for that area of the lesion to a depth of 500 μm. The integrated reflectivity correlated well with the matching mineral density profiles taken from the microradiographs. PS-OCT images also accurately represent the highly convoluted internal structure of caries lesions. Longitudinal studies have demonstrated that PS-OCT can be used for monitoring erosion and demineralization [46, 89, 90]. The progression of artificially produced caries lesions in the pit and fissure systems of extracted molars can be monitored nondestructively [89].

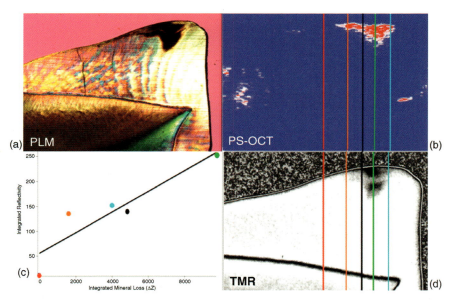

Figure 68.9 Pigmented lesion: (a) reflected light image, (b) ⊥-axis PS-OCT image, and (c) plot of the integrated reflectivity versus the integrated mineral loss to a depth of 500 μm. Each color coded symbol corresponds to the line profiles of (b) and (d). (d) Microradiograph. From ref. [170].

It is not sufficient simply to detect a carious lesion or areas of demineralization – it is also necessary to assess the state of the carious lesion and determine whether it is active and progressing or whether it is arrested and has remineralized. One can take two approaches for determining the state of activity of carious lesions. The first approach is based on monitoring the lesion progression over time to see if it changes. The second approach is to examine the structure of the lesion to determine if there are any definitive features, such as a weakly scattering surface zone, that are indicative of remineralization. Ismail presented a review of visual caries detection criteria [91]. Surfaces of arrested lesions are typically hard and shiny with less light scattering, in contrast to the soft and chalky surface of active lesions. It is likely that if PS-OCT can resolve changes in lesion structure and demineralization, for example the presence of a surface zone of reduced light scattering indicative of remineralization, then this method has potential as a more quantitative diagnostic of the activity of the lesion. Figure 68.10 shows PS-OCT, PLM, and TMR images of human dentin that was exposed to a demineralization solution (left side) and a demineralization solution followed by a remineralization solution (right side). The lower reflectivity surface layer of remineralized dentin is clearly visible in the PS-OCT image.

PS-OCT can also be used to image secondary caries under sealants where the advantage of having the ability to directly integrate the PS-OCT fast axis (⊥-axis) images to quantify lesion severity is even more apparent [92]. The imaging depth of PS-OCT through composite is sufficient to resolve early demineralization under the sealant or restoration. Studies also suggest that polarization sensitivity can be used to

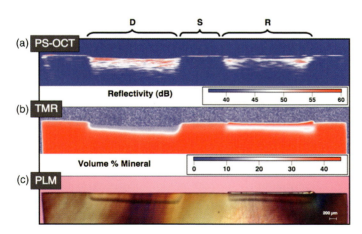

Figure 68.10 (a) (⊥-axis) PS-OCT scan taken along long axis of a sample before sectioning. Intensity scale is shown in red–white–blue in units of decibels (dB). Remineralized layer is visible as a blue gap above the remineralized lesion. (b) The corresponding TMR of a 260 μm thick section in a similar orientation to the PS-OCT image. The intensity scale is also shown in red–white–blue with units in volume-% mineral. (c) The PLM image of the same section is shown at 15× magnification. From ref. [171].

differentiate between composites and tooth structure and even differentiate between different composites [92]. The composite reflectivity, depolarization, and penetration depth is not influenced by the composition of the filler; however, the reflectivity was markedly increased in sealants with an added optical opacifier such as titanium dioxide. Therefore, polarization sensitivity can also be advantageous for differentiating demineralized enamel under composite sealants and restorations.

In summary, OCT is an excellent tool for acquiring images of dental caries. However, OCT images are fairly complicated and considerable experience is required to interpret the images. PS-OCT is advantageous for providing a quantitative measure of the lesion severity for monitoring lesion progression and for assessing the success of chemical intervention on arresting and remineralizing lesions.

68.6
Fluorescence Imaging

68.6.1
Quantitative Light Fluorescence (QLF)

Teeth fluoresce naturally upon irradiation with UV and visible light. Alfano et al. [93] and Bjelkhagen and Sundstrom [94] demonstrated that laser-induced fluorescence (LIF) of endogenous fluorophores in human teeth could be used as a basis for discrimination between carious and noncarious tissue.

Quantitative light fluorescence (QLF) is the most extensively investigated optical technique for the measurement of surface demineralization. ten Bosch [95] described the mechanism for the QLF phenomenon, that is, the loss of yellow–green fluorescence under illumination with blue light, and it can be solely explained by light scattering effects. Lesion areas appear dark due to increased light scattering in the lesion area that prevents the fluorescence from the underlying collagen in dentin or unknown fluorophores in enamel that are deeper in the tooth from reaching the tooth surface. Excitation wavelengths have varied from 370 nm to 488 nm and blue laser diodes are more typically employed today. Emission spectra are typically independent of excitation wavelengths – Kasha's rule [96]. A study by Amaechi et al. [84] compared the loss of reflectivity with lesion depth as measured with OCT with the loss of fluorescence measured with QLF and achieved a very high correlation, suggesting that both experiments measure the same thing, namely increased light attenuation due to an increase in light scattering in the lesion. Fluorescence images provide increased contrast between sound and demineralized tooth structure and avoid the interference caused by specular reflection or high glare from the tooth surface which can interfere with visual detection of white spot lesions. Figure 68.4c shows an LIF image of the buccal surface of a tooth with a demineralized area.

Hafström-Björkman et al. [97] established an experimental relationship between the loss of fluorescence intensity and extent of enamel demineralization [97]. The method was subsequently labeled the QLF (quantitative laser fluorescence) method. An empirical relationship between overall mineral loss (ΔZ) and loss of fluorescence was established that can be used to monitor lesion progression on enamel surfaces [98–100]. The gold standard for quantifying lesion severity and tooth surface and subsurface demineralization is microradiography. The lesion severity is typically reported as the product of the volume-percent mineral loss and the lesion depth, ΔZ (vol.% × μm). Therefore, it is advantageous to be able to report a similar measure using optical methods. It is important to point out that QLF researchers report changes in fluorescence radiance, ΔF (%), calculated as

$$\Delta F_{\text{Ref}} = \frac{F_{\text{Ref}}(\text{demin})}{F_{\text{Ref}}(\text{sound})} \times 100 \tag{68.3}$$

for comparison with the ΔZ value measured with microradiography. Excellent correlation has been established between ΔF and ΔZ for shallow, uniform, artificial lesions [97, 101]. This approach requires that the user has prior knowledge that the $F_{\text{Ref}}(\text{sound})$ measurement is actually in a sound area of the tooth, free of any surface defects. The ΔF value is a ratio of reflectivity of two regions of the tooth; therefore, if there is some complexity to the lesion structure, ΔF may not be meaningful. This is a serious limitation for high-risk areas such as the occlusal pits and fissures. QLF also manifests a strong dependence on the imaging angle and the remaining enamel thickness of the sample. Ando and co-workers [102, 103] established that the ΔF_{Ref} intensity for similar lesions depends on the actual enamel thickness. That result suggests a very serious limitation for clinical implementation since the enamel thickness varies markedly between positions on each tooth and from tooth to tooth. Even though acquiring a reference sound area needed for calculation of ΔF would be challenging in such areas, this method still provides increased contrast for better visualization of the lesion area.

Another complication is that stains and plaque fluoresce strongly, greatly confounding detection. Therefore, QLF has not been successfully validated for quantifying surface demineralization in the pits and fissures of the occlusal surfaces where most lesions are found, and has manifested relatively low specificity in clinical studies on smooth surfaces [101].

QLF has also been used to quantify and measure remineralization both *in vitro* and *in vivo* [104]. Studies have shown optical changes in the fluorescence intensity after exposure of *in vivo* white spot lesions around orthodontic brackets to a remineralization solution or removal of the plaque retention device [105, 106]. However, since QLF cannot produce an image of the structure of the lesion, it cannot be used to determine whether the lesion simply eroded away after removal of the bracket from abrasive action, whether there was a change in the surface roughness of the lesion, or confirm that actual mineral repair has taken place.

In summary, QLF performs well on carefully produced, shallow, uniform lesions on smooth surfaces that are consistent with the proposed fluorescence loss mechanism. However, such performance cannot be expected on highly convoluted occlusal surfaces or for the complex structures of natural lesions. Moreover, stains, plaque, and developmental defects interfere with QLF, which explains the somewhat disappointing clinical performance to date.

68.6.2
DIAGNOdent (Porphyrin Fluorescence)

Bacteria produce significant amounts of porphyrins and dental plaque fluoresces upon excitation with red light [107]. The earliest fluorescence measurements of dental caries showed the distinctive salmon red fluorescence due to porphyrins [93, 94]. A novel caries detection system, the DIAGNOdent (KaVo, Biberach, Germany) has received FDA approval in the United States. This device, shown in Figure 68.11, uses a red diode laser and a fiber-optic probe designed to detect the fluorescence emitted from porphyrins at longer wavelengths. The probe is designed to be inserted in an occlusal pit and fissure and an electronic reading is generated representing the amount of fluorescence from the lesion. Bacteria produce significant amounts of porphyrins and dental plaque fluoresces upon excitation with red light [107]. This device is designed to detect hidden occlusal lesions that have penetrated into the dentin, where the high porosity concentrates porphyrins from bacteria. It is important to note that the primary microorganism responsible for dental decay, *Streptococcus mutans*, does not contain porphyrins and that this method is not an effective means of monitoring cariogenic bacteria. Moreover, this device is designed to detect lesions in the later stage of development after the lesion has penetrated into the dentin and accumulated a considerable amount of bacterial

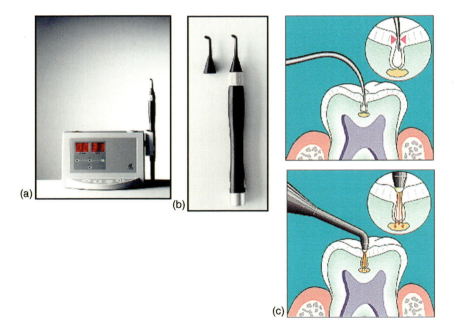

Figure 68.11 Images of (a) the KaVo DIAGNOdent instrument and (b) probes for fluorescence-based detection of caries. The diagrams in (c) compare the use of the DIAGNOdent with the dental explorer for probing the occlusal pits and fissures. Images taken from KaVo web site (http://www.kavo.com/Default.aspx?navid=40&oid=002&lid=En).

byproducts. The DIAGNOdent system has poor sensitivity (~0.4) for lesions confined to enamel [108], since porphyrins have not accumulated in those lesions. Laboratory studies have shown a very good relationship between lesion extent measured histologically after sectioning and the original DIAGNOdent readings [109]. An interproximal caries probe designed to reach proximal surfaces has also been introduced more recently [110]. The DIAGNOdent is well designed for the detection of "hidden" dentinal caries but it does not provide quantitative measurements of demineralization.

68.6.3
Other Fluorescence-Based Imaging Methods

Other fluorescence imaging techniques that have been employed for caries detection include time-resolved fluorescence [111], multiphoton fluorescence [112], dye-enhanced fluorescence [113], confocal fluorescence [100, 114], and modulated fluorescence or luminescence. Confocal fluorescence has considerable promise for studying very early incipient caries lesions.

68.7
Thermal Imaging

Thermal imaging has been used as a tool to measure enamel demineralization. Kaneko *et al.* [115] used a thermal imaging camera to measure the thermal emission from the surface of shallow enamel lesions. The increasing water content of highly porous lesion areas leads to a lower temperature due to evaporative cooling that was clearly discernable in the 3.6–4.6 µm wavelength range.

At highly absorbed wavelengths at which the absorption coefficient greatly exceeds the scattering coefficient, the magnitude of the absorption coefficient can be determined by measurement of the rate of thermal emission from the tooth surface. This is called pulsed photothermal radiometry [116]. Chebotareva *et al.* [117] and Zuerlein *et al.* [42] utilized this approach to determine the magnitude of the absorption coefficient of enamel and dentin at highly absorbed laser wavelengths in the IR region. This approach can also be used to monitor changes in the hard tissues as the absorption coefficient changes with increasing temperature or by modification during laser irradiation [43].

At wavelengths at which there is deep optical penetration such as the visible and NIR region for dental enamel, there can be increased absorption of light by subsurface structures. In theory, the depth of those structures can be calculated by analyzing the temporal profile of the thermal emission from the sample. The incident thermal radiation can be temporally modulated for frequency-domain photothermal radiometry (FD-PTR) and Mandelis and co-workers [118–121] have applied FD-PTR to detect and analyze dental caries, cracks, and other subsurface structures in teeth. Analysis of the phase and amplitude of the PTR signals at different modulation frequencies provides depth-specific information. Mandelis and co-workers indicated

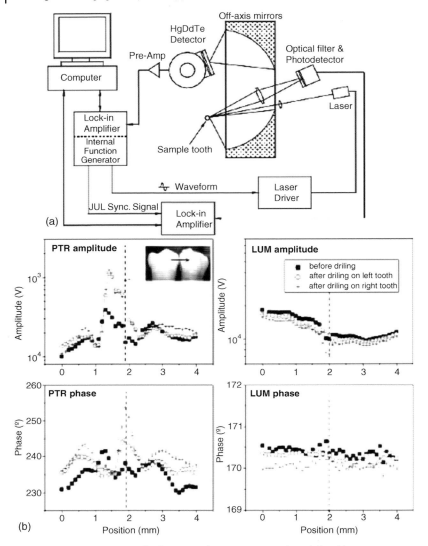

Figure 68.12 (a) Apparatus for frequency-domain IR PTR and modulated luminescence (LUM) imaging. (b) PTR and LUM phase and amplitude measurements at 5 Hz across the proximal contact point before and after drilling holes in each tooth. From reference [118].

that this method can be used potentially to monitor defects up to 3 mm below the surface. The apparatus used to acquire simultaneous FD-PTR signals and modulated fluorescence or luminescence signals is shown in Figure 68.12. Studies have been carried out looking at natural and artificial lesions in enamel and dentin and holes and cracks in teeth. A sample image of holes drilled in teeth positioned at the proximal contact points is also shown in Figure 68.12, indicating that the PTR signals change after drilling holes in the teeth at the appropriate positions. The FD-PTR signals acquired from artificial lesions show that the PTR signals change with increasing lesion severity, which is consistent to what would be expected with an increase in the scattering coefficient as is observed with other methods.

68.8
Infrared Spectroscopy

Enamel and dentin have distinctive molecular absorption bands in the IR region due to the water, mineral, protein, and lipid components. IR spectra can be used to monitor chemical and crystalline orientational changes in the mineral and organic components of dental hard tissues. The surfaces of mineralized tissues are well suited for probing by light scattering, reflectance, attenuated total reflectance, and photoacoustic IR methods. Transmission measurements are more difficult owing to the high water absorption across the mid-IR region and the extremely high absorption near 3 μm (water) and the mineral phase from 9 to 11 μm. Spectra are useful for determining changes to the organic components, crystalline changes to the mineral phase that are indicative of the formation of either more acid resistance phases, or the disproportionation of hydroxyapatite to form other calcium phosphate phases that are more susceptible to acid dissolution.

IR spectroscopy has been used for half a century to study the structure of bony tissues [122–125]. The most intense absorption bands are due to the phosphate group between 1000 and 1200 cm^{-1}. The spectral properties of the carbonate bands [126, 127] located near 1500 and 800–900 cm^{-1} are of particular importance since it is the carbonate inclusions in the hydroxyapatite mineral phase that make dental enamel highly susceptible to acid dissolution. Fowler and Kuroda [128, 129] used IR transmission spectroscopy to show the chemical changes induced in laser-irradiated dental enamel. Featherstone et al. [130] showed that the amount of carbonate in synthetic carbonated hydroxyapatites could be quantified using IR transmission measurements. These early transmission measurements employed the very destructive KBr pellet method, in which the surface of the irradiated enamel is ground away and mixed with KBr to produce a pellet. Fourier transform infrared (FTIR) spectroscopy in specular reflectance mode can provide the same information without destruction of the sample [131]. The principal advantage of this technique is that the tissue reflectance is only influenced by a surface layer with a thickness on the order of the wavelength of the light. FTIR reflectance spectra showing the influence of laser irradiation on the chemical composition of tooth enamel are shown at several irradiation intensities in Figure 68.13. Thermal transformation of the carbonated hydroxyapatite mineral in enamel to a purer phase hydroxyapatite is indicated by the decrease in the area of the carbonate bands between 6 and 8 μm. High-brightness synchrotron radiation sources such as the advanced light source (ALS) provide a means to probe nondestructively specific areas such as laser ablation craters and resolve the laser-induced chemical changes that occur in the mineral phase. During high-intensity laser irradiation, marked chemical and physical changes may be induced in the irradiated dental enamel. Thermal decomposition of the mineral

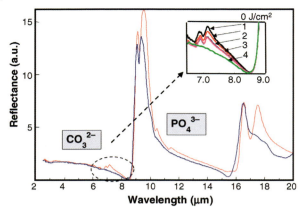

Figure 68.13 IR reflectance spectra of enamel before and after irradiation by a carbon dioxide laser at various irradiation intensities (fluence). The inset shows the area of the carbonate peaks as the carbonate is removed with increasing fluence. From ref. [131].

can lead to changes in the susceptibility of the modified mineral to organic acids in the oral environment. The mineral hydroxyapatite, found in bone and teeth, contains carbonate inclusions that render it highly susceptible to acid dissolution by organic acids generated from bacteria in dental plaque. Upon heating to temperatures in excess of 400 °C, the mineral decomposes to form a new mineral phase that has increased resistance to acid dissolution [132]. Studies have suggested that as a side effect of laser ablation, the walls around the periphery of a cavity preparation will be transformed through laser heating into a more acid-resistant phase with an enhanced resistance to future decay [133–135]. However, poorly crystalline non-apatite phases of calcium phosphate may have an opposite effect on plaque acid resistance [128] and may increase the quantity of poorly attached grains associated with delamination failures.

Polarized IR reflectance measurements can also show orientation changes in the enamel crystals [136]. Such information can be used to show changes to the enamel crystal structure as a result of thermal modification by laser irradiation and chemical changes through acid dissolution and fluoride-enhanced remineralization.

68.9
Raman Spectroscopy and Imaging

Raman spectroscopy has been employed in several studies to measure changes in the mineral phase of demineralized and remineralized enamel and dentin, aid in the discrimination of carious enamel, measure the mineral content of dental calculus, and measure the presence of CaF_2 after fluoride treatments. Raman spectra have been obtained of hydroxyapatite, enamel, dentin, and calculus [137–141]. High-resolution micro-Raman spectroscopy has been used to analyze the chemical composition of the internal interfaces between enamel/dentin and cementum [142].

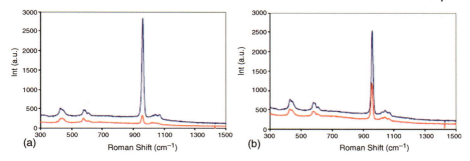

Figure 68.14 Representative parallel- (upper trace) and cross- (lower trace) polarized Raman spectra of (a) sound enamel and (b) caries lesions acquired with the portable Raman spectrometer system. From ref. [144].

Tsuda and Arends [138] used polarization-resolved Raman scattering to gain additional information about changes in the crystal structure of enamel. In a similar manner to IR reflectance spectroscopy [136], polarized Raman analysis can reveal the orientation of single crystals of hydroxyapatite. Such polarized Raman spectra can show the orientation of newly deposited mineral crystals during the repair of the enamel crystals via remineralization and orientational loss during demineralization. It has been well established that caries lesions rapidly depolarize incident light, and Ko and co-workers [143, 144] used polarized Raman spectroscopy to help discriminate caries lesions from sound enamel. Figure 68.14 shows polarized Raman spectra of sound enamel and caries lesions acquired with a portable Raman spectrometer.

Hill and Petrou [145] showed that NIR Raman spectroscopy could be used to differentiate between sound and carious enamel both *in vitro* and *in vivo*. NIR Raman spectroscopy is advantageous for the examination of biological tissues due to the reduced fluorescence background. They used the intensity ratio of the intense 960 cm^{-1} phosphate peak and the background luminescence (fluorescence) at 930 cm^{-1}. Yokoyama *et al.* [146] examined carious lesions with a hollow optical fiber probe that was cemented to a glass ball lens allowing the same fiber to be used for both light delivery and collection. They used the ratio of the 433/446 cm^{-1} peaks, which are peaks due to phosphate that manifest intensity differences in polarized Raman spectra, to indicate caries along with the fluorescence ratio of the 962/930 cm^{-1} peaks. Even though fluorescence is typically considered a major source of unwanted noise in Raman spectroscopy of biological tissues, it can also be exploited in the case of dental caries. Ko *et al.* used polarization ratios of the 959 cm^{-1} band to indicate caries [144]. They also used Raman spectroscopy in conjunction with OCT to help differentiate sound and demineralized enamel [143].

Raman spectroscopy can also be used to monitor changes in the carbonate content of enamel and dentin in a similar fashion to IR spectroscopy [147–150]. Carbonate is preferentially lost during demineralization since those sites on the enamel crystal have a higher solubility than the rest of the mineral phase. Liu and Hsu [151] and Klocke *et al.* [152] showed that changes in the carbonate content could be measured after laser irradiation.

68.10
Ultrasound and Terahertz Imaging

Ultrasound uses high-frequency mechanical pressure waves that are reflected in tissues upon encountering changes in tissue density. Different materials such as enamel dentin and carious tissues manifest different acoustic velocities [153]. Although ultrasound has been highly successful in medicine, bulky probes, the large acoustic impedance mismatch (low coupling efficiency) at the tooth surface due to the high density of the enamel of the tooth, and the relatively low resolution of ultrasound has greatly hindered its use for caries imaging. Ghorayeb et al. recently provided a review of ultrasonography for dentistry [154]. Lees and Barber first used ultrasound to measure the internal interfaces in the tooth, namely the enamel–dentin junction and the dentin–pulp junction over 40 years ago [155]. Huysmans and co-workers [156, 157] demonstrated that the a conventional ultrasonic probe could be used with a glycerin coupling agent to measure the remaining enamel thickness for monitoring erosion on extracted human incisors. Culijat et al. [158] acquired an entire circumferential scan of a tooth accurately showing the enamel thickness using water as a coupling agent. Blodgett [159] used laser-produced ultrasonic waves to image internal tooth interfaces. A pulsed CO_2 laser was used to produce the ultrasonic waves directly in the tooth, which overcomes the problem of low coupling efficiency.

Several studies have demonstrated that ultrasound can be used to measure tooth demineralization [160, 161]. Ng et al. [162] showed that 18 MHz pulse echo ultrasound could be used to differentiate shallow lesions artificially induced on extracted teeth. Teeth were immersed in a water-bath for measurement. Tagtekin et al. [160] used an ultrasound probe with glycerin as a coupling agent to measure early white spot lesions on the proximal surfaces of extracted third molar teeth.

Matalon et al. [163] demonstrated in both *in vitro* and *in vivo* studies that a hand-held probe that utilizes ultrasonic surface waves (Rayleigh) could be used to detect cavitated interproximal lesions. This device yielded higher sensitivity to cavitated interproximal lesions than bitewing radiographs, but the specificity was lower.

Terahertz radiation is located at wavelengths of ∼75 μm–7.5 mm between the far-IR and microwave regions. Biological tissues are semi-transparent and have terahertz fingerprints, permitting them to be imaged, identified, and analyzed. High-intensity synchrotron radiation sources have been used to generate terahertz radiation for imaging. Schade et al. [164] acquired multispectral images of teeth in the range 0.06–0.6 THz using a synchrotron. Near-field imaging techniques were employed to increase the resolution. An image of a 2.7 mm thick tooth slab with a lesion through the enamel is shown in Figure 68.15. More portable terahertz imaging systems are now available that utilize pulsed femtosecond lasers and semiconductors [165]. One of the most promising implementations involves terahertz pulse imaging (TPI). This method can be used to measure the remaining enamel thickness to provide a measure of enamel erosion [166]. The large refractive index mismatch at the dentin–enamel junction provides a strong reflective signal in TPI. Crawley et al. [165] showed that areas of demineralized and hypomineralized enamel can

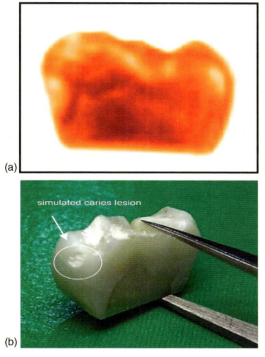

Figure 68.15 A near-field THz image (a) of a 2.7 mm thick human tooth section with an embedded small lesion (b). From ref. [164].

be discriminated from sound tissue in TPI images. The depth of artificial "surface-softened" lesions created in enamel using an acid gel measured by TPI compared well with measurements [167]. These methods appear best suited for measurement of the lesion depth and the remaining enamel thickness for erosion measurements.

References

1 National Institutes of Health (2001) Diagnosis and Management of Dental Caries Throughout Life 1–24, NIH Consensus Statement. National Institutes of Health, Wahington, DC.
2 Lussi, A. (1991) Validity of diagnostic and treatment decisions of fissure caries. *Caries Res.*, **25**, 296–303.
3 Featherstone, J.D.B. and Young, D. (1999) The need for new caries detection methods. *Proc. SPIE*, **3593**, 134–140.
4 Hume, W.R. (1996) Need for change in dental caries diagnosis, in *Early Detection of Dental Caries, Proceedings of the 1st Annual Indiana Conference* (ed. G.K. Stookey), Indiana University School of Dentistry, Indianapolis, IN, pp. 1–10.
5 Featherstone, J.D.B. (1996) Clinical implications: new strategies for caries prevention, in *Early Detection of Dental Caries, Proceedings of the 1st Annual Indiana Conference* (ed. G.K. Stookey), Indiana University School of Dentistry, Indianapolis, IN, pp. 287–295.

6 ten Cate, J.M. and van Amerongen, J.P. (1996) Caries diagnosis: conventional methods, in *Early Detection of Dental Caries, Proceedings of the 1st Annual Indiana Conference* (ed. G.K. Stookey), Indiana University School of Dentistry, Indianapolis, IN, pp. 27–37.

7 Featherstone, J.D.B. (1999) Prevention and reversal of dental caries:role of low level fluoride. *Community Dent. Oral. Epidemiol.*, **27**, 31–40.

8 Zijp, J.R. and ten Bosch, J.J. (1991) Angular dependence of HeNe laser light scattering by bovine and human dentine. *Arch. Oral Biol.*, **36**, 283–289.

9 Zijp, J.R. and ten Bosch, J.J. (1993) Theoretical model for the scattering of light by dentin and comparison with measurements. *Appl. Opt.*, **32**, 411–415.

10 Fried, D., Featherstone, J.D.B., Glena, R.E., and Seka, W. (1995) The nature of light scattering in dental enamel and dentin at visible and near-IR wavelengths. *Appl. Opt.*, **34**, 1278–1285.

11 Zijp, J.R., ten Bosch, J.J., and Groenhuis, R.A. (1995) HeNe laser light scattering by human dental enamel. *J. Dent. Res.*, **74**, 1891–1898.

12 Schmidt, W.J. and Keil, A. (1971) *Polarizing Microscopy of Dental Tissues*, Pergamon Press, Oxfordk.

13 Theuns, H.M., Shellis, R.P., Groeneveld, A., Dijk, J.W.E. and Poole, D.F.G. (1993) Relationships between birefringence and mineral content in artificial caries lesions in enamel. *Caries Res.*, **27**, 9–14.

14 Cheong, W., Prahl, S.A. and Welch, A.J. (1990) A review of the optical properties of biological tissues. *IEEE J. Quantum Electron.*, **26**, 2166–2185.

15 Wilson, B.C., Patterson, M.S., and Flock, S.T. (1987) Indirect versus direct techniques for the measurement of the optical properties of tissues. *Photochem. Photobiol.*, **46**, 601–608.

16 Izatt, J.A., Hee, M.R., Huang, D., Fujimoto, J.G., Swanson, E.A., Lin, C.P., Schuman, J.S., and Puliafito, C.A. (1993) Optical coherence tomography for medical applications, in *Medical Optical Tomography:Functional Imaging and Monitoring* (ed. G. Muller), SPIE, Bellingham, WA, pp. 450–472.

17 Tuchin, V. (2000) *Tissue Optics: Light Scattering Methods and Instruments for Medical Diagnostics*, SPIE, Bellingham, WA.

18 Spitzer, D. and ten Bosch, J.J. (1975) The absorption and scattering of light in bovine and human dental enamel. *Calcif. Tissue Res.*, **17**, 129–137.

19 ten Bosch, J.J. and Zijp, J.R. (1987) Optical properties of dentin, in *Dentine and Dentine Research in the Oral Cavity* (eds. A. Thylstrup, S.A. Leach, and V. Qvist), IRL Press, Oxford, pp. 59–65.

20 Jones R.S. and Fried, D. (2002) Attenuation of 1310 nm and 1550 nm laser light through sound dental enamel. *Proc. SPIE*, **4610**, 187–190.

21 Hale, G.M. and Querry, M.R. (1973) Optical constants of water in the 200 nm to 200-μm wavelength region. *Appl. Opt.*, **12**, 555–563.

22 Wang, L. and Jacques, S.L. (1992) *Monte Carlo Modeling of Light Transport in Multi-Layer Tissues in Standard C*, University of Texas M.D. Anderson Cancer Center, Houston, TX.

23 Wilson, B.C. and Adam, G. (1989) A Monte Carlo model for the absorption and flux distribution of light in tissue. *Med. Phys.*, **10**, 824–830.

24 van der Zee, P. (1993) Methods for measuring the optical properties of tissue in the visible and near-IR wavelength range, in *Medical Optical Tomography: Functional Imaging and Monitoring* (eds. G.H. Mueller et al..), vol. **IS11**, SPIE, Bellingham, WA, pp. 450–472.

25 Altshuler, G.B. (1995) Optical model of the tissues of the human tooth. *J. Opt. Technol.*, **62**, 516–521.

26 Altshuler, G.B. and Grisimov, V.N. (1990) New optical model in the human hard tooth tissues. *Proc. SPIE*, **1353**, 97–102.

27 Kienle, A., Forster, F.K., Diebolder, R., and Hibst, R. (2003) Light propagation in dentin: influence of microstructure on anisotropy. *Phys. Med. Biol.*, **48**, N7–N14.

28 Kienle, A., Forster, F.K. and Hibst, R. (2004) Anisotropy of light propagation in

biological tissue. *Opt. Lett.*, **29**, 2617–2619

29 Kienle, A. and Hibst, R. (2006) Light guiding in biological tissue due to scattering. *Phys. Rev. Lett.*, **97**, 018104.

30 Kienle, A., Michels, R. and Hibst, R. (2005) Light propagation in a cubic biological tissue having anisotropic optical properties. *Proc. SPIE*, **5859**, 585917-1–585917-7.

31 Kienle, A., Michels, R., and Hibst, R. (2006) Magnification – a new look at a long-known optical property of dentin. *J. Dent. Res.*, **85**, 955–959.

32 Brinkman, J., ten Bosch, J.J., and Borsboom, P.C.F. (1988) Optical quantification of natural caries in smooth surfaces of extracted teeth. *Caries Res.*, **22**, 257–262.

33 ten Bosch, J.J., van der Mei, H.C. and Borsboom, P.C.F. (1984) Optical monitor of *in vitro* caries. *Caries Res.*, **18**, 540–547.

34 Angmar-Månsson, B. and ten Bosch, J.J. (1987) Optical methods for the detection and quantification of caries. *Adv. Dent. Res.*, **1**, 14–20.

35 ten Bosch, J.J. and Coops, J.C. (1995) Tooth color and reflectance as related to light scattering and enamel hardness. *J. Dent. Res.*, **74**, 374–380.

36 Ko, C.C., Tantbirojn, D., Wang, T., and Douglas, W.H. (2000) Optical scattering power for characterization of mineral loss. *J. Dent. Res.*, **79**, 1584–1589.

37 Wefel, J.S., Clarkson, B.H. and Heilman, J.R. (1985) Natural root caries: a histologic and microradiographic evaluation. *J. Oral Pathol.*, **14**, 615–623.

38 Vaarkamp, J., ten Bosch, J.J., and Verdonschot, E.H. (1995) Light propagation through teeth containing simulated caries lesions. *Phys. Med. Biol.*, **40**, 1375–1387.

39 Darling, C.L., Huynh, G.D., and Fried, D. (2006) Light scattering properties of natural and artificially demineralized dental enamel at 1310 nm. *J. Biomed. Opt.*, **11**, 034023 1–11.

40 Fried, D., Glena, R.E., Featherstone, J.D.B., and Seka, W. (1997) Permanent and transient changes in the reflectance of CO_2 laser irradiated dental hard tissues at $\lambda = 9.3$, 9.6, 10.3, and 10.6 μm and at fluences of 1–20 J/cm^2. *Lasers Surg. Med.*, **20**, 22–31.

41 Duplain, G., Boulay, R. and Belanger, P.A. (1987) Complex index of refraction of dental enamel at CO_2 wavelengths. *Appl. Opt.*, **26**, 4447–4451.

42 Zuerlein, M., Fried, D., Featherstone, J.D.B., and Seka, W. (1999) Absorption coefficients of dental enamel at CO_2 wavelengths. *Spec. Top. J. Quantum Electron.*, **5**, 1083–1089.

43 Zuerlein, M., Fried, D., and Featherstone, J.D.B. (1999) Modeling the modification depth of carbon dioxide laser treated enamel. *Lasers Surg. Med.*, **25**, 335–347.

44 Benson, P.E., Ali Shah, A., and Robert Willmot, D. (2008) Polarized versus nonpolarized digital images for the measurement of demineralization surrounding orthodontic brackets. *Angle Orthod.*, **78**, 288–293.

45 Everett, M.J., Colston, B.W., Sathyam, U.S., Silva, L.B.D., Fried, D., and Featherstone, J.D.B. (1999) Non-invasive diagnosis of early caries with polarization sensitive optical coherence tomography (PS-OCT). *Proc. SPIE*, **3593**, 177–183.

46 Fried, D., Xie, J., Shafi, S., Featherstone, J.D.B., Breunig, T., and Lee, C.Q. (2002) Early detection of dental caries and lesion progression with polarization sensitive optical coherence tomography. *J. Biomed. Opt.*, **7**, 618–627.

47 Fried, D., Featherstone, J.D.B., Darling, C.L., Jones, R.S., Ngoatheppitak, P., and Buehler, C.M. (2005) *Early Caries Imaging and Monitoring with Near-IR Light*, Saunders, Philadelphia, pp. 771–794.

48 Blodgett, D.W. and Webb, S.C. (2001) Optical BRDF and BSDF measurements of human incisors from visible to mid-infrared wavelengths. *Proc. SPIE*, **4257**, 448–454.

49 Analoui, M., Ando, M., and Stookey, G.K. (2000) Comparison of reflectance spectra of sound and carious enamel. *Proc. SPIE*, **3910**, 1017–2661.

50 Wu, J.I. and Fried, D. (2009) High contrast near-infrared polarized reflectance images of demineralization

on tooth buccal and occlusal surfaces at $\lambda = 1310$ nm. *Lasers Surg. Med.*, **41** (3), 208–213.
51 Barenie, J., Leske, G., and Ripa, L.W. (1973) The use of fiber optic transillumination for the detection of proximal caries. *Oral Surg.*, **36**, 891–897.
52 Pine, C.M. (1996) Fiber-optic transillumination (FOTI) in caries diagnosis, in *Early Detection of Dental Caries, Proceedings of the 1st Annual Indiana Conference* (ed. G.K. Stookey), Indiana University School of Dentistry, Indianapolis, IN, pp. 51–66.
53 Peltola, J. and Wolf, J. (1981) Fiber optics transillumination in caries diagnosis. *Proc. Finn. Dent. Soc.*, **77**, 240–244.
54 Holt, R.D. and Azeevedo, M.R. (1989) Fiber optic transillumination abnd radiographs in diagnosis of approximal cariesin primary teeth. *Community Dent. Health*, **6**, 239–247.
55 Mitropoulis, C.M. (1985) The use of fiber optic transillumination in the diagnosis of posterior approximal caries in clinical trials. *Caries Res.*, **19**, 379–384.
56 Hintze, H., Wenzel, A., Danielsen, B., and Nyvad, B. (1998) Reliability of visual examination, fibre-optic transillumination, and bite-wing radiography, and reproducibility of direct visual examination following tooth separation for the identification of cavitated carious lesions in contacting approximal surfaces. *Caries Res.*, **32**, 204–209.
57 Schneiderman, A., Elbaum, M., Schultz, T., Keem, S., Greenebaum, M., and Driller, J. (1997) Assessment of dental caries with digital imaging fiber-optic transillumination (DIFOTI): in vitro study. *Caries Res.*, **31**, 103–110.
58 Bin-Shuwaish, M., Yaman, P., Dennison, J., and Neiva, G. (2008) The correlation of DIFOTI to clinical and radiographic images in Class II carious lesions. *J. Am. Dent. Assoc.*, **139**, 1374–1381.
59 Young, D.A. and Featherstone, J.D. (2005) Digital imaging fiber-optic trans-illumination, F-speed radiographic film and depth of approximal lesions. *J. Am. Dent. Assoc.*, **136**, 1682–1687.
60 Young, D.A. (2002) New caries detection technologies and modern caries management: merging the strategies. *Gen. Dent.*, **50**, 320–331.
61 Yang, J. and Dutra, V. (2005) Utility of radiology, laser fluorescence, and transillumination. *Dent. Clin. North Am.*, **49**, 739–752, vi.
62 Keem, S. and Elbaum, M. (1997) Wavelet representations for monitoring changes in teeth imaged with digital imaging fiber-optic transillumination. *IEEE Trans. Med. Imaging*, **16**, 653–663.
63 Jones, R. and Fried, D. (2001) Attenuation of 1310 nm and 1550 nm laser light through dental enamel. *J. Dent. Res.*, **80**, 737.
64 Huynh, G.D., Darling, C.L., and Fried, D. (2004) Changes in the optical properties of dental enamel at 1310 nm after demineralization. *Proc. SPIE*, **5313**, 118–124.
65 Darling, C.L. and Fried, D. (2005) Optical properties of natural caries lesions in dental enamel at 1310 nm. *Proc. SPIE*, **5687**, 34–41.
66 Bühler, C.M., Ngaotheppitak, P., and Fried, D. (2005) Imaging of occlusal dental caries (decay) with near-IR light at 1310 nm. *Opt. Express*, **13**, 573–582.
67 Jones, R.S., Huynh, G.D., Jones, G.C., and Fried, D. (2003) Near-IR transillumination at 1310 nm for the imaging of early dental caries. *Opt. Express*, **11**, 2259–2265.
68 Jones, G., Jones, R.S., and Fried, D. (2004) Transillumination of interproximal caries lesions with 830 nm light. *Proc. SPIE*, **5313**, 17–22.
69 Bouma, B.E. and Tearney, G.J. (eds.) (2002) *Handbook of Optical Coherence Tomography*, Marcel Dekker, New York.
70 Derickson, D. (1998) *Fiber Optic Test and Measurement*, Prentice Hall, Upper Saddle River, NJ.
71 Youngquist, R.C., Carr, S., and Davies, D.E.N. (1987) Optical coherence-domain reflectometry. *Appl. Opt.*, **12**, 158–160.
72 Huang, D., Swanson, E.A., Lin, C.P., Schuman, J.S., Stinson, W.G., Chang, W., Hee, M.R., Flotte, T., Gregory, K., Puliafito, C.A., and Fujimoto, J.G. (1991)

Optical coherence tomography. *Science*, **254**, 1178–1181.
73 Brezinski, M. (2006) *Optical Coherence Tomography: Principles and Applications*, Academic Press, Burlington, MA.
74 Colston, B., Everett, M., Da Silva, L., Otis, L., Stroeve, P., and Nathel, H. (1998) Imaging of hard and soft tissue structure in the oral cavity by optical coherence tomography. *Appl. Opt.*, **37**, 3582–3585.
75 Colston, B.W., Everett, M.J., Da Silva, L.B., and Otis, L.L. (1998) Optical coherence tomography for diagnosis of periodontal diseases. *Proc. SPIE*, **3251**, 52–58.
76 Colston, B.W., Sathyam, U.S., DaSilva, L.B., Everett, M.J., and Stroeve, P. (1998) Dental OCT. *Opt. Express*, **3**, 230–238.
77 Feldchtein, F.I., Gelikonov, G.V., Gelikonov, V.M., Iksanov, R.R., Kuranov, R.V., Sergeev, A.M., Gladkova, N.D., Ourutina, M.N., Warren, J.A., and Reitze, D.H. (1998) In vivo OCT imaging of hard and soft tissue of the oral cavity. *Opt. Express*, **3**, 239–251.
78 Otis, L.L., Al-Sadhan, R.I., Meiers, J., and Redford-Badwal, D. (2000) Identification of occlusal sealants using optical coherence tomography. *J. Clin. Dent.*, **14**, 7–10.
79 Madjarova, V.D., Yasuno, Y., Makita, S., Hori, Y., Voeffray, J.B., Itoh, M., Yatagai, T., Tamura, M., and Nanbu, T. (2006) Investigations of soft and hard tissues in oral cavity by spectral domain optical coherence tomography. *Proc. SPIE*, **6079**, 60790N-1–60790N-7.
80 Seon, Y.R., Jihoon, N., Hae, Y.C., Woo, J.C., Byeong, H.L., and Gil-Ho, Y. (2006) Realization of fiber-based OCT system with broadband photonic crystal fiber coupler. *Proc. SPIE*, **6079**, 60791N-1–60791N-7.
81 Yamanari, M., Makita, S., Violeta, D.M., Yatagai, T., and Yasuno, Y. (2006) Fiber-based polarization-sensitive Fourier domain optical coherence tomography using B-scan-oriented polarization modulation method. *Opt. Express*, **14**, 6502–6515.

82 Furukawa, H., Hiro-Oka, H., Amano, T., DongHak, C., Miyazawa, T., Yoshimura, R., Shimizu, K., and Ohbayashi, K. (2006) Reconstruction of three-dimensional structure of an extracted tooth by OFDR-OCT. *Proc. SPIE*, **6079**, 60790T-1–60790T-7.
83 Amaechi, B.T., Higham, S.M., Podoleanu, A.G., Rodgers, J.A., and Jackson, D.A. (2001) Use of optical coherence tomography for assessment of dental caries. *J. Oral. Rehab.*, **28**, 1092–1093.
84 Amaechi, B.T., Podoleanu, A., Higham, S.M., and Jackson, D.A. (2003) Correlation of quantitative light-induced fluorescence and optical coherence tomography applied for detection and quantification of early dental caries. *J. Biomed. Opt.*, **8**, 642–647.
85 Baumgartner, A., Hitzenberger, C.K., Dicht, S., Sattmann, H., Moritz, A., Sperr, W., and Fercher, A.F. (1998) Optical coherence tomography for dental structures. *Proc. SPIE*, **3248**, 130–136.
86 Dicht, S., Baumgartner, A., Hitzenberger, C.K., Sattmann, H., Robi, B., Moritz, A., Sperr, W., and Fercher, A.F. (1999) Polarization-sensitive optical optical coherence tomography of dental structures. *Proc. SPIE*, **3593**, 169–176.
87 Baumgartner, A., Dicht, S., Hitzenberger, C.K., Sattmann, H., Robi, B., Moritz, A., Sperr, W., and Fercher, A.F. (2000) Polarization-sensitive optical optical coherence tomography of dental structures. *Caries Res.*, **34**, 59–69.
88 Wang, X.J., Zhang, J.Y., Milner, T.E., Boer, J.F.D., Zhang, Y., Pashley, D.H., and Nelson, J.S. (1999) Characterization of dentin and enamel by use of optical coherence tomography. *Appl. Opt.*, **38**, 586–590.
89 Jones, R.S., Darling, C.L., Featherstone, J.D.B., and Fried, D. (2004) Imaging artificial caries on occlusal surfaces with polarization sensitive optical coherence tomography. *Caries Res.*, **40**, 81–89.
90 Ngaotheppitak, P., Darling, C.L., and Fried, D. (2006) PS-OCT of natural

91 Ismail, A.I. (2004) Visual and visuo-tactile detection of dental caries. *J. Dent. Res.*, **83**, C56–C66.

92 Jones, R.S., Staninec, M., and Fried, D. (2004) Imaging artificial caries under composite sealants and restorations. *J. Biomed. Opt.*, **9**, 1297–1304.

93 Alfano, R.R., Lam, W., Zarrabi, H.J., Alfano, M.A., Cordero, J., and Tata, D.B. (1984) Human teeth with and without caries studied by laser scattering, fluorescence and absorption spectroscopy. *IEEE J. Quantum Electron.*, **20**, 1512–1515.

94 Bjelkhagen, H. and Sundstrom, F. (1981) A clinically applicable laser luminescence for the early detection of dental caries. *IEEE J. Quantum Electron.*, **17**, 266–268.

95 ten Bosch, J.J. (1999) Summary of research of quantitative light fluorescence, in *Early Detection of Dental Caries II*, vol. **4**, Indiana University School of Dentistry, Indianapolis, IN, pp. 261–278.

96 Lakowicz, J.R. (1999) *Principles of Fluorecence Spectroscopy*, Kluwer Academic, New York.

97 Hafström-Björkman, U., Sundström, F., de Josselin de Jong, E., Oliveby, A., and Angmar-Månsson, B. (1992) Comparison of laser fluorescence and longitudinal microradiography for quantitative assessment of *in vitro* enamel caries. *Caries Res.*, **26**, 241–247.

98 Ando, M., Gonzalez-Cabezas, C., Isaacs, R.L., Eckert, A.F., and Stookey, G.K. (2004) Evaluation of several techniques for the detection of secondary caries adjacent to amalgam restorations. *Caries Res.*, **38**, 350–356.

99 de Josselin de Jong, E., Sundström, F., Westerling, H., Tranaeus, S., ten Bosch, J.J., and Angmar-Månsson, B. (1995) A new method for *in vivo* quantification of changes in initial enamel caries with laser fluorescence. *Caries Res.*, **29**, 2–7.

100 Fontana, M., Li, Y., Dunipace, A.J., Noblitt, T.W., Fischer, G., Katz, B.P., and Stookey, G.K. (1996) Measurement of enamel demineralization using microradiography and confocal microscopy. *Caries Res.*, **30**, 317–325.

101 Stookey, G.K. (2005) *Quantitative Light Fluorescence: a Technology for Early Monitoring of the Caries Process*, Saunders, Philadelphia, pp. 753–770.

102 Ando, M., Eckert, G.J., Stookey, G.K., and Zero, D.T. (2004) Effect of imaging geometry on evaluating natural white-spot lesions using quantitative light-induced fluorescence. *Caries Res.*, **38**, 39–44.

103 Ando, M., Schemehorn, B.R., Eckert, G.J., Zero, D.T., and Stookey, G.K. (2003) Influence of enamel thickness on quantification of mineral loss in enamel using laser-induced fluorescence. *Caries Res.*, **37**, 24–28.

104 al-Khateeb, S., Oliveby, A., de Josselin de Jong, E., and Angmar-Månsson, B. (1997) Laser fluorescence quantification of remineralisation *in situ* of incipient enamel lesions: influence of fluoride supplements. *Caries Res.*, **31**, 132–140.

105 Tranaeus, S., Al-Khateeb, S., Björkman, S., Twetman, S., and Angmar-Månsson, B. (2001) Application of quantitative light-induced fluorescence to monitor incipient lesions in caries-active children. A comparative study of remineralisation by fluoride varnish and professional cleaning. *Eur. J. Oral Sci.*, **109**, 71–75.

106 al-Khateeb, S., ten Cate, J.M., Angmar-Månsson, B., de Josselin de Jong, E., Sundström, G., Exterkate, R.A., and Oliveby, A. (1997) Quantification of formation and remineralization of artificial enamel lesions with a new portable fluorescence device. *Adv. Dent. Res.*, **11**, 502–506.

107 Koenig, K., Schneckenburger, H., Hemmer, J., Tromberg, B.J., Steiner, R.W., and Rudolf, W. (1994) In-vivo fluorescence detection and imaging of porphyrin-producing bacteria in the human skin and in the oral cavity for diagnosis of acne vulgaris, caries, and squamous cell carcinoma. *Proc. SPIE*, **2135**, 129–138.

108 Shi, X.Q., Welander, U., and Angmar-Månsson, B. (2000) Occlusal caries detection with KaVo DIAGNOdent and radiography: an *in vitro* comparison. *Caries Res.*, **34**, 151–158.

109 Lussi, A., Imwinkelreid, S., Pitts, N.B., Longbottom, C., and Reich, E. (1999) Performance and reproducibility of a laser fluorescence system for detection of occlusal caries *in vitro*. *Caries Res.*, **33**, 261–266.

110 Lussi, A., Hack, A., Hug, I., Heckenberger, H., Megert, B., and Stich, H. (2006) Detection of approximal caries with a new laser fluorescence device. *Caries Res.*, **40**, 97–103.

111 Konig, K., Schneckenburger, H., and Hibst, R. (1999) Time-gated *in vivo* autofluorescence imaging of dental caries. *Cell Mol. Biol. (Noisy-le-Grand)*, **45**, 233–239.

112 Hall, A. and Girkin, J.M. (2004) A review of potential new diagnostic modalities for caries lesions. *J. Dent. Res.*, **83** (Spec No. C), C89–C94.

113 Eggertsson, H., Analoui, M., Veen, M.H.V.D., Gonzalez-Cabezas, C., Eckert, G.J., and Stookey, G.K. (1999) Detection of early interproximal caries *in vitro* using laser fluorescence, dye-enhanced laser fluorescence and direct visual examination. *Caries Res.*, **33**, 227–233.

114 Ando, M., Hall, A.F., Eckert, G.J., Schemehorn, B.R., Analoui, M., and Stookey, G.K. (1997) Relative ability of laser fluorescence techniques to quantitate early mineral loss *in vitro*. *Caries Res.*, **31**, 125–131.

115 Kaneko, K., Matsuyama, K., and Nakashima, S. (1999) Quantification of early carious enamel lesions by using an infrared camera, in *Early Detection of Dental Caries II*, vol. 4, Indiana University School of Dentistry, Indianapolis, IN, pp. 83–99.

116 Tam, A.C. (1985) Pulsed photothermal radiometry for noncontact spectroscopy, material testing and inspection measurements. *Infrared Phys.*, **25**, 305–313.

117 Chebotareva, G.P., Nikitin, A.P., and Zubov, B.V. (1995) Comparative study of CO_2 and Er:YAG laser heating of tissue using the pulsed photothermal radiometry technique. *Proc. SPIE*, **2394**, 243–251.

118 Jeon, R.J., Matvienko, A., Mandelis, A., Abrams, S.H., Amaechi, B.T., and Kulkarni, G. (2007) Detection of interproximal demineralized lesions on human teeth *in vitro* using frequency-domain infrared photothermal radiometry and modulated luminescence. *J. Biomed. Opt.*, **12**, 034028.

119 Lena, N., Andreas, M., and Stephen, H.A. (2000) Novel dental dynamic depth profilometric imaging using simultaneous frequency-domain infrared photothermal radiometry and laser luminescence. *J. Biomed. Opt.*, **5**, 31–39.

120 Raymond, J.J., Adam, H., Anna, M., Andreas, M., Stephen, H.A., and Bennett, T.A. (2008) *In vitro* detection and quantification of enamel and root caries using infrared photothermal radiometry and modulated luminescence. *J. Biomed. Opt.*, **13**, 034025.

121 Raymond, J.J., Andreas, M., Victor, S., and Stephen, H.A. (2004) Nonintrusive, noncontacting frequency-domain photothermal radiometry and luminescence depth profilometry of carious and artificial subsurface lesions in human teeth. *J. Biomed. Opt.*, **9**, 804–819.

122 Carden, A. and Morris, M.D. (2000) Application of vibrational spectroscopy to the study of mineralized tissues. *J. Biomed. Opt.*, **5**, 259–268.

123 Brown, P.W. and Constantz, B. (1994) *Hydroxyapatite and Related Materials*, CRC Press, Boca Raton, FL.

124 Elliot, J.C. (1994) *Structure and Chemistry of the Apatites and Other Calcium Orthophosphates*, Elsevier, Amsterdam.

125 Pleshko, N., Boskey, A., and Mendelsohn, R. (1991) Novel infrared spectroscopic method for the determination of crystallinity of hydroxyapatite minerals. *Biophys. J.*, **60**, 786–793.

126 LeGeros, R.Z., LeGeros, J.P., Trautz, O.T., and Klein, E. (1970) *Spectral Properties of Carbonate in Carbonate-Containing*

Apatites, Plenum Press, New York, vol. **7b**, pp. 3–13.

127 LeGeros, R.Z., Trautz, O.R., LeGeros, J.P., and Klein, E. (1968) Carbonate substitution in the apatite structure. *Bull. Soc. Chim. Fr.*, 1712–1718.

128 Fowler, B. and Kuroda, S. (1986) Changes in heated and in laser-irradiated human tooth enamel and their probable effects on solubility. *Calcif. Tissue Int.*, **38**, 197–208.

129 Kuroda, S. and Fowler, B.O. (1984) Compositional, structural and phase changes in *in vitro* laser-irradiated human tooth enamel. *Calcif. Tissue Int.*, **36**, 361–369.

130 Featherstone, J.D.B., Pearson, S., and LeGeros, R.Z. (1984) An IR method for quantification of carbonate in carbonated-apatites. *Caries Res.*, **18**, 63–66.

131 Featherstone, J.D.B., Fried, D., and Duhn, C. (1998) Surface dissolution kinetics of dental hard tissue irradiated over a fluence range of 1–8 J/cm^2 at a wavelength of 9.3 μm. *Proc. SPIE*, **3248**, 146–151.

132 Featherstone, J.D.B. and Nelson, D.G.A. (1987) Laser effects on dental hard tissue. *Adv. Dent. Res*, **1**, 21–26.

133 Konishi, N., Fried, D., Featherstone, J.D.B., and Staninec, M. (1999) Inhibition of secondary caries by CO_2 laser treatment. *Am. J. Dent.*, **12**, 213–216.

134 Young, D.A., Fried, D. and Featherstone, J.D.B. (2000) Ablation and caries inhibition of pits and fissures by IR laser irradiation. *Proc. SPIE*, **3910**, 247–253.

135 Fried, D. (2000) IR ablation of dental enamel. *Proc. SPIE*, **3910**, 136–148.

136 Klein, E., LeGeros, J.P., Trautz, O.T., and LeGeros, R.Z. (1970) *Polarized Infrared Reflectance of Single Crystals of Apatites*, Plenum Press, New York, vol. **7b**, pp. 3–13.

137 Nelson, D.G.A. and Williamson, B.E. (1985) Raman spectra of phosphate and monophosphate ions in several dentally-relevant materials. *Caries Res.*, **19**, 113–121.

138 Tsuda, H. and Arends, J. (1994) Orientational micro-Raman spectroscopy on hydroxyapatite single crystals and human enamel crystallites. *J. Dent. Res.*, **73**, 1703–1710.

139 Tsuda, H. and Arends, J. (1997) Raman spectroscopy in dental research: a short review of recent studies. *Adv. Dent. Res.*, **11**, 539–547.

140 O'Shea, D.C., Bartlett, M.L., and Young, R.A. (1974) Compositional analysis of apatites with laser-Raman spectroscopy(OH, F, Cl). *Arch. Oral Biol.*, **19**, 995–1006.

141 Tsuda, H., Ruben, J., and Arends, J. (1996) Raman spectra of human dentin mineral. *Eur. J. Oral Sci.*, **104**, 123–131.

142 Schulze, K.A., Balooch, M., Balooch, G., Marshall, G.W., and Marshall, S.J. (2004) Micro-Raman spectroscopic investigation of dental calcified tissues. *J. Biomed. Mater. Res. A*, **69**, 286–293.

143 Ko, A.C., Choo-Smith, L.P., Hewko, M., Leonardi, L., Sowa, M.G., Dong, C.C., Williams, P., and Cleghorn, B. (2005) *Ex vivo* detection and characterization of early dental caries by optical coherence tomography and Raman spectroscopy. *J. Biomed. Opt.*, **10**, 031118.

144 Ko, A.C., Hewko, M., Sowa, M.G., Dong, C.C., Cleghorn, B., and Choo-Smith, L.P. (2008) Early dental caries detection using a fibre-optic coupled polarization-resolved Raman spectroscopic system. *Opt. Express*, **16**, 6274–6284.

145 Hill, W. and Petrou, V. (2000) Caries detection by diode laser raman spectroscopy. *Appl. Spectrosc.*, **54**, 795–799.

146 Yokoyama, E., Kakino, S., and Matsuura, Y. (2008) Raman imaging of carious lesions using a hollow optical fiber probe. *Appl. Opt.*, **47**, 4227–4230.

147 Nishino, M., Yamashita, S., Aoba, T., Okazaki, M., and Moriwaki, Y. (1981) The laser-Raman spectroscopic studies on human enamel and precipitated carbonate-containing apatites. *J. Dent. Res.*, **60**, 751–755.

148 Penel, G., Leroy, G., Leroy, N., Behin, P., Langlois, J.M., Libersa, J.C., and Dupas, P.H. (2000) Raman spectrometry applied to calcified tissue and calcium-phosphorus biomaterials. *Bull. Group Int. Rech. Sci. Stomatol. Odontol.*, **42**, 55–63.

149 Penel, G., Leroy, G., Rey, C., and Bres, E. (1998) MicroRaman spectral study of the PO_4 and CO_3 vibrational modes in synthetic and biological apatites. *Calcif. Tissue Int.*, **63**, 475–481.

150 de Mul, F.F.M., Hottenhuis, M.H.J., Bouter, P., Greve, J., Arends, J., and ten Bosch, J.J. (1986) Micro-Raman line broadening in synthetic carbonated hydroxyapatite. *J. Dent. Res.*, **65**, 437–440.

151 Liu, Y. and Hsu, C.Y. (2007) Laser-induced compositional changes on enamel: a FT-Raman study. *J. Dent.*, **35**, 226–230.

152 Klocke, A., Mihailova, B., Zhang, S., Gasharova, B., Stosch, R., Guttler, B., Kahl-Nieke, B., Henriot, P., Ritschel, B., and Bismayer, U. (2007) CO_2 laser-induced zonation in dental enamel: a Raman and IR microspectroscopic study. *J. Biomed. Mater. Res. B Appl. Biomater.*, **81**, 499–507.

153 Ng, S.Y., Payne, P.A., Cartledge, N.A. and Ferguson, M.W. (1989) Determination of ultrasonic velocity in human enamel and dentine. *Arch. Oral Biol.*, **34**, 341–345.

154 Ghorayeb, S.R., Bertoncini, C.A., and Hinders, M.K. (2008) Ultrasonography in dentistry. *IEEE Trans. Ultrason. Ferroelectr. Freq. Control*, **55**, 1256–1266.

155 Lees, S. and Barber, F.E. (1968) Looking into teeth with ultrasound. *Science*, **161**, 477–478.

156 Huysmans, M.C. and Thijssen, J.M. (2000) Ultrasonic measurement of enamel thickness: a tool for monitoring dental erosion? *J. Dent.*, **28**, 187–191.

157 Louwerse, C., Kjaeldgaard, M., and Huysmans, M.C. (2004) The reproducibility of ultrasonic enamel thickness measurements: an *in vitro* study. *J. Dent.*, **32**, 83–89.

158 Culjat, M., Singh, R.S., Yoon, D.C. and Brown, E.R. (2003) Imaging of human tooth enamel using ultrasound. *IEEE Trans. Med. Imaging*, **22**, 526–529.

159 Blodgett, D.W. (2003) Applications of laser-based ultrasonics to the characterization of the internal structure of teeth. *J. Acoust. Soc. Am.*, **114**, 542–549.

160 Tagtekin, D.A., Ozyoney, G., Baseren, M., Ando, M., Hayran, O., Alpar, R., Gokalp, S., Yanikoglu, F.C. and Stookey, G.K. (2008) Caries detection with DIAGNOdent and ultrasound. *Oral Surg. Oral Med. Oral Pathol. Oral Radiol. Endodontol.*, **106**, 729–735.

161 Lees, S., Gerhard, F.B. Jr., and Oppenheim, F.G. (1973) Ultrasonic measurement of dental enamel demineralization. *Ultrasonics*, **11**, 269–273.

162 Ng, S.Y., Ferguson, M.W., Payne, P.A., and Slater, P. (1988) Ultrasound studies of unblemished and artificially demineralized enamel in extracted human teeth: a new method for detecting early caries. *J. Dent.*, **16**, 201–209.

163 Matalon, S., Feuerstein, O., Calderon, S., Mittleman, A., and Kaffe, I. (2007) Detection of cavitated carious lesions in approximal tooth surfaces by ultrasonic caries detector. *Oral Surg. Oral Med. Oral Pathol. Oral Radiol. Endodontol.*, **103**, 109–113.

164 Schade, U., Holldack, K., Martin, M., and Fried, D. (2005) THz near-field imaging of biological tissues employing synchrotron radiation. *Proc. SPIE*, **5687**, 46–52.

165 Crawley, D.A., Longbottom, C., Cole, B.E., Ciesla, C.M., Arnone, D., Wallace, V.P. and Pepper, M. (2003) Terahertz pulse imaging: a pilot study of potential applications in dentistry. *Caries Res.*, **37**, 352–359.

166 Crawley, D., Longbottom, C., Wallace, V.P., Cole, B., Arnone, D. and Pepper, M. (2003) Three-dimensional terahertz pulse imaging of dental tissue. *J. Biomed. Opt.*, **8**, 303–307.

167 Pickwell, E., Wallace, V.P., Cole, B.E., Ali, S., Longbottom, C., Lynch, R.J., and Pepper, M. (2007) A comparison of terahertz pulsed imaging with transmission microradiography for depth measurement of enamel demineralisation *in vitro*. *Caries Res.*, **41**, 49–55.

168 Fried, D., Zuerlein, M., Featherstone, J.D.B., Seka, W.D., and McCormack, S.M. (1997) IR laser ablation of dental enamel: mechanistic dependence on the primary absorber. *Appl. Surf. Sci.*, **128**, 852–856.

169 Zuerlein, M., Fried, D., Featherstone, J., and Seka, W. (1999) Optical properties

of dental enamel at 9–11 μm derived from time-resolved radiometry. *Spec. Top. IEEE J. Quantum Electron.*, **5**, 1083–1089.

170 Ngaotheppitak, P., Darling, C.L., and Fried, D. (2005) Polarization optical coherence tomography for measuring the severity of caries lesions. *Lasers Surg. Med.*, **37**, 78–88.

171 Manesh, S.K., Darling, C.L., and Fried, D. (2009) Nondestructive assessment of dentin demineralization using polarization sensitive optical coherence tomography after exposure to fluoride and laser irradiation. *J. Biomed. Mater. Res. B Appl. Biomater.*, **90** (2), 802–812.

69
Optical Coherence Tomography in Dentistry

Natalia D. Gladkova, Yulia.V. Fomina, Elena B. Kiseleva, Maria M. Karabut, Natalia S. Robakidze, Alexander A. Muraev, Stefka G. Radenska-Lopovok, and Anna V. Maslennikova

69.1
Introduction

Studies on the application of optical coherence tomography (OCT) in dentistry were published in 1998 simultaneously by two research teams: one at the Lawrence Livermore Laboratory (Livermore, CA, USA) [1–10] and the other at the Institute of Applied Physics of the Russian Academy of Sciences (Nizhny Novgorod, Russia) [11–14]. Later, a series of studies were launched at the Beckman Laser Institute [15–21]. The majority of dental OCT studies were performed using time domain OCT and later frequency domain OCT systems [22, 23]. There are many publications on time domain polarization-sensitive optical coherence tomography (PS-OCT) [24–31]. The available experience in using OCT for oral cavity examination is sufficiently significant to discuss its clinical value.

In this chapter, we consider the application of OCT in dentistry and present the most significant recent clinical, results obtained by the Russian team with the use of a standard time domain OCT system [32] and a cross-polarization optical coherence tomography (CP-OCT) system described by Gelikonov and Gelikonov [33].

CP-OCT detects light reflected in two polarizations: the polarization of probing light (referred to as co-polarization) and orthogonal polarization (cross-polarization) [34]. The CP-OCT device used by our group acquires two conjugate images (see Figure 69.2b). The image in co-polarization (at the bottom) contains information about the backscattering characteristics of biotissue. The cross-polarized image (at the top) gives an insight not only into the backscattering features but also into the depolarizing properties of collagen, keratin, and other anisotropic structures of biotissue, which can increase the clinical value of OCT substantially [35].

Dental applications of OCT are traditionally focused on three areas: (i) examination of hard tissues (teeth), (ii) periodontal tissues, and (iii) oral cavity mucosa.

Handbook of Biophotonics. Vol.2: Photonics for Health Care, First Edition. Edited by Jürgen Popp,
Valery V. Tuchin, Arthur Chiou, and Stefan Heinemann.
© 2012 Wiley-VCH Verlag GmbH & Co. KGaA. Published 2012 by Wiley-VCH Verlag GmbH & Co. KGaA.

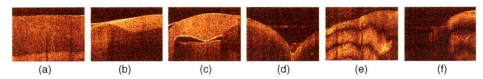

Figure 69.1 OCT images of 11th (a, b, c), 24th (d) and 4th (e, f) teeth. The dentinoenamel junction is seen in a healthy tooth (a); composite material is well distinguished from tooth tissue (b); well seen are small defects of the composite (c); surface caries has no visual manifestations (d). In healthy enamel, the birefringent bands coincide with the orientation of the enamel rods (e); early caries deprives enamel of birefringent properties (f). The cross-section of the OCT images is 1.7 mm.

69.2
Hard Tooth Tissues

The importance of early diagnosis of caries cannot be overestimated as preventive therapy and restoratives may be used at this stage. It is common knowledge that X-ray imaging detects contact caries with moderate sensitivity and high specificity, whereas it is much less efficient in detecting occlusive caries [36, 37]. A diagnostic method of choice for detecting carious cavity is visual examination [38].

Hard tissues are good objects for OCT because of their transparency. Technical characteristics of OCT allow the diagnosis of early carious and noncarious lesions, demineralization, early stages of caries of different localizations (Figure 69.1d and f), and also efficiency of sealant and composite restoration (Figure 69.1b and c) [1–3, 11, 13, 39–41]. Enamel possesses pronounced birefringent properties, which permits additional information to be obtained using PS-OCT. It was found, in particular, that the birefringence and depolarizing properties of healthy enamel and enamel even with early caries are different (Figure 69.1e and f). This phenomenon may be used for preclinical detection of enamel pathology [26, 29].

OCT imaging of subsurface changes in enamel properties and microdefects of composite restorations have no analogs in dental practice and may enhance both diagnosis and quality of dental therapy [2, 11, 19].

69.3
Periodontal Tissues

Periodontal diseases are the main cause of tooth loss. Exact assessment not only of tooth pockets but also of the state of soft tissues is of great significance in periodontology [42]. The OCT technique is a good tool for periodontal diagnosis as it has sufficient penetration depth and high resolution. OCT reliably images borders of soft and hard periodontal tissues [10] and ensures fast, reproducible imaging of surface tissue structure of gingiva, reproducing its specific features in different sections of oral cavity and pocket morphology [22]. OCT gives quantitative information about the thickness and level of gingiva attachment, and also about the position of alveolar crest. Comparison of OCT images taken at different times enables clinicians to reveal promptly potential sections of disease progression, probably even before clinical manifestations of recession [43]. OCT may be extremely useful for assessing the efficiency of periodontal therapy [19]. Electron recording of clinical data allows the avoidance of subjective and erroneous interpretations inherent in available metrical methods of diagnosis [42].

Dental implants are used to replace missing teeth. An implant is considered to be successful if it does not move, has no periimplantitis, clinical symptoms of infection, discomfort, or pain, and is esthetically pleasant [44–46]. Available diagnostic techniques are based on visual assessment of clinical symptoms of inflammation and are inefficient in detecting periimplantitis at an early stage. Further, it is difficult to control loss of alveolar bone mass around implants. Information provided by OCT may be useful for controlling the volume of attached keratinized gingiva, and also for monitoring gingival inflammatory processes [19, 23].

We studied in a clinical environment healthy volunteers and patients with implants. We analyzed CP-OCT images in co-polarization and cross-polarization of different gingival sections with different degrees of epithelial keratinization and subepithelial collagen organization. For the investigation of specific features of the histomorphology of gum above each tooth, we independently studied tissues postmortem of people without symptoms of gingival pathology. In our histologic analyses we used, in addition to standard hematoxylin–eosin staining, Picrosirius Red (PSR) staining of collagen with subsequent assessment by means of polarization microscopy. This technique allows the detection and differentiation of type I collagen (red and yellow color) from type III collagen (green). The color intensity shows the degree of collagen organization that is directly related to its polarization characteristics [47]. Collagen that has been disorganized in the course of inflammation or neoplasia development produces no color [48].

In a CP-OCT image of healthy gingiva in co-polarization (Figure 69.2b), keratinizing epithelium is not differentiated from connective-tissue stroma because stromal

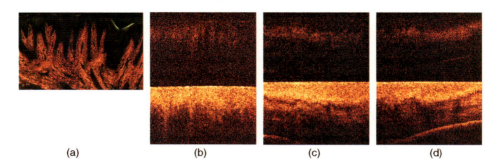

Figure 69.2 Attached gingiva above 11th tooth in norm (a, b): histologic specimen, PSR staining in polarized light, size of histologic frame 1.7 × 1 mm (a); OCT image in co-polarization (bottom) and in cross-polarization (top) (b). Attached gingiva above implant of 11th tooth (c, d): CP-OCT image with signs of gingivitis (c); CP-OCT image illustrating thin gingival biotype. Gingiva thickness 0.7 mm (d). The cross-section of OCT frames is 1.7 mm.

papillae are oriented vertically (Figure 69.2a); hence the image has no layered structure typical of mucosa. In cross-polarization (Figure 69.2b), there is signal from stromal papillae that include vertically oriented fibers, primarily of type I that depolarize light. This is confirmed by histologic data obtained with PSR staining of the specimens and visualized in polarized light (Figure 69.2a).

The CP-OCT image of gingiva above implant clearly demonstrates signs of gingivitis (Figure 69.2c). In co-polarization, inflammation appears as horizontally arranged areas of low signal (signs of edema), but in cross-polarization as regions in which the signal from collagen becomes weaker due to its inflammatory organization. CP-OCT (Figure 69.2d) permits the measurement of the thickness of gingiva above implant thanks to a strong signal from metal gingiva former. Figure 69.2d demonstrates a thin gingival biotype 0.7 mm thick.

69.4
Oral Cavity Mucosa

One of the traditional tasks of OCT is the search for dysplasia and malignization of the epithelium of oral cavity mucosa in patients with leukoplakia and erythroplakia [49–51]. The diagnostic efficiency of OCT and its combinations with various kinds of spectroscopy have been shown in many studies on animal [16] and human [17, 20] models. In all cases, diagnostics is based on the fact that images lose stratification and contrast between epithelium with dysplasia and underlying connective tissue. *In vivo* OCT images demonstrate the excellent capability of OCT for detecting and diagnosing oral premalignancy and malignancy in humans [12, 14, 18].

Our recent unpublished results on the application of CP-OCT for the solution of this important clinical problem demonstrated that, whereas in the case of flat leukoplakia image stratification is still preserved (Figure 69.3c), verrucous leukoplakia without malignization features in co-polarization a low-contrasted image with high-level signal from the keratin layer that obscures the underlying layers (Figure 69.3e). Such an image without contrast also corresponds to carcinoma *in situ* (Figure 69.3g), which reduces the diagnostic value of traditional OCT. In contrast, CP-OCT provides a cross-polarized conjugate image, thus expanding substantially the diagnostic potential of the technique. Fibrillar, highly structured protein keratin, and also organized collagen, depolarizes radiation, leading to the appearance of a signal in the cross-polarized image (Figure 69.3e). Collagen of lamina propria of mucosa immediately under malignant epithelium responds to disturbances of epithelial cell differentiation by collagen destruction, and hence by degradation of its depolarization properties. In this case, there is no signal in the cross-polarized image (Figure 69.3g). The level of signal in cross-polarization agrees well with the intensity of collagen color revealed with PSR staining in polarized light (Figure 69.3b, d, f, and k).

We extended the scope of CP-OCT clinical tasks to the evaluation of microstructural alterations of oral mucosa (in the cheek area, in particular) in patients with inflammatory intestinal diseases, such as Crohn disease and nonspecific ulcerative

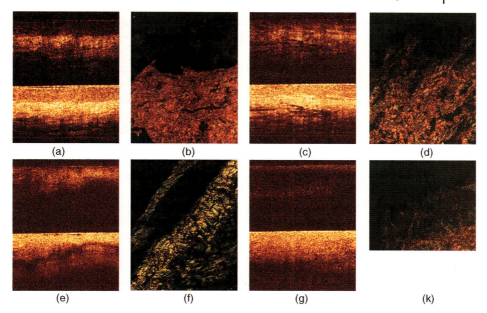

Figure 69.3 Cheek mucosa in norm (a, b), in patients with leukoplakia (c–f) and with cancer (g, k). Healthy cheek mucosa: CP-OCT image (a); the corresponding histologic specimen, PSR staining (b). Flat cheek leukoplakia: co-polarized (bottom) and cross-polarized (top) OCT images (c); histologic specimen, PSR staining in polarized light (d). Verrucous cheek leukoplakia: co-polarized (bottom) and cross-polarized (top) CP-OCT images (e); histological specimen, PSR staining in polarized light (f). Carcinoma *in situ* in cheek area: co-polarized (bottom) and cross-polarized (top) OCT images (g); histological specimen, PSR staining in polarized light (k); histologic frame size, 1.7 × 1 mm.

colitis. Differential diagnosis of Crohn disease and nonspecific ulcerative colitis is difficult because even "gold standard" histology does not necessarily offer definitive distinguishing features. Diagnosis is usually based on a combination of clinical, endoscopic, and radiologic features [52, 53]. However, a full spectrum of well-known diagnostic techniques is not always sufficient for differential diagnosis of diseases. The role of the oral cavity in a unified model of diagnostic algorithms is currently under active discussion [54]. Some researchers adhere to the opinion that oral mucosa is involved in the inflammatory process in Crohn disease, which is attributed to systemic manifestations of the disease. The susceptibility of the gastrointestinal tract to fibrosis in Crohn disease is regarded as a proven fact [55–57].

We analyzed CP-OCT images of 15 patients and revealed a recurrent phenomenon of a high-intensity signal from fibrous collagen of lamina propria of mucosa in patients with Crohn disease (Figure 69.4a), weak signal intensity in patients with nonspecific ulcerative colitis (Figure 69.4c), and no signal in the aphthae area in patients with stomatitis (Figure 69.4e). The signal level in cross-polarization agrees well with the intensity of collagen color revealed with PSR staining in polarized light (Figure 69.4b, d, and f).

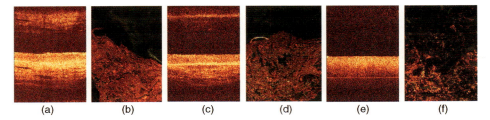

Figure 69.4 Cheek mucosa of a patient with Crohn disease (a, b), with nonspecific ulcerative colitis (c, d), and with aphthous stomatitis (e, f). Co-polarized (bottom) and cross-polarized (top) OCT images (a, c, e); the corresponding histological specimens (b, d, f), PSR staining; histologic frame size, 1.7 × 1 mm.

Such an approach was used by Lee and co-workers for diagnosing oral submucous fibrosis (OSF) [58, 59]. By analyzing OCT scanning results for OSF cases, two indicators, epithelium thickness and standard deviation (SD) of the A-mode scan intensity in the lamina propria layer, were found useful for real-time OSF diagnosis. The statistics showed that the sensitivity and specificity of lamina propria SD can reach 84.1 and 95.5%, respectively. Also, both the sensitivity and specificity of epithelium thickness can reach 100%.

OCT is an attractive tool for noninvasive monitoring of mucositis [15, 21, 60]. OCT proved to be efficient in assessing the dynamics of cheek mucosa microstructure during radio- and radio(chemo)therapy of mucositis in patients with oropharyngeal cancer [61, 62]. Typical OCT manifestations of mucositis were a gradual reduction of contrast of optical layers in OCT images up to their complete disappearance and a decrease in the epithelium layer thickness. The maximum of clinical manifestations of mucositis corresponded to the most pronounced alterations in the OCT images. It was found that the image dynamics depend strongly on mucositis grade, and prognostic criteria of individual radiosensitivity of mucosa were formulated.

For prognosis of mucositis grade in the course of radio- and radio(chemo)therapy of cancer of the oral cavity and pharynx in a specific patient, OCT images of cheek mucosa taken at the same site before treatment and on the day of appearance of the first clinical signs of radioreaction should be compared. Grade 3–4 mucositis is predicted by disappearance in the OCT image of the border between the epithelium and underlying connective tissue (Figure 69.5a and b), and grade 1–2 mucositis is predicted if this border is preserved (Figure 69.5c and d). Numerical analysis of images revealed statistically significant differences in the rates of image contrast reduction as a function of developed mucositis. This information may be extremely important for choosing mucositic prevention and therapy [63].

69.5
OCT Images of Labial Salivary Glands in Norm and Pathology

The main objects of attention for OCT are usually surface layers of superficial tissues: epithelium and subepithelial layers of connective tissue. Recent research [64, 65]

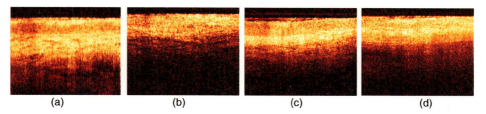

Figure 69.5 OCT images of cheek during radio (chemo)therapy. OCT images showing the alterations caused by mucositis in a 70-year-old male patient with oropharyngeal cancer, stage III (squamous cell keratinizing carcinoma), treated with radio(chemo)therapy resulting in grade 3 mucositis (a, b). Before irradiation, interface between layers is shown by the arrow (a) and after a total dose of 12 Gy with clinically manifested hyperemia and edema (b). The OCT contrast is absent, and no layers can be differentiated. White scale bars measure 1 mm. OCT images showing the alterations caused by mucositis in a 47-year-old male patient with oropharyngeal cancer, stage III (squamous cell non-keratinizing carcinoma), treated with radio (chemo)therapy resulting in a grade 2 mucositis (c, d). Before irradiation (c); after a total dose of 14 Gy with clinically manifested hyperemia and edema (d). The OCT contrast is reduced, but epithelium is still visible. White scale bars measure 1 mm. Interface between the layers is shown by an arrow [63].

demonstrated that labial glands (LGs) located at a depth of 1–1.5 mm in oral mucosa may be visualized by means of OCT. LGs in normal condition have an average size of 1.8 ± 0.8 mm (ranging from 1 to 7 mm, depending on saliva filling and salivation phase). LGs in norm are imaged by OCT in connective tissue stroma as homogeneous regions with low signal and clear contours. When LGs are moderately filled with saliva, lobulous gland structure can be visualized (Figure 69.6a); when LG are well filled with saliva, their size may reach 5–7 mm (Figure 69.6b and c). Ducts of healthy LGs can easily be measured (Figure 69.6d).

Salivary gland diseases in autoimmune processes (primary and secondary Sjögren syndrome), lymphoproliferative lesions, and lymphomas have different clinical courses, in spite of their similarity; therefore, long-term case monitoring is

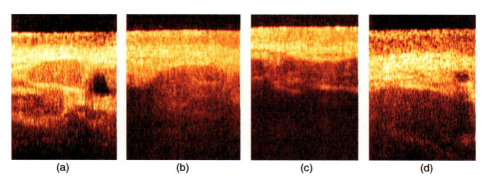

Figure 69.6 Variants of OCT images of LGs and ducts in norm: LG consisting of two lobes and duct, horizontal gland size 0.85 mm (a); LG consisting of two lobes, horizontal gland size 0.85 mm (b). LG with substantial secretion, horizontal gland size 6 mm, only part of the gland is visualized (c). OCT image of LG with excretory duct (d) [33].

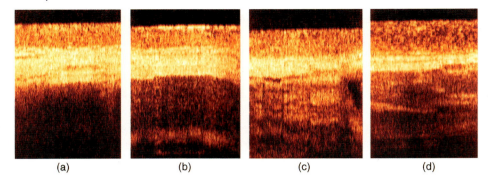

Figure 69.7 OCT images of LGs in a patient with sialadenosis (a, b) obtained with the probe slightly pressed to mucosa (a), and when the probe is strongly pressed to the glandular area so that the bottom of the gland appears (the gland is readily compressed) (b). OCT images of minor salivary glands in a patient with sialadenitis with strongly reduced glandular (c, d) acquired with the probe slightly pressed to mucosa (c); when the probe is strongly pressed, the gland remains almost uncompressed (d) [43].

demanded [66, 67]. The only method of verification is recurrent biopsy. Consequently, objective noninvasive high-resolution techniques that would detect structural changes in salivary glands are highly desirable [68].

We have studied OCT images of LGs in patients with sialadenosis, sialadenitis, and lymphoproliferative alteration lesions. Morphologically, sialadenosis manifests itself as dystrophic alterations of salivary glands without focal or diffuse lymphocytic infiltrations in stroma and with retained functional gland activity. The determining OCT features of sialadenosis are preserved contrast and structure of LGs, and good elasticity ("compressibility") on the OCT probe pressure (Figure 69.7a and b). Being an inflammatory disease, morphologically sialadenitis is characterized by the presence in LG lobes or in periductal space of massive (more than 50 cells in the field of view) focal or diffuse lymphocytic infiltrations (as a manifestation of active immune infiltration inflammation), periductal and interlobular sclerosis (as a consequence of immune inflammation), and disturbance of saliva drainage from small-sized ducts (duct sialadenitis) or disturbance of saliva production (parenchymatous sialedenitis). The OCT features of these structural alterations are a nonuniform signal from the gland with sections of high intensity due to expressed cellular infiltration and stroma sclerosis. The gland remains almost uncompressed in this case (Figure 69.7c and d). Thus, we have shown that OCT adequately identifies the glandular system of oral micosa. The histologic features of sialadenosis, sialadenitis, and benign lymphoproliferative lesion have well-defined OCT equivalents. OCT allows the assessment not only of the structure but also of the function of glands, which is no less important. Noninvasive imaging of LGs by means of OCT may be useful for differential diagnosis of various diseases and also for their monitoring in the course of therapy.

To conclude, OCT, especially its polarization modalities, provides objective information about the properties of biotissues, thus broadening the clinical capabilities of early diagnosis of dental, periodontium, and soft oral tissue diseases and ensuring their efficient monitoring.

Acknowledgments

This work was supported by the State Contract of the Russian Federation No. 02.740.11.5149 and the Russian Foundation for Basic Research (project 10-02-01175). The authors greatly appreciate the assistance of their colleagues G.V. Gelikonov, V.M. Gelikonov, F.I. Feldchtein, N.B. Krivatkina, V.M. Gubova, V.S. Kozlovsky, L.B. Snopova, E. Yunusova, O.B. Shchukina, A.V. Botina, and S.Yu. Tytyuk.

References

1 Colston, B.W., Everett, M.J., Da Silva, L.B., Otis, L.L. *et al.* (1998) Imaging of hard- and soft-tissue structure in the oral cavity by optical coherence tomography. *Appl. Opt.*, **37** (16), 3582–3584.

2 Colston, B.W., Everett, M.J., Sathyam, U.S., and Da Silva, L.B. (2002) Optical coherence tomography in dentistry, in *Handbook of Optical Coherence Tomography* (eds. B.E. Bouma and G.J. Tearney), Marcel Dekker, New York, pp. 591–612.

3 Colston, B.W., Sathyam, U.S., Da Silva, L.B., Everett, M.J. *et al.* (1998) Dental OCT. *Opt. Express*, **3** (6), 230–238.

4 Otis, L.L., Al-Sadhan, R.I., and Meiers, J. (2003) Identification of occlusal sealants using optical coherence tomography. *J. Clin. Dent.*, **14**, 7–10.

5 Otis, L.L., Al-Sadhan, R.I., Redford-Badwal, D., and Meiers, J. (2000) Identification of occlusal sealants using optical coherence tomography. *J. Dent. Res.*, **79**, 456.

6 Otis, L.L., Colston, B.W., and Everett, M.J. (1998) Dental optical coherence tomography: a novel assessment of oral tissues. *J. Dent. Res.*, **77**, 228.

7 Otis, L.L., Colston, B.W. Jr., Everett, M.J., and Nathel, H. (2000) Dental optical coherence tomography: a comparison of two *in vitro* systems. *Dentomaxillofac. Radiol.*, **29**, 85–89.

8 Otis, L.L., Everett, M.J., and Colston, B.W. (1998) Optical coherence tomography: a novel assessment of coronal restorations. *J. Dent. Res.*, **77**, 824.

9 Otis, L.L., Everett, M.J., Sathyam, U.S., and Colston, B.W. (2000) Optical coherence tomography: a new imaging technology for dentistry. *J. Am. Dent. Assoc.*, **131** (4), 511–514.

10 Otis, L.L. (2005) Optical coherence tomography of periodontal tissues. *Lasers Surg. Med.*, **3**, 19.

11 Feldchtein, F.I., Gelikonov, G.V., Gelikonov, V.M., Iksanov, R.R. *et al.* (1998) In vivo OCT imaging of hard and soft tissue of the oral cavity. *Opt. Express*, **3** (6), 239–250.

12 Gladkova, N.D., Fomina, J.V., Shakhov, A.V., Terentieva, A.B. *et al.* (2003) Optical coherence tomography as a visualization method for oral and laryngeal cancer, in *Oral Oncology*, vol. **9** (eds. A.K. Varma and P. Reade), Macmillan Press, New Delhi, pp. 272–281.

13 Ourutina, M.N., Gladkova, N.D., Feldchtein, F.I. and Gelikonov, G.V. *et al.* (eds) (1999) In-vivo optical coherent tomography of teeth and oral mucosa. *Proc. SPIE*, **3567**, 97–107.

14 Fomina, J.V., Gladkova, N.D., Snopova, L.B., Shakhova, N.M. *et al.* (2004) *In vivo* OCT study of neoplastic alterations of the oral cavity mucosa. *Proc. SPIE*, **5316**, 41–47.

15 Wilder-Smith, P., Hammer-Wilson, M.J., Zhang, J., Wang, Q. *et al.* (2007) *In vivo*

imaging of oral mucositis in an animal model using optical coherence tomography and optical Doppler tomography. *Clin. Cancer Res.*, **13** (8), 2449–2454.

16 Wilder-Smith, P., Jung, W.G., Brenner, M., Osann, K. et al. (2004) In vivo optical coherence tomography for the diagnosis of oral malignancy. *Lasers Surg. Med.*, **35** (4), 269–275.

17 Wilder-Smith, P., Krasieva, T., Jung, W.G., Zhang, J. et al. (2005) Noninvasive imaging of oral premalignancy and malignancy. *J. Biomed. Opt.*, **10** (5), 51601.

18 Wilder-Smith, P., Lee, K., Guo, S., Zhang, J. et al. (2009) In vivo diagnosis of oral dysplasia and malignancy using optical coherence tomography: preliminary studies in 50 patients. *Lasers Surg. Med.*, **41** (5), 353–357.

19 Wilder-Smith, P., Otis, L., Zhang, J., and Chen, Z. (2009) Dental OCT, in *Optical Coherence Tomography: Technology and Applications* (eds. W. Drexler and J. Fujimoto), Springer, Berlin, pp. 1151–1182.

20 Matheny, E.S., Hanna, N.M., Jung, W.G., Chen, Z. et al. (2004) Optical coherence tomography of malignancy in hamster cheek pouches. *J. Biomed. Opt.*, **9** (5), 978–981.

21 Kawakami-Wong, H., Gu, S., Hammer-Wilson, M.J., Epstein, J.B. et al. (2007) In vivo optical coherence tomography-based scoring of oral mucositis in human subjects: a pilot study. *J. Biomed. Opt.*, **12** (5), 051702.

22 Xiang, X., Sowa, M.G., Iacopino, A.M., Maev, R.G. et al. (2010) An update on novel non-invasive approaches for periodontal diagnosis. *J. Periodontol.*, **81** (2), 186–198.

23 Ionita, I. and Reisen, P. (2009) Imaging of dental implant osseointegration using optical coherent tomography. *Proc. SPIE*, **7168**, 71682F–71687F.

24 Wang, X.J., Milner, T.E., de Boer, J.F., Zhang, Y. et al. (1999) Characterization of dentin and enamel by use of optical coherence tomography. *Appl. Opt.*, **38** (10), 2092–2096.

25 Baumgartner, A., Dichtl, S., Hitzenberger, C.K., Sattmann, H. et al. (2000) Polarization-sensitive optical coherence tomography of dental structures. *Caries Res.*, **34** (1), 59–69.

26 Fried, D., Xie, J., Shafi, S., Featherstone, J.D. et al. (2002) Imaging caries lesions and lesion progression with polarization sensitive optical coherence tomography. *J. Biomed. Opt.*, **7** (4), 618–627.

27 Chen, Y., Otis, L., Piao, D., and Zhu, Q. (2005) Characterization of dentin, enamel, and carious lesions by a polarization-sensitive optical coherence tomography system. *Appl. Opt.*, **44** (11), 2041–2048.

28 Ngaotheppitak, P., Darling, C.L., and Fried, D. (2005) Measurement of the severity of natural smooth surface (interproximal) caries lesions with polarization sensitive optical coherence tomography. *Lasers Surg. Med.*, **37** (1), 78–88.

29 Kang, H., Jiao, J.J., Lee, C., and Fried, D. (2010) Imaging early demineralization with PS-OCT. *Proc. SPIE*, **7549**, 75490M.

30 Le, M.H., Darling, C.L., and Fried, D. (2010) Automated analysis of lesion depth and integrated reflectivity in PS-OCT scans of tooth demineralization. *Lasers Surg. Med.*, **42**, 62–68.

31 Stahl, J., Kang, H., and Fried, D. (2010) Imaging simulated secondary caries lesions with cross polarization OCT. *Proc. SPIE*, **7549**, 754905.

32 Gelikonov, V.M., Gelikonov, G.V., Dolin, L.S., Kamensky, V.A. et al. (2003) Optical coherence tomography: physical principles and applications. *Laser Phys.*, **13** (5), 692–702.

33 Gelikonov, V.M. and Gelikonov, G.V. (2006) New approach to cross-polarized optical coherence tomography based on orthogonal arbitrarily polarized modes. *Laser Phys. Lett.*, **3** (9), 445–451.

34 Schmitt, J.M. and Xiang, S.H. (1998) Cross-polarized backscatter in optical coherence tomography of biological tissue. *Opt. Lett.*, **23** (13), 1060–1062.

35 Zagaynova, E.V., Gladkova, N.D., Shakhova, N.M., Streltsova, O.S. et al. (2011) Interoperative OCT monitoring, in *Handbook of Biophotonics*, vol. **2** (ed. J. Popp), Wiley-VCH Verlag GmbH, Weinheim.

36 Lussi, A. (1991) Validity of diagnostic and treatment decisions of fissure caries. *Caries Res.*, **25**, 296–303.

37 Ricketts, D.N.J., Ekstrand, K.R., Martignon, S., Ellwood, R. *et al.* (2007) Accuracy and reproducibility of conventional radiographic assessment and subtraction radiography in detecting demineralization in occlusal surfaces. *Caries Res.*, **41**, 121–128.

38 Versteeg, C.H., Sanderink, G.C.H., and van der Stelt, P.F. (1997) Efficacy of digital intra-oral radiography. *J. Dent.*, **25** (3–4), 215–224.

39 Douglas, S.M., Fried, D., and Darling, C.L. (2010) Imaging natural occlusal caries lesions with optical coherence tomography. *Proc. SPIE*, **7549**, 75490N.

40 Jablow, M. (2009) Caries detection in the 21st century – sharpening our diagnostic abilities. *DentistryIQ*, 11 December 2009.

41 Negrutiu, M.L., Sinescu, C., Rominu, M., Hughes, M. *et al.* (2009) Optical coherence tomography and confocal microscopy investigations of dental structures and restoration materials. *Proc. SPIE*, **7258**, 72584N.

42 Lindhe, J., Karring, T., and Lang, N.P. (2003) *Clinical Periodontology and Implant Dentistry*, 4th edn, Blackwell, Oxford.

43 Baek, J.H., Na, J., Lee, B.H., Choi, E. *et al.* (2009) Optical approach to the periodontal ligament under orthodontic tooth movement: a preliminary study with optical coherence tomography. *Am. J. Orthod. Dentofacial Orthop.*, **135** (2), 252–259.

44 National Institutes of Health (1978) Dental Implants: Benefit and Risk. NIH Consensus Statement Online, National Institutes of Health, Washington, DC, 13–14 June, 1 (3), pp. 13–19.

45 Albrektsson, T., Zarb, G., Worthington, P., and Eriksson, R.A. (1986) The long-term efficacy of currently used dental implants: a review and proposed criteria of success. *Int. J. Oral Maxillofac. Implants*, **1**, 11–25.

46 Iacono, V.J. and Committee on Research, Science and Therapy, the American Academy of Periodontology (2000) Dental Implants in Periodontal Therapy. Academy Report. *J. Periodontol.*, **71** (12), 1934–1942.

47 Junqueira, L., Bignolas, G., and Brentani, R. (1979) Picrosirius staining plus polarization microscopy, a specific method for collagen detection. *Histochem. J.*, **11** (4), 447–455.

48 Borges, L., Gutierrez, P., Marana, H., and Taboga, S. (2007) Picrosirius-polarization staining method as an efficient histopathological tool for collagenolysis detection in vesical prolapse lesions. *Micron*, **38**, 580–583.

49 Clark, A.L., Gillenwater, A., Alizadeh-Naderi, R., El-Naggar, A.K. *et al.* (2004) Detection and diagnosis of oral neoplasia with an optical coherence microscope. *J. Biomed. Opt.*, **9** (6), 1271–1280.

50 Tsai, M.T., Lee, C.K., Lee, H.C., Chen, H.M. *et al.* (2009) Differentiating oral lesions in different carcinogenesis stages with optical coherence tomography. *J. Biomed. Opt.*, **14** (4), 044028.

51 Tsai, M.T., Lee, H.C., Lee, C.K., Yu, C.H. *et al.* (2008) Effective indicators for diagnosis of oral cancer using optical coherence tomography. *Opt. Express*, **16** (20), 15847–15862.

52 Zonderman, J. and Vender, R. (2000) *Understanding Crohn Disease and Ulcerative Colitis*, University Press of Mississippi, Jackson, MS.

53 Sklar, J. (2002) *The First Year: Crohn's Disease and Ulcerative Colitis: An Essential Guide for the Newly Diagnosed*, Marlowe, New York.

54 Bricker, S.L., Landlais, R.P., and Miller, C.S. (eds.) (2001) *Oral Diagnosis, Oral Medicine and Treatment Planning*, BC Decker, Hamilton, ON.

55 Graham, M.F., Diegelmann, R.F., Elson, C.O., Lindblad, W.J. *et al.* (1988) Collagen content and types in the intestinal strictures of Crohn's disease. *Gastroenterology*, **94**, 257–265.

56 Stumpf, M., Krones, C.J., Klinge, U., Rosch, R. *et al.* (2006) Collagen in colon disease. *Hernia*, **10** (6), 498–501.

57 Floer, M., Binion, D.G., Nelson, V.M., Manley, S. *et al.* (2008) Role of MutS homolog 2 (MSH2) in intestinal myofibroblast proliferation during Crohn's disease stricture formation. *Am. J. Physiol. Gastrointest. Liver Physiol.*, **295** (3), G581–G590.

58 Lee, C.-K., Tsai, M.-T., Yang, C.C., and Chiang, C.-P. (2009) Clinical diagnosis of oral submucous fibrosis with optical coherence tomography. *Proc. SPIE*, **7634**, 763405.

59 Lee, C.-K., Tsai, M.-T., Yang, C.C., Lee, H.-C. *et al.* (2009) Diagnosis of oral submucous fibrosis with optical coherence tomography. *J. Biomed. Opt.*, **14**, 054008.

60 Muanza, T.M., Cotrim, A.P., McAuliffe, M., Sowers, A.L. *et al.* (2005) Evaluation of radiation-induced oral mucositis by optical coherence tomography. *Clin. Cancer Res.*, **11** (14), 5121–5127.

61 Gladkova, N., Maslennikova, A., Balalaeva, I., Vyseltseva, Y. *et al.* (2006) OCT visualization of mucosal radiation damage in patients with head and neck cancer: pilot study. *Proc. SPIE*, **6078**, 6071J–6078J.

62 Gladkova, N., Maslennikova, A., Terentieva, A., Fomina, Y. *et al.* (2006) OCT visualization of acute radiation mucositis: pilot study. *Proc. SPIE*, **5861**, 58610T.

63 Gladkova, N., Maslennikova, A., Balalaeva, I., Feldchtein, F. *et al.* (2008) Application of optical coherence tomography in the diagnosis of mucositis in patients with head and neck cancer during a course of radio(chemo)therapy. *Med. Laser Appl.*, **23**, 186–195.

64 Gladkova, N.D., Radenska-Lopovok, S.G., Fomina, Yu.V., Gaydyk, I.V. *et al.* (2006) In vivo imaging of labial salivary glands using optical coherence tomography. *Clin. Stomatol.*, **2**, 30–35.

65 Ozawa, N., Sumi, Y., Shimozato, K., Chong, C. *et al.* (2009) In vivo imaging of human labial glands using advanced optical coherence tomography. *Oral Surg. Oral Med. Oral Pathol. Oral Radiol. Endodontol.*, **108** (3), 425–429.

66 Fox, R.I., (2005) Sjögren's syndrome. *Lancet*, **366**, 321–331.

67 Vitali, C., Bombardieri, S., Jonsson, R., Moutsopoulos, H.M. *et al.* (2002) Classification criteria for Sjögren's syndrome: a revised version of the European criteria proposed by the American–European Consensus Group. *Ann. Rheum. Dis.*, **61**, 554–558.

68 Xu, K.P., Katagiri, S., Takeuchi, T., and Tsubota, K. (1996) Biopsy of labial salivary glands and lacrimal glands in the diagnosis of Sjogren's syndrome. *J. Rheumatol.*, **23**, 76–82.

70
Fluorescence Spectroscopy
Adrian Lussi

70.1
Introduction

Dental caries is a bacteria-associated progressive process of the hard tissues of the coronal and root surfaces of teeth. Net demineralization may begin soon after tooth eruption in caries-susceptible children without being recognized. This process may progress further, resulting in a caries lesion that is the sign and/or the symptom of the carious process. Caries is, in other words, a continuum which may by assessed falsely when only a certain time point is considered.

Clinical–visual diagnosis may be amenable to longitudinal monitoring even though the assessment is qualitative. It would be easier to have a device that would not only detect caries but also quantify it. Then monitoring progression or arrest would be simple: use the device again and see in what direction the numbers change. This concept is persuading so it is little wonder that researchers have made such efforts to develop, test, and perfect such devices. All of these methods for caries detection are based on the interpretation of one or more physical signals. These are causally related to one or more features of a caries lesion. First, the signals must be received using a receptor device and classified. The classification of a signal is part of the diagnostic decision-making process. However, none of the methods is capable of processing all of these signals to a status that could be called diagnosis. "The art of identifying a disease from its signs and symptoms" [1] is a process that cannot be replaced by a machine or a device. Nevertheless, such devices may be of help especially in the early stage of caries, which may be difficult to diagnose.

The aim of this chapter is to provide insight in today's methods for caries detection based on laser fluorescence.

70.2
Laser/Light-Induced Fluorescence

Fluorescence is a phenomenon by which the wavelength of the emitted light (coming from the light source) is changed into a longer wavelength as it travels back for detection. The larger wavelength is caused by some loss of energy to the surrounding tissue and therefore will have a different color than the emitting light. By using a filter through which only the fluorescent light may pass, the intensity of the fluorescent light can be measured. The fluorescence of dental hard tissues has been known for a very long time [2]. The chromophores causing the fluorescence of dental hard tissues, however, are not clearly identified. The blue fluorescence was assigned to dityrosine [3]. It seems likely that most of the yellow fluorescence stems from proteinic chromophores, probably cross-links between chains of structural proteins [4]. It has also been discussed whether or not the apatite of dental hard tissues would also contribute [5, 6]. Caries lesions, plaque, and microorganisms also contain fluorescent substances. The difference between the fluorescence of sound tooth tissues and that of a caries lesion can be made visible by the DIAGNOdent method and by quantitative laser- or light-induced fluorescence (QLF).

70.3
DIAGNOdent

The setup and function of the DIAGNOdent (KaVo, Biberach/Riss, Germany) (Figure 70.1) are as follows. Light from a laser diode (655 nm) is coupled into an optical fiber and transmitted to the tooth. The backscattered excitation and short-wavelength ambient light is absorbed by a filter in front of the photodiode detector. To discriminate the fluorescence from the ambient light in the same spectral region, the laser diode is modulated. Owing to its relatively short lifetime, fluorescence follows this modulation. By amplifying only the modulated portion of the signal, the ambient

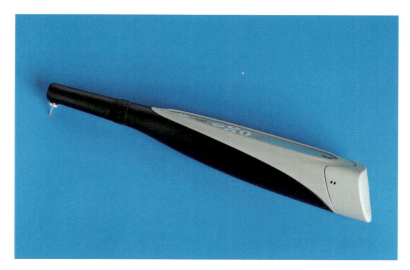

Figure 70.1 The DIAGNOdent is available with two tips, one for occlusal caries and the other for interproximal caries.

light is suppressed. The remaining signal is proportional to the detected fluorescence intensity and displayed as a number (0–99, in arbitrary units). In order to compensate for potential variations of the system (e.g., laser diode output power), it can be calibrated by a standard of known and stable fluorescence yield. This makes the measurement absolute (although in arbitrary units) and allows comparisons of fluorescing tooth spots over time. A systematic review revealed a high sensitivity of the DIAGNOdent compared with traditional diagnostic methods. However, the method was less specific in that the DIAGNOdent also identified a larger proportion of sound sites as being carious than did the visual method [7]. We recommend this tool as a second opinion in the diagnostic process. Figure 70.2 shows its use on (a) occlusal and (b) interproximal surfaces.

70.3.1
Quantitative Light-Induced Fluorescence (QLF)

Demineralization of a dental hard tissue results in loss of its autofluorescence, the natural fluorescence. As early as the 1920s, this phenomenon was suggested to be useful as a tool for diagnosing dental caries [8]. More recently, laser light was used to induce fluorescence of enamel in a sensitive, nondestructive, diagnostic method for detection of caries [9, 10]. The tooth was illuminated with a broad beam of blue–green light from an argon laser, producing diffuse monochromatic light with a wavelength (λ) of 488 nm. The fluorescence of the enamel occurring in the yellow region was observed through a yellow high-pass filter ($\lambda \geq 520$ nm), which filters out all reflected and backscattered light. Demineralized areas appeared as dark areas because the fluorescence of a caries lesion viewed by QLF is lower than that of sound enamel (Figure 70.3).

Several effects may contribute to the decreased fluorescence of white spot caries lesions, and most probable ones being as follows:

- The light scattering in the lesion, which is much stronger than in sound enamel [11], causes the light path in the lesion to be much shorter than in sound enamel. Therefore, the absorption per unit volume is much smaller in the lesion and the fluorescence is less strong.
- The light scattering in the lesion acts as a barrier for excitation light to reach the underlying fluorescent sound tooth tissues, and as a barrier for fluorescent light from the layers below the lesion to reach the tooth surface. This is one of the reasons for limited depth detection with QLF (up to 400 μm).

The laser fluorescence method was developed further for *in vivo* quantification of mineral loss in natural enamel lesions using a color microvideo charge-coupled device (CCD) camera and computed image analysis [12]. To allow calculation of fluorescence loss in the caries lesion, the fluorescence of the lesion is subtracted from the fluorescence of the surrounding sound tissue. The difference between the actual values and the reconstructed ones gives the resulting fluorescence loss.

QLF is a sensitive, reproducible method for quantification of enamel lesions limited to a depth of only 400 μm. A relationship between mineral loss and fluorescence has been found for artificial and natural caries lesions with correlation

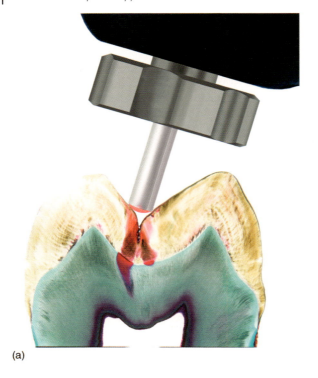

(a)

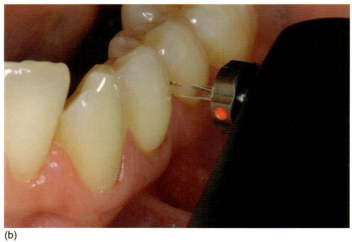

(b)

Figure 70.2 The DIAGNOdent has to be tilted in order to detect the occlusal caries (a) and has to be inserted between two teeth in order to measure fluorescence from interproximal surfaces (b).

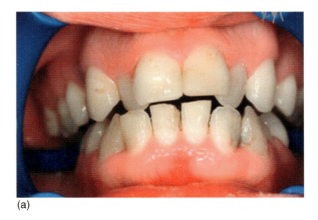
(a)

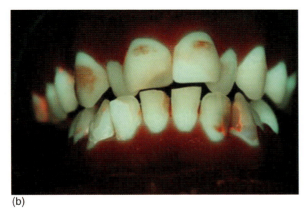
(b)

Figure 70.3 Appearance of teeth with normal white light (a) and its fluorescence image (b); caries appears as a dark spot (Inspektor Research Systems BV, Amsterdam Netherlands).

coefficients of $r = 0.73-0.86$ [6, 13–15]. Hence a measure of fluorescence may indicate the amount of mineral loss. This device may be used primarily for research purposes where small changes in the outer surface of caries may be of great interest.

References

1 Merriam-Webster (2003) http://www.merriam-webster.com/dictionary/diagnosis.
2 Benedict, HC. (1928) Note on the fluorescence of teeth in ultra-violet rays. *Science*, **67**, 442.
3 Booij, M. and ten Bosch, J.J. (1982) A fluorescent compound in bovine dental enamel matrix compared with synthetic dityrosine. *Arch. Oral Biol.*, **27**, 417–421.
4 Scharf, F. (1971) Über die natürliche Lumineszenz der Zahnhartgewebe 'Schmelz und Dentin'. *Stoma*, **24**, 11–25.
5 Spitzer, D. and ten Bosch, J.J. (1976) The total luminescence of bovine and human dental enamel. *Calcif. Tissue Res.*, **20**, 201–208.

6 Hafström-Björkman, U., Sundström, F., and ten Bosch, J.J. (1991) Fluorescence of dissolved fractions of human enamel. *Acta Odontol. Scand.*, **49**, 133–138.

7 Bader, J.D. and Shugars, D.A. (2004) A systematic review of the performance of a laser fluorescence device for detecting caries. *J. Am. Dent. Assoc.*, **135**, 1413–1426.

8 Benedict, H.C. (1929) The fluorescence of teeth as another method of attack on the problem of dental caries. *J. Dent. Res.*, **9**, 274–275.

9 Bjelkhagen, H. and Sundström, F. (1981) A clinically applicable laser luminescence method for the early detection of dental caries. *IEEE J. Quantum Electron.*, **17**, 266–286.

10 Bjelkhagen, H., Sundström, F., Angmar-Månsson, B., and Rydén, H. (1982) Early detection of enamel caries by the luminescence excited by visible laser light. *Swed. Dent. J.*, **6**, 1–7.

11 ten Bosch, J.J. (1996) Light scattering and related methods, in *Early Detection of Dental Caries: Proceedings of the 1st Annual Indiana Conference* (ed. G.K. Stookey), Indiana University School of Dentistry, Indianapolis, IN, pp. 81–90.

12 de Josselin de Jong, E., Sundström, F., Westerling, H., Tranaeus, S., ten Bosch, J.J., and Angmar-Månsson, B. (1995) A new method for *in vivo* quantification of mineral loss in enamel with laser fluorescence. *Caries Res.*, **29**, 2–7.

13 Hafström-Björkman, U., Sundström, F., de Josselin de Jong, E., Oliveby, A., and Angmar-Månsson, B. (1992) Comparison of laser fluorescence and longitudinal microradiography for quantitative assessment of *in vitro* enamel caries. *Caries Res.*, **26**, 241–247.

14 Emami, Z., Al-Khateeb, S., de Josselin de Jong, E., Sundström, F., Trollsås, K., and Angmar-Månsson, B. (1996) Mineral loss in incipient caries lesions quantified with laser fluorescence and longitudinal microradiography. *Acta Odontol. Scand.*, **54**, 8–13.

15 Al-Khateeb, S., Oliveby, A., de Josselin de Jong, E., and Angmar-Månsson, B. (1997) Laser fluorescence quantification of remineralisation *in situ* of incipient enamel lesions: influence of fluoride supplements. *Caries Res.*, **31**, 132–140.

71
Photothermal Radiometry and Modulated Luminescence: Applications for Dental Caries Detection

Jose A. Garcia, Andreas Mandelis, Stephen H. Abrams, and Anna Matvienko

71.1
Introduction

The traditional methods for dental caries detection, visual examination and radiography, are not effective in detecting early caries lesions [1]. If dental caries is detected early, before a substantial amount of the tooth is destroyed, the area can be remineralized [2], which can prevent invasive procedures, radiation exposure, and create a sound surface devoid of restorations. Recently, a new group of light-based dental diagnostic techniques has been proposed [3]. The group includes techniques such as light-induced fluorescence, digital imaging fiber-optic transillumination (DIFOTI), electrical caries monitoring (ECM), optical coherence tomography (OCT), alternating current impedance spectroscopy, and photothermal radiometry (PTR) with modulated luminescence (LUM).

PTR–LUM combines PTR and LUM for detection and monitoring of carious lesions. In PTR, a beam of energy (typically a laser), intensity modulated at a certain frequency, is focused on the sample surface. The resulting periodic heat flow due to the absorbed optical energy in the material is a diffusive process, producing a periodic temperature rise (spatial distribution) which is called a "thermal wave." This temperature distribution, in turn, causes a modulated thermal infrared (IR) (black-body or Planck radiation) emission which is used to monitor the material under examination. PTR has the ability to penetrate, and yield information about, an opaque medium well beyond the range of optical imaging. Specifically, the frequency dependence of the penetration depth of thermal waves makes it possible to perform depth profiling of materials [4]. In PTR applications to turbid media, such as hard dental tissue, depth information is obtained following optical-to-thermal energy conversion and transport of the incident laser power in two distinct modes: conductively, from a near-surface distance controlled by the thermal diffusivity of enamel (50–500 µm); and radiatively, through blackbody IR emissions from considerably deeper regions commensurate with optical penetration of the diffusely scattered laser-induced optical field (several millimeters) and with the existence of IR spectral "windows" for IR photon transmission.

Handbook of Biophotonics. Vol.2: Photonics for Health Care, First Edition. Edited by Jürgen Popp,
Valery V. Tuchin, Arthur Chiou, and Stefan Heinemann.
© 2012 Wiley-VCH Verlag GmbH & Co. KGaA. Published 2012 by Wiley-VCH Verlag GmbH & Co. KGaA.

The term luminescence is used to describe a process in which light is produced when an energy source causes a transition of electrons or other energetic particles from their lowest ("ground") energy state to a higher ("excited") state; subsequent return to the ground state releases the excess energy in the form of light photons of longer wavelength than the optical source (Stokes shift). There are several types of luminescence, each named according to the source of energy, or the trigger for the luminescence, for example, photoluminescence (caused by electromagnetic radiation), or fluorescence (caused by ultraviolet or visible excitation). However, these terms are often used interchangeably in the literature. Many important biological objects containing fluorescing components (fluorophores) exhibit intrinsic fluorescence (or autofluorescence). The reduction in fluorescence when enamel demineralizes has been attributed to the increase in porosity of carious lesions when compared with sound enamel. There is an associated uptake of water and decrease in the refractive index of the lesion resulting in increased scattering and a decrease in light-path length, absorption, and autofluorescence [5]. Additional sources of luminescence decrease are associated with the destruction of the enamel crystalline structure as a result of demineralization. In summary, a combination of PTR and LUM provides valuable complementary information about the conversion pathways of optical energy in a sample.

It has been a decade since the first publications on PTR–LUM for examining hard dental tissue appeared. In this chapter, we review the development of and future outlook for this novel dental diagnostic technology, including the clinical PTR-LUM device (The Canary System).

71.2
Theoretical Modeling

To appreciate better the capabilities of PTR–LUM, it is important to understand how the laser light interacts with turbid media generating two energy fields: optical and thermal. A coupled diffuse photon density and thermal wave model was developed for theoretical analysis of the photothermal field in demineralized teeth [6, 7]. The solution of the radiative transport equation in the limit of diffuse photon density field is considered as a source term in the thermal wave field equation. The chromophore (hydroxyapatite) relaxation lifetimes, optical absorption, scattering, and spectrally averaged IR emission coefficients were determined by fitting the model to the measured PTR–LUM data [6]. The influence of the foregoing optical and thermal parameters (thermal diffusivity and conductivity) of each layer of a demineralized tooth on the diffuse photon density and therma wave depth profiles has recently been analyzed using computer simulations that allow the optimized simultaneous measurement of these major optical and thermal properties of teeth from experimental PTR scans [7].

71.3
Depth Profilometry

One of the main advantages of PTR–LUM, intrinsic to diffusion wave methods, is the ability to perform depth profiling through scanning of the excitation source modulation frequency. By selecting several modulation frequencies, PTR measurements at different depths in the enamel can be obtained. Studies on the depth profilometric capability of PTR–LUM in dental applications [8] showed that the radiometric amplitude exhibited a superior dynamic range (two orders of magnitude signal resolution) to luminescence (a factor of two only) in distinguishing between intact and cracked subsurface structures in the enamel. Further studies [9] assessed the feasibility of PTR–LUM to detect deep lesions. PTR frequency scans over the surface of a fissure into demineralized enamel and dentin exhibited higher amplitude than those for healthy teeth, and also a pronounced curvature in both amplitude and phase signal channels. These can be excellent markers for the diagnosis of subsurface carious lesions. In addition, PTR exhibited superior sensitivity to the presence of sharp boundaries, and also to changes in natural demineralized regions of the tooth.

71.4
Pits and Fissures Caries Detection

Jeon *et al.* [10] examined pit and fissure caries in 52 extracted human teeth with simultaneous measurements of PTR and LUM. These measurements were compared with conventional diagnostic methods including continuous (DC) luminescence (DIAGNOdent), visual inspection, and radiographs. Sensitivities and specificities were calculated by using histologic observations as the gold standard to compare all examined methods. With the combined criteria of four PTR and LUM signals (two amplitudes and two phases), it was found that the sensitivity of this method was much higher than those of any of the other methods used in this study, whereas the specificity was comparable to that of the DC luminescence. Therefore, the combination of PTR and LUM has the potential to be a reliable tool to diagnose pit and fissure caries and could provide detailed information about deep lesions.

71.5
Interproximal Caries Detection

The interproximal contact area of teeth is a common location for dental caries. Here plaque (containing bacteria in a biofilm) and food particles can become trapped, leading to the creation of carious lesions. Jeon *et al.* used PTR–LUM to examine interproximal caries in extracted human teeth [11]. Three types of lesions were created: with dental burs in a high-speed handpiece, with 37% phosphoric acid, and with demineralization in artificial caries media. Each sample pair was examined before and after bur application, or sequential treatment with acid (etched for 20 s) and gel (time periods spanning from 6 h to 30 days). The results showed distinct differences in the signal for these types of lesions. Dental bitewing radiographs showed no sign of demineralized lesions even for the samples treated for 30 days. Only scanning electron microscopy and micro-computed tomographic imaging, which are destructive, showed visible signs of treatment. PTR further exhibited excellent reproducibility and consistent signal changes in the presence of interproximal demineralized lesions. The results suggested that PTR can be a reliable probe to detect early interproximal lesions. The LUM channel was also measured simultaneously, but it showed a lower ability than PTR to detect these lesions.

71.6
Early Lesion Detection

There is a great deal of value in detecting carious lesions early. PTR–LUM research in the laboratory setting [12] and observations from an early investigational trial [13] using the clinical version (Canary System), indicated that lesions can remineralize or stabilize if exposed to remineralizing agents such as fluoride and casein phosphopeptide–amorphous calcium phosphate (CCP–ACP). In laboratory studies, PTR–LUM was used to monitor early stages of tooth demineralization and remineralization [12]. Extracted teeth were treated with an artificial demineralization gel to simulate controlled mineral loss on the enamel surface and then exposed to a remineralizing agent. The treated region of the tooth was monitored with PTR–LUM before and after treatment. The results showed that PTR–LUM has very high sensitivity to incipient changes in the enamel structure. The PTR and LUM amplitudes and phases showed gradual and consistent changes with treatment time. There was good correlation of PTR–LUM results with the mineral loss or the lesion depth measured with transverse microradiography (TMR). This indicated that PTR–LUM is capable of monitoring artificially created carious lesions, their evolution during demineralization, and the reversal of the lesions during remineralization.

71.7
Recent Developments

In 2009, The Canary System developed by Quantum Dental Technologies (Toronto, ON, Canada) was used in a Health Canada-approved human investigational trial [13]. In that study, amplitude (A) and phase (P) responses at various modulation frequencies from healthy and carious dental enamel (ICDAS 0–6) were measured. Over 500 regions on healthy tooth surfaces of 50 subjects were used to construct a healthy baseline for each output channel (1-PTR-A, 2-PTR-P, 3-LUM-A, and 4-LUM-P). PTR-A and PTR-P were used to detect near-surface and subsurface lesions, whereas LUM-A and LUM-P were used to detect near-surface lesions. The results indicated that The Canary System did not cause any adverse events or soft or hard tissue trauma. There was no difference in signal from anterior and posterior healthy tooth surfaces, and the presence of surface stain and biofilm did not affect the signal from healthy tooth surfaces. A clear shift from the baseline in both PTR and LUM in carious enamel was observed depending on the type, depth, and nature of the lesion. The results showed that The Canary System is safe and discriminates between healthy and carious enamel. In 2010, a second Health Canada-approved investigational trial has shown the ability of The Canary System to detect carious lesions on smooth surfaces, occlusal surfaces and around restoration margins [14]. Preliminary results show that the system is safe and provided repeatable measurements that allowed it to monitor lesion growth and response to various remineralization therapies [14].

71.8
Conclusion

This review of the principles of PTR–LUM and its applications to dentistry has highlighted the significant progress in bringing this novel technology to clinical applications. PTR–LUM offers features beyond what is currently available in traditional dental detection methods. These features include the ability to perform depth profilometry and very early caries detection and monitoring on various tooth surfaces. It was also shown that The Canary System, a portable PTR–LUM instrument for clinical applications, is safe and can discriminate healthy and carious tooth tissue.

References

1 Dove, S.B. (2001) Radiographic diagnosis of dental caries. *J. Dent. Educ.*, **65** (10), 985–990.
2 Kidd, E.A. and Fejerskov, O. (2004) What constitutes dental caries? Histopathology of carious enamel and dentin related to the action of cariogenic biofilms. *J. Dent. Res.*, **83** (Spec. No. C), C35–C38.
3 Amaechi, B.T. (2009) Emerging technologies for diagnosis of dental caries: the road so far. *J Appl. Phys.*, **105**, 102047.
4 Munidasa, M. and Mandelis, A. (1992) Photothermal imaging and microscopy, in *Principles and Perspectives of Photothermal and Photoacoustic Phenomena*, vol. **1** (ed. A. Mandelis), Elsevier, New York, pp. 299–367.
5 Bjelkhagen, H., Sundstrom, F., Angmar-Mansson, B., and Ryden, H. (1982) Early detection of enamel caries by the luminescence excited by visible laser light. *Swed. Dent. J.*, **6** (1), 1–7.
6 Nicolaides, L., Feng, C., Mandelis, A., and Abrams, S.H. (2002) Quantitative dental measurements by use of simultaneous frequency-domain laser infrared radiometry and luminescence. *Appl. Opt.*, **41** (4), 768–777.
7 Matvienko, A., Mandelis, A., Jeon, R.J., and Abrams, S.H. (2009) Theoretical analysis of coupled diffuse-photon-density and thermal-wave field depth profiles photothermally generated in layered turbid dental structures. *J. Appl. Phys., Special Issue Appl. Biophys.*, **105**, 102022.
8 Jeon, R.J., Mandelis, A., and Abrams, S.H. (2003) Depth profilometric case studies in caries diagnostics of human teeth using modulated laser radiometry and luminescence. *Rev. Sci. Instrum.*, **74** (1), 380–383.
9 Jeon, R.J., Han, C., Mandelis, A., Sanchez, V., and Abrams, S.H. (2004) Non-intrusive, non-contacting frequency-domain photothermal radiometry and luminescence depth profilometry of carious and artificial sub-surface lesions in human teeth. *J. Biomed. Opt.*, **9** (4), 809–819.
10 Jeon, R.J., Han, C., Mandelis, A., Sanchez, V., and Abrams, S.H. (2004) Diagnosis of pit and fissure caries using frequency-domain infrared photothermal radiometry and modulated laser luminescence. *Caries Res.*, **38**, 497–513.
11 Jeon, R.J., Matvienko, A., Mandelis, A., Abrams, S.H., Amaechi, B.T., and Kulkarni, G. (2007) Detection of interproximal demineralized lesions on human teeth *in vitro* using frequency-

domain infrared photothermal radiometry and modulated luminescence. *J. Biomed. Opt.*, **12** (3), 034028.

12 Jeon, R.J., Hellen, A., Matvienko, A., Mandelis, A., Abrams, S.H., and Amaechi, B.T. (2008) *In vitro* detection and quantification of enamel and root caries using infrared photothermal radiometry and modulated luminescence. *J. Biomed. Opt.*, **13** (3), 048803.

13 Sivagurunathan, K., Abrams, S.H., Garcia, J., Mandelis, A., Amaechi, B.T., Finer, Y., Hellen, W.M.P., and Elman, G. (2010) Using PTR–LUM ("The Canary System") for *in vivo* detection of dental caries: clinical trial results. *Caries Res.*, **44**, 229.

14 Abrams, S., Sivagurunathan, K., Jeon, R.J., Mandelis, A., Silvertown, J.D., Hellen, A., Hellen, W.M.P., Elman, G.I., Ehrlich, B.R., Chouljian, Finer, Y., and Amaechi, B.T. (2011) Multi-center clinical study to evaluate the safety and effectivness of "The canary system" (PTR-LUM Technology), *Caries Res*, **45**, 174–242.

72
Lasers in Restorative Dentistry
Anil Kishen

72.1
Introduction

Current developments in laser technology and laser delivery systems in tandem with an improved understanding of laser–dental tissue interactions have broadened the scope of the application of lasers in dentistry. There has been growing interest among clinicians to apply lasers in different clinical procedures in dentistry. However, the main concern is the lack of sufficient clinical studies that significantly support the advantage of lasers over conventional treatment methods. Dental lasers for clinical procedures operate in the infrared, visible, or ultraviolet range of the electromagnetic spectrum. These lasers are delivered as either a continuous, pulsed (gated), or running pulse waveform. It is imperative for the clinician to determine the specific treatment goals and then select the laser technology best suited to achieve the desired effect(s).

The photons in a laser beam are emitted as a coherent, unidirectional, monochromatic light beam that can be collimated into an intensely focused ray of energy. This focused ray of energy will interact with a target tissue or material by absorption, reflection, transmission, or scattering. The absorbed light energy may result in heating, coagulation, or vaporization of tissues, depending on the laser parameters such as wavelength, power, waveform (continuous or pulsed), pulse duration, energy/pulse, energy density, duration of exposure, angulation of the delivery tip to the target surface, and so on. The optical properties of the target tissue also dictate to a great extent the nature and intensity of interaction with specific laser wavelengths. Pigmentation, water content, mineral content, thermal conductivity, tissue density, and latent heat of transformation are factors that would influence the nature of light–tissue interaction. Furthermore, physiological processes, for instance, tissue vascularity, degree of tissue inflammation, and the presence of progenitor cells to participate in the healing process, will also influence the tissue response to laser radiation.

Handbook of Biophotonics. Vol.2: Photonics for Health Care, First Edition. Edited by Jürgen Popp, Valery V. Tuchin, Arthur Chiou, and Stefan Heinemann.
© 2012 Wiley-VCH Verlag GmbH & Co. KGaA. Published 2012 by Wiley-VCH Verlag GmbH & Co. KGaA.

72.2
Classification of Lasers in Restorative Dentistry

Lasers can be classified as follows:

i. Based on the type of laser medium used:
   - *Gas*
   - *Solid*
   - *Liquid*
ii. Based on the type of delivery system:
   - *Flexible hollow waveguide*
   - *Glass fiber optic cable*
iii. Based on type of interaction with tissue:
   - *Contact lasers*
   - *Noncontact lasers*
iv. Based on the type of application:
   - **Soft tissue lasers (athermic) (~1000 m)**: *The three main types of soft tissue lasers are the helium–neon (He–Ne) diode, gallium arsenide (GaAs), and gallium aluminum arsenide (GaAlAs) lasers.*
   - **Hard tissue lasers (thermic) (~ 3 W or more)**: *The three main types of hard tissue lasers are the argon (Ar), CO_2, and Nd:YAG lasers.*
v. Based on the type by source:
   - **Infrared example**: CO_2, *Ho:YAG, and Nd:YAG lasers*
   - **Visible example**: - *He–Ne and argon lasers*
   - **Ultraviolet example**: - *XeCl, XeF, KrF, and ArF (excimer) lasers*
vi. Based on the type by wavelength:
   - **Long-wavelength**: – *Infrared laser*
   - **Short-wavelength**: – *Ultraviolet laser*
vii. Based on safety (American National Standards Institute Laser Classification)
   - **Class 1**: *Exempt lasers or laser systems that cannot, under normal operation conditions, produce a hazard.*
   - **Class 2**: *Low-power visible lasers or laser systems which, because of normal human aversion responses, do not normally present a hazard, but may present some potential for hazard if viewed directly for extended periods of time.*
   - **Class 3a**: *Lasers or laser systems that normally would not produce a hazard if viewed for only momentary periods with the unaided eye. They may present a hazard if viewed using collecting optics.*
   - **Class 3b**: *Lasers or laser systems that can produce a hazard if viewed directly. This includes intrabeam viewing or specular reflections. Except for the high-power Class 3b lasers, this class of laser will not produce a hazardous diffuse reflection.*
   - **Class 4**: *Lasers or laser systems that can produce a hazard not only from direct or specular reflection, but also from diffuse reflection. In addition, such lasers may produce fire hazards and skin hazards.*

72.3 Lasers and Delivery Systems

The construction of a laser source requires an active medium (gas, liquid, or solid) and a collection of atoms or molecules. The active medium is excited to emit the photons by stimulated emission. The light produced by the laser source is usually coupled to a delivery system and is used for therapeutic purposes. The active medium producing the beam identifies and distinguishes one laser from another. Different types of lasers used in dentistry, such as the carbon dioxide (CO_2), erbium (Er), and neodymium (Nd) lasers, various other substances used in the medium [e.g., yttrium aluminum garnet (YAG) and yttrium scandium gallium, garnet (YSGG), erbium, chromium-doped yttrium scandium gallium garnet (Er,Cr:YSGG)], argon, diode, and excimer types, all produce light of a specific wavelength. The CO_2, Er:YAG, Er,Cr:YSGG, and Nd:YAG lasers emit invisible beams in the infrared range. The argon laser emits a visible light beam at 488 or 514 nm, whereas the excimer lasers, which operate on rare gas monohalides, emit invisible ultraviolet light beams at predetermined wavelengths (ArF, 193 nm; KrF, 248 nm, XeCl, 308 nm).

In clinical applications, it is important to deliver laser energy precisely to the target tissue and ergonomically to the operator. Laser energy is delivered to the target tissue by means of two common delivery systems: (1) a flexible hollow waveguide with mirror finish interior and (2) a glass fiber optic cable. In a flexible hollow waveguide, the laser energy is reflected along this tube and exits through a delivery handpiece with the beam striking the target tissue in noncontact mode. If the delivery probe has to be used in contact mode, then an accessory tip of sapphire or hollow metal may be connected to the end of the waveguide. A glass fiber optic cable can be more flexible and fits snugly into a handpiece with a bare fiber end or attached to a sapphire or quartz tip. A fiber optic delivery system allows greater accessibility to different areas in the oral cavity. Infrared dental lasers are equipped with a separate aiming visible beam that is coaxially superimposed on the infrared beam. This will allow the operator to spot where the infrared laser energy will be focused [1]. Shorter wavelength instruments such as Ar, diode, and Nd:YAG lasers have a small, flexible optical fiber system with bare glass fibers that delivers the laser energy to the target tissue. Er and CO_2 laser devices are constructed with more rigid glass fibers, semiflexible hollow waveguides, or articulated arms to deliver the laser energy to the surgical site [1].

The fiber optic delivery system can be used in either contact or noncontact mode (a few millimeters away from the target). For the optic fiber, the focal point is at or near the delivery tip, and has the greatest energy (focused mode of delivery). The laser beam is divergent and defocused as the handpiece is moved away from the focal point (defocused mode of delivery) (Figure 72.1). At a small divergent distance, the laser light can cover a wider area. At a greater distance, the beam effectiveness decreases because the energy is dissipated. In the defocused mode, there is a wider zone of tissue removal but a shallower depth of penetration. Lasers with shorter emission wavelengths, such as argon, diode, and Nd: YAG lasers, can be designed with small, flexible glass fibers. The Er,Cr:YSGG and Er:YAG devices present challenges to fiber

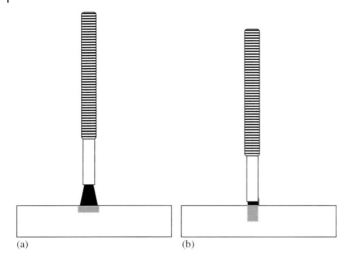

Figure 72.1 Schematic diagram showing the application of laser in (a) defocused and (b) focused modes of delivery. The defocused mode produces a wider zone of tissue removal but shallower depth of penetration compared with the focused mode.

manufacture because their wavelengths are large and do not easily fit into the crystalline molecules of the conducting glass. Further, they are highly absorbed by water, hence a special expensive fiber design with minimal hydroxyl content, peripheral cooling air, and water spray for the handpiece is necessary [1, 2].

72.4
Laser–Dental Tissue Interaction

The nature of laser–tissue interaction is influenced by the properties of the laser (e.g., wavelength, energy density, and pulse duration of the laser radiation) and the characteristics of the tissue (e.g., absorption, reflection, transmission, and scattering). Different types of lasers may produce different effects on the same tissue, and the same laser can have varying effects on different tissues. The nature of absorption and transmission of laser light is primarily wavelength dependent and it is important to note that the intensity of light will not remain constant throughout a definite volume of tissue. Consequently, laser effects will change depending upon their depth of penetration. The clinician controls four parameters while operating a laser system: (a) the level of applied power (power density), (b) the total energy delivered over a given surface area of tissue (energy density), (c) the rate and duration of the exposure to laser energy (pulse repetition), and (d) the mode of energy delivery to the target tissue (i.e., continuous or pulsed energy; direct contact, or no target tissue contact). Four different types of interactions with the target dental tissue have been suggested: (1) photobiological interactions, (2) photochemical interactions, (3) photothermal interactions, and (4) photomechanical and photoelectrical interactions.

1) **Photobiological Interactions:** Transmission of light conducts light energy through the tissue without any interaction and therefore does not have any effect on the tissue. Reflection is another process that does not produce any thermal effect on the tissue. The absorption of laser energy by a tissue is an indication of its ability to produce tissue changes and generation of heat. The nature of the absorption of laser energy by a tissue generally depends upon the laser wavelength and the optical characteristics of the target tissue. In scattering, the light energy is absorbed over a greater surface area, resulting in a less intense and less precise tissue or thermal effect. Scattering is observed to be the most pronounced effect in a tissue, and most tissues behave as a highly scattering turbid medium. The particular properties of each type of laser and the specific target tissue render them suitable for various procedures.

    The CO_2 laser has a high absorption coefficient in water, and is well absorbed by dental hard and soft tissue components. Consequently, it produces effects on most biological soft and hard tissues. However, because of its high thermal absorption, the enamel and dentin surfaces reach very high temperatures on application of a CO_2 laser. The Er:YAG laser is the most efficient for cutting enamel and dentin as its energy is also well absorbed by water and by hydroxyapatite. The argon laser is effective on pigmented or highly vascular tissues, whereas the Nd:YAG laser energy is transmitted through tissues by water and interacts well with dark pigmented tissue. The excimer lasers functions by breaking molecular bonds and reducing the tissue to its atomic constituents before their energy is dissipated as heat.

2) **Photochemical Interactions:** A photochemical effect occurs when there is absorption of laser light by chromophores (natural or artificial) or cellular components to induce biochemical reactions without any thermal effect. Photochemical effects may lead to alterations in the physicochemical properties of the irradiated tissue by photoaddition, photofragmentation, photo-oxidation, photohydration, *cis–trans* isomerization, or photorearrangement. The chemical reaction that leads to the curing of dental composite resin is an example of a photochemical process. Another example of photochemical reactions is photodynamic therapy (PDT). PDT is produced by the absorption of light by the chemicals introduced into a cell or a tissue. The principle of PDT is applied to destroy tumor cells and microbial cells. In PDT, the applied chemical, called a photosensitizer, acts as an exogenous chromophore that performs the function of photosensitization. Photosensitization is the process of sensitizing a photoprocess. The mechanism of most photosensitization reactions involves photoaddition and photo-oxidation.

3) **Photothermal Interactions:** In a photothermal interaction, the laser energy is absorbed by the tissue and the light energy is transformed into heat energy, which is the cause of the final tissue effect. Photothermal interactions can lead to heating with denaturation (45–60 °C), and rapid absorption of laser energy leading to removal of tissue (photoablation) by vaporization or burning away of the tissue by carbonization (photopyrolysis). This interaction forms the basis for many surgical applications of lasers. The photothermal effect on a tissue largely depends on the

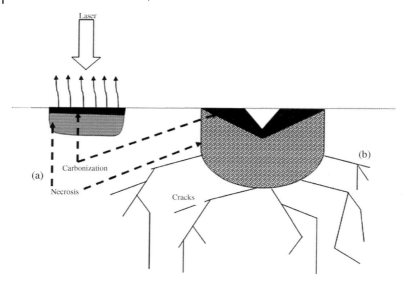

Figure 72.2 Illustration of laser-induced photothermal effects. The laser light energy is absorbed and converted to thermal energy by the lattice vibration of the hard tissue molecules (mainly water). This leads to the heating of surrounding tissues, boiling of water, and drying of tissue. Drying of tissue will cause rapid increase in absorption, carbonization and tissue removal (a). Usually, the surrounding hard tissue damage after treatment is manifested by massive zones of carbonization, necrosis, and cracks in the hard tissue (b) [2].

wavelength of the laser, since the amount of heat generated in the tissue is determined by the extent of absorption of the light. In a photothermal interaction, the laser light is absorbed and converted into thermal energy by stimulating the lattice vibrations of the tissue molecules. The lattice vibrations lead to heating of surrounding tissues and boiling of tissue water. On subsequent to drying, there will be a rapid increase in absorption leading to carbonization and tissue removal. The level of tissue dehydration plays a significant role in the photothermal interaction of lasers with tissue (Figure 72.2).

4) **Photomechanical Interactions:** The photomechanical effect is characteristic of ultra-short pulses of very high energy density. Even a small energy of 1 µJ in 1 ps corresponds to a peak power of 10^6 W and when focused to a 10 m spot would lead to an intensity of about 10^{12} W cm^{-2}. This localized absorption of intense laser irradiation can lead to rapid heating and very large temperature gradients in dental tissues, resulting in enormous pressure waves and photodisruption (or photodisassociation). Photomechanical disruption occurs in three phases: (1) ionization, (2) plasma formation, and (3) shock wave generation.

The energy levels resulting in shock waves are capable of rupturing intermolecular and atomic bonds due to the conversion of light energy into kinetic energy. In dental hard tissues, a photomechanical shock wave produced by the rapid photovaporization of water leads to a volumetric change of water within the tooth. This change creates high pressures, removing and destroying selective areas of adjacent tissue. Therefore, it is important to maintain the maximum laser energy density of all pulsed lasers below a certain threshold to avoid microcracks in dental tissues. Er:YAG and Er,Cr:YSGG lasers are known to cause photomechanical effects. This principle is also used to ablate infected dental hard tissue in dental caries (bactericidal effect). Furthermore, the short pulse width associated with high fluence rates in lasers such as the Ho:YAG laser may also induce a photoacoustic interaction, which involves the removal of tissue with shock wave generation. Photoelectrical interactions include photoplasmolysis, which involves the removal of tissue through the formation of electrically charged ions and particles that exist in a semi-gaseous, high-energy state.

5) **Effect of Target Tissue Properties on Laser–Tissue Interaction:** The response of tissues to laser energy is complex and different thermal properties of the target tissue are considered to influence the tissue response. The different thermal properties of interest are the following:
   - **Thermal diffusivity:** This is the ability of the tissue to conduct heat upon exposure to a transient temperature gradient.
   - **Thermal coefficient of expansion:** This property measures the expansion that is experienced by the tissue upon heating.
   - **Heat capacity:** This is the amount of heat energy needed to raise the temperature of $1\,cm^3$ of tissue by $1\,°C$.
   - **Phase transformation temperature:** This temperature indicates the melting and vaporization temperatures of the tissue components.
   - **Latent heat of transformation:** This is the characteristic amount of heat absorbed or released by a substance undergoing a change in physical state.

72.5
Laser Effects on Dental Tissues

72.5.1
Laser Effect on Enamel

The laser effect on enamel tissue is primarily due to a temperature effect, which leads to melting and fusion of enamel, vaporization, and crater formation. A pulse mode of application produces less heat than the continuous mode, and therefore melting of the enamel surface can be achieved without vaporization. CO_2 laser radiation is reasonably well absorbed by the enamel, since human enamel has its absorption peak close to the wavelength of the CO_2 laser. Laser energy causes a rapid local temperature rise and prompts melting and cooling of apatite crystals up to a depth of $5\,\mu m$. The melting point of human enamel is above $1280\,°C$. It forms a mixture of β-tricalcium phosphate and tetracalcium phosphate above $1450\,°C$. These changes were shown either to decrease or increase the solubility of the enamel depending on the calcium: phosphate ratio in the newly formed minerals. Depending on the energy density of the laser irradiation, the chemical and solubility characteristics of enamel tissue can be modified. Low-energy lasing in particular has been shown to increase the acid resistance of enamel, while high-energy density lasing may decrease the acid resistance. Additionally, lasing of enamel enhances the fluoride uptake, resulting in an increased acid resistance. The laser effect on enamel seems to depend on the structure and orientation of enamel prisms. It was concluded based on this study that the observed resistance of laser-exposed enamel to subsurface demineralization was due to a physical alteration in permeability rather than to a chemical change in solubility [3, 4]. Enamel surfaces exposed to Nd:YAG laser energy show roughness, fused crystallites, charring, and cracking, whereas Er:YAG-lased enamel shows characteristics of cavitated and flaky rough surfaces. An argon laser has a minimal effect on the surface due to reflection of the beam.

72.5.2
Laser Effects on Dentin

The exposure of sound dentin to a CO_2 laser beam causes high absorption of laser radiation by the minerals in the dentin, resulting in an increase in temperature, which after evaporation of water leads to combustion of the organic material and fusion of the hydroxyapatite. Lased dentin recrystallizes while cooling to form a structure that is similar to that of normal enamel [5]. This recrystallized dentin can form a sealed layer [6]. The depth of the sealed layer depends on the energy density used. The recrystallized sealed layer showed a higher mineral content and this layer seemed to be partially acid resistant. However, demineralization of dentin was observed underlying the sealed layer, which was attributed to the acid diffusion that occurs through the unsealed dentin in the laser-treated area and also through the radiation-induced micro-cracking. The level of dentin dehydration plays a significant role in the thermal interaction of lasers with dentin. CO_2 laser energy is easily absorbed by water and therefore results in less carbonization on or heat penetration into the tissue surface that contains water. If the tissue contains no water, carbonization and crack formation occur on the surface of the tissue. It is important to note that the continuous-wave CO_2 laser, when used at energies necessary to ablate the dentin structure, can cause pulpal damage.

Laser irradiation renders the dentin surface harder and more brittle. Different wavelengths of irradiation could provide different surface morphology with less brittle dentin. Laser irradiation of dentin produced a surface morphology characterized by localized melting and recrystallization of dentin which resulted in projections covered with a glaze-like surface that enhanced the mechanical bond of composite restorations. Unfortunately, failures occurred most often within the dentin, indicating that the bond strength of the composite was stronger than the mechanical properties of the dentinal projections [7]. Studies have shown that dentin irradiated with the Nd:YAG laser prior to an adhesive procedure results in a reduction in bond strength with composite resin. This effect is due to the obliteration of the dentinal tubules due to the melting and resolidification of irradiated dentin. In the red and near-infrared electromagnetic spectral region (600 nm–1.5 μm), the tissues have a transmission window so that the absorption of neodymium laser (1064 nm) energy by water and hydroxyapatite is low. Even though the energy is only partially absorbed by dentin, the laser is still able to heat the tissue to the point of carbonization

(600–800 °C). During carbonization, the organic and inorganic materials can melt and vaporize, leading to micro-explosions and ejection of the molten mineral phase, which subsequently resolidifies at the surface. Therefore, the Nd:YAG laser is not indicated for cutting hard tissues or for tissue modification prior to adhesive restoration. Excessive thermal damage has been one of the major problems associated with lasers such as the Nd:YAG and CO_2 lasers when used for cavity preparation due to lower absorption coefficients. Er:YAG laser energy is highly absorbed by water and hydroxyapatite. Er lasers offer the ability to remove enamel, dentin, and carious tissue with minimal tissue disruption of the residual tooth. The absorption of the ArF (193 nm) excimer laser radiation by dentin is considerably higher than that of the XeCl (308 nm) laser. As a result, a much thinner melted layer is formed following Ar:F ablation. It was found that when pulses of extremely short duration are used with this laser, heat accumulation does not occur and thermal damage to the surrounding tissues may be avoided.

72.5.3
Laser Effects on Dental Pulp

The effect of lasers on dental pulp is an important factor that determines the nature of their clinical application. Normally, pulp tissue cannot survive in an environment of elevated temperature for extended periods when the tooth structure is irradiated with lasers [8]. Lasers that produce thermal damage to the pulp are not particularly suitable for cavity preparation. Semiconductor diode laser irradiation has been reported to produce no dental pulp damage, and these lasers having a power of 3–30 W are considered to have potential applications for dental pulp treatment. He–Ne lasers with powers of more than 15 mW are also known not to damage dental pulp. The Nd:YAG laser has a wide energy emission range and its effects should be carefully considered before application. Nd:YAG and CO_2 lasers have the greatest potential for causing damage to the pulp and surrounding tissue. Nd:YAG laser treatment results in total disruption of the normal architecture of the pulp, including destruction of the odontoblastic cell layer and vasculature. There is extravasation of red blood cells and inflammatory cell infiltration consisting of neutrophils, lymphocytes, and plasma cells. The Er:YAG laser has been used to ablate hard and soft tissues under copious water spray without conspicuous damage to dental pulp [9]. Laser preparation of teeth should include short intervals with continuous water application to allow dissipation of heat. This strategy of fractionating the energy and continuous water spray is suggested to minimize the risk of damage to the dental pulp. The Er,Cr:YSGG laser can also be used to ablate hard and soft tissues with a water coolant. The argon laser, which has been used for laser polymerization of dental composite resin, has been reported to cause no significant temperature increase in the pulp tissue [10].

The parameters of the laser that affect the dental pulp tissue are waveform, power, and duration of exposure. Different methods have been utilized to reduce heat transfer from the enamel and dentin surface to the pulp. Pulsing of lasers has been used as an effective method to prevent collateral tissue damage. However, this method was noted to be effective during soft tissue ablation and less effective when dental hard tissue was involved. The difference is attributed to the difference in the thermal diffusion rate and relaxation time between the hard calcified and soft oral tissues. The use of a combination of air and water spray before, during, and immediately after laser irradiation of enamel and dentin is suggested to be a more effective means of temperature control and reduction of heat transfer to the pulp and surrounding vital structures. Precooling with an air–water spray prior to lasing may be used with laser systems with wavelengths that are more readily absorbed by water (e.g., CO_2, Ho, and Er).

72.6
Lasers in Operative Dentistry

Lasers have been employed for a wide variety of operative procedures on teeth such as bleaching of teeth, adhesive restorations, and soft tissue procedures such as crown lengthening. The treatment goal of applying lasers in operative dentistry is to reduce pain and perform operative procedures as atraumatically as possible.

72.6.1
Laser-Assisted Tooth Bleaching

Laser-assisted tooth bleaching, also called as power bleaching, is carried out for cosmetic purpose and in most cases does not need restoration. Argon, KTP (potassium titanyl phosphate), and diode lasers are most commonly used for this purpose. Laser-assisted tooth bleaching is a method of removing extrinsic and intrinsic stains from teeth. In laser-assisted tooth bleaching, laser activation energy is utilized to trigger faster degradation of the hydrogen peroxide bleaching agent into reactive oxygen free radicals. This in turn increases the rate of tooth bleaching and bleaching efficiency. Certain dental lasers are also used with bleaching products containing a specific chromophore that will absorb the laser wavelength. Studies have recommended minimum laser exposure times to minimize the rise in pulpal temperature. Generally, the rise in laser-induced intra-pulpal temperature is directly proportional to the power of the laser and irradiance and inversely related to tooth thickness [11, 12]. Bleaching gel (absorption agent) is indicated to reduce surface thermal changes and intra-pulpal temperature during laser-assisted bleaching [e.g., titanium dioxide (TiO_2)] [13].

Systems used for laser-assisted tooth bleaching include BritSmile Teeth Whitening System (Discus Dental), LaserSmile Teeth Whitening System (Biolase Technology), LumaArch Teeth Whitening System (Lumalite), and Opus 10 Laser System (Lumenis). In addition to laser systems, light-emitting diode (LED) and halogen light sources are also used for tooth bleaching. The advantages of laser-assisted tooth bleaching are that this approach aids in distributing light energy more uniformly throughout the tooth surface. It has been found that the photochemical activation provides a higher intrinsic overall radical yield than thermal activation, and that the rate at which reactive radicals are generated is higher than with thermal activation. However, several studies have suggested that laser-assisted tooth bleaching is not more effective than many professional teeth bleaching systems and further research is indicated in this area [14, 15].

72.6.2
Laser-Assisted Cavity Preparation and Caries Removal

In laser-assisted cavity preparation, intense laser energy is used for the ablation of dental hard tissues. The main advantage of laser-assisted cavity preparation is the ability to be specific to the carious lesion and conserve maximum healthy tooth structure. Depending on the wavelength of the light applied, the ablative effect can produce photochemical or photothermal effects on the target tissue. The nature of absorption of light energy is different for different tissues with different textures and water contents [16]. Subsequently, different laser parameters are required for the ablation of enamel, dentin, and carious tissue. Different lasers such as CO_2 and pulsed Nd:YAG lasers have been applied for this caries removal.

The Er-based laser system has been widely used for the preparation of dental hard tissues. During ablation, the main constituent that absorbs the laser energy is the bound water in the tissue. The maximum absorption peak in the infrared range (2.9 μm) matches well with the wavelengths of Er:YAG (2.94 μm) and Er,Cr:YSGG (2.79 μm) lasers. The absorption of water in this range is so high that scattering and absorption of light contributed by other tissue components become insignificant. Since the Er wavelength has an affinity for the water content of hard tissues, (a) it has a small tissue penetration depth and (b) less energy is required to ablate caries (greater water content) than dentin or enamel (lesser water content). When irradiating dentin or enamel with Er-based laser pulses with sufficient pulse energy, the tissue water is heated very rapidly above the boiling point. This causes a micro-explosion, leading to blowing out of the generated small particles (spallation) and mechanical removal of target tissue. Since certain amount of heat conduction cannot be avoided, even though most of the radiation is absorbed by the water, a water-spray has to be used for cooling to prevent damage to pulp tissue. During cavity preparation, a water–air stream is directed on the cutting tip and on to the target tissue. The laser tip should be held perpendicular to the cutting surface to maximize the cutting efficiency. The tip end should be moved constantly to provide effective ablation and better tissue cooling. Once the enamel has been removed, the energy settings must be reduced because dentin and caries have a higher water content than enamel and cut more readily. An alternative approach to changing the operating parameters is to reposition the laser tip to a noncontact mode to decrease the energy density. For deep cutting, the tip is moved up and down as in a pumping action. The operator can also recognize different tooth structures by the sound of ablation (popping sound), which differentiates the tissue types. Both the pitch and resonance of this popping sound are related to the propagation of acoustic shock waves within tooth, and vary according to the presence or absence of caries. This feature assists the user to confirm if the caries has been removed completely. During interproximal tooth preparation, adjacent teeth should be isolated and protected with the use of a rubber dam or a metal matrix. Laser-based cavity preparation usually results in a cavity with a rough or irregular surface, which is ideal for the placement of a composite resin or glass ionomer restoration.

Lasers have also been applied to prevent enamel and dental caries. The mechanism by which laser irradiation enhances enamel resistance to artificial caries ranges from a physical seal achieved by melting the surface through partial fusion and recrystallization to compositional alteration of enamel [17]. It has been suggested that the thermal treatment converts carbonated hydroxyapatite mineral to a less soluble mineral. Scanning electron microscopic studies have shown that enamel and dentin surfaces subjected to continuous-wave (cw) CO_2 laser radiation (1, 2, 3 W) were sufficiently melted and the smear layer was solidified. This lased region was unchanged even after acid demineralization. This supported the hypothesis that the lased enamel surface is less permeable to diffusion of ions and is resistant to solubility [18]. The formation of a micropore system within the mineral phase in the enamel, dentin, and cementum following laser treatment has also been suggested. The micropore system provides a means for trapping calcium, phosphate, and fluoride ions released during demineralization and these subsequently act as sites for reprecipitation. During the process of demineralization, dissolution of mineral occurs with the mobilization of ions from the affected dental hard tissue. As mineral phases are released from the deeper layers of hard tissue, reprecipitation in the more superficial layer occurs and the surface remains intact [19–21]. It has been shown that irradiation with an argon laser alters the surface layer of enamel by producing microporosities admixed with globular surface coating. This surface is rich in calcium, phosphate, and fluoride. The presence of fine porosities with a confluent globular surface coating suggests that the effect of the argon laser at relatively low fluences (11.5 and 100 J cm^{-2}) may alter or eliminate the organic material overlying and within the surface enamel. Further, it appears as though the mineral phases previously embedded in this organic matrix may form globular precipitates, resembling a calcium fluoride confluent surface [22]. These globular deposits may provide a reservoir for mineral phases during a cariogenic attack.

Nd:YAG, CO_2, and argon lasers have been applied to induce remineralization of subsurface lesions of enamel and dentin, by combining the effect of lasers with fluorides. This treatment approach renders the tooth surface less susceptible to caries. An argon laser has been found to alter the surface characteristics resulting in a significant reduction in its acid solubility. The Nd:YAG laser in combination with acidulated phosphate fluoride application increased the acid resistance of enamel whereas CO_2 laser irradiation created a sealed tissue layer that delayed diffusion of acids into the underlying sound dentin. The lattice strain was decreased after laser irradiation and this resulted in a better arrangement of ions in the crystals of enamel. A slight contraction of the A-axis was observed but the C-axis was not altered by laser irradiation at an optimal energy density. Reductions of water, carbonate, and organic substances were also revealed in lased enamel. The reduction of carbonate and bound water from the enamel crystals has been suggested to contribute to caries resistance. Even the uptake of fluoride was found to be very high in the lased enamel [23, 24].

72.6.3
Laser-Assisted Adhesion in Tooth Color Restoration

Laser etching is an alternative option to acid etching to bond tooth colored restoration to the tooth structure. Argon, CO_2, Nd:YAG, and excimer lasers are used for this purpose. Laser etching is due to the continuous vaporization of water trapped within the inorganic matrix of enamel and dentin. The laser etching procedure is considered to result in an etched tooth surface, similarly to the acid etching technique. However, conflicting results on bond strength to enamel have been reported after different laser etching. CO_2 laser energy was reported to increase the shear bond strength [7], which exceeded the value obtained for acid etching. These studies supported CO_2 laser etching to improve the bonding of resin composite to enamel [25, 26]. When the bond strengths of porcelain laminate veneers to tooth surfaces after etching with acid and after conditioning with an Er,Cr:YSGG laser was compared, laser-etched bonding produced a similar bond strength to acid-etched bonding of laminate veneers to tooth surfaces [27]. The effect of etching with an Er:YAG laser was also evaluated *in vitro*. There was a significant positive correlation between the etching fluence and the shear bond strength, but pitting of the enamel surface at fluences above 25 J cm^{-2} limited the maximum fluence for etching purposes. It was observed that the shear bond strengths were significantly inferior after Er:YAG laser etching than those obtained using conventional acid etching [28]. In another study, it was shown that laser pretreatment of Ni–Cr alloy increased the bond strength to composite resin compared with sandblasting techniques. Laser pretreatment in combination with sandblasting further increased the bond strengths [29].

72.6.4
Laser-Activated Polymerization of Composite Resin

The argon laser has been found to be useful for polymerizing composite resin. The argon laser emits specific light with wavelengths that correspond to the absorption peak of camphorquinone, the initiator of polymerization for the efficient photopolymerization of resin composite. An *in vitro* experiment has shown that the argon laser is a possible alternative for photopolymerizing composite resin, providing the same quality of polymerization as the halogen lamp. However, none of the photocured units tested in that study completely eliminated microleakage [30]. It is also important to note that laser-induced polymerization can produce an increase in pulpal temperature. However, the magnitude of the heat increase produced would depend upon the polymerization modes and remaining dentin thicknesses. It was reported that thicker the remaining dentin and the lower the polymerization mode energy, the lower will be the temperature rise [31].

72.6.5
Laser-Assisted Removal of Restorative Materials and Metal Dowels

Lasers are used to remove different restorative materials such as zinc phosphate cement, polycarboxylate cement, glass ionomer cement, polyketone, composites, and silver amalgam. The Er:YAG laser has been shown to be effective in removing these restorative materials with a total energy delivered that was slightly lower than that used on enamel and dentin tissue [32]. The removal of silver amalgam using lasers is not recommended, mainly owing to the high concentration of mercury vapor released during ablation. There are also disadvantages when lasers are applied to remove dowels or endodontic posts: (a) it is difficult to manipulate the fiber optic illumination tip deep within the root canal; (b) some lasers could, directly or indirectly, induce damaging effects on surrounding tissues; and (c) tiny metallic particles are generated during lasing and these particles could be aspirated by the patient or the operator [16]. Further research is required before lasers can be applied for dowel removal. Experiments have also demonstrated that Nd:YAG laser irradiation can be useful for removing obturation materials from the root canal. It has been suggested that these lasing procedures require less time than the conventional method with drills and files. Furthermore, following laser irradiation, the dentin wall was found to be free of debris and the smear layer [33].

72.6.6
Laser-Based Management of Dentin Hypersensitivity

Dentinal hypersensitivity is characterized by pain of short duration arising from the exposed dentin in response to stimuli, typically thermal, evaporative, tactile, osmotic, or chemical. Pain due to dentin hypersensitivity cannot be attributed to any other form of dental defect or pathology. Dental erosion, abrasion, attrition, gingival recession, periodontal treatment, and anatomic defects are risk factors for dentinal hypersensitivity. Hypersensitive tooth surfaces are mostly located at the cervical margin on the buccal surface of the teeth [34]. Brännström *et al.* proposed that the nerve endings in the dentin–pulp interface are activated by the rapid dentinal fluid flow in response to the dentinal stimulation (hydrodynamic theory) [35]. Patent dentinal tubules are the primary factor that leads to sensitivity of exposed dentin and clinical management of dentin hypersensitivity is aimed at blocking the fluid flow through the exposed dentinal tubules. Most in-office treatments for dentin hypersensitivity employ some form of "barrier," either a topical solution or gel or an adhesive-based restorative material to block the dentinal fluid movements. The application of desensitizing agents to reduce neuronal responsiveness to dentinal stimuli has also been investigated. Potassium-containing dentifrices, fluoride-containing medicaments, and agents containing 10% strontium chloride were found to be partially effective in reducing dentinal hypersensitivity. However, to date most treatment procedures have failed to produce satisfactory long-term treatment outcome in patients with dentin hypersensitivity [36].

The application of lasers in the management of dentinal hypersensitivity is based on two mechanisms. (1) The first is the direct effect of laser irradiation on the neurogenic activity of the nerve fibers. The GaAlAs (diode) laser wavelengths (780, 830, and 900 nm) were considered to produce an analgesic effect by depressing the nerve transmission by blocking the depolarization of C-fiber afferents. (2) The second is sealing of the dentinal tubule by the melting and fusing of dentin tissue/smear layer. The different lasers used for the management of dentin hypersensitivity are He–Ne (632.8 nm), GaAlAs (780 nm), GaAlAs (830 nm), Nd:YAG (1.064 µm), and CO_2 (10.6 µm). The CO_2 laser has been applied to fuse or recrystallize the dentin surface [37]. However, the possibility of carbonization of organic material together with the melting of dentin cannot be avoided. The CO_2 laser at moderate energies produces mainly sealing of dentinal tubules, and also a reduction in permeability [38]. Nd:YAG laser energy (3 W) was also used to manage dentin hypersensitivity [39] by occluding the open dentinal tubules. The sealing depth of Nd:YAG laser irradiation on dentinal tubules measured less than 4 µm [40]. Direct nerve analgesia and a suppressive effect achieved by blocking the depolarization of A-delta and C-fibers were also considered possible mechanisms of action for Nd:YAG laser irradiation in the treatment of dentin hypersensitivity [41]. The Er:YAG laser at a low power has also been used for the treatment of dentin hypersensitivity. However, addition of sodium fluoride was found to enhance markedly the occlusion produced by CO_2 and Er:YAG lasers. Excimer lasers have also been highlighted as a potential treatment option to manage dentin hypersensitivity and prevent bacterial penetration through dentinal tubules. A possible advantage of using excimer lasers could be the lack of thermal damage to the surrounding tissues [42, 43].

72.7
Application of Lasers in Endodontics

The goals of endodontic therapy are to (1) debride and disinfect the infected root canal, (2) shape the root canal, and (3) completely seal the root canal space. Currently, lasers are applied for the diagnosis of pulpal health, pulp capping, pulpotomy, root canal disinfection, root canal obturation, endodontic retreatment, and apical surgery.

72.7.1
Application of Lasers to Diagnose Dental Pulp Health

The selection of a tooth for root canal treatment depends on the pulp health or vitality. Pulp vitality depends on the presence of intact blood supply in dental pulp microvasculature. Nevertheless, the location of dental pulp within a rigid and calcified dental hard tissue has made measuring this parameter very difficult clinically. The current vitality tests do not aid in assessing the pulp vitality. Laser Doppler flowmetry (LDF) was developed to assess the blood flow in pulpal microvasculature, and subsequently the health of dental pulp. This method is useful for diagnosing the health of pulp in clinical cases of pulp revascularization 3–4 months ahead of electric pulp testing [44].

LDF is a noninvasive method, which assess the blood flow in the microvasculature of dental pulp. This system uses He–Ne and diode lasers at a low power of 1–2 mW. The dentinal tubules in the dentin are said to act as light guides and direct the light incident on the tooth surface into the pulp. Moving red blood cells cause the frequency of the laser beam to be Doppler shifted and some of the light to be backscattered out of the tooth. The scattered light is detected by a photocell that is placed on the surface of the tooth. The measured output is proportional to the number and velocity of the blood cells. The mean blood flux level in teeth with vital pulp tissue is significantly higher than for teeth with non-vital pulp. However, in teeth with vital pulp with impaired blood supply, the flux level can be low and in such cases the presence of pulsation is the only indication of pulp vitality.

Laser light can be directed to the tissue surface via an optical fiber or as a light beam. If an optical fiber-based delivery system is used then the optical fiber can terminate as a probe, which can be placed directly on the tissue surface. One or more light-collecting optical fibers also terminate in the probe head (all-in-one bundle probe). In this case, the collecting optical fibers would transmit a proportion of the scattered light to a photodetector and the signal/image processing systems. Generally, this system will measure blood flow in a tissue volume of typically $1\ mm^3$ or smaller. The measured blood flow change in a small tissue volume is generally considered to be a representation of the larger volume. Optical probe stability on the tooth surface is very important during laser Doppler measurements. Comparison of a flux trace measured from the contralateral healthy tooth often aids the diagnosis of vitality. Improved signal processing techniques permit diagnosis with a sensitivity and specificity of over 90%.

The advantages of LDF are that (1) it is an objective measurement of the pulpal vitality, (2) in comparison with other vitality tests, it does not rely on the sensation of pain to determine the vitality of a tooth, and (3) this method can be useful in patients, especially young children, who have difficulty in communicating and teeth that have experienced recent trauma. The disadvantages of LDF are that (1) it is difficult to calibrate the measurements in absolute units and their output may not be linearly related to blood flow, (2) the anterior teeth, in which the enamel and dentin are thin, do not present a problem, but molar teeth, with their thicker enamel and dentin and the variability in the position of the pulp within the tooth, may cause deviations in pulpal blood flow and blood flow measurements, (3) differences in the sensor output and inadequate calibration by the manufacturer may indicate the use of multiple probes for accurate measurement, and (4) the system is expensive [44, 45].

72.7.2
Laser-Assisted Pulp Capping and Pulpotomy

Pulp capping is defined as a procedure in which a biomaterial is placed over an exposed or nearly exposed pulp tissue to encourage the formation of an irritational dentin bridge at the location of exposure. Pulpotomy also involves the surgical removal of a small portion of vital pulp tissue as a means of preserving the remaining coronal and radicular pulp tissues. Pulp capping is indicated when the pulp exposure occurs during the removal of healthy dentin tissue and when the pulp exposure size is very small (≤ 1.0 mm). Pulp capping is recommended only on uninflamed pulp tissue since the pulp tissue should possess the capacity to form tertiary dentin.

Pulpotomy is indicated when the young pulp is exposed to caries and the root formation is not complete. Restorative materials such as calcium hydroxide [$Ca(OH)_2$], bioactive molecules, or mineral trioxide aggregate (MTA) is used during pulp capping to encourage the formation of reparative dentin. Traditionally, $Ca(OH)_2$ is used widely as a pulp capping agent. When $Ca(OH)_2$ is applied to pulp tissue, a necrotic layer is formed subjacent to which a dentin bridge is expected to form. The disadvantage of this process is that the necrotic zone may allow bacterial growth if the restoration leaks. MTA shows favorable results when applied to exposed pulp. It produces more dentinal bridge in a shorter time with significantly less inflammation than $Ca(OH)_2$. However, a long setting time (3–4 h) is required for the complete setting of MTA [46, 47].

CO_2, Nd:YAG, and Er:YAG lasers have been used to manage exposed pulp tissue. The most important effect of CO_2 laser irradiation is the ability to sterilize and form scar tissue in the irradiated area due to thermal effects, which in turn would aid in protecting the pulp from bacterial invasion. A bloodless field would be easier to achieve owing to the ability of the laser to vaporize tissue, coagulate, and seal small blood vessels. Further, laser treatment may directly stimulate dentin formation [48]. The CO_2 laser was observed to produce new mineralized dentin without cellular modification of pulp tissue when tooth cavities were irradiated in beagles and primates [49, 50]. Jerkit *et al.* [51] studied the histological response of the direct pulp capping after laser irradiation for pulpotomy in the premolars and molars of dogs. An energy level of 1 W with a 0.1 s exposure time and 1 s pulse intervals was applied until the exposed pulp area was completely irradiated. They were then dressed with $Ca(OH)_2$. This study reported more predictable results (90%) with pulp capping performed using laser of different wavelengths than traditional procedures that report a success rate of ~60%. The long intervals (1 s) between the pulses (0.1 s), the relatively low power setting (1 W), and the wavelength (10.6 μm), which is absorbed within 100 μm by water, seem to be sufficient to avoid any thermal damage to the pulp. The final restoration of the cavity is done after 6 months. The high success rate is thought to be due to control of hemorrhage, disinfection, sterilization, carbonization, and stimulation effects on the dental pulp cells. It causes scar tissue formation in the irradiated area due to thermal effects, which may help to preserve the pulp from bacterial invasion. In addition, the laser minimizes the formation of hematoma between the pulp tissue and the $Ca(OH)_2$ dressing, allowing close contact between the dressing and the exposed pulp.

The Nd:YAG laser has been applied in laser-assisted pulp capping procedures. A study showed that the success rate after direct pulp capping with an Nd:YAG laser and Vitrebond are significantly higher than in conventional $Ca(OH)_2$ treatment [52]. However, another study revealed that Nd:YAG and CO_2 laser irradiation of exposed pulp tissue caused carbonization, necrosis, infiltration of inflammatory cells, edema, and hemorrhage in the pulp tissue [53]. Under the conditions of this experiment, there was little histologic evidence of repair on treated pulp, which was in contrast to the control samples treated with $Ca(OH)_2$-containing cement. Studies have also examined the application of argon, semiconductor diode and Er:YAG lasers for direct pulp capping. Interestingly, more dentin formation and better healing capacity were observed in the Er:YAG laser group than in the conventional control treatment group. Nevertheless, further investigations were suggested to study the effect of the blood extravasation, which appeared near the Er:YAG laser exposure sites [54].

72.7.3
Laser-Assisted Root Canal Disinfection and Shaping

Most current lasers, such as Nd:YAG, diode, and Er:YAG lasers, have bactericidal effects *in vivo* and *in vitro*. Laser radiation produces a bactericidal effect by causing alterations to the bacterial cell wall. Delivery of laser energy through an extremely thin, flexible fiber optic system is important in endodontics. However, the bactericidal effect of laser radiation deep within the dentin differs because of the different absorption of the different wavelengths of lasers. The disadvantage of endodontic irrigants (disinfectants) is that their bactericidal effect is limited to the main root canal lumen. The penetration depth of chemical disinfectants into the dentinal tubules is suggested to be 100 µm [55]. However, laser light is shown to penetrate more than 1000 µm into the dentin. Although different studies have highlighted that the energy from a laser declines as it penetrates into the dentin, the bactericidal effect is found to be effective even to a depth of 1000 µm or more [56]. It was noted that Gram-negative bacteria showed a higher resistance against laser irradiation than Gram-positive bacteria. This higher resistance of Gram-negative bacteria was attributed to their cell wall characteristics.

The laser is an effective tool for killing microorganisms because of the laser energy and wavelength characteristics. Nd:YAG, semiconductor diode, and Er:YAG lasers have commonly been used for endodontic disinfection. However, complete disinfection of the root canal system is very difficult. Additionally, a smear layer is always formed on the instrumented root canal wall. The smear layer, which forms a superficial layer on the surface of the root canal wall, is ∼1–2 µm thick. A deeper smear layer packed into the dentinal tubules is also observed to a depth of up to 40 µm (smear plug). The smear layer contains inorganic and organic substances that include microorganisms and necrotic pulp debris. In addition to the possibility that the smear layer itself may be infected, it also can protect the bacteria already present in the dentinal tubules. A smear layer containing bacteria or bacterial products might provide or serve as a reservoir of irritants. During endodontic disinfection, it is vital not only to remove microorganisms from the root canals, but also remove the smear layer formed on the root canal wall and dentinal tubules. In most cases, the effect is directly related to the amount of irradiation and to its energy level. It has also been documented in numerous studies that CO_2 [57], Nd:YAG [58], argon [59], Er,Cr:YSGG [60], and Er:YAG [61] laser irradiation has the ability to remove debris and the smear layer from the root canal walls following instrumentation. KTP laser irradiation was able to remove smear layer and debris from the root canals. At specific fluences, the XeCl laser (wavelength of 308 nm) can melt dentin and seal exposed dentinal tubules [63]. The ArF excimer laser emitting at 193 nm caused significant removal of peritubular dentin at relatively high fluence (10–15 J cm^{-2}) [64]. Inspite of this tremendous interest of using laser in endodontics, studies have stressed the possible limitations of the use of lasers in the root canal system.

The task of cleaning and disinfecting a root canal system that contain microorganisms gathered in a biofilm becomes very difficult. Certain bacterial species become more virulent when harbored in a biofilm demonstrating stronger pathogenic potential and increased resistance to antimicrobial agents as biofilms have the ability to prevent the entry and action of such agents. To increase the effect of disinfection of the infected root canal, black Indian ink or 38% silver ammonium solution was placed in the root canal before irradiating with pulsed Nd:YAG. Rooney *et al.* [65] reported disinfection rates of 80% to 90% with Nd:YAG, whereas others have reported rates of 60%, depending on the condition of the root canals, the type of laser device, the application parameters and the techniques. Bergmans *et al.* [66] tried to define the role of the laser as a disinfecting tool by using Nd:YAG laser irradiation on some endodontic pathogens *ex vivo*. They concluded that Nd:YAG laser irradiation is not an alternative but a possible supplement to existing protocols for canal disinfection, as the properties of laser light may allow a bactericidal effect beyond 1 mm of dentin. Endodontic pathogens that grow as biofilms are difficult to eradicate even upon direct laser exposure.

There are several limitations that may be associated with the intracanal use of lasers that cannot be overlooked. (1) The emission of laser energy from the tip of the optical fiber or the laser guide is directed along the root canal and not necessary laterally to the root canal walls. Thus it is almost impossible to obtain uniform coverage of the canal surface using a laser [67, 68]. (2) The safety of such a procedure is another limitation because potential thermal damage to the periapical tissues is possible. (3) Direct emission of laser irradiation from the tip of the optical fiber in the vicinity of the apical foramen of a tooth may result in transmission of the irradiation beyond the foramen. This transmission of irradiation, in turn, may adversely affect the supporting periradicular tissues of the tooth and can be hazardous in teeth with close proximity to the mental foramen or to the mandibular nerve [68].

The Er:YAG laser is mostly applied for root canal preparation. It has the ability to remove smear layer and open the dentinal tubules. Harashima *et al.* [58, 60] compared 17% EDTA with 6% phosphoric acid and Er:YAG laser energy. The results showed that the Er:YAG laser was the most effective at removal of the smear layer from the root canal walls. When Er:YAG lasers were compared with other lasers, notably argon and Nd:YAG lasers, the results showed that the Er:YAG laser had the most effective wavelength and was more effective than 17% EDTA in removal of the smear layer from the root canal walls. It is suggested that the removal of the smear layer and debris by lasers is possible; however, it is difficult to clean all root canal walls. Therefore, further research is necessary to improve the endodontic tip to permit irradiation of the entire root canal walls. A modified beam delivery system has been tried for the Er:YAG laser. This system consists of a hollow tube allowing lateral emission of the radiation (side firing), rather than direct emission through a single opening at its terminal end [69].

This new endodontic side-firing spiral tip (RCLase, Opus Dent, Tel Aviv, Israel) was designed to fit the shape and volume of root canals prepared by nickel–titanium rotary instrumentation. It emits the Er:YAG laser irradiation laterally to the walls of the root canal through a spiral slit located all along the tip. The tip is sealed at its far end, preventing the transmission of irradiation to and through the apical foramen of the tooth. The prototype of the RCLase side-firing spiral tip is shown in the root canal of an extracted maxillary canine in which the side wall of the root was removed to allow visualization of the tip. Studies have shown that bacteria and their byproducts, present in infected root canals, may invade the dentinal tubules. The presence of bacteria in the dentinal tubules of infected teeth at approximately half the distance between the root canal walls and the cementodentinal junction was also reported. These findings justify the rationale and need for developing effective means of removing the smear layer from root canal walls following biomechanical instrumentation. This removal would allow disinfectants and laser irradiation to reach and destroy microorganisms in the dentinal tubules.

72.7.4
Laser-Assisted Root Canal Obturation

The thermoplastic properties of gutta-percha are utilized during laser-assisted root canal obturation. The laser energy is used to melt the gutta-percha and the heat-softened gutta-percha is vertically compacted into the root canals. The first laser-assisted root canal filling procedure involved using the argon 488 nm laser. This wavelength, which can be transmitted through dentin, was used to polymerize a resin that was placed in the main root canal. The ability of this obturating material to penetrate into the accessory root canals was tested and it was shown that the resin in the lateral canals was readily polymerized at a low energy level (30 mW). Argon, CO_2, and Nd:YAG lasers have been used to soften gutta-percha [70], and the results indicate that the argon laser can be used for this purpose to produce a good apical seal. The photopolymerization of camphorquinone-activated resins for obturation is possible using an argon laser (477 and 488 nm) [71]. The results indicate that an argon laser coupled to an optical fiber could become a useful modality in endodontic therapy. Er:YAG lasers have shown a remarkable ability to enhance the results of the obturation process. Application of the Er:YAG laser energy to the root canal wall has been shown to increase the adhesion of epoxy-based root canal sealers (AH-26, AH plus, Topseal, Sealer 26, and Sealer Plus) to the canal wall [72].

Studies have shown that laser irradiation using Nd:YAG (100 mJ per pulse, 1 W, 10 Hz) and Er:YAG (170–250 mJ, 2 Hz) lasers improves the apical seal by hindering apical leakage [61, 73]. Since the Nd:YAG laser is very well absorbed by black color, an absorbent paper point soaked with black ink was introduced to the working length and the apical root canal was painted [74]. Maden *et al.* [75] used the dye penetration method to measure apical leakage by comparing lateral condensation, System B technique, and Nd:YAG-softened gutta-percha. No statistically significant differences were reported. In another study, Anić and Matsumoto [70] demonstrated that the temperature elevation induced on the outer root surface when Nd:YAG and argon lasers were used ranged from 12.9 °C (argon laser) to 14.4 °C (Nd:YAG laser). Such an increase in temperature may be detrimental to the tissues of the attachment apparatus of the teeth. The implication of such methodologies remains questionable. Experiments also evaluated the degree of apical leakage *in vitro* after root canal preparation using Er:YAG laser irradiation. Er:YAG laser irradiation at parameters of 2 Hz and 170–230 mJ per pulse was employed. After obturation, the teeth were immersed in a vacuum flask containing 0.6% rhodamine for 48 h, longitudinally bisected, and observed by stereoscopy and scanning electron microscopy (SEM). It was observed that the degree of apical leakage from the teeth prepared by the laser was not significantly less than that from control teeth. Morphologic findings showed that contact between the root canal walls and obturated materials was hermetic in both groups, but canal walls prepared by laser irradiation were rough and irregular. These results show that root canal preparation by laser irradiation does not affect apical leakage after obturation compared with leakage in canals prepared using the conventional obturation methods [74].

Eriksson and Albrektsson [76] found that the threshold level for bone survival was 47 °C for 1 min. The effects of Nd:YAG laser irradiation on periradicular tissues was examined using a mongrel dog model. The results showed that Nd:YAG laser-treated teeth exhibited ankylosis, cemental lysis, and major bone remodeling 30 days after treatment. The parameters used in this study (3 W and 25 pps for 30 s) were excessive. Since that time, many other studies on periodontal effects of lasers in dogs and rats have been published [77]. According to Kimura *et al.*, the effect on the periodontal ligament when using Er:YAG laser energy is minimal, and no discernible effects on the periodontal ligament were noted [78]. They suggested that the Er:YAG laser can be used for root canal preparation if appropriate parameters are selected.

72.7.5
Laser-Assisted Root Canal Retreatment

The rationale for using laser irradiation in nonsurgical retreatment may be ascribed to the need to remove foreign material from the root canal that may be otherwise difficult to remove using conventional methods. However, studies conducted in this line to evaluate the efficacy of Nd:YAP laser radiation to remove gutta-percha, zinc oxide–eugenol sealer, silver cones, and broken instruments noted that laser radiation alone would not completely remove debris and obturating materials from the root canal. When used at 200 mJ with a pulse duration of 150 ms, an exposure time of 1 s, and a frequency of 10 Hz, the Nd:YAP laser preserved the dentinal walls of the root canal and permitted root canal retreatment without thermal damage to the periodontal tissue. It was concluded that, in combination with hand instruments, the Nd:YAP laser is an effective device for root canal preparation in endodontic retreatment [79]. In another study, an Nd:YAG laser was used at three power outputs (1, 2, and 3 W) to remove gutta-percha and broken files from the root canal. It was suggested that they were able to remove the obturating materials in more than 70% of the samples, whereas broken files were removed in 55% of the samples [80]. In straight root canals, laser irradiation at appropriate parameters was useful in removing root canal filling materials. The time required to remove any root canal filling material using laser ablation was shorter than that required for conventional methods. However, following laser irradiation, some orifices of the dentinal tubules were noted to be blocked following the melting or ledging of the root canal dentin.

72.7.6
Laser-Assisted Apical Surgery

Endodontic surgery (apicoectomy) is indicated when teeth do not respond to conventional nonsurgical endodontic treatment or when they cannot be treated appropriately by nonsurgical methods. Persistence of irritants in the root canal system zinc oxide eugenol sealer has been observed to be the cause of inflammation of the periapical tissue. The CO_2 laser was first used to irradiate the apex of a tooth during apicoectomy. The advantages of using a laser for this application are (1) improved hemostasis and concurrent visualization of the operating field, (2) reduction in the permeability of the lased dentin, (3) potential disinfection of the contaminated root apex, and (4) ability to achieve recrystallization of the apical root dentin which appeared smooth and ideal for the placement of apical filling material. However, results from an *in vivo* study on dogs showed that the success rate following apicoectomy using the CO_2 laser was not superior and failed to support the use of the CO_2 laser [81]. In another prospective clinical study of two apical preparations with and without a CO_2 laser, in which 320 cases were evaluated, the results did not show that a CO_2 laser improved the healing process [82]. *In vitro* studies with Nd:YAG lasers have shown a reduction in penetration of dye or bacteria through resected roots [83, 84, 85]. It was suggested that the reduced permeability in the lased specimens was probably the result of structural changes in the dentin following laser application [85]. Although SEM examination showed melting, solidification, and recrystallization of hard tissue, the structural changes were not uniform and the melted areas appeared connected by areas that looked like those in the non-lased specimens. It was postulated that this was the reason why the permeability of the dentin was reduced but not completely eliminated.

The Er:YAG laser has been applied for apical cavity preparation of extracted teeth [86]. No significant difference in dye penetration was found between the laser-treated groups and those in which ultrasonic tips were used. This finding was attributed to the inability of the Er:YAG laser to melt or seal dentinal tubules. Further, it has been reported that when using an Er:YAG laser with a low output power in apical surgery, it was possible to resect the apex of extracted teeth. Smooth and clean resected surfaces devoid of charring were observed [87, 88]. It was suggested that although the cutting speed of the Er:YAG laser was slightly slower than that of a conventional high-speed bur, the absence of vibration and discomfort, the smaller chance of contamination of the surgical site, and reduced risk of trauma to adjacent tissues may compensate for the additional time [89]. In a 3 year clinical study [90], a new protocol for lasers in apical surgery was reported. An Er:YAG laser was used for osteotomy and root resection, whereas Nd:YAG laser irradiation served to seal the dentinal tubules to reduce possible bacterial contamination of the surgical cavity. Improvement in healing was achieved with the use of a LILT GaAlAs diode laser. It has been suggested that after the appropriate wavelength to melt the hard tissues of the tooth has been established, the main contribution of laser technology to surgical endodontics is to convert the apical dentin and cementum structure into a uniformly glazed area that does not allow egress of microorganisms through the dentinal tubules or other openings in the apex of the tooth. Hemostasis and sterilization of the contaminated root apex will have an additional significance. Further, laser treatment in endodontics appears to potentiate spreading of bacterial contamination from the root canal to the patient and the clinicians via the laser plume produced by the laser. This may lead to bacterial dissemination unless precautions are taken to protect against spreading infections while using lasers.

72.8
Lasers in Periodontal Therapy

The applications of lasers in periodontics can be categorized as follows:

1) periodontal soft tissue management
2) removal of periodontal pocket lining
3) disinfection of periodontal pocket
4) removal of calculus, root surface planning
5) periodontal hard tissue management.

The advantages of applying lasers in periodontal treatment are (a) effective and efficient soft and hard tissue ablation with a greater degree of hemostasis, (b) their bactericidal effect, (c) minimal wound contraction, (d) minimal collateral damage with reduced use of local analgesia, and (e) better patient compliance. The disadvantages are that (a) if appropriate power settings are not used, the irradiation of root surfaces can cause detrimental effects, (b) precautions need to be taken during clinical application, (c) laser-based treatment is not very cost-effective, (d) the size of the laser device is cumbersome in the setup involved, (e) application of lasers requires trained personnel, and (f) there is still a need for more sound evidence-based clinical studies.

72.8.1
Laser-Assisted Periodontal Soft Tissue Management

Lasers are mostly used to prepare soft incisions in periodontics. Three methods are commonly used to prepare soft tissue incisions in dentistry: scalpel, electrosurgery, and laser. Each of these methods is effective; however, they differ with respect to hemostasis, healing time, cost of instruments, width of the cut, anesthetic requirement, and disagreeable characteristics such as production of smoke, odor, and undesirable taste. Many reports have confirmed the safety and efficacy of CO_2 and Nd:YAG lasers, commonly for soft tissue application in periodontics [91]. These lasers can be used for procedures such as frenectomy, gingivectomy, and gingivoplasty, de-epithelization of reflected periodontal flaps, removal of granulation tissue, second-stage exposure of dental implants, lesion ablation, incisional and excisional biopsies of both benign and malignant lesions, coagulation of free gingival graft donor sites, and gingival depigmentation.

The use of surgical lasers in periodontology has been explored in three areas of treatment: (1) removal of diseased pocket lining epithelium, (2) bactericidal effect of lasers on pocket organisms, and (3) removal of calculus deposits and root surface disinfection. Current dental lasers such as diode laser, Nd:YAG, Er,Cr:YSGG, Er:YAG, and CO_2 lasers are advocated for the treatment of sulcular debridement, elimination of bacteria, and removal of calculus. In addition to the current wavelengths, the recently developed frequency-doubled Nd:YAG laser at 532 nm, termed the KTP laser, which has a range of actions similar to the diode laser, has been used for periodontal therapy. It is important to realize that many local and systemic factors influence the treatment outcome in the management of periodontal diseases. The application of most laser delivery systems depend on an axial, end-on emission of light energy, which predisposes the target tissue to a potential build-up of direct and conductive heat effects. Therefore, the applied laser power has to be as low as possible to obtain the desired effect without any unwanted interaction with the tooth or the supporting tissues. In view of the above, blind treatment procedures that lack tactile feedback must be carried out with great caution.

72.8.2
Laser-Assisted Management of Periodontal Pocket

The development of the quartz optic fiber delivery system associated with the diode and Nd:YAG group of lasers, with diameters of 200–320 μm, makes access into the periodontal pocket easy. Longer wavelengths usually rely on fine-bore waveguide probes and sapphire handpiece tips, which are designed for treatment purposes. Following the removal of all hard and soft tissue deposits, the pocket depth is reassessed. The laser fiber is measured to a distance of 1–2 mm short of the pocket depth and inserted at an angle to maintain contact with the soft tissue walls at all times. Using power values sufficient to ablate the epithelial lining ($\sim$0.8 W cw diode, 100 mJ/20 pps, 2.0 W Nd:YAG, and Er:YAG/YSGG, 1.0 W cw CO_2), the laser probe is used in a light contact, sweeping mode to cover the entire soft tissue lining. Ablation should commence near the base of the pocket and proceed upwards by slowly removing the probe. Some bleeding of the pocket site may occur due to the disruption of the fragile inflamed pocket epithelium. Each pocket site should be treated for 20–30 s, amounting to possibly 2 min per tooth site with retreatment at approximately weekly intervals during any 4 week period. If the pocket is infra-bony, a number of procedures have been advocated, including laser-ENAP (excisional new attachment procedure), where the Nd:YAG (1064 nm) laser is used in a non-flap procedure to reduce the pocket depth by several millimeters, through a succession of treatment appointments.

72.8.3
Bactericidal Effect of Lasers on Periodontal Pocket

Bacteria have been implicated as one of the major causative factors in periodontal disease. Many studies have demonstrated the ability of laser energy to disinfect periodontal pockets [92–94]. The additional role of lasers in disinfecting infected tissue when used in conjunction with scaling and root planning was also noted [93]. Studies have addressed some of the difficulties of using wavelengths in the range 810–830 nm in the periodontal pocket. The build-up of char and denatured protein material on the delivery fiber of the diode laser results in the development of a carbonized tip, with the temperature rising to $>700\,°C$. If it is not removed, it can lead to attenuation of the subsequent laser beam, replaced by the secondary emission of radiant thermal energy from the carbonized deposits (the "hot-tip effect"). The conductive heat would lead to unwanted damage to the delicate tissues [94].

72.8.4
Laser-Assisted Calculus Removal

Er:YAG and Er,Cr:YAG lasers together with innovative near-ultraviolet wavelengths such as frequency-doubled alexandrite (wavelength 377 nm) have been safely applied to remove calculus. Access to the calculus deposits was achieved by utilizing specific laser handpiece tips. The poorly calcified deposits together with a higher water content rendered supra- and sub-gingival calculus susceptible to defragmentation through photomechanical ablation with the erbium group [95]. Potentially this enables deposits to be removed using laser energy levels lower than those required to ablate dental hard tissues. In a study by Aoki *et al.* [96], laser power levels as low as 0.3 W were shown to be sufficient to ablate calculus.

72.8.5
Laser-Assisted Management of Oral Hard Tissues

Since laser–hard tissue interactions are photothermal events which are wavelength dependent, with the possible exception of two wavelengths (Er:YAG and Er,Cr:YSGG), the effect of most dental lasers on bone is generally detrimental. Bone tissue heated to $>47\,°C$ is known to undergo cellular damage, leading to osseous resorption, while heating bone tissue to $>60\,°C$ would result in tissue necrosis [76]. However, few investigations have examined bone surface temperatures while the overlying soft tissues are being irradiated by a dental laser. A study by Fontana *et al.* [97] examined the temperature increase at the bone surface while using an 810 nm diode laser applied within the periodontal pockets in rats. Following 9 s of irradiation using laser powers of 800 mW, 1.0 W, and 1.2 W delivered through a 300 μm optical fiber, they reported 10 and $11\,°C$ increases in bone surface temperature. Only at a 600 mW setting was the bone surface temperature below the threshold of inducing cellular damage. If the exposure time was shortened to 3 s, all power selections resulted in temperature increases that remained below the critical threshold. In another *in vitro* study [98] CO_2 and Nd:YAG lasers using comparable energy densities were compared for their effects on bone surface temperature while ablating overlying soft tissues. The results showed that the bone surface temperature increase ranged from 1.4 to 2.1 °C for the CO_2 laser and from 8.0 to 11.1 °C for the Nd:YAG laser. These results indicated that when ablating relatively thin, soft tissues supported by subjacent bone, the Nd:YAG laser should be used at lower energy densities for short intervals; otherwise there is a risk of irreversible bone damage. Severe secondary tissue damage has been identified as a major factor in delayed healing of laser-induced bone incisions. Studies have been carried out to compare the osteotomies created by the Er:YAG laser with those created by rotary burs and the CO_2 laser [99, 100]. Overall, the two studies indicated that the Er:YAG laser when used at a peak pulse energy of 100 mJ per pulse and 10 Hz produced well-defined intra-bony incisions without any evidence of melting and carbonization, whereas CO_2 laser-induced osteotomies exhibited extensive charring, melting of the mineral phase, and delayed healing.

72.8.6
Laser-Assisted Management of Dental Hard Tissue

Surface modifications of cementum and dentin have been studied using a variety of laser wavelengths, primarily CO_2, Nd:YAG, Er:YAG, and to a lesser extent diode lasers [101]. A major consideration is the selection of a wavelength that will effectively remove calculus while suppressing both thermal damage to the pulp tissue and undesired removal of sound root structure. The mineral phase of both dentin and cementum is a carbonated hydroxyapatite that has intense absorption bands in the mid-infrared region. Consequently, an Er:YAG laser would appear to be the optimum choice for the effective removal of calculus, for root etching, and for creating a biocompatible surface for cell or tissue reattachment [101, 102]. However, studies on the biocompatibility of CO_2-treated laser surfaces, even when applied at low energy densities, have yielded conflicting results. Some studies [103, 104] reported increased *in vitro* attachment of fibroblasts to laser-treated surfaces compared with controls or chemically treated surfaces. On the other hand, some studies [105, 106] reported a total lack of fibroblast attachment to irradiated surfaces. However, owing to the larger diameter of the hollow delivery tip that is required to transmit CO_2 laser radiation, it has only restricted application in subgingival periodontal therapy. Owing to its high absorption in both water and hydroxyapatite, the bulk of recent research concerning laser-induced root surface modification has involved the Er:YAG laser. The wavelength of this laser has been shown to remove effectively smear layers cementum, and cementum-bound endotoxins [107]. When used at low energy densities with a water spray surface coolant, the majority of studies have reported little or no heat-induced tissue damage and production of smooth root surfaces. In addition, *in vitro* fibroblast adhesion studies have shown that the resultant root surface appears to be at least as biocompatible as that produced by scaling and root planning [108].

72.9
Conclusion

The rapid maturity of laser systems, the availability of improved detection systems and sensitive delivery systems, and high information-processing capabilities of present-day computers are facilitating the use of lasers in noninvasive and sensitive diagnostic and selective therapeutic applications. With the continued developments in the field of lasers and the associated technologies, the application of lasers in dentistry is expected to grow even more rapidly in the coming years.

References

1 Coluzzi, D.J. and Convissar, R.A. (2004) Lasers in clinical dentistry. *Dent. Clin. North Am.*, **48** (4), xi–xii.
2 Miserendino, L. and Robert, P.M. (1995) *Lasers in Dentistry*, Quintessence Publishing, Hanover Park, IL, Q18 3.
3 Brugnera, A. Jr., Garrini dos Santos, A.E.C., Bologna, E.D., and Pinheiro Ladalardo, T.C.C.G. *Atlas of Laser Therapy Applied to Clinical Dentistry*, Quintessence Publishing, Hanover Park, IL.
4 Oho, T. and Morioka, T. (1990) A possible mechanism of acquired acid resistance of human dental enamel by laser irradiation. *Caries Res.*, **24** (2), 86–92.
5 Stern, R.H., Vahl, J., and Sognnaes, R.F. (1972) Lased enamel: ultrastructural observations of pulsed carbon dioxide laser effects. *J. Dent. Res.*, **51** (2), 455–460.
6 Kantola, S. (1973) Laser-induced effects on tooth structure. VII. X-ray diffraction study of dentine exposed to a CO2 laser. *Acta Odontol. Scand.*, **31** (6), 381–386.
7 Nammour, S., Renneboog-Squilbin, C., and Nyssen-Behets, C. (1992) Increased resistance to artificial caries-like lesions in dentin treated with CO2 laser. *Caries Res.*, **26** (3), 170–175.
8 Cooper, L.F., Myers, M.L., Nelson, D.G., and Mowery, A.S. (1988) Shear strength of composite bonded to laser-pretreated dentin. *J. Prosthet. Dent.*, **60** (1), 45–49.
9 Keller, U. and Hibst, R. (1989) Experimental studies of the application of the Er:YAG laser on dental hard substances: II. Light microscopic and SEM investigations. *Lasers Surg. Med.*, **9** (4), 345–351.
10 Anic, I., Pavelic, B., Peric, B., and Matsumoto, K. (1996) In vitro pulp chamber temperature rises associated with the argon laser polymerization of composite resin. *Lasers Surg. Med.*, **19** (4), 438–444.
11 Baik, J.W., Rueggeberg, F.A., and Liewehr, F.R. (2001) Effect of light-enhanced bleaching on in vitro surface and intrapulpal temperature rise. *J. Esthet. Restor. Dent.*, **13** (6), 370–378.
12 Sulieman, M., Addy, M., and Rees, J.S. (2005) Surface and intra-pulpal temperature rises during tooth bleaching: an in vitro study. *Br. Dent. J.*, **199** (1), 37–40.
13 Yazici, A.R., Muftu, A., and Kugel, G. (2007) Temperature rise produced by different light-curing units through dentin. *J. Contemp. Dent. Pract.*, **8** (7), 21–28.
14 Strobl, A., Gutknecht, N., Franzen, R., Hilgers, R.D., Lampert, F., and Meister, J. (2010) Laser-assisted in-office bleaching using a neodymium:yttrium-aluminum garnet laser: an in vivo study. *Lasers Med. Sci.*, **25** (4), 503–509.
15 ADA Council on Scientific Affairs (1998) Laser-assisted bleaching: an update. *J. Am. Dent. Assoc.*, **129** (10), 1484–1487.
16 Moritz, A. (2006) *Oral Laser Application*, Quintessence Publications, Berlin.
17 Wigdor, H.A., Walsh, J.T., Jr. Featherstone, J.D., Visuri, S.R., Fried, D., and Waldvogel, J.L. (1995) Lasers in

dentistry. *Lasers Surg. Med.*, **16** (2), 103–133.
18 Azevedo Rodrigues, L., Nobre dos Santos, M., Pereira, D., Videira Assaf, A., and Pardi, V. (2004) Carbon dioxide laser in dental caries prevention. *J. Dent.*, **32** (7), 531–540.
19 Westerman, G.H., Hicks, M.J., Flaitz, C.M., Blankenau, R.J., Powell, G.L., and Berg, J.H. (1994) Argon laser irradiation in root surface caries: in vitro study examines laserís effects. *J. Am. Dent. Assoc.*, **125** (4), 401–407.
20 Hicks, M.J., Flaitz, C.M., Westerman, G.H., Berg, J.H., Blankenau, R.L., and Powell, G.L. (1993) Caries-like lesion initiation and progression in sound enamel following argon laser irradiation: an in vitro study. *ASDC J. Dent. Child.*, **60** (3), 201–206.
21 Oho, T. and Morioka, T. (1990) A possible mechanism of acquired acid resistance of human dental enamel by laser irradiation. *Caries Res.*, **24** (2), 86–92.
22 Silverstone, L.M., Hicks, M.J., and Featherstone, M.J. (1988) Dynamic factors affecting lesion initiation and progression in human dental enamel. Part I. The dynamic nature of enamel caries. *Quintessence Int.*, **19** (10), 683–711.
23 Silverstone, L.M., Hicks, M.J., and Featherstone, M.J. (1988) Dynamic factors affecting lesion initiation and progression in human dental enamel. II. Surface morphology of sound enamel and caries like lesions of enamel. *Quintessence Int.*, **19** (11), 773–785.
24 Liu, Y. and Hsu, C. (2007) Laser-induced compositional changes on enamel: a FT-Raman study. *J. Dent.*, **35** (3), 226–230.
25 Tagomori, S. and Morioka, T. (1989) Combined effects of laser and fluoride on acid resistance of human dental enamel. *Caries Res.*, **23** (4), 225–231.
26 Walsh, L.J., Abood, D., and Brockhurst, P.J. (1994) Bonding of resin composite to carbon dioxide laser-modified human enamel. *Dent. Mater.*, **10** (3), 162–166.
27 Usumez, A. and Aykent, F. (2003) Bond strengths of porcelain laminate veneers to tooth surfaces prepared with acid and Er, Cr:YSGG laser etching. *J. Prosthet. Dent.*, **90** (1), 24–30.
28 Martınez-Insua, A., Da Silva Dominguez, L., Rivera, F.G., and Santana- Penın, U.A. (2000) Differences in bonding to acid-etched or Er:YAG-laser-treated enamel and dentin surfaces. *J. Prosthet. Dent.*, **84** (3), 280–288.
29 Usumez, A. and Aykent, F. (2003) Bond strengths of porcelain laminate veneers to tooth surfaces prepared with acid and Er,Cr:YSGG laser etching. *J. Prosthet. Dent.*, **90** (1), 24–30.
30 Ramos Lloret, P., Lacalle Turbino, M., Kawano, Y., Sanchez Aguilera, F., Osorio, R., and Toledano, M. (2008) Flexural properties, microleakage, and degree of conversion of a resin polymerized with conventional light and argon laser. *Quintessence Int.*, **39** (7), 581–586.
31 Aguiar, F.H., Barros, G.K., Lima, D.A., Ambrosano, G.M., and Lovadino, J.R. *Biomed. Mater.*, **1** (3), 140–143.
32 Hibst, R. and Keller, U. (1991) Removal of dental filling materials by Er: YAG laser radiation. *Proc. SPIE*, **1424**, 120–126.
33 Takashina, M., Ebihara, A., Sunakawa, M., Anjo, T., Takeda, A., and Suda, H. (2002) The possibility of dowel removal by pulsed Nd:YAG laser irradiation. *Lasers Surg. Med.*, **31** (4), 268–274.
34 Orchardson, R. and Collins, W.J. (1987) Clinical features of hypersensitive teeth. *Br. Dent. J.*, **162** (7), 253–256.
35 Brannstrom, M., Lind_en, L.A., and Astrom, A. (1967) The hydrodynamics of the dental tubule and of pulp fluid. A discussion of its significance in relation to dentinal sensitivity. *Caries Res.*, **1** (4), 310–317.
36 Orchardson, R. and Gillam, D.G. (2006) Managing dentin hypersensitivity. *J. Am. Dent. Assoc.*, **137** (7), 990–998.
37 Kantola, S. (1972) Laser-induced effects on tooth structure. IV. A study of changes in the calcium and phosphorus contents in dentine by electron probe microanalysis. *Acta Odontol. Scand.*, **30** (4), 463–474.
38 Bonin, P., Boivin, R., and Poulard, J. (1991) Dentinal permeability of the dog canine after exposure of a cervical cavity to the beam of a CO_2 laser. *J. Endod.*, **17** (3), 116–118.

39 Liu, H.C., Lin, C.P., and Lan, W.H. (1997) Sealing depth of Nd:YAG laser on human dentinal tubules. *J. Endod.*, **23** (11), 691–693.

40 Renton-Harper, P. and Midda, M. (1992) Nd:YAG laser treatment of dentinal hypersensitivity. *Br. Dent. J.*, **172** (1), 13–16.

41 Cakar, G., Kuru, B., Ipci, S.D., Aksoy, Z.M., Okar, I., and Yilmaz, S. (2008) Effect of Er:YAG and CO_2 lasers with and without sodium fluoride gel on dentinal tubules: a scanning electron microscope examination photomedicine and laser surgery. *Photomed. Laser Surg.*, **26** (6), 565–571.

42 Ipci, S.D., Cakar, G., Kuru, B., and Yilmaz, S. (2009) Clinical evaluation of lasers and sodium fluoride gel in the treatment of dentine hypersensitivity. *Photomed Laser Surg.*, **27** (1), 85–91.

43 Andreasen, J.O., Andreasen, F.M., and Andersson L. (eds) (2007) *Textbook and Color Atlas of Traumatic Injuries of the Teeth*, 4th edn, Blackwell Munksgaard, Copenhagen.

44 Stabholz, A., Sahar-Helft, S., and Moshonov, J. (2004) Lasers in endodontics. *Dent. Clin. North Am.*, **48** (4), 809–832.

45 Matsumoto, K. (2000) Lasers in endodontics. *Dent. Clin. North Am.*, **44** (4), 889–906.

46 Torabinejad, M. and Chivian, N. (1999) Clinical applications of mineral trioxide aggregate. *J. Endod.*, **25** (3), 197–205.

47 Stark, M.M., Myers, H.M., Morris, M., and Gardiner, R. (1964) The localization of radioactive calcium hydroxide Ca45 over exposed pulps in rhesus monkey teeth: a preliminary report. *J. Oral Ther. Pharmacol.*, **54**, 290–297.

48 Paschoud, Y. and Holz, J. (1988) Effect of the soft laser on the neoformation of a dentin bridge following direct pulp capping of human teeth with calcium hydroxide. I. Histological study with the scanning electron microscope. *Schweiz. Monatsschr. Zahnmed.*, **98** (4), 345–356.

49 Melcer, J., Chaumette, M.T., and Melcer, F. (1987) Dental pulp exposed to the CO_2 laser beam. *Lasers Surg. Med.*, **7** (4), 347–352.

50 Santucci, P.J. (1999) Dycal versus Nd:YAG laser and Vitrebond for direct pulp capping in permanent teeth. *J. Clin. Laser Med. Surg.*, **17** (2), 69–75. NdYAG laser treatment of dentinal hypersensitivity.

51 Jukic, S., Anic, I., Koba, K., Najzar-Fleger, D., and Matsumoto, K. (1997) The effect of pulpotomy using CO_2 and Nd:YAG lasers on dental pulp tissue. *Int. Endod. J.*, **30** (3), 175–180.

52 Olivi, G., Genovese, MD., Maturo, P., and Docimo, R. (2007) Pulp capping: advantages of using laser technology. *Eur. J. Paediatr. Dent.*, **8** (2), 89–95.

53 Suzuki, M., Ogisu, T., Kato, C., Shinkai, K., and Katoh, Y. (2011) Effect of CO_2 laser irradiation on wound healing of exposed rat pulp. *Odontology.*, **99** (1), 34–44.

54 Ipci, S.D., Cakar, G., Kuru, B., and Yilmaz, S. (2009) Clinical evaluation of lasers and sodium fluoride gel in the treatment of dentine hypersensitivity. *Photomed Laser Surg.*, **27** (1), 85–91.

55 Berutti, E., Marini, R., and Angeretti, A. (1997) Penetration ability of different irrigants into dentinal tubules. *J. Endod.*, **23** (12), 725–727.

56 Klinke, T., Klimm, W., and Gutknecht, N. (1997) Antibacterial effects of Nd:YAG laser irradiation within root canal dentin. *J. Clin. Laser Med. Surg.*, **15** (1), 29–31.

57 Onal, B., Ertl, T., Siebert, G., and Müller, G. (1993) Preliminary report on the application of pulsed CO_2 laser radiation on root canals with AgCl fibers: a scanning and transmission electron microscopic study. *J. Endod.*, **19** (6), 272–276.

58 Harashima, T., Takeda, F.H., Kimura, Y., and Matsumoto, K. (1997) Effect of Nd:YAG laser irradiation for removal of intracanal debris and smear layer in extracted human teeth. *J. Clin. Laser Med. Surg.*, **15** (3), 131–135.

59 Moshonov, J., Orstavik, D., Yamauchi, S., Pettiette, M., and Trope, M. (1995) Nd:YAG laser irradiation in root canal disinfection. *Endod. Dent. Traumatol.*, **11** (5), 220–224.

60 Harashima, T., Takeda, F.H., Zhang, C., Kimura, Y., and Matsumoto, K. (1998) Effect of argon laser irradiation on

instrumented root canal walls. *Endod Dent Traumatol.*, **14** (1), 26–30.

61 Biedma, B.M., Varela Patino, P., Park, S.A., Barciela Castro, N., Magan Munoz, F., Gonzalez Bahillo, J.D., and Cantatore, G. (2005) Comparative study of root canals instrumented manually and mechanically, with and without Er:YAG laser. *Photomed. Laser Surg.*, **23** (5), 465–469.

62 Kesler, G., Gal, R., Kesler, A., and Koren, R. (2002) Histological and scanning electron microscope examination of root canal after preparation with Er:YAG laser microprobe: a preliminary in vitro study. *J. Clin. Laser Med. Surg.*, **20** (5), 269–277.

63 Stabholz, A., Neev, J., Liaw, L.H., Stabholz, A., Khayat, A., and Torabinejad, M. (1993) Sealing of human dentinal tubules by XeCl 308-nm excimer laser. *J. Endod.*, **19** (6), 267–271.

64 Stabholz, A., Neev, J., Liaw, L.H., Stabholz, A., Khayat, A., and Torabinejad, M. (1993) Effect of ArF-193nm excimer laser on human dentinal tubules. A scanning electron microscopic study. *Oral Surg. Oral Med. Oral Pathol.*, **75** (1), 90–94.

65 Rooney, J., Midda, M., and Leeming, J. (1994) A laboratory investigation of the bactericidal effect of a Nd:YAG laser. *Br. Dent. J.*, **176** (2), 61–64.

66 Bergmans, L., Moisiadis, P., Teughels, W., Van Meerbeek, B., Quirynen, M., and Lambrechts, P. (2006) Bactericidal effect of Nd:YAG laser irradiation on some endodontic pathogens ex vivo. *Int. Endod. J.*, **39** (7), 547–557.

67 Goodis, H.E., Pashley, D., and Stabholz, A. (2002) Pulpal effects of thermal and mechanical irritant, in *Seltzer and Bender's Dental Pulp* (eds K.M. Hargreaves and H.E. Goodis), Quintessence Publishing, Hanover Park, IL, pp. 371–410.

68 Stabholz A., Zeltser, R., Sela, M., Peretz, B., Moshonov, J., Ziskind, D., and Stabholz, A. (2003) The use of lasers in dentistry: principles of operation and clinical applications. *Compend. Contin. Educ. Dent.*, **24** (12), 935–948.

69 Cozean, C., Arcoria, C.J., Pelagalli, J., and Powell, G.L. (1997) Dentistry for the 21st century? Erbium:YAG laser for teeth. *J. Am. Dent. Assoc.*, **128** (8), 1080–1087.

70 Anic, I. and Matsumoto, K. (1995) Dentinal heat transmission induced by a laser-softened gutta-percha obturation technique. *J. Endod.*, **21** (9), 470–474.

71 Potts, T.V. and Petrou, A. (1991) Argon laser initiated resin photopolymerization for the filling of root canals in human teeth. *Lasers Surg. Med.*, **11** (3), 257–262.

72 Pecora, J.D., Cussioli, A.L., Gueri, D.M., Marchesan, M.A., Sousa-Neto, M.D., and Brugnera Junior, A. (2001) Evaluation of Er:YAG laser and EDTACon dentin adhesion of six endodontic sealers. *Braz. Dent. J.*, **12** (1), 27–30.

73 Gekelman, D., Prokopowitsch, I., and Eduardo, C.P. (2002) In vitro study of the effects of Nd:YAG laser irradiation on the apical sealing of endodontic fillings performed with and without dentin plugs. *J. Clin. Laser Med. Surg.*, **20** (3), 117–121.

74 Kimura, Y., Yonaga, K., Yokoyama, K., Matsuoka, E., Sakai, K., and Matsumoto, K. (2001) Apical leakage of obturated canals prepared by Er:YAG laser. *J. Endod.*, **27** (9), 567–570.

75 Maden, M., Gorgul, G., and Tinaz, A.C. (2002) Evaluation of apical leakage of root canals obturated with Nd:YAG laser softened gutta-percha, System-B, and lateral condensation techniques. *J. Contemp. Dent. Pract.*, **3** (1), 16–26.

76 Eriksson, A.R. and Albrektsson, T. (1983) Temperature threshold levels for heat-induced bone tissue injury: a vital-microscopic study in the rabbit. *J. Prosthet. Dent.*, **50** (1), 101–107.

77 Koba, K., Kimura, Y., Matsumoto, K., Takeuchi, T., Ikarugi, T., and Shimizu, T. (1998) A histopathological study of the morphological changes at the apical seat and in the periapical region after irradiation with a pulsed Nd:YAG laser. *Int. Endod. J.*, **31** (6), 415–420.

78 Kimura, Y., Yonaga, K., Yokoyama, K., Matsuoka, E., Sakai, K., and Matsumoto, K. (2001) Apical leakage of obturated canals prepared by Er:YAG laser. *J. Endod.*, **27** (9), 567–570.

79 Blum, J.Y., Peli, J.F., and Abadie, M.J. (2000) Effects of the Nd:YAP laser on

coronal restorative materials: implications for endodontic retreatment. *J Endod.*, **26** (10), 588–592.

80 Yu, D.G., Kimura, Y., Tomita, Y., Nakamura, Y., Watanabe, H., and Matsumoto, K. (2000) Study on removal effects of filling materials and broken files from root canals using pulsed Nd:YAG laser. *J. Clin. Laser Med. Surg.*, **18** (1), 23–28.

81 Miserendino, L.J. (1988) The laser apicoectomy: endodontic application of the CO_2 laser for periapical surgery. *Oral Surg. Oral Med. Oral Pathol.*, **66** (5), 615–619.

82 Bader, G., and Lejeune, S. (1998) Prospective study of two retrograde endodontic apical preparations with and without the use of CO_2 laser. *Endod. Dent. Traumatol.*, **14** (2), 75–78.

83 Stabholz, A., Khayat, A., Ravanshad, S.H., McCarthy, D.W., Neev, J., and Torabinejad, M. (1992) Effects of Nd:YAG laser on apical seal of teeth after apicoectomy and retrofill. *J. Endod.*, **18** (8), 371–375.

84 Arens, D.L., Levy, G.C., and Rizoiu, I.M. (1993) A comparison of dentin permeability after bur and laser apicoectomies. *Compend. Contin. Educ. Dent.*, **14** (10), 1290–1298.

85 Stabholz, A., Khayat, A., Weeks, D.A., Neev, J., and Torabinejad, M. (1992) Scanning electron microscopic study of the apical dentine surfaces lased with Nd:YAG laser following apicectomy and retrofill. *Int. Endod. J.*, **25** (6), 288–291.

86 Ebihara, A., Majaron, B., Liaw, L.H., Krasieva, T.B., and Wilder-Smith, P. (2002) Er:YAG laser modification of root canal dentine: influence of pulse duration, repetitive irradiation and water spray. *Lasers Med. Sci.*, **17** (3), 198–207.

87 Paghdiwala, A.F. (1993) Root resection of endodontically treated teeth by erbium:YAG laser radiation. *J. Endod.*, **19** (2), 91–94.

88 Komori, T., Yokoyama, K., Matsumoto, Y., and Matsumoto, K. (1997) Erbium:YAG and holmium:YAG laser root resection of extracted human teeth. *J. Clin. Laser Med. Surg.*, **15** (1), 9–13.

89 Komori, T., Yokoyama, K., Takato, T., and Matsumoto, K. (1997) Clinical application of the erbium:YAG laser for apicoectomy. *J. Endod.*, **23** (12), 748–750.

90 Gouw-Soares, S., Tanji, E., Haypek, P., Cardoso, W., and Eduardo, C.P. (2001) The use of Er:YAG, Nd:YAG and Ga-Al-As lasers in periapical surgery: a 3-year clinical study. *J. Clin. Laser Med. Surg.*, **19** (4), 193–198.

91 Coleton, S. (2004) Lasers in surgical periodontics and oral medicine. *Dent Clin North Am.*, **48** (4), 937–962 vii. Review.

92 Midda, M. (1992) Lasers in periodontics. *Periodontal Clin Investig.*, (1992) **14** (1), 14–20.

93 Moritz, A., Gutknecht, N., Doertbudak, O., Goharkhay, K., Schoop, U., Schauer, P., and Sperr, W. (1997) Bacterial reduction in periodontal pockets through irradiation with a diode laser: a pilot study. *J. Clin. Laser Med. Surg.*, **15** (1), 33–37.

94 Cobb, C.M. (2006) Lasers in periodontics: a review of the literature. *J. Periodontol.*, **77** (4), 545–564.

95 Aoki, A., Ando, Y., Watanabe, H., and Ishikawa, I. (1994) In vitro studies on laser scaling of subgingival calculus with an erbium:YAG laser. *J. Periodontol.*, **65** (12), 1097–1106.

96 Aoki, A., Ishikawa, I., Yamada, T., Otsuki, M., Watanabe, H., Tagami, J., Ando, Y., and Yamamoto, H. (1998) Comparison between Er:YAG laser and conventional technique for root caries treatment in vitro. *J. Dent. Res.*, **77** (6), 1404–1414.

97 Fontana, C.R., Kurachi, C., Mendonça, C.R., and Bagnato, V.S. (2004) Temperature variation at soft periodontal and rat bone tissues during a medium-power diode laser exposure. *Photomed. Laser Surg.*, **22** (6), 519–522.

98 Spencer, P., Cobb, C.M., Wieliczka, D.M., Glaros, A.G., and Morris, P.J. (1998) Change in temperature of subjacent bone during soft tissue laser ablation. *J. Periodontol.*, **69** (11), 1278–1282.

99 Friesen, L.R., Cobb, C.M., Rapley, J.W., and Forgas-Brockman, L. (1999) *and Spencer P.*, Laser irradiation of bone: II. Healing response following treatment by CO_2 and Nd:YAG lasers. *J. Periodontol.*, **70** (1), 75–83.

100 van As, G. (2004) Erbium lasers in dentistry. *Dent. Clin. North Am.*, **48** (4), 1017–1059.

101 Sasaki, K.M., Aoki, A., Ichinose, S., and Ishikawa, I. (2002) Morphological analysis of cementum and root dentin after Er:YAG laser irradiation. *Lasers Surg. Med.*, **31** (2), 79–85.

102 Chanthaboury, R., and Irinakis, T. (2005) The use of lasers for periodontal debridement: marketing tool or proven therapy? *J. Can. Dent. Assoc.*, **71** (9), 653–658.

103 Pant, V., Dixit, J., Agrawal, A.K., Seth, P.K., and Pant, A.B. (2004) Behavior of human periodontal ligament cells on CO_2 laser irradiated dentinal root surfaces: an in vitro study. *J. Periodontal. Res.*, **39** (6), 373–379.

104 Crespi, R., Barone, A., Covani, U., Ciaglia, R.N., and Romanos, G.E. (2002) Effects of CO_2 laser treatment on fibroblast attachment to root surfaces. A scanning electron microscopy analysis. *J. Periodontol.*, **73** (11), 1308–1312.

105 Belal, M.H., Watanabe, H., Ichinose, S., and Ishikawa, I. (2007) Effect of Er:YAG laser combined with rhPDGF-BB on attachment of cultured fibroblasts to periodontally involved root surfaces. *J. Periodontol.*, **78** (7), 1329–1341.

106 Dilsiz, A., Aydin, T., and Yavuz, M.S. (2010) Root surface biomodification with an Er:YAG laser for the treatment of gingival recession with subepithelial connective tissue grafts. *Photomed Laser Surg.*, **28** (4), 511–517.

107 Liu, C.M., Shyu, Y.C., Pei, S.C., Lan, W.H., and Hou, LT. (2002) In vitro effect of laser irradiation on cementum-bound endotoxin isolated from periodontally diseased roots. *J. Periodontol.*, **73** (11), 1260–1266.

108 Hakki, S.S., Korkusuz, P., Berk, G., Dundar, N., Saglam, M., Bozkurt, B., and Purali, N. (2010) Comparison of Er, Cr: YSGG laser and hand instrumentation on the attachment of periodontal ligament fibroblasts to periodontally diseased root surfaces: an in vitro study. *J. Periodontol.*, **81** (8), 1216–1225.

73
Laser Ablation of Dental Restorative Materials
Ernst Wintner and Verena Wieger

73.1
Introduction

The purpose of the complete study [1] reported in this chapter was to test the applicability of ultra-short pulse laser (USPL) systems for biological hard tissue removal, that is, dental hard tissues including dental restorative materials, and bone structures. The focus of the investigations was restorative or filling materials together with selective ablation in comparison with dentin and enamel. In dental practice, composite filling material often has to be removed to treat secondary caries appearing at the rims of restorations or underneath them. Hence ablation of composites had to be investigated, particularly because hardly any research has been conducted so far on this topic. Therefore, throughout this chapter a comparison between erbium and USPL ablation is made, involving different erbium and USPL systems.

Ablation by erbium lasers has already become a dental treatment standard, despite the fact that only a few percent of the hard tissue treatments involve lasers at all. Employing USPL systems represents a potential future perspective in dentistry, which at present is severely inhibited by the extremely high costs. An estimation by the author yielded a current cost of around €250 000 and more for a USPL system dedicated to dental applications. On the other hand, in materials processing technology, USPL systems are being progressively employed, yielding innovative and superior results. This, together with substantial cost reductions due to series production of such tools, can be expected to lead to applications in dentistry also.

In general, and as an outstanding advantage of USPL ablation, the morphologic characteristics of the cavity boundaries of remaining tissue and/or dental restorative materials are of superior quality. In general, the advantages of scanned [2] USPL ablation are (i) precise cavity preparation resulting in smooth cavity rims; (ii) no molten or resolidified zones; (iii) no carbonization; (iv) no microcracks as there is no collateral damage due to thermal load and shock waves; (v) almost no temperature increase and therefore no heat transport to the adjacent tissue; (vi) gentle tissue removal; and (vii) selective ablation.

Handbook of Biophotonics. Vol.2: Photonics for Health Care, First Edition. Edited by Jürgen Popp,
Valery V. Tuchin, Arthur Chiou, and Stefan Heinemann.
© 2012 Wiley-VCH Verlag GmbH & Co. KGaA. Published 2012 by Wiley-VCH Verlag GmbH & Co. KGaA.

73.2
Dental Restorative Materials

Esthetic dental restorative materials generally are composites which initially are pasty to enhance cavity preparations, including plastic shaping. After the restoration, chemical or physical hardening techniques, depending on the specific composite, have to be applied. Thereby a stable polymer network is built. As the term "composites" suggests, they are composed of several chemical substances that determine their characteristics. In the following, a brief overview is provided [3–9].

The main constituents of composites can be arranged in three groups: (i) organic composite matrix; (ii) dispersive phase; and (iii) compound phase. The first group contains monomers, comonomers, initiators, stabilizers, and other additives. The dispersive phase comprises the inorganic filler particles. The compound phase is made of silanes that act as intermediates between the first two phases mentioned.

73.3
Ablation by Erbium Laser

The erbium laser systems used in ablation rate measurements were a Fotona Fidelis Er:YAG laser ($\lambda = 2940$ nm, $\tau \approx 100\,\mu$s) [10] and a Biolase Waterlase Er, Cr:YSGG laser ($\lambda = 2780$ nm, $\tau \approx 50\,\mu$s) [11]. Cavities were generated with front panel power settings ranging from 2 to 6 W. Ablations were always accompanied by an air–water spray (60% water, 65% air, each out of 100% max). With each laser setting, six cavities were prepared for a defined period of time (15 s). The experiments were performed on dentin, enamel, and different composite materials. Concerning teeth, extracted caries-free human permanent third molars were collected and stored in pure water until use. For dentin ablation, the occlusal enamel was removed using a diamond saw under water cooling (Accutom-2, Struers). Before and after laser ablation, the samples were weighed employing a microbalance to determine the total mass loss. The average of all six total mass values was included to calculate the mean total ablated volume according to $V = m/\varrho_{d/e}$, where the densities are given by $\varrho_d = 2.14\,\mathrm{g\,cm^{-3}}$ for dentin and $\varrho_e = 2.97\,\mathrm{g\,cm^{-3}}$ for enamel [12]. In a final step, the mean ablated volume per pulse was established by dividing the averaged total volumes by the number of applied laser pulses N, which is the product of pulse repetition rate PRR times the duration of ablation D: $N = PRR \times D$.

As the density of the composite samples generally was not known, another method was applied to determine the ablation rates in terms of volume per pulse. For that, imprint material commonly used in dentistry was considered. As this material is capable of reproducing tooth structures exactly, it was assumed that it fits the demands of the experiment, which are to attach closely to the surface of the cavities, to be able to fill even very small holes, and to imprint the cavities accurately without any excess material. Because all of these issues could be confirmed, the laser-ablated cavities were weighed, filled with imprint material, and weighed again. The difference between the mass values gave the mass $m_{i,c}$ of the imprint model of the cavity.

By producing blocks of imprint material with defined volumes V_i and measuring their masses m_i, the density of the material was determined as $\varrho_i = m_i/V_i$, so that at least the total ablated volume of the cavity could be calculated: $V_c = m_{i,c}/\varrho_i$. This procedure was applied for all cavities per sample. Finally, the mean ablated volumes per pulse were retrieved in a similar way as for dentin and enamel.

73.3.1
Results for Dentin and Enamel

First, results on the basis of the front panel power settings of the lasers were compared. As both erbium systems were in use for dental applications, it was expected that due to regular maintenance, the pre-settings and output parameters would apply. Thereby, a direct comparison of the Fotona Fidelis Er:YAG and the Biolase Er, Cr:YSGG ablation rates was possible by choosing the same powers of 2, 3, 4, 5, and 6 W and a fixed pulse repetition rate of 20 Hz. Table 73.1 (left) gives the results obtained.

In general, the dentin ablation rates obtained with the Fotona system are about 42% lower than those with the Biolase system. Concerning enamel, the situation is similar, as deviations of ~33% can be observed. Because the absorption coefficient of water ($\mu_a = 7000$ cm^{-1}) for Er, Cr:YSGG laser radiation at room temperature is only ~55% of that of the Er:YAG laser ($\mu_a = 13\,000$ cm^{-1}) [13], the ablation performance of the Er:YAG laser apparently is superior. This discrepancy motivated the measurement of the output parameters of the laser systems. It turned out that relying just on the front panel power settings is insufficient to develop useful data. Actually, the measured average output powers deviate significantly from the front panel settings for each laser system. The output powers of the Biolase system are around 13% higher than the displayed values, whereas the Fotona laser emits on average 31% less power than indicated. Using the measured laser parameters, a second attempt was made to analyze the ablation rates. The basis for the renewed discussion is the radiant exposure or fluence. Table 73.1 (right) reports the ablation rates in this new context to be considered as reliable data.

Owing to the different measured pulse energy values and focal spot diameters of the two erbium laser systems, the single fluence values are not directly comparable. Nevertheless, by drawing regression lines, interpretation of the results is permitted. Obviously, in a wide fluence range the ablation rates of the Er:YAG system are higher than those of the Er,Cr:YSGG laser. This finding is in agreement with the above-mentioned absorption coefficients of the two laser wavelengths. For higher laser fluences the ablation behavior is reversed, with the Er,Cr:YSGG laser performing slightly better. This also corresponds to the results of Vogel and Venugopalan [14], who reported on the changes in the absorption coefficients for both erbium laser wavelengths with rising fluence. The high absorption coefficient of the Er:YAG laser wavelength declines monotonically with increasing radiant exposure and even becomes lower than that of the Er,Cr:YSGG laser. At fluences where laser ablation takes place, the performances of both systems should at least be comparable.

Table 73.1 Mean ablated dentin and enamel volumes per pulse by two different erbium lasers for increasing (imprecise) front panel power settings (left) and for varying laser fluences (right).

Dental tissue	Power (W) (front panel settings)	Fotona Er:YAG	Biolase Er,Cr:YSGG	Fotona Er:YAG		Biolase Er,Cr:YSGG	
		Ablated volume per pulse (mm³)	Ablated volume per pulse (mm³)	Fluence (J cm⁻²)	Ablated volume per pulse (mm³)	Fluence (J cm⁻²)	Ablated volume per pulse (mm³)
Dentine	2	0.0043	0.0062	13.5	0.0043	24.9	0.0062
	3	0.0067	0.0125	16.2	0.0067	38.0	0.0125
	4	0.0101	0.0181	21.7	0.0101	51.4	0.0181
	5	0.0121	0.0234	26.9	0.0121	64.1	0.0234
	6	0.0161	0.0275	32.5	0.0161	77.9	0.0275
Enamel	4	0.0044	0.0063	21.7	0.0044	51.4	0.0063
	5	0.0059	0.0093	26.9	0.0059	64.1	0.0093
	6	0.0079	0.0115	32.5	0.0079	77.9	0.0115

Table 73.2 Ablation rates of dental composite materials with the Biolase Er, Cr:YSGG laser.

Composite	Ablation volume per pulse (mm^3)	
	Fluence 19 J cm^{-2}	Fluence 51 J cm^{-2}
Tetric Ceram	0.0026	0.0161
Tetric Flow	0.0036	0.0187
Heliomar	0.0041	0.0213
Point 4	0.0047	0.0233
Premise	0.0031	0.0176
Z100	0.0021	0.0156
XRV Herculite	0.0031	0.0176

73.3.2
Results for Composites

Composite ablation rates were determined with the Biolase Er,Cr:YSGG laser as it showed the best laser characteristics among Er systems. In Table 73.2, different types of dental restorative materials and their ablation rates for 1.7 and 4.5 W laser ablation, corresponding to fluences of 19 and 51 J cm^{-2}, respectively, are reported. The ablated volumes per pulse were determined with the imprint method.

Although the ablation rates of all composites are on the same scale, deviations among different filling materials are apparent. Ranking the composite samples according to their material removal per laser pulse at 19 J cm^{-2} from the highest to the lowest amount, the Point 4 composite occupies the first position with 4.7×10^{-3} mm^3 per pulse and Z100 occupies the last place with 2.1×10^{-3} mm^3 per pulse. The ranking remains consistent for higher applied laser powers. On average, the ablated volume per pulse is about 5.8 times higher for the 51 J cm^{-2} ablation than the 19 J cm^{-2} ablation.

The effective removal of composite material is determined by its chemical composition and the related absorption of erbium laser wavelengths. On the one hand, there is sufficient water present in all composites for strong absorption around 3 μm, but other components such as quartz and poly(methyl methacrylate) (PMMA) resin also absorb substantially in the mid-infrared region [15]. Kalachandra et al. [16] investigated the transmittance of BIS-GMA (bisphenol A- glycidyl methacrylate) depending on the wavenumber and reported a decrease in transmittance at a stretching frequency of 3458.32 cm^{-1} corresponding to a wavelength of ~2.89 μm. This broad absorption band belongs to an OH group of BIS-GMA. As the mentioned components apparently are contained in some of the above-listed composites, although in different concentrations, the observed deviations of the ablation rates can be explained. For example, the concentration of BIS-GMA in Tetric Ceram is 8.3 wt% whereas Tetric Flow contains 13.6 wt% [17]. On this basis, the higher ablation rates of Tetric Flow may be understood.

Comparison of the ablation rates of composites with those of dentin indicates that the selectivity of erbium laser ablation is not very pronounced. A dentin volume of 18.1 $\times$ 10^{-3} mm^3 per pulse ablated with the Er,Cr:YSGG laser at 51 J cm^{-2} fits fairly well into the range achieved in composites for the same fluence. Whereas for Z100, Tetric Ceram, Premise, and XRV Herculite the tissue removal procedure is slower compared with dentin, Tetric Flow, Heliomar, and Point 4 can be removed faster. A clear tendency concerning selectivity valid for all tested materials could therefore not be found.

Erbium laser ablation thresholds were derived by fitting a regression line (using a plot of ablation rates versus fluence) to measured data and extrapolating it to the abscissa. Dentin threshold values of 5 and 8 J cm^{-2} for Er:YAG and Er, Cr:YSGG lasers, respectively, were obtained. Wannop et al. [18] investigated 250 μs Er:YAG laser performance, yielding a threshold value of 10 ± 2 J cm^{-2}, which is somewhat higher than the value we calculated. Several groups have already reported on the erbium laser ablation threshold in enamel: Wannop et al. [18] 25 ± 8 J cm^{-2} for Er:YAG (250 μs); Apel et al. [19] 9–11 J cm^{-2} for Er:YAG (150 μs) and in identical experiments 10–14 J cm^{-2} for Er,Cr:YSGG; in another report by Apel et al. [20] for varying pulse durations of a Fotona Fidelis Er:YAG system, 9–10 J cm^{-2} for 700 μs, 8 J cm^{-2} for 350 μs, and f7 J cm^{-2} or 150 and 100 μs; Fried et al. [21] 7–9 J cm^{-2}; Hibst and Keller [22] $\sim$10 J cm^{-2}; and Blinkov et al. [23] 8 J cm^{-2}. There are no comparative reports on ablation thresholds of restorative materials.

73.4
Ablation by Ultra-Short Pulse Laser

Ablation rates generated by ultra-short laser pulses were determined using either a High Q [24] IC-10000 REG AMP Microproc 12 ps Nd:vanadate laser ($\lambda = 1064$ nm, $P_{av} = 5$ W) under $r - \varphi$ scanning, or a Spectra-Physics Hurrican-i Ti:sapphire laser with variable pulse duration. The aim was to determine the ablated volume per laser pulse to judge the effectiveness of USPL compared with erbium systems (Tables 72.3 and 72.4). The 12 ps Nd:vanadate laser was operated for a duration of 15 s at a pulse energy of

Table 73.3 Ablation rates (left) and ablated volume per second of dental composite materials obtained with an $r-\varphi$-scanned Nd:vanadate laser (center); $\tau = 12$ ps, $PRR = 50$ kHz, $E_p = 100 \mu$J, $\Phi = 11.2$ J cm^{-2}; for comparison, corresponding values for an Er,Cr:YSGG laser are given (right).

Composite	Nd:vanadate 5 W	Er,Cr:YSGG 4.5 W	
	Ablation rate (10^{-6} mm^3 per pulse)	Ablated volume (mm^3 s^{-1})	
Tetric Ceram	5.18	0.259	0.322
Tetric Flow	4.15	0.208	0.374
Heliomar	5.53	0.277	0.426
Point 4	5.81	0.291	0.466
Premise	6.08	0.304	0.352
Z100	8.71	0.436	0.312
XRV Herculite	7.88	0.394	0.352
Dentin	0.32	0.016	0.362

Table 73.4 Ti:sapphire laser ablation thresholds of composite materials for 150 fs – 7 ps pulses.

τ (fs)	Ablation threshold (J cm^{-2})								
	Compoglass F	Premise Enamel	Premise Body A2	Heliomar A3	XRV Herculite	XRV Herculite Dentin	XRV Herculite Enamel	Point 4	Z100
150	0.27	0.30	0.35	0.24	0.25	0.25	0.30	0.36	0.27
500	0.44	0.38	0.43	0.35	0.39	0.34	0.33	0.48	0.44
2000	0.56	0.58	0.51	0.48	0.57	0.44	0.53	0.51	0.46
7000	0.60	0.63	0.58	0.59	0.64	0.54	0.56	0.56	0.59

100 µJ and a *PRR* of 50 kHz yielding 5 W. As the focal spot diameter was 33.7 µm, this setting corresponds to a fluence of 11.2 J cm^{-2}. Dentin and composite materials were ablated and, again, the imprint method was used to reconstruct the cavity volume from which ablation rates were calculated.

Ablation rates using the Hurrican-i Ti:sapphire via line scans ($\lambda \approx 800$ nm, $\tau = 130$ fs–7 ps, $E_{pulse} > 1$ mJ, *PRR* 0–10 kHz [25]) have been published [26]. The depths of the grooves were measured by means of digital light microscopy with implemented software. The motivation for these experiments was a comparison with dental tissue ablation rates increasing under rising pulse duration [27].

For sound dentin and enamel and also for carious dentin, ablation thresholds obtained with a Ti:sapphire laser can be derived from the literature. Serbin *et al.* [27] focused on laser ablation with pulse durations from 120 fs to >2 ps. For 120 fs Ti:sapphire pulses, the highest threshold was found for sound enamel at >0.5 J cm^{-2}, followed by sound dentin at >0.3 J cm^{-2}, whereas for carious dentin the threshold was only <0.12 J cm^{-2}. Similar values were found by Lubatschowski *et al.* [28]. For all applied pulse durations, the composite thresholds of this study are lower than the enamel thresholds reported by Serbin *et al.* [27]. Pulse durations of 150 and 500 fs yield comparable values for composite and dentin, whereas for longer pulse widths the dentin thresholds are situated well above composite thresholds (~2 ps corresponding to a threshold of 0.75 J cm^{-2} in dentin). That implies that the selectivity at longer pulse duration is very pronounced for enamel and composite ablation.

73.5 Conclusion

The following conclusions can be drawn:

- The etch depths per pulse achieved by USPL in dental hard tissue and composite restorations are in the micrometer region and therefore much smaller than for erbium ablation rates. The USPL ablation procedure can be enhanced by applying *PRR* of some tens of kiloherz. Ablation rates per second are then comparable to erbium laser ablation rates.

- Selective ablation is very pronounced for USPL as dentin revealed lower ablation rates than composites, which is beneficial for minimum invasive secondary caries treatment. This is not generally so for erbium systems. Depending on the composite, sometimes a controversial effect can be observed, that is, a composite becomes equally or even less ablated than dentin.
- Ablation thresholds in dentin amount to $5\,\mathrm{J\,cm^{-2}}$ for Er:YAG and $8\,\mathrm{J\,cm^{-2}}$ for Er,Cr:YSGG lasers; enamel ablation starts at even higher fluences, whereas femtosecond laser ablation thresholds of these tissues are below $1\,\mathrm{J\,cm^{-2}}$. USPL thresholds of composites determined in this study are lower than those of dental structures. Again, this fact contributes to selective ablation. With rising pulse duration, thresholds increase and ablation rates decline at constant fluence.
- The remaining dentin surface after erbium laser treatment is fissured and scaly. It depends strongly on the air–water spray, that is, insufficient cooling and hydration, leading to microcracks, melting, or carbonization. The morphology of USPL-processed cavities in dentin involving a scanner shows superior tissue quality. An appropriate set of laser and scanner parameters leads to surfaces with no evidence of melting, carbonization, or microcracks. In contrast to cavities excavated by erbium lasers, the cavity rims after scanned USPL treatment are well defined and smooth.
- Scanned USPL ablation leaves a fine micro-retentive pattern. Although its appearance is distinct from that of an etched surface, this regular structure bears the potential for good adhesion of compound to filling material without additional etching. On the other hand, supplementary etching is recommended after erbium conditioning because of the scaly appearance of the tooth surface.

References

1 Wieger, V. (2006) Ultra-Short Pulse Laser Ablation of Biological Hard Tissue and Dental Restorative Materials, PhD thesis, Vienna University of Technology, Photonics Institute.

2 Strassl, M., Yousif, A., and Wintner, E. (2007) Scanning of ultra-short laser pulses in dental applications. *J. Laser Micro/Nanoeng.*, **2** (3), 206–211.

3 Ivoclar Vivadent (1990) Composite-Füllungsmaterial, Report No. 5, Ivoclar Vivadent, Schaan, Liechtenstein.

4 Ivoclar Vivadent (1992) Der gefüllte Zahn – ein komplexes Verbundsystem, Report No. 7, Ivoclar Vivadent, Schaan, Liechtenstein.

5 Ernst, C.P. and Willsershausen, B. (2000) Quo vadis Komposit? Eine aktuelle Standortbestimmung zahnärztlicher Füllungskomposite. Polyklinik für Zahnerhaltungskunde des Klinikums der Johannes Gutenberg-Universität Mainz, Mainz.

6 Kultermann, G. (2001) Moderne Adhäsivsysteme – Fortschritt oder Marketing?, *Der Freie Zahnarzt*, **5** (1), 36–41.

7 Lonnroth, E. and Shahnavaz, H. (1997) Use of polymer materials in dental clinics, case study. *Swed. Dent. J.*, **21** (4), 149–159.

8 Schärer, P. and Chen, L. (1998) Komposite-Zemente und Dentinhaftmittel. *Philips J.*, **11/12**, 326–334.

9 Ivoclar Vivadent (2004) SR Adoro im Fokus: Indirekte Komposite – Werkstoffkunde und Entwicklung, Report

No. 15, Ivoclar Vivadent, Schaan, Liechtenstein.
10. Fotona (2010) www.fotona.si (last accessed 4 June 2011).
11. Biolase (2010) www.biolase.com (last accessed 4 June 2011).
12. Ciba Found. Symp. (1997) **205**, 54–67; discussion 67–72.
13. Stock, K., Hibst, R., and Keller, U. (1997) Comparison of Er:YAG and Er:YSGG laser ablation of dental hard tissues. *Proc. SPIE*, **3192**, 88–95.
14. Vogel, A. and Venugopalan, V. (2003) Mechanisms of pulsed laser ablation of biological tissues. *Chem. Rev.*, **103**, 577–644.
15. Dumore, T. and Fried, D. (2000) Selective ablation of orthodontic composite by using sub-microsecond IR laser pulses with optical feedback. *Lasers Surg. Med.*, **27**, 103–110.
16. Kalachandra, S., Sankarapandian, M., Shobha, H.K., Taylor, D.F., and McGrath, J.E. (1997) Influence of hydrogen bonding on properties of BIS-GMA analogues. *J. Mater. Sci. Mater. Med.*, **8**, 283–286.
17. *Information sheet of Ivoclar Vivadent*: Tetric Ceram Familie.pdf; www.ivoclarvivadent.com/zoolu-website/media/ (last accessed Aug. 11, 2011).
18. Wannop, N.M., Dickinson, M.R., and King, T.A. (1993) Erbium:YAG laser radiation interaction with dental tissue. *Proc. SPIE*, **2080**, 33–43.
19. Apel, C., Meister, J., Ioana, R.S., Franzen, R., Hering, P., and Gutknecht, N. (2002) The ablation threshold of Er:YAG and Er:YSGG laser radiation in dental enamel. *Lasers Med. Sci.*, **17**, 246–252.
20. Apel, C., Franzen, R., Meister, J., Sarrafzadegan, H., Thelen, S., and Gutknecht, N. (2002) Influence of the pulse duration of an Er:YAG laser system on the ablation threshold of dental enamel. *Lasers Med. Sci.*, **17**, 253–257.
21. Fried, D., Featherstone, J.D.B., Visuri, S.R., Seka, W., and Walsh, J.T. (1997) The caries inhibition potential of Er:YAG and Er:YSGG laser radiation. *Proc. SPIE*, **3593**, 73–78.
22. Hibst, R. and Keller, U. (1989) Experimental studies of the application of the Er:YAG laser on dental hard substances: I. measurement of the ablation rates. *Lasers Surg. Med.*, **9**, 338–344.
23. Belinkov, A.V., Erofeev, A.V., Shumilin, V.V., and Tkachuk, A.M. (1993) Comparative Study of the 3 μm Laser Action on Different Tooth Tissue Samples Using Free-Running Er-Doped YAG, YSGG, YSP and YLF Lasers. *Proc. SPIE*, **2080**, 60–67.
24. High Q Laser (2010) www.highQlaser.com (last accessed 4 June 2011).
25. Spectra-Physics (2003) Manual of the Hurricane-i Laser. Spectra-Physics, Irvine, CA.
26. Strassl, M., Wieger, V., Brodoceanu, D., Beer, F., Moritz, A., and Wintner, E. (2008) Ultra-short pulse laser ablation of biological hard tissue and biocompatibles. *J. Laser Micro/Nanoeng.*, **3** (1), 30–40.
27. Serbin, J., Bauer, T., Fallnich, C., Kasenbacher, A., and Arnold, W.H. (2002) Femtosecond lasers as novel tool in dental surgery. *Appl. Surf. Sci.*, **197–198**, 737–740.
28. Lubatschowski, H., Heisterkamp, A., Will, F., Singh, A.I., Serbin, J., Ostendorf, A., Kermani, O., Heermann, R., Willing, H., and Ertmer, W. (2002) Medical applications of ultrafast laser pulses. *RIKEN Rev.*, **50**, 113–118.

74
Laser Ablation of Hard Tissues

Gregory B. Altshuler, Andrey V. Belikov, and Felix I. Feldchtein

74.1
Introduction

Bone tissue, cartilage, and dental hard tissue (enamel, dentin, cementum) are traditionally classified as hard biological tissues of the human body. This chapter is focused on hard tissues such as enamel and dentin of human teeth. Interest in lasers for the treatment of hard tissues in dentistry has increased because such treatment is localized, comfortable, and less painful (due to the lack of vibration), relatively bloodless, and antiseptic.

74.2
Tooth Structure

A tooth [1] consist of the crown, neck, and root. The root and neck of the tooth are covered with cement. The tooth crown is covered with enamel – the most durable tissue of the human body. The underlying tissue of the tooth is dentin. The pulp chamber is located in the dentin interior. In the root of the tooth, the pulp chamber extends into the root canal. The pulp filling the chamber and the root canal is the only soft and the most sensitive tissue of the tooth, which consists of connective tissue, plexus of nerve fibers, and blood vessels.

Tooth enamel [1] consists of about 96% inorganic material, 1% organic material, and 3% water by weight. The volume water content is about 10%. The inorganic material is mostly a hydroxyapatite (HA), a substance also found in bone and dentin. The basic structural components of enamel are HA-like nanocrystals with the formula $A_{10}(BO_4)_6X_2$, where A is Ca, Cr, Ba, or Cd, B is P, As, or Si, and X is F, OH, or $ClCO_2$. The dominant formula of enamel apatite is an ideal HA $Ca_{10}(PO_4)_6(OH)_2$ with a Ca:P ratio of 1.67. In addition to HA (~75%), carbide apatite (~3–20%), chlorine apatite (~4%), and fluorine apatite (~0.5%) are also present in enamel. HA nanocrystals form a hexagonal structure with a size of $14–46 \times 27–78$ nm. These crystals have the typical crystal defects in the lattice arrangement, including shifted, disrupted, and curved lattice planes. Defective lattices in the boundary between crystals are fused with each other. The crystals in enamel are surrounded by a water shell, with a typical width of about 10 nm. A tightly packed mass of apatite crystals forms the basic structural unit of enamel, called the "enamel rod" or "enamel prism." It is shaped like a keyhole and has an average width of 5 μm. The rod length is determined by the local enamel thickness, with a maximum of ∼2.5 mm. Rods run from the dentinoenamel junction perpendicularly to the outer enamel surface and are organized in rows. Neighboring rods are surrounded by 0.1–0.5 μm wide rod sheaths and separated from each other by inter-rod enamel. The enamel rod consists almost entirely of inorganic material, whereas the rod sheaths are made up largely of organic material comprised of amelogenin polypeptide and non-amelogenin proteins impregnated by water.

Dentin comprises [1] the bulk of the tooth crown. The mature dentin consists (by weight) of up to 70% inorganic substances, about 20% organic components (mainly proteins that form collagen fibers), and up to 10% water (about 22% water by volume). The basic inorganic components are HA, calcium carbonate, and a small amount of calcium fluoride. Two main structural units of dentin are ground substance and dentinal tubules. The ground substance of dentin is permeated with numerous dentinal tubules with density ranging from 15 000 to 75 000 per mm^2 of dentin. The diameter of dentinal tubules depends on the location and ranges from 0.5 μm near the dentinoenamel junction to 5 μm near the pulp.

74.3
Laser Ablation

The main application of lasers for hard tissue treatment is laser ablation. The mechanism and parameters of ablation are determined by the optical, thermal, and mechanical properties of treated tissue. The most important characteristic is the coefficient of absorption of a tissue and its components.

Investigation of the absorption spectra of intact tissues allows one to determine wavelength regions with maximum absorption and minimal energy for ablation. Based on this analysis, one can choose a laser with radiation that is best absorbed by tissue and is the most destructive.

Absorption spectra of hard tissues of human teeth have been studied by several groups [2–6]. The spectrum of the transport coefficient for dental enamel and dentin $\mu_{tr} = \mu_a + \mu'_s$ [5], where μ_a is coefficient of absorption and μ'_s is reduced coefficient of scattering, is presented in Figure 74.1a. One can recognize three ranges of strongest absorption of enamel and dentin: in the UV region for $\lambda < 0.3$ μm and in the IR region for 2.7 μm $< \lambda <$ 3.5 μm and 9 μm $< \lambda <$ 11 μm. The UV absorption peak corresponds to electron absorption of all hard tissue components. The 3 μm region is related to free water (2.96 μm peak) and OH group of HA (2.8 μm peak). The 10 μm region mostly corresponds to PO$_4$ groups of HA (9.6 μm peak) and free water (Figure 74.1b). For these ranges, the absorption is several orders of magnitude higher than the scattering ($\mu_a > \mu'_s$) and penetration in the hard tissue is determined by absorption.

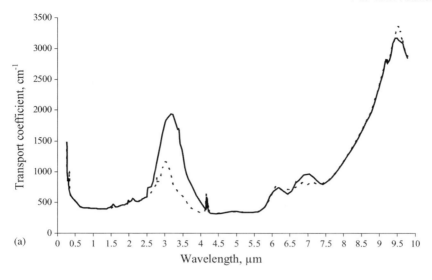

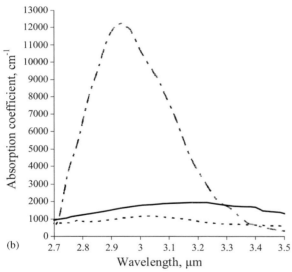

Figure 74.1 Transport coefficient (a) and absorption (b) spectra of intact enamel (dotted line) [5], dentin (solid line) [5], and water (dashed-dotted line) [6].

Due to a very high absorption coefficient in the 3 and 10 µm ranges, the penetration of laser radiation into the tissue is about $1/\mu_a \approx 2\text{--}20\,\mu\text{m}$ for enamel and 1–10 µm for dentin.

The highest absorption in the spectrum is observed for free water in hard tissue at about 2.96 µm. The water absorption at this wavelength is about $12\,000\,\text{cm}^{-1}$ [7], whereas the enamel absorption does not exceed $1000\,\text{cm}^{-1}$ [5, 8]. Hence the water absorption is more than an order of magnitude higher (Figure 74.1b). The absorption coefficients of hard dental tissues may change during laser irradiation, for example due to changes in the optical properties of water when it is heated, so although the absorption coefficient at normal body temperature is higher at $\lambda = 2.94\,\mu m$ than at $\lambda = 2.79\,\mu m$, the ratio may change to the opposite with temperature increase because of water absorption changes substantially with temperature. In particular, the absorption coefficient of free water decreases by about an order of magnitude upon heating from room temperature to the critical point of 374 °C [9, 10].

From the analysis of the absorption spectrum, it can be concluded that the most effective and low threshold mechanism of laser ablation can be achieved through selective absorption of free water in the rod sheaths, pores, and surface microcracks for enamel or in free water in dentinal tubules and ground substance of dentin.

The destruction mechanism of dental hard tissues associated with selective absorption was described by Altshuler et al. [11]. Selective absorption of laser radiation by water contained in the rod sheaths, pores, and surface microcracks for enamel or in dentinal tubules and ground substance of dentin heats water much faster than HA. As noted by Lee et al. [9], at 130 °C the pressure developing in closed pores may reach 1000 bar and, as a consequence, microcracks localized primarily as clusters in inter-rod enamel start to appear and propagate in the material surrounding the pores (Figure 74.2).

Cracks will most likely develop through the inter-rod enamel, where HA crystals are disordered and as a result the inter-rod enamel has a lower mechanical strength. The propagation of cracks through the inter-rod enamel will lead to the separation of individual enamel rods or groups from the ground enamel substance. Such a mechanism of laser ablation will be referred to as "thermomechanical". Thus, radiation with a wavelength of 3 µm does not penetrate deeply into the dental hard tissue owing to the high absorption [12]. Ablation of dental hard tissues occurs in microexplosions, when tissue does not evaporate completely, but breaks into small

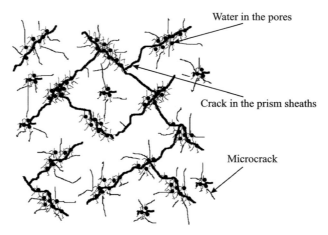

Figure 74.2 Microcrack cluster formation in enamel.

fragments and is ejected from the bulk of the tissue [13]. The role of melting and vaporization of HA and carbonization of organic components is minimal. The rate of tissue removal depends on the amount of water contained in the tissue. Carious dentin contains the largest amount of water among the hard dental tissues [14]. Hence, for this mechanism of laser ablation carious dentin has the highest and sound enamel has the lowest ablation rate.

Selective absorption of the free water content within structural components with size r is possible if the laser pulse duration is shorter than thermal relaxation time (TRT) of these components [15]:

$$TRT = \frac{r^2}{k\alpha} \qquad (74.1)$$

where α is the thermal diffusivity of the tissue and k is the form-factor coefficient, equal to 8, 16, and 24 for layer, cylindrical, and spherical structural components, respectively. For enamel $\alpha = 4.7 \times 10^{-7}\,m^2\,s^{-1}$ and for dentin $\alpha = 1.8 \times 10^{-7}\,m^2\,s^{-1}$ [16]. A laser pulse duration τ less than the thermal relaxation time TRT is considered optimum for tissue layer destruction because in this case the heat absorbed by the tissue components is not transferred to the surrounding tissues due to the heat conductivity mechanism, providing local and most effective laser action. Thus, in addition to optical and thermophysical properties of tissue, the optimum laser pulse duration is determined by the size of its structural component which is instrumental for ablation. For the thermomechanical mechanism, r is the size of components which contain free water: the rod sheaths ($<0.5\,\mu m$), pores ($<1\,\mu m$), and surface microcracks ($<10\,\mu m$) for enamel and free water in dentinal tubules ($<5\,\mu m$) and ground substance of dentin. The optimum pulse duration for selective heating of rod sheaths should be shorter than $0.1\,\mu s$, for enamel microcracks shorter than $100\,\mu s$, and for dentinal tubules shorter than $20\,\mu s$. The most effective lasers for thermomechanical ablation are Er:YAG ($\lambda = 2.94\,\mu m$), Er:YLF ($\lambda = 2.81\,\mu m$), and Er:YSGG ($\lambda = 2.79\,\mu m$) lasers.

For the $10\,\mu m$ wavelength region, the dominant mechanism of laser ablation is melting with subsequent vaporization ("thermal vaporization" mechanism) of HA or removal of the molten phase in the form of individual drops ("hydrodynamic ejection" mechanism) of HA [3]. Both mechanisms require heating of the enamel surface to a temperature higher than melting temperature of HA (about $1280\,°C$). The optimum pulse duration for both mechanisms should be shorter than the TRT of the hard tissue layer equal to the penetration depth $1/\mu_a$ of the laser radiation. The same basic equation (Eq. 74.1) for the TRT of such a layer for a wavelength of $9.6\,\mu m$ provides values of about $2\,\mu s$ for enamel and $4\,\mu s$ for dentin.

As can be seen for both the thermomechanical and thermal vaporization mechanisms of ablation, a period of nanoseconds provides the most efficient laser ablation. However, the delivery of such pulses to the tissue may be problematic. In addition, plasma formation on the hard tissue surface may be induced by a short pulse with high power density and the laser pulse will be blocked by the plasma due to reflection and absorption. Therefore, the optimum pulse duration for clinical use should be in the submicrosecond or microsecond range.

One known risk of laser dental tissue treatment is overheating of the pulp [17], which results in pulp injury [18]. The temperature on the hard tissue surface during an ablation pulse can be higher than 1300 °C, but the amount of heat left after ablation is very low because of the very small ablation volume per pulse. This amount of heat can be characterized as residual energy factor R, which is the ratio of the energy left in a tissue after ablation to the total energy of the laser pulse. R is critical to parameters of the laser pulse. For optimum conditions of ablation, $R = 0.4$ for the thermomechanical mechanism [19] and 0.25 for the thermal vaporization mechanism [3]. Residual heat from ablation propagates to the pulp and spreads over a large volume. Finally, the temperature rise in the pulp depends on the average laser power and exposure time. The safe temperature limit for pulp is about 42 °C [20]. The temperature rise in the pulp chamber depends on the average laser power and treatment time. Water cooling is used to reduce the likelihood of pulp overheating [18, 21]. Water cooling removes part of the residual heat and allows more laser power to be delivered and the drilling speed to be increased. In the optimum regime, about 50% of residual heat may be removed by water cooling. Heating of the pulp chamber during laser treatment has been simulated and measured *in vitro* in several studies [22–31]. The safety limit for laser ablation with an erbium laser and water cooling is about 6 W, which corresponds to a residual power dissipation of about 1.2 W. The water flow can be pulsed or continuous, in the form of a jet or spray. As a result, a water film is created on the tooth surface. The film absorbs laser light and can decrease the ablation efficiency. However, pulse laser radiation absorption by a water film is normally accompanied by the formation of shock waves. These waves contribute to removing water from the exposure zone on the tissue, if the initial thickness of the water film is small [32], and increase the ablation efficiency [33]. Water can soak (impregnate) a metamorphosed layer formed on the walls of the cavity under laser irradiation, creating additional absorption centers for laser radiation and enhancing the ablation efficiency due to the thermomechanical mechanism.

The efficacy of laser ablation of hard dental tissues by laser irradiation depends on many factors, such as the laser radiation wavelength, laser pulse structure and duration, water cooling, contact or noncontact mode of treatment, laser radiation energy density, distribution of power density in laser beam, repetition rate, and so on. Multimodal free-running erbium lasers with flashlamp pumping at wavelengths of 2.79 and 2.94 μm are the most widely used for hard dental tissue ablation. Pulses of these lasers have durations from 50 to 1000 μs and consist of a set of random spikes with durations from 1 to 10 μs [34].

In experiments on hard tissue laser ablation with an Er:YAG laser [35], the laser damage threshold of hard dental tissues was measured with a photoacoustic method. The threshold value was 4–7 J cm^{-2} for enamel for contact mode irradiation, which is approximately two times lower than for noncontact exposure. The results of measurements of dental hard tissue removal efficiency by the first Er:YAG laser pulse incident on the intact sample surface in contact and noncontact modes are shown in Figures 73.3 and 73.4.

Clearly, the removal efficiency in the contact mode of laser ablation without water cooling for enamel at energy densities up to 100 J cm^{-2} is almost 1.5 times higher

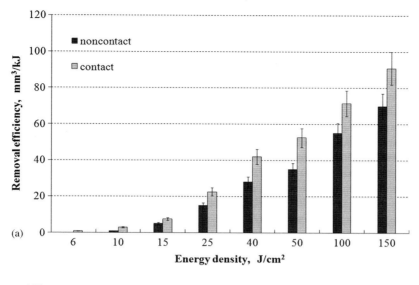
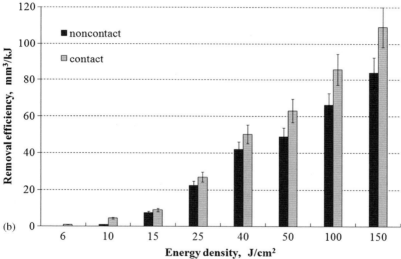

Figure 74.3 Enamel removal efficiency versus energy density of an Er:YAG laser in contact and noncontact modes of treatment, with no water (a) and with water (b) cooling (water flow rate 0.2 ml min^{-1}).

than that in the noncontact mode, and almost 1.3 times higher at energy densities up to 100–150 J cm^{-2}.

The efficiency of dentin removal at energy densities up to 100 J cm^{-2} for laser ablation without water cooling is 1.2 times higher in the contact mode than in the noncontact mode, whereas at energy densities up to 100–150 J cm^{-2} the efficiencies are almost the same in the contact and noncontact modes. In the case of laser

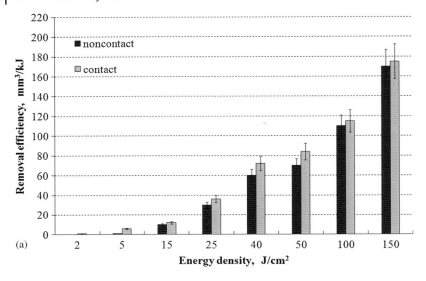

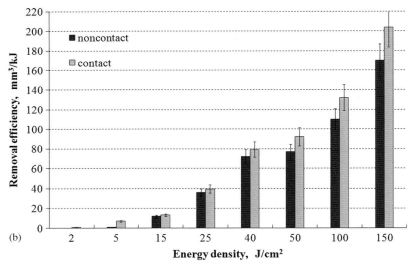

Figure 74.4 Dentin removal efficiency versus energy density of an Er:YAG laser in contact and noncontact modes of treatment, with no water (a) and with water (b) cooling (water flow rate 0.2 ml min^{-1}).

ablation with water cooling, the efficiency of enamel removal at energy densities up to 50 J cm^{-2} is almost 1.2 times higher in the contact mode than the noncontact mode and 1.3 times higher in the contact mode at energy densities of 50–150 J cm^{-2}.

In the contact mode of laser ablation with water cooling (the water flow rate in the experiment was 0.2 ml min^{-1}), the efficiency of dentin removal at energy densities up to 50 J cm^{-2} was nearly 1.1 times higher in the contact mode, and 1.2 times higher at energy densities of 50–150 J cm^{-2}.

It is worth noting that the presence of water in the noncontact mode enhances the enamel removal efficiency nearly 1.5-fold at energy densities up to 50 J cm^{-2}, and 1.2-fold at energy densities of 50–150 J cm^{-2}. For dentin, the presence of water in the noncontact mode increases the removal efficiency almost 1.2-fold at energy densities up to 50 J cm^{-2} and does not change the removal efficiency at energy densities of 50–150 J cm^{-2}.

Water cooling increases the enamel removal efficiency in the contact mode almost 1.2-fold at energy densities up to 150 J cm^{-2} and does not change the efficiency of dentin removal.

The difference in removal efficiency for the contact and noncontact modes arises from differences in the mechanisms of laser damage. Rapid heating and microexplosions associated with the absorption of light by structural elements containing free water are characteristic for the noncontact mode. The same processes occur in the contact mode, but they are supplemented by a mechanism associated with the motion of particles produced by ablation in a closed space between a laser crater and the distal end of the contact tip (sapphire fiber). Particles of enamel are reflected from the distal end of the tip and produce additional tissue damage due to the high kinetic energy. This effect is similar to hard tissue damage by high-speed sapphire particles (air abrasive). This laser abrasive mechanism is related to the bombardment of the tissue surface by hard particles of the same tissue produced during laser ablation [36].

The efficiency of laser removal of enamel and dentin may be increased by adding hard abrasive particles (sapphire, diamond) to the treatment area [36]. It has been demonstrated [36] that the abrasive particles are accelerated by the laser radiation. The most likely mechanisms of hard particle acceleration by laser irradiation are the reactive mechanism and the mechanism related to microexplosions of particle material. Particles accelerated by laser irradiation reach speeds exceeding 1000 m s^{-1}. The kinetic energy of laser-accelerated particles is sufficient to destroy the hard tooth tissues. When fast abrasive particles collide with the enamel surface, the most likely location of destruction is inter-rod enamel having less mechanical cohesion and hardness than enamel rods. As has been shown [36], delivery of sapphire particles with a diameter of 10–30 μm as a powder or aqueous suspension into the treatment area synchronized with the Er:YAG laser radiation can improve the enamel removal efficiency almost twofold compared with enamel treatment by laser ablation alone.

The impact of erbium and CO_2 laser radiation on hard dental tissues has been investigated [37–41] using multimodal lasers with a 300–1000 μm beam size in the treatment area. In clinical practice, a beam with a diameter of 0.5 − 1 mm is normally used in laser systems for hard tissue drilling. Cavities created by such radiation are characterized by a low aspect ratio and intended for traditional tooth filling. Such a large spot size does not allow the benefits of laser beams with extremely small size, comparable to the wavelength of laser light, to be realized. Apparently, the smaller the beam size, the more local and potentially safe material processing will be. Using smaller beam sizes provides cavity formation with high accuracy, while the lateral size of these cavities may be significantly less than the lateral size of the cavities formed by mechanical tools. Using microbeams provides cavity formation with a unique profile. Another advantage of microbeams is their ability to form a relief structure on the treated surface with a preset profile and microscopic dimensions comparable to the size of the microbeams, and also extended microchannels. Another characteristic of laser radiation is the ability of lasers to form short solitary spikes and their reproducible sequences – micropulses. These capabilities have not been fully utilized in modern laser dentistry. The process of laser hard tissue removal may be significantly optimized by controlling the spike duration, their duty cycle in a micropulse, and the micropulse repetition rate. Optimization may be achieved by delivering laser energy when it can reach the treated material with minimal loss, and will not be reduced by a water cooling system or by products of laser damage that have not yet left the laser cavity. We call the simultaneous use of micropulses and microbeams "M^2 technology." Single-mode radiation should be used to create optical beams with a small diameter of <0.2 mm [42–44]. Craters formed in human tooth dentin by different modes of YAG:Er laser ablation are shown in Figure 74.5.

The crater depth created by a single-mode laser beam with a diameter of ∼0.1 mm greatly exceeds the crater depth created by a traditional multimode laser beam with a large diameter (∼1.0 mm) for the same total laser energy. The crater diameter created by a single-mode laser beam is smaller than the crater diameter created by the traditional multimode laser beam in this case.

The possibility of extended cavity formation with a high aspect ratio [42] in dental hard tissues may create a possibility for new technologies in dentistry, minimally invasive microdrilling for drug delivery to dentin and pulp, dentinal bleaching [45], creating cavities of programmable shape, and so on.

Another possibility is associated with surface texturing to increase the adhesion of filling materials [46]. The M^2 beam can be used to increase the mechanical strength of compounds and reduce microleakage for bonding dental material to hard tissue [46], the essence of which is to increase the contact area of adhesive materials through the creation of textures on the hard material surface by laser radiation. Textures are represented by a sequence of microcraters (Figure 74.6).

In a recent study [47], the composite material Revolution (Kerr, Orange, CA, USA) was placed on a surface containing the texture and polymerized with light radiation produced by a Rembrandt Rembrandt® Allegro™ (Den-Mat® Corporation, Santa Maria, CA, USA) for 30 s. Conventional technology was used as a control, when a smooth enamel surface is first covered with Nano-Bond Self-Etch Primer, then Nano-Bond Adhesive, and after that Revolution, with subsequent polymerization. The shear-bond strength was investigated. It was shown that the bond strength of Revolution with the textured surface was almost three times higher.

In another recent study [48], the adhesion of bonding resin (Tetric EvoCeram; Ivoclar Vivadent, Schaan, Liechtenstein) to dentin of an extracted tooth after texturing was investigated. In control samples, G-Bond (GC America, Alsip, IL, USA) was first applied to the smooth dentin surface, then bonding resin, followed by polymerization. It was shown that the shear-bond strength with the textured dental surface was 24.5 ± 9.6 MPa and for the control surface 14.9 ± 4.4 MPa.

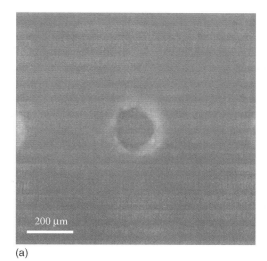
(a)

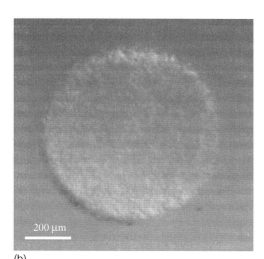
(b)

Figure 74.5 Photographs of craters formed in the dentin by single-mode (a, c) and multimode (b, d) Er:YAG lasers for the same exposure (0.1 J).

74.4
Conclusion

Significant efforts have been made by numerous groups and companies in the period 1985–1995 to understand and optimize laser ablation of dental hard tissue. As a result, in last 15 years the erbium laser for hard tissue was introduced as a commercial instrument for laser drilling. Laser drilling in clinical practice has been proven to be a safe, minimally invasive method for cavity preparation. The speed of

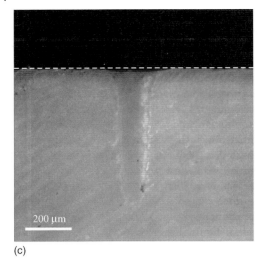

(c)

(d)

Figure 74.5 (Continued)

laser drilling for the best lasers is comparable to that of mechanical drilling. Laser drilling is less painful than a high-speed turbine but anesthesia is still necessary in about 50% of cases. A laser drill is mostly used for class II and V cavity preparation, selective caries removal, and calculus removal. Future applications may include laser "etching" of hard tissue surfaces before bonding. However, the hard dental tissue laser drill has not yet become a standard instrument in dental practice because of the very high price and complexity of a standard flashlamp-pumped erbium laser and insufficient advantages compared with alternative instruments. Hence further

(a)

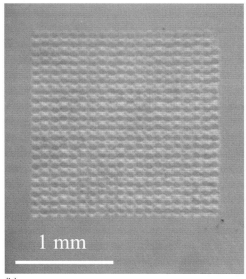

(b)

Figure 74.6 Photograph of human tooth enamel fragment textured by an Er:YAG laser with laser beam diameter $D \approx 100\,\mu m$ and period $dx = 40\,\mu m$ (a) and $D \approx 100\,\mu m$ and $dx = 100\,\mu m$ (b). Scanning electron micrograph of texture with $D \approx 70\,\mu m$ and $dx = 100\,\mu m$ (c).

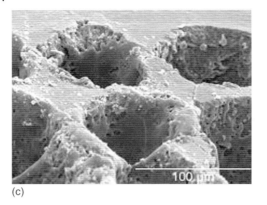

(c)

Figure 74.6 (Continued)

improvements in the laser ablation technique, finding more cost-effective lasers and discovering unique and important laser applications in the dental clinic are the new challenges for future research.

References

1 Nanci, A. and Ten Cate, A.R. (2003) *Ten Cate's Oral Histology: Development, Structure, and Function*, Mosby, St. Louis, MO.
2 Tasev, E., Delacretaz, G.P., and Woeste, L.H. (1990) Drilling in human enamel and dentin with lasers: a comparative study. *Proc. SPIE*, **1200**, 437–445.
3 Fried, D. (2000) IR laser ablation of dental enamel. *Proc. SPIE*, **3910**, 136–148.
4 Ragadio, J.N., Lee, C.K., and Fried, D. (2000) Residual energy deposition in dental enamel during IR laser ablation at 2.79, 2.94, 9.6, and 10.6 μm. *Proc. SPIE*, **3910**, 204–209.
5 Belikov, A.V., Skripnik, A.V., and Shatilova, K.V. (2010) Study of the dynamics of the absorption spectra of human tooth enamel and dentine under heating and ablation by submillisecond pulse radiation of an erbium laser with a generation wavelength of 2.79 μm. *Opt. Spectrosc.*, **109** (2), 211–216.
6 Wieliczka, D.M., Weng, S., and Querry, M.R. (1989) Wedge shaped cell for highly absorbent liquids: infrared optical constants of water. *Appl. Opt.*, **28** (9), 1714–1719.
7 Zolotarev, V.M., Mikhailov, B.A., Alperovich, L.I., and Popov, S.I. (1969) Dispersion and absorption of liquid water in the infrared and radio regions of the spectrum. *Opt. Spectrosc.*, **27**, 430–432.
8 Fried, D. (2005) Laser processing of dental hard tissues. *Proc. SPIE*, **5713**, 259–269.
9 Lee, C., Ragadio, J.N., and Fried, D. (2000) Influence of wavelength and pulse duration on peripheral thermal and mechanical damage to dentin and alveolar bone during IR laser ablation. *Proc. SPIE*, **3910**, 193–203.
10 Cummings, J. and Walsh, J. (2009) Erbium laser ablation: the effect of dynamic optical properties. *Appl. Phys. Lett.*, **62** (16), 1988–1990.
11 Altshuler, G.B., Belikov, A.V., Erofeev, A.V., and Egorov, V.I. (1993) Simulation of laser destruction of hard tooth tissues. *Proc. SPIE*, **2080**, 10–19.
12 Majaron, B., Sustercic, D., Lukac, M., Skaleric, U., and Funduk, N. (1998) Heat diffusion and debris screening in Er:YAG

laser ablation of hard biological tissues. *Appl. Phys. B Lasers Opt.*, **66** (4), 479–487.

13 Altshuler, G.B., Belikov, A.V., and Erofeev, A.V. (1994) Laser treatment of enamel and dentine by different Er lasers. *Proc. SPIE*, **2128**, 273–281.

14 Keller, U. and Hibst, R. (1995) Er:YAG laser effects on oral hard and soft tissues in *Lasers in Dentistry* (eds. L.J. Miserendino and R.M. Pick), Quintessence Publishing, Hanover Park, IL, pp. 161–172.

15 Altshuler G.B., Anderson, R.R., Manstein, D., Zenzie, H.H., and Smirnov, M.Z. (2001) Extended theory of selective photothermolysis. *Lasers Surg. Med.*, **29** (5), 416–432.

16 Brown, W.S., Dewey, W.A., and Jacobs, H.R. (1970) Thermal properties of teeth. *J. Dent. Res.*, **49** (4), 752–755.

17 Belikov, A.V., Erofeev, A.V., and Skripnik, A.V. (1995) Thermal response of pulp during laser treatment of enamel and dentine. *Tech. Phys. Lett.*, **21**, 330–331.

18 Visuri, S.R., Walsh, J.T. Jr., and Wigdor, H.A. (1996) Erbium laser ablation of dental hard tissue: effect of water cooling. *Lasers Surg. Med.*, **18** (3), 294–300.

19 Altschuler, G. and Erofeev, A. (1995) New concepts in laser-tissue interaction in *Lasers in Dentistry* (eds. L.J. Miserendino and R.M. Pick), Quintessence Publishing, Hanover Park, IL, pp. 299–315.

20 Zach, L. and Cohen, G. (1965) Pulp response to externally applied heat. *Oral Surg. Oral Med. Oral Pathol.*, **19**, 515–530.

21 Mir, M., Meister, J., Franzen, R., Sabounchi, S.S. et al. (2008) Influence of water-layer thickness on Er:YAG laser ablation of enamel of bovine anterior teeth. *Lasers Med. Sci.*, **23** (4), 451–457.

22 White, J.M., Fagan, M.C., and Goodis, H.E. (1994) Intrapulpal temperatures during pulsed Nd:YAG laser treatment of dentin, *in vitro*. *J. Periodontol.*, **65** (3), 255–259.

23 Goodis, H.E., Schein, B., and Stauffer, P. (1988) Temperature changes measured *in vivo* at the dentinoenamel junction and pulpodentin junction during cavity preparation in the *Macaca fascicularis* monkey. *J. Endod.*, **14** (7), 336–339.

24 Mollica, F.B., Camargo, F.P., Zamboni, S.C., Pereira, S.M. et al. (2008) Pulpal temperature increase with high-speed handpiece, Er:YAG laser and ultrasound tips. *J. Appl. Oral Sci.*, **16** (3), 209–213.

25 Krmek, S.J., Miletic, I., Simeon, P., Mehicic, G.P. et al. (2009) The temperature changes in the pulp chamber during cavity preparation with the Er:YAG laser using a very short pulse. *Photomed. Laser Surg.*, **27** (2), 351–355.

26 Staninec, M., Darling, C.L., Goodis, H.E., Pierre, D. et al. (2009) Pulpal effects of enamel ablation with a microsecond pulsed lambda=9.3-microm CO_2 laser. *Lasers Surg. Med.*, **41** (4), 256–263.

27 Glockner, K., Rumpler, J., Ebeleseder, K., and Stadtler, P. (1998) Intrapulpal temperature during preparation with the Er:YAG laser compared to the conventional burr: an *in vitro* study. *J. Clin. Laser Med. Surg.*, **16** (3), 153–157.

28 Attrill, D.C., Davies, R.M., King, T.A., Dickinson, M.R., and Blinkhorn, A.S. (2004) Thermal effects of the Er:YAG laser on a simulated dental pulp: a quantitative evaluation of the effects of a water spray. *J. Dent.*, **32** (1), 35.

29 Park, N.S., Kim, K.S., Kim, M.E., Kim, Y.S., and Ahn, S.W. (2007) Changes in intrapulpal temperature after Er:YAG laser irradiation. *Photomed. Laser Surg.*, **25** (3), 229–232.

30 Geraldo-Martins, V.R., Tanji, E.Y., Wetter, N.U., Nogueira, R.D., and Eduardo, C.P. (2005) Intrapulpal temperature during preparation with the Er:YAG laser: an *in vitro* study. *Photomed. Laser Surg.*, **23** (2), 182–186.

31 Cavalcanti, B.N., Lage-Marques, J.L., and Rode, S.M. (2003) Pulpal temperature increases with Er:YAG laser and high-speed handpieces. *J. Prosthet. Dent.*, **90** (5), 447–451.

32 Hibst, R. and Keller, U. (1993) Mechanism of Er:YAG laser-induced ablation of dental hard substances. *Proc. SPIE*, **1880**, 156–162.

33 Fried F D., Ashouri, N., Breunig, T., and Shori, R. (2002) Mechanism of water augmentation during IR laser ablation of dental enamel. *Lasers Surg. Med.*, **31** (3), 186–193.

34 Altshuler, G.B., Belikov, A.V., Gagarskiy, S.V., Erofeev, A.V., and

Parakhuda, S.E. (1995) Peculiarities of temporal structure of erbium lasers. *Proc. SPIE*, **1984**, 190–200.

35 Belikov, A.V., Erofeev, A.V., Shumilin, V.V., and Tkachuk, A.M. (1993) Comparative study of the 3 μm laser action on different hard tooth tissue samples using free running pulsed Er-doped YAG, YSGG, YAP and YLF lasers. *Proc. SPIE*, **2080**, 60–67.

36 Altshuler, G.B., Belikov, A.V., and Sinelnik, Y.A. (2001) A laser-abrasive method for the cutting of enamel and dentin. *Lasers Surg. Med.*, **28** (5), 435–444.

37 Hibst, R. and Keller, U. (1989) Experimental studies of the application of the Er:YAG laser on dental hard substances: I. Measurement of the ablation rate. *Lasers Surg. Med.*, **9** (4), 338–344.

38 Dostalova, T., Jelinkova, H., Hamal, K., Krejsa, O. *et al.* (1996) Evaluation of depth and profile cavity after laser ablation with different energy of Er:YAG laser radiation. *Proc. SPIE*, **2623**, 88–93.

39 Nishimoto, Y., Otsuki, M., Yamauti, M., Eguchi, T. *et al.* (2008) Effect of pulse duration of Er:YAG laser on dentin ablation. *Dent. Mater. J.*, **27** (3), 433–439.

40 Stock, K., Hibst, R., and Keller, U. (1997) Comparison of Er:YAG and Er:YSGG laser ablation of dental hard tissues. *Proc. SPIE*, **3192**, 88–95.

41 Lizarelli, R., Moriyama, L.T., and Bagnato, V.S. (2002) Ablation rate and micromorphological aspects with Nd:YAG picosecond pulsed laser on primary teeth. *Lasers Surg. Med.*, **31** (3), 177–185.

42 Tokarev, V.N. (2006) Mechanism of laser drilling superhigh-aspect-ratio holes in polymers. *Quantum Electron.*, **36** (7), 624–637.

43 Meister, J., Franzen, R., Apel, C., and Gutknecht, N. (2004) Influence of the spatial beam profile on hard tissue ablation. Part II: pulse energy and energy density distribution in simple beams. *Lasers Med. Sci.*, **19** (2), 112–118.

44 Meister, J., Apel, C., Franzen, R., and Gutknecht, N. (2003) Influence of the spatial beam profile on hard tissue ablation. Part I: multimode emitting Er:YAG lasers. *Lasers Med. Sci.*, **18** (2), 112–118.

45 Belikov, A.V. and Feldchtein, F.I. (2009) New method for intrinsic whitening of vital teeth using Er,Cr:YSGG laser microperforation: first *in vivo* cases. Presented at the Academy of Laser Dentistry 16th Annual Conference, Las Vegas, NV, SA-51.

46 Belikov, A.V., Skrypnik, A.V., and Shatilova, K.V. (2009) YAG:Er laser texturing of human teeth hard tissue surface. *Proc. SPIE*, **7547**, 754705–754706.

47 Belikov, A.V., Pushkareva, A., Skrypnik, A.V., Strunina, T., and Shatilova, K.V. (2009) Laser texturing of material surfaces. *Sci. Instrum.*, **53** (4), 52–56.

48 Samad-Zadeh, A., Kugel, G., Harsono, M., Defuria, C. *et al.* (2009) The influence of laser-textured dentinal surface on the bond strength. Presented at the International Association of Dental Research 87th General Session, Miami, FL.

Index

1-methyl-4-phenylpyridinium (MPP) 958
3D image 202, 230, 231, 422, 427, 722
4-hydroxyphenylpyruvate dioxygenase (HPPD) 483
5-ALA 329, 331, 531, 727

a

ABCD rule 818
aberrometer 631, 636, 687
– wavefront analysis 636
ablation
– laser ablation 378, 384, 393, 511–514, 689, 924, 927, 931, 963, 967, 969, 970, 1015, 1016, 1074, 1085–1092, 1095–1106, 1108
– photoablation 609, 610, 675, 676, 678, 917, 924, 926–928, 930, 1057
– thermal ablation 945
absorption
– cross-section 207
– spectroscopy 8, 105–107, 119, 437, 527, 839, 848
– spectrum 50, 106, 263, 323, 324, 411, 446, 453, 590, 836, 839, 841, 844, 906, 1098
acceptor 13, 18, 23–25, 50, 178, 331, 838
acetylcholine 575
acne 271, 457, 837, 855, 856, 858, 860, 862–865, 933
actin 98
actinic keratosis (AK) 271, 272, 386, 387, 806–810, 812
action potential 720
active nanoparticles 323, 325
adenoma 355–357
adenosine triphosphate (ATP) 445, 638, 730, 835, 873, 956–958
adhesion 117, 119, 384, 574, 971, 1065, 1072, 1078, 1092, 1104
advanced light source (ALS) 1015

aerosol 327, 448, 469
aerosol-OT (AOT) 327
AFB. *see* autofluorescence bronchoscopy
AFM. *see* atomic force microscopy
AFS. *see* autofluorescence spectroscopy
Age-Related Eye Disease Study (AREDS) 631
age-related macular degeneration (AMD) 321, 597, 598, 600–602, 654, 988
AH. *see* assisted hatching
AIDS 298, 653, 1068
AK. *see* actinic keratosis
ALA. *see* aminolevulinic acid
ALS. *see* advanced light source
Alzheimer's disease 55, 58, 716
AMD. *see* age-related macular degeneration
amino acids 129, 131, 207, 211, 482, 527, 675, 837
aminolevulinic acid (ALA) 192, 266, 269, 272–275, 282, 292, 316, 329, 331, 345, 346, 469, 477, 480, 528, 530–533, 548, 601, 727–729, 789, 859–865
angiogenesis 230, 467, 552, 726, 855, 870, 871
animal model 192, 419, 420, 452, 548, 550, 551, 553, 555, 572, 637, 702, 767, 769, 782, 797, 870, 958, 959
anisotropy 605, 997, 998, 1000
ANN. *see* artificial neural network
anterior segment (AS) 244, 302, 384, 585, 586, 606, 608, 620, 638, 679, 680
antibiotic resistance 122
antibiotics 106, 126, 128–130, 136, 267, 862
antibodies 36, 38, 61, 134, 145, 163, 164, 208, 220, 239, 291, 292, 295, 322, 729
– monoclonal antibodies 292, 295, 322
antigen 38, 44, 45, 239, 274, 789
anti-Stokes Raman scattering 78, 220, 235, 244, 719, 724, 732

AOT. *see* aerosol-OT
APD. *see* avalanche photodiode
apoptosis   56, 164, 165, 230, 263, 265, 299, 322, 382, 552, 730, 861, 873, 908, 957, 958
aptamer   290, 324, 332
arachidonic acid   956
AREDS. *see* Age-Related Eye Disease Study
Argon ion laser   190
argon laser trabeculoplasty   596, 614, 661, 975
argon plasma coagulation   477, 479, 524, 529
arteriovenous passage time (AVP)   666
arthritis   388, 415, 547–557, 561, 566, 569, 571, 572, 786
artificial neural network (ANN)   108, 113, 114, 116, 122, 241, 282
AS. *see* anterior segment
aspect ratio   605, 1103, 1104
assisted hatching (AH)   895–898, 1072
asthma   432
atherosclerosis   182, 188, 189, 237, 246, 548, 574, 575
atomic force microscopy   106
atoms   6, 210, 324, 635, 637, 842, 924, 927, 1055
ATP. *see* adenosine triphosphate
ATR. *see* attenuated total reflection
atrial fibrillation   457–460, 531
attenuated total reflection (ATR)   7, 117–120, 132, 135, 1015
autofluorescence bronchoscopy (AFB)   468, 477, 491, 492, 494, 495
autofluorescence spectroscopy (AFS)   526, 528
autoimmune diseases   780
automation   63
autophagy   322
avalanche photodiode   227, 771
AVP. *see* arteriovenous passage time
axon   716

b

background fluorescence   299, 471, 593, 594
bacteria   27, 107, 113, 116, 118, 122, 123, 127–136, 301, 837, 1012, 1016, 1041, 1049, 1070, 1072, 1074, 1076, 1077
bacterial infection   872
Barrett's esophagus (BE)   332, 337, 338, 348–355, 381, 387, 523–530, 532, 534–536
basement membrane   391, 695, 697–699, 707, 708, 968, 969
BCC. *see* Burkholderia cepacia complex
BCVA. *see* best-corrected visual acuity
BE. *see* Barrett's esophagus

beam quality   934
bending   500, 839
bending vibrations   839
benzoporphyrin derivative (BPD)   266, 859
best-corrected visual acuity (BCVA)   650, 652–654
bidirectional scattering distribution functions (BSDF)   1002
bilayer membrane   327
binning   186
bioassay   24, 25, 47, 56
biochip   35–63
biodegradable nanoparticles   326, 330, 332
biological samples   5, 8, 38, 106, 235
bioluminescence   725, 730
biophotonic imaging   715, 725, 726, 729, 735
biophotonics
– methods   77, 777
– networks   241
– technologies   77, 94
biosensor   14, 15, 19, 838, 847
biotechnology   11, 194, 211
biotin-streptavidin   61
birefringence   340, 363, 364, 379, 380, 389, 421, 605, 620, 621, 702, 845, 919, 1006, 1007, 1030
bladder   15, 191, 270, 281, 337, 341–346, 348, 364, 366–368, 370, 379, 387, 388, 505–507, 511, 913
blood cells   57, 147, 237, 238, 240, 406, 410, 434, 448, 571, 667, 721, 746, 768, 790, 827, 844, 1061, 1068
blood flow   308, 309, 403–438, 449, 450, 548, 571, 572, 574, 576, 593, 610, 611, 665–669, 719–721, 732, 734, 743, 745–748, 761, 767–772, 784–787, 812, 824, 827–830, 841, 844, 861, 870, 871, 955, 1067, 1068
– cerebral (CBF)   416, 419, 428, 429, 745–749, 761, 767–772
– measurement   610, 665, 666, 668, 745, 1068
– microvascular   403–438
blood oxygen level-dependent (BOLD)   227, 716, 727, 747
blood oxygen saturation   839
blood–brain barrier   728, 729
BOLD. *see* blood oxygen level-dependent
bond strength   1060, 1065, 1104
bone   161, 164, 165, 388, 465, 552, 564, 565, 567, 874, 1016, 1031, 1073, 1077, 1078, 1085, 1095
BPD. *see* benzoporphyrin derivative
brain development   724, 743, 744, 746, 748–750

breast   37, 52–54, 57, 135, 160, 163–165, 168–171, 192, 215–219, 221, 225, 230, 232, 263, 306, 325, 327, 328, 381, 390, 391, 427
breast cancer   37, 52, 54, 57, 160, 163–165, 169–171, 192, 216, 217, 230, 263, 325, 327, 390, 391
breast cancer diagnostics   160, 163, 164, 192
bright field   3, 4, 119, 126, 166, 176, 179, 215
– illumination   176, 179
– microscopy   215
bronchoscopy   465–486, 491–499
BSDF. see bidirectional scattering distribution functions
Burkholderia cepacia complex (BCC)   116, 122, 123, 193, 271, 272, 387, 805–809, 811, 813

c
Caenorhabditis elegans   247
calcium   17, 183, 238, 720, 956, 1015, 1016, 1050, 1059, 1064, 1069, 1096
calcium imaging   718
cancer diagnosis   57, 58, 146, 163, 192, 880, 882
– CT   57, 58, 146, 163, 192, 880, 882
– sensitivity   192, 882
– specificity   192, 882
– techniques   163, 882
cancer incidence   261, 271, 327, 337, 531, 698, 777, 817, 879
cancer mortality   337
capillaroscopy   406, 407, 409, 410, 437, 785, 786
CAR. see carotenoid
carbohydrates   6, 105, 128, 164, 290, 839, 848
carbon monoxide   448
carboxyfluorescein   674, 675
carcinogen   220, 862
carcinogenesis   344, 526, 862
carcinoma in situ (CIS)   191, 219, 220, 267, 268, 337, 342–344, 466, 474–476, 480, 492, 497, 505, 506, 697, 1032, 1033
cardiac myocyte   906
cardiovascular disease   54, 591, 669
cardiovascular surgery   385
carotenoid (CAR)   125, 298, 299, 302, 798
CARS microscopy   243–248
CARS. see coherent anti-Stokes Raman scattering
cartilage   189, 363, 378, 381, 388, 389, 551, 552, 561, 695, 699, 873, 1095
caspases   299
cataract   598–600, 604, 607, 609, 629–640, 662, 663

cataract detection   629–640
cathepsin B   301, 553
cavernous nerve (CN)   388
cavitation   515, 599, 600, 914, 924, 928, 929, 964, 971
CBF. see cerebral blood flow
CCD. see charge-coupled device
cDNA   36, 37, 39, 41, 44, 52
– arrays   36, 37, 39, 41, 52, 53
– microarrays   36, 37, 39, 41, 52, 53
cell biology   61, 89, 96, 159, 181, 211, 292, 892
cell characterization   146, 236, 237, 239
cell cycle   55, 57, 165, 236, 730
cell death   164, 299, 321, 322, 328, 445, 483, 495, 548, 552, 596, 861, 959
cell differentiation   164, 292, 1032
cell sorting   43, 145–148
cell wall   128, 129, 131, 835, 1070
cell-penetrating peptide (CPP)   290
cellular activation   292
cellulose   20, 43
central nervous system (CNS)   381, 729, 732, 734, 872, 955
central retinal vein occlusion (CRVO)   600, 987
cerebral blood flow (CBF)   416, 419, 428, 429, 745–749, 761, 767–772
cerebral hemodynamic activity   232
cerebral ischemia   955–958, 960
cerebrospinal fluid (CSF)   126, 127, 451, 669, 735
cervical intraepithelial neoplasia   879
cervical neoplasia   338, 360, 392, 879–883
cervix   267, 306, 341, 358–362, 364–367, 379, 879, 880, 882
CGH. see comparative genome hybridization
charge-coupled device (CCD)
– camera   15, 20, 49, 217, 416, 417, 468, 474, 476, 548, 720, 721, 768, 888, 1043
– sensor   492
– technology   49
chemical bonds   824
chemiluminescence   50
chemometric methods   112
chemometrics   111, 135
chemotherapy   52, 230, 289, 321, 477, 707
chiral molecules   845
chitin   327
chlorophyll   207, 262, 482, 530, 531
cholesterol   731, 796
choroidal melanoma   381, 990
choroidal neovascularization (CNV)   597, 985, 988, 989
chromatin   525

chromophores 106, 155, 182, 207, 209, 229, 267, 268, 324, 404, 405, 422, 479, 715, 717, 718, 720, 771, 787, 821, 830, 836, 957, 1042, 1057
chromosomes 46
CIA. *see* collagen-induced arthritis
CIN. *see* cervical intraepithelial neoplasia
CIS. *see* carcinoma *in situ*
classifier 111, 113, 388
CLE. *see* confocal laser endomicroscopy
clinical diagnostics 36, 37, 46, 52, 56–59, 63, 388, 552
clinical imaging 182, 189, 388, 715
clinical study 124, 346, 468, 511, 526, 880, 1074
clinical trial 277, 280, 322, 530, 534, 574, 598, 727, 728, 864, 898, 944, 956, 959, 960, 995
clinically significant macula edema (CSME) 645
CLSM. *see* confocal laser scanning microscopy
CME. *see* cystoid macular edema
CMOS. *see* complementary metal-oxide semiconductor
CN. *see* cavernous nerve
CNS. *see* central nervous system
CNV. *see* choroidal neovascularization
co-oximetry 446, 447
cofactors 27, 46, 527, 790
cognition function 760
cognitive development 229, 750
coherence length 154, 217, 389, 721, 781, 819, 846, 848
coherent anti-Stokes Raman scattering (CARS) 8, 78, 80, 90–94, 220, 221, 235, 243–248, 719, 724, 732, 735
– microscope 235, 732
collagen
– type I 363–366, 368, 873, 919, 1031, 1032
– type II 873
– type III 363, 368, 1031
collagen-induced arthritis (CIA) 549, 550, 552–554
colon 191, 337, 348, 355, 370, 497, 526, 528
colonoscopy 355
colorectal cancer 53, 54, 337
colposcopy 338, 359–361, 364, 879–883
combination bands 7
comparative genome hybridization (CGH) 52
complementary metal-oxide semiconductor (CMOS) 203, 414, 435, 828, 831
computed tomography (CT) 466, 547, 561, 716, 725, 726, 743, 759
condenser 4, 5

confocal fluorescence microscopy 486, 494, 495
confocal laser endomicroscopy (CLE) 523, 529
confocal laser scanning microscopy (CLSM) 17, 786–790, 798, 808–810, 821
confocal microscopy 5, 163, 186, 245, 338, 404, 405, 718, 722, 723, 732, 788, 789, 795
conformations 6, 919
conventional microscopy 737
cornea 155, 380, 383, 384, 445, 583–586, 588, 600, 607, 608, 610, 620, 633, 634, 636, 677, 686, 689, 690, 781, 846, 914, 921, 927–930, 932, 963, 964, 966, 967, 971
corneal ablation 972
corneal flap 384, 600, 610, 914, 929
corneal topography 607, 608, 690, 972
covalent bond 298, 592
CP-OCT. *see* cross-polarization optical coherence tomography
CPP. *see* cell-penetrating peptide
CRC. *see* colorectal cancer creatinine
critical angle 51
cross-correlation frequency-resolved optical gating (XFROG) 91, 92
cross-polarization optical coherence tomography (CP-OCT) 340, 363–367, 369, 370, 1029, 1031–1033
CRVO. *see* central retinal vein occlusion
CSF. *see* cerebrospinal fluid
CSME. *see* clinically significant macula edema
CT. *see* computed tomography
cyclins 165
cyclophotocoagulation 614, 661, 663
cystoid macular edema 651, 652, 654
cystoscope 343, 345, 346
cytochrome c 957
cytoplasm 163, 164, 169, 218, 237, 238, 639, 787, 790, 811, 907
cytoskeleton 719

d

d-aminolevulinic acid (ALA) 266, 269, 272–275, 282, 292, 316, 329, 331, 345, 346, 469, 477, 480, 528, 531–533, 548, 601, 727–729, 859–865
DAFE. *see* diagnostic autofluorescence endoscopy
dark field 3, 4, 167, 201
– illumination 201
– image 201
DCS. *see* diffuse correlation spectroscopy
denaturation 44, 45, 161, 919, 920, 950, 1057
dendritic cells 38, 193, 821–823, 855

dental caries detection 1047, 1048, 1049
dental hard tissues 996, 1002, 1005, 1015, 1042, 1058, 1063, 1077, 1085, 1098, 1104
dental restorative materials 1085, 1086, 1088–1090, 1092
dentin 995–998, 1000, 1001, 1005, 1006, 1009, 1010, 1012–1018, 1049, 1057, 1060–1072, 1074, 1075, 1078, 1085–1092, 1095–1099, 1101–1105
deoxyribonucleic acid (DNA) 525
deposition 41, 47, 252, 609, 735, 789, 872, 903, 926, 975, 1001
depth profilometry 1048, 1051
dermoscopy 777–780, 806, 817–819, 821
dextran 16, 23–25, 27, 28, 327, 732, 838
DFB laser. *see* distributed feedback laser
DFS. *see* drug-induced fluorescence spectroscopy
DHM. *see* digital holographic microscopy
diabetes 11, 53, 574, 635, 645, 649, 650, 669, 835, 869, 986, 988
diabetic macular edema (DME) 602, 645–657, 985
diabetic retinopathy 588, 590, 595, 645–648, 650, 652, 654, 656, 985, 986
diagnostic autofluorescence endoscopy (DAFE) 474, 475
diagnostic marker 813
DIC. *see* differential interference contrast
differential interference contrast 4, 5, 167, 904
diffraction 5, 81, 82, 87, 151, 212, 217, 235, 427, 685, 715, 717, 736, 796, 929
– disc 212, 235
– limit 151, 212, 217, 235, 427, 685, 715, 717, 736, 929
diffuse correlation spectroscopy (DCS) 743, 744, 746, 748, 749
diffuse optical imaging (DOI) 225–232
diffuse optical tomography (DOT) 216, 561–568, 717, 767
diffuse reflection 727, 831, 1054
DIFOTI. *see* digital imaging fiber-optic transillumination
DIG. *see* digoxigenin
digital imaging fiber-optic transillumination (DIFOTI) 1002, 1003, 1047
digital micromirror device (DMD) 767, 771
digital microscopy 167, 171
digital slide 168
digitalis 945
digoxigenin (DIG) 44, 45
diode laser trabeculoplasty (DLT) 976, 977
diploid 525

dipole moment 842
discriminant partial least-squares (DPLS) 123
dispersion 77, 78, 89, 94–96, 98, 190, 291, 448, 639, 700, 795
DLPFC. *see* dorsolateral prefrontal cortex
DLS. *see* dynamic light scattering
DLT. *see* diode laser trabeculoplasty
DMD. *see* digital micromirror device
DME. *see* diabetic macular edema
DNA 6, 7, 35–45, 49–61, 78, 106, 121, 125, 129, 130, 147, 162, 163, 165, 208, 209, 211, 236, 238, 240, 252, 254, 256, 271, 301, 303, 309, 525–527, 535, 536, 861, 898
DNA microarrays 36, 37, 41, 42, 50, 54, 55, 57
DNA sequencing 6, 121, 125, 208, 209
DNA. *see also* deoxyribonucleic acid
DOI. *see* diffuse optical imaging
dorsolateral prefrontal cortex (DLPFC) 760, 761, 763
DOT. *see* diffuse optical tomography
double-stranded DNA 59
DPLS. *see* discriminant partial least-squares
drug delivery 90, 292, 293, 327–331, 422, 530, 789, 794, 798, 1104
drug development 52, 55, 404
drug-induced fluorescence spectroscopy (DFS) 528
dye laser 90, 269, 270, 419, 480, 496, 860, 870, 930, 983
dynamic light scattering (DLS) 631, 633, 639, 640
dynamic range 38, 49, 590, 839, 1049
dysplasia 267, 337, 338, 343, 344, 346, 349, 353–355, 360, 392, 466, 468, 471, 473, 476, 492, 523–526, 528–530, 532, 535, 536, 697, 698, 705–707, 778, 1032

e
E. coli 114, 115, 116, 122, 128, 131–134
ear, nose, and throat (ENT) 202, 270, 380, 381, 383, 391, 393, 465, 693, 697, 699
early central lung cancer (ECLC) 495
early diagnosis 614, 697, 777, 1030, 1037
early increase in blood supply (EIBS) 526
EB. *see* Epstein–Barr virus
EBUS. *see* endobronchial ultrasound
ECC. *see* enhanced corneal compensation
ECG. *see* electrocardiogram
ECLC. *see* early central lung cancer
ECM. *see* extracellular matrix *or* electrical caries monitoring

EDHF. *see* endothelium-derived hyperpolarizing factor
EEM. *see* excitation-emission matrix
EFE. *see* endovenous fluence equivalent
EGF. *see* epidermal growth factor
EGFP. *see* enhanced green fluorescent protein
EIBS. *see* early increase in blood supply
elastic light scattering   191, 448, 633, 717, 723, 830, 843
elastic scattering spectroscopy (ESS)   525, 526
elastin   182, 193, 194, 207, 267, 268, 471, 495, 527, 790, 792–794, 797, 822, 837
electric field   4, 5, 145, 759
electrical caries monitoring (ECM)   1047
electrocardiogram (ECG)   418, 736
electromagnetic radiation   261, 534, 592, 1048
electromagnetic wave   201
electron density   100, 928, 929
electron multiplied CCD   203
electronic excitation   49
ELISA. *see* enzyme-linked immunosorbent assay
ellipticity   340
ELM. *see* external limiting membrane
ELT. *see* endovenous laser therapy
EMCCD. *see* electron multiplied CCD
emission wavelength   49, 190, 208, 291, 722, 723, 838, 1055
EMR. *see* endoscopic mucosal resection
enamel   116, 995–998, 1000–1011, 1013–1019, 1030, 1043, 1047–1050, 1057, 1059–1066, 1068, 1085–1088, 1090–1092, 1095–1104, 1107
endobronchial ultrasound (EBUS)   465, 466, 485, 486
endocrine   161, 164
endocytosis   324
endodontics   1067, 1070, 1071, 1073, 1075
endomicroscopy   388, 485, 523
endoplasmic reticulum   290, 322
endoscope   99, 187, 203, 339, 341, 354, 390, 418, 460, 461, 466, 468, 492, 493, 512, 513, 662, 698–700
endoscopic mucosal resection   531
endoscopic optical coherence tomography (EOCT)   355, 357, 370
endoscopy   80, 90, 189, 190, 201–203, 337, 338, 357, 370, 387, 391, 392, 474, 505, 509, 510, 512, 514, 516, 523, 526, 529, 532, 535, 697, 698, 934
– advantages   387
– application fields   934
– components   80
– types   337, 370, 505, 934

endothelial dysfunction   574–576
endothelium   574, 575, 966
endothelium-derived hyperpolarizing factor (EDHF)   575
endovenous fluence equivalent (EFE)   941, 942
endovenous laser therapy (ELT)   661–663, 939–946, 948–951
enhanced corneal compensation (ECC)   621
enhanced green fluorescent protein (EGFP)   97, 98, 100, 212
enhanced permeability and retention (EPR)   291, 293, 295, 301, 322, 326
ENT. *see* ear, nose, and throat
enzyme-linked immunosorbent assay   46
enzymes   11–14, 27, 43, 46, 50, 55, 57, 58, 61–63, 162, 263, 292, 295, 296, 298, 302, 330, 480, 482, 553, 554, 836, 844, 956
EOCT. *see* endoscopic optical coherence tomography
epidermal growth factor (EGF)   164, 165, 219, 290
epidermal growth factor receptor   164, 219
epidermis   193, 387, 421, 422, 425, 781–783, 787, 788, 795, 796, 807, 808, 811–813, 821, 822, 923
epiretinal membrane (ERM)   385, 651, 653, 655
epithelial cells   251, 252, 254, 256, 595, 599, 969
epithelial layer   387, 495, 695, 707, 1034
epitopes   729
EPR. *see* enhanced permeability and retention
Epstein–Barr (EB) virus   253, 256
Er:YAG laser   387, 1057, 1061, 1065–1067, 1070–1075, 1078, 1085–1092, 1100–1103, 1105–1107
ERM. *see* epiretinal membrane
erythrocytes   23, 238, 445, 611, 665, 666, 732, 784, 923
ESD. *see* endoscopic submucosal dissection
esophagus   270, 337, 338, 348, 350, 351, 357, 358, 364, 369, 370, 381, 387, 497, 498, 523, 524, 526–528, 530, 532, 534, 536
ESS. *see* elastic scattering spectroscopy
etiology   898
eukaryotic cells   957
evanescent field   78, 81, 85, 89
excimer laser   383, 384, 609, 661, 662, 686, 687, 689, 858, 916, 927, 930, 963–965, 967, 968, 970, 971, 1055, 1057, 1061, 1065, 1067, 1071
excimer laser trabeculotomy   661, 662

excitation wavelength 106, 107, 114, 184, 186, 189–191, 204, 237, 238, 245, 325, 338, 481, 484, 485, 527, 528, 722, 732, 795, 796, 1010
excitation-emission matrix (EEM) 189
exocytosis 788, 789
external limiting membrane (ELM) 651, 652, 654–657
extinction coefficients 47, 209, 210, 322, 446, 447
extracellular matrix (ECM) 321, 552, 554, 790, 811, 822, 873, 895, 976, 1047
extraction 42, 43, 53, 134, 178, 446, 746, 748, 749, 841, 845, 928, 995
eye
– correction of wavefront aberrations 685–691
– examination 586
– surgery 383

f

FA. *see* fluorescein angiography
Fabry-Pérot interferometer 78, 184
FACS. *see* fluorescence-activated cell sorting
FAD. *see* flavin adenine dinucleotide
FAP. *see* fibroblast activation protein
far-infrared region 6
fast Fourier transform 431
fatty acids 238, 675, 796
fatty tissue 381
FCFM. *see* fibered confocal fluorescence microscopy
FD. *see* frequency domain
feature space 108, 111, 380
femtosecond laser 90, 154, 384, 600, 611, 727, 914, 924, 928, 929, 963, 964, 966, 970, 971, 1018, 1092
femtosecond laser pulse 154, 727, 929
FFPE. *see* formalin-fixed, paraffin-embedded tissues
FIA. *see* flow injection analysis
fiber laser 77, 91, 189, 246, 509, 889
fiber photocatheter (FPC) 457–460
fiber-based imaging 77, 82, 83, 461
fibered confocal fluorescence microscopy (FCFM) 494, 495
fiber-optic transillumination (FOTI) 1002, 1003, 1047
fibroblast activation protein (FAP) 299
fibroblasts 382, 673, 675, 678, 944, 1078
fibronectin 552, 904
field microscopy 3, 4, 80, 215, 718
– bright field 3, 4, 215
– dark field 3

field of view 165, 166, 379, 393, 438, 506, 586, 618, 718, 768, 791, 888, 1036
filaggrin gene (FLG) 797
filtration surgery 610, 673–680
fine needle aspiration (FNA) 160, 381
fingerprint region 105, 106, 135, 221, 839, 843
FIR. *see* far-infrared
FISH. *see* fluorescence *in situ* hybridization
FITC. *see* fluorescein isothiocyanate
fixation 39, 126, 152, 160, 162, 252, 392, 585, 690
FLA. *see* fluorescein angiography
flavin adenine dinucleotide (FAD) 13, 27, 182, 194, 207, 471, 837
FLDF. *see* functional laser Doppler flowmetry
flexible endoscope 203, 466, 698
FLG. *see* filaggrin gene
FLIM. *see* fluorescence lifetime imaging microscopy
flow cytometry 215, 525
fluence rate (FR) 273–276, 279, 280, 307, 308, 321, 483, 512, 946–948, 950
fluorescein angiography (FA/FLA) 602, 604, 646–649, 652, 653, 989
– applications 602
fluorescein isothiocyanate (FITC) 20, 23, 27, 789
fluorescence
– decay time 183
– endoscopy 190, 338, 474
– energy transfer 207
– (Förster) resonance energy transfer (FRET) 13, 16, 18, 23–30, 49, 50, 61, 62, 187, 208, 209, 212, 296, 323, 325, 838
– imaging 6, 181, 184, 187, 189, 190, 192, 194, 212, 246, 338, 345, 346, 387, 388, 393, 461, 476, 506, 528, 547–550, 552–557, 592, 723, 733, 879, 1010, 1011, 1013
– *in situ* hybridization (FISH) 39, 62
– intensity 15, 20, 21, 27, 193, 204, 268, 298, 301, 549–553, 555–557, 906, 1010, 1011, 1043
– labeling 212, 236
– lifetime 6, 24, 50, 61, 181, 183–185, 192–194, 204, 207, 209, 210, 727, 794, 824
– lifetime imaging microscopy (FLIM) 6, 181, 184–189, 191, 193, 204, 727
– measurements 6, 201, 473, 486, 837, 1012
– microscopy 6, 89, 94, 96, 163, 184, 186, 187, 189, 194, 201, 202, 235, 243, 248, 338, 486, 494, 495, 718, 727, 887
– quantum yield 49, 207–210, 212
– quenching 208, 326, 554, 838

– spectroscopy   5, 189, 190, 192, 207, 208, 275, 276, 460, 526–528, 837, 1041, 1042, 1044
– standard   207–212
– techniques   194, 201, 469, 471, 838
fluorescence-activated cell sorting   145–147
fluorescent dye   21–23, 44–46, 49, 57, 201, 207, 208, 210, 290, 330, 423, 548, 550, 595, 715, 732, 787
fluorescent imaging agents   547–549, 551, 553, 555
fluorescent nanoparticles   210
fluorescent probe   11, 12, 14, 16, 18, 20, 22, 24, 26, 28, 30, 179, 202, 207–212, 235, 677, 891
fluorescent proteins   29, 96, 98, 212, 636
fluorescent sensing   14, 15, 17–21, 23, 25–27, 29
fluorochromes   461, 548, 549, 554, 729, 903, 905
fluorophore   17, 20, 28, 45, 49, 50, 166, 182–184, 208, 210–212, 460, 469, 491, 495, 593, 675, 719, 722, 723, 728, 729, 787, 790, 822, 838
– endogenous   181–183, 194, 268, 495, 528, 727, 790, 1010
– – collagen   182, 194, 528, 790, 1010
– – elastin   182, 194, 495, 790
flux   183, 235, 278, 280, 432, 436, 571, 572, 599, 722, 785, 790, 1068
fMRI. *see* functional magnetic resonance imaging
FNA. *see* fine needle aspiration
focal plane array   732
focal volume   186
food-borne pathogens   106, 131–135
formalin-fixed, paraffin-embedded tissues (FFPE)   162
Förster (fluorescence) resonance energy transfer (FRET)   13, 16, 18, 23–30, 49, 50, 61, 62, 187, 208, 209, 296, 323, 325, 838
Fourier transform infrared (FTIR)   6, 105, 117, 119, 122–124, 126, 128, 129, 131, 132, 134, 135, 527, 732, 1015
FPC. *see* fiber photocatheter
FR. *see* fluence rate
frame rate   187, 391, 392
frequency domain (FD)   186, 187, 378, 701, 724, 745, 748, 749, 1005, 1013, 1014, 1029
frequency doubling   723
FRET. *see* fluorescence resonance energy transfer (also Förster resonance energy transfer)
frozen section   162, 189, 215, 216
– assessment   215, 216

FTIR. *see* Fourier transform infrared
functional imaging   725, 727, 729, 731, 835
functional laser Doppler flowmetry (FLDF)   668
functional magnetic resonance imaging (fMRI)   227, 452, 453, 716, 717, 724, 727, 743, 746, 747, 759, 760
functionalization   40, 47, 49, 61
fundus photography   587–591, 603, 605, 606, 646, 648
fusion protein   211

g
gains   452
galactose   21, 27
gas lasers   795
gastroesophageal reflux disease   523, 524, 526, 528, 530, 532, 534, 536
gastrointestinal (GI)   161, 190, 341, 348, 353, 357, 369, 370, 379, 381, 383, 387, 523, 529, 535, 1033
gastrointestinal tract   161, 341, 353, 369, 370, 1033
gating   91, 421–423, 690
gel electrophoresis   38, 120, 127
gene   36, 37, 39, 41, 50, 52–58, 63, 89, 90, 98, 122, 125, 165, 171, 204, 301, 321, 573, 730, 797, 873, 903, 906, 928
– expression analysis   39, 54, 78
– profiling   37, 38, 41, 54, 57, 63, 180
generalized nonlinear Schrödinger equation (GNSE)   95
genomics   42, 78, 80, 90
genus   116, 124, 131, 135
GFP. *see* green fluorescent protein
GI. *see* gastrointestinal
GIP. *see* glucose indicator protein
glandular mucosa (GM)   100, 342, 351, 352
glaucoma   340, 592, 596–598, 602, 603, 604, 613, 614, 616–620, 653, 661, 662, 665, 668, 669, 673–680, 975, 978–980, 988
glaucoma therapy   978
glucose   11–16, 18–30, 46, 62, 445, 716, 835, 836, 838–842, 844–849, 956
glucose determination   11, 12, 14, 16, 18, 20, 22, 24–26, 28, 30
glucose indicator protein (GIP)   28, 29
glucose sensing   11–14, 16, 18–20, 22, 24–26, 28, 30, 840
glutamate   956
glutathione   331
glycoproteins   23, 895
GM. *see* glandular mucosa

GNSE. *see* generalized nonlinear Schrödinger equation
Golgi apparatus 322
gradient refractive index (GRIN) 583
Gram-negative bacteria 27, 116, 122, 128, 1070
Gram-positive bacteria 122, 128, 1070
granulocytes 239
green fluorescent protein (GFP) 28, 29, 89, 98, 100, 211, 212, 905–908
green protein technology 207–212
GRIN. *see* gradient refractive index
group velocity dispersion (GVD) 78, 88, 95, 98
GVD. *see* group velocity dispersion

h

HA. *see* hyaluronic acid *or* hydroxyapatite
halogen lamp 888, 1065
hard tooth tissues 1030, 1103
HCA. *see* hierarchical cluster analysis
head and neck squamous cell carcinoma (HNSCC) 254, 255
heart rate 418, 718
heat capacity 842, 922, 1059
heat shock proteins 958
Heidelberg retina tomograph (HRT) 614, 617–620
helium-neon laser 190, 572
heme 219, 238, 240, 263, 266, 272, 448, 469, 471, 480, 530, 531, 836, 837
hemoglobin oxygenation 745, 746, 760, 767, 769, 770, 827, 828, 830
Henyey–Greenstein (HG) function 997, 998
hexose monophosphate shunt (HMPS) 638
HFUS. *see* high-frequency ultrasound
HG. *see* Henyey–Greenstein function
hierarchical cluster analysis (HCA) 108, 112–116, 121, 122, 124, 129, 130
high-frequency ultrasound (HFUS) 338, 404, 426, 806, 808, 879
high-power laser 509
high-resolution imaging 80, 81, 201, 203, 217, 377, 391, 431, 718
HMPS. *see* hexose monophosphate shunt
HNSCC. *see* head and neck squamous cell carcinoma
hormone 57, 164, 274, 835
horseradish peroxidase (HRP) 16, 46, 62
HPPD. *see* 4-hydroxyphenylpyruvate dioxygenase
HPV. *see* human papillomavirus
HRP. *see* horseradish peroxidase
HRT. *see* Heidelberg retina tomograph

HSA. *see* human serum albumin
human genome 256
human papillomavirus (HPV) 252–256, 338, 359, 879
human serum albumin (HSA) 328, 550, 836, 842
hyaluronic acid (HA) 554
hybridization 39, 40, 47, 50, 52, 57, 59–62, 163, 165, 171, 177, 302
hydrogen bond 919
hydroxyapatite (HA) 1095, 1096, 1098, 1099
hyperbilirubinemia 836, 844
hypermetropia 610
hyperplasia 274, 338, 343, 345, 509, 707, 855
hyperspectral imaging 184, 186–188, 452, 453, 720
hypertension 298, 574, 602, 645, 668, 669, 978, 987, 988

i

ICA. *see* independent component analysis
ICD. *see* irritant contact dermatitis
ICG. *see* indocyanine green
ICSI. *see* intracytoplasmic sperm injection
identification of microorganisms 105–136
ILM. *see* inner limitans membrane
IMAC. *see* intramucosal adenocarcinoma
image-guided optical spectroscopy 567
imaging
– fiber-based 77, 82, 83, 461
– high-resolution 80, 81, 201, 203, 217, 377, 391, 431, 718
– optical 80, 203, 216, 218, 226, 229, 230, 292, 379, 405, 453, 461, 547, 554, 638, 715–719, 724, 729, 734–736, 743, 744, 750, 769, 772, 835, 874, 1047
imaging techniques 84, 90, 181, 182, 185, 187, 189, 203, 216, 222, 419, 437, 449, 477, 547, 548, 557, 668, 718, 725, 726, 735, 743, 750, 763, 1013, 1018
IMC. *see* intramucosal adenocarcinoma
immersion 117, 237, 409, 858, 887
immobilization 20, 23, 35, 36, 41, 42, 54, 63
immune system 256, 263, 265, 321
immunofluorescence 39
immunohistochemistry (IHC) 39, 163, 164, 171, 871
in situ hybridization (ISH) 39, 61, 62, 163, 165, 171, 177
in vitro fertilization 895
in vitro instrumentation 181, 182, 184, 186, 188, 190, 192, 194
in vivo brain imaging 94, 715, 716, 718–720, 722, 724, 726, 728, 730, 732, 734, 736

in vivo instrumentation 201, 202, 204
in vivo optical imaging 734, 736, 743
incubation 46, 115, 124, 307, 322, 328
independent component analysis (ICA) 110, 111
indocyanine green (ICG) 202, 548–551, 555, 557, 593, 595, 666, 727, 732, 733, 746, 863, 864
inelastic collision 635
inelastic light scattering 723
infant brain 230, 743, 744, 746, 748–750
infectious diseases 52, 321
infrared spectroscopy 6, 251, 449, 451, 527, 724, 743, 760, 796, 1015
inner limitans membrane (ILM) 654
inner mitochondrial membrane 837, 957
inner segments (IS) 382, 651, 652, 654
integrating sphere 831, 844, 1001
intense pulsed light (IPL) 651, 864, 865
intensified charge-coupled device 203
interactions
– intermolecular 479
– laser–tissue 914, 915, 917, 919, 921, 923, 925, 927, 929
– protein–protein 37, 50, 61
interference 4, 6, 84, 92, 93, 147, 154, 167, 201, 301, 340, 416, 458, 601, 666, 721, 762, 763, 781, 838, 890, 1010
interferometer 78, 184, 217, 339–341, 391, 615, 846, 848
interferometric techniques 78
intermolecular interactions 479
internal conversion 208
internal reflection element (IRE) 117, 118
interstitial fluid (ISF) 840, 846
interstitial illumination 261
intracytoplasmic sperm injection (ICSI) 896–898
intramucosal adenocarcinoma (IMAC) 349, 352, 353, 355
intramucosal adenocarcinoma (IMC) 349, 350, 352–354
intraocular pressure (IOP) 596–598, 602, 610, 613, 614, 619, 661–663, 665, 666, 669, 673, 674, 676, 677, 975–980
intraoperative ultrasound (IOUS) 726
intraretinal fluid (IRF) 651
INVAC. *see* invasive adenocarcinoma
invasive adenocarcinoma (INVAC) 349, 352, 353, 355
inversion 447, 567
ion channels 186, 461
ion pumps 445
iontophoresis 574, 575, 789

IOP. *see* intraocular pressure
IOUS. *see* intraoperative ultrasound
IPL. *see* intense pulsed light
IR absorption 8, 105–107, 119, 123, 130, 135, 208, 209, 449, 839, 843, 847, 848, 1001
– spectroscopy 8, 105–107, 119, 123, 130, 135, 208, 449, 839, 843, 847, 848
– spectrum 106
IR laser 147, 325, 514, 515, 724, 934, 959, 960
IR microscopy 107, 243
IR spectrum 6, 105, 839, 841
IRE. *see* internal reflection element
IRF. *see* intraretinal fluid
irradiance 1062
irritant contact dermatitis (ICD) 781, 788, 789
IS. *see* inner segments
ISF. *see* interstitial fluid
ISH. *see in situ* hybridization
isosbestic point 422, 446, 719

k

Kaposi's sarcoma (KS) 788, 827
Kasha's rule 1010
keratin 182, 707, 790, 794, 797, 807, 811, 1029, 1032
keratinocytes 193, 194, 329, 811, 821, 824, 837, 861, 865
keratoconus 384, 607, 608
keratomileusis 384, 929, 963, 967
keratoplasty 384, 385, 610, 963, 964, 966
keratosis 271, 272, 386, 806–810, 812
KS. *see* Kaposi's sarcoma
Kubelka–Munk theory 424

l

lab-on-a-chip 56, 80, 146
label-free techniques 236
labial salivary glands 1034, 1035
Lambert–Beer law 915
laryngeal cancer 695–698
larynx 190, 379, 391, 695–702
LASCA. *see* laser speckle contrast analysis
LASCI. *see* laser speckle contrast imaging
LASEK. *see* laser-assisted subepithelial keratectomy
laser ablation 378, 384, 393, 511–514, 689, 924, 927, 931, 963, 967, 969, 970, 1015, 1016, 1074, 1085–1092, 1095–1106, 1108
laser bronchoscopy 498
laser diode 227, 411, 430, 449, 451, 459, 473, 492, 721, 1010, 1042, 1043
laser Doppler flowmetry (LDF) 407, 571, 572, 576, 611, 665, 666, 668, 767, 769, 784, 785, 1067, 1068

laser Doppler imaging (LDI)   413, 414, 418, 420, 430, 436, 437, 571–576, 667, 721, 784–786, 812, 824
laser Doppler perfusion imaging (LDPI)   404, 405, 411, 412, 414, 415, 418, 422, 425, 429, 431–435, 437, 824, 828
laser emission   918, 923, 924
laser endoscopy   509, 510, 512, 514, 516
laser line   413, 431, 435, 1001
laser lithotripsy   514, 515, 913, 927, 930
laser microdissection and pressure catapulting   147
laser microdissection and pressure catapulting (LMPC)   147
laser microtome   151–155, 157
laser penetration   914
laser photocoagulation   593, 647, 654, 983
laser scanning microscopy   202, 717, 786, 787, 789, 808, 809, 821
laser speckle contrast analysis (LASCA)   417, 419, 420, 829
laser speckle contrast imaging (LASCI)   417, 767–769, 829, 830
laser speckle perfusion imaging (LSPI)   404, 405, 415, 417–420, 424, 425, 429, 431, 433, 435, 437
laser surgery   391, 509, 614, 661, 702, 911, 913, 914, 916, 918, 920–922, 924, 926, 928–930, 932, 934, 943
laser therapy   481, 491, 492, 494–498, 500, 512, 513, 524, 654, 661, 662, 860, 870, 872, 921, 939, 940, 942, 944, 946, 948–950, 957–960, 977
laser thinning   895, 896, 898
laser trabeculoplasty   457, 596, 597, 614, 661, 931, 975–980
laser vision correction   963, 964, 966, 968, 970, 972
laser–tissue interactions   914, 915, 917, 919, 921, 923, 925, 927, 929
laser-assisted *in situ* keratomileusis (LASIK)   384, 600, 610, 634, 685, 688, 689, 929, 963, 964, 966–971
laser-assisted intracytoplasmic sperm injection   896
laser-assisted subepithelial keratectomy (LASEK)   967, 969
laser-induced fluorescence (LIF)   189, 207, 267–272, 528, 1010
LASIK. *see* laser-assisted *in situ* keratomileusis
lateral area of prefrontal cortex (LPFC)   762, 763
lattice vibrations   1058
layered media   422

LCTF. *see* liquid crystal tunable filter
LDA. *see* linear discriminant analysis
LDF. *see* laser Doppler flowmetry
LDI. *see* laser Doppler imaging
LDL. *see* low-density lipoprotein
LDPI. *see* laser Doppler perfusion imaging
LED. *see* light-emitting diode
leukemias   165
leukocytes   53, 127, 237, 554, 611
LIF. *see* laser-induced fluorescence
LIFE. *see* light-induced fluorescence endoscopy or lung imaging fluorescence endoscope
ligands   27, 38, 47, 295, 322, 551
light microscope   49, 239
light microscopy   47, 159, 161, 166, 363, 404, 405, 717, 736, 1008, 1091
light penetration   265, 266, 275, 279, 480, 724, 915, 916, 918
light scattering
– dynamic (DLS)   631, 633, 639, 640
– elastic   191, 448, 633, 717, 723, 830, 843
– inelastic   723
– quasi-elastic (QELS)   631, 633, 635, 830
light-emitting diode (LED)   166, 265, 270, 316, 958, 1062
light-induced fluorescence endoscopy (LIFE)   190, 471, 476
light-scattering spectroscopy (LSS)   525, 526
linear discriminant analysis   21, 241, 525
linearly polarized light   5, 421, 845, 846, 1007
lipid bilayer   327
lipid distribution   246, 247
lipopolysaccharide (LPS)   552
liquid crystal tunable filter (LCTF)   831
lithotripsy   514, 515, 913, 927, 930, 933
LITT. *see* laser interstitial thermal therapy
live cell imaging   246
LLLT. *see* low-level laser therapy
LMPC. *see* laser microdissection and pressure catapulting
LOC. *see* lab-on-a-chip
LOH. *see* loss of heterozygosity
long-term memory (LTM)   760, 763–765
loss of heterozygosity (LOH)   53
low-density lipoprotein (LDL)   322
low-level laser therapy (LLLT)   872–874, 955, 957–960
LPFC. *see* lateral area of prefrontal cortex
LSM. *see* laser scanning microscopy
LSPI. *see* laser speckle perfusion imaging
LSS. *see* light-scattering spectroscopy
LTM. *see* long-term memory

luminescence 6, 12, 14, 19, 21, 22, 24, 26, 84, 85, 87, 89, 94, 298, 323, 326, 368, 788, 873, 1013, 1014, 1017, 1047–1049
luminescence decay time 24
lung imaging fluorescence endoscope (LIFE) 492, 493
lymphocytes 855, 861, 1061
lymphomas 165, 1035
lysosomes 265, 322, 329, 482, 675

m

Mach–Zehnder interferometer 78
macrophages 237, 385, 552, 596, 729
MACS. *see* magnetic-activated cell sorting
magnetic nanoparticles 18, 134, 219, 220, 331
magnetic resonance imaging (MRI) 216, 230, 232, 275, 276, 279, 292, 331, 404, 452, 453, 553, 555, 561, 631, 633, 637, 638, 668, 716, 726, 729, 735, 743, 747, 759
magnetic-activated cell sorting (MACS) 145, 147
MAL. *see* methyl aminolevulinate
MALDI. *see* matrix assisted laser desorption/ionization
malignant lesions 190, 192, 218, 254, 271, 272, 345, 382, 476, 707, 1076
malignant melanoma (MM) 219, 220, 271, 272, 306, 350, 386, 732, 778, 791, 795, 812, 817, 818, 820–824, 837
malignant tissue 190, 192–194, 267, 363, 470
marker molecule 94, 98
matrix, extracellular (ECM) 321, 552, 554, 790, 811, 822, 873, 895, 976
matrix-assisted laser desorption/ionization (MALDI) 179
mediator 11, 305, 675
medicine, regenerative 157
melanin 155, 404, 405, 411, 422, 425, 595–597, 732, 787, 790, 806, 808, 821, 822, 824, 830, 837, 844, 975, 983
melanocytes 194, 791, 821, 823, 824, 837
membrane potential 720, 891
membranes 15, 42, 43, 46, 164, 265, 322, 330, 370, 384, 448, 479, 640, 675, 835, 891, 906, 919, 985, 987
memory 94, 229, 282, 759–765
MEMS. *see* microelectromechanical systems
meso-tetra(4-carboxyphenyl)porphyrin (mTCP) 326
meso-tetraphenylporpholactol (mTPPL) 328
metabolism 56, 127, 159, 329, 472, 474, 480, 549, 567, 590, 637, 715, 730, 733, 734, 743, 746, 749, 787, 790, 796

metal nanoparticles 47, 48, 50, 51, 58, 59, 61–63
metal-oxide-semiconductor 414
metastasis 466, 495, 731, 732
methicillin-resistant Staphylococcus aureus (MRSA) 121, 302
methyl aminolevulinate (MAL) 860, 864
micelles 291, 295, 322, 323, 327, 331
micro-Raman set-up 126
micro-Raman spectroscopy 106, 107, 114, 120, 126, 1016
microarrays 35–39, 41, 42, 49, 50, 52–58, 61, 178
microcirculation imaging 403–406, 408, 410, 412, 414, 416, 418, 420, 422, 424, 426, 428, 430, 432, 435–438
microcutting 913
microelectromechanical systems (MEMS) 339, 345, 461
microfluidic system 63, 146
microlenses 721
microorganisms 4, 56, 63, 105–108, 113, 116, 117, 120–124, 126, 127–136, 1042, 1070–1072, 1075
microscopic analysis 173
microscopy
– classical 159–173
– phase contrast 4, 5, 201, 887
microstructure fibers 77–100
microsurgical endoscopy 697, 698
microvascular blood flow 403–438, 666, 827
mid-infrared (MIR) 6–8, 457, 715, 732, 836, 1078, 1089
migration 225, 380, 548, 554, 556, 566, 567, 787
miniaturization 56, 63, 339, 514
minimally invasive surgery 934
MIR. *see* mid-infrared
mitochondria 265, 290, 322, 445, 446, 479, 525, 675, 790, 837, 891, 957
mitochondrial membrane potential (MMP) 299, 891
mitochondrial permeability transition pore (mPTP) 903, 906
mitomycin C (MMC) 872, 968
MIX. *see* Multiple Image Xposition
MLST. *see* multilocus sequence typing
MLT. *see* multiphoton laser tomography
MM. *see* malignant melanoma
MMC. *see* mitomycin C
MMP. *see* mitochondrial membrane potential
mode locking 964
modulated luminescence 1014, 1047, 1048, 1050

molecular
- beacons 295–303
- contrast imaging 218, 219, 221
- diagnostics 7, 37, 39, 53, 57, 59, 63, 90, 167, 461, 479, 813
- fingerprints 54, 732, 839
- imaging 219, 220
- targeting 289, 290, 292
- targets 38, 39, 55, 57, 129, 290, 298, 302, 861
monocytes 554
monomethoxy poly(ethylene glycol) (MPEG) 300, 554
Monte Carlo technique 411
Moorfields regression analysis (MRA) 618, 619
MP. see muscularis propria
MPA. see multiphoton absorption
MPEG. see monomethoxy poly(ethylene glycol)
MPM. see multiphoton microscopy
MPP. see 1-methyl-4-phenylpyridinium
MPT. see multiphoton tomography
mPTP. see mitochondrial permeability transition pore
MRA. see Moorfields regression analysis
MRI. see magnetic resonance imaging
mRNA 37, 43, 44, 55, 302
MRSA. see methicillin-resistant Staphylococcus aureus
MSRI. see multispectral reflectance imaging
mTCP. see meso-tetra(4-carboxyphenyl) porphyrin
mTetrahydroxyphenylchlorin (Foscan) (mTHPC) 266, 275, 330, 481, 530, 533
mTHPC. see mTetrahydroxyphenylchlorin (Foscan)
MTPE. see multispectral two-photon emission
mTPPL. see meso-tetraphenylporpholactol
multi-photon excitation 186, 422
multicolor imaging 98, 99
multidimensional fluorescence imaging 181, 189, 194
multilocus sequence typing (MLST) 120, 127
multiphoton absorption (MPA) 92, 94, 928
multiphoton imaging 790, 795, 811, 822
multiphoton laser imaging 790–795
multiphoton laser tomography (MLT) 791–794, 822, 823
multiphoton microscopy 78, 80, 186, 380, 718, 732, 734, 736
multiphoton tomography (MPT) 794, 810–812, 822, 823
Multiple Image Xposition (MIX) 476

multispectral reflectance imaging (MSRI) 767, 769, 770
multispectral two-photon emission (MTPE) 191
muscle 98, 135, 164, 357, 387, 419, 450, 451, 453, 457, 574, 575, 588, 695, 723
muscular dystrophy 55
muscularis propria (MP) 350, 351, 506
mutations 27, 29, 44, 52–54, 165, 797
mycobacteria 124, 125
myelin sheath 93
myosin 98

n

NAD. see nicotinamide adenine dinucleotide
nano-surgery 913, 928, 929
nanocarriers 293, 295, 327, 329, 331
nanoparticles
- active 323, 325
- magnetic 18, 135, 219, 220, 331
- metal 47, 48, 50, 51, 58, 59, 61–63
- passive 322, 323
- polymeric 323, 327, 330
narrowband imaging (NBI) 467, 468, 477, 486
National Institute for Health and Clinical Excellence (NICE) 530, 535
National Institutes of Health Stroke Scale (NIHSS) 960
natural moisturizer factor (NMF) 796, 797
NBI. see narrowband imaging
Nd:YAG laser 477–479, 495, 497, 498, 501, 513, 514, 524, 534, 599, 600, 662, 663, 795, 864, 873, 930, 975, 983, 1054, 1055, 1057, 1060, 1061, 1063–1067, 1069, 1071–1076, 1078
near-infrared imaging (NIRI) 452, 827
near-infrared spectroscopy (NIRS) 449–453, 724, 726, 727, 733, 743–750, 759–765
necrosis 38, 263, 265, 275, 322, 327, 382, 409, 479, 480, 484, 511, 552, 574, 726, 730, 919, 1058, 1069, 1077
- zone 730, 1058, 1069
neoplasia 182, 337, 338, 342, 344, 345, 353–355, 359, 360, 363, 365, 368, 370, 379, 392, 526, 707, 708, 879–883, 1031
nerve fiber indicator (NFI) 620
neuro-oncology 716, 719, 725, 727, 729–731
neurology 389, 713, 718
neuronal damage 246, 956
neurons 94, 98, 113, 229, 716, 720, 956–958
neurophotonics 94, 95, 97
NFI. see nerve fiber indicator

NICE. *see* National Institute for Health and Clinical Excellence
nicotinamide adenine dinucleotide (NAD)   12, 13, 27, 46, 182, 190, 194, 202, 207, 267, 268, 471, 528, 790, 837
NIHSS. *see* National Institutes of Health Stroke Scale
NIRI. *see* near-infrared imaging
NIRS. *see* near-infrared spectroscopy
nitric oxide (NO)   574, 575, 906, 956
NIVI. *see* nonlinear interferometric vibrational imaging
NLOM. *see* nonlinear optical microscopy
NMF. *see* natural moisturizer factor
NO. *see* nitric oxide
non-melanoma skin cancer   282, 386, 779, 780, 788, 789, 805, 806, 808, 810, 812, 814, 862
nonbiodegradable nanoparticles   322, 326, 330, 332
nonlinear interferometric vibrational imaging (NIVI)   220, 221
nonlinear optical microscopy (NLOM)   874
nonlinear optical microspectroscopy   90, 91, 93
nonradiative energy transfer   16, 50
nonresonant background   92, 93, 245
nontuberculous Mycobacteria (NTM)   124, 125
normal-tension glaucoma (NTG)   668, 669
normalization   108, 110, 112, 423, 604
NSOM. *see* scanning near field optical microscopy
NTG. *see* normal-tension glaucoma
NTM. *see* nontuberculous Mycobacteria
nuclear receptor   164
nucleases   43
nucleolus   525
nucleus   28, 164, 170, 238, 252, 290, 301, 309, 324, 525, 585, 632–634, 637, 811, 821
numerical aperture   77, 85, 86, 94, 151, 152, 165, 176, 202, 203, 245, 588, 700, 717, 890

o

OA. *see* osteoarthritis
objective lens   5, 201, 202, 414
observable   129, 646, 648, 652, 655, 919
OCM. *see* optical coherence microscopy
OCT. *see* optical coherence tomography
OD. *see* optical density
ODMS. *see* ocular-free digital microscope system
OEF. *see* oxygen extraction fraction
OIS. *see* optical imaging of intrinsic signals
OISI. *see* optical intrinsic signal imaging
OLED. *see* organic light-emitting diode
oligonucleotides   36, 37, 41, 42, 44, 45, 50, 53, 59, 302, 303
oncogenes   53
online monitoring   90
OP. *see* oropharynx
open-angle glaucoma   661, 665, 975, 978–980
operator   175, 338, 388, 498, 561, 599, 618, 619, 1055, 1063, 1066
ophthalmology   202, 218, 379, 383, 384, 581, 600, 609, 633, 656, 781, 983, 991
OPS. *see* orthogonal polarization spectroscopy
optical breakdown   151, 927–929
optical coherence imaging   215, 216, 218, 220, 222
optical coherence microscopy   380
optical density (OD)   5, 50, 447, 451, 591, 632, 744, 844, 906
optical diagnostics   529
optical imaging   80, 203, 216, 218, 225–232, 292, 379, 405, 453, 461, 547, 554, 583, 638, 715–719, 724, 729, 734–736, 743, 744, 750, 769, 772, 835, 874, 1047
– diffuse (DOI)   225–232
– *in vivo*   734, 736, 743
optical imaging of intrinsic signals (OIS)   719, 720, 769, 770
optical intrinsic signal imaging (OISI)   731
optical mammography   230
optical microangiography   427
optical oximetry   445, 446, 448, 450, 452
optical resolution   168, 904
optical rotation   845, 846, 848
optical sectioning   5, 186, 188, 718, 723
optical sensing   49, 78, 79, 836
optical sensor   12, 15, 30, 80, 84
optical tomography   216, 225, 229, 230, 232, 390, 561–566, 568, 717, 750, 767
optical topography   225–229, 232, 452, 762, 763
optical trap   146, 147, 891, 892
optical tweezer   146
optical window   370
optoporation   903–908
oral cancer   382, 383, 707
oral cavity   190, 253, 270, 699, 705, 706, 708, 1005, 1029, 1030, 1032–1034, 1055
oral cavity mucosa   1029, 1032, 1033
organelles   162, 201, 290, 525, 919
organic fluorophores   49, 60, 61, 207–209, 211, 212
organic light-emitting diode (OLED)   315–320
oropharynx (OP)   699, 700, 705–709

orthogonal polarization spectroscopy (OPS)   422, 423
OS. *see* outer segments
oscillator strength   208
osteoarthritis (OA)   182, 388, 389, 561–569, 573, 668
outer mitochondrial membrane   164
outer segments (OS)   651, 652, 654–657, 843, 844
overtones   7
oxygen extraction fraction (OEF)   746, 749

p

p53   58, 165
PA. *see* photoacoustic
PAM. *see* potential acuity meter
PAMAM. *see* poly(amidoamine)
PAOD. *see* peripheral arterial occlusive disease
pap smear   338, 359
Parkinson's disease   729
partial least-squares regression (PLSR)   135, 841
passive nanoparticles   322, 323
pathogen detection   36, 37, 56, 57, 63
pathogenesis   52, 654, 785, 797
pathogenic microorganisms   113, 131, 133, 134
pathology   5, 38, 143–256, 281, 340, 341, 345, 358, 359, 370, 377, 378, 385, 389, 391, 393, 525, 561, 648, 668, 702, 728, 732, 880, 985, 1030, 1031, 1034, 1035, 1066
patient monitoring   861
pattern recognition   131, 393, 468
PBS. *see* phosphate-buffered saline
PCA. *see* principal component analysis
PCNL. *see* percutaneous nephrolithotomy
PCR. *see* polymerase chain reaction
PCT. *see* photochemotherapy
PDA. *see* photodiode array
PDD. *see* photodynamic diagnosis
PDMS. *see* polydimethylsiloxane
PDT. *see* photodynamic therapy
PDVP. *see* proliferative diabetic vitreoretinopathy
PEG. *see* poly(ethylene glycol)
penetration depth   117, 154, 218, 309, 324, 379, 411, 420, 452, 480, 483, 485, 514, 548, 557, 715, 717, 722, 726, 732, 781, 791, 795, 805, 806, 810, 864, 914–916, 925, 983, 1010, 1030, 1047, 1063, 1070, 1099
penicillin   904
peptide nucleic acid   53
peptides   290, 292, 298, 324, 325
percutaneous nephrolithotomy (PCNL)   514, 515
perfusion monitoring   410, 411, 413, 828
perimetry   604
periodontal therapy   1031, 1075–1078
periodontal tissues   1029–1031
peripheral nervous system   734
permeability   25, 164, 291, 295, 322, 326, 548, 647, 903, 904, 908, 1060, 1067, 1074
personalized medicine   52, 57
PET. *see* photoinduced electron transfer *or* positron emission tomography
PFGE. *see* pulsed-field gel electrophoresis
phagocytes   237
phagocytosis   549
phase contrast   4, 5, 167, 201, 887, 888
phase matching   89, 96
phase modulation   93, 99
phosphate groups   129
phosphate-buffered saline   161
phosphatidylcholine   238, 731, 796
phosphatidylethanolamine   796
phosphatidylserine   552, 796
phosphorescence   6, 715, 767, 770–772
phosphorescence quenching   767, 770, 771
phosphorylation   445, 449, 956, 957
photoablation   609, 610, 675, 676, 678, 917, 924, 926–928, 930, 1057
photoacoustic (PA)   212, 404, 405, 426, 452, 453, 485, 836, 839, 841, 842, 848, 914, 925, 956, 1015, 1059, 1100
photoacoustic spectroscopy   841
photoacoustic tomography   404, 405, 426
photobleaching   46, 49, 98, 186, 235, 243, 280, 307, 309, 325, 483, 728
photochemotherapy (PCT)   261, 262, 855, 856, 858, 859
photodamage   98, 100, 248, 298, 299, 722, 723, 870
photodiode   227, 413, 563, 771, 846, 1042
photodiode array   413
photodynamic diagnosis (PDD)   181, 486
photodynamic molecular beacons   295–302
photodynamic therapy (PDT)   84, 90, 261–266, 268–283, 289–293, 295, 297–299, 302, 305–310, 315–332, 382, 415, 419, 420, 457, 458, 472, 477, 479–486, 491, 495–497, 524, 529–536, 600, 601, 673–675, 678–681, 736, 789, 827, 855, 856, 859–865, 989, 991, 1057
photoinduced electron transfer (PET)   208, 296, 297
photoluminescence   26, 1048
photomultiplier tube   771
photon scattering   795

photoreceptor outer segment (PROS) 655, 656
photosensitization 262, 295, 305, 528, 863, 865, 1057
photosensitizer 263–267, 276, 280, 289–292, 295–299, 301, 302, 305–309, 315, 316, 321–332, 345, 470, 478, 480, 483, 495, 529–531, 533, 674, 675, 855, 856, 859, 860, 862, 863, 1057
photosensitizer carriers 326
photosynthesis 262
phototherapy (PT) 261, 262, 320, 788, 842, 855–859, 861, 863–865, 870
photothermal radiometry 1013, 1047, 1048
Picrosirius Red (PSR) 363–368, 1031–1034
pigmented skin lesions 778, 779, 817, 818, 820–822, 824
pigments 790
plasma membrane 321, 322, 675
PLDD. see percutaneous laser disc decompression
PLM. see polarized light microscopy
PLSR. see partial least-squares regression
PMMA. see poly(methyl methacrylate)
PMT. see photomultiplier tube
POAG. see primary open-angle glaucoma
POBF. see pulsatile ocular blood flow
POC diagnostics. see point-of-care diagnostics
POCT. see point of care testing
point mutations 54
point spread function 382, 451
polarimetry 614, 620, 845–848
polarizability 106, 842
polarization contrast 167
polarization spectroscopy 421–423, 425, 430, 431
polarization-sensitive optical coherence tomography (PS-OCT) 340, 363, 371, 380, 388, 422, 698, 702, 782, 1005–1010, 1029, 1030
polarized light 51, 192, 363–368, 421–423, 437, 525, 620, 778, 845, 846, 1001, 1006–1008, 1031–1033
polarized light microscopy (PLM) 1008, 1009
polarizers 831, 1002, 1003
poly(amidoamine) (PAMAM) 329
poly(ethylene glycol) 300, 324, 729
poly(ethylene glycol) (PEG) 26, 324, 325, 327, 329, 331
poly(methyl methacrylate) (PMMA) 686, 1089
poly(propylenimine) (PPI) 329
polydimethylsiloxane (PDMS) 41, 385
polymerase chain reaction (PCR) 37, 39, 44, 45, 52, 57, 122, 147, 215, 873
polymeric nanoparticles 323, 327, 330
polymerization 41, 1061, 1065, 1066, 1104
polymers 41, 42, 326–328, 330, 609
port-wine stain (PWS) 870, 871, 923
positive predictive value (PPV) 338, 344–346, 353, 354
positron emission tomography (PET) 292, 404, 472, 547, 716, 743, 759
post-voiding residual urine (PVR) 511
potential acuity meter (PAM) 329, 631
Pourcelot's index of resistivity 665
PPG. see photoplethysmography
PPI. see poly(propylenimine) or proton pump inhibitor
PPV. see positive predictive value
PR. see progesterone receptor
precision 57, 153, 167, 310, 384, 609, 619, 840, 929, 930, 964
prefrontal cortex 759–764
prevalence 645, 748
primary open-angle glaucoma (POAG) 661, 662, 975, 978
principal component analysis (PCA) 108, 110, 111, 129, 130, 132, 133, 135, 184, 185, 192, 241, 252, 526
probe design 771, 1012, 1013
progenitor cells 734, 957, 958, 1053
progesterone receptor (PR) 164, 170
proliferation 56, 165, 380, 536, 548, 574, 653, 675, 797, 861, 871, 872
proliferative diabetic vitreoretinopathy (PDVP) 986
PROS. see photoreceptor outer segment
prostate cancer 38, 52, 239, 261, 274–277, 279–282, 342, 387
prostate vaporization 509, 510, 512, 514, 516
prostate-specific antigen (PSA) 239, 274, 276
proteases 293, 298, 553, 797, 897
protein domains 221
protein–protein interaction 37, 50, 61
proteoglycan 920
proteomics 58, 78, 80, 90
proton pump inhibitor (PPI) 530, 535
PS-OCT. see polarization-sensitive optical coherence tomography
PSA. see prostate-specific antigen
PSF. see point spread function
psoralen with ultraviolet A (PUVA) 858, 859, 861–863
psoriasis 262, 409, 457, 601, 781, 784, 789, 795, 796–798, 855–862, 864, 931
PSR. see Picrosirius Red

PT. *see* phototherapy
pulsatile ocular blood flow (POBF)　666
pulse oximetry　448, 449, 498, 746, 828
pulsed laser　151, 245, 514, 599, 719, 722, 732, 861, 869, 871, 913, 925, 928, 1059
pulsed-field gel electrophoresis (PFGE)　120, 121, 127
PUVA. *see* psoralen with ultraviolet A
PVP. *see* photoselective vaporization of the prostatectomy
PVR. *see* post-voiding residual urine
PWS. *see* port-wine stain

q

QELS. *see* quasi-elastic light scattering
QLF. *see* quantitative light fluorescence
quantitative light fluorescence (QLF)　1006, 1010, 1011, 1042, 1043
quantum dots　16, 19, 58–63, 323, 725, 729, 838
quantum efficiency　183, 241, 282
quasi-elastic light scattering (QELS)　631, 633, 635, 830
quencher　21–23, 296–300, 302

r

RA. *see* radial artery
radial artery (RA)　385, 386, 547, 548, 552, 554, 555, 557, 571–577
radial fiber　946–951
radiance　1011
radiative transport equation (RTE)　279, 566, 567, 1048
radiofrequency ablation　524, 534, 535, 939, 943–945, 949, 951
radioimmunoassay　49
radiotherapy　275, 315, 496, 707
Raman effect　96, 99
Raman microscopy　220, 235, 236, 238, 240, 241, 719
Raman optical activity (ROA)　8
Raman scattering　78, 80, 90, 105, 106, 109, 121, 210, 220, 235, 236, 243, 244, 635, 719, 724, 732, 843, 1017
Raman spectroscopy　7, 8, 50, 59, 105–107, 114, 117, 119–121, 124–126, 132, 134, 135, 219–221, 235, 236, 240, 422, 527, 631, 633, 635, 636, 723, 731, 732, 735, 795, 813, 824, 842, 843, 847, 848, 1016, 1017
– micro-Raman set-up　126
– micro-Raman spectroscopy　106, 107, 114, 120, 126, 1016
– Raman set-up　121, 126
– resonance Raman spectroscopy　8, 106, 842
– surface-enhanced　8, 50, 106
RATTS. *see* real-time automated tracking and trapping system
Rayleigh scattering　210
RBC. *see* red blood cell
reactive oxygen species (ROS)　295, 321, 323–325, 327, 331, 332, 495, 554, 903, 906, 956
real-time automated tracking and trapping system (RATTS)　888, 889
receptor tyrosine kinase　164
receptors　12, 27, 55, 164, 165, 170, 290, 291, 293, 322, 324, 705, 835, 956
reconstruction　202, 230, 231, 239, 279, 280, 426, 562, 564–567, 602, 725, 727
red blood cell (RBC)　147, 164, 238, 240, 406–408, 417, 418, 420, 423–425, 434, 435, 437, 448, 571, 667, 721, 790, 827, 844, 1061, 1068
redox state　471, 638, 760
reflectance spectroscopy　182, 844, 848, 1002, 1017
refraction　377, 606, 607, 685–688, 963
refractive surgery　384, 606, 607, 636, 686–690, 924, 929, 963–967, 969, 971, 972
regeneration　24, 321, 524, 532, 535, 719, 734, 782
regenerative medicine　157
region of interest (ROI)　167, 175, 359, 379, 411, 416, 425, 549, 555, 556, 769, 771
relaxation time　922, 923, 926, 975, 1001, 1062, 1099
remote microscopy　175, 176
resonance Raman scattering (RRS)　236, 238, 724
respiration　265, 309, 705, 736, 830, 957
respiratory disease　298
restorative dentistry　1053, 1054, 1056, 1060, 1066, 1068, 1072, 1074, 1076, 1078
restricted fragment length polymorphism (RFLP)　122
retention　38, 210, 262, 263, 291, 295, 322, 324, 326, 482, 511, 528, 549, 905, 906, 908, 1011
retinal degeneration　602, 984
retinal nerve fiber layer (RNFL)　590–592, 605, 614–620, 669
retinal pigment epithelium (RPE)　584, 585, 592, 595, 597, 605, 651–655, 657, 989
retinal vein occlusions　987
reverse transcriptase　44
reverse transcription　873
RFBR. *see* Russian Foundation for Basic Research

RFLP. *see* restricted fragment length polymorphism
rheumatoid arthritis   547, 548, 550, 552, 554, 556, 557, 561, 569, 571, 786
RIA. *see* radioimmunoassay
ribonucleic acid   525
RIfS. *see* reflectometric interference spectroscopy
RifS. *see* reflectometric interference spectroscopy
rigid endoscope   203, 390
RNA   36–40, 42–44, 53, 54, 129, 131, 162, 163, 208, 236, 861
RNA. *see* nucleic acid
RNAi   39
RNFL. *see* retinal nerve fiber layer
ROA. *see* Raman optical activity
ROI. *see* region of interest
ROS. *see* reactive oxygen species
rotation   345, 354, 355, 638, 845, 846, 848
RPE. *see* retinal pigment epithelium
RRS. *see* resonance Raman scattering
RTA. *see* retinal thickness analyzer
RTE. *see* radiative transport equation
Russian Foundation for Basic Research (RFBR)   371, 849, 883, 1037

s

saccharides   12, 20, 21, 26, 27
sample preparation   7, 40, 42, 46, 57, 106–108, 116, 117, 132, 134, 135, 843
sampling volume   340, 605
sarcomas   306
scaffold   98
scanning
– A-scan   615, 651, 846
– en-face   370, 371
scanning electron microscopy (SEM)   79, 328, 389, 1049, 1073, 1074
scanning laser ophthalmoscope (SLO)   593, 594, 615, 616, 648, 651, 666
scanning laser polarimetry (SLP)   620
scanning laser tomography   617–619, 666
scattering process   47
SCC. *see* squamous cell carcinoma
scotoma   988
SCP. *see* spectral cytopathology
screening   121, 189, 190, 215, 216, 230, 254, 338, 379, 384, 386, 390, 466, 506, 588, 590, 646, 707, 879, 880
SDD. *see* sequential digital dermoscopy
SDOCT. *see* spectral domain optical coherence tomography

second-harmonic generation (SHG)   92, 191, 244, 718, 719, 723, 790–792, 795, 811, 822
secondary structure   635
segmentation   189, 241, 388, 605
selective laser trabeculoplasty (SLT)   597, 614, 661, 662, 975–980
self-phase modulation   93
SEM. *see* scanning electron microscopy
semiconductor nanocrystals   210, 324, 725, 729
sensitization   263, 272, 282, 859
sequential digital dermoscopy (SDD)   819
serous macular detachment (SMD)   652, 656
SERS. *see* surface-enhanced Raman spectroscopy
SHG. *see* second-harmonic generation
side chain   46, 237, 238, 290, 300, 554
signal transduction   55
signal-to-noise ratio   227, 237, 302, 386, 417, 427, 549, 562, 621, 736, 743, 796, 843
signaling pathway   165, 322, 957
single molecule   6, 61, 236
single nucleotide polymorphism (SNP)   52–54, 575
single-cell sorting   145, 147
single-point laser trapping   887
singlet state   208
siRNA   39
skeletal muscle   450, 451, 453
skin cancer   262, 266, 271, 273, 274, 282, 306, 315–320, 386, 601, 673, 777–780, 788, 789, 805, 806, 808, 810, 812, 814, 859, 862
skin diagnostics   777, 778, 780, 782, 784, 786, 788, 790, 792, 794, 796, 798
slit-lamp examination   630
SLO. *see* scanning laser ophthalmoscope
SLP. *see* scanning laser polarimetry
SLT. *see* selective laser trabeculoplasty
SM. *see* submucosa
SMD. *see* serous macular detachment
smooth muscle   387, 574
SNP. *see* single nucleotide polymorphism
solid-state lasers   509
spatial coherence   491
spatially resolved spectroscopy (SRS)   451
speckle contrast   416–420, 430, 767–769, 771, 829, 830
spectral cytopathology (SCP)   251–254, 256
spectral domain optical coherence tomography (SDOCT)   460, 605, 614, 648
spectral intensity   88
spectral unmixing   184, 239, 241
spectrograph   184, 186, 188
spectroscopic fingerprint   238

speed of light 599, 887
sperm energetics 889, 891
sperm motility 887–890, 892
spherical harmonics 566
sphingolipids 796
spotting 42
SPR. see surface plasmon resonance
squamous cell carcinoma (SCC) 190, 193, 254, 271, 273, 328, 358, 366, 386, 387, 533, 695, 696, 705, 707, 795, 805, 807, 809, 810, 814
SRF. see subretinal fluid
SRS. see spatially resolved spectroscopy or stimulated Raman scattering
SS. see swept-source
SSF. see Swedish Strategic Research Foundation
SSIM. see subsquamous specialized intestinal metaplasia
staining 4, 107, 127, 147, 160, 163, 164, 166–172, 209, 221, 239, 338, 358, 363–368, 906, 1031–1034
Staphylococcus aureus 113, 121, 302, 872
STED. see stimulated emission depletion
stem cell 54, 146, 554, 555, 734, 735, 871, 874, 955
– embryonic 54
– mesenchymal 554, 871, 874
– monitoring 734
– therapy 871, 955
stereo fundus photography 646, 648
steroid 164, 170, 653, 783, 786
stimulated emission 80, 208, 491, 1055
stimulated emission depletion (STED) 80
stimulated Raman scattering (SRS) 80, 90, 235, 724
stimulation 227, 419, 727, 731, 732, 747, 873, 898, 955, 957, 1066, 1069
– electrical 419, 731
Stokes scattering 843
Stokes shift 14, 25, 209, 592, 1048
stray light 601
stretching 118, 129, 130, 590, 796, 839, 1089
stroke 231, 548, 716, 727, 732, 733, 767, 769–771, 872, 873, 955–960
structural contrast imaging 217
structural proteins 194, 207, 1042
subcellular structures 791
submucosa (SM) 350–352, 527, 531, 534
subretinal fluid (SRF) 651, 989
subsquamous specialized intestinal metaplasia (SSIM) 535, 536
subventricular zone (SVZ) 958
sugars 27, 128

superficial illumination 264, 266, 270
superposition 109
support vector machine 241
surface modification 42, 1001, 1078
surface plasmon resonance (SPR) 12, 50, 51
surface plasmons 51, 239
surface-enhanced Raman scattering 236
surface-enhanced Raman spectroscopy (SERS) 8, 50, 57, 59, 60, 106, 107, 120, 127, 128, 134, 236, 238, 239
surfactant 327, 906
surgical microscope 379, 380, 384, 386, 700, 934
surgical pathology 159, 161, 167, 215, 216, 218, 220, 222
SVZ. see subventricular zone
Swedish Strategic Research Foundation (SSF) 282
swept-source (SS) 378, 382, 383, 701, 708, 709

t

target validation 55
targeted fluorescence 552
taxonomy 54
TCA. see topographic change analysis
TCSPC. see time-correlated single-photon counting
TD. see time domain
TE. see trabeculectomy
telepathology 175–177, 179
temperature distribution 921, 929, 1047
temporal coherence 491
terahertz pulse imaging (TPI) 1018, 1019
TERS. see tip-enhanced Raman spectroscopy
tertiary structure 251
tetramethylrhodamine isothiocyanate (TRITC) 27
tetrasulfocyanine (TSC) 552
therapeutics 57
thermal conductivity 922, 1053
thermal imaging 1013
thermal process 918
third harmonic generation 244
third-order polarization 244
time domain (TD) 96, 185, 187, 378, 379, 614, 615, 701, 709, 724, 730, 745, 748, 749, 1029
time gate 187, 204
time-correlated single-photon counting (TCSPC) 186, 187
time-resolved fluorescence spectroscopy technique (TRST) 460
time-resolved spectroscopy (TRS) 451, 761

tip-enhanced Raman spectroscopy (TERS) 106, 120
tissue autofluorescence 181, 527, 548
tissue cutting 913, 929, 930, 932
tissue damage 278, 307, 308, 382, 480, 510, 511, 771, 790, 975, 1058, 1061, 1078, 1103
– irreversible 1078
– reversible 1078
tissue microarrays 38, 178
tissue perfusion 411, 548, 572, 733, 734, 746
tissue plasminogen activator (tPA) 872, 956, 959
tissue viability imaging 405
TL. see tube lens
TM. see trabecular meshwork
TMP. see trimethylpsoralen
TMR. see transverse microradiography
TOI. see total oxygenation index
tooth decay 996
topographic change analysis (TCA) 619
total internal reflection 80, 459, 934, 935
total oxygenation index (TOI) 745, 746, 749
total reflectance 7, 117, 135, 916, 1015
toxicology studies 177
tPA. see tissue plasminogen activator
TPI. see terahertz pulse imaging
TPL. see two-photon-excited luminescence
TPM. see two-photon microscopy
trabecular meshwork (TM) 380, 596, 597, 613, 614, 661, 662, 673, 975–977, 980
trabeculectomy (TE) 384, 613, 661, 673, 677
training 108, 111, 113, 116, 178, 184, 222, 252, 780, 845, 873, 1005
transcription 44, 98, 129, 164, 165, 873
transcription factors 164
transcriptional regulation 861
transducer 15, 18, 26, 36, 427, 841, 848
transfection 39, 903–908
transformation, malignant 267, 468, 525
transillumination 201, 225, 446, 447, 999, 1002–1004, 1047
transplantation 385, 734
transporters 790
transpupillary thermotherapy (TTT) 598, 988, 989
transurethral resection of bladder tumor (TURBT) 346, 347
transverse microradiography (TMR) 1008, 1009, 1050
trimethylpsoralen (TMP) 211, 862
triplet state 208, 305
TRITC. see tetramethylrhodamine isothiocyanate

TRS. see time-resolved spectroscopy
TRST. see time-resolved fluorescence spectroscopy technique
TSC. see tetrasulfocyanine
TTT. see transpupillary thermotherapy
tube lens (TL) 890
tumor
– deoxygenation 308
– diagnosis 159
– marker 169, 178
– suppressor gene 53, 57, 58, 204
tumorigenesis 165, 805
TURBT. see transurethral resection of bladder tumor
tweezers 146, 154, 887–889, 891
two-photon absorption 84
two-photon excitation 6, 100, 191, 207, 244, 722
two-photon microscopy (TPM) 84, 193, 194, 718, 719, 722, 723
two-photon-excited luminescence (TPL) 84–90, 94–96, 98

u

UCTD. see undifferentiated connective tissue disease
ultra-short pulse laser (USPL) 1085, 1090–1092
ultrashort pulses 96, 244, 964
ultraviolet A (UVA) 324, 636, 856, 858, 859, 861, 862
ultraviolet B (UVB) 424, 856–859, 861, 862
ultraviolet light 383, 528, 1055
ultraviolet phototherapy 858, 863
undifferentiated connective tissue disease (UCTD) 409
upper urinary tract transitional cell carcinoma (UUTT) 512
urinary bladder 270, 281, 337, 341, 343–345, 364, 366–368, 370, 379, 388
urinary tract infection (UTI) 121, 122
urology 342, 343, 345, 347, 387, 503, 509, 510, 512, 514, 516, 933
USPL. see ultra-short pulse laser
uterine cervix 306, 358–362, 364–367, 379, 879, 880, 882
UTI. see urinary tract infection
UUTT. see upper urinary tract transitional cell carcinoma
UV-resonance 106
UVA. see ultraviolet A
UVB. see ultraviolet B

v

varicose veins 944, 949, 951
vascular surgery 385, 931
vascularization 467, 603, 726, 732, 812, 824, 942
vasculartargeted photodynamic therapy (VTP) 276
VCA. *see* vertex component analysis
vertex component analysis (VCA) 239, 241
vesicles 327, 388
vibrational frequency 245, 246
vibrational spectroscopy 105–136, 219, 221, 732
video endoscope 528
viral infection 251–256, 789
virtual microscopy 175–180
virtual slide 176
virulence factors 119
virus 252, 253, 256, 338
viscosity 749, 829
visible wavelengths 190, 722, 916
vitamin A 798
vitamin D 292
vitreo-macular traction (VMT) 652–654
VMT. *see* vitreo-macular traction
voltage-sensitive dyes 720
volume data set 735
VTP. *see* vasculartargeted photodynamic therapy

w

wavefront analysis 606, 610, 636, 972
– applications 636
wearable light sources 315, 317, 319, 320
Western blot 39, 47
wound healing 673–680, 782, 783, 797, 798, 869, 870, 872–874, 950, 968, 969, 971

x

X-ray mammography 216
xenon lamp 202, 471, 474, 492, 528
XFROG. *see* cross-correlation frequency-resolved optical gating

y

yeast 55, 107, 124, 247, 789
yellow fluorescent protein (YFP) 28, 29
YFP. *see* yellow fluorescent protein
YSGG. *see* yttrium scandium gallium, garnet
yttrium scandium gallium, garnet (YSGG) 1001, 1055, 1059, 1061, 1063, 1065, 1070, 1076, 1077, 1086–1090, 1092, 1099

z

zona pellucida 895, 896, 898